HIGH MAGNETIC FIELDS IN THE PHYSICS OF SEMICONDUCTORS

11*th* International Conference

HIGH MAGNETIC FIELDS IN THE PHYSICS OF SEMICONDUCTORS

August 8 - 12, 1994
Francis Bitter National Magnet Laboratory
Massachusetts Institute of Technology

Editor

DONALD HEIMAN

Francis Bitter National Magnet Laboratory
Massachusetts Institute of Technology
Cambridge, Massachusetts, USA

World Scientific

Published by

World Scientific Publishing Co. Pte. Ltd.

P O Box 128, Farrer Road, Singapore 9128

USA office: Suite 1B, 1060 Main Street, River Edge, NJ 07661

UK office: 57 Shelton Street, Covent Garden, London WC2H 9HE

HIGH MAGNETIC FIELDS IN THE PHYSICS OF SEMICONDUCTORS

ISBN 981-02-2181-9

Printed in Singapore.

Preface

These Proceedings contain the manuscripts of research presented at The Eleventh International Conference *High Magnetic Fields in the Physics of Semiconductor* (SEMIMAG-94). The Conference was organized by the Francis Bitter National Magnet Laboratory and held on the MIT campus from August 8-12, 1994. As in the past, it was a satellite of the *International Conference on the Physics of Semiconductors*, which was held in Vancouver the following week. This marks the first time the Conference has been sited in the U.S. It was attended by 211 researchers from 24 countries, who presented 27 invited talks, 42 regular talks, 106 poster papers, and 3 summary talks.

The purpose of this meeting was to bring together an international group of researchers to discuss new results and current problems in the area of high magnetic field effects in semiconducting and related materials. In 1972 the first conference on this subject was organized by Professor G. Landwehr at Würzburg University. In this initial meeting two concepts were emphasized – all talks were invited and tutorial in nature, and participants were provided with workbooks containing full manuscripts. This approach promoted lengthy and intense discussions during and after the lectures. The first conference was so successful that it led to a succession of meetings held every two years. Subsequent meetings were held in Würzburg in 1974 and 1976, Oxford in 1978, Hakone in 1980, Grenoble in 1982, Würzburg in 1986, 1988 and 1990, and Chiba in 1992.

In keeping with the previous traditions we supplied participants with a workbook containing unrefereed manuscripts of all papers. Unfortunately, we had to limit the number of invited tutorial talks to about two per session because of the overwhelming number of abstracts received. To make up for this, a summary session was arranged for Friday afternoon. But instead of having one person summarize the whole Conference, three speakers were chosen to review timely subjects. Günther Bauer gave a thorough overview of the materials aspects, Sankar Das Sarma eloquently reviewed localization effects, and Bertrand Halperin gave an insightful summary of the current status of composite fermion research. Initially, I felt that it was not reasonable to ask the summary speakers to prepare a manuscript for the Proceedings, but after seeing what an outstanding job they did, I was sorry that their presentations were not recorded. Perhaps in a future Conference a video recording could be made of the summary session.

There is no doubt that high magnetic fields have been crucial for the advancement of semiconductor research. In the past, the application of high magnetic fields has been an indispensable tool for determining the properties of semiconductors and understanding their underlying basic physics. One might call this "semiconductor physics." Now, the dominant research focuses on the effects of electron-electron interactions — the semiconductor structure is more commonly a simple container for the electrons. Thus the majority of the research reported at the present Conference involves "physics *in* semiconductors," which is reflected in the title of these Proceedings.

Support for the Conference was kindly provided by the Office of Naval Research, the International Science Foundation/Soros, the National Science Foundation, NEC Research Institute, Oxford Instruments, Keithley Instruments, and Taylor & Francis Publishers. This support allowed a number of people to attend the Conference and present their work, who would otherwise have missed the opportunity to share their important results.

As with organizing any conference, many people gave their unselfish support to make the Conference a success. At the top of this list is Margaret O'Meara who contributed innumerable hours communicating with the authors and function organizers; and Tomasz Dietl who spent several weeks at MIT helping to organize the initial program, coordinated ISF/SOROS support, and ensured that researchers from Eastern Europe were adequately represented. I would also sincerely like to thank: Gottfried Landwehr for his support of MIT as the Conference site; Larry Rubin for important suggestions, especially in the early stages of planning; Simon Foner for acting as a sounding board, providing valuable advice, and keeping us to a reasonable schedule by making sure the fire was always going under my feet; Robin Elices for her suggestions for siting various functions and dealing with several bureaucratic snags; Zachary Lee for putting the workbook together and providing the sounds signalling the end of the coffee breaks; Gervais Favrot for editing much of the written material and compiling the restaurant list; Steve Klepper for organizing the onsite sponsors; Steve and Thang Vu for setting up email communications; Judy Haberl for judging the artistic merits of the poster presentations; Noboru Miura for providing mailing lists and information on the previous High-Field Conference; Bruce McCombe and Susan Masoian for general advice on conference organizing and information about outside support; Rose Godinho and her staff at Joanies Kitchen for keeping the participants well-fed and charged up with strong coffee; numerous students and staff at the Magnet Lab; and finally, Gayle Fitzgerald and her staff at the MIT Conference Services Office for putting together the events so smoothly that left many people complaining that they had no complaints. Of course the ultimate success of any conference depends on the scientific program and the program committee did an outstanding job carving out an interesting program from the nominations and abstracts.

Finally, I was delighted to see the active participation and enthusiastic discussion between the speakers and participants. I thank you all for contributing to the success of this Conference.

Don Heiman, Chairman

INTERNATIONAL ADVISORY COMMITTEE

J.S. Brooks	(Boston University)
P.M. Chaikin	(Princeton University)
R.G. Clark	(U New South Wales)
M.S. Dresselhaus	(MIT)
M. Grynberg	(Warsaw University)
B.I. Halperin	(Harvard University)
F. Herlach	(Katholieke U Belgium)
M.A. Kastner	(MIT)
K.v. Klitzing	(Max Planck Stuttgart)
G. Landwehr	(Wurzburg University)
B. Lax	(MIT)
P.A. Lee	(MIT)
J.C. Maan	(Nijmegen High Field Lab)
G. Martinez	(Grenoble High Field Lab)
E.E. Mendez	(IBM, Watson Lab)
N. Miura	(ISSP Tokyo)
R.J. Nicholas	(University Oxford)
E.I. Rashba	(University Utah)
D.G. Seiler	(NIST Maryland)
L.J. Sham	(U Cal San Diego)
H.L. Stormer	(AT&T Bell Labs)
P.A. Wolff	(NEC Research Lab)
P. Wyder	(Grenoble High Field Lab)

ORGANIZING AND PROGRAM COMMITTEE

T. Dietl	(Institute Physics Warsaw)
S. Foner	(MIT)
D. Heiman (Chairman)	(MIT)
A.H. MacDonald	(Indiana University)
B.D. McCombe	(SUNY Buffalo)
A. Pinczuk	(AT&T Bell Labs)
D.C. Tsui	(Princeton University)

LOCAL COMMITTEE

G. Dresselhaus	(MIT)
G. Favrot	(MIT)
B.B. Goldberg	(Boston University)
S.J. Klepper	(MIT)
C.H. Perry	(Northeastern University)
L.G. Rubin	(MIT)

LOCAL STAFF

Margaret O'Meara	FBNML
Robin Elices	FBNML
Zachary Lee	FBNML
Gayle Fitzgerald	Conf. Service Office
Trish Ezekiel	Conf. Service Office

SPONSORS

U.S. Office of Naval Research
National Science Foundation
NEC Research Institute
Oxford Instruments
Keithley Instruments
Taylor & Francis Publishers

CONTENTS

List of Sections

List of Papers

I. COMPOSITE FERMIONS

A. Experiment

B. Theory

II. COUPLED LAYERS

III. EDGE STATES

A. Experiment

B. Theory

IV. LOCALIZATION

A. Metal/Insulator Transition

B. Coulomb Gap

C. Scaling

V. QUANTUM HALL EFFECTS

VI. MAGNETO-OPTICS

A. Quantum Hall Regime

B. Visible and Near-visible Region

C. Infrared Region

D. D^- and X^-

VII. ELECTRON SOLID

VIII. NANOSTRUCTURES

A. Quantum dots

B. Quantum wires

IX. TUNNELING

X. DELTA-DOPED LAYERS

XI. Si-Ge

XII. MAGNETIC SEMICONDUCTORS

XIII. 3D SYSTEMS

XIV. ORGANIC CONDUCTORS

XV. EXPERIMENTAL METHODS

Geometrical Resonance of Composite Fermions in an Anti-dot Superlattice

W. Kang, H.L. Stormer, L.N. Pfeiffer, K.W. Baldwin, and K.W. West

AT&T Bell Laboratories, Murray Hill, New Jersey 07974

ABSTRACT

According to recent theories, a system of electrons at half-filled Landau level can be transformed to an equivalent system of composite fermions at zero effective magnetic field. The fractional quantum Hall effect then arises as the integral quantum Hall effect of composite fermions. In order to test for these new particles, we have studied transport of anti-dot superlattices in a two-dimensional electron gas. At low magnetic fields electron transport exhibits well-known resonances at fields where the classical cyclotron orbit becomes commensurate with the anti-dot lattice. At half-filling we observe the same dimensional resonances. This establishes the "semi-classical" behavior of composite fermions.

1. Introduction

When all electrons reside in the lowest Landau level, the kinetic energy is quenched and electron-electron interaction dominates the collective dynamics. The quantum liquid states of fractional quantum Hall effect (FQHE)[1-3] represent such a demonstration of strong electron correlation at high magnetic fields. Recently the nature of electronic interaction at half-filled Landau level has received much attention. Several experiments provide strong support for a novel particle called "composite fermion" that plays a crucial role in the physics of two-dimensional (2D) electron systems in the lowest Landau level. In this paper we present results of an experiment that demonstrates the semi-classical motion of such a particle.

The FQHE, occurring in high-quality samples, is considered a quantum fluids of electrons condensed due to electron-electron interaction[4]. The primary Laughlin states occur at $\nu = 1/m$ and $\nu = 1 - 1/m$, and the higher order FQHE states are derived from lower order states as Laughlin states of fractionally charged quasi-particles[5,6] . Thus, in the hierarchical model, the two prominent series of FQHE states observed at $\nu = p/(2p \pm 1)$ represent a succession of parental and daughter states starting from $\nu = 1/3$ and $\nu = 2/3$ and converging towards $\nu = 1/2$. However, little can be stated regarding the $\nu = 1/2$ state.

The significance of the physics at half-filled Landau level was foreshadowed in exceptional electrical transport and surface acoustic wave anomalies exactly at $\nu = 1/2$. At this filling fraction Jiang *et al.*[7] observed a deep minimum in the magnetoresistivity ρ_{xx} that persisted to unusually high temperatures, exhibiting a temperature dependence distinctly different from the neighboring FQHE-states. Surface acoustic wave experiments by Willett *et al.*[8] at $\nu = 1/2$ revealed attenuation and velocity changes that were opposite to those observed in the regime of the FQHE liquids.

Halperin, Lee and Read[9] used a Chern-Simons gauge field construction to transform the state at exactly $\nu = 1/2$ to a mathematically equivalent state of "composite fermions" with a well defined Fermi surface. The magnetic field is "incorporated" into the particles themselves and the resulting composite fermions move in an apparently vanishing external magnetic field. Away from $\nu = 1/2$ the composite fermion experiences an effective magnetic field and they form Landau levels similar to electrons in a magnetic field. This is the essence of the Jain's model. The series of FQHE liquids are derived as the integral quantum Hall effect of composite fermions[10].

The composite particles of these theories have been able to account for several of the previously puzzling features in the vicinity of $\nu = 1/2$. The electronic transport anomaly at $\nu = 1/2$ is explained in terms of the appearance of a metallic state and the suppression of electron localization[11]. The surface acoustic wave data at this filling factor are interpreted as wavevector dependent relaxation of the composites that make up the Fermi sea[9]. Finally, the composite fermion picture provides a very natural interpretation of the observation by Du *et al.*[12] that the size of the energy gaps of the main sequence of FQHE states at $\nu = p/(2p \pm 1)$ increases linearly with the magnetic field deviation from $B_{1/2}$ at $\nu = 1/2$. It simply reflects the linearly increasing cyclotron gap or "Landau-level splitting" of composite fermions exposed to an effective magnetic field $B_{eff} = B - B_{1/2}$.

Since the $\nu = 1/2$ state is proposed to be largely equivalent to a metal at zero magnetic field, one wonders whether experiments that usually reveal the semi-classical motion of electrons could not be performed on these new particles. One such experiment is to investigate dimensional resonance in anti-dot superlattice. We have studied magnetotransport in anti-dot superlattice to first establish the semi-classical behavior of electrons around $B = 0$ and then probe the equivalent semi-classical behavior of the composite particles around $\nu = 1/2$. One of the particularly telling signatures for the composite fermions we expect to find is their modified resonant field as compared to the resonances of the usual electrons. Not only should the peaks around $B = 0$ reoccur symmetrically around $\nu = 1/2$ but their spacing should differ by exactly a factor of $\sqrt{2}$ between electrons and composites. This is due to the spin alignment of the fermions which increases their Fermi velocity v_f by a factor of $\sqrt{2}$ as compared to the unpolarized electron systems[9].

2. Experimental

The anti-dot superlattice was fabricated on a high quality modulation-doped GaAs/AlGaAs heterostructure with electron density $n_e = 1.45\times10^{11}$ cm^{-2} and mobility $\mu = 7.8\times10^6$cm^2/Vsec prior to processing. It was initially defined as an array of holes by standard electron-beam lithography. These holes were transferred to the 2D electron gas via reactive ion etching, producing cylindrical holes with minimal undercutting. The period of anti-dots ranged from d = 500-700nm with the dot sizes of 100-200nm. The electron-beam exposure and the etch depth were varied to optimize the size of the well established anti-dot oscillations for electrons around B = 0.

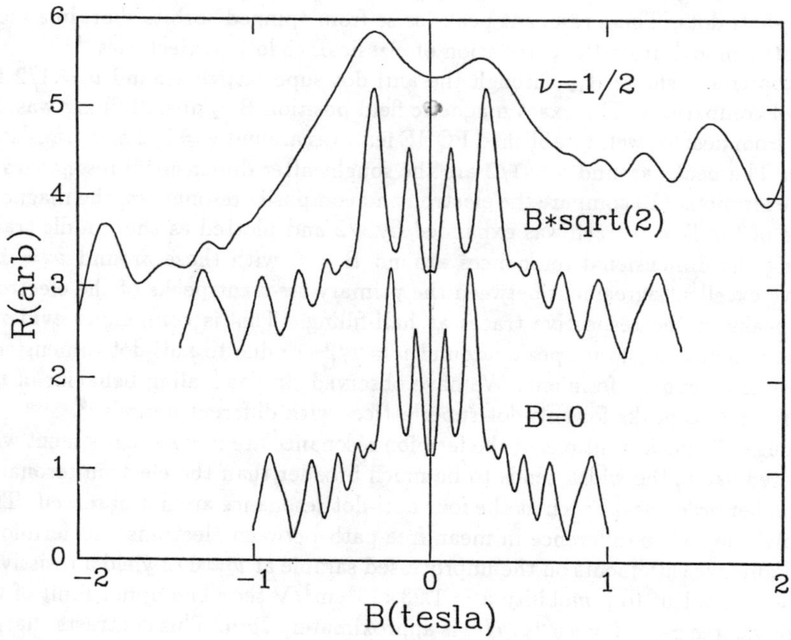

FIG. 1. Comparison of the magnetoresistance R_{xx} of the 600 nm period anti-dot superlattice near B = 0 (bottom), the R_{xx} of near $\nu = 1/2$ (top), and the B = 0 electron trace with magnetic field scale expanded by $\sqrt{2}$ (middle). For the sake of comparison, the $\nu = 1/2$ trace was shifted down to zero magnetic field by determining the position of $\nu = 1/2$ from neighboring fractions.

3. Results and Discussion

Fig. 1 shows the magnetoresistance R_{xx} of a d = 600 nm period anti-dot superlattice near B=0 (bottom), around $\nu = 1/2$ (top), and expanded view of the electron trace (middle). For the bottom trace, the origin, B = 0, resides at the center of the figure and the electron transport data was taken for positive and negative magnetic fields. The sharp features around B = 0 are clearly identifiable as the well-known geometrical resonances of the electrons [13] followed by the appearance of Shubnikov-de Haas oscillations in higher magnetic field. A simple geometrical construct reveals that the resonances occur when the classical cyclotron orbit, $r_c = m^* v_f / eB = \hbar k_f / eB$, encircles a specific number of anti-dots (m^* is the effective mass, v_f is the Fermi velocity, $k_f = (2\pi n_e)^{1/2}$ is the Fermi wavevector, and n_e is the electron density). The lower (higher) magnetic field resonance peak corresponds to the resonance around four (one) anti-dots. These resonant peaks arise from "pinned" orbits [13] or, in a more sophisticated model, from the correlation of classical, chaotic trajectories [14].

The top traces shows R_{xx} through the anti-dot superlattice around $\nu = 1/2$ for the sake of comparison. The exact magnetic field position $B_{1/2}$ at half-filling was determined from nearby, well-established FQHE features around $\nu = 1/2$ and translated to B = 0. The peaks around $\nu = 1/2$ are the sought after dimensional resonances of composite fermions. To compare the electron and composite resonances, the magnetic field scale of the B = 0 trace was expanded by $\sqrt{2}$ and plotted as the middle trace. Comparing the dimensional resonances around B = 0 with those around $\nu = 1/2$ we observe excellent agreement between the primary resonant peaks of the electrons and the peaks in the respective traces at half-filling. This is compelling evidence that these magnetoresistance peaks around $\nu = 1/2$ are due to anti-dot dimensional resonances of composite fermions. We have observed similar scaling behavior of the anti-dot resonance peaks for anti-dot superlattices with different periods [15].

Although the peak positions of the fermion resonance are in good agreement with the expected value, the width tends to be much broader than the electron resonance and the higher order peaks such as the four anti-dot resonance are not observed. This is probably due to the difference in mean free path between electrons and fermions. Van der Pauw measurements on the unprocessed sample at $\nu = 1/2$ yield a resistivity of $350\Omega/\square$ equivalent to a mobility $\mu = 1.23\times10^5 \mathrm{cm}^2/\mathrm{V\,sec}$. The upper limit of the fermion mean free path, $l = \hbar\sqrt{2}k_f\mu/e$, is approximately 2μm. This contrasts sharply with the electronic mean free path of $\sim 50\mu$m at zero magnetic field. Thus, it is not surprising that the fermion features are so much broader than the electron features and that higher order peaks are absent.

We have extended the transport study of anti-dot superlattice to higher magnetic field to the regime around $\nu = 1/4$ [16]. Though strong resonances were observed around $\nu = 1/2$, no resonances were observed around $\nu = 1/4$. Since the resistivity

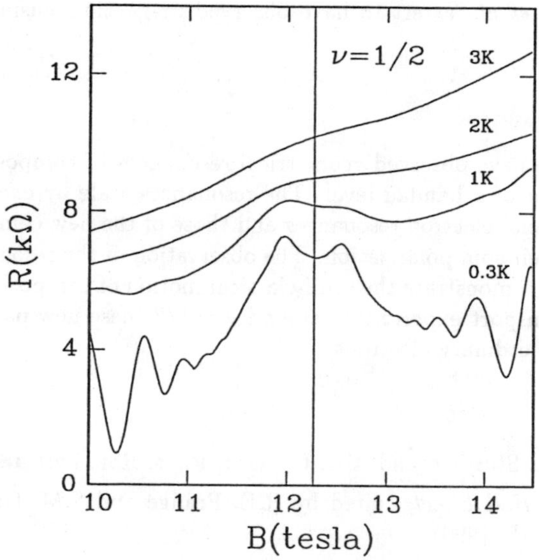

FIG. 2. Magnetoresistance R_{xx} of anti-dot superlattice around $\nu = 1/2$ for different temperatures.

at $\nu = 1/4$ is about six times larger than that at $\nu = 1/2$, the mean free path at $\nu = 1/4$ is expected to be six times smaller compared to $\nu = 1/2$. Consequently, observing similar resonance at $\nu = 1/4$ would require a superlattice period on the order of 100 nm. At present we are unable to fabricate such small anti-dot structures.

Fig. 2 shows the R_{xx} of anti-dot superlattice around $\nu = 1/2$ for different temperatures. Although strong geometrical resonances are observed at 0.3 K, the composite fermion resonances become much broader at 1K. Upon raising the temperature even higher, the resonance around $\nu = 1/2$ becomes barely observable. This behavior is in a sharp contrast to the electron resonances which are known to persist up to temperature as high as 40K [13]. One noticeable feature of the data is that the background resistivity at $\nu = 1/2$ increases with increasing temperature. This suggests that the mean free path of the composite fermions decreases with increasing temperature and that the broadening of the resonance simply reflects such reduction of the mean free path.

While our experiments were performed in a static geometry, related geometrical resonances in surface acoustic wave experiments have been observed by Willett *et al.*[17] and Goldman *et al.* assert to have observed magnetic focusing of composite fermions.[18].

4. Conclusion

In summary, we have observed geometrical resonances of composite fermions at $\nu = 1/2$ filling factor of a Landau level. The resonances scale by exactly a factor of $\sqrt{2}$ between traditional electron resonances and those of the new composite particles as expected from their spin polarization. The observation of the resonances and their appropriate scaling demonstrate the semi-classical motion of composite fermions and suggests that in transport experiments around $\nu = 1/2$ these new particles, in many aspects, behave like ordinary electrons.

References

[1] D.C. Tsui, H.L. Stormer, and A.C. Gossard, Phys. Rev. Lett **48**, 1559 (1982).

[2] *The Quantum Hall Effect,* edited by R.E. Prange and S.M. Girvin (Springer-Verlag, New York, 1990).

[3] T. Chakraborty and P. Pietilainen, *The Fractional Quantum Hall Effect,* (Springer-Verlag, New York, 1988).

[4] R.B. Laughlin, Phys. Rev. Lett **50**, 1395 (1983).

[5] F.D.M. Haldane, Phys. Rev. Lett. **51**, 605 (1983).

[6] B.I. Halperin, Phys. Rev. Lett. **52**, 1583 (1984); **52**, 2390(E) (1984).

[7] H.W. Jiang, H.L. Stormer, D.C. Tsui, L.N. Pfeiffer, and K.W. West, Phys. Rev. **B 40**, 12013 (1989).

[8] R.L. Willett, M.A. Paalanen, R.R. Ruel, K.W. West, L.N. Pfeiffer, and D.J. Bishop, Phys. Rev. Lett. **65**, 112 (1990).

[9] B.I. Halperin, P.A. Lee, and N. Read, Phys. Rev. **B 47**, 7312 (1993).

[10] J.K. Jain, Phys. Rev. Lett. **63**, 199 (1989); Phys. Rev. **B 40**, 8079 (1989); **41**, 7653 (1990).

[11] V. Kalmeyer and S.C. Zhang, Phys. Rev. **B 46**, 9889 (1992).

[12] R.R. Du, H.L. Stormer, D.C. Tsui, L.N. Pfeiffer, and K.W. West, Phys. Rev. Lett. **70**, 2944 (1993).

[13] D. Weiss, M.L. Roukes, A. Menschig, P. Grambow, K. von Klitzing, and G. Weimann, Phys. Rev. Lett. **66**, 2790 (1991).

[14] R. Fleischmann, T. Geisel, and R. Ketzmerick, Phys. Rev. Lett. **68**, 1367 (1992).

[15] W. Kang, H.L. Stormer, K.W. Baldwin, L.N. Pfeiffer, K.W. West, Phys. Rev. Lett. **71**, 3850 (1993).

[16] W. Kang, R.R. Du, A.S. Yeh, H.L. Stormer, L.N. Pfeiffer, K.W. Baldwin, and K.W. West, unpublished.

[17] R.L. Willett, R.R. Ruel, K.W. West, and L.N. Pfeiffer, Phys. Rev. Lett. **71**, 3846 (1993).

[18] V.J. Goldman, B. Su, and J.K. Jain, Phys. Rev. Lett. **72**, 2065 (1994).

EXPERIMENTAL DEMONSTRATION OF A FERMI SURFACE AT ONE-HALF FILLING FACTOR

R. L. WILLETT
AT&T Bell Laboratories, 600 Mountain Avenue, Murray Hill, New Jersey 07974, U.S.A.

ABSTRACT

Experiments using surface acoustic waves (SAW) to probe the 2D electron system conductivity over small dimensions are described. These experiments reveal two new length scales in the 2DES apparent near filling factor 1/2: the mean free path and Fermi wavevector of the gauge transformed fermion. These results provide definitive evidence for Fermi surface formation at half filling of the lowest Landau level.

The 2D electron system (2DES) in sufficiently low disorder GaAs/AlGaAs heterostructures is able to support electron correlation effects at low temperatures, and it is the objective of our experiments to understand these effects. Conventional d.c. transport measurements probe the 2DES over macroscopic dimensions, and as such have demonstrated electron correlations in the form of the fractional quantum Hall effect (FQHE)[1]. We describe here a series of experiments[2-4] using surface acoustic waves (SAW) that instead probe the 2DES conductivity over microscopic dimensions. By measuring the conductivity at microscopic dimensions we reveal conductivity effects more fundamental than the FQHE. Our results demonstrate two new length scales in the magnetic field spectrum of the 2DES; these length scales correspond to the mean free path and Fermi wavevector of new quasiparticles. By experimentally describing the quasiparticle behavior, the magnetic field spectrum of the 2DES can be understood, including the origin of the FQHE, through recent theoretical work.

Before examining the findings of the SAW measurements, the theory which provides a framework for their interpretation will be reviewed. In the theory of the half-filled Landau level[5], Halperin, Lee, and Read (HLR) have used the fermion Chern-Simons method to examine the properties of a single layer 2D electron system at $\nu = 1/2$ and at various other even denominator fractions. In this method a singular gauge transformation converts the electrons to a system of particles interacting with a "fictitious" or Chern-Simons gauge field. A flux tube containing an integer number n of quanta of Chern-Simons magnetic field is attached to each particle. Choosing $n=2$, at $\nu = 1/2$ in their construction the average Chern-Simons field just cancels the applied external magnetic field. The resultant Hartree ground state is a filled Fermi sea of particles, gauge transformed fermions, in zero magnetic field with a distinct Fermi wavevector $k_F = (4\pi n_e)^{1/2}$, with n_e the electron areal density. Away from 1/2 filling, the particles are expected to execute cyclotron motion with radius $R_c = \hbar k_F / e\Delta B$ in an effective magnetic field $\Delta B = B - B(\nu = 1/2)$.

The consequences of this Fermi surface are significant in that they may explain the major observed properties of the 2DES in the extreme quantum limit. For large effective B fields, the Shubnikov-deHaas oscillations of these gauge transformed fermions are

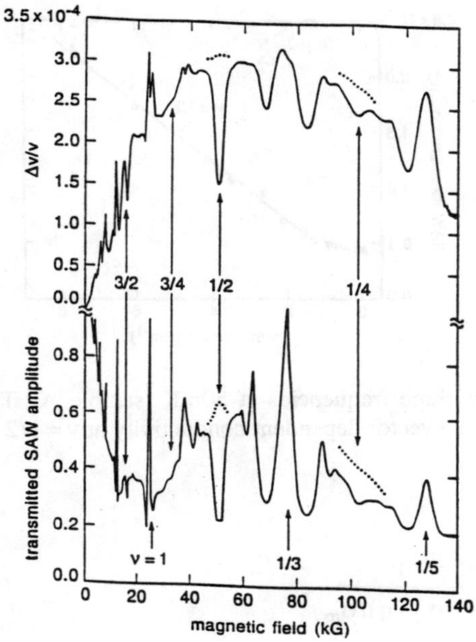

Fig. 1. SAW velocity shift and transmitted amplitude versus magnetic field at 3.4GHz, 120mK. The dotted lines represent $\Delta v/v$ and amplitude calculated from the DC conductivity (offset for clarity).

precisely the series of observed odd denominator FQHE states at $\nu = p/(2p+1)$, p an integer. In the vicinity of $\nu = 1/2$, where dc transport exhibits no quantum Hall effect, HLR provide explicit predictions for surface acoustic wave propagation results to be described below.

SAW Technique and Early Results

These experiments were performed on two sets of GaAs/AlGaAs heterostructures cut from two different wafers (A and B): electron mobility is $\sim 1-4 \times 10^6 \, cm^2/V-sec$ with areal densities $\sim 6-7 \times 10^{10} \, cm^{-2}$. Surface acoustic waves were generated and detected using lithographically defined interdigitated transducers evaporated onto the heterostructures. The 2DES was etched away in the vicinity of each transducer and they were placed roughly 1mm apart with a mesa containing the 2DES between. The fundamental spatial period of the transducer is the sound wavelength. Frequencies ranged from 0.1 to 8.5GHz, corresponding to wavelengths of approximately $28 \mu m$ to $0.3 \mu m$.

The SAW is launched from one transducer at frequency $f = v_s/\lambda = v_s q/2\pi$ where v_s is the velocity of sound, λ is the SAW wavelength, and q is the SAW wavevector. The sound wave traverses the 2DES where it is attenuated and slowed by the interaction of its piezoelectric field and the 2D electrons, according to a simple relaxation process. The sound velocity shift monotonically decreases with increasing wavevector dependent sheet conductivity, as follows:

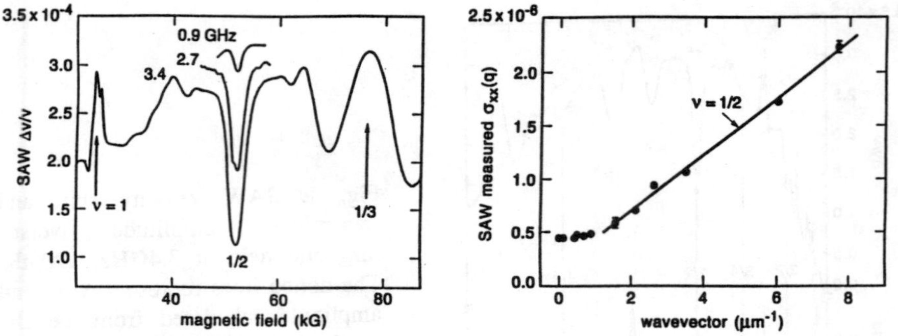

Fig. 2. Left panel: SAW velocity shift for three frequencies at 50mK, sample A. The traces are offset for clarity. Right panel: wavevector dependent conductivity at $\nu=1/2$ as derived from the SAW velocity shift.

$$\frac{\Delta v}{v} = \left(\frac{\alpha^2}{2}\right) \frac{1}{1+(\sigma_{xx}(q)/\sigma_m)^2} \tag{1}$$

where the piezoelectric coupling constant $(\alpha^2/2)=0.0032$ for GaAs, and $\sigma_m = v_s(\varepsilon+\varepsilon_o)$; see ref. 4. By measuring the sound velocity shift the wavevector dependent conductivity of the 2DES is measured.

Figure 1 shows the low-temperature transmitted SAW amplitude and sound velocity shift for sample A at 3.4 GHz as a function of magnetic field. The relaxation model was used to calculate the expected SAW values from the dc transport $\sigma_{xx}(q=0)$. Direct comparison can be made between the calculated and measured SAW parameters to look for discrepancies, with these differences demonstrating that $\sigma_{xx}(q=0) \neq \sigma_{xx}(q_{SAW})$. Little to no discrepancy is seen between the dc conductivity and the σ_{xx} at large q for FQHE states.

However, a remarkable difference between $\sigma_{xx}(q=0)$ and $\sigma_{xx}(q\neq 0)$ occurs at $\nu=1/2$, with a marked minimum in both amplitude and $\Delta v/v$. This feature at 1/2 represents additional or enhanced conductivity $\sigma_{xx}(q)$ for $q=2\pi/(0.8\mu m)$. The sheet conductivity is well parametrized by a single σ_{xx} rather than by invoking a complex conductivity. A distinct minimum at $\nu=1/4$ is also demonstrated in both amplitude and $\Delta v/v$, in addition to a small effect at $\nu=3/4$. At $\nu=3/2$ a deep minimum similar to the 1/2 result is present.

The wavevector dependence of the SAW response is examined in Fig. 2, with comparison of raw $\Delta v/v$ data for three SAW frequencies displayed in the left panel. The effect gets both broader and deeper with increasing wavevector. Precisely at $\nu=1/2$ measuring $\Delta v/v$ gives $\sigma_{xx}(q)$, and this can be plotted for the full range of q used on this sample (right panel). At small q, σ_{xx} is q independent as seen for the rest of the B-field

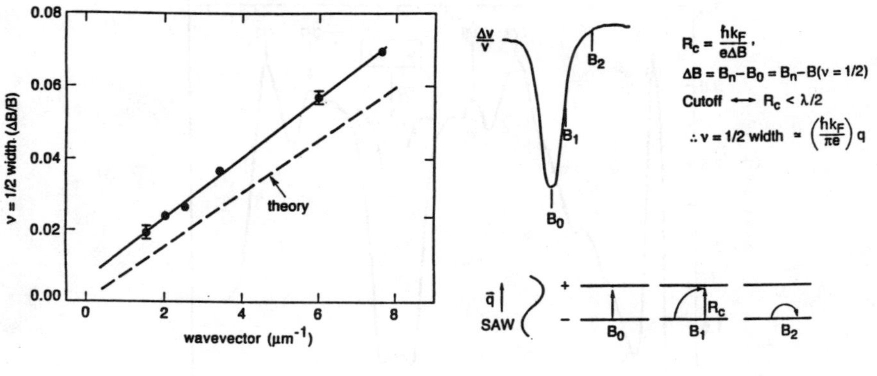

Fig. 3. Left panel: $\nu = 1/2$ width $(\Delta B/B)$ versus SAW wavevector at 50mK, sample A. The width is defined as the full B field extent at the half maximum. Right panel: map of $\Delta v/v$ near 1/2 and schematic of gauge transformed fermion cyclotron radius for ΔB.

range for all q. For $q > 2\mu m^{-1}$, $\sigma_{xx}(q)$ increases linearly with q. The width of the enhanced conductivity clearly increases with increased SAW q. Fig. 3 shows the full width in $\Delta B/B$ at half maximum of the effect at $\nu = 1/2$, demonstrating a distinct linear dependence in q.

The results described above are explicity predicted[5] for the Chern-Simons gauge transformed fermion at $\nu = 1/2$. Enhanced conductivity occurs at 1/2 because the quasiparticle will have a component of linear motion in the direction of the sound propagation. At 1/2 there is no magnetic field apparent to the gauge transformed fermion, and therefore no Lorentz force. If the quasiparticle does not scatter, then this motion in the SAW direction will short the SAW piezoelectric field and slow the wave, resulting in enhanced conductivity. For SAW wavelengths small compared to the quasiparticle mean free path ℓ $(q >> 2/\ell)$, the quasiparticle conductivity is proportional to q: $\sigma_{xx}(q) = (q/k_F)(e^2/8\pi\hbar)$. This behavior is qualitatively just what is seen in Fig. 2. The inflection in the data at $q \sim 2\mu m^{-1}$ demonstrates a transition from mean free path limited to q dependent conductivity, and therefore determines the quasiparticle mean free path to be $\sim 1\mu m$. The predicted linear dependence of the conductivity on q is confirmed, but a discrepancy between theory and experiment in overall conductivity scale of about a factor of two exists and is unexplained.

The width of the enhanced conductivity at 1/2 is also consistent with the picture of Fermi surface formation and provides a scaling of the gauge transformed fermion Fermi wavevector, k_F. As shown schematically in the right panel of Fig. 4 the quasiparticles will move in a cyclotron orbit of radius $R_c = \hbar k_F/e\Delta B$. We assume here that SAW $\lambda < \ell$. One expects that for small ΔB the quasiparticle will move in a sufficiently large orbit R_c that the particle will still short-out the SAW electric field, but less than that occuring at $\Delta B = 0$. However, as ΔB increases, the cyclotron radius is reduced to the point that

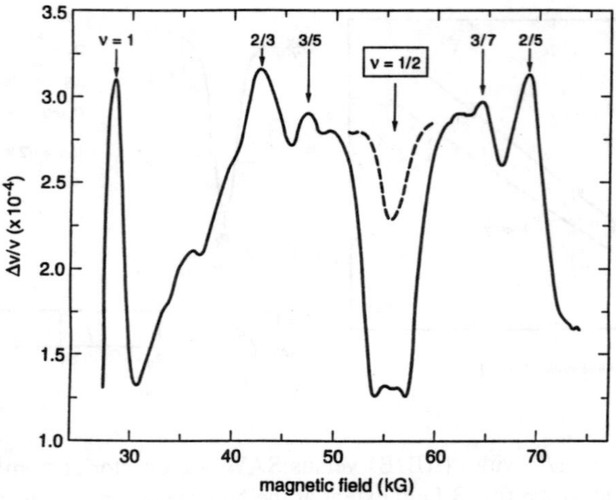

Fig. 4. Sound velocity shift versus magnetic field at T~400mK at 5.4GHz for sample B. The dashed line is a similar measurement at 2.4GHz.

movement of the quasiparticle is essentially transverse to the SAW propagation direction and as such the quasiparticle motion will not enhance the $\sigma_{xx}(q)$. Therefore, the width of the effect at 1/2 provides a scale of the gauge transformed fermion Fermi wavevector. Using this mechanism it is expected that the width of the effect will be proportional to the SAW wavevector, $\Delta B = \hbar k_F q/e\pi$, where the enhanced conductivity is halved for $R_c = \lambda/2$. This line is shown in the figure. Note that the data form a line parallel to this theoretical construct that extrapolates back to a zero wavevector width of about 0.01. This represents the density inhomogeneity of the sample over the sound path. This inhomogeneity value will be used later.

Geometric Resonances

While the early experimental results described above show consistency with the field theoretic construction elucidated by HLR, the expected conductivity as a function of magnetic field is actually more complicated than that described above. This is due to the commensurability of the cyclotron orbit with a wavelength, or wavelengths, of the surface acoustic wave. It is this commensurability that allows us to precisely measure the gauge transformed fermion k_F, and therefore demonstrate the presence of a Fermi surface.

In a semiclassical sense, if the quasiparticle can traverse an entire cyclotron orbit without scattering (e.g., $2\pi R_c < \ell$) and if $\lambda_{SAW} \sim 2R_c$, then a geometric resonance of the quasiparticle motion and the sound wave can occur. This principle resonance manifests itself as a peak in the conductivity (minimum in the sound velocity) versus magnetic field

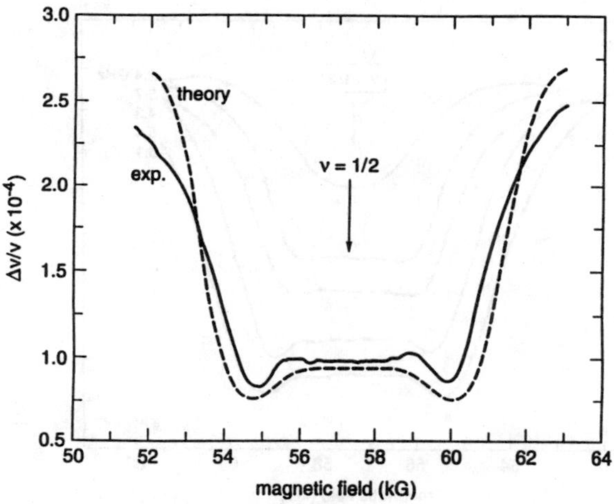

Fig. 5. Sound velocity shift versus magnetic field for SAW frequency 8.5GHz at T~200mK, sample B. The dashed line is the theoretical curve described in the text including sample density inhomogeneity.

as the orbit is commensurate with the dimension of the wavelength. As the λ_{SAW} is made smaller, higher order resonances can occur if the mean free path is sufficiently large.

While we have described the general concept of the geometric resonance above the actual conductivity may be approximated (see ref. 5) for appropriate ℓ as

$$\sigma_{xx}(q) = \frac{\rho_{yy}(q)}{\rho_{xy}^2}, \; \rho_{yy}(q) \approx \frac{1}{\tilde{\sigma}_{yy}} \text{ near } 1/2 \tag{2}$$

and

$$\tilde{\sigma}_{yy}(q) = \frac{2}{\rho_o} \sum_{n=-\infty}^{\infty} \frac{[dJ_n(X)/dX]^2}{1+n^2(q\ell/X)^2}; \tag{3}$$

$J_n(X)$ is the Bessel function and $X = qR_c$. In the limit $q\ell >> 1$, the semiclassical approximation predicts oscillations in $\sigma_{xx}(q)$, $(\Delta v/v)$, with maxima occurring close to the zeroes of $J_1(qR_c)$, or at $\Delta B/B = q/K_{1,n}k_F$ where $K_{1,n}$ is the position of the nth zero of J_1; $K_{1,1}$=3.83. Therefore the peak position of the principal resonance in B as a function of SAW q gives a measure of the gauge transformed fermion k_F.

A large product of mean free path and SAW wavevector is needed to generate a large commensurability effect. Therefore, in order to observe these geometric resonances it was necessary to substantially increase the SAW q, since the mean free path of the samples cannot be increased significantly. Figure 4 demonstrates the sound velocity shift versus magnetic field for larger SAW q in sample B. New structure is present in the $\Delta v/v$ minimum at $\nu = 1/2$: these are the geometric reasonances that demonstrate the presence of a Fermi surface. Lower q SAW have shown only a simple minimum in the

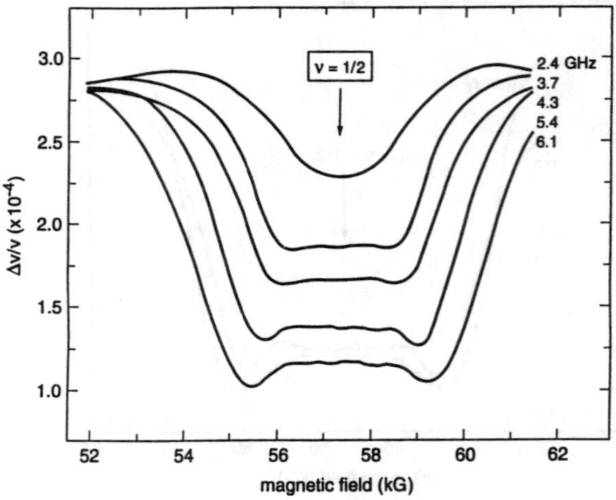

Fig. 6. Development of the gauge transformed fermion geometric resonance with increasing SAW frequency in sound velocity shift versus magnetic field, T~200mK.

sound velocity at 1/2, as demonstrated by the dashed trace in the figure. For the higher frequencies featured here, one clearly sees local minima, or resonances, in the sound velocity on each side of 1/2. The two minima are symmetrically spaced about $\nu=1/2$ and correspond to commmensurability of the sound wave and motion of charge carriers.

Figure 5 shows a comparison of the theoretical prediction for the sound velocity shift near 1/2 and that measured at 8.5GHz. The parameters for this theory curve are extracted from the data as follows. We use the dc resistivity value at $\nu=1/2$ to give an approximation to the gauge transformed fermion mean free path $\ell \sim 0.4\mu m$, resulting in $q\ell \sim 8$. This $q\ell$ can be used in eqs. 2 and 3 to produce theoretical $\Delta v/v$ versus B. Note that we adjust the minimum and background theoretical $\Delta v/v$ to match that of the data: the relative magnitude of the resonances is set by $q\ell$. To compare this to the data we must impose the density inhomogeneity which we know to be present in the samples. This inhomogeneity was measured above in Fig. 3; it is the offset from the origin at $q=0$ of the width versus q plot for this sample. A value of about 1.5% was derived from data for this sample, which is independently observed in these heterostructures in d.c. transport. We convoluted the original $\Delta v/v$ with a Gaussian distribution of densities whose FWHM is 1.5%. The resultant theoretical curve is shown in the Fig. 5 to closely mimic the actual measured sound velocity.

Figure 6 shows how these resonances develop as the SAW q increases. In general, structure becomes more prominent as the SAW frequency (wavevector) increases. At low frequency (2.4GHz) no resonances are apparent in $\Delta v/v$. As SAW q is increased the side minima appear. Using B=0 of the cyclotron orbit at $\nu=1/2$ is dramatically supported by plotting the resonance positions in $\Delta B=B-B(1/2)$ versus SAW q, as shown in Fig. 7. Note that the data are consistent with a line extrapolating through the

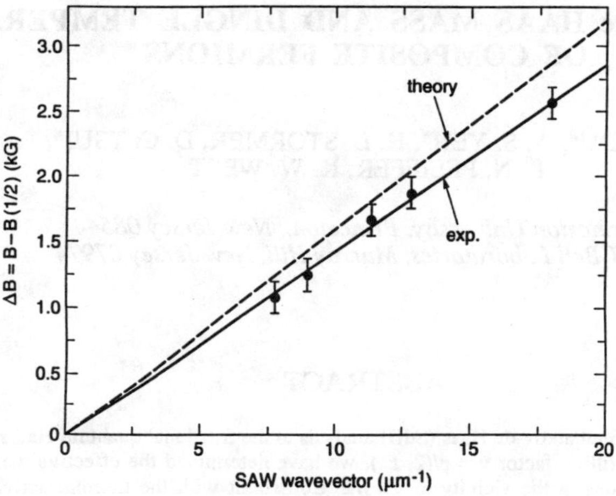

Fig. 7. Principal resonance position in ΔB vs. SAW wavevector, with theoretical $k_F = (4\pi n_e)^{1/2} = 93 \mu m^{-1}$ for the gauge transformed fermion (dashed line).

origin, as expected for $\Delta B = 0$ at $\nu = 1/2$. From the resonance positions the gauge transformed fermion Fermi wavevector is extracted. Using $k_F = q/(3.83 \Delta B/B)$ for the data and $k_F = (4\pi n_e)^{1/2}$ for the theoretical line, the figure shows the data to be just below the theoretical prediction, almost within experimental error.

In conclusion, our early results demonstrating enhanced conductivity at 1/2 provided the first indications of novel length scale dependent behavior in the correlated 2DES: these results revealed the gauge transformed fermion mean free path. Our higher q SAW results showed distinct resonances in the sheet conductivity near 1/2: these findings allowed measurement of the gauge transformed fermion Fermi wavevector, k_F. These resonances emphatically demonstrate the presence of a Fermi surface at one-half filling.

It is my pleasure to acknowledge my collaborators M.A. Paalanen and L.N. Pfeiffer at A.T.&T. Bell Laboratories.

References

1. D.C. Tsui, H.L. Stormer, and A.C. Gossard, Phys.Rev.Lett. 48,1559(1982).
2. R.L. Willett, M.A. Paalanen, R.R. Ruel, K.W. West, L.N. Pfeiffer, Phys.Rev.Lett. 65, 112(1990).
3. R.L. Willett, R.R. Ruel, M.A. Paalanen, K.W. West, L.N. Pfeiffer, Phys.Rev.B 47, 7344 (1993).
4. R.L. Willett, R.R. Ruel, K.W. West, L.N. Pfeiffer, Phys.Rev.Lett. 71, 3846 (1993).
5. B.I. Halperin, P.A. Lee, and N. Read, Phys.Rev.B 47, 7312 (1993).

SHUBNIKOV-de HAAS MASS AND DINGLE TEMPERATURE OF COMPOSITE FERMIONS

R. R. DU#, A. S. YEH#, H. L. STORMER, D. C. TSUI#,
L. N. PFEIFFER, K. W. WEST

 # Princeton University, Princeton,, New Jersey 08544
AT&T Bell Laboratories, Murray Hill, New Jersey 07974

ABSTRACT

Applying a Shubnikov-de Haas (SdH) analysis to the fractional quantum Hall series at Landau level filling factor $\nu = p/(2p\pm1)$, we have determined the effective mass of composite fermions in the vicinity of $\nu = 1/2$. Consistent with the thermal activation results, the SdH masses of the new particles exceed the electron mass in GaAs by about a factor of 10. An enhancement of masses is observed as the filling factor approaches $\nu = 1/2$ from both sides, indicating that the state at exactly half-filling is not an ordinary fermi liquid.

1. Introduction

The fractional quantum Hall effect (FQHE) [1] is well understood as the condensation of two-dimensional electrons in high magnetic field into a sequence of quantum liquids. Traditionally, the relationship between the different liquids was viewed as a hierarchy in which the higher-order fractional states develop from the condensation of the elementary excitation of the next lower-order fractional states [2]. Recent theory [3, 4] has shed a new light on these relationships and provided us with a new framework for the FQHE. The central ingredients of this new picture are new quasiparticles, often termed composite fermions. Such particles result from the highly correlated motion of the carriers which effectively attaches flux quanta to each electron. The state at exactly $\nu = 1/2$ can be regarded as a degenerate system of fermions, in the absence of magnetic field, populating a fermi sea up to the fermi wavevector k_F. The FQHE sequence $\nu = p/(2p\pm1)$ is viewed then as the quantization of composite fermions in the presence of B_{eff}, the deviation of magnetic field from $\nu = 1/2$. The validity of this view has recently been supported by activation energy measurements [5] and Shubnikov-de Haas (SdH) oscillation analyses [6,7].

The conceptual background for the gap energy measurements is the traditional quantum liquid picture in which thermally activated fractionally-charged quasiparticles are generated across an energy gap. The energy gap Δ can be deduced from the exponential temperature dependence of the magneto-resistivity $\rho_{xx} \propto \exp(-\Delta / 2kT)$. Our data have established that the excitation gap of the quantum liquid can be interpreted as the cyclotron energies, $\hbar\omega_c = \hbar e B_{eff} / m^*$, of the composite fermions [5]. The energy gap can thus be associated with an effective mass, m*. The existence of a well defined effective mass, originating purely from electron-electron interactions, is a remarkable property of the composite fermion system.

By employing the standard SdH formalism to the oscillations around ν = 1/2, one implicitly invokes quite a different picture. It implies a fermi-sea with a fixed fermi-energy E_f and an initially smooth density of states (DOS). The application of a magnetic field introduces periodic oscillations in the DOS whose amplitude at E_f is probed by magneto-transport.

Our interpretation of the magneto-transport data around ν =1/2 in terms of a standard SdH effect yields a carrier mass and a Dingle temperature [6], consistent with the activation energy results. Moreover, the SdH analyses yield effective mass at filling factor close to ν =1/2, where the system is not accessible to the activation measurements. An interesting and puzzling feature of our SdH analyses is the evidence of mass enhancement of composite fermions as the filling factor approaches 1/2. In fact, there exist theoretical considerations which imply a divergence of the composite fermion mass as $\nu \to 1/2$ as a result of gauge fluctuations [4].

2. Experiment

We have performed magneto-transport measurements on a series of GaAs-AlGaAs heterostructures. The electron density ranges 1.1 to 2.3 x 10^{11} cm^{-2} and mobility 6.8 to 12.8 x 10^6 cm^2 / Vsec. For all these samples the distance between electrons and the Si-doping layer is d_s = 800 Å. The sample is a 5mm x 5mm square with eight Indium contacts. The electron density and mobility were established after a brief illumination of light-emitting diode at low temperature. Standard low-frequency (3-11 Hz) lock-in technique was employed. The measurements were performed using a dilution refrigerator and a combination of superconducting and Bitter magnets up to 30 T.

3. Results and Discussions

For a sample of electron density n = 1.1 x 10^{11} cm^{-2}, the gap energies as determined by the thermal activation measurement are plotted versus the applied magnetic field, in Fig. 1. The simplest relationship consistent with the data were two linear dependence on B_{eff} with a common negative intercept at ν = 1/2. One can characterize our findings in very simple terms once one establishes $B_{1/2}$ at ν = 1/2 as a new origin for B and adopts its deviation B_{eff} from this value as the *effective magnetic field*. Just like in a simple fermion system of density n, the minima in ρ_{xx} due to the primary FQHE sequences occur at $B_{eff}(p) = B_{p/(2p\pm1)} - B_{1/2} = \pm nh/ep$. Since the gap energies are linear in B_{eff}, they can be characterized by an effective

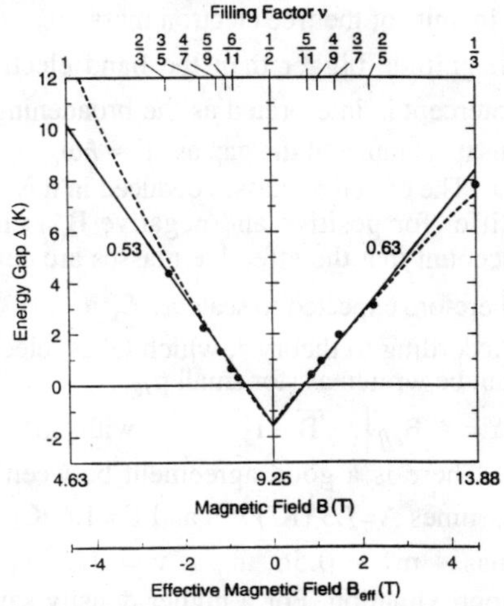

Fig.1 Gap Energies for varies filling factors in the vicinity of ν =1/2 plotted vs magnetic field. The solid lines are a linear approximation of the gaps, and the deduced masses are indicated. The dashed line takes the $\sqrt{B}$ dependence of the masses into account.

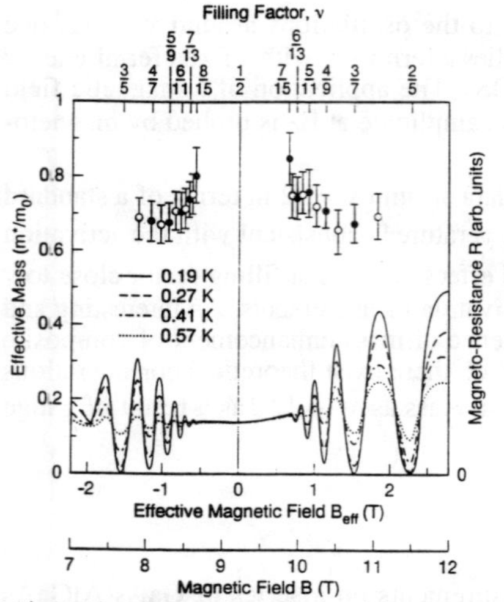

Fig. 2 Temperature dependence of the magneto-resistance, R (bottom), and deduced effective masses, m* (in unites of the electron mass m_o, top) as a function of magnetic field, B, and effective magnetic field, B_{eff}, around $\nu = 1/2$ Landau level filling factor.

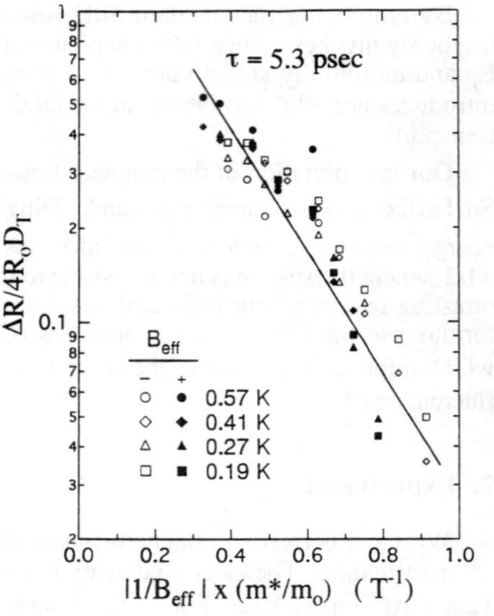

Fig. 3 Dingle plot of magneto-resistance data.

mass m* via $\hbar\omega_c = \hbar e B_{eff} / m^*$, assuming e to be the electron charge. The values for m* (in units of the free electron mass, m_o) are indicated in Fig. 1. They are about one order of magnitude bigger than the band electron mass $m_b \approx 0.07\ m_o$ of GaAs. The negative intercept is interpreted as the broadening Γ of the fermion "Landau levels" which causes disappearance of the gap as $\Gamma \approx \hbar\omega_c$.

The effective masses deduced in this simple manner from the linear dependence on $|B_{eff}|$ differ for positive and negative B_{eff}; in a more accurate approach one must take into account that the effective masses are determined by electron-electron interaction and are therefore expected to scale as ℓ_o^{-1} ($\ell_o = \sqrt{\hbar / eB}$ is the magnetic length) and hence as $\sqrt{B}$. According to theory [4], which takes electron-hole symmetry into account, the gap energy can be written as (for small p)

$$\Delta = A\left|B_{eff}\right| / \sqrt{B} - \Gamma \qquad \text{with } m^* = (\hbar e / A)\sqrt{B}$$

There is a good agreement between the experimental data and this equation if one assumes A=7.3 ($K/T^{1/2}$) and Γ =1.6 K (dashed line in Fig. 1). This results in an effective mass $m^* \approx 0.56\ m_o$ at $\nu = 1/2$, about the average value deduced from linear approximation. For a higher density sample ($n = 2.3 \times 10^{11} cm^{-2}$, see Ref. 5), a similar agreement can be found assuming the same value of A and Γ = 2.2 K.

The temperature dependence of the amplitude ΔR (Fig. 2) follows extraordinarily well the traditional sinh-dependence of the SdH formalism usually denoted as :

$$\Delta R / 4R_0 = A_T / \sinh A_T \times \exp(-\pi / \omega_c \tau), \qquad D_T = A_T / \sinh A_T$$

with $A_T = 2\pi^2 kT / \hbar\omega_c$,

once the magnetic field is replaced by B_{eff}. The scattering rate $1/\tau$ is often expressed in terms of the Dingle temperature, usually defined as $kT_D = \hbar / \tau\pi$.

The effective mass values and the uncertainties deduced form our data are shown in the upper part of Fig. 2. Over the range of magnetic fields we can explore, and to within the accuracy of our data, the effective mass is ~ 0.7 m_o with a slight upturn toward $\nu = 1/2$. In a higher density sample ($n = 2.25 \times 10^{11}$ cm $^{-2}$), we observe a mass enhancement by as much as 40% as the filling factor approaches $\nu = 1/2$, with a tendency to diverge further in closer proximity.

A modified Dingle plot is shown in Fig. 3, in which the mass variation is taken into account by modifying the horizontal axis from $1/|B_{eff}|$ to $1/|B_{eff}| \times (m^*/m_o)$ and including the varying mass into D_T on the vertical axis. The plot reflects a scattering time of $\tau \approx 5.3$ psec equivalent to a broadening $\Gamma = \hbar / \tau$ of 1.4 K. This value is in good agreement with the value of Γ derived in Fig. 1.

4. Conclusions

We have determined the effective mass and the Dingle temperature of the composite fermions near $\nu = 1/2$. The sum of our findings reemphasizes the internal consistency of the SdH formalism when applied to the FQHE data.

5. Acknowledgments

We thank P. Coleridge for useful conversations regarding SdH analyses in electron system. The experiments were performed at the Francis Bitter National Magnet Lab, Cambridge, USA. RRD, ASY, and DCT are supported by the AFOSR and a grant from NEC Corporation.

6. References

1. D. C. Tsui, H. L. Stormer, and A. C. Gossard, *Phys. Rev. Lett.* **48** (1982)1559.
2. *The Quantum Hall Effect*, edited by R. E. Prange and S. M. Girvin (Springer-Verlag, New York, 1990).
3. J. K. Jain, *Phys. Rev. Lett.* **63** (1989) 199; *Phys. Rev.* **B40** (1989) 8079; **B41** (1990) 7653.
4. B. I. Halperin, P. A. Lee, and N. Read, *Phys. Rev.* **B47** (1993) 7312.
5. R. R. Du, H. L. Stormer, D. C. Tsui, L. N. Pfeiffer, and K. W. West, *Phys. Rev. Lett.* **70** (1993) 2944.
6. R. R. Du, H. L. Stormer, D. C. Tsui, L. N. Pfeiffer, and K. W. West, *Solid State Communication* **90** (1994) 71.
7. D. R. Leadley, R. J. Nicholas, C. T. Foxon, and J. J. Harris, *Phys. Rev. Lett.* **72** (1994) 1906.

THE QUANTUM HALL EFFECT OF COMPOSITE FERMIONS

*D.R. Leadley, *R.J. Nicholas, *P.J. Gee, *J. Singleton, #J.J. Harris and +C.T. Foxon
*Department of Physics, Clarendon Laboratory, Parks Road, Oxford, OX1 3PU, UK
#Department of Electronics and Electrical Engineering, University College London. UK
+Department of Physics, The University, Nottingham NG7 2RD, UK

The Fractional Quantum Hall Effect is described as Shubnikov-de Haas oscillations of the composite Fermions (CF) formed by a flux attachment transformation. We find a CF effective mass of $0.5m_e$ at zero effective field which increases weakly with effective magnetic field at a rate independent of carrier density. By applying successive flux attachments we may describe all the observed features in ρ_{xx}.

The Fractional Quantum Hall Effect (FQHE) has been observed in the resistivity of high mobility heterojunctions for many years[1,2] but a full description has remained elusive due to the complexity of the electron-electron interactions. Now, by using weakly interacting composite Fermions (CF)[3,4] in an effective magnetic field B^*, it is possible to provide a complete and consistent picture. We present experimental data which show the similarity between the Integer Quantum Hall Effect (IQHE) and each series of FQHE features, and measure the effective mass of the CF particles.

Consider first the IQHE of bare electrons which forms the gross structure of the resistivity. Quantized plateaux appear in ρ_{xy} whenever the Fermi energy lies outside regions of extended states, i.e. all the current carrying states are filled. In the CF picture we consider a new basis of current carrying states by making a gauge transformation, equivalent to attaching 2 flux quanta of $\Phi_0=h/e$ to each electron. At $\nu=1/2$ this results in a metallic system of independent CFs at zero effective field,[4] replacing the system of interacting electrons in high field. Increasing (or decreasing) the external field from $\nu=1/2$ quantises the CF energies into Landau levels (LL) in the effective field $B^* = B - 2n_e\Phi_0$, gaps open in the density of states and hence plateaux appear in ρ_{xy}. For CFs the filling factor ν^* is simply related to ν for the electrons by $\nu = \nu^*/(1+2\nu^*)$, which for

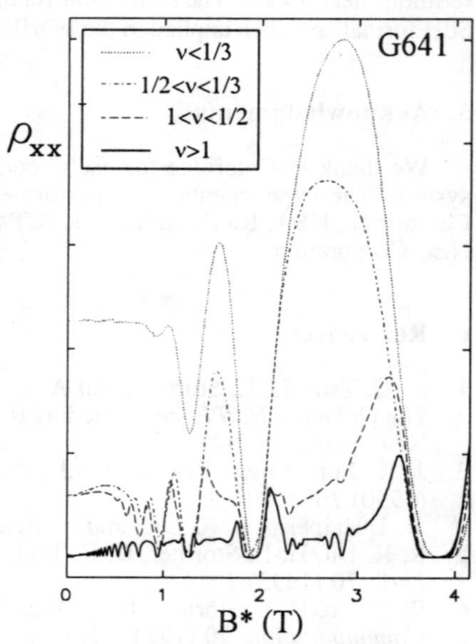

Figure 1. ρ_{xx} at 90mK plotted as a function of effective field, showing how the FQHE may be regarded as SdHO of composite Fermions.

Table 1 Sample data after illumination at 300mK

Sample	Spacer Width Å	n_e $10^{15}m^{-2}$	Mobility m^2/Vs
G650	400	2.3	1100
G640	1200	1.2	680
G641	1600	0.9	400
G646	2400	0.6	200

positive (negative) integral ν^* generates plateaux at $\nu = 1/3, 2/5, 3/7...$ $(1, 2/3, 3/5...)$. These principal FQHE features *look* identical to the IQHE and indeed *are* the IQHE of first generation (m=1) CFs.

Even if the LLs are not resolved sufficiently to produce plateaux in ρ_{xy}, Shubnikov-de Haas oscillations (SdHO) are seen in ρ_{xx}. In Fig. 1 we have divided a single recording of ρ_{xx} for sample G641 into the four regions: $\nu>1$; $1>\nu>1/2$; $1/2>\nu>1/3$ and $\nu<1/3$, plotting each in terms of effective field with the origin at B=0; $\nu=1/2$ (twice) and $\nu=1/4$ respectively. For each region an exponentially damped harmonic series can be seen, typical of SdHO. From the temperature and magnetic field variation of these oscillations values of effective mass m^* and scattering time τ_q may be deduced for the current carrying particles in each series.[5] We use the conventional Ando formula[6]:

$$\Delta\rho_{xx} \propto \frac{X}{\sinh X}\exp\left(-\frac{\pi}{\omega_c\tau_q}\right)\cos 2\pi(\nu-\tfrac{1}{2}) \tag{1}$$

where $X=2\pi^2 k_B T/\hbar\omega_c$ and $\omega_c=eB/m^*$ is the cyclotron frequency. This is the first term in a harmonic series, with higher terms that account for the increased peak sharpening at low temperatures.[7] For CFs B is replaced by B^*, ν by ν^*, m^* by M^* and τ_q by T_q.

The measurements were performed on lithographed Hall bars of high quality GaAs-GaAlAs heterojunctions grown by MBE at Philips Research Laboratories[8], with the characteristics after photoexcitation given in Table 1. The resistivity components ρ_{xx} and ρ_{xy} were measured between 40mK and 1.6K in magnetic fields up to 15T. Eq. (1) was found to give a very good description of the temperature and field dependence of the oscillation amplitudes, providing $\Delta\rho_{xx}/\rho_{xx}<0.5$. All the oscillations at small B^* are in this range where they are predominately of cosine form, but the analysis was restricted to higher temperatures for peaks at larger B^* (e.g. 1/3). A similar analysis of the single particle SdHO at low fields using this condition gives $m^*=0.070m_e$. The amplitude of the oscillations in ρ_{xx} were fitted to Eq. (1) for both the maxima and minima between $\nu=1$ and 1/3 (i.e. integral and half integral filling of the m=1 CF LLs) and values of the CF effective mass M^* deduced for each sample. These are shown in Fig. 2, plotted as a function of the effective field, for both positive (B+) and negative (B-) values of ν^*. The remarkable result is that the composite Fermions display a consistent value of M^*, dependent only on the single parameter B^*, and independent of both the sign of this field and n_e (which determines the total B). Whilst the error associated with an individual data point is up to $\pm 0.05m_e$ the overall trend from all the samples is very clear. The field dependence of the mass only becomes significant for features at high effective fields, where it can be described by:

$$M^* = 0.51 + 0.074B^* \tag{2}$$

in units of m_e with B^* in tesla. Using these mass values we calculate an effective cyclotron gap $E_{CR}^* = eB^*/M^*$ as shown in Fig. 3. The increase of mass with B^* now translates to a sub-linear increase of E_{CR}^* with field. This energy gap is equivalent to what was previously described as the FQHE quasi-particle gap, and studied by activation measurements where again a sub-linear increase with field is found. However, by showing

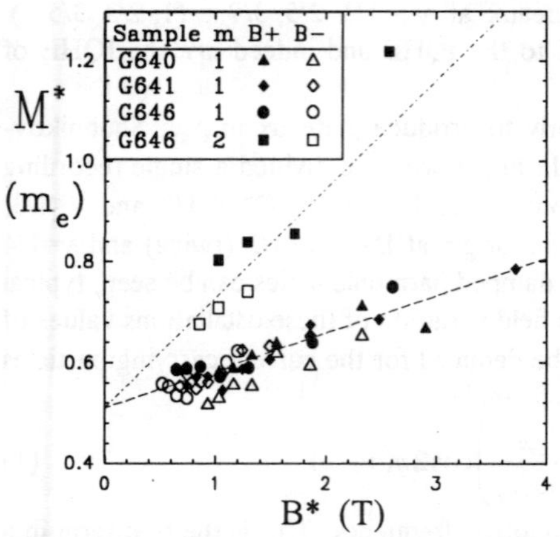

Figure 2 Effective mass of first & second generation CF. Line through points as Eq. (2).

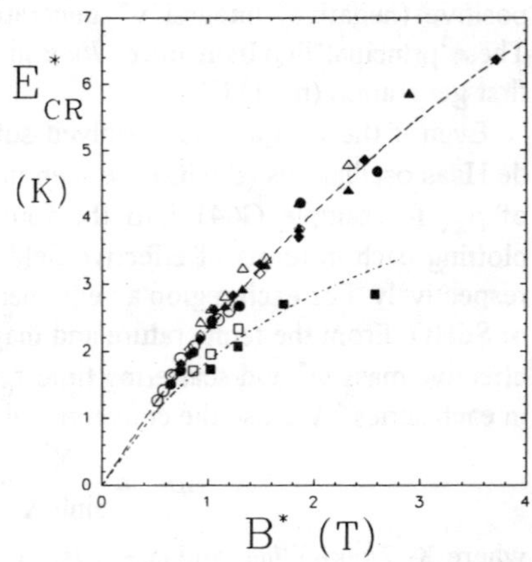

Figure 3. The cyclotron energy gap of the composite Fermion LLs. Key as Fig. 2.

the data as a function of effective field values from a large number of fractions and several samples can be included simultaneously on a single smooth curve. Furthermore, values deduced from oscillation maxima give similar results to data from minima, confirming that the CF picture provides a consistent description for all filling factors. By contrast only FQHE minima are well described in the hierarchical model.

By also fitting the field dependence of the oscillation amplitudes the CF lifetime T_q, can be evaluated. We find[5] that for each sample T_q is the same for both directions of B^* and for the m=1 CFs is also identical to τ_q deduced for the (m=0) electrons at low field. These findings support the notion of new particles in an effective field, which may be either parallel or anti-parallel to the external field, and show that in high fields where short range scattering is dominant there is no additional scattering of the CFs.

A similar series of oscillations occurs around the metallic state at $\nu=1/4$ where a second generation of CFs (m=2) have been formed by attaching a second pair of flux quanta to each m=1 CF, i.e. now electrons with 4 (=2m) flux quanta. In this case the fits to Eq. (1) are not quite as good due to the strong underlying increase in resistivity in the region $\nu<1/3$. However, the masses shown in Fig. 2 demonstrate a much larger dependence of M^* on B^* for the m=2 generation. Indeed at the 1/5 minimum ($\nu^*=1$) the value of $M^*=1.2m_e$ is exceptionally large. The data are also shown as $E_{CR}{}^*$ in Fig. 3 where the energy gap is seen to saturate more quickly with B^* than for the m=1 data.

Now consider higher LLs. At $\nu=N+1/2$, N LL are fully occupied with electrons and one is half occupied. Again a metallic state of first generation CF states can be formed by attaching 2 flux quanta to those electrons in the partially filled LL, while those in the full levels do not contribute to the conductivity and may still be considered as bare electrons. As the external field is changed so does the LL degeneracy. Accounting for the number of particles remaining in the partially filled level, the electron and CF filling factors are related by $\nu=N+\nu^*/(1+2\nu^*)$ as expected, but now the effective field is

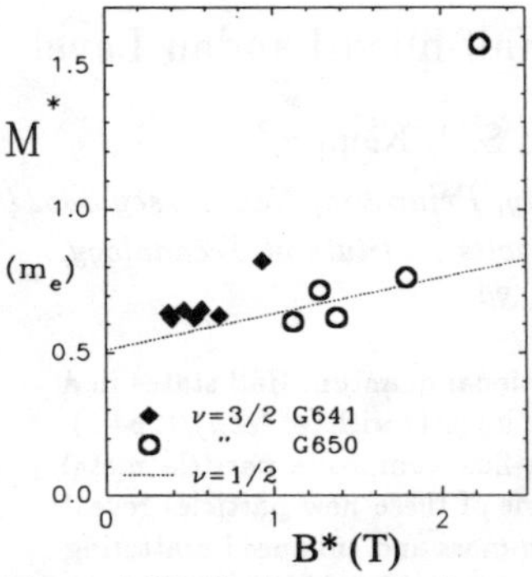

Figure 4 CF effective mass in the region of ν=3/2. Line shows ν=1/2 data from Fig. 2. The deviation at ν=4/3 is due to mixed spin states.

$B^* = (2N+1)B - 2n_e\Phi_0$, much larger than simply the difference from the field of the even fraction. We have examined ρ_{xx} in the region of ν=3/2 using the above effective field in Eq. (1). We find M^* has the same value and dependence on the appropriate B^* as for the oscillations around ν=1/2 (Fig. 4). If this larger value of B^* is not used a consistent set of values for M^* cannot be found by fitting to Eq. (1). The result demonstrates that the current carrying particles around ν=3/2 are again first generation CFs. It also provides strong support for the CF model and the fractal picture of the QHE.

Whenever the filling factor for the m=1 CFs is fractional (e.g. ν<1/3, 1>ν>2/3, etc.) a second flux attachment transformation may be applied to each m=1 CF in the partially filled level, producing m=2 CFs as the current carrying states and full m=1 levels that do not contribute to the conductivity. Thus metallic states can be formed at ν=1/4, 3/4 etc. and more FQHE features appear as the IQHE of the second generation CFs. Since the QHE is fractal this process can be repeated *ad infinitum* producing series of oscillations centred on every even denominator fraction. This can be seen by comparing the regions around ν=3/2 and 3/4 at B^*~2.7T in the two lowest traces of Fig. 1. However, as each higher generation requires the attachment of more pairs of flux quanta the oscillations become weaker. Our measurements for m=2 CFs around ν=1/4 and 3/4 suggest this is due to the higher mass of the resulting particles.

So far we have only considered spinless particles. While this is a good approximation at high magnetic fields, spin must be included for a complete description when the Zeeman splitting is small. Fig. 4 shows that, in both samples, M^* is exceptionally large for the mixed spin state at ν=4/3 (B^*=0.95T and 2.2T) corresponding to a reduced energy gap.

In conclusion we have shown that the FQHE can be analysed accurately in terms of the Shubnikov-de Haas oscillations of a series of renormalised composite Fermion states. Their effective masses are weakly dependent only on the single parameter of effective field B^*. The conductivity can now be explained in terms of the density of states and scattering properties of composite Fermions.[9]. Only the physics of the IQHE is required and it is not necessary to invoke fractional charge and fractional statistics to describes the quasi-particles used in the hierarchical model of the FQHE.

[1] D.C. Tsui, H.L. Stormer and A.C. Gossard, *Phys. Rev. Lett.* **48**, 1559 (1982)

[2] R.L. Willett *et al.*, *Phys. Rev. Lett.* **59**, 1776 (1987); J.R. Mallett *et al.*, *Phys. Rev. B* **38**, 2200 (1988)

[3] J.K. Jain, *Adv. Phys.* **41**, 105 (1992)

[4] B.I. Halperin, P.A. Lee and N. Reed, *Phys. Rev. B* 47, 7312 (1993)

[5] D.R. Leadley, R.J. Nicholas, C.T. Foxon and J.J. Harris, *Phys. Rev. Lett.* **72** 1906 (1994)

[6] T. Ando, *J. Phys. Soc. Jpn.* **37**, 1233 (1974)

[7] Including higher harmonics does not significantly change Eq. 2 L. Smrcka, *private communication.*

[8] C.T. Foxon *et al.*, Semicond. Sci. Tech. **4**, 582 (1989)

[9] D.R. Leadley *et al.*, Proc ICPS-22, Vancouver (1994)

Hole-flux Composite Particles around the Half-filled Landau Level

H. C. Manoharan,[1,2] M. Shayegan,[1,2] and S. J. Klepper[2]

[1]*Department of Electrical Engineering, Princeton University, Princeton, New Jersey 08544*
[2]*Francis Bitter National Magnet Laboratory, Massachusetts Institute of Technology, Cambridge, Massachusetts 02139*

Quasiparticle excitation gaps of the $\nu = p/(2p+1)$ fractional quantum Hall states in a low-disorder two-dimensional hole system are observed to scale linearly with $(e^2/\epsilon\ell_B)/(2p+1)$, predicted by the Chern-Simons gauge field theory if a hole-flux composite particle metal forms around $\nu = \frac{1}{2}$. The effective mass and quantum lifetime of these new particles reveal a remarkable renormalization structure, including a diverging mass and enhanced scattering close to the Fermi surface, suggesting a novel class of *marginal* Fermi liquid behavior.

Recent theories[1,2] have exploited a singular gauge transformation that maps the fractional quantum Hall effect (FQHE) at Landau level (LL) filling factor $\nu = p/(2mp+1)$ onto the integral quantum Hall effect (IQHE) at $\nu^* = |p|$ (p integer), where ν^* is the quasi-LL filling factor of particle-flux "composite fermions" (CFs). These unusual particles experience an effective magnetic field $B^* = B - 2mn\Phi_0 = B - B_{\nu=1/2m}$ (n is the carrier density) and therefore experience zero net magnetic field at exactly half filling. Here they are expected to form a Fermi surface and have a renormalized mass dependent only on electron-electron interaction.

We concentrate on the temperature (T) dependence of R_{xx} extrema around $\nu = \frac{1}{2}$ in a low-disorder 2D *hole* system (2DHS) of density $n_h = 1.6 \times 10^{11}$ cm^{-2} and low-T mobility $\mu \simeq 1 \times 10^6$ cm^2/V s, in the range $0.04 \lesssim T \lesssim 1.1$ K. The sample, grown by molecular beam epitaxy on an undoped (311)A substrate, consists of a 200 Å GaAs quantum well surrounded by $Al_{0.3}Ga_{0.7}As$ spacer layers and Si-doped regions. Eutectic In:Zn was alloyed in a van der Pauw geometry for ohmic contacts. Together with a calibrated RuO_2 resistor with known magnetoresistance corrections used for thermometry, the sample was immersed in the dilute phase of a $^3He/^4He$ dilution refrigerator inside a superconducting/Bitter hybrid magnet system.

Figure 1 shows R_{xx} vs B around $\nu = \frac{1}{2}$ at $T \simeq 40$ mK. The density of the sample places $B_{\nu=1/2}$ at 13.2 T, the deviation from which is labeled by the B^* axis. Minima (maxima) are marked by integer (half-integer) p. The exceptional quality of this 2DHS is manifested in the high-order FQH states up to $p \simeq \pm 5$ around $\nu = \frac{1}{2}$ as well as states at $\nu = \frac{2}{7}$ and $\frac{3}{11}$ at higher field. Our measurements focus on the simplest form of CF renormalization, $m = 1$, corresponding to two flux quanta per hole. We begin by examining the T dependence of the R_{xx} minima corresponding to FQH states at $\nu = p/(2p+1)$ (p integer), and extract the quasiparticle excitation gaps Δ_ν of these incompressible states from fits to the activated regions, using $R_{xx} \propto \exp(-\Delta_\nu/2T)$. As shown in Fig. 2, these energy gaps obey a remarkable scaling law

$$\Delta_\nu = \frac{\kappa}{|2p+1|}\frac{e^2}{\epsilon\ell_B} - \Gamma, \tag{1}$$

where ϵ is the dielectric constant, $\ell_B \equiv \sqrt{\hbar/eB}$ is the magnetic length, and κ and Γ

are constants. In the plot of Δ_ν vs $(e^2/\epsilon\ell_B)/(2p+1)$, linear fits from both positive and negative B^* yield nearly identical slopes and intercepts, from which we obtain $\kappa \simeq 0.07$ and $\Gamma \simeq 1.4$ K.

The observation of the energy gap scaling (1) is an impressive finding for the following reasons. In the simplest interpretation of the CF theory, the incompressibility gaps of the FQH liquids map precisely onto the "single-particle" cyclotron energy gaps of the corresponding IQH states of CFs. One would then expect $\Delta_\nu \propto \hbar\omega_c^*$, where $\omega_c^* \equiv eB^*/m^*$ and m^* is the effective mass of the CFs. This interpretation was used in Ref. 3 to extract m^* for CFs in 2DESs, assuming a constant m^* for each sign of B^*. The Chern-Simons gauge field theory of HLR,[2] however, explicitly postulates modifications to this simple interpretation. First, since m^* is a result of electron-electron interaction, it cannot remain rigorously constant as B varies. In fact, the theory predicts from dimensionality arguments that

$$m^* = \left(\frac{\hbar}{e}\right)^2 \frac{\epsilon}{\kappa\ell_B} = \left(\frac{\Phi_0}{2\pi}\right)^{3/2} \frac{\epsilon}{\kappa}\sqrt{B}. \tag{2}$$

Furthermore, the requirements of particle-hole symmetry in the FQHE posit the equivalency of states at p and $-(p+1)$. These two considerations predict directly the observed scaling (1). We therefore interpret this observation as strong evidence that a hole-flux composite particle Fermi liquid forms around $\nu = \frac{1}{2}$.

We can make several inferences about these renormalized fermions. Our value of κ, using Eq. (2) and $\epsilon = 13.1$, gives a CF effective mass at $B^* = 0$ of $m_0^* \simeq 1.4 m_e$ for this density. The observation of the gap scaling behavior then provides direct experimental evidence for a field-dependent mass $m^* = m_0^*(2\nu)^{-1/2} \propto \sqrt{B}$, *unlike any conventional Fermi liquid.* One may still think of the energy gaps (1) as CF cyclotron gaps $\hbar\omega_c^*$ corrected for disorder, $\Delta_\nu = \hbar eB^*/m^* - \Gamma$, as long as m^* is recognized to be field-dependent in the form of Eq. (2). Interpreted in this way, Γ is the collision broadening of the CF LLs.[3] We may then estimate the quantum lifetime τ_q^* of CFs at $B^* = 0$ as $\tau_q^* = \hbar/2\Gamma \simeq 2.7$ ps.

Assuming the presence of a Fermi surface, we have analyzed the R_{xx} oscillations around $\nu = \frac{1}{2}$ by treating them as the Shubnikov-de Haas (SdH) oscillations originating from the B-field-fractured density of states of the CF metal.[4,5] The analysis, based on 2D carriers at $B = 0$, utilizes the first Fourier component of the oscillatory amplitude ΔR_{xx} normalized to the zero-field resistance R_0,

$$\frac{\Delta R_{xx}}{R_0} = 4\exp\left(-\frac{\pi}{\omega_c\tau_q}\right)\frac{\xi}{\sinh\xi}\cos\left[\pi(2\nu-1)\right], \tag{3}$$

where $\xi \equiv 2\pi^2 k_B T/\hbar\omega_c$ appears in the Dingle factor $D(\xi) \equiv \xi/\sinh\xi$ describing temperature-induced damping.[6] We apply it to the metallic state of hole-flux composite particles around $\nu = \frac{1}{2}$ by evaluating R_0 at $B^* = 0$ and making the transformations $\omega_c \to \omega_c^* \equiv eB^*/m^*$, $\tau_q \to \tau_q^*$, and $\nu \to \nu^* \equiv |p|$.

From the T dependence of the R_{xx} extrema, where the cosine term of (3) is ± 1, we can extract m^* of the CFs from the Dingle factor. This was done for each extremum via a nonlinear least-squares fit. As manifest in Fig. 3, we find outstanding agreement to the hypothesis of metallic behavior over more than two orders of magnitude in the fitted amplitude variable within our entire temperature range, giving further credence to Fermi liquid formation around $\nu = \frac{1}{2}$. The renormalized mass, plotted vs $p^{-1} \propto B^*$ in Fig. 4(a), exhibits extraordinary features. We note that m^* very roughly follows the $\sqrt{B}$ dependence (dashed line) expected from the energy gap scaling of Fig. 2, but demonstrates strong symmetrical features on both sides of $\nu = \frac{1}{2}$ as well as an apparent divergence as $B^* \to 0$. This is incongruous with the experimental electron-flux CF behavior propounded by Leadley *et al.*[4] The overall behavior also differs significantly from the observations of Du *et al.*,[5] who observe a constant SdH mass with a small enhancement near $\nu = \frac{1}{2}$ that is stated to be within their experimental error bars. In contrast, we observe significantly more pronounced m^* variations both near $\nu = \frac{1}{2}$ as well as at large B^*.

It is also enlightening to examine the quantum lifetime of the hole-flux composites. Typical Dingle plots of $\log[\Delta R_{xx}/R_0 D(\xi)]$ vs $1/B^*$ or m^*/B^*, however, turn out to be nonlinear, betraying a B dependence of the CF lifetimes. With confidence that Eq. (3) provides an accurate description of the CF SdH oscillations, we use it to extract τ_q^* *as a function of* B^* from the m^* fits. We plot this renormalized lifetime in Fig. 4(b) and note its striking behavior: in the low-B^* limit, scattering appears to be strongly enhanced, while away from $\nu = \frac{1}{2}$, τ_q^* recovers and then drops again.

Our observations have several implications relating to the fundamental nature of the CF metallic state. The energy gap scaling alone indicates that m^* is *not* constant; hence, conventional Fermi liquid (FL) behavior is ruled out. Rather, the SdH-derived m^* is strongly suggestive of a *marginal* FL,[7,8] characterized by a diverging m^* close to the Fermi surface. While the gauge field approach[2] in fact entails a logarithmic mass divergence close to the Fermi surface, our observed divergence appears to be faster than any power law in $(B^*)^{-1} \propto p$. Our best estimate, based on fitting the data of Fig. 4(a) for $p \leq -3$ and $p \geq 2.5$, is an exponential divergence $m^* \sim \exp(|p|^\gamma)$, where $\gamma \simeq 1.4 \pm 0.3$. The nature of this divergence suggests a novel universality class of marginal FL behavior. The bulk of the observed renormalization structure of this exotic metal, however, awaits a comprehensive theoretical explanation.

We thank D. V. Khveshchenko, X. G. Wen, I. A. McDonald, B. I. Halperin, and A. H. MacDonald for illuminating discussions, and P. W. Emery and J. J. Heremans for technical assistance. H. C. Manoharan is grateful to the Fannie and John Hertz Foundation for fellowship support. This work was supported by the NSF.

1. J. K. Jain, *Phys. Rev. Lett.* **63**, 199 (1989).
2. B. I. Halperin, P. A. Lee, and N. Read, *Phys. Rev. B* **47**, 7312 (1993).
3. R. R. Du *et al.*, *Phys. Rev. Lett.* **70**, 2944 (1993).
4. D. R. Leadley *et al.*, *Phys. Rev. Lett.* **72**, 1906 (1994).
5. R. R. Du *et al.*, *Solid State Commun.* **90**, 71 (1994).
6. R. B. Dingle, *Proc. Roy. Soc. A* **211**, 517 (1952); E. N. Adams and T. D. Holstein, *J. Phys. Chem. Solids* **10**, 254 (1959); A. Isihara and L. Smrčka, *J. Phys. C* **19**, 6777 (1986).

7. C. M. Varma *et al.*, *Phys. Rev. Lett.* **63**, 1996 (1989).
8. P. W. Anderson, *Phys. Rev. Lett.* **64**, 1839 (1990); **65**, 2306 (1990).

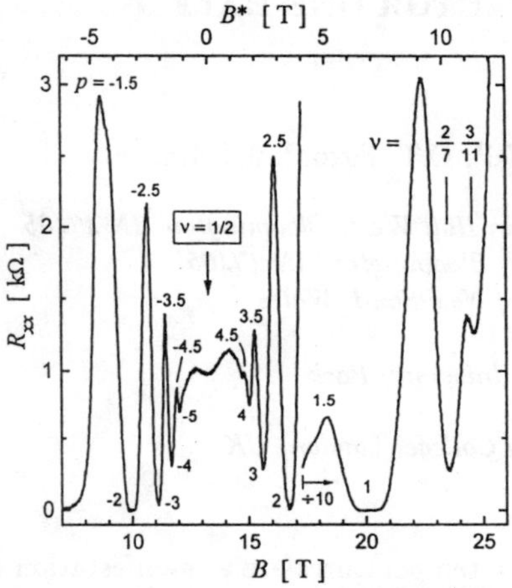

Fig. 1. R_{xx} **of the 2DHS vs applied magnetic field** B **and CF effective field** $B^* = B - B_{\nu=1/2}$.

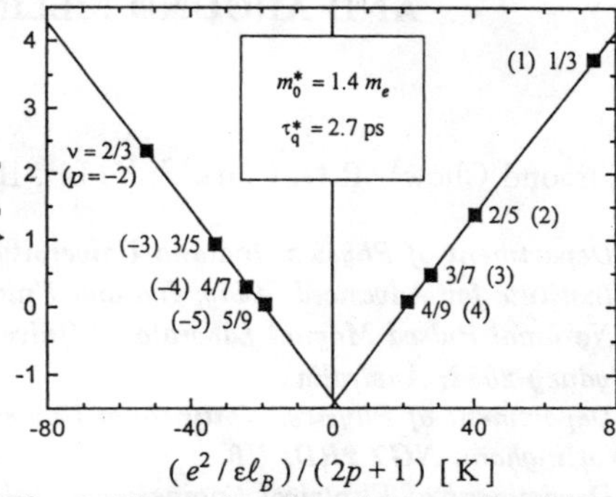

Fig. 2. **Observed scaling of the quasiparticle excitation gaps of the FQH states around** $\nu = \frac{1}{2}$. **Solid lines are linear fits to the data.**

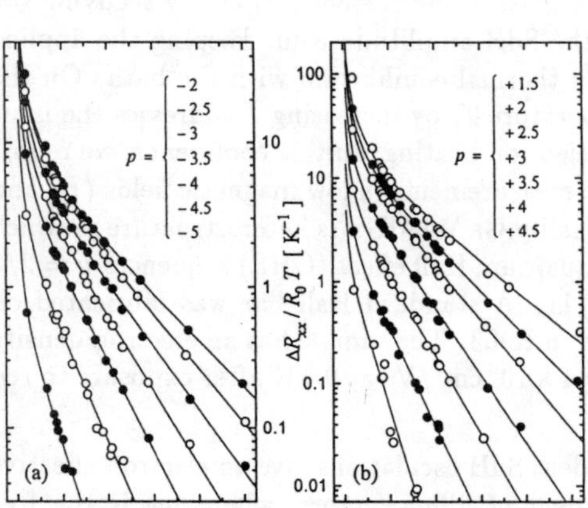

Fig. 3. **SdH analysis for (a)** $p < 0$ **and (b)** $p > 0$ **(**$p = -1.5$ **and +1 omitted for clarity). Solid lines are fits to Eq. (3) for minima (closed circles) and maxima (open circles).**

Fig. 4. **The renormalized mass (a) and quantum lifetime (b) of the CFs. Open squares and dashed line are expected values from the scaling in Fig. 2. Lines are guides to the eye.** ⟶

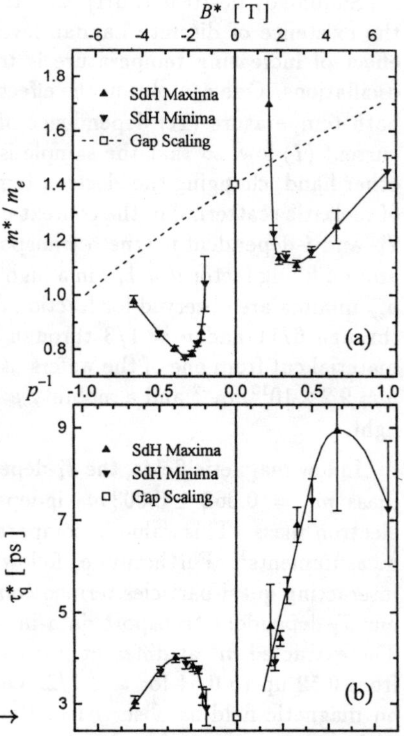

TEMPERATURE AND CURRENT DEPENDENT MAGNETO-TRANSPORT IN LOW MAGNETIC FIELDS AND AROUND FILLING FACTOR ONE HALF

Edmond Chow[1], R.G. Clark[2,3], Y. Li[1], H.P. Wei[1], C.T. Foxon[4], J.J. Harris[5]

[1]*Department of Physics, Indiana University, Swain Hall West, Bloomington, IN 47405*
[2]*Institute for Advanced Study, Indiana University, Bloomington, IN 47405*
[3]*National Pulsed Magnet Laboratory, University of New South Wales Sydney 2052, Australia*
[4]*Department of Physics, Nottingham University, University Park Nottingham, NG7 2RD, UK*
[5]*Department of Electrical Engineering, University College, London, UK*

Shubnikov de-Haas (SdH) oscillations at low temperatures are a manifestation of the existence of discrete Landau levels and a well defined Fermi energy (E_F). The effect of increasing temperature is to smear E_F and thus reduce the amplitude of oscillations. One can obtain the effective mass of the current carriers by studying the bath temperature (T_b) dependence of the SdH amplitude while keeping the applied current (I) low, so that the sample is in thermal equilibrium with the bath. On the other hand, changing the electron temperature T_e by increasing I addresses the issue of inelastic scattering in the context of electron heating. In this conference, we report T_b- and I-dependent magneto-transport measurements in low magnetic fields (B) and around filling factor $\nu = 1/2$ in a high quality GaAs/AlGaAs heterostructure in which ρ_{xx} minima are observed for fractional quantum Hall effect (QHE) sequences $\nu = 2/3$ through $6/11$ and $\nu = 1/3$ through $5/11$. A standard Hall bar was fabricated on material cut from one of the wafers used in Ref.3. The sample has an electron density $n = 9.2 \times 10^{10}\,\mathrm{cm}^{-2}$ and a mobility $\mu = 4 \times 10^{6}\,\mathrm{cm}^2/\mathrm{Vs}$ at 0.3 K after exposure to red light.

In low magnetic fields, the T_b-dependent SdH oscillations give an electron effective mass $m^* = 0.062 \pm 0.003\, m_0$ independent of filling factors, where m_0 is the free electron mass. This value is comparable to that obtained from cyclotron resonance measurements[1]. Furthermore, following a recent proposal[2] that there exist weakly interacting quasi-particles termed composite fermions (CF) near $\nu = 1/2$, we analyze our T_b-dependent transport data in terms of SdH oscillations of CF for $T_b > 0.3$ K. The extracted m^* at different ν is constant, $0.69 \pm 0.08\, m_0$, for $\nu > 1/2$, and ranges from 0.52 up to 0.64 for $\nu < 1/2$. Our data does not have a systematic dependence on magnetic field as observed in Ref.3 and is more in line with results reported in Ref.4, where m^* was found to be approximately constant.

The decreasing amplitude of SdH oscillations with increasing T_b has been usually used as a thermometer for measuring the electron temperature T_e[5]. By increasing the applied current, T_e can be increased. Combining the T_b- and I-dependent magneto-transport data measured in a ^{3}He cryostat, we find that, in low magnetic fields ($B < 0.2\,\mathrm{T}$), there is a characteristic current I_0 at each T_b, below which T_e remains constant, and above which it merges into a single power law $T_e \sim I^{0.5}$. This result, obtained from semi-classical SdH oscillations, is the same as that observed in the integer QHE regime[6] indicative of a unique inelastic scattering process controlling the energy dissipation at low temperatures.

In analogy to the electron heating model for thin metal films[7] in zero magnetic field, the smearing of the Fermi distribution function is characterized by T_e and is given by the energy acquired by an electron diffusing a distance l_{in} in an electric field E. Here, l_{in} is the inelastic scattering length and $l_{in} \sim T_e{}^{-p/2}$. Therefore, $k_B T_e \sim eEl_{in}$ and $T_e \sim I^{2/(2+p)}$. Our data then gives $p = 2$.

We also extend our study to the region near $\nu = 1/2$ in the context of the CF model, in which there exists a well defined Fermi energy for the CF energy distribution. Within our limited temperature range, our data is not inconsistent with the relation $T_e \sim I^{0.5}$. Measurement at lower temperatures ($< 0.3\,\mathrm{K}$) is currently in progress to extend the range of T_b and I for both the low B measurements and the $\nu = 1/2$ regime.

The work at Indiana University is supported by the IU Foundation and an NSF grant DMR-9311091.

1. M.J. Chow, D.C. Tsui, G. Weimann, *Phys. Rev.* **B 37**, 848 (1988), and references therein.

2. J.K. Jain, *Adv. Phys.* **41**, 105 (1992). B.I. Halperin, P.A. Lee and N. Reed, *Phys. Rev.* **B 47**, 7312 (1993).

3. D.R. Leadley, R.J. Nicholas, C.T. Foxon, J.J. Harris, *Phys. Rev. Lett.* **72**, 1906 (1994).

4. R.R. Du, H.L. Stormer, D.C. Tsui, L.N. Pfeiffer, K.W. West, (Preprint).

5. See, for example, K. Hirakawa, H. Sakaki, *Appl. Phys. Lett.* **49**, 889 (1986), and references therein.

6. H.P. Wei, L.W. Engel, D.C. Tsui, (submitted to *Phys. Rev. Lett.*).

7. P.W. Anderson, E. Abrahams, T.V. Ramakrishnan, *Phys. Rev. Lett.* **43**, 718 (1979). M.R. Arai, *Appl. Phys. Lett.* **42**, 906 (1983). M.L. Roukes, M.R. Freeman, R.S. Germain, R.C. Richardson, *Phys. Rev. Lett.* **55**, 422 (1985).

A STUDY OF THE HIERARCHY OF THE FQHE BY TIME-RESOLVED LUMINESCENCE OR COMPOSITE FERMION EFFECTIVE MASS FROM LUMINESCENCE MEASUREMENTS

I.V. KUKUSHKIN[a], R.J. HAUG, K. v. KLITZING, K.PlOOG[b], K. EBERL
Max-Planck-Institut für Festkörperforschung
D-70569 Stuttgart, Germany

ABSTRACT

In time-resolved magnetoluminescence experiments the hierarchy of the FQHE states have been studied. A striking symmetry in the dependence of the chemical potential discontinuity on filling factor is found inside the families of the FQHE states. The measurements could be interpreted as a a drastic increase of the composite fermion effective mass in approaching filling factor $\nu = 1/2$.

Recently, an alternative attempt to unify Laughlin's theory of the fractional quantum Hall effect (FQHE) was proposed by Jain[1] who modified Laughlin's wavefunctions and used Jastrow-Slater functions instead of pure Jastrow functions. Similar to the Landau Fermi-liquid theory, the FQHE is described in terms of non-interacting quasiparticles (called composite fermions) with peculiar properties (due to renormalization) - the quasiparticle consists of an electron to which an even number of flux quanta is attached. A well defined Fermi-surface for composite fermions is proposed[2] to exist at ν=1/2 and a linear dependence of the FQHE energy gap on magnetic field inside the FQHE families ending at ν=1/2 is predicted.

We used time-resolved magnetoluminescence to study the hierarchy of FQHE states and observed a striking symmetry in the dependence of chemical potential discontinuity on filling factor, which would correspond to a divergence of the mass of the composite fermions in approaching $\nu = 1/2$

We studied high quality GaAs/AlGaAs single heterojunctions with 1000 nm GaAs buffer layer in which a monolayer of acceptors (with concentration $5 \cdot 10^9 cm^{-2}$) was created at a distance of 40 nm away from the interface. It was possible to change the concentration of 2D-electrons in the range of $(0.6-3.5) \cdot 10^{11} cm^{-2}$ by the intensity of illumination. The transport mobility of the 2D-electrons at the lowest concentration was $3 \cdot 10^6 cm^2/Vs$. For the photoexcitation we used pulses of an Ar^+ -laser with

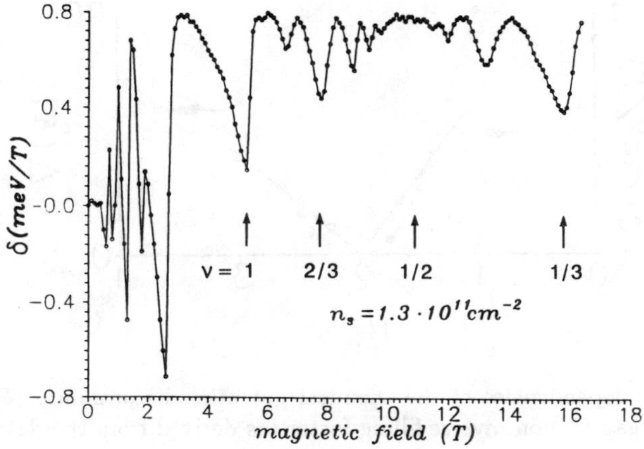

Figure 1: Magnetic field dependence of the first derivative of the spectral position of the luminescence line (first moment) measured at $\Delta t = 500ns$.

a duration of 20 ns, a peak power of $(10^{-5} - 10^{-3})W/cm^2$ and a frequency of 0.1 MHz[3]. The luminescence was detected by a gatable photon counting system. The time resolution of the system was about 30 ns with a spectral resolution of 0.01 meV. The samples were mounted in the mixing chamber of a He^3/He^4 dilution refrigerator.

To obtain the spectral position of the luminescence line on magnetic field, we calculated the first momentum M_1 for the total luminescence spectrum measured for the recombination of electrons from the two-dimensional electron gas with holes bound to the acceptors:

$$M_1 = \int I(E)EdE / \int I(E)dE \tag{1}$$

where I(E) is the energy distribution of the luminescence signal. At filling factors $\nu > 2$ the luminescence line splits into several peaks due to Landau quantization, but we always took the total spectrum. In the case of holes bound to remote acceptors the spectral position of the recombination line reflects the dependence of the mean energy of interacting electrons on filling factor[4]. Therefore, the investigation of radiative recombination of 2D-electrons with holes bound to acceptors located far away from the interface is a direct method of the determination of the FQHE energy characteristics. The exact correspondence[4] between the spectral position of the luminescence line and the mean energy of 2D-electrons exists for the case when the acceptor is located at an infinite distance from the 2D-channel. However, in real situations we deal with finite distances and therefore it is important that the corrections

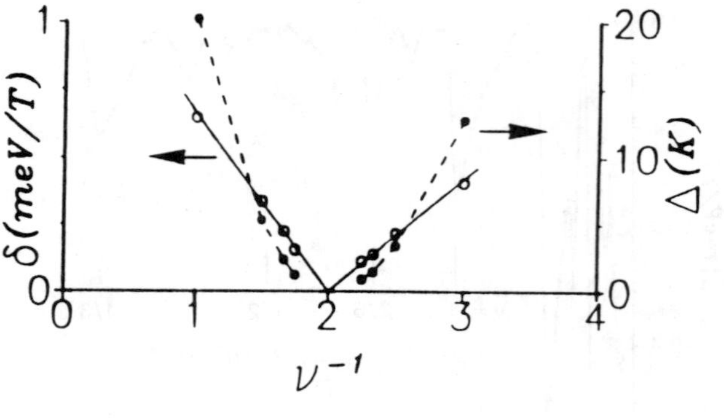

Figure 2: The dependencies of the chemical potential discontinuity (δ) and of the FQHE energy gap (Δ) on inverse filling factor (as derived from the data of Fig. 1).

to this simple expression are sufficiently small. These corrections are defined by the third power of the ratio l_c/d (d - is the distance to the acceptor, l_c - magnetic length) and therefore, for magnetic fields B in the range of 5-20T, d has to be at least 40 nm to keep these corrections small, i.e. less than 5% at ν=1/3. [5]

Due to the correspondence between M_1 and the mean energy of 2D-electrons, the derivative of the first moment ($\delta = dM_1/dB$) is proportional to the chemical potential of 2D-electrons. Its dependence on magnetic field is shown in Fig. 1 for a carrier concentration of $1.3 \cdot 10^{11}$ cm^{-2}. Some filling factors are marked in the figure. At filling factor ν=2 the value of dM_1/dB shows an abrupt change from the value -$(e\hbar/2m_e)$ to $+(e\hbar/2m_e)$, as one expects for non-interacting 2D-electrons. The reduction in amplitude of dM_1/dB at ν=4,6,8 is mainly due to the finite step size in magnetic field which was used in our measurements. Jumps in dM_1/dB at fractional filling factors are also visible. They are smaller in amplitude than those observed for the integer filling factors and their amplitudes clearly change with filling factor. One can see a monotonic reduction of the amplitude of dM_1/dB in the sequence of the FQHE states 2/3, 3/5, 4/7 and 5/9. Much smaller amplitudes of dM_1/dB are measured for the states 4/5 and 5/7 which is consistent with the picture of composite fermions where 4/5 and 5/7 belong to other FQHE families in comparison to 2/3, 3/5. The amplitudes of the discontinuities in δ depend linearly on magnetic field within the same family and these linear dependencies end at filling factor 1/2 (for the shown carrier concentration). Only special sequences of FQHE states are visible: ν=n/(2mn+1) and symmetrical ones ν=n/(2mn-1) with m=1 and n=1,2,3,4 and 5 in different FQHE states. One can consider here the state ν=1 as FQHE state of the family n/(2n-1) for n=1. In terms of chemical potential discontinuity a symmetry between the states n/(2mn+1) and n/(2n(m+1)-1) is visible, however, it does not mean that the energy gaps are the same for these FQHE states because the denominators of these fractions are different. We derived the FQHE energy gap values (Δ) from the amplitude δ of the discontinuities in (dM_1/dH) using the relationship for ν=p/q: $\delta = 2q\ \Delta/B$. The corresponding data are presented in Fig.2 together with the dependence of the amplitude of the discontinuities δ on inverse of the filling factor. Similar dependencies have also been observed for other families of the FQHE states[3]. It is seen from this plot that for the results presented in terms of FQHE energy gaps a linear dependence is less obvious. A linear extrapolation of the magnetic field dependence of the FQHE gaps[6] would give negative values of the gap at ν=1/2, but one can see from the figure that this dependence is closer to the quadratic dependence. Interpreting these measurements in terms of the effective mass of the composite fermions[2] gives a large increase in mass in approaching $\nu = 1/2$, which was shown to possibly originate from the nature of the interaction. [7] However, the detailed dependence near $\nu = 1/2$ cannot be extracted from our measurements.

In conclusion, the hierarchy of the FQHE states has been investigated by time-resolved magnetoluminescence. The observed linear dependence of the chemical potential discontinuity on magnetic field starts at ν=1/2 for the corresponding family of the FQHE states.

This work has partly been supported by the Volkswagen Stiftung, Germany, and by the Bundesministerium für Forschung und Technologie, Germany.

[a] Permanent address: Institute of Solid State Physics, Russian Academy of Science, 142432 Chernogolovka, Moscow district, Russia

[b] Present address: Paul-Drude-Institut fur Festkörperelektronik, Hausvogteiplatz 5-7, 10117 Berlin, F.R.Germany

References

1. J.K.Jain, Phys.Rev.Lett. **63**, 199 (1989); J.K.Jain, S.A. Kivelson, and N. Trivedi, Phys.Rev.Lett. **64**, 1297 (1990).

2. B.I.Halperin, P.A. Lee, and N. Read, Phys.Rev.B **47**, 7312 (1993).

3. I.V.Kukushkin, R.J. Haug, K. v. Klitzing and K. Ploog, Phys. Rev. Lett. **72**, 736 (1994).

4. V.M.Apal'cov and E.I.Rashba, JETP Lett. **53**, 442 (1991).

5. I.V. Kukushkin, R.J. Haug, K. v. Klitzing, K. Eberl, and K. Tötemeyer, Phys. Rev. B (in print).

6. R.R. Du, H.L. Störmer, D.C. Tsui, L.N. Pfeiffer, and K. W. West, Phys.Rev. Lett. **37**, 2944 (1993).

7. D.V. Khveshchenko, Phys. Rev. B **49**, 10514 (1994).

RADIATIVE RECOMBINATION FROM COMPOSITE FERMIONS

A.J. Turberfield, I.N. Harris, H.D.M. Davies, J.F. Ryan
Department of Physics, University of Oxford, Clarendon Laboratory, Parks Road, Oxford OX1 3PU U.K.

C.T. Foxon
Physics Department, University of Nottingham, Nottingham NG7 2RD U.K.

and

J.J. Harris
Semiconductor I.R.C., Imperial College, London SW7 2BZ U.K.

ABSTRACT

The composite fermion model is used as the basis of a discussion of photoluminescence from a two-dimensional electron system with both free and bound holes. Estimates of the composite fermion effective mass are obtained.

The composite fermion (CF) model[1] provides a intuitive description of the hierarchies of incompressible liquid states responsible for the fractional quantum Hall effect (FQHE). A change of gauge corresponding to threading $2m$ flux quanta through each electron (where m is an integer) transforms electrons into particles (composite fermions) that obey Fermi statistics and experience an effective mean magnetic field B^* that is reduced by $2m$ flux quanta per electron. The FQHE is the integer quantum Hall effect of CFs. (A topologically equivalent description has been proposed in which a CF consists of an electron bound to a vortex in the many–electron wavefunction[2]). At filling factor $\nu = 1/2m$ B^* vanishes and the existence of a Fermi surface has been predicted[3]. The CF effective mass, which sets the energy scale for correlation phenomena, is relatively well behaved (it diverges logarithmically at the Fermi surface) which leads to the further prediction[3] that the FQHE excitation gap 2Δ measured in activated transport experiments is equivalent to the gap between CF Landau levels $\hbar\omega_c^*$ which is proportional to effective field; this prediction has been verified[4].

In this paper we use a constant CF mass and the mean gauge field to provide the lowest level of description of the effects of electron correlation on the photoluminescence spectrum of a two–dimensional electron system (2DES) (long–range interaction of CFs via fluctuations in the gauge field[3] are neglected). We consider both symmetry between FQHE hierarchies observed in optical experiments and the energies of optical transitions in the range $1/3 < \nu < 2/3$ in which it is not necessary to invoke mixed spin or particle–hole symmetry to explain the observed incompressible states.

The electron Landau level filling factor ν is related to the CF filling factor ν^* by $\nu = \nu^*/(2m\nu^* + \alpha)$ $(\alpha = \pm 1)$[1]; incompressible states of the FQHE hierarchies form where $\nu^* = p$ (p integer). A success of CF theory was the observation that ρ_{xx} oscillations in the $\nu = 1/3$ and $2/3$ hierarchies (corresponding to $m = 1, \alpha = \pm 1$) may

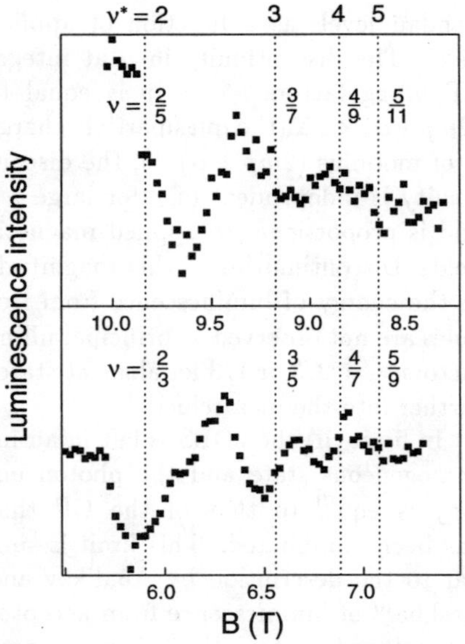

Fig. 1. Oscillations in the peak intensity of photoluminescence from the 2DES in (a) 1/3 hierarchy; (b) 2/3 hierarchy. Measurements made in single heterojunction G641[6]: $n_s = 10^{11}\text{cm}^{-2}$, $\mu = 10^7\text{cm}^2\text{V}^{-1}\text{s}^{-1}$, T = 100 mK.

be understood by considering ρ_{xx} as a function of ν^* [5]. In an optical experiment we expect this symmetry between hierarchies to be broken. When an electron recombines with a valence band hole the gauge flux attached to it disappears from the system and the total effective flux experienced by the remaining carriers increases by $2m$ quanta. In the 1/3 hierarchy this augments B^* whereas in the 2/3 hierarchy it reduces the magnitude of the (negative) effective field. Figure 1 shows minima in the peak intensity of luminescence from a 2DES at a GaAs single heterojunction that identify incompressible states[6,7]. Data from the 2/3 and 1/3 hierarchies are presented separately (with field axes reversed in the top section) such that states with the same CF filling factor $\nu^* = p$ are aligned. Asymmetry between the hierarchies is visible in the systematic displacement of intensity minima to low applied field.

The CF density of states consists of Landau levels whose energies and degeneracy are proportional to B^*; it is tempting to assume that optical spectra should display the same simple structure. To calculate optical transition energies, however, we must account for the change in gauge flux associated with recombination. There are two simple limits: (i) effective flux is redistributed evenly in the final state, affecting all electrons in the system slightly and changing the effective filling factor ν^* and field B^*; (ii) the change in effective flux remains as a local defect and the effective field experienced by other electrons is unchanged.

Limit (i) is a natural choice to describe recombination with free holes[6,7] where there is no perturbation of the 2DES in the final state. In (i) both initial and final states are homogeneous and the photon energy is equal to the electron chemical potential (μ). The CF density of states suggests the possibility of transitions from filled Landau levels below the CF Fermi energy or from thermally excited particles in higher levels; these processes occur at energies displaced by $\hbar\omega_c^*$ and are equivalent to the recombination channels involving creation or annihilation of quasielectron–quasihole pairs that have been invoked to explain multiplet structure in photoluminescence spectra[8,9]. Figure 2(a) shows the energies of transitions at μ and from lower CF

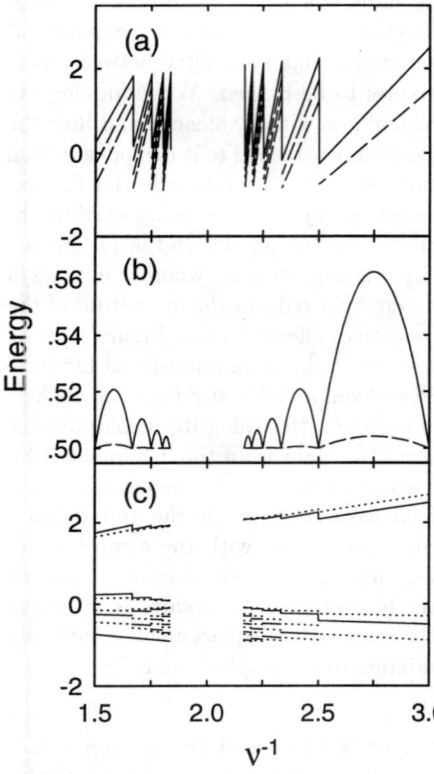

Fig. 2. (a) Chemical potential of a uniform 2DES (solid) and optical recombination from filled CF Landau levels (dashed); (b) mean energy of 2DES when homogeneous (solid) and after domain formation (dotted); (c) optical transitions from domains of $\nu^* = p$ (solid), $p+1$ (dashed). The upper pair of transitions are optical absorption processes. The unit of energy for all sections is the CF Fermi energy at $\nu = 1/2$.

Landau levels as a function of applied field. The discontinuity in μ at integer CF filling factors $\nu^* = p$ is equal to $(2mp+\alpha)\times 2\Delta$ (the quasiparticle charge is of modulus $(2mp+\alpha)^{-1}$); the discontinuity is independent of p for large p [3] and is proportional to applied magnetic field. Discontinuities of this magnitude in the energy of luminescence from free holes are not observed at principal filling factors $\nu = 2/3$ or $1/3$ let alone at states further into the hierarchies[6,7].

In limit (ii) the 2DES is left in an inhomogeneous state and the photon energy is equal to that of the CF that has been annihilated. This limit is similar to the description by Apal'kov and Rashba[10] of luminescence from acceptor delta–doped systems[11]. If we assume that transitions occur from all occupied CF Landau levels then the mean photon energy is equal to the mean CF energy shown in Figure 2(b). Discontinuities in the magnetic–field *derivative* of the mean transition energy at CF filling factors $\nu^* = p$ are independent of p and equal to the CF Fermi energy at $\nu = 1/2$. Measured discontinuities in the field derivative of the mean luminescence energy from acceptor–doped samples are *inversely proportional* to p, however[12].

An interesting consequence of the assumption of constant CF mass is that the homogeneous 2DES is unstable everywhere except at exact fractional filling (note in Figure 2 that in the homogeneous system μ may be less than the mean energy). The derivative of the chemical potential w.r.t. electron number is negative so density fluctuations are amplified until domains of exact integer CF filling factor are formed (at which filling factors μ is discontinuous)[3]. A 2DES with filling factor $\nu = 0.65$ will spontaneously separate into domains with $\nu = 2/3$ and $3/5$, i.e. CF filling factors $\nu^* = 2,\ 3$. It is likely that the disorder potential will nucleate, stabilise and localise domains. Figure 2(b) shows the mean CF energy assuming domain formation (neglecting domain wall and electrostatic contributions). The discontinuities in the field derivative of the mean energy do decrease to zero as ν^* increases but comparison with discontinuities deduced from the luminescence of acceptor-doped samples[12] leads to an estimate of the CF effective mass that is of the order of the GaAs band mass, an order of magnitude smaller than the CF mass deduced from transport measurements[4].

If we assume that the change in gauge flux on recombination is redistributed evenly within the domain to give a relaxed final state then we expect optical transitions in each domain at its chemical potential and displaced from μ by the value of $\hbar\omega_c^*$ within that domain. Because μ is pinned at a discontinuity where $\nu^* = p$ we move ν^* to $p \pm \epsilon, \epsilon \ll 1$ (depending on whether a decrease in electron density increases or decreases ν^*) to determine the emission and absorption energies shown in Figure 2(c). Discontinuities in these transition energies are much smaller than the discontinuity in μ. At every filling factor some of these transitions are continuous as observed in free-hole systems[6,7]. There is a gap between emission and absorption energies that is approximately equal to discontinuities in μ at neighbouring filling factors in the homogeneous system. The width of the band of emission energies is always approximately equal to the CF Fermi energy; by measuring the total width of free-hole photoluminescence spectra between $\nu = 2/3$ and $1/3$ [6,9] we obtain an estimate of the CF mass of order 10 $\sim$15m^*_{GaAs}.

Conclusions

If we neglect the possibility of density instability this simple treatment of the recombination process from a homogeneous system does not account for experimental results. When the formation of domains of exact fractional filling is permitted then general features of luminescence involving both free and bound holes are reproduced (although different estimates of the CF effective mass are obtained for these two cases). The instability of the 2DES merits further study.

References

1. J. K. Jain, *Phys. Rev. Lett.* **63**, 199 (1989)
2. E. Rezayi and N. Read, *Phys. Rev. Lett.* **72**, 900 (1994)
3. B. I. Halperin et al., *Phys. Rev.* **B47**, 7312 (1993)
4. R. R. Du et al., *Phys. Rev. Lett.* **70**, 2944 (1993)
5. V. J. Goldman et al., *Phys. Rev. Lett.* **65**, 907 (1990)
6. A. J. Turberfield et al., *Phys. Rev. Lett.* **65**, 637 (1990)
7. B. B. Goldberg et al., *Phys. Rev. Lett.* **65**, 641 (1990)
8. A. H. MacDonald et al., *Phys. Rev. Lett.* **68**, 1939 (1992)
9. I. N. Harris et al., *Surf. Sci.* **305**, 61 (1994)
10. V. M. Apal'kov and E. I. Rashba, Pis'ma Zh. Eksp. Teor. Fiz. **53**, 420 (1991) [*JETP Lett.* **53**, 442 (1991)]
11. H. Buhmann et al., *Phys. Rev. Lett.* **65**, 1056 (1990)
12. I. V. Kukushkin et al., *Phys. Rev. Lett.* **72**, 736 (1994)

The quantum-limit thermopower in two-dimensional electron systems

U. Zeitler[a,b], B. Tieke[a], S.A.J. Wiegers[a], J.C. Maan[a],
R. Fletcher[c], V.I. Fal'ko[d], C.T. Foxon[e], and J.J. Harris[f]

[a] *High Field Magnet Laboratory, Research Institute for Materials, University of Nijmegen, Toernooiveld 1, NL-6525 ED Nijmegen*

[b] *High Magnetic Field Laboratory, MPIF-CNRS, B.P. 166, F-38042 Grenoble Cedex 9*

[c] *Physics Department, Queen's University, Kingston, Ontario, Canada K7L 3N6*

[d] *Max-Planck-Institut für Festkörperforschung, Heisenberstr. 1, D-70569 Stuttgart*

[e] *Department of Physics, University of Nottingham, Nottingham MG7 2RD, UK*

[f] *Semiconductor Interdisciplinary Research Centre, Imperial College, London SW7 2BZ, UK*

ABSTRACT

We have measured the phonon-drag thermopower of two-dimensional electron gases in GaAs-$Ga_{1-x}Al_xAs$ heterojunctions at low temperatures and high magnetic fields. In the quantum limit two distinct plateaus are seen in the thermopower which can be related to the composite fermions near ν =1/2 and ν =1/4. The large magnitude of the TEP implies that the coupling between phonons and composite fermions is strongly enhanced as compared to non-interacting electrons.

1. Introduction

The fractional quantum Hall (FQH) effect in two-dimensional electron gases (2DEGs) is characterized by the appearance of plateaus in the Hall resistivity ρ_{xy} and minima in the transverse magnetoresistivity ρ_{xx} at rational fractional filling factors p/q with q odd.[1,2] The FQH effect is well explained for filling factors $\nu = 1/m$ where m is odd[3] but the explanation of higher fractions has been controversial.[2,4] A new model appropriate to filling factors near $\nu = 1/p$ with *even* denominator p has been proposed recently[5,6] and introduces a new particle, the "composite fermion" (CF). Experimental evidence for its validity has been found[7,8] in resistivity measurements. The FQH effect is then regarded as the equivalent of the integer quantum Hall effect for CFs around $\nu = 1/p$. The particularity of even-denominator filling factors has also been pointed out be means of sound acoustic wave studies in 2DEGs.[9]

Whereas resistivity is mainly sensitive to the electron-impurity scattering, the thermoelectric power (TEP) of the 2DEG in GaAs-$Ga_{1-x}Al_xAs$ heterojunctions is normally dominated by the electron-phonon interaction and can therefore give important complementary information about the FQH effect. Previously, experimental data on the TEP of 2DEGs (and 2D hole gases) were reported at zero field and in the integer quantum Hall regime[10] and, recently, also in the FQH regime.[11,12]

In this paper we will present our experimental investigations concerning the TEP of the 2DEG in GaAs-$Ga_{1-x}Al_xAs$ heterojunctions at temperatures between 200 mK and 2 K and in magnetic fields up to 20 T corresponding to the regime where features related to the FQH effect become visible. We will show from zero field data that the TEP in our samples is dominated by phonon drag in this temperature range and the arguments are extended to the case of the quantum limit. We will give a phenomenological interpretation of the general behaviour of the TEP as a function of temperature and magnetic field. Finally, the strong change observed in the experimental TEP in moving from filling factor $\nu > 1/3$ to $\nu < 1/3$ will be interpreted within the framework of the CF model[5,6].

2. Basic principles for thermoelectric effects

Thermoelectric effects are caused by a temperature gradient ∇T which leads to a locally induced electric current density $\mathbf{j}_{th} = \epsilon\nabla T$ where ϵ is the thermoelectric tensor. In the experimental situation, where there is no net electric current, an electric field $\mathbf{E} = \mathbf{S}\nabla T$ is built-up which in turn induces a drift current $\mathbf{j}_{el} = \sigma\mathbf{E}$ (σ is the conductivity tensor) to compensate the thermally induced current. The thermopower tensor $\mathbf{S}$ is defined by the relation $\mathbf{S} = \sigma^{-1}\epsilon$. In a 2DEG, with a magnetic field applied in the confinement direction, only two independent components of the thermopower tensor $\mathbf{S}$ exist: the thermopower S_{xx} parallel to ∇T and the Nernst-Ettingshausen coefficient S_{yx} perpendicular to it.

In general the TEP of a 2DEG system is composed of two contributions, diffusion and phonon drag. In GaAs-$Ga_{1-x}Al_xAs$ heterojunctions grown on undoped substrates, the diffusion contribution, which can be related to the thermodynamics of the system and the dominant scattering mechanisms (normally impurity scattering), can normally be neglected compared to phonon-drag TEP.[10] This is the case in this work.

3. Experimental

3.1. Samples

We have studied three high-mobility ($\mu \approx 200$ m^2/Vs) MBE grown GaAs-$Ga_{1-x}Al_xAs$ heterojunction samples (Philips Research Laboratory, Redhill, England) labeled G645 ($n = 0.87 \times 10^{11}$ cm^{-2}), G647 ($n = 0.43 \times 10^{11}$ cm^{-2}), and G650 ($n = 1.0 \times 10^{11}$ cm^{-2}). They consist of a semi-insulating GaAs substrate with a 2DEG in form of a Hall bar with 8 AuGeNi contacts.

3.2. Thermopower measurements

All the experiments were performed at low temperatures; between 350 mK and 2 K we used a ^{3}He system, and for lower temperatures a ^{3}He-^{4}He dilution refrigerator. Magnetic fields were produced with a Bitter coil (up to 20 T) or a superconducting magnet (12 T).

The thermopower measurements were carried out in a vacuum. One end of the free-standing sample was thermally anchored to the ^{3}He bath or the mixing chamber. In order to establish a temperature gradient, a strain gauge resistance was glued to the other end of the sample and used as a heater. The temperature gradient was measured with two calibrated film resistors glued on the rear side of the sample. Electric fields in the sample were measured by means of 25μm Au wires glued to each contact (using silver-filled epoxy), which were in turn In-soldered to 100μm manganin wires of about 1 cm length: these latter being there to eliminate any extraneous heat loss from the sample. For similar reasons, 50μm manganin wires were used to contact the heater and thermometers.

The measured heat conductance of the system should be determined by the GaAs substrate of the samples, and we have checked this experimentally. The measured thermal conductivity of the semi-insulating GaAs substrates, as determined from the applied heating power, the measured temperature gradient and the known dimensions of the samples, follows roughly the expected T^3 dependence appropriate to a constant phonon mean free path (limited by boundary scattering) $L \approx 1$ mm.

In order to measure the TEP, a temperature difference of 10 mK to 100 mK was established with 2-20 Hz pulsed heating. We verified that the thermal response of the system was fast enough to follow the heating frequency by obtaining the same results (within 10%) for higher and lower frequencies. The thermopower S_{xx} and the Nernst-Ettingshausen coefficient S_{yx} were obtained by measuring the voltages on the Hall bar along and perpendicular to the temperature gradient with a lock-in technique.

4. Experimental results and discussion

4.1. Results at zero magnetic field

In Fig. 1 we show the zero-field TEP for the three samples investigated. As expected for phonon drag, the TEP decreases strongly (much stronger than linearly) with decreasing temperature. The variation of the phonon drag with temperature can be understood phenomenologically in either the force-balance approach or the Π approach[13]. These models emphasize either the phonon scattering rate by electrons τ_{pe}^{-1}, or the electron scattering rate by phonons τ_{ep}^{-1}, respectively. We will use the Π approach to interpret our experimental zero-field data and later extend it to the case with a magnetic field.

In the Π approach[14] the temperature dependence of the phonon drag TEP S^g at zero magnetic field can be expressed in terms of τ_{ep}^{-1} as

$$S^g = -\frac{vLm^*}{eT\tau_{ep}} = -\rho^{ph}\frac{nevL}{T} . \tag{1}$$

Here $v \approx 3000$ m/s is the sound velocity, m^* the electron effective mass, n the 2D electron density and $\rho^{ph} = m^*/ne^2\tau_{ep}$ the electronic resistivity due to phonon scattering. In contrast to the measured resistivity ρ which is dominated by impurity scattering in our temperature range, the phonon drag TEP is only sensitive to the electron-phonon scattering described by the very small contribution of ρ^{ph} to ρ.

The temperature dependence of S^g arises from the temperature dependence of $(T\tau_{ep})^{-1}$. For very low temperatures, where the dominant phonon wave-vector for momentum transfer $q = 4.96kT/\hbar v$ is smaller than $2k_F$ (k_F is the Fermi wave vector)[15], S^g varies as T^4 implying a T^5 variation of τ_{ep}^{-1}. This can be crudely understood as arising from a T^3 dependence from the number of phonons and an additional T^2 dependence arising from the predominance of forward scattering for small phonon wave vectors. At higher temperatures (typically $T > 2$ K) the TEP levels off mainly because the phonons are less effective in imparting momentum to the 2DEG because of the small diameter of the Fermi circle.

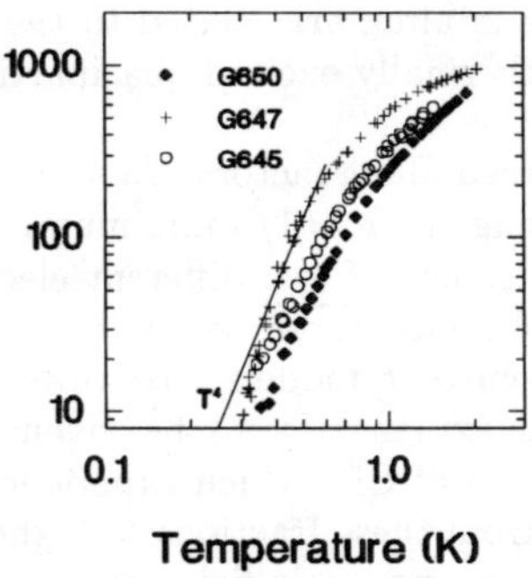

Fig. 1. The zero field TEP as a function of temperature.

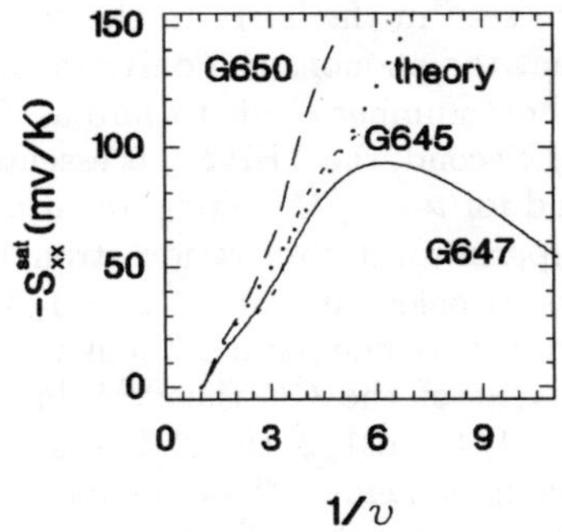

Fig. 2. The field dependence of the saturation TEP for the three different samples compared to theory.

4.2. The TEP in the fractional quantum Hall regime

Two regions of temperature can be distinguished in considering TEP in the quantum limit. First, at high temperatures (typically above 1K) where the FQH effect is not resolved, the system can be described as a non-interacting 2DEG. We will only briefly mention the most important features of this case; more details can be found in Refs. 12 and 16. The temperature dependence (Fig. 2 in Ref. 16) of the drag current, and thus of the TEP, is expected to saturate when the thermal phonon wavelength $\lambda = \hbar v/kT$ becomes comparable to the magnetic length and this is confirmed experimentally.[12] For GaAs heterojunctions, the field dependence of the TEP in the saturation regime is found to be

$$S_{xx}^{sat}(\nu) = 28\,\sqrt{n/n_0}\,\nu^{1/2}(1-\nu)(1+\delta(B))\ \ \mathrm{mV/K}. \qquad (2)$$

n is the 2DEG density (with $n_0 = 10^{15}$ m^{-2}) and $\nu = nh/eB$ is the filling factor. The quantity $\delta(B)$ depends on the ratio between the deformation potential and the polar interaction.[16]

In Fig. 2 the saturation thermopower of the three different samples is shown and compared to the theoretically expected curve for $n = 10^{15}$ m^{-2}. The experimental curves have been extrapolated from the actual results to a temperature where the magnetic length is larger than the characteristic thermal phonon wavelength by using the theoretically expected temperature dependence below saturation. For $\nu > 1/4$, theory and experiment agree reasonably well with a rapid increase of the TEP when entering into the quantum limit and a tendency to saturation for higher fields. However, we cannot yet explain the observed decrease of the TEP for higher fields corresponding to $\nu < 1/5$.

At low temperatures ($T < 1$ K) two features appear which can be related to the FQH effect. The results for G645 and G647 are shown in Fig. 3.

First, minima at fractional filling factors, which are also visible in resistivity, are seen in the TEP. In the case of the TEP these minima are related to the fact that the phonons can only transfer momentum to thermally excited quasiparticles, whose number tends to zero as $T \to 0$.

Second, the TEP S_{xx} is essentially constant between filling factors $1/3 < \nu < 2/3$ and for $\nu < 1/3$, making a distinct jump after passing the $\nu = 1/3$ minimum. This experimental observation strongly supports the existence of two different electron states below and above $\nu = 1/3$. When we first obtained these results[12] we were unable to understand them in terms of the then-current models. However, the advent of the CF theory[5,6] has allowed a natural explanation to be given. At $\nu = 1/4$ and $1/2$, the 2DEG is best described in terms of CFs which experience no magnetic field at these points. As one moves away from these fractions to higher or lower fields, the CF experiences only the difference in magnetic field, and the CF gas can be considered to be at low fields for ν not too far from these points. The structure observed in the resistivity and TEP at the fractions p/q is then described in terms of the "low field" Shubnikov-de Haas oscillations of the 2D CF gas. Strong evidence to support this interpretation has been seen in resistivity measurements[7,8].

We believe the two plateaus in the TEP correspond to phonons interacting with different coupling factors to the CFs centered at the two fractions 1/2 and 1/4. Thus we regard the two plateaus in the TEP as the semi-classical "low field" TEP of the $\nu = 1/2$ and the $\nu = 1/4$ CF gas where quantum effects do not play a major role (which is confirmed by the fact that at these temperatures the fractions are only just visible). We note that when $\hbar\omega_c < kT$, the phonon drag TEP of the usual non-interacting 2DEG is experimentally found to be independent of magnetic fields, i.e. the presence of the Landau levels has no visible effect on the magnitude in this case. Using semiclassical arguments, the magnitude is also found to be independent of $\omega_c\tau$[17] in a manner rather similar to the resistivity. In this picture one views the weak FQH oscillatory structure in resistance and TEP as a small quantum effect in a semiclassical regime.

However, in contrast to TEP, resistivity does not display any distinct discontinuity around $\nu = 1/3$. This appears to have an explanation in the fact that the latter is controlled by electron-impurity scattering, for which the quantum lifetimes of the non-interacting electron and CFs are found to remain similar,[8] while the former is sensitive to CF-phonon coupling which we anticipate will be different in the two cases. Indeed, the ratio of the electron-phonon scattering times in the two distinct plateaus of the TEP can be estimated quantitatively from the TEP data within the framework of the CF model.

Eqn. 1 can be extended to the case where a magnetic field is applied.[16] Using the same argument as at zero field and taking into account the Hall effect we find in the low-disorder limit

$$S^g_{xx} = -\rho^{ph}_{xx} \frac{nevL}{T} . \tag{3}$$

ρ^{ph}_{xx} is the contribution to the resistivity due to phonon absorption. We assume that this relation is still valid for CFs and rewrite it in analogy to Eqn. 1 as

$$S^g_{cf} = -\frac{vLm^*_{cf}}{eT\tau_{cfp}} \quad = \quad -\rho^{ph}_{cf} \frac{nevL}{T} . \tag{4}$$

In this equation, ρ^{ph}_{cf} is the resistivity of the CF gas at $\nu = 1/2$ or $\nu = 1/4$ due to phonon scattering and contains all the information about the internal structure of the particles and their interaction with the phonons. We have defined a CF-phonon scattering rate τ^{-1}_{cfp} by $1/\rho^{ph}_{cf} = ne^2\tau_{cfp}/m^*_{cf}$ where m^*_{cf} is the effective mass of the CF at $\nu = 1/2$ or $\nu = 1/4$.

The magnitude of the experimental S_{xx} thus gives information about the CF-phonon scattering. In particular, we can compare the quantity $m^*_{cf}\tau^{-1}_{cfp}$ of the CF with the zero-field value $m^*\tau^{-1}_{ep}$ for non-interacting electrons.

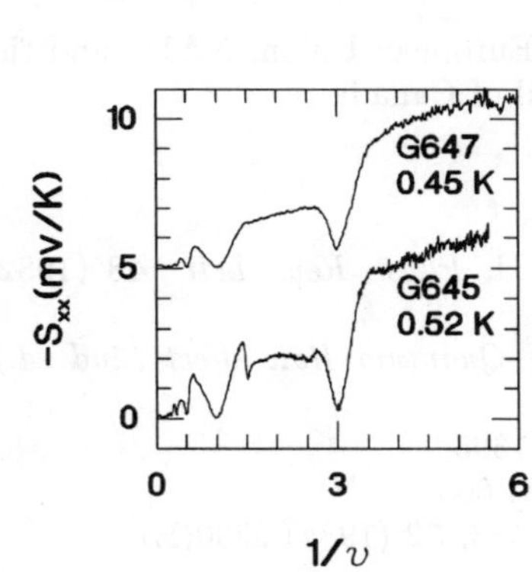

Fig. 3. The low-T TEP of G645 and G647 (shifted by 5 mV/K) as a function of the inverse filling factor.

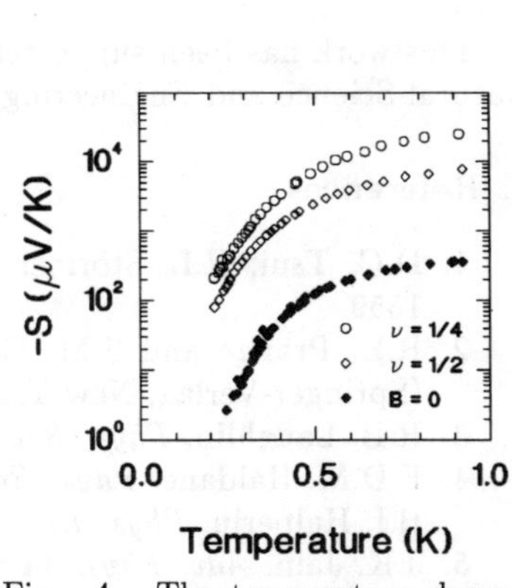

Fig. 4. The temperature dependence of S_{xx} at filling factors 1/2 and 1/4 compared to the zero field $S(T)$ of G647.

In Fig. 4 the temperature dependence of S_{xx} for G647 at $\nu = 1/2$ and $\nu = 1/4$ is shown. It is very similar to the zero field TEP S which is also shown in the same figure for comparison. We see that $m^*_{cf}\tau^{-1}_{cfp}$ is about 20 times larger for the $\nu = 1/2$ CF than the zero field value and about 50 times larger for $\nu = 1/4$. Most of this enhancement can be attributed to a larger effective mass for CFs of $\sim 0.5m_e$ at both $\nu = 1/2$ and $1/4$ [8], as compared to non-interacting electrons of $\sim 0.067m_e$. Leadley *et al.*[8] have suggested a dependence of m^*_{cf} on the magnetic field but we are unable to extract a variation from our data at present. Using $m^*_{cf} = 0.5m_e$ at $\nu = 1/2$, the observed enhancement of $m^*_{cf}\tau^{-1}_{cfp}$ by a factor of 20 implies an about three times higher CF-phonon scattering rate compared to non-interacting electrons. For $\nu = 1/4$ CF, τ^{-1}_{cfp} is another factor 2-3 higher if the same effective mass is used.

5. Conclusions

We have measured the thermoelectric power of 2DEGs in the FQH regime. At high temperatures ($T > 1$ K) the TEP can be described by a model of phonon drag appropriate to a non-interacting quantized 2DEG. At lower temperatures two distinct plateaus appear in the field dependence of S_{xx} which we relate to two different electronic states. In the CF model these two plateaus are associated with the low field TEP of the CFs near 1/2 and 1/4. The enhanced CF phonon drag is believed to be mainly due to a higher effective mass compared to non-interacting electrons, but detailed calculations are clearly required.

6. Acknowledgements

This work has been supported in part by the European Union, NATO and the Natural Science and Engineering Research Council of Canada.

7. References

1. D.C. Tsui, H.L. Störmer, and A.C. Gossard, *Phys. Rev. Lett.* **48** (1982) 1559.
2. R.E. Prange and S.M. Girvin (eds.), *The Quantum Hall effect (2nd ed.)*, (Springer-Verlag, New York, 1990).
3. R.B. Laughlin, *Phys. Rev. Lett.* **50** (1983) 1395.
4. F.D.M. Haldane, *Phys. Rev. Lett.* **51** (1983) 605;
   B.I. Halperin, *Phys. Rev. Lett.* **52** (1984) 1583, **52** (1984) 2390(E).
5. J.K. Jain, *Adv. Phys.* **41** (1992) 105.
6. B.I. Halperin, P.A. Lee, and N. Reed, *Phys. Rev.* **B47** (1993) 7312
7. R.R. Du, H.L. Störmer, D.C. Tsui, L.N. Pfeiffer, K.W. West, *Phys. Rev. Lett.* **70** (1993) 2944.
8. D.R. Leadley, R.J. Nicholas, C.T. Foxon, and J.J. Harris, *Phys. Rev. Lett.* **72** (1994) 1906.
9. R.L. Willet, *Surf. Sci.* **305** (1993) 76;
   R.L. Willet, R.R. Ruel, M.A. Paalanen, K.W. West, and L.N. Pfeiffer, *Phys. Rev.* **B47** (1993) 7344.
10. for a review see B.L. Gallagher and P.N. Butcher, in *Handbook on Semiconductors*, ed. T.S. Moss and P.T. Landsberg, (Elsevier Science Publishers

B.V., Amsterdam, 1992) Vol. 1, p. 817.

11. V. Bayot, X. Ying, M.B. Santos and M. Shayegan, *Europhys. Lett.* **25** (1994) 613.
12. U. Zeitler, J.C. Maan, P. Wyder, R. Fletcher, C.T. Foxon, and J.J. Harris, *Phys. Rev.* **B47** (1993) 16008;
    U. Zeitler, R. Fletcher, J.C. Maan, C.T. Foxon, J.J. Harris, and P. Wyder, *Surf. Sci.* **305** (1994) 91;
    U. Zeitler, PhD thesis, Universität Konstanz (1994, to be published).
13. M.J. Smith and P.N. Butcher, *J. Phys.: Condens. Matter* **2** (1990) 2375.
14. C. Herring, *Phys. Rev.* **96**, (1954) 1163.
15. G.M. Gusev, N.V. Zavaritskii, Z.D. Kvon, and A.A. Yurgens, *Pis'ma Zh. Eksp. Teor. Fiz.* **40** (1984) 275, [*JETP Lett.* **40** (1984) 1056].
16. V.I. Fal'ko and S.V. Iordanskii, *J. Phys.: Condens. Matter* **4** (1992) 9201.
17. X. Zianni, P.N. Butcher, and M.J. Kearney, *Phys. Rev.* **B49** (1994) 7520.

Thermopower of a dilute 2D hole system at half-filled Landau levels and in the reentrant insulating phase

X. Ying,[†] V. Bayot,[†,*] M. B. Santos,[†] and M. Shayegan[†]

[†]*Department of Electrical Engineering, Princeton University, Princeton, NJ 08544, USA*
[*]*Laboratoire PCPM, Université Catholique de Louvain*
Place Croix du Sud 1, B-1348 Louvain-la-Neuve, Belgium

Measurements of the low temperature diffusion thermopower (S_{xx}) of a dilute two-dimensional hole system reveal that the magnitude and temperature dependence of S_{xx} at filling factors ν=1/2 and 3/2 can be explained well in terms of a recently proposed particle-flux composite fermion picture. We also observe that S_{xx} in the reentrant insulating phase at low fillings diverges as $T \to 0$, consistent with hole condensation into a pinned Wigner solid.

Two novel ground states of a two-dimensional (2D) system of interacting particles in a strong magnetic field are of particular current interest: the Fermi-liquid-like ground state at half-integer filling factors (ν) and its associated composite fermions (CFs),[1-2] and the Wigner solid state at sufficiently small ν. We have measured the diagonal thermopower S_{xx} of high-quality 2D hole systems (2DHS) at GaAs/AlGaAs interfaces down to $T \approx 40$mK at these fillings. In the diffusion thermopower regime (typically $T<100$mK), we observe: (1) The magnitude of S_{xx} at ν=1/2 is *less* than that at ν=3/2, and neither the temperature dependence nor the magnitude of S_{xx} is consistent with the universal quantized values expected in a model of non-interacting particles. Instead, the data are remarkably consistent with the CF picture. (2) S_{xx} of the insulating phases reentrant around ν=1/3 and 2/7 diverges as $T \to 0$, suggesting an energy gap separates the ground state from its excitations; this is consistent with the insulating phases being pinned Wigner solid states.

The samples used in the measurements were GaAs/GaAlAs heterojunctions grown by molecular beam epitaxy on undoped (311)A substrates. We concentrate here on a 2DHS with a density of $6.5\times10^{10}/\text{cm}^2$ and low temperature mobility of $7\times10^5\text{cm}^2/\text{Vs}$. Details of our S_{xx} measurements are discussed elsewhere.[3]

The diagonal (R_{xx}) and Hall (R_{xy}) resistances of the 2DHS as a function of applied magnetic field (B) are shown in Fig.1 (left panel). The high quality of the 2DHS is evident by the set of the fractional quantum Hall (FQH) states at $\nu=j/(2j\pm1)$ (j is a positive integer) which, in the CF picture,[1,2] are just the integer quantum Hall (IQH) states of CFs in an effective magnetic field measured from ν=1/2; these include the strong 1/3 and 2/3 states and the high order 2/5, 3/5, 3/7 and 4/7 states. R_{xx} near ν=0.29 and 0.37, on the other hand, diverges as $T \to 0$; these insulating phases reentrant around ν=1/3 and 2/7 FQH liquids can be associated with the localization of the 2DHS into a pinned Wigner solid.[4]

Typical traces of S_{xx} as a function of B are shown in Fig. 1 (right panel). A positive S_{xx} is observed, consistent with hole conduction. Despite the presence of noise (primarily because of the small signal), several features of the data are very clear. At IQH states, S_{xx}, which is a measure of heat or entropy (per carrier), vanishes as $T \rightarrow 0$, reflecting vanishing entropy of the filled Landau levels.[5] By analogy, one also expects the incompressible FQH liquid to have zero entropy and therefore vanishing thermopower as observed. As shown in Fig. 2, the temperature dependences of S_{xx} of the half-filled Landau levels and of the insulating phase are distinctly different. For $T > 100$mK S_{xx} at these fillings follows approximately a T^3 dependence. For $T < 100$mK, S_{xx} at $\nu=1/2$, $3/2$ decreases roughly linearly with T, similar to S_{xx} at $B\approx 0$,[6] and $S_{xx}(\frac{1}{2})<S_{xx}(\frac{3}{2})$ even though $R_{xx}(\frac{1}{2})$ is larger than $R_{xx}(\frac{3}{2})$ by nearly an order of magnitude. On the other hand, as $T \rightarrow 0$, S_{xx} in the insulating phase ($\nu\approx 0.29$) diverges after going through a minimum.

At low temperatures primarily two mechanisms, the diffusion and the phonon-drag, are expected to contribute to the thermopower. In zero magnetic field, the diffusion thermopower depends linearly on T: $S_d = aT$, with $a = \pi^2(p+1)k_B{}^2 / 3eE^f$, where E^f is the Fermi energy of the carriers, and p is a constant (≈ 1 for GaAs/AlGaAs heterojunctions).[7] On the other hand, the phonon drag thermopower (S_{ph}) varies approximately as T^3. In our 2DHS, the temperature dependence of S_{xx} at $B\approx 0$ (Fig. 2) suggests that S_{xx} is dominated by S_{ph} for T>100mK. As T is reduced below 100mK, S_d starts to dominate, exhibiting a roughly linear T dependence from which we obtain $a\approx 80\mu V/K^2$. From this value and using p=1, we obtain a carrier effective mass which is consistent with the weighted hole effective mass of the 2DHS ($\approx 0.3m_o$, where m_o is the free electron mass).[8]

Now we turn to a discussion of S_d in a magnetic field. In a model where electron-electron interactions are ignored, S_d vanishes for filled Landau levels, and reaches maximum values at half-filled Landau levels. For $k_BT \ll \hbar\omega_c$, where $\hbar\omega_c$ is the cyclotron energy, and in the limit of zero impurity scattering, S_{xx} at half-filled Landau levels has the universal quantized values: $S_{xx}(\nu)=k_B \ln 2/e\nu \approx 60/\nu(\mu V/K)$.[5] Clearly this model cannot account for our data in the FQH regime where electron-electron interactions are important. For example the diffusion thermopower $S_{xx}(\frac{1}{2})$ is found to be temperature dependent; its value (~10μV/K at T=100mK) is much less than the predicted universal quantized value (≈120μV/K), and $S_{xx}(\frac{1}{2})\approx 0.7S_{xx}(\frac{3}{2})$ while the model predicts $S_{xx}(\frac{1}{2})=3S_{xx}(\frac{3}{2})$. The magnitude of S_{xx} maxima may be reduced significantly by impurity scattering if $k_BT/\Gamma \ll 1$, where Γ is the Landau level broadening.[5] However, $\Gamma(\frac{1}{2})$ has to be about four times larger than $\Gamma(\frac{3}{2})$ to account for the ratio between $S_{xx}(\frac{1}{2})$ and $S_{xx}(\frac{3}{2})$. Although in principle this is possible, we believe it is unlikely: e.g., specific heat measurements on 2DESs, which were performed for ν>1, have indicated that Γ has nearly the same value at half-integer fillings.[9] Moreover, there is no obvious reason why S_{xx} should vary linearly with T when $k_BT<\Gamma$. We conclude that the non-interacting particle model is unlikely to account for our data. Instead, we believe that electron-electron interactions are dominant and should be considered.

Our data of $S_{xx}(\frac{1}{2})$ and $S_{xx}(\frac{3}{2})$ can find a natural and satisfactory explanation in the CF picture.[8] First, if S_{xx} at these fillings is due to the CFs, its temperature dependence should be similar to that at zero field, as observed (Fig. 2). Second, the ratio of the diffusion thermopower at ν=1/2 and 3/2 should be consistent with the CF Fermi energy: $E^f_{CF}=\hbar^2 k_{CF}^2/2m_{CF}$, where k_{CF} is the CF Fermi wavevector. The CF effective mass, m_{CF}, which is determined by electron-electron interactions, is expected to roughly scale with ℓ_B^{-1}, where $\ell_B=(\hbar/eB)^{1/2}$ is the magnetic length.[2] Since a filled Landau level does not contribute to conduction, we have $k^2_{CF}(\frac{1}{2}) = 3k^2_{CF}(\frac{3}{2})$ and therefore $S_{xx}(\frac{1}{2})/S_{xx}(\frac{3}{2}) = [m_{CF}(\frac{1}{2})/m_{CF}(\frac{3}{2})][k_{CF}(\frac{3}{2})/k_{CF}(\frac{1}{2})]^2 = 1/\sqrt{3} \cong 0.58$. This ratio is close to the value measured in the experiment (≈0.7). Finally, m_{CF} can be estimated from the magnitude of the thermopower. Assuming p≈1 is also valid for CFs, we obtain $m_{CF}(\frac{1}{2}) \approx 0.7m_o$ for B=5.6T with an accuracy of about ±20%. This is a reasonable value for m_{CF} compared to other measurements.[10]

Finally, we consider the divergence of S_{xx} in the insulating phase. In the case of an insulator with an energy gap E_g, the diffusion thermopower is ~ E_g/k_BT~$1/T$.[11] In contrast to this diverging behavior, in an insulating system with a finite density of states at E^f (i.e., with a mobility gap), the thermopower is expected to vanish as $T\rightarrow 0$. The diverging behavior of $S_{xx}(0.29)$ shown in Fig. 2 suggests that an energy gap, rather than a mobility gap, separates the ground state from its excitations. The energy gap can be associated with the energy required to thermally activate defects for conduction at finite temperature in the Wigner solid phase.[3]

In conclusion, we find that the diagonal thermopower S_{xx} of a dilute, high-quality 2DHS is dominated by the diffusion thermopower at low temperatures. Both the approximately linear temperature dependence and the magnitude of the low-temperature S_{xx} at ν=1/2 and 3/2 can be explained well in terms of the CF picture. The divergence of S_{xx} in the reentrant insulating phase can be associated with an energy gap separating the pinned Wigner solid ground state from its thermal excitations.

We thank P. M. Chaikin, B. I. Halperin, S. M. Girvin and H. C. Manoharan for useful discussions. This work is supported by the National Science Foundation through grant DMR-9222418 and the Materials Research Group grant DMR-9224077.

---

1. J. K. Jain, *Phys. Rev. Lett.* **63**, 199 (1989).
2. B. I. Halperin, P. A. Lee, and N. Read, *Phys. Rev.* **B47**, 7312 (1993).
3. V. Bayot, X. Ying, M. B. Santos, and M. Shayegan, *Europhys. Lett.* **25**, 613 (1994).
4. M. B. Santos, et al., *Phys. Rev. Lett.* **68**, 1188(1992); M. B. Santos, et al., *Phys. Rev.* **B46**, 13639(1992).
5. See, for example, B. L. Gallagher and P. N. Butcher, in *Handbook on Semiconductors*, Vol. 1, ed. P. T. Landsberg (Elsevier, Amsterdam, 1992), p. 721.
6. In our experiments we could not reliably measure S_{xx} at B=0. Here we use B≈0 to refer to data taken at B slightly larger than zero (≈0.2T).
7. V. C. Karavolas and P. N. Butcher, *J. Phys.: Condens. Matter* **3**, 2597(1991).
8. X. Ying, V. Bayot, M. B. Santos, and M. Shayegan, unpublished.

9. J. K. Wang, D. C. Tusi, M. Santos, and M. Shayegan, *Phys. Rev.* **B45**, 4384 (1992).
10. R. R. Du, H. L. Stormer, D. C. Tsui, L. N. Pfeiffer, and K. W. West, *Phys. Rev. Lett.* **70**, 2944(1993); H. C. Manoharan, M. Shayegan, and S. J. Klepper, unpublished.
11. P. M. Chaikin, in *Organic Superconductivity*, ed. V. Z. Kresin and W. A. Little, (Plenum Press, New York, 1990), p. 101.

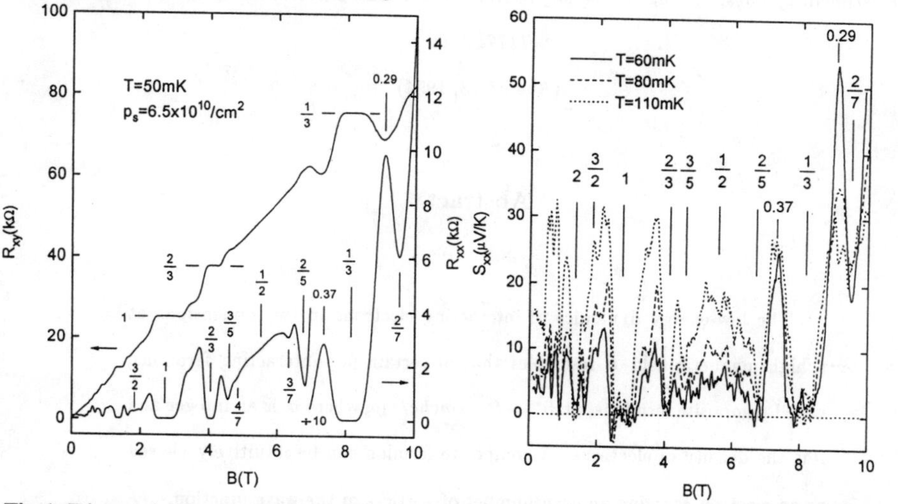

Fig.1. Diagonal (R_{xx}) and Hall (R_{xy}) resistances (left panel) and diagonal thermopower S_{xx} (right panel) versus applied magnetic field. Filling factors are marked with vertical lines.

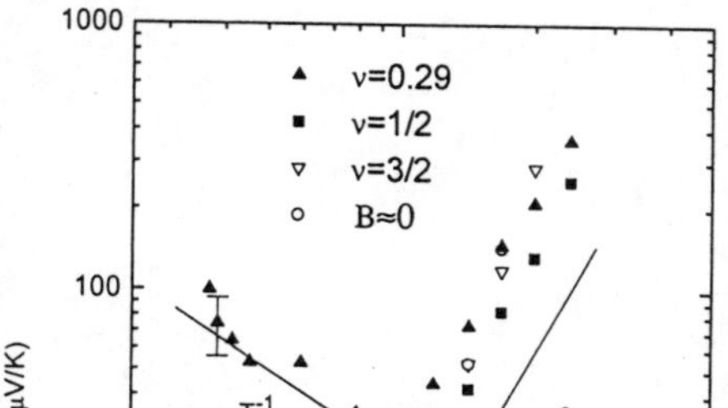

Fig. 2. Temperature dependence of S_{xx} at ν =0.29, 1/2 , 3/2 and at $B \approx 0$. Typical error bars for the low temperature S_{xx} are indicated. Both the magnitude and the temperature dependence of S_{xx} at the half fillings are consistent with the CF picture. The low temperature divergence of S_{xx} in the insulating phase (ν=0.29) suggests an energy gap separates the ground state from its excitations.

Composite Fermions

J.K. Jain

Department of Physics, State University of New York at Stony Brook, Stony Brook, New York

11794-3800

(August 18, 1994)

Abstract

The low-energy dynamics of interacting electrons in two dimensions at a high magnetic field B resembles that of certain non-interacting fermions, called composite fermions, at $B^* = B - 2m(hc/e)\rho$, where m is an integer and ρ is the density of electrons. A composite fermion can be intuitively viewed as an electron carrying an even number of vortices of the wave function.

Spectacular phenomena are observed when a two-dimensional electron system (2DES) is subjected to a high magnetic field (B) [1]. This has led to an intense theoretical study of this system over the last decade. The objective is to provide a simple model which is close to the reality, but at the same time allows an intuitive understanding of the relevant physics. It has now become clear that the formation of a new kind of particles, called composite fermions [2], is at the root of much of the interesting physics exhibited by 2DES at high B. This tutorial lecture briefly discusses some simple features of the composite fermion (CF) theory. For a detailed discussion, the reader is referred to a review article [3].

The composite fermions were proposed in order to provide a unified description of the fractional quantum Hall effect (FQHE) and the integer quantum Hall effect (IQHE). The line of argument that establishes a connection between the two is the following [2].

(i) Consider a system of non-interacting electrons at magnetic field B^*, or filling factor $\nu^* = \rho\phi_0/B^*$ (ρ = density; $\phi_0 = hc/e$ = quantum of flux).

(ii) Its spectrum has bands of degenerate states separated by the cyclotron energy, $\hbar\omega_c$. Start with the many-body (Slater-determinant) states of the lowest band, and imagine attaching to each electron an infinitesimally thin massless solenoid carrying an even number ($2m$) of flux quanta. This is similar to the way "anyons" are described, except that here the resulting particle, electron carrying an even number of flux quanta, obeys ordinary fermionic statistics. We will call this particle a composite fermion. The system of composite fermions at B^* (or filling factor ν^*) obtained in this manner is indistinguishable from the electron system we started with. Therefore, composite fermions form Landau levels (LL's), and exhibit IQHE. Their IQHE will later be recognized as being related to the FQHE of electrons.

(iii) Now we perform a mean-field approximation (MFA), in which the flux, which is a part of the composite fermion, is smeared. This type of MFA was first used by Laughlin in the context of many anyon systems [4]. At the end of the MFA, we obtain electrons at a different magnetic field, which is easily seen to be:

$$B = B^* + 2m\phi_0\rho \tag{1}$$

since $2m$ flux quanta were added per electron. In terms of filling factors, this relationship can be expressed as

$$\nu = \frac{\nu^*}{2m\nu^* + 1} \tag{2}$$

where $\nu = \rho\phi_0/B$. The MFA is reasonable provided the gap survives, in which case, the states below the gap *after* the MFA have a one-to-one correspondence with the states below the gap *before* the MFA. In this way, the problem of interacting electrons at B is related to that of non-interacting composite fermions at B^*.

The fundamental defining property of composite fermions is that they experience an effective field $B^* = B - 2m\rho\phi_0$. Exact diagonalization studies of few electron systems [5,6] have indeed shown a remarkable similarity between the low-energy spectra of interacting electrons at B and non-interacting fermions at B^*. These studies, which are exact in the limit of large B, establish that composite fermions are formed in two-dimensional electron systems in a wide range of filling factors. We emphasize that these studies demonstrate that it is a good approximation to treat composite fermions as non-interacting.

Now we would like to obtain a more microscopic picture for composite fermions. We note that the wave function of the CF system after we add flux quanta is

$$\Phi'_{CF} = \prod_{j<k} \frac{(z_j - z_k)^{2m}}{|z_j - z_k|^{2m}} \Phi_{el} \ , \tag{3}$$

where Φ_{el} is the non-interacting-electron wave function, and $z_j = x_j - iy_j$ denotes the position of the jth electron. This is clearly not a good wave function of the actual state of interacting electrons. In particular, it involves all Landau levels, even when Φ_{el} is restricted to a finite number of lowest LL's, which is not good, since the FQHE occurs at high magnetic fields. Motivated by the Laughlin states [7], the denominator in Eq. (3) is thrown away [2] to obtain

$$\Phi_{CF} = \prod_{j<k} (z_j - z_k)^{2m} \Phi_{el} \ . \tag{4}$$

The wave function Φ_{CF} is superior to Φ'_{CF} in two important respects. First, it involves no more LL's than Φ_{el}, since multiplication by an analytic function cannot result in mixing with higher LL's. In fact, multiplication by $\prod_{j<k}(z_j - z_k)^{2m}$ actually brings a lot of electrons in the higher LL's of Φ_{el} down to the lowest LL. For example, if we multiply the state with two filled LL's (which has only 50% electrons in the lowest LL) by the Jastrow factor, the product state Φ_{CF} (at $\nu = 2/5$) has 96% electrons in the lowest LL [8]. Secondly, Φ_{CF} is more efficient than Φ'_{CF} in keeping electrons apart; the former vanishes as r^{2m+1} as two electrons are brought close to one another, while the latter only as r. Thus, Φ_{CF} explicitly brings out the role of interactions in making composite fermions.

Finally, we would like to formulate an adiabatic scheme which allows us to connect Φ_{CF} to the "true" state. The true state has, in general, certain amount of mixing with higher LL's, depending on B. For simplicity, we consider the theoretical limit of $B \to \infty$, since here, the true wave function is strictly confined to the lowest LL. It is a reasonable guess that the adiabatic continuation of Φ_{CF} to the lowest LL can be accomplished simply by throwing away its part which involves higher LL's, i.e., by projecting it onto the lowest LL. We denote the projected wave function by $\mathcal{P}\Phi_{CF}$. Such projection is not expected to alter Φ_{CF} in any drastic manner, since Φ_{CF} is already predominantly in the lowest LL. In several detailed studies on few electron systems, we have compared $\mathcal{P}\Phi_{CF}$ to the true wave function obtained by brute force numerical diagonalization, and found them to be almost identical [5,6]. Table I shows a comparison of the true electron ground states at the special filling factors $\nu = n/(2n+1)$ with the CF states constructed from the n filled LL states of electrons. We note that the projected wave functions are rather complicated: they do not bind vortices to electrons in any transparent manner, and there is no know way of getting to them without going through the steps described above. Their closeness to the true wave functions, however, 'proves' that the intuitively much simpler wave functions Φ_{CF}, from which they were derived, contain the right physics.

Eqs. (3) and (4) provide two microscopic ways of thinking about the composite fermions. In both, composite fermion is an electron carrying $2m$ vortices of the wave function, but the Jastrow factors, which attach vortices to electrons, are different. While $\prod_{j<k}[(z_j - z_k)/|z_j -$

$z_k|]^{2m}$ is a pure phase factor, $\prod_{j<k}(z_j - z_k)^{2m}$ also serves to keep electrons further apart. Φ_{CF} is closer to the reality, and a better microscopic model for composite fermions.

There is another way to see that composite fermions experience an effective field B^*. Consider a composite fermion taken around in a closed loop of area A. The phase associated with this loop is

$$2\pi BA/\phi_0 - 2\pi 2mN_{enc} \tag{5}$$

where the first term is the Aharonov-Bohm phase and the second term comes from taking the composite fermion around N_{enc} composite fermions enclosed in the loop (since that each vortex produces a phase of 2π). Now make a mean-field approximation by replacing N_{enc} with its average value ρA (where ρ is the density). The effective field experienced by the composite fermions is obtained by equating the phase to $2\pi B^* A/\phi_0$, which gives

$$B^* = B - 2m\phi_0\rho \ . \tag{6}$$

This again shows explicitly how binding of vortices to electrons (i.e., formation of composite fermions) affects the low-energy dynamics in a profound manner.

The residual interaction between composite fermions depends on which microscopic state, Φ'_{CF} or Φ_{CF}, is used as the reference. Since Φ_{CF} already keeps electrons apart, and is only a small perturbation away from the true state, the residual interaction is expected to be weak. This is not the case when composite fermions are modeled as Φ'_{CF}. Recently, a Chern-Simons (CS) field theoretical approach has been developed for calculating with composite fermions [9,10]. In simple language, the Hamiltonian is written as $H_0 + H'$, where H_0 is chosen so that Φ'_{CF} is obtained as its solution. A systematic perturbation theory is then set up in the residual interaction H', which contains, in addition to the Coulomb interaction, certain two and three body interaction terms, called 'gauge interaction'. Perturbation theory is usually carried out at the RPA level. It has been speculated that RPA on Φ'_{CF} produces Φ_{CF} [9], but no proof exists. It should be useful to formulate a theory that performs perturbation around Φ_{CF} (instead of Φ'_{CF}), which is closer to the desired exact solution.

In summary, the following overall picture has emerged. In a range of filling factors, electrons in the lowest LL capture an even number of vortices of the many-particle wave function to transform into composite fermions. The composite fermions experience an effective field B^* and are weakly interacting. They form quasi-LL's, analogous to the LL's of electrons. (We use the term "quasi-LL" for the energy levels of composite fermions, since these are completely within the *lowest real* LL of electrons.) Many experimental observations can be understood straightforwardly from this picture.

The FQHE is explained as follows. There is a gap in the excitation spectrum when composite fermions fill an integer number of quasi-LL's, i.e., when $\nu^* = n$. This corresponds to electron filling factors

$$\nu = \frac{n}{2mn+1} \, , \tag{7}$$

which are precisely the observed fractions. (Both positive and negative integer values are allowed for n [2,5].) An interesting limit is $n \to \infty$, which corresponds to $\nu = 1/2$. Here, if composite fermions exist, they fill an infinite number of quasi-LL's, i.e., form a quasi-Fermi sea. However, *a priori*, the applicability of the composite fermion theory is not obvious here, since there is no gap in the electron state Φ_{el}. Motivated by certain experimental anomalies, Halperin *et al.* [10] proposed that the compressible state at $\nu = 1/2$ is indeed a Fermi sea of composite fermions. Several beautiful experiments [11] have confirmed this view.

In conclusion, composite fermions are defined by the simple property that they experience an effective magnetic field $B^* = B - 2m\phi_0\rho$. In the $B \to \infty$ limit, the most accurate description is provided by $\mathcal{P}\Phi_{CF}$, which, however, does not lend itself to any simple intuitive understanding. The physics of formation of composite fermions and their quasi-LL structure is more apparent in Φ_{CF}', which is closely related to $\mathcal{P}\Phi_{CF}$.

ν	n	N	overlap
$\frac{1}{3}$	1	7	0.9964
		8	0.9954
		9	0.9941
$\frac{2}{5}$	2	6	0.9998
		8	0.9996
$\frac{3}{7}$	3	9	0.9994
$\frac{2}{3}$	-2	6	0.9965
		8	0.9982
		10	0.9940

Table I. This table gives the overlap between the true ground state at $\nu = n/(2n+1)$, obtained numerically for Coulomb interaction, with the CF wave function constructed from the n filled LL state. (Negative values of n imply opposite direction of the magnetic field.) The CF wave function contains no fitting parameter. N is the number of electrons. The wave functions have been properly normalized for the calculation of the overlap. The CF state for $\nu = 1/3$ is identical to the state proposed earlier by Laughlin [7]. The overlaps are taken from: G. Fano, F. Ortolani, and E. Colombo, [Phys. Rev. B **34**, 2670 (1986)] for $\nu = 1/3$; Ref. [5] for 2/5; and Ref. [6] for 3/7 and 2/3.

REFERENCES

[1] D.C. Tsui, H.L. Stormer, and A.C. Gossard, Phys. Rev. Lett. **48**, 1559 (1982).

[2] J.K. Jain, Phys. Rev. Lett. **63**, 199 (1989); Phys. Rev. B **41**, 7653 (1990).

[3] J.K. Jain, Adv. Phys. **41**, 105 (1992).

[4] R.B. Laughlin, Phys. Rev. Lett. **60**, 2677 (1988).

[5] G. Dev and J.K. Jain, Phys. Rev. Lett. **69**, 2843 (1992).

[6] X.G. Wu, G. Dev, and J.K. Jain, Phys. Rev. Lett. **71**, 153 (1993).

[7] R.B. Laughlin, Phys. Rev. Lett. **50**, 1395 (1983).

[8] N. Trivedi and J.K. Jain, Mod. Phys. Lett. B **5**, 503 (1991).

[9] A. Lopez and E. Fradkin, Phys. Rev. B **47**, 7080 (1993).

[10] B.I. Halperin, P.A. Lee, and N. Read, Phys. Rev. B **47**, 7312 (1993).

[11] W. Kang *et al.*, Phys. Rev. Lett. **71**, 3850 (1993); V.J. Goldman *et al.*, Phys. Rev. Lett. **72**, 2065 (1994); R.R. Du *et al.*, Phys. Rev. Lett. **70**, 2944 (1993); D.R. Leadley *et al.*, Phys. Rev. Lett. **72**, 1906 (1994); R.R. Du *et al.*, Solid State Commun. **90**, 71 (1994); R.L. Willett *et al.*, Phys. Rev. Lett. **71**, 3846 (1993).

COMPOSITE-FERMION ANALYSIS OF THE FERROMAGNETIC SPIN WAVE IN THE FRACTIONAL QUANTUM HALL SYSTEM

H. AOKI and T. NAKAJIMA*

Department of Physics, University of Tokyo, Hongo, Tokyo 113, Japan

Abstract : We give a composite-fermion theory for the ferromagnetic spin-wave excitations in the spin-polarized fractional quantum Hall liquid with odd fractions in the spherical geometry. The result is shown to accurately describe the exact numerical result for finite systems, which indicates that the composite picture extends to excited states for wavenumbers smaller than the inverse magnetic length.

1 Introduction

The enduring fascination of the fractional quantum Hall (FQH) effect[1] comes largely from its gauge field theoretic aspects. The composite-fermion picture, initiated by Jain,[2] asserts that a quantum Hall liquid of electrons in an external magnetic field B of $\nu^{-1} = 2m+1$ flux quanta per electron (ν: Landau level filling, m: an integer), may be regarded, in a mean-field sense, as a liquid of composite fermions (electron + $2m$ flux quanta) with $\nu^{-1} = 1$. Although there is a mounting evidence [3] that the composite-particle picture, or the Chern-Simons (CS) field theory,[4,5] is surprisingly adequate, the attention has been primarily focussed on the ground state and transport properties. Here we explore the ability of the composite picture to describe excitations, which we believe is a true test for a many-body theory. Since the Laughlin liquid is incompressible, the low-lying collective excitation is the ferromagnetic spin wave that restores the broken SU(2) symmetry[7].

Role of spins in strongly correlated systems is interesting in its own right, but the peculiarity in the FQH system is that winding up the spin state costs an energy because the spins tend to allign due to the electron correlation. In the composite-boson picture, the FQH liquid has an off-diagonal long-range order similar to a Bose condensation. Unlike the superfluid ^{3}He where the nuclear spin is relevant[6], the question in the FQH system is the electron spin itself.

2 Formulation for the spin wave of composite fermions

We start from the spin-wave theory of Kallin and Halperin[8] exactly given for $\nu = 1$. If we re-formulate this in the spherical geometry to exploit the rotational symmetry, everything is written in terms of angular momentum quantum numbers, and the wavenumber k is related to the total angular momentum L via $k = L/R$, where R is the radius of the sphere. When the total number of magnetic flux quanta going out of the sphere is $2S$ (an integer due to Dirac's condition), ν is given by $2S = \nu^{-1}N-$ integer with N being the number of electrons.

The spin-wave excitation spectrum in the spherical geometry requires a tedious calculation, but the result is elegantly given in terms of Wigner's $6j$-symbol, familiar in the nuclear physics, as

$$\omega_L - g\mu_B B = \sum_{J=0}^{2S} (2J+1)\,(-1)^{2S-J}\, V_J \left[\frac{1}{2S+1} - (-1)^{2S-J} \begin{Bmatrix} S & S & L \\ S & S & J \end{Bmatrix} \right], \quad (1)$$

for $\nu = 1$, where $L = 0, 1, \cdots, 2S$, $\{ {SSL \atop SSJ} \}$ is the $6j$-symbol, and V_J is the Haldane pseudopotential for the relative angular momentum $2S - J$ [9]. Since $\{ {SS0 \atop SSJ} \} = (-1)^{2S-J}/(2S+1)$, Larmor's theorem, $\omega_0 = g\mu_B B$ at $k = 0$, is guaranteed.

Now, for $\nu = 1/(2m+1)$ we make $2m$ flux quanta coalesce to each electron to translate the system back into the $\nu = 1$ state, which amounts to a shift of the relative angular momentum by $2m$. Then the advantage of working in the spherical geometry is that the transformation from $\nu = 1/(2m+1)$ into $\nu = 1$ composite system is simply given by

$$2\tilde{S} = 2S/(2m+1) = N - 1, \quad \tilde{V}_{2\tilde{S}-(n-2m)} / \frac{e^2}{\tilde{\ell}} = V_{2S-n} / \frac{e^2}{\ell}. \quad (2)$$

The field is reduced, in a mean-field sense, to $\tilde{B} = B/(2m+1)$, so that the magnetic length $\ell = \sqrt{c\hbar/eB}$ changes into $\tilde{\ell} = \sqrt{2m+1}\,\ell$.

If we plug this transformation into the '$6j$-formula', eqn(1), we arrive at the desired expression for $\nu = 1/(2m+1)$ as

$$\omega_L - g\mu_B B = \sum_{J=0}^{2\tilde{S}} (2J+1)\,(-1)^{2\tilde{S}-J}\, \tilde{V}_J \left[\frac{1}{2\tilde{S}+1} - (-1)^{2\tilde{S}-J} \begin{Bmatrix} \tilde{S} & \tilde{S} & L \\ \tilde{S} & \tilde{S} & J \end{Bmatrix} \right], \quad (3)$$

where the range of L now reduces to $L = 0, 1, \cdots, 2\tilde{S}$.

3 Comparison with numerical results and discussions

We now introduce the numerical results for the low-lying excitations in the spherical geometry for $\nu = 1/3, 1/5$ in Fig.1. In the lowest branch (spin-wave mode) of one-spin-flip excitations (with the change in the total spin $\Delta S_{tot} = -1$), we immediately recognize that the spin mode in these finite systems appear only in the predicted range $0 \leq L \leq 2\tilde{S} = N - 1$, while naively there is no reason why the mode should not extend to $0 \leq L \leq 2S = (2m+1)(N-1)$. In fact, higher-energy spin excitations do indeed exist for larger L in Fig.1.

The spin-wave dispersion curve predicted from the composite-fermion theory, eqn(3), also exhibits an excellent agreement with the exact result in Fig.1. A deviation starts for $k > \ell^{-1}$, implying that the effect of fluctuations from the mean CS field becomes appreciable for large-wavenumber excitations.

To check the sample-size dependence, we have applied the composite picture to a larger 51-electron system in Fig.2. For comparison we also plot the result of the single-mode approximation (SMA) devised for long-wavelength spin waves given in terms

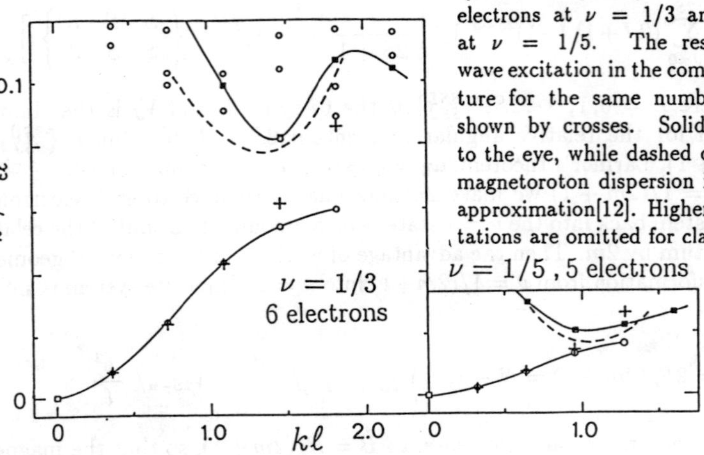

Fig.1 One-spin-flip excitations with $\Delta S_{\rm tot} = -1$ (open circles) and charge-excitations with $\Delta S_{\rm tot} = 0$ (solid squares) for a FQH system of spin 1/2 electrons with (a) 6 electrons at $\nu = 1/3$ and (b) 5 electrons at $\nu = 1/5$. The result for the spin-wave excitation in the composite-fermion picture for the same number of electrons is shown by crosses. Solid curves are guide to the eye, while dashed curves indicate the magnetoroton dispersion in the single-mode approximation[12]. Higher $\Delta S_{\rm tot} = -1$ excitations are omitted for clarity in (b).

of the radial distribution function by Rasolt *et al*[10]. All of the exact, composite and SMA results are seen to agree with each other for $k < \ell^{-1}$ up to a finite-size correction in the Haldane pseudopotential.

The more we attach the flux to the fermion to decrease $\nu = 1 \rightarrow 1/3 \rightarrow 1/5 ...$, the softer the spin wave becomes as numerically depicted in Fig.3. This may be seen by transforming the Kallin-Halperin formula via eqn(2), where $\tilde{V}$ becomes sharply reduced.

Crudely the reason why the composite picture is good may be traced back to the 'correlation hole', i.e., electrons tend to avoid each other over the magnetic length scale. Still, the agreement is rather surprising, since there is no apriori reason why the CS gauge field, which is highly nonlocal, should conform to the mean field. As for the multi-spin flips, all the low-lying excitations at $\nu = 1$ can be entirely interpreted in terms of the multiplets of weakly-interacting magnons where the interaction is attractive at short distances.[11] It is interesting to ask the applicability of the composite picture to these excitations. We also notice in Fig.1 that the roton-like charge-excitation mode (with $\Delta S_{\rm tot} = 0$) exists as well. Extension of the composite picture to these modes is another interesting problem.[13]

This work was in part supported by a Grant-in-Aid from the Ministry of Education, Science and Culture, Japan.

References

* Present address: Semiconductor Research Lab, Mitsubishi Electric Co., Amagasaki, Hyogo 661, Japan.

[1] *The Quantum Hall Effect* 2nd ed., edited by R.E. Prange and S.M. Girvin (Springer,

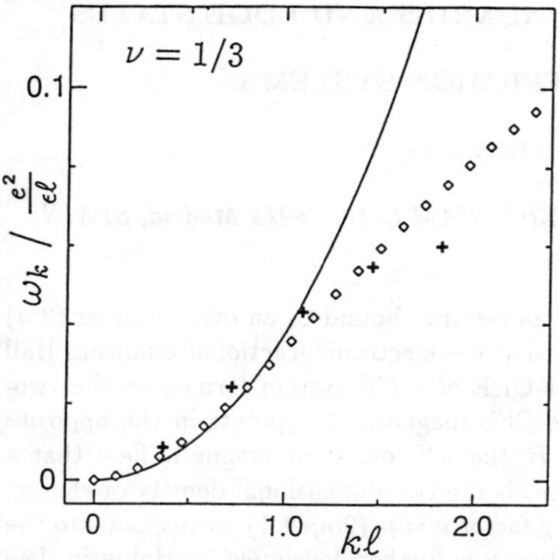

Fig.2 The spin-wave excitation spectrum for a FQH system of spin 1/2 electrons at $\nu = 1/3$ from the composite-fermion mean-field approximation in a spherical system with 51 electrons (open diamonds), the single-mode approximation for an infinite (flat) system (solid curve) and the exact result for a 6-electron system (crosses).

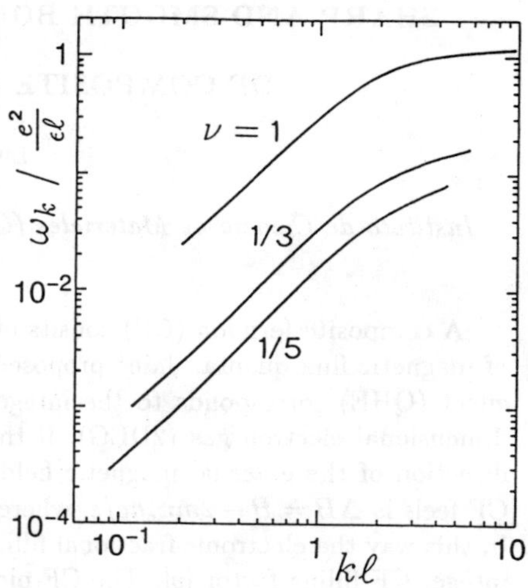

Fig.3 The composite-fermion mean-field result for the spin-wave dispersion for a FQH system of 51 spin 1/2 electrons for various fractions is shown in double logarithmic scale along with the $\nu = 1$ result of Kallin and Halperin.

New York, 1990); T. Chakraborty and P. Pietiläinen, *The Fractional Quantum Hall Effect* (Springer, Berlin, 1988).

[2] J.K. Jain, *Phys. Rev. Lett.* **63** (1989) 199.

[3] R.R. Du et al, *Phys. Rev. Lett.* **70** (1993) 2944; R.L. Willett et al, *Phys. Rev. Lett.* **71** (1993) 3846; W. Kang et al, *ibid* (1993) 3850; D.R. Leadley et al, *Phys. Rev. Lett.* **72** (1994) 1906; V.J. Goldman et al, *ibid* (1994) 2065.

[4] S.C. Zhang, T.H. Hansson and S. Kivelson, *Phys. Rev. Lett.* **62** (1989) 82; N. Read, *ibid* **62** (1989) 86; Z.F. Ezawa, M. Hotta and A. Iwazaki, *Phys. Rev.* B **46** (1992) 7765.

[5] B.I. Halperin, P.A. Lee and N. Read, *Phys. Rev.* B **47** (1993) 7312; E. Rezayi and N. Read, *Phys. Rev. Lett.* **72** (1994) 900.

[6] A.S. Borovik-Romanov et al, *JETP Lett.* **40** (1984) 1033.

[7] A. Pinczuk et al, *Phys. Rev. Lett.* **70** (1993) 3983.

[8] C. Kallin and B.I. Halperin, *Phys. Rev.* B **30** (1984) 5655.

[9] F.D.M. Haldane, *Phys. Rev. Lett.* **51** (1983) 605.

[10] M. Rasolt and A.H. MacDonald, *Phys. Rev.* B **34** (1986) 5530.

[11] T. Nakajima and H. Aoki, Physica B, in the press.

[12] S.M. Girvin, A.H. MacDonald and P.M. Platzman, *Phys. Rev.* B **33** (1986) 2481.

[13] A.H. MacDonald and S.C. Zhang, preprint.

SHARP AND SMOOTH BOUNDARIES AND EDGE STATES OF COMPOSITE FERMION SYSTEMS.

Luis Brey

Instituto de Ciencia de Materiales (CSIC), UAM C-12, 28049 Madrid, SPAIN.

A composite fermion (CF) consits of an electron bound to an even number ($2m$) of magnetic flux quanta. Jain[1] proposed that the electronic fractional quantum Hall effect (QHE) corresponds to the integer QHE of a CF system formed in the two-dimensional electron gas (2DEG). If the CF's magnetic flux points in the opposite direction of the external magnetic field B, the self-consistent magnetic field that a CF feels is $\Delta B = B - 2mn_e hc/e$, where n_e is the two-dimensional density of charge. In this way the electronic fractional filling factor $\nu = p/(2mp+1)$ corresponds to the integer CF filling factor $|p|$. The CF picture was further developed by Halperin, Lee and Read[2], who obtain that a 2DEG, with Coulomb interactions between electrons, has a strong Fermi surface at filling factor 1/2.

In this work we study the edge of the CF system in the Hartree approximation. We consider that the CF move in the $x-y$ plane, and the external potential v_0 depends only on x. Using the Landau gauge for describing the external magnetic field $\vec{A} = (0, Bx, 0)$, in the Hartree aproximation[3,4] the one-CF eigenfunction takes the form, $\varphi_{j,k}(\mathbf{r}) = e^{iky}\phi_{j,k}(x)$. The one-dimesional eigen-function, $\phi_{j,k}$, and their corresponding eigenenergies $\varepsilon_{j,k}$ are obtained by solving an one-dimensional Schrödinger equation, with the following Hamiltonian,

$$\frac{\hbar\omega_c}{2}\ell^2\left[-\frac{d^2}{dx^2}+\left(k+\frac{x}{\ell^2}-\tilde{\phi}\pi\int dx'\rho(x')\frac{x-x'}{|x-x'|}\right)^2\right]+\hbar\pi\tilde{\phi}\int dx' J(x')\frac{x-x'}{|x-x'|}+V_H+v_0 . \tag{1}$$

Here V_H is the usual Hartree potential, $\omega_c = eB/m_b c$ is the cyclotron frequency and $\ell = \sqrt{\hbar c/eB}$ is the magnetic length. In this equation $\rho(x) = \sum_{k,j}^{occ} |\phi_{k,j}(x)|^2$ is the charge density and $J(x)$ has the expression,

$$J(x) = \frac{\omega_c}{L_y}\sum_{j,k}^{occ}|\phi_{j,k}(x)|^2\ell^2\left(k+\frac{x}{\ell^2}-\tilde{\phi}\pi\int dx'\rho(x')\mathrm{sgn}(x-x')\right) . \tag{2}$$

Here *occ* means that the sum is just over the occupied states. In our case, where the CF's are confined in a wire, there is a continuum of eigen-states and therefore the higher energy of all occupied states defines the Fermi energy of the system. For studying the edge structure of the CF at different edge potentials we take the external potential v_0 as the created by a fictitious distribution of positive charge with the trapezoidal shape,

$$\rho_+(x) = \frac{n_+}{L} \begin{cases} x, & 0 < x < L \\ L, & L < x < W - L \\ W - x & W - L < x < W \, . \end{cases} \tag{3}$$

We choose the parameters which characterized $\rho_+(x)$ in order to have overall charge neutrallity in the system. In presence of B the electronic filling factor at the center of this system is $\nu = 2\pi n_+ \ell^2$. With this choice for v_0 the sum of the Hartree and the external potential has the form,

$$V_H + v_0 = 2\frac{e^2}{\epsilon} \int dx' \left[\rho_+(x') - \rho(x')\right] \log|x - x'| \, . \tag{4}$$

In order to perform the calculations we use the typical values corresponding to GaAs, ϵ=12.5, m_b=0.067 and B=5T.

We have previously[3] applied this method to the study of sharp boundaries (L=0) of CF systems corresponding to different electronic filling factors. In the case of the 2/3 quantum Hall state we have obtained that a droplet of charge with filling factor 1 appears at the edge of the system. In the case of the 1/3 quantum Hall state the droplet of charge corresponds to filling factor 2/5. For an 2DEG at filling factor 1/2 we have found[3] Friedel oscillations, with period $\sim \pi\ell$, at the edge of the system, these oscillations reveals the existence, at ν=1/2, of a Fermi surface in the system.

In this work we are to show results for the case of a smooth edge and for the case of a sharp barrier separating two 2DEG in the fractional QHE regime. The existence of gaps for the CF at magnetic fields corresponding to electronic fractional filling factor $\nu_p = p/(2p+1)$ implies discontinuities of the chemical potential of the CF at densities $\nu_p Be/h$. Then, at the edge of the system one can expect a stepwise decrease of density from the center of the wire or a setwise increase of the self-consistent potential[5]. In Figure 1 we study the case of a very smooth edge. Figure 1 shows the self-consistent charge density profile and the band structure for the case $L = 70\ell$, $W = 2\ell$ and n_+ corresponding to ν=1/2 at the center of the system. We see in Fig.1a that at the edge of the system there is a wide plateau with charge density corresponding to ν=1/3. Also there are some droplets in the charge density profile. Comparing figures 1a and 1b we associate the plateau and the droplets in the charge density profile to regions in the reciprocal space where the CF filling factor is 1, which corresponds to the electronic filing factor 1/3. The existence of this plateau and these droplets means that there is not metallic regions at the edge of the CF system. This behaviour is because in the CF system, the relation between the charge density and the total self-consistent magnetic field prevents the formation of regions with dispersionless band structure and nonuniform density. The CF system prefers forming droplets instead of having a thicker plateau at ν=1/3 because in this way, it minimize electrostatic energy We see also the appearance of a plateau at ν=3/7, but not at $\nu = 2/5$.

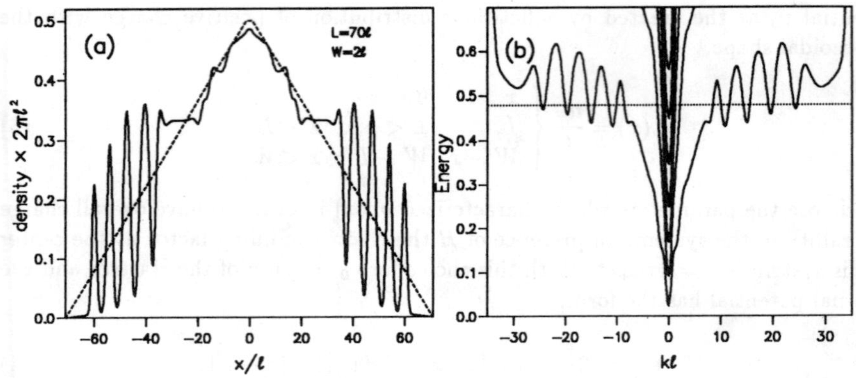

Figure 1. (a) Self-consistent charge density profile and (b) band structure and Fermi energy (dotted line), in units of $\hbar\omega_c$, for the case $L = 70\ell$, $W = 2\ell$ and n_+ corresponding to a filling factor 1/2 at the center of the well.

This is because the energy gap at $\nu = 1/3$ is much large than the the energy gap at $\nu = 2/5$ and then the CF's prefer being in the ν=1/3 state than creating another plateau at electronic filling factor 2/5. On the other hand, the difference in energy between the self-consistent confining potential at the center of the plateaus $\nu = 1/3$ and $\nu = 3/7$ is bigger than the difference of the corresponding energy gaps. Therefore the system prefers to form a plateau at $\nu = 3/7$. There are other features in the density profile but they are difficult to assign to other fractional filling factor as they are mixed with confinement effects.

The method developed here can also be applied in order to study the electronic structure of two 2DEG in the QHE regime separated by a potential barrier of thickness d and height V_b. In figure 2 we show the results for the case of a square potential barrier, 4ℓ thick and with a height of $0.25\hbar\omega_c$, separating two 2DEG at filling factor 1/2. In Fig. 2a it is plotted the self-consistent charge density profile. Observe that a big amount of charge have penetrated in the barrier region. Note also that the potential barrier induce the existence of Friedel oscillations in the 2DEG. The sum of V_H, v_0 and the potential barrier is plotted in Fig. 2(b). Note that due to the self-consistent Hartree potential the effective potential barrier is just $\sim 0.1\hbar\omega_c$ instead of its bare value $0.25\hbar\omega_c$. In Fig. 2(b) it is also plotted the total self-consistent vector potential that the CF feel. In the regions where the density does not correspond to the filling factor 1/2, there is a residual magnetic field and the vector potential in these regions is not constant. Therefore, far from the barrier, in the *bulk* part, the vector potential is constant. The difference between the vector potential at $x = \pm\infty$ (ΔA) is related with the integral of the magnetic field in the perturbed region. In this form, if there is not charge in the barrier region, so that in this region the total magnetic field is the external one, ΔA=d (Δ and d in magnetic units).

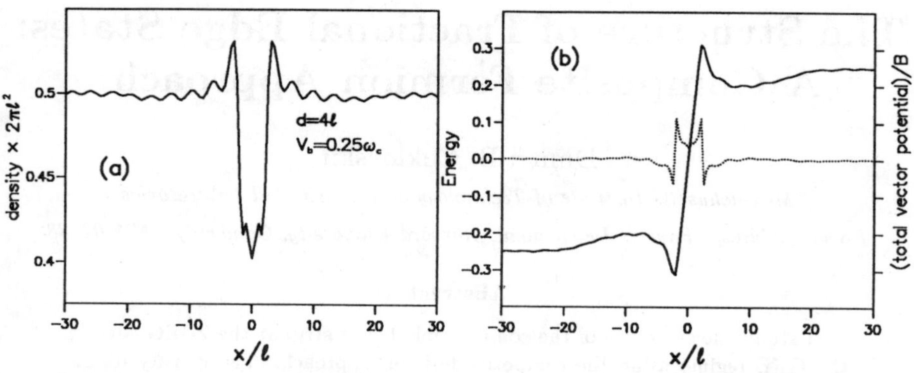

Figure 2. (a) Self-consistent charge density profile and (b) self-consistent total vector potential (continuous line) and the sum of V_H, v_0 and the barrier potential (dotted line) for the case of a barrier (d=4ℓ and V_b=0.25$\hbar\omega_c$) separating two 2DEG with filling factor 1/2. These magnitudes are plotted in magnetic units.

In Fig.2(b) we observe that the self-consistent calculation has reduced ΔA from 4 to 0.5. The precise value of ΔA is very important in order to study the transport through the barrier because the Lorentz force produces that only particles with velocities v_x and v_y verifying the condition (in magnetic units), $(\Delta A)^2 - 2\Delta A v_y - v_x^2 < 0$, can pass from the left to the right part of the barrier[6]. In the case of linear response and in our case where the Fermi energy in magnetic units is 1/2, the above condition implies the opacity of barriers when $\Delta A > 2$.

In conclussion we have presented a method based in the CF picture, for calculating the electronic edge structure of a 2DEG in the fractional QHE regime. We have applied this method to the case of smooth boundaries and we have obtained the existence of plateaus and droplets of charge at fractional filling factors. Also we have studied the electronic structure of two 2DEG separated by a potential barriers. In this case we have discussed the importance of obtaining a precise self-consistent total vector potential.

Acknowledgements This work has been supported by CICyT of Spain under contract No. MAT 91-0201.

1. J.K.Jain, Phys. Rev. Lett. **63**, 199 (1989); Phys.Rev.B **40**, 8079 (1989); Adv. Phys. **41**, 105 (1992).
2. B.I.Halperin, P.A. Lee and N. Read, Phys.Rev. B **47**, 7312 (1993).
3. Luis Brey *preprint*.
4. D.B.Chklovskii, this conference.
5. C.W.L.Beenaker, Phys.Rev.Lett. **64**, 216 (1990); D.B.Chklovskii *et al*, Phys.Rev.B **46**, 4026 (1992);, L.Brey *et al*, Phys.Rev.B **47**, 13884 (1993).
6. A.Matulis *et al* , Phys.Rev.Lett. **72**, 1518 (1994).

The Structure of Fractional Edge States: A Composite Fermion Approach.

Dmitri B. Chklovskii

Massachusetts Institute of Technology and AT&T Bell Laboratories

Present address: Physics Department, Harvard University, Cambridge, MA 02138

Abstract

I study the structure of the compressible liquid strip at the 2DEG edge in the QHE regime using the composite fermion approach. The density distribution and the fermion energy spectrum are obtained numerically in Hartree approximation for the $\nu = 1$ state. As the strip width a is increased fractional channels appear in pairs. For a wide strip I find quasiclassically the number of channels $p \sim \sqrt{a/l_H}$, where l_H is the magnetic length.

In the single–electron theory of edge states[1, 2] the 2DEG edge is viewed as a set of one–dimensional chiral channels. Each channel can be assigned a chemical potential and their number is just the Landau level filling factor ν in the bulk. This simple picture has been very useful in interpreting a large number of experiments.[3] However, in the case of a realistic confining potential edge channels are known to be compressible liquid strips of finite width.[4] This explains a number of experimental observations which were unresolved within the single–electron picture. The existence of the finite width channels brings up the problem of their internal structure. In particular, one may wonder about the number of one–dimensional channels within a compressible strip of a given width. These channels could possibly be probed in non-local transport, edge magnetoplasmon, and tunneling experiments. The complexity of the problem is due to a strongly correlated nature of the electron state. Any theory that solves this problem should also describe the formation of the incompressible FQHE strips, which must exist in the limit of a very smooth edge. In this paper I present such a theory based on the composite fermion approach.[5, 7, 6] The advantage of this approach is that the important correlations are automatically built in in the single–particle (Hartree) approximation for composite fermions.

Without loss of generality I consider the edge of the $\nu = 1$ quantum Hall state and restrict myself to the fractions of the principal sequence $\nu = p/(2p+1)$.[5] I assume that the system is translationally invariant along the edge.[1] The edge confinement is modeled by introducing a positive background charge density which is uniform in the bulk and goes linearly to zero over the distance a. Since any unbalanced charge

[1]It is possible that in reality the translational invariance is broken at the edge because of the Wigner solid formation or by the presence of disorder.

costs electrostatic energy electron density must roughly follow that of the positive background. Thus lengthscale a determines the smoothness of the 2DEG edge.

The case of the sharp edge ($a \sim l_H$) can be treated numerically in Hartree approximation for composite fermions. The self-consistent Hamiltonian for the gauge field problem has been derived in the context of anyon superconductivity.[8, 9] Density distribution and composite fermion energy spectrum obtained from the calculation for various values of a are presented in Fig.1. A similar calculation has been performed by Brey.[10] Next I give a simple physical interpretation of the results.

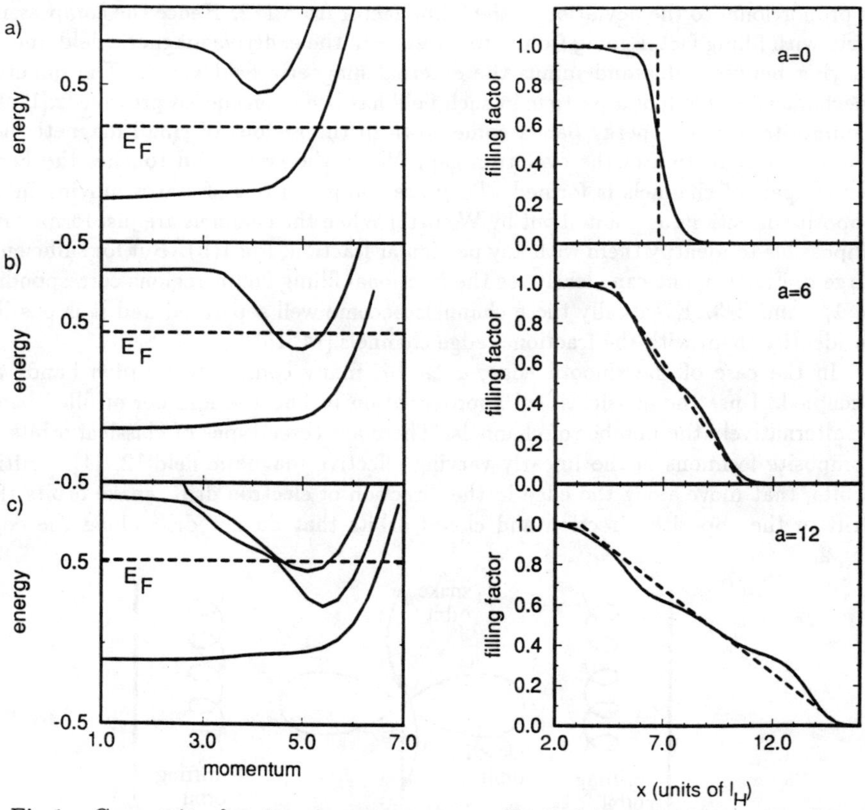

Fig.1. Composite fermion energy spectrum (dashed line designates the Fermi level) and the density profile (dashed line designates the positive background charge) for the sharp and smooth edge. (a) $a = 0$, (b) $a = 6l_H$, (c) $a = 12l_H$.

When $a = 0$ confinement potential dominates over electron–electron interactions. Hence I expect the single–electron theory to valid, implying that there is just a single one–dimensional channel. This can be also understood in terms of composite fermions, Fig.1(a). Attachment of two flux quanta maps the $\nu = 1$ electron state to the single filled Landau level for composite fermions with the magnetic field direction reversed. This level bends up and crosses the Fermi level at the edge forming a single channel. Higher Landau levels for composite fermions remain empty. As the confinement gets smoother (larger a) a finite width region with the partially filled Landau level must emerge. Single–electron theory is useless in this case since it predicts a huge degeneracy. The composite fermion approach, on the other hand, yields an interesting picture. The effective magnetic field seen by composite fermions is proportional to the deviation of the filling factor from 1/2. Hence the compressible strip with filling factor going from 0 to 1 results in the effective magnetic field linearly varying between plus and minus the external magnetic field value. The quantum mechanical problem of a particle in such field has been considered previously.[11, 12] Composite fermion energy bands come down in the regions of small magnetic field due to the reduction in the cyclotron gap. When the next band touches the Fermi level a pair of channels is formed. They are composed out of states moving in the opposite directions as pointed out by Wen.[13] When the channels are just formed it is impossible to identify them with any particular fraction, Fig.1(b). But for sufficiently large a, Fig.1(c), one can clearly see the fractional filling factor regions corresponding to 1/3 and 2/3. Eventually these channels become well separated and it is possible to identify them with the fractional edge channels.[14, 15]

In the case of the smooth edge, $a \gg l_H$, many composite fermion bands are occupied. I use the quasiclassical approximation to find the number of filled bands or alternatively the number of channels. There are three types of classical orbits for composite fermions in the linearly varying effective magnetic field[12, 11]: drifting orbits, that move along the edge in the direction of electron drift, snake orbits that drift in the opposite direction and closed orbits that do not drift along the edge, Fig.2.

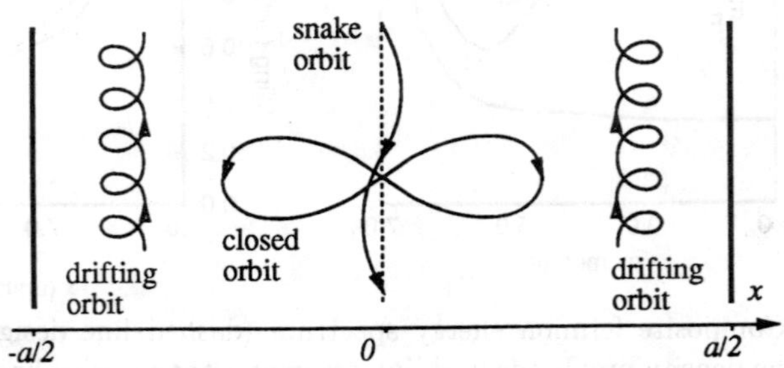

Fig.2. Classical orbits of composite fermions in a varying effective magnetic field neglecting the variation in the Fermi velocity.

In order to determine the total number of channels I have to consider a closed orbit as it corresponds to the bottom of the last occupied band. Cyclotron radius as a function of the distance from the zero effective field line x is:

$$R_c \sim l_H a/|x| \tag{1}$$

where I have used the fact that the cyclotron radius at the boundary of the compressible strip ($x = -a$) is l_H. A periodic orbit should have the extent in x direction such that R_c at x is equal to x. From Eq.(1) I find that $R_c \sim \sqrt{l_H a}$.

The number of channels is equal to the number of the effective flux quanta enclosed by the orbit. The area of the orbit is proportional to R_c^2. Area corresponding to one flux quantum is $l_H^2 R_c/a$. Then the number of flux quanta enclosed is

$$p \sim \frac{R_c^2}{l_H^2 R_c/a} \sim \sqrt{\frac{a}{l_H}} \tag{2}$$

This gives the number of channels at the smooth 2DEG edge.

In the case of the sufficiently smooth edge composite fermion channels must become wide. Then one should make a transformation to the composite fermions of the next generation and follow the same line of argument. Eq.(2) only counts the channels for composite fermions of the principal sequence.

I am grateful to B.I. Halperin, P.A. Lee, and X.-G. Wen for many helpful discussions and to F.G. Pikus for advice on computer simulation. I would like to thank L. Brey for sending me the preprint of his work prior to publication. This work has been supported by NSF Grant No. DMR92-16007.

References

[1] B.I. Halperin, Phys. Rev. B **25**, 2185 (1982).

[2] M. Büttiker, Phys. Rev. Lett. **57**, 1761 (1986).

[3] C.W.J. Beenakker *et al.*, in *Solid State Physics*, ed. H. Ehrenreich *et al.*, Vol.44

[4] D.B. Chklovskii *et al.*, Phys. Rev. B **46**, 4026 (1992); **46**, 15606(E) (1992).

[5] J.K. Jain, Phys. Rev. Lett. **63**, 199 (1989).

[6] B.I. Halperin, P.A. Lee, and N. Read, Phys. Rev. B **47**, 7312 (1993).

[7] V. Kalmeyer and S.-C. Zhang, Phys. Rev. B **46**, 9889 (1992).

[8] C.B. Hanna, R.B. Laughlin and A.L. Fetter, Phys. Rev. B **40**, 8745 (1989).

[9] B.I. Halperin, Phys. Rev. B **45**, 5504 (1992).

[10] L. Brey, preprint; L.Brey, this conference.

[11] D.B. Chklovskii and P.A. Lee, Phys. Rev. B **48**, 18060 (1993).

[12] J.E. Müller, Phys. Rev. Lett. **68**, 385 (1992).

[13] X.-G. Wen, Int. J. Mod. Phys. B **6**, 1711 (1992).

[14] C. W. J. Beenakker, Phys. Rev. Lett. **64**, 216 (1990).

[15] A. M. Chang, Solid State Commun. **74**, 871 (1990).

ELECTROMAGNETIC RESPONSE OF THE FRACTIONAL QUANTIZED HALL STATE

Steven H. Simon and Bertrand I. Halperin
Department of Physics, Harvard University, Cambridge, MA 02138

and

Song He
AT&T Bell Labs, Murray Hill, NJ 07974

A fractional quantized Hall state with filling fraction $\nu = p/(2mp+1)$ can be modeled as an integer quantized Hall state of transformed fermions, interacting with a Chern-Simons field. The electromagnetic response function for these states at arbitrary frequency and wavevector can then be calculated using the Random Phase Approximation. However, such calculations do not properly take into account the large effective mass renormalization which is present in the Chern-Simons theory. We show how the mass renormalization can be incorporated in a calculation of the response function within a Landau Fermi liquid theory approach such that Kohn's theorem and the f-sum rules are properly satisfied. We present results of such calculations for planar and spherical geometries. We compare our calculations with the results of exact numerical diagonalizations for finite size systems.

1. Definition and Features of Response Function and Structure Factor

The electromagnetic response function $K_{\mu\nu}(q,\omega)$ at wavevector q and frequency ω is defined by the equation $J_\mu(q,\omega) = K_{\mu\nu}(q,\omega)A_\nu^{ext}(q,\omega)$ where A^{ext} is an externally applied vector potential (with A_0 the scalar potential) and J is the induced current (with J_0 the induced change in density). In a recent series of papers,[1-3] we have studied the response function for fractional quantized Hall states. In particular we have been interested in the imaginary part of $K_{00}(q,\omega)$, which is related to dynamical structure factor $S(q,\omega)$ by

$$S(q,\omega) = -\mathrm{Im}[K_{00}(q,\omega)]/\pi = \sum_j \ |\langle\Phi_j|\rho_q|\Phi_0\rangle|^2\delta(\omega+\epsilon_0-\epsilon_j) \tag{1}$$

where Φ_i is the ith many-particle eigenstate of the system whose energy is given by ϵ_i, and ρ_q is the qth Fourier mode of the density operator. Thus, the poles occur at the excitation energies, and the residues of the poles (given by the matrix element) tell us how easy it is to create this excitation from the ground state Φ_0 by applying a density perturbation at a given wavevector.

There are two rules that we will require our response function to satisfy: Kohn's theorem and the f-sum rule. Kohn's theorem is the statement that in the $q \to 0$ limit a mode at the bare cyclotron frequency $\omega_c = eB/m_b c$, with m_b the bare electron band mass, should be the only mode with any residue. This is simply the statement that in the $q \to 0$ limit, all of the electrons oscillate together at the cyclotron frequency. Similarly, the f-sum rule is simply the statement that in the $\omega \to \infty$ limit, all of the electrons oscillate in place and again the response of the any interacting system should be identical to the response of a free electron system. By using a Kramers-Krönig relation, this can be expressed as an "f-sum" : $\int S(q,\omega)\omega d\omega = q^2\rho/m_b$ where ρ is the electron density.

2. Composite Fermion RPA

In order to model the fractional quantized Hall state, we use the composite fermion approach.[4–6] The system of electrons in magnetic field B is transformed exactly into a system of fermions interacting with a Chern-Simons gauge field. Loosely, we can think of this transformed system as fermions bound to an even number $2m$ of flux quanta (ϕ_0). Within a mean field approximation the fermions see an effective field $\Delta B = B - 2m\phi_0\rho$ which is just the externally applied field minus the average field due to the flux quanta bound to the fermions. When the filling fraction for the original electron system is given by $\nu = \rho\phi_0/B = p/(2mp+1)$, then the effective filling fraction for the transformed fermions is $\nu_f = \rho\phi_0/\Delta B = p$. Thus we see that the fractional quantized Hall states at filling fraction $\nu = p/(2mp+1)$ are just integer quantized Hall states $\nu_f = p$ for the transformed fermions.

The corrections to this mean field model can be considered as an interaction between the fermions due to the fact that the Chern-Simons field seen by a given fermion depends on the positions of all of the other fermions. We can treat this interaction, as well as the direct Coulomb interaction between the fermions, in the Random Phase Approximation (RPA). The RPA equation is given as $K^{-1} = K_0^{-1} + V$, where K is the response matrix of the full system, K_0 is the (mean field) response matrix for a system of noninteracting fermions in a magnetic field ΔB, and V is an interaction matrix that includes both the Chern-Simons and the Coulomb interactions. It can then be shown that K calculated in the RPA satisfies Kohn's theorem and the f-sum rule if and only if K_0 satisfies these rules with respect to the mean field ΔB.

The problem with this RPA calculation is that the low energy excitations are on the wrong energy scale. The excitation spectrum of the mean field system is just the usual Landau levels with spacing $\Delta\omega_c = e\Delta B/m_b c$, and this effective cyclotron frequency sets the energy scale for the excitations of the original electron system calculated in the RPA. However, as pointed out in previous work,[6,1] the composite fermion effective mass m^* can be renormalized significantly by interactions ($m^* \approx 4m_b$ is a reasonable estimate[6] for typical experimental parameters). Thus in an exact treatment of the system, the scale of the low energy excitations should be set by a renormalized effective cyclotron frequency $\Delta\omega_c^* = e\Delta B/m^*c$. A naive approach to this problem would be to simply repeat the RPA calculation replacing m_b with m^* everywhere it occurs. Although this approach correctly models the energy scale of the low energy excitations, it would incorrectly predict the Kohn mode to occur at eB/m^*c instead of eB/m_bc. Such an approach would similarly fail to satisfy the f-sum rule.

3. Mass Renormalization - Modified RPA

In order to repair this naive RPA calculation, we use a Landau Fermi liquid theory approach and write and energy functional E of a phase space density $n(\mathbf{p}, \mathbf{r})$:

$$E[n(\mathbf{p},\mathbf{r})] = E^0_{m^*}[n(\mathbf{p},\mathbf{r})] + \frac{f_1}{2\rho m^*}\sum_{\mathbf{p},\mathbf{p}'}(\mathbf{p}-\mathbf{A})\cdot(\mathbf{p}'-\mathbf{A}')n(\mathbf{p},\mathbf{r})n(\mathbf{p}',\mathbf{r}')+\ldots \quad (2)$$

where $E^0_{m^*}$ is an energy functional for noninteracting fermions of effective mass m^*, and f_1 is the first Landau coefficient (a phenomenological parameter) which (by Galilean invariance) relates the effective mass to the bare mass via $m^* = (1 + f_1)m_b$. We then treat this mass renormalizing interaction as an additional interaction that is not properly included in the above RPA. Thus, we propose a "Modified RPA" that includes this mass renormalizing Landau Fermi liquid interaction. The modified RPA equation is given by $K^{-1} = K^{-1}_{0,m^*} + V + M$ where K_{0,m^*} is the response of noninteracting fermions of mass m^* in the mean field ΔB, and M is an interaction matrix that represents the above mass renormalizing interaction. The modified RPA satisfies Kohn's theorem and the f-sum rule as well as correctly predicting the energy scale of the low energy excitations to be given by $\Delta\omega^*_c$.

Results of calculations[1] of the response function using the modified RPA are shown in Fig. 1. In the case of $\nu = \frac{1}{3}$ the modified RPA does not predict much of a minimum

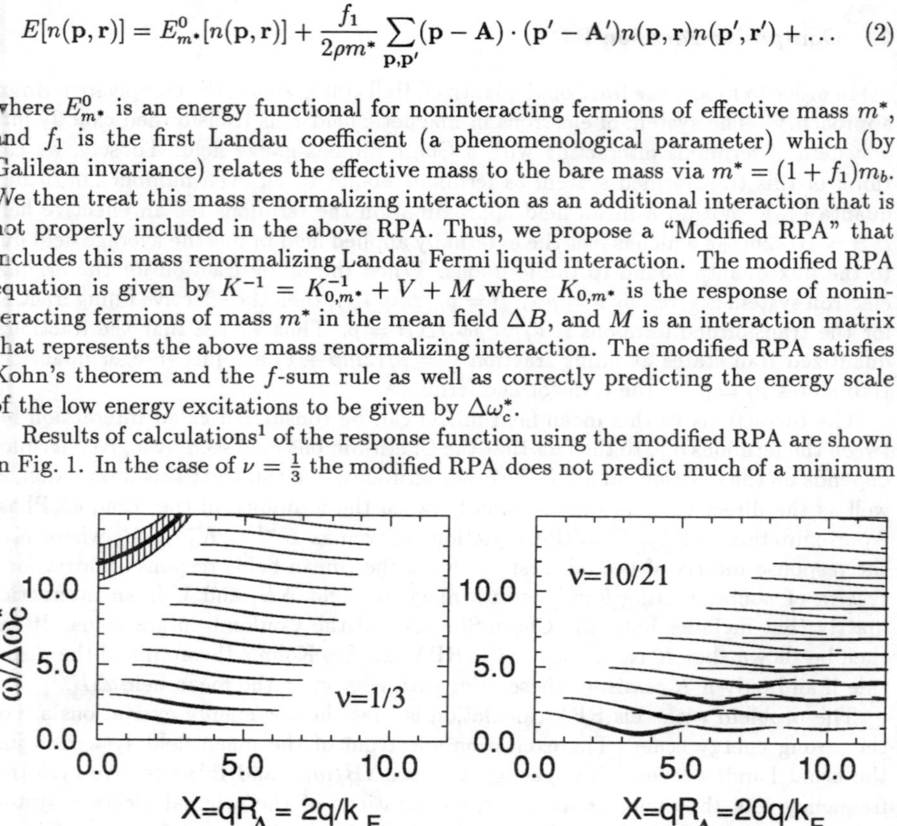

Figure 1: Response function calculated with the modified RPA at $\nu = \frac{1}{3}$ and $\nu = \frac{10}{21}$. The lines are the location of the poles of K_{00}, and the width of the lines indicate the residue of the poles. Here we have used $m^* = 3.9m_b$. For $\nu = \frac{1}{3}$, the cyclotron mode at high frequency has all of the weight at $q \to 0$ in accordance with Kohn's theorem. (For $\nu = \frac{10}{21}$, the cyclotron mode is off the top of the graph.)

in the dispersion curve, but as we approach $\nu = \frac{1}{2}$ (as shown in the $\nu = \frac{10}{21}$ curve), a pronounced minimum forms in the lowest branch of the spectrum, and a series of additional minima start to form at higher wavevector. The apparent q-periodicity evident in the $\nu = \frac{10}{21}$ spectrum is a reflection of the length scale $R_\Delta = \hbar k_F c/e\Delta B$ associated with the composite-fermion cyclotron radius. Although the modified RPA

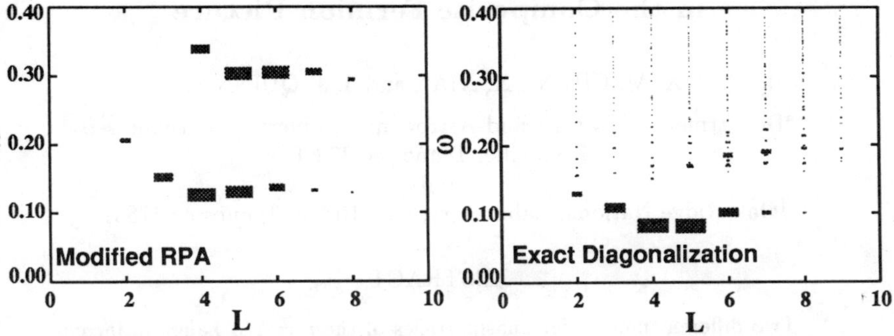

Figure 2: Response function for 8 electrons on a sphere at $\nu = \frac{1}{3}$. The boxes are the location of the poles in the response function, and the size of each box gives the residue of the corresponding pole. The frequency is given in units of e^2/l_0. (We have actually used $m^* = 50m_b$ here instead of $m^* \to \infty$). L here is the angular momentum which is analogous to the wavevector q in the planar system.

predicts a series of modes at higher energy corresponding to excitations of the fermions by more than one effective Landau level (with spacing $\Delta\omega_c^*$), we would expect these higher modes to be smeared out when scattering is taken into account.

4. Finite Sized Spherical Systems and Exact Diagonalizations

The modified RPA can be reformulated for a finite spherical system[2] such that comparison can be made with the results of exact diagonalizations.[3] Such a comparison is shown in Fig. 2. Since the exact diagonalizations are restricted to the lowest Landau level, we must take the limit $m^* \to \infty$ while fixing the ratio of the effective fermion cyclotron energy $\Delta\omega_c^*$ to the Coulomb energy e^2/l_0 where l_0 is the magnetic length. For the results shown in Fig. 2, we have used an estimate of this ratio made in Ref. 6. The discrepancy in energy scales between the modified RPA and the exact calculation may be caused by an error in this estimate. We see that there is a reasonably good agreement between the shapes of the two curves at low energy. As expected, we see that the modified RPA does not accurately predict the higher energy modes.

References

[1] S. H. Simon and B. I. Halperin, *Phys. Rev.* **B48** (1993) 17386.
[2] S. H. Simon and B. I. Halperin, *Phys. Rev.* **B50** (1994) To Be Published.
[3] Song He, S. H. Simon, and B. I. Halperin, *Phys. Rev.* **B50** (1994) To Be Published.
[4] J. K. Jain, *Adv. Phys.* **41**, 105 (1992); and references therein.
[5] A. Lopez and E. Fradkin, *Phys. Rev.* **B44**, 5246 (1991); **B47**, 7080 (1993).
[6] B. I. Halperin, P. A. Lee, and N. Read, *Phys. Rev.* **B47** (1993) 7312.

Charge Density of Fractional Charged Quasiparticles in the Composite Fermion Picture

X. M. CHEN*, X. XIA*, and J. J. QUINN*†

*Department of Physics and Astronomy, University of Tennessee
Knoxville, Tennessee 37996
and
†Oak Ridge National Laboratory, Oak Ridge, Tennessee 37831

ABSTRACT

Two different models for quasiparticles of the $\nu = 1/3$ Laughlin incompressible state are investigated in the Fermion-Chern-Simons picture. In the first model a quasielectron (quasihole) is induced by adiabatically introducing an infinitesimally thin flux tube parallel (antiparallel) to the external magnetic field. The charge density induced is calculated in linear response theory. The quasielectron (quasihole) of the second model is obtained by creating (annihilating) a composite fermion consisting of a real electron with two attached magnetic flux quanta. Both models give a same total charge $\pm\frac{1}{3}e$ for quasiparticles; but they produce different charge density distributions $\rho(r)$. The self-energies (Coulomb) of quasiparticles are discussed.

It is now known that the physics of fractional quantum Hall effects of a two-dimensional electron system can be understood in terms of that of integer quantum Hall effects[1]. By attaching to each electron an even number of magnetic flux quanta and employing the mean field approximation, an electron system with a filling factor $\nu = p/(2mp+1)$ (m and p are integers) can be transformed into a composite fermion system[2,3] with a integer filling factor $\nu = p$. The Chern-Simons composite-fermion picture has been quite successful recently in explaining several intriguing properties of fractional quantum Hall systems[4]. Two of the most distinctive features associated with fractional quantum Hall states are the fractionalizing of quasiparticle charges and the finiteness of the quasiparticle creation energies. There are two ways to create a quasiparticle in the Chern-Simons composite-fermion picture, namely, the Laughlin approach and the composite-fermion approach. In the Laughlin approach[5] a single flux quanta confined in a thin tube is adiabatically introduced at the origin. The response of the fermion system to the external flux leads to the accumulation of a fractional charge at the origin of the sample, thus the creation of a quasiparticle. In the composite-fermion approach, on the other hand, a composite fermion with a charge $-e$ is either added or removed from the system. The response to the magnetic flux and electric charge carried by the composite-fermion also produces a correct fractional charge for quasiparticles. The purpose of this paper is to study and compare the charge density distributions of quasiparticles created in the two different models, and to calculate their self (Coulomb) energies associated with the charge distributions.

We use the perturbation method developed by Chiu and Quinn[6] to calculate the response functions. The method is similar to that employed by Halperin, Lee and

Reed in Ref. (3). The Hamiltonian of the composite-fermion system after the Chern-Simons transformation is given by

$$\hat{H} = \sum_{i=1}^{N} \frac{1}{2m_b}[\vec{p}_i + \frac{e}{c}\vec{A}_0(\vec{r}_i) - \frac{e}{c}\vec{A}_{cs}(\vec{r}_i)]^2 + e\sum_{i<j} v(|\vec{r}_i - \vec{r}_j|), \tag{1}$$

where $\vec{A}_0(\vec{r})$ is the vector potential corresponding to the uniform magnetic field $B\hat{z}$, and $v(\vec{r})$ is the Coulomb interaction between fermions. As a result of the Chern-Simons gauge interaction, $\vec{A}_{cs}(\vec{r}) = \sum_i \vec{V}_{cs}(\vec{r} - \vec{r}_i)$ with $\vec{V}_{cs}(\vec{r}) \equiv \tilde{\phi}(\hbar c/e)(\hat{z} \times \vec{r}/r^2)$. Here $\tilde{\phi}$ $(= 2)$ is the number of flux quanta attached to each electron.

In the mean field approximation, the Hamiltonian of Eq. (1) is reduced to

$$\hat{H}_0 = \sum_{i=1}^{N} \frac{1}{2m_b}[\vec{p}_i + \frac{e}{c}\Delta\vec{A}(\vec{r}_i)]^2 + e\sum_{i<j} v(|\vec{r}_i - \vec{r}_j|), \tag{2}$$

where the effective vector potential $\Delta\vec{A}(\vec{r}) = \vec{A}_0(\vec{r}) - \bar{\vec{A}}_{cs}(\vec{r})$ and the corresponding effective magnetic field is $\Delta B = B - \frac{2\pi\hbar c}{e}\tilde{\phi}n_e$ (n_e is the average electron density).

The current and charge response of the composite-fermion system to an external vector potential $\vec{A}_{ext}(\vec{r})$ and scalar potential $\varphi_{ext}(\vec{r})$ can be written, after the Fourier transformation, as

$$\vec{j}_{ind}(\vec{q},\omega) = \frac{n_e e^2}{m_b c}[-\vec{A}(\vec{q},\omega) - \vec{I}(\vec{q},\omega)\cdot\vec{A}(\vec{q},\omega) + \vec{K}(\vec{q},\omega)\varphi(\vec{q},\omega)] \tag{3}$$

and

$$\rho_{ind}(\vec{q},\omega) = \frac{n_e e^2}{m_b c}[-\frac{\omega}{c}\vec{K}^*(\vec{q},\omega)\cdot\vec{A}(\vec{q},\omega) + \vec{q}\cdot\vec{K}(\vec{q},\omega)\varphi(\vec{q},\omega)] \tag{4}$$

where the vector $\vec{K}(\vec{q},\omega)$ and the 2×2 tensor $\vec{I}(\vec{q},\omega)$ are the same as those defined in Ref. (6). $\vec{A}(\vec{q},\omega)$ and $\varphi(\vec{q},\omega)$ are the total perturbative potentials, given by[7]

$$\vec{A}(\vec{q},\omega) = \vec{A}_{ext}(\vec{q},\omega) + \frac{L^2}{e}\vec{V}_{cs}(\vec{q})\cdot\rho_{ind}(\vec{q},\omega) \tag{5}$$

and

$$\varphi(\vec{q},\omega) = \varphi_{ext}(\vec{q},\omega) + v(\vec{q})\rho_{ind}(\vec{q},\omega) - \frac{L^2}{ec}\vec{V}_{cs}(\vec{q})\cdot\vec{j}_{ind}(\vec{q},\omega) \tag{6}$$

Here $v(q)$ and $\vec{V}_{cs}(\vec{q})$ are the Fourier transforms of $v(\vec{r})$ and $\vec{V}_{cs}(\vec{r})$. By solving the coupled equations (3)~(6) self-consistently, one gets the induced current and charge density $\vec{j}_{ind}(\vec{q},\omega)$ and $\rho_{ind}(\vec{q},\omega)$. The charge density distribution in real space is obtained from the Fourier tranform of $\rho(\vec{q})$.

Laughlin Quasiparticles—— By introducing a infinitesimally thin flux at the origin, we have $\vec{A}_{ext}(\vec{r}) = \pm(\hbar c/e)(\hat{z}\times\vec{r}/r^2)$ with $\varphi_{ext}(\vec{r}) = 0$, where the plus (minus) sign corresponds to a flux parallel (antiparallel) to the uniform magnetic field and the creation of a quasielectron (quasihole). The induced charge distribution depends on the ratio of the Coulomb energy and the Landau level separation, i.e., $\gamma_c = (e^2/\epsilon l)/\hbar\omega_c$ ($\omega_c = eB/m_b c$, $l = \sqrt{\hbar c/eB}$). In our calculations we take $\gamma_c = 2/3$. The charge density $\rho(r)$ of a Laughlin quasielectron at $\nu = 1/3$ is plotted as a function of r in Fig. 1. The structure of $\rho(r)$ resembles very much to that obtained by Laughlin from his trial wave function[5]. Most of the induced charge is localized in the area $r < l$; $\rho(r)$ decreases as a power (instead of an exponential) of r when $r \longrightarrow \infty$.

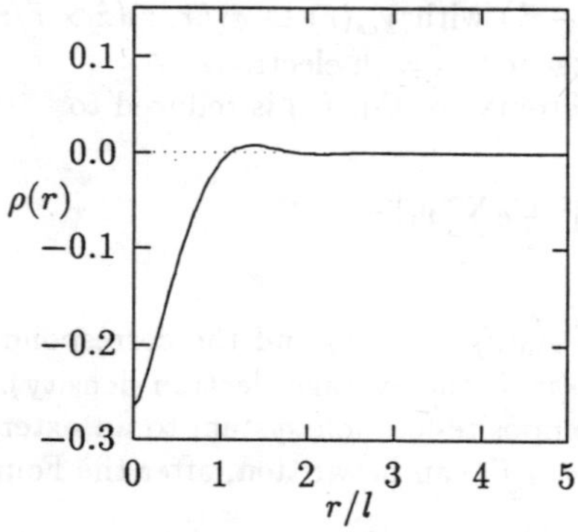

Fig. 1. Charge density of a Laughlin quasielectron at $\nu = 1/3$ (in units of $e/2\pi l^2$).

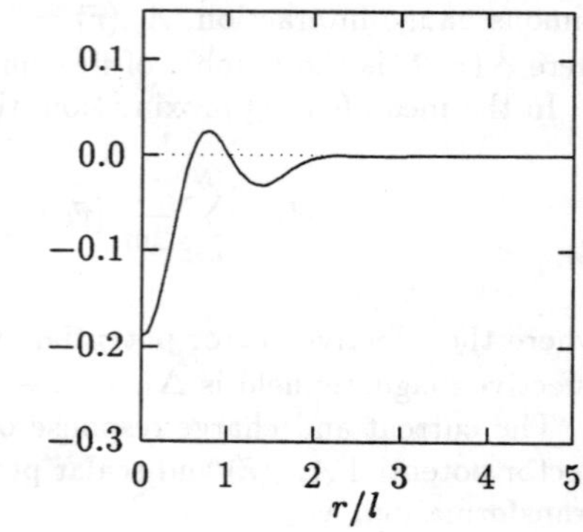

Fig. 2. Charge density of a quasielectron created by the composite-fermion approach with n=1 and m=0.

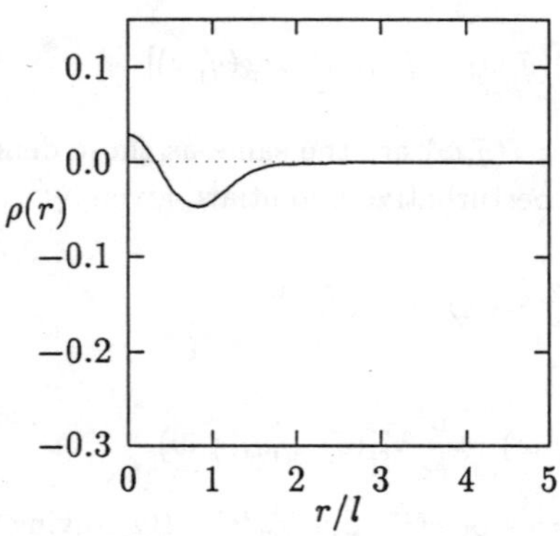

Fig. 3. Charge density of a quasielectron created by the composite-fermion approach with n=0 and m=1.

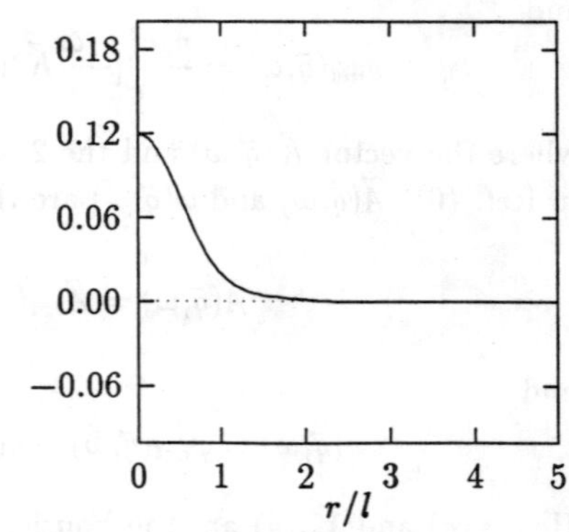

Fig. 4. Charge density of a quasihole created by the composite-fermion approach with n=0 and m=0.

Composite-Fermion Approach —— We can also create a quasiparticle of $\nu = 1/3$ state by adding or removing one composite-fermion from the fermion system. In this case we not only introduce an external vector potential but also a scalar potential since each fermion has a charge $-e$. $\vec{A}_{ext}(\vec{q})$ and $\varphi_{ext}(\vec{q})$ are given by $\vec{A}_{ext}(\vec{q}) = \pm\frac{L^2}{e}\vec{V}_{cs}(\vec{q})\rho_{nm}(q)$ and $\varphi_{ext}(\vec{q}) = \pm v(q)\rho_{nm}(q)$, where the plus (minus) sign corresponds to the creation of a quasielectron (quasihole). $\rho_{nm}(q)$ is the charge density of the composite-fermion that is added to or removed from the single particle state $\mid nm\rangle$. (n and m are the principle and angular momentum quantum numbers in the symmetric gauge). The total charge distributions $\rho_{ind}(r) \pm \rho_{nm}(r)$ are plotted in Fig. 2 and 3 for two quasielectrons (n=1, m=0; and n=0, m=1), and in Fig. 4 for a quasihole created by extracting one composite-fermion from the origin (the n=0 and m=0 state). Note the different structures in Fig. 2 and 3 in comparison with the Laughlin quasiparticle of Fig. 1. $\rho(r)$ in Fig. 3 resembles to the quasielectron densities obtained from several finite-size simulations[8]

Self-energy of Quasiparticles —— As is pointed out by Laughlin[5], the quasiparticle creation energy is primarily the Coulomb repulsion of the accumulated charge with itself. From the charge distribution obtained above we can easily estimate the self-energy of a quasiparticle by a simple integration, i.e., $E_c = \frac{1}{2}(L/2\pi)^2 \int d^2\vec{q}\, v(q)\rho(\vec{q})\rho(-\vec{q})$. For Laughlin quasiparticles, we get $E_c = 0.0443e^2/\epsilon l$ at $\gamma_c = 2/3$ and $E_c = 0.0584e^2/\epsilon l$ at $\gamma_c = 0$ for both quasielectrons and quasiholes. These values give roughly a correct quasielectron-quasihole pair creation energy ($\sim 0.1e^2/\epsilon l$). In the composite-fermion model, the self-energies of quasielectrons are different from that of quasiholes because of the difference in charge distributions. At $\gamma_c = 2/3$, a quasielectron has a energy of $E_c = 0.0192$ for the case of n=1 and m=0, and $E_c = 0.0189$ for n=0 and m=1. A quasihole with n=0 and m=0 has a higher energy of $E_c = 0.0265$, which is in disagreement with results from finite-size numerical calculations[8]. The inclusion of the exchange interaction may therefore be necessary in an accurate calculation of quasiparticle creation energies.

Acknowledgments —— We would like to thank Lizheng Zhang for helpful discussions. This research was supported in part by the Division of Materials Sciences of the U. S. Department of Energy under Contract No. DE-AC05-84OR21400 with Martin Marietta Energy Systems, Inc.

REFERENCES

[1] J. K. Jain, *Phys. Rev. Lett.* **63** (1989) 199; *Adv. Phys.* **41** (1992) 105.
[2] Ana Lopez and E. Fradkin, *Phys. Rev.* **B47** (1993) 7080.
[3] B. I. Halperin, P. A. Lee, and N. Reed, *Phys. Rev.* **B47** (1993) 7312.
[4] R. Willet *et al.*, *Phys. Rev.* **B47** (1993) 7344; R. R. Du *et al.*, *Phys. Rev. Lett.* **70** (1993) 2944.
[5] R. Laughlin, in *The Quantum Hall Effect*, ed. R. E. Prange and S. M. Girvin (Springer-Verlag, New York, 1990).
[6] K. W. Chiu and J. J. Quinn, *Phys. Rev.* **B9** (1974) 4724.
[7] Q. Dai *et al.*, *Phys. Rev.* **B46** (1992) 5642.
[8] R. Morf and B. I. Halperin, *Phys. Rev.* **B33** (1986) 2221; F. Haldane and E. Rezayi, *Phys. Rev. Lett.* **54** (1985) 237.

NEW PHYSICS IN DOUBLE LAYER 2D ELECTRON SYSTEMS

J.P. Eisenstein

AT&T Bell Laboratories, Murray Hill, NJ 07974

ABSTRACT

Recent work on bilayer two-dimensional electron systems (2DES) in GaAs/AlGaAs double quantum wells (DQW) is reviewed. Three sets of experiments which illustrate some of the new physics obtained from such systems are briefly described. These include decoupled DQWs used to make quantitative measurements of the compressibility of the 2DES in the extreme quantum limit, weakly coupled bilayers for examining the tunneling density of states of the 2DES, and lastly strongly coupled systems which exhibit incompressible quantum Hall states not found in single layer systems.

1. Introduction

A sufficient motivation for examining the properties of bilayer two dimensional electron systems (2DES) is the potential they hold, by virtue of the controlled re-introduction of the third spatial dimension, for revealing new and interesting many body physics. The recent observations of a fractional quantum Hall plateau at one-half filling of the lowest Landau level in strongly coupled systems provide a vivid first example of this potential[1]. But beyond this, bilayer structures which are coupled weakly or not at all offer numerous possibilities as well. The recent work on frictional drag between two 2D layers is one such example[2]. In this paper I will very briefly review a selection of our own work on double layer 2D systems. Three topics will be discussed: the use of decoupled double layer systems to make accurate determinations of the compressibility of a single layer 2DES; measurements of the tunneling of electrons between weakly coupled 2D layers; and lastly studies of the unusual quantized Hall state at unit filling factor ($\nu=1$) in strongly coupled structures. The discussion is at a superficial level; hopefully interested readers will consult the literature for the details.

2. Samples and Contacts

These experiments rely heavily on the reproducible MBE growth of double quantum wells containing high mobility 2DES's with controlled sheet densities. Both of these requirements can now be met using MBE-grown double quantum wells (DQW) in the GaAs/AlGaAs family. Such structures are now routinely grown with mobilities well above $1\times10^6\,\mathrm{cm}^2/\mathrm{Vs}$. While not as high as in the best ordinary heterostructures, this level of disorder is adequate for the study of the electron correlation phenomena at high magnetic fields. Our typical sample consists of two GaAs quantum wells $100-200\,\mathrm{\AA}$-wide separated by an $Al_xGa_{1-x}As$ barrier. Depending upon the experiment the barrier width can range from 20 to $5000\,\mathrm{\AA}$ with an Al concentration between x=0.1-1.0. Above and below the GaAs wells are thick AlGaAs cladding layers containing Si doping layers.

Typical 2D densities range from N=0.4 to $1.5\times10^{11}\ \mathrm{cm}^{-2}$ in each layer.

In order to fully realize the potential of bilayer 2D systems separate Ohmic contacts to the individual layers are essential. Without these we would be restricted to parallel transport studies and the tunneling, frictional drag and compressibility experiments that we have done would not have been possible. As described in detail elsewhere[3] we have satisfied this requirement by employing a selective depletion scheme in which one or the other 2DES is depleted in the immediate vicinity of an ordinary Ohmic contact by the action of a Schottky gate on the top or backside of the sample. This allows the contact to "see" the rest of the sample only through the one undepleted 2DES.

3. Compressibility of the 2DES

When a perpendicular electric field E_0 is applied, with a gate, to a grounded bilayer 2DES some of the field penetrates the "top" layer and impinges upon the lower layer. This effect is due entirely to the finite density of states, or compressibility, of the upper 2DES. It can be shown[4] that for two thin decoupled 2D systems the equilibrium ratio of the differential penetrating electric field δE_p to the differential applied gate field δE_0 is

$$\delta E_p/\delta E_0 = d_t/(d_t+d_b+d) \tag{1}$$

where d is the layer spacing and d_j (j=t,b) are distance parameters directly proportional to the inverse thermodynamic density of states of the top and bottom 2D gases: $d_j=(\kappa\varepsilon_0/e^2)\,\partial\mu_j/\partial N_j$. This derivative of chemical potential with respect to density is also closely related to the *compressibility* K and total energy E_{tot} (per unit area) of the 2D system:

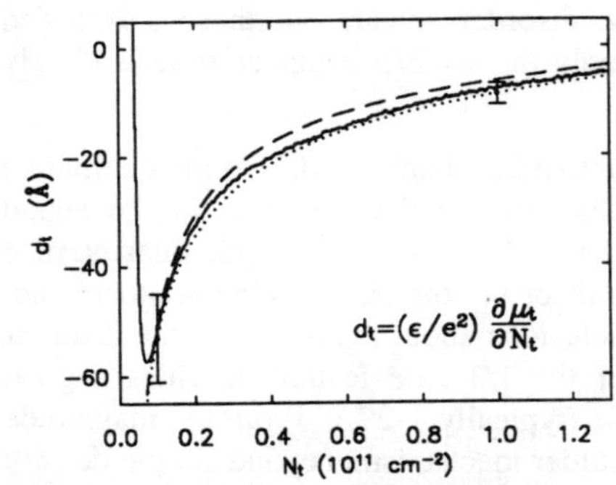

Fig. 1. Compressibility parameter d_t vs. upper 2DES density at B=0. The truncated negative divergence at low N_t is due to the exchange energy. Dashed and dotted lines are two different theoretical results.

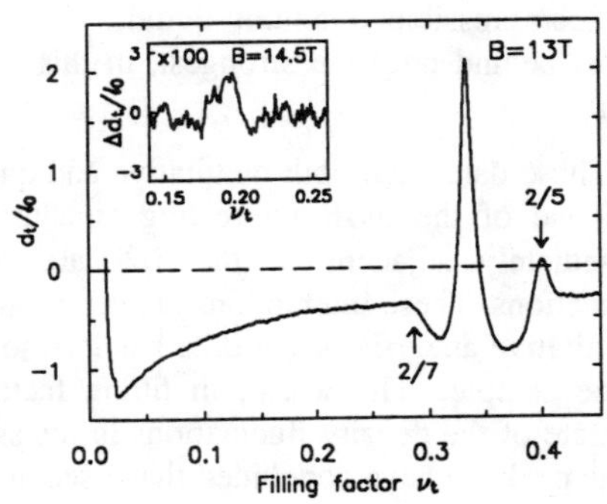

Fig. 2. Compressibility parameter d_t at B=13T and T=0.4K showing $\nu=1/3$ and 2/5 FQHE states. Inset shows $\nu=1/5$ state at B=14.5T.

$$\partial\mu/\partial N = K^{-1}/N^2 = \partial^2 E_{tot}/\partial^2 N \qquad (2)$$

Thus, a measurement of the differential penetration field provides direct access to $\partial\mu/\partial N$.

Given separate contacts to the individual 2D layers in a DQW, measuring the differential penetration field is easy. In essence the technique[4] consists of measuring the capacitance between top-side gate and the lower 2D electron layer, with the upper layer grounded. Any electric field which penetrates the upper layer causes currents to flow to the lower layer and these are detected. A dc bias is applied to the gate in addition to a small ac voltage, this allows measurements of the differential penetration as a function of the density of the upper layer. Since the penetration is usually weak, the lower layer density remains near its nominal value, and δE_p is determined almost entirely by the top layer compressibility.

Our experiments have shown that $\partial\mu/\partial N$ in typical 2DES's in GaAs is dominated by electron-electron interaction effects[4]. As Fig. 1 proves, this is true even at zero magnetic field where the exchange energy makes $\partial\mu/\partial N$ diverge *negatively* at low density. (In the figure the $\partial\mu/\partial N$ values have been converted into the equivalent distance parameter d_t.) At high magnetic field we have used this technique to measure the compressibility in the fractional quantum Hall regime[4]. Figure 2 shows data taken at 13T. At this field the Fermi level is in the lowest Landau level over the accessible density range. Hence the total energy is determined *only* by interactions, the kinetic energy being quenched by the field. This leads to a generally negative compressibility punctuated by structure associated with the fractional quantum Hall effect (FQHE). For the figure the compressibility parameter d_t has been normalized by the magnetic length ℓ_0. The primary feature, of course, is the strong upward peak associated with the $\nu=1/3$ FQHE. Were the sample ideal, i.e. no disorder, no inhomogeneity, etc, the FQHE would create a dense family of upward δ-functions in K^{-1} owing to the formation of a hierarchy of incompressible quantum liquids. In reality, disorder smears out these spikes and leaves behind only the strongest, in this case only the $\nu=2/5$ daughter state is clearly seen.

These data have both qualitative and quantitative implications[4]. On the qualitative side one of the most interesting conclusions is that the deep minima in the signal immediately adjacent to the 1/3-state peak are due to quasiparticle-quasiparticle interactions. These interactions are central to the theory of the FQHE. On the other hand, quantitative analysis of the data has lead to conclusions about the nature of the disorder in the sample. The width, in filling factor, of the 1/3-state feature provides a good estimate of the density fluctuations in the sample (typically 1-2%). From the magnitude of the peak we have concluded that a second disorder mechanism, beyond simple density smearing, also exists. This latter effect is at least consistent with homogeneous broadening of the quasiparticle states due to scattering and reasonable values for the scattering rates are found.

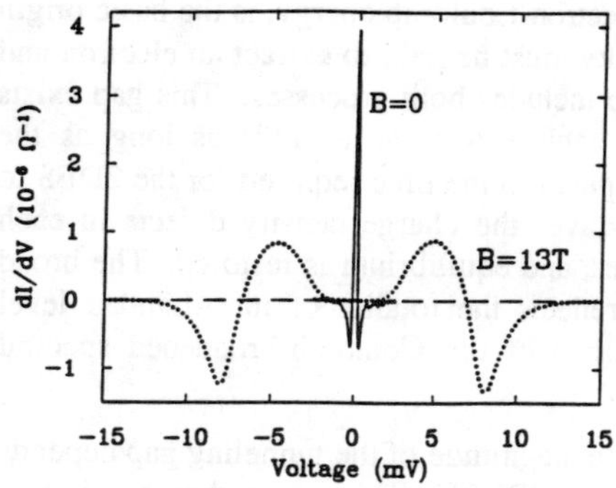

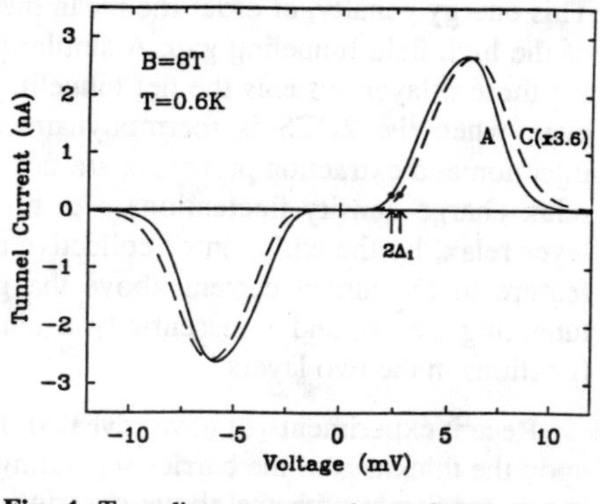

Fig. 3. Comparison of B=0 and B=13T tunneling spectra at low temperature. The sharp peak at B=0 is due to energy-momentum conservation while the suppressed tunneling around V=0 at 13T is a purely many-body effect.

Fig. 4. Tunneling current-voltage characteristics at B=13T in two samples with different barrier thicknesses: 175Å in sample A and 340Å in sample B. The different magnitude of the gap around V=0 is an interlayer excitonic effect.

4. Tunneling

We now turn to tunneling experiments on double layer 2DES's. Our studies so far have been confined to very weakly coupled systems in which the interlayer tunneling resistance dominates all other relevant resistances. With the separate contacts to the layers in hand we can directly measure the tunneling I-V characteristics. At zero magnetic field sharp resonances in the tunneling conductance are observed[5]. This is a consequence of the twin constraints of energy and momentum conservation. Simply put, for a weakly coupled DQW these constraints imply that tunneling between the ground subbands in the two wells can only occur when the subband edges are almost exactly lined up. In fact, the observed width of the resonances provides a new measure of the disorder in the 2DES[5]. While the zero field tunneling results are consistent with single electron physics, the results obtained at high magnetic field[6] are clearly not and offer a new perspective on the strongly interacting 2DES in the quantum limit. The dramatic difference between the zero and high field cases is shown in Fig. 3. Over a significant energy range around zero bias the tunneling is strongly suppressed at high magnetic field, to less than 1/4000 of the B=0 value. Unlike the zero field resonance this high field suppression always occurs at V=0, i.e. at the Fermi level, even when the densities in the two wells are unequal. Above this gap around V=0 are broad features where significant tunneling is evident. This basic tunneling signature persists over wide ranges of magnetic field and is only slightly perturbed when the 2DES's enter FQHE states.

These results are explained as follows[6,7]. At high field electrons are localized on the scale of the magnetic length ℓ_0 and can therefore avoid their neighbors much more effectively than at B=0. At short distances the electron liquid is strongly correlated and is not much different than a Wigner crystal. An electron tunneling into such a system must have enough energy to be thrust into an "interstitial" position within the liquid.

This energy penalty, of order the mean inter-electron Coulomb energy, is the basic origin of the high field tunneling gap. A similar penalty must be paid to extract an electron and for these bilayer systems the net tunneling gap includes both processes. This gap exists even when the 2DES is thermodynamically gapless (e.g. at $\nu = 1/2$) as long as the injection and extraction processes are fast compared to the time required for the 2DES to relax charge density fluctuations. As time evolves the charge density defects in each layer relax, by the emission of collective modes, and equilibrium is restored. The broad feature in the tunnel current above the gap reflects the totality of intra-Landau level tunneling events and is essentially a convolution of the Coulomb-broadened spectral functions in the two layers.

Recent experiments[8] have revealed that the magnitude of the tunneling gap depends upon the thickness of the barrier separating the two 2DES's. This result, shown in Fig. 4, is not consistent with the above description of the tunneling in which the observed I-V characteristics were assumed to reflect a convolution of independent spectral functions for each 2DES and interlayer correlations were ignored. We believe that data like that in Fig. 4 can be neatly explained by including the attractive interaction between a tunneled electron and the hole which it leaves behind. This excitonic effect *reduces* the net energy cost for tunneling and hence when the barrier thickness is increased the tunneling gap also increases. The magnitude of observed change in the tunneling gap is consistent with this model, being about 1/3 of the Coulomb energy between a point-like electron and hole separated by the center-to-center quantum well spacing. To our knowledge, this is the first observation of an interlayer excitonic interaction in a tunneling experiment. We believe that two-dimensionality and high magnetic fields are critical to the detection of this effect. In tunnel junctions made of ordinary 3D metals the enormous electron density not only makes the screening length extremely small (~1 Å) but also eliminates the possibility of reaching the lowest Landau level with available magnetic fields.

5. Bilayer Quantized Hall Effect

In the previous section the double layer systems were weakly coupled by both tunneling and Coulomb interactions. In this final section we discuss the quantized Hall effect in strongly coupled DQWs having narrow barriers ($d_b \sim 30$ Å). The discussion is limited to the case of total filling fraction $\nu_{tot} = 1/2 + 1/2 = 1$.

For uncoupled 2D layers each at 1/2-filling there is, of course, no QHE. If the tunneling between the layers is significant a QHE does, however, appear. This is the result of the energy gap between the ground symmetric and antisymmetric subbands of the DQW and is a purely single particle effect. But even in the absence of tunneling, theory predicts[9] that a QHE at $\nu = 1$ would occur in a bilayer system owing to interlayer Coulomb interactions. The problem for experiment was whether it was possible to cleanly distinguish between these two cases in real DQW samples where a $\nu = 1$ state was observed rather early on[1].

In order to examine the interplay of tunneling and Coulomb interactions at $\nu = 1$ Murphy, *et al.*[10] examined a large number of samples with varying barrier heights and

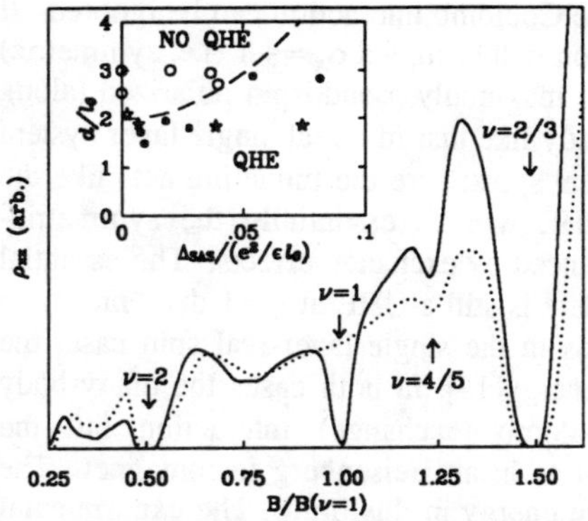

Fig. 5. Transport data and phase diagram for the $\nu=1$ QHE in strongly coupled DQW's. The ρ_{xx} data is from the two samples denoted by the left-most open and solid stars in the phase diagram.

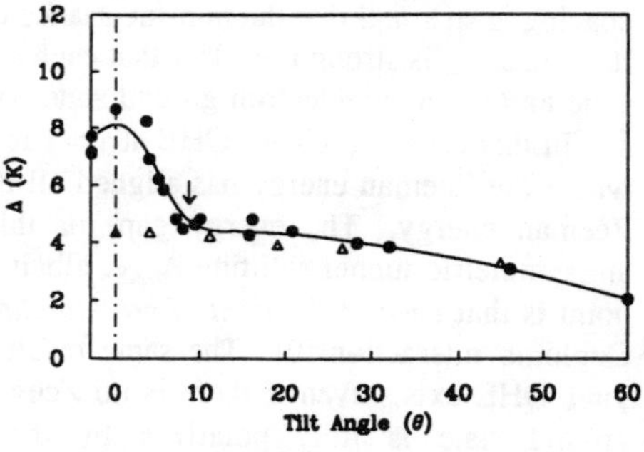

Fig. 6. Tilt dependence of the energy gap for the $\nu=1$ QHE state in a weakly tunneling sample (the leftmost solid star on the phase diagram in Fig. 5.) The open triangles are for the $\nu=2/3$ state and show no anomalous angular dependence.

widths and electron densities. Figure 5 shows the resultant $\nu=1$ phase diagram along with representative ρ_{xx} traces from two samples. The horizontal axis parameterizes the tunneling strength by the ratio of the calculated symmetric-antisymmetric gap Δ_{SAS} to the mean *intra*layer Coulomb energy $e^2/\varepsilon\ell_0$. The vertical axis is the ratio of the center-to-center quantum well separation to the magnetic length: d/ℓ_0. This is the ratio of intra- to interlayer Coulomb energies. The solid symbols denote samples for which a $\nu=1$ QHE was observed while the open symbols denote samples in which it was absent. This diagram shows that the $\nu=1$ QHE evolves continuously between a regime where single particle tunneling is dominant to one where interlayer Coulomb interactions stabilize the state in the absence of tunneling. Thus a clear distinction between the two cases is not possible. At large layer separation the QHE is destroyed. This is obviously expected in the weak tunneling limit where the state is stabilized by interlayer Coulomb interactions. For the strong tunneling case the phase boundary signals the collapse[11] of the tunneling gap (at finite wavevector) due to interlayer interactions first observed by Boebinger, *et al.*[12].

An attractive qualitative picture of the bilayer $\nu=1$ QHE can be constructed by making use of an analogy to quantum ferromagnetism[13]. The bilayer system can be modeled as a single 2D layer possessing an extra internal quantum number. This new quantum number, the pseudospin, replaces the layer index in the DQW. (We emphasize that DQW FQH physics is not merely equivalent to that of a single layer 2DES with real spin since the Coulomb interaction is not pseudospin invariant: electrons in different layers repel one another less than electrons in the same layer. For the present discussion the true spin is assumed to be fully polarized and is thus irrelevant.) The pseudospin axes are conventionally chosen so that $\sigma_z=\pm1$ states correspond to electrons in either one layer or the other while $\sigma_x=\pm1$ denotes electrons in either the symmetric or the antisymmetric DQW state. Consider for the moment a structure in which the layer spacing is so small that the non-invariance of the Coulomb interaction can be ignored. If the tunneling is strong it is clear that each electron will occupy a $\sigma_x = +1$ (i.e. symmetric) state and the many-electron ground state will be maximally pseudospin polarized (along $\hat{x}$). In this case the bilayer QHE at $\nu = 1$ is exactly like that in a real single layer system where the Zeeman energy has aligned all the real spins; here the tunneling acts like the Zeeman energy. The energy gap for this QHE will be essentially the symmetric-antisymmetric tunnel splitting Δ_{SAS}, albeit enhanced by exchange effects. The essential point is that *even in the limit of no tunneling* there is still a QHE at $\nu = 1$ due entirely to Coulomb interactions[9]. The same result holds in the single layer-real spin case: the $\nu = 1$ QHE exists even if there is no Zeeman energy[14]. In both cases the many-body ground state is fully polarized by the Coulomb (exchange) interaction but the "magnetization" can point in any direction, just as in an Heisenberg ferromagnet. The QHE energy gap is due entirely to the exchange energy in this limit. The experimental finding that the DQW $\nu = 1$ QHE evolves continuously between the weak and strong tunneling limit now appears natural. Turning on a small amount of tunneling merely breaks the symmetry of the ferromagnetic state, forces the pseudomagnetization to lie along the x-axis, and further lowers the ground state energy. The existence of this broken symmetry[15] is essential to the remarkable properties that the $\nu = 1$ state is predicted[16] to possess.

The analogy with true spin in a single layer system breaks down as soon as the interlayer separation is large enough that the pseudospin dependence of the Coulomb interaction cannot be ignored. This is the case in our actual samples where $d/\ell_0 > 1$. Fortunately, however, the broken symmetry of the ground state is not lost. The difference now is that the pseudomagnetization is confined to lie in a disk in the x-y plane since any z-component would imply unequal densities in the two layers and this is energetically unfavorable. Instead of a Heisenberg ferromagnet we now have an easy-plane ferromagnet with 2D XY symmetry[13].

This model provides a pretty explanation for the puzzling tilted field behavior of the $\nu = 1$ DQW QHE observed by Murphy, *et al.*[10]. Figure 6 shows the energy gap Δ (determined from the ρ_{xx} temperature dependence) at $\nu = 1$ as a function of the angle between the total magnetic field and the normal to the 2D planes. (The total field is always adjusted to keep the perpendicular component fixed at the $\nu = 1$ value.) The rapid variation of Δ at small angles was completely unexpected. Simple explanations for this behavior based on real spin effects or changes in the tunneling matrix elements have not been found. Yang, *et al.*[13] have proposed that the data in Fig. 6 is the signature of a novel textural phase transition in the pseudospin field. They point out that the added in-plane magnetic field $B_{||}$ adds a spatially-dependent phase to the tunneling matrix element. The effect of this phase is to cause the local direction (in pseudospin space) of symmetric DQW eigenstates to "tumble" along an axis normal to the in-plane field. Thus if the ground state is to maintain the energetic advantage of tunneling the pseudospin field must twist in the xy-plane. The wavelength of the twist is proportional to $B_{||}^{-1}$. On the other hand, the twisting costs exchange energy since "adjacent" pseudospins are not precisely parallel. As $B_{||}$ increases this cost becomes prohibitive and the pseudospin field snaps back into an aligned state. Obviously, the critical tilt angle depends on the strength of the tunneling, and Yang, *et al.*[13] find that $\theta_c \propto \Delta_{SAS}^{1/2}$, at least for small $B_{||}$. The calculated critical angle comes within about a factor of 2 of the experimental result.

6. Conclusion

This paper has described, in very abbreviated form, three experiments that we are pursuing using double layer 2D systems. There are many other experiments on such structures that would be worthwhile. Possibilities include extending the interlayer Coulomb drag studies to high magnetic field, making direct tunneling measurements on strongly coupled systems where coherence effects are important, looking for other unusual many-body states (e.g. a bilayer Wigner crystal), etc. All of these are very challenging experimentally but they represent one of the most exciting frontiers in low dimensional electronic systems.

7. Acknowledgements

This work represents an active collaboration with S.Q. Murphy, L.N. Pfeiffer, K.W. West, and G.S. Boebinger.

8. References

1. Y.W. Suen, *et al., Phys. Rev. Lett.* **68,** 1379 (1992); J.P. Eisenstein, *et al., Phys. Rev. Lett.* **68,** 1383 (1992).
2. T.J. Gramila, *et al., Phys. Rev. Lett.* **66**, 1216 (1991); U. Sivan, *et al., Phys. Rev. Lett.* **68**, 1196 (1992).
3. J.P. Eisenstein, L.N. Pfeiffer, and K.W. West, *Appl. Phys. Lett.* **57**, 2324 (1990).
4. J.P. Eisenstein, L.N. Pfeiffer, and K.W. West, *Phys. Rev. Lett.* **68**, 674 (1992); and *Phys. Rev. B*, in press.
5. J.P. Eisenstein, L.N. Pfeiffer, and K.W. West, *Appl. Phys. Lett.* **58**, 1497 (1991); and S.Q. Murphy, *et al., Bull. Am. Phys. Soc.* **38,** 591 (1993).
6. J.P. Eisenstein, L.N. Pfeiffer, and K.W. West, *Phys. Rev. Lett.* **69**, 3804 (1992) and *Surf. Sci.* **305**, 393 (1994).
7. Y. Hatsugai, P.-A. Bares, and X.G. Wen, *Phys. Rev. Lett.* **71,** 424 (1993); Song He, P.M. Platzman, and B.I. Halperin, *Phys. Rev. Lett.* **71,** 777 (1993); Peter Johansson and Jari M. Kinaret, *Phys. Rev. Lett.*, **71,** 1435 (1993) and *Phys. Rev. B*, in press; A.L. Efros and F.G. Pikus, *Phys. Rev. B***48**, 14694 (1993) and to be published.
8. J.P. Eisenstein, L.N. Pfeiffer, and K.W. West, submitted for publication.
9. T. Chakraborty and P. Pietilainen, *Phys. Rev. Lett.* **59**, 2784 (1987); D. Yoshioka, A.H. MacDonald, and S.M. Girvin, *Phys. Rev. B***39**, 1932 (1989).
10. S.Q. Murphy, *et al., Phys. Rev. Lett.* **72,** 728 (1994).
11. A.H. MacDonald, P.M. Platzman, and G.S. Boebinger, *Phys. Rev. Lett.* **65,** 775 (1990); Luis Brey, *Phys. Rev. Lett.* **65,** 903 (1990).
12. G.S. Boebinger, H.W. Jiang, L.N. Pfeiffer, and K.W. West, *Phys. Rev. Lett.* **64,** 1793 (1990).
13. Kun Yang, *et al., Phys. Rev. Lett.* **72,** 732 (1994).
14. S.L. Sondhi, *et al., Phys. Rev. B***47,** 16419 (1993).
15. X.G. Wen and A. Zee, *Phys. Rev. Lett.* **69,** 1811 (1992); *Phys. Rev. B***47**, 2265 (1993).
16. See Refs. 13 and 15 and Z.F. Ezawa and A. Iwazaki, *Phys. Rev. B***47**, 7295 (1993); *Phys. Rev. Lett.* **70**, 3119 (1993); *Phys. Rev. B***48**, 15189 (1993).

SPONTANEOUS INTERLAYER COHERENCE IN DOUBLE-LAYER QUANTUM HALL SYSTEMS

K. MOON, H. MORI, KUN YANG, S.M. GIRVIN, A.H. MACDONALD
Department of Physics, Indiana University
Bloomington, IN 47405

and

L. ZHENG
Department of Physics, University of Kentucky
Lexington, KY 40506

and

D. YOSHIOKA
Institute of Physics, College of Arts and Sciences
University of Tokyo, Komaba, Meguroku Tokyo 153, Japan

and

SHOU-CHENG ZHANG
Department of Physics, Stanford University
Palo Alto, CA 94305

ABSTRACT

At strong magnetic fields double-layer two-dimensional-electron-gas systems can form an unusual broken symmetry state with spontaneous inter-layer phase coherence. The system can be mapped to an equivalent system of spin $1/2$ particles with spin-dependent interactions and easy-plane XY magnetic order. We estimate the critical temperature for the Kosterlitz-Thouless phase transition, below which the system exhibits 'pseudospin' superconductivity. Above the KT transition, the system exhibits unconfined vortices each of which carry fractional charge $e^* = \pm e/2$. In the presence of weak inter-layer tunneling, a parallel magnetic field induces a zero-temperature commensurate-incommensurate phase transition. The latter has been recently observed by Murphy et al.

1. Introduction

The study of correlated electron systems in fewer than three dimensions continues to be an important theme in condensed matter physics. In particular the study of strongly correlated two-dimensional electron systems on a lattice has been motivated by high-temperature superconductivity while the study of strong correlations in continuum two-dimensional systems has been motivated by the fractional quantum Hall effect[1]. Properties of high-temperature superconductors are thought by some to be strongly influenced by the weak coupling which exists between the superconducting planes. In the fractional quantum Hall effect, early work by Halperin[2] anticipated novel fractional quantum Hall effects due to interlayer correlations[3] in multi-layer systems.

Recent technological progress has made it possible to produce double-layer two dimensional electron gas systems of extremely high mobility.[4] As illustrated schematically in Fig.(1), these systems consist of a pair of 2D electron gases separated by a distance d so small ($d \sim 100\text{Å}$) as to be comparable to the typical spacing between electrons in the same layer. A second type of system has also recently been developed to a high degree of perfection.[5] These systems consist of single wide quantum wells in which strong mixing of the two lowest electric subbands allows the electrons to localize themselves on opposites sides of the well to reduce their correlation energy. We will take the point of view that these systems can also be approximately viewed as double well systems with some effective layer separation and tunnel barrier height.

In a large magnetic field, strong correlations between the layers have long been expected to lead to novel fractional quantum Hall effects. Correlations are especially important in the strong magnetic field regime because all electrons can be accommodated within the lowest Landau level and execute cyclotron orbits with a common kinetic energy. The fractional quantum Hall effect occurs when the system has a gap for making charged excitations, *i.e.* when the system is incompressible, and theory has predicted[2,3,6] that at some Landau level filling factors, gaps occur in double-layer systems only if interlayer interactions are sufficiently strong. These theoretical predictions have recently been confirmed[7]. More recently work from several different points of view[8,9,10,11,12] has suggested that inter-layer correlations can also lead to unusual broken symmetry states with spontaneous phase coherence between layers which are isolated except for inter-layer Coulomb interactions. We have recently argued[13,14] that it is spontaneous interlayer phase coherence which is responsible for the recently discovered[4] extreme sensitivity of the fractional quantum Hall effect at total Landau level filling factor $\nu = 1$ to small tilts of the magnetic field away from the normal to the layers. ($\nu \equiv N/N_\phi$ where N is the number of electrons and N_ϕ is the number of single-particle levels per Landau level.)

2. Experimental Background

In this section we review the experimental indications that the system is ordered and exhibits excitations which are highly collective in nature. We focus here and throughout the paper on the case of Landau level filling factor $\nu = 1$ (1/2 in each layer) and assume for simplicity that the electronic spin degrees of freedom are frozen out by the Zeeman energy. The schematic energy level diagram for the growth direction degree-of-freedom in the double-layer system is shown in Fig.(1). For simplicity we assume that electrons can only occupy the lowest electric subband in each quantum well. If the barrier between the wells is not too strong, tunneling from one side to the other is allowed. The lowest energy eigenstates split into symmetric and antisymmetric combinations separated by an energy gap Δ_{SAS} which can, depending on the sample vary from essentially zero to many hundreds of

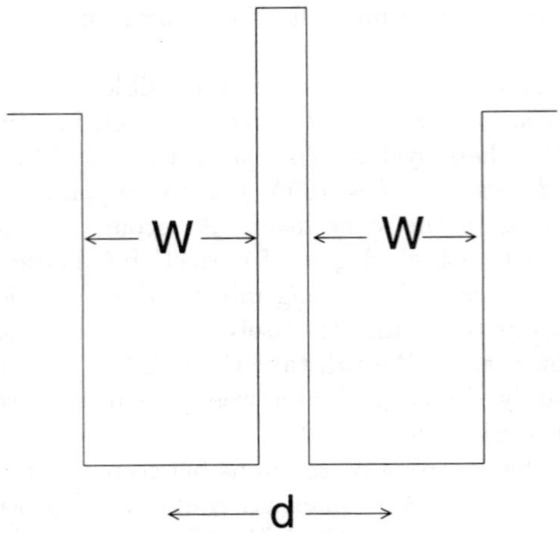

Figure 1: Schematic conduction band edge profile for a double-layer two-dimensional electron gas system. Typical widths and separations are $W \sim d \sim 100\text{Å}$ and are comparable to the spacing between electrons within each inversion layer.

Kelvins. The splitting can therefore be much less than or greater than the interlayer interaction energy scale, $E_c \equiv e^2/\epsilon d$.

When the layers are widely separated, there will be no correlations between them and we expect no dissipationless quantum Hall state since each layer has[15] $\nu = 1/2$. For smaller separations, it is observed experimentally that there is an excitation gap and a quantized Hall plateau.[16,5,4] This has either a trivial or a highly non-trivial explanation, depending on the ratio Δ_{SAS}/E_c. For large Δ_{SAS} the electrons tunnel back and forth so rapidly that it is as if there is only a single quantum well. The tunnel splitting Δ_{SAS} is then analogous to the electric subband splitting in a (wide) single well. All symmetric states are occupied and all antisymmetric states are empty and we simply have the ordinary $\nu = 1$ integer Hall effect. Correlations are irrelevant in this limit and the excitation gap is close to the single-particle gap Δ_{SAS} (or $\hbar\omega_c$, whichever is smaller). What is highly non-trivial about this system is the fact that the $\nu = 1$ quantum Hall plateau survives even when $\Delta_{SAS} \ll E_c$. In this limit the excitation gap has clearly changed to become highly collective in nature since the observed[5,4] gap can be on the scale of 20K even when $\Delta_{SAS} \sim 1$K. Because of a spontaneous broken symmetry,[8,9,12,13] the excitation gap actually survives the limit $\Delta_{SAS} \longrightarrow 0$! This cross-over from single-particle to collective gap is quite analogous to the result that for spin polarized single layers, the excitation gap survives the limit of zero Landé g factor and hence '$\nu = 1$ is a fraction too.'[17]

A second indication of the highly collective nature of the excitations can be seen in the Arrhenius plots showing thermally activated dissipation.[4] The low temperature activation energy Δ is, as already noted, much larger than Δ_{SAS}. If Δ were

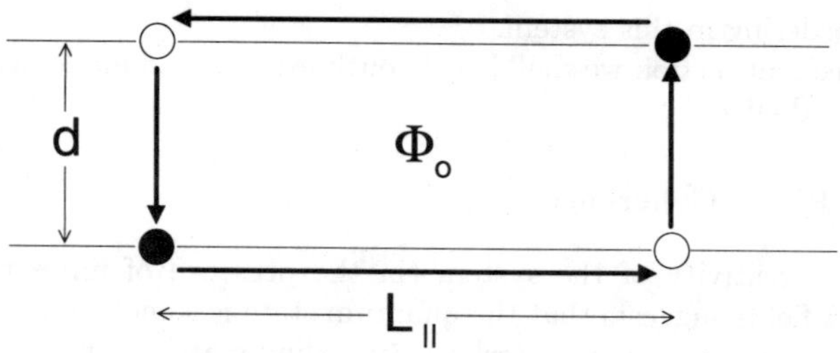

Figure 2: A process in double-layer two-dimensional electron gas systems which encloses flux from the parallel component of the magnetic field. The quantum amplitude for such paths is sensitive to the parallel component of the field.

nevertheless somehow a single-particle gap, one would expect the Arrhenius law to be valid up to temperatures of order Δ. Instead one observes a fairly sharp leveling off in the dissipation as the temperature increases past values as low as $\sim 0.1\Delta$. This is consistent with the notion of a thermally induced collapse of the order that had been producing the collective gap.

The third significant feature of the experimental data pointing to a highly-ordered collective state is the strong response of the the system to relatively weak magnetic fields $B_{\parallel}$ applied in the plane of the 2D electron gases. Within a model that neglects higher electric subbands, we can treat the electron gases as strictly two-dimensional. This is important since $B_{\parallel}$ can affect the system only if there are processes that carry electrons around closed loops containing flux. A prototypical such process is illustrated in Fig.(2). An electron tunnels from one layer to the other at point A, and travels to point B. Then it (or another indistinguishable electron) tunnels back and returns to the starting point. The parallel field contributes to the quantum amplitude for this process (in the 2D gas limit) a gauge-invariant Aharonov-Bohm phase factor $\exp(2\pi i\Phi/\Phi_0)$ where Φ is the enclosed flux and Φ_0 is the quantum of flux.

Such loop paths evidently contribute significantly to correlations in the system since the activation energy gap is observed to decrease very rapidly with $B_{\parallel}$, falling by factors of order two or more until a critical field, $B_{\parallel}^* \sim 0.8$T, is reached at which the gap essentially ceases changing.[4] To understand how remarkably small $B_{\parallel}^*$ is, consider the following. We can define a length $L_{\parallel}$ from the size of the loop needed to enclose one quantum of flux: $L_{\parallel}B_{\parallel}^*d = \Phi_0$. ($L_{\parallel}[\text{Å}] = 4.137 \times 10^5/d[\text{Å}]B_{\parallel}^*[\text{T}]$.) For $B_{\parallel}^* = 0.8$T and $d = 150$Å, $L_{\parallel} = 2700$Å which is approximately twenty times the spacing between electrons in a given layer and thirty times larger than the quantized cyclotron orbit radius $\ell \equiv (\hbar c/eB_{\perp})^{1/2}$ within an individual layer. Significant drops in the excitation gap are already seen at fields of 0.1T implying enormous phase coherent correlation lengths must exist. Again this shows the highly collective nature of the ordering in this system.

In the subsequent section we shall briefly outline a detailed model which explains all these observed effects.

3. Interlayer Phase Coherence

The strong sensitivity of the system (in the presence of finite tunneling) to small parallel B fields suggests that the quantum state is somehow very sensitive to changes in the local phase of the interlayer tunneling matrix element. This in turn suggests that we consider the possibility of a broken symmetry corresponding to the following order parameter condensing (even in the absence of tunneling between the layers)

$$\Psi(\mathbf{r}) \equiv \langle \psi_\uparrow^\dagger(\mathbf{r})\psi_\downarrow(\mathbf{r})\rangle,$$

where the arrows indicate layer index. This complex order parameter has an amplitude and a phase like that of a superconductor but in many ways is more reminiscent of an excitonic insulator since it is charge neutral and represents a particle in one layer bound to a hole in the other layer.

In the pseudo-spin formulation of the problem[13] this spontaneous interlayer phase coherence corresponds to a finite expectation value of the pseudospin raising operator

$$\langle S^+(\mathbf{r})\rangle = \langle S^x(\mathbf{r}) + iS^y(\mathbf{r})\rangle \sim e^{i\varphi(\mathbf{r})},$$

where $\varphi(\mathbf{r})$ is the local orientation of the pseudospin in the XY plane. The z component of the pseudospin represents the local charge density difference between the layers

$$S^z(\mathbf{r}) = \frac{1}{2}\left[\psi_\uparrow^\dagger(\mathbf{r})\psi_\uparrow(\mathbf{r}) - \psi_\downarrow^\dagger(\mathbf{r})\psi_\downarrow(\mathbf{r})\right].$$

The charging energy introduces an $(S^z)^2$ term in the action which quenches fluctuations of the magnetic order parameter out of the plane, leaving us with an XY model whose classical action is of the form

$$H = \frac{1}{2}\rho_s \int d^2r |\nabla\varphi|^2.$$

It may seem impossible that in the absence of tunneling, the system energy could depend on the relative phase φ for finding the particle in the upper or lower layer. Indeed the energy is unchanged when a constant is added to φ. However the energy does depend on gradients of φ. The 'spin stiffness' ρ_s is non-zero because of the loss of exchange energy which occurs when φ varies with position. Imagine that two particles approach each other. They are in a linear superposition of states in each of the layers (even though there is no tunneling!). If they are characterized by the same phase, then the wave function is symmetric under pseudospin exchange and so must the spatial wave function is antisymmetric and must vanish as the particles approach each other. This lowers the Coulomb energy. If a phase gradient exists then there is some amplitude for the particles to be in the same place and hence the energy is higher. This loss of exchange energy is the source of the finite spin stiffness and is what causes the system to spontaneously 'magnetize'.

Hartree-Fock estimates indicate that ρ_s and hence the Kosterlitz-Thouless critical temperature are on the scale of 1K in typical samples. Vortices in the φ field are reminiscent of Laughlin's fractionally charged quasiparticles but in this case carry charges $\pm\frac{1}{2}e$ and can be left- or right-handed for a total of four 'flavors'.[13] It is also possible to show that the presence spontaneous magnetization due to the finite spin stiffness means that the charge excitation gap is finite (even though the tunnel splitting is zero).

Since the 'charge' conjugate to the phase φ is S^z, the pseudospin 'supercurrent'

$$J = \rho_s \nabla \varphi \tag{1}$$

represents oppositely directed charge currents in each layer. Below the KT transition temperature, such current flow will be dissipationless (in linear response) just as in an ordinary superfluid.

A finite tunneling amplitude t between the layers breaks the $U(1)$ symmetry

$$H = \int d^2r \left[\frac{1}{2}\rho_s|\nabla\varphi|^2 - \frac{t}{2\pi\ell^2}\cos\varphi\right]$$

by giving a preference to symmetric tunneling states. This can be seen from the tunneling Hamiltonian

$$H_{\rm T} = -t \int d^2r \left\{\psi_\uparrow^\dagger(\mathbf{r})\psi_\downarrow(\mathbf{r}) + \psi_\downarrow^\dagger(\mathbf{r})\psi_\uparrow(\mathbf{r})\right\} \tag{2}$$

which can be written in the spin representation as

$$H_{\rm T} = -2t \int d^2r S^x(\mathbf{r}).$$

(Recall that the eigenstates of S^x are symmetric and antisymmetric combinations of up and down.)

The presence of parallel field $B_{||}$ field modulates the phase of t with position (in the direction perpindicular to $B_{||}$)

$$t \to t\, e^{iQx}$$

and the energy becomes

$$H = \int d^2r \left[\frac{1}{2}\rho_s|\vec{\nabla}\varphi|^2 - \frac{t}{2\pi\ell^2}\cos(\varphi - Qx)\right]$$

where $Q = 2\pi/L_{||}$ and $L_{||}$ is the length associated with one quantum of flux for the loops shown in Fig. 2. This is the so-called Pokrovsky-Talopov model which exhibits a commensurate-incommensurate phase transition. At low $B_{||}$, Q is small and the low energy state has $\varphi \approx Qx$; i.e. the local spin orientation 'tumbles'. In contrast, at large $B_{||}$ the gradient cost is too large and we have $\varphi \approx$ constant. It is possible to show that this phase transition semiquantitatively explains the rapid drop and subsequent leveling off of the activation energy vs. $B_{||}$ seen in experiment.[4]

4. Acknowledgements

It is a pleasure to acknowledge useful conversations with Daniel Arovas, Greg Boebinger, Nick Bonesteel, Luis Brey, René Côté, Jim Eisenstein, Herb Fertig, Matthew Fisher, Jason Ho, Jun Hu, David Huse, Sheena Murphy, Phil Platzman, Scott Renn, Nick Read, Ed Rezayi, Mansour Shayegan, Shivaji Sondhi, and Mats Wallin. The work at Indiana University was supported by NSF DMR91-13911. The work at the University of Kentucky was supported by NSF DMR92-02255. We are pleased to acknowledge the Aspen Center for Physics where part of this work was performed.

5. References

1. *The Quantum Hall Effect, 2nd Ed.*, edited by Richard E. Prange and Steven M. Girvin (Springer-Verlag, New York, 1987); T. Chakraborty and P. Pietiläinen, *The Fractional Quantum Hall Effect* (Springer-Verlag, Berlin, New York, 1988); *Quantum Hall Effect*, edited by Michael Stone (World Scientific, Singapore, 1992); Allan H. MacDonald, *Quantum Hall Effect: A Perspective*, (Kluwer Academic Publishers, 1989).
2. B. I. Halperin, *Helv. Phys. Acta* **56**, 75(1983).
3. T. Chakraborty and P. Pietiläinen, Phys. Rev. Lett., **59**, 2784 (1987); E.H. Rezayi and F.D.M. Haldane, *Bull. Am. Phys. Soc.* **32**, 892 (1987); Song He, S. Das Sarma and X.C. Xie, *Phys. Rev. B* **47**, 4394 (1993); D. Yoshioka, A.H. MacDonald, and S.M. Girvin, *Phys. Rev. B* **39**, 1932 (1989).
4. S.Q. Murphy, J.P. Eisenstein, G.S. Boebinger, L.N. Pfeiffer, and K.W. West, *Phys. Rev. Lett.* **72**, 728 (1994).
5. M. B. Santos, L. W. Engel, S. W. Hwang, and M. Shayegan, *Phys. Rev. B* **44**, 5947 (1991); T. S. Lay, Y. W. Suen, H. C. Manoharan, X. Ying, M. B. Santos, and M. Shayegan, Princeton University preprint (1994).
6. For a brief review of the fractional quantum Hall effect in double-layer systems see A.H. MacDonald, *Surface Science* **229**, 1 (1990).
7. Y. W. Suen *et al.*, *Phys. Rev. Lett.* **68**, 1379 (1992); J. P. Eisenstein *et al.*, *Phys. Rev. Lett.* **68**, 1383 (1992). preprint (1994).
8. X.G. Wen and A. Zee, *Phys. Rev. Lett.* **69**, 1811 (1992); X.G. Wen and A. Zee, *Phys. Rev. B* **47**, 2265 (1993).
9. Z.F. Ezawa and A. Iwazaki, *Int. J. of Mod. Phys. B*, **19**, 3205 (1992); Z.F. Ezawa and A. Iwazaki, *Phys. Rev. B* **47**, 7295 (1993); Z.F. Ezawa, A. Iwazaki, *Phys. Rev. B* **48**, 15189 (1993).
10. A.H. MacDonald, P.M. Platzman, and G.S. Boebinger, *Phys. Rev. Lett.* **65**,

775 (1990)

11. Luis Brey, *Phys. Rev. Lett.* **65**, 903 (1990); H.A. Fertig, *Phys. Rev. B* **40**, 1087 (1989).
12. R. Côté, L. Brey, and A.H. MacDonald, *Phys. Rev. B* **46**, 10239 (1992); X.M. Chen and J.J. Quinn, *Phys. Rev. B* **45**, 11054 (1992).
13. Kun Yang, K. Moon, L. Zheng, A.H. MacDonald, S.M. Girvin, D. Yoshioka, and Shou-Cheng Zhang, *Phys. Rev. Lett.* **72**, 732 (1994), and *Phys. Rev. B* (submitted).
14. K. Moon, H. Mori, Kun Yang, S.M. Girvin, A.H. MacDonald, L. Zheng, D. Yoshioka, and Shou-Cheng Zhang, in preparation (1994).
15. A single-layer system at Landau level filling factor $\nu = 1/2$ has no charge gap but does show interesting anomolies which may indicate that it forms a liquid of composite fermions. For a discussion of recent work see B.I. Halperin, Patrick A. Lee, and Nicholas Read, *Phys. Rev. B* **47**, 7312 (1993) and work cited therin.
16. G. S. Boebinger, H. W. Jiang, L. N. Pfeiffer, and K. W. West, *Phys. Rev. Lett.* **64**, 1793 (1990); G. S. Boebinger, L. N. Pfeiffer, and K. W. West, *Phys. Rev. B* **45**, 11391 (1992).
17. S. L. Sondhi, A. Karlhede, S. A. Kivelson, and E. H. Rezayi, *Phys. Rev. B* **47**, 16419 (1993).

THE UNQUANTIZED QUANTUM HALL EFFECT IN COULOMB DRAG

EFRAT SHIMSHONI
Beckman Institute, 405 North Mathews Avenue,
and Department of Physics, 1110 West Green Street
University of Illinois at Urbana-Champaign
Urbana, IL 61801, U.S.A

and

SHIVAJI SONDHI
Department of Physics, 1110 West Green Street
University of Illinois at Urbana-Champaign
Urbana, IL 61801, U.S.A

ABSTRACT

We show that the Coulomb drag between two weakly coupled quantum Hall systems is a novel probe of their internal dynamics, particularly of the compressible states. The temperature dependence of the drag is sensitive to a frequency and wave-vector regime ($\omega << q$) inaccessible to transport measurements in a single layer. Most strikingly, at the transitions between plateaux it directly measures an exponent η, first discussed by Chalker, that is characteristic of the fractal structure of the critical eigenstates. In contrast, in the $\nu = 1/2$ state in a clean system the drag is identical to that in the zero field Fermi liquid.

1. Introduction

Recently, much interest has focussed[1] on states in compressible regions of the phase diagram of the two-dimensional electron system (2DES) in a strong magnetic field B. In these states, which are observed at the transitions between the plateaux and even denominator fillings, the 2DES exhibits novel correlated behavior *without* quantized values of the conductivites — a state of affairs that Halperin has dubbed the "unquantized" quantum Hall effect (QHE). The dynamics of these states is of considerable interest and indeed is richer than that of the ideal (odd denominator) QH states whose response is dominated by the gap in the excitation spectrum. In this work[2] we show that the phenomenon of Coulomb drag can be used to probe these dynamical properties in an unusual frequency and wave-vector regime, where existing theories can be used to make some striking predictions.

Coulomb drag, first predicted by Pogrebinskii and by Price[3], is the phenomenon whereby a current flowing in one system induces a current/voltage in another, nearby system, even though the two do not exchange particles (a typical geometry is depicted in Fig. 1). The drag is measured by the trans-resistivity ρ_t,

$$\rho_t = \mathcal{E}_u / j_d \,, \tag{1}$$

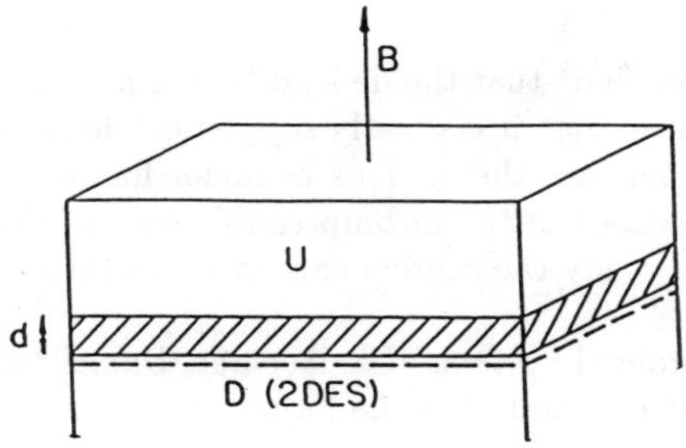

Fig. 1. Schematics of the heterostructure; The shaded area is an insulator.

where $\mathcal{E}_u$ is the parallel electric field induced in U (in open circuit) in response to a current density j_d established in D. Recent experiments, all in $B = 0$, have successfully measured the drag in heterostructures involving a 2DES, separated from another electron or hole system[4]. These were accompanied by a number of theoretical studies[5,6]. The physics of the drag is quite intuitive for it is much like ordinary friction. While a perfectly uniform current carrying state does not exert a Coulomb force on its neighbor, inhomogeneities in the density creates a "corrugated" potential which pushes charge carriers in the other layer in the direction of net current flow. The drag is therefore sensitive to the ability of the electronic states in the layers to support long lived density fluctuations, and thus to their internal dynamics. More formally, Zheng and MacDonald[6] have shown that, to leading order in the interaction V between two 2DESs, ρ_t can be expressed as

$$\rho_t = \frac{\beta \hbar^2}{\pi n^{(u)} n^{(d)} e^2} \int \frac{d^2 q}{(2\pi)^2} q^2 |V(q)|^2 \int_0^\infty d\omega \frac{\text{Im}\chi_d(q,\omega)\text{Im}\chi_u(q,\omega)}{4\sinh^2(\beta\hbar\omega/2)} \,, \tag{2}$$

where $\text{Im}\chi_{d(u)}(q,\omega)$ are the dissipative density-density response functions of layer D (U); $\beta = 1/k_B T$, where T is the temperature, and $n^{(d(u))}$ the density in layer D (U).

In the physics of the QHE, density fluctuations indeed play a priviliged role. The QH plateaux typically reflect the presence of incompressible states at odd denominator fillings. Near even denominator fillings one observes compressible behavior, either arising from critical behavior between neighboring plateaux in "dirty" samples[7,8], or from Fermi liquid (FL) states in the cleanest samples[9]. In both cases, problems such as the relative significance of interactions, disorder and their interplay is still a subject of active research. The former case has been extensively studied, within non-interacting models, in terms of delocalization transitions to "metallic" states[7]. Following this approach, Chalker has predicted the possibility of anomalously slow diffusion, characterized by an exponent η (see below), which reflects the fractal structure of the critical eigenstates[8]. The dynamical properties of these critical states are substentially distinct from the FL behavior predicted by ref. 9; however, they are inaccessible to conventional transport measurements. Below we show that interlayer drag can be potentially used as such a probe.

2. Principal results

At a fixed temperature, we find[2] that the dependence of ρ_t on the filling factor ν is qualitatively similar to that of σ_{xx}: it is greatly suppressed deep in the QH phases, and peaks at the transition between them. This behavior follows directly from the suppression of density fluctuations in the incompressible states. Below we focus on the compressible regions, and study the consequence of conjectured properties of the corresponding states.

For non-interacting, disordered systems, Chalker and Daniell[8] have shown that the low frequency response at criticality has the form:

$$\mathrm{Im}\chi(q,\omega) = \frac{(dn/d\mu)\omega\mathcal{D}(q,\omega)q^2}{[\omega^2 + (\mathcal{D}(q,\omega)q^2)^2]}, \quad \mathcal{D}(q,\omega) = D\left(\frac{\mathcal{C}\omega}{q^2}\right)^{\eta/2} \tag{3}$$

for $\mathcal{C}\omega < q^2$; here $D = 0.087/\hbar(dn/d\mu)$, $\mathcal{C} \approx 60\hbar(dn^{(d)}/d\mu)$, where $dn/d\mu$ is the density of states at the band center, and the universal exponent $\eta \approx 0.38$. For $\mathcal{C}\omega > q^2$, Eq. (3) is replaced by a conventional diffusive form, i.e. $\mathcal{D}(q,\omega) \approx D$. Inserting this for $\mathrm{Im}\chi_{d(u)}$ in Eq. (2) (assuming two identical layers), we obtain

$$\rho_t \approx 3.1 \times 10^{-4}\frac{h}{e^2}\left(\frac{k_BT}{\hbar Dn}\right)^2\left(\frac{\epsilon}{e^2d(dn/d\mu)}\right)^2\left(\frac{\mathcal{C}d^2k_BT}{\hbar}\right)^{-\eta}, \tag{4}$$

where n is the density, ϵ the background dielectric constant, and d the interlayer distance. Note that we made use of an important property of Eq. (2): for $T \to 0$, the T dependence of the denominator forces the relevant range of ω to vanish, and hence the behavior of ρ_t becomes sensitive to the form of $\mathrm{Im}\chi_{d(u)}(q,\omega)$ for $\omega << q$. This is *in contrast* with dc transport measurements in a single layer, which probe the opposite limit, and consequently is not sensitive to the anomalous diffusion regime. Eq. (4) implies that ρ_t in the critical region is *parametrically* enhanced at low T over non-critical filling factors (at which[2] $\rho_t \sim O(T^2)$), and its T dependence directly measures η. In fact[2], at any arbitrary critical point ρ_t probes an equivalent exponent, that parametrizes anomalous low-frequency dissipation. For a sample of mobility $\sim 10^6\,\mathrm{cm}^2/\mathrm{Vs}$, $n \sim 10^{11}\,\mathrm{cm}^{-2}$, $\epsilon \sim 10$, $d = 200\,\AA$ and $B = 10T$, we estimate $\rho_t \approx 10\,\mathrm{m}\Omega$ at $T \approx 0.1K$.

To compare the above result to the extremely clean limit, in which the transition regions are believed to contain FL states[9], we use the form $\mathrm{Im}\chi \sim \omega/q$ in Eq. (2). We then find $\rho_t \sim T^2$, as in the $B = 0$ case!

3. Discussion

The enhancement of interlayer drag by anomalous diffusion, reflected by the low T behavior $\rho_t \sim T^{2-\eta}$, is in accord with our earlier description of the underlaying physics. The anomalously slow dynamics implied by Eq. (3) ensures that large amplitude fluctuations in the density are maintained for long times ($\mathcal{D}(q,\omega)$ describes the rate of relaxation of such fluctuations). In comparison, ballistic dynamics (such as in a free electron gas) implies $\rho_t \sim T^2$, while regular diffusion enhances the drag at low T by a ln$T$ factor[6]. As a consequence, at $T \to 0$ the expression in Eq. (4) is expected to overwhelm any contribution to ρ_t from 'conventional' dynamical processes.

We are grateful to A.H. MacDonald for much useful advice and for informing us of his work. We thank J. Eisenstein, E. Fradkin, S. Girvin, S. Kivelson and B. Shklovskii for helpful discussions. This work was supported in part by the Beckman Foundation (ES), and NSF grants No. DMR 91–22385, DMR 91–57018 and the Aspen Center for Physics (SS).

References

1. For extensive references see, e.g., B. McCombe and A. Nurmikko eds, *EP2DS-10 Conference Proceedings* in *Surface Science* **305** No. 1-7 (1994).

2. E. Shimshoni and S. L. Sondhi, *Phys. Rev.* **B49**, 11484 (1994).

3. M. B. Pogrebinskii, *Fiz. Tekh. Poluprovodn.* **11**, 637 (1977) [Sov. Phys. Semicond. **11**, 372 (1977)]; P. M. Price, *Physica* (Amsterdam) **117B**, 750 (1983).

4. P. M. Solomon, P. J. Price, D. J. Frank and D. C. La Tulipe, *Phys. Rev. Lett.* **63**, 2508 (1989); T. J. Gramila, J. P. Eisenstein, A. H. MacDonald, L. N. Pfeiffer and K. W. West, *Phys. Rev. Lett.* **66**, 1216 (1991); U. Sivan, P. M. Solomon and H. Shtrikman, *Phys. Rev. Lett.* **68**, 1196 (1992).

5. B. Laikhtman and P. M. Solomon, *Phys. Rev.* **B41**, 9921 (1990); A. P. Jauho and H. Smith, *Phys. Rev.* **B47**, 4420 (1993); H. C. Tso and P. Vasilopolous, *Phys. Rev.* B **B45**,1333 (1992).

6. L. Zheng and A. H. MacDonald, *Phys. Rev.* **B48**, 8203 (1993).

7. H. Levine, S. B. Libby and A. M. M. Pruisken, *Phys. Rev. Lett.* **53**, 1915 (1983); D. E. Khmelnitiskii, *JETP Lett.* **38**, 552 (1983).

8. J. T. Chalker, *J. Phys.* **C20**, L493 (1987); J. T. Chalker and G. J. Daniell, *Phys. Rev. Lett.* **61**, 593 (1988); J. T. Chalker, *Physica* **167A**, 253 (1990); B. Huckestein and L. Schweitzer, *Phys. Rev. Lett.* **72**, 713 (1994).

9. B. I. Halperin, P. A. Lee and N. Read, *Phys. Rev.* **B47**, 7312 (1993).

Correlated Quantum Hall Effect in Bilayer Electron Systems

T. S. Lay, Y. W. Suen, H. C. Manoharan, X. Ying, M. B. Santos and M. Shayegan
Department of Electrical Engineering, Princeton University, Princeton, NJ 08544, USA

We present experimental phase diagrams for many-body quantum Hall effects at $\nu = 1/2$ and $\nu = 1$ in bilayer electron systems in wide single quantum wells. The $\nu = 1$ quantum Hall state exhibits an unusual temperature dependence suggestive of a finite temperature compressible/ incompressible phase transition.

Low-disorder bilayer electron systems (BLESs) exhibit novel quantum Hall phenomena such as a $\nu = 1/2$ even-denominator fractional quantum Hall effect (QHE) which has no counterpart in single-layer two-dimensional electron systems (ν is the *total* filling factor for the BLES).[1-3] This unique QHE has been associated with a correlated, two-component Ψ_{331} Laughlin-like state stabilized by interlayer Coulomb interaction.[2-4] In BLESs with appropriate parameters, the interlayer interaction can also lead to new, *correlated* QHE at certain *integral* fillings such as $\nu = 1$.[2, 4-7] In contrast to the single-particle $\nu = 1$ state associated with the tunneling gap (Δ_{SAS}) separating the symmetric and antisymmetric states, the many-body $\nu = 1$ QHE state characterized by the two-component Ψ_{111} wavefunction has been predicted to exhibit novel properties such as neutral superfluid modes and a Kosterlitz-Thouless transition.[5-7]

In this paper, we first present experimental phase diagrams for the presence or absence of the $\nu = 1/2$ and $\nu = 1$ QHE in BLESs in wide single quantum wells (WSQWs) establishing that in WSQWs with appropriate parameters the $\nu = 1$ QHE indeed has a many-body origin. We then concentrate on a striking temperature dependence observed for the $\nu = 1$ state which is suggestive of a finite temperature QHE to metallic transition.

The samples are modulation-doped GaAs wells grown by molecular beam epitaxy, have a width (W) ranging from 600 Å to 1200 Å, and typical low-temperature mobility ~ 1 x 10^6 cm^2/Vs. The electron density in the well, N_S, and the charge distribution symmetry were controlled via front- and back-side gates and Δ_{SAS} was measured from the Fourier transform of the low-field magnetoresistance oscillations. The measured Δ_{SAS} were in excellent agreement with those determined from self-consistent Hartree-Fock calculations, an example of which is shown in Fig. 1(c).

Figure 1 shows the magnetotransport data of a 770 Å-wide well. In addition to the strong integer QHE state at $\nu = 1$, we observe well-developed fractional QHE states at $\nu = 1/2$ and 3/2. To understand the origin of these even-denominator states, we made a careful and detailed study of their strength by measuring the quasiparticle excitation gaps as a function of N_S, W, and the asymmetry of the charge distribution.[3] A summary of this study is shown in the d/l_B vs α phase diagram of Fig. 2(a) for the presence or absence of the 1/2 QHE state. Here d is the distance between electron layers and $l_B = (\hbar/eB)^{1/2}$ is the magnetic length; d/l_B is therefore a measure of the competition between the intralayer and interlayer interaction while $\alpha = \Delta_{SAS} / (e^2/\varepsilon l_B)$ is proportional to the ratio of the tunneling gap and intralayer electron correlation energy. In Fig. 2(a) the range of α studied for each sample is indicated by the thin solid curve, while the presence of the 1/2 QHE state is marked by a symbol whose size qualitatively indicates the strength of the observed 1/2 QHE state (for clarity, we have used two symbols, circles and squares, to show data for six WSQWs with $600 \leq W \leq 960$ Å and variable N_S). The $\nu = 1/2$ state is stable only in the region of $0.04 \leq \alpha \leq 0.1$ between the dashed lines. Outside this region the BLES is metallic except for $\alpha < 0.04$ where it becomes insulating; the insulating phase possibly being a pinned bilayer Wigner crystal.[8] The collapse of the $\nu = 1/2$ QHE at large α where the system becomes more single-layer- (one-component-) like, together with the fact that it also collapses when the charge distribution becomes asymmetric, strongly suggests the two-component origin of this state.[3]

Figure 2(b) shows the corresponding d/l_B vs α phase diagram for the presence (solid symbols) or absence (open symbols) of a $\nu = 1$ QHE in WSQWs at our experimental base temperature of 25 mK. In the range of $4 < d/l_B < 8$, the $\nu = 1$ QHE collapses into a metallic state at the boundary $\alpha \approx 0.035$, which is very close to the boundary $\alpha \approx 0.04$ where the $\nu = 1/2$ QHE state makes a transition to an insulating phase. As α increases, the $\nu = 1$ QHE state remains stable and becomes of single-particle origin, stabilized by the tunneling gap. Our measured quasiparticle excitation gaps for the $\nu = 1$ QHE state near the experimental phase boundary $\alpha \approx 0.035$ are much larger than the Δ_{SAS} gap and slowly decrease to values close to Δ_{SAS} at large α (≥ 0.15). These observations strongly suggest that the $\nu = 1$ QHE states observed near the phase boundary of Fig. 2(b), in approximately the same range of α and d/l_B where the many-body $\nu = 1/2$ QHE state is stable, also have a many-body origin. The data also provide clear evidence that, as α increases, these many-body $\nu = 1$ QHE states *continuously* evolve to single-particle QHE states stabilized by substantial tunneling.

Further credence to the many-body origin of the $\nu = 1$ QHE states observed near the $\alpha \approx 0.035$ boundary is provided by their unusual temperature dependence. Figure 3(a) presents Arrhenius plots of the diagonal resistance (R_{xx}) at $\nu = 1$ for a 770 Å-wide

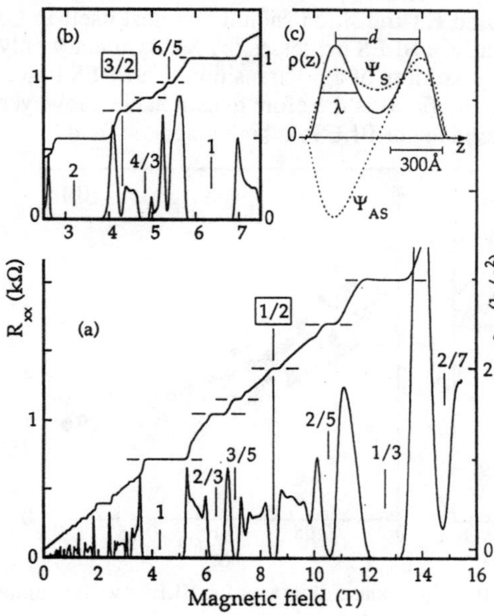

Fig. 1. Magnetotransport data for a 770 Å-wide well with (a) $N_S = 1.03 \times 10^{11}$ cm^{-2} and (b) $N_S = 1.55 \times 10^{11}$ cm^{-2}, showing well-developed even-denominator fractional QHE at $\nu = 1/2$ and $3/2$. In (c), the calculated electron wavefunctions (dotted curves denoted Ψ_S and Ψ_{AS}) and distribution function [solid curve denoted $\rho(z)$] are shown for this well with $N_S = 1.03 \times 10^{11}$ cm^{-2} at $B = 0$.

quantum well with electron density $5.0 \leq N_S \leq 16.4 \times 10^{10}$ cm^{-2} ($0.25 \geq \alpha \geq 0.05$). The quasiparticle excitation gaps, $^1\Delta$, determined from the slopes of the activated regions of these plots according to $R_{xx} \sim \exp(-^1\Delta/2T)$, are shown in Fig. 3(b) as a function of α. Also shown are the measured Δ_{SAS}. Several features of the data of Fig. 3 are particularly noteworthy: (1) While Δ_{SAS} decreases with decreasing α, $^1\Delta$ increases and exceeds Δ_{SAS} by more than a factor of three near the phase boundary of $\alpha \approx 0.035$. (2) For $N_S \geq 1.0 \times 10^{11}$ cm^{-2} ($\alpha \leq 0.11$), the activated behavior of R_{xx} vs 1/T ends rather abruptly above an N_S-dependent temperature T^*. Above T^*, R_{xx} is nearly independent of temperature and magnetic field. For $N_S \leq 1.0 \times 10^{11}$ cm^{-2}, the Arrhenius plots show a smoother behavior and $^1\Delta$ gaps start to decrease with decreasing N_S. (3) The measured $^1\Delta$ for $N_S \geq 1.0 \times 10^{11}$ cm^{-2} are approximately constant ($\approx$ 20 K) and exceed T^* by more than an order of magnitude. (4) Plot of the measured T^* vs α presented in Fig. 3(c) shows that T^* decreases with α and extrapolates to zero at $\alpha \approx 0.035$, consistent with the phase diagram of Fig. 2(b). These features clearly point to the many-body origin of $\nu = 1$ QHE states in the region of the d/l_B vs α phase diagram near the phase boundary.

The temperature dependence data of Fig. 3 are collectively very unusual and qualitatively different from what is observed for the integral or fractional QHE in standard single-layer 2D systems. The data may suggest a finite-temperature QHE to metallic phase transition, and it is tempting to examine them in light of the Kosterlitz-Thouless (KT) transition recently proposed for BLESs.[5,7] Associating the layer index with an "isospin," the theories argue that in the limit of zero tunneling but finite interlayer separation, the system has an XY symmetry and thus may support a KT transition at a critical temperature T_{KT}. The data of Fig. 3 are qualitatively consistent with the crude estimates given in Ref. 7 for T_{KT} and its dependence on d and l_B: our measured T^* are of the order of the estimated T_{KT} and both T^* and T_{KT} decrease with increasing d/l_B. It is not obvious, however, how the proposed KT transition should manifest itself in QHE transport data of a BLES. More importantly, a BLES possesses the XY symmetry only in the absence of interlayer tunneling. The possibility of a KT transition in a BLES like ours with substantial tunneling (at zero magnetic field) is therefore questionable. However, it is possible that such tunneling is suppressed in our BLESs at high magnetic fields.

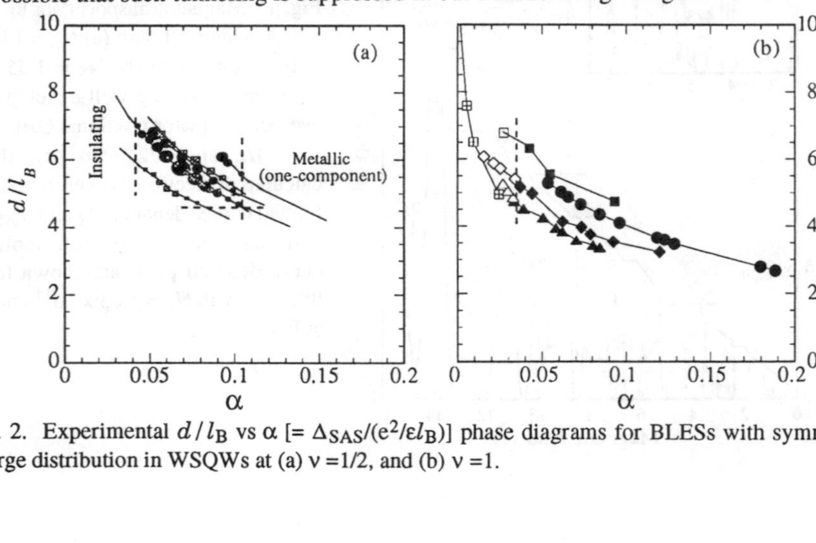

Fig. 2. Experimental d/l_B vs α [$= \Delta_{SAS}/(e^2/\varepsilon l_B)$] phase diagrams for BLESs with symmetric charge distribution in WSQWs at (a) $\nu = 1/2$, and (b) $\nu = 1$.

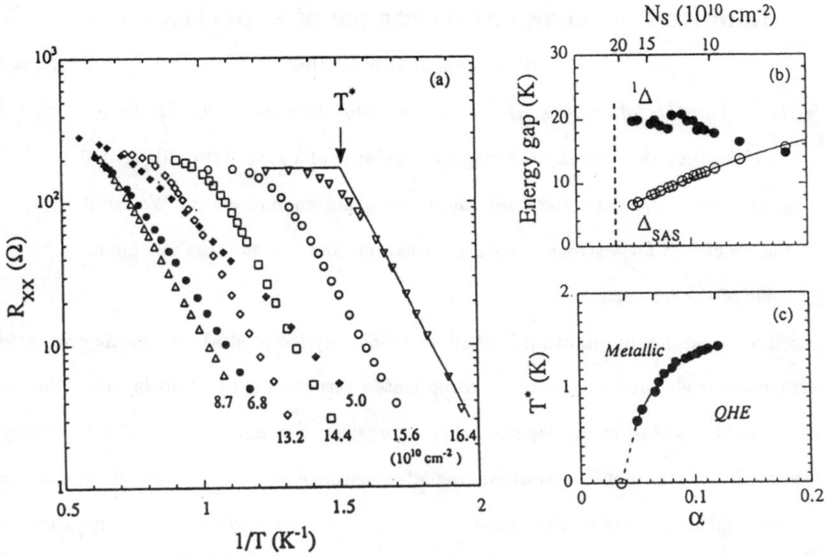

Fig. 3. (a) Arrhenius plots at $\nu = 1$ for a 770 Å-wide WSQW with different N_S. The temperature above which the activated behavior ends is marked T^*. (b) The measured energy gaps $^1\Delta$ for the $\nu = 1$ QHE and Δ_{SAS} as a function of α. The vertical dashed line marks the $\alpha = 0.035$ boundary. (c) Plot of T^* vs α.

An alternative explanation for our data is the following. Consider the BLES at α greater than but close to 0.035. Two possible states of the system are a correlated (Ψ_{111}-like) QHE state, and an uncorrelated metallic state with $\nu = 1/2$ in each layer. The correlated state, energetically more favorable, is the ground state. The metallic state, however, has larger entropy. It is therefore possible that the BLES makes a transition from the correlated QHE to the metallic state at finite temperature. This scenario is qualitatively consistent with the observation that the closer the BLES is to the $\alpha \approx 0.035$ boundary the smaller the temperature needed to destroy the correlated QHE state (Fig. 3).

We thank F.D.M. Haldane, A. MacDonald, I.A. McDonald, D.C. Tsui, X.G. Wen and K. Yang for useful discussions and D. Shahar and L.W. Engel for technical assistance. H.C.M. thanks the Fannie and John Hertz Foundation for fellowship support. This work was supported in part by the NSF DMR-9222418 and DMR-9224077 grants.

1. Y. W. Suen *et al., Phys. Rev. Lett.* **68**, 1379 (1992); **69**, 3551 (1992).
2. J. P. Eisenstein *et al., Phys. Rev. Lett.* **68**, 1383 (1992).
3. Y. W. Suen *et al., Phys. Rev. Lett.* **72**, 3405 (1994).
4. S. He, S. Das Sarma, and X. C. Xie, *Phys. Rev.* B **47**, 4394 (1993).
5. X. G. Wen and A. Zee, *Phys. Rev. Lett.* **69**, 1811 (1992).
6. S. Q. Murphy *et al., Phys. Rev. Lett.* **72**, 728 (1994).
7. K. Yang *et al., Phys. Rev. Lett.* **72**, 732 (1994).
8. Y. W. Suen, H. C. Manoharan, and M. Shayegan, unpublished.

Longitudinal magnetoresistance of superlattices

B.Laikhtman and D.Menashe

Racah Institute of Physics, Hebrew University, Jerusalem, 91904, Israel

We study theoretically the magnetoresistance of a superlattice along the growth direction under the condition of the quantum Hall effect. We show the conditions for a strong quantum oscillations and estimate the maximum value of the conductivity.

Very active researches of quantum Hall effect eventually led to study magnetic properties of two dimensional electron gas in more complicated structures, like two layers. One can expect that further extension can lead to other interesting new effects. One of natural ways of the research is the study of longitudinal magnetoresistance in superlattices. We mean that the magnetic field is applied in the growth direction and the conductivity is measured in the same direction. In classical mechanics magnetic field does not affect an electron motion in the direction of the field and any longitudinal magnetoresistance can result only from the quantization of electron states. Few experimental measurements[1-3] did not report any significant effect of magnetic field on the longitudinal resistance. In this paper we discuss the conditions for the observation of the effect and make some estimates of the amplitude and the shape of the magnetoresistance oscillations.

In superlattices the quantization of the electron motion in the growth directions results in the formation of minibands. Magnetic field in the same direction leads to the quantization of the in-plane electron motion. The electron spectrum in a strong enough magnetic field ($\bar{\Omega} \gg \Gamma$, where Ω is the cyclotron frequency and Γ is the width of the level due to scattering) is characterized by the Landau quantum number, the number of the miniband, and the wave vector in the growth direction. The resulting spectrum can be understood in two ways. One is the splitting of each miniband into Landau levels. The other is the spreading of Landau levels in quantum wells into Landau minibands with the width equal to the original miniband without magnetic field, 4Λ (in the case of a narrow band Λ is an overlap integral between two adjacent quantum wells). As long as the separation between Landau levels, $\hbar\Omega$, is smaller than 4Λ the Landau minibands overlap with each other and the density of states of the superlattice miniband remains continuous (Fig.1). When $\hbar\Omega$ grows larger than 4Λ the density of states of the original miniband splits into the densities corresponding to different Landau minibands (Fig.2).

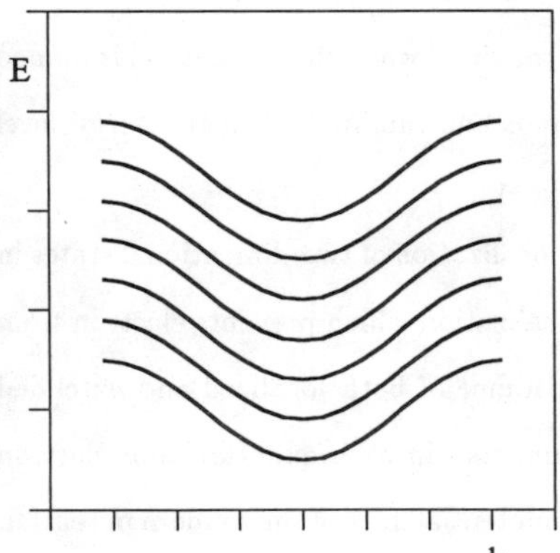

Fig.1

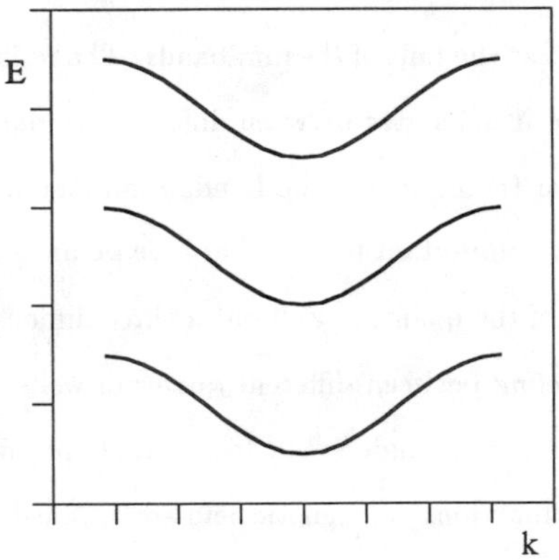

Fig.2

In doped superlattices the electron concentration is controlled by the doping and remains constant when the magnetic field is changing. That means that with the increase of the magnetic field the Fermi level passes from one Landau miniband to another so that more and more Landau minibands appear above the Fermi level. One can expect stong oscillations of the longitudinal magnetoresistance if the condition $\hbar\Omega > 4\Lambda$ is satisfied for the magnetic field when $\hbar\Omega$ is still smaller than the Fermi energy, E_F, so that a few Landau minibands are occupied and are depopulated with the further increase of the magnetic field. For this condition a superlattice with narrow minibands is necessary. Actually, it is enough if the width of the first miniband is a few times smaller than E_F and the gap between the first and the second miniband is so large that the Fermi level is below the bottom of the second miniband.

Another necessary condition is that the inequality $\hbar\Omega \gg \Gamma$ also has to be satisfied for the magnetic field when $\hbar\Omega < E_F$.

The physical reason for the oscillations of the longitudinal magnetoresistance is the same as for the oscillations of the magnetoresistance in the quantum Hall effect. A random potential resulting from impurities or from the roughness of the interfaces between quantum wells and barriers leads to the smearing of the Landau minibands and the localization of states at the tails of the minibands. The resistance is maximal when the Fermi level is pinned by localized states between different Landau minibands and minimal when the Fermi level is near the middle of the Landau miniband.

It is important to note that here we mean not the localization of two-dimentional states in each of the quantum well but a three dimensional localization which prevents electron from tunneling between different quantum wells. The existance of both localized and extended states can be understood from simple physical arguments. In each quantum well electron wave functions in magnetic field are confined near equipotential lines of the random potential. The width of the confinement steep around equipotential lines is of the order of $l\sqrt{N}$, where $l = \sqrt(c\hbar/eB)$ is the magnetic length and N is the Landau quantum number. The energy of the state deviates from $\hbar\Omega(N+1/2)$ by the value of the random potential. (One has to keep in mind that the relevant random potential equals the original random potnetial averaged over the cyclotron orbit[4].) Random potentials in adjacent wells are not correlated because there is hardly any correlation between the rughnesses of different well/barrier interfaces. An electron can tunnel from an equipotnetial line in one well to an equipotential line in an adjacent well in the region of their spacial overlap. Tunneling is an elastic process and is possible only if energies of the initial and final states are different no more than by the overlap integral Λ. The width of the random potential distribution is of the order of Γ. If $\Lambda \ll \Gamma$ then the area where the random potential is different from a given value no more than by Λ with the probability about 1 is of the order of $a^2\Gamma/\Lambda$ (a is the correlation length of the random potential). The area of the electron confinement is of the order of $tl\sqrt{N}$. That is the tunneling is possible if $tl\sqrt{N} \gtrsim a^2\Lambda/\Gamma$. This condition is a limitation for the perimeter of the equipotential line. It is satisfied for states near the center of the Landau level. The deviation of the potential very far from its average value is a rear occasion and typical perimeters of corresonding regions are small. So that the probability of electron tunneling between the wells is small and such states are localized. This consideration shows, by the way, that in the case $\Lambda \ll \Gamma$ not only Landau levels in separate quantum wells are smeared but also that the wave vector in the growth direction is not a good quantum number.

In the case of $\Lambda \gg \Gamma$ the probability to find a state in one well which is not overlapped in energy with another state in an adjacent well is exponentially small. That is tails of localized states are very narrow compared to the width of the extended states.

Comparing these two situations we see that if $\Lambda \ll \Gamma$ the widths of regions of both extended states near a Landau level, where the longitudinal resistance is minimal, and localized states, where the longitudinal resistance is maximal, are of the order of Γ. In the opposite case, when $\Lambda \gg \Gamma$, the regions of a large resistance are very narrow comared to the regions with a small resistance.

For the conductivity in the center of the Landau band at zero temperature the self-consistent Born approximation in the case $\Lambda \gg \Gamma$ gives the following result,

$$\sigma = \frac{2e^2\Lambda^2 d}{\pi^2 \hbar l^2 u}\,, \tag{1}$$

where u is the mean square random potential and d is the period of the superlattice.

References

1. K.K.Choi, B.F.Levine, N.Jarosik, J.Walker, and R.Malik, Phys.Rev.B **38**, 12362 (1988).

2. A.Sibille, J.F.Palmier, A.Celeste, J.C.Portal, and F.Mollot, Europhys.Lett. **13**, 279 (1990).

3. W.Müller, H.T.Grahn, R.J.Haug, and K.Ploog, Phys.Rev.B **46**, 9800 (1992).

4. M.E.Raikh and T.V.Shahbazyan, Phys.Rev.B **47**, 1552 (1993).

MAGNETIC FIELD INDUCED MINIGAP
IN DOUBLE QUANTUM WELLS

J. A. SIMMONS, N. E. HARFF,* S. K. LYO, AND J. F. KLEM
Sandia National Laboratories, P. O. Box 5800,
Albuquerque, New Mexico 87185 USA

ABSTRACT

We report the discovery of a partial energy gap, or minigap, in strongly coupled double quantum wells (QWs), due to an anticrossing of the two QW dispersion curves. The anticrossing and minigap are induced by an in-plane magnetic field $B_{||}$, and give rise to large distortions in the Fermi surface and density of states, including a Van Hove singularity. Sweeping $B_{||}$ moves the minigap through the Fermi level, with the upper and lower gap edges producing a sharp maximum and minimum in the low-temperature in-plane conductance, in agreement with theoretical calculations. The gap energy may be directly determined from the data.

1. Introduction

Applying an in plane magnetic field $B_{||}$ to a double quantum well (QW) causes the canonical momenta **k** of electrons in one QW to be shifted relative to those in the other. The shift is given by $\Delta k_y = edB/\hbar$, where d is the distance between the two electron layers, and direction y is normal to both the growth and $B_{||}$ directions. Recent studies[1,2] have shown this to have a strong effect on interwell tunneling, due to the conservation of **k**. Little attention, however, has been paid to the effect of this shift on the in-plane conductance $G_{||}$.[3] This is of interest since the shifted dispersion curves of two strongly coupled QWs can be expected to produce an anticrossing, resulting in a large minigap. At sufficiently high $B_{||}$ the minigap can be made to pass through the Fermi level. Under these conditions the DQW has a multi-component Fermi surface whose shape and topology is tunable by $B_{||}$.[4] Our data show the presence of this minigap, and demonstrate that the electronic density of states and group velocities are dramatically distorted near its edges, producing the observed sharp features in $G_{||}(B_{||})$. By plotting the $B_{||}$-positions of the features as a function of a surface gate voltage V_G, the gap energy can be extracted.

2. Experimental Set-up

Three MBE-grown structures were studied, each consisting of a modulation-doped pair of GaAs QWs of equal width w separated by an $Al_{0.3}Ga_{0.7}As$ barrier of thickness t. Samples A, B, and C had w=150 Å, t = 25 Å; w = 100 Å, t = 35Å; and w = 150 Å, t =

* Also at Oregon State University, Corvallis, Oregon 97331

15 Å, respectively. Standard four terminal low-frequency lock-in measurements were performed at $T \geq 0.3$ K on 0.2 mm wide Hall bars with Cr/Au surface gates. At V_G=0, samples A, B, and C had top QW densities $n_1 \approx 1.4$, 1.2 and 0.7 x 10^{11} cm^{-2}, and bottom QW densities $n_2 \approx 1.5$, 1.2, and 0.9 x 10^{11} cm^{-2}, respectively. The V_G=0 mobilities of the top QW were $\mu_1 \approx 2.7$, 1.2, and 0.6 x 10^5 cm^2/Vs, and of the bottom QW were $\mu_2 \approx 2.2$, 0.6, and 0.2 x 10^5 cm^2/Vs, respectively. n_1 and n_2 were determined from quantum Hall and Shubnikov-de Haas (SdH) measurements in $B_\perp$, followed by a rotation *in situ* for measurement in $B_\parallel$, with the residual component of $B_\perp$ less than 0.005 T. The applied electric field was nominally parallel to $B_\parallel$.

3. Effect of Anticrossing on the In-Plane Conductance

Fig. 1 shows the normalized $G_\parallel(B_\parallel)$ for several V_G. For V_G=0, a sharp maximum appears at $B_{max} \approx 5.8$ T, followed by a sharp minimum at $B_{min} \approx 6.4$ T. As V_G decreases, both features move to lower $B_\parallel$, with B_{max} moving more rapidly than B_{min}. The features disappear once the top QW depletes at V_G = -0.29 V, since their origin is due solely to interactions between the two QWs. The maximum and minimum correspond respectively to the upper and lower edges of the minigap crossing the Fermi level.

Fig. 2 shows the results of a tight-binding calculation[5] for a double QW structure having w = 140 Å, t = 60 Å, and unequal QW depths of 280 and 277 meV. (These

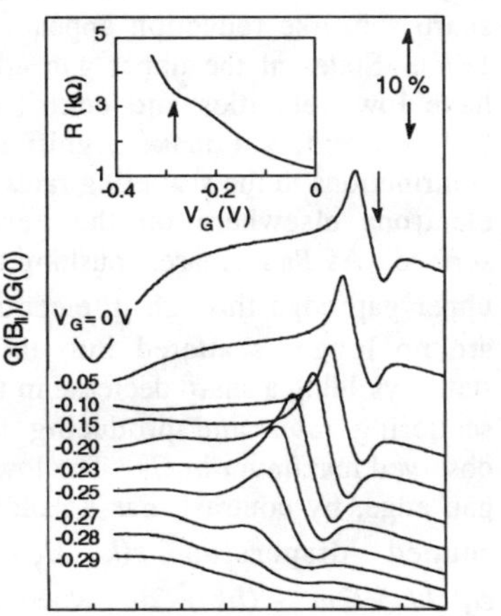

FIG. 1. Normalized $G_\parallel$ vs. $B_\parallel$ of sample A for different V_G, offset from one another by 2 %. Arrow indicates $\hbar(k_1+k_2)/ed$ for V_G=0, with d from a Hartree self-consistent calculation. Inset: $B_\parallel$=0 resistance vs. V_G.

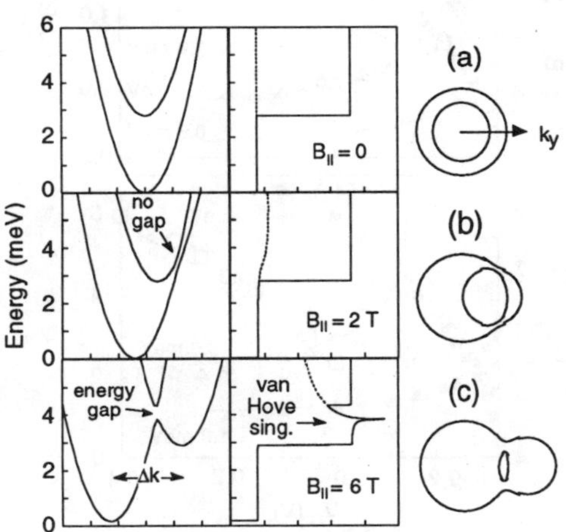

FIG. 2. Calculated dispersion $\varepsilon(k_y)$ (left); calculated density of states $D(E)$ (middle) for lower energy branch (dotted line) and both energy branches (solid line); and sketch of Fermi surface for $\mu \approx 5$ meV (right); for $B_\parallel$ = (a) 0 T, (b) 2 T, and (c) 6 T.

parameters, while somewhat different from the experiment, illustrate the essential physics nicely.) The left column shows the two lowest eigenenergies $\varepsilon_{\pm}(k_y)$, the middle column the density of states D(E), and the right column a sketch of the Fermi surface for chemical potential $\mu = 5$ meV. The three rows are (a) $B_{\|} = 0$ T, (b) 2 T, and (c) 6 T. At $B_{\|} = 0$ the $\varepsilon_{\pm}(k_y)$ curves are parabolic and $D(E)$ is constant for each energy branch, with the energy splitting mainly due to the unequal QW depths. The other two $B_{\|}$ values were chosen so that with no coupling the Fermi circles touch tangentially on the inside and outside, respectively, at the same $\mu \approx 4$ meV. This corresponds to $\Delta k_y = k_1 \pm k_2$, where k_1 and k_2 are the Fermi wavevectors of the top and bottom QWs. When coupling is present, these two cases exhibit two types of anticrossing: In the first, at 2 T, the slopes of $\varepsilon_{\pm}(k_y)$ deviate but do not change sign, changing the group velocity only slightly. No minigap appears, and the slight changes in $D(E)$ of each energy branch cancel one another. Thus the effect on $G_{\|}$ of this type of anticrossing passing through μ is negligible, in accord with the data of Fig. 1, for which no sharp features appear at the low fields $B_{\|} = \hbar(k_1 - k_2)/ed$.

In the second type of anticrossing, at 6 T, the slopes of $\varepsilon_{\pm}(k_y)$ change sign, and a minigap of width E_G appears. E_G is insensitive to $B_{\|}$ and $\approx \Delta_{SAS}$, the $B_{\|}=0$ symmetric-antisymmetric gap existing when the QW depths (densities) are equal.[5] At the upper gap edge the dispersion is nearly parabolic and a sharp step-like reduction appears in $D(E)$. States at the upper gap edge have low velocities and contribute little current, yet make significant contributions to the scattering rates of electrons elsewhere on the Fermi surface. As $B_{\|}$ increases, pushing the upper gap edge through μ, electrons are no longer scattered into these states, yielding a sharp decrease in the scattering rates and producing the observed maximum in $G_{\|}$. The lower gap edge, by contrast, has a saddle-shaped dispersion $\varepsilon(k_x, k_y) = \varepsilon_0 + (\hbar k_x)^2/2m^* - (\hbar k_y)^2/2m'$, where ε_0 is the saddle-point energy, and m' is determined by the saddle-point curvature. This type of dispersion produces a Van Hove singularity at ε_0 of the form $D(E) \propto -\ln(\varepsilon - \varepsilon_0)$. Since the saddle point states have zero group

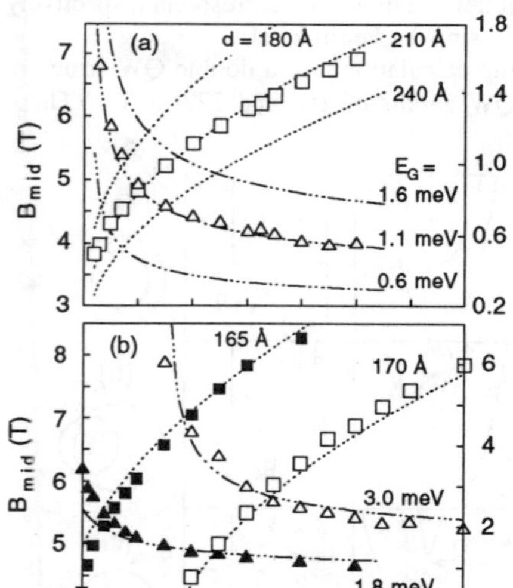

FIG. 3. (a) Plot of B_{mid} (squares) and ΔB (triangles) for sample A. Using sample A parameters, dotted lines are plots of Eq. (1) for various d, while dash-dotted lines are plots of Eq. (2) (using the d obtained from the fit to Eq. 3) for various E_G. (b) Same as in (a), but for samples B (solid symbols) and C (open symbols), and only showing the best fits.

velocity, when $B_{\parallel}$ is swept and the lower gap edge passes through μ, electrons are divergently scattered into states that carry no current, yielding the sharp minimum in $G_{\parallel}$.

4. Gate Bias and Minigap Energy

Negatively biasing the gate causes the features in $G_{\parallel}$ to move to lower $B_{\parallel}$. To a first approximation, V_G linearly changes n_1 while leaving n_2 unchanged. Assuming $B_{mid} = (B_{min}+B_{max})/2$ corresponds to μ lying in mid-gap, or equivalently, to $\hbar(k_1+k_2)/ed$, then

$$B_{mid} = \left(\sqrt{\frac{C_G}{e}(V_G - V_{depl})} + \sqrt{n_2}\right)\frac{\hbar\sqrt{2\pi}}{ed} \tag{1}$$

where C_G is the calculated gate capacitance and n_2 and V_{depl} are known from measurements in $B_{\perp}$. In Fig. 3 we show fits of Eq. (1) to the B_{mid} vs V_G data, with d as the only adjustable parameter. Hartree self-consistent calculations (HSCCs) give values for d of 195, 140, and 180 Å for samples A, B, and C, respectively, in relatively good agreement with the data.

As seen in Fig. 1, $\Delta B = B_{min}\text{-}B_{max}$ increases as V_G is decreased Using the same model and assuming that B_{max} and B_{min} occur at $\hbar[k_1+k_2\pm(\Delta k_1+\Delta k_2)]/ed$, where $\Delta k_{1,2} = E_G/(2\partial E/\partial k_{1,2})$ and $\partial E/\partial k_{1,2}$ is evaluated as $(2\pi n_{1,2})^{1/2}\hbar^2/m^*$, we obtain

$$\Delta B = \left[\left(\frac{C_G}{e}(V_G - V_{depl})\right)^{-1/2} + n_2^{-1/2}\right]\frac{E_G m^*}{ed\hbar\sqrt{2\pi}} \tag{2}$$

Fig. 3 shows fits of Eq. (2) to the ΔB data, using the values for d obtained from the fits of Eq. (1). E_G is the only adjustable parameter. The fits yield E_G = 1.1, 1.8, and 3.0 meV for samples A, B, and C, in fair agreement with the values of 1.4, 2.0, and 3.4 meV obtained from HSCCs, and demonstrating that E_G can be directly obtained from the data. The model breaks down as E_G becomes large, so the discrepancy is largest for sample C.

We thank T. R. Castillo for technical assistance. This work was supported by the U.S. Dept. of Energy under Contract DE-AC04-94AL85000.

5. References

1. J. P. Eisenstein *et al.*, Phys. Rev. **B44** (1991) 6511.
2. J. A. Simmons *et al.*, Phys. Rev. **B47** (1993) 15741; S. K. Lyo and J. A. Simmons, J. Phys.: Condens. Matter **5** (1993) L299.
3. While G. S. Boebinger *et al.*, Phys. Rev. B **43** (1991) 12673 studied Fermi-surface distortions through measurements of $B_{\parallel}$-induced changes in the Shubnikov-de Haas oscillations in a $B_{\perp}$, the $B_{\parallel}$ component was too small to produce an anticrossing-type energy gap, and $G_{\parallel}(B_{\parallel})$ for $B_{\perp}$=0 was not investigated.
4. Many years ago vicinal planes of Si were shown to have a similarly shaped Fermi surface arising by an entirely different mechanism, intervally coupling. See D. C. Tsui *et al.*, Phys. Rev. Lett. **25** (1978) 1667, T. Ando, Phys. Rev. **B19**, (1979) 3089, and references therein.
5. S. K. Lyo, unpublished.

Magnetotransport and Electric Subband Studies of $In_{0.52}Al_{0.48}As/In_xGa_{1-x}As$ Asymmetric Coupled Quantum Wells

T. W. Kim, M. Jung, T. H. Park, and J. W. Cho
Department of Physics, Kwangwoon University, 441-1 Wolgye-dong, Nowon-ku, Seoul 139-701, Korea
and
Keon-Ho Yoo
Department of Physics, Kyung Hee University, Seoul 130-701, Korea
and
Kyung-Hwa Yoo
Superconducting Laboratory, Korea Reserch Institute of Standard and Science, Taejon 305-606, Korea
and
G. Ihm
Department of Physics, Chungnam National University, Taejeon 305-764, Korea
and
S. J. Lee
Department of Physics, Korea Military Academy, Seoul 139-799, Korea

Abstract

Shubnikov-de Haas (S-dH) and Van der Pauw Hall-effect measurements on $In_{0.52}Al_{0.48}As/In_xGa_{1-x}As$ coupled quantum wells grown by metalorganic chemical vapor deposition have been carried out to investigate the magnetotransport properties of an electron gas in the coupled quantum wells. Transmission electron microscopy measurements showed that an $In_{0.8}Ga_{0.2}As$ and an $In_{0.53}Ga_{0.47}As$ quantum wells are separated by an $In_{0.25}Ga_{0.75}As$ potential barrier in an active region. The measurements S-dH and Hall-effect at 1.5 K demonstrated clearly the existence of a quasi-two-dimensional electron gas in the quantum wells. The fast Fourier transformation results for the S-dH data indicate clearly the occupation of three subbands in the $In_{0.52}Al_{0.48}As/In_xGa_{1-x}As$ asymmetric coupled quantum wells.

I. Introduction

Electrical and optical measurements and theoretical calculations of the two-dimensional electron gas (2DEG) at the heterointerfaces in various structures have been attempted for many years[1-5]. Among these structures, coupled quantum wells have been attractive because of the interest in both an investigation of the fundamental physical properties[6-8] and tunable coherent light sources for optical communications[9-11]. Recently, although few works concerning the coupled quantum wells were investigated[10,11], to the best of our knowledge, the results for the optical application devices only have been reported. However, the 2DEG studies in the $In_{0.52}Al_{0.48}As/In_xGa_{1-x}As$ asymmetric coupled quantum wells have not been reported yet.

This paper reports transmission electron microscopy (TEM) measurements which were carried out to investigate the structure of $In_{0.52}Al_{0.48}As/In_xGa_{1-x}As$ asymmetric coupled quantum wells and Shubnikov-de Haas (S-dH) and Van der Pauw Hall-effect measurements which were performed in order to investigate the electric properties of the subbands for the $In_{0.52}Al_{0.48}As/In_xGa_{1-x}As$ coupled quantum wells grown by the metalorganic chemical vapor deposition (MOCVD).

II. Experimental Details

The samples used in this work were grown on Fe-doped semi-insulating (100)-oriented InP substrates by MOCVD and consisted of the following structures: a 90-Å Si-doped (3×10^{19} cm^{-3}) $In_{0.53}Ga_{0.47}As$ capping layer for ohmic contacts, a 50-Å undoped $In_{0.53}Ga_{0.47}As$ layer, a 150-Å undoped $In_{0.52}Al_{0.48}As$ layer, a 50-Å Si-doped (2×10^{19} cm^{-3}) $In_{0.52}Al_{0.48}As$ modulation layer, a 30-Å undoped $In_{0.52}Al_{0.48}As$ spacer layer, a 100-Å $In_{0.8}Ga_{0.2}As$ quantum well, a 35-Å $In_{0.25}Ga_{0.75}As$ potental barrier, a 100-Å $In_{0.53}Ga_{0.47}As$ quantum well, a 500-Å undoped $In_{0.52}Al_{0.48}As$ buffer layer, and a 30 period of [$In_{0.52}Al_{0.48}As/In_{0.53}Ga_{0.47}As$] superlattice buffer layer. The compositions of the samples were measured by the double-crystal X-ray diffraction, and the thickness of the each layer was determined by the cross-sectional TEM. Ohmic contacts to the samples were made by diffusing a small amount of indium through several layers at 450 ℃ in a hydrogen atmosphere for approximately 10 min. The S-dH and Hall effect measurements were carried out at a temperature of 1.5 K in magnetic fields up to 12 T in an Oxford superconducting magnet system using a 181 nanovoltmeter. After ohmic contacts were formed, the $In_{0.53}Ga_{0.47}As$ cap layer was removed to get rid of parallel conductance for measurements.

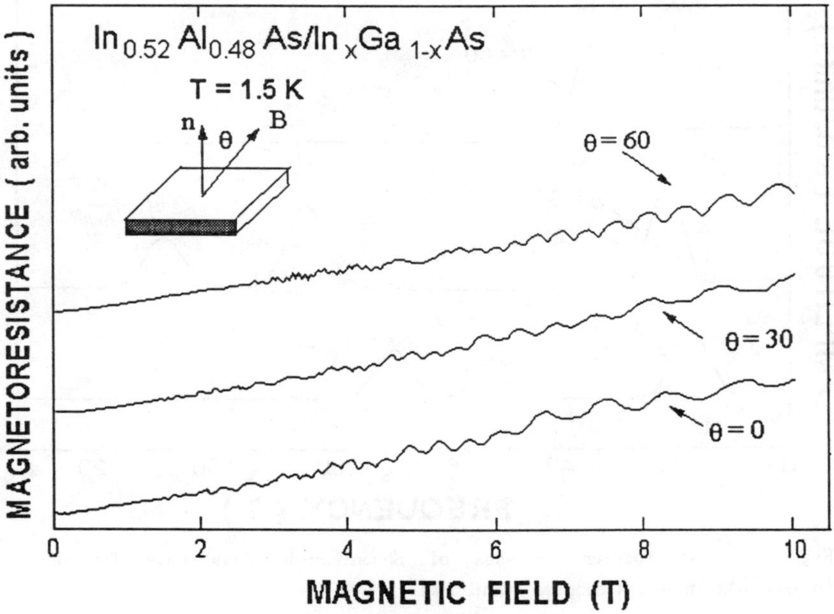

Fig. 1. Shubnikov-de Haas results at 1.5 K for the $In_{0.52}Al_{0.48}As/In_xGa_{1-x}As$ coupled quantum well with three different orintations of the magnetic field relative to heterointerface normal.

III. Results and Discussion

A high-resolution TEM image of a cross-sectional structure shows that a 100-Å $In_{0.8}Ga_{0.2}As$ and 100-Å $In_{0.53}Ga_{0.47}As$ quantum wells were separated by a 35-Å $In_{0.25}Ga_{0.75}As$ potential barrier in its active region. A TEM image shows mifit dislocations at the $In_{0.8}Ga_{0.2}As/In_{0.25}Ga_{0.75}As$ and $In_{0.25}Ga_{0.75}As/In_{0.53}Ga_{0.47}As$ heterointerfaces due to the lattice mismatch. The $In_{0.25}Ga_{0.75}As$ barrier was partially strained with misfit dislocations due to the upper $In_{0.53}Ga_{0.47}As$ and the lower $In_{0.8}Ga_{0.2}As$ layers, and the spatial roughness at heterointerfaces was believed to be the result of the balance between a strain energy and a surface step energy of the growing surface.

The S-dH oscillations in Fig. 1 show multiple periods, indicative of the occupation of several subbands by a 2DEG in the $In_{0.52}Al_{0.48}As/In_xGa_{1-x}As$ asymmetric coupled quantum wells. The quasi-2DEG nature giving rise to the S-dH oscillations was substantiated by studies with the magnetic field oriented 30° and 60° to the normal to the surface as shown in the middle and top curves of Fig. 1.

The oscillation frequencies of the subbands were obtained by digitizing the data linearly in terms of 1/B and by performing high-pass filtering and a fast Fourier transformation (FFT) by computer. The frequency peaks determined from the S-dH data were observed at 16.8 T, 51.6 T, and 69.5 T as shown in Fig. 2. The oscillation

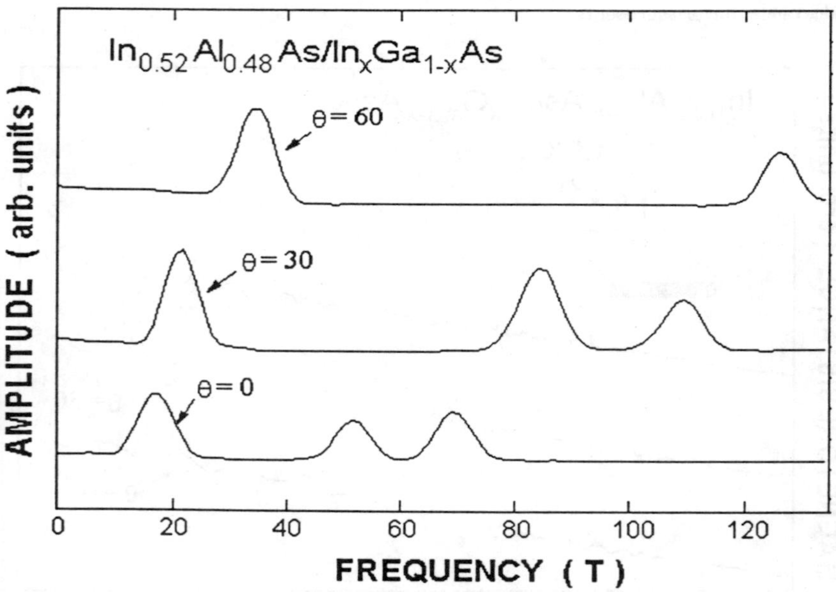

Fig. 2. Fast Fourier analyses of Shubnikov-de Haas data for the $In_{0.52}Al_{0.48}As/In_xGa_{1-x}As$ coupled quantum well.

frequencies of 16.8 T, 51.6 T and 69.5 T correspond to electron densities of 8.13×10^{11} cm^{-2}, 2.5×10^{12} cm^{-2}, and 3.364×10^{12} cm^{-2} for the second, the first, and the zeroth subbands, respectively. The total electron density determined from the S-dH measurements is $(6.674 \pm 0.4) \times 10^{12}$ cm^{-2}. The Hall-effect measurements at 1.5 K in a magnetic field of 2 T showed that the mobility and carrier concentration were 6×10^5 cm^2/V-sec and $(6.1 \pm 0.6) \times 10^{12}$ cm^{-2}, respectively. Even though there were combined uncertainties resulting from the non-ideal geometry of the sample contacts and the complex pattern of the S-dH oscillations, the results of these two kinds of measurements were in reasonable agreement with each other. The oscillation frequencies of the subbands determined from the S-dH measurements at 1.5 K were shifted for measurements with the magnetic field oriented at 30° and 60° to the normal to the sample surface as shown in the upper curves of Fig. 2.

IV. Summary and Conclusions

The results of the S-dH measurements at 1.5 K demonstrate clearly the existence of 2DEG in the $In_{0.52}Al_{0.48}As/In_xGa_{1-x}As$ coupled quantum well grown by MOCVD. The FFT results for the data show three oscillation frequencies, indicative the occupation of three subbands in the $In_{0.52}Al_{0.48}As/In_xGa_{1-x}As$ coupled quantum wells. Although more detailed studies on basic physical properties, including optical measurements, remain to be carried out, the present observations can help improve our understanding of this complicate quantum-well.

Acknowledgments

This work was supported in part by the Korean Science and Engineering Foundation (Contract No. 941 - 0200 - 044 - 2) and the Korea Telecommunication Authority. The authors also would like to thank Dr. W-P. Hong for his useful discussions.

References

1. T. Ando, A. B. Fowler, and F. Stern, Rev. Mod. Phys. **54**, (1982) 437.
2. F. Capasso, **Physics of quantum Electron Devices** (Springer-Verlag, Heidelberg, 1990).
3. T. W. Kim, Y. H. Chang, Y. D. Zheng, A. A. Reeder, B. D. McCombe, R. F. C. Farrow, T. Temofonte, F. A. Shirland, and A. Noreika, J. Vac. Sci. Technol. B **5**, (1987) 980.
4. G. Ihm, S. K. Noh, J. I. Lee, J-S Hwang, and T. .W. Kim, Phys. Rev. B **44**,(1991) 6266 .
5. T. W. Kim, J. I. Lee, K. N. Kang, K-S. Lee, and K-H. Yoo, Phys. Rev. B **44** , (1991) 12891.
6. J. Feldman, G. Peter, E. O. Gobel, P. Dawson, K. Moore, C. Foxon, and R. J. Elliot, Phys. Rev. Lett. **59**, (1987) 2337.
7. S. L. Wong, R. J. Warburton, R. J. Nicholas, N. J. Mason, and P. J. Walker, Physica B **184**, (1993) 106.
8. P. Yuh and K. L. Wang, Phys. Rev. B **38**, (1988) 8377.
9. L. Y. Liu, E. E. Mendez, and H. Meier, Apl. Phys. Lett. **60**, (1992) 2971.
10. M. Ogawa and E. E. Mendez, Appl. Phys. Lett. **63**, (1993) 2189.
11. J. H. Smet, C. G. Fonstad, and Q. Hu, Appl. Phys. Lett. **63**, (1993) 2225.

THE BREAKDOWN OF THE QUANTUM HALL EFFECT AS A PROBE OF CURRENT DISTRIBUTION

U. Meirav, N. Q. Balaban, and Hadas Shtrikman

Braun Center for Submicron Research, Weizmann Institute of Science, Rehovot, Israel

ABSTRACT

We study the relation between the critical current I_c for the breakdown of the Quantum Hall effect (QHE), and the sample dimensions. In high quality material I_c is found to scale logarithmically with the channel width, while in lower quality material I_c is proportional to the width. Furthermore, a transition between the two types of behavior can be induced in a given sample, by in-situ illumination. These are manifestations of different patterns of current distribution which can exist in the QHE.

1. Introduction

1.1 Current Distribution in the QHE

The Quantum Hall Effect (QHE)[1] can be viewed as a manifestation of dissipationless current flow in a bulk 2-dimensional electron gas (2DEG), produced by the combination of a large magnetic field B, perpendicular to the 2DEG, and a moderate electric field E transverse to the current - the Hall field - which act together to impart a longitudinal velocity $v = E/B$ to the electrons. Scattering and dissipation are suppressed due to the energy gap between the 2D Landau levels (LL). We refer to this picture as the *bulk model* of the QHE.

An alternative view, known as the *edge state model*, views the effect as current flowing in quantum edge channels,[2] which are virtually immune to backscattering, hence free of dissipation. The two parallel edges carry current in opposite directions, and the net current is a result of a chemical potential difference between the opposite edges, which is the Hall voltage.

In most experimental situations concerning the QHE, current is constrained to flow through a Hall bar - namely, a straight 2D channel - of width W. Such a Hall bar is depicted in the inset of Fig. 2. Two questions are of primary interest with respect to the spatial distribution of the dissipationless current in such channel of width W:

(a) Does some, or all, of this current flow in the bulk of the channel, as opposed to the edges ?

(b) If so, what is the distribution of current *density* $j(x)$ within the channel ?

The former question seems to have been the subject of debate and confusion, especially in light of the simplicity and remarkable success of the edge state model in accounting for many experimental results.[2] Nevertheless, it is widely recognized that a significant part of the current must flow in the bulk, particularly when sizable currents are forced through a channel.

The latter question has been considered theoretically by several workers.[3-5] Experimental attempts to study the problem electrically and, perhaps most notably, by optical methods, have shown interesting results, but so far many issues remain open.

1.2 The Breakdown of the QHE

Consider an experimental situation when the magnetic field is fixed at a value corresponding to a dissipationless QHE state, in other words an integer filling factor $\nu \equiv nh/eB$, where n is the density of the 2DEG, h being Planck's constant and e the electronic charge. If the current forced through the channel is gradually increased beyond a certain threshold I_c, a sudden onset of dissipation - namely, a rise in the resistance - occurs.[1,6] This phenomenon is called the *breakdown* of the QHE. Its characteristic abruptness allows one to identify a well defined value of I_c, known as the *critical current.*

The quantitative experimental picture for I_c is somewhat hazy. The reported values of the *critical current density* j_c, obtained simply by dividing I_c by the width of the channel W, vary over more than two orders of magnitude, a typical value for j_c being ~ 1 A/m. Furthermore, I_c is highly sensitive to small changes in the magnetic field, for a given sample and filling factor. It is also known to depend in a nontrivial way on the measurement geometry.

The mechanism causing breakdown is not fully known. Among the proposed theories,[7] one could mention the model of a runaway heating instability, inter-LL transitions, injection of hot electrons at the current contacts, and emission of acoustical phonons. Each of the mechanisms has a characteristic threshold electric field, which should determine j_c. The large experimental variability of j_c seems to limit to the applicability of any such theory.

1.3 The question

The motivation of our work was to study the relation between the I_c and W, since it is linked to the question of the distribution of current density and electric fields in the QHE. The widespread assumption, $j_c = I_c/W$, had not been systematically checked when this work began.

The few attempts to address this issue in the recent past led to contradictory conclusions; in particular, j_c was reported to decrease[8] with W or to be independent[9] of W, with opposite conclusions as to the possible role of edge currents. We will argue that these contradictory results may be reconciled. At any rate, the primary objective was to clarify the experimental relation between a channel's width and its capacity to carry dissipationless-current, and to understand this relation as an indicator of that current's distribution.

2. The Experiments

2.1 Preview of the Results

The experiments we performed[10] showed that in high-quality samples, I_c scales very sublinearly - apparently *logarithmically* - with W, for a variety of integer filling factors. This observation has at least two significant implications. First, it sheds new light on previous quantitative discussions of the breakdown of the QHE, which have often made the implicit assumption that I_c scales linearly with W; in particular, we conclude that the experimental value of j_c had often been underestimated, leading to a premature rejection of certain theories.

Secondly, and perhaps most interestingly, this result is strong, albeit indirect, evidence for an intrinsically non-uniform distribution of *bulk* current in the QHE. We show a model for the current distribution across a homogeneous 2D channel in the QHE,, which can account for our observations. While this model suggests that the current tends to concentrate towards the channel's boundaries, it should be emphasized that this is a macroscopically distributed current rather than quantum edge channel transport.

The case of lower quality samples is remarkably different. Such samples show a linear relation between I_c and W, suggestive of uniform bulk current density. We find that in-situ illumination, which causes a (persistent) increase in the samples' mobility, causes a *qualitative* change in the current distribution. This will be discussed in Section 4.

2.2 Samples and Measurements

In a sequence of experiments a large number of samples were measured. Different material samples were used; though all were GaAs-based heterojunctions with a 2DEG, they differed in design and growth conditions, and had different values of density n and mobility μ. Mobilities ranged from 10^5 to 10^6 cm^2/Vsec, and densities varied from 1×10^{11} - 3×10^{11} cm^{-2}.

Using standard wet-etch, photolithography and liftoff techniques, each sample was patterned into multi-contact Hall bar devices of various widths, all in the range of 5-100 μm, with otherwise-similar geometry, namely length and contact positions. Details of the geometry and the measurement method, and their importance for reliably determining the critical current, have been reported before.[10] Measurements were performed at temperatures of ~ 40 mK. A light-emitting diode was installed inside the refrigerator to allow in-situ illumination of the samples.

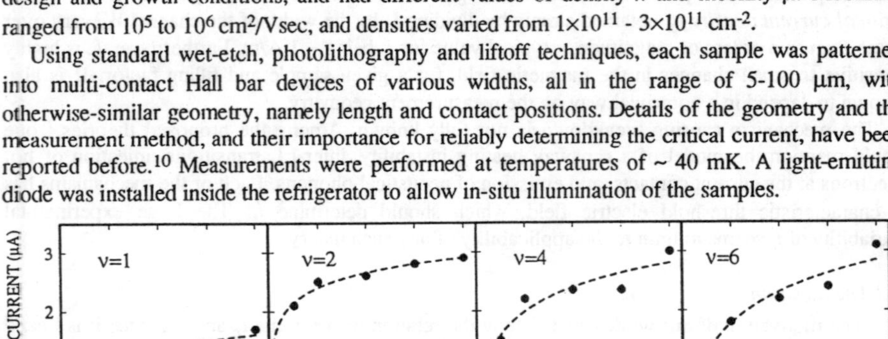

FIG. 1. Critical currents of 5 Hall bars in the same piece of material, with widths ranging between 8 and 78 μm. Each plot corresponds to a different filling factor, namely, ν = 1, 2, 4 and 6. Note the clear sublinearity. The solid line is a logarithmic fit, which is explained below.

2.3 Critical Current vs Width

Fig. 1 shows I_c vs W in a moderately-high-mobility sample, with $\mu \approx 7\times10^5$ cm^2/Vsec. Each box corresponds to a different filling factor. A distinct *sublinear* dependence can be seen for all filling factors. The solid line fit, of the form $I_c(W) = I_0 \ln(W/W_0)$, will be discussed later. This behavior has been repeatedly observed in numerous different high-mobility samples, and appears to be rather general.

Note the values of I_c, which seem to hover around a few micro amperes, even for relatively wide channels. Not only would this correspond to a very low *average* velocity of the electrons, it is almost 2 orders of magnitude below some of the values reported in the literature.[6]

3. Discussion

3.1 A model for current distribution in the QHE

To interpret these findings, we begin by pointing out that theoretical models of breakdown determine a *critical electric field* E_c, transverse to the direction of current flow, which causes an abrupt rise in dissipation. Moreover, unlike the case of the Hall effect in 3-dimensional conductors, in the QHE the transverse (Hall) field $E(x)$ in the 2DEG is *not* uniform. This implies a non-uniform current distribution as well, since the current density in the QHE at any point is just $j(x) = R_H E(x)$, where $R_H = \nu e^2/h$ is the quantized Hall resistance.

The analysis we apply has been presented before,[10] and therefore we only recapitulate the main assumptions and results. It stems from a model for the potential and current distribution in the QHE first studied by MacDonald *et al.*[3] and subsequently by many other workers.[4]

We are considering a uniform channel, where the current flows in the *y* direction, and the transverse coordinate *x* spans from $-W/2$ to $W/2$, with no disorder and abrupt (hard-wall) boundaries. It is assumed that a Landau level is completely filled while the next one is completely empty. The transverse potential distribution $V = V(x)$ is studied under the constraints of satisfying both the electrostatics and the quantum mechanical density redistribution (of LL electrons) in the presence of a slowly varying electric field.

The resulting equation has a mathematical divergence near the edges which is of no physical meaning, and there are various ways to deal with it.[4,5] A thorough theoretical treatment must seriously address the details of the boundary at equilibrium,[11] which is by no means a hardwall potential. Our approach was to treat the edge by means of an extrapolation,[10] which results in a potential profile of the form

$$V(x) = V_0\left(\ln\frac{W}{W_0}\right)^{-1} \ln\left|\frac{x+W/2}{x-W/2}\right| \tag{1}$$

where $2V_0 = IR_H$ is the total Hall voltage and W_0 is a microscopic length of order 10 nm, which depends on the magnetic field and filling factor but also on the details of the edge-extrapolation scheme. Eq. (1) is valid throughout most of the channel, except for some vicinity of the edges.

3.2 Critical current vs width

Such a potential distribution would have a rather definite implication to the question of breakdown, and specifically to the relation between critical current and channel width. The essential assumption is that once a critical electric field E_c is attained at *some* point in the channel, dissipation will commence. By differentiation of Eq. (1) we obtain the transverse electric field $E(x)$ at any point, and consider its highest relevant value E_{max}, namely at the position of the outermost electrons, which is roughly a magnetic length, $l_B \equiv (\hbar/eB)^{1/2}$, away from the edge. Its value will depend on W, W_0 and on the total current I, through V_0. By definition, I_c is the value of I where $E_{max}=E_c$. Thus we can express I_c in terms of E_c

$$I_c \approx \frac{2d}{R_H}\left(\ln\frac{W}{W_0}\right)E_c(B) \tag{2}$$

where d is the effective position of the outermost electrons with respect to the mathematical edge. While some of the quantities on the right hand side of (2) depend on the material parameters and on B, it is clear that for a *given* material and magnetic field this can be written as

$$I_c = I_0 \ln \frac{W}{W_0} \tag{3}$$

namely, the critical current depends *logarithmically* on the sample width. This result qualitatively motivates the line fits shown in Fig. 1, which shows I_c vs W for several different filling factors. We now turn to a quantitative discussion, which helps substantiate the picture.

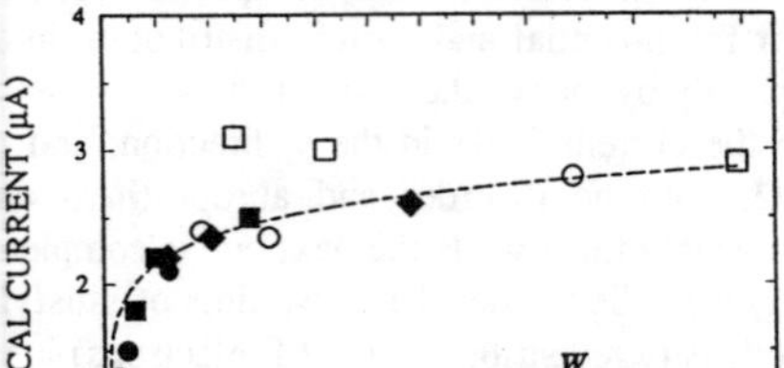

FIG. 2. Measured critical currents for five different widths (● 8μm, ■ 18μm, ◆ 38μm, O 58μm, □ 78μm) and three even filling factors ν = 2, 4 and 6, versus the normalized width W/W_0, where $W_0 \sim 10$ nm, and depends on ν. The dashed line shows the theoretical result $I_c = I_0 \ln(W/W_0)$ for $I_0 = 0.3$ μA, which corresponds to inter-LL transitions with $\gamma \approx 0.2$. Inset: A schematic drawing of the Hall bars.

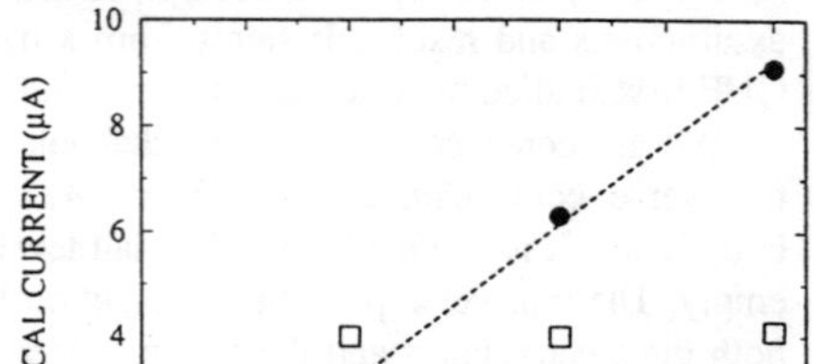

FIG. 3. Critical current vs width in a low-quality, low-mobility material: (●) Samples cooled in the dark show a linear relation between width and critical current, indicative of microscopic inhomogeneities; (□) after illumination, which increases the density, mobility and homogeneity of the 2DEG, we partially recover the sub-linear relation, characteristic of "perfect" samples.

3.3 *Filling factor dependence*

In order to make a quantitative comparison between the data and Eqs. (2) and (3), we need a model for $E_c(B)$, which is really the microscopic physics of the breakdown mechanism. Inter-LL transitions, possibly including the emission of a phonon, are a likely mechanism to consider. Indeed it has been studied theoretically,[12] but was considered to be in discrepancy with experiments, predicting relatively high values of E_c. However, as said above, the latter is not measured directly and its experimental value had been underestimated by the assumption of uniform current distribution. Therefore, the inter-LL mechanism merits re-evaluation.

The rate of such transitions becomes significant only when the electric field becomes high enough to produce spatial and energetic overlap between filled and empty LLs; in other words, E_c must be comparable to a potential drop of $\hbar\omega_c$ - the cyclotron energy - falling over a distance corresponding to the extent of the wavefunction. The latter is, for the Nth LL, just $N^{1/2} l_B$. Thus E_c is determined to within a multiplicative constant $\gamma \sim 1$.

The final result we obtain is $I_0 = \gamma\ (2e\hbar n/m)$, where m is the effective mass. Note the somewhat surprising result that I_0 is independent of B. For example, for the samples of Fig. 1, which had a density of $n = 2.6 \times 10^{11}\ cm^{-2}$, this model predicts $I_0 = 1.4\ \gamma$ micro amperes.

This prediction can be put to a rather stringent test, by trying to fit the critical current for all widths *and* all even-integer filling factors on the same, *single* curve. This is done by plotting all the measured I_c's vs their appropriate normalized width, W/W_0. The quantity W_0 is a well defined function of B within our analysis.[10] Such a plot is shown in Fig. 3. The 15 data points result from 5 widths and three filling factors: 2, 4 and 6. The only fitting parameter is the multiplicative constant $\gamma \equiv I_0/1.4\ \mu A$. The fit yields a value of $\gamma \approx 0.2$, which is a reasonable value[12].

3.4 Odd integer filling factor

Odd filling factors correspond to a half-filled LL such that one spin polarization is fully populated while the other is empty. The electrostatics and the current distribution analysis presented above apply in this case, essentially unmodified. However, the energy gap, which is the Zeeman spin energy, is smaller by a factor $gm/2m_0$, where g is the effective gyromagnetic ratio (the *g-factor*) and m_0 is the bare electron mass, which is $\approx 15m$. In GaAs $g \sim 0.4$, which means that the Zeeman energy splitting can be almost two orders of magnitude smaller than the LL gap.

However, it is well known that g can be enhanced by more than an order of magnitude in the regime of odd-integer QHE, due to electron interactions.[13] It is interesting to note that this enhanced g-factor is manifested, e.g., in the temperature dependence of the conductivity, but is not observed in electromagnetic absorption, where the "bare" g-factor determines the spin-flip excitation energy.

At any rate, the analysis of current distribution given in the previous sections should apply equally well to the case of odd filling factors, only the gap for excitations is smaller. Thus we would expect the critical current to have the same logarithmic width dependence, only with a different prefactor I_0, corresponding to the onset of electric-field-assisted spin-flip excitations.

As seen in Fig. 1, the qualitative picture for odd filling factors is the same, namely a distinct sublinear dependence of critical current on channel width, which fits equally well to a logarithmic curve. Quantitatively, if we apply the most straightforward extension of the previous analysis, we obtain a value for the effective energy gap, which turns out to be smaller than, but comparable to, the LL gap. This can be interpreted as another, new manifestation of the enhanced g-factor.

However, the justification for any ***quantitative*** conclusions can be questioned because the matrix elements of spin-flip transitions, as well as the energy-momentum constraints,[12,14] are different. A study of this topic has yet to be completed.

3.5 Bulk vs edge currents: a comment

The above analysis of current distribution makes no explicit reference to the edge states. The validity of the edge state picture is questionable at the high current and voltage levels corresponding to the breakdown phenomenon.[2] The fact that dissipationless current is observed beyond the small-signal limits of the Buttiker-Landauer formalism suggests the QHE is more general than the edge model itself would suggest.

The unusual logarithmic scaling of I_c with W can be viewed, in particular, as a manifestation of non uniform fields and dissipationless bulk currents in the QHE. However, there may well be additional edge state currents. In fact, several recent theoretical works[5] discuss the coexistence of bulk and edge currents in the QHE. They conclude that a potential imbalance between the edges must be accompanied by bulk fields and currents similar to Eq. (1). Our results, rather than distinguish between two opposite pictures, can be seen to support a unified view of the current distribution in the QHE along the lines of these works.[5]

4. Low quality vs high quality samples

4.1 Linear scaling in low quality samples

The high values of critical currents reported by previous workers, and the recent results of Kawaji *et al.*,[9] who observe linearity or, at most, a weak sublinearity in I_c vs W, raised doubts as to the universality of the logarithmic scaling of I_c. Therefore we repeated the experiments on low-quality material, with $n = 1.3 \times 10^{11}$ cm^{-2} and $\mu = 1.2\times10^5$ cm^2/Vsec.

The result was a clear *linear* relation between I_c and W, in contrast to the higher mobility samples we had tested before. This is shown in Fig. 3. This qualitative difference not only opens the possibility of much larger critical currents in sufficiently wide samples,[6] but also suggests an altogether different pattern of current distribution, which we discuss below.

4.2 Effect of illumination

Using the well-known persistent-photoconductivity property of GaAs modulation doped heterojunctions, we gradually changed these samples' characteristics by a sequence of short exposures to light, repeating the breakdown measurements after each illumination. The results, seen in Fig. 3, show how the linear width-dependence evolved, after sufficient illumination, into the already familiar sublinear dependence. Note the remarkable fact that in narrow channels the critical current increased after illumination, while in wide channels it decreased.

4.3 Interpretation

This behavior, while not understood in full, may be related to the effect of illumination on the *homogeneity* of the 2DEG. It is believed[1] that, due to inherent inhomogeneity of the 2DEG material, the current in the QHE permeates the sample via a network of microscopic channels. In such a case, the breakdown would occur only when the current exceeds the combined current capacity of all the internal channels, which is proportional to the number of such channels. With sufficient density fluctuations, these channels are numerous and randomly distributed, and their number is proportional to the samples' macroscopic dimensions, therefore we expect $I_c \propto W$.

On the other hand, for relatively homogeneous samples, there is only a small number of wide channels, ultimately just one, and its current capacity is only logarithmic in its width, as discussed above. In fact, once the characteristic length-scale for the inhomogeneity fluctuations is larger than the sample width, we expect logarithmic scaling of the critical current with width.

We believe that this explains our observation, namely, that the illumination, by releasing trapped electrons in the modulation doped layer, smoothens out the microscopic inhomogeneity and increases the characteristic length scale of density fluctuations to a value comparable to the width of our samples. As a result the entire width of the Hall bars forms a single channel with the ensuing logarithmic critical current. Note that this is not the only effect of illumination: The increase in density, hence the energy gap, and the possible changes in the density of states, also modify the critical current, and must be taken into account for a detailed picture.

5. Conclusions

The distribution of currents in the QHE is a complex issue. In a single, perfect channel, current flows throughout the entire bulk of the sample, although its density is largest near the edge, and it decreases slowly towards the center. This is primarily a result of 2D electrostatics, but it is a peculiar property of the non-metallic QHE state. One important manifestation of this is the fact that the total current capacity of a channel is essentially logarithmic in its width.

However, most samples have a certain level of inhomogeneity, which, if sufficiently high, changes the current distribution in a fundamental way. Somewhat paradoxically, this inhomogeneity tends to promote a more uniform current distribution which, among other things, may also increase the critical current. This suggests, incidentally, that "bad" samples are in a sense good for the QHE, especially when large dissipationless currents are desired.

6. Acknowledgements

Y. Levinson made important contributions to our understanding of the subject. We also acknowledge useful discussions with K. von Klitzing, R. Haug, and J. Hajdu. This work was supported by the Basic Research Foundation administered by the Israeli Academy of Sciences and Humanities.

7. References

1. *The Quantum Hall Effect*, ed. R. E. Prange and S. M. Girvin (Springer, New York, 1987).
2. For a recent review see R. J. Haug, *Semicond. Sci. Technol.* **8** (1993) 131.
3. A. H. MacDonald, T. M. Rice, and W. F. Brinkman, *Phys. Rev.* **B28** (1983) 3648.
4. C. W. J. Beenakker and H. van Houten in *Solid State Physics* vol. 44, ed. H. Ehrenreich and D. Turnbull (Academic Press, San Diego, 1991), p. 177-181, and references therein.
5. D. J. Thouless, *Phys. Rev. Lett.* **71** (1993) 1879; C. Wexler and D. J. Thouless, *Phys. Rev.* **B49** (1994) 4815; H. Hirai and S. Komiyama, *ibid.*, 14012.
6. G. Ebert *et al.*, *J. Phys.* C **16** (1983) 5441; M. E. Cage *et al.*, *Phys. Rev. Lett.* **51** (1983) 1374; F. Kuchar *et al.*, *Surf. Sci.* **142**, 196 (1984).
7. See, for example, P. C. van Son, G. H. Kruithof, and T. M. Klapwijk, *Phys. Rev.* B **42** (1990) 11267.
8. R. Haug, K. von Klitzing, and K. Ploog, in *High Magnetic Fields in Semiconductor Physics II*, ed. G. Landwehr (Springer, Berlin, 1989) p. 185.
9. S. Kawaji, K. Hirakawa, and M. Nagata, *Physica* **B184** (1993) 17; S. Kawaji *et al.* (unpublished).
10. N. Q. Balaban *et al.*, *Phys. Rev. Lett.* **71** (1993) 1443.
11. D. B. Chklovskii, B. I. Shklovskii, and L. I. Glazman, *Phys. Rev.* **B46** (1992) 4026.
12. O. Heinonen, P. L. Taylor, and S. M. Girvin, *Phys. Rev.* **B30** (1984) 3016.
13. T. Ando and Y. Uemura, *J. Phys. Soc. Jap.* **37** (1974) 1044.
14. P. Streda and K. von Klitzing, J. Phys. C **17** (1984) L483; L. Smrcka, *ibid.* **18** (1985) 2897.

ELECTROOPTIC IMAGING OF ELECTRIC FIELDS IN TWO-DIMENSIONAL ELECTRON SYSTEMS

R. KNOTT, W. DIETSCHE, K. VON KLITZING, K. EBERL, AND K. PLOOG[a]

Max-Planck-Institut für Festkörperforschung, Heisenbergstraße 1, 70569 Stuttgart, Germany

The potential distribution in Hall effect experiments on two-dimensional electron systems (2DES) is imaged with a scanning polarization optical microscope utilizing the internal electrooptic effect. Outside the quantum Hall regime (QHR), Hall potential profiles vary linearly between the two edges. In the QHR, the observed Hall potential profile is linear in low-mobility but non-linear in high-mobility samples. This nonlinear shape depends on magnetic field and current, indicating inhomogeneous current transport in the interior of the 2DES. In the plateau center, the Hall potential drop is concentrated in the middle of the 2DES structure. The results are interpreted in terms of Hall-voltage induced electron density inhomogeneities. These lead to the observed potential distributions because of the very strong dependence of the conductivity on the local electron density.

1. Introduction

The electronic transport properties of two-dimensional electron systems (2DES) are generally studied by passing currents through contacts alloyed at the edges of the samples and measuring voltages via other contacts. If the 2DES is exposed to a quantizing magnet field in which phenomena like the quantum Hall effect (QHE) are observed then this technique is not sufficient because local current and potential distributions are then expected to be very inhomogeneous.[1] This is accentuated in physical situations like the edge-channel transport regime or in the one of the breakdown of the QHE which occurs at larger currents. Consequently, contactless measurements are desirable to provide a 2D mapping of the potential distributions in the 2DES.

The electrooptic effect (Pockels effect) in III-V compound semiconductors makes contactless potential measurements possible even at liquid helium temperatures and in high magnetic fields. An electric field applied to the crystal causes birefringence, which changes the ellipticity of subbandgap light which is passed through the sample.[2] Recently, this technique has been applied to study the Hall potential profile in a modulation doped heterostructure subjected to a quantizing magnetic field.[3] In the quantum Hall regime (QHR) around filling factor 4, the profile was found to be significantly non-linear, with the major potential drop split symmetrically within approximately 150μm of the system boundaries.

The profile of Ref. 3 agreed qualitatively with an early theoretical model[4] which assumes a vanishing dissipative conductivity everywhere in the sample which is unrealistic for a real sample. In the case of a homogeneous but nonzero conductivity a linear profile was predicted, however.[5] In a realistic 2DES sample the conductivity varies laterally because of its strong dependence on carrier concentrations. For such cases, the main potential drop was expected to occur in the interior of the 2DES. Clearly, further experimental studies are necessary.

In this contribution we report new imaging experiments based on the electrooptic technique. The lateral potential distributions built up by a transport current are mapped. Significant differences in the potential landscapes were found between samples having high and low mobilities. The dependence on current and on magnetic field has been studied in detail. Most effects can be interpreted by models taking int account local conductivities which depend on the local carrier density.

2. Experimental Details

For the experiments a scanning polarization microscope was attached to an optical cryostat. External galvanometer scanners allow for raster scanning of a laser focus over the sample placed in the bore of a superconducting magnet. A cryogenic lens yields a fine focus (≈10μm). As the light source we choose a collimated diode laser (850nm, 2mW). This light is made circularly polarized with a calcite polarizing prism and an appropriately set Soleil-Babinet compensator. In the scanning experiments we operate in a front side probing geometry, where we analyze the light reflected from the gold-coated polished back side ((001) surface) of the sample. Its intensity is detected with a Si photodiode. Changes in this intensity are proportional to the potential difference between the sample surface and the backside reference electrode, laterally averaged over the probing beam profile.[6] As the transport current through the sample is modulated, an intensity change is synchronously detected. With the laser focus raster scanned across the sample under computer control, the potential signal measured at each point is displayed color coded on the screen. Thus, the potential distribution in the plane of the 2DES is sequentially probed to yield an image. The potential resolution (1 mV/$\sqrt{\text{Hz}}$ at 10 μA photocurrent) is limited only by shot noise.

3. Low mobility sample

With this method, we have investigated the potential distributions in delta-doping layers in GaAs.[7] These structures do not show undesired persistent bypass effects due to their single minimum vertical bandstructure. Despite of their low mobility ($\mu \approx 1.5*10^3 cm^2/Vs$), they showed quantized Hall resistance around filling factor 2. Simultaneously, the longitudinal resistance drops to ≈500 Ω at T=1.6K. From zero to quantizing magnetic field, we have obtained the potential distributions (Fig. 1). The boundary of the 2DES area is indicated by the black line. The width and length of the sample were 0.5 and 1.5 mm.Source and drain contacts are visible at the left and at the right of the image. At the top and at the bottom additional voltage probes are visible. In the top panel the magnetic field is zero. The original color coding is not reproduced in this figure but it is obvious that the equipotential lines run perpendicular to the current flow. The potential increases linearly along the sample as expected.

In the case of the magnetic field being 7 T (center panel) the equipotential lines have tilted by 45° to the edges of the sample. At this field the Hall resistance is equal to the dissipative (longitudinal) resistance. This is the case where a Hall angle of 45° is realized. Towards the current contacts the equipotential bend to match the potential distribution in the low resistivity contacts.

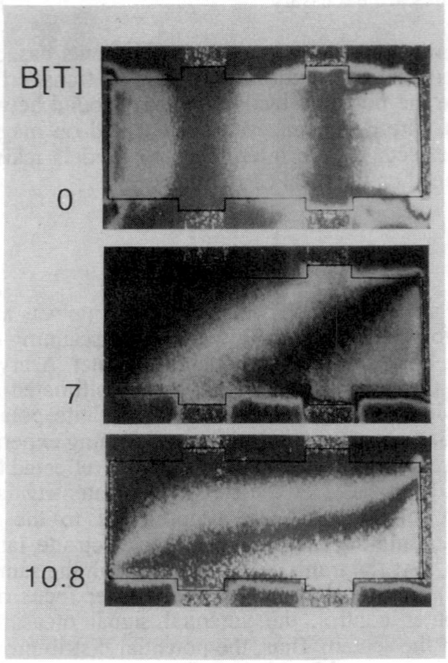

Fig. 1. Potentials measured in a low-mobility δ- doping sample in three magnetic field values.

In the bottom image, the magnetic field is 10.8 T, which is just at the center of the quantized Hall plateau at filling factor 2. Indeed, here we observe that the equipotentials are nearly parallel to the edges in the center of the sample. This Hall angle of approximately 90° agrees with the low dissipative resistance measured galvanically (500Ω). The potential drops linearly across the sample as expected for a uniform conductor.[1] This agreement between galvanic and electrooptic measurements holds also in the nonohmic transport regime, where electron heating further increases the dissipative conductivity.[7]

4. High mobility sample

In addition to this low-mobility sample we investigated also a high mobility one. To avoid the bypass effects which are detrimental to our technique, we have prepared a weakly doped heterostructure by molecular beam epitaxy. The two-terminal magnetoresistance of this sample shows the patterns typical for high quality samples. It is well quantized around low integer filling factors. Three-terminal I-V characteristics around filling factor 2 show a pronounced breakdown behaviour, with a maximum critical current of 100µA at T=1.6K.

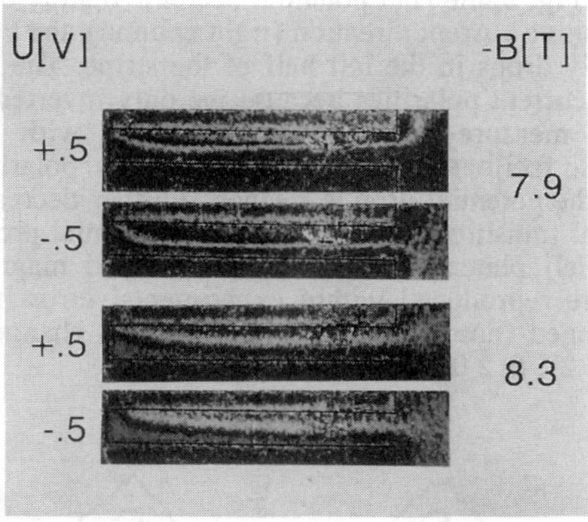

Fig. 2. Electrostatic potential in a high mobility sample. At B=7.9T (upper two panels) the center of a QHE plateau is reached. The field of 8.3 T corresponds to the edge of the QHE plateau.

The potential distribution in this structure is shown in Fig. 2. The four panels correspond to two Hall voltages (± 0.5V) at two different magnetic fields. The outline of the structure is again indicated by the black line. Current is fed at the left side via an alloyed contact. At the right hand side the channel merges in a wide 2DES area with the other current contact (not shown). The voltage of 0.5V corresponds to a transport current of 38 µA which is well below the breakdown current of the QHE (110 µA). Again the original color coding is not reproduced in this figure.

At the edge of the plateau (B=-8.3T), the Hall potential drop is uniformly linear and independent on current direction. Towards the plateau center (B=-7.9T), the potential profile in the direction perpendicular to the current becomes markedly nonlinear. It aquires a steplike shape, with the step position depending on the current direction and the shape varying with magnetic field. In the images, the step is approximately marked by the white stripe. At the opposite edge of the plateau (B=-7.5T), the slope of the step reduces again to lead over to the linear profile outside the plateau region (not shown in the figure).

The potential profiles across the 2DES channel are much better visible in Fig. 3. Here, Hall potential profiles taken across the 2DES channel of Fig. 2, 400µm away from the source contact, are presented. The transport current is switched on and off at a frequency of 320Hz with a duty cycle of 50%. This is indicated at the bottom of the figure. The amplitude (p-p) of the two-terminal voltage of 0.3 V corresponds to a current value of 23 µA. The positions of the stripe edges are indicated with dashed vertical lines. Zero electrooptical signal is indicated by the short horizontal line at the left of each trace. It corresponds to the potential of the reference electrode at the back of the sample.

At the high field edge of the plateau (B=-8.3T), the Hall potential profile is linear within the channel independent of polarity. This changes upon entering the QHR with decreasing magnetic field. Now the profiles deviate from linearity. In addition the evolution of the profile with magnetic field across the plateau is different for both current polarities. For negative current direction (left column), i.e. electrons flowing from left to right in Fig. 2, the Hall potential decrease mainly occurs in the right half of the channel. For positive current direction (right column), the behaviour is opposite: the Hall potential mainly drops in the left half of the stripe. The signals plotted have the same sign for both current polarities because we only inverted the DC bias to reverse polarity. Thus we measure an averaged derivative with respect to the current modulation. This also facilitates comparison between the polarities. At the center of the plateau (B≈-7.7T), the potential drop is steepest. Further decrease of the magnetic field then yields a gradual transition towards the linear potential profile observed at the low field side of the Hall plateau (B=-7.1T). At reversed magnetic field direction the potential profiles are reproduced within experimental error if the current polarity is simultaneously changed (not shown in the figure). No changes with the helium bath temperatures from 1.2K to 2.0K were observed.

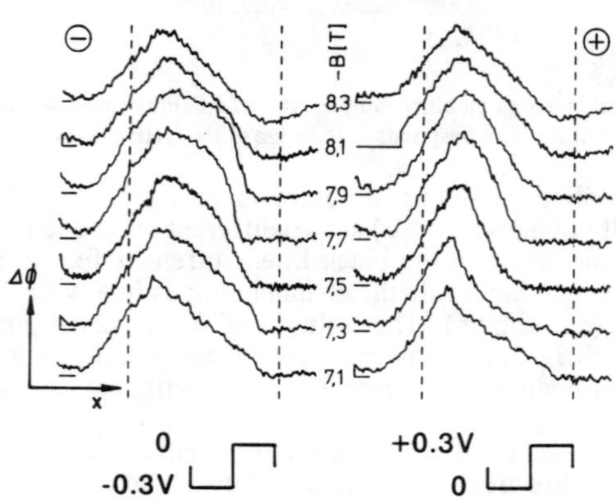

Fig. 3. Potential profiles taken across the 2DES channel of Fig. 2 at different magnetic fields.

The nonlinear shape remains down to the lowest currents investigated (I=2.3μA). At large current levels which just exceed the breakdown threshold the potential profile becomes linear again. This is visible in Fig. 4 for the center panel where the Hall voltage is modulated between .5 and 1.5 V. At even higher Hall voltages (bottom panel) the potential distribution is both remarkable nonlinear and asymmetric with respect to current polarity. These effects are not yet further studied by us.

Qualitatively potential distributions can be understood models considering the strong density dependence of the dissipative conductivity in the QHR.[8-11] Carrier density inhomogeneities will then enforce nonlinear potential profiles. Physically spoken, the current will flow where the dissipation is minimum. At elevated transport currents, the space charge necessary to set up the Hall voltage will locally change the filling factor dynamically. Quantitative evaluation yields that a density inhomogeneity around 1% is required to explain the effect. Such a Hall-voltage-induced inhomogeneity is possible if a screening electrode (gate electrode) is located approximately 1μm from the 2DES plane. This distance would approximately correspond to the thickness of the buffer layer. Possibly a conductive layer forms at the substrate-buffer interface and acts as a virtual gate electrode. By contrast, in the low mobility sample the minimum of the dissipative conductivity as function of carrier concentration is not as pronounced. Furthermore, a virtual gate electrode is probably absent. Consequently, the potential distribution remains linear.

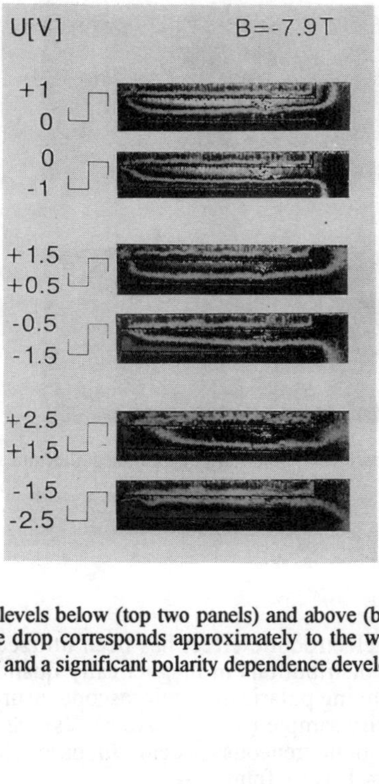

Fig. 4. Potential at current levels below (top two panels) and above (bottom panels) the QHE breakdown threshold. The width of the drop corresponds approximately to the white stripe. With increasing current this width becomes broader and a significant polarity dependence develops.

5. Nonlocal transport

In the QHR, the local description of transport is not necessarily valid. This is demonstrated in a special sample geometry, Fig. 5. The current enters the 2DES on the left side of the narrow stripe at the bottom of the images. This is the stripe which was used for the previous studies. The exit contact is on the right outside the frame. Above the narrow stripe, there is a dead end part of the 2DES. Off plateau (B=10T) this part is at a homogenous potential as expected since no net current leaves this stripe. Simultaneously, the potential distribution in the current-carrying stripe is described by the Hall angle less than 90° because of the finite dissipative conductivity.

At the center of the quantized Hall plateau at filling factor 2 (B=7.9T), the potential distribution in the lower stripe is as described previously. It shows the remarkable sharp potential drop in the interior of the channel. But now the upper 2DES area is no longer an equipotential. Instead of that, the Hall potential is concentrated at the system edges, while the center of the dead end part is at a lower potential. Clearly, this behavior is different from any expected from classical electrodynamics. Possibly, it is a dynamic phenomenon: The edges may reach equilibrium much faster than our measuring time (100 Hz to 10 kHz) while the interior attains equilibrium only after much longer times.

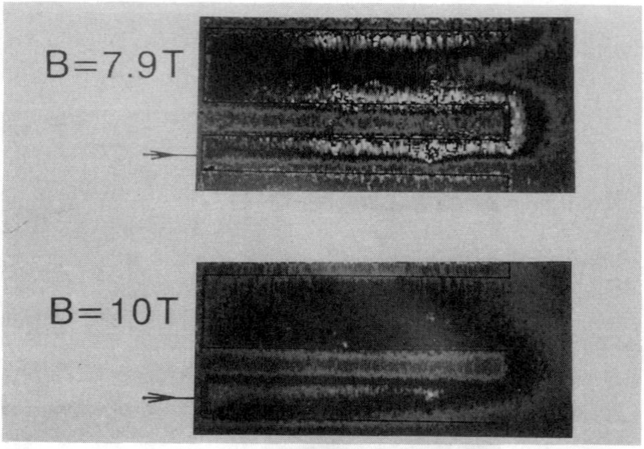

Fig 5. Potential profile in a 'dead end' (the upper stripe). A nonequilibrium situation exists in the QHR (B=7.9T).

6. Conclusion

In conclusion, the linear electrooptic effect has been utilized to map transport-current induced lateral potential distributions in magnetically quantized 2DES. Local sub-bandgap illumination in a scanning polarization microscope insures minimal disturbance of the system. In a low mobility sample ($\mu \approx 1.5*10^3 cm^2/Vs$) we observed potential distributions as expected for a homogeneous system. In particular no inhomogeneities caused by the Hall potential itself were found.

In a high mobility sample ($\mu \approx 5.0*10^5 cm^2/Vs$) the observed potential distributions show: Outside the QHR, the Hall potential drops linearly over the width of the stripe. Inside the QHR, the potential profile is distinctly non-linear with a pronounced maximum of the Hall electric field in the interior of the sample. This indicates current transport in the interior of the sample. Details of the potential profile depend continuously on filling factor and current value.

These nonlinear potential profiles are a consequence of a strong dependence of the transport coefficients on the carrier concentrations. Since the conductivity depends strongly on filling factor in the QHR, a pronounced bunching of the Hall potential drop will result. In our case both a carrier concentration gradient in the original sample and density inhomogeneities set up by the Hall potential itself were necessary to explain all effects.

Effects related to edge channel or incompressible stripes near the edges were not necessary to describe the potential profiles. They will become important only on length scales much shorter than available with the present technique. Even so, a rich spatial response was encountered which reflects the quantum structure in the conductivity of the 2DES.

7. References

(a) now at: Paul-Drude-Institut für Festkörperelektronik, Hausvogteiplatz 5-7, 10117 Berlin, Germany

1. J. Wakabayashi, and S.Kawaji, J. Phys. Soc. Japan **44**, 1839 (1978)
2. J. M. Wiesenfeld, and R. K. Jain, in: Measurement of High Speed Signals in Solid State Devices, Semiconductors and Semimetals **28** (Academic Press, Boston 1990)
3. P. F. Fontein, P. Hendriks, F. A. P. Blom, J. H. Wolter, H. G. M. Lochs, F. A. J. M. Driessen, L. J. Giling, and C. W. J. Beenakker, Phys. Rev. **B43**, 12090 (1991)
4. A. H. MacDonald, T. M. Rice, and W. F. Brinkman, Phys. Rev. **B28**, 3648 (1983)
5. D. J Thouless, J. Phys. **C 18**, 6211 (1985)
6. J. L. Freeman, S.R. Jefferies, and B. A. Auld, Optics Letters **12**, 795 (1987)
7. R. Knott, W. Dietsche, K. von Klitzing, K. Eberl, and K. Ploog, Solid State Electronics **37**, 689 (1994). Please note that Figure 2 in this publication was accidentally turned upside down.
8. V. M. Pudalov, and S. G. Semenchinskii, JETP Letter **42**, 232 (1985)
9. R. Woltjer, R. Eppenga, J. Mooren, and C. E. Timmering, Europhys. Lett. **2**, 2 (1986)
10. A. A. Shashkin, V. T. Dolgopolov, and S. I. Dorozhkin, Sov. Phys. JETP **64**, 1124 (1986)
11. G. Diener, and J. Collazo, J. Phys. **C 21**, 3305 (1988)
12. R. J. Haug, Semicond. Sci. Technol. **8**, 131 (1993)
13. D. B. Chklovskii, B. I. Shklovskii, and L. I. Glazman, Phys. Rev. **B46**, 4026 (1992)

MAGNETIC FIELD SYMMETRY PROPERTIES OF THE CAPACITANCE TENSOR OF A TWO-DIMENSIONAL ELECTRON SYSTEM

W. Chen, T.P. Smith III, and M. Büttiker
IBM T. J. Watson Research Center, P.O. Box 218
Yorktown Heights, New York 10598

and

M. Shayegan
Department of Electrical Engineering, Princeton University
Princeton, New Jersey 08540

ABSTRACT

Large *asymmetries* in the capacitance tensor elements are observed under field reversal in a 2-dimensional electron system with a multi-lead measurement configuration. The saturation fields for the asymmetry at low field are typically less than 0.05 Tesla at $T < 4K$. In quantizing magnetic fields, we observed almost complete asymmetry at integer filling factors, and it is very sensitive to temperature. We show that this asymmetry is a manifestation of the quantum Hall effect in the capacitance tensor elements. A physical model is presented to account for the observed asymmetry. Results showing the general symmetry relations of the capacitance tensor elements are also presented.

1. Introduction

Capacitance measurements in two-dimensional (2D) electron systems, quantum dots and semiconductor nanostructures have revealed the *electro-chemical* nature of a capacitance [1-9]. The capacitances in these systems are usually measured two-terminally with the metal gate connected to one electron reservoir and the 2DES connected to another. In such a two-terminal configuration the capacitance is an even function of magnetic field. In a multi-terminal measurement where the 2DES is connected to two or more electron reservoirs, there are, however, capacitance matrix elements which are asymmetric under field reversal [6]. We investigate these asymmetries in both low and quantizing magnetic fields, and show that they are in accordance with the general symmetry relations for the capacitance tensor elements.

2. Capacitance and Emittance Tensor

For an assembly of N-conductors, each connected to a single contact with electro-chemical potential μ_k, $k = 1, 2, ..N$ the capacitance matrix is defined as

[7,8]

$$C_{kl} = edQ_k/d\mu_l, \tag{1}$$

where $d\mu_l$ is the change of the electro-chemical potential at contact l and dQ_k is the change of the charge on conductor k. In this purely capacitive configuration it is the full equilibrium electron density which matters. Consequently, all capacitance coefficients in such a configuration are *even* functions of magnetic field, $C_{kl}(B) = C_{kl}(-B)$ and the capacitance tensor is symmetric $C_{kl}(B) = C_{lk}(B)$. Conservation of the total charge on the N conductors requires $\sum_k C_{kl}(B) = 0$ and $\sum_l C_{kl}(B) = 0$.

The situation changes dramatically if one or more of the N-conductors has at least two or more contacts connected to the electron reservoirs. The low-frequency behavior of such a configuration of conductors is now characterized not by a capacitance matrix but by an admittance matrix. The current $I_k(\omega)$ measured at contact k in response to an oscillating potential $V_l(\omega)$ applied at contact l is given by [8]

$$g_{kl}(\omega) = g_{kl}(0) - i\omega E_{kl} + O(\omega^2). \tag{2}$$

Here $g_{kl}(0)$ are the dc-conductances which are non-vanishing only if both contacts k and l belong to the same conductor. The leading order frequency dependence is determined by an emittance[8] E_{kl}. If both k and l denote contacts of the same conductor this coefficient describes a response which is either capacitive or inductive or a combination of both depending on the properties of the conductor. If k and l denote contacts belonging to different conductors, the response is purely capacitive and $E_{kl} = C_{kl}$. Thus for the system shown in Fig. 1a where contact 1 denotes the gate and probes 2 and 3 connect to the 2DES, the emittance matrix $\mathbf{E}$ is

$$\mathbf{E} = \begin{pmatrix} C_{11} & C_{12} & C_{13} \\ C_{21} & E_{22} & E_{23} \\ C_{31} & E_{32} & E_{33} \end{pmatrix}. \tag{3}$$

Because of the occurence of emittance elements in this matrix, the off-diagonal capacitance elements can be *uneven* under magnetic field reversal [8]. Just like in the case of a capacitance tensor, the sum of a row or a column of an emittance tensor is equal to zero. The diagonal elements are even functions of magnetic field but for the off-diagonal elements the symmetry requires $C_{kl}(B) = C_{lk}(-B)$ and $E_{kl}(B) = E_{lk}(-B)$. In the following we focus on our investigation of the dramatic asymmetry observed in the capacitance tensor elements such as C_{12} and C_{13}.

3. Experimental

The samples used in the experiment are modulation-doped GaAs/AlGaAs heterostructures grown by molecular beam epitaxy. The low-temperature electron mobilities are better than $1.5 \times 10^6 cm^2/Vsec$ and the sheet densities are in the range of $1.5 - 3.5 \times 10^{11}/cm^2$, as deduced from van der Pauw measurements. Typical sample geometry is schematically shown in Fig. 1a. The size of the sample is about 3 mm × 3 mm and the gate size 300 μm × 300 μm. Note in the figure that the 2DES is connected to *two* potential reservoirs through contacts 2 and 3 located at the *edge* of the sample, and the gate covers the *edge*. These are key conditions for observing complete asymmetry in the capacitance matrix elements [6].

The capacitance matrix element C_{kl} were measured using standard lock-in techniques with a modulation voltage ($V_l e^{-i\omega t}$) applied to contact l and a phase-sensitive AC current preamplifier connected to gate k while other contacts are connected to their potential reservoirs [6]. The excitation voltage is typically of the order of 1 mV, and the frequency is 1 KHz. Because there is no dc conductance between the gate and the 2D gas, the capacitance coefficients can be obtained directly from the current-voltage ratio: for example, $-i\omega C_{12} = I_1(\omega)/V_2(\omega)$.

4. Results and Discussion

4.1 Asymmetries at Low Fields

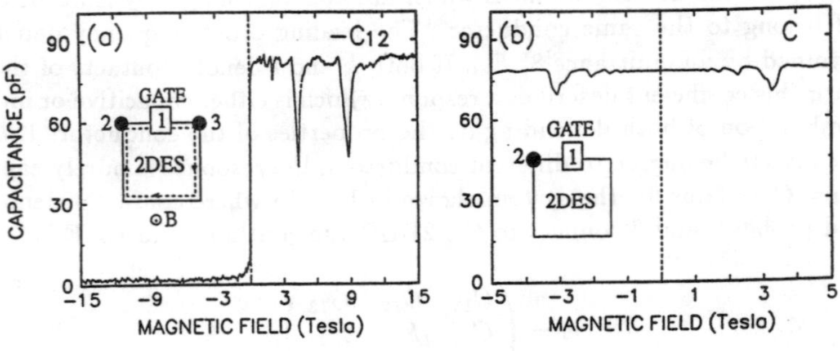

Fig. 1: (a) The capacitance tensor element $-C_{12}$ in a magnetic field. Inset: the three-terminal measurement geometry. The arrows indicate the direction in which electrons tend to flow in a positive field (pointing out of paper). (b) The two-terminal capacitance C obtained from a similar 2DES sample at T=1.3 K. Inset: the two-terminal measurement geometry.

Very large asymmetry of C_{12} under field reversal is clearly displayed in Fig. 1a. C_{12} is small in a negative magnetic field, while it increases quickly and saturates at approximately its geometric value in a positive field,. The magnetic-field range for this switching is less than 0.05 Tesla. The deep dips occurring at higher fields corresponds to the integer quantum Hall effect of the 2DES (e.g. ν=1 at B=8.8 T and ν =2 at B=4.4 T). Except in the field regions that are close to these filling factors, the measured signal is purely capacitive (within 1 %).

In stark contrast, the two-terminal capacitance C under similar conditions (see Fig. 1b) is essentially symmetric upon field reversal. No sharp transition is present at low fields. The tensor properties are completely lost and density of states oscillations are seen in both positive and negative fields.

4.2 General Symmetry Relations

The results for other capacitance matrix elements in this three-terminal measurement are summarized in Fig. 3. A pronounced asymmetry is also observed for C_{13} (see Fig. 3a). When we add C_{13} to C_{12}, we obtain a symmetric trace of $C_{13} + C_{12}$ with no sharp transition present at low fields. Experimentally, we can also directly measure the sum by introducing the AC modulation voltage ($\Delta V e^{i\omega t}$) to both contacts 2 and 3, and measure the total AC current (ΔI_{tot}) at gate 1. The result is shown in Fig. 2b. It is symmetric within experimental accuracy. Because of the total charge conservation, $C_{11} = -(C_{12} + C_{13})$. Thus, we obtain that C_{11} is symmetric in magnetic field. We verified that this symmetry also holds for E_{22} and E_{33} within measurement accuracy. When we compare C_{21} (Fig. 2c) with C_{12} under the same experimental conditions (Fig. 1a), we obtain that $C_{12}(B) = C_{21}(-B)$ within the experimental uncertainty. The Onsager-Casimir symmetry relation for capacitance [8], $C_{kl}(B) = C_{lk}(-B)$, holds in every case that we studied.

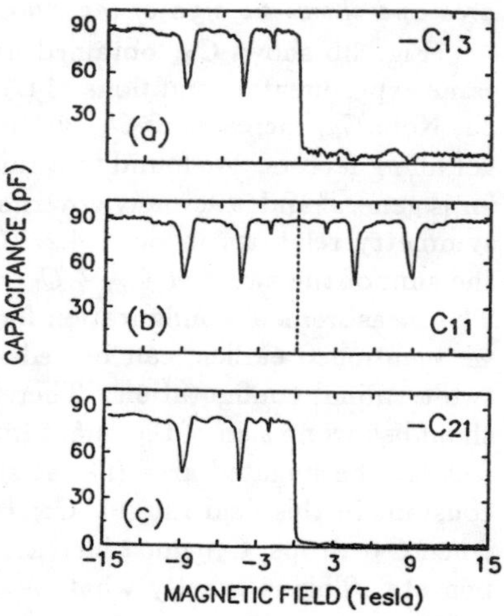

Fig. 2: Capacitance tensor elements in the three-terminal measurement: (a) $-C_{13}$; (b) $C_{11} = -(C_{12} + C_{13})$; (c) $-C_{21}$.

4.3 Asymmetries in the Quantum Hall Regime

Even though the asymmetry observed at low magnetic fields is quite large, it is not complete. There exists a small offset. As clearly shown in Fig. 2a, this offset does drop at integer filling factors ($\nu = 1, 2$). But, because in this experiment the sheet densities of the 2DES underneath the gate and in the rest area of the sample was kept to be very close, it is hard to pin-point the origin of the drop. Furthermore, the accuracy of measurement in this field region generally suffers from a noticeable phase shift.

To clarify this point, we apply a suitable bias to the gate so that the sheet densities in the two areas become different. Fig. 3a shows the result of such a measurement for C_{21} in one of the samples (where ν_g and ν denote the filling factors of the 2DES in the gated and ungated area, respectively). The valleys occurring at $\nu = 1, 2...$ clearly result from the quantum Hall effect in the ungated region of the 2DES; their positions remained constant as we changed the gate bias. The signal in these valleys is purely capacitive (within 0.5 %). Thus the capacitance tensor elements in a multi-terminal configuration contains information of the quantum Hall effect in the ungated region of the 2DES. *In stark contrast, a two-terminal capacitance shows only siguratures of the density of states oscillations of the gated area and shows no sign of the quantum Hall effect for the ungated area.*

Fig. 3b shows C_{31} obtained under the same experimental conditions as C_{21} in Fig. 3a. Note C_{31} increases as C_{21} drops at integer filling factors. We found that this behavior is general and is actually governed by the symmetry relations in Sec. 4.2. Consider the sum of the two, i.e. $C_{21} + C_{31} = -C_{11}$. The measurement configuration for C_{11}, as we mentioned earlier, can be reduced to a two-terminal configuration. Therefore C_{11} should show no sign of the quantum Hall effect for the ungated area (i.e. it should be constant in this field range). Consequently, when C_{21} drops, C_{31} must increase to compensate. This is exactly what we observed in the experiment.

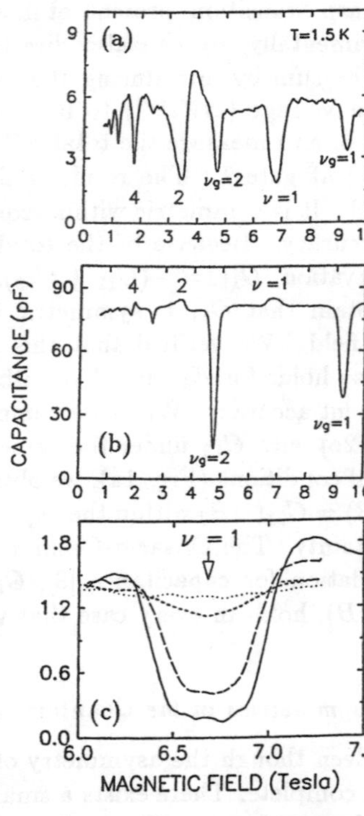

Fig. 3: Capacitance tensor elements $-C_{21}$ (a) and C_{31} (b) measured with a bias voltage $V_g = 0.4V$. (c) Temperature dependence of the $\nu = 1$ valley in $-C_{21}$: T = 0.5 K (solid), 3 K (long dash), 7 K (short dash) and 10 K (dot).

The valleys in C_{21} are very sensitive to temperature. As an example, the temperature dependence of the $\nu = 1$ valley in C_{21} obtained from one of the samples is shown in Fig. 3c. The valley disappears at T < 10 K. The dc magneto-resistance R_{xx} measured on the same type of sample shows quite similar temperature activation behavior. The non-zero offset of C_{12} at $\nu = 1$ at T=0.5 K, as we have identified, is largely due to the contact resistance (with possible mixing of the Hall resistance R_{xy}). Because in such a capacitance experiment we measure AC current that flows through a contact, we unavoidably include the contact resistance. As the sample magneto-resistance

R_{xx} diminishes, this effective resistance becomes the limiting factor. We have applied a load resistor to simulate the function of the contact resistance, and we found that we could artificially change the offset. Because of the presence of this finite offset the determination of the activation energy from such a measurement becomes rather difficult and inaccurate. For samples with ideal contacts of *zero* resistance, we would expect *complete asymmetry* in the capacitance tensor elements at integer filling factors.

4.4 Physical Model

To account for the observed asymmetry, we present a model based on the discussion of Ref. 7 and 8. Fig. 4 depicts schematically the areas in the gate and in the sample over which we have charge accumulation (or depletion) when a voltage dV_1 is applied to the gate. The easiest way to understand this effect is under the conditions when the 2DES is in a quantized Hall state with ideal contacts [9]. At the Fermi surface electron motion from one contact to another is then via edge states [9,10]. Carriers accumulated underneath the gate will eventually flow along the edge state to contact 3 (but not to contact 2). Similarly, a depletion of the carrier density at gate 1 will reduce the number of carriers along the edge state into contact 3 but will not affect the injection and absorption of carriers at contact 1.

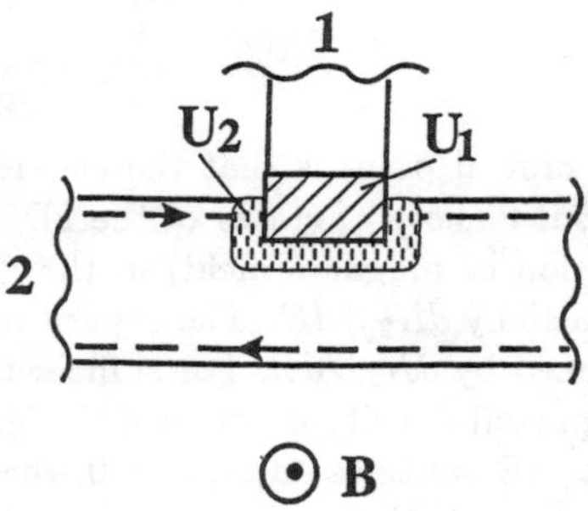

Fig. 4: A schematic diagram of the model for the three-terminal measurement configuration (described in the text). The arrows and dashed lines indicate the edge-state electron-current flow in a quantizing magnetic field.

An increase in the electro-chemical potential at the gate can be viewed first as an increase of the chemical potential which injects an additional charge $e(dN_{11}/dE)edV_1$ onto the gate. Here $e(dN_{11}/dE)$ is the density of states of the shaded gate area in Fig. 4. The index pair 11 indicates that this density of states is due to carriers which are injected by contact 1 and are reflected back into contact 1. This additional charge creates, via Coulomb forces, an electrostatic potential dU_1 which counteracts departures from a charge neutral state. The net result is a charge $dQ_1 = e(dN_{11}/dE)(edV_1 - edU_1)$ on the gate. Similarly the additional charge injected onto the gate will cause an induced voltage dU_2 in the 2DES over an area indicated in Fig. 4. This voltage induces a charge $dQ_3 = -e(dN_{3g}/dE)edU_2$ where the indices pair $3g$ indicates that the density of states is due to carriers which reach contact 3 from the area of the gate which overlaps the sample. Any charge imbalance must be a consequence of the long range Coulomb coupling. For small changes in the electro-chemical potential dV_1 the induced voltages are also small and are linearly related. Since our system must remain charge neutral overall, we have

$$e(dN_{11}/dE)(edV_1 - edU_1) = C(dU_1 - dU_2), \quad (4)$$

$$-e(dN_{3g}/dE)edU_2 = C(dU_2 - dU_1), \quad (5)$$

where C plays the role of the electrostatic capacitance. In our samples the gate is actually a macroscopic conductor and the density of states dN_{11}/dE is very large compared to C. This implies $dU_1 = dV_1$, and simple algebra gives for the capacitance $C_{31} = dQ_3/dV_1$,

$$\frac{1}{C_{31}} = -(\frac{1}{C} + \frac{1}{e^2(dN_{3g}/dE)}). \quad (6)$$

The crucial point is that the electro-chemical capacitance C_{31} is determined by a *partial* density of states of the 2DES. The total density of states (that is an even function of magnetic field) in the gated area is determined not only by dN_{3g}/dE but also by dN_{2g}/dE. The capacitance C_{21} is determined by Eq. (6) with dN_{3g}/dE replaced by dN_{2g}/dE. For sufficiently large fields with polarity as shown in Fig. 4, the partial density of states dN_{2g}/dE vanishes and $C_{21} = 0$. If the field is reversed dN_{3g}/dE vanishes and $C_{31} = 0$ whereas C_{21} is non-vanishing. This is in very good agreement with the experiment.

5. Summary

We observed large asymmetries in the capacitance matrix elements of 2DES with a multi-terminal configuration. This observation is in accordance with the microreversibility principle. The asymmetry in the quantizing magnetic fields clearly shows the manifestation of the integer quantum Hall effect in the capacitance tensor elements. A model based on the edge-state conduction is used to understand the origin of the asymmetry. Future work will aim at analyzing the asymmetry in the fractional quantum Hall effect regime.

6. Acknowledgements

The authors thank M.B. Santos for expertise in molecular beam epitaxy, and J.P. Cheng, L. Rubin, and the staff at the Francis Bitter National Magnet Lab for their assistance. The work is supported at IBM by the Office of Naval Research under contract No. N0014-92-c-0017 and at Princeton University by the National Science Foundation.

7. References

1. T.P. Smith III, W.I. Wang, and P.J. Stiles, *Phys. Rev.* **B34**, (1986) 2995.

2. A.M. Voshchenkov, and J.N. Zemel, *Phys. Rev.* **B9**, (1974) 4410; J. Lambe and R.C. Jaklevic, *Phys. Rev. Lett.* **22**, (1969) 1371.

3. W. Hansen, T.P. Smith III, K.Y. Lee, J.A. Brum, C.M. Knoedler, J.M. Hong, and D.P. Kern, *Phys. Rev. Lett.* **62**, (1989) 2168.
4. R.C. Ashoori, H.L. Stormer, J.S. Weiner, L.N. Pfeiffer, S.J. Pearton, K.W. Baldwin, and K.W. West, *Phys. Rev. Lett.* **71**, (1993) 613; P. Lafarge et al., *Z. Phys.* **B85**, (1991) 327.
5. S. Takaoka, K. Oto, H. Kurimoto, and K. Murase, K. Gamo, and S. Nishi, *Phys. Rev. Lett.* **72**, (1994) 3080.
6. W. Chen, T.P. Smith III, M. Büttiker, and M. Shayegan, *Phys. Rev. Lett.* **73**, 146, (1994).
7. M. Büttiker, H. Thomas, and A. Prêtre, *Phys. Lett.* **A180**, (1993) 364; M. Büttiker, A. Pretre, H. Thomas, *Phys. Rev. Lett.* **70**, (1993) 4114.
8. Büttiker, *J. Phys.* **C5**, (1993) 9361.
9. M. Büttiker, *Phys. Rev.* **B38**, (1988) 9375.
10. D. B. Chklovskii, K. A. Matveev, and B. I. Shklovskii, *Phys. Rev.* **B47**, (1992) 12605; C. W. J. Beenakker, *Phys. Rev. Lett.* **64**, (1990) 216; N. R. Cooper and J. T. Chalker, *Phys. Rev.* **B48**, (1993) 4530.

PHONON EMISSION FROM THE CONTACT REGIONS OF A MAGNETICALLY QUANTIZED TWO-DIMENSIONAL ELECTRON GAS (2DEG)

Z Xin, F F Ouali, L J Challis, A F Jezierski and M Henini

Department of Physics, University of Nottingham, NG7 2RD UK.

ABSTRACT

Differences in the nature of the phonon emission from the current entry and exit corners of 2DEGs in the form of Hall bars in the quantum Hall regime are not yet understood. We report measurements showing differences in size of the regions of phonon emission at a corner for electron entry and exit.

1: Introduction

When a current I passes through a 2DEG in the form of a Hall bar in the quantum Hall regime (QHR), the energy dissipation I^2R_H is believed to be confined to regions around the two diagonally opposite corners, called "hotspots", where the current enters and leaves. The energy relaxation is predominantly as phonons and from observations of the phonon emission it appears that in Si MOSFETS the dissipation is equally shared between the two corners[1]. The situation is less clear in 2DEGs[2,3] and 2DHGs[3] in GaAs/(AlGa)As heterojunctions: there are evident differences in the phonon emission but it has not yet been established to what degree these are due to differences in the location of the emission, in the angular distribution of the phonons emitted or in the magnitude of the dissipation at the two corners. It is also not yet known in any of the 2D systems how much of the dissipation and subsequent relaxation at each corner occurs within the 3D region, how much at the interface between the 2D and 3D regions and how much within the 2D region as a result of quantum Hall breakdown. We report measurements which provide information on the size of the regions where the emission is taking place. Complimentary studies of FIR emission are also in progress[4].

2: Experimental

Measurements have been made on a 2DEG in a 12 nm wide GaAs/(AlGa)As quantum well in the form of a Hall bar of dimensions 3×1 mm^2. The electron density n_s is ~2.4×10^{15} m^{-2} and the mobility is 44 m^{2}V^{-1}s^{-1}. Two phonon detectors each of dimensions ~350×350 μm^2 are attached on opposite sides of the substrate surface facing one of the corners as shown in Fig 1. The detectors, D_N and D_F are at distances ~1 and ~400 μm from the plane of the 2DEG respectively so that, if the phonon emission occurs from a region of area small compared with that of the detector,

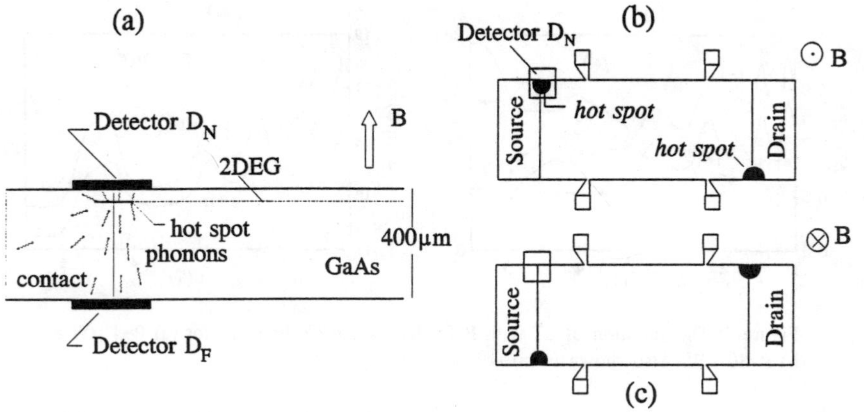

Figure 1: The experimental arrangement

essentially all the flux emitted above the 2DEG should strike the nearer detector D_N while the further detector D_F should only receive phonons whose group velocity makes an angle <25° to the normal. The phonons incident on a detector produce a rise in its temperature proportional to the flux incident on it. The rise in temperature ΔT is measured relative to a reference contact and so includes a small contribution from the thermal resistance of a 4.5 mm length of substrate.

3: Results

Measurements have been made for power inputs P in the range 0.5-100 μW which are maintained constant while the magnetic field B is varied from 1-6T. Fig 2(a) and (b) show the temperature change $\Delta T(B)$ of D_N for P=4 and 80 μW respectively for both current directions with B in the ⊙ direction so that D_N lies above a hotspot (Fig 1(b)). Fig 2(a) also includes the source-drain resistance for this power showing the quantized plateaux of the quantum Hall regime. $\Delta T(B)$ oscillates as the dissipation moves from the bulk of the 2DEG to the contact regions[1,3] and at 4 μW $\Delta T(B)$ is always larger when the electrons are entering the 2DEG. This changes however at powers between 5 and 10 μW depending on B and at 20μW and above ΔT is significantly greater when the electrons are leaving the 2DEG as seen in Fig 2(b). At all powers and for both current directions, there is a steady decrease in the value of ΔT and so in the proportion of total phonon flux striking D_N as B increases.

Fig 3 shows $\Delta T(B)$ for D_N for P=80 μW for both current directions when B is in the ⊗ direction so that the hotspot lies on the opposite side of the Hall bar to D_N (Fig 1(c)). ΔT is appreciably smaller as expected and for electrons leaving the 2DEG is

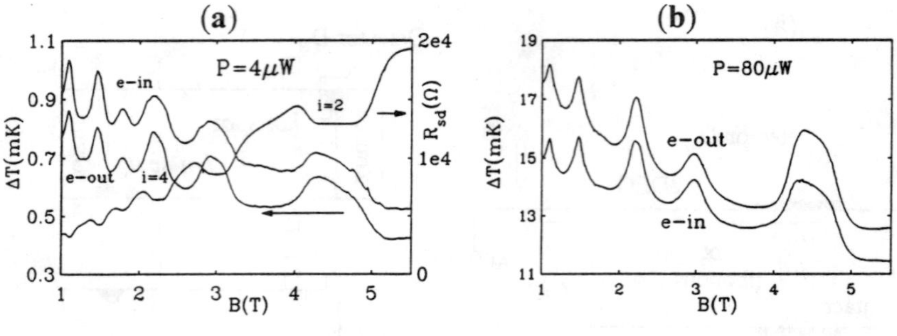

Figure 2: The variation of ΔT with B for D_N for the ⊙ direction for (a) P=4 μW and (b) P=80 μW. Also shown in (a) is $R_{sd}(B)$.

predominantly due to the thermal resistance of the substrate. Oscillations due to changes in the flux falling on the detector are still apparent however. Of particular note is the change in phase of the oscillations when the current is reversed (e.g i=3). When the electrons are leaving the 2DEG ΔT has minima in the QHR but, for P≳ 15 μW, these become maxima when the electrons are entering. This behaviour, including the pronounced maximum at i=2, is seen over a range of powers. Minima in the QHR have also been seen in Si[1] for this field direction and are consistent with the movement of dissipation from the detector to the hotspots. However the maxima that occur in the QHR when the electrons are entering the 2DEG indicate that for P≳ 15 μW the tail of the hotspots now extends ~1 mm across the width of the Hall bar although the relative sizes of ΔT suggest that most of the emission still occurs within the area of the detector and only ~ 5% lies within a similar area on the opposite side of the bar. This presumably is also the reason for the behaviour seen for the ⊙ direction. The flux on D_N is smaller when the electrons are entering the 2DEG because the hotspot is so large that some of the emission from it hits the surface outside the detector.

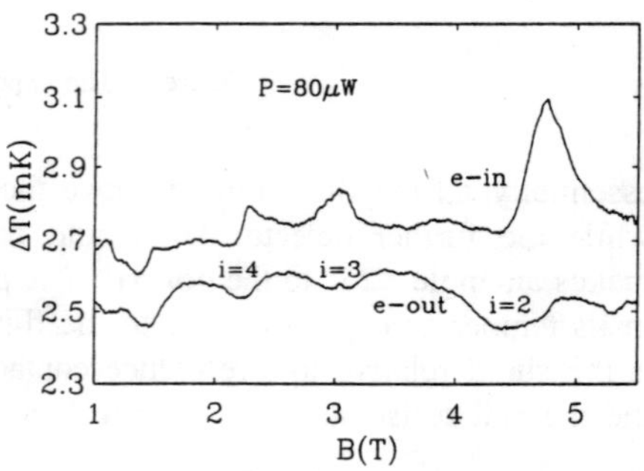

Figure 3: Variation of ΔT with B for D_N for the ⊗ direction for P=80 μW.

Observations on a detector D_F on the opposite surface to a hotspot have been reported previously for GaAs/(AlGa)As heterostructures[3] and the present work on a quantum well shows broadly similar behaviour with ΔT being greater at all powers when the electrons enter the 2DEG.

4: Discussion

The main conclusion from this work is that, for powers greater than ~10 μW, the tail of a hotspot extends appreciably further across the width of contact when the electrons are entering the 2DEG than when they are leaving. The decrease in flux striking D_N as the field is increased might also suggest that the hotspot size increases with field for both directions of electron current. The contrasting results, except at low powers, for the flux falling on D_F suggests this is strongly affected by the angular distribution of the phonons; this should evidently be much less important for D_N. The wavevectors of phonons emitted through relaxation in the 2DEG should mainly occur at angles close to the normal while those emitted within the 3D contact seem likely to lie in all directions. The angular distribution of the TA flux is focussed strongly along [100] directions so the emission from the 2DEG will be further channelled towards the normal and hence towards D_F while that from the 3D contact will also occur in the 4 [100] directions parallel to the surface. It seems likely therefore that the phonon flux striking D_F mainly consists of phonons emitted from the 2DEG. The proportion of these is likely to be greater when hot electrons are moving away from the 3D region than towards it as previously discussed[3] and the results suggest that this effect outweighs effects associated with the difference in size of the hotspots for the two current directions. Differences in the size of the hot spot may also be related to relaxation: when the electrons are entering the 2DEG and relaxing relatively slowly, the hotspot should be larger than when the electrons are leaving and should increase in size with field because of the decrease in relaxation rate[5]. However, whether this is the main reason for the present result or whether the electrical propoerties of the 2D/3D interface also play a part is not yet clear. Further work is in progress.

5: Acknowledgements

We are grateful to Dr N N Zinov'ev and Prof A Y Shik for useful discussions and to the Science and Engineering Research Council and the British Council for financial support.

6: References

1. P A Russell, F F Ouali, N P Hewett and L J Challis, Surface Science, **54** 229 (1990); F F Ouali, L J Challis and J Cooper, J Semicond Sci and Technol, **7**, 608 (1992)
2. U Klaß, W Dietsche, K von Klitzing and K Ploog, Z Phys **B82**, 351 (1991)
3. Z Xin, F F Ouali, L J Challis, J Cooper, A F Jezierski and M Henini, Semicond Sci and Technol, **9**, 800 (1994)
4 N N Zinov'ev, A V Akimov, L J Challis, R Fletcher, B Sujak-Cyrul, A F Jezierski, and M Henini, Semicond Sci and Technol, **9**, 831 (1994)
5 V I Fal'ko and L J Challis, J Phys C, Condensed Matter, **5**, 3945 (1993)

Evaluation of Edge Channel Width through Temperature Dependent Capacitance Minima in Quantum Hall Regime

K.Oto, S.Takaoka, H.Kurimoto and K.Murase

Department of Physics, Faculty of Science, Osaka University, Toyonaka, 560, Japan

The temperature dependence of the magneto-capacitance between two dimensional electron systems and the gate has been studied at low temperatures (80 mK - 10 K). It is found the minimum capacitance at the quantum Hall plateau regime is determined by area of the edge channels and a temperature dependent bulk channel conductivity $\sigma_{xx}(T)$, where σ_{xx} is expressed as an activation type at higher temperatures and a variable range hopping type at lower temperatures. The width of the edge channels is evaluated for different Landau occupations.

1. Introduction

The (differential) capacitance between a two dimensional electron system (2DES) and a metal gate shows U valley minima at the quantum Hall plateau (QH plateau). The origin of the magneto-capacitance oscillations was explained by the oscillations of the density of states at the Fermi level[1,2], or those of the bulk conductivity σ_{xx}[3,4]. But the contribution of the edge channels to the capacitance was overlooked. Recently, we reported for the first time that the minimum values of the capacitance are mostly determined by the effective area of edge channels at low temperatures[5]. In this paper, to evaluate more accurate area of the edge channels, the contribution of the bulk channel to the capacitance is estimated by measuring the magneto-capacitance of a GaAs/AlGaAs heterostructure in the quantum Hall regime very precisely at temperatures between 10 K and 80 mK.

2. Experimental

The samples were made from the GaAs/AlGaAs heterostructure wafer of the mobility μ=3.7×10^5 cm^2/Vs and electron density N_S=2.8×10^{11} cm^{-2} at 0.5 K. The thicknesses of a nondoped AlGaAs spacer, Si-doped AlGaAs, and the GaAs cap layer were 20, 60, and 20 nm, respectively. Since the capacitance measurement is very much sensitive to the existence of parallel conduction in the Si-doped layer, such a contribution should be completely excluded. We tested the gate voltage (Vg) dependence of capacitance at low temperatures as follows; (1) capacitance became zero when the 2DES was depleted (Vg< −0.4 V), (2) measured capacitance at Vg= 0 V agreed with the calculated value from the sample geometry, (3) no capacitance increase due to the parallel conduction was detected up to Vg= +0.05 V.

The samples were fabricated by the uv lithograph and wet mesa etching (etching depth was 250 nm). The schottky gate was formed by thermally evaporated Au (thickness 100 nm) on the GaAs cap layer. There was only one ohmic contact region in the sample to minimize a stray capacitance. The differential complex capacitance between the gate and 2DES through the ohmic contact was measured by a capacitance bridge operating at 1 KHz with the modulation voltage 1 mV with the zero dc bias. The real and imaginary (loss) parts of the capacitance were measured simultaneously. All lead wires to the sample are coaxial cables to reduce the stray capacitance. The stray capacitance 0.2 - 0.35 pF, mostly due to the sample geometry, was estimated from the value at a gate bias Vg= −0.5 V, where the 2DES under the gate was completely depleted. It was confirmed that the stray capacitance was independent of the magnetic fields. The samples were cooled by a ^{3}He/^{4}He dilution refrigerator, and the temperature was monitored by a calibrated RuO_2 thermometer. All measurements were made in a dark condition.

3. Results and Discussion

3.1. The capacitance of the edge channels

The temperature dependence of the capacitance minima at each filling factor ν in a sample(A), whose gate covers the 2DES edge, is plotted as shown in Fig.1. The size of the gated 2DES region is 1.4 mm × 0.4 mm. With lowering temperatures, the capacitance shows an abrupt reduction, because the bulk conductivity σ_{xx} decreases. At the lower temperature regime, the capacitance at each ν tends to be gently reduced to a certain value, that is determined by the effective area of the edge channels[5].

To reasonably evaluate the area of edge channels, we take the contribution of the bulk state to the capacitance into account by a model as shown in Fig.2. The bulk channel capacitance C_b (complex) is treated as a distributed circuit of the resistive plate[3] with the conductivity σ_{xx}. The total complex capacitance C is given by;

$$C = C_e + C_b(\sigma_{xx}) = C_e + (C_0 - C_e)\frac{\tanh(\alpha)}{\alpha}, \quad (1)$$

where C_e is the edge channel capacitance (real), C_0 the capacitance at zero field, $\alpha = (i\,\omega\, C_0\, K / \sigma_{xx})^{1/2}$, ω the angular frequency of the ac modulated voltage, K a constant determined by the sample shape, respectively. It is supposed that the $\sigma_{xx}(T)$ is expressed by two kinds of terms as;

$$\sigma_{xx}(T) = \sigma_A \exp(-\Delta E / k_B T) + \sigma_H \exp[-(T_0/T)^{1/2}], \quad (2)$$

the first term is due to an activated conduction (an activation energy ΔE) and the second is due to a variable range hopping conduction[6] with a Coulomb gap in the 2DES. The calculated curves of the real part of the capacitance with a stray capacitance 0.2 pF for sample(A) can be fitted very well to the measured one as shown in Fig.1.

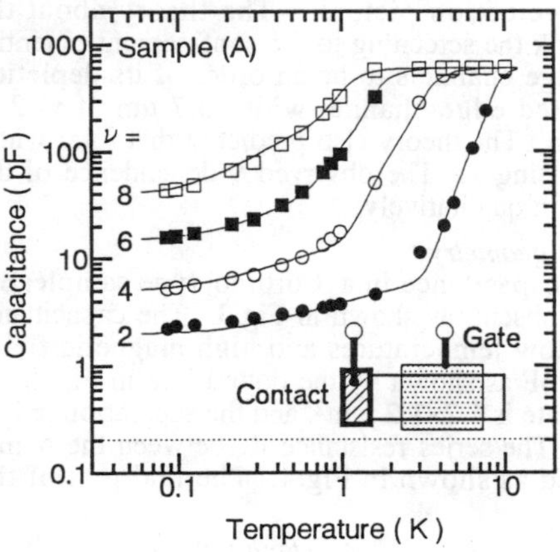

FIG.1 Temperature dependence of the capacitance minima at each ν for the sample(A). The dotted line shows a stray capacitance 0.2 pF of the sample. Solid curves are calculated by using Eqs.(1) and (2) (see text). Inset: The schematic view of the sample(A), whose gate covers the sample edge.

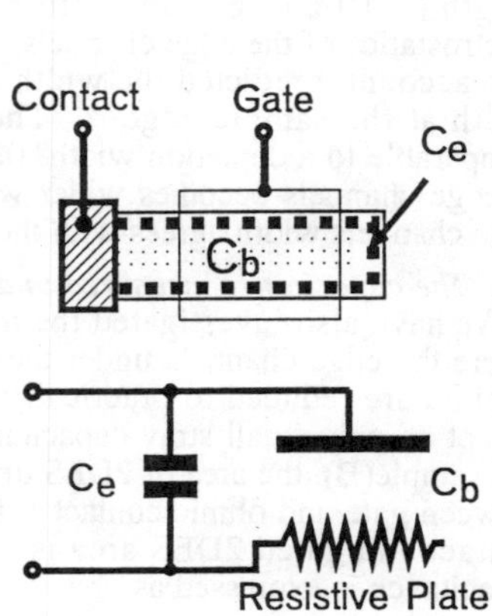

FIG.2 The equivalent circuit for the sample(A). The complex capacitance C_b is treated as a distributed system. At the QH plateau regime, C_b decreases with decreasing σ_{xx}, the residual capacitance is C_e.

Table I The parameters used in the fitting of Eqs. (1) and (2) in the sample(A). The activation energy ΔE, characteristic temperatures of the hopping conduction T_0, capacitance of the edge channels C_e, total width of the edge channels W are listed. A half energy of Landau level separation for each ν are also listed. The evaluated total edge channel width W(ν) from C_e are calculated by the assumption of the two-plate capacitor between edge channels and gate.

Filling factor (ν)	2	4	6	8
$\hbar\omega_c / 2k_B$ (K)	48	24	16	12
$\Delta E / k_B$ (K)	45	11	5.5	4.4
T_0 (K)	11	11	11	11
C_e (pF)	2	5	15	40
W(ν) (μm)	0.7	1.7	5.2	13

Thus obtained parameters ΔE, T_0 in Eqs.(1) and (2) are listed in Table I. The activation energy ΔE is expected to be $\approx(\hbar\omega_c - \Gamma)/2$, where $\hbar\omega_c$ is the Landau level separation, Γ the width of the extended state in the Landau level which is much smaller than $\hbar\omega_c$ at ν=2. With decreasing magnetic field (or increasing filling factor ν), ΔE deviates from $\hbar\omega_c/2$, since Γ becomes significantly large. The characteristic temperature T_0 of the hopping conduction is determined to be 11 K, which is independent of the filling factor ν as predicted by a theory[6]. The pre-exponential factor of the activation term σ_A is of the order of 10^{-6} S, while that of the hopping term σ_H varies 10^{-11} to 10^{-7} S with increasing ν from 2 to 8.

Note that the parameters C_e in Table I are finite. The total width of edge channels evaluated from C_e at ν=2 is 0.7 μm, which is much larger than the order of a magnetic length (~10 nm) expected from the one-electron picture. The theory about the electrostatics of the edge channels, which took the screening to the confinement potential into account, predicted the width of the edge channels to be an order of its depletion width at the sample edge[7,8]. The evaluated edge channel width 0.7 μm at ν=2 is comparable to a depletion width[9] 0.2–0.5 μm. The theory also predicted that total width of edge channels becomes wider with increasing ν. The observed ν dependence of the edge channels width agrees with the prediction qualitatively.

3.2. The capacitance measured in a Corbino geometry

We have also investigated the magneto-capacitance in a 'Corbino' type sample(B), where the edge channels under the gate are absent, as shown in Fig.3. The capacitance minima are reduced to practically zero at low temperatures and high magnetic fields except an only small stray capacitance 0.35 pF as shown by the dotted line in Fig.3. In this sample(B), the area of 2DES under the gate is 1.2×1.2 mm², and the separation length between gate and ohmic contact is 0.1 mm. The series resistance R_S between the ohmic contact and gated 2DES area is considered as shown in Fig.4. The real part of the capacitance is expressed as;

$$C = \frac{C_b(\sigma_{XX})}{1 + [\,\omega C_b(\sigma_{XX}) R_S(\sigma_{XX})\,]^2}\,, \quad C_b(\sigma_{XX}) = C_0 \frac{\tanh(\alpha)}{\alpha}\,. \tag{3}$$

The curves in Fig.3 are calculated by using Eq.(3) with the same parameters, σ_A, ΔE, σ_H, T_0, for each filling factor ν in the sample(A), and a stray capacitance 0.35 pF is added. It should be noted that the curves with the fitting parameters used in the sample(A) almost reproduce the measured temperature dependence of the capacitance in the sample(B). This is a strong evidence that the fitting parameters used in the sample(A), including the capacitance of the edge channels C_e, are quite reasonable.

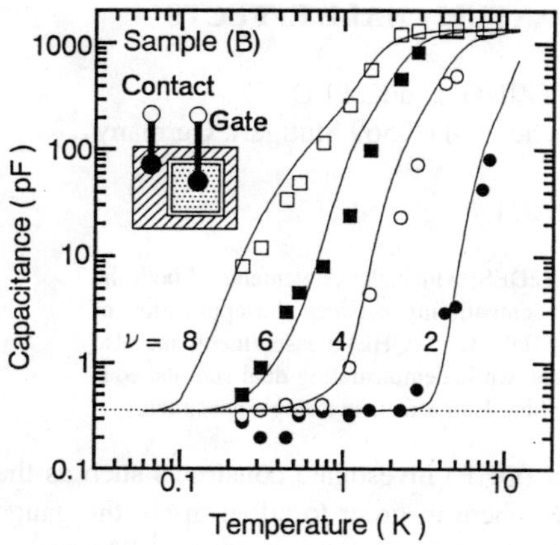

FIG.3 Temperature dependence of the capacitance minima for the Corbino sample(B), where the edge channels are absent under the gate. At low temperatures, the measured values tend to the value of a stray capacitance 0.35 pF. Solid curves are the capacitance calculated by Eq.(3) using the parameters in the sample(A). The schematic view of the sample(B) is also shown in the inset.

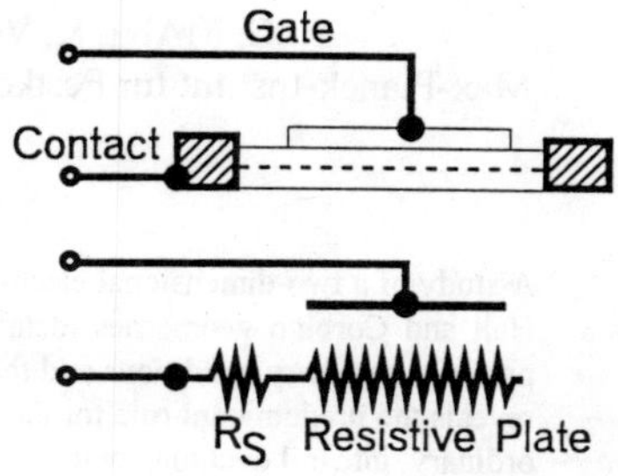

FIG.4 The equivalent circuit for the sample(B). The series resistance R_S between the ohmic contact around 2DES and 2DES under the gate is considered. The complex capacitance C_b is treated by a resistive plate model. At the QH plateau regime, C_b decreases with decreasing σ_{xx}, the capacitance tends to be zero.

4. Summary

In conclusion, the temperature dependence of the magneto–capacitance minima is analyzed by using a temperature dependent bulk channel conductivity $\sigma_{xx}(T)$, where the conduction mechanisms are of an activation type at higher temperatures and a variable range hopping type at lower temperatures. It is found that the area of the edge channels can be unambiguously evaluated by the magneto–capacitance measurement at the QH plateau at low temperatures.

5. Acknowledgements

We would like to thank S.Nishi of Oki Electric Industry Co. Ltd. for providing the GaAs/AlGaAs wafers. We also acknowledge Prof. Gamo and Prof. Namba at Research Center for Extreme Materials, Osaka University for providing microfabrication facilities. This work is partially supported by Grant–in–Aid for Scientific Research (B), and for Scientific Research on Priority Area, both from the Ministry of Education, Science and Culture (Japan). One of us (K.O) acknowledges the support by the Foundation Advanced Technology Institute and the Murata Science Foundation. Another (S.T) thanks the support by the Casio Science Promotion Foundation and the Shimazu Science Foundation.

6. References

1. T.P.Smith, B.B.Goldberg, P.J.Stiles, and J.P.Harrang, *Phys. Rev.* **B32** (1985) 2696.
2. V.Mosser, D.Weiss, K.v.Klitzing, K.Ploog, and G.Weimann, *Solid State Commun.* **58** (1986) 5.
3. R.K.Goodall, R.J.Higgins, and J.P.Harrang, *Phys. Rev.* **B31** (1985) 6597.
4. J.I.Lee, P.J.Stiles, and M.Heiblum, *Surf. Sci.* **263** (1992) 120.
5. S.Takaoka, K.Oto, H.Kurimoto, K.Murase, K.Gamo, and S.Nishi, *Phys. Rev. Lett.* **72** (1994) 3080.
6. D.G.Polyakov, and B.I.Shklovskii, *Phys. Rev. Lett.* **70** (1993) 3796.
7. D.B.Chklovskii, B.I.Shklovskii, and L.I.Glazman, *Phys. Rev.* **B46** (1992) 4026.
8. D.B.Chklovskii, K.A.Matveev, and B.I.Shklovskii, *Phys. Rev.* **B47** (1993) 12605.
9. K.K.Choi, D.C.Tsui, and K.Alavi, *Appl. Phys. Lett.* **50** (1987) 110.

DUAL ORDINARY - AND QUANTUM - HALL EFFECTS

R. G. MANI, K. VON KLITZING, and K. PLOOG
Max-Planck-Institut für Festkörperforschung, D70569 Stuttgart, Germany

ABSTRACT

A study of a two-dimensional electron system (2DES) which shares elements of both the Hall and Corbino geometries identifies the incompatibility between the application of periodic boundary conditions and the quantum Hall effect (QHE) measurement, and also reveals the predominant role for the bulk current, while demonstrating dual simultaneous ordinary, integral quantum, and fractional quantum, Hall effects in a single specimen.

Theoretical studies of the quantum Hall effect (QHE) investigate constructs such as the short cylinder, the annulus, the torus, and the sphere in order to either apply the gauge argument or examine numerically the ground state of few electron systems.[1] The use of such configurations follows from the premise that disturbing the ends of a long Hall bar has little influence on the QHE, and therefore, one may modify the boundary conditions (geometry) as convenient, and draw general conclusions. Hence, the Hall bar is bent, and the ends joined together, in order to form a short (Laughlin's) cylinder. Further, the boundaries of the cylinder are eliminated by applying periodic boundary conditions to obtain a torus, or even a sphere. Such simplifications based on "topological invariance" lead, however, to measurement-complications which include the use of magnetic monopoles to bias magnetically the 2DES. In the gauge argument, one also threads the cylinder with flux quanta in order to relate a preexisting current to the QHE, which involves a voltage measurement between the boundaries of the cylinder; this argument, involving quasiparticles with charge $e^* = e/m$, is also invoked to relate the (correlated) Laughlin ground states to the (1/m) fractional QHE's, i.e., $R_{xy} \equiv \rho_{xy} = mh/e^2 = h/e^*e$.[1]

The edge-current picture relates the IQHE to one dimensional resistance quantization by asserting that an insulating bulk sends the current to the quasi-1D edges; this suggestion has fueled much interest in the nature of the current distribution at high magnetic fields, B.[1] Thus, we address the questions: (1) How does the application of periodic boundary

Fig. 1) (top) The "anti-Hall bar within a Hall bar". (A) & (B): The Hall voltage on a boundary depends exclusively upon the current injected via the same boundary.

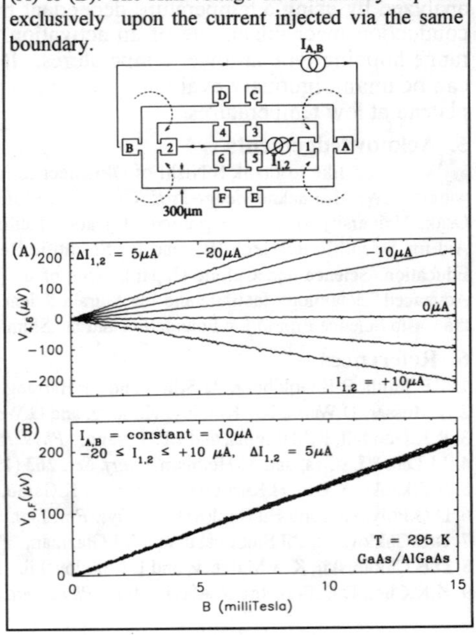

conditions modify the QHE measurement? (2) How is the current distributed at high B?

These experiments examine transport in inversion symmetric homogeneous, and (gated) inhomogeneous, doubly connected structures including several small contacts in the interior and exterior of the specimen, see Figs. 1(top) & 4(top).[2] Here, the configuration (Hall or Corbino-type) is poorly defined until a source is connected to the sample, and it remains ambiguous even after engaging a current supply between, e.g., the contact pair (A,B), until one observes that a Hall bar is obtained by making the hole vanishingly small. Upon noting this point, the hole may be replaced, and the sample turned inside-out. Thus, the "Hall configuration" is moved into the interior where, by symmetry, one expects also to measure a Hall effect $V_{4,6}$ when the current is injected via (1,2). As this transformation effectively reverses orientation with respect to the external B-field, one expects sign reversal in the interior Hall effect and thus one views the interior configuration as the complement of the Hall bar or the "anti-Hall bar." The measurements shown in Fig. 1 include configuration-superposition, which is realized by injecting current simultaneously from two disconnected current sources (see Fig. 1 (top)). The results, obtained with relatively large currents in milliTesla magnetic fields at 295K, shows that the interior Hall effect $V_{4,6}$ varies in proportion to $I_{1,2}$, while the exterior Hall effect $V_{D,F}$ is independent of $I_{1,2}$. The results of a experiment with $I_{1,2}$ = constant and varying $I_{A,B}$ are shown in Fig. 2. Here, one obtains complementary results although the opposite limit of low-temperature (T = 1.6K) and high-B (B < 8 Tesla) is examined utilizing small currents. The observed characteristics also show immediately that dual Hall resistances, e.g., $R_{D,F;A,B} = |V_{D,F}/I_{A,B}|$ and $R_{4,6;1,2} = |V_{4,6}/I_{1,2}|$, are

Fig. 2) (left) (a) At T = 1.6K, the Hall effect $V_{3,5}$ is insensitive to changes in $I_{A,B}$. (b) The Hall effect $V_{C,E}$ varies in proportion to $I_{A,B}$. Note sign reversal, i.e., $V_{3,5} = -V_{C,E}$ for $I_{1,2} = I_{A,B}$.
Fig. 3) (right) (a) $V_{3,4}$ reflects also the exterior injected current $I_{A,B}$, unlike the Hall effect (Fig. 2). (b) The exterior magnetoresistive voltage $V_{C,D}$. Note the vanishing signal when $I_{A,B} = -I_{1,2}$.

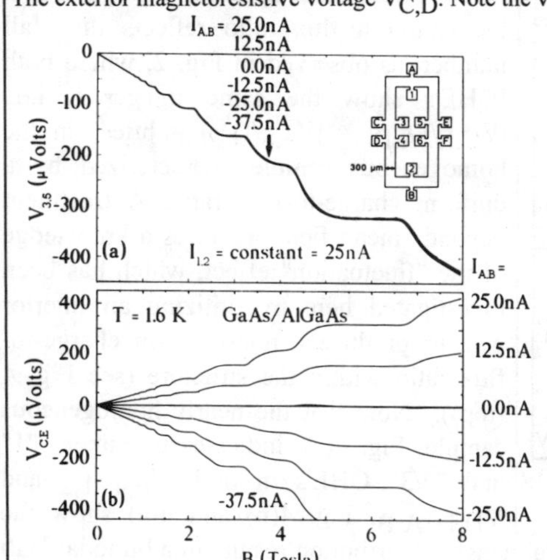

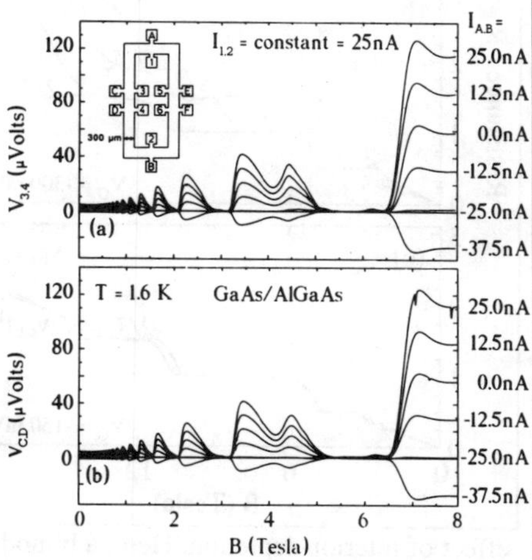

Fig. 4) (top) Partially gated configuration which exhibits simultaneous dual distinct QHE's. (a) $V_{6,4}/I_{1,2}$ and $V_{D,F}/I_{A,B}$, in units of R_K = 25.813 kOhm, are shown for V_G = -450mV. Here, an i = 1 interior QHE coincides with an i = 2 exterior QHE for 3 < B < 4.5 Tesla. At B ~ 12 Tesla, one observes an interior "1/3" FQHE coinciding with an exterior "4/7". Quantum numbers i corresponding to marked resistances are indicated in italics. (b) V_G = -300mV. (c) The nearly homogeneous condition at V_G = -150mV.

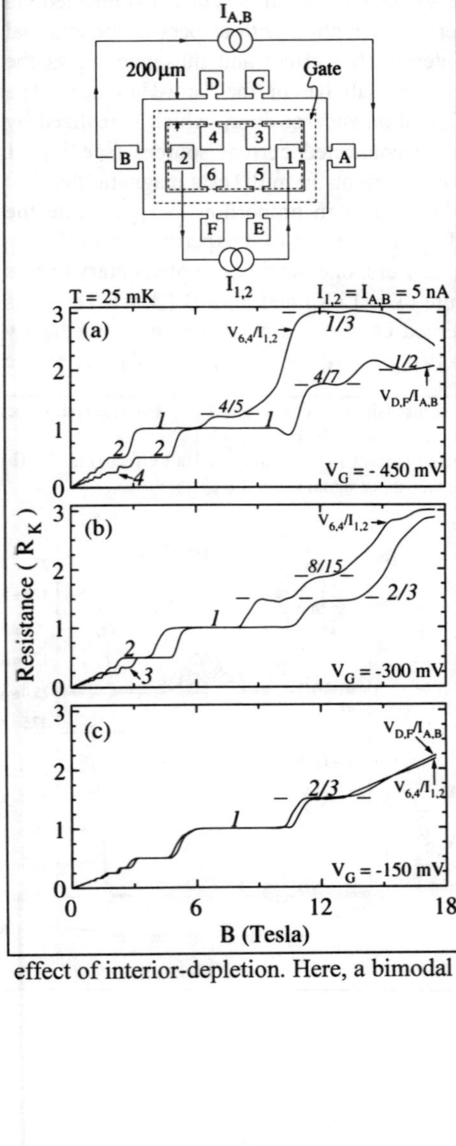

possible in a single specimen.

A description of these Hall effects does not require a knowledge of the current distribution. Yet, the unchanging $V_{3,5}$ in Fig. 2(a), which goes with the constant $I_{1,2}$, does not imply an invariant current distribution along the interior. This point is clarified upon examining the magneto-resistive voltages under the same experimental conditions (see Fig. 3), which show that $V_{3,4}$ and $V_{C,D}$ are insensitive to the origin of the current, unlike the global Hall effects. Indeed, the characteristics of Figs. 2 and 3 indicate a progressive reduction in the branch currents for $-25 < I_{A,B} \leq 0$ nA, and <u>null branch current</u> for $I_{A,B} = -I_{1,2}$ = -25nA. The null branch current seems, initially, to be inconsistent with non-vanishing interior, and exterior, Hall effects (Fig. 2), and finite currents injected by the two current sources. One resolves the paradox, however, by observing that a <u>bulk current</u> is driven from one supply to the other, *via* the the ends of the device (see Fig. 5(f)).

"Mean field" approaches suggest that the charge-to-flux ratio reflects the Hall number, as observed in Fig. 2, where both IQHE's show the same integer i, i.e., $|V_{C,E}/I_{A,B}| = |V_{3,5}/I_{1,2}| = h/ie^2$, in the homogeneous sample characterized by a uniform charge-to-flux ratio. A treatment beyond "mean field" requires a knowledge of the "fluctuation" effect, which has been investigated here by utilizing an interior gate to produce a non-uniform charge-to-flux ratio within the structure (see Fig. 4 (top)). Note that the nearly homogeneous sample, Fig. 4(c), indicates prominent "1" and "2/3" QHE's in both $V_{6,4}/I_{1,2}$ and $V_{D,F}/I_{A,B}$. Fig. 4(b) and 4(a) show the effect of interior-depletion. Here, a bimodal density distribution results in a bimodal Hall

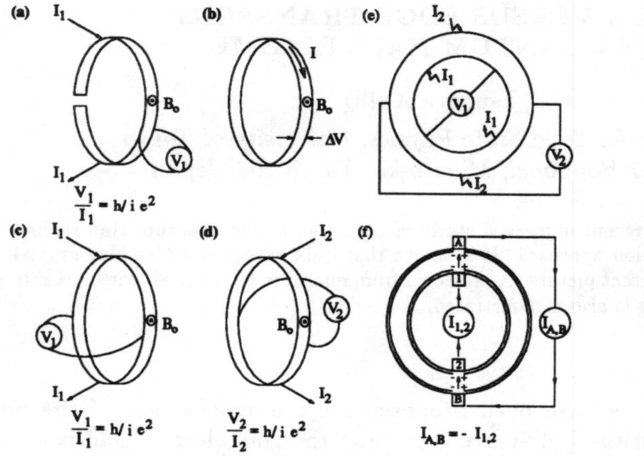

Fig. 5) (a) The QHE measurement in a bent Hall bar. (b) A change in the end conditions in (a) results in the cylinder (Laughlin's geometry). (c) & (d) QHE measurements in "anti-Hall bar" type configurations. Superposition of (c) and (d) allows for simultaneous dual QHE's in the cylinder. (e) Dual QHE's in the annular configuration. (f) A bulk current configuration in the quasi-Corbino geometry.

response that differs quite significantly from expectations based on the mean charge-to-flux ratio; this result suggests an important role for "fluctuations" in the QHE especially when connectivity is preserved between the current contacts. The data (Fig. 4(a)) also show that inducing a "1/3" along the interior shifts the exterior "2/3" (Fig. 4(c)) towards "4/7" (Fig. 4(a)); this is another bulk current effect, originating from a shunt of the exterior injected current via the gated ("1/3") interior, in the FQHE regime.

The effect of periodic boundary conditions upon the QHE measurement is examined in Fig. 5. Fig. 5(a) shows a long, bent Hall bar. If one applies periodic boundary conditions and, further, assumes a sourceless current, then one obtains Fig. 5(b).[1] Steady state experiments demonstrate, however, the existence of a large two-terminal resistance and corresponding heat dissipation under QHE conditions, which indicates the necessity of the "power supply". These experiments also show that the source serves to define the effective geometry of the specimen. Together, these points suggest that viable four-point QHE measurements in the cylinder are measurements in the "anti-Hall bar" type configuration (see Fig. 5(c) and 5(d)), where the gauge argument does not help to relate the voltage to the current *via* elementary constants.[2]

In conclusion, this study suggests that (a) typical applications of periodic boundary conditions are incompatible with the QHE measurement, and (b) the bulk current predominates at high B, under QHE conditions. Both conclusions originate from the identification of peculiarities in the Hall measurement, which are revealed through configuration-superposition studies in the so-called "anti-Hall bar within a Hall bar".

We thank T. Murphy, E. Palm, B. Brandt, and J. Crow for facilitating visits to the National High Magnetic Field Laboratory in Tallahassee, Florida.

1) The quantum Hall effect, eds. R. E. Prange and S. M. Girvin (Springer, New York, 1987); M. Buttiker, Phys. Rev. B 38, 9375 (1988).

2) R. G. Mani and K. von Klitzing, Z. Phys. B92, 335 (1993); Appl. Phys. Lett. 64, 1262 (1994); Appl. Phys. Lett. 64, 3121 (1994); (unpublished).

BULK VERSUS EDGE TRANSPORT IN QUANTUM HALL REGIME

Tsuneya ANDO

Institute for Solid State Physics, University of Tokyo
7-22-1 Roppongi, Minato-ku, Tokyo 106, Japan

A review is given of recent numerical study of transport in the quantum Hall regime based on a transmission approach. It is shown that bulk extended states play crucial roles even in edge-current picture. A crossover from quantum to classical transport due to inelastic scattering is also demonstrated.

1. Introduction

Two different pictures have been proposed for the explanation of the integer quantum Hall effect. In the bulk-current picture, the Hall electric field is essential and the current is carried by extended states existing in the vicinity of the center of each broadened bulk Landau level. In the edge-current picture, on the other hand, the current is carried by edge states localized along system boundaries and the Hall field is not required. The relationship between these two pictures remains still unclear. The purpose of the present paper is to give a brief review of some of theoretical works performed for the purpose of shedding some light on this problem.

This paper is organized as follows. In Sec. 2 a short review is given of explanations of the quantum Hall effect. In Sec. 3 transport quantities calculated in a transmission approach are compared with those obtained in the bulk-current picture.[1-3] Effects of inelastic scattering are studied in Sec. 4, where a crossover from edge transport to classical transport based on local conductivities is demonstrated.

2. Bulk versus Edge Current Picture

In the bulk-current picture, a key ingredient of the quantum Hall effect is the localization of Landau states in the presence of random potential fluctuations. It has been shown in various different ways that the Hall conductivity is quantized into integer multiples of $-e^2/h$ when states around the Fermi level are all localized. Several reviews have been published already.[4-6]

A Thouless-number study has shown that states are all exponentially localized except those in the vicinity of the center of the broadened Landau levels.[8] The critical exponent s in the presence of dominant short-range scatterers defined by $\alpha(E)\propto|E|^s$ is $s\sim2$ for the lowest Landau level $N=0$, where energy E is measured from the center of the Landau level. A more powerful finite-size scaling confirmed the result of the Thouless number study that $s\sim2$ for the lowest Landau level.[9,10] Huckestein-Kramer obtained $s\approx2.34\pm0.4$ based on a random-matrix model.[11]

Although the localization problem in the lowest Landau level is quite well-understood, the situation is not so good for higher Landau levels. Various numerical calculations have been reported,[12–14] but they have not been successful in supporting the universality predicted based on a field-theoretic nonlinear σ model.[15]

The quantized Hall effect can also be explained by the transmission approach based on the Büttiker-Landauer formula.[16] In fact, if we consider a four terminal geometry used in Hall-effect experiments and assume that $T_L=1$, $R=T_F=T_R=0$, then we immediately obtain the quantized Hall resistance $R_H = h/e^2$, where R, T_L, T_F, and T_R are transmission probabilities for being reflected back, turning left, going straight, and turning right at the junction of two wires.

In a two-terminal geometry, the conductance of a wire is given by Landauer's formula,[17]

$$G = \frac{e^2}{\pi\hbar}\sum_{\mu\nu} T_{\mu\nu}, \tag{2.1}$$

where G is the conductance, ν is the index for incident channels, and μ is that for out-going channels. This transmission probability is related to the diagonal conductivity σ_{xx} of the wire[18–21] and does not depends on the Hall conductivity σ_{xy}. This is quite in contrast to the classical galvanomagnetic effect in which σ_{xy} plays a decisive role.

Consider a wire with width W in the y direction and infinitely long in the x direction. The usual textbook procedure is to start with the transport relations, $j_x=\sigma_{xx}E_x+\sigma_{xy}E_y$ and $j_y=\sigma_{yx}E_x+\sigma_{yy}E_y$, and impose the condition $j_y=0$, where $\mathbf{j}=(j_x,j_y)$ is the current density and $\mathbf{E}=(E_x,E_y)$ is the electric field. Then, we have $E_y=-(\sigma_{yx}/\sigma_{yy})E_x$ and $j_x=(\sigma_{xx}-\sigma_{xy}\sigma_{yx}/\sigma_{yy})E_x$. In high magnetic fields, the diagonal elements of the conductivity tensor is much smaller than the offdiagonal and the total current is dominated by the Hall current determined by the Hall field E_y.

A quantum analogue of the classical procedure is to consider the linear relation:

$$\begin{pmatrix} j_x(y) \\ j_y(y) \end{pmatrix} = \int dy' \begin{pmatrix} \tilde{\sigma}_{xx}(y,y') & \tilde{\sigma}_{xy}(y,y') \\ \tilde{\sigma}_{yx}(y,y') & \tilde{\sigma}_{yy}(y,y') \end{pmatrix} \begin{pmatrix} E_x(y') \\ E_x(y') \end{pmatrix}, \tag{2.2}$$

with nonlocal conductivity tensor $\tilde{\sigma}_{\mu\nu}$. Unfortunately, this procedure does not work because we usually get vanishing $\tilde{\sigma}_{yy}$ and nonvanishing $\tilde{\sigma}_{yx}$ for wires with finite width. This leads to the conclusion that the current flows in the y direction perpendicular to the wire direction.

This unphysical result is due to the fact that the current calculated in the conventional linear response theory does not represent observable quantity and can be avoided by introduction of a flux average in the cylinder geometry.[22] The flux average of $\bar{j}_y$ vanishes identically and also leads to the conclusion that the flux average of $\bar{j}_x$ is determined by σ_{xx} alone as has been given by Landauer's formula, where $\bar{j}_x$ and $\bar{j}_y$ are the average of j_x and j_y in the y direction.

It should be noted that the above does not necessarily mean that the current is carried only by edge states in actual systems. It may just show that the total current is independent of the Hall field and/or the Hall conductivity σ_{xy} in quantum wires.

Note also that the Hall polarization field due to charges accumulated spatially[23–30] has not been included in the above discussion. It is expected that this polarization field does not influence the total Hall current although it may strongly modify the local current distribution.

3. Edge Transport and Extended Bulk States

We consider both two- and four-terminal geometries, schematically illustrated in Fig. 1. For the two-terminal case, a wire with width W in the y direction and length L in the x direction is connected to an infinitely long ideal wire which is eventually connected to a reservoir. For the four-terminal geometry, we shall consider a junction of two wires with length L and width W. Ideal wires are connected at all ends.

The strength of scatterers with short-range delta-potentials is characterized by broadening Γ of the Landau level. In order to calculate transmission probability for the entire energy range of the broadened 2D Landau levels, we shall lower the bottom energy in the ideal wires by ΔE ($\Gamma < \Delta E < \hbar\omega_c - \Gamma$ with $\omega_c = eH/mc$ and H being a magnetic field) as shown in Fig. 1. The transmission probabilities are calculated using the Green function techniquebased on a lattice model[31] and the conductance can easily be given by the Büttiker-Landauer formula.[16]

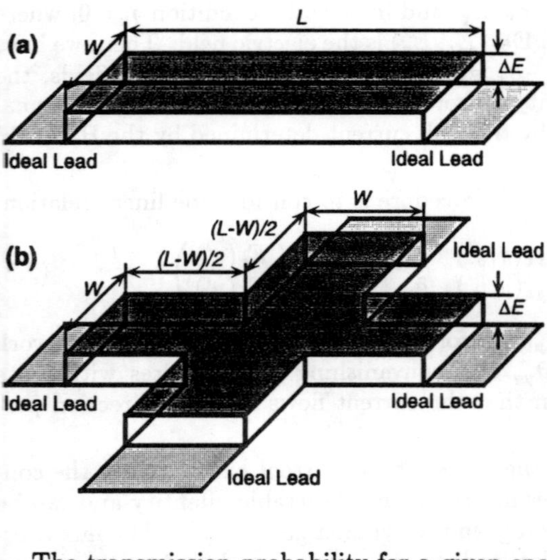

Fig. 1 Schematic illustration of a quantum wire (a) and its crossing (b) with width W and length L. The wire is confined by infinitely high barriers in one direction and connected to an infinitely long ideal wire which is eventually connected to a reservoir. The bottom energy is raised by ΔE from that in the ideal wire.

The transmission probability for a given energy is calculated by averaging over those of typically a few times 10^3 different samples with width W lying between $W/l = 5$ and 50. The broadening of the Landau level is $\Gamma/\hbar\omega_c = 0.2$ and the energy shift is chosen as $\Delta E = \hbar\omega_c/2$. The results do not depend on a slight shift in the choice of ΔE. Numerical calculations are performed for $L/W = 1/2$, 1, and 2 in the two-terminal case and for $L/W = 3$ in the crossed-wires geometry.

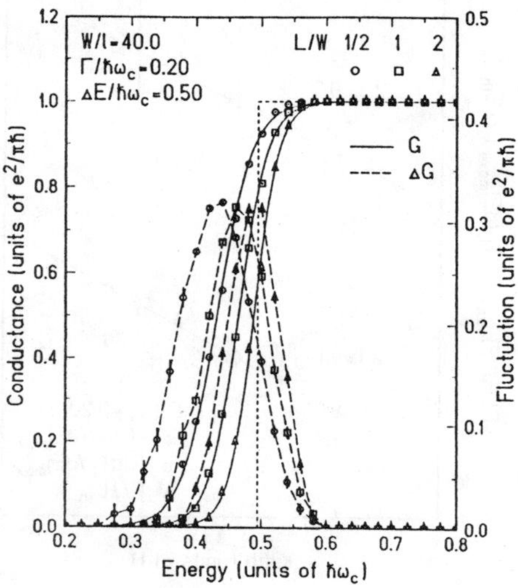

Fig. 2 Calculated conductance and its fluctuation (the open symbol) as a function of energy for the lowest Landau level. The circles denote the results for a short rectangular system ($L/W = 1/2$), squares those for a square ($L/W=1$), and triangles those for a long rectangular system ($L/W = 2$). The dotted line represents the conductance in the absence of randomness. The solid lines represent a fitted curve and the dashed lines are guides to the eye. $W/l=20$.

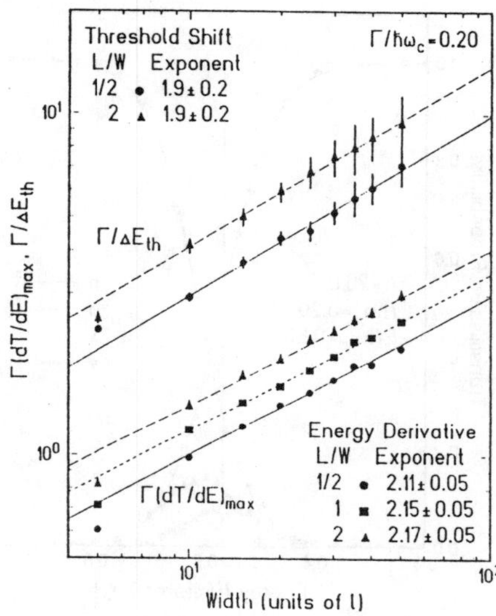

Fig. 3 Log-log plot of the maximum energy derivative of the two-terminal conductance as a function of the wire width. The straight lines represent $\Gamma(d\langle T\rangle/dE)_{\max} \propto W^{1/s}$ with exponent s listed at the bottom-right corner. The dots denoted as $\Gamma/\Delta E_{\text{th}}$ represent the threshold energy difference between $L/W = 1$ and 1/2, and the squares the difference between $L/W = 2$ and 1. The exponent defined by $\Delta E_{\text{th}} \propto W^{-1/s}$ are listed at the top-left corner.

3.1. Two-Terminal Geometry

Figure 2 gives some examples of calculated energy dependence of the averaged conductance given by $\langle G\rangle = (e^2/\pi\hbar)\langle T\rangle$ for the lowest Landau level, where $\langle\cdots\rangle$ denotes a sample average and T is the transmission probability. The dotted lines represent the transmission probability for square systems with $L/W = 1$ in the absence of randomness. It is a step function rising at the energy of the Landau level in two dimensions, which is slightly smaller than $\hbar\omega_c/2$ because of the small nonparabolicity present in the present lattice model. The energy dependence of the transition probability in the presence of randomness becomes steeper with increase of the system size.

In order to deduce the effective energy width over which the transmission probability increases from zero to unity, we fit the results to the function $\langle T(E)\rangle = \{\exp[\beta(E-E_{\text{th}})]+1\}^{-1}$. The parameter β or $\Gamma(d\langle T\rangle/dE)_{\max}$ represents the inverse of the effective energy width and E_{th} the effective threshold energy. The fitted results are shown by solid lines in Fig. 2.

Figure 3 gives the steepness parameter as a function of the width in logarithmic

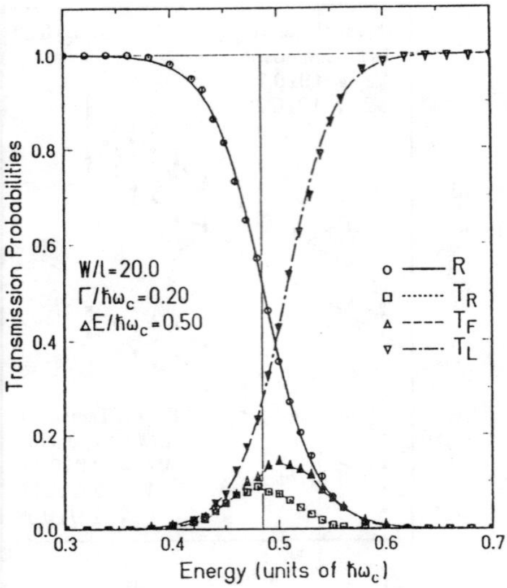

Fig. 4 Averaged reflection and transmission probabilities at wire junctions with width $W/l = 20$. The lines for $\langle R\rangle$ and $\langle T_L\rangle$ represent a fitted result.

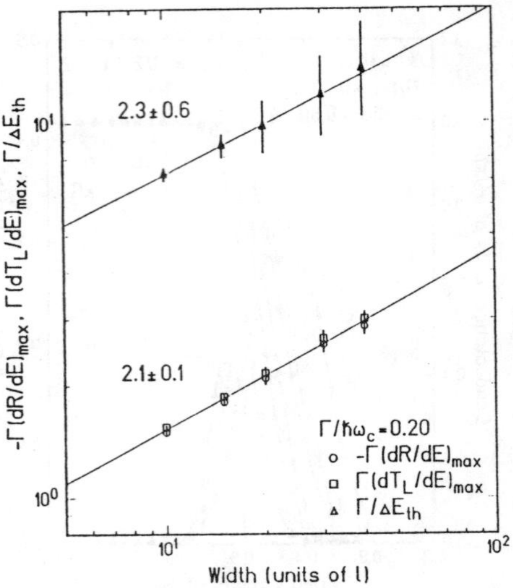

Fig. 5 Log-log plot of the maximum derivatives of the averaged reflection $\langle R\rangle$ and transmission probability $\langle T_L\rangle$ and the inverse of the difference of the threshold energy for $\langle R\rangle$ and $\langle T_L\rangle$ as a function of the wire width for crossed-quantum wires. The straight lines represent $W^{1/s}$ with critical exponent s listed.

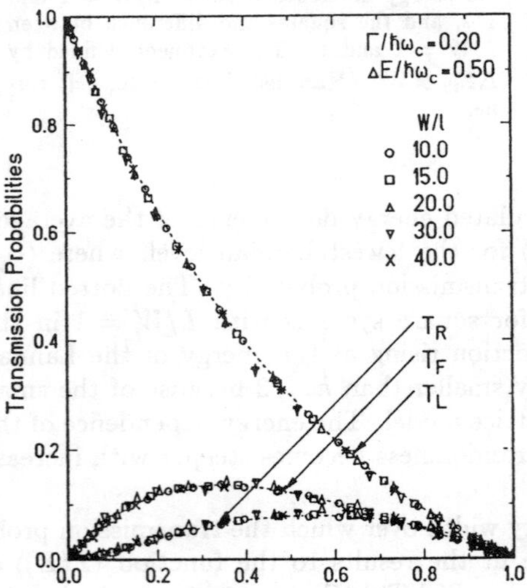

Fig. 6 Average transmission probabilities as a function of average reflection probability at wire junctions with width $W/l = 10$, 15, 20, 30, and 40 for different energies. The dotted lines are guides to the eye. It is evident that the average transmission probabilities are determined by $\langle R\rangle$ alone within the present numerical accuracy.

scales. It is clear that the results can be fitted to a straight line, except those for $W/l = 5$, showing that $\Gamma(d\langle T\rangle/dE)_{\max} \propto W^{1/s}$. The exponent is almost independent of the form of the system and is obtained as $s=2.2\pm0.1$. This exponent is consistent with that of almost all the existing numerical study discussed in the previous section because actual errors can be larger than those estimated statistically.

Figure 2 contains also the conductance fluctuation $\sqrt{\langle(G-\langle G\rangle)^2\rangle}$. It has been demonstrated that the fluctuation is a universal function of the average conductance and that the distribution function of the conductance is also universal.

3.2. Four-Terminal Geometry

In the four-terminal geometry we calculate reflection (R) and transmission probabilities, turning right (T_R), turning left (T_F), and going straight (T_F), as a function of energy for the lowest Landau level. Figure 5 shows an example for $W/l=20$. With the increase of the wire width, the transition region, where $\langle R\rangle$ and $\langle T_L\rangle$ change between zero and unity and where $\langle T_R\rangle$ and $\langle T_F\rangle$ are nonzero, becomes narrower. The effective width of this transition region can be deduced by fitting the results to $\langle R\rangle=\{\exp[-\beta_R(E-E_{\text{th}}^R)]+1\}^{-1}$ and $\langle T_L\rangle=\{\exp[\beta_T(E-E_{\text{th}}^T)]+1\}^{-1}$, with two parameters β_R and E_{th}^R for $\langle R\rangle$ and β_T and E_{th}^T for $\langle T_L\rangle$. The fitted results are shown by solid lines for $\langle R\rangle$ and by dot-dashed lines for $\langle T_L\rangle$ in Fig. 5.

Figure 4 shows the maximum derivative $-\Gamma(d\langle R\rangle/dE)_{\text{max}}$ and $\Gamma(d\langle T_L\rangle/dE)_{\text{max}}$ as a function of W together with $\Delta E_{\text{th}}^{TR}=E_{\text{th}}^T-E_{\text{th}}^R$. Within the numerical accuracy, the critical exponents obtained for the crossed-wires geometry agree with those obtained for the two-terminal case discussed in the previous section.

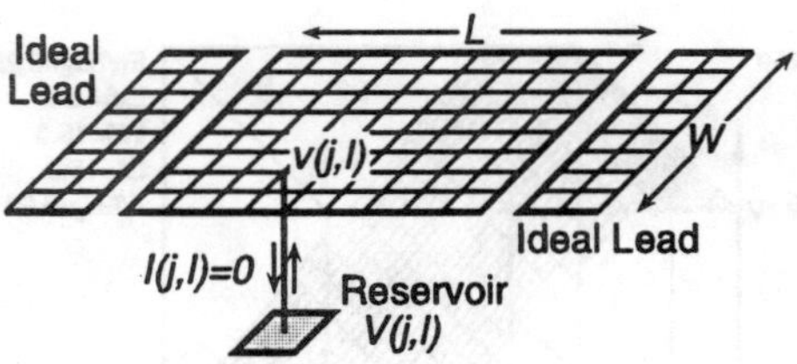

Fig. 7 A schematic illustration of a quantum wire in the presence of inelastic scattering. Each lattice point is connected to a reservoir through an ideal wire characterized by the transfer integral $-\hbar/2\tau_\phi$. The voltage of each reservoir is determined in such a way that the current along the wire should vanish.

Figure 6 gives the average transmission probabilities as a function of the average reflection probability for systems with different widths. It is evident that the average transmission probabilities are determined by $\langle R\rangle$ alone within the present numerical accuracy. The universality prevails also for the fluctuations and correlation functions although not explicitly shown here. This is consistent with a scaling relation between σ_{xx} and σ_{xy} obtained numerically for systems without a boundary having a size lying in the similar range.[35] The universality demonstrated in the above numerical study strongly suggests that $\alpha(E)$ is the only relevant parameter in the present system.

4. Inelastic Scattering and Voltage Distribution

In order to introduce inelastic scattering and calculate the voltage distribution in the system we shall extend Büttiker's method.[36–39] The system is schematically

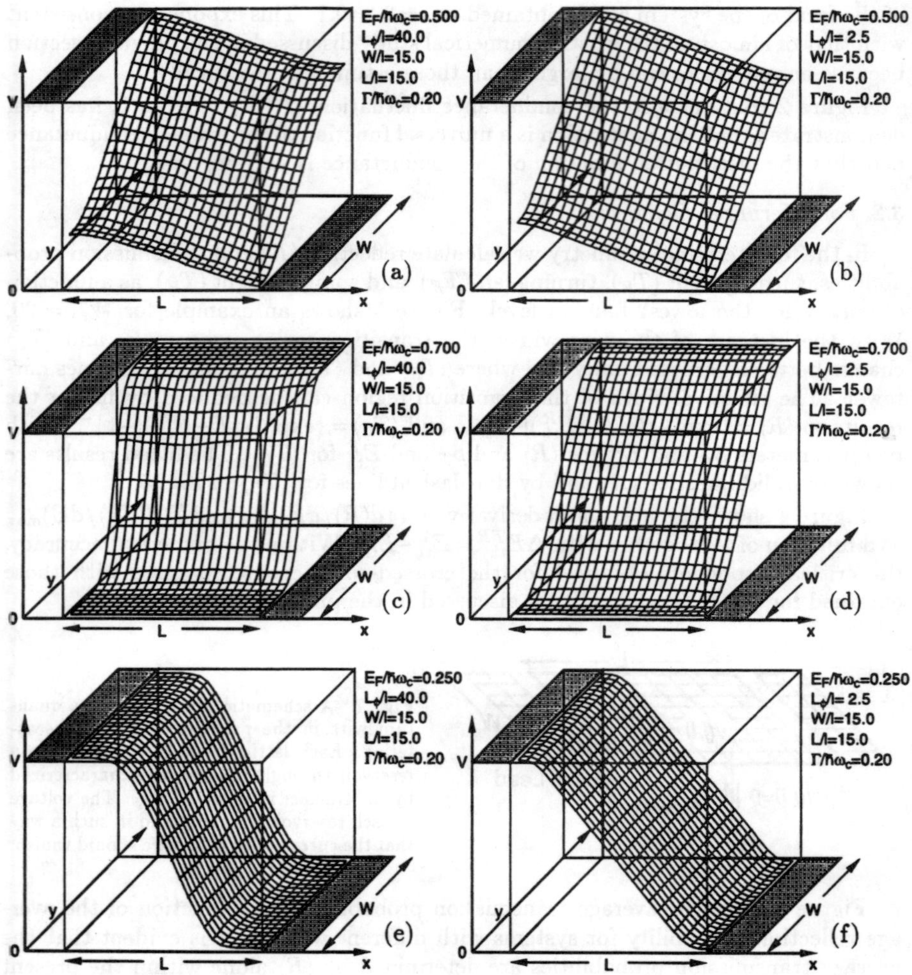

Fig. 8 Calculated voltage distribution for $E_F/\hbar\omega_c = 0.5$ [(a) and (b)], in 0.7 [(c) and (d)], and 0.25 [(e) and (f)]. $L/l = W/l = 15$ and $L_\phi/l = 40$ on the left hand side [(a), (c), and (e)] and 2.5 on the right hand side [(b), (d), and (f)], respectively.

illustrated in Fig. 7, in which a wire with a single current channel is attached and the wire is connected to a reservoir with a certain voltage. The voltage is determined by the condition that the net current passing through the wire vanishes. This voltage probe introduces inelastic scattering which can be adjusted freely by the change in the coupling of the wire to the system. The coupling is characterized by the imaginary part $-i\hbar/2\tau_\phi$ in the diagonal site energy with τ_ϕ being the inelastic scattering time, which is related to the phase coherence length L_ϕ through $L_\phi = \sqrt{2\tau_\phi D^*}$ with D^* being the diffusion coefficient given by $l^2\Gamma/\hbar$.

The two cases are considered: $L_\phi = 40l > L, W$ and $K_\phi = 2.5l < L, W$. In the former case the phase coherence length is larger than the system size and therefore the system remains coherent, while in the latter case the coherence is totally broken and the system becomes classical. Figure 8 shows some examples of calculated voltage distribution. When the Fermi level E_F lies near the center of the broadened lowest Landau level, i.e. (a) and (b), the voltage distribution is independent of the strength of inelastic scattering and is approximated well by a flat plane determined by the local conductivity tensor σ_{xx} and σ_{xy} except in the region close to the system boundaries. The situation changes drastically when the Fermi level moves away from the center.

When E_F lies in the region corresponding to the high-energy tail of the broadened Landau level, i.e. (c) and (d), the voltage exhibits a step-function-like change in the y direction in the quantum case, $L_\phi \gg L, W$, while the voltage decays linearly in the y direction in the classical case, $L_\phi \ll L, W$. In the quantum case, the width of the region where the voltage changes suddenly is highly likely to be determined by the localization length of localized bulk Landau states. In the classical case, on the other hand, the usual description based on local σ_{xx} and σ_{xy} is valid because of the presence of the Hall field almost independent of the position. The same is applicable in the case that E_F lies in the region of the low-energy tail, i.e. (e) and (f), except that the voltage change now occurs in the x direction instead of the y direction.

5. Summary and Conclusion

The numerical study of the transport in two-dimensional systems in high magnetic fields has been performed based on an edge-current picture for the purpose of understanding the relation between bulk- and edge-current pictures of the quantum Hall effect. The results are all consistent with those predicted in the bulk-current picture, including the critical exponent of localization of bulk states and the universal behavior of transport coefficients. Effects of inelastic scattering are also studied. The voltage distribution calculated shows a clear crossover from the quantum case where the phase coherence length is much larger than the system size to the classical case where the phase coherence length is much smaller and the conventional description based on local σ_{xx} and σ_{xy} is valid.

Acknowledgments

This work is supported in part by Grant-in-Aid for Scientific Research on Priority Area "Quantum Coherent Electronics: Physics and Technology" from Ministry of Education, Science and Culture, Japan.

References

1. T. Ando, *J. Phys. Soc. Jpn.* **61** (1992) 415.
2. T. Ando, *Physica B* **183** (1993) 361.
3. T. Ando, *Phys. Rev. B* **49** (1994) 4679.
4. T. Ando, *Prog. Theor. Phys. Suppl.* **84** (1985) 69.
5. *The Quantum Hall Effect*, edited by R.E. Prange and S.M. Girvin (Springer, New York, 1987).
6. H. Aoki, *Rep. Prog. Phys.* **50** (1987) 655.
7. H. Aoki, in *High Magnetic Fields in Semiconductor Physics III*, ed. G. Landwehr (Springer, Heidelberg, 1992), p. 17.
8. T. Ando, *J. Phys. Soc. Jpn.* **52** (1983) 1893; **53** (1983) 310 and 3126; *Phys. Rev. B* **40** (1989) 9965.
9. H. Aoki and T. Ando, *Phys. Rev. Lett.* **54** (1985) 831.
10. T. Ando and H. Aoki, *J. Phys. Soc. Jpn.* **54** (1985) 2238.
11. B. Huckestein and B. Kramer, *Phys. Rev. Lett.* **64** (1990) 1437; *Solid State Commun.* **71** (1989) 445.
12. B. Huckestein, *Europhys. Lett.* **20** (1992) 451
13. B. Mieck, *Eutophys. Lett.* **13** (1990) 453; *Z. Phys. B* **90** (1993) 427
14. D. Liu and S. Das Sarma, *Phys. Rev. B* **49** (1994) 2677
15. H. Levine, S.B. Libby, and A.M.M. Pruisken, *Phys. Rev. Lett.* **51** (1983) 1915; Nucl. Phys. B **240**, 30 (1984); A.M.M. Pruisken, *Phys. Rev. B* **32** (1985) 2636.
16. M. Büttiker, *Phys. Rev. Lett.* **57** (1986) 1761.
17. R. Landauer, *IBM J. Res. Dev.* **1** (1957) 223; *Philos. Mag.* **21** (1970) 863.
18. E.N. Economou and C.M. Soukoulis, *Phys. Rev. Lett.* **46** (1981) 618.
19. D.S. Fisher and P.A. Lee, *Phys. Rev. B* **23** (1981) 6851.
20. D.C. Langreth and E. Abrahams, *Phys. Rev. B* **24** (1981) 2978.
21. D.J. Thouless, *Phys. Rev. Lett.* **47** (1981) 972.
22. T. Ando and H. Aoki, *Physica B* **184** (1983) 365.
23. A.H. MacDonald, T.M. Rice, and W.F. Brinkman, *Phys. Rev. B* **28** (1983) 3648.
24. J. Riess, *J. Phys. C* (1984) 17,L847.
25. D.J. Thouless, *J. Phys. C* **18** (1985) 6211.
26. O. Heinonen and P.L. Taylor, *Phys. Rev. B* **32** (1985) 633.
27. D. Pfannkuche and J. Hajdu, *Phys. Rev. B* **46** (1992) 7032.
28. D.J. Thouless, **Phys. Rev. Lett.** (1993) 71,1879.
29. C. Wexler and D.J. Thouless, *Phys. Rev. B* **49** (1994) 4815.
30. H. Hirai and S. Komiyama, *Phys. Rev. B* **49** (1994) 14012.
31. T. Ando, *Phys. Rev. B* **44** (1991) 8017.
32. B.L. Altshuler, *Pis'ma Zh. Eksp. Theor. Fiz.* **41** (1985) 530 [*JETP Lett.* **41** (1985) 648].

33. P.A. Lee and A.D. Stone, *Phys. Rev. Lett.* **55** (1985) 1622.
34. P.A. Lee, A.D. Stone, and H. Fukuyama, *Phys. Rev. B* **35** (1987) 1039.
35. T. Ando, *J. Phys. Soc. Jpn.* **35** (1986) 3199
36. M. Büttiker, *Phys. Rev. B* **32** (1985) 1846.
37. M. Büttiker, *Phys. Rev. B* **33** (1986) 3020.
38. M. Büttiker, *IBM J. Res. Develop.* **32** (1988) 63.
39. J.L. D'Amato and H.M. Pastawski, *Phys. Rev. B* **41** (1990) 7411.

The Effect of Magnetic Field on the Edge Plasmons of Two-dimensional Electron Gas Systems

X. XIA*, X. M. CHEN*, and J. J. QUINN*†
*Department of Physics and Astronomy, University of Tennessee
Knoxville, Tennessee 37996
and
†Oak Ridge National Laboratory, Oak Ridge, Tennessee 17831

ABSTRACT

Edge magnetoplasmons of two-dimensional electron gas (2DEG) systems with density profiles that fall from their bulk value to zero in a non-abrupt manner are investigated. Higher multipole modes, which can be thought of as bulk plasmons of the low density edge region, exist as localized edge modes only for finite values of wavevector along the edge. In addition, a sequence of very low frequency edge magnetoplasmons displaying linear dispersion are found for only one direction of propagation along the edge for a given applied magnetic field. Some insight is gained from the analysis of a simple double step density profile.

In 2DEG systems with an abrupt edge, localized edge plasmons with a frequency about 90% of a bulk 2D plasmon of the same wavevector are well known[2~4]. For a non-abrupt edge profile additional edge modes[4], analogous to 3D higher multipole surface modes[5], occur. These 2D higher multipole edge modes can be thought of as bulk 2D plasmons of the low density edge region which are trapped near the edge. The presence of a perpendicular magnetic field plays an important role in determining the allowed edge mode frequencies. At the boundary between two 2DEG systems of different density Mikhailov and Volkov[6] found an inter-edge magnetoplasmon which could propagate only in one direction for a given magnetic field. For a particular non-abrupt edge electron density profile, Nazin and Shikin[7] found an infinite number of low frequency magnetoplasma modes for a classical 2DEG on the surface of liquid helium. These acoustic magnetoplasma modes were recently investigated in the high magnetic field limit by Aleiner and Glazman[8]. In this paper we study the edge magnetoplasmons of a 2DEG with a nonabrupt edge density profile. Analytic results for a simple double step density profile, which give some insight into the problem are also presented.

Consider a 2DEG confined in the x-y plane with an equilibrium density profile $n(x)$ with a magnetic field $\mathbf{B}$ along the positive z-direction. The collective electronic modes of the system correspond to self-sustaining density fluctuations of the form of $\delta n(\mathbf{r},t) = \delta n(x) exp(iqy - i\omega t)$. In the simple local approximation for the conductivity tensor, the plasmon modes are determined by requiring non-trivial solutions of the equation [9]

$$\phi(x)=\frac{4\pi e^2}{m_e\epsilon(\omega^2-\omega_c^2)}\int dx' L_q(x-x')[n(x')(q^2-\frac{d^2}{d(x')^2})-\frac{dn(x')}{dx'}(\frac{d}{dx'}-\frac{\omega_c}{\omega}q)]\phi(x'), \tag{1}$$

for the electric potential, where the integration kernel $L_q(x)=\frac{1}{4\pi}\int dp\ \exp(ipx)/(p^2+q^2)^{1/2}$, and ω_c $(=eB/m_ec)$ is the electron cyclotron frequency. Here m_e is the electron mass, and ϵ is the background dielectric constant.

For an electron density profile described by $n(x)=n_B\Theta(-x-a)+n_S\Theta(-x)\Theta(x+a)$, where Θ is the single step function, we can solve Eq. (1) analytically using the Wiener-Hopf method if L_q is replaced by an approximate kernel $L_q(x)=2^{-3/2}exp(-\sqrt{2}|qx|)$. The plasmon modes are found to be solutions of

$$(\alpha_S-\alpha_B)\cosh(\gamma_S qa)+(\alpha_S\alpha_B-1)\sinh(\gamma_S qa)=0, \tag{2}$$

where $\gamma_\nu^2=2[\omega_\nu^2-(\omega^2-\omega_c^2)]/[2\omega_\nu^2-(\omega^2-\omega_c^2)]$ with $\nu=B$ or S, $\alpha_B=\{\gamma_B[2\omega_B^2-(\omega^2-\omega_c^2)]-2(\omega_B^2-\omega_S^2)\omega_c/\omega\}/\gamma_S[2\omega_S^2-(\omega^2-\omega_c^2)]$, and $\alpha_S=(\sqrt{2}(\omega^2-\omega_c^2)+2\omega_S^2\omega_c/\omega)/\gamma_S[2\omega_S^2-(\omega^2-\omega_c^2)]$, and $\omega_\nu^2=2\pi n_\nu e^2q/\epsilon m$ is the bulk 2D plasmon frequency for a system with electron density n_ν. In the limiting case of $n_S=n_B$ or $a=0$, Eq. (2) reduces to the expected abrupt single step result[2] . On the other hand, when $a\to\infty$ and $q\neq 0$, Eq. (2) reduces to $(\alpha_B+1)(\alpha_S-1)=0$. Setting the first factor equal to zero gives abrupt single step result with ω_B replaced by ω_S. This is just the edge magnetoplasmon of the lower density edge region. Setting the second factor to zero yields

$$\gamma_B(2\omega_B^2-\omega^2+\omega_c^2)+\gamma_S(2\omega_S^2-\omega^2+\omega_c^2)=2(\omega_B^2-\omega_S^2)\frac{\omega_c}{\omega}. \tag{3}$$

Eq. (3) determines the frequency of the inter-edge plasmon[3] localized near the edge $x=-a$. In other words, as expected the system sustains two well-defined magnetoplasma modes, an edge mode of the lower density edge region and a well separated inter-edge magnetoplasmon localized at the inner edge $x=-a$. The inter-edge mode propagating in the positive y-direction exists only for a finite magnetic field along the positive z-direction, in agreement with the result of Mikhailov and Volkov[6].

In the absence of an applied magnetic field the dispersion relation for the higher multipole edge modes is approximately $\omega_n^2=(2\pi n_S e^2/\epsilon m)[q^2+(n\pi/a)^2]^{1/2}$ in the limit of $n_B\gg n_S$ and $qa\gg 1$, where $n=1,2,3,\ldots$. It is clear from this result that physically the these edge modes are standing wave plasmons with an integral number of half wavelengths in the x-direction fitting into the low electron density edge region. Only for $q>q_n^*\sim(n\pi/a)n_S(n_B^2-n_S^2)^{-1/2}$ will these modes exist as well defined edge excitations.

For an arbitrary edge electron density profile Eq. (1) cannot be solved analytically. To facilitate numerical calculations we transform the integral equation into a matrix equation by expanding the potential $\phi(x)$ in Laguerre polynomials, $\phi(x)=exp(|q|x)\sum_{n=0}^{\infty}c_nL_n(-2|q|x)$. We find that if the edge density profile takes the form of a polynomial in x, the computation is simplified considerably. The plasmon dispersion relations are obtained by requiring that non-trivial solutions exist for the matrix equation truncated at a finite order N. We find that the numerical results converge quickly with increasing N, and that $N \simeq 20$ gives the desired numerical accuracy for many density profiles.

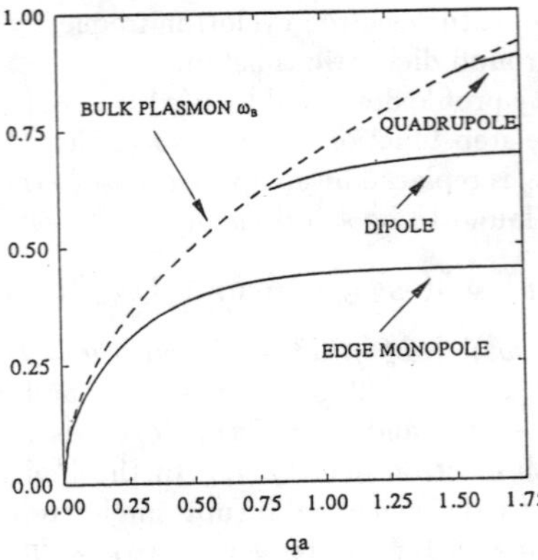

Fig. 1. Dispersion of edge plasmon modes for an electron density profile decreasing linearly (see text) near the edge. Here a is the width of the edge region and $\Omega = (4\pi n_0 e^2/\epsilon m a)^{1/2}$. The 2D bulk plasmon mode with an electron density n_0 is shown as the dashed line.

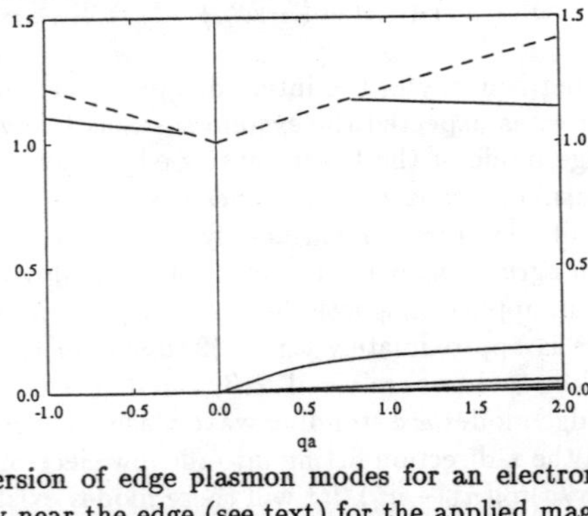

Fig. 2. Dispersion of edge plasmon modes for an electron density profile decreasing smoothly near the edge (see text) for the applied magnetic field along the positive z-direction. Here $|\omega_c| = \Omega$, $\Omega = (4\pi n_0 e^2/\epsilon m a)^{1/2}$. The 2D bulk plasmon dispersion with an electron density n_0 is shown as the dashed line.

We have studied several models for the electron density profile in which the density drops from its bulk value to zero smoothly. Although the details may differ a little, the basic features remain the same. In addition to the regular edge monopole, a sequence of higher multipole modes emerges from the bulk plasmon as well-defined edge modes at finite values of the wavevector q. In Fig. 1 we plot for $B = 0$, the plasmon dispersion for the density profile $n(x)/n_0 = 0$ for $x > 0$, $= -x/a$ for $0 > x > -a$, and $= 1$ for $x < -a$. The figure is unchanged for any value of a when the unit of frequency is taken as $\Omega = (4\pi n_0 e^2/\epsilon m a)^{1/2}$, so that the nonlocal effect can be neglected for sufficiently large value of a. When q is small the system can support only the monopole mode. When the value of q increases additional higher multipole modes begin to appear below the bulk plasma frequency.

In the presence of a perpendicular magnetic field, the edge magnetoplasmon modes form two distinct groups. The modes in the high frequency (greater than ω_c) are the higher multipole modes trapped in the low density edge region. The low frequency gapless group exists only for propagation along the positive y-direction; they appear to be infinite in number if damping is ignored. The usual monopole mode is one of this group. The modes in the low frequency group are all acoustic modes in the long wavelength limit. Furthermore, at a given q, as the magnetic field increases, the frequency of the monopole mode decreases while that of the other modes increases. In Fig. 2, we plot the edge magnetoplasmon dispersion for the density profile $n(x)/n_0 = 0$, for $x > 0$, $= 2(x/a)^2 + 2(x/a)^3$, for $0 > x > -a$, and $= 1$, for $x < -a$. From our numerical study, we find that the number of nodes in x-direction for the electron density fluctuation associated with the high frequency excitations increase by unity with each successive higher mode. Surprisingly, the density variation associated with the low frequency group exhibit the same behavior with each successive lower frequency mode. In other words, with similar density variations, we find two charge density excitations with entirely different frequencies for a fixed q. The reason for this appears to be the highly frequency dependent nature of the magneto-conductivity tensor. The very different response in the high and low frequency regions appears to give rise to two modes with similar spatial variation and quite different frequency.

This research was supported in part by the Division of Materials Sciences of the U. S. Department of Energy under Contract No. DE-AC05-84OR21400 with Martin Marietta Energy Systems, Inc.

REFERENCES

[1] F. Stern, *Phys. Rev. Lett.* **18**, (1967) 546.
[2] D. B. Mast, A. J. Dahm, and A. L. Fetter, *Phys. Rev. Lett.* **54**, (1985) 1706; D. C. Glattli *et al.*, *Phys. Rev. Lett.* **54**, (1985) 1710; J.-W. Wu, P. Hawrylak, and J. J. Quinn, *Phys. Rev. Lett.* **55**, (1985) 879; A. L. Fetter, *Phys. Rev.* **B33**, (1986) 3717; J. J. Quinn, *Solid State Comm.* **84**, (1992) 139,
[3] V. A. Volkov and S. A. Mikhailov, *JETP Lett.* **42**, (1986) 556; *JETP* **67**, (1989) 1639.
[4] X. Xia and J. J. Quinn, *Phys. Rev.* B, to be published.
[5] S. S. Nazin and V. B. Shikin, *JETP* **67**, (1988) 288.
[6] I. L. Aleiner and L. I. Glazman, *Phys. Rev. Lett.*, **72** (1994) 2935.
[7] K.-D. Tsuei *et al.*, *Phys. Rev. Lett.* **64** (1990) 44.
[8] V. A. Mikhailov and V. A. Volkov, *J. Phys.: Condensed Matter* **4** (1992) 6523;
[9] X. Xia and J. J. Quinn, unpublished

Two-channel Resonant Tunneling between Edge States in a Quantum Hall Sample

M. E. RAIKH and T. V. SHAHBAZYAN

Department of Physics, University of Utah,
Salt Lake City, Utah 84112

Abstract

We study two-channel resonant tunneling between the edge states in the quantum Hall geometry. We show that for the perfect resonant conditions the coupling between the edge states is non-zero only within a narrow interval of magnetic fields.

Tunneling between edge states of 2D electron gas in a strong magnetic field is widely studied since for narrow samples this is the process responsible the dissipation and, hence, for a finite value of ρ_{xx}. Due to the exponentially small probability of direct tunneling, the resonant tunneling processes involving localized states (LS) in the bulk can strongly influence the coupling between the edge states. The resonant tunneling through a single LS was first considered by Jain and Kivelson.[1]

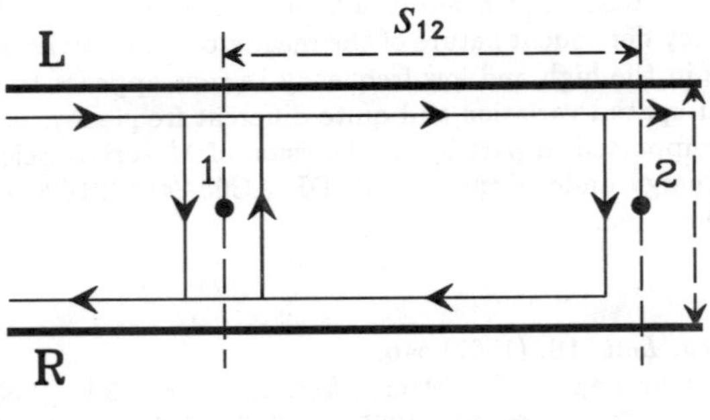

Figure 1: Schematic representation of the tunneling junction with two resonant impurities (top view).

In this paper we study the resonant tunneling through a two parallel LS, 1 and 2, separated by a distance s_{12} along the edges (see Fig. 1). We assume $s_{12} \gg D$, D being the sample width, so that each LS is coupled to the both edges while the direct overlap between LS is vanishingly small. The specifics of the process we study compared to a single-channel resonant tunneling[1] is the following. First, an electron propagating, say, in the left edge state can be resonantly transmitted to the right edge via either LS 1 or LS 2. As a result, the reflection coefficient R is a periodic function of the magnetic flux through the area $s_{12}D$. More importantly, the electron resonantly transmitted through the LS 2 from the left to the right edge can be then transmitted back to the left through the LS 1. This means that a single act of two-channel resonant tunneling involves multiple circulations of the electron within the loop formed by the two LS and edges. As a result, the reflection coefficient instead of the usual Breit-Wigner form $R = \Gamma^2/(\mathcal{E}^2 + \Gamma^2)$ takes the form

$$R = \frac{4\mathcal{E}^2\Gamma^2 \cos^2\theta}{[\mathcal{E}^2 + \Gamma^2(1+\cos\theta)][\mathcal{E}^2 + \Gamma^2(1-\cos\theta)]}, \tag{1}$$

where $\theta = s_{12}D/2l^2$ (l stands for the magnetic length), $\mathcal{E}$ is the deviation of the Fermi energy from the resonance and Γ is the width associated with the coupling of the LS to the edges. The dependences of R on $\mathcal{E}$ and θ are shown in Fig. 2.

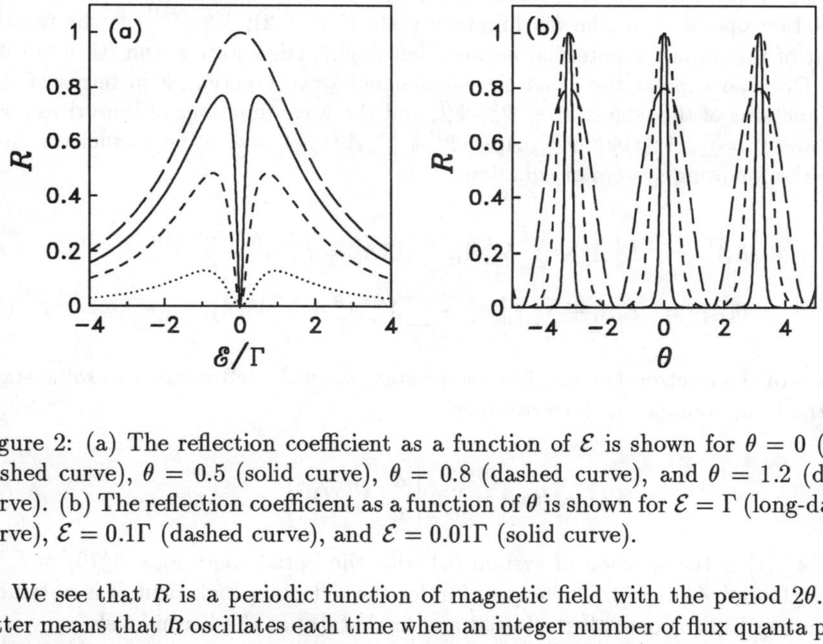

Figure 2: (a) The reflection coefficient as a function of $\mathcal{E}$ is shown for $\theta = 0$ (long-dashed curve), $\theta = 0.5$ (solid curve), $\theta = 0.8$ (dashed curve), and $\theta = 1.2$ (dotted curve). (b) The reflection coefficient as a function of θ is shown for $\mathcal{E} = \Gamma$ (long-dashed curve), $\mathcal{E} = 0.1\Gamma$ (dashed curve), and $\mathcal{E} = 0.01\Gamma$ (solid curve).

We see that R is a periodic function of magnetic field with the period 2θ. The latter means that R oscillates each time when an integer number of flux quanta passes through the area $s_{12}D$. For $\theta = \pi n$ (n being an integer) the reflection coefficient reaches its maximum value $R = 1$ at exact resonance, $\mathcal{E} = 0$, while for $\theta = \pi/2 + \pi n$ we have $R = 0$ in the entire region of $\mathcal{E}$. For all other θ the reflection coefficient as a function of $\mathcal{E}$ exhibits a deep minimum (antiresonance). The width of this antiresonance, $\Gamma(1 - \cos\theta)$, depends in oscillatory way on magnetic field. It can be seen that at small $\mathcal{E}$ there are wide intervals in magnetic field within which the reflection coefficient is precisely *zero* ($R = 0$). The latter means that the quantum Hall plateau in this interval is undisturbed. One should emphasize that the reflection coefficient is extremely sensitive to the deviation $\mathcal{E}$ of the Fermi energy from the resonance. Namely, at $\mathcal{E} = 0$ the periodic peaks in $R(\theta)$ have zero width, while for $\mathcal{E} \sim \Gamma$ the widths of the minima and the maxima are of the same order (see Fig. 2b).

In the rest of the paper we derive the expression (1) for the reflection coefficient R following the approach of Larkin and Matveev.[2] We adopt the following Hamiltonian of the tunneling junction

$$\begin{aligned} H = \quad & \sum_\nu E_\nu^L \mathbf{C}_{L\nu}^\dagger \mathbf{C}_{L\nu} + \sum_\nu E_\nu^R \mathbf{C}_{R\nu}^\dagger \mathbf{C}_{R\nu} + \sum_i E_i \mathbf{C}_i^\dagger \mathbf{C}_i \\ & + \sum_{\nu,i} (V_{\nu i}^L \mathbf{C}_{L\nu}^\dagger \mathbf{C}_i + V_{i\nu}^L \mathbf{C}_i^\dagger \mathbf{C}_{L\nu}) + \sum_{\nu,i} (V_{\nu i}^R \mathbf{C}_{R\nu}^\dagger \mathbf{C}_i + V_{i\nu}^R \mathbf{C}_i^\dagger \mathbf{C}_{R\nu}), \end{aligned} \tag{2}$$

where E_ν^L (E_ν^R), $\mathbf{C}_{L\nu}^\dagger$ ($\mathbf{C}_{R\nu}^\dagger$), and $\mathbf{C}_{L\nu}$ ($\mathbf{C}_{R\nu}$) are the energy, creation, and annihilation operators of the left (right) edge state ν; E_i, $\mathbf{C}_i^\dagger$, and $\mathbf{C}_i$ are the energy, creation, and annihilation operators of the ith impurity state ($i = 1, 2$); $V_{\nu i}^L$ ($V_{\nu i}^R$) is the matrix element of the impurity potential between left (right) edge state ν and ith impurity state. One can express the exact time-dependent wave function Ψ in terms of the wave functions of the edge states, Ψ_ν^L, Ψ_ν^R, and the wave functions of impurities, Ψ_i, as follows $\Psi = \sum_\nu A_\nu^L(t)\Psi_\nu^L + \sum_\nu A_\nu^R(t)\Psi_\nu^R + \sum_i A_i(t)\Psi_i$, where the coefficients $A(t)$ satisfy the following system of equations

$$\begin{aligned} i\hbar \dot{A}_\nu^L &= E_\nu^L A_\nu^L + \sum_i V_{\nu i}^L A_i, \qquad i\hbar \dot{A}_\nu^R = E_\nu^R A_\nu^R + \sum_i V_{\nu i}^R A_i, \\ i\hbar \dot{A}_i &= E_i A_i + \sum_\nu V_{i\nu}^L A_\nu^L + \sum_\nu V_{i\nu}^R A_\nu^R + \sum_{j\neq i} V_{ij} A_j. \end{aligned} \tag{3}$$

The rate of the electron transfer from some state ν_0 in the left contact to some state μ_0 in the right contact can be written as

$$W(L\nu_0|R\mu_0) = \lim_{t\to\infty} \left[\frac{d}{dt}|A_{\mu_0}^R(t)|^2\right], \tag{4}$$

where $A_{\mu_0}^R(t)$ is the solution of system (4) with the initial conditions $A_\nu^L(0) = \delta_{\nu\nu_0}$, $A_\nu^R(0) = 0$, and $A_i(0) = 0$. The expression for the reflection coefficient R is obtained by summing Eq. (4) over all the initial and final states at the Fermi level

$$R = 2\pi\hbar \sum_{\mu,\nu} W(L\nu|R\mu)\delta(E_F - E_\nu^L). \tag{5}$$

The solution of system (3) can be easily found. With this solution, the expression (4) for the rate takes the usual "golden rule" form

$$W(L\nu|R\mu) = \frac{2\pi}{\hbar}|T_{\nu\mu}|^2\delta(E_\nu^L - E_\mu^R), \tag{6}$$

where $T_{\nu\mu}$ is the amplitude of transition from state ν in the left contact to state μ in the right contact. It can be presented in the following form

$$T_{\nu\mu} = V_{\nu 1}^L S_{11} V_{1\mu}^R + V_{\nu 1}^L S_{12} V_{2\mu}^R + V_{\nu 2}^L S_{21} V_{1\mu}^R + V_{\nu 2}^L S_{22} V_{2\mu}^R = \sum_{ij} V_{\nu i}^L S_{ij} V_{j\mu}^R. \qquad (7)$$

We see that the amplitude (7) represents a sum of four terms. The first term describes a transition from the left edge state ν to the impurity state 1 and, then, the transition from the impurity state 1 to the right edge state μ; the second term describes a transition from the left edge state ν to the impurity state 1 and, then, the transition from the impurity state 2 to the right edge state μ, *etc.* The matrix elements S_{ij}, which are defined below, can be viewed as propagators of an electron in the state i (S_{11} and S_{22}) and between the states i and j (S_{12} and S_{21}). These propagators describe all the virtual transitions between the impurity and edge states.

Substituting Eq. (7) into Eq. (5), we obtain

$$R = \mathrm{Tr}(\hat{\Gamma}^L \hat{S} \hat{\Gamma}^R \hat{S}^\dagger), \qquad (8)$$

where $\hat{\Gamma}^{L,R}$ and $\hat{S}$ are 2×2 matrices, defined as

$$\Gamma_{ij}^{L,R} = 2\pi \sum_{\nu} V_{i\nu}^{L,R} V_{\nu j}^{L,R} \delta(E_F - E_\nu^L), \qquad \hat{S} = \left[E_F - \hat{E} + \frac{i}{2}(\hat{\Gamma}^L + \hat{\Gamma}^R)\right]^{-1}, \qquad (9)$$

where we also introduced an energy matrix $E_{ij} = E_i \delta_{ij}$. Chosing the y-axis along the edges the wave function of the edge state in the Landau gauge represents a plane wave $e^{ip_y y}$, where $p_y = x_0/l^2$, x_0 being the x-coordinate of the center of cyclotron orbit. Then it can be easily shown that the diagonal and non-diagonal elements of $\hat{\Gamma}$ differ only by a phase factor. Namely,

$$\hat{\Gamma}^L = \Gamma \begin{pmatrix} 1 & e^{i\theta} \\ e^{-i\theta} & 1 \end{pmatrix}, \qquad \hat{\Gamma}^R = \Gamma \begin{pmatrix} 1 & e^{-i\theta} \\ e^{i\theta} & 1 \end{pmatrix}, \qquad (10)$$

where the phase $\theta = s_{12}D/2l^2$ has a meaning of a number of the magnetic flux quanta through the area $s_{12}D/2$. Note that $\hat{\Gamma}^L$ and $\hat{\Gamma}^R$ are complex conjugates. The appearance of a phase factors $e^{i\theta}$ in Eq. (10) is a consequence of a "chiral" structure of the edge states. Substituting Eq. (10) into Eq. (9) and then into Eq. (8) yields Eq. (1).

1. J. K. Jain and S. A. Kivelson, Phys. Rev. Lett. **60**, 1542 (1988).
2. A. I. Larkin and K. A. Matveev, Zh. Eksp. Teor. Fiz. **93**, 1030 (1987) [Sov. Phys. JETP **66**, 580 (1987)].

Magnetic domains in spin-polarized edge channels

Lex Rijkels and Gerrit E.W. Bauer

Department of Applied Physics
Delft Institute for Microelectronics and Submicrontechnology (DIMES)
Delft University of Technology
Lorentzweg 1, 2628 CJ Delft, The Netherlands

Abstract

The magnetic configuration of edge channels near the boundary of a quasi-two-dimensional electron gas in the integer quantum Hall regime is studied in the Hartree-Fock approximation. At a critical non-equilibrium occupation of spin levels a reversal of edge channel positions at the Fermi energy is predicted. The crossings which arise when edge channels switch position in open systems are identified as soliton-like, one-dimensional magnetic domain walls. The interaction between these magnetic defects is investigated.

In mesoscopic structures of the quasi-two-dimensional electron gas the concept of edge channels[1,2] can be used to describe magnetotransport along the boundaries in the integer quantum Hall regime. Recently, the influence of electron-electron interactions on the ground state of edge channels has been studied in the Hartree-Fock approximation,[3] in which single-particle wave functions have an integral occupation at zero temperature. The direct interactions cause a spin-polarizing transition of the outermost edge states, whereas the exchange interaction is believed to stabilize the spin-polarized, integer filling configuration relative to states with a fractional filling. It is noted that the validity of the Hartree-Fock approximation in the integer quantum Hall regime is still controversial.[4,5]

In the present paper we extend the work of Dempsey *et al.*[3] on the magnetic configuration of edge channels to situations where spin-states of a Landau level have a non-equilibrium occupation, taking th finite width of the quasi-two-dimensional electron gas into account.[6] We follow Ref. 3 in separating the Hartree energy into spatially slowly and rapidly varying contributions. In the absence of microscopic information on sample geometry, gate potential and the distribution of screening charges, the external confinement $V_c^{\text{eff}}(x)$ is linearized and treated as an adjustable electric field at the edges. Assuming that the exchange sufficiently favors integral filling, we concentrate on spin-polarization of the lowest Landau level, and solve the $\nu_{\text{bulk}} = 2$ case for zero temperature.

The edge channel separation at the Fermi energy ΔX^* as a function of the scaled electric field $\alpha = a_1\ell/(\frac{e^2}{\epsilon\ell})$ is displayed in Fig. 1 for a strictly two-dimensional and a quasi-two-dimensional system, both with and without a bare Zeeman splitting. The values have been calculated for $B = 7.2$ T, corresponding to a zero field density of 3.5×10^{11} cm^{-2}. In GaAs-Al$_x$Ga$_{1-x}$As the static dielectric constant $\epsilon = 12.5$, and the bare Landé factor $|g| = 0.44$. The inset depicts the self-consistent single-particle

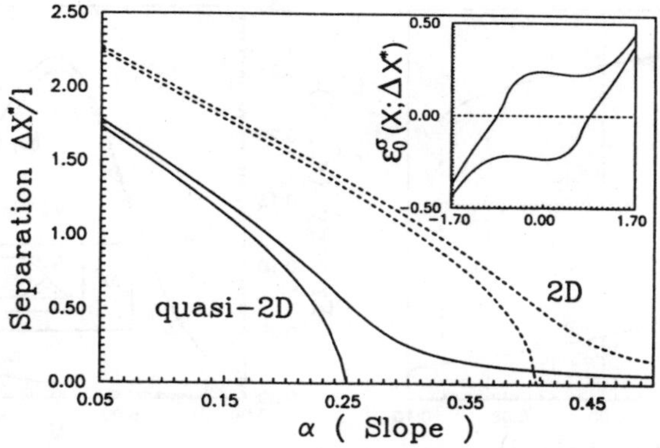

Figure 1: Equilibrium separation $\Delta X^*/\ell$ of edge states as function of the scaled slope $\alpha = a_1\ell/(\frac{e^2}{\epsilon\ell})$ of the confining potential $V_c^{\rm eff}$, with and without Zeeman energy for $B = 7.2$ T. Results are shown for a strictly two-dimensional (dashed lines), and a quasi-two-dimensional (solid lines) system. The upper curves of each set contain the Zeeman energy. Inset: self-consistent single-particle energies $\varepsilon_0^\sigma(X;\Delta X^*)$ for $\Delta X^* \approx 2\ell$ (solid lines); Fermi level (dashed line).

energies for a value of α corresponding to $\Delta X^* \approx 2\ell$. Quantitatively, the results for the quasi-2D systems differ from those of the strictly 2D systems obtained earlier by Dempsey *et al.*.[3] The Coulomb interaction form factor significantly weakens the effect of the interaction.

When the spin-up and spin-down levels are populated up to different chemical potentials $\mu_\uparrow = \bar{\mu} - \Delta\mu$, and $\mu_\downarrow = \bar{\mu} + \Delta\mu$, the separation ΔX^* continuously decreases with increasing $\Delta\mu$. At a critical value $\Delta\mu_{cr}^+ > 0$, the edge channels switch position and the absolute value of the separation increases. When the applied bias is reduced again, an energy barrier between the edge channels prohibits the relaxation of the system, until a value $\Delta\mu_{cr}^-$ is reached. In Fig. 2 the separation ΔX^* is plotted versus $2\Delta\mu$, for $B = 7.2$ T and $\alpha = 0.05$. The value of $\Delta\mu_{cr}^+$ is of the order $\sim 8\varepsilon_Z$. The edge channels switch position at values $2\Delta\mu$ of about $\frac{1}{6}$ of the enhanced energy difference, which is again caused by the direct, and not the exchange interaction. A chemical potential difference can be realized by irradiation of a quantum dot with polarized light or by applying a voltage difference to the contacts of an open system.[7] Here it is assumed that the small amount of scattering between the spin channels leaves the value of $\Delta\mu$ unaltered over the length of their trajectories through the wire. This is not unreasonable, considering the long spin-flip relaxation times,[7] but it will limit the length of the channels over which the potential difference can be maintained.

Channel switching in an open system causes the topological problem of how to connect the channels to the contacts. Relative to the original configuration, a strip

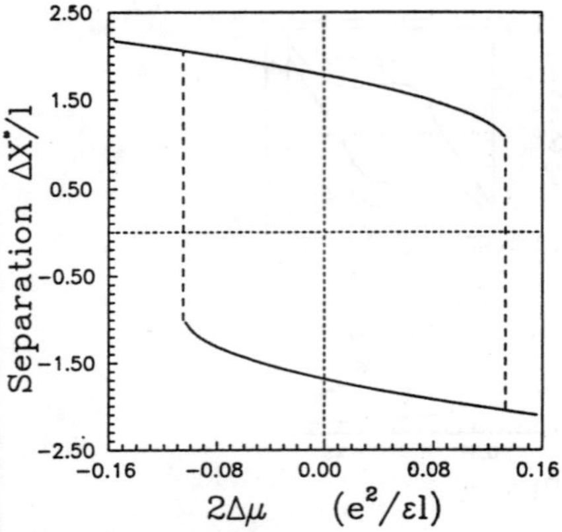

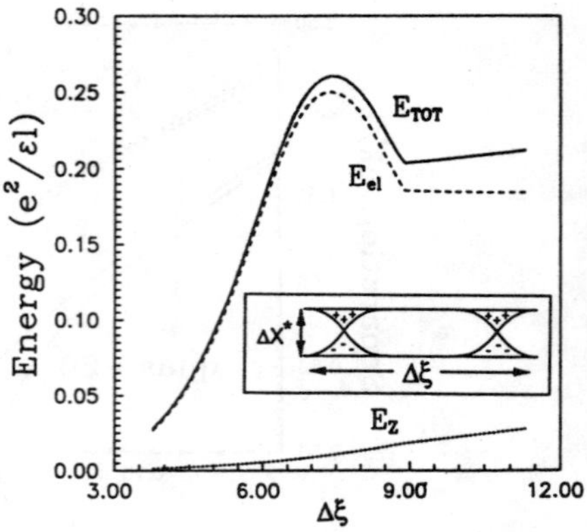

Figure 2: (Left) Equilibrium separation $\Delta X^\star/\ell$ of edge states as function of a chemical potential difference $2\Delta\mu$ between the spin-up and spin-down levels (units $\frac{e^2}{\epsilon\ell}$) for $B = 7.2$ T and $\alpha = 0.05$. Negative $\Delta X^\star$ correspond to reversed channel positions.

Figure 3: (Right) Energy (units $\frac{e^2}{\epsilon\ell}$) of interacting defects as function of their mutual distance $\Delta\xi$ for an equilibrium separation of edge channels $\Delta X^\star = 2\ell$. The energy E_{TOT} consists of electrostatic (E_{el}, dashed) and Zeeman (E_Z, dotted) energy ($B =$ 7.2 T). The inset shows the dipolar charge at a crossing.

with a reversed spin direction is formed. However, the contacts itself are supposed to be at thermal equilibrium with normal ordering of spin-states, which means that the channels have to switch again somewhere. The magnetic regions are separated by non-dispersive defects, corresponding to soliton-like "Bloch points" between one-dimensional, magnetic domains. The defect and anti-defect repel eachother as long as the chemical potential difference is maintained to be larger than the critical one. When the bias is removed, the Zeeman energy difference between these domains exerts an attractive force between the defect-anti-defect pair. Between these extremes the interaction can be freely tuned. The annihilation of the defects in the attractive regime can be prevented by impurity pinning or the short-range electrostatic repulsion due to the dipolar charges (see below). We have estimated the electrostatic energy of the dipolar charges defined by the crossing edge channels for a strictly two-dimensional system.[6] In Fig. 3 this energy is plotted as function of the distance $\Delta\xi$ between the defects, for an equilibrium separation of edge channels $\Delta X^\star = 2\ell$ and a minimum radius of curvature $r_c^{\min} = 2\ell$, beyond which the kinetic energy strongly increases. The results depend only weakly on the precise values of these parameters. An order of magnitude estimate of the lifetime τ of a double-defect ("bipolaron") system without a chemical potential difference indicates $\tau \sim 10^{-10}$ s. The electrostatic repulsion is not strong enough to suppress the recombination of the two defects for an appreciable time. The reader is reminded, however, that an applied bias does provide for a long-range repulsion, which can be tuned to create a system of two defects with only a short-range repulsive interaction and a long recombination lifetime.

At a crossing, enhanced scattering between spin channels is expected.[8] Although this forward scattering will not influence the total conductance, the amount of current per edge channel will be affected. The relaxation of a non-equilibrium current distribution depends exponentially on the distance between the channels. In contrast to the non-interacting picture, the present results show that the separation in typical samples is always appreciable, except at the singular defects. This implies a low scattering rate at all applied voltages, which is suddenly enhanced when a crossing appears. Selective detection of the contribution per edge state to the conductance is possible by the use of quantum point contacts as voltage probes.

An intriguing possibility is the spin-blockade of transport through a quantum dot connected to the leads by tunnel-barriers. In a system which contains two Landau levels, tunneling occurs selectively from the reservoirs to the outer spin level.[9] When the polarization of the spin levels inside the dot is reversed, electrons have to undergo a spin-flip in addition to the tunneling process, in order to be able to pass the barrier, which blocks transport. In a quantum dot containing 300 electrons, reversal of the polarization requires a spin-flip of about six electrons, which may be achieved by a pulse of polarized microwave radiation. Relaxation of the system out of the metastable spin-blockade state is an example of "macroscopic quantum tunneling of spin", which is expected to be slow.

The authors wish to thank A. (Sasha) V. Khaetskii and Theo Stoof for many helpful discussions. This work is part of the research program of the "Stichting Fundamenteel Onderzoek der Materie (FOM)"

---

1. B.I. Halperin, Phys. Rev. B **25**, 2185 (1982).
2. See for instance: C.W.J. Beenakker and H. van Houten, Solid State Phys. **44**, 1 (1991), and references therein.
3. J. Dempsey, B.Y. Gelfand, and B.I. Halperin, Phys. Rev. Lett. **70**, 3639 (1993); B.Y. Gelfand and B.I. Halperin, Phys. Rev. B **49**, 1862 (1994).
4. D.B. Chklovskii, B.I. Shklovskii, and L.I. Glazman, Phys. Rev. B **46**, 4026 (1992); D.B. Chklovskii, K.A. Matveev, and B.I. Shklovskii, *ibid.* **47**, 12605 (1993); A.V. Khaetskii, V. Fal'ko, and G.E.W. Bauer, *ibid.*, in press.
5. C. de C. Chamon and X.G. Wen, Phys. Rev. B **49**, 8227 (1994); T.H. Stoof and G.E.W. Bauer, these Proceedings.
6. A. Rijkels, Diploma Thesis, TU Delft, 1994; A. Rijkels and G.E.W. Bauer, Phys. Rev. B, in press.
7. G. Müller *et al.*, Phys. Rev. B **45**, 3932 (1992).
8. A.V. Khaetskii, Phys. Rev. B **45**, 13777 (1992).
9. N.C. van der Vaart *et al.*, to be published.

The Influence of Exchange on Edge States in the Integer Quantum Hall Effect Regime

T.H. Stoof and G.E.W. Bauer

Faculty of Applied Physics and DIMES, Delft University of Technology, 2628 CJ Delft, The Netherlands

The competition between the exchange and the direct Coulomb interaction near the edge of a strip of two-dimensional electron gas (2DEG) in a strong magnetic field is discussed using density-functional theory and a local approximation for the exchange energy functional. We find that exchange strongly reduces the spatial extent of the compressible strips as compared to a purely electrostatic model, which has important consequences for the transport properties in the integer quantum Hall regime.

In recent years a large effort has been devoted to incorporate the effects of electron-electron interactions into the description of edge states,[1,2] both in the integer [3–5] and fractional [6–8] quantum Hall regime.

At present, two apparently incompatible pictures for the electronic ground state of edge channels in the integer quantum Hall effect regime exist. In the Hartree-Fock approximation the ground state wave function is a single Slater determinant, which corresponds to occupation numbers zero and one, [3] *i.e.* the edge state is incompressible and the electron density drops rapidly at the edges. An electrostatic description, however, favors a more gradual electron density variation which is achieved by fractional occupation at the edges,[4] corresponding to compressible edge states. Although the Hartree-Fock approach has been challenged on the grounds that it does not take global electrostatics into account properly[4] it has also been pointed out[9] that a purely electrostatic model of a parabolically confined quantum dot in high magnetic field seriously underestimates the stability of the ground state.

In the present work we interpolate between the different regimes of applicability of both theories by extending the Thomas-Fermi approach of Ref. 4. This comes down to a Thomas-Fermi-Dirac like treatment of the exchange effects in strong magnetic fields. Recently the transition between abruptly en smoothly varying density distributions has been studied by Chamon and Wen.[5] They predict a formation of compressible edge states when the confinement potential is decreased beyond a certain point.

Our system is a quantum wire of the strictly two-dimensional electron gas along the y axis in the xy plane with a uniform positive background charge to ensure global neutrality. The wire is confined in the x-direction by a parabolic confinement potential $V_c(x) = \alpha \frac{e^2}{\epsilon l^3} x^2$, where α is a dimensionless parameter. We use periodic boundary conditions in the y direction and adopt the Landau gauge so that $\mathbf{A}_0 = B_0 x\hat{\mathbf{y}}$ and $\mathbf{B}_0 = B_0\hat{\mathbf{z}}$.

We use density-functional theory in strong magnetic fields[10] to find the ground state of the system. In general, the total energy of the system is a functional of the density $n(\mathbf{r})$, the spin density $\mathbf{s}(\mathbf{r})$ and paramagnetic current density $\mathbf{j}_p(\mathbf{r})$. The energy functional contains a small kinetic part which is disregarded in our treatment. We also disregard all correlation effects, *e.g.* fractional quantization of the Hall conductance, and consider only the exchange energy functional. The latter depends on density and spin density only and reads in the local approximation:[11] $E_x[n(\mathbf{r}), \mathbf{s}(\mathbf{r})] = \int d^2r\ n(\mathbf{r})\ \epsilon_x[n(\mathbf{r}), \mathbf{s}(\mathbf{r})]$, where ϵ_x is the exchange energy per electron of a uniform electron gas with density $n(\mathbf{r})$ and spin density $\mathbf{s}(\mathbf{r})$. We write the total-energy functional as:[12]

$$\begin{aligned} E[n^\uparrow(x), n^\downarrow(x)] &= \sum_\sigma \int_{-\infty}^{\infty} dx\ n^\sigma(x) \left\{ V_c(x) + \frac{1}{2} V_H[n^\uparrow(x), n^\downarrow(x)] + \epsilon_x^\sigma[n^\sigma(x)] \right\} \\ &+ \frac{ge\hbar}{2mc} B_0 \int_{-\infty}^{\infty} dx\ \{n_\uparrow(x) - n_\downarrow(x)\}. \end{aligned} \quad (1)$$

where g is the Landé factor and $\sigma = \uparrow, \downarrow$ denotes up or down spin respectively. The Hartree electrostatic potential is:

$$V_H(x) = -2\, \frac{e^2\, l}{\epsilon} \int_{-\infty}^{\infty} dx'\ n(x')\ \ln \left| \frac{x - x'}{l} \right|, \quad (2)$$

where the electron density $n = n^\uparrow + n^\downarrow$ is the sum of the densities of the spin up and spin down electrons and $l = \sqrt{\hbar c / eB}$ is the magnetic length. The one-particle exchange energy of the homogenous 2DEG in the absence of a magnetic field is proportional to $\sqrt{n}$.[13] However, in the presence of a strong magnetic field it is given by:

$$\epsilon_x^\sigma(x) = -\sqrt{2\pi^3}\, \frac{e^2\, l}{\epsilon}\, n^\sigma(x). \quad (3)$$

The solution which minimizes the total energy now follows from:

$$\frac{\delta E[n^\uparrow(x), n^\downarrow(x)]}{\delta n^\uparrow(x)} = \frac{\delta E[n^\uparrow(x), n^\downarrow(x)]}{\delta n^\downarrow(x)} = \mu, \quad (4)$$

where μ is the chemical potential.

We consider the high magnetic field limit, where the lowest Landau level is occupied and mixing with higher Landau levels is negligible. We use the filling factor ν as the variational function from which the density is obtained by a convolution with the square of the wave functions

$$n(x) = \frac{1}{2l^3\sqrt{\pi^3}} \int_{-\infty}^{\infty} dX\ \nu(X)\ e^{-(\frac{x-X}{l})^2}. \quad (5)$$

We numerically minimize the energy with respect to the filling factor, while keeping the total number of electrons constant. With our approach we can obtain both the electrostatic solution and, by forcing the filling factor to be an integer, the Hartree-Fock solution obtained by Dempsey, Gelfand and Halperin[3]. We therefore expect

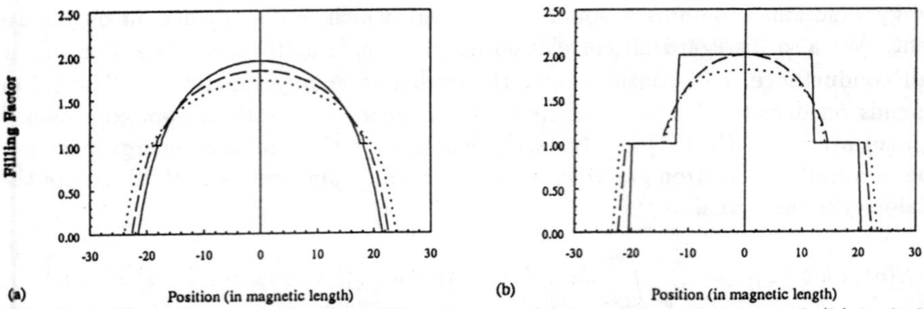

Figure 1: Electronic ground state filling factors, (a) ignoring exchange and (b) including exchange, for confinement strengths $\alpha = 0.045$ (solid line), 0.040 (dashed line) and 0.035 (dotted line).

our method to give a good description of the intermediate regime between these two extremes. In Fig. 1 ground state filling factors are plotted as a function of the strength of the confinement. Fig. 1(a) shows purely electrostatic solutions whereas in Fig. 1(b) solutions with exchange taken into account are plotted. The effect of exchange in favoring integer filling and reducing the extent of the compressible regions is obvious from the figure. The small incompressible regions occuring in the electrostatic solutions are due to the Zeeman splitting.

We have also carried out calculations for a system where a constant potential at a distance d from the 2DEG is assumed, which may represent a real back gate or a plane of not fully ionized donors parallel to the 2DEG. This gate causes the potential of a mirror charge distribution of opposite sign located a distance $2d$ from the 2DEG:

$$V_{bg}(x) = +2\,\frac{e^2 l}{\epsilon} \int_{-\infty}^{\infty} dx'\; n(x')\; \ln\sqrt{\frac{(x-x')^2\; +\; (2d)^2}{l^2}}. \tag{6}$$

In Fig. 2. solutions that minimize the total energy of the system including the back

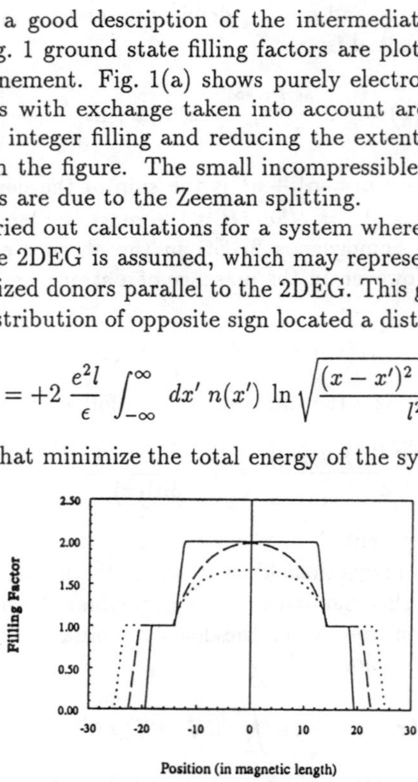

Figure 2: Electronic occupation of the system in the presence of a back gate at a distance $d = \infty$ (dotted line), $d = 20l$ (dashed line) and $d = 10l$ (solid line).

gate are plotted as a function of the distance of the gate to the 2DEG, while keeping the number of electrons and the confinement potential fixed. From Fig. 2 it can be seen that moving the gate closer to the 2DEG has the effect of reducing the width of the compressible regions. This can be understood by noting that the gate screens the long-range Coulomb interaction in the 2DEG, thus effectively reducing the importance of the direct interaction relative to the exchange interaction.

It is clear from Figs. 1(b) and 2 that the width of the incompressible region is of the same order of magnitude as that of the outermost compressible strip, even for relatively wide strips like the ones we consider (approximately 4000Å for a magnetic field of 5 T). This in contrast to the electrostatic description of edge states given in Fig. 1(a) where, for wide strips, the width of the incompressible region is always much smaller than that of the compressible one. However, for soft confinement potentials, the solution including exchange does not deviate dramatically from the purely electrostatic one, which implies that in this regime the Hartree-Fock approach breaks down.[5] Only a strong confinement potential can cause the edge channels to become incompressible, although in Fig. 1(b) it is still not strong enough to destroy spin splitting altogether. However, in all cases exchange causes the formation of relatively wide incompressible strips which strongly reduces the back scattering of spin split edge channels and enhances the equilibration length.

The authors thank Sasha Khaetskii, Yuli Nazarov and Lex Rijkels for many helpful discussions and A.H. MacDonald for a reprint of Ref. 9. This work is part of the research program of the "Stichting Fundamenteel Onderzoek der Materie (FOM)".

---

1. B.I. Halperin, Phys. Rev. B **25**, 2185 (1982).
2. See, e.g., C.W.J. Beenakker and H. van Houten, in *Solid State Physics*, edited by H. Ehrenreich and D. Turnbull, (Academic, New York, 1991), Vol. 44, and references therein.
3. J. Dempsey, B.Y. Gelfand, and B.I. Halperin, Phys. Rev. Lett. **70**, 3639 (1993); B.Y. Gelfand and B.I. Halperin, Phys. Rev. B **49**, 1862 (1994).
4. D.B. Chklovskii, B.I. Shklovskii, and L.I. Glazman, Phys. Rev. B **46**, 4026 (1992); D.B. Chklovskii, K.A. Matveev, and B.I. Shklovskii, *ibid.* **47**, 12605 (1993)
5. C. de C. Chamon and X.G. Wen, Phys. Rev. B **49**, 8227 (1994).
6. C.W.J. Beenakker, Phys. Rev. Lett. **64**, 216 (1990).
7. A.H. MacDonald, Phys. Rev. Lett. **64**, 220 (1990).
8. X.G. Wen, Phys. Rev. Lett. **64**, 2206 (1990).
9. A.H. MacDonald, S.R. Eric Yang, and M.D. Johnson, Aust. J. Phys. **46**, 345 (1993).
10. G. Vignale and M. Rasolt, Phys. Rev. Lett. **59**, 2360, (1987); Phys. Rev. B **37**, 10685, (1988).
11. G. Vignale and P. Skudlarski, Phys. Rev. B **46**, 10232 (1992).
12. T.H. Stoof and G.E.W. Bauer, unpublished.
13. F. Stern, Phys. Rev. Lett. **30**, 278 (1973).

Electrostatics of Inter - Landau - Level Diodes

A.V. Khaetskii [1)], V.I. Fal'ko [2)] and G.E.W. Bauer [1)]
1) Faculty of Applied Physics and DIMES, Delft University of Technology, 2628 CJ Delft, The Netherlands ,
2) Max-Planck-Institut für Festkörperforschung, 70569 Stuttgart, Germany

Abstract

We present the exact solution of the electrostatic problem of the incompletely depleted two-dimensional electron gas in a heterostructure covered by a half-infinite gate. Using this solution we calculate configurations of compressible and incompressible strips which are formed in the presence of a quantizing magnetic field. We discuss the application of our results to transport experiments on the inter-Landau-level and spin-diodes, as well as edge-magnetoplasmon propagation.

As it is known the chemical potential of a *homogeneous* two-dimensional electron gas (2DEG) in a quantizing magnetic field experiences a jump each time when the density n reaches the value at which an integer number of Landau levels are filled. As a result, the details of the screening are highly dependent on the filling factor $\nu = n/n_L (n_L = 1/2\pi\lambda^2, \lambda$ being the magnetic length) and the *inhomogeneous* electronic system is split into compressible and incompressible regions.[1,2]

In experiments, the inhomogeneity is frequently created by applying a negative potential to a gate, the electron gas under the gate being partially depleted. In this case, far from the gate edge we deal with half-infinite compressible (incompressible) regions separated by a transition region . This transition region nearby the gate edge is the subject of the present study. Its structure plays a crucial role in the formation of an electron transport in the inter - Landau - level and spin - diodes [3,4] and may be important for the quantitative description of the edge-magnetoplasmons which propagate along the border of two 2DEG's with different densities. [5,6]

1 Electrostatics of the incompletely depleted 2DEG

The structure we consider below is shown in Fig.1 (a). The inhomogeneity in the 2DEG is created by applying a negative voltage, $-V_g$, to the metal half-plane serving as a gate and lying on top of the heterostructure. The half-space, $z < 0$, is occupied by a semiconductor with dielectric constant $\epsilon \gg 1$. The 2DEG plane is buried at a distance d below the semiconductor surface. We assume that, at $V_g = 0$, the electron concentration is uniform and equal to n_0. The boundary conditions at $z = 0, x < 0$ (uncovered surface of a heterostructure) are:

$$\frac{d\varphi}{dz}|_{z=+0} = \epsilon\frac{d\varphi}{dz}|_{z=-0}, \; \frac{d\varphi}{dx}|_{z=+0} = \frac{d\varphi}{dx}|_{z=-0}. \quad (1)$$

Since $\epsilon \gg 1$, we replace Eq. (1) by the simplified one, condition $d\varphi/dz|_{z=-0} = 0$ (for $x < 0$). This makes it easier to apply methods of complex analysis and to obtain

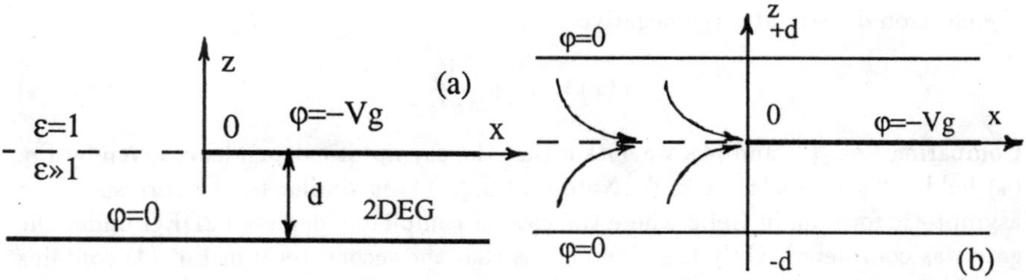

Figure 1: (a) Geometry of the device considered.(b) Plane of the complex variable $\xi = x + iz$. Curves with arrows show the direction of the electric field lines.

an exact analytical solution of the electrostatic problem. (Below we will discuss the character of the solution obtained beyond this approximation).

Consideration of the electrostatic potential alone is not sufficient for solving the problem of splitting of the electronic system into compressible and incompressible regions. The jump $\Delta\mu$ in the chemical potential at some filling factors should be taken into account. But the problem is appreciably simplified by the fact that $eV_g \gg \Delta\mu$. For this reason we can find a zero-order electron density distribution neglecting the internal energy μ of the 2DEG and then improve the solution using perturbation theory with respect to the small parameter μ/eV_g. Thus, we first solve the purely electrostatic problem, when everywhere in the 2DEG plane $\mu(n) \equiv 0$.

The potential distribution $\varphi(x,z)$ is obtained from the solution of the Laplace equation $\nabla^2\varphi = 0$ in the strip $-d \le z \le 0$ with the following boundary conditions:

$$\varphi(x,-d) = 0,\ \varphi(x \ge 0,\ 0) = -V_g,\ \frac{d\varphi}{dz}|_{z=-0}(x<0) = 0. \tag{2}$$

This problem is equivalent to searching for the symmetric solution of the Laplace equation in the strip of a width $2d$ (see Fig. 1(b)) with the boundary conditions $\varphi(x,\pm d) = 0$ and $\varphi(x>0, z=0) = -V_g$. Using methods of analytical functions we got exact solution of this problem and found the main approximation for the density profile in the system:

$$n(x) = n_0 - \frac{CV_g}{e}\frac{1}{\sqrt{1+\exp(-\pi x/d)}}, \tag{3}$$

where $C = \frac{\epsilon}{4\pi d}$ is the capacitance per unit area and n_0 is the electron density at $V_g = 0$. Note that, due to the assumed incomplete depletion of the 2DEG under the gate, Eq. (3) differs from the corresponding result by Chklovskii *et al.* [2]: the variation of the 2D density is localized within the width $\sim d$, whereas far from the gate edge the density variation is exponentially small. (The last statement, strictly speaking, is true only for the region under the gate, as discussed below).

Solution (3) was obtained in the lowest order in small parameter $1/\epsilon$. Account of the terms of the next order in this parameter [7] changes the asymptotic behaviour of the electron density at large negative x :

$$n(x) = n_0 + \frac{V_g}{4\pi^2 e x} \tag{4}$$

Comparing Eqs. (3) and (4), we obtain that the asymptotic dependence given by Eq. (4) holds at $|x| \gg d \ln \epsilon, x < 0$. Note that Eq. (4) is similar to the corresponding asymptotic formula in Ref.2 where the case of completely depleted 2DEG under the gate was considered. Only the difference is that the second term in Eq. (4) contains an additional small factor $1/\epsilon$ compared to the corresponding term in Ref.2. Thus, at large distances to the left of the gate edge, the characteristic scale of the potential felt by the electrons is V_g/ϵ rather than V_g. As it follows from the foregoing considerations this is due to the incomplete depletion of the 2DEG under the gate (the electric field can penetrate into the 2DEG plane only from above).

2 Incompressible strips formation in high magnetic field

Now we take into account the influence of the internal energy by perturbation theory. [2] Since eV_g is much larger than the energy gap $\Delta\mu$, the magnetic field does not change the electrostatic density profile significantly. In the case of compressible regions on both sides of the gate edge (for instance, $\nu(-\infty) = 5/2$, $\nu(+\infty) = 3/2$), the only effect of the internal energy is the formation of a narrow strip of incompressible phase near point x_0, where the electron density corresponds to the filling factor $\nu = 2$. Within this strip, the electron concentration is constant and equal to $2n_L$ and the width W_{inc} of the strip can be found following Chklovskii *et al.* [2]:

$$W_{inc}^2 = \frac{4\epsilon\hbar\omega_c}{\pi^2 e^2 (dn/dx)|_{x=x_0}}. \tag{5}$$

The derivative $dn(x)/dx$ of function (3) is taken at point x_0 , and Eq. (5) was derived for the limit of $W/d \ll 1$. This is, indeed, the case if the values of the filling factor at $x \to \pm\infty$ are not anomalously close to integer ones. On the other hand, W can be varied easily with gate voltage up to $W \sim d$.

The experimental situation where regions of incompressible liquid exist on both sides of the gate edge (at $x \to \pm\infty$) is also frequently encountered, for instance, $\nu = 2$ under the gate and $\nu = 4$ far from the gate. In the region under the gate for the boundary x_R between the compressible and incompressible regions we got :

$$x_R = +\frac{d}{\pi} \ln(2\frac{eV_g}{\Delta\mu_R}) \tag{6}$$

where $\Delta\mu_R$ is the magnitude of the jump of the chemical potential at an integer value of the filling factor under the gate.

Concerning the left side of the structure, the details of incompressible strip formation in this region depend strongly on the value of the quantity $\Delta\nu(-\infty) \equiv \nu(-\infty) - 4$

(for the case considered). We discuss only the case when this quantity, being positive, tends to zero and $\Delta\nu(-\infty) \ll (\epsilon \ln \epsilon)^{-1}$. It means that the point x_0 where filling factor equals 4 lies to the left of the crossover point $\sim d \ln \epsilon$ and the variation of electron density is described by Eq.(4). Thus we have for x_0:

$$x_0 = -\frac{V_g}{4\pi^2 e n_L \Delta\nu(-\infty)} \tag{7}$$

and a distance between the gate edge and the nearest edge of the incompressible strip x_L is slightly smaller, but of the same order [7].

The above calculated value of W_{inc} (see Eq.(5)) can be measured experimentally, [3] since the incompressible region is insulating and at low temperatures the current in the transverse direction is caused by electron tunneling. At a high magnetic field the tunneling is mainly determined by the effective magnetic barrier. When $W_{inc} \gg \lambda$, the tunneling probability is exponentially small. Hence, the value of W_{inc} can be derived from experiments with logarithmic accuracy when a formula for the conductance is available. The foregoing reasons can be applied to the case when the jump of the chemical potential occurs at $\nu = 1$.[4] The incompressible state corresponding to $\nu = 1$ is assumed to be spin-polarized and the elementary charged excitation is an electron with an opposite orientation of a spin. The value of the jump in the chemical potential for this case can be experimentally verified with the use of Eqs. (3,5), where in Eq. (5) $\Delta\mu$ should be substituted for $\hbar\omega_c$.

The propagation of edge magneto-plasmons along the border of a two 2DEG [5,6] is another physical effect which is sensitive to the structure of a transitional region. The experimental conditions for observation of edge magneto-plasma excitations are mainly restricted to the integer filling factors in both (left and right) semi-infinite planes which provides minimal friction and slowest relaxations. Therefore, it deals with the second of the examples considered above (Section 2.). The width of the compressible region, in such a case, determines the width of a strip where the magneto-plasmon charge can be accumulated, and therefore is an important quantity determining the dispersion of this mode.

Acknowledgement. We thank D. Chklovslkii, R. Gerhardts, I. Grodnensky, Yu. Nazarov, B. Shklovskii and especially B. Kane for useful discussions.

---

1. C.W.J. Beenakker, Phys. Rev. Lett. **64**, 216 (1990); A.M. Chang, Solid State Commun. **74**, 871 (1990);A.L. Efros, Solid State Commun. **67**, 1019 (1988); Phys.Rev.B **45**, 11354 (1992).
2. D.B. Chklovskii, B.I. Shklovskii and L.I. Glazman, Phys. Rev. B **46**, 4026 (1992)
3. B.E. Kane, D.C. Tsui and G. Weimann, Phys. Rev. Lett. **61**, 1123 (1988)
4. B.E. Kane, L.N. Pfeiffer and K.W. West, Phys. Rev. B **46**, 7264 (1992)
5. S.A. Mikhailov and V.A. Volkov, Journ. Phys. Cond. Matter **4**, 6523 (1992)
6. V.I.Talyanskii *et al.*, Journ. Phys.Cond.Matter **5**, 7643 (1993)
7. A.V.Khaetskii,V.I.Fal'ko and G.E.W.Bauer, Phys. Rev. B **50**, N 31 (1994).

CURRENT INDUCED POTENTIAL AND CHARGE DISTRIBUTION IN A TWO DIMENSIONAL ELECTRON SYSTEM IN A MAGNETIC FIELD

HANS NIELSEN

Niels Bohr Institute, Ørsted Laboratory
Universitetsparken 5, DK-2100 Copenhagen Ø, Denmark

ABSTRACT

It is shown that under steady state conditions a current may induce a constant electric field in a two dimensional electron system (2DES) in a magnetic field. This happens when the conductivities are independent of space coordinates. If, however, there is a local position-dependent conductivity tensor then equal positive and negative charge densities are induced in the system. Outside the 2DES the electric field disappears. This is possible because the 2DES becomes distorted and a quadrupole density arises.

Introduction

In the usual 3 dimensional Hall effect a current induces positive and negative surface charge densities on the confining sides. The sample then acts as a parallel plate capacitor and the electric field $\vec{E}$ is constant. To observe the quantum Hall effect (QHE) it is, however, necessary to use a 2 dimensional electron system (2DES). The corresponding capacitor is then assumed to be confined by two long strips with line charge densities of opposite sign. In this case the field $\vec{E}$ from one of the strips is assumed to vary as $\frac{1}{r}$ where r is the distance to the strip. The result is a logarithmic potential[1]. We will, however, show that under steady state conditions $\vec{E}$ is constant in the bulk region if there is a local relation $\vec{J}(\vec{r}) = \bar{\bar{\sigma}}\vec{E}(\vec{r})$ where the conductivity tensor $\bar{\bar{\sigma}}$ is independent of $\vec{r}$ in bulk. $\vec{J}(\vec{r})$ is the local current density. It is essential for the argument that also the diagonal components of $\bar{\bar{\sigma}}$ are non-zero. Because of hopping conductivity this will always be the case for T ≠ 0.

Considerations on a long Hall bar

For a long Hall bar with the applied current in the y-direction and the constant magnetic field $\vec{B}$ in the z-direction we have the following equations for the relation between $\vec{J}$ and $\vec{E}$ in the 2DES.

$$J_x(x,y) = \sigma_{xx}E_x - \sigma_{yx}E_y$$

$$J_y(x,y) = \sigma_{yx}E_x + \sigma_{xx}E_y \qquad \text{Eq. (1)}$$

where $\sigma_{yy} = \sigma_{xx}$ *and* $\sigma_{yx} = -\sigma_{xy}$ (Onsager relations) have been used.

For σ_{ik} independent of (x,y) we get by differentiation and addition:

$$\frac{\partial J_x}{\partial x} + \frac{\partial J_y}{\partial y} = \sigma_{xx}\left(\frac{\partial E_x}{\partial x} + \frac{\partial E_y}{\partial y}\right) - \sigma_{yx}\left(\frac{\partial E_y}{\partial x} - \frac{\partial E_x}{\partial y}\right)$$

Because of the continuity equation $\frac{\partial J_i}{\partial x_i} = 0$ *and* $(\nabla \times \vec{E})_z = -\frac{\partial B_z}{\partial t} = 0$ we then have the Laplace equation $\frac{\partial E_x}{\partial x} + \frac{\partial E_y}{\partial y} = 0$. We consider a system confined in the x-direction. Under steady state conditions $J_x(x,y) = 0$ must hold. Otherwise an increasingly accumulation of charge at the edges of the 2DES will appear. This gives the condition $E_y = \frac{\sigma_{xx}}{\sigma_{yx}} E_x$ from (1). By differentiation with respect to x and y we get two further equations for the 4 components $\frac{\partial E_i}{\partial x_k}$. All together we get 4 linear independent equations:

$$\frac{\partial E_x}{\partial y} - \frac{\partial E_y}{\partial x} = 0, \quad \frac{\partial E_x}{\partial x} + \frac{\partial E_y}{\partial y} = 0,$$

$$\frac{\sigma_{xx}}{\sigma_{yx}} \frac{\partial E_x}{\partial x} - \frac{\partial E_y}{\partial x} = 0 \; and \; \frac{\sigma_{xx}}{\sigma_{yx}} \frac{\partial E_x}{\partial y} - \frac{\partial E_y}{\partial y} = 0.$$

Obviously the only solution is that $\frac{\partial E_i}{\partial x_k} = 0$ for all 4 components. Thus E_x and E_y are constants. The current induced potential φ is then

$$\varphi(x,y) = -\rho_{xx} J_y \cdot y - \rho_{yx} J_y \cdot x \qquad \text{Eq. (2)}$$

where the components of the resistivity tensor $\rho_{xx} = \frac{\sigma_{xx}}{\sigma_{xx}^2+\sigma_{yx}^2}$ *and* $\rho_{yx} = \frac{\sigma_{yx}}{\sigma_{xx}^2+\sigma_{yx}^2}$ are introduced.

The expected line charge densities at the edges has so far not been introduced. It can be done by assuming that σ_{xx} and σ_{yx} varies in the transverse direction especially near the edges of the 2DES. Then from (1) we get $E_x = \frac{\sigma_{yx}(x)}{\sigma_{xx}(x)} E_y$, $J_y(x) = \sigma_{yx}E_x + \sigma_{xx}E_y$. This gives $J_y(x) = \left(\sigma_{yx}\frac{\sigma_{yx}}{\sigma_{xx}} + \sigma_{xx}\right)E_y = \rho_{xx}^{-1}\left(-\frac{\partial\varphi}{\partial y}\right)$. In the resulting equation $-\frac{\partial\varphi}{\partial y} = \rho_{xx}(x)J_y(x)$ the right hand side is independent of y. Integration gives $\varphi = -\rho_{xx}(x)J_y(x)y - g(x)$ where the unknown function $g(x)$ is found from the relation between E_x and E_y, which gives

$$-\frac{\partial\varphi}{\partial x} = y \cdot \frac{\partial}{\partial x}\left(\rho_{xx}(x)J_y(x)\right) + \frac{dg}{dx} = \frac{\sigma_{yx}(x)}{\sigma_{xx}(x)} \cdot \left(\rho_{xx}(x)J_y(x)\right) \text{ where } E_y = \rho_{xx}(x)\,J_y(x)$$

has been used. Also here the right hand side is independent of y and the factor to y on the left side must be zero. Therefore $\frac{\partial}{\partial x}\left(\rho_{xx}(x)J_y(x)\right) = 0$ which gives

$$\rho_{xx}J_y(x) = \rho_{xx}J_y = constant\ (=E_y) \; and \; g(x) = \rho_{xx}J_y \cdot \int_{x_0}^{x} \frac{\sigma_{yx}(x')}{\sigma_{xx}(x')}\,dx'.$$

The resulting potential and field is

$$\varphi = -\rho_{xx}J_y \cdot \left(y + \int_{x_0}^{x} \frac{\sigma_{yx}(x')}{\sigma_{xx}(x')}\, dx' \right)$$

$$E_x(x) = \rho_{xx}J_y \frac{\sigma_{yx}(x)}{\sigma_{xx}(x)}, \quad E_y = \rho_{xx}J_y = constant \qquad \text{Eq. (3)}$$

From symmetry it follows that $\frac{\partial}{\partial x}\left(\frac{\sigma_{yx}(x)}{\sigma_{xx}(x)}\right)$ has opposite signs at the left and the right edge of the 2DES. The Poisson equation applied on (3) gives the current induced charge density $e \cdot n^i(x)$.

$$\frac{\partial E_x}{\partial x} + \frac{\partial E_y}{\partial y} = \rho_{xx}J_y \frac{\partial}{\partial x}\left(\frac{\sigma_{yx}(x)}{\sigma_{xx}(x)}\right) = \frac{en^i(x)}{\epsilon_b} \qquad \text{Eq. (4)}$$

where $\epsilon_b = \kappa\epsilon_0$ is the background dielectric constant ($\epsilon_b \sim 11\ \epsilon_0$ *for GaAs*). From (4) we have two charge densities with opposite signs at the edges. However, from (4) it seems that $en^i(x) = \epsilon_b\rho_{xx}J_y\frac{\partial}{\partial x}\left(\frac{\sigma_{yx}(x)}{\sigma_{xx}(x)}\right)$ is independent of z. That would correspond to surface charge densities and give a trivial explanation of the constant $\vec{E}$ in bulk. The system is simply a plate capacitor. However, $n^i(x)$ must disappear outside the quantum well (QW) where the 2DES is embedded. It would cost a large energy if the wave function for the electrons were allowed to expand into the potential walls confining the 2DES in the z-direction. Therefore $n^i \sim |\zeta(z)|^2 \to 0$ outside the QW. Here $\zeta(z)$ is the z-dependent part of the electron wavefunction which has the form $\psi(x,y) \cdot \zeta(z)$. With this correction the current induced potential becomes

$$\phi(x,y,z) = -\rho_{xx}J_y \cdot |\zeta(z)|^2 \cdot \left(y + \int_{x_0}^{x} \frac{\sigma_{yx}(x')}{\sigma_{xx}(x')}\, dx' \right) \qquad \text{Eq. (5)}$$

where $\int |\zeta(z)|^2\, dz = 1$ *and* $\zeta(z) = 0$ outside the quantum well. As a simple example we may consider the rectangular well where

$$\zeta^2(z) = \frac{2}{b}\sin^2\frac{\pi z}{b} \quad 0 \le z \le b$$

$$\zeta(z) = 0 \ \ for\ z < 0\ and\ z < b \qquad \text{Eq. (6)}$$

The triangular well at the interface between AlGaAs and GaAs is more complicated because the width of this well depends on the electron density. It will not be treated here.

The distorted 2DES

The expression (5) explains why it is possible to have a constant $\vec{E}$ between two oppositely charged strips. The 2DES is distorted in the z-direction. The charge density distribution for the distortion $en^d(x,z)$ and for the strips $en^i(x,z)$ are found from the Poisson equation as follows:

$$\nabla^2\phi = -\rho_{xx}J_y \cdot \left[\frac{d}{dx}\left(\frac{\sigma_{yx}(x)}{\sigma_{xx}(x)}\right) \cdot \zeta^2(z) + \int_{x_0}^{x} \frac{\sigma_{yx}(x')}{\sigma_{xx}(x')}\, dx' \cdot \frac{d^2\zeta^2(z)}{dz^2} \right]$$

$$= -\frac{e}{\epsilon_b}\left(n^i(x,z) + n^d(x,z) \right)$$

Eq. (7)

From (6) it follows that
$\int \frac{d^2\zeta^2(z)}{dz^2} \cdot dz = 0$ $\int z \frac{d^2\zeta^2(z)}{dz^2}\, dz = 0$ and $\int z^2 \frac{d^2\zeta^2(z)}{dz^2}\, dz \neq 0$. This shows that it is a density of quadrupoles of the form $Q_{zz}(x,z)$ that makes $\vec{E}$ constant inside the 2DES and zero outside.

Two dimensions and three dimensions

A quantum system may be called two dimensional if the Hamiltonian is of the form $H = H_2\,(x,y) + H_1\,(z)$ which corresponds to wavefunctions that can be written $\psi\,(x,y) \cdot \zeta\,(z)$. This is usually the situation for theoretical treatments of the QHE. For $J = 0$ it is also the case here. However, the current induced potential (5) is a 3 dimensional potential, and if it is added to the original Hamiltonian as a first step in a selfconsistent treatment then it follows that for a large current the system is not longer two dimensional and the QHE may break down. From (5) it follows that breakdown starts near the edges where the variation of $\frac{\sigma_{yx}(x)}{\sigma_{xx}(x)}$ may be large. In fact recent investigations on edge current have been analyzed by using a model[2] where bulk is weakly coupled to the edge regions. If $\sigma_{xy} \sim \frac{e^2}{h}$ and σ_{xx} is determined by tunnelling through a perfect filled region as in [2] then $\frac{\sigma_{yx}}{\sigma_{xx}}$ may vary very much.

1. D.J. Thouless, *Phys. Rev. Lett.* **71** (1993), 1879.
2. D.B. Chlovskii, B.I. Shklovskii, L.I. Glazman, *Phys. Rev. B* **46** (1992) 4026.

THE NONLOCAL BREAKDOWN OF THE QUANTUM HALL EFFECT

M.V.Budantsev, G.M.Gusev*, Z.D.Kvon, A.G.Pogosov

Institute of Semiconductor Physics,
Russian Academy of Sciences, Siberian Branch.
Prosp. Lavrent'eva, 13, 6300090 Novosibirsk, RUSSIA.

The differential nonlocal resistance of two-dimensional (2D) electron gas at magnetic fields up to 7 T at the temperature T = 1.3 K at high direct currents is investigated and the breakdown of quantum Hall effect at the filling factor $\nu = 4$ is found. The critical currents I_c for breakdown of the nonlocal resistance are much higher than that of the local resistance. It is found, that I_c for the nonlocal resistance strongly depends on the sign of the magnetic field **B** , normal to the plane of 2D electron gas, and does not depend on the sign of direct current.

The experimental samples were fabricated on the basis of 2D electron gas in GaAs/AlAsGa heterojunction with 2D-electron density $n_s = 4.8\times10^{11}cm^{-2}$ and mobility $\mu = 4\times10^5 cm^2/Vs$. The geometry and relevant dimensions of the samples are shown at figure 1.

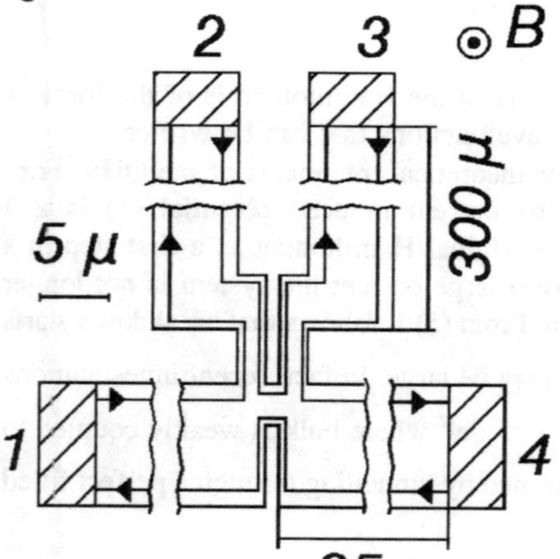

FIG. 1 Geometry of experimental samples.

In the work the differential resistance $\mathbf{R_{ijkl} = dV_{kl}/dI_{ij}}$ was measured. The direct current $\mathbf{I_{ij}^{DC}}$ modulated by small alternating current $\mathbf{I_{ij}^{AC}}$ ($\mathbf{I_{ij} = I_{ij}^{DC} + I_{ij}^{AC}}$) was passed through the contacts **i , j** . The voltage $\mathbf{V_{kl}}$ between the contacts **k** and **l** was measured with a lock-in amplifier at the frequency of ac current f = 70 Hz as a function of $\mathbf{I_{ij}^{DC}}$. The amplitude of the alternating current was 0,02 μA. Figure 2 shows the dependence of the resistance of the sample on the magnetic field for local $\mathbf{R_{1423}}$ and nonlocal $\mathbf{R_{1243}}$ cases (the positive sign of the magnetic field corresponds to the direction, shown at the figure). The local resistance at the high magnetic fields has a form of the Shubnikov-de Haas oscillations and it vanished at **B** ~ 5T, which corresponds to filling factor $\nu = 4$.

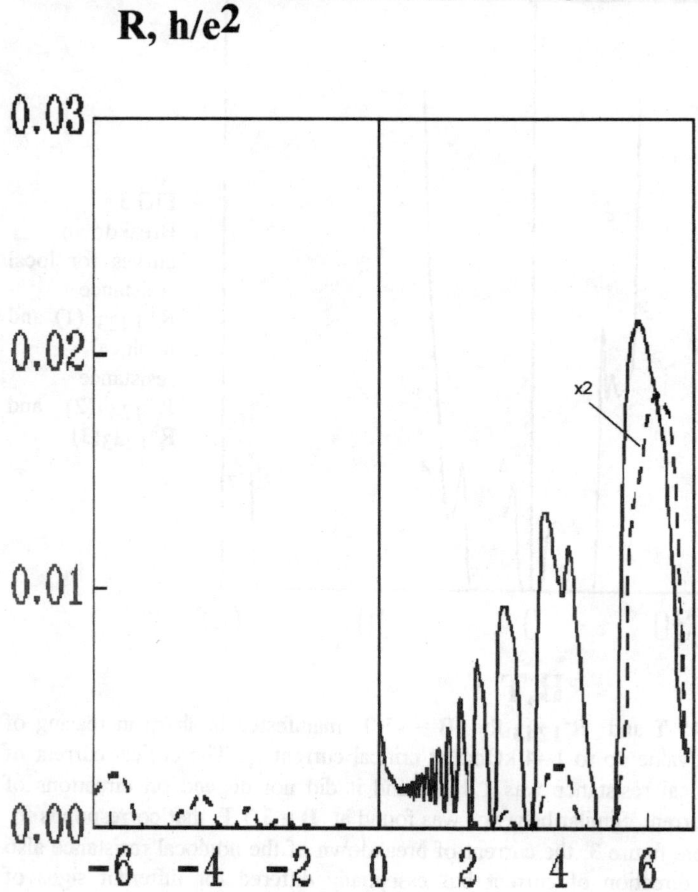

FIG. 2 Local R_{1423}(solid line) and nonlocal R_{1243} (dashed line) resistances as a function of magnetic field.

As is seen from fig.2 the nonlocal resistance at weak magnetic fields up to 2 T is equal to zero. Maxima in the nonlocal resistance correspond to half-integer filling factors, i.e. dissipative regime of QHE. Also one can see from this figure, that the nonlocal resistance has essentially asymmetrical form about **B** = 0 T axis, that connected with the absence of symmetry in geometry of the sample. At integer filling factor $\nu = 4$ the experimental behavior of the local and nonlocal resistances had breakdown characteristics (fig. 3).

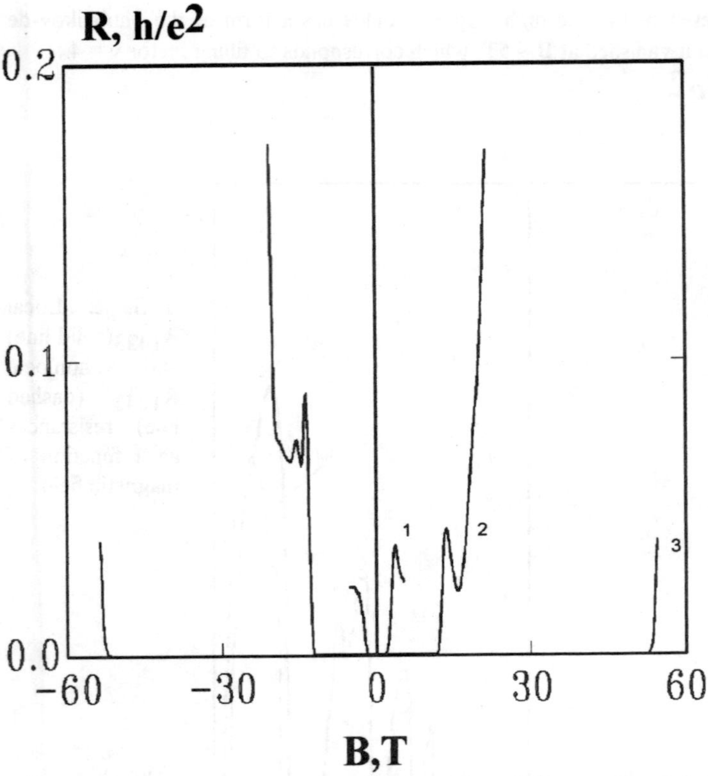

FIG.3 Breakdown curves for local resistance R^{+}_{1423} (1) and nonlocal resistance R^{+}_{1243}(2) and R^{-}_{1243}(3).

$\mathbf{R^{+}_{1243}}$ for $\mathbf{B}$ = +5 T and $\mathbf{R^{-}_{1243}}$ for $\mathbf{B}$ = -5 T. manifested in sharp increasing of resistance from zero value up to 1--4 kOhm at critical current $\mathbf{I_c}$. The critical current of breakdown of the local resistance was 1.8 μA and it did not depend on directions of magnetic field and current. Similar behavior was found at $\mathbf{B}$ = 3.3 T, that corresponds to $\nu = 6$. As is seen from figure 3, the current of breakdown of the nonlocal resistance also did not depend on direction of current but essentially differed for different signs of magnetic field, changing from 11μA at $\mathbf{B}$ = +5 T to 54 μA at $\mathbf{B}$ = -5 T. It should be noted, that the critical currents for nonlocal resistance far exceed those of local. The breakdown of local resistance is usually explained by the strong Hall field stimulating the backscattering of electrons. This picture is not valid in nonlocal case due to the absence of Hall field in nonlocal lead, for the current flows through the other lead. We believe that in this case the breakdown originates from the spread out of the breakdown region from the local lead through the constriction to the nonlocal lead (this is possible for the finite length of edge states coupling). Therefore, the experimentally measured critical breakdown currents in the nonlocal case are sufficiently larger than those in the local one. The value of the breakdown current in the nonlocal case depends on the distance the edge current goes from the contact to the constriction. For the different signs of magnetic field the edge current entering the constriction starts from the different ends of the lead. Since the geometry of the samples was asymmetrical, this distance was different for different directions of magnetic field, resulting in the noticeable asymmetry of measured critical currents.

G.M.G. is supported by FAPESP (Brazil)

*Present address :Instituto de Física de São Carlos, Universidade de São Paulo, SP, Brasil

Transition from Quantum Hall Metal to Insulator Induced by Floating-up of Extended-State Bands in Si-MOSFET's

Tohru Okamoto, Yoshinori Shinohara, and Shinji Kawaji
Department of Physics, Gakushuin University, Mejiro Toshima-ku, Tokyo 171, Japan
and
Atsuo Yagi
NPC Ltd., Shiobara-cho, Tochigi 329-28, Japan

ABSTRACT

Diagonal and Hall conductivities, σ_{xx} and σ_{xy}, have been experimentally investigated through the disorder-induced transition from the quantum Hall metal to the insulating phase in Si-MOSFET's. The transition occurs when a dimensionless parameter $\omega_c\tau$ decreases to the order of unity. These observations are consistent with the theory that extended-state bands float up from the energy center of the Landau levels with increasing disorder.

1. Introduction

In order to understand the quantum Hall effect, which observed in a two-dimensional electron system (2DES) in a magnetic field, existence of disorder should be taken into account. In a strong magnetic field, small disorder localizes the major part of Landau states.[1] Only the extended-state bands at the center of the Landau levels below the Fermi level E_F are responsible for the Hall conductivity in the system. Thus, if E_F lies on a localized state, the Hall conductivity σ_{xy} takes a quantized value ie^2/h and does not depend on E_F within the plateau region. Here, i indicates the number of the extended-state bands lying below E_F. With increasing disorder, how does the conducting behavior of the 2DES change? Khmelnitskii[2] and Laughlin[3] proposed that the extended-state bands should float up from the center of the Landau level with decreasing scattering time. The transition into insulator occurs when the lowest extended-state band goes up through E_F. Experimentally, giant magnetoresistance oscillation caused by the magnetic-field-induced metal-insulator transition was observed in Si-MOSFET's[4] and in GaAs-(AlGa)As heterostructures.[5] In the present work, disorder dependence of σ_{xx} and σ_{xy} in a high-mobility Si-MOSFET has been investigated for a fixed filling factor ν by changing the electron density N_s.

2. Experimental

We used a Si-MOSFET Hall bar sample (3 mm x 0.3 mm with eight potential probes) with a high-peak-mobility (μ_{peak} = 2.7 m^2/ V s). The sample was cooled using a dilution refrigerator through a thermal link to a mixing chamber. A magnetic field perpendicular to the 2DES was applied by a Nb_3Sn superconducting magnet up to 15.6 T. Four-wire measurements both on the diagonal and Hall resistivities, ρ_{xx} and ρ_{xy}, were carefully performed in the linear region of the *I-V* relations using lock-in amplifiers (PAR model-124A) operated at 7 Hz or a source-and-measure-unit (Keithley model-236) with an ultra-high input resistance of 10^{14} Ω. The source-drain voltage and current were limited below 50 μV and 1 nA, respectively, in the whole region of the resistance of the sample. The diagonal and Hall conductivities were calculated from ρ_{xx} and ρ_{xy} data. All the transport data reported in this work were measured at the temperature of 60 mK. The temperature dependence of ρ_{xx} and ρ_{xy} is almost saturated below 100 mK except for very high resistivity regions. The zero-magnetic-field mobility μ has a strong N_s dependence due to screening effect by electrons.[6] The scattering time τ $(= \mu m^*/e)$ was determined from μ as a function of N_s.

3. Results and Discussions

3.1. Diagonal and Hall conductivities through the metal-insulator transition at $\nu = 1.7$

In Fig. 1, diagonal and Hall conductivities are shown against a dimensionless parameter $\omega_c\tau = \mu B$ at a constant filling factor $\nu = 1.7$. Here, $\omega_c = eB/m^*$ is the cyclotron frequency and m^* is the effective mass. At $\omega_c\tau \gg 1$, σ_{xy} has a quantized value of $2e^2/h$ with $\sigma_{xx} = 0$. The result shows that E_F lies at a localized state

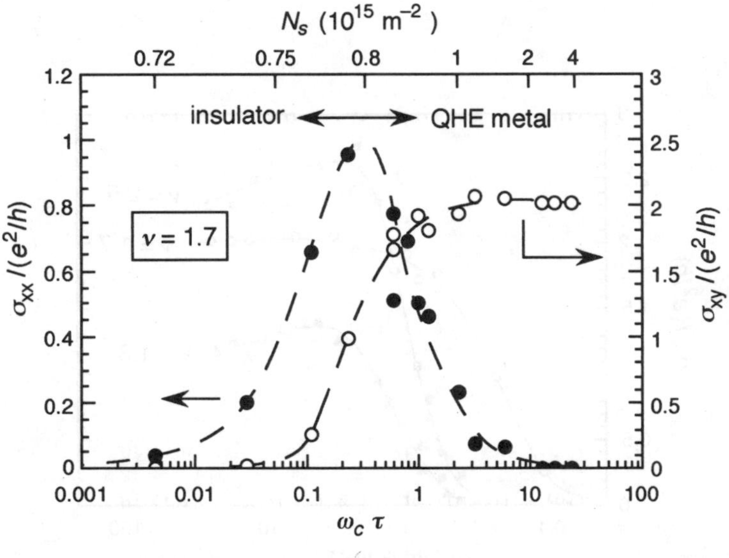

Fig. 1. Diagonal (closed circles) and Hall (open circles) conductivities through the QHE-metal↔insulator transitions at the constant filling factor of $\nu = 1.7$. These data were obtained at $T = 60$ mK.

between the Landau level centers separated by the Zeeman effect. As $\omega_c\tau$ is reduced to the order of unity, σ_{xy} decreases monotonically, while σ_{xx} increases first and decreases to zero after reaching the maximum value of $\sigma_{xx} \approx e^2/h$. These behaviors show that the extended states escape up through E_F and the fraction of the extended states below E_F which contribute to the conduction decreases. Finally, both σ_{xy} and σ_{xx} go to zero when $\omega_c\tau \rightarrow 0$. The transition is finished then no extended states exist below E_F. In the theories by Khmelnitskii[2] and Laughlin,[3] the energy of the nth extended-state band E_n should rise with decreasing $\omega_c\tau$ as

$$E_n = \left(n + \frac{1}{2}\right)\hbar\omega_c\left[1 + \left(\omega_c\tau\right)^{-2}\right]. \tag{1}$$

Although τ in Fig. 1 is obtained in the absence of magnetic field, the results can be well represented by eq. (1) in which considerable floating-up of E_n occurs at $\omega_c\tau \sim 1$.

3.2. Metal-insulator transition at other filling factors

The transition region shifts with the filling factor ν. Figure 2 shows typical σ_{xy}–$\omega_c\tau$ data for three filling factors. With decreasing $\omega_c\tau$, the σ_{xy} at $\nu = 1.3$ and 2.5 decrease faster than that at $\nu = 1.7$. In other words, the metallic phase is more stabilized at $\nu = 1.7$ than those at $\nu = 1.3$ and 2.5. The $\rho_{xx}(\nu)$ shows a large oscillation and its clear minimum at $\nu \sim 1.7$ as well as at $\nu \sim 0.9$ as a result of the

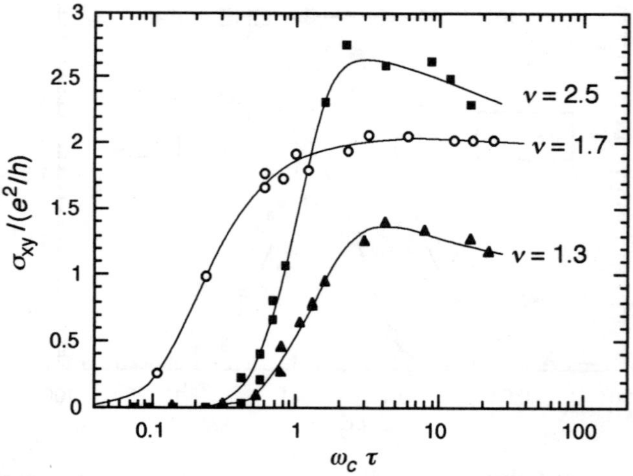

Fig. 2. Hall conductivity through the metal-insulator transitions for three filling factors.

metal-insulator transitions at an appreciate density. The ρ_{xx} minima at $\nu = 0.9$ and 1.7 change into those at $\nu = 1$ and $\nu = 2$ in the QHE region continuously with increasing N_S.[7] In a simple picture, we expect that the valley splitting at $\nu = 1$ and the Zeeman splitting at $\nu = 2$ cannot be observed at $\omega_C\tau \leq 1$. They should be so small in comparison with the Landau level separation and its broadening by disorder. The enhancement caused by the many-body effects can explain the resolved splittings.[8,9,10] The many-body exchange interactions give rise to the maximum separation of the Landau levels when the Fermi level lies between the concerned levels. Simultaneously, the energy difference between E_F and the center of the lower Landau level should also have the maximum value. Since the metal-insulator transition should occur when the energy of the lowest extended-state band–$E_{n=0}$ in eq. (1)–is equal to E_F, it is reasonable that the metallic phase is stable at $\nu = 1$ and $\nu = 2$ even for lower $\omega_C\tau$.

4. Conclusions

We have studied the transition from the quantum Hall metal to the insulator in Si-MOSFET's. Our σ_{xx} and σ_{xy} data at $\nu = 1.7$ demonstrate the floating-up of the extended-state bands induced by disorder theoretically proposed by Khmelnitskii and Laughlin. The oscillating behavior of the ν-dependence on the transition can be well understood if the many-body effects make the enhancement of the spin and valley separations.

5. Acknowledgments

The authors wish to thank K. Hayashi and T. Okuno for their experimental assistance. This work was supported in part by the Grants-in-Aid for Scientific Research from the Ministry of Education, Science and Culture, Japan.

6. References

1. T. Ando, Prog. Theor. Phys. Suppl. **84** (1985) 69.
2. D. E. Khmelnitskii, Phys. Lett. **106A** (1984) 182.
3. R. B. Laughlin, Phys. Rev. Lett. **52** (1984) 2304.
4. S. V. Kravchenko, J. A. A. J. Perenboom, and V. M. Pudalov, Phys. Rev. **B44** (1991) 13513; M. D'Iorio, V. M. Pudalov, and S. G. Semenchinsky, Phys. Rev. **B46** (1992) 15992.
5. H. W. Jiang, C. E. Johnson, K. L. Wang, and S. T. Hannahs, Phys. Rev. Lett. **71** (1993) 1439; T. Wang, K. P. Clark, G. F. Spencer, A. M. Mack, and W. P. Kirk, Phys. Rev. Lett. **72** (1994) 709.
6. F. Stern, and W. E. Howard, Phys. Rev. **163** (1967) 816.
7. T. Okamoto, Y. Shinohara, and S. Kawaji, (in preparation).
8. T. Ando, Y. Uemura, J. Phys. Soc. Jpn. **37** (1974) 1044.
9. Y. Uemura, Surf. Sci. **58** (1976) 1.
10. F. J. Ohkawa, and Y. Uemura, Surf. Sci. **58** (1976) 254.

TRANSFORMATION OF THE QHE INTO THE QUANTUM OSCILLATIONS OF THE 2D METAL TO INSULATOR CRITICAL BOUNDARY

V. M. PUDALOV

Institute for High Pressure Physics, Troitsk, Moscow district, 142092, Russia

and

M. D' IORIO

National Research Council Canada, IMS, Ottawa, Ont., K1A 0R6, Canada

ABSTRACT

In the high mobility 2D electron system in Si, the Landau levels were found to remain resolved down to the metal – insulator critical density. We studied the location of the resistance minima on the Landau fan diagram, through the reentrant M-I transition. We found that the R_{xx} minima deviate towards high field contrary to what is expected for the 2D metal phase. It reflects the transformation of the regular QHE into the quantum oscillations of the M-I critical boundary. The capacitance minima, in contrast, remain at their location linked to the minima of the density of states.

1. Introduction

The 2D electron system in high mobility Si-MOSFETs, subjected to a perpendicular quantizing magnetic field, behaves in a spectacular fashion upon decreasing density[1]. It was found experimentally that, in high mobility samples, the Landau levels remain well resolved down to the critical density for the metal to insulator (M-I) transition[2,3]. When approaching and further crossing the critical density of the M-I transition, the regular quantized Hall effect (QHE) transforms into a reentrant insulating state[1,2], as illustrated in Figs. 1 (a) and (b).

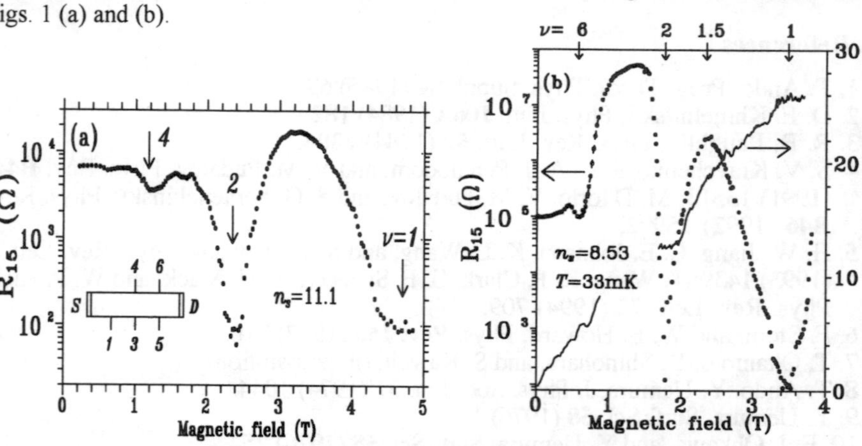

Fig. 1. Diagonal and Hall resistances vs magnetic field at two different densities: (a) in the metallic range, and (b) - in the insulating range. Density is shown in units of 10^{10}cm^{-2}.

The insulating state is interrupted at integer filling factors by the QHE state which is characterized by the quantized Hall resistance[3] and low diagonal resistivity values, $\rho_{xx} \ll h/e^2$. The diagonal resistivity, modulated by the reentrant M-I transitions, looks as in the QHE but is scaled to $10^4\times(h/e^2)$ or higher peak values.

The aim of this work is to clarify the nature of the reentrant insulating transition, and in particular, the low-resistivity phase in the regions of integer filling factors. We present here data that provide evidence for the transformation of the QHE low-resistance phase (related to the energy gap in the electron spectrum) into the metal phase (related to the minima of the chemical potential).

2. Experimental

To explore the reentrant transition in details we have selected two high mobility (100)-Si MOSFET samples: Si-11 with peak mobility 3.6×10^4cm^2/Vs and Si-5 with $\mu^{peak}=4.3\times10^4$cm^2/Vs. We have measured the magnetic field positions of the ρ_{xx} minima and compared them to those expected in the 2D metal or electron liquid phase. Most of the measurements were done by a 4-terminal dc technique described in detail in Ref.[2]. The results for the major R_{xx} minima are shown on the Landau fan diagram in Fig. 2. The empty triangles designate the location of the minima, while the full dots show the oscillatory boundary for the M-I transition [2,4]. Two of the R_{xx} minima, at $\nu=3$ and $\nu=4$, are seen only in the metallic phase at high n_s. This is because the corresponding energy levels intersect and the energy gaps vanish[4], in the magnetic field of ~1–1.5 T.

In the high density range (true metallic phase) the ρ_{xx} minima, as expected, follow the straight lines $n_s = i(hH/ec)$ where i is the number of occupied Landau levels. Below the critical density $n_c \sim 10^{11}$cm^{-2}, the resistivity minima start to deviate from the straight lines

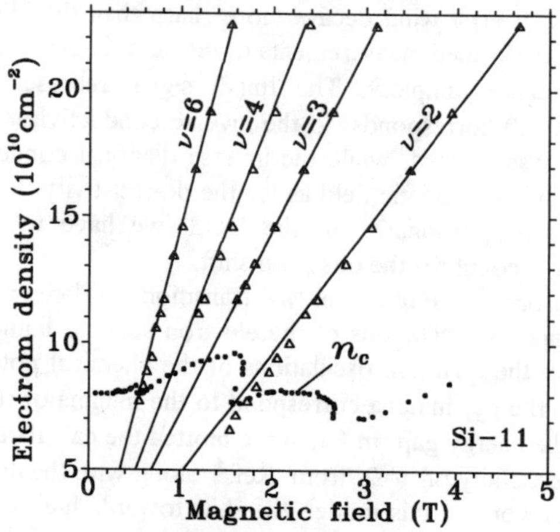

Fig. 2. Landau fan diagram for the four different R_{xx} minima (empty triangles). Full dots show the M-I boundary at T=30mK.

on the fan diagram. In Fig. 3 we plotted the deviation of the measured R_{xx} minima from the calculated positions for all three observed minima $\nu = 1$, 2 and 6. The deviation starts to appear at a carrier density of $10.5 \times 10^{10} cm^{-2}$, which is close to the maxima of the M-I critical boundary, plotted in Fig. 2 by full dots.

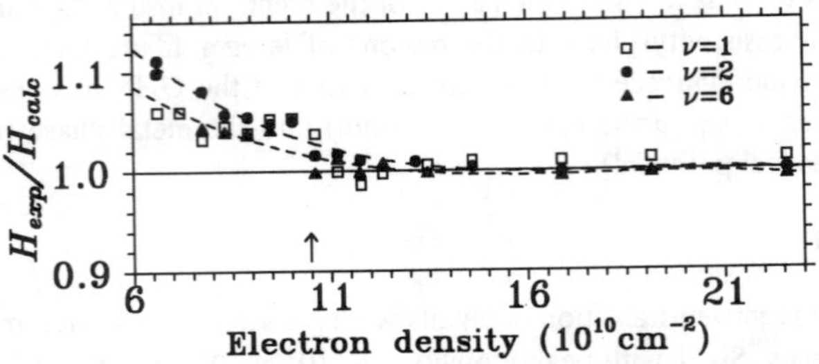

Fig. 3. Normalized deviation in the measured positions of the R_{xx} minima relative to the expected ones.

3. Discussion

The deviation from the regular $n(H)$ straight lines in Fig. 2 occurs towards the high field side, i.e. in the direction opposite to that expected for single-particle localization. In other words, there is no decrease in the density of delocalized carriers in the 2D layer down to at least 5×10^{10} cm^{-2}. On the other hand, the excess carriers can not appear in the 2D system because of the high cost in the corresponding Coulomb energy. This implies that the deep ρ_{xx} minima may not be associated with the position of the Fermi energy in the energy gap between neighboring Landau levels.

The asymmetry in the density of states (DOS) might cause a shift of the minima to the high field side, if the low energy wing decays more sharp than the high energy one. To probe this possibility we performed measurements of the ac-impedance Z of the "2D layer –gate" system on the same samples[6]. The Im(Z) signal represents changes in the capacitance δC, while Re(Z) corresponds to the inverse conductivity $1/\sigma_{xx}$ No shift in the capacitance minima was noticed, while the inverse diagonal conductivity did exhibit the same shift towards higher magnetic field as for the dc-resistivity. Since the changes in capacitance are directly proportional[5,7] to the DOS, we have to conclude that the asymmetric DOS cannot account for the observed shift.

In the alternative approach[4], the reentrant M-I transition has been modelled quantitatively in terms of the quantum oscillations of the electron solid to liquid phase boundary. The latter originate from the quantum oscillations of the chemical potential in the liquid phase. In this approach, the ρ_{xx} minima correspond to the minima of the chemical potential μ_L, rather than to the energy gap. In Fig. 4 we plotted the calculated variations of the chemical potential in the vicinity of $\nu=2$, from Ref.[4] along with the measured shift in the R_{xx} minima. In the framework of this model, the shift towards high magnetic field has a natural explanation and corresponds roughly to half of the Landau level broadening. The energy gap, DOS minima, and capacitance minima all remain at their true positions on the

Landau fan diagram, while the resistance minima do shift to higher magnetic field. The suggested interpretation agrees qualitatively with the experimental data, as seen in Fig. 4.

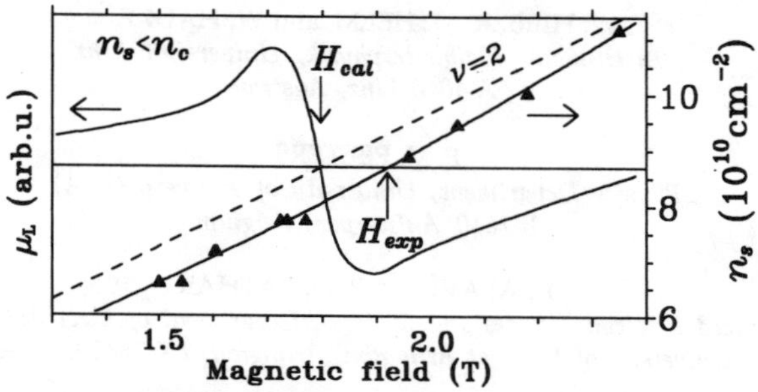

Fig. 4. Chemical potential vs magnetic field in the vicinity of ν=2 (full curve).and the fragment of the measured diagram (triangles). Dashed straight line corresponds to two fully filled Landau levels. Vertical arrows indicate the calculated (expected) and measured positions of the R_{xx} minima.

In summary, we found the diagonal resistance minima to shift from the expected positions on the Landau fan diagram, towards *higher magnetic field*. This occurs when the carrier density is decreased down to the M-I transition. The experimental data are explained by a model[4] where the oscillatory M-I transitions originate from the quantum oscillations of the electron solid to liquid boundary.

4. Acknowledgements.

Authors acknowledge support from the Russian Foundation for Basic Research by the grant N° 93-0214235, from the International Science Foundation by the grant MUG000, from Ministry for Science and Technical Policy, and from the Nederlandse Organisatie voor Wetenschappelijk Onderzoek (N.W.O) by the grant N° 07-13-217.

5. References

1. M. D' Iorio, V. M. Pudalov, and S. G. Semenchinsky. *Phys. Lett.* **A150** (1990) 422.
2. M. D' Iorio, V. M. Pudalov, and S. G. Semenchinsky, *Phys. Rev.* **B46** (1992) 15992.
3. V. M. Pudalov, M. D' Iorio, and J. W. Campbell. *JETP Lett* **57** (1993) 608.
4. V. M. Pudalov. *Bull. Amer. Phys. Soc.* **39** (1994) 805. Will be published elsewhere.
5. S. V. Kravchenko, J. A. A. J. Perenboom, and V. M. Pudalov. *Phys. Rev.* **B44** (1991) 13513.
6. V. M. Pudalov and M. D' Iorio. Will be published elsewhere.
7. A. M. Voshchenkov and J. N. Zemel. *Phys. Rev.* **B9** (1974) 4410.

MAGNETIC-FIELD-INDUCED METAL-INSULATOR TRANSITION IN GaAs/AlGaAs QUANTUM WELLS AND SUPERLATTICES

W. HILBER, M. HELM, and W. HAUKE
Institut für Halbleiterphysik, Universität Linz
A-4040 Linz, Austria

F. M. PEETERS
Physics Department, University of Antwerp (UIA)
B-2610 Antwerpen, Belgium

K. ALAVI and R. N. PATHAK
Center for Advanced Electron Devices and Systems, Department of Electrical Engineering, The University of Texas at Arlington, Arlington, TX 76019, USA

ABSTRACT

The in-plane magnetotransport of a homogeneously doped, metallic GaAs/AlGaAs superlattice and a multi-quantum well system is investigated. In the superlattice, a metal-insulator transition can be induced by a magnetic field above 7 T, with a similar behavior (apart from the anisotropy) as bulk GaAs. The multi-quantum well system shows a more complex behavior, exhibiting features related to weak localization, the quantum Hall effect, and the magnetic-field-induced metal-insulator transition. The transport studies are complemented by preliminary infrared absorption measurements, which yield information on the position of the Fermi energy.

1. Introduction

If a semiconductor is doped higher than a critical density, the conductivity remains finite at low temperature and the semiconductor behaves like a metal. The critical density, n_c, is approximately given by the Mott criterion, $n_c^{1/3}a_0 = 0.25$, where a_0 is the effective Bohr radius. A sufficiently large external magnetic field diminishes the extent of the electronic wavefunctions, and their reduced overlap leads to a metal-insulator transition (MIT). In order to achieve the transition with readily available magnetic fields, materials with a relatively small effective mass are required, which then also exhibit a small donor binding energy and a large effective Bohr radius. In the past, the magnetic-field-induced MIT has been studied mainly in InSb, InP, and GaAs. While initial studies focused on the magnetic freeze-out behavior on the insulating side of the transition, more recent studies dealt with the scaling behavior in the close vicinity of the transition[1,2]. Another interesting question relates to the position of the Fermi energy at the MIT - whether it is located in the impurity band or in the conduction band. Recent far-infrared absorption experiments[3] in GaAs indicate that the transition occurs in the impurity band.[4]

The advent of semiconductor heterostructures opens up the possibility to study the influence of lowering the dimensionality on the MIT. Previous experiments addressing this issue were performed by successively depleting a GaAs MESFET structure with a gate voltage[5] or using δ-doped systems.[6] Very recently a magnetic-field-induced delocalization has been reported in a low-mobility two-dimensional electron gas.[7]

In the present paper we report first results on the magnetic-field-induced MIT for a GaAs/AlGaAs superlattice as well as a multi-quantum well system consisting of low-mobility two-dimensional electron systems .

2. Samples

The multi-quantum well sample (MQW, grown by MOCVD) consists of 180 periods of 270 Å GaAs quantum wells separated by 60 Å $Al_{0.3}Ga_{0.7}As$ barriers and is homogeneously doped to $n=4x10^{16}cm^{-3}$. The barriers are thick enough so that the electrons are confined to two-dimensional layers with an areal electron concentration of $n_s=1.3x10^{11}cm^{-2}$. The superlattice sample (SL, grown by MBE) consists of 500 periods of 75 Å GaAs wells and 25 Å $Al_{0.3}Ga_{0.7}As$ barriers and is homogeneously doped to $n=5x10^{16}cm^{-3}$. Here the thin barriers lead to a finite width of the first miniband of 18 meV. The Fermi energy at low temperature lies approximately in the middle of the first miniband, so the system behaves three-dimensional, but strongly anisotropic.

Both samples remain metallic at low temperature, in agreement with an estimate using the Mott critical density for bulk GaAs, which is $n_c= 1.5x10^{16}cm^{-3}$. The mobility of the MQW sample reaches a maximum of 9000 cm^2/Vs at 100 K and drops to 6000 cm^2/Vs at 10 K. The mobility of the SL sample has a maximum of 6000 cm^2/Vs at 100 K, but drops more strongly to 1000 cm^2/Vs at low temperature. It is not clear at present, whether this different behavior is due to the different degree of compensation in the two samples or a result of stronger interface roughness scattering in the SL sample, where the wavefunction is extended and therefore quite large at the interfaces.

3. Transport

Magnetotransport measurements for both samples were performed at temperatures down to 0.4 K and in magnetic fields up to 14 T. Fig. 1 shows the tranverse magneto-resistivity, ρ_{xx}, of the SL sample in a magnetic field perpendicular to the layers for different temperatures. A last minimum (ν=2) in the Shubnikov-de Haas (SdH) oscillations is observed at 4.5 T followed by a steep increase of the resistance. Above 7.2 T the resistance consistently increases on lowering the temperature. In particular, the resistance curves for all temperatures $T \leq 4.2K$ have a common crossing point at B=7.2 T (Fig. 1b), which can apply as a criterion for the critical field of the MIT. Note, however, that the conductivity σ_{xx} (calculated from ρ_{xx} and ρ_{xy}) extrapolates to zero[2] for $T \rightarrow 0$ only above 11 T. The Hall resistivity (not shown) does not exhibit any special features except a superlinear rise beyond 7 T.

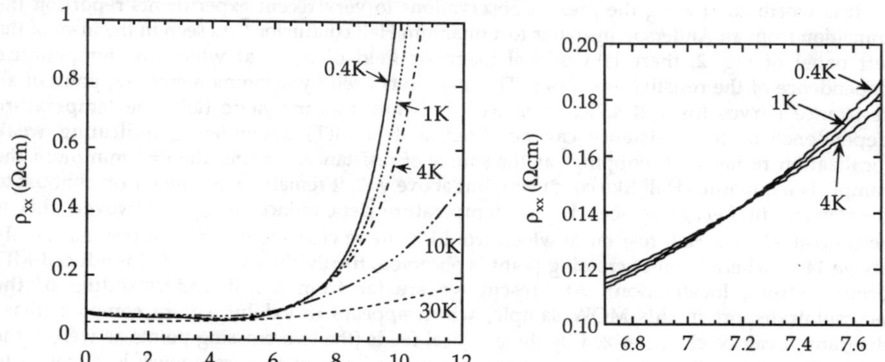

Fig. 1: Magnetoresistivity of the SL sample for different temperatures (left). Crossing point at 7.2 T (right).

The MQW sample is a system of 180 parallel, "dirty" two-dimensional electron gases. Fig. 2 shows ρ_{xx} and ρ_{xy} for different temperatures. At magnetic fields close to zero a significant resistance increase due to weak localization is observed, consistent with the two-dimensional nature of the system (of course also the SL sample exhibits weak localization, but much less pronounced). Then the "last" SdH minimum (ν=2) appears and is followed by a sharp kink in the Hall resistance, corresponding to a decrease of the electron concentration. Also ρ_{xx} rises quite sharply (note the log scale), however upon cooling down further, the rise is less steep and instead, a plateau-like structure develops. Only above 10 T, the strong resistance increase is recovered.

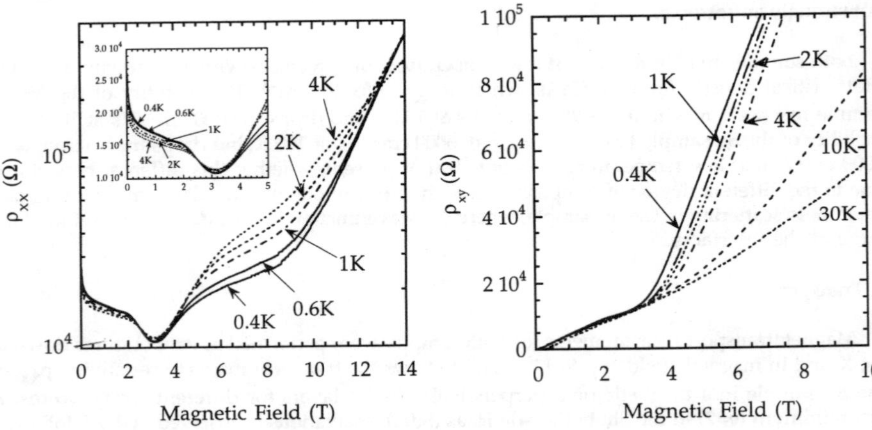

Fig. 2: Magnetoresistivity (left) and Hall resistivity (right) - both per 2D layer - of the MQW sample for different temperatures. The inset in the left panel shows the crossing point at 2.5 T.

It is worth comparing the present observations to very recent experiments reporting the transition from an Anderson insulator to a quantum Hall conductor.[7] As seen in the inset of the left panel of Fig. 2, there is a critical magnetic field of 2.5 T at which the temperature dependence of the resistivity reverses. This manifests itself by a common crossing point of all resistance curves for $T \leq 4.2K$, as in Ref. 7. Below this magnetic field the temperature dependence of the resistance can be fitted with a ln(T) dependence, indicating weak localization rather than hopping[7] as the source of resistance. Around the ν=2 minimum the sample is a quantum-Hall-like conductor, but above ν=2, it remains a conductor (in contrast to the reports in Ref. 7) as seen by the temperature dependence of ρ_{xx}. However, this is accompanied by a Hall resistance which would be more characteristic for an insulator. Only above 14 T, where another crossing point is observed, finally the magnetic-field-induced MIT occurs (strong localization). At present we are far from a full understanding of the magnetotransport in this MQW sample, which appears to exhibit a very complex phase diagram. It can be characterized by three critical fields (the two crossing points as well as the field where the kink in the Hall resistance occurs) and a transition from weak localization to quantum-Hall-like conduction and strong localization, with an intermediate, possibly new phase, where ρ_{xx} has the opposite temperature dependence than ρ_{xy}.

4. Infrared absorption

Infrared absorption spectroscopy can be employed to determine the position of the Fermi energy relative to the impurity and conduction bands, namely by monitoring the relative strength of impurity (1s-2p_+ or 1s-2p_0) and conduction band (cyclotron resonance or intersubband) absorption. In this way it has been proven recently that the magnetic field induced MIT in bulk GaAs occurs in the impurity band.[3] In the present SL sample, both interminiband absorption and 1s-2p_0 absorption are visible at B=0 (see Ref. 8). With an applied magnetic field, we have measured the far-infrared absorption, which already at 5T exhibits only the 1s-2p_+ transition and no sign of cyclotron resonance. This confirms the picture of the Fermi energy dropping into the impurity band already before the MIT occurs. For the MQW sample we observe the intersubband absorption at low temperature (see Ref. 9), but no clear sign of any impurity absorption (at least of impurities located in the wells; unfortunately the theoretical position of the barrier-impurity absorption coincides with the intersubband absorption, so no definite statement can be made). No data are available yet for finite magnetic fields, but they will most likely provide additional information allowing a full interpretation of the transport properties.

To summarize, the SL sample exhibits magnetotransport properties very similar to bulk GaAs, which is not surprising, since it represents essentially a 3D, though ansiotropic system. The magnetotransport of the MQW sample exhibits a rather intricate behavior reaching from weak to strong localization with an intermediate range of metallic conduction, whose proper interpretation requires further investigations.

This work was supported by the Fonds zur Förderung der wissenschaftlichen Forschung (Vienna), under project number P10211. We are grateful to E. Colas for the growth of the MQW sample and to G. Brunthaler and G. Bauer for helpful discussions.

References

1. D. J. Newson and M. Pepper, *J. Phys. C: Solid State Phys.* **19** (1986) 3983
2. M. C. Maliepaard, M. Pepper, R. Newbury, and G. Hill, *Phys. Rev. Lett.* **61** (1988) 369; M. C. Maliepaard, M. Pepper, R. Newbury, J. E. F. Frost, D. C. Peacock, D. A. Ritchie, G. A. C. Jones, and G. Hill, *Phys. Rev.* **B 39** (1989) 1430
3. M.-W. Lee, D. Romero, H. D. Drew, M. Shayegan, and B. S. Elman, *Solid State Commun.* **66** (1988) 23; D. Romero, S. Liu, H. D. Drew, and K. Ploog, *Phys. Rev.* **B 42** (1990) 3179; Y. Wang, B. D. McCombe, and W. Schaff, in *Localization and Confinement in Semiconductors*, Springer, Berlin (1990), p. 314
4. S. Liu, H. D. Drew, A. Illiades, and S. Hadjipanteli, *Phys. Rev.* **B 45** (1992) 1155; S. Liu, K. Karrai, F. Dunmore, H. D. Drew, R. Wilson, and G. A. Thomas, *Phys. Rev.* **B 48** (1993) 11394
5. D. J. Newson, C. M. McFadden, and M. Pepper, *Philos. Mag.* **52** (1985) 437
6. A. Zrenner, F. Koch, J. Leotin, M. Goiran, and K. Ploog, *Semicond. Sci. Technol.* **3** (1988) 1132
7. H. W. Jiang, C. E. Johnson, K. L. Wang, S. T. Hannahs, *Phys. Rev. Lett.* **71** (1993) 1439; T. Wang, K. P. Clark, G. F. Spencer, A. M. Mack, and W. P. Kirk, *Phys. Rev. Lett.* **72** (1994) 709
8. M. Helm, W. Hilber, T. Fromherz, F. M. Peeters, K. Alavi, and R. N. Pathak, *Phys. Rev.* **B 48** (1993) 1601
9. M. Helm et al., *Appl. Phys. Lett.* **63** (1993) 3315

TEMPERATURE-DRIVEN TRANSITIONS FROM INSULATING BEHAVIOR AT $\nu = 1$

Whitney Mason, S. V. Kravchenko, and J. E. Furneaux
Laboratory for Electronic Properties of Materials and Department of Physics and Astronomy, University of Oklahoma, Norman, Oklahoma 73019

J. M. Caulfield and J. Singleton
Clarendon Laboratory, Department of Physics, University of Oxford, Parks Road, Oxford OX1 3PU, United Kingdom

V. M. Pudalov
Institute for High Pressure Physics, Troitsk, 142092 Moscow District, Russia

In the study of two-dimensional electron systems (2DES), transitions between insulating and quantum Hall (QH) groundstates thus far have been observed only as the magnetic field and/or the electron density is changed[1–3]. Here we show experimental evidence that transitions from insulating to metallic and from thence to QH types of behavior of transport coefficients can be observed at fixed field and carrier density as the temperature is changed. We define behaviors as follows:

(i) insulator-like: $d\sigma_{xx}/dT$, $d\sigma_{xy}/dT > 0$; $\sigma_{xy} < \sigma_{xx} < e^2/h$;

(ii) metallic: $d\sigma_{xx}/dT$, $d\sigma_{xy}/dT < 0$;

(iii) QH-like: $\sigma_{xy} \to e^2/h$, $\sigma_{xx} \to 0$ as $T \to 0$.

Transitions between these behaviors at $\nu = 1$ can be explained as follows: the lowest band of extended states, which lies above the Fermi energy (E_F) at high temperatures (making the system insulating), decreases in energy as the temperature is decreased. As this band of extended states passes through E_F, the system behaves as a metal. Finally, as the energy sinks below E_F, the system is in the QH regime.

We studied 2DES in silicon at constant filling factor ν at very low electron density n_s, at the border of existence of the QH effect. Two samples from wafers with different mobilities, μ, were studied: Si-14 with a maximum $\mu = 1.9 \times 10^4$ cm^2/Vs, and Si-22 with a maximum $\mu = 3.5 \times 10^4$ cm^2/Vs. The samples and experimental technique are described elsewhere[4].

The dependencies of ρ_{xx} and ρ_{xy}, both in units of h/e^2, on magnetic field for three different temperatures are shown in figure 1. Note that as the temperature is

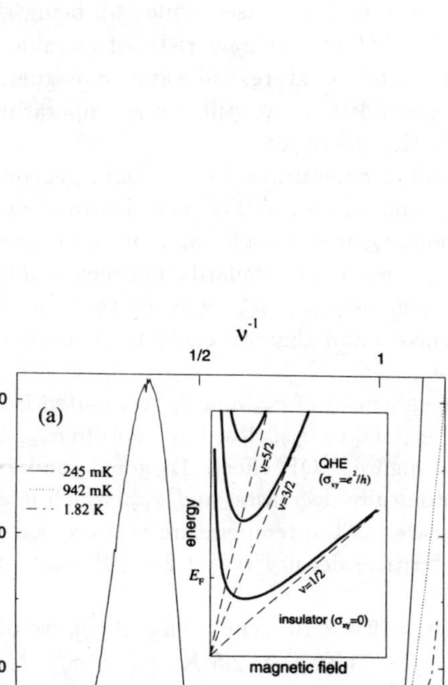

ν^{-1}
1/2
1
(a)
245 mK
942 mK
1.82 K
QHE
$(\sigma_{xy}=e^2/h)$
ν=5/2
ν=3/2
ν=1/2
energy
E_F
insulator $(\sigma_{xy}=0)$
magnetic field
ρ_{xx}, ρ_{xy} (h/e^2)
ρ_{xx}
ρ_{xy}
B (T)

Fig. 1

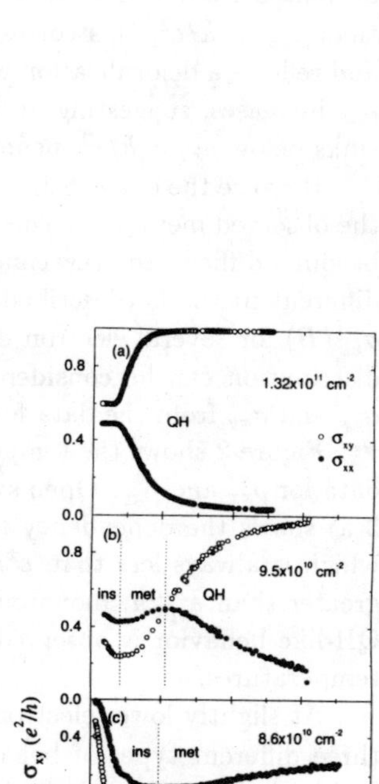

(a)
1.32×10^{11} cm^{-2}
QH
σ_{xy}
σ_{xx}
(b)
9.5×10^{10} cm^{-2}
ins
met
QH
(c)
8.6×10^{10} cm^{-2}
ins
met
(d)
8.1×10^{10} cm^{-2}
insulator
σ_{xx}, σ_{xy} (e^2/h)
$1/T$ (K^{-1})

Fig. 2

decreased from 1.82 K to 942 mK, ρ_{xx} at $\nu = 1$ decreases while still being greater than $\rho_{xy} = h/e^2$; this corresponds to $d\sigma_{xx}/dT < 0$ characteristic of metallic states and reflects a delocalization with decreasing temperature, while at zero magnetic field ρ_{xx} increases, suggesting an insulating groundstate. At still lower temperature, ρ_{xx} sinks below $\rho_{xy} = h/e^2$, penetrating into the QH region.

Because the current distribution at low temperatures can be inhomogeneous, and the observed metallic-like decrease in ρ_{xx} could be a result of special current paths, it becomes difficult to draw conclusions about σ_{xx} from data for ρ_{xx}. However, using two different methods (described elsewhere [4]), we found similarity between $\rho_{xx}(B)$ and $\sigma_{xx}^{-1}(B)$ for several electron densities as long as $\rho_{xx} > \rho_{xy}$, showing that the current distribution can be considered homogeneous and that we could therefore calculate σ_{xx} and σ_{xy} from the data for ρ_{xx} and ρ_{xy}.

Figure 2 shows the temperature dependencies of σ_{xx} and σ_{xy} computed from the data for ρ_{xx} and ρ_{xy}. Open symbols correspond to σ_{xy}, closed symbols to σ_{xx}. Figure 2(a) shows the dependency for "normal high-n_s" QH effect. Diagonal conductivity, which is always less than $e^2/2h$, monotonically decreases, and σ_{xy}, which is always greater than $e^2/2h$, monotonically increases as the temperature is decreased. Only QH-like behavior is observed above a "critical density"$n \geq 1.2 \times 10^{11}$ cm^{-2} for all temperatures.

At slightly lower electron density ($n_s = 9.5 \times 10^{10}$ cm^{-2})[fig. 2(b)], we observe three different types of behavior at $\nu = 1$: (i) at $T > 2.5$ K, $\sigma_{xy} < \sigma_{xx} < e^2/2h$, and both diminish with diminishing T which is characteristic for an insulating state. Furthermore, according to Khmelnitskii[5], σ_{xy} in units of e^2/h is a "counter" of the number of bands of extended states below E_F, and because $\sigma_{xy} << e^2/h$ at this temperature, there are not extended states below E_F in this temperature region, which confirms an insulating state; (ii) at $T < 2.5$ K, both σ_{xx} and σ_{xy} increase with decreasing temperature, showing "metallic" behavior and both reach $e^2/2h$, the value expected for an extended state; (iii) for $T < 1$ K, σ_{xx} again tends to zero as $T \to 0$, while σ_{xy} approaches e^2/h. The value of σ_{xy} now corresponds to one band of extended states below E_F; this is a QH state.

At even lower $n_s = 8.6 \times 10^{10}$ cm^{-2}) [fig. 2(c)], both σ_{xx} and σ_{xy} change their slope from "insulator-like" to "metallic" at $T = 1$ K, but σ_{xy} never reaches e^2/h, and the QH conditions are never met for this electron density.

Finally, for the lowest $n_s = 8.1 \times 10^{10}$ cm^{-2}) [fig. 2(d)], both σ_{xx} and σ_{xy} decrease monotonically down to the lowest temperatures while $\sigma_{xy} << \sigma_{xx} << e^2/2h$; for this n_s, the system remains insulating regardless of temperature.

According to references [5,6], the energy of the lowest extended state at T = 0 is

$$E_c = \frac{1}{2}\hbar\Omega_c[1 + (\Omega_c\tau)^{-2}]$$

where Ω_c is the cyclotron frequency and τ is the relaxation time. Therefore, at constant B, E_c decreases with increasing τ. Our data shows that at $\nu = 1$, the effect of temperature is qualitatively similar to the effect of disorder: with decreasing temperature, the energy of the lowest band of extended states decreases. At high T, all extended states are above the Fermi energy, and conductivity is due to temperature activation to the nearest extended state. At lower temperatures, the band of extended states crosses the Fermi level, and we observe a metallic state with characteristic metallic temperature dependencies for both diagonal and Hall conductivities. At still lower temperatures, the band of extended states sinks below the Fermi energy, σ_{xy} approaches e^2/h, and the system enters the QH regime.

In conclusion, we have studied the temperature dependent behavior of the very dilute 2DES in silicon at constant filling factor ν. We have found that at $\nu = 1$ for n_s below some critical value, the energy of the lowest band of extended states decreases as the temperature is decreased. This gives rise to transitions from insulator-like to metallic, and from metallic to QH-like behaviors. The transition from insulator to metal shows delocalization as temperature is lowered, whereas lowering the temperature usually makes 2DES more localized.

We acknowledge useful discussions with E. I. Rashba, L. Zhang, B. A. Mason, X. C. Xie, and D. Shahar. This work was supported by grants from the National Science Foundation, DMR-89-22222 and Oklahoma EPSCoR via LEPM, and the SERC (UK).

REFERENCES

[1] M. D'Iorio, V. M. Pudalov, and S. G. Semenchinsky, *Phys. Lett. A* **150** (1990) 422.
[2] H. W. Jiang, C. E. Johnson, K. L. Wang, and S. T. Hannahs, *Phys. Rev. Lett.* **71** (1993) 1439.
[3] T. Wang, K. P. Clark, G. F. Spencer, A. M. Mack, and W. P. Kirk, *Phys. Rev. Lett.* **72** (1994) 709.
[4] S. V. Kravchenko, Whitney Mason, J. E. Furneaux, J. M. Caulfield, J. Singleton and V. M. Pudalov, submitted to *Phys. Rev. B*, and references therein.
[5] D. E. Khmelnitskii, *Phys. Lett.* **106A** (1984) 182; *Helvetica Physica Acta* **65** (1992) 164.
[6] R. B. Laughlin, *Phys. Rev. Lett.* **52** (1984) 2304.

HALL INSULATOR IN SILICON INVERSION LAYERS IN THE HIGH-FIELD EXTREME QUANTUM LIMIT

S. V. Kravchenko and J. E. Furneaux

Laboratory for Electronic Properties of Materials and Department of Physics and Astronomy, University of Oklahoma, Norman, OK 73019

V. M. Pudalov

Institute for High Pressure Physics, Troitsk, 142092 Moscow District, Russia

A new insulating state for two-dimensional (2D) electron systems was predicted recently: the Hall insulator.[1,2] In this state, both diagonal and Hall conductivities, σ_{xx} and σ_{xy}, tend to zero with decreasing temperature such that $\sigma_{xy} \propto \sigma_{xx}^2$. This gives a diverging diagonal resistivity, $\rho_{xx} \to \infty$, and a *finite* Hall resistivity, ρ_{xy}, close to its classical value $B/n_s ec$ (here B is magnetic field, n_s is electron density, e is the electron charge, and c is the speed of light). Therefore, this state differs qualitatively from a Mott insulator, in which both ρ_{xx} and ρ_{xy} tend to infinity as $T \to 0$. According to the global phase diagram for the metal/insulator (M/I) transition proposed in [2], this behavior is generic for transitions into an insulating state involving 2D systems, exhibiting the quantized Hall effect (QHE). For GaAs/AlGaAs heterojunctions, the Hall insulator state is expected for Landau level filling factor, $\nu = n_s ch/eB$, around 1/5 for high mobility samples or for ν around 1/3 for samples with lower mobility. For the even more disordered 2D systems in silicon, exhibiting no fractional quantum Hall effect, the transition to the Hall insulator state is expected for filling factors below $\nu = 1$.

Experimental data on GaAs/AlGaAs confirmed that the Hall resistance remains nearly equal to its classical value in the insulating states around $\nu = 1/7$,[3] below $\nu = 1/3$,[4] and around $\nu = 1/5$ [5] (the data in [3] had been obtained before the theory [1,2]). Results are in a good agreement with predictions of theory [2]; however, it was pointed out [4,5] that a finite Hall resistance is also characteristic of transport in the presence of a pinned Wigner solid previously reported for the same conditions (see, e. g., [6–8]). For the 2D system in silicon, the Hall resistance has been studied in low magnetic fields,[9,10] in the presence of a reentrant insulating state. This insulating state is also likely to be caused by the strong Coulomb interaction and was well explained in terms of the formation of a pinned electron solid (see [11] and references therein). In fact, the M/I phase diagram experimentally found in [10] is inconsistent with the global phase diagram,[2] and, therefore, the classical value of the Hall resistance observed in [10] was explained in terms of dislocation motion within the electron solid.

Here we concentrate on the essencially single-particle situation, *i.e.*, on the transition from the lowest integer QHE-state, $\nu = 1$, to an insulator, which occurs with increasing magnetic field. In the high B extreme quantum limit, the existence of the electron solid is doubtful even at low temperatures [12,13] and can be excluded at higher

T. We used two Si MOSFET's from different wafers with peak mobilities of 1.9×10^4 cm^2/Vs (Si-1) and 3.1×10^4 cm^2/Vs (Si-2). Both samples have Hall bar geometry and dimensions 5×0.8 mm^2.

Figure 1 shows a typical dependence of ρ_{xx} and ρ_{xy} on electron density, n_s, in the high magnetic field limit (at Landau level filling factor $\nu < 1$). The dashed curve corresponds to the calculated classical Hall resistivity for free electrons. The data at $n_s > 2.20 \times 10^{11}$ cm^{-2} is at the edge of the $\nu = 1$ Hall plateau with $\rho_{xx} = 0$ within our experimental accuracy. When electron density is lowered, ρ_{xx} increases approximately exponentially to about 1 $M\Omega$ while ρ_{xy} remains at approximately its classical value, B/n_sec (dashed curve). We were unable to take data at $n_s < 1.78 \times 10^{11}$ cm^{-2} because contact resistance became very high resulting in large noise and irreproducible measurements.

The temperature dependence for both components of the resistivity tensor at two different electron densities is shown in Fig. 2. Dashed lines show the classical Hall resistivities. Again, ρ_{xy} remains nearly constant while ρ_{xx} increases by orders of magnitude as the temperature is lowered. The deviations of ρ_{xy} from its classical value, especially pronounced in Fig. 2 (a), were irreproducible. These deviations are probably caused by increasing contact noise mentioned above. High contact resistance made it impossible to perform measurements at lower temperatures.

At low temperatures we observe that transport in the insulating phase in the extreme quantum limit is strongly nonlinear similar to measurements reported in [12,13]; therefore, all the measurements reported here were carried out in the linear regime with electric fields below the nonlinearity threshold. To eliminate the admixture of the diagonal voltage into the Hall voltage, we extracted the antisymmetric contribution from measurements taken in two opposite magnetic field directions.

The observed independence of the Hall resistance on temperature and the observed consistency with the classical value in the extreme quantum limit insulator region are consistent with theoretical predictions [1,2] for the Hall insulator. Our results agree also with the global phase diagram proposed for this state in:[2] We see that in our MOSFET's, where the 2D systems have a relatively large disorder potential, the transition to the Hall insulator state proceeds directly from the $\nu = 1$ quantum Hall liquid with increasing magnetic field. Recent transport measurements [13] show that a second possible state of 2D electrons in the extreme quantum limit, an electron solid, is very unlikely even at low temperatures. At higher temperatures, it can be definitely rulled out because estimates for the melting temperature of the electron solid in Si MOSFET's at infinite magnetic field [14] give $T_m \sim e^2(\pi n_s)^{1/2}/\Gamma_m k_B \epsilon \sim 900$ mK for $n_s \sim 10^{11}$ cm^{-2} (here ϵ is the dielectric constant and $\Gamma_m = 127 \pm 3$; [6] this sets an upper limit for solid formation at T_m.

In summary, we have experimentally shown that in the extreme quantum limit, the Hall resistance of two-dimensional electron gas of Si MOSFET's remains independent of temperature and close to its classical value even when ρ_{xx} reaches $\sim 10^8$ Ω. It is very unlikely that the electron solid formation could be responsible for the observed

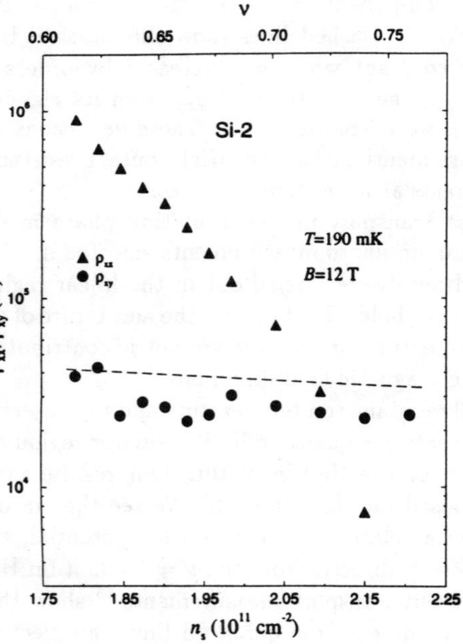
ν
0.60
0.65
0.70
0.75
Si-2
ρxx
ρxy
T=190 mK
B=12 T
ρxx, ρxy (Ω)
10^6
10^5
10^4
1.75
1.85
1.95
2.05
2.15
2.25
ns (10^11 cm^-2)

Fig. 1

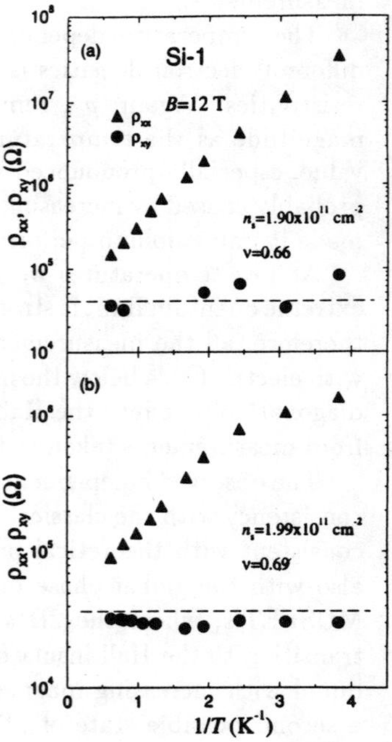
(a)
Si-1
B=12 T
ρxx
ρxy
ns=1.90x10^11 cm^-2
ν=0.66
(b)
ns=1.99x10^11 cm^-2
ν=0.69
ρxx, ρxy (Ω)
10^8
10^7
10^6
10^5
10^4
0
1
2
3
4
1/T (K^-1)

Fig. 2

constancy of ρ_{xy} and diverging ρ_{xx} in the conditions of our experiment. Therefore, our data indicate development of the single-particle Hall insulating state in accordance with theory [1,2].

We acknowledge useful discussions with Bruce Mason, Ryan Doezema, and George Kravchenko, and experimental assistance from Alison Japikse and Jason Dotter. This work was supported by grants from the National Science Foundation, DMR 89-22222 and Oklahoma EPSCoR via LEPM.

1. O. Viehweger and K. B. Efetov, J. Phys. Cond. Mat. **2** (1990) 7049.

2. S. Kivelson, D.-H. Lee, and S.-C. Zhang, Phys. Rev. B **46** (1992) 2223.

3. V. J. Goldman, M. Shayegan, and D. C. Tsui, Phys. Rev. Lett. **61** (1988) 881; J. Wakabayashi, A. Fukano, S. Kawaji, K. Hirakawa, H. Sakaki, Y. Koike, and T. Fukase, J. Phys. Soc. Jpn. **57** (1988) 3678.

4. S. I. Dorozhkin, A. A. Shashkin, G. V. Kravchenko, V. T. Dolgopolov, R. J. Haug, K. von Klitzing, and K. Ploog, Pis'ma Zh. Eksp. Teor. Fiz. **57** (1993) 55 [JETP Lett. **57** (1993) 58].

5. V. J. Goldman, J. K. Wang, Bo Şu, and M. Shayegan, Phys. Rev. Lett. **70** (1993) 647.

6. D. C. Glattli, G. Deville, V. Duburcq, F. I. B. Williams, E. Paris, B. Etienne, and E. Y. Andrei, Surf. Sci. **229** (1990) 344.

7. V. J. Goldman, M. Santos, M. Shayegan, and J. E. Cunningham, Phys. Rev. Lett. **65** (1990) 2189.

8. H. W. Jiang, H. L. Stormer, D. C. Tsui, L. N. Pfeiffer, and K. W. West, Phys. Rev. B **44** (1991) 8107.

9. S. V. Kravchenko, J. A. A. J. Perenboom, and V. M. Pudalov, Phys. Rev. B **44** (1991) 13513.

10. V. M. Pudalov, M. D'Iorio, and J.W. Campbell, Pis'ma Zh. Eksp. Teor. Fiz. **57** (1993) 592) [JETP Lett. **57** (1993) 608].

11. M. D'Iorio, V. M. Pudalov, and S. G. Semenchinsky, Phys. Rev. B **46** (1992) 15992.

12. V. T. Dolgopolov, G. V. Kravchenko, A. A. Shashkin, and S. V. Kravchenko, Phys. Rev. B **46** (1992) 13303; Surf. Sci. **305** (1994) 96.

13. A. A. Shashkin, V. T. Dolgopolov, and G. V. Kravchenko, Phys. Rev. B **49** (1994) 14486.

14. H. Fukuyama and D. Yoshioka, J. Phys. Soc. Japan **48** (1980) 1853.

MAGNETOTRANSPORT IN SEMIMETALLIC AND SEMICONDUCTING InAs/$Ga_{1-x}In_xSb$ HETEROSTRUCTURES

K.S.H. Dalton, D.M. Symons, M. van der Burgt, M. Lakrimi, R.J. Nicholas, N.J. Mason and P.J. Walker.

Physics Dept., Oxford University: Clarendon Laboratory, Parks Rd., Oxford OX1 3PU, UK

Abstract

We have studied the magnetotransport characteristics of semimetallic and semiconducting superlattices of the crossed gap system InAs/$Ga_{1-x}In_xSb$ in magnetic fields up to 45 T. In the case of intrinsic semimetallic structures at low temperatures, we observe large oscillations in ρ_{xy}, corresponding to strong minima in σ_{xx}. These features become much more pronounced as the temperature is decreased to 0.6 K. We relate these features to the movement of the Fermi level, E_F, between localised and extended states. We have used high temperature magnetotransport to measure the electron and hole densities for both a semiconducting and a semimetallic superlattice, and thus determine their energy gaps. By calculating the positions of the electron and hole confinement energies and using the measured energy gaps, we have deduced that the band overlap increases by ~30 meV when the temperature is increased from 4 K to 300 K.

InAs/$Ga_{1-x}In_xSb$ is a crossed gap type II system with the conduction band minimum of InAs lying below the valence band maximum of the $Ga_{1-x}In_xSb$. For sufficiently thin layers, the confinement energy of the electrons in the InAs, E_0^e, lies above the confinement energy of the holes in the GaSb, E_0^h, and the system is semiconducting. For structures with thicker layers, E_0^e lies below E_0^h and the system is semimetallic, with electrons transferred from the valence band of the $Ga_{1-x}In_xSb$ into the conduction band of the InAs.[1] In the structures grown to date, there is usually an additional contribution to the electron density due to donors[2] and pinning of the Fermi level.[3] However, the samples studied here have been grown by MOVPE and have only a low extrinsic doping level,[4] with electron to hole density ratios as low as 1.1.

Firstly, we describe magnetotransport measurements up to 45 T, which have been made on semimetallic structures at temperatures down to 0.6 K. The Hall resistivity, ρ_{xy}, shows large oscillations, as reported in our previous work,[5,6] and we have now found that these become much stronger at low temperatures. ρ_{xx} and ρ_{xy} for a 20 period InAs/$Ga_{0.9}In_{0.1}Sb$ superlattice are shown in Figure 1, along with σ_{xx} and σ_{xy} deduced using the tensor relationship between the conductivity and resistivity. Maxima in ρ_{xx} and ρ_{xy} coincide with minima in σ_{xx} and σ_{xy}. As the temperature is decreased, the maxima in ρ_{xy} become higher and widen, and the conductivity minima, particularly in σ_{xx}, become broader and flatter, with an almost square minimum at the lowest temperature.

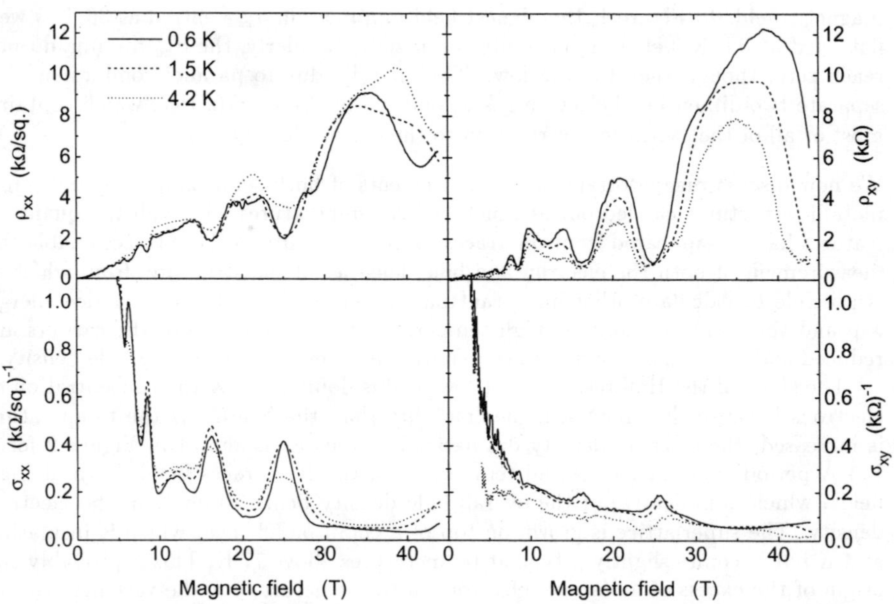

Figure 1. The resistivity and conductivity components ρ_{xx} and σ_{xx}, ρ_{xy} and σ_{xy}, for a 20-period semimetallic superlattice with InAs($Ga_{0.9}In_{0.1}Sb$) layer thicknesses of 440(150) Å. A fit to low-field data gives $n_e(n_h) = 9.2(8.5)\times10^{15}$ m^{-2} and $\mu_e(\mu_h) = 10.9(0.8)$ m^2(Vs)$^{-1}$. As the temperature is decreased from 4.2 to 0.6 K, the maxima in ρ_{xy} become larger, while the minima in σ_{xy} and particularly σ_{xx} become wider and flatter.

This strong temperature dependence leads us to interpret the minima in σ_{xx} as being in the regions of field for which the Fermi level, E_F, is within localised states for both electrons and holes. These regions would be expected to become wider as the temperature is decreased and thermal excitation to extended states in the system becomes less probable. We find that when σ_{xx} is at a minimum, σ_{xy} is also at a minimum, tending to the quantised value $\sigma_{xy} = (\nu_e - \nu_h)e^2/h = 0$. This is because the net carrier density is small enough that, at high and moderate fields, the Landau level filling factors ν_e and ν_h are effectively equal. The excess electrons are only enough to populate localised states in the tail of a Landau level and so do not alter the effective filling factor, i.e., the number of Landau levels with the extended states filled.

Between these regions there are peaks in σ_{xx}, corresponding to regions where the centres of Landau levels cross and E_F is within extended states. Both σ_{xx} and σ_{xy} show peaks and ρ_{xx} and ρ_{xy} fall to non-zero minima. These features narrow with decreasing temperature, as localisation becomes more important.

For a superlattice with almost equal electron and hole densities, we would expect the minima in σ_{xx} and σ_{xy} to become zero at sufficiently low temperature and high magnetic field. In Figure 1, the highest-field minimum in σ_{xx} centred at 36 T is well flattened at 0.6 K, yet does not quite reach zero. Similarly, the σ_{xy} minima do not reach zero, though they too are low. This may be due to parallel conduction of a separate two-dimensional electron gas at one end of the superlattice, which contains most or all of the extrinsic contribution to the carrier density.

We now discuss magnetotransport measurements of both semiconducting and semimetallic structures carried out at much higher temperatures where all the quantum features have disappeared from the traces. Two-carrier fits to these traces enable the measurement of both the electron and hole densities of the structure, from which it is possible to calculate other important parameters such as the superlattice energy gap and the band overlap.[7] At high temperatures, the mobilities of the carriers are reduced and high magnetic fields are required in order to measure the hole density.

The slope of the Hall resistance at low field is dominated by the concentration of electrons because they have a higher mobility than the holes. As the temperature is increased, the electron density dramatically increases as shown in Figure 2 for a 145 Å period superlattice. At all temperatures, the Hall traces bend down at high fields, which is indicative of the overall hole density being larger than the electron density. The superlattice is grown on top of a GaSb buffer layer which is insulating at 4 K but becomes slightly p-type at temperatures above 77 K. This is probably the origin of the excess of holes over electrons in the structure, but the vast majority of the carriers are intrinsic.

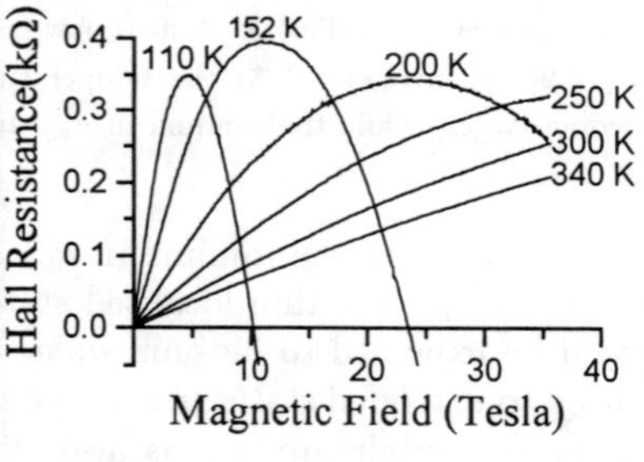

Figure 2. The measured Hall resistance traces for the semiconducting 100 period superlattice.

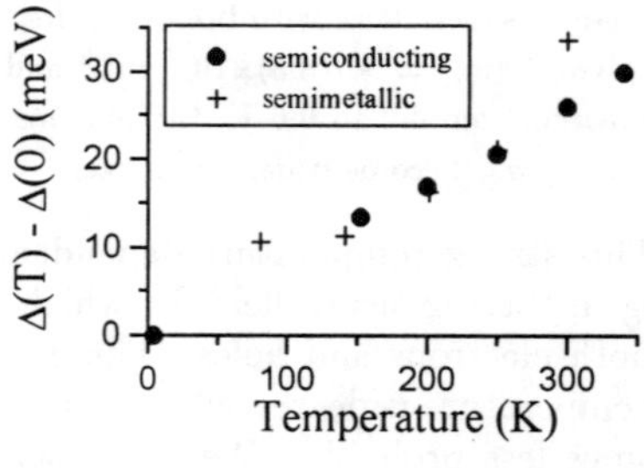

Figure 3. The band overlap, $\Delta(T)$, as a function of temperature for both samples.

When a magnetic field is applied to the structure, the effective energy gap of the superlattice increases due to the increase in zero point motion, $\frac{1}{2}\hbar\omega_c$, of both the electrons and the holes. We have included the field-dependence of the gap, and modelled the carrier densities using two-dimensional Fermi statistics for the electrons and three-dimensional Fermi statistics for the holes. We include the field-dependence of the electron and hole densities in the classical two-carrier expression for the Hall resistance giving a fit to the experimental traces. We can deduce the energy gap of the superlattice at each temperature by fitting the zero-field densities of electrons and holes and using the known band structure parameters such as electron and hole effective masses.

We find that the energy gap of the semiconducting sample is 20 meV at 340 K, increasing to 50 meV at 150 K, and it has been measured to be 65 meV at 4 K by photoconductivity measurements.[8] Figure 3 shows the temperature dependence of the band overlap which is calculated from the measured energy gap, taking into account the band bending produced by the carriers confined in the wells. The band overlap is seen to increase by about 30 meV when the temperature is increased from 4 K to 300 K. Using values for the deformation potentials of the InAs conduction band and the GaSb valence band and the thermal expansion coefficients of the bulk semiconductors,[9] the model solid theory of Van de Walle[10] predicts that the band overlap should be 25 meV larger at 300 K than at 0 K, which agrees very well with our experimental results.

We have performed similar measurements on a thicker period semimetallic sample (with InAs(GaSb) layer thicknesses of 160(105) Å). The higher temperature Hall resistance traces are very similar to those measured for the semiconducting sample, indicating that nearly all the carriers are intrinsic but that there is a small excess of holes over electrons. When the temperature is reduced, the sample remains conducting all the way to zero temperature because the energy gap of this superlattice is negative. This sample can be modelled in a very similar way to that described above for the short period sample and yields the temperature dependence of the band overlap which is shown in Figure 3. This agrees very well with both the results for the semiconducting sample and the value predicted by theory.

References

1. L.L. Chang *J. Phys. Soc. Japan* **49** (1980) 997.
2. J. Luo, H. Munekata, F.F. Fang and P.J. Stiles *Phys. Rev.* **B 38** (1988) 10142.
3. M. Altarelli, J.C. Maan, LL. Chang and L. Esaki *Phys. Rev.* **B 35** (1987) 9867.
4. M. Lakrimi, R.W. Martin, N.J. Mason R.J. Nicholas, and P.J. Walker *J. Cryst. Growth* **110** (1991) 677.
5. K.S.H. Dalton, R. Bogaerts, R.W. Martin, M. Lakrimi, D.M. Symons, R.J. Warburton, R.J. Nicholas, F. Herlach, N.J. Mason and P.J. Walker *Physica B* **184** (1993) 202.
6. K.S.H. Dalton, M. van der Burgt, M. Lakrimi, R.J. Warburton, M.S. Daly, W. Lubczyński, R.W. Martin, D.M. Symons, D.J. Barnes, N. Miura, R.J. Nicholas, N.J. Mason and P.J. Walker *Surf. Sci.* **305** (1994) 156
7. J. P. Omaggio, J. R. Meyer, R. J. Wagner, C. A. Hoffman, M. J. Yang, D. H. Chow and R. H. Miles *Appl. Phys. Lett.* **61** (1992) 207.
8. T.A. Vaughan, M. Lakrimi, R.J. Nicholas, N.J. Mason and P.J. Walker *Proc. 22nd International Conference on the Physics of Semiconductors* (1994) to be published.
9. Landolt-Börnstein *Numerical Data and Functional Relationships in Science and Technology* (Springer-Verlag, New York 1982), Vol. **17a**.
10. C. G. Van de Walle *Phys. Rev.* **B39** (1989) 1871.

LOCALIZATION AND AN INCREASE OF THE EFFICIENCY OF IMPACT IONIZATION IN SEMI-INSULATING GaAs NEAR THE ULTRA QUANTUM LIMIT REGIME

J. ŁUSAKOWSKI*, S. HUANT** AND M. GRYNBERG*
*Institute of Experimental Physics, University of Warsaw
ul. Hoża 69, 00-681 Warsaw, Poland
**High Magnetic Field Laboratory, MPIF and CNRS
25, Avenue des Martyrs, 38042 Grenoble, France

ABSTRACT

Measurements of conductivity (σ) of optically excited electrons in semi-insulating GaAs were performed for the magnetic field (B) up to 21T at 4.2K, for the electric field up to $E_{max} = 65$V/cm. Impact ionization of metastably populated shallow bound states was observed and its efficiency was estimated by $\gamma = \sigma(E_{max})/\sigma(E \to 0)$. γ decreased by about 10^3 times with B growing from 0T to 8T, then remained constant up to 13T and rose again by a factor of a few with B increasing to 21T. A decrease of γ for $B < 8$T was caused by magnetic field induced localization of electrons in a tail of the density of states of the conduction band. It is proposed that the increase of γ for strongest B was caused by quasi-resonant optical phonon assisted excitations of electrons from $N = 0$ to $N = 1$ Landau level. This process increased a number of weakly localized electrons which caused a more effective impact ionization.

1. Introduction

Impact ionization of shallow impurities has been a widely investigated hot-electron phenomenon of the physics of semiconductors. In the last decade much attention was devoted to an influence of the magnetic field (B) on this process[1]. In a typical experiment a high quality n- or p-type material was studied and a value of B did not exceed a few tenths of Tesla. On the other hand, studies of impact ionization at magnetic fields up to a few Tesla were carried out on undoped semi-insulating (SI) GaAs[2]. In this material, at liquid helium temperatures (LHT), the Fermi level is pinned to the mid-gap dominant level, the EL2[3], which overcompensates residual shallow donors and shallow acceptors. A random spatial distribution of charged centers is a source of long-range fluctuations of the electrostatic potential, which lead to a tail of localized states at the bottom of the conduction band. A lack of free carriers disables conductivity measurements at LHT at a thermal equilibrium. A permanent sub-band illumination of a sample leads to a non-zero concentration of conducting electrons, which occupy both localized and extended states in the conduction band, and to neutralization of some of shallow donors. The illumination makes the conductivity measurements possible, and an avalanche breakdown caused by impact ionization of shallow bound states takes place for a sufficiently high applied electric field. It was shown[2] that application of the magnetic field of a few Tesla led to a rapid decrease of an efficiency of impact ionization due to magnetic field induced localization of electrons.

In the present study, investigations of impact ionization in SI-GaAs are extended to the magnetic field up to 21T. It is shown that the drop of an efficiency of impact ionization caused by localization is followed by its increase at magnetic fields corresponding to the ultra quantum limit regime. It is proposed that this is caused by a quasi-resonant, optical phonon assisted excitation of electrons from localized states at the bottom of $N = 0$ Landau level (LL) to weakly localized states at the bottom of $N = 1$ LL.

2. Experiment and Results

The measurements were carried out at 4.2K on a sample of undoped SI-GaAs with the EL2 concentration equal to $5.3 \cdot 10^{16}\text{cm}^{-3}$, supplied with Au/Ge/Ni contacts. Before the measurements the EL2 centers were bleached to the metastable state[3]. The sample was illuminated with a sub-band (0.846μm) laser light. A foton flux absorbed by the sample was estimated to be of the order of 10^{13}s^{-1}. The electrical circuit consisted of the sample connected in series with a voltage source and a ammeter. A Keithley 617 electrometer was used for measuring a voltage drop between two side contacts.

Current-voltage characteristics for a few values of the magnetic field are shown in Fig.1. For each characteristic a step of the current is observed. A threshold electric field (E_{th}) corresponding to a step increases from 17V/cm for $B = 0$T to 45V/cm for $B = 7$T, then decreases by about 15V/cm for B increasing to 8T and then increases again to reach 55V/cm for $B = 21$T. The height of the step decreases about 10^3 times for the magnetic field growing form 0T to 7T and then increases again by a factor of a few.

To evaluate an efficiency of impact ionization a parameter $\gamma = \sigma(E_{max})/\sigma(E \to 0)$ was calculated. Fig.2 shows the logarithm of γ as a function of the magnetic field. The plateau of γ for 9T$< B <$13T results from a similar slope of both $\sigma(B)$ curves at this range of B (see the inset of Fig. 2).

3. Discussion

Results of measurements for low magnetic fields are in agreement with previous studies[2]. The avalanche breakdown observed as a step on a current – voltage characteristics is interpreted as impact ionization of shallow donors and states created

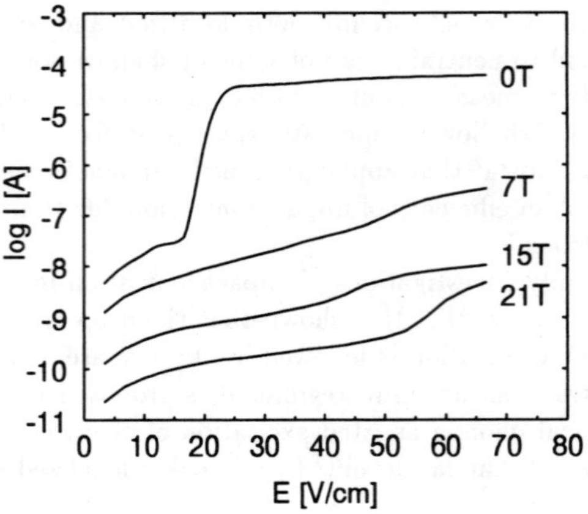

Figure 1: Current – voltage characteristics of SI-GaAs at 4.2K for the magnetic field equal to 0T, 7T, 15T and 21T.

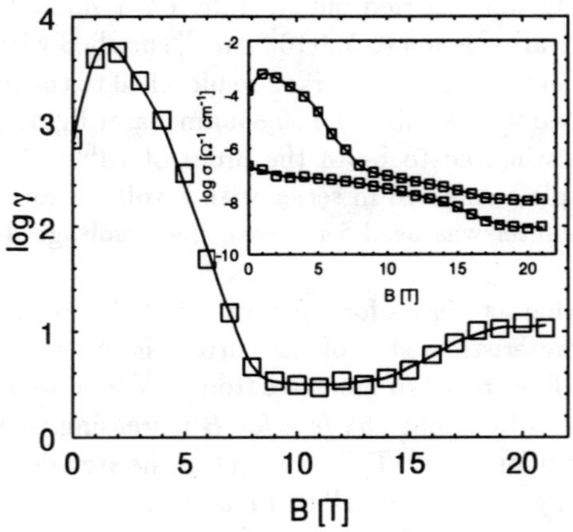

Figure 2: A magnetic field (B) dependence of $\gamma = \sigma(E_{max})/\sigma(E \to 0)$. The inset shows $\sigma(B)$ dependence for $E = E_{max}$ (the upper curve) and $E \to 0$ (the lower one).

by potential fluctuations. A rapid decrease of γ for B growing from 0T to 8T results from a shift of the quasi-Fermi level below the mobility edge[2]. This is caused by an increase of the density of states at the bottom of the conduction band with an increase of B and leads to magnetic field induced localization of hot electrons. The mechanism leading to the maximum of $\sigma(B)$ dependence for $B \approx 1$T (the upper curve in the inset of Fig. 2) will be discussed elsewhere.

The localization does not cause a total suppression of impact ionization. However, a decrease of E_{th} for the magnetic field increasing from 7T to 8T suggests that in the localization regime ($B > 7$T) one observes impact ionization of states with a smaller ionization energy than that of states which were ionized for $B < 8$T. It is proposed that the weakly bounding states are localized states in a tail of $N = 1$ LL. This is based on the fact that an influence of potential fluctuations on a motion of electrons is weaker for $N = 1$ LL than for $N = 0$ LL [4].

Optically excited electrons thermalize by a rapid emission of optical phonons. The energy distribution of hot electrons is expected to be limited by the energy of optical phonons $\hbar\omega_{ph}$, which is equal to about 34meV at the Γ point[5]. Thus, for 10T ¡ B ¡ 20T only two lowest Landau levels ($N = 0$ and $N = 1$) are occupied and for $B \approx 20$T the ultra quantum limit is reached. Electrons are excited from $N = 0$ LL to $N = 1$ LL by absorption of non-equilibrium optical phonons. If a cyclotron resonans separation between the bottoms of the two LL is less than $\hbar\omega_{ph}$ then electrons are excited mainly to extended states of $N = 1$ LL. However, when this two energies are close to each other, the final states of the phonon absorption process belong to localized states in a tail of the $N = 1$ LL. This results in a kink on $\sigma(B)$ curve for $E \to 0$ at $B \approx 17$T. However, there is not a kink on $\sigma(B)$ curve for $E = E_{max}$, because this electric field is high enough to cause impact ionization of states at the bottom of $N = 1$ LL. Thus, the efficiency of impact ionization increases near the ultra-quantum-limit regime due to an increased population of weakly bounding states. This increase of γ is not sharp because the density of states of the Landau levels is broadened by potential fluctuations and the energy spectrum of optical non-equilibrium phonons is expected to be smeared.

4. Acknowledgments

The authors are thankful to Dr. E. Kamińska and Dr. A. Piotrowska, Institute of Electron Technology, Warsaw, for preparing contacts to the samples. A financial support to J. Ł. from the Polish SCSR grant No. 2 P302 204 06 is kindly acknowledged.

5. References

[1] A. Brandl, M. Vöckler and W. Prettl, *Appl. Phys. Lett.* **55** (1989) 238.
[2] J. Łusakowski, R. Merten and M. Grynberg, *Acta Phys. Polonica* **82** (1992) 551.
[3] M. Kamińska, *Rev. Phys. Appl.* **23** (1988) 793
[4] B. I. Shklovskii and A. L. Efros, *Zurnal Eksperim. Teoret. Fiz.* **64** (1973) 2222.
[5] J. S. Blakemore, *J. Appl. Phys.* **53** (1982) R123.

ELECTRIC FIELD DEPENDENT PARTICLE DYNAMICS AND CONDITIONS FOR THE INTEGER QUANTUM HALL EFFECT

J. RIESS

Centre de Recherches sur les Très Basses Températures,
Centre National de la Recherche Scientifique, B.P. 166,
38042 Grenoble Cedex 9, France

ABSTRACT

The basic experimental facts of the integer quantum Hall effect (IQHE), including electric field dependent mobility edges and breakdown (onset of dissipation), reflect general properties of the solutions of the time-dependent Schrödinger equation, where the macroscopic electric field **E** is properly included. In particular, it is not only the disorder but also the value of |**E**| which determines whether a state is insulating or conducting. This leads to two conditions for the occurrence of the IQHE, which must be simultaneously fulfilled. The first condition is violated at sufficiently high values of |**E**|, which leads to dissipation. The second condition is violated at sufficiently low |**E**|. In this case one expects the plateau value of σ_{xy} to fall below (e^2/h)xinteger. This might be observable at extremely low currents in special systems with strong disorder.

1. Introduction

An important feature of the integer quantum Hall effect[1,2] (IQHE) is the following: If the total current increases, the width of the conductance plateaux shrinks until they disappear completely at currents corresponding to a critical electric field of the order of 100 Vcm^{-1}. This shows that the strength |**E**| of the electric field is a crucial parameter in the system.

The starting point of a microscopic theory of the IQHE is the Hamiltonian

$$H = [1/(2m)]\{(\hbar/i)\nabla - (e/c)\mathbf{A}(\mathbf{r})\}^2 + V(\mathbf{r}) - e\mathbf{E}\mathbf{r} \qquad (1)$$

Here V(**r**), **r** = (x,y), is a static substrate potential which contains homogeneous disorder. **B** = (0,0,B) = curl **A**(**r**) is the magnetic field perpendicular to the x-y-plane. **E** is a homogeneous electric field, i. e., H describes subdomains of a macroscopic sample, over which the macroscopic electric field **E**(**r**) can be considered constant.

The major difficulty comes from the fact that the Schrödinger equation

$$[H - i\hbar\partial_t]\psi = 0 \qquad (2)$$

cannot be solved exactly because of the disorder potential V(**r**). Therefore, various approximations and assumptions have been used in order to obtain a Hall conductance that agrees with the measured quantized value σ_{xy} =

(e^2/h)xinteger in the center of the Hall conductance plateaux. Most theories are based on the linear response approximation. However, the electric field dependent conductivities outside the conductance plateaux and in the breakdown regime indicate that such an assumption may not always be justified.

In this paper we will not make such assumptions. We will investigate some general properties of the exact solutions of the time-dependent Schrödinger equation (2), where the macroscopic electric field **E** is properly included. An important result is, that the dynamic behaviour (insulating or conducting) of each state in the disordered system depends on the strength of the electric field **E**.

2. Properties of wave functions in the presence of disorder and of a macroscopic electric field.

First we remark that in the presence of a homogeneous electric field **E** and of a disorder potential V(x,y) (which varies in *both* spatial directions) no stationary solutions of (2) exist[3] (although (1) does not contain the time explicitly). This important result indicates, that the process of electric conductance in the IQHE is intrinsically time-dependent.

This result can also be understood in another way by performing the time-dependent gauge transformation

$$\mathbf{A}'(\mathbf{r},t) = \mathbf{A}(\mathbf{r}) + \nabla f(\mathbf{r},t), \quad \psi' = \psi \exp[(ief/(\hbar c)\,], \quad f(\mathbf{r},t) = -c\mathbf{E}\mathbf{r}t, \tag{3}$$

whence

$$(H' - i\hbar\partial_t)\psi' = 0. \tag{4}$$

The Hamiltonian

$$H' = [1/(2m)]\{(\hbar/i)\nabla - (e/c)A'(\mathbf{r},t)\}^2 + V(\mathbf{r}) \tag{5}$$

depends explicitly on time. (Note that H' coincides with H if **E** = 0).

The energies ε_s in the broadened Landau bands are determined by the stationary solutions of (1) or (4) in the absence of **E**: $H[\mathbf{E}=0, V(x,y)\,]\psi_s = \varepsilon_s\psi_s$. We consider that this problem has been solved and we ask, how the states ψ_s change when **E** is different from zero, and what are the corresponding velocity components parallel and perpendicular to **E**. The former determine the dissipative conductivity σ_{xx} and the latter the Hall conductivity σ_{xy}.

It has been found[3] that for each state ψ_s two limits exist: 1. For sufficiently low $|\mathbf{E}|$ the macroscopic velocity parallel to **E** vanishes. 2. For sufficiently high $|\mathbf{E}|$ the Hall velocity of the state tends towards the value $v_d = c|\mathbf{E}|/B$ (i. e., to the classical value in the absence of the disorder potential V(x,y).

The low $|\mathbf{E}|$-limit corresponds to the adiabatic[4] limit of the time-dependent Schrödinger equation (4). The adiabatic solutions are momentary eigenfunctions of H'[V(x,y), **E**, t]. Due to the absence of symmetry (presence of disorder) the corresponding energies $\varepsilon_s(t)$ do not intersect[5]. In addition, in a macroscopic system the levels $\varepsilon_s(t)$ are closely spaced. It follows then from the general relation

$$\partial_t\langle\psi'|H'\psi'\rangle = e\mathbf{E}\langle\psi'|\mathbf{v}'_{op}|\psi'\rangle = e\mathbf{E}\langle\psi|\mathbf{v}_{op}|\psi\rangle \qquad (6)$$

that the velocity component parallel to **E** of an adiabatic state rapidly oscillates around zero and therefore does not contribute to the macroscopic current.

The existence of the high **E**-limit can be proved[3, 6] by transforming Eq. (2) to a frame of reference which moves with the drift velocity $\mathbf{v}_d = (c/B^2)\mathbf{E}\mathbf{x}\mathbf{B}$. In the transformed equation the electric field **E** vanishes and the disorder potential moves with the constant velocity $-\mathbf{v}_d$. The high $|\mathbf{E}|$-limit is reached when the solution of the transformed Schrödinger equation is in the sudden approximation limit[4]. This limit occurs whenever $v_d = c|\mathbf{E}|/B$ is sufficiently high (the solution feels then only some time average of the moving disorder potential). The Hall velocity of the back transformed sudden approximation solution is just $\mathbf{v}_d$.

When the Fermi level is in a range of adiabatic levels the total energy (in the representation H') of the system does not increase with time. Hence, according to (6) no net current parallel to **E** is possible, i. e., the dissipative conductivity vanishes. On the other hand, when **E** is increased such that the states near ε_F are no longer in the adiabatic limit, i. e., when nonadiabatic transitions between the levels near ε_F occur, the energy (in the representation H') of the system increases with time. This is equivalent to a current parallel to **E**, hence to a nonvanishing dissipative conductivity. (Dissipation occurs because the accumulated excess energy above ε_F is periodically lost to the surrounding heat bath by inelastic scattering events).

In order to calculate the left hand side of Eq. (6) for each state, one needs to know the ($|\mathbf{E}|$-dependent) nonadiabatic transition probabilities between the disorder perturbed Landau levels ε_s. However, at present these transition probabilities can only be calculated in special cases, but not in the presence of an arbitrary disorder potential. Nevertheless, a qualitative measure of the transition probabilities can be obtained from an estimation of the electric "threshold" fields $E^{th}(s)$, which characterize the crossover from the low **E**-limit to the high **E**-limit of a particular state ψ_s. $E^{th}(s)$ is defined by the following inequality: a state ψ_s is in the low (high) field limit if $|\mathbf{E}| \ll (\gg) E^{th}(s)$.

It is important to know how the values $E^{th}(s)$ are distributed among the states ψ_s in a broadened Landau band. It has been found[6] that states at the band edges have the highest fields $E^{th}(s)$, for which an estimation gave values of the order of 100 Vcm^{-1} (for typical disorder potentials). This is consistent with experimental values of the breakdown fields of the integer quantum Hall effect (IQHE). Further, the values of $E^{th}(s)$ rapidly deacrease from the edges towards the center of the bands, where they are estimated to be many orders of magnitude smaller[6]. This leads to an electric field dependent mobility edge, corresponding to a decrease of the conductance plateaux when the total current is increased, as observed in experiments.

3. Conditions for the IQHE

1. In order to have $\sigma_{xx} = 0$ it is necessary that all states ψ_s near the Fermi level ε_F are in the adiabatic limit, i. e., that the condition $E^{th}(\varepsilon_F) >> |\mathbf{E}|$ is fulfilled. (This is equivalent to the statement that these states do not contribute to the macroscopic current at the given value of $|\mathbf{E}|$). At sufficiently high currents (i. e., high $|\mathbf{E}|$) this condition is no longer fullfilled, hence dissipation occurs (see sec. 2). This represents the usual breakdown of the IQHE.

2. In order to observe $\sigma_{xy} = (e^2/h)n$ it is necessary that conducting (i. e., nonadiabatic) states exist in the band center (in addition to the fulfilment of condition 1.) But this is not sufficient. A further requirement is that among these states there must be states which are in the high $|\mathbf{E}|$-limit (sudden approximation) at the given experimental value of $|\mathbf{E}|$ (i. e., states ψ_s for which $E^{th}(s) << |\mathbf{E}|$). Further, there must exist at least one path P across the sample (in a direction parallel $\mathbf{E}$) along which such states percolate. (Finally, in a small neighbourhood of P the particle density must be that of n fully occupied Landau levels. This last condition is always fulfilled if ε_F is sufficiently above the band center, since the states with the lowest fields $E^{th}(s)$ have been shown[6] to lie along zero equi-potential lines of V(x,y), which corresponds to the center of a broadened Landau band). From conditions 1. and 2. together it can be proved[7] that $\sigma_{xy} = (e^2/h)n$. The proof (and the reason for condition 2.) becomes particularly transparent in model systems, where the electric field induced nonadiabatic transition probabilities between the levels ε_s are known and the velocity of each conducting state can be calculated explicitly[8].

Condition 2. is no longer fulfilled if $|\mathbf{E}|$ is so low that even the states in the band center tend towards the adiabatic limit. In this case the value of σ_{xy} at integer filling factors is expected to be below (e^2/h)xinteger. This could be called the low current breakdown of the IQHE. This effect might perhaps be observable at extremely low currents in specially prepared samples with strong disorder.

4. References

1. K. von Klitzing, G. Dorda and M. Pepper, Phys. Rev. Lett. **45** (1980) 494.
2. R. E. Prange and S. M. Girvin, Eds. "*The Quantum Hall Effect*", (Springer Verlag, New York) 1990.
3. J. Riess and P. Magyar, Physica B**182** (1992) 149.
4. A. Messiah, *Quantum Mechanics*, North Holland, Amsterdam, (John Wiley & Sons, INC., New York-London-Sidney 1966) Vol. II, ch. XVII.
5. J. von Neumann and E. Wigner, Phys. Z. **30**, (1929) 467.
6. J. Riess, Physica B**190** (1993) 366.
7. J. Riess, in "*High Magnetic Fields in Semiconductor Physics III*", (G. Landwehr, ed.), Springer Ser. Solid State Sc. Vol. **101** (Springer-Verlag, Berlin, Heidelberg,1992); J.Riess, Physica B, in press.
8. J. Riess, Z. Phys. B77 (1989) 69; D. Bicout, P.-A. Hervieux and J. Riess, Phys. Rev. B**45** (1992) 9577 and references quoted therein.

ELECTRONIC PROPERTIES OF TYPE II PbTe/PbS SUPERLATTICES

F.F.Sizov, J.V.Gumenjuk-Sichevskaya, V.V.Tetyorkin, and V.V.Zabudsky

Institute of Semiconductor Physics, 252028, pr.Nauki 45, Kiev, Ukraine

Strong magnetic field Hall coefficient and magnetoresistivity dependencies in PbTe/PbS semimetallic SLs with periods of 12.0÷100.0 nm were observed. From the analysis of these dependencies by Monte-Carlo fitting procedure the band-offset ΔE_v=0.32±0.05 eV was obtained (T= 77 K) and it was proved that these SLs are type II "misaligned "structures. The band structure calculations of PbTe/PbS SLs in the envelope wave function approximation showed that in such SLs a semimetal-semiconductor transition should occur for layer thicknesses of ≈6.0 nm.

1. *Introduction.*

Investigated up to now IV-VI superlattices (SLs) as a rule belong to type I SLs.[1] At the same time type II SLs offer some interests for infrared (*IR*) detector applications as it is possible to realize in such kind of systems favorable optical properties for long-wavelength detector applications through the valence to conduction band optical transitions between the states in alternative layers due to overlap of the envelope wave functions. Type II IV-VI SLs can be realized in PbTe/PbS system as these semiconductors, with large part of ion bond, differ much by the affinity energies (ϕ(PbTe) = 4.6 eV[2], ϕ(PbS) = 3.9 - 4.2 eV[3]).

2. *Preparation of PbTe/PbS SLs.*

PbTe and PbS compounds differ substantially by lattice constants (a_0(PbTe) = 0.6460 nm and a_0(PbS) = 0.594 nm) that is a reason of misfit dislocations at the interface, which are generated in a thin layer at the interface (d≈1÷2 nm).[4] Beyond this thin highly deteriorated layer the films are unstrained and one can suppose to obtain the perfect layers in SLs with good electrical characteristics.

PbTe/PbS SLs with number of periods up to N=100 were prepared by "hot wall" epitaxy technique in a vacuum chamber with gas pressure of 10^{-5}-10^{-6} Torr at substrate temperatures T≤350 C. At these temperatures the metallurgical boundary between PbTe and PbS layers should be very pronounced due to the absence of the constituents intermixture at temperatures T<400C,[5] which is strengthened by the presence of misfit dislocations.[4]

SLs grown exhibit several regularly spaced θ-2θ scans of X-ray diffraction spectra satellites (Fig.1). From angle distance $\Delta(2\theta)$ between the satellites the SL period can be obtained: $d=\lambda/\Delta(2\theta)\cos\theta$. Appearance of numerous sharp, regularly spaced X-ray satellites is an indication of high structural quality SLs with little constituents interdiffusion.

3. *Electrical characteristics and band offset in PbTe/PbS SLs.*

Studies of the angle dependencies of transverse magnetoresistance to reveal the quasi-two-dimensional carriers conductivity, showed very pronounced $\rho_\perp/\rho_0$ dependencies in

PbTe/PbS SLs. To obtain the quantitative values of the band offset in these objects the studies of the temperature and magnetic field dependencies of Hall coefficient R_H(H) and transverse magnetoresistance $\rho_\perp$(H) were carried out. To analyze the experimental dependencies in PbTe/PbS SLs the multilayer conductivity model was applied[6] and Monte-Carlo fitting procedure was used to find the equations solutions for this multilayer conductivity model.

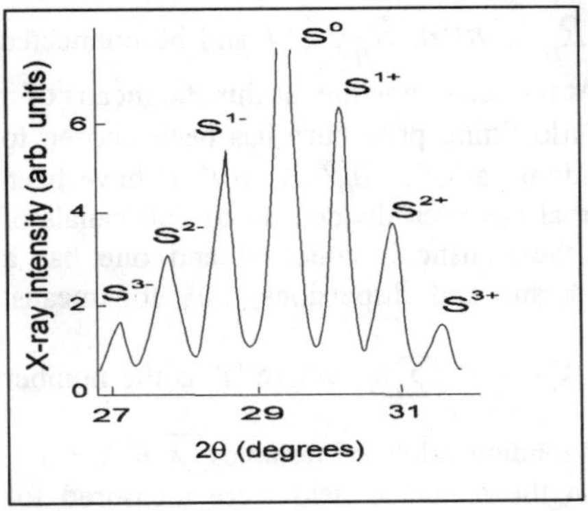

Fig.1. θ-2θ scans of X-ray diffraction spectra (CuK_α -radiation, λ = 0.154 nm) of 100-period PbTe/PbS (6.0 nm/6.0 nm) SL.

PbTe/PbS SLs can be considered as the structures with a sequence of alternative layers of semiconductors. For multilayer structure for $\rho_\perp$(H) and R_H(H) dependencies it follows[6]:

$$\frac{\rho_\perp}{\rho_o} = \frac{\sum_i d_i N_i \mu_i(0)}{\sum_i d_i N_i \mu_i(H)} \left[1 + (\mu_H H)^2\right], \tag{1}$$

$$R_H = \rho_\perp(H)\mu_H, \tag{2}$$

where the Hall mobility of a structure with different types of carriers in the layers is determined by the expression:

$$\mu_H = \frac{\sum_i d_i N_i \mu_i^2(H)}{\sum_i d_i N_i \mu_i(H)}. \tag{3}$$

and the $\mu_i(H)$ and $\mu_i^2(H)$ mobility values in "i"-layer in a magnetic field are defined by the ordinary expressions (see, e.g.,[6]). Determining the layer thicknesses and the resistivity ρ_0, it is possible to obtain the basic electrical characteristics of each separate layer in multilayer structure, analyzing R_H(H) and $\rho_\perp$(H) dependencies on the magnetic field strength.

For experimental results on R_H(H) and $\rho_\perp$(H) dependencies it is possible to find the solution of the system of Eqs. (1) and (2) with some inaccuracy ε.

$$R_j^{theor}(\mu_n^0, N, \mu_p^0, P) = R_j^{exp}(H^j) \pm \varepsilon^R; \rho_0^{theor}(\mu_n^0, N, \mu_p^0, P) = \rho_0^{exp}(H^j) \pm \varepsilon^\rho, \tag{4}$$

where $\mu_n{}^0$, N, $\mu_p{}^0$, P are unknown values of mobilities, electron and hole concentrations respectively, R_j^{exp}, $\rho_0{}^{exp}$, $\mu_H{}^{exp}$ are the experimental data at some magnetic field. The index "j" references to the variables parameters in a magnetic field and ε is the error that should satisfy the relationship: $\varepsilon << [max(R_H)/min(R_H)-1]$ and be connected with the dispersion values of system solutions. At the same time the arithmetic means of it should have weak dependence on ε. Monte-Carlo fitting procedure has been chosen to find the solution of equations system. The random variables μ_n^0, N, μ_p^0, P have been generated and substituted into Eq. (4). The interval has been divided up to 200 canals of measurements. After some quantities of run, the statistic is collected and one has a possibility to recover the value of average means and dispersions, i.e., for means: $\overline{X} = \sum_i n_i X_i \Big/ \sum_i n_i$, for dispersion: $\sigma^2 = \sum_i n_i (X_i - \overline{X})^2 \Big/ \sum_i n_i$, where "$i$" is the number of canal, n_i is its weight coefficient, X_i are corresponding values of variables, $\overline{X} = X \pm \sigma$.

The $R_H(H)$ and $\rho_\perp(H)/\rho_0$ dependencies on the magnetic field were measured for PbTe/PbS SLs, PbS films and PbTe/PbS heterostructures (Fig.2). An essential difference exists in the magnetic field dependencies for SLs, film and heterostructure. For last two objects there is almost no $R_H(H)$ and $\rho_\perp(H)/\rho_0$ dependencies, which is characteristic for films and heterostructures in the case of in-plane transport with one type of carriers. For SLs these dependencies are very strong even in weak magnetic fields which is characteristic for systems with two types of carriers with different mobilities.

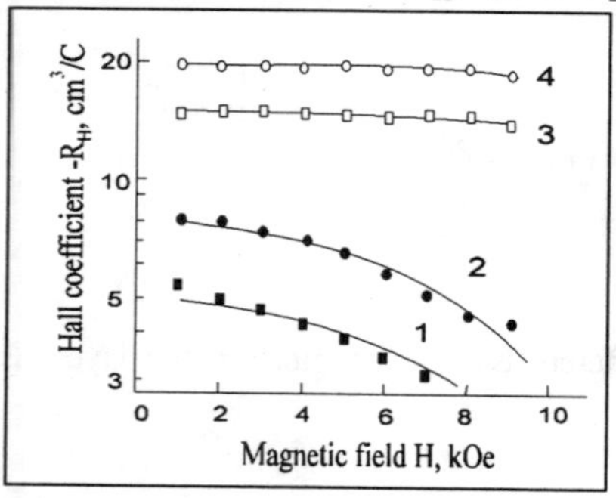

Fig.2. Magnetic field dependencies of Hall coefficient in PbTe/PbS SLs (1 - 6.0/6.0 nm, 2 - 20/50 nm), PbS film (3) and PbTe/PbS heterostructure (4 - 1.5 μm/1.5 μm). The thicknesses of all the structures are about 3 μm.

The carrier mobilities calculated in the SLs layers correspond to that observed in perfect bulk or epitaxial PbTe and PbS samples that confirms the results of X-ray diffraction analysis and assumption that outside the PbTe/PbS interface regions the layers are unstrained and misfit dislocations weakly influence the electrical properties of the layers.

The Monte-Carlo simulation procedure can be fitted with the experimental data only in the case, when in PbS layers there are holes and in PbTe layers there are electrons. In the growth conditions used PbS or PbTe layers were grown only of n-type conductivity. Thus, the change of the conductivity type in PbS layers, when they form PbTe/PbS SL, can only be in a case when the electrons from PbS layers overflow to PbTe layers. It is possible when these materials form type II "misaligned" structure, when the top of PbS valence band is above the bottom of the PbTe conduction band. Typical results for hole concentrations in PbS layers are of $P \approx 1.5\ 10^{18}$ cm^{-3} with hole mobilities of $\mu_p \approx 1.0\ 10^4$ cm^2/Vs and for electrons in PbTe buffer or SL layers there are $N \approx 6.0\ 10^{17}$ cm^{-3} and $\mu_n \approx$

3.0 10^4 cm^2/Vs. Than, calculating the position of the Fermi level in the SL, one can obtain the band-offset in this system: ΔE_v=0.32±0.05 eV (E_g(PbTe)=0.21 eV, E_g(PbS)=0.31 eV).

4. Theory of the band structure of PbTe/PbS SLs.

Using the ΔE_v-values obtained PbTe/PbS SLs band structure was calculated for [100] growth direction. The calculations were made in the envelope-function approximation[7] taking into account the bands nonparabolicity at *L*-points of the Brillouin zone.

In VI-VI semiconductors four isoenergetical ellipsoids directed along <111> axes form the same angle with respect to the SL axis. When writing a two-band Kane Hamiltonian in the geometry connected with the SL [100] axis one can apply a unitary transformation[8] to the standard Hamiltonian describing the <111> directed ellipsoids.

In calculations it was taken into account that the temperature change of the band gaps is the same as for bulk crystals (= +4.5 10^{-4} eV/K). In deriving the equation for eigenenergies there were used the boundary conditions of continuity and periodicity of the wave functions and their derivatives.

The band spectra of PbTe/PbS SLs for different layer thicknesses and for temperatures T=77 and 300 K were calculated. For layer thicknesses greater than 6.0 nm the PbTe/PbS SLs without the buffer layers should exhibit semimetallic conductivity due to hybridization of *s*-type valence band states of PbS and *p*-type conduction band states of PbTe at T=77 K. At higher temperatures the band states overlap decreases, and the transition to the state with the semimetallic conductivity should occur at higher layer thicknesses for SLs without the buffer layers. The presence of the buffer layer changes the situation, as for any PbTe layer thicknesses and for PbS thicknesses ≥3.5 nm there always would be *n*-type conductivity in PbTe buffer layer and *p*-type conductivity in PbS SL layers. Thus, such kind of a system will be a semimetallic one.

For the layer thicknesses a≤6.0 nm the PbTe/PbS SLs should go over to the semiconductor one and can be used for the design of *IR*-photosensitive structures. Estimations of the interband absorption coefficient α have shown that $\alpha \approx 10^3$ cm^{-1} at 10 μm wavelength cutoff, which is a sufficient value for *IR*-photodetector applications.

5. References.

1. F.F.Sizov and A.Rogalski, *Progr. Quantum Electron.*, **17**, 93 (1993).
2. B.Lai, G.M.Wells, F.Cerrina, *J.Vac.Sci.Technol.*, **A4**, 977 (1986).
3. V.S.Fomenko. *Properties of Matters and Materials* (Atomizdat, Moscow, 1975) (in Russian).
4. S.S.Borisova, I.F.Michailov, L.S.Palatnik, A.Yu.Sipatov, A.I.Fedorenko and L.P.Shpa kovskaya, *Kristallographia*, **34**, 716 (1989) (*in Russian*).
5. A.Kh.Abrikosov and L.E.Shelimova, *Semiconductor materials based on IV-VI Compounds* (Nauka, Moscow, 1970) (*in Russian*).
6. F.F.Sizov and S.D.Darchuk, *Phys. Stat. Sol.* (a), **113,** 107 (1989).
7. G.Bastard, *Phys. Rev.*, **B24**, 5693 (1981).
8. M. de Dios Leyva, V.Lopez-Gonar, *Phys. Stat.Sol.*(b), **138.** 253 (1986) .

COULOMB GAP IN TWO-DIMENSIONAL ELECTRON SYSTEM IN A STRONG MAGNETIC FIELD

A.L. EFROS and F.G. PIKUS
Department of Physics, University of Utah,
Salt Lake City, Utah 84112

ABSTRACT

We argue that recently observed magnetic field suppression of the tunneling current from 2D electron gas to another 2D gas or to a metal can be explained within the framework of the classical Coulomb gap.

Interest in the Coulomb gap in the single-particle density of states (DS) of two-dimensional localized electrons has been recently stimulated by observations of a tunneling gap in heterostructures[1] and quantum wells.[2] The experiments have shown that the tunneling conductivity at small bias is significantly suppressed by low temperature and by a strong magnetic field. Several theoretical groups have offered both classical[3] and quantum-mechanical[4] explanations for this phenomenon. The data obtained in heterostructures (Fig. 1) show that the gap at zero bias develops only in a strong magnetic field, and it saturates at high fields. The gap for tunneling between two quantum wells[2] also has a kind of saturation in a strong magnetic field. We think that magnetic field creates a localization of electronic states, and the nature of the gap is classical. Relatively high temperatures, disorder and small filling factors are in favor of this description.

The aim of this talk is to show that a good understanding of the experimental data on the tunneling gap can be achieved in terms of the classical Coulomb gap. The problem of the Coulomb gap was originally associated with strongly disordered systems.[5,6] The following classical Hamiltonian has been studied traditionally

$$H = \sum_i \phi_i n_i + \frac{1}{2}\sum_{i\neq j} \frac{e^2}{r_{ij}}(n_i - \nu)(n_j - \nu). \tag{1}$$

The problem is formualted on a lattice, $n_i = 0, 1$ are occupation numbers of lattice sites, and ν is the average occupation number, i.e. the filling factor of the lattice. To make the system neutral each site has a positive background charge νe. The external disorder is introduced by a set of random energies ϕ_i distributed uniformly within the interval $(-A, A)$. Efros and Shklovskii[5] have predicted that the DS $G(\varepsilon)$ of the single particle energies ε_i has a soft Coulomb gap near the chemical potential E_F. Efros[7] has shown that the Coulomb gap exists also without any external disorder. In this case the system is ordered at $T = 0$. However, this ordered phase melts at a very low temperature, and an internal disorder provides a gap near E_F. A crossover from weak disorder ($A \ll 1$) to strong disorder ($A \gg 1$) has been studied recently.[8]

To describe the tunneling between two quantum wells we have performed computer modeling, where the structure has been simulated by two parallel plane arrays.[3] The distance S between the planes has been chosen in such a way that the capacitance between them is equal to the experimental one. The thermal motion of electrons is described by the Hamiltonian in Eq. (1) with both intra- and interplane Coulomb interactions but without interplane particle exchange. The lattice constant l is equal to magnetic length $\sqrt{hc/eB}$ and ν is the filling factor of the lowest Landau level. In this case we have not assumed any external disorder ($A = 0$). To simulate an applied voltage we make the number of electrons in one plane larger than in the other, while the background charge of the sites is the same. To calculate the tunneling current between the planes we assume that a tunneling probability is non-zero for pairs of sites with the same lateral coordinates and only if the work required to transfer electron from one plane to the other is smaller than some energy Δ:

$$\left|\varepsilon_i^{(e)} - \varepsilon_i^{(o)} - e^2/\kappa S\right| < \Delta. \tag{2}$$

Here $\varepsilon_i^{(o)}, \varepsilon_i^{(e)}$ are the energies of the occupied site in one plane and of the empty site in the other plane, S is the distance between the planes. (See details in Ref. 3.)

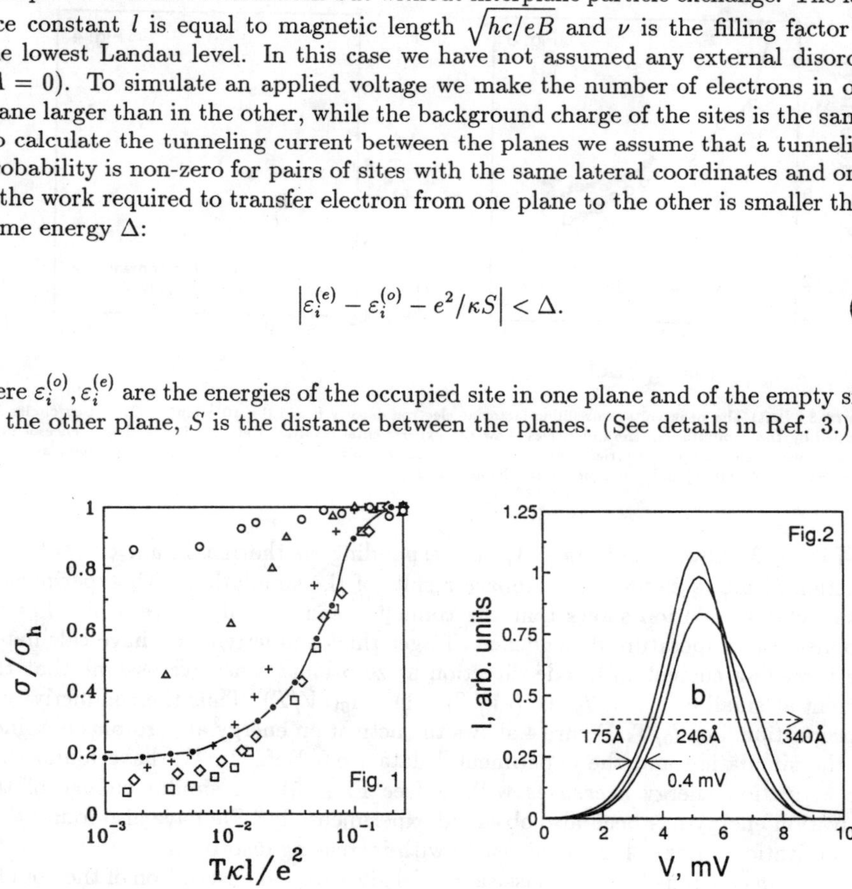

Figure 1. (left) Ratio of the conductivity to its high-temperature value. The markers show the data from Fig. 7 of Ref. 1 for different magnetic fields: B = 1.0 T — circles; B = 2.0 T — triangles; B = 4.0 T — plus sign; B = 6.5 T — diamonds; B = 8.5 T — squares. Solid line shows our results at $\nu = 0.3$, $A = 0.4\, e^2/\kappa l$, and $d = 0.4\, l$.
Figure 2. (right) I-V characteristics for tunneling between two quantum wells as obtained by computer modeling at different barrier thicknesses b between the wells. B = 9.4 T, electron density $n = 1.6 \times 10^{11}\,\mathrm{cm}^{-2}$, T = 2 K.

The current-voltage characteristics, as obtained by computer modeling, are shown in Fig. 2. They are close to the experimental results including the observed shift due to change of the distance between quantum wells. This shift can be qualitatively understood from Eq. (2). The third term in this equation increases the current with the decreasing S, because due to this term the energies of electronic states participating in tunneling become further from E_F. This effect has been confirmed experimentally.[9]

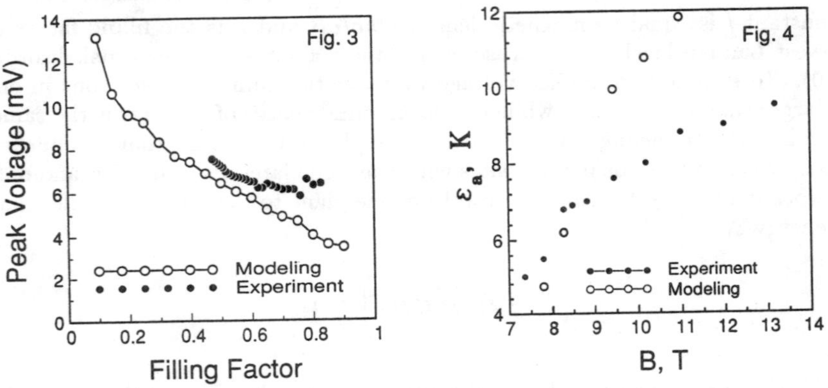

Figure 3. (left) The peak voltage *vs* filling factor at electron density $n = 1.6 \times 10^{11}$ cm^{-2}. The open circles are obtained by the computation, the full circles show the experimental result of Ref. 2. The T-dependence of peak voltage is very weak. **Figure 4.** (right) Activation energy at zero bias as obtained by computer modeling at a very small disorder ($A = 0.01$) and experimental results of Ref. 2.

Figure 3 shows the voltage V_m corresponding to the maximum current as a function of filling factor. We compare results of the simulation with experimental data. Our simulation shows that the tunneling conductivity at zero bias has an exponential temperature dependence. To get this conductivity we have calculated the tunneling current I_0 in one direction at zero bias. Then we assume that the current at small bias V is $I_0(\exp(eV/T) - 1) \approx I_0(eV/T)$. Thus the conductivity σ is proportional to I_0/T. Figure 4 shows the activation energy at zero bias obtained by the simulation and the experimental data from Ref. 2. We have found that the activation energy *decreases* with S (see Fig. 5). A similar change of the activation energy has been also observed experimentally.[9] We have also found that the activation energy sharply decreases with increasing disorder.

One can conclude that the classical model gives a good description of the double-well tunneling experiment. Note that our theory has *no fitting parameters*. The only relevant characteristic energy is the Coulomb interaction between electrons. Nevertheless, the theory gives correct values for both the peak voltage and the activation energy, which differ by a factor of 10 from each other. Moreover, the theory gives such tiny effects as the shift of I-V characteristics and the change of the activation energy due to the variation of the distance between wells.

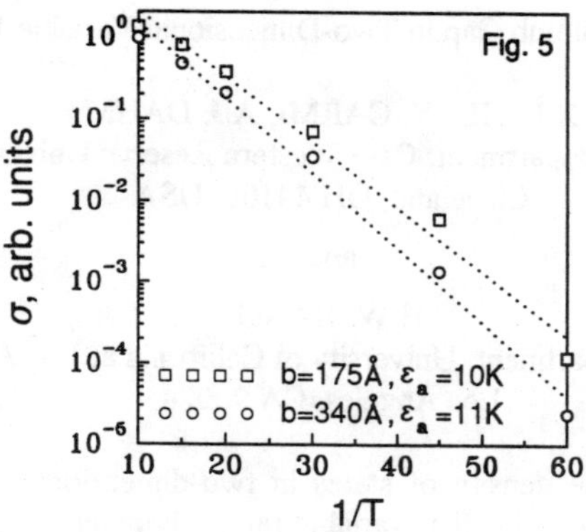

Figure 5. Tunneling conductivity between two wells at zero bias *vs* inverse temperature at $n = 1.6 \times 10^{11}$ cm^{-2}, $B = 9.4$ T, and at different barrier thicknesses b, as obtained by simulation.

The experimental data of Ref. 1 are quite different: in this case a highly doped material is pretty close to the electron layer and, therefore, the disorder potential may be large. In addition, the proximity of the metal contact leads to a strong screening of the electron-electron interaction. The striking feature of the gap observed in Ref. 1 is that it is independent of the electron density at a given magnetic field. We have performed computations taking into account the screening by a metallic plane. The result is that the conductivity at zero bias becomes independent of the filling factor at moderate disorder $A > 0.4 - 0.6$, and its temperature dependence is in good agreement with the experimental data (Fig. 1).

Acknowledgement

We acknowledge support from NSF Grant No. DMR-9116748 and from the San Diego Supercomputer Center.

References

1. R.C. Ashoori, J.A. Lebens, N.P. Bigelow, and R.H. Silsbee, *Phys. Rev. B* **48** (1992) 4616.
2. J.P. Eisentstein, L.N. Pfeiffer, and K.W. West, *Phys. Rev. Lett.* **69** (1992) 3804.
3. A.L. Efros and F.G. Pikus, *Phys. Rev. B* **48** (1993) 14694.
4. Song He, P.M. Platzmann, and B.I. Halperin, *Phys. Rev. Lett.* **71** (1993) 777; Y. Hatsugai, P. Bares, and X.G. Wen, *Phys. Rev. Lett.* **71** (1993) 424; C.M. Varma, A.I. Larkin, and E. Abrahamas, *Phys. Rev. B* **49** (1994) 13999.
5. A.L. Efros and B.I. Shklovskii, *J. Phys. C*8, **49** (1975).
6. J.H. Davies, P.A. Lee, and T.M. Rice, *Phys. Rev. B* **49** (1982) 758.
7. A.L. Efros, *Phys. Rev. Lett.* **68** (1992) 2208.
8. F.G. Pikus and A.L. Efros, unpublished.
9. Jim Eisenstein, private communication.

Screening of the Coulomb Gap in Two-Dimensional Variable-Range Hopping

X.L. HU, Y. CARMI, A.J. DAHM
Physics Department, Case Western Reserve University
Cleveland, OH 44106, USA

and

H.W. JIANG
Physics Department, University of California at Los Angeles
Los Angeles, CA 90024

A Coulomb gap in the density of states in two-dimensions is confirmed from resistivity measurements in the variable-range hopping regime in a gated GaAs/$Al_xGa_{1-x}As$ heterostructure. The Coulomb interaction is screened by the metallic gate for large hopping lengths. A universal crossover to Mott hopping at low temperatures is observed. Internal consistency of the data and excellent agreement with theory is achieved by adjusting the dielectric constant to be 4.8 times the lattice value.

A soft Coulomb gap in the density of states at the Fermi surface has been predicted by Efros and Shklovskii[1]. This gap, which results from the long-range nature of the Coulomb interactions, has been observed in three dimensions by many authors. In two-dimensional samples the Coulomb gap has been suggested as a possible explanation for a suppression in the tunneling conductivity at low temperatures[2] and for the temperature dependence of the resistivity at the resistivity minima in the quantum Hall regime[3,4] although an alternate explanation was offered.

A metallic gate, separated a distance d from the electron layer, screens the Coulomb interaction for $r \geq 2d$. This leads to a finite density of low energy states, $|E| < e^2/2\varepsilon d$, which lie at a distance greater than 2d from occupied states at the Fermi level. Here E is measured from E_F, and ε is the dielectric constant.. The density of states is predicted to be given by the following expressions[5,6]

$$g(E) = 2\varepsilon^2 |E| / \pi e^4; \qquad |E| > \pi\alpha\varepsilon^2/2\varepsilon d \tag{1}$$

$$g = g_s = \alpha\varepsilon/e^2 d; \qquad |E| < \pi\alpha\varepsilon^2/2\varepsilon d \tag{2}$$

where[7] $\alpha = 0.1$. The screened density of states, g_s, is two orders of magnitude less than the free electron density of states.

The longitudinal resistivity in the presence of a soft Coulomb gap is given by

$$\rho_{xx} = \rho_1(T) \exp(r_M/\xi) = \rho_1(T) \exp(T_1/T)^{1/2} \tag{3}$$

where r_M is the Mott hopping length and ξ is the localization length. In the presence of screening the longitudinal resistivity is given by the Mott hopping formula[8]

$$\rho_{xx} = \rho_0(T) \exp(r_M/\xi) = \rho_0(T) \exp(T_0/T)^{1/3} \tag{4}$$

The localization lengths are obtained in the Coulomb gap and Mott hopping regimes,

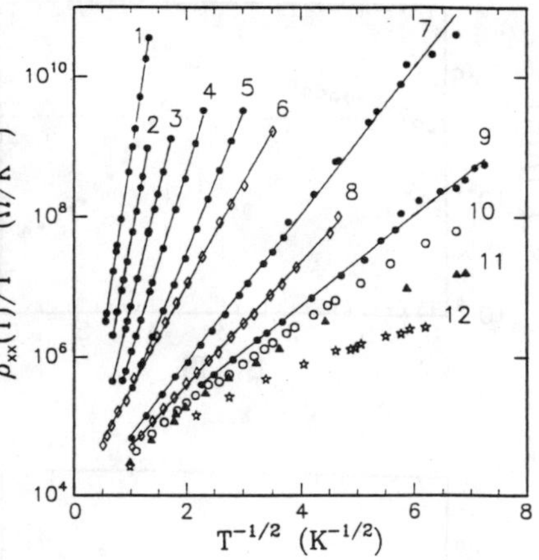

Fig. 1. A plot of ρ_{xx}/T versus $T^{1\ 2}$. The curves in increasing number correspond to n = [0.61, 0.72, 0.83, 0.94, 1.05, 1.16, 1.27, 1.33, 1.38, 1.41, 1.44] 1.49]×10^{15} m^{-2}. Solid lines are best fits to the data.

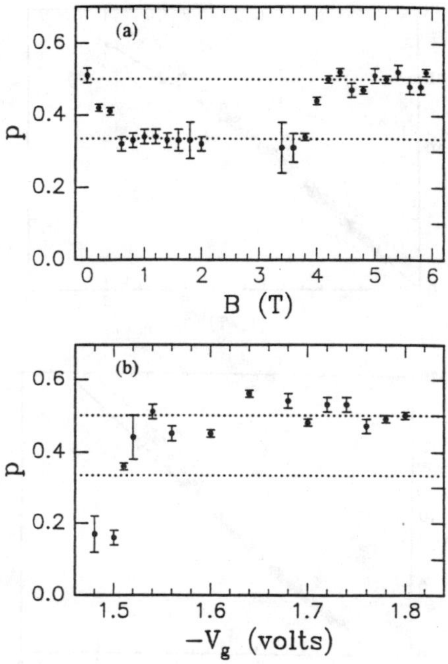

Fig. 2. The exponent p from Eq. (9). (a) p vs B for V_g = -1.54 V. (b) p vs V_g for B = 0. Dotted lines are drawn at p 1/2 and 1/3.

respectively, from measurements of $\rho(T)$ as[5]

$$\xi = Ce^2/\varepsilon k_B T_1 \tag{5}$$

$$\xi = (\beta/g_s k_B T_0)^{1/2} \tag{6}$$

where C = 6.2 and β = 13.8.

Aleiner and Shklovskii[6] examined screening by a metallic gate and found a universal function f(x) for the temperature dependence of r_M. Here $x = T/T_*$, $T_* = e^2\xi/\varepsilon k_B d^2$, and

$$r_M/d = [Cf(x)/x]^{1/2}. \tag{7}$$

This experiment explores the temperature dependence of the resistance of a gated heterostructure. The localization length is controlled by a voltage applied to the gate which alters the electron density and hence the screening of the donors. The density is given by

$$n = (5.56 + 2.73\ V_g)\times 10^{15}\ m^{-2} \tag{8}$$

The localization length is also varied with an applied field. It increases in small fields and is reduced again at large fields. By adjusting the localization length we are able to explore both the Coulomb gap and Mott hopping regimes, and the universal transition between the two regimes as a function of temperature.

The prefactors in Eqs. (3) and (4) and their temperature dependence have not been determined theoretically. We find empirically that ρ_0 and ρ_1 vary as T^m, where m = 0.8 ± 0.2. We present in Fig. 1 a plot of $\rho_{xx}/T^{0.8}$ vs $T^{-1/2}$ to show evidence of the Coulomb gap

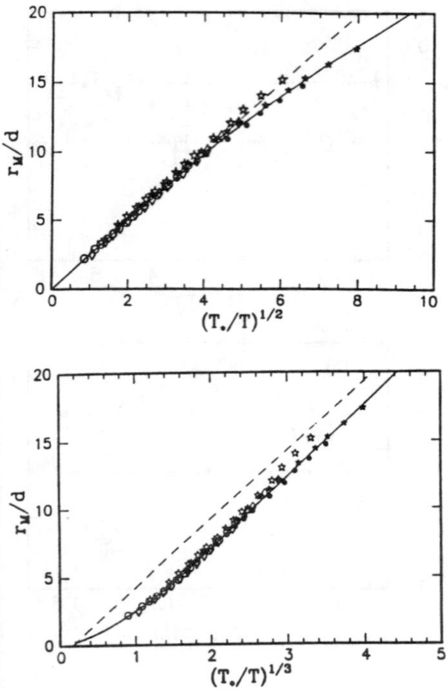

Fig. 3. The ratio r_M/d vs (a) $(T*/T)^{1/2}$; (b) $(T*/T)^{1/3}$. The solid line is given by Eq. (10) with $\varepsilon = 4.8\ \varepsilon_\ell$. Dashed lines are extrapolations from (a) high and (b) low T limits. Symbols represent: ○ξ = 0.23; ◊ ξ = 0.35; ☆ξ = 0.53; ● ξ = 0.63; ★ ξ = 0.93 μm.

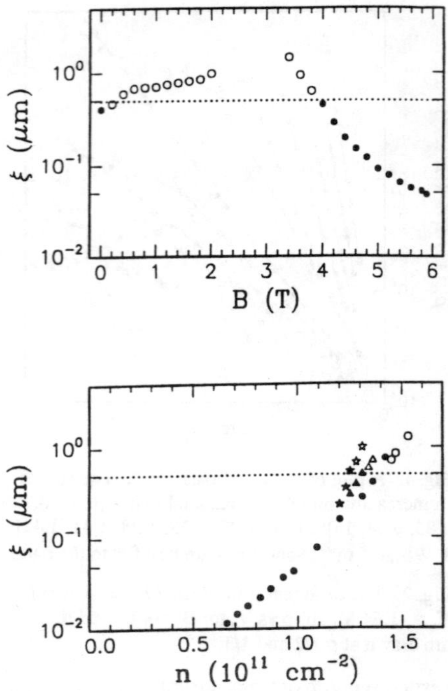

Fig. 4. Localization length vs n and B. (a) ξ vs B for n = 1.33×10^{15} m^{-2}. (b) ξ vs V$_g$ for B = 0. Symbols represent: ○ B = 0; △ B = 1 T.; ☆ B = 3 T. Solid and open symbols are taken from the Coulomb gap and Mott hopping regimes, respectively.

in resistance measurements in zero magnetic field. The strictly linear behavior of the data for $n < 1.38\times10^{15}$ m^{-1} verifies the existence of the Coulomb gap at small electron densities.

In order to demonstrate the transition from Coulomb gap behavior to Mott hopping we fit our data to

$$\rho_{xx} = A\, T^{0.8} \exp(T'/T)^p \tag{9}$$

where T'(n,B) is a constant for each set of data. The best fit value of the exponent p is plotted as a function of B and V_g, respectively in Figs. 2a and 2b. Figure 2a shows a transition from p = 1/2 to p = 1/3 at small B and another transition back to 1/2 at B ≅ 4 T. A transition at high densities for B = 0 from p = 1/2 to a smaller value occurs at $V_g \cong 1.52$ V.

In Fig. 3 we demonstrate the universal screening crossover from the Coulomb gap behavior to Mott hopping in plots of r_M/d vs $(T_*/T)^{1/2}$ and $(T_*/T)^{1/3}$. Data were taken at different gate voltages at B = 3 T. The experimental value of r_M/d is defined as

$$r_M/d = [(\xi/d) \ln(\rho_{xx}/T^{0.8})/ \ln[\lim_{T\to\infty} (\rho_{xx}/T^{0.8})] \tag{10}$$

The $\lim_{T\to\infty} (\rho_{xx}/T^{0.8})$ was taken from the Coulomb gap regime. The symbols represent different localization lengths. The solid lines are a single parameter, ε, fit to Eq. (7) with the dielectric constant set equal to 4.8 ε_ℓ, where ε_ℓ is the lattice dielectric constant. The data shown in Fig. 3b lie below the low temperature $T^{-1/3}$ screened limit represented by the dashed line. Most of the data shown as p = 1/3 in Fig. 2 lie in this "apparent 1/3" region where the slope is less than in the totally screened regime.

The density and field dependencies of the localization length are shown in Fig. 4 with data taken from both the Coulomb gap and Mott hopping regimes. The localization lengths were determined from Eqs. (5) and (6) using the screened density of states in Eq. (6) and $\varepsilon = 4.8\ \varepsilon_\ell$, determined from the fit to Fig. 3. The value of T_0, obtained from the slope of $\rho_{xx}/T^{0.8}$ vs $T^{-1/3}$, was corrected for its deviation from the low temperature slope. In the temperature range investigated in this experiment the transition between regimes occurred for localization lengths near 500 nm. The good match of the localization lengths confirms the correctness of the screened density of states given in Eq. (2).

The scaling theory of localization[9] predicts that $\xi \sim \exp(2\pi^2\sigma h/e^2) \sim \exp[cn^{1/2}\ell(n)]$, where σ is the conductivity, ℓ is the elastic mean free path, c is a constant, and $\sigma h/e^2 = k_F\ell$. From Fig. 4b, we observe that ξ varies as exp[f(n)], where f(n) is approximately linear in n. This is in reasonable agreement with expectations of the scaling theory.

In summary, we have confirmed the existence of the Coulomb gap in two-dimensions and the screening of the gap by a metallic electrode located 300 nm from the electron layer. The data are in excellent agreement with the predicted universal crossover by Aleiner and Shklovskii with one adjustable parameter, the dielectric constant.

The authors are indebted to I. L. Aleiner and B. I. Shklovskii for many stimulating conversations, and to F. Van Keuls for valuable technical assistance. This work was supported in part by NSF grants DMR-93-13786 and DMR-94-02647.

References

1. A. L. Efros and B. I. Shklovskii, . hys. **C8**, L49 (1975).
2. R. D. Ashoori, J. A. Levens, N. P. Bigelow, and R. H. Silsbee, Phys. Rev. **B 48**, 4616, (1993).
3. G. Ebert, K. von Klitzing, C. Probst, E. Schuberth, K. Ploog, and G. Weiman, Solid State Comm. **45**, 625 (1983).
4. The behavior was also seen by A. Briggs, Y. Guldner, J.P. Vieren, M. Voos, J.P. Hirtz, and M. Razeghi, Phys. Rev. **B 27**, 6547 (1983).
5. B. I. Shklovskii, and A. L. Efros, Electronic Properties of Doped Semiconductors, (Springer, Berlin, 1984).
6. I. L. Aleiner and B. I. Shklovskii, Phys. Rev. **B 49**, 13721 (1994).
7. A. A. Mogilyanskii and M. E. Raikh (unpublished).
8. N. F. Mott, J. Non-Crystalline Solids **1**, 1, (1968).
9. E. Abrahams, P. W. Anderson, D. C. Licciardello, and T. V. Ramakrishnan, Phys. Rev. Lett. **42**, 673 (1979).

INTERACTIONS AND DISORDER IN THE INTEGER QUANTUM HALL EFFECT

S.-R. ERIC YANG
Institute for Microstructural Sciences
National Research Council of Canada
Ottawa , Ontario, Canada, K1A 0R6

ABSTRACT

The interplay between Coulomb interactions and localization properties in the integer quantum Hall regime is investigated extensively by performing finite-size Hartree-Fock calculations in the presence of disorder. Pseudogaps are found in the tunneling density of states. Investigations of participation ratios and the Thouless numbers show no evidence that Coulomb interactions change either the localization length or the critical conductivity. A strong cancellation between self-energy and vertex corrections is responsible for the unchanged critical conductivity. It is also shown that in the strongly localized regime spin-splitting is not due to the exhange interactions but is rather due to the formation of a Hubbard-like band.

In the integer quantum Hall effect (IQHE) regime it is usually assumed that Coulomb interactions can be ignored compared to disorder[1]. Under this assumption the localization length is found to diverge[2] as $|E-E_c|^{-7/3}$ for energies near the center of the lowest Landau level E_c. Such a system is like an insulator when the Fermi level E_F is in the Landau level tail regime and like a metal when E_F is near the middle of the Landau level. As E_F approaches E_c a metal-insulator transition takes place and both the diagonal- and off-diagonal conductivities[3] assume the critical value $e^2/(2h)$. This paper investigates the properties of the tunneling DOS, localization length, critical conductivity, and spin-splitting in the presence of Coulomb interactions.

In this paper a Hartree-Fock calculation is implemented on a finite-size system. This method treats localization physics exactly and includes self-consistently disorder and Coulomb effects. As long as the strong correlations leading to the fractional QHE are absent the present method is expected to be qualitatively correct. In the model used in the numerical work N electrons in the lowest Landau level are confined to move on a square of size L^2. The strong magnetic field limit is assumed and the number of applied flux quanta is N_φ. The impurity potential is $V_I(\vec{r})=\sum_i V_i\delta(\vec{r}-\vec{r}_i)$ where the potential strength V_i varies randomly between $-\lambda$ and λ. The characteristic disorder energy is $\Gamma=(\lambda^2 N_I/l^2L^2)^{1/2}$ where the number of impurities is $N_I=5N_\varphi$ and l is the magnetic length. The relative strength of Coulomb interactions and disorder is specified by the parameter $\gamma=(e^2/\varepsilon l)/\Gamma$ and the filling factor is $\nu=N/N_\varphi$ (ε is the dielectric constant). Calculations are performed in the Landau gauge $\vec{A}=(0,Bx,0)$ and quasiperiodic boundary conditions are applied to the HF single-particle orbitals inside the square. The basis states used to represent the HF Hamiltonian are related to elliptic theta functions and can be labeled by a set of guiding centers $|X_i\rangle$, i=1,...,N_φ inside the fundamental cell of the finite size system. The eigenstates φ_α and eigenvalues ε_α satisfy the following matrix equation

$$\Sigma_i\left[\langle X_j|V_I|X_i\rangle+\langle X_j|V_{HF}|X_i\rangle\right]\langle X_i|\alpha\rangle=\varepsilon_\alpha\langle X_j|\alpha\rangle \tag{1}$$

where the matrix elements of the HF potential are

$$\langle X_j|V_{HF}|X_i\rangle = \Sigma_{\vec{q}}\Delta(\vec{q},\{\varphi_\alpha\})\exp[iq_x(X_i+X_j)/2]\delta(j,i+m_y)U_{HF}(\vec{q}) \tag{2}$$

Here the sum is over the discrete set of wave vectors $\vec{q}$ (or equivalently over m_x and m_y) consistent with the boundary conditions and δ(n',n) is 1 if n'=n(mod N_φ) and 0 otherwise. The quantity $\Delta(\vec{q},\{\varphi_\alpha\})$ is proportional to the Fourier component of the charge density at wavevector $\vec{q}$ and depends on the eigenstates φ_α. The quantity U_{HF} is independent of $\{\varphi_\alpha\}$. The Eqs. (1) and (2) must be solved self-consistently for each disorder realization. The number of disorder realization used in the computation is N_D.

This method has been used to calculate the tunneling DOS, D(E), of a spin polarized system[4]. Calculations have been completed for ν=1/5 and for ν=1/2. For these filling factors electrons are in a quantum regime since already at ν=1/5 the states near the Fermi level have a localization length comparable to the distance between electrons.

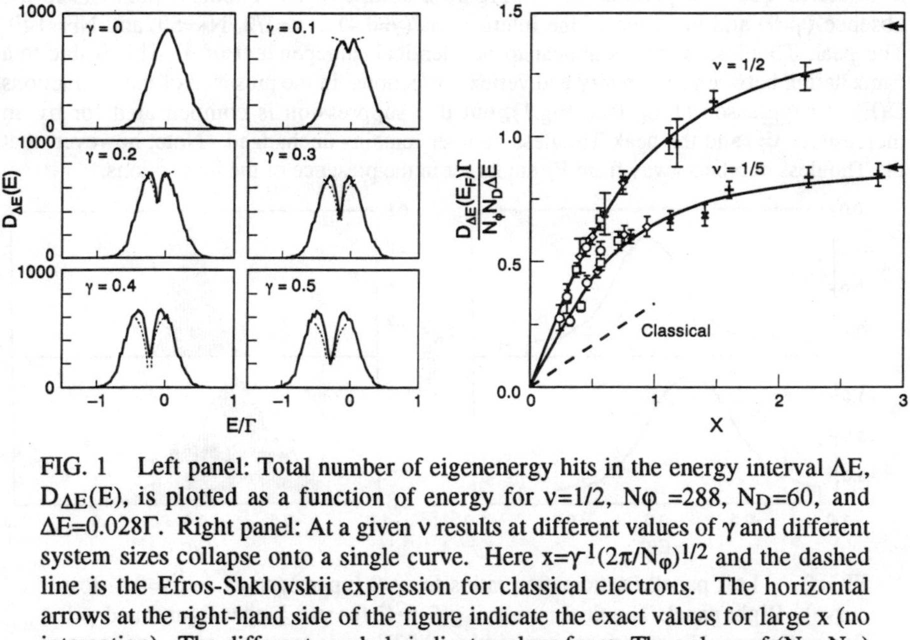

FIG. 1 Left panel: Total number of eigenenergy hits in the energy interval ΔE, $D_{\Delta E}(E)$, is plotted as a function of energy for ν=1/2, Nφ =288, N_D=60, and ΔE=0.028Γ. Right panel: At a given ν results at different values of γ and different system sizes collapse onto a single curve. Here $x=\gamma^{-1}(2\pi/N_\varphi)^{1/2}$ and the dashed line is the Efros-Shklovskii expression for classical electrons. The horizontal arrows at the right-hand side of the figure indicate the exact values for large x (no interaction). The different symbols indicate values for γ. The values of (N_φ, N_D) for each symbol are given in Ref.4.

Clear evidence is found for the formation of a pseudogap where the DOS at the Fermi level vanishes in the thermodynamic limit. Figure 1 displays tunneling DOS for 6 different values of γ. A finite size scaling analysis has been performed by investigating the system size dependence of the DOS at E_F. It provides strong evidence that the tunneling DOS goes to zero in the thermodynamic limit. Recently Ashoori et al[5] and Eisenstein et al[6] have observed a gap in the tunneling DOS. The present results apply to samples which are sufficiently disordered that the fractional QHE does not occur.

The localization properties are investigated by calculating the participation ratio $P_\alpha = \int d\vec{r}|\varphi_\alpha(\vec{r})|^2 / L^2 \int d\vec{r}|\varphi_\alpha(\vec{r})|^4$ of the HF wavefunctions φ_α. Figure 2 displays $\ln(P_\alpha)$ vs. E/Γ for a spin polarized system with ν=1/2, N_D=11, and N_ϕ=128. One can see that the results for γ=0 and 0.4 show identical energy dependence of $\ln(P_\alpha)$ on E/Γ except for a constant shift. It is inferred from this that the interactions do not change the exponent ν=7/3, in agreement with recent experiments[7]. Also the Thouless number, which is expected to be proportional to the dissipative conductivity[8], is investigated. The definition of the Thouless number is taken to be $g(E) = \overline{\Delta E} D(E) L^2$ where $\overline{\Delta E}$ is the geometric mean of energy level shifts when the boundary condition is changed from periodic to anti-periodic. It will be argued elsewhere that, within the HFA[9], this is indeed a reasonable definition even in the presence of the interactions. It is found that the critical conductivity is unchanged by the inclusion of the interactions. Lee et al reached the same conclusion using a model of quantum percolation[10]. Figure 2 compares the Thouless numbers in the absence (γ=0) and presence of the interactions (γ=0.4) at ν=1/2, N_ϕ=50, and N_D=140. The peak Thouless numbers appear to be identical, independent of γ. This is due to a cancellation between self energy and vertex corrections: in the presence of the interactions D(E) is suppressed at E_F (see Fig.1), but this suppression is compensated for by an increase in $\overline{\Delta E}$ and the peak Thouless number remains unchanged. Note, however, that the Thouless numbers away from E_F are larger in the presence of the interactions.

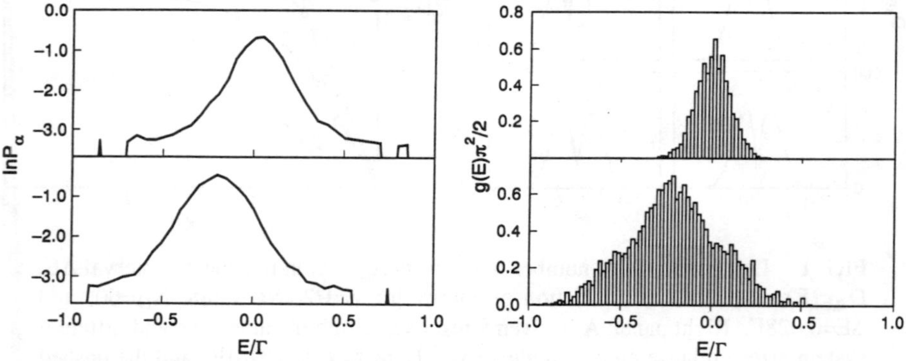

FIG. 2. Left panel: Participation ratios for γ=0 (upper curve) and γ=0.4 (lower curve). Right panel: Thouless numbers $g(E)\pi^2/2$ for γ=0 (upper figure) and γ=0.4 (lower figure). Histogram energy width is 0.02Γ.

Spin-splitting in a two-dimensional electron gas in a strong magnetic field is usually attributed to the exchange interactions. The amount of the exchange spin-splitting in the lowest Landau level without localization effects is $E_x = -\sqrt{\pi/2}\nu e^2/\varepsilon l$. When the electrons are localized the exchange interaction is smaller since their wavefunctions overlap little and the corresponding exchange enhanced spin-splitting is smaller. However when two localized electrons occupy a single site their mutual Coulomb repulsion can give rise to a Hubbard-like band splitting. For a simple model of square density of states the condition for full spin polarization is $E_F < g\mu B + U$ where $U \approx e^2/\varepsilon\xi(E_F)$ (here E_F is measured from the bottom of the spin-up band, $g\mu B$ is the bare spin-splitting, and $\xi(E_F)$ is the localization length at the Fermi level). The finite size HFA provides a quantitative theory of the effect of the on-site Coulomb interaction in the IQHE regime. It is necessary to introduce two different single-particle HF equations, one for each spin species. The effective HF potentials for spin-up and down electrons will be different since the electron numbers are different (details of this theory will be presented in Ref.11). The DOS is plotted in Fig.3 for $\nu=1/5$, $N_\varphi=250$, $\gamma=0.7$, $N_D=10$ and $g\mu B=0.15\Gamma$. A gap is observed between spin-up and -down bands. Note that a Coulomb gap appears in the spin-up band. Figure 3 also shows the DOS for $\nu=1$, $N_\varphi=98$, $\gamma=0.7$, $N_D=10$ and $g\mu B=0.15\Gamma$. No matter how strong the Coulomb interactions are, as long as γ remains unchanged, complete spin polarization is absent .

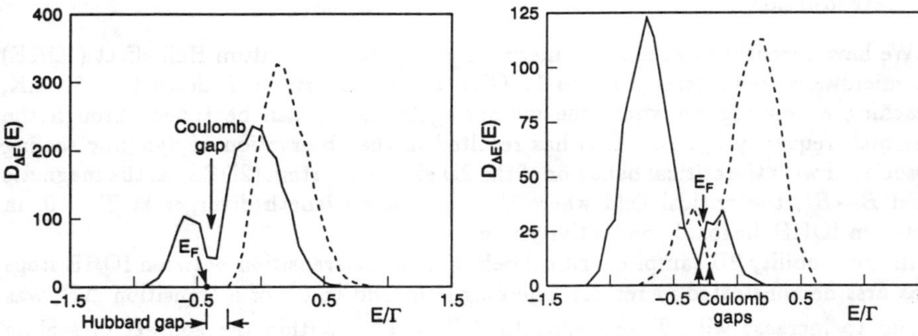

FIG. 3. $D_{\Delta E}(E)$ for $\nu=1/5$ (left panel) and $\nu=1$(right panel). Here $\Delta E=0.06\Gamma$.

Acknowledgments
The work presented in this paper was done in close collaboration with A.H. MacDonald. The author is grateful to him for a critical reading of this paper.

References

1. For a review see, for example, T. Ando, Prog. Theor. Phys. Suppl., **84**, 69 (1985).
2. G. V. Mil'nikov and I. M. Sokolov, Pis'ma Zh. Eksp. Teor. Fiz, **48**, 494 (1988) [JETP Lett. **48**, 536 (1988)].
3. Y. Huo, R. E. Hetzel, and R. N. Bhatt, Phys. Rev. Lett., **70**, 481 (1993).
4. S.-R. Eric Yang and A. H. MacDonald, Phys. Rev. Lett., **70**, 4110 (1993).
5. R. C. Ashoori et al, Phys. Rev. Lett., **64**, 681 (1990).
6. J. P. Eisenstein et al, Phys. Rev. Lett., **69**, 3804 (1992).
7. H. P. Wei et al, Phys. Rev. Lett., **61**, 1294 (1988); S. Koch et al , Phys. Rev. Lett., **67**, 883 (1991).
8. E. Akkermans and G. Montambaux, Phys. Rev. Lett., **68**, 642 (1992).
9. S.-R. Eric Yang and A. H. MacDonald, to be published.
10. D. H. Lee, Z. Wang, and S. Kivelson, Phys. Rev. Lett.**70**, 4130 (1993).
11. S.-R. Eric Yang and A. H. MacDonald, to be published.

MICROWAVE FREQUENCY SCALING IN THE INTEGER QUANTUM HALL EFFECT

L.W. ENGEL, D. SHAHAR, Ç. KURDAK and D.C. TSUI
Dept. of Electrical Engineering, Princeton University, Princeton NJ 08544

ABSTRACT

We describe a method for measuring the conductivity $\mathrm{Re}(\sigma_{xx})$ at microwave frequencies to 14 GHz, magnetic fields to 15.5 Tesla and temperatures down to 50 mK. Results on dynamic scaling in the integer quantum Hall effect are reviewed, and discussed in the context of recent theories.

1. Introduction

We have recently succeeded[1] in measuring the integer quantum Hall effect (IQHE) at microwave frequencies f up to 14 GHz and temperatures T down to ~50 mK, reaching a new regime where the measuring frequency can be tuned through the thermal frequency k_BT/h. This has resulted in the observation of dynamic scaling associated with the critical behavior of the 2D electron system (2DES) as the magnetic field $B \rightarrow B^*$, the critical field where the localization length diverges at $T = 0$, in between IQHE diagonal conductivity minima.

In low mobility 2D samples, critical behavior in the transition between IQHE steps was first demonstrated in ref. 2. The magnetic field width of a transition ΔB, was found to increase with T according to $\Delta B \sim T^{\kappa}$. Within the context of scaling theory[3], and assuming that inelastic scattering controls the effective size of the sample, this power law was related to the divergence of the localization length ξ as the magnetic field $B \rightarrow B^*$, with the scale of the system determined by ξ/L, where L is the inelastic length.

Quite generally for continuous phase transitions, a divergent characteristic length is accompanied by a divergent characteristic time[4]. Dynamic scaling describes the time-dependent behavior of the system relative to this correlation time. If the transitions between IQHE steps are critical phenomena, as ref. 2 indicates, with ξ diverging as $\xi \sim |B - B^*|^{-\nu}$ we can expect that there is a concomitantly diverging correlation time[5] $\tau_c \sim |B - B^*|^{-y} \sim \xi^z$, where $z = \nu/y$ is the dynamic exponent.

In this paper we present an especially detailed description of our unusual experimental method, emphasizing the use of a transmission line to measure $\mathrm{Re}[\sigma_{xx}(f)]$. We then review ac measurements of $\mathrm{Re}(\sigma_{xx})$ in a sample that exhibits the IQHE, for a frequency (f) range of 0.2 to 14 GHz. The width of transitions between IQHE minima in $\mathrm{Re}(\sigma_{xx})$ increases with f, for $f \gtrsim 2$ GHz and temperature $T \sim 50$ mK, roughly

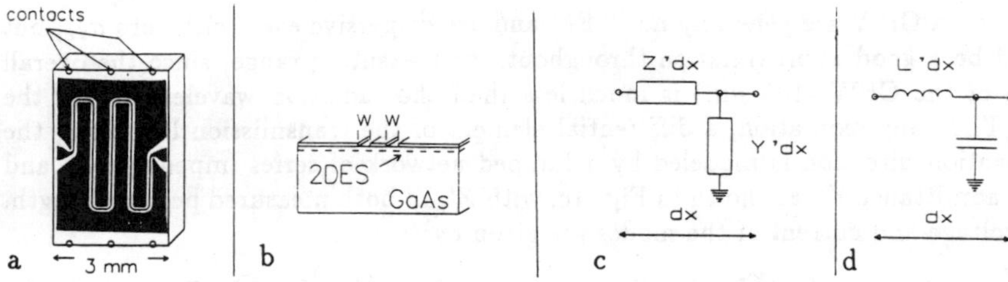

Fig. 1 *a.* Top view of sample with coplanar waveguide transmission line. Al film shown as black. *b.* Magnified cross section of sample near the transmission line (not to scale). Slotwidth W=30 μm. *c.* Equivalent circuit of transmission line differential element, general case. *d.* Differential element equivalent circuit with $\sigma_{xx} = 0$.

as f^γ. This power-law behavior is interpreted as due to dynamic scaling, where the exponent is related to the τ_c exponent as $\gamma = 1/y$. We measure $\gamma \approx 0.41$, implying $z = 1$ if $\nu = 7/3$. Our results on transition widths are discussed in the context of recent theories[5-9], that do not rely on a T-dependent inelastic length in setting the scale of the system.

2. Experimental Method

We obtain Re(σ_{xx}) from the rf loss of a transmission line patterned onto the front surface of the sample, $\lesssim 1\mu$m away from the 2DES. Coupling between the 2DES and the transmission line is capacitive. Since this technique for measuring transport has not been widely used, and should be applicable to measurement of other high resistivity systems, we describe our transmission line method in detail here.

Figs.1a and 1b show the sample with the Al transmission line on its surface. This three-conductor transmission line is of a well-known type called a coplanar waveguide (CPW) [10], though it is operated more like a coaxial cable than a waveguide, since rf is applied to the narrow center conducting strip relative to the two side planes. These "ground" planes are shorted together at each end of the sample. The line meanders to make it long enough to give adequate sensitivity with a reasonably sized sample. The periodicity of the CPW plays no role in our experiments, and coupling between successive meanders is minimal due to the shielding effects of the ground planes. The total length of the line l is 14 mm, and the width W of one of the gaps between the center conductor and the ground plane is 30 μm. Use of CPW instead of the more conventional microstrip transmission line allows easy fabrication of the long, thin line, and permits a simple calculation of the attenuation due to the 2DES conductivity σ_{xx}, since for our experimental conditions the in-plane electric field is concentrated mainly under the 30 μm gaps. As shown in Fig. 1a, we can place ohmic contacts on the edges of the sample, allowing measurement of dc and microwave transport at the same time. To calculate 2DES conductivity from the observed loss of the transmission line, we use the quasi-TEM approximation[11]. This is a low frequency approximation since the modes on a CPW are generally not TEM and are dispersive even with zero σ_{xx}, but should be a good approximation throughout our measuring range, since the overall width of the CPW (105 μm) is much less than the radiation wavelength. In the quasi-TEM approximation, a differential element of the transmission line along the propagation direction is modeled by a lumped network of series impedance Z' and shunt admittance Y', as shown in Fig. 1c, with Z', Y' both measured per unit length. The voltage and current of the modes are given by[12]:

$$V(x) = V_+ e^{-Kx} + V_- e^{Kx}; \qquad I(x) = (V_+ e^{-Kx} - V_- e^{Kx})/Z_c \tag{1}$$

where the phase constant K and the characteristic impedance Z_c are

$$K = (Y'Z')^{1/2}; \qquad Z_c = (Z'/Y')^{1/2}. \tag{2}$$

The starting point for our analysis is the lossless ($\sigma_{xx} = 0$) transmission line whose differential element network is shown in Fig. 1d, in which $Y' = i\omega C'$ and $Z' = \omega L'$, so that the phase constant and characteristic impedance become

$$K_0 = i\omega(L'C')^{1/2} = i\omega/v_{p0}; \qquad Z_0 = (L'/C')^{1/2} \tag{3}$$

where v_{p0} is the phase velocity. The geometry of our CPW is designed such that $Z_0 = 50\Omega$, so as to match the coaxial cable used to make connections in our measuring system, when $|\sigma_{xx}|$ is small. For our CPW $v_{p0} = 1.2 \times 10^{11}$ mm/sec.

A few assumptions, true for large f and small $|\sigma_{xx}|$ result in a simple, f-independent formula for the transmission line attenuation A as a function of Re(σ_{xx}). First, f must be high enough that the metal of the transmission line effectively confines the induced current in the 2DES to the region under the gaps, which means $(2|\sigma_{xx}|/\omega C_g)^{1/2} \ll W$, where C_g is the capacitance per unit area between the 2DES and a front gate. With this assumption the admittance Y' of Fig. 1c becomes $\sigma_{xx}/2W + i\omega C'$. Second, we take $|\sigma_{xx}|/2W \ll \omega C'$, so we can expand the phase constant K to first order in σ_{xx}. With these two assumptions, and also neglecting the losses of the metal transmission line conductors, the attenuation of the line becomes

$$A = e^{-2l\mathrm{Re}K} = e^{-\sigma_{xx} l/Z_0 W}. \tag{4}$$

We have analyzed our measuring system in detail, and find that under our experimental conditions, Eq. (4) needs only slight correction to allow calculation of Re(σ_{xx}) from the measured, transmitted power. In our detailed analysis, we keep the quasi-TEM approximation to describe the CPW, but calculate Z', Y', and hence K, Z_c, with the assumptions of the previous paragraph relaxed. The distributed nature of the capacitance between the CPW conductors and the 2DES is taken into account, assuming only that σ_{xx} can be handled classically on length scales much smaller than W. Loss in the CPW conductors, and the possiblity that σ_{xx} may have a large imaginary component are included, as well. In practice, we calculate Re(σ_{xx}) from P, the measured transmitted power divided by the transmitted power when $\sigma_{xx} = 0$ (as is nearly the case on IQHE σ_{xx} minima). Our detailed analysis to determine the relationship between P and Re(σ_{xx}) requires us to use Eqs. (1), given K, Z_c and

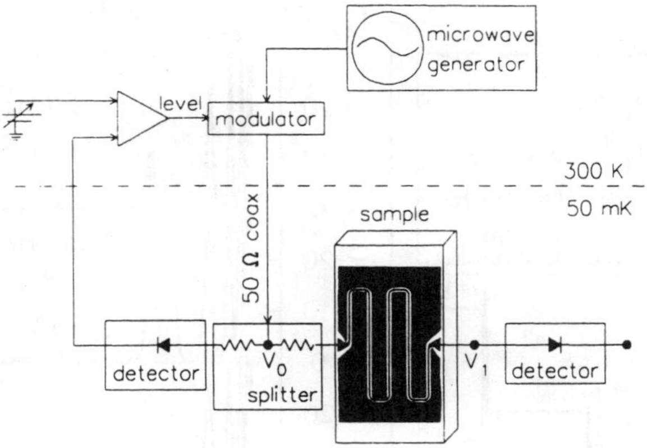

Fig. 2 Simplified schematic of our experimental setup. Voltage V_0 is fixed by the servo loop.

source and detector impedances, thus including effects of reflections at the ends of the transmission line, as is done in ref. 13. The result of the analysis is the correction factor C in the expression we use to obtain $\mathrm{Re}(\sigma_{xx})$:

$$\mathrm{Re}[\sigma_{xx}(f)] = C(f,P)W|\ln P|/Z_0 \tag{5}$$

With $C = 1$ this becomes the same as Eq. (4). For all the $\mathrm{Re}(\sigma_{xx})$ data shown here $1 \leq C(f,P) \leq 1.4$, and for $\mathrm{Re}(\sigma_{xx})$ at the value used to determine the widths of peaks between IQHE minima, $1 \leq C(f,P) \leq 1.14$. Spurious reflections due to interconnections and to departure of source and detector impedances from 50 Ω cause an oscillatory part of the transmitted power of about ±25 percent when $\sigma_{xx} = 0$, and so result in an error of similar magnitude between $\mathrm{Re}(\sigma_{xx})$ measured at different frequencies. Typically our measurements are $\sim$ 10 times more sensitive to the real part of σ_{xx} than to the imaginary part.

Fig. 2 shows a schematic of the microwave measuring setup. Difficulties with making connections between room-temperature instruments and the sample at T =50 mK are largely avoided by using detectors that operate at low T and high B, only about 1 cm away from the sample. Microwaves are fed from room temperature down a single coaxial cable to a splitter consisting of two 50 Ω resistors. One arm of the splitter goes directly into a "reference" detector whose output is fed back to a modulator to regulate V_0, the voltage at the point where the two resistors of the splitter meet. The feedback eliminates any frequency response due to the feed cable, allowing us to keep the cable compatible with the cryogenics. The other arm of the splitter goes through the sample into another detector, which is matched to the first one and measures the transmitted power. The detectors are square-law Ge back diodes, matched to 50Ω with an R-L network.

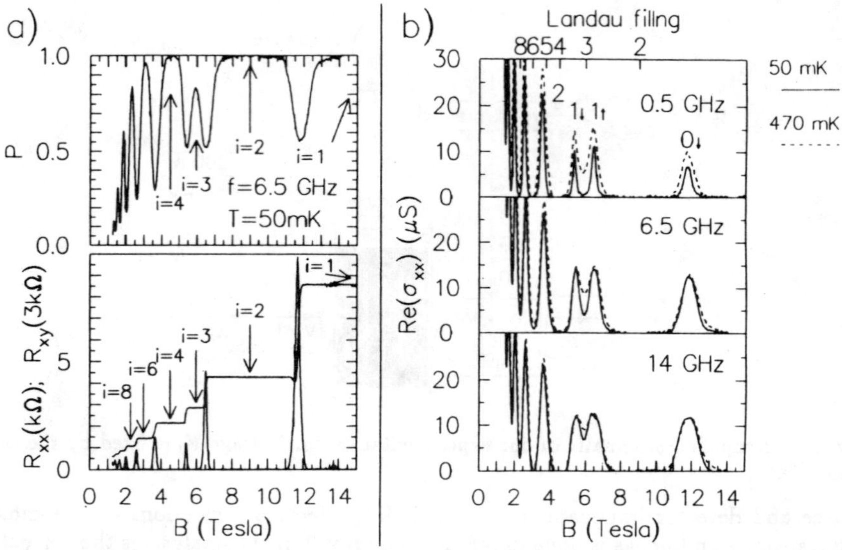

Fig. 3 *a.* Microwave relative transmission, $f = 6.5$ GHz, T=50 mK, and dc resistances, $T = 50$ mK. *b.* Re(σ_{xx}) vs B at three frequencies and two temperatures.

3. Experimental Results

The sample for which we observe the effects of dynamic scaling between IQHE σ_{xx} minima was a GaAs/Al$_x$Ga$_{1-x}$As heterojunction grown by liquid phase epitaxy, with density $n \approx 4.2\times10^{11}$ cm^{-2}, and mobility $\mu \approx 4\times10^4$ cm^2/V-sec, similar to that of other samples used to study scaling in the IQHE[2,14].

We present data for a nominal lowest T of 50 mK, which is the T of the bath in which the sample was immersed. Varying the microwave power indicated that the applied microwaves were not causing heating. However, the dc ΔB for IQHE transitions ceased to narrow with decreasing T at around 65 mK, so the sample T was apparently above the bath T of 50 mK.

Fig. 3a shows the dc transport and 6.5 GHz P vs B. The dc transport is typical of low-mobility samples at low T, with wide IQHE plateaus and narrower transition regions between them. Odd-integer (i) plateaus with $i > 3$ do not appear. Microwave transmission decreases as Re(σ_{xx}) increases, so the 6.5 GHz Re(σ_{xx}) is mimimal in the IQHE regions, just as the dc σ_{xx} is. The $i = 1, 2, 4$ maxima all saturated at the same transmitted power indicating our data are unaffected by σ_{xy}. The transitions between the IQHE P maxima at 6.5 GHz are clearly wider than the corresponding transitions in the dc transport traces.

Fig. 3b shows the calculated Re(σ_{xx}) vs B for $f = 0.5, 6.5$ and 14 GHz and $T = 50$ and 470 mK. The peaks of the $N = 0\downarrow, 1\uparrow, 1\downarrow, 2, 3$ Landau levels broaden with increasing f, as is seen most clearly for the $0\downarrow$ peak. The 6.5 GHz and 14 GHz data

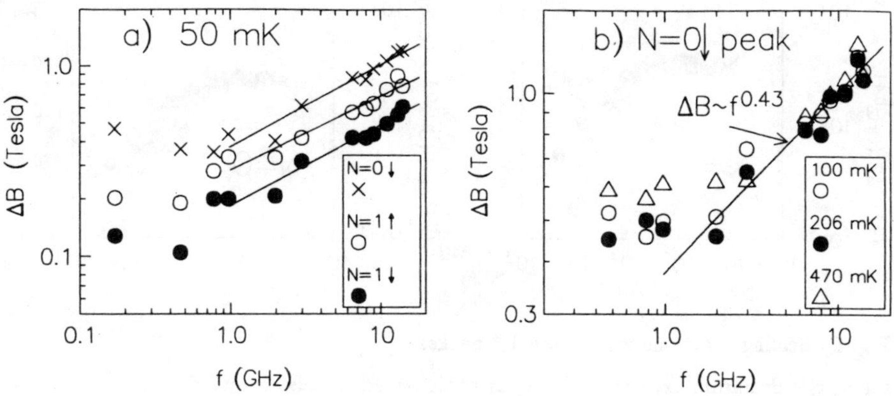

Fig. 4 *a.* **Peak width vs frequency, for three different Landau levels. Lines are least squares fits to data with $f \geq 0.97$GHz.** *b.* **Peak width vs frequency at three different temperatures, for $N = 0 \downarrow$ Landau level.**

barely differ at the two T's, but the 0.5 GHz data are much broader at 470 mK than they are at 50 mK.

The widths ΔB of $\mathrm{Re}(\sigma_{xx})$ peaks between different sets of IQHE minima, as a function of f, at $T = 50$ mK are plotted in Fig. 4a. ΔB are measured between extremal derivatives of $\mathrm{Re}(\sigma_{xx})$ vs B. The systematic errors appearing for different peaks at the same f's are due to microwave reflections affecting the σ_{xx} data, as mentioned above. The ΔB's shown increase with f for $f \gtrsim 2$GHz, but at lower f appear to be f-independent. The lines in the figure are least-squares fits to the data with $f \geq 0.97$ GHz, to the power law $\Delta B \propto f^{\gamma}$. For the $N = 0\downarrow, 1\uparrow, 1\downarrow$ Landau level peaks the fits give respectively $\gamma = 0.43,\ 0.38,\ 0.42$. Errors estimated from scatter are $\pm \sim 10$ percent, and the average $\gamma = 0.41$.

The T- and f-dependence of ΔB for the $0\downarrow$ Landau level are shown in Fig. 4b. The figure suggests that at low f ΔB is f-independent and determined by T, while at high f ΔB is T-independent and determined by f. f_c, the crossover frequency between the two regions, appears to be of order $k_B T/h$.

4. Discussion

We interpret the high f regions of Fig. 4, where $\Delta B \sim f^{\gamma}$, as manifestations of dynamic scaling. Though the f and ΔB ranges are small, the γ's from the fits match each other for different Landau levels, consistent with γ being a universal critical exponent.

Our value of 0.41 for γ agrees with the T exponent κ of ref. 2. We identify γ as $\gamma = 1/y = 1/\nu z$. The localization length exponent ν has been extensively studied theoretically [6,15] and has also been derived from experimental results[14,16]. There is general agreement that $\nu \approx 7/3$. With this value of ν, and our experimental value

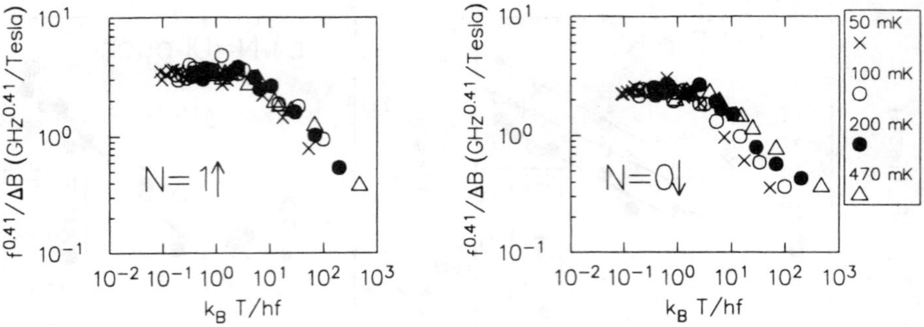

Fig. 5 Scaling functions for 0 ↓ and 1 ↑ peaks.

for γ, the dynamic exponent $z = 1$, as anticipated in refs. 5 and 6.

Our measured ΔB's increase with f as f^γ in the high-f regions of Fig. 4, but also increase with T in the low-f regions of Fig. 4b. Our dc T-dependence of ΔB goes as $\Delta B \sim T^\kappa$, where κ is the same as our γ to within error. With $\kappa = \gamma$, factoring out the f divergence of $1/\Delta B$ results in

$$F\left(\frac{k_B T}{hf}\right) = \frac{f^\gamma}{\Delta B} \tag{6}$$

where F is a scaling function, and is plotted in Fig. 5 for our $\Delta B(f,T)$ data on the 0 ↓ and 1 ↑ peaks. For each peak, the data fall on a function of k_BT/hf to within a factor of 2, and the function exhibits a crossover around $k_BT/hf = 3$.

Finding $\gamma = \kappa$, and the crossover from f- to T-determined ΔB at $k_BT \sim hf$ would be natural consequences of the T scaling observed at low f being due to to dynamic scaling. In this picture[5,7], the maximum of $(f\tau_c)$, $(k_BT\tau_c/h)$ determines the scale of the system, giving $f_c \sim k_BT/h$, in rough agreement with the crossover seen in the scaling function. κ becomes a universal critical exponent, equal to $\gamma = 1/\nu z$, and is unaffected by the relationship of the inelastic length and T.

A unity value for the dynamic exponent has been identified as a consequence of electron-electron interaction[5-7], and was predicted for the superconductor-insulator transition by Fisher, Grinstein and Girvin[17]. The Coulomb interaction energy $e^2/\epsilon\xi$, (where ϵ is the dielectric constant) is the characteristic energy for electron transport, and is related to the characteristic time τ_c by $\hbar/\tau_c \sim e^2/\epsilon\xi$, so that $z = 1$. If this is indeed the case, an experiment on the f and T dependence of the transport in a 2DES whose electron-electron interaction is modified by a nearby metallic "screening" layer would no longer necessarily have $z = 1$. We are presently pursuing experiments on this sort of system, with a 2DES-screening layer distance of $d \sim 100$ Å.

Finally, we relate our results to the theory of Polyakov and Shklovskii[8], who consider transport in the IQHE regime with a variable range hopping model incorporating the Coulomb gap and a diverging ξ. Polyakov and Shklovskii obtain $\Delta B \propto f^\gamma$ with

$\gamma = 1/\nu$, consistent with our observations. Aleiner and Shklovskii[9] also considered the variable range hopping model for a 2DES with a nearby metallic layer.

To summarize, we have presented a detailed description of an experimental method for measuring $\text{Re}(\sigma_{xx})$ in the quantum Hall effect regime, at microwave frequencies, and dilution refrigerator temperatures. We review our results on peak widths ΔB as a function of f and T, finding a power law $\Delta B \sim f^\gamma$ at high f, which is interpreted as dynamic scaling, only for $f \gtrsim k_B T/h$. The measured $\gamma = 0.41$ implies $z = 1$ if $\nu = 7/3$. It is possible to interpret our results with T entering the scaling picture through the thermal frequency $k_B T/h$.

ACKNOWLEDGEMENTS: We thank Prof. R. Bhatt for discussions, and acknowledge support of NSF grant DMR-9319264.

1. L. W. Engel, D. Shahar, Ç. Kurdak, and D. C. Tsui, *Phys. Rev. Lett.* **71** (1993) 2638.
2. H. P. Wei, D. C. Tsui, M. Paalenen, and A. M. M. Pruisken, *Phys. Rev. Lett.* **61** (1988) 1294.
3. A. M. M. Pruisken, *Phys. Rev. Lett.* **61** (1988) 1297.
4. P. C. Hohenberg and B. I. Halperin, *Rev. Mod. Phys.* **49** (1977) 435.
5. S. Kivelson, D.-H. Lee, S.-C. Zhang, *Phys. Rev. B* **46** (1992) 2223.
6. D.-H. Lee, Z. Wang, and S. Kivelson, *Phys. Rev. Lett.* **70** (1993) 4130.
7. Sondhi and Kivelson, unpublished.
8. D. G. Polyakov and B. I. Shklovskii, *Phys. Rev. Lett.* **70** (1993) 3796; D. G. Polyakov and B. I. Shklovskii, *Phys. Rev. B* **48** (1993) 11167.
9. I. L. Aleiner and B. I. Shklovskii, *Phys. Rev. B* **49** (1994) 13721.
10. C. P. Wen, *IEEE Trans. Microwave Theory Tech.*, **MTT-17**, (1969) 1087.
11. T. Itoh, *Planar Transmission Line Structures* (IEEE Press, New York, 1987) Parts 1 and 8.
12. S. Y. Liao, *Microwave Devices and Circuits* (Prentice Hall, Englewood Cliffs, 1980) p. 70.
13. Hewlett-Packard Application note 183, *High Frequency Swept Measurements*(1978); S. F. Adam, *Microwave Theory and Applications*, (Prentice Hall, Englewood Cliffs, 1969) p. 101.
14. S. Koch, R. J. Haug, K. von Klitzing, and K. Ploog, *Phys. Rev. Lett.* **67** (1991) 883.
15. J. T. Chalker and P. D. Coddington, *J. Phys. C* **21** (1988) 2665; G. V. Mil'nikov and I. M. Sokolov, *JETP Lett.* **48** (1988) 536; B. Huckestein and B. Kramer, *Phys. Rev. Lett.* **64** (1990) 1437; Y. Huo and R. N. Bhatt, *Phys. Rev. Lett.* **68** (1992) 1375.
16. H. P. Wei, L. W. Engel, and D. C. Tsui, submitted for publication in *Phys. Rev. Lett.*
17. M. P. A. Fisher, G. Grinstein and S. M. Girvin, *Phys. Rev. Lett.* **64** (1990) 587.

UNIVERSALITY AND CORRECTIONS TO SCALING IN THE INTEGER QUANTUM HALL EFFECT

BODO HUCKESTEIN
Max-Planck-Institut für Kernphysik, Postfach 10 39 80, 69029 Heidelberg, Germany

ABSTRACT

We study the scaling behavior of the localization length in Landau levels $n = 0$ and $n = 1$. We establish universality in the framework of a two-parameter scaling law. The universal scaling indices are the localization length exponent $\nu = 2.35 \pm 0.03$ and an irrelevant scaling index $y_{\rm irr} = -0.38 \pm 0.04$ associated with an irrelevant scaling field that depends strongly on the Landau level index and the correlation length of the disorder potential. This strong dependence is explained, using renormalization group arguments, by the behavior of the conductivity on short length scales.

Since numerical calculations are available for the localization length exponent ν an apparent discrepancy is observed between the values in different Landau levels if the disorder potential is short-ranged.[1] While the value obtained by Aoki and Ando in 1985 for the lowest Landau level and δ-function scatterers, $\nu \approx 2$, is close to the best numerical estimate $\nu = 2.35 \pm 0.03$[2,3], it deviates quite significantly from the value for the first Landau level $\nu \approx 4$. Subsequently,[3] it was shown numerically that the scaling behavior is independent of the correlation length of the disorder in the lowest Landau level and is also the same for Landau levels $n = 0$ and $n = 1$ if the correlation length σ is equal to the magnetic length $l_c = \sqrt{\hbar/eB}$. For shorter correlation length no universal scaling behavior was obtained.[3,4]

Experimentally, no indication for a non-universal behavior of ν was found. On the contrary, the measured values of 2.3 ± 0.1 for spin-split Landau levels agree well with the numerical value of 2.35 ± 0.03 independent of Landau level index and sample parameters.[5]

Chalker and Eastmond observed that deviations from scaling behavior in an extension of the network model[6–8] can be analysed in terms of an irrelevant scaling field.[7] The same analysis can be used in the first Landau level to show that the apparent nonuniversality can be explained by an irrelevant scaling field that generates a large length scale.[9] The idea is to replace the single-parameter scaling law for the localization length $\lambda_M(E)$ of a cylinder of circumference M,

$$\lambda_M(E) = M\Lambda\left(M/\xi(E)\right), \tag{1}$$

where $\xi(E) = \xi_0|E - E_c|^{-\nu}$ is the localization length of the infinite system and $E_c = (n + 1/2)\hbar\omega_c$ is the critical energy in the center of the n-th Landau level, by a

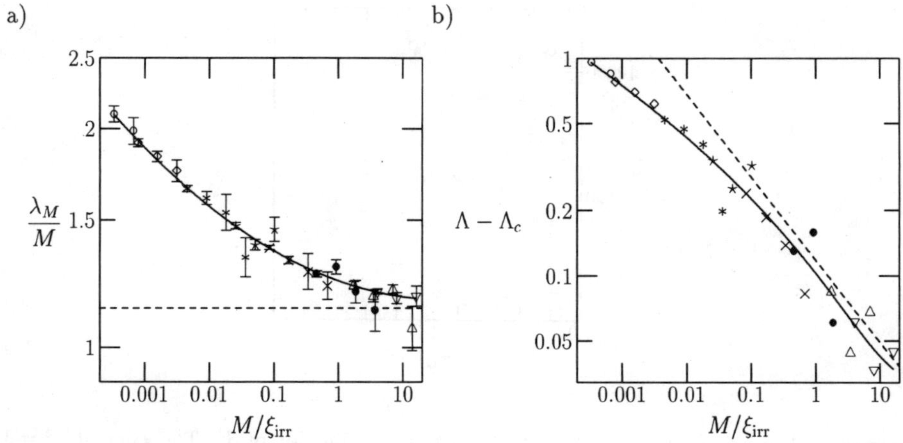

Figure 1: a) The renormalized localization length λ_M/M for β^2 between 1 and 1.7. The dashed line represents the asymptotic value $\Lambda_c = 1/\ln(1+\sqrt{2})$. b) Deviation of the scaling function Λ from its asymptotic value Λ_c. The dashed (shifted) line with slope $y_{\rm irr} = -0.38$ serves as a guide to the eye.

two-parameter scaling law

$$\lambda_M(E,\sigma) = M\Lambda\left(M/\xi(E), M/\xi_{\rm irr}\right). \tag{2}$$

The irrelevant length scale $\xi_{\rm irr}$ is related to an irrelevant scaling field $\zeta_{\rm irr}$ and an irrelevant scaling index $y_{\rm irr} < 0$ by

$$\xi_{\rm irr} \propto \zeta_{\rm irr}^{-1/y_{\rm irr}}. \tag{3}$$

The unknown scaling field $\zeta_{\rm irr}$ depends on the correlation length σ of the disorder potential. The relation $y_{\rm irr} < 0$ reflects the fact that it is irrelevant in a renormalization group sense and $\xi(E)$ is the only diverging length scale at the transition.

At the critical energy $E = E_c$ the two-parameter scaling law (2) reduces to a single-parameter scaling relation in the irrelevant scaling field,

$$\lambda_M(E_c,\sigma) = M\Lambda\left(0, M/\xi_{\rm irr}\right). \tag{4}$$

The irrelevant scaling index $y_{\rm irr}$ can be extracted from this relation by noting that the scaling function Λ has to be analytic in $M^{y_{\rm irr}}\zeta_{\rm irr}$. To leading order, Eq. (4) becomes

$$\lambda_M(E_c,\sigma) = M\left(\Lambda_c + aM^{y_{\rm irr}}\zeta_{\rm irr}\right). \tag{5}$$

$\lambda_M(E_c)$ was calculated for $\beta^2 = (\sigma^2 + l_c^2)/l_c^2$ ranging from 1 to 2 while M varied between 16 and 128 (in multiples of $\sqrt{2\pi}l_c$). For every value of β^2 the length $\xi_{\rm irr}$ was adjusted so that $\lambda_M/M(E_c)$ becomes a function of a single variable $M/\xi_{\rm irr}(\beta^2)$

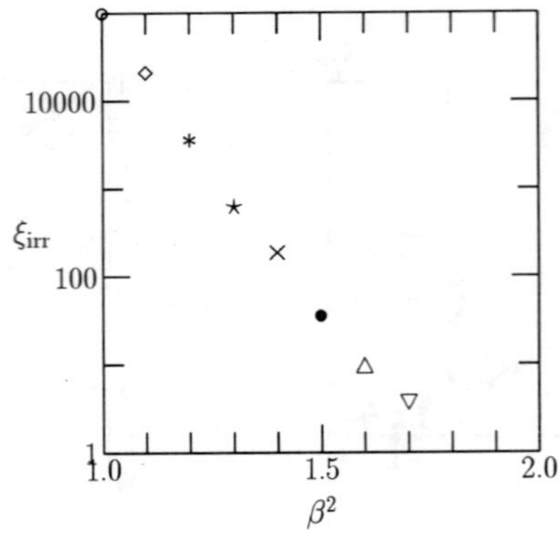

Figure 2: The length scale ξ_{irr} (in units of $\sqrt{2\pi}l_c$) as a function of β^2. The value at $\beta^2 = 1$ has been arbitrarily fixed to $\xi_{irr} = 50\,000$.

as shown in Fig. 1a). By performing this fit the overall scale of ξ_{irr} cannot be fixed and was taken to be of the order of l_c for $\sigma = l_c$. The observation of the corrections to scaling implies that not only the fixed point value Λ_c of the scaling function but also the localization length exponent ν are universal and independent of Landau level index and microscopic details of the disorder. In addition, the irrelevant scaling index can be extracted from the doubly logarithmic plot of $\Lambda - \Lambda_c$ vs. M/ξ_{irr} in Fig. 1b). $\Lambda_c = 1/\ln(1+\sqrt{2}) = 1.13459\ldots$ was used which is close to the best fit estimate of $\Lambda_c = 1.14 \pm 0.02$.[8] The slope of the dashed line is given by the irrelevant scaling index $y_{irr} = -0.38 \pm 0.04$. This value agrees well with the value $y_{irr} = -0.38 \pm 0.02$ for the network model.[7] An additional periodic potential leads to the same corrections to scaling and is hence an irrelevant perturbation at the fixed point in the center of the Landau levels.[9]

The irrelevant length scale ξ_{irr} in Fig. 2 shows an extremely strong dependence on β^2. It grows by more than four orders of magnitude when the correlation length σ is decreased from $0.8l_c$ to 0. This large increase in the cut-off length scale for finite-size corrections is the reason why previous finite-size-scaling studies were unable to observe the true asymptotic scaling behavior.[3,4] The behavior of the irrelevant length scale can be understood from renormalization group arguments. The critical fluctuations that drive the phase transition can only be observed on length scales larger than the localization length ξ_0 of the underlying mean-field theory. The mean-field theory for the integer quantum Hall transition is given by the unitary nonlinear sigma model and the self-consistent Born approximation (SCBA).[10,11] The localization length ξ_0 can be calculated by integrating the β-function of the unitary nonlinear sigma model and is given by

$$\xi_0 = l_c \exp\left(\pi^2 {\sigma^0_{xx}}^2\right), \tag{6}$$

where σ^0_{xx} is the bare conductivity in units of e^2/h given by the SCBA. Note the quadratic dependence on σ^0_{xx} in the exponent due to the absence of weak-localization corrections in the magnetic field. For δ-function scatterers the SCBA value for the bare conductivity in the center of the n-th Landau level is $\sigma^0_{xx} = (2n+1)/\pi$.[11] Thus, while the mean-field localization length ξ_0 is of the order of the magnetic length in the lowest Landau level it is larger by a factor of $e^8 \approx 3 \times 10^3$ in the first Landau level. The dependence of the bare conductivity on the range of the disorder potential also explains the strong range dependence of ξ_{irr}. In the first Landau level the conductivity decreases by about a factor of 6 between $\sigma = 0$ and $\sigma = l_c$ leading to a reduction of ξ_{irr} to microscopic values.

In conclusion, the observation of corrections to scaling according to Eq. (2) strongly supports the notion of *universal* metal-insulator-transitions at the centers of Landau levels in the integer Quantum Hall Effect. The occurrence of a very large irrelevant length scale related to the bare conductivity of the system explains why the universality of the localization length exponent ν could not be observed directly in previous calculations. Irrelevant perturbations seem to be described by a scaling index $y_{\text{irr}} = -0.38 \pm 0.04$, irrespective of the type of perturbation.

I thank J.T. Chalker for telling me about the results of Ref. 7 before publication.

References

1. H. Aoki and T. Ando, *Phys. Rev. Lett.* **54** (1985) 831.
2. B. Huckestein and B. Kramer, *Phys. Rev. Lett.* **64** (1990) 1437.
3. B. Huckestein, *Europhysics Lett.* **20** (1992) 451.
4. B. Mieck, *Europhysics Lett.* **13** (1990) 453; *Z. Phys.* **B90** (1993) 427; D. Liu and S. Das Sarma, *Mod. Phys. Lett.* **B7** (1993) 449.
5. S. Koch, R. J. Haug, K. von Klitzing, and K. Ploog, *Phys. Rev. Lett.* **67** (1991) 883; H. P. Wei, L. W. Engel, and D. C. Tsui, (1994) unpublished.
6. J. T. Chalker and P. D. Coddington, *J. Phys.* **C21** (1988) 2665.
7. J. T. Chalker and J. P. G. Eastmond, unpublished; J.F.G. Eastmond, Ph.D. thesis, Oxford University (1992).
8. D.-H. Lee, Z. Wang, and S. Kivelson, *Phys. Rev. Lett.* **70** (1993) 4130.
9. B. Huckestein, *Phys. Rev. Lett.* **72** (1994) 1080.
10. A. M. M. Pruisken, in *The Quantum Hall Effect*, eds. R. E. Prange and S. M. Girvin (Springer, Berlin, 1987).
11. T. Ando and Y. Uemura, *J. Phys. Soc. Jpn.* **36** (1974) 959.

LOCALIZATION IN A SPIN-DEGENERATE LANDAU LEVEL IN THE SEMICLASSICAL LIMIT

D.K.K. LEE
Department of Physics, Massachusetts Institute of Technology,
77 Massachusetts Ave, Cambridge MA02139

J.T. CHALKER
Theoretical Physics, University of Oxford, 1 Keble Rd, Oxford OX1 3NP, United Kingdom

We study localization in a spin-degenerate Landau level in the semiclassical limit, reached when the correlation length of the disorder is large compared to the magnetic length. It is argued that the localization length diverges at two distinct energies, with the same critical behavior as in a spin-split Landau level, and that states at the Landau band center are localized. This is supported by a numerical study of a random network model. We also discuss semiclassical approaches to two other related localization problems: localization in a random magnetic field, and localization in a pair of Landau levels with disorder only in the inter-Landau-level scattering.

The *spin-split* Landau level in the presence of a random potential represents the best characterized example to date of critical behavior at a metal-insulator transition. Within each disorder-broadened Landau level, almost all eigenstates are Anderson localized, but the localization length, ξ, diverges at a critical energy, E_c (which lies at the band center in the case of a symmetric potential): $\xi(E) \sim |E-E_c|^{-\nu}$. This scaling behavior has been observed experimentally and the critical exponent has been estimated to be $\nu = 2.3 \pm 0.1$ (*e.g.* see Koch *et al.* [1]). The same value for the critical exponent has also been obtained from numerical simulations, using a variety of models and techniques.[2–4]

By contrast, different critical behavior has been reported in sufficiently disordered samples at low magnetic field. Under these conditions, it is possible for the disorder-broadening of Landau levels to exceed the Zeeman energy, so that each orbital level becomes *spin-degenerate.* Assuming, as for spin-split Landau levels, that the localization length diverges only at one energy, the experimental results imply that the critical exponent ν is roughly twice that of the conventional value: $\nu \simeq 6.5$ (*e.g.* see Hwang *et al.* [5] and references therein).

We present below results[6–7] supporting an alternative scenario for the spin-degenerate system, first proposed by Khmelnitskii[8]. We study an extension of a "network model"[2] which is based on a semiclassical picture of electron motion in a smooth, two-dimensional random potential, under a strong, perpendicular magnetic field. Taking the magnetic length to be much shorter than the correlation length of the potential, electron motion can be separated into two components: a rapid cyclotron orbit and a slow drift of the guiding center along equipotential lines.

Formally, the model consists of a square network of links, representing a simplified view of the equipotentials of the random potential. Each link is characterized by a direction for the flow of probability flux, corresponding to the guiding-center drift. The electron acquires a phase shift on traversing the link. In addition, the current amplitudes for the two spin states ($s = 1, 2$) experience a random U(2) rotation in spin space. This models the effects of strong spin-orbit scattering, and is responsible for the difference between the localization properties of this two-channel system and that of the single-channel (spin-split) one.

The nodes of this network correspond to saddles points in the random potential where equipotentials approach each other. Here, a particle may tunnel across the saddle point, from left (L) to right (R) say, between two disjoint portions of an equipotential. Electron propagate coherently in this network so that interference between different tunneling paths is automatically included. This tunneling process is described at each node by a transfer matrix which, after allowing for the constraints of unitarity, contains only one important quantity, the "node parameter", θ_s, for each spin species. Current amplitudes, $\{\psi_s\}$, on the four links meeting at a given node are related by (for each s)

$$\begin{pmatrix} \psi_{\mathrm{out,R}} \\ \psi_{\mathrm{in,R}} \end{pmatrix} = \begin{pmatrix} \cosh\theta & \sinh\theta \\ \sinh\theta & \cosh\theta \end{pmatrix} \begin{pmatrix} \psi_{\mathrm{in,L}} \\ \psi_{\mathrm{out,L}} \end{pmatrix} \tag{1}$$

The transmission probability is $1/\cosh^2\theta$. Thus, tunneling is turned off in two limits: $\theta = 0$ and ∞, corresponding, respectively, to particle trajectories in the valleys and hills of the potential landscape, so that the energy of the particle lies in the tails of the Landau band. At $\sinh\theta = 1$, tunneling is maximal. This occurs at the percolation threshold of the equipotentials which, in terms of energy, corresponds to the center of the Landau band. In this study, we take tunneling at every node in the network to be the same. We can then discuss behavior of the system in terms of a two-dimensional phase diagram in the (θ_1, θ_2) plane. We will focus on the spin-degenerate limit, in which the Zeeman energy vanishes, so that the scattering of the trajectories of the two spin species should be described by the same node parameters: *i.e.* $\theta_1 = \theta_2 \equiv \theta$ line in the parameter space. On sweeping the Fermi level through the Landau band, the node parameter θ increases from 0 to ∞.

We are now in a position to discuss localization in this network model. In the absence of scattering between channels, the model would consist of two uncoupled networks, each as studied in Chalker and Coddington[2], with delocalized states on the lines $\sinh\theta_1 = 1$ and $\sinh\theta_2 = 1$. We are concerned with the result of coupling the two channels. We have studied numerically networks which are up to 128 links wide and 1.2×10^6 links long. The bulk localization lengths, ξ_∞, are extracted as a function of θ by a conventional one-parameter finite-size scaling analysis. Our results are consistent with Khmelnitskii's hypothesis[8] that there are *two* distinct mobility edges, situated away from the center of the Landau band, with the *same* critical exponent as in the spin-split case. Our results are presented in Fig. 1a, where we show that there exists a choice for a critical $\theta = \theta_c$ for which the data is compatible with the exponent value $\nu = 2.3$. Wang, Lee and Wen[9] have obtained similar results. (Polyakov and Shklovskii[10] have also discussed this scenario.)

It is of interest to explore the rest of the parameter space for this network model. In particular, we have studied the line $\sinh\theta_1 = \sinh\theta_2$, labeled "rf" in Fig. 1b. This represents the center in energy of a spin-pair of Landau levels, with varying Zeeman splitting along the line. These networks can be shown also to describe motion in a smooth random magnetic field [6–7]. We obtain a finite localization length everywhere on the line. Since this *includes the point* $\sinh\theta_1 = \sinh\theta_2 = 1$, we have further evidence that states at the center of the spin-degenerate Landau band are localized, as are states in a random magnetic field.

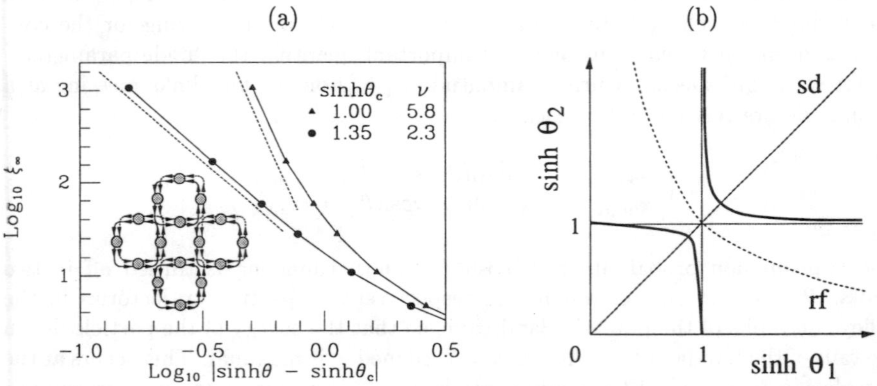

Figure 1. (a) Spin-degenerate Landau level. A better power-law fit to the form $\xi_\infty \sim |\sinh\theta - \sinh\theta_c|^{-\nu}$ is obtained for $\sinh\theta_c = 1.35$. Inset: schematic diagram of network. (b) Phase diagram of the network model (sd: spin-degenerate Landau level, rf: random magnetic field). Bold line denotes location of delocalization transition.

Finally, we apply the semiclassical approach to a model of theoretical interest, recently proposed by Hikami, Shirai and Wegner[11]. Details are presented elsewhere[12]. Consider a pair of degenerate Landau levels in a uniform magnetic field B, with random inter-level transitions (*cf.* the random U(2) mixing in the network model), but without disorder within each level. Regarding the level index a $S = 1/2$ pseudospin degree of freedom, the Hamiltonian can be written as:

$$H = \frac{1}{2m}(-i\boldsymbol{\nabla} + e\mathbf{A})^2 + \mathbf{h}(\mathbf{r}) \cdot \boldsymbol{\sigma} \tag{2}$$

where $\mathbf{A}$ is the vector potential, $\boldsymbol{\sigma}$ are the Pauli matrices and $\mathbf{h}$ is a *random* pseudofield in the xy-plane with a Gaussian distribution of zero mean: $\overline{h_{x,y}} = 0$. We consider here the semiclassical limit where the spatial fluctuation of $\mathbf{h}$ is smooth on the scale of the magnetic length. In this limit, the particle performs cyclotron motion along contours of $|\mathbf{h}(\mathbf{r})|$ with energy $\pm|\mathbf{h}|$ (as measured from the center of the Landau level). By analogy with the single Landau level in scalar disorder, one expects that this system has a pair of delocalization transitions at energies corresponding to the percolation of the $|\mathbf{h}|$ contours, as discussed by Wang, Lee and Wen[9]. This is reminiscent of the transitions discussed above for the spin-degenerate Landau level.

Hikami *et al.* also found a finite conductance at the band center (for the case of short-ranged disorder) and suggested a new universality class. We can recover such a delocalization transition in the limit of smooth disorder. In the semiclassical picture, band-center states are concentrated in the vicinity of $\mathbf{h} = 0$ points. Here, the pseudofield typically has a vortex-like configuration. It can be shown that there is a special symmetry in this model: the spectrum has reflection symmetry in energy about the band center. A consequence of this symmetry is that, around an isolated vortex, there is always an eigenstate at the band center. In general, there is more than one zero in the pseudofield. The associated band-center states, initially degenerate, will hybridize, and it appears that this gives rise to a delocalization transition. It can be argued further that these delocalized states are removed by a random scalar potential, which destroys the special symmetry of the spectrum. Therefore, we believe that the pair of delocalization transitions away from the band center should be robust to additional forms of disorder in the Hamiltonian, but that the band-center transition is special to this model.

In summary, we have presented evidence for the existence of a pair of mobility edges in spin-degenerate Landau levels. We hope that future experimental data can be analyzed in the context of this scenario. This work was supported by the EPSRC (GR/GO2727) and the European Community (SCC CT90 0020). DKKL acknowledges the support of the Lindemann Trust and an EPSRC/NATO fellowship.

References

[1] S. Koch *et al.* , *Phys. Rev. Lett.* **67** 883 (1991)
[2] J.T. Chalker and P.D. Coddington, *J. Phys. C: Solid State Phys.* **21** 2665 (1988)
[3] B. Huckestein and B. Kramer, *Phys. Rev. Lett.* **64** 1437 (1990)
[4] Y. Huo and R.N. Bhatt, *Phys. Rev. Lett.* **68** 1375 (1992)
[5] S.W. Hwang *et al.* , *Phys. Rev.* B **48** 11416 (1993)
[6] D.K.K. Lee and J.T. Chalker, *Phys. Rev. Lett.* **72** 1510 (1994)
[7] D.K.K. Lee, J.T. Chalker and D.Y.K. Ko, *Phys. Rev.* B **50** , in press (1994)
[8] D.E. Khmelnitskii, *Helv. Phys. Act.* **65** 164 (1992)
[9] Z. Wang, D.H. Lee and X.G. Wen, *Phys. Rev. Lett.* **72** 2454 (1994)
[10] D.G. Polyakov and B.I Shklovskii, *Phys. Rev. Lett.* **70** 3796 (1993)
[11] S. Hikami, M. Shirai and F. Wegner, *Nucl. Phys.* **B408** 415 (1993)
[12] D.K.K. Lee, *Phys. Rev.* B **50** , in press (1994)

TEMPERATURE DEPENDENCE OF QUANTUM HALL PLATEAUX WIDTH FOR LOW AND HIGH MOBILITY HETEROJUNCTIONS.

P.J. Gee, D.R. Leadley, S.O Hill, J. Singleton, R.J. Nicholas, #J.J. Harris, +C.T. Foxon
Department of Physics, Clarendon Laboratory, Parks Road, Oxford, OX1 3PU, UK
#Department of Electronics and Electrical Engineering, University College London, UK
+Department of Physics, The University, Nottingham, NG7 2RD, UK

We have performed magnetotransport measurements to search for possible universal behaviour of the conductance of a two dimensional system in high magnetic fields and very low temperatures using wide spacer layer high mobility GaAs-(Al,Ga)As heterojunctions.

Electrons in two-dimensional semiconductor heterostructures give rise to the Integer and Fractional Quantum Hall effects (IQHE and FQHE)[1,2] in high magnetic fields and at low temperature. In a single particle picture,[3] IQHE plateaux are thought to arise when the Fermi energy lies in a region of localised states. Recent theory[4] describes how disorder localises almost all the single particle states with the result that states at the Fermi energy are extended only at one energy within each Landau level. Consequently, the width of risers between QH plateaux and associated conductivity peaks should become infinitely narrow at T=0. Our motivation is to examine the IQHE for GaAs-(Ga,Al)As heterojunctions with varying amounts of disorder, including very high mobility (μ_e) structures with wide spacer layers where the disorder potential is expected to be smooth on the scale of the spacer layer thickness.[5] The QHE is expected to be qualitatively different from that observed in low mobility samples.[6-9]

GaAs-(Al,Ga)As heterojunctions with μ_e from 3 to 1200 m²/Vs (see Table 1) were studied in a dilution refrigerator in magnetic fields up to 17 tesla. Resistivity components ρ_{xx} and ρ_{xy} were recorded simultaneously at temperatures in the range 25mK $\leq T \leq$ 2K both in the dark and then after illumination with an LED. The temperatures were recorded by calibrated Ge and Ru_2O_3 sensors.

G378L and G590D (Figs. 1 and 2) possess similar carrier densities (n_e) so the ρ_{xx} minima occur at similar magnetic fields, but G378L has a lower mobility due to the deliberate introduction of an equal number of acceptors and donors in the channel. The relatively low μ_e of these samples gives rise to wide integer minima and sharp resistivity maxima, as opposed to the higher μ_e samples, where fractions appear and the ρ_{xx} maxima are broader. A comparison of the evolution of the ρ_{xx} maximum between filling factors ν = 1 and 2 in these samples is interesting. For G590D, the height of the

Table 1. Sample characteristics. Suffixes D or L indicate samples measured in dark or after illumination. μ_e is the transport mobility; μ_q is the quantum mobility measured from Dingle analysis.

Sample	Spacer Å	n_e 10^{15} m⁻²	μ_e m²/Vs	μ_q m²/Vs
G378D	100	2.76	3.7	0.59
G378L	100	3.52	5.8	0.88
G590D	100	3.19	11.5	0.73
G590L	100	6.14	64.9	1.81
G650L	400	2.18	1186	4.76
G640L	1200	1.15	684	3.81
G137L	1600	0.97	98.9	4.73
G641L	1600	0.90	451	4.32

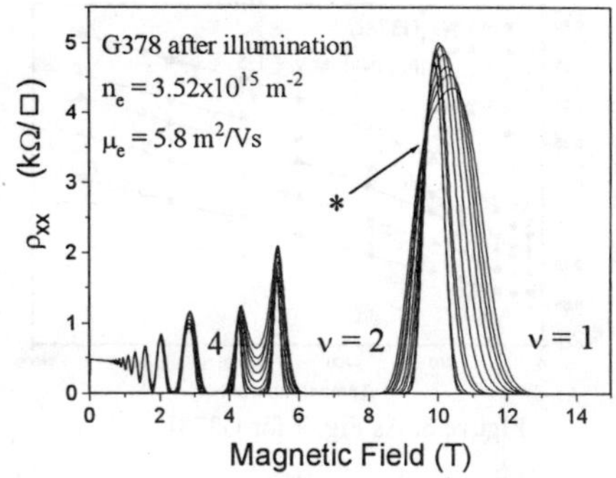

Figure 1. Resistivity of G378L for 25mK≤T≤2K

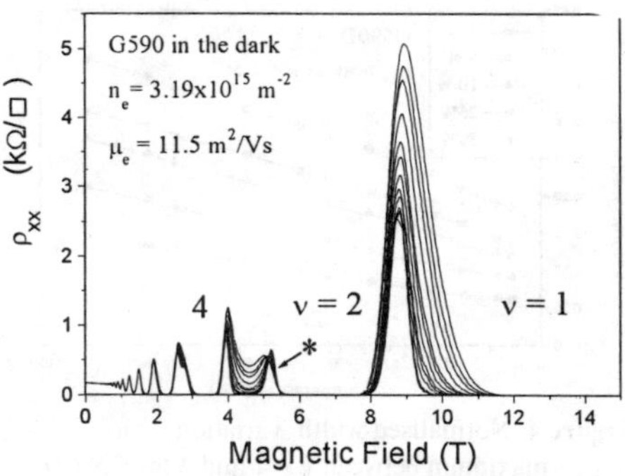

Figure 2. Resistivity of G590D for 25mK≤T≤2K

maximum is seen to increase with increasing T and its field position is fairly constant. However, after illumination, the height of the corresponding ρ_{xx} maximum of G378L is seen at first to increase and then decrease with increasing T; its mean position in field also increases. The broadening of this G378L ρ_{xx} maximum occurs chiefly on the spin split ($g\mu_B B$) rather than the cyclotron split ($h\omega_c/2\pi$) side. Similar behaviour occurs in G650L after light (higher mobility) between ν = 4 and 5 (Fig. 3), illustrating the general point that characteristic features appearing in high disorder samples at high fields are reproduced in low disorder samples at low fields.

It is thought that the different T-dependent behaviours of ρ_{xx} maxima in unilluminated and illuminated samples (Figs. 1 and 2) may reflect inhomogeneities on different length scales. Before exposure to light, the samples may contain regions of different carrier density, which would act as a series of small samples. The mesa as a whole would therefore only show finite ρ_{xx} when all of the regions had $\rho_{xx} \neq 0$,[10] i.e. when an extended state could propagate through the whole sample. As more states become extended across the whole sample at higher T the peak height would increase. This effect might give rise to ρ_{xx} peaks which do not shift with T, as seen in most of the unilluminated samples. After light, the samples will have a roughly homogeneous carrier density on a large length scale, and potential fluctuations will occur on a length scale approximately equal to the spacer layer thickness; the asymmetric shifting of ρ_{xx} peaks with T (Fig. 1) is thus likely to represent the "true" behaviour of ρ_{xx} due to potential fluctuations on the length scale of the spacer layer.

Note that the data (Figs. 1-3) exhibits apparent fixed points (starred) where ρ_{xx} does not vary within the given temperature range.[3,6,11] Additionally, comparing the evolution of the spin-splitting around ν = 3 in G590D, the ρ_{xx} maximum at ν = 2½ is lower than that at ν = 3½ whereas the opposite occurs in G378L.

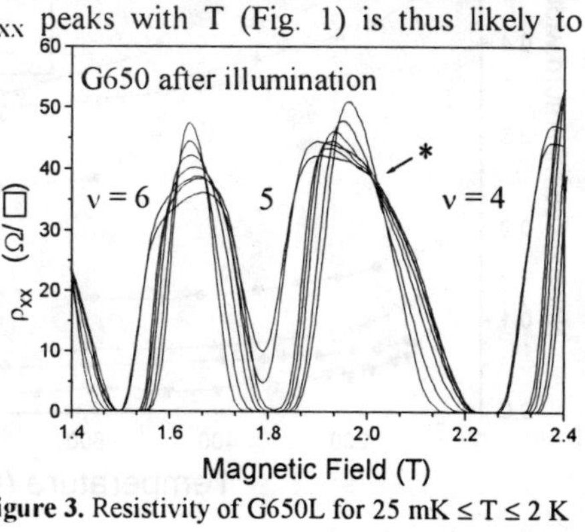

Figure 3. Resistivity of G650L for 25 mK ≤ T ≤ 2 K

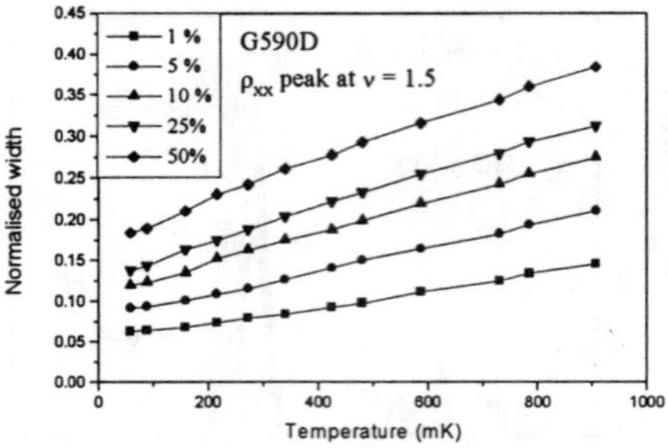

Figure 4. Normalised width variation with T of ρ_{xx} maximum between ν = 1 and 2 for G590D.

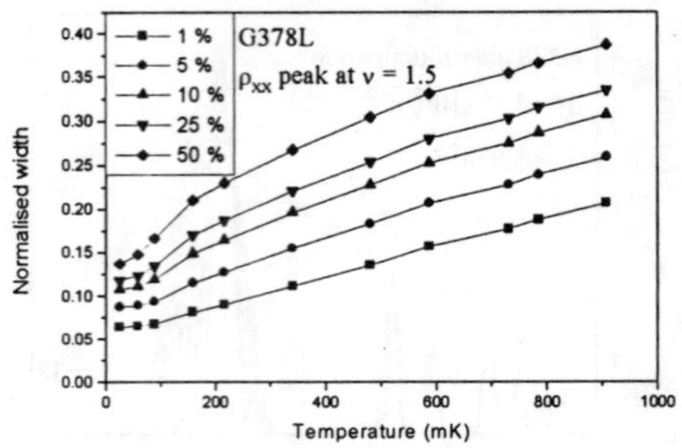

Figure 5. As Fig. 4 for G378L.

Our method of analysis has examined the T-dependent features of both localisation and delocalisation in order to obtain a picture of the evolution of ρ_{xx} and ρ_{xy} as T→0. The temperature variation of the ρ_{xx} maxima has been measured by calculating the widths in field at a set of fixed percentage heights of the maxima. These widths were then normalised by dividing by the value of field at which the maximum occurred. This method is only appropriate for the lower mobility samples where there are simple peaks. Samples with $\mu_e > 7$ m²/Vs also exhibit FQHE features within the maxima, so it is more appropriate to observe how the widths of the minima vary with T. The widths of the ρ_{xx} minima are calculated at constant ρ_{xx}, and normalised by dividing by the field at the corresponding integer filling factor. Both normalisation procedures compensate for the different n_e in the samples, allowing direct comparisons to be made.

Fig. 4 shows the T-dependence of the normalised width of the ρ_{xx} maximum at ν = 1½ for G590D, and Fig. 5 shows the same width variation for G378L. In Fig. 4, the width variation can be seen to be linear within the given T range (slight curvature below 200 mK), whereas in Fig. 5, in the low T region, the rate at which the width varies with T levels off. This effect becomes more evident with increasing sample quality. Additionally,

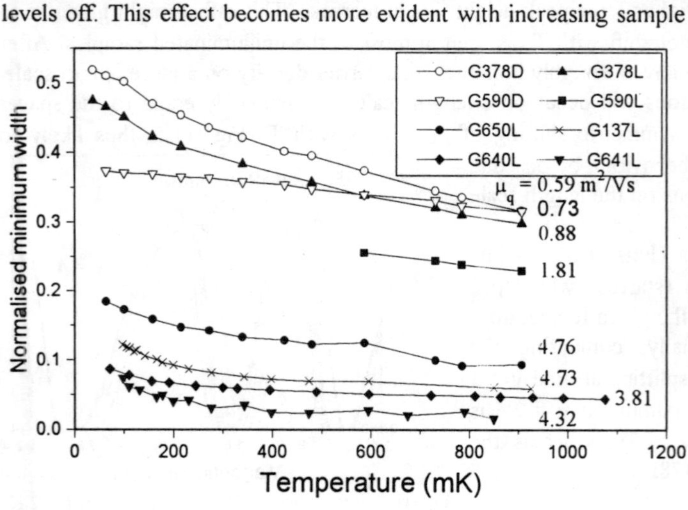

Figure 6. Normalised minimum width variation with temperature at υ = 2 for all samples in the dark and after illumination at ρ_{xx} = 10 Ω/□. The samples are labelled by their quantum mobility.

the linear regions of Fig. 4 and 5 do not extrapolate to zero at T=0 i.e., they possess finite ρ_{xx} maximum width at T=0, indicating that the theory of Ref. 4 is not appropriate. In each of the lower μ_e (3 to 7 m²/Vs) samples in this study, the gradient of the linear region of the ρ_{xx} maximum width versus T plot is greater at higher fields than at lower fields. The low T curvature is also more prominent for ρ_{xx} maxima at higher fields.

Fig. 6 shows the width variation with T for all the samples in this study for the $\nu = 2$ minimum, at the constant height $\rho_{xx} = 10\Omega/\square$. It gives the impression of similar behaviour in all samples with the possible exception of G590D. As discussed previously, this may result from the large potential fluctuation length scales in this case. It also appears that all the curves are predominately ordered by increasing spacer layer width, such that the higher mobility samples possess lower normalised widths. The curvature is also found to be stronger in the integer minima than in the fractional minima (in samples which demonstrate the FQHE) and increases with increasing field (decreasing ν) for any given sample such that it is strongest at $\upsilon = 1$.

For all samples in Fig. 6, the normalised width increases linearly with decreasing T until ~300 mK, at which point it tends to increase more rapidly. It appears therefore that there are two terms which govern the behaviour of all samples over the given T range. The first term appears to be linear at high T and is sample (spacer layer) dependent. The second term leads to the sample independent curvature at lower T. This suggests a possible "universal" behaviour; however this is not the power law suggested in studies of low mobility samples.[8]

We conclude that for the lower mobility disordered samples, where the behaviour of ρ_{xx} maxima are investigated, there is a finite width to the ρ_{xx} maxima as $T \to 0$, showing that more than one populated state is extended at T=0. However, comparisons of the experimental T dependence of ρ_{xx} minima with theory is perhaps less meaningful for the higher mobility samples, due to the absence of many body interactions in the theory.[4] Using this theory, the T=0 normalised width (W) of the ρ_{xx} minimum at filling factor υ is expected to be $W = \upsilon / (\upsilon^2 - \frac{1}{4})$. If the ρ_{xx} maxima approach δ-functions as $T \to 0$, then all curves of Fig. 6 should extrapolate to this value (for $\nu = 2$, W = 0.53). It appears unlikely that any of the samples save G378D will do this, illustrating a fundamental difference in behaviour between these samples and those of low mobility. Further studies are under way using samples in which μ_q and n_e can be adjusted by means of gates whilst keeping the sample geometry constant.

1 K.von Klitzing, G. Dorda and M. Pepper, Phys. Rev. Lett. **45**, 494 (1980).
2 D. C. Tsui, H. L. Störmer and A. C. Gossard, Phys. Rev. Lett. **48**, 1562 (1982).
3 The Quantum Hall Effect, edited by R. E. Prange and S. M. Girvin (Springer-Verlag, Heidelberg, 1990).
4 N. R. Cooper and J. T. Chalker, Phys. Rev. B **48**, 4530 (1993).
5 B. Shapiro, Phys. Rev. B **33**, 8447 (1986).
6 H. P. Wei, D. C. Tsui, M. Paalanen and A. M. M. Pruisken, Phys. Rev. Lett **61**, 1294 (1988).
7 H. P. Wei, S. Y. Lin, D. C. Tsui A. M. M. Pruisken, Phys. Rev. B **45**, 3926 (1992)
8 S. Koch, R. J. Haug, K. von Klitzing and K. Ploog, Phys. Rev. B **43**, 6828 (1991).
9 R. J. Haug, K. von Klitzing and K . Ploog, Phys. Rev. B, **35**, 5933 (1987)
10 Reinout Woltjer, Ph.D thesis, University of Utrecht (1988).
11 L. Engel, H. P. Wei, D. C. Tsui and M. Shayegan, Surf. Sci. **229** (1990)

MICROWAVE CONDUCTIVITY OF THE IQHE-TO-HALL INSULATOR TRANSITION

D. SHAHAR, L.W. ENGEL and D.C. TSUI
Dept. of Electrical Engineering, Princeton University, Princeton NJ 08544

ABSTRACT

We study the Integer Quantum Hall Effect-to-Hall Insulator transition at microwave frequencies f between 0.16 and 14.5 GHz and temperatures $T \gtrsim$ 25 mK. The transition is manifested by a diagonal conductivity peak near filling factor $\nu = 1/2$. For $f \gtrsim 1$ GHz, the peak width (ΔB) follows $\Delta B \sim f^{\gamma}$, with $\gamma = 0.41 \pm 0.05$.

1. Introduction

In the language of the scaling theory of localization[1,2] the Integer Quantum Hall Effect (IQHE)[3] is described as a series of continuous phase transitions separating adjacent quantum Hall liquid states. In low temperature (T) magnetotransport measurements these transitions manifest themselves as peaks in the diagonal resistivity (ρ_{xx}) accompanied by steps in the Hall resistance (ρ_{xy}) between IQHE plateaus. In the theoretical limits of an infinite sample and of $T = 0$, these peaks (and steps) are expected to be infinitely sharp. The experiments done by Wei *et al*[4] showed that the widths of the ρ_{xx} peaks (ΔB, where B is the magnetic field) and the inverse slopes of the ρ_{xy} steps $(d\rho_{xy}/dB)^{-1}$ diminish upon lowering T and both follow the same critical behavior

$$\Delta B, (d\rho_{xy}/dB)^{-1} \sim T^{\kappa}. \tag{1}$$

κ was found to be universal ($\kappa = 0.42 \pm 0.02$ for spin polarized states) and independent of the Landau level. These results demonstrate the scaling nature of the conductivity near these transitions, and indicate the existence of a correlation length, ξ, that diverges at the transition as $\xi \sim |B - B^*|^{-x}$, where B^* is the magnetic field at the transition. The best estimate[5] currently available for x is $\approx 7/3$, which is consistent with the results discussed above.

Scaling behavior of 2DES in B field is not limited to inter-QHE transitions. Wang *et. al.*[6] investigated a disordered sample that exhibited a B field driven transition from a weak-localization type behavior to the $\nu = 2$ IQHE state (ν is the Landau level filling factor). The critical exponent measured for that transition indicates that it belongs to the same universality class as the inter-IQHE transitions.

An important aspect of phase transitions is their dynamics[7]. The theory of dynamic scaling postulates the existence of a time scale that diverges as $\tau \sim \xi^z$ as a phase transition is approached. This relation defines z, the dynamic exponent.

Recently we[8] investigated the dynamics of the inter-IQHE transitions. Utilizing a new microwave (MW) technique to measure the real part of the diagonal conductivity, $Re[\sigma_{xx}(f)]$, at frequencies (f) from 0.16 to 14.5 GHz and $T \geq 50$ mK, we found that ΔB (referring to the width of the $Re[\sigma_{xx}(f)]$ transition peaks) increases with increasing f and follows $\Delta B \sim f^{0.41\pm0.04}$. Considering this result in the framework of dynamic scaling, and taking $x = 7/3$, we obtained $z = 1$.

Here we report on an implementation of the same MW technique to study an IQHE-to-Hall insulator transition. This transition is observed for samples with low enough mobility ($\mu \simeq 10^4$ cm^2/V·sec) such that the Fractional Quantum Hall Effect (FQHE) is not resolved. The term Hall insulator (HI) was introduced by Kivelson, Lee and Zhang (KLZ) [9], referring to the insulating phase that borders, at high B and high disorder, on the QHE. They proposed a global phase diagram in the disorder-magnetic field plane pertaining to both the IQHE and the FQHE. The topology of this phase diagram restricts the allowed continuous phase transitions into this HI to be only from a principal QHE state such as $\nu = 1$, $1/3, 1/5$ etc. The experiment presented here is a systematic study of the transition from the $\nu = 1$ IQHE to this HI.

We will focus, in this presentation, on the dynamic scaling aspect of this transition. Our main result is the observation, near $\nu = 1/2$, of a $Re[\sigma_{xx}(f)]$ transition peak (c.f. fig. 1), that, at $T = 25$ mK and above $f \approx 1$ GHz, becomes wider with increasing f and follows a scaling law, $\Delta B \sim f^\gamma$ with $\gamma = 0.41 \pm 0.05$.

2. Experimental

Our experimental method is described in detail elsewhere in this volume[10]. In short, we use a balanced bridge method to measure the MW absorption of a 2DES. Both bridge arms are terminated with matched MW square law power detector (Ge backward diode). We use one arm as the sensor of a feedback loop that maintains a constant MW power at the input of the bridge, and place the sample on the other. A coplanar transmission line deposited on top of the semiconductor heterostructure couples capacitively to the 2DES. The difference between the two detectors represents the power loss in the sample. The major contribution to that loss is absorption in the 2DES, from which $Re[\sigma_{xx}(f)]$ can be deduced in a straightforward manner using a standard transmission line model[11]. The sample and detectors are immersed in the mixing chamber of a top loading dilution refrigerator (capable of a base T of 25 mK) equipped with a superconducting solenoid that generates B field up to 15.5 T.

The sample used was a MOCVD grown lattice matched InP/$In_{0.53}Ga_{0.47}As$/InP single quantum well similar to those used in ref. 4. The electron density $n = 1.2\times10^{11}$ cm^{-2} and mobility 30,000 cm^2/V·sec at $T = 1.2$ K. We have chosen this density such that the $\nu = 1$ IQHE-to-Hall-insulator transition will fit comfortably in the available B range.

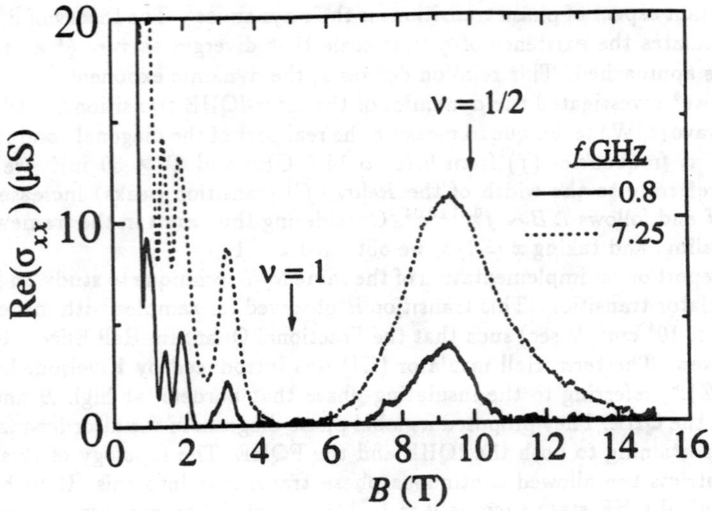

Fig. 1 $Re[\sigma_{xx}(f)]$ *vs.* B at two different f's at T = 25 mK. At f = 0.8 GHz, the ν = 1 and 2 IQHE states are fully developed. The peak near ν = 1/2 separates the IQHE and the Hall insulator phases. Note that the peak is at $\nu > 1/2$.

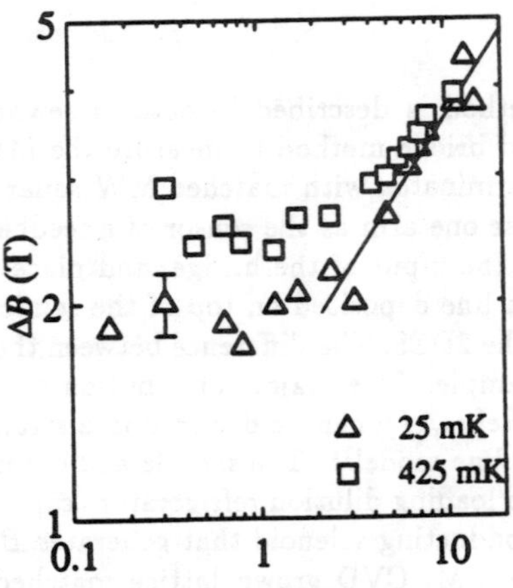

Fig. 2 The IQHE-to-Hall insulator transition peak width, ΔB, *vs.* f for two temperatures. The straight line is a power law fit to the 25 mK data above 2 GHz. The slope is 0.41 ± 0.05.

3. Results and Discussion

Typical $Re[\sigma_{xx}(f)]$ traces taken at $T = 25$ mK and $f = 0.8$ and 7.25 GHz are shown in fig. 1. The usual set of IQH states separated by peaks is clearly observed. A new feature in these data is the appearance of an additional peak near $\nu = 1/2$. This peak is a consequence of the $\nu = 1$ IQHE-to-HI transition since, at the lowest f and T measured, $Re[\sigma_{xx}] \rightarrow 0$ both for the $\nu = 1$ IQHE state in the low B side and for the HI on the high B side of the transition. It is noted that the transition peak in fig. 1 is at ν close to, but at *lower* B than $\nu = 1/2$, in agreement with the discussion in ref. 9, for samples with very large disorder.

The evolution of this peak as a function of frequency is summarized in fig. 2 where we plot ΔB, the full width at half maximum of the $Re[\sigma_{xx}(f)]$ peak. Above ~ 1 GHz, and at 25 mK, ΔB increases with f and follows $\Delta B \sim f^{\gamma}$ (solid line in fig. 2), with $\gamma = 0.41 \pm 0.05$. The 425 mK data converge to the same slope above ~ 5 GHz. This establishes that the $\nu = 1$ IQHE-to-HI transition follows dynamic scaling, with the critical exponent γ similar to the one obtained for the spin polarized, inter-IQHE plateau-to-plateau transitions.

To summarize, we observed and characterized the $\nu = 1$ IQHE-to-Hall-insulator transition as it appears in the MW $Re[\sigma_{xx}(f)]$. The transition follows similar dynamic scaling behavior as reported for the IQHE plateau-to-plateau transitions.

4. Acknowledgements

We would like to acknowledge discussions with N. Sivan. This work was supported by the NSF (DMR-9319264).

5. References

1. E. Abrahams, P.W. Anderson, D.C. Licciardello, and T.V. Ramakrishnan, *Phys. Rev. Lett.* **42** (1979) 673.
2. A.M.M. Pruisken, *Phys. Rev. Lett.* **61** (1988) 1297.
3. K. Von Klitzing, G. Dorda, and M. Pepper, *Phys. Rev. Lett.* **45** (1980) 494.
4. H.P. Wei, D.C. Tsui, and A.M.M. Pruisken, *Phys. Rev. Lett.* **61** (1988) 1294.
5. B. Huckenstein and B. Kramer, *Phys. Rev. Lett.* **64** (1990) 1437.
6. T. Wang, K.P. Clark, G.F. Spencer, A.M. Mack, and W.P. Kirk, *Phys. Rev. Lett.* **72** (1994) 709.
7. P.C. Hohenberg, and B.I. Halperin, *Phys. Rev.* **177** (1969) 952.
8. L.W. Engel, D. Shahar, C. Kurdak, and D.C. Tsui, *Phys. Rev. Lett.* **71** (1993) 2638.
9. S. Kivelson, D.H. Lee, and S.C. Zhang, *Phys. Rev. B* **46** (1992) 2223.
10. L.W. Engel, D.Shahar, C. Kurdak and D.C. Tsui, (In this volume).
11. C.P. Wen,*IEEE Trans. Microwave Theory Tech.* **17** (1969) 1087.
12. M.A. Paalanen, D.C. Tsui, A.C. Gossard and J.C.M. Hwang, *State Comm.* **50** (1984) 841.

ELECTRONIC STATES AND LEVEL STATISTICS IN 2D ELECTRONS IN STRONG MAGNETIC FIELDS

Yoshiyuki ONO

Department of Physics, Toho University, Funabashi, Chiba 274, Japan

The application of level statistics analysis along the random matrix theory to the study of localization in disordered systems is considered. Particularly the energy dependence of the level repulsion parameter is analyzed in the Landau bands of 2D disordered electrons in strong magnetic fields. The scaling behavior of the level repulsion parameter can yield the exponent of the localization length. The dependence of the exponent on the Landau band index and on the potential correlation length is discussed. The results are compared with other type of studies such as the real space scaling method. Possibility of extending the present method to the study of other disordered systems with different basic symmetries is pointed out.

1 Introduction

In order to explain the electronic properties of disordered systems, the understanding of quantum mechanical electronic states is inevitable. The electronic states are most directly studied by looking at the corresponding wave functions. An alternative method is to analyze the statistics of energy spectra. Since the energy and the wave function are both determined by the eigenvalue problem of the Hamiltonian describing the systems, it will be reasonable that the characteristics of one of them are reflected in the other. In this paper it will be discussed how useful the analysis of level statistics in disordered system is in understanding the localization-delocalization property of electronic states.

Since the energy levels are obtained as eigenvalues of the "random" Hamiltonian matrix reflecting random properties of the disordered potential, it is quite natural to apply the random matrix theory (RMT) which has a long history in the fields of the nuclear physics, the metallic fine particles, and the chaotic dynamics.[1-3] The RMT was originally proposed by Wigner to explain complicated energy spectra of heavy nuclei in terms of eigenvalues of random matrices with uncorrelated elements. According to RMT, the level statistics, such as the level spacing distribution, have quite universal properties depending only on the basic symmetry of the system. If the system has no time reversal symmetry due to the presence of magnetic fields or magnetic impurities, it is said to belong to the *unitary* universality class. If the system satisfy the time reversal symmetry, there are two universality classes; *orthogonal* when there is no spin-orbit interaction or the spin of the particle is integer and *symplectic* when the spin of the particle is half odd-integer and there exists a spin-orbit interaction. The most prominent difference between the usual RMT and the disordered electron systems will be the existence of correlations among the Hamiltonian matrix elements in the latter. This will be clear if we remind the Hamiltonian matrix in the Anderson model where the diagonal elements are uncorrelated random numbers while the off-diagonal elements take 0 or a non-zero value (= the nearest-neighbor transfer integral) in a regular way. It is not an easy task to relate the correlations in matrix elements to the localization-delocalization properties of wave functions, since the apparent correlations of matrix elements depend on the choice of basis functions.

2 Universality of Level Statistics

The level statistics are characterized by various quantities such as the level spacing distribution, the two level correlations, the Δ_3-statistics (or the rigidity of the level distribution),[3] and the fluctuations of the number of levels within an energy band of a given width.[4,5] Among them the level spacing distribution seems to reflect most directly the difference in the universality classes. In the case of Gaussian ensembles where the random elements of Hamiltonian matrices obeys the Gaussian distribution, the distribution of the spacing t of neighboring levels are believed to be given by the following expressions according to three different universality classes,[3]

$$P_{\mathrm{GOE}}(t) = (\pi/2)t\exp(-\pi t^2/4), \tag{1}$$
$$P_{\mathrm{GUE}}(t) = (32/\pi^2)t^2\exp(-4t^2/\pi), \tag{2}$$
$$P_{\mathrm{GSE}}(t) = (2^{18}/3^6\pi^3)t^4\exp(-64t^2/9\pi), \tag{3}$$

where GOE, GUE, and GSE are the abbreviations of Gaussian orthogonal, unitary, and symplectic ensembles, respectively, and the numerical coefficients are determined so as to normalize the distribution function in the region $t > 0$ and to make the average spacing unity. The expression for P_{GOE} is known as Wigner's surmise and obtained by diagonalizing a 2×2 real symmetric Gaussian random matrix. Other two expressions are its straightforward extension. As far as the level statistics in the delocalized regime of disordered electron systems are concerned, the level spacing distribution is considered to be well described by the above expressions in each universality class. In the strongly localized regime, however, the levels are uncorrelated because of negligible overlap of localized wave functions. Therefore the level statistics are subject to Poissonian, and the spacing distribution is given by

$$P_{\mathrm{Poisson}}(t) = \mathrm{e}^{-t}, \tag{4}$$

irrespectively of the universality class. If we look at the spacing distribution, the transition from the delocalized to the localized regime is expected to reveal itself as the transition from one of the distributions, (1) to (3), to the Poissonian (4) when we change the energy or the strength of disorder.

The behavior of $P(t)$ in the small t region is expressed as t^β, where $\beta = 1, 2, 4$, and 0 for GOE, GUE, GSE, and Poissonian. Physically the parameter β represents the level repulsion strength, i.e. the larger value of β means stronger exclusion of vanishing level spacing. It will be quite natural to see the localization-delocalization transition in terms of the level repulsion parameter. In the following we discuss the behavior of the level repulsion parameter as a function of energy and system size in the case of the quantum Hall effect(QHE) regime and show how fruitful conclusions can be derived from the analyses of level statistics.

3 Level Spacing Distribution in QHE Regime

In many of recent investigations the localization-delocalization transitions are studied by level statistics.[4-8] They are concerned mainly with the orthogonal case, i.e. the case satisfying the time reversal symmetry and having no spin-orbit interaction. We have studied the level statistics in the lowest Landau band (LLB) of 2D disordered electron systems subject to strong magnetic fields.[9,10] Such 2D systems are the stage of the QHE and have been studied quite precisely. Nevertheless the detailed analysis of level statistics was not done until recently. Since the delocalization occurs only at the LB centers irrespectively of the strength of disorder, this system is very convenient to study the relation between the level statistics and the localization-delocalization transition. Furthermore the appearance of the magnetic field allows us to study the effect of the violation of time reversal symmetry.

We assume the system is described by the following 2D Hamiltonian,

$$H = (\boldsymbol{p} + e\boldsymbol{A})^2/2m + V(\boldsymbol{r}), \tag{5}$$

where $\boldsymbol{A} = (0, Bx)$ is the vector potential giving rise to the perpendicular magnetic field, and the disordered potential $V(\boldsymbol{r})$ is characterized by ensemble averages,

$$\overline{V(\boldsymbol{r})} = 0, \tag{6}$$

$$\overline{V(\boldsymbol{r}+\boldsymbol{b})V(\boldsymbol{r})} = (u^2/2\pi s^2)\exp(-b^2/2s^2). \tag{7}$$

Here u and s represent the strength and the correlation length of the potential.* In order to discuss the bulk effect, the periodic boundary conditions (p.b.c.'s) are imposed in both directions.[11,12] If we express the linear dimension of the system by L, the p.b.c.'s require that L is integer times ΔX, the spacing of the center coordinate of the Landau function, i.e. $L = N\Delta X = N \times 2\pi\ell^2/L$ with $\ell = \hbar/eB$ the magnetic length. The energy levels are obtained by diagonalizing the Hamiltonian matrix calculated by using properly chosen basis functions,[18] in the present case the Landau functions extended to satisfy the p.b.c.'s in both directions.[12] The matrix dimension is N per one LB.

The density of states of a LB, $D(E)$, is expressed schematically as in Fig. 1. Since the density of states is inversely proportional to the local average of the level spacing, the shape of $D(E)$ shown in Fig. 1 indicates that the average spacing depends strongly on the energy region. In order to normalize the average spacing, energy eigenvalues $\{E_i\}$ are mapped to $\{x_i\}$ through the so-called unfolding procedure defined by $x_i = N(E_i)$, where $N(E)$ is the integrated density of states, $N(E) = \int^E \mathrm{d}ED(E)$. A simple argument[9] leads to the fact that the level density for $\{x_i\}$ is unity within the

*As far as we consider the case where the LB mixing may be neglected, the band width can be the unit of energy in the system, and therefore the value of u need not be specified.

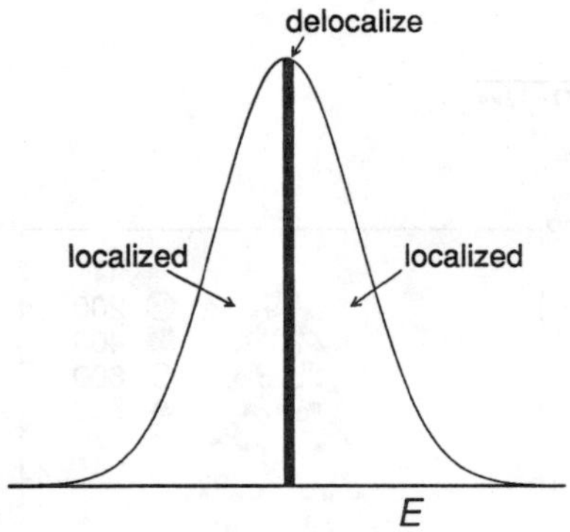

Fig. 1: Density of states in a LB (schematic)

"band", $0 < x < N$. For the later convenience we shift x_i by $N/2$ so that the band center corresponds to $x = 0$. We discuss the statistics of levels $\{x_i\}$. In order to study the energy region dependent level statistics, we take out the levels within a width of $0.1N$ centered at x and discuss e.g. the x-dependence of the level repulsion parameter. The number of samples are chosen so that the total number of levels may be about 300,000.

Since the numerical data for the level spacing distribution function cannot be described by any of the expressions (1) to (4), we employ Brody's interpolation formula[13] extended to the region $\beta > 1$,

$$P_\beta(t) = (1+\beta)at^\beta \exp(-at^{1+\beta}), \qquad \text{for } 0 \le \beta \le 1, \tag{8}$$

$$= ct^\beta \exp(-b^2t^2), \qquad \text{for } \beta > 1, \tag{9}$$

with the coefficients a, b and c determined so as to normalize the distribution function and the average spacing.[9] It is straightforward to verify that the above interpolation formula reduces to P_{Poisson}, P_{GOE}, P_{GUE}, and P_{GSE} for $\beta = 0, 1, 2$, and 4, respectively.

In Fig. 2 an example of fitting is shown for $x = 0$ in the case of the LLB with the matrix size $N = 800$ and the δ-correlated potential. For comparison, GUE and GOE curves are also shown. It should be noted that the level repulsion parameter β is smaller than 2 although the localization length exceeds the system size in this region. Similar fittings are carried out for different values of x and for different values of N. The result for β is summarized in Fig. 3.

4 Scaling Property of Level Repulsion Parameter

The size dependence of the function $\beta(x)$ shown in Fig. 3 suggests the existence of scaling behavior. Since whether the level statistic obeys the Gaussian ensemble property or the Poissonian depends on whether the states are delocalized or localized, it will be plausible to assume that the energy and system size dependences of β is scaled as $\beta(E, L) = f_1(L/\xi(E))$, where $\xi(E)$ is the energy dependent localization length and $f_1(z)$ is a proper scaling function. In the case of a 2D LB, ξ diverges as $|E|^{-\nu}$ with energy E measured from the band center. Many numerical and analytic works have shown that the exponent ν is around 2.4 in the case of the LLB irrespectively of the correlation length of the disordered potential.[18,14-21] In order to replace E and L dependences by x and N dependences, we note that $x = \int^E D(E)\mathrm{d}E - N/2 \simeq D(0)E$, and $D(0) \propto N \propto L^d$ with d the system dimensionality. As a result, the scaling property of β is expressed as

$$\beta(x, N) = f_2\left(\frac{x}{N^{1-1/\nu d}}\right). \tag{10}$$

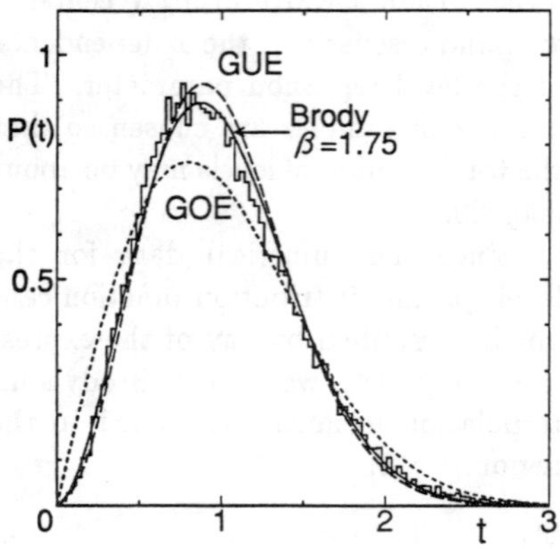

Fig. 2: Brody fitting at the center of the LLB. The matrix size is 800, and the potential is δ-correlated. The best fit value of β is indicated in the figure.

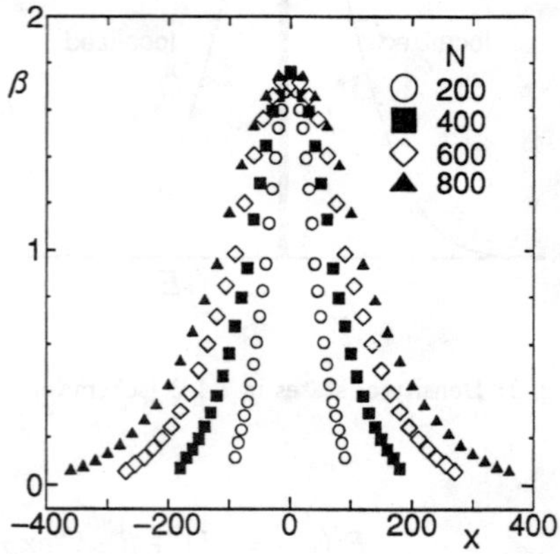

Fig. 3: $\beta(x)$ for four different values of the matrix size, $N = 200, 400, 600$, and 800 in the LLB with δ-correlated potential.

If we employ $\nu = 2.34$ (the most reliable value)[18] and $d = 2$, x should be scaled by $N^{0.786}$. In fact the data in Fig. 3 show a single curve when plotted as a function of $x/N^{0.786}$ (see Fig. 4). The result in Fig. 4 indicates that the localization length behaves as $|x|^{-\nu}$ in a rather wide range of x. According to the above scaling Ansatz, $\beta_c \equiv \beta(x = 0)$ is independent of N (or L). This is clearly seen in Figs. 3 and 4. Similar analyses have been done for different values of the potential correlation length and for a different LB (the 2nd LLB). In Fig. 5, the result for the LLB with $s = 2\ell$ is summarized. Again this result shows that the localization length behaves as $\xi \sim |x|^{-1.56}$ in a wide range of x. The corresponding exponent of the localization length contradicts with previous calculations.[19,21] If we note that the present analysis yields overall behavior of ξ instead of the very critical behavior, this contradiction will not be so serious. In order to investigate the very critical behavior, it will be necessary to study larger system sizes and many more samples and we have to concentrate ourselves to a very narrow region of x around $x = 0$. This has not yet been done. From the comparison of Figs. 4 and 5, we find that β_c is decreasing function of the potential correlation length.[†] This seems reasonable since in the long

[†]In the case with $s = \ell$, we find $\beta_c \simeq 1.5$.

range limit each state is represented by a classical locus which is nothing but an equipotential line. In this limit the quantum level repulsion is expected to play no essential role.

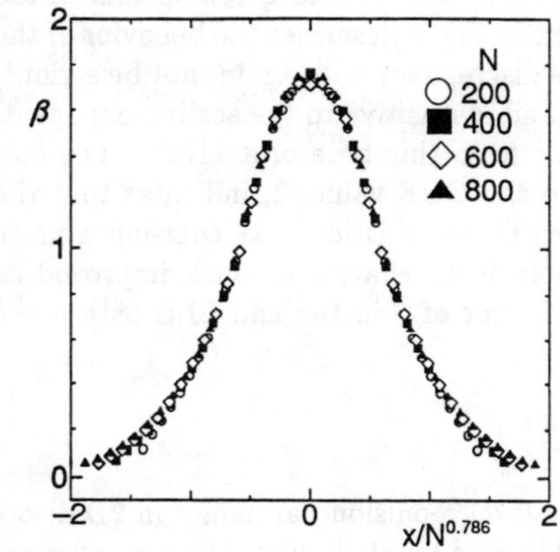

Fig. 4: Scaling behavior of β in the LLB with δ-correlated potential. The localization length exponent is $\nu = 2.34$.

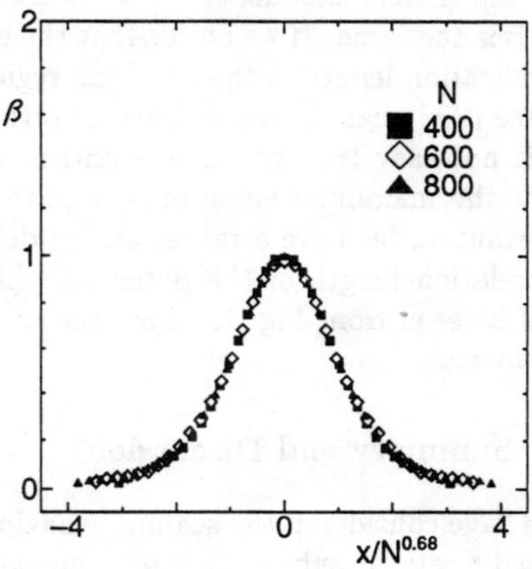

Fig. 5: Scaling behavior of β in the LLB with $s = 2\ell$. The corresponding exponent of the localization length is $\nu = 1.56$.

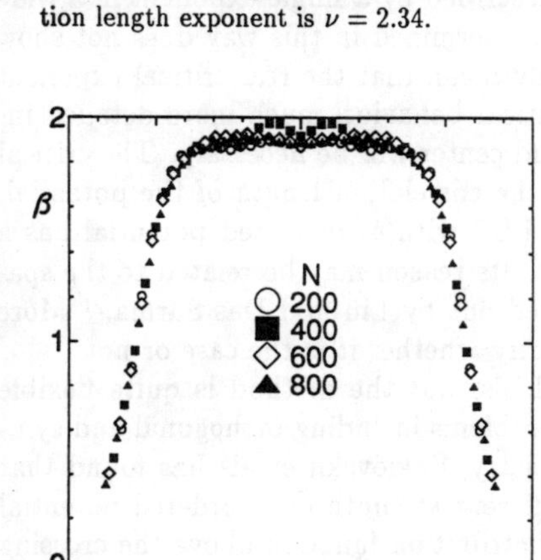

Fig. 6: Scaling behavior of β in the 2nd LLB with δ-correlated potential. The localization length exponent is $\nu = 7.1$.

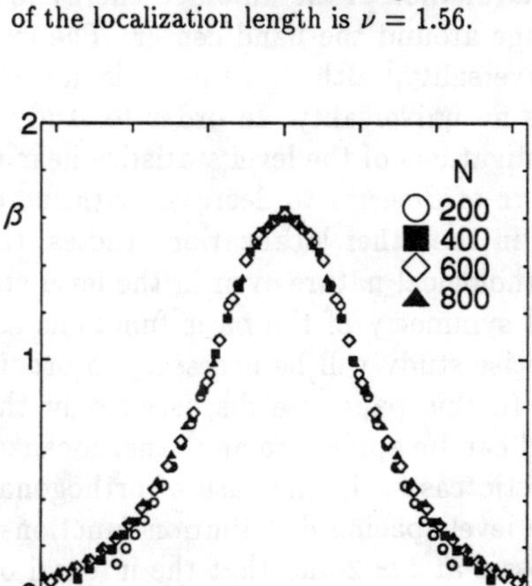

Fig. 7: Scaling behavior of β in the 2nd LLB with $s = \ell$. The corresponding localization length exponent is $\nu = 1.67$.

The result of analyses for the case of the 2nd LLB with $s = 0$ is summarized in Fig. 6 in a similar way as in Figs. 4 and 5. As was pointed out by previous works, the localization behavior in this case is known to be pathological. Figure 6 shows that this pathological behavior is reflected also in the level statistics. The corresponding exponent of the localization length is 7.1 and somewhat larger than reported values for the critical exponent,[19,21] although the tendency that it is larger than that of the LLB is the same. If we notice that the exponent in Fig. 6 describes the behavior of the localization length in the band tail region, this discrepancy will, again, not be serious. Since $\beta(x)$ near the band center is rather flat and insensitive to the scaling exponent, it is not easy to derive a real critical behavior from this type of analysis. The fact that the maximum value of β is quite near to the GUE value, 2, indicates that the relevant states have a rather strong delocalized characteristic. If we consider a finite correlation length of the potential, this pathological behavior is much improved as will be seen from Fig. 7, where the scaling behavior of β in the 2nd LLB with $s = \ell$ is shown.

5 Summary and Discussion

We have considered the scaling behavior of the level repulsion parameter in 2D disordered electrons subject to strong magnetic fields, and pointed out that some informations about the localization length can be derived from the analysis. The existence of the scaling behavior in a wide range of x suggests that the localization length viewed as a function of the unfolded energy may be described by a single exponent in a wide range around the band center. The exponent determined in this way does not show universality,[‡] although this does not necessarily mean that the real critical exponent has no universality. In order to study the critical behavior, much more detailed investigations of the level statistics near the band center will be necessary. The critical value of β seems to decrease with increasing the correlation length of the potential. As in the other localization studies, the 2nd LLB with δ-correlated potential has a pathological nature even in the level statistics. Its reason may be related to the spatial symmetry of the basis functions as pointed out by Liu and Das Sarma.[21] More precise study will be necessary in order to clarify whether it is the case or not.

In this paper we discussed only the 2D LB's, but the method is quite flexible and can be applied to any other localization problems including orthogonal and symplectic cases. In the case of orthogonal symmetry, Shklovskii et al[6] has found that the level spacing distribution functions for different strength of disordered potential crosses at $t \simeq 2$ and that the integral of the distribution function above the crossing point obeys the scaling law. From this scaling they found the localization length exponent consistent with other numerical studies.[22] Hofstetter and Schreiber[8] used an interpolation formula different from Brody's one in order to quantify the transition of the level repulsion parameter from unity to zero. They also found a scaling behavior and the proper exponent of the localization length.

These successes suggest that the quantity, from which we derive informations about the localization, need not be the level repulsion parameter. If the quantity re-

[‡]The exponent is decreasing function of s, for example.

flects the change of the level statistics properly, it may be used as a scaling parameter. In fact the level spacing distribution as shown in Fig. 2 can be regarded as GUE in the small t region and GOE in the large t region as far as $\beta > 1$. This crossover can be explained by considering semiclassical orbits in magnetic fields[23] and characterized by a single parameter α.[9] This α as a function of x is found to obey the scaling law. The same exponent as used for β can explain the scaling behavior of α as far as this analysis is possible. This method is, however, not applicable e.g. to the case shown in Fig. 5. Nevertheless these analyses indicate that the scaling property governs the level statistics as a whole. The scaling analysis of the level statistics will become a powerful method for the study of localization.

Acknowledgements

Useful and fruitful comments from B. Kramer and T. Ohtsuki are acknowledged. Discussions with K. Slevin are also appreciated.

References

1. E.P. Wigner: Ann. Math. **53**, 36 (1951), **62**, 548 (1955), **65**, 203 (1957), **67**, 325 (1958).
2. F.J. Dyson: J. Math. Phys. **3**, 140, 157, 166 (1962).
3. M.L. Mehta: *Random Matrices* 2nd ed. (Academic Press, 1991).
4. B.L. Al'tshuler and B.I. Shklovskii: Sov. Phys. JETP **64**, 127 (1986).
5. B.L. Al'tshuler, I.K. Zharekeshev, S.A. Kotochigova, and B.I. Shklovskii: Sov. Phys. JETP **67**, 625 (1988).
6. B.I. Shklovskii, B. Shapiro, B.R. Sears, P. Lambrianides, and H.B. Shore: Phys. Rev. **B 47**, 11487 (1993).
7. V.E. Kravtsov, I.V. Lerner, B.L. Al'tshuler, and A.G. Aronov: Phys. Rev. Lett. **72**, 888 (1994).
8. E. Hofstetter and M. Schreiber: Phys. Rev. B **48**, 16979 (1993) and preprint.
9. Y. Ono, H. Kuwano, K. Slevin, T. Ohtsuki, and B. Kramer: J. Phys. Soc. Jpn. **62**, 2762 (1993).
10. Y. Ono and T. Ohtsuki: J. Phys. Soc. Jpn. **62**, 3813 (1993), **63**, Suppl. A 158 (1994).
11. T. Ando: J. Phys. Soc. Jpn. **52**, 1740 (1983), **53**, 3101, 3126 (1984).
12. Y. Ono, T. Ohtsuki and B. Kramer: J. Phys. Soc. Jpn. **60**, 270 (1991).
13. B. Huckestein and B. Kramer: Solid State Commun. **71**, 445 (1989), Phys. Rev. Lett. **64**, 1437 (1990).
14. T.A. Brody: Nuvo Cimento Lett. **7**, 482 (1973).
15. H. Aoki and T. Ando: Phys. Rev. Lett. **54**, 831 (1985), T. Ando and H. Aoki: J. Phys. Soc. Jpn. **54**, 2238 (1985).
16. S. Hikami: Prog. Theor. Phys. **76**, 1210 (1986).
17. J.T. Chalker and P.D. Coddington: J. Phys. **C21**, 2665 (1988).
18. G.V. Mil'nikov and I.M. Sokolov: JETP Lett. **48**, 536 (1988).
19. B. Huckestein: Europhys. Lett. **13**, 453 (1992).
20. Y. Huo and R.N. Bhatt: Phys. Rev. Lett. **68**, 1375 (1992).
21. Dongzi Liu and S. Das Sarma: Phys. Rev. B **49**, 2677 (1994).
22. B. Kramer, K. Broderix, A. MacKinnon, and M. Schreiber: Physica **A 167**, 163 (1990).
23. N. Argaman, Y. Imry, and U. Smilansky: Phys. Rev. **B 47**, 4440 (1993).

RESONANT SCATTERING BY IMPURITIES AND UNIVERSAL LOCALIZATION SCALING IN THE INTEGER QHE

S.A. GURVITZ
Department of Particle Physics, Weizmann Institute of Science, Rehovot 76100, Israel
and TRIUMF, Vancouver. B.C., Canada V6T 2A3

ABSTRACT

We investigate the inter-edge transport due to resonant scattering on randomly distributed impurities in the vicinity of Landau levels. It is shown that the transport takes place along resonant-percolation trajectories. The problem is related to one of localization in one-dimensional systems with off-diagonal disorder. It is shown that for identical impurities each Landau band has one extended state $\bar{E}_N$, while all other states are localized. Neglecting self-crossing resonant-percolation trajectories, one finds that the localization length $L(E) \sim (E - \bar{E}_N)^{-2}$. The exponential depends neither on the type of impurity not on the number of the Landau band.

1. Introduction

It is known that the electronic states are localized for a two-dimensional system with any amount of disorder. However, in the presence of a uniform magnetic field perpendicular to the system, both the localized and the extended states are necessary in order to explain the integer QHE. In this paper we investigate the mechanism of a delocalization by taking into account only the resonance component of the scattering amplitude, which is the dominant one if the scattering goes via intermediate quasi-stationary states near impurity sites (tunneling). Such a process always takes place in the vicinity of Landau levels, since any weak impurity (attractive or repulsive), irrespective of its strength will generate quasi-stationary states (resonances) in that region[1].

It was found[1] that in the case of a strong magnetic field the resonance energy does not depend on the impurity podesition, provided that the impurity is not in the region of the sample boundaries. In contrast, the resonance width (Γ) drops down very rapidly with the distance from the edge. Thus, the resonances generated by impurities in the bulk can be considered as stationary states. Due to the overlap of these states, localized near different impurity sites, an electron can propagate across a sample via resonant scattering (resonant tunneling) when the Fermi energy is near the Landau levels. In the following we are going to analyze this process. For the definiteness we restrict our discussion to the $N = 0$ Landau band, although the results are valid for any N.

2. Resonant-scattering via different impurities

Consider the inter-edge electron transport across a sample as a sequence of resonant transitions via n impurities located at the points $\boldsymbol{r}_1, \ldots, \boldsymbol{r}_n$ inside a sample. We label the impurities in such a way that the first one is closest to the upper edge, and the last one is closest to the lower edge. The probability of the inter-edge transition $P_{1n}(E)$ can be written as[1] $P_{1n}(E) = 4\Gamma_1 |b_n(E)|^2 \Gamma_n$, where $\Gamma_{1,n}$ are the widths of the states near the impurities 1 and n sites respectively and are due to the coupling with the edge states, and $b_n(E)$ is the amplitude of the inter-edge transition via n impurities. The amplitude b_n is determined from the Schroedinger equation

$$\sum_{j'=1}^{n} (E\delta_{jj'} - \mathcal{H}_{jj'}) b_{j'}(E) = i\hbar \delta_{1j} \tag{1}$$

The diagonal matrix elements in the effective Hamiltonian, except for $\mathcal{H}_{11}$ and $\mathcal{H}_{nn}$, are the same: $\mathcal{H}_{jj} \equiv \bar{E}_0 = E_0 + \epsilon_0$, where E_0 is the energy of the $N = 0$ Landau level and ϵ_0 is the energy of the resonance state generated by the j-th impurity. It does not depend on the impurity position. (The energy of the state near the first and the last impurity sites obtains imaginary parts, so that $\mathcal{H}_{11} = \bar{E}_0 - i\Gamma_1$ and $\mathcal{H}_{nn} = \bar{E}_0 - i\Gamma_n$). However, the non-diagonal matrix elements $\Omega_{ij} \equiv \mathcal{H}_{ij}$ are strongly dependent on impurity coordinates. For instance, one finds for short range impurities[1]:

$$\Omega_{ij} = \epsilon_0 e^{-(\boldsymbol{r}_i - \boldsymbol{r}_j)^2/4\ell^2} e^{-i(x_i - x_j)(y_i + y_j)/2\ell^2} \tag{2}$$

where $\boldsymbol{r}_i = (x_i, y_i)$ and ℓ is the magnetic length. Since Ω_{ij} drops down very fast for $|\boldsymbol{r}_i - \boldsymbol{r}_j| > 2\ell$, one expects that the inter-edge transport takes place along the most probable paths of resonant transitions ("resonant percolation" trajectories). Therefore it is not necessary to keep all the amplitudes Ω_{ij} in the effective Hamiltonian, but only those which are associated with the most probable path of transitions.

Let us assume that there exists only one resonant-percolation trajectory connecting the impurities 1 and n. Then the effective Hamiltonian $\mathcal{H}$ can be re-arranged in a form of the nearest-neighbour couplings tunneling Hamiltonian for a one-dimensional system with off-diagonal disorder[2]. In this case Eqs. (1) can be solved thus obtaining for the probability of the inter-edge transition $P_{1n}(E) \equiv \mathcal{P}_{1n}(\mathcal{E})$:

$$\mathcal{P}_{1n}(\mathcal{E}) = 4\Gamma_1 \frac{|\Omega_{12}\Omega_{23} \cdots \Omega_{n-1,n}|^2}{|\det[E - \mathcal{H}]|^2} \Gamma_n \tag{3}$$

where $\mathcal{E} = E - \bar{E}_0$. Correspondingly, the localization length is defined as

$$L^{-1}(\mathcal{E}) = -\lim_{n\to\infty} \frac{1}{n} \ln |\mathcal{P}_{1n}(\mathcal{E})| \tag{4}$$

3. Extended state in the center of band.

Rewriting the determinant in (3) in terms of minors and cofactors, one can easily find the probability of inter-edge transition at the resonance energy, $\mathcal{E} = 0$.

For instance, if $n = 2k+1$ one gets

$$\mathcal{P}_{1n}(0) = \frac{4\Gamma_1\Gamma_n}{(\Gamma_1\kappa_{1n}^{-1} + \Gamma_n\kappa_{1n})^2} \tag{5}$$

where

$$\kappa_{1n} = \left|\frac{\Omega_{12}\Omega_{34}\cdots\Omega_{n-2,n-1}}{\Omega_{23}\Omega_{45}\cdots\Omega_{n-1,n}}\right| = \exp[(-\boldsymbol{r}_{12}^2 + \boldsymbol{r}_{23}^2 - \boldsymbol{r}_{34}^2 + \cdots + \boldsymbol{r}_{n-1,n}^2)/4\ell^2] \tag{6}$$

and $\boldsymbol{r}_{ij}^2 \equiv (\boldsymbol{r}_i - \boldsymbol{r}_j)^2$.

It follows from Eqs. (5),(6) that the state at $E = \bar{E}_0$ is an extended one as far as $\kappa_{1n} \sim 1$, or $|\sum_j(-1)^j\boldsymbol{r}_{j-1,j}^2| \lesssim 4\ell^2$. Hence, all possible resonant-percolation trajectories of electrons across a sample have to satisfy the above condition, which can also be considered as a definition of these trajectories. One can prove that the resonant-percolation trajectories would always appear by increasing the width of a sample, but by maintaining the impurity density as a constant, no matter how small the density is. Notice that the described mechanism of a delocalization is similar to that of Lifshitz and Kirpichenkov[3] (LK) in their analysis of tunneling transparency of disordered systems (with no magnetic field). However our requirement for the resonant-percolation trajectories is much more loose in comparison with that of LK. The reason is that LK looked for a band of extended states, and therefore they required approximately the same distance between impurities in the chain. In contrast, our condition ($\kappa_{1n} \sim 1$) is sufficient only for a delocalization of one level in the band ($\mathcal{E} = 0$), where all other states remain localized.

Let us demonstrate that it is indeed the case. The most simple way to show that randomly distributed impurities (but obeying the constraint $\kappa_{1n} \sim 1$) localizes all states with $\mathcal{E} \neq 0$, is to start with aligned impurities. Thus, we assume that the non-diagonal amplitudes in the Hamiltonian $\mathcal{H}$ are the same, i.e. $\Omega_{ij} = \Omega$. Then the conducting band would appear around $\mathcal{E} = 0$. Now let us displace the impurity 'k' from its position, so $\Omega_{k-1,k} \neq \Omega$ and $\Omega_{k,k+1} \neq \Omega$. As a result the impurity 'k' would generate the reflection of the Bloch waves inside the band. For the energy $\mathcal{E} \ll \Omega$, one finds that the situation resembles resonant tunneling through a double-barrier structure, namely the Bloch waves inside the band can pass the impurity 'k' only via the virtual state at $\mathcal{E} = 0$. Then the probability of inter-edge transition can be written as

$$\mathcal{P}_{1n}(\mathcal{E}) = \mathcal{P}_{1n}^{(k)}(\mathcal{E}) = \frac{4\gamma_{k-1}\gamma_{k+1}}{\mathcal{E}^2 + (\gamma_{k-1} + \gamma_{k+1})^2} \tag{7}$$

where $\gamma_{k\pm1} = |\Omega_{k\pm1,k}|^2/|\Omega|$

The above result clarifies the meaning of the constraint $\kappa_{1n} \sim 1$ for a delocalization of the state at $\mathcal{E} = 0$. Indeed, as it always happens in the case of resonant tunnelling, the maximal value of the transition probability is reached for the same partial widths of the resonant state, $\gamma_{k-1} = \gamma_{k+1}$ i.e. for $\Omega_{k-1,k}/\Omega_{k,k+1} = 1$. Since all other $\Omega_{ij} = \Omega$, the above condition is equivalent to $\kappa_{1n} = 1$. The displacement of an another impurity would create a second double barrier structure along the chain. By approximating the total transmission probability as a product[2] $\mathcal{P}_{1n}(\mathcal{E}) \simeq \mathcal{P}_{1n}^{(k_1)}\mathcal{P}_{1n}^{(k_2)}$, one finds that $\mathcal{P}_{1n}(0) = 1$ for $\Omega_{k_1-1,k_1}/\Omega_{k_1,k_1+1} = 1$ and $\Omega_{k_2-1,k_2}/\Omega_{k_2,k_2+1} = 1$. By continuing this procedure of increasing disorder one finds that the absence of attenuation along the chain is achieved by keeping $\Omega_{k-1,k}/\Omega_{k,k+1} = 1$, which corresponds to a particular case of the requirement $\kappa_{1n} \sim 1$.

Now let us estimate the transmission probability for $\mathcal{E} \neq 0$, by imposing $\Omega_{k-1,k} = \Omega_{k,k+1}$. Using Eq. (7) and approximating the total transmission by a product form, we get

$$\mathcal{P}_{1n}(\mathcal{E}) \simeq \prod_k \frac{1}{\mathcal{E}^2 \frac{|\Omega|^2}{4|\Omega_k|^4} + 1} \tag{8}$$

where $\Omega_{k-1,k} = \Omega_{k,k+1} \equiv \Omega_k$. Therefore, the electron states for $\mathcal{E} \neq 0$ are localized with the localization length, Eq. (4)

$$L^{-1} = C \ln\left(1 + \frac{\mathcal{E}^2}{4\langle\Omega\rangle^2}\right) \sim \frac{\mathcal{E}^2}{\langle\Omega\rangle^2}, \tag{9}$$

so that the transmission probability across a sample decreases exponentially with n, i.e. $\mathcal{P}_{1n}(\mathcal{E}) \sim \exp[-(\mathcal{E}^2/\langle\Omega\rangle^2)n]$.

It is clear that the above results remain the same for any of the Landau bands, since our analysis of the localization properties was based only on the off-diagonal disorder in the effective Hamiltonian $\mathcal{H}$, and subject to the constraint $\kappa_{1n} \sim 1$, Eq. (6). Hence, the exponential is expected to be the same for any of the Landau bands. Therefore the localization length near the Landau level N can be written as

$$L_N^{-1}(E) \sim (E - \bar{E}_N)^2 \tag{10}$$

where $\bar{E}_N = E_N + \epsilon_N$ and ϵ_N is the energy of the resonance near the Landau level E_N.

Finally we would like to mention that in the case of self-crossing resonant-percolation trajectories (i.e. when the same initial and final impurity sites can be connected by more than one trajectories), the Aharonov-Bohm interference-effects may play an essential role and the value of the exponential in Eq. (10) might be modified. This effect needs special investigation and is not considered in this work.

4. Acknowledgements

I am indebted to A. Aronov for useful discussions and important suggestions.

5. References

1. S.A. Gurvitz, Preprint TRI-PP-94-33, *cond-mat@babbage.sissa.it* 9406022.
2. G.T. Theodorou and M. Cohen, *Phys. Rev.* **B13** (1976) 4597.
3. I.M. Lifshitz and V.Ya. Kirpichenkov, *Sov. Phys. JETP* **50** (1979) 499.

UNIVERSAL PREFACTOR OF ACTIVATED CONDUCTIVITY IN THE QUANTUM HALL EFFECT

D. G. Polyakov[(a)] and B. I. Shklovskii

Theoretical Physics Institute, School of Physics and Astronomy, University of Minnesota
116 Church St. SE, Minneapolis, MN 55455

ABSTRACT

The prefactor of the activated dissipative conductivity in the quantum Hall regime is studied in the case of a long-range random potential. It is shown that due to long time it takes for an electron to drift along the perimeter of a large percolation cluster, phonons are able to maintain quasi-equilibrium inside the cluster. The saddle points separating such clusters may then be viewed as ballistic point contacts between electron reservoirs with different electrochemical potentials. In spite of important role of phonons, the activation conductivity does not depend on the constant of electron-phonon interaction. The prefactor is universal and equal to $2e^2/h$.

When the Fermi level lies in the middle between two Landau levels and the temperature T is not too low the conductivity σ_{xx} has a form $\sigma_{xx} = \sigma_0 \exp(-\Delta/T)$.The prefactor σ_0 has attracted a great deal of interest since it was claimed [1,2] that σ_0 is equal to e^2/h and does not depend on the Landau level number. This puzzling universality was reported to be valid in the fractional quantum Hall effect (FQHE) as well: $\sigma_0 = (e/q)^2/h$ at filling $\nu = p/q$. There have been several interesting attempts to calculate the prefactor universally, for any type of disorder, using the idea that at $T = 0$ the extended state exists at only one energy and phase-breaking processes are responsible for the delocalization of electrons within a narrow band of states near the Landau level center [2,3,4,5,6,7]. Nevertheless, it has not been shown yet that the range of temperatures exists where $\sigma_0 = e^2/h$.

We suggest a more specific approach and consider separately the cases of a short and a long range random potential. In our previous work [8] we evaluated explicitly two major contributions to the conductivity in the case of a white noise potential: the contribution of the narrow band of delocalized states which appear near the Landau level center at finite temperatures and the other one, related to the variable range hopping (VRH) between localized states in the tail of the density of states. It was shown that the interplay of these contributions is responsible for the inflection point in the dependence of $\ln \sigma_{xx}$ on $1/T$ (Fig. 1a). We demonstrated that approximation of σ_{xx} by $\sigma_0 \exp(-\Delta/T)$ in the vicinity of the inflection point yields $\sigma_0 \lesssim e^2/h$ being a very slow function of the rate of phase breaking processes. So we concluded that, strictly speaking, no universality can be obtained for the case of a short range potential.

In this paper, we calculate the prefactor σ_0 in the case of a long range random potential. Such a potential exists in heterojunctions with large spacer $d >> \lambda$, λ being the magnetic length. We show below that in this case, due to the suppression of tunneling, the conductivity may be described by a completely *classical* theory which yields a range of temperatures, growing as a power of the ratio d/λ, where the prefactor is constant and universal (Fig. 1b). We obtain $\sigma_0 = e^2/h$ for the contribution of one Landau level to the conductivity in this range. The Plank constant appears in this expression only through the density of states of the Landau level.

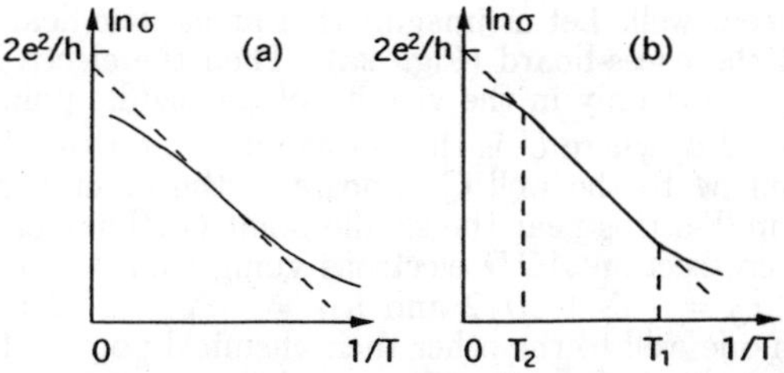

Fig. 1. Schematical plot of $\ln\sigma$ vs $1/T$ when the Fermi level is in the middle between two Landau levels: (a) as calculated for a short range potential[8], (b) as obtained in this work for a long range potential.

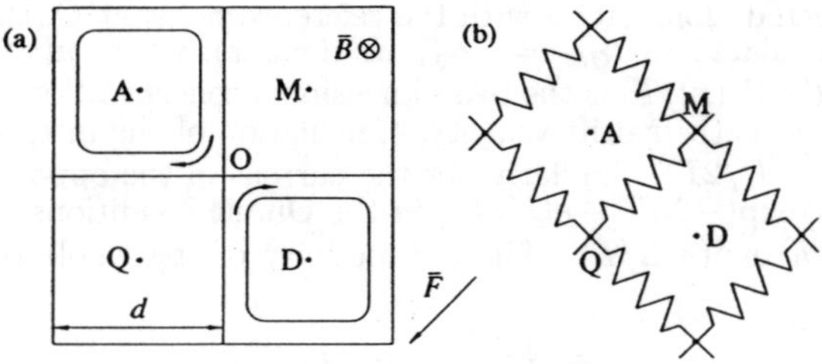

Fig. 2. (a) Equipotential lines $W(x,y)=0$ (square lattice) and $W(x,y)=T$ for a chess-board potential. Saddle point O separates two hills A, D and two wells Q, M. Arrows show directions of the drift current. (b) Equivalent curcuit.

We start with a demonstration of the universality of σ_0 for a simple model problem. We consider a periodic "chess-board" potential $W(x,y)=W\cos(\pi x/d)\cos(\pi y/d)$. It is assumed that the Fermi level μ is located in the middle between two Landau levels separated by the gap 2Δ and that $W \lesssim \Delta$. At $T \ll \Delta$ the concentration of electrons activated to the upper Landau level is exponentially small so that we can neglect the interaction between them. The inequality $d \gg \lambda$ allows us to treat the electrons semiclassically. Electron guiding centers drift along equipotential lines $W(x,y)=E$, all of which except one at $E=0$ are closed loops (Fig. 2a). The probability of tunneling between two loops falls off with increasing $|E|$ as $\exp(-|E|/T_1)$, where $T_1 \sim W(\lambda/d)^2$ [9]. Below we consider the wide temperature range $T_1 \ll T \ll W$ in which tunneling can be neglected.

In the absence of inelastic processes $\sigma_0 = 0$. To find the conductivity of electrons due to inelastic collisions, we should know their energy distribution in the presence of an external electric field. Low energy electrons are in equilibrium with each other within a given well because they circulate many times around the well and experience many inellastic collisions before leaving the well due to activation to trajectories with $E>0$. Therefore, their distribution is characterized by an electrochemical potential which is constant inside the well. We show below that electrons with energies $0<E\lesssim T$ play a very important role in transport. They occupy "T-strips" surrounding each potential hill (Fig. 2a). The distribution of these electrons depends on the parameter $v\tau/d$, where v is the typical drift velocity and τ is the time necessary to change the energy by T. Our crucial assumption is that $v\tau \ll d$. This means that electrons of T-strips "tune" their electrochemical potential so that it equals the electrochemical

potential in the adjacent well. Let us imagine that an electric field F is applied along a diagonal MOQ of the chess-board (Fig. 2a). Then the electrochemical potential drops along the line QOM only in the vicinity of the saddle point O. The value of this drop is $eU = \sqrt{2}eFd$, where U is the voltage drop between M and Q. The net current from the well M to the well Q appears as the difference between the two opposite drift flows in T-strips near the saddle point O. They are shown by arrows in Fig. 2a. In the crossection AOD electrons going from Q to M and back have chemical potentials $\mu_Q = -\Delta + eU/2$ and $\mu_M = -\Delta - eU/2$ respectively. When electrons arrive from one well to the other their chemical potential quickly relaxes to the new value.

Thus we have arrived at the concept of a network of reservoirs (potential wells) connected via ballistic contacts (saddle points) shown in Fig. 2b. Let us show that the conductance G of each of these contacts has the form $G = (e^2/h)\exp(-\Delta/T)$. If the x-axis is directed along AOD with the reference point at O, the current from Q to M can be calculated as $I_{QM} = -e\int_0^\infty dx n(x)v(x)$, where $n(x) = (1/2\pi\lambda^2) \times \exp(-\Delta/T + eU/2T - W(x)/T)$ is the two-dimensional concentration of electrons and $v(x) = (c/eB)\partial W/\partial x$ is their drift velocity. Calculation of the integral gives $I_{QM} = (eT/h)\exp(-\Delta/T + eU/2T)$. Similarly, for the current in the opposite direction we get $I_{MQ} = -(eT/h)\exp(-\Delta/T - eU/2T)$. In the Ohmic conditions ($eU/T \ll 1$) we arrive at $G = (e^2/h)\exp(-\Delta/T)$. The conductivity of the whole resistor network shown in Fig. 2b is

$$\sigma_{xx} = G = \frac{e^2}{h}\exp(-\Delta/T). \tag{1}$$

Note that this calculation gives contribution to the conductivity only of electrons on the upper Landau level. Holes on the lower Landau level form their own resistor network.Thus for the total conductivity of both parallel networks $\sigma_0 = 2e^2/h$.

Let us now turn to the random long range potential. The simplest way to imagine it, departing from the chess-board potential, is to assume that heights of the saddle points W_i are randomly distributed in an interval of energies $(-W_1, W_1)$, where $W_1 \lesssim W$. Assuming then that $v\tau \ll d$ and repeating the same arguments as for the chess-board case, we get the network of random conductances $G_i = (e^2/h)\exp(-\Delta/T - W_i/T)$. To calculate the conductivity of such a network, we use the Dykhne theorem [10]. According to this theorem a two-dimensional network of random conductances with a symmetrical distribution of $\ln G_i$ around the average value $< \ln G_i >$ has the conductivity $\sigma_{xx} = \exp(< \ln G_i >)$. In our case $< \ln G_i >= -\Delta/T$ because $< W_i >= 0$, and thus we again arrive at Eq. (1). It is important that the limits of applicability of Eq. (1) are much broader for the random potential than for the periodic one with the same values of W and d. According to percolation theory, randomness of the saddle point heights generates two new large scales at $|E| \ll W$: the "diameter" $\xi = d\,(W/|E|)^{\nu_p}$ and the perimeter $p = d\,(W/|E|)^\gamma$ of critical equipotential loops at a given energy E ($\nu_p = 4/3$ and $\gamma = \nu_p + 1 = 7/3$). In the above derivation of Eq. (1) we considered all of the saddle points of the random chess-board potential. Actually, the conductivity is determined only by those of them for which $|W_i| \sim T$. At $T \ll W$ these saddle points separate critical loops corresponding to $|E| \sim T$ which have very long perimeters $p_T \sim d\,(W/T)^\gamma$. Such "T-loops" together with saddle points with heights $|W_i| \sim T$ form a network which is topologically similar to the chess-board. We can consider it as a network of ballistic conductances G_i and obtain Eq. (1) if the electrons of a T-loop are in equilibrium with the cluster of potential wells they circulate alongside, i.e. if $v\tau \ll p_T$. This inequality is a much weaker than $v\tau \ll d$.

We discuss now the width of the interval of temperatures where $v\tau \ll p_T$ and $\sigma_0 = 2e^2/h$. This requires an explicit discussion of the nature of the inelastic processes. As we mentioned above, activated electrons are far from each other and thus interaction between them is negligible. Therefore, we assume below that inelastic processes are only due to electron-phonon interactions. One can easily verify that under realistic conditions one-phonon processes are permitted by the conservation laws. They give $\tau \sim (\hbar T/\alpha)(d/W\lambda)^2$ for $T \gtrsim W\lambda/d$ and $\tau \sim \hbar d/\alpha\lambda W$ for $T \lesssim W\lambda/d$. Here $\alpha = \hbar C^2/2\rho s^3\lambda^2$ is the electron-phonon coupling constant, C, s and ρ are the deformation potential constant, the sound velocity and the crystal density respectively. For $GaAs$ $\alpha \simeq 0.1(100\AA/\lambda)^2$. Using these results we find that for a large enough spacer $d \gg d_c$, the inequality $v\tau \ll p_T$ is valid for T-loops and, correspondingly, $\sigma_0 = 2e^2/h$ in a wide range of temperatures $T_1 \ll T \ll T_2$. Here $d_c = \lambda/\alpha^{1/(\gamma+1)}$ and $T_2 = W\alpha^{1/(\gamma+1)}$. For $\lambda \simeq$10 nm one gets $d_c \simeq$20 nm. Thus for samples with $d >> 20$ nm we arrive at the universal prefactor $\sigma_0 = 2e^2/h$. Let us briefly discuss what happens away from the temperature range $T_1 \ll T \ll T_2$. If $T \lesssim T_1$ VRH becomes important and leads to deviation upwards from the straight line on the plot of $\ln\sigma$ vs $1/T$ (Fig. 1b). On the other side, when $T \gtrsim T_2$ the perimeter p_T becomes smaller than $v\tau$ and only the low energy part of the T-strip is still in equilibrium with the adjacent potential well. Finding the width of this part selfconsistently one can show that in this case $\sigma_0 \sim (e^2/h)(T_2/T)^\beta$, where $\beta = (\gamma+1)/(\gamma+2) = 10/13$. This expression is responsible for deviation downwards from the straight line on the plot of $\ln\sigma$ vs $1/T$ at $T \gtrsim T_2$ (Fig. 1b).

In conclusion, we have shown that in a heterostructure with a large spacer there is a range of temperatures in which the prefactor σ_0 is universal. It is equal to $2e^2/h$ if the Fermi level is in the middle between two Landau levels. Similarly, it should be expected that in the FQHE $\sigma_0 = 2(e/q)^2/h$ at $\nu = p/q$.

We are grateful to I. L. Aleiner, M. M. Fogler and V. I. Perel for valuable discussions. This work was supported by the NSF Grant No. DMR-9321417.

1. References

[a] Present address: Institut für Theoretische Physik, Universität zu Köln, 50937 Köln, Germany.

1. R. G. Clark, J. R. Mallett, S. R. Haynes, J. J. Harris, and C. T. Foxon, Phys. Rev. Lett. **60**, 1747 (1988).
2. R. G. Clark, Phys. Scr. T**39**, 45 (1991).
3. Y. Katayama, D. C. Tsui, and M. Shayegan, Phys. Rev. B**49**, 7400 (1994)
4. D.-H. Lee, S. Kivelson, and S.-C. Zhang, Phys. Rev. Lett. **68**, 2386 (1992).
5. S. Das Sarma and D. Liu, Phys. Rev. B**48**, 9166 (1993).
6. D.-H. Lee, S. Kivelson, and S.-C. Zhang, Phys. Rev. Lett. **71**, 2679 (1993).
7. R. N. Bhatt, N. Read, and B. Huckestein, Phys. Rev. Lett. **71**, 2678 (1993).
8. D. G. Polyakov and B. I. Shklovskii, unpublished
9. H. A. Fertig and B. I. Halperin, Phys. Rev. B**36**, 7969 (1987).
10. A. M. Dykhne, Zh. Eksp. Teor. Fiz. **59**, 110 (1971) [Sov. Phys. JETP **32**, 63 (1971)].

PHONON ABSORPTION AT THE MAGNETO-ROTON MINIMUM IN THE FRACTIONAL QUANTUM HALL EFFECT

J.E. DIGBY, C.J. MELLOR, R.H. EYLES, A.J. KENT, K.A. BENEDICT, L.J. CHALLIS, M. HENINI, C.T. FOXON, J.J. HARRIS*

Physics Department, Nottingham University, Nottingham, NG7 2RD, UK.
* Semiconductor Materials I.R.C., University College London, London, UK.

ABSTRACT

The energy of the magneto-roton minimum has been measured by non-equilibrium phonon absorption. Experiments have been conducted on two samples which have similar electron densities but greatly differing mobilities. The energy gaps as measured by activation studies of the longitudinal resistance differ by a factor of two. Phonon absorption measurements give almost identical values for the energy gap demonstrating that this method of measuring the gap is rather insensitive to disorder. The values obtained at ν=2/3 are in good agreement with theory.

1. Introduction

The fractional quantum Hall effect (FQHE) that occurs in high mobility two-dimensional electron systems (2DES) in the presence of a high magnetic field B, results from the formation of an incompressible quantum fluid. Collective excitations of this fluid can be excited across a wavevector dependent energy gap $\Delta(q)$. A theory of these collective excitations was developed by Girvin, MacDonald and Platzman[1]. They predict that the excitations have finite energy at zero wavevector. This then falls to a deep minimum known as the magneto-roton minimum at the wavevector corresponding to the the peak in the structure factor of the quantum fluid. This minimum is analogous to the roton minimum in superfluid helium.

Ballistic acoustic phonons have proved to be a unique probe of low dimensional systems[2]. The typical energies and wavevectors are well-matched to those of the 2DES. For example the dispersion curve of a longitudinal acoustic phonon incident at 60° to the normal on a 2DES in the FQHE regime crosses the excitation curve close to the magneto-roton minimum.

2. Experimental details

We have studied two high mobility GaAs-AlGaAs heterostructures with similar electron densities but with greatly differing mobilities. NU409 has a mobility of 1.0×10^6 $cm^2V^{-1}s^{-1}$ whereas G635 has a mobility of 8.0×10^6 $cm^2V^{-1}s^{-1}$. After illumination with red light the electron densities were both within a few per cent of 1.5×10^{11} cm^{-2}.

The phonon absorption is measured from the change in longitudinal resistance produced by a burst of non-equilibrium ballistic phonons. The 3×1 mm Hall device is supplied with a constant bias current and the phonons are generated by applying a 100ns electrical pulse to a 60×600 μm constantan heater evaporated on the back face of the sample. The spectral distribution of the phonons is assumed to correspond to that of a black body at the temperature of the heater, T_h. This temperature is calculated from the total power dissipated in it using acoustic mismatch theory[3] and confirmed to within ten per cent by measuring the energy gaps at ν=12 and 14 on G635. To obtain the the magneto-roton gap we measure the increase in longitudinal voltage in the tail of the ballistic phonon response. We assume that this voltage is proportional to the absorption and measure it as a function of heater temperature. Care was taken to ensure that the results are independent of bias current, length of heater pulse, repetition rate of the heater pulses and substrate temperature. Preliminary time-resolved measurements show that the dissipation caused by the phonon pulse rises to its maximum value as soon as the phonon pulse reaches the 2DES. This is consistent with direct absorption of the phonons by the 2DES.

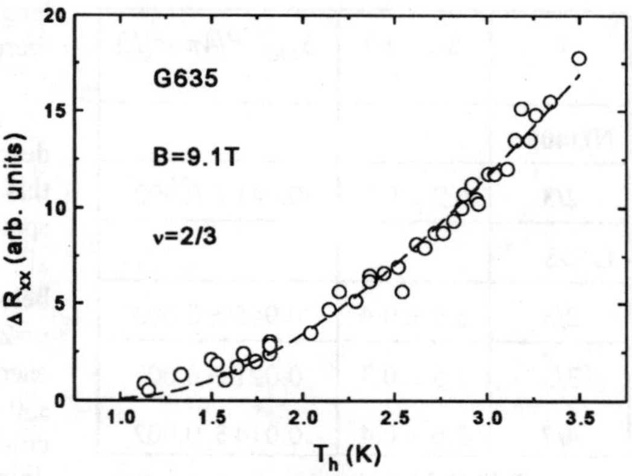

Figure 1. Phonon absorption at ν=2/3. Dashed line is theoretical curve for c = 0.045.

3. Results and Analysis

The energy gaps at ν=2/3 for NU409 and G635 determined by activated magneto-transport are 2.7 ± 0.2K and 5.5 ± 0.5K respectively. The factor of two difference is attributable to differing amounts of disorder in the samples. The phonon absorption results for G635 are shown in Fig. 1. The energy gaps for NU409 and G635, obtained by phonon absorption are 6.2 ± 0.2 K and 6.9 ± 0.4 K respectively. This striking result strongly suggests that this method probes an energy gap that is largely independent of the disorder in the sample and that disorder may have a slight effect on the activated magneto-transport values even in G635. Values for these and other filling factors are summarised in Table 1.

The predominant phonon absorption is expected to occur at the magneto-roton minimum where there is a peak in the density of states of the collective excitations. Therefore to a good approximation the gap, Δ_{MR}, can be obtained by assuming the absorption is proportional to $1/[\exp(\Delta_{MR}/k_B T)-1]$. A fit of this curve to the data at ν=2/3 for G635 is shown on figure 1. The curve fits the data extremely well and gives a value of Δ_{MR} = 6.4 ± 0.2 K or c=0.045±0.03 where $\Delta_{MR}=c.(e^2/4\pi\epsilon\epsilon_0 l_c)$, and l_c is the magnetic

ν	Δ_{MR} (K)	$\Delta_{MR}/(e^2/4\pi\epsilon\epsilon_0 l_c)$
NU409		
2/3	6.2 ± 0.2	0.041 ± 0.002
G635		
2/3	6.9 ± 0.4	0.045 ± 0.003
3/5	3.5 ± 0.3	0.021 ± 0.002
4/7	2.6 ± 0.4	0.014 ± 0.002

Table 1. Magneto-roton energy gaps measured by phonon absorption.

length. The results summarised in Table 1 were obtained using this analysis.

A more complete analysis has been described previously[4]. We have extended this model to include the excitation spectrum for $\nu=1/3$ calculated by Girvin et al[1]. This has improved the agreement between the two models. For instance at $\nu=2/3$ on G635, this model gives an energy gap of 6.2 ± 0.3K, compared to 6.9 ± 0.4K from the simplified model. It also confirms that the phonon absorption is dominated by absorption close to the magneto-roton minimum.

4. Discussion

The energy gaps determined from the phonon experiment are nearly equal for the two samples but those determined from activated transport measurements differ by a factor of more than two, in line with the difference in their zero field mobilities. Clearly disorder affects the two measurements in different ways. We interpret this difference as being due to large scale ($>>l_c$) inhomogeneity in the lower mobility sample leading to a variation in the local conduction band edge. The transport measurements are taken at thermal equilibrium so that the measured gap will be dominated by the largest negative fluctuation in the conduction band edge energy. The temperature of the ballistic phonon pulse is far in excess of that of the 2DES and our time-resolved experiments show that the non-equilibrium phonons are absorbed directly by the 2DES. Hence we conclude that the phonon experiment is a genuine *local* spectroscopic probe, measuring the excitation gap at a specific point in the 2DES which should only be affected by short length scale disorder.

The energy gap obtained at $\nu=2/3$ on G635 can be compared to the theoretically predicted value for $\nu=1/3$. Two effects that must be accounted for are the finite thickness of the 2DES and Landau level mixing. In G635, at $\nu=2/3$, $bl_c=1.75$, where b is the Fang-Howard parameter. For this value of bl_c theoretical studies[5] predict that the energy gap is reduced by a factor of 0.7 from the ideal case of a 2DES with zero thickness. An estimate of the effect of Landau level mixing can be made following the work of Yoshioka[6]. When these two effects are combined the theoretical gap is found to be c=0.045 in excellent agreement with our experimentally obtained value of c=0.045±0.002. The small difference between the NU409 and G635 may be due to the differences in layer structure and doping levels ; this may alter the vertical extent of the wavefunction. Another possibility is that it is due to the higher disorder in NU409. Pinzcuk et al[7] have measured the excitation curve in the limit of low wavevector by inelastic light scattering in a quantum well. Allowing for the differences between the samples our results are in good agreement.

The mechanism by which phonon absorption causes a change in the longitudinal voltage is as yet unknown. One possibility is that there is a mutual friction between the magneto-rotons and the Laughlin liquid (possibly due to the dipole moment of the former) which causes dissipation[8]. Another possibility is that once created by a high energy phonon, a magneto-roton can absorb many low energy phonons and so dissociate into an unbound quasiparticle and quasihole[9]. Research is in progress to determine which of these is the dominant effect.

The variation of the energy gaps at $\nu=2/3$, 3/5 and 4/7 with magnetic field is consistent with the composite fermion theory[10] of the FQHE. The value of the composite fermion effective mass is similar to that obtained by magneto-transport[11], at $0.6 \pm 0.1m_e$, where m_e is the mass of an electron.

5. Conclusions

The energy of the magneto-roton minimum for several filling factors of a high mobility heterojunction has been measured by phonon absorption. The value at $\nu=2/3$ is in excellent agreement with theory. The values for $\nu=2/3$, 3/5 and 4/7 are consistent with the composite fermion theory of the FQHE although detailed comparison with theory has not yet been made.

6. Acknowledgements

We are grateful for support from the E.P.S.R.C. and E.U. We would like to thank W. Dietsche, and V. Fal'ko for helpful discussions.

7. References

1. S. M. Girvin, A. H. MacDonald and P. M. Platzman, *Phys. Rev. B***33** (1986) 2481.
2. For a review see L.J. Challis and A.J. Kent in *Aspects of Electronic Nanostructures*, ed. G. Bauer, F. Kuchar and H. Heinrich (Springer, Heidelberg, 1992) p. 31.
3. W. Kappus and O. Weiss, *J. Appl. Phys.* **44** (1973) 1947.
4. R. H. Eyles, et al, *Surf. Sci.* **305** (1994) 87.
5. S.C. Zhang and S. Das Sarma, *Phys. Rev. B***33** (1986) 2903.
6. D. Yoshioka, *J. Phys. Soc. Jpn.* **55** (1986) 885.
7. A. Pinczuk, B. S. Dennis, L. N. Pfeiffer, and K. West, *Phys. Rev. Lett.* **70**, (1993) 3983.
8. P.M. Platzman, S.M. Girvin and A. H. MacDonald, *Phys. Rev. B***32** (1985) 8458.
9. M. Brownlie and K.A. Benedict, in preparation.
10. B.I. Halperin, P.A. Lee and N. Read, *Phys.Rev. B***47** (1993) 7312.

Magnetisation measurements on two-dimensional electron systems in the magnetic quantum limit

C.L.Jones, A.Usher
Department of Physics, Stocker Road, University of Exeter, Exeter, EX4 4QL, UK

M.Elliott, W.G.Herrenden-Harker, A.Potts, R.Shepherd
Department of Physics and Astronomy, UWCC , PO Box 913, Cardiff, CF1 9HQ, UK

G.A.C Jones, D.A. Ritchie, E.H.Linfield, M.Grimshaw
Department of Physics, University of Cambridge, Cambridge, CB3 OHE, UK

T.S.Cheng, C.T.Foxon
Department of Physics, University of Nottingham, Nottingham, NG7 2RD, UK

We have performed magnetisation measurements at magnetic fields up to 17T and temperatures as low as 30mK on a variety of two-dimensional electron systems with different electron concentrations. De Haas-van Alphen oscillations in the highest concentration sample have been fitted to various model densities of states, a Gaussian with a $B^{1/2}$ dependence of Landau level width and a Lorentzian with no such field dependence both fitting the data well. At the lowest temperatures, large peaks appear at integer Landau level filling factors, which are associated with long-lived eddy currents induced in the zero resistance state. These peaks are the dominant features in the lower concentration samples. A study of these peaks reveals a strongly non-linear temperature dependence and decay times greater than ten hours at T = 30mK.

A measurement of the magnetisation of a two-dimensional electron system (2DES) has the advantage that the system is always in its equilibrium state and is not disturbed in any way during measurement. The equilibrium density of states (DOS) may then be determined directly, without the complication of electron scattering and the distinction between localised and extended states. A measurement of magnetisation in the extreme magnetic quantum limit could also potentially provide valuable information about the exotic phase changes that are expected in this limit, most notably the transition to the Laughlin liquid ground state[1] and its quasiparticle excitations.

The most sensitive method of measuring the weak magnetisation of such systems is torque magnetometry .[2,3] In this method the sample to be studied is fixed

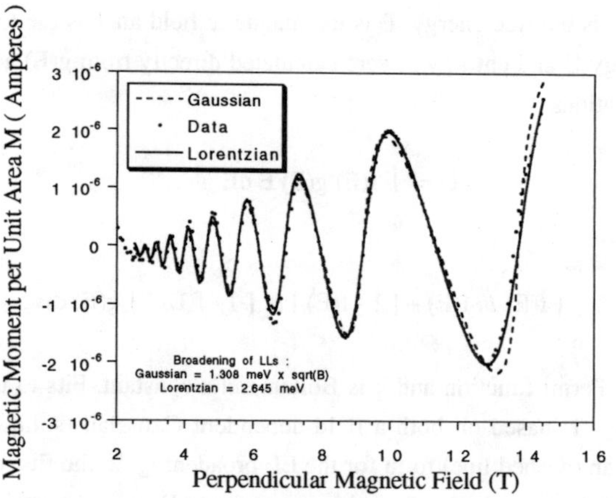

Fig. 1. Magnetisation data from a 10 layer MQW at 2.2K. Theoretical fits to the data are also shown, as described in the text.

to a plate with a square electrode evaporated on its underside. The plate is held by a fine suspension wire with its normal at a slight angle to the magnetic field direction. The magnetic moment of the 2DES causes a torque which turns the sample with respect to a pair of fixed rectangular electrodes . This enables the rotation of the plate to be detected capacitatively with the use of a high stability A.C. capacitance bridge.

Two different magnetometers were used to obtain the results. The first was based on Templeton's design[3] and was mounted vertically in a ^{4}He cryostat with a 6T magnet. The second was a slightly modified version in which both the split and single electrodes were on rotating plates, the sample on one plate and a piece of substrate of the same size on the other, with the aim of minimising anisotropic background moments coming from the substrate and the plate itself. This magnetometer was mounted with the plates almost horizontal in the mixing chamber of a ^{3}He-^{4}He dilution refrigerator, in a field of up to 17T .

Fig. 1 shows de Haas - van Alphen (dHvA) oscillations in a 10-layer multiple quantum well sample with a sheet carrier density n_s (per layer) of 1.2 x 10^{16} m^{-2} and Hall mobility of 4.0 m^2 V^{-1} s^{-1} , after removing a field-dependent background signal and converting to magnetic moment per unit area. The magnetometer was calibrated as described previously.[3]

Theoretical fits are made based on a trial electron DOS, g(E), consisting of a sum of broadened Landau Levels (LLs). The magnetisation is then calculated using

$$m = - (\partial F/\partial B)_{T,\, n_s}$$

where F = U - TS is the free energy, B is the magnetic field an T is the temperature. The internal energy U and entropy S were calculated directly from g(E) according to the following equations

$$U = \int_0^\infty f(E)\, g(E)\, E\, dE$$

$$S = -k \int_0^\infty \{ f(E)\, ln\, f(E) + [\, 1 - f(E)\,]\, ln\, [\, 1 - f(E)\,] \}\, g(E)\, dE\,.$$

where f(E) is the Fermi function and k is Boltzmann's constant. Fits to the data are also shown in Fig. 1, based on both a field-dependent Gaussian (solid line) and a constant Lorentzian (dashed line) form for the LL broadening. In the fits, the electron density and broadening are both adjustable parameters. Due to uncertainties in the magnetometer calibration, the amplitude of the fits is also multiplied by an adjustable fitting parameter. It is evident that the data are well described by either model DOS. However, a constant Gaussian broadening does not fit the data, in contrast to Templeton's findings.

At temperatures below a few Kelvin, sharp spikes are observed superimposed upon the dHvA oscillations, which are many orders of magnitude larger than the oscillations. The spikes reverse direction when the field sweep direction is reversed and can therefore be attributed to long-lived eddy currents induced in the 2DES in the zero resistance state at integer Landau level filling factors.[4]

The eddy currents are more clearly seen in a second sample with a lower concentration (3.5×10^{15} m^{-2}), which shows no visible dHvA oscillations (Fig. 2). The temperature dependence of the peak at 7.3T (LL filling factor $\nu = 4$) is shown in Fig. 3. The axes have been chosen as in Ref. 4. For temperatures between 1 and 2 Kelvin the size of the peak is dependent on sweep-rate and the magnetisation increases rapidly as the temperature decreases (see inset). This is suggestive of a variable-range hopping mechanism for electron conduction, but a more detailed study over a wider temperature range would be required to provide a quantitative test. The eddy current decay time is of the order of a few seconds at these temperatures and no eddy currents are observed above 2 Kelvin. At bath temperatures below 1K the size of the peak is largely independent of the bath temperature, and does not vary with sweep rate. We believe that an increasing sweep rate increases the eddy currents in the sample but at the same time increases the electron temperature, and hence the resistance. These two effects balance each other leading to eddy currents whose magnitude is independent of sweep rate. The decay time of the peak at 30mK is observed to be at least 10 hours , and about 3 hours at 0.4 K. Various estimates of the 2D resistivity may be made, depending on the model used for conduction and dissipation of the eddy current, and these estimates give results in the range 10^{-11} to $10^{-13}\,\Omega/\square$ at 30mK.

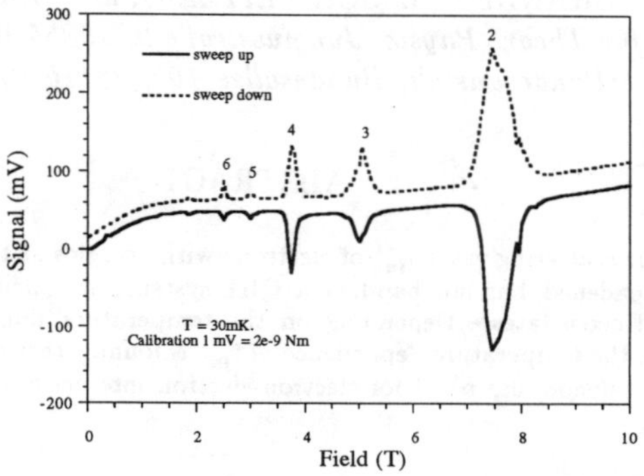

Fig. 2. Raw magnetometer output showing eddy currents as described in the text.

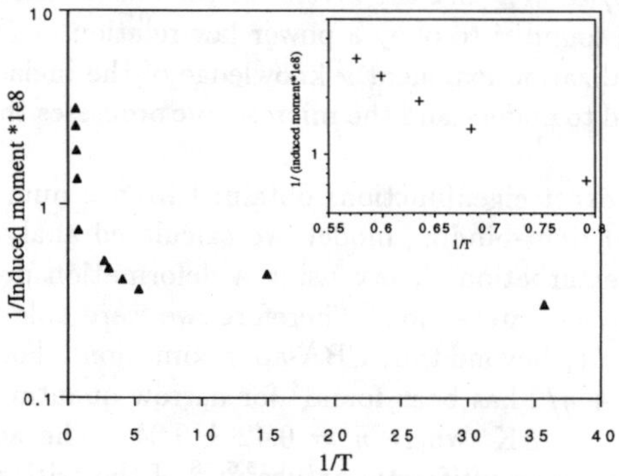

Fig. 3. Temperature dependence of ν = 4 eddy current peak. Inset expands the higher temperature data.

References

1. R.B.Laughlin, Phys. Rev. Lett , **50**, 18, 1395 (1983).

2. I. M. Templeton, J. Appl. Phys., **64**, 3570 (1988).

3. M. Elliott, M.G.M. Harris, W.G. Herrenden-Harker, R. Shepherd, G.A.C.Jones, D.A. Ritchie, E.H. Linfield and M. Grimshaw, Proceedings of the 4th International Conference on the Formation of Semiconductor Interfaces, Juelich 1993, World Scientific (1994).

4. J. P. Eisenstein, H. L. Stormer, V. Narayanamurti and A. C. Gossard, Superlattices Microstruct., **1** ,11 (1985).

INELASTIC SCATTERING AND TEMPERATURE–DEPENDENT BROADENING OF ρ_{xx}–PEAKS IN THE IQHE

T. BRANDES[1], L. SCHWEITZER[2], and B. KRAMER[1]

[1]*I. Inst. für Theor. Physik, Jungiusstraße 9, 20355 Hamburg, Germany*

[2]*Phys.-Techn. Bundesanstalt, Bundesallee 100, 38116 Braunschweig, Germany*

ABSTRACT

The inelastic scattering rates τ_{in}^{-1} of electrons with energies at the center of the lowest disorder broadened Landau band of a QHE-system are calculated using the exact multifractal eigenstates. Depending on the temperature range and on the system parameters, the temperature dependence of τ_{in}^{-1} is found to scale $\sim T^p$, with $1 \leq p \leq 4$ for electron-phonon and $p = 1$ for electron-electron interactions.

1. Introduction

The temperature dependence of the maximal slope of the derivative of the Hall-resistivity $(d\rho_{xy}/dB)_{max}$ and the inverse of the half-width of the $\rho_{xx}(B)$-peaks in the IQHE has been found[1,2] to obey a power law relation $\sim T^{-\kappa}$, with $\kappa = p/2\nu$. Thus, besides the localization exponent ν knowledge of the inelastic scattering length $l_{in} \sim T^{-p/2}$ is required to understand the microscopic processes that govern the temperature scaling.

Taking the exact eigenfunctions obtained from a numerical diagonalization of a two-dimensional tight-binding model, we calculated analytically[3] the exponent p in second order perturbation theory using a deformation potential approximation for the electron-phonon interaction. Therefore, we were able to determine the inelastic scattering length l_{in} beyond the SCBA-approximation[4]. For scattering with acoustical phonons, $p = 2 + \eta/2$ has been found[3] for narrow quantum wells in the temperature range $1\,\text{K} \leq T \leq 10\,\text{K}$, where $\eta = 0.42 \pm 0.04$. The additional term $\eta/2$ in the exponent reflects the multifractal nature[5,6,7,8] of the critical IQHE eigenstates in the center of the disorder broadened Landau band. For temperatures below 2 K, piezo-electric processes are believed to dominate the electron-phonon interaction. The corresponding temperature dependence of the inelastic scattering rates is $\sim T^2$ in the narrow and $\sim T$ in the wide well limit, respectively.

The inelastic scattering rate due to electron-electron interaction shows a linear temperature dependence[9,3], $p = 1$. This, however, seems to be at variance with considerations of hopping transport in the presence of a Coulomb gap[10] and with arguments for a dynamical scaling exponent[11], $z = 1/(\kappa\nu) = 2/p = 1$, from which $p = 2$ is concluded. Either electron-phonon processes unexpectedly dominate the inelastic scattering even at very low temperatures ($T < 0.5\,\text{K}$) or the electron-electron inter-

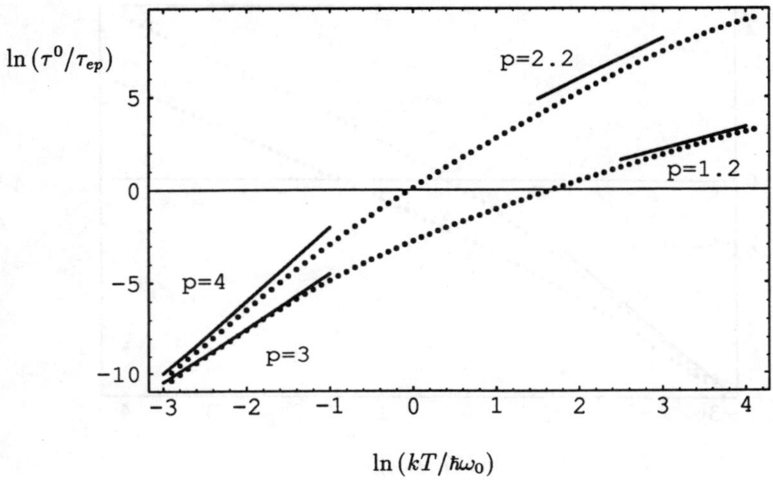

Figure 1: The *e-p* scattering rate for deformation potential coupling. The upper curve corresponds to the narrow ($\omega_z = 100\,\omega_0$) and the lower to the wide well case ($\omega_z = 0.1\,\omega_0$). Indicated is the asymptotic behaviour obtained analytically and $l_z = c_S/\omega_z$ is the effective well thickness.

action has to be treated non-perturbatively, taking into account possible Coulomb gap[12,10,13] effects.

2. Electron-phonon interactions

We assume the *e-p* interaction to be sufficiently weak to be treated by perturbation theory. In GaAs/AlGaAs this approximation is certainly justified for acoustical phonons. The electronic self energy is calculated to second order in the *e-p* coupling matrix element, V_Q. The inelastic lifetime of an electron of energy E is derived from the imaginary part of the self energy

$$\tau_{ep}^{-1}(E) = \frac{2\pi}{\hbar\Omega} \sum_{\vec{Q},E'} |V_Q|^2 \, |M_{E,E'}(\vec{q})|^2 \, |F(Q_z)|^2 \qquad (1)$$
$$\times \quad \{(n(\omega_Q) + f(E'))\, \delta(E - E' + \hbar\omega_Q) - (\omega_Q \to -\omega_Q)\}$$

Here, $n(\omega_Q)$ denotes the Bose function at phonon frequency ω_Q and $f(E)$ the Fermi function. The terms in the curly brackets correspond to phonon emission ($+\hbar\omega_Q$) and absorption ($-\hbar\omega_Q$), respectively. The E'-sum runs over all intermediate eigenstates of energy E' which also appear in the momentum matrix element $M_{E,E'}(\vec{q})$ which has been calculated numerically[3]. We assume that the wavefunction $|\chi\rangle$ in z-direction

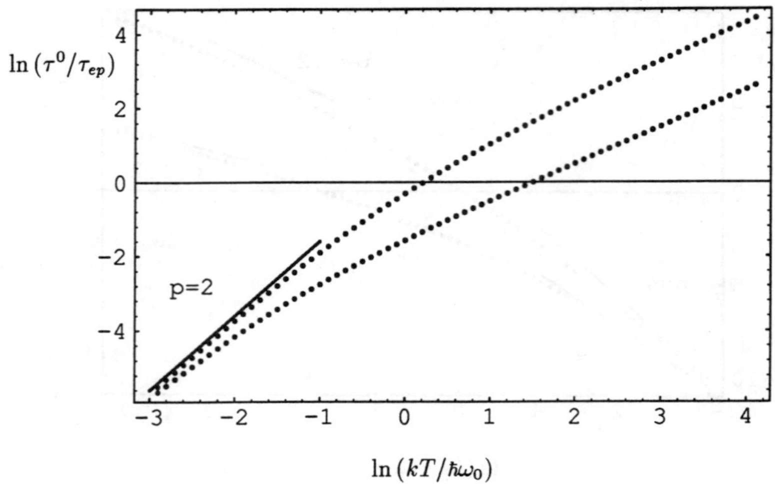

Figure 2: The *e-p* scattering rate for piezo-electric coupling. The curves correspond to the thin ($\omega_z = 100\,\omega_0$, upper) and the thick well case ($\omega_z = 0.1\,\omega_0$, lower). Indicated is the asymptotic behavior obtained analytically for the narrow well limit at low T.

decouples from the motion in the x-y-plane and so the momentum matrix element separates yielding the formfactor $F(Q_z) = |\langle\chi|e^{iQ_z}|\chi\rangle|^2$, where $\vec{Q} = (\vec{q}, Q_z)$ is the phonon wave vector.

For higher temperatures the deformation potential approximation is sufficient, $V_Q = \Xi(\hbar Q/2\rho_M c_S)^{1/2}$ (Ξ, ρ_M, and c_S are the deformation potential, the mass density and the speed of sound, respectively), whereas piezo-electric processes dominate the electron-phonon scattering at $T \lesssim 2\,\mathrm{K}$.

In Fig. 1 we show the dependence of the *e-p* scattering rate for $E = E_F$ as a function of $kT/\hbar\omega_0$ according to Eq. (1), where $\hbar\omega_0 := \hbar c_S^2/D$ with diffusion constant D and $1/\tau^0 := \Xi^2\omega_0^3/(\pi^2\hbar\rho_M c_S^5)$ is a nominal scattering rate.

The low-temperature behavior is $\tau_{ep}^{-1} \sim T^p$ with $p = 3$ for the wide well case ($\omega_z = 0.1\,\omega_0$) and $p = 4$ for the narrow well case ($\omega_z = 100\,\omega_0$). In the temperature range $1\,\mathrm{K} \leq T \leq 10\,\mathrm{K}$ (well below the Debye energy, which is taken as $\hbar\omega_D = 100\,\hbar\omega_0$) the exponents are $p = 1.2$ and $p = 2.2$, respectively. For T approaching ω_D one also notices the slow crossover to the regime linear in T in the plot for the thin well.

At very low temperatures the piezo-electric phonon coupling with a different Q dependence and piezo-electric constant P, $|V(Q)|^2 = \hbar P/(2\rho_M c_S Q)$, should dominate over the deformation potential coupling. Therefore, in the low-temperature regime the T-dependence of the scattering rate is reduced from T^4 to T^2 in the narrow well and from T^3 to T in the wide well limit. For higher temperatures we refer to τ_{ep}^{-1} shown in Fig. 2, where again we introduced a nominal scattering rate $1/\tau^0 := P/(\pi^2\hbar\rho_M c_s^3)$. The figure shows the crossover from the T^2–behavior to a nearly linear behavior of $1/\tau_{ep}$ for $kT \gg \hbar\omega_0$.

3. Electron-electron interaction

The inelastic scattering rate, τ_{ee}^{-1}, due to electron-electron interaction was calculated as the response of the system to a dynamically screened Coulomb potential. In the exact eigenstate formalism[3] this rate is due to Nyquist noise of electrons near the mobility edge. Astonishingly, the temperature dependence is linear for low T, both in 2 and 3 dimensions in agreement with previous results[9,14], with only the prefactor depending on the quantity η describing the correlations of the multifractal eigenstates. It remains to be seen, how an attempt to treat the electron-electron interaction non-perturbatively will change this result.

4. Acknowledgements

L.S. thanks the Institut for Scientific Interchange (ISI) foundation at Turin for hospitality during the ESPRIT workshop and euroconference *The quantum Hall effect* and gratefully acknowledges useful discussions with J. Eisenstein, B. Gallagher, R. Morf, and B. Shklovskii.

5. References

1. H. P. Wei, D. C. Tsui, M. Paalanen, and A. M. M. Pruisken, Phys. Rev. Lett. **61**, 1294 (1988).
2. S. Koch, R. J. Haug, K. v. Klitzing, and K. Ploog, Phys. Rev. Lett. **67**, 883 (1991).
3. T. Brandes, L. Schweitzer, and B. Kramer, Phys. Rev. Lett. **72**, 3582 (1994).
4. B. Tanatar, M. Singh, and A. H. MacDonald, Phys. Rev. B **43**, 4308 (1991).
5. J. T. Chalker and G. J. Daniell, Phys. Rev. Lett. **61**, 593 (1988).
6. W. Pook and M. Janßen, Z. Phys. B **82**, 295 (1991).
7. B. Huckestein, B. Kramer, and L. Schweitzer, Surf. Science **263**, 125 (1992).
8. B. Huckestein and L. Schweitzer, Phys. Rev. Lett. **72**, 713 (1994).
9. S. C. Siak, J. Phys.: Condensed Matter **3**, 4213 (1991).
10. D. G. Polyakov and B. I. Shklovskii, Phys. Rev. Lett. **70**, 3796 (1993).
11. D.-H. Lee, Z. Wang, and S. Kivelson, Phys. Rev. Lett. **70**, 4130 (1993).
12. A. L. Efros and B. I. Shklovskii, J. Phys. C **8**, L49 (1975).
13. S. R. E. Yang and A. H. MacDonald, Phys. Rev. Lett. **70**, 4110 (1993).
14. D. Belitz and K. I. Wysokinski, Phys. Rev. B **36**, 9333 (1987).

SUBMILLIMETER PHOTOCONDUCTIVITY IN MAGNETICALLY QUANTIZED TWO-DIMENSIONAL ELECTRON GAS

S.D.SUCHALKIN, Yu.B.VASILYEV*, Yu.L.IVANOV
S.V.IVANOV, P.S.KOP'EV, and I.G.SAVEL'EV
Ioffe Physical Technical Institute,194021 St.Petersburg, Russia
and
T.OHYAMA
Department of Physics, Faculty of Sciences, Osaka University, Toyonaka, Japan

ABSTRACT

We report on magnetic field and probe frequency dependencies of submillimeter photoconductivity in the quantum Hall effect regime. The measurements were made on samples of Hall and Corbino shapes. The results of our experiments permit the conclusion that there are two mechanisms of photoconductivity, namely, resonant inter-Landau level transitions induced by submillimeter emission and non-resonant absorption of radiation power giving rise to bolometric heating. The comparative contribution of them are strongly dependent on both the filling factor and the shape of the sample. We experimentally demonstrate that the photoresponse in the Hall bar is enhanced by over two orders in the center of the resistivity oscillation minimum when the current through the structure increases from 100 μA to 300 μA.

1. Introduction

In spite of a large number of papers on submillimeter photoconductivity (PC) of two dimensional electron gas (2DEG) in GaAs-AlGaAs heterojunctions[1-6] the detail mechanism of PC is still unclear. Most of the previous experiments were performed on samples of Hall geometry at fixed radiation frequencies by sweeping the magnetic field of the sample. This techniques has some drawbacks since such important parameters of the 2DEG as the density of states and filling factor are changing during the experiment. Here we report oh experimental studies of PC by means of tunable germanium cyclotron resonance lasers.[7] This method allows us to measure PC at fixed filling factors by sweeping the frequency of the probe emission. The experimental results on Hall bar and Corbino disk samples are presented.

2. Experimental

The emission source was a continuously tunable germanium cyclotron resonance laser with the pulse duration of 0.2μs and the pulse cut-off of less than 50ns. The experiments were done on standard Hall bars of 2.5mm long, 0.3mm wide, and 0.55mm between the potential probes and Corbino disks. The Hall bar samples were prepared from the MBE grown AlGaAs-GaAs heterojunction wafer with the sheet concentration of $2.6\times10^{11}cm^{-2}$ and the mobility of $8.2\times10^{4}cm^{2}/Vs$ at T=77K. The Corbino disk samples were prepared from the similar wafer with $n_s=2.6\times10^{11}cm^{-2}$ but with higher mobility of $1.2\times10^{5}cm^{2}/Vs$ at T=77K. We measured the change of the longitudinal resistivity (conductivity) due to submillimeter radiation fallen normally to the layer at a fixed current. All the measurements were performed at T=4.2 K.

3. Results

Figures 1a and 1b show magnetic field dependences of the PC and resistivity and conductivity oscillations for the Hall bar and Corbino disk samples, respectively.

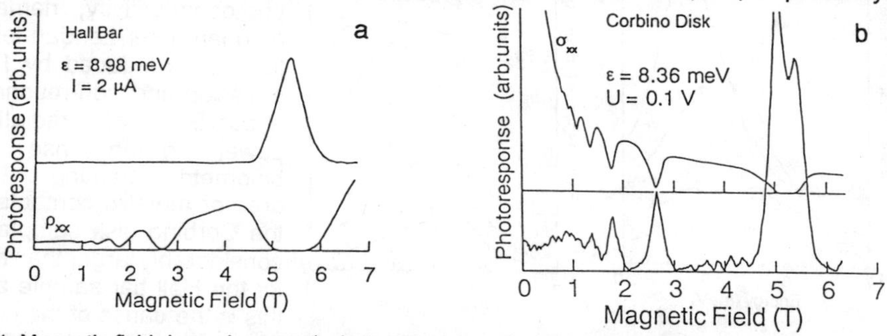

Fig.1. Magnetic field dependences: a) photoresponse and resistivity oscillation for the Hall bar, b) photoresponse and conductivity oscillations for the Corbino disk.

One can clearly see the cyclotron resonance (CR) peaks which measure approximately the same effective masses for the Hall bar and Corbino disk. The variation of the CR shape with the radiation energy is shown in Fig.2a and 2b. The most striking difference in magnetic field characteristic of the both geometries is the strong oscillations of PC observed for the Corbino disk which is absent in the case of the Hall bar. The non-resonant oscillation maxima coincide with the conductivity oscillation minima and has the representative decay time of the signal of about 10 μs which much longer than the typical decay time of the resonant signal (about 0.1 μs). Figure 3a and 3b represent the photoconductivity signal as a function of the emission frequency at different fixed filling factors near ν=2 and a current I=2 μA for the Hall bar and a voltage U=0.1 V for the Corbino disk The PC of Corbino and Hall geometries bear similar features: a sharp

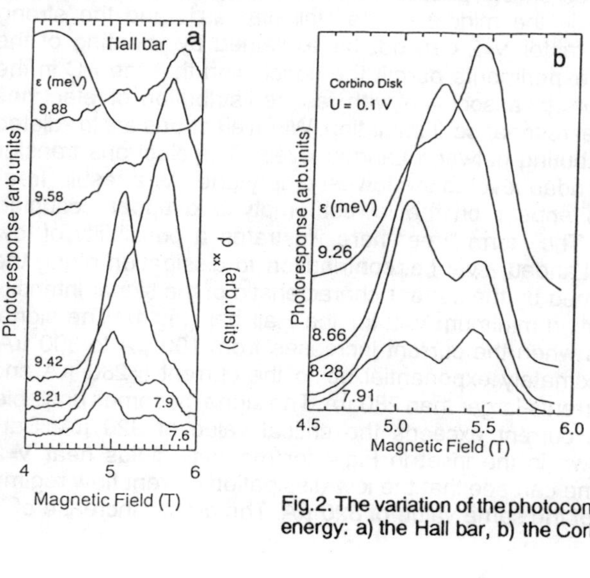

Fig.2. The variation of the photoconductivity signal with the radiation energy: a) the Hall bar, b) the Corbino disk.

peak near the center of the Shubnikov-de Haas (SdH) oscillation minimum and a broad smooth spectrum at the magnetic field corresponding to the edge of the Hall plateau. Besides, it is clearly seen that the peaks split into two and the relative intensity of the splitting lines depends on the filling factor.

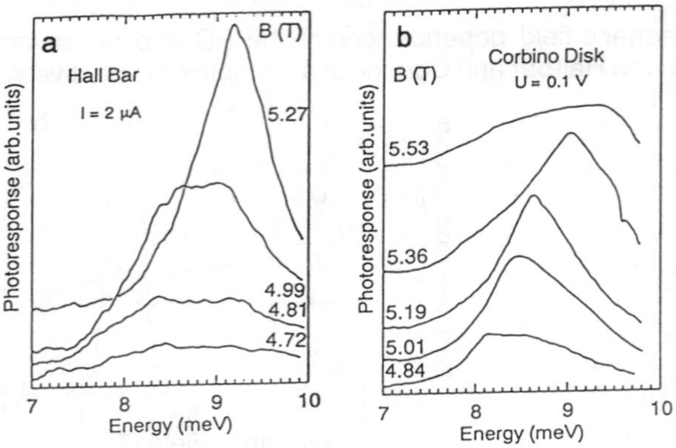

Fig.3 Spectra of the photoresponse: a) at I = 2μA for the Hall bar, b) at U=0.1V for the Corbino disk.

Our results provide strong evidence that there are two different mechanisms of photoconductivity, namely, resonant inter-Landau level transitions induced by FIR emission and non-resonant absorption of radiation power giving rise to bolometric heating. The area of metallic contacts in the Corbino disk sample is considerably larger than that for the Hall bar sample and this is the cause of the non-resonant oscillation structure in the Corbino disk sample which does not depend on the probe emission frequency. This structure is attributed to the electron heating, namely, bolometric behavior of the 2D system which was put forward to explain signals in some experiments.[5,8] Bolometric behavior means that a change of photoconductivity is due to an increase of the system temperature $\Delta\rho=(\delta\rho/\delta T)\Delta T$. This is clearly seen for the Corbino disk samples where the heating is stronger. For the Hall bars the bolometric behavior of the 2D system is observed on the plateau edge where $\delta\rho/\delta T$ is maximum. In contrast $\delta\rho/\delta T$ is minimum in the middle of the Hall plateau[3] and the strong signal at the integer filling factor $\nu=2$ can not be explained by heating of the system. The results of our experiments permit the conclusion that the PC in the vicinity of even filling factors is associated with fast redistribution of electrons between Landau levels under resonance illumination. When electrons are irradiated resonantly, they are redistributing between Landau levels. The electrons transfer from the upper occupied Landau level to the lowest empty one. As a result "free" electrons and "free" holes appear on the lowest empty and upper occupied Landau level respectively. The term "free" here illustrates a possibility of low energy scattering within a Landau level i.e. contribution to dissipation along the Hall bar. This is also confirmed by the current characteristic of the signal intensity in the center of the oscillation minimum $\nu=2$ for the Hall bar (Fig.4). The signal increases by over 2 orders when the current increases from 100 μA to 300 μA. The dependence is approximately exponential up to the current I=280 μA and becomes more abrupt at currents larger than 280 μA. The signal becomes unstable and sharply decreases as current exceeds the critical value of 320 μA. SdH oscillation minima are shown in the inset in Fig.4 for magnetic fields near $\nu=2$ and at different currents. One can see that the low-dissipation current flow regime is destroyed in the vicinity of the same value of 320 μA. The drastic increase of the signal at currents more than 100 μA can be explained by enhancement of the ρ_{xx}(illum) when the filling factor deviates from the integer value ν=2 under illuminations. It is known that at non-integer filling factors the breakdown occurs at weaker currents and at ν(illum) it is likely to be near I=280 μA as can be concluded from Fig.4. Before breakdown ρ_{xx}(illum) increases exponentially which results in exponential increase of the photoresponse (Fig.4). Thus the incident submillimeter emission changes the electron distribution quantitatively making possible the scattering which leads to the disruption of the low-dissipation state. This process is most apparent when the Fermi level lies in the middle of the mobility gap, in another word at the integer filling factor. We call this effect photoinduced breakdown of the QHE because it reflects the main point of the phenomenon which is the breakdown of the low-dissipation state under illumination. When the current is so strong that the low-dissipation state is destroyed, the carriers are distributed among Landau levels lying above the Fermi level without illumination. In this case the incident submillimeter emission does not change the resistivity and the signal disappears.

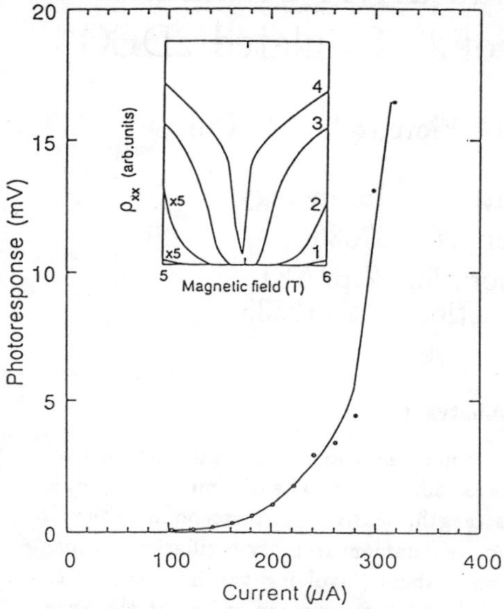

Fig. 4. The photoresponse as a function of the bias current at B= 5.45 T. The inset : SdH oscillation minima at currents: 1) 5μA , 2) 90μA , 3) 280μA, 4) 350 μA .

4. Acknowledgments

This work was supported by the Russian Fund of Fundamental Researches, project N 93-02-2838 and the Japan Society for the Promotion of Sciences.

5. References

*Also *Department of Physics, Faculty of Sciences, Osaka University, Toyonaka, Japan*

[1]J.C.Maan, et al., *Appl.Phys.Lett.* **40**, 609,(1982).
[2]D.Stein, G.Ebert, K.v.Klitzing, and G.Weimann, *Surf.Sci.* **142**, 406 (1984).
[3]R.E.Horstman, et al., *Solid State Commun.* **50**, 753 (1984).
[4]M.J.Chou, D.C.Tsui, and A.Y.Cho, *Proc. 18th ICPS*, p.437 (1986).
[5]F.Thiele, et al., *Phys.Rev. B* **40**, 1414 (1989).
[6]Yu.B.Vasilyev, et al., *JETP Lett.* , **56**, 377 (1992).
[7]Yu.B.Vasilyev, and Yu.L.Ivanov, *Proc. 18th ICPS*, p.1659 (1986).
[8]F.Neppl, J.P.Kotthaus, and J.F.Koch, *Phys.Rev. B* **19**, 5240 (1979).

Commensurability and dHvA Oscillations of Exchange Energy of a Modulated 2DEG

N.J.M. Horing, G. Fiorenza*, H.L. Cui

Stevens Institute of Technology
Hoboken, N.J. 07030
*IBM Corporation, Zip/AE1
Hopewell Junction, N.Y. 12533

Abstract

The exchange energy of a weakly modulated two dimensional electron gas in a normal quantizing magnetic field is examined here. The parameters employed-modulation period, magnetic field strength, density, etc.-correspond to the circumstances attending the observation of commensurability oscillations in magnetoconductivity by Weiss et.al. and others (involving the interplay of the length scales of the modulation period and the cyclotron radius at the Fermi level). Such commensurability oscillations are known to occur in a variety of other phenomena, including the dynamic nonlocal polarizability of noninteracting 2D electrons subject to weak density modulation and the associated magnetoplasmon spectrum, energy loss to a modulated 2DEG and magnetocapacitance. The polarizability provides the basis for our determination of the exchange energy as a function of inverse magnetic field and modulation period. In this analysis, we find novel commensurability oscillatory structure, as well as dHvA oscillations, in exchange phenomena for a weak periodic density modulation. In particular, these features are both evident in our result for the exchange energy $U_{EX}(meV)$ as a function of inverse field 1/H(Tesla).

Commensurability oscillatory phenomena due to a weak periodic density/potential modulation imposed on a 2D electron plasma(x,y plane) in a normal quantizing magnetic field($\vec{H} = H\hat{z}$) have been examined in several contexts recently. Arising from the interplay of the length scales of the cyclotron diameter(at the Fermi energy) and the modulation period(x-direction), they have been discussed in magnetoconductivity[1–5], magnetocapacitance[6], charged particle energy loss to a density modulated 2DEG[7,8], and the dynamic, nonlocal polarizability and magnetoplasmon spectrum[9]. In this work, we examine the role of commensurability oscillations in the exchange energy of a density modulated 2DEG and find that they form an envelope within which one also has deHaas-van Alphen oscillations of the exchange energy as a function of inverse magnetic field. In this analysis, the parameters employed(modulation period, magnetic field strength, density, etc.) correspond to the circumstances attending the observation of commensurability oscillations in magnetoconductivity by Weiss et al[1].

The exchange energy U_{EX}(per particle) of a modulated 2DEG in a quantizing magnetic field may be written in terms of the free-electron polarizability[10,11]. Considering a weak modulation which provides a first order energy correction to Landau levels spreading them into Landau minibands, and neglecting second order corrections which would modify spatial wavefunctions in accordance with Bragg reflection phenomena, the polarizability is spatially translationally invariant to this level of approximation. Consequently, it may be Fourier analyzed in terms of a single 2D wavevector $\vec{q}$, having the form $\alpha(\vec{q},\omega)$, and U_{EX} is given by (N is the 2D particle density and ε_0 is the background dielectric constant)

$$U_{EX} = -\frac{\hbar}{32\pi^4\varepsilon_0^3 N}\int d^2q \int d\omega\ \alpha(\vec{q},i\omega) \tag{1}$$

The noninteracting 2D electron polarizability determined in accordance with the approximation indicated above has been given by Cui, Fessatidis, and Horing[9] as

$$\alpha(\vec{q},i\omega) = \frac{2me^2\omega_c}{\varepsilon_0 a\hbar q}\sum_{n,n'} C_{nn'}\left(\frac{\hbar q^2}{2m\omega_c}\right)\int_0^a dx_0\,\frac{f\left[E_n\left(x_0+x_0'\right)\right]-f\left[E_{n'}(x_0)\right]}{E_n\left(x_0+x_0'\right)-E_{n'}(x_0)+\omega+i\delta} \tag{2}$$

where ω_c is the cyclotron frequency, m is the effective electron mass,
a is the modulation period in the x-direction, and $f(E)$ is Fermi-Dirac distribution. Also,

$E_n(x_0) = \left(n+\frac{1}{2}\right)\hbar\omega_c + V_n\cos(2\pi x_0/a)$is the modulated Landau miniband energy where $V_n = V_0\exp(-x/2)\,\mathrm{L}_n(x), x = \dfrac{\hbar(2\pi x_0/a)^2}{2m\omega_c}$ and $\mathrm{L}_n(x)$ is a Laguerre polynomial of order n.

Furthermore, $x_0' = -q_y\hbar/(m\omega_c)$ and V_0 is the modulation potential amplitude.
In Eq. (2),

$$C_{nn'}(x) = \frac{n_2!}{n_1!}x^{n_1-n_2}\exp(-x)\left[\mathrm{L}_{n_1}^{n_1-n_2}(x)\right]^2$$

where $n_1 = \max(n,n'), n_2 = \min(n,n')$ and $\mathrm{L}_n^l(x)$ is the Associated Laguerre polynomial. Finally,

$N = \dfrac{m\omega_c}{\pi a\hbar}\sum_{n'=1}^{n_L}\int_0^a dx_0\ f\left[E_{n'}(x_0)\right]$ is the 2D density, where $n_L$is the number of occupied Landau levels at zero temperature.

The resulting form of the exchange energy U_{EX} is

$$U_{EX} = -\frac{m\omega_c^2}{2\pi N}\sqrt{\frac{R^*}{\hbar\omega_c}}\sum_{n,n'}\int_0^1 d\phi'\int_0^\infty dz\, C_{nn'}\left(z^2\right)\int_0^1 dy \bullet$$

$$\bullet\frac{f\left[E_n\left(y+\frac{x_0'}{a}\right)\right]-f\left[E_{n'}(y)\right]}{n-n'+\bar{V}_n\frac{V_0}{\hbar\omega_c}\cos 2\pi\left(y+\frac{x_0'}{a}\right)-\bar{V}_{n'}\frac{V_0}{\hbar\omega_c}\cos 2\pi y} \quad (3)$$

where the effective Rydberg constant R^* is related to the usual Rydberg unit R by $R^* = mR/(m_0\varepsilon_0^2)$ (m_0 is the bare electron mass) and $\bar{V}_n = \exp(-x/2)\,\mathrm{L}_n(x)$. The integrals in Eq.(3) were evaluated numerically for a 2DEG at a GaAs-AlGaAs heterojunction with the parameter values,

$m = 0.07m_0$ =effective electron mass
$\kappa = 12.9$ = background dielectric constant
$N = 3.16\times10^{15}m^{-2}$ =2D electron density
$a = 382\ nm$=modulation period
$V_0 = 1.0\ meV$ =modulation potential amplitude

The results for the density modulated 2DEG exchange energy $U_{EX}(meV)$ are plotted as function of inverse magnetic field $1/H(Tesla)$ in the figure below. The rapid oscillations in this figure are of the deHaas-van Alphen(dHvA) type, and they fall within a commensurability oscillatory envelope which is clearly evident as a much slower oscillation.

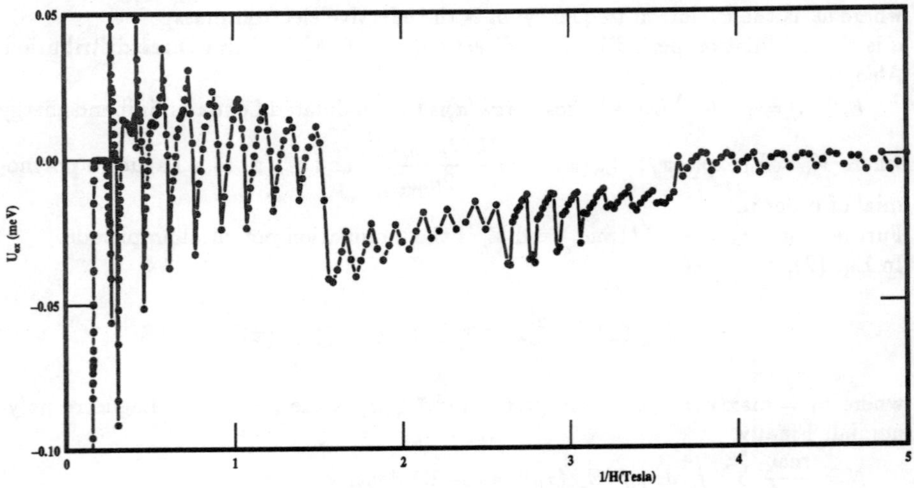

References

[1]D. Weiss,K.v. Klitzing,K. Ploog, and G. Weimann, *Europhys. Lett.* **8**, 179 (1989).

[2]R.R. Gerhardts, D. Weiss, and K.v. Klitzing, *Phys. Rev. Lett.*, **62**, 1173 (1989).

[3]R.W. Winkler, J.P. Kotthaus, and K. Ploog, *Phys. Rev. Lett.*, **62**, 1177 (1989).

[4]P. Vasilopoulos and F.M. Peters, *Phys. Rev. Lett.*, **63**, 2120 (1989).

[5]C.W.J. Beenakker, *Phys. Rev. Lett.*, **62**, 2020 (1989).

[6]D. Weiss *et al.*, *Phys. Rev. B* **39**, 13020 (1989).

[7]N.J. M. Horing, V. Fessatidis, and H.L. Cui, *Europhys. Lett.* **18**, 613 (1992).

[8]N.J. M. Horing, V. Fessatidis, and H.L. Cui, *Superlatt. & Microstru.* **10**, 447 (1991).

[9]H.L. Cui, V. Fessatidis, and N.J. M. Horing,*Phys. Rev. Lett.* **63**, 2598 (1989).

[10]J.J. Labonney, *Ph.D. Thesis*, "*Exchange and Correlation Effects in Two-Dimensional Quantum Plasmas in Magnetic Field*", Stevens Institute of Technology,1979.

[11]G. Fiorenza, *Ph.D. Thesis*, "*Superlattice Correlation and Self Energy Phenomena*", Stevens Institute of Technology,1992.

Quantum Magnetic Field Oscillations in GaAs Superlattice Correlation Energy

G. Fiorenza* and N.J.M. Horing

Stevens Institute of Technology
Hoboken, N.J. 07030
*IBM Corporation, Zip/AE1
1580 Route 52
Hopewell Junction, N.Y. 12533

Abstract

Quantum magnetic field deHaas-vanAlphen (DHVA) oscillatory effects in the correlation energy E^{corr} of a GaAs-AlGaAs superlattice are analyzed in terms of the two-dimensional polarizability $\alpha(q,\omega)$ of the constituent thin superlattice planes. In the range of low to intermediate magnetic field strengths, the DHVA oscillatory part $\alpha_{dhva}(q,\omega)$ is small compared to the monotonic part $\alpha_{mon}(q,\omega)$ ($\alpha(q,\omega) = \alpha_{mon}(q,\omega) + \alpha_{dhva}(q,\omega)$; $E^{corr} = E^{corr}_{mon} + E^{corr}_{dhva}$),

$$\alpha_{dhva}(q,\omega)/\alpha_{mon}(q,\omega) \sim \hbar\omega_c/\zeta < 1,$$

(ω_c is the cyclotron frequency and ζ is the chemical potential) and this facilitates an expansion of E^{corr}_{dhva} to linear order as a functional of $\alpha_{dhva}(q,\omega)$. In this, the DHVA oscillatory part E^{corr}_{dhva} is explicitly exhibited to vary as a linear sawtooth function of $\zeta/(\hbar\omega_c)$. The coefficient of this closed form DHVA dependence involves the monotonic polarizability, and we analyze this coefficient using a 2-D quantum mechanical random phase approximation (RPA) model for $\alpha_{mon}(q,\omega)$ in the limit of zero magnetic field. Detailed numerical evaluation of the coefficient of the linear sawtooth DHVA oscillatory part of the correlation energy, E^{corr}_{dhva}, is carried out for typical superlattice period and density parameters under conditions of low-intermediate magnetic field strength for a GaAs based superlattice.

Correlation energy calculations have played a central role in the historical development of many-electron theory with long range electron-electron Coulomb interactions $v(\vec{r}_1 - \vec{r}_2)$ manifested in terms of dynamic plasmon phenomenology and decay into electron-hole pairs(Landau damping), as well as statically shielded short range electron-electron scattering. Although originally intended for a homogenous solid state plasma, the use of quantum field theory techniques readily accommodates a spatially inhomogenous system with the grand potential W(logarithm of the grand partition function) represented in terms of the density-density correlation function[1-2] $G_2(1,2;1^+,2^+)_{t_1=t_2}$ as ($\beta = 1/kT$ = inverse thermal energy)

$$W = W(0) + \frac{\beta}{2}\int d^3r_1 \int d^3r_2 \int_0^1 \frac{d\lambda}{\lambda} v(\vec{r}_1 - \vec{r}_2) G_2(1,2;1^+,2^+)_{t_1=t_2}. \tag{1}$$

Here, $W(0)$ is the value in the absence of interelectron interactions($v \to 0$) and λ is a strength parameter attached to $v(\vec{r}_1 - \vec{r}_2) \to \lambda e^2/(\varepsilon_0 |\vec{r}_1 - \vec{r}_2|)$ ranging from no interaction($\lambda = 0$) to full interaction strength($\lambda = 1$)[ε_0 is the background dielectric constant]. As the inverse dielectric function is intimately related to the density-density correlation function[3](even for inhomogenous systems), we may write[4](in this, we invoke spatial translational invariance of the superlattice in the x-y plane and Fourier transform $(x_1 - x_2, y_1 - y_2) \to (q_x, q_y)$ in that plane and $(t_1 - t_2 \to \omega)$

$$W - W(0) = \frac{i\beta A}{2}\int \frac{d\omega}{2\pi}\int \frac{d^2q}{(2\pi)^2}\int dz \int \frac{d\lambda}{\lambda}\left[K(z,z,q,\omega) - \delta(z-z)\right] + \frac{\beta}{2}\int d^3r_1 \int d^3r_2 \int \frac{d\lambda}{\lambda} v(\vec{r}_1 - \vec{r}_2) G_1\left(1,1^+\right) G_1\left(2,2^+\right), \tag{2}$$

where $G_1(1,2)$ is a one-electron thermal Green's function. Subtracting from Eq.(2) its Hartree-Fock approximation to get the correlated part W^{corr} and executing standard ω-contour integral manipulations[2], we employ earlier results for the causal inverse dielectric function of a type I superlattice of period a in the random phase approximation(RPA)[5]

$$K(z_1,z_2,q,\omega) \to K_{rpa}(z_1,z_2,q,\omega - i\varepsilon) = \delta(z_1 - z_2) - \frac{a}{2\pi}\sum_{n,n'} \exp\left(-q\left|z_1 - n'a\right|\right)\int_{-\pi/a}^{\pi/a} dk_z \exp\left(-ik_z\left(n - n'\right)\right)\frac{\delta(z_2 - na)\alpha_0(q,\omega - i\varepsilon)}{1 + \mathrm{S}(q,k_z)\alpha_0(q,\omega - i\varepsilon)} \tag{3}$$

to obtain the correlation energy $E^{corr} = -\partial W^{corr}/\partial\beta$ at zero temperature Kelvin as

$$E^{corr} = \frac{\hbar N_p A a}{8\pi}\int \frac{d\omega}{2\pi}\int \frac{d^2q}{(2\pi)^2}\int_0^1 \frac{d\lambda}{\lambda}\int_{-\pi/a}^{\pi/a} dk_z \bullet \left[\frac{\lambda\,\mathrm{S}(q,k_z)\alpha_0(q,i\omega)}{1 + \lambda\,\mathrm{S}(q,k_z)\alpha_0(q,i\omega)} - \lambda\alpha_0(q,i\omega)\right]. \tag{4}$$

Here $\alpha_0(q,i\omega)$ is the 2D polarizability with quantizing magnetic field[6] in the present case, N_p is the infinite number of superlattice planes, A is the infinite planar area and $\mathrm{S}(q,k_z)$ is the superlattice structure factor,

$$\mathrm{S}(q,k_z) = \frac{\sinh(qa)}{\cosh(qa) - \cos(k_z a)}. \tag{5}$$

Executing the coupling strength parameter λ-integral, the correlation energy per particle is(ρ is the 2D electron density and ζ is the chemical potential),

$$\frac{E^{corr}}{N_p \rho A} = \frac{2\zeta}{\pi^2}\int_0^\infty dz\, z^2 \int_{-\infty}^{+\infty} du \int_{-\pi}^{+\pi} dx \left\{\ln\left[1 + \mathrm{S}(z,x)\alpha_0(z,iu)\right] - \alpha_0(z,iu)\right\}, \tag{6}$$

where we have set $z = q/(2q_f)$ and $u = m\omega/(\hbar q_f q)$ and $x = k_z a$, where q_f is Fermi wavevector and m is the effective electron mass(an evaluation of Eq.(6) for null magnetic field has been carried out by Gumbs[7]). Carrying out the x-integral we finally obtain

$$\frac{E^{corr}}{N_p \rho A} = \frac{4\zeta}{\pi} \int_0^{\infty} dz\ z^2 \int_{-\infty}^{+\infty} du \left\{ \ln\left[1 + A_0(z, iu)\right] - \alpha_0(z, iu) \right\}, \tag{7}$$

where

$$A_0(z, iu) = \frac{\alpha_0(z, iu) - 1 + \sqrt{1 + \alpha_0^2(z, iu) + 2\alpha_0(z, iu)\coth\left(a\sqrt{2\pi\rho z}\right)}}{1 + \coth\left(a\sqrt{2\pi\rho z}\right)}. \tag{8}$$

The deHaas-vanAlphen oscillatory part of superlattice correlation energy $E^{corr} = E^{corr}_{mon} + E^{corr}_{dhva}$ may be extracted from the structure of the 2D polarizability[6] $\alpha_0(q, \omega) = \alpha_{mon}(q, \omega) + \alpha_{dhva}(q, \omega)$. Here "*mon*" designates the part which is monotonic as a function of magnetic field strength and "*dhva*" designates the quantum magnetic field oscillatory part. In the range of low to intermediate magnetic fields, $\alpha_{dhva}(q, \omega)$ is much smaller than $\alpha_{mon}(q, \omega)$(ω_c is the cyclotron frequency)

$$\alpha_{dhva}(q, \omega)/\alpha_{mon}(q, \omega) \sim \hbar\omega_c/\zeta < 1, \tag{9}$$

and this facilitates an expansion of E^{corr}_{dhva} to linear order as functional of $\alpha_{dhva}(q, \omega)$. In this, the dHvA oscillatory part E^{corr}_{dhva} is explicitly exhibited to vary as a linear sawtooth function of $(\zeta/(\hbar\omega_c))$ for the superlattice as in earlier 2D calculations[8]. The coefficient of this closed form dHvA dependence involves the monotonic polarizability as follows:

$$\frac{E^{corr}_{dhva}}{N_p \rho A} = \left[\frac{\zeta}{\hbar\omega_c} - \frac{1}{2}\right]_{per} \frac{4m\omega_c\zeta^2}{\pi^2\hbar\rho} \int_0^{\infty} dz\ z^2 \int_{-\infty}^{+\infty} du \bullet$$

$$\bullet \left[1 - \left(1 + \alpha^2_{mon}(z, iu) + 2\alpha_{mon}(z, iu)\coth\left(a\sqrt{2\pi\rho z}\right)\right)^{-1/2}\right] \frac{\partial\alpha_{mon}}{\partial\zeta}. \tag{10}$$

All dHvA oscillatory dependence is embodied in the periodic linear sawtooth function[6,8] $[y]_{per} = y$ in the fundamental interval $-1/2 \le y \le +1/2$ with periodic extension outside. The coefficient[6,8] of this closed form dHvA oscillatory function $\left[\frac{\zeta}{\hbar\omega_c} - \frac{1}{2}\right]_{per}$ on the right hand side of Eq.(10) for E^{corr}_{dhva} involves α_{mon} and $\partial\alpha_{mon}/\partial\zeta$. In this regard we treat α_{mon} approximately, replacing it by its zero field random phase approximation(RPA) limit as given by Stern[9], with monotonic quantum magnetic field corrections taken to be small. On this basis, we carried out a numerical evaluation of E^{corr}_{dhva} as given by Eq.(10), using parameter values appropriate to a GaAs-AlGaAs superlattice as follows[10]: $m = 0.07m_0$ is the effective electron mass, $\kappa = 12.9$ is the background dielectric constant, and $\rho = 3.16 \times 10^{12} cm^{-2}$ is the 2D electron density.

The resulting dependence of E^{corr}_{dhva} on$(\zeta/(\hbar\omega_c))$ is shown in Figure 1[10], with superlattice period $a = 9\ 7.5\AA$.

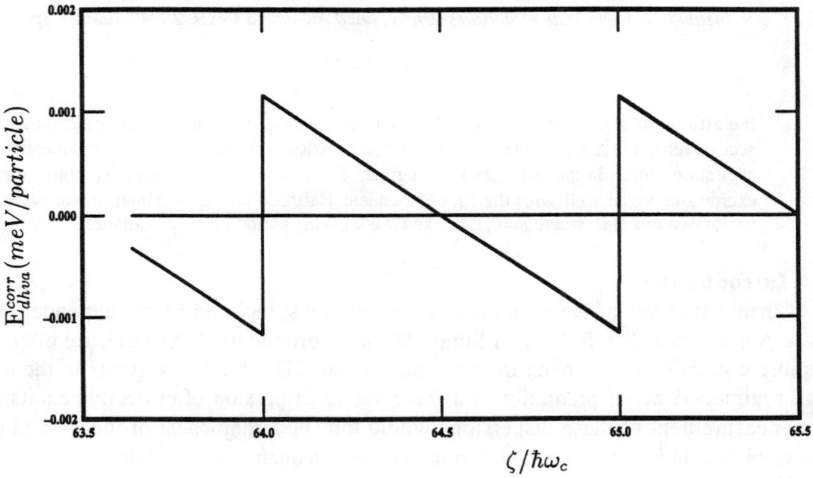

Figure 1: dHvA oscillatory part of GaAs-based superlattice correlation energy

References

[1] P.C. Martin and J. Schwinger, *Phys. Rev.* **115**, 1342 (1959).
[2] D. Pines, *Elementary Excitations in Solids*(The Benjamin/Cummings Publishing Company, London, 1963)(Appendix C for contour integration).
[3] N.J.M. Horing, *Ann. Phys.(N.Y.)* **31**, 1 (1965).
[4] N.J.M. Horing, R.W. Danz, and M.L. Glasser, *Phys. Rev.* **6**, 2391 (1972).
[5] N.J.M. Horing, G. Fiorenza, and H. Cui, *Phys. Rev. B* **31**, 6349 (1985).
[6] N.J.M. Horing and M.M. Yildiz, *Ann. Phys.(N.Y.)* **97**, 216 (1976).
[7] G. Gumbs, *Phys. Rev. B* **18**, 546 (1967).
[8] N.J.M. Horing and J.J. Labonney, *J. Mag. and Mag. Materials* **11**, 73 (1979); J.J. Labonney and N.J.M. Horing, *Surf. Sci.* **64**, 437 (1977); J.J. Labonney, *Ph.D. Thesis*, '*Exchange and Correlation Effects in Two-Dimensional Quantum Plasmas in Magnetic Field*', Stevens Institute of Technology,1979.
[9] F. Stern, *Phys. Rev. Lett.* **18**, 546 (1967).
[10] G. Fiorenza, *Ph.D. Thesis*,'*Superlattice Correlation and Self Energy Phenomena*', Stevens Institute of Technology,1992.

WAVE VECTOR-INDUCED COLLECTIVE EXCITATIONS IN THE QUANTUM HALL REGIME

L. L. SOHN,[(1)] A. PINCZUK,[(1)] B. DENNIS,[(1)] L. N. PFEIFFER,[(1)] K. W. WEST,[(1)] and L. BREY[(2)]
[(1)]*AT&T Bell Laboratories, Murray Hill, NJ 07974--0636 U.S.A.*
[(2)]*Instituto de Ciencia de Materiales ,Universidad Autonoma C-XII, 28049 Madrid, Spain*

ABSTRACT

Inelastic light scattering measurements on a weakly density-modulated 2D electron gas access, for the first time, the large wave vectors needed to probe the dispersive collective excitation modes in the quantum Hall regime. For $\nu=1$, measured dispersive collective excitations agree well with the time-dependent Hartree-Fock approximation for wave vectors $q_m=m(\pi/d)$, where $m=0,1,2,...$ and d is a partial period of the modulation.

1. Introduction

Current theories of electron-electron interactions, including time-dependent Hartree Fock Approximation (HFA)[1] and Single Mode Approximation (SMA)[2], are often used to explore collective phenomena in two dimensional (2D) electron systems in the quantum Hall regime. A major prediction is a wave vector dispersion of collective excitations.[1,2] The measurement of these dispersions would thus be a major test of theory and provide powerful insight into electron-electron interactions in quantum Hall states.

This paper considers recent observations of the wave vector dispersion of collective excitations for a 2D electron system in the quantum Hall regime.[3,4] We performed inelastic light scattering measurements on samples with a weak in-plane electron-density modulation. For $\nu=1$, we observe collective excitations whose dispersions are in agreement with those predicted by HFA. Observed modes disperse with wave vectors $q_m{}^*=m(\pi/d)$, where $m=0,1,2,...$ and d is the partial period of the modulation. The wave vector dependence is the result of the decoupling between alternating compressible and incompressible strips within the system which consequently *confines* the optical exciton states and their corresponding electron wave functions to the compressible strips.

2. Materials and Methods

The measured samples are asymmetric GaAs-Al_x-$Ga_{1-x}As$ single quantum wells with x~0.1 and width 250Å. Each quantum well is capped with 1600Å Si delta-doped Al_x-$Ga_{1-x}As$ and 100Å GaAs. All samples prior to processing have an electron density $n_o=1.33 \times 10^{11}$ cm^{-2} and mobility $\mu \geq 3 \times 10^6$ cm^2/Vsec. Several arrays of 65nm wide lines are e-beam written onto each sample. Each array has a period a =450nm and is 84μm x 84μm large. 30nm of each sample's unprotected surface is reactively-ion-etched with a 3:50 BCl_3:Ar mixture to create a weak in-plane electron density modulation. A detailed outline of experimental parameters and procedures can be found in Ref. 3.

Photoluminescence spectra confirm the creation of a density-modulated system with two *distinct* densities: n_o which corresponds to the region (of width d=385nm) *between* the etched lines and $n_1=1.43 \times 10^{11}$ cm^{-2}, to the region (of width a-d=65nm) *below* the etched lines. That the etched regions show higher electron density may result from the shallow-etching process which modifies only the surface states and not the Si dopant layer. Regardless, wave vectors $q_m=m(2\pi/a)$ should be accessed[5,6] and indeed they are: observed 2D plasmon modes in zero field disperse with q_m. For n_o, the well known $\nu=1$ optical anomaly[7] occurs over the range B=[5.3T, 5.8T], for n_1 over B=[5.8T-6.2T].

3. Results and Discussion

Figure 1a shows the inelastic light scattering measurements performed at B=5.9T. The spectra show a sharp peak at 23.75meV which corresponds to the q=0 charge-density intersubband excitation of the 250Å quantum well.[8] Also present are peaks, which correspond to induced wave vector components m=1,2, ...,6, in the range 22.8-23.6meV. For comparison, Fig. 1b shows the spectra obtained from the unmodulated part of the very same sample: only the q=0 intersubband peak is present.

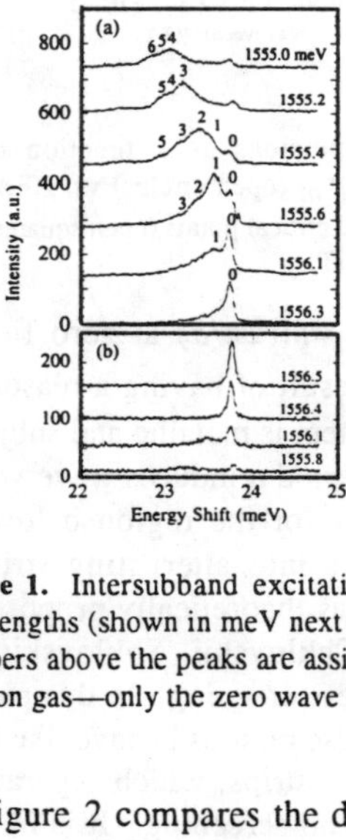

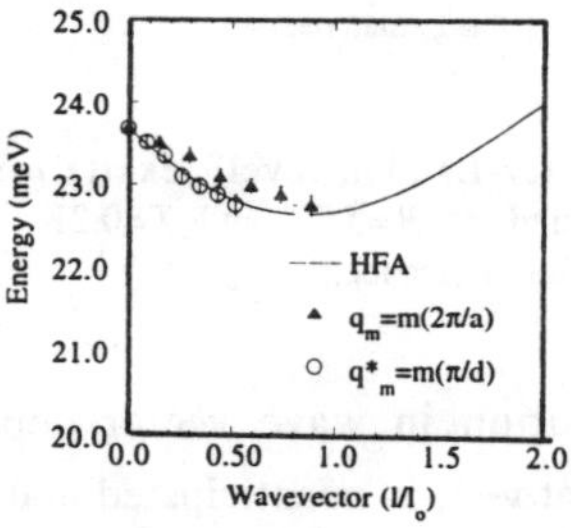

Figure 2. Dispersion as a function of q_m (triangles) and q^*_m (open circles) vs. HFA.

Figure 1. Intersubband excitations. **(a)** Inelastic light scattering spectra taken for a range of laser wavelengths (shown in meV next to the individual spectrum). All spectra obtained at B=5.9T and T=0.2K. Numbers above the peaks are assignments to wave vector components m. **(b)** Spectra of unmodulated 2D electron gas—only the zero wave vector mode is visible. The smaller satellite peak is luminescence.

Figure 2 compares the dispersion derived from the observed multiple intersubband modes to that predicted by HFA.[9] As a function of $q_m=m(2\pi/a)$, the dispersion extrapolates to a roton minimum of ql_o~1.5, as a function of $q^*_m=m(\pi/d)$ to ql_o~1.0. Since HFA theory accurately accounts for critical-point energies[10-12] and predicts the roton to occur at $ql_o \leq 1.0$, we conclude that the induced wave vector is q^*_m. We comment that the q_m modes may still exist—given that a is only slightly larger than d, these modes would coincide with the even q^*_m ones.

The inter-Landau level excitation spectra of the modulated system at B=5.7T is shown in Fig. 3. The q=0 excitation occurs at the cyclotron energy ω_c=9.65meV. Well-defined higher-order dispersion modes, m=1,2,...,6, occur in the range 9.9-11.5meV. The dispersion critical points are weak and near modes m=4 and 6—they become strongly disorder-activated at B=5.9T and 5.5T.[10,11] Fig. 4 compares the obtained dispersion as a function of q_m and q^*_m to that of HFA.[12] Again, only the q_m^* dispersion shows quantitative agreement with theory.

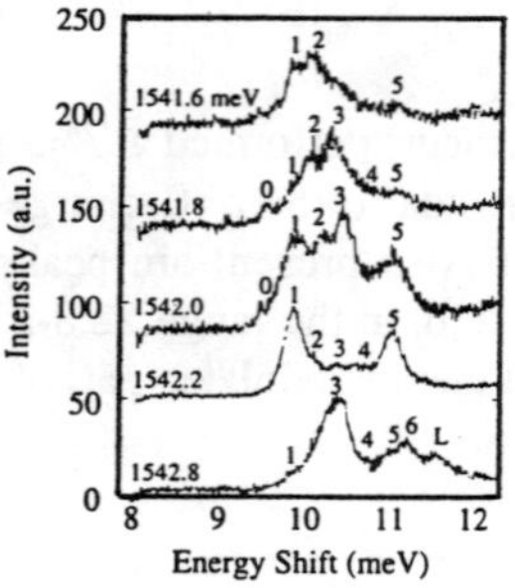

Figure 3. Inter-Landau-level excitations. Spectra obtained at B=5.7T and T=0.2K. L indicates weak luminescence.

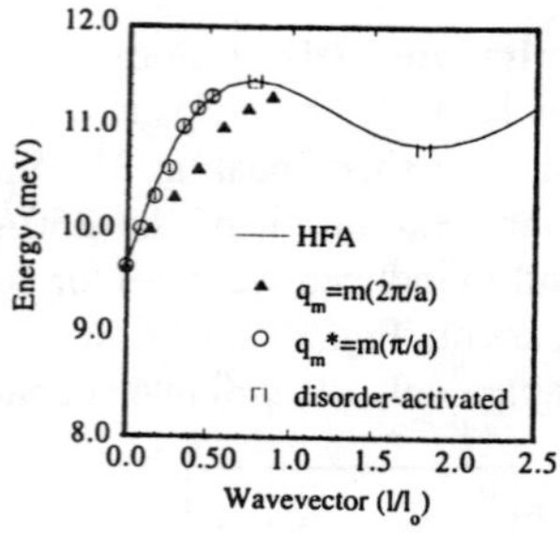

Figure 4. Dispersions as a function of q_m (triangles) and q^*_m (open circles) vs. HFA. The disorder-activated critical points (open squares) were obtained at B=5.9T.

The transition in wave vector dependence, from $q_m=m(2\pi/a)$ at zero field to $q^*_m=m(\pi/d)$ at $\nu\sim1$, is unanticipated and can be seen as a result of having a reasonably well-defined density-modulated system. In zero field, the system is metallic and subject to screening. Consequently, the *weak* periodic potential with period a induces wave vectors q_m. In high magnetic fields, B=5.9T and 5.7T where ν=1 for the region below and between the etched lines, respectively, the system divides into alternating strips of incompressible and compressible liquids. Such a system was theoretically proposed by Beenakker[13] and Chang[14] and analytically studied by Chklovskii, Shklovskii, and Glazman [15] to provide a picture of edge states which includes screening. In this picture, screening is nearly perfect within the compressible strips—these regions behave like metal strips at constant electrostatic potential. The incompressible strips, which separate the compressible ones, behave as insulators and therefore have no screening. It is in these regions where all the electrostatic potential drops occur. A barrier is effectively created between the compressible and incompressible strips and the two thus become *decoupled*. As recent tunneling measurements[16] demonstrate, this barrier increases with magnetic field. For our system, the barrier *confines* the optical exciton states to the compressible regions of width d.

4. Conclusion

In conclusion, a novel technique of employing a density-modulated 2D system successfully accesses large wave vectors. Consequently, the dispersive collective excitation modes that reflect the novel electron-electron interactions of the quantum Hall regime can be directly observed.

5. Acknowledgements

We would like to thank A. M. Chang, L. P. Kouwenhoven, D. B. Chklovskii, and L. Glazman for insightful discussions. We would also like to thank K. Baldwin for assisting us with sample etching. One of us, L. B., acknowledges financial support from the CICYT of Spain, No. MAT 91 0201.

6. References

1. C. Kallin and B. I. Halperin, Phys. Rev. **B 30**, 5655 (1984); A. H. MacDonald, J. Phys. C **18**, 1003 (1985).
2. S. M. Girvin, A. H. MacDonald, and P. M. Platzman, Phys. Rev. Lett. **B 54**, 581 (1985); and A. H. MacDonald, H. C. A. Oji, and S. M. Girvin, Phys. Rev. Lett. **55**, 2208 (1985).
3. L. L. Sohn, A. Pinczuk, B. S. Dennis, L. N. Pfeiffer, K. W. West, and L. Brey, submitted for publication (1994).
4. L. L. Sohn, A. Pinczuk, B. S. Dennis, L.N. Pfeiffer, and K. W. West, in preparation (1994).
5. G. Eliasson, Ji.-Wei Wu, P. Hawrylak, and J. J. Quinn, Solid State Comm. **60**, 41 (1986); W. Lai, Akiko Kobayashi, and S. Das Sarma, Phys. Rev. **B 34**, 7380 (1986).
6. U. Mackens, D. Heitmann, L. Prager, J. P. Kotthaus and W. Beinvogl, Phys. Rev. Lett. **53**, 1485 (1984).
7. D. Heiman et al., Phys. Rev. Lett. **61**, 605 (1988); A. J. Tuberfield, ibid., 637 (1990); and B. B. Goldberg et al., ibid., **65**, 6416 (1990).
8. A. Pinczuk, S. Schmitt-Rink, G. Danan, J. P. Valladares, L. N. Pfeiffer, and K. W. West, Phys. Rev. Lett., **63**, 1633 (1989); D. Gammon, B. V. Shanabrook, J. C. Ryan, and D. S. Katzer, ibid., **68**, 1884 (1992).
9. L. Brey, Phys. Rev. **B 44**, 3772 (1991).
10. A. Pinczuk, J. P. Valladares, D. Heiman, A. C. Gossard, J. H. English, C. W. Tu, L. N. Pfeiffer, and K. W. West, Phys. Rev. Lett. **61**, 2701 (1988); A. Pinczuk, B. S. Dennis, D. Heiman, C. Kallin, L. Brey, C. Tejedor, S. Schmitt-Rink, L. N. Pfeiffer, and K. W. West, ibid. **68**, 3623 (1992).
11. T. Zettler, C. Peters, J. P. Kotthaus, and K. Ploog, Phys. Rev. **B 39**, 3931 (1989).
10. The time-dependent HFA calculation includes the effect of the extent of the electron wave function along the perpendicular direction. C. Kallin, private communication.
12. C. W. J. Beenakker, Phys. Rev. Lett. **64**, 215 (1990).
13. A. M. Chang; Solid State Comm. **74**, 871 (1990).
14. D. B. Chklovskii, B. I. Shklovskii, and L. I. Glazman, Phys. Rev. **B 46**, 4026 (1992). See also L. Brey et al., Phys. Rev. **B 47**, 13884 (1993).
15. N. C. van der Vaart, M. P. de Ruyter van Steveninck, L. P. Kouwenhoven, A. T. Johnson, Y. V. Nazarov, and C. J. P. M. Harmans, preprint.

MAGNETO-PHOTOLUMINESCENCE OF THE 2DHS IN P-TYPE GaAs

A G Davies,[1] E E Mitchell,[1] R G Clark,[1] P E Simmonds,[1,2] D A Ritchie,[3] M Y Simmons,[3] M Pepper[3] and G A C Jones[3]

[1] *National Pulsed Magnet Laboratory, University of New South Wales, Sydney 2052, Australia.*
[2] *Department of Physics, University of Wollongong, PO Box 1144, Wollongong 2500, Australia.*
[3] *The Cavendish Laboratory, University of Cambridge, Madingley Road, Cambridge CB3 0HE, UK.*

Bandgap magneto-photoluminescence (PL) has been used to study the optical signature of a high quality p-type GaAs-$Al_xGa_{1-x}As$ single quantum well in the extreme quantum regime. Two PL lines associated with recombination in the quantum well are observed which undergo intensity modulations at integer and fractional quantum Hall effect ground states. We relate the optical signature to recent PL studies of n-type single heterojunctions and quantum wells.

1. Introduction

Bandgap magneto-photoluminescence (PL) has recently been used to investigate the low temperature, high magnetic field ground states of the two-dimensional hole system (2DHS) formed in p-type GaAs-$Al_xGa_{1-x}As$ single heterojunctions (SHJs)[1] and narrow single quantum wells (SQWs)[2]. In contrast to optical spectroscopy of 2D electron systems (2DESs) which is now well-established (for a review, see Ref. 3), the use of PL to probe the 2DHS is only just being addressed.

In a previous paper,[1] we presented a detailed PL study of a high quality, low-density p-type SHJ at 75mK. One sharp spectral line which we associated with recombination at the heterointerface showed an intensity maximum which correlated with the $\nu=1$ quantum Hall effect (QHE). Overall, the results pointed to a general conclusion that progress towards observing sharp PL signatures of integer and fractional QHE states and the magnetically-induced Wigner solid (MIWS), comparable to those obtained in n-type experiments, requires characterised flat band SHJs or for attention to shift to high quality SQW structures, in which problems associated with strong band-bending are avoided. In this paper we report a PL investigation of a 300Å wide p-type SQW at temperatures down to 80mK.

2. P-Type Single Quantum Well PL Data

PL measurements at an excitation wavelength of 737nm (below $Al_xGa_{1-x}As$ bandgap) have been carried out on the low density ($n=2.7\times10^{10}cm^{-2}$) p-type modulation-doped SQW sample T128 grown by MBE on the (311)A GaAs surface. Ref. 1 contains the experimental details. InZn contacts were annealed into the corners of the $2\times2mm^2$ sample and transport data revealed structure at IQHE states and $\nu=2/3$, 1/3 FQHE states, although the sample geometry caused some admixture of the ρ_{xx} and ρ_{xy} components.

A 300Å wide well was selected for this study to reduce deleterious effects associated with the interface. To demonstrate that the well width is sufficient to separate the quantum well signal from bulk recombination, a comparison of the 9T PL spectra obtained for p-type SHJ A325 ($n=6\times10^{10}cm^{-2}$)[1] and SQW T128 is presented in Fig. 1. Fig. 1(a) shows the 75mK, 1.75μW spectrum of SHJ A325 (see Ref. 1 for a complete discussion). The peak labelled X is assigned to GaAs bulk free exciton recombination, confirmed by photoluminescence excitation studies, the A^0, X emission is due to recombination of acceptor-bound excitons and A^0, X-D to acceptor-defect bound excitons. We associated the arrowed peak with emission at the heterointerface.[1] PL spectra from SQW sample T128 are shown in Figs. 1(b) and 1(c) where structure associated with recombination in the quantum well is labelled QW; Fig. 1(b) was obtained at 4.2K with an enhanced excitation power of 100μW, Fig. 1(c) shows the 80mK, 5μW spectrum. Under high illumination conditions (Fig. 1(b)), several bulk features resulting from recombination in the GaAs buffer layer below the quantum well are brought up and the bulk free exciton peak X dominates the spectrum. Under our normal mK illumination conditions of 5μW (Fig. 1(c)), PL structure associated with recombination in the quantum well is resolved together, at this high magnetic field, with a weak free exciton signal ~2meV lower in energy. Care must be taken when assigning PL structure lying at energies below the main quantum well emission owing to the many bulk PL lines present in this region; note for

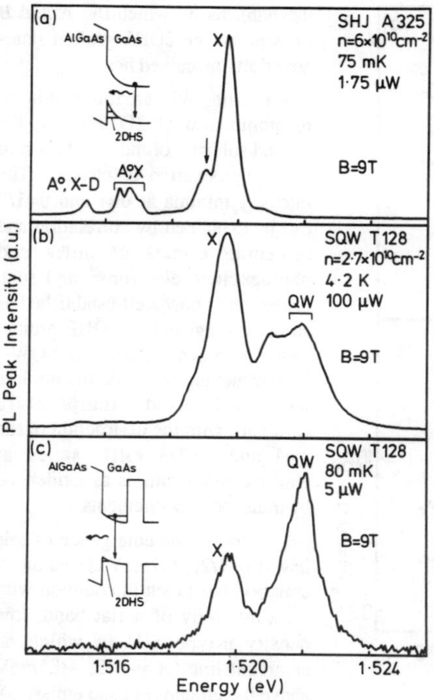

Fig. 1 Comparison of 9T PL spectra for (a) SHJ A325 at 75mK and SQW T128 at (b) 4.2K and (c) 80mK.

example the line between the QW emission and X brought up at high laser power in Fig. 1(b) also seen in Fig. 1(a). Schematic diagrams showing the band-edge profiles and 2DHS recombination with photoexcited electrons for the p-type SHJ and SQW are shown in Figs. 1(a) and 1(c), respectively.

The magnetic field dependencies of the PL peak energy positions and intensities of T128 at 80mK are shown in Fig. 2. The inset to Fig. 2(a) shows the 3.0T PL spectrum – the quantum well emission is characterised by a clear doublet at this field with no bulk emission evident. Fig. 2(b) also shows the ρ_{xy} data around $\upsilon=1$ (offset for clarity) and the ρ_{xx} out-of-phase component ($\rho_{xx}(\theta=90°)$) which increases sharply at $\upsilon=1/3$ – such behaviour has been associated with the onset of a MIWS phase in n-type material.[3] The T-dependence of the PL structure is addressed in Figs. 3 and 4. Figs. 3(a) and 3(b) present the 80mK and 4.2K intensity data in the regions around $\upsilon=1$ and $\upsilon=1/3$. Figs. 4(a) and 4(b) show the T-dependence of the raw PL spectra at 1.1T ($\upsilon=1$) and 3.5T ($\upsilon=1/3$).

PL emission associated with recombination in the well is characterised by two spectral lines A and B. The A line is present over the entire magnetic field range studied whereas the B line is not observed for low fields, emerging at $\upsilon\sim1/2$ in the FQHE regime, ~0.7meV higher in energy than the A emission (Fig. 2). Emission associated with the bulk free exciton X is observed for fields beyond 5T – the energy trajectory of this line exactly overlays that of the bulk free exciton in p-type SHJ A325.[1]

In the IQHE regime at 80mK, the A line intensity shows a sharp minimum at the $\upsilon=1$ QHE state (Figs. 2(b) and 3(a)). The width of the minimum correlates with the transport structure (Fig. 2(b)). The association of this PL feature with the $\upsilon=1$ QHE is reinforced by its temperature dependence; at 4.2K the A line intensity minimum is quenched (Fig. 3(a)). Owing to the lower overall spectral intensity at elevated temperatures, the 4.2K data was obtained under illumination conditions of 10µW which additionally serves to normalise the 4.2K and 80mK A line intensities at B=0T for clearer comparison. The quenching of the $\upsilon=1$ A line intensity minimum can also be seen in the raw PL data in Fig. 4(a).

In the FQHE regime at 80mK, the A line intensity drops markedly concomitant with the emergence and strengthening of B emission (Fig. 2(b)). At $\upsilon=1/3$, the emergent B line has an intensity maximum together with an intensity minimum in the A line emission (Figs. 2(b) and 3(b)). At 4.2K, the collapse of the A intensity and the increase in B intensity are very much weakened and both the $\upsilon=1/3$ A line minimum and the B line maximum are quenched (Fig. 3(b)). This can be seen in the raw PL data at $\upsilon=1/3$ shown in Fig. 4(b); the higher energy B line is quenched relative to the A line with increased temperature.

3. Discussion

The magneto-PL of p-type SQW T128 fits in well with the overall picture that has emerged from spectroscopic studies of 2D electron to free valence hole (e-h) recombination in the 2DESs formed in both n-type SHJs and SQWs.[3,4] A recent study[2] of a 200Å p-type SQW shows some similar features to those reported here, but important differences exist between this study and our data (see below). The

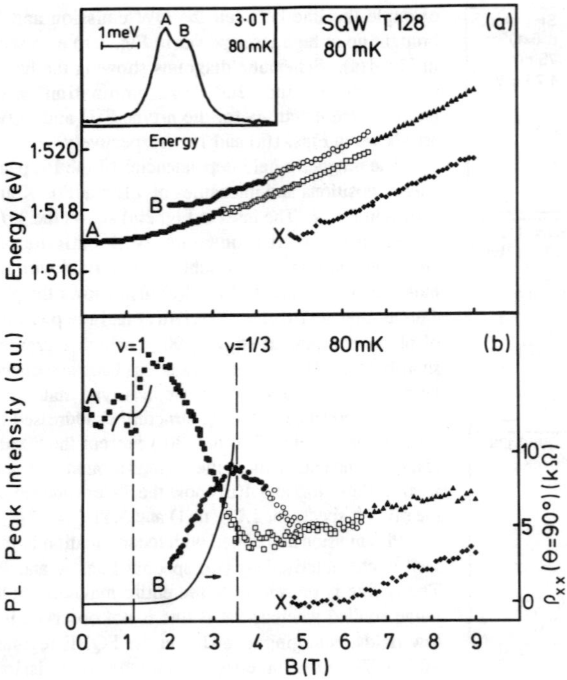

Fig. 2 (a) Field dependence of PL peak energy data for T128 at 80mK. (b) Corresponding PL peak intensity data at 80mK together with ρ_{xy} data around υ=1 (offset for clarity) and the ρ_{xx} out-of-phase component. Open symbols indicate A, B components were determined by overlay with 4.2K spectra where they were more clearly resolved. Beyond 7T the doublet is not resolved and so the mid-point of the spectrum is plotted.

mechanisms by which the A and B lines probe the 2DHS ground states are briefly discussed here.

(i) **A line** We attribute this to recombination of 2D holes in the ground subband of the quantum well with photoexcited electrons. The intensity minima at υ=1 and υ=1/3 can be explained by correlation and screening effects of holes and photoexcited electrons[3] and such behaviour is now well-established as an optical signature of QHE ground states in n-type SHJs and SQWs. The quenching of the A line minima with increased temperature correlates with the destruction of the υ=1 and υ=1/3 QHE states at elevated temperatures as evidenced by transport measurements.

(ii) **B line** The emergence of this line at υ~1/2, ~0.7meV above the A emission has much in common with a recent study of a flat band, low density n-type SHJ in which an emergent line (at υ~1/2, ~0.5meV above the ground subband emission), reported as structure additional to the first excited subband emission, was attributed to recombination of 'interface excitons' bound (in z) at the heterointerface and not directly involving electrons in the 2DES.[4] Such a mechanism is consistent with B emission in T128. In a SQW, excited subband recombination is expected to be enhanced over the ground subband emission with increased temperature and excitation power – in T128, however, the A/B intensity ratio increases with increasing temperature and power. The B line intensity maximum at υ=1/3 can be explained as an indirect result of reduction in A emission due to the FQHE ground state. Ref. 4 considered an exciton created in the bulk GaAs buffer layer diffusing quickly to the heterointerface to form an interface exciton; whilst the details may differ, recombination of single-electron excitons is also possible in our wide SQW. The appearance of PL lines to higher energy than the main PL emission is also a common feature of n-type SQWs.[5] The emergence of a *lower* energy line at υ=1/3 reported in Ref. 2 attributed to the formation of a MIWS phase is not observed in our sample nor in intrinsic recombination in n-type SQWs.[5] The Ref. 2 data are reminiscent of the PL signature associated with electron to neutral acceptor recombination (e-A^0) in n-type SHJs incorporating a deliberately introduced layer of beryllium impurities ~200Å below the heterointerface (see Ref. 3 for a review).

(iii) **MIWS regime** The emergence of the B line and collapse of the A line intensity at mK temperatures correlates with the onset at υ~1/2 of the ρ_{xx} dephasing which increases asymptotically at υ=1/3 (Fig. 2(b)). Dephasing of ρ_{xx} has been related to the formation of a MIWS phase which is anticipated beyond υ=1/3 in a 2DHS.[6] For υ<1/3 our PL spectrum remains relatively unchanged with field and the A, B line intensities

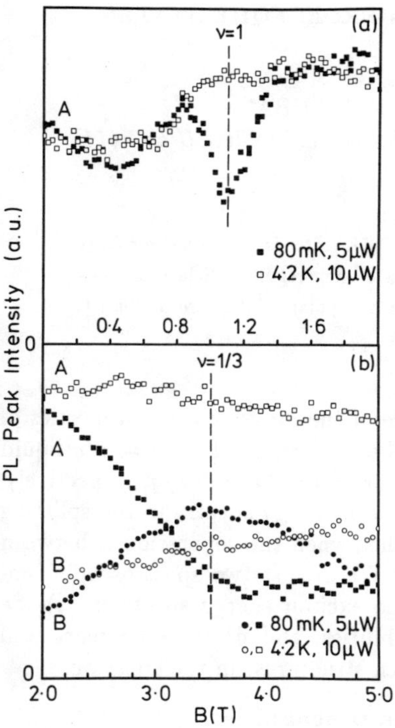

Fig. 3 Comparison of the 80mK (closed symbols) and 4.2K (open symbols) A, B PL peak intensities as a function of field around (a) υ=1 and (b) υ=1/3.

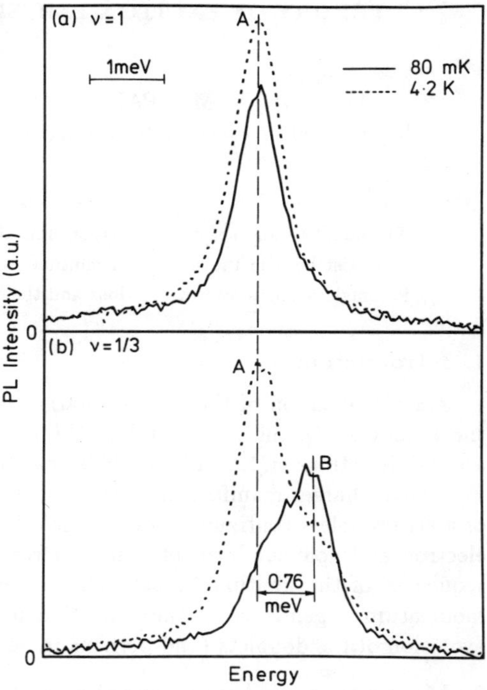

Fig. 4 Comparison of the 80mK and 4.2K raw PL spectra at (a) υ=1 and (b) υ=1/3.

become equal. Results reported in Ref. 4 for an n-type SHJ, showed that the emergent line attributed to interface exciton recombination gave way to two new lines in the MIWS regime (υ<1/5 for a 2DES). The lines were associated with interface excitons which either became localised (in x-y) by the formation of an electron lattice at the heterointerface or which remained unlocalised; the field and temperature dependencies of this PL structure were used to map a MIWS phase diagram. This behaviour is not observed in our p-type SQW which could be related to the spatial offset of the single-electron exciton from the 2D plane. Results for n-type SQWs also showed no clear emergence of prominent additional structure at the anticipated MIWS boundary.[5]

4. References

1. A G Davies *et al.*, Proc. of the 4th ISSP Int. Symp. *'Frontiers in High Magnetic Fields'*, Tokyo, November 1993; *Physica B* (in press).
2. L V Butov *et al.*, *Phys. Rev.* **B49** (1994) 14054.
3. R G Clark in *Springer Series in Solid State Sciences* **111**, 239 (Springer-Verlag, Berlin, 1992).
4. E M Goldys *et al.*, *Phys. Rev.* **B46** (1992) 7957; S A Brown *et al.*, *Surf. Sci.* **305** (1994) 42.
5. D Heiman *et al.*, *Surf. Sci.* **305** (1994) 50.
6. M B Santos *et al.*, *Phys. Rev. Lett.* **68** (1992) 1188; P J Rodgers *et al.*, *Physica B* **184** (1993) 95.

THEORY OF EXCITON EMISSION FROM FQHE STATES

V. M. APALKOV* and E. I. RASHBA*
Department of Physics, University of Utah, Salt Lake City, Utah 84112

ABSTRACT

The doublet structure of the emission from the FQHE states is shown to be robust and exist both in the single and multiple exciton branch regimes. Selection rules for single-magnetoroton transitions and the effect of z-polarizability are discussed.

1. Introduction

Fractionalization of the electric charge[1] is one of the most exciting properties of the Fractional Quantum Hall Effect (FQHE). Charge carriers of an $\nu = p/q$ liquid are quasielectrons (QE) and quasiholes having charges $(-e/q)$ and e/q, respectively. Fractional charges manifest themselves in the exciton energy spectra in the splitting of a structureless continuum into multiple branches when the separation h between electron and hole confinement planes increases.[2] We discuss two spectroscopic consequences of the multiple-branch structure of the exciton energy spectrum: i) the momentum dependence of the transition probabilities, and ii) the emergence and evolution of the doublets (and perhaps more reach structures) in the emission.

2. Momentum dependence of the transition probabilities

In Fig. 1 the lower part of the energy spectrum of a system of 7 electrons and a hole found in a spherical geometry is shown. It describes an exciton against a background of an incompressible $\nu = 1/3$–liquid of 6 electrons. The separation between electron and hole confinement planes equals $h = 1.5l$; l is the magnetic length. The branches are assigned quantum numbers L_m, for each branch m equals the minimum angular momentum the branch starts with. For $h = 0$ the spectrum comprises the L_0 branch and a continuum. Other branches appear gradually when h increases; see Ref. 2.

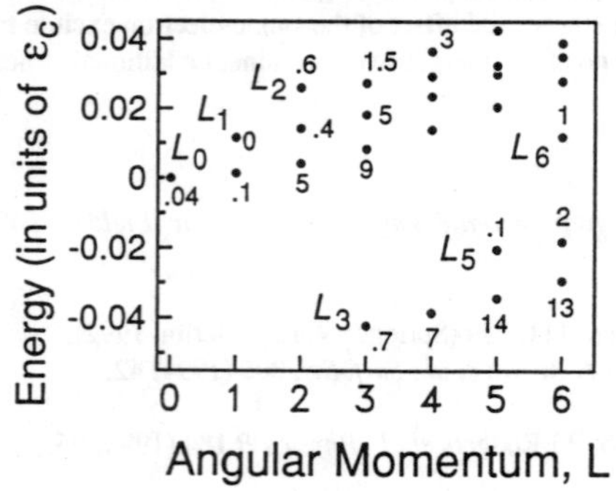

Fig. 1. Energy spectrum of a $N_e = 7, N_h = 1$ system, the flux $2S = 15$, $h/l = 1.5$. Numbers show intensities of single-MR transitions in percents to the transition probability from multiplicative states, for $h = 0$). L_m numerate exciton branches.

The branches L_m with $m \geq 3$ can be described in terms of bound states of a valence hole and three QEs. The prescription for mapping the spectrum of a composite particle like an exciton from the spherical into the plane geometry is as follows. A projection of the angular momentum on the normal to the confinement plane, M, equals $M = -L_m < 0$, and a quasimomentum $k = (L - L_m)/R$, where R is the sphere radius.

Optical transitions to the ground (Laughlin) state are the only allowed from the $L = 0$ state of the L_0 branch due to the L (or k) conservation law in the spherical (or plane) geometry. Selection rules for transitions into excited states, to be specific, single magnetoroton (MR) states, are less obvious. Since the starting point of any L_m branch is identified as the $k = 0$ point, single-MR transitions are allowed only if $k = 0$ MRs possess the same M. MRs are composite particles, hence, we expect them to obey the same mapping prescription as excitons. MRs with $L \leq 1$ are absent in the charge density model,[3,4] which implies that MRs should be classified as L_2 excitations. However, $L = 1$ states were found in the quasiexciton model;[5] therefore, L_1 assignment can not be ruled out. If the transition is forbidden at $k = 0$, its intensity should be small at the starting point of the branch (it is an artifact of the spherical model) and should show a fast increase with L. For allowed transitions only a slow L dependence is expected. In Fig. 1 single-MR transition probabilities are shown for all points having an unambiguous assignment. The data indicate the branch L_2 as a candidate for allowed single-MR transitions (the transition is weak, but the intensity shows a slow L dependence). Therefore, they imply that MRs are L_2 excitations (L_1 assignment was proposed recently from different arguments[6]). A strong and regular L dependence of the intensity along the L_0, L_1, L_3, and L_5 branches is an additional argument for the generalized mapping prescription.[2]

3. Doublets in emission spectra

The shape of the emission spectra strongly depends on h/l. Owing to a hidden

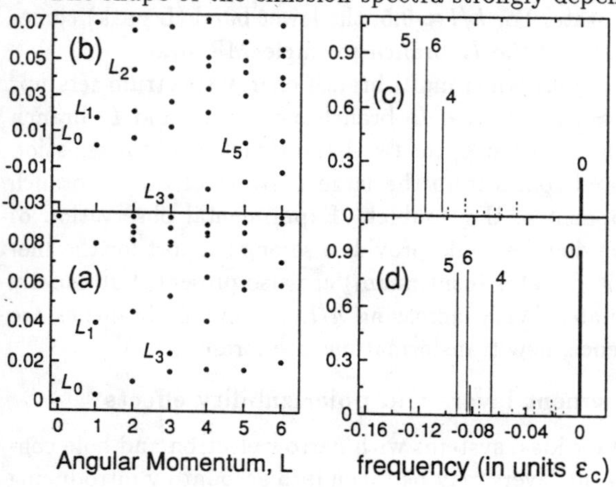

Fig. 2. Energy (left, in units of $\varepsilon_C = e^2/\epsilon l$) and emission spectra (right, arbitrary units) for two values of h/l: 0.8 – (a) and (d), and 1.2 – (b) and (c). Solid lines – single–MR transitions, dotted lines – transitions to the continuum. Numbers show L values. The temperature $T = 0.025\varepsilon_C$.

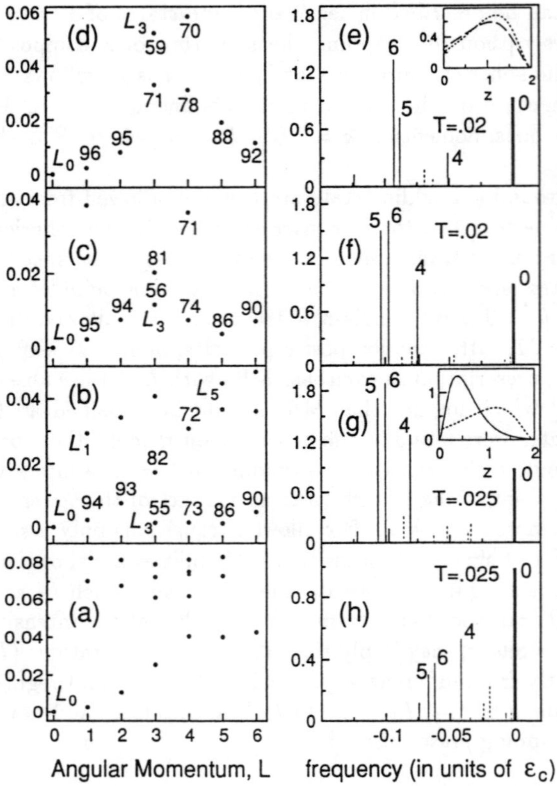

Fig. 3. Energy (left, in units of ε_C) and emission spectra (right, arbitrary units) for finite confinement layer widths. The ratio h/l is constant in each row: 1.5 for (a), 2.0 for (b)-(d); $(b_e l)^{-1} = 0.2$ and $(b_h l)^{-1} = 0.4$. Inset to (g): electron (solid line) and hole (dashed line) z-distributions for $\alpha = 0$. Polarizability effect is shown for $\alpha = 0.02$ in (c) and (f), and $\alpha = 0.1$ in (d) and (e). Inset to (e): change in the hole z-distribution for $L = 0, \alpha = 0.1$. T in units of ε_C.

symmetry, optical transitions occure only from multiplicative states if $h = 0$. In the lower part of the spectrum only the $L = 0$ state of the L_0 branch belongs to this group. For larger values of h/l the oscillator strength is partly transferred to the lower part of the energy spectrum. As a result, a two-band structure of the emission spectrum forms. The upper band (U peak) comes from the direct transition from the $L = 0$ exciton state (L_0 branch) to the ground state. For $h/l \approx 0.5$, the lower band (L peak) comes from transitions from large L states of the L_0 branch to single–MR states.[7]

Formation of doublets for $h/l \geq 0.7$ when multi-branch energy spectrum sets up[2] is illustrated by Fig. 2. For $h/l \approx 1$ the L_0 and L_3 branches compete, and L_3 branch is becoming the spectrum bottom. As a result, the two-band structure of the emission spectrum survives, but the L band comes from the large L states of the L_3 branch and turns into the low-T component of the doublet. Experimental observation of this transformation of the exciton doublet could provide a strong support for the theory. Transitions to different excited states (continuum) are also present, but usually the single-MR transitions dominate. With increasing h/l L_3 and L_5 branches are expected to interchange, etc., hence, new transformations are foreseen.

4. Finite widths of confinement layers and polarizability effects

The above results were obtained for ideal systems with narrow electron and hole confinement layers. Finite widths of the layers may be taken into account by introducing effective electron-electron[8] and electron-hole interactions. We used z-functions of the Fang-Howard type both for electrons and holes: $\zeta_e(z) = (b_e^3/2)^{1/2}\, z \exp(-b_e z/2)$, $\zeta_h(z) = (b_h^3/2)^{1/2}\,(h-z)\exp[-b_h(h-z)/2]$. The results are presented in Fig. 3. The energy and emission spectra strongly resemble the spectra of an ideal system but with one important difference. Similar spectra arise for different h values. h is larger for systems with a z-distributed charge density, for the parameter values we used, Fig. 3, more than twice. In such systems the maxima of the electron and hole density distributions draw together (inset to Fig. 3g), and the mean separation between them decreases.

In systems with z-distributed densities the electron-hole attraction modulates the mean particle z-separation. Since it depends on the quantum number L, one can anticipate some changes in the exciton dispersion law. We accepted that the force is proportional to the electron density on the hole, d, averaged over the quantum state, and that the parameter b_h shows a linear dependence on the density, $b_h(d,\alpha) = b_h + \alpha(d/d_F - 1/3)$, where $d_F = 1/2\pi l^2$ is the Fermi limit. Figs. 3b-d show that the effect exists and is unexpectedly strong. The spectrum changes drastically, though the density (inset to Fig. 3e) changes slightly. The states with large d draw down, Figs. 3c,d. The emission spectrum, Figs. 3e-g, is more stable than the dispersion law. In general, the doublet structure of the emission spectrum is robust, it exist in a wide range of the parameter values. Optical transition probabilities strongly depend on $d(L)$ and may be found sensitive to the angular momentum L of the initial state, i.e., be different for different emission bands. These anticipations are in agreement with the data and discussion of the Ref. 9.

5. References

* Also: L. D. Landau Institute for Theoretical Physics, Moscow 117940.
1. R. B. Laughlin, Phys. Rev. Lett. **50**, 13 (1983).
2. E. I. Rashba and V. M. Apalkov, these Proceedings.
3. G. Dav and J. K. Jain, Phys. Rev. Lett. **69**, 2843 (1992).
4. S. He, S. H. Simon, and B. I. Halperin, preprint, 1994.
5. J. Yang, Phys. Rev. B **49**, 5443 (1994).
6. D.-H. Lee and X.-G. Wen, Phys. Rev. B **49**, 11066 (1994).
7. V. M. Apalkov and E. I. Rashba, Phys. Rev. B **48**, 18312 (1993).
8. F. C. Zhang and S. Das Sarma, Phys. Rev. B **33**, 2903 (1986).
9. D. Heiman, A Pinczuk, M. Dahl *et al.*, Surf. Sci. **305**, 50 (1994).

6. Acknowledgments

We are grateful fo Prof. A. L. Efros for numerous stimulating discussions, and to Dr. F. G. Pikus for advice in developing computational programs. The support by NSF Grant No. DMR-9116748 and San Diego Supercomputer Center where the computations were performed is acknowledged.

EXCITON SPECTRA UNDER THE FQHE REGIME: FROM A CONTINUUM TO MULTIPLE BRANCHES

E. I. RASHBA* and V. M. APALKOV*
Department of Physics, University of Utah, Salt Lake City, Utah 84112

ABSTRACT

Exciton continuum splits into multiple exciton branches when the separation between electron and hole confinement planes increases. Symmetry classification of the branches is proposed, and the connection between the exciton spectrum and the properties of the quasielectron Hilbert space is established.

1. Introduction

Elementary excitations of Fractional Quantum Hall systems are quasielectrons and quasiholes. They have fractional charges which for a $\nu = p/q$ incompressible liquid are equal to $(-e/q)$ and e/q, respectively. These excitations obey a fractional statistics, i.e., they are anyons. From this end an exciton may be considered as a composite particle consisting of a valence hole and q quasielectrons (QE); it can be termed an anyon exciton. This approach must be valid when a separation h between electron and hole confinement planes is large, $h > l$, where l is the magnetic length; under these conditions QEs having the size l retain their individuality. In a strong magnetic field every charged particle possesses a single quantum number. Therefore, an exciton possesses $q+1$ quantum numbers. Two of them form a 2D quasimomentum $\mathbf{k}$ inherent in neutral particles, and one more may be chosen as a projection $M < 0$ of the angular momentum on the normal to the confinement plane. When $\mathbf{k}=0$, M is the internal angular momentum, other $q-2$ quantum numbers also are internal quantum numbers of an exciton. Such a view on the exciton problem opens a new approach to the low energy physics of the FQHE. A hole acts as a probe which feels through the Hilbert space of QEs. Quantum numbers of an anyon exciton and the dependence of its spectrum on external parameters reflect intimate properties of this space.

2. Multi-Branch Spectra

The interplain distance h is the main parameter governing the exciton spectra.[1] In Fig. 1 the energy spectrum of an exciton against a background of a six-electron incompressible liquid (a system of 7 electrons and a single hole, $\nu = 1/3$) is shown. It was found by exact diagonalization in a spherical geometry. The spectrum consists of a single exciton branch and a structureless continuum above it. They come very close together in the region right from the magnetoroton minimum. Only for $h/l \approx 0.5$ a noticeable gap is seen between them. For comparison, in Fig. 2 the energy spectrum

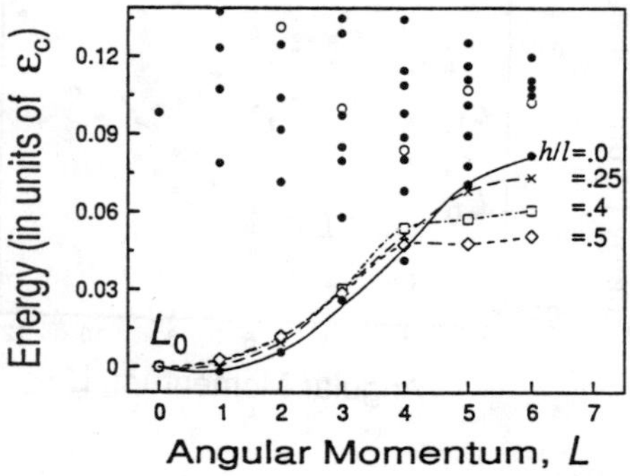

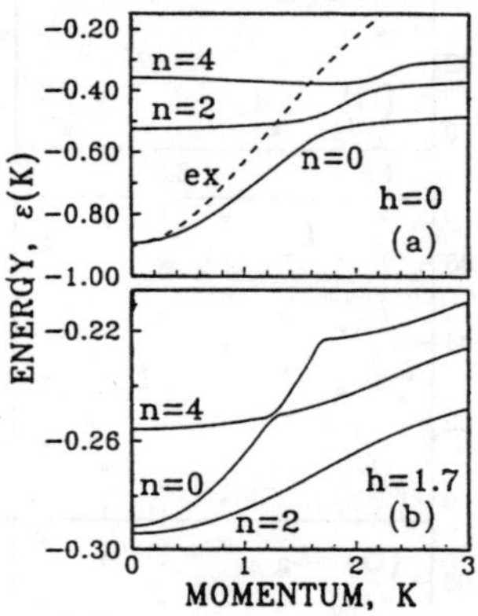

Fig. 1. Energy spectrum of a system of $N_e = 7, N_h = 1$, the flux $2S = 15$. Dispersion law, $\varepsilon(k)$, is shown for four values of h/l, the continuum for $h = 0$; $\varepsilon_C = e^2/\epsilon l$. Empty circles show both the multiplicative exciton states and the magnetoroton dispersion law.

Fig. 2. Energy spectrum of an anyon exciton in a two-anyon model, n is the internal angular momentum; h in units of l. The curve 'ex' shows the exciton dispersion law in an empty crystal. From Ref. 3.

of an anyon exciton found using Halperin's[2] anyon pseudo (envelope) wave functions is shown. This approach results in a multiplicity of branches for any value of h/l. For $h = 0$ the criterion of the applicability of the model is violated, but for $h/l = 1.7$ (Fig. 2b) it is hopefully satisfied. The question is whether the spectrum of Fig. 1 acquires a multi-branch structure when h increases, and how this transformation occurs. In Fig. 3 the spectrum of the same system is shown for larger values of h/l. Beginning from $h/l = 0.7$ the emergence of new branches is seen distinctly, they split off of the bottom of the continuum. Every branch L_m is characterized by the value of the angular momentum, $m = L$, it starts from. L_m do not depend on the system size, i.e., on the flux S, at a given ν. Therefore, L_m are quantum numbers which survive in the plane limit, $S \to \infty$. It is seen that the branch L_3 plays a special role, it is the first to appear, and it divides all the branches onto two groups. The branches with $m > 3$ drove down relative to L_0 with increasing h, while the branches L_1 and L_2 remain above L_0. L_3 also is the critical branch for a $\nu = 2/5$ liquid.

3. Quantum Numbers and Hilbert Space of Quasielectrons

One can show, e.g., using the Jain's intermapping projection, that for a $\nu = 1/3$ liquid the maximal value of the angular mmentum of three QEs equals $(L_{QE})_{\max} = 3N_e/2 - 6$. Angular momentum of a hole in the presence of three QEs is equal to $L_h = 3N_e/2 - 3$. Therefore,

$$L_h - (L_{QE})_{\max} = 3$$

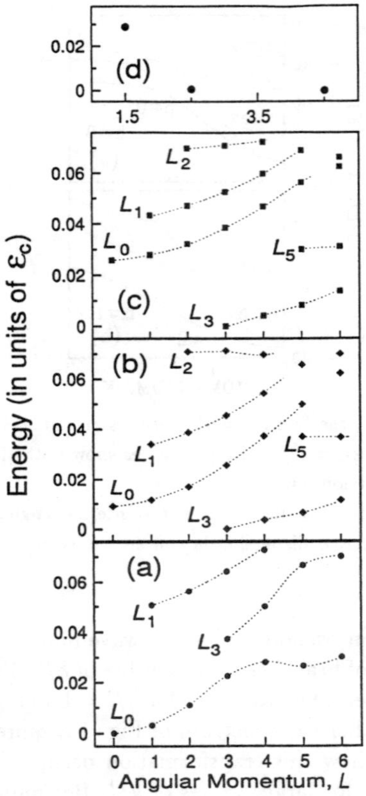

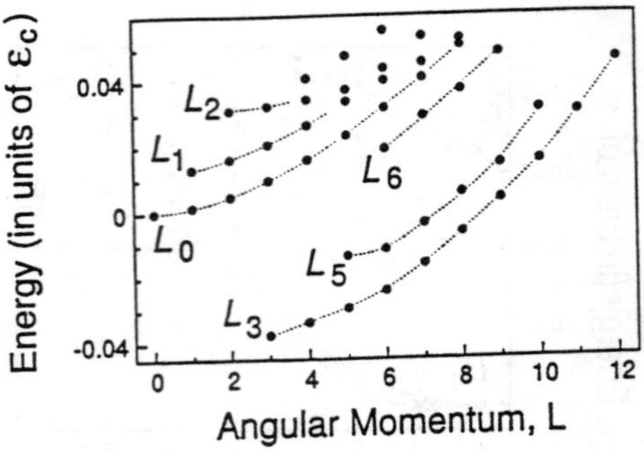

Fig. 3. Energy spectrum of the system shown in Fig. 1 for $h/l \geq 0.7$; $h/l = 0.7$ (a), 1.0 (b), and 1.2 (c). In (d) is shown the low-energy spectrum of a $N_e = 7$ system.

Fig. 4. Energy spectrum of the same system for $h/l = 1.4$ shown with a wide L-window. The number of states in the three lowest branches agrees with the $\{QE\}h$–coupling scheme.

for arbitrary N_e. The standard procedure of the addition of the angular momenta suggests that the minimum angular momentum which may possess an exciton built of a hole and three QEs equals $L = 3$. It is just the angular momentum of the L_3 branch. Therefore, only the branches L_m with $m \geq 3$ may be described in the quasiparticle representation. Decomposition of the Hilbert space of QEs onto subspaces with different momenta is $3/2 \oplus 5/2 \oplus 9/2$, as it is seen from Fig. 3d. Minimal momenta resulting from the addition of these three momenta with $L_h = 15/2$ are L_m=6, 5, and 3, respectively. L_3 appears first in Fig. 3a, L_5 in Fig. 3b, appearance of L_6 can be observed in Fig. 3c. One can see that first emerge the excitons which originate from subspaces having lower energies, and between the excitons originating of subspaces with comparable energies first come those which have lower values of m, i.e., a lesser exciton size. These regularities are supported by data for a $\nu = 1/3$ system of $N_e = 8, N_h = 1$, and a $\nu = 2/5$ system of $N_e = 9, N_h = 1$. The above results establish the momentum coupling scheme for low-energy anyon excitons: angular momenta of QEs form the quasielectron angular momentum L_{QE} which couples to the hole momentum L_h ($\{QE\}h$–coupling). Therefore, exciton spectrum reflect the basic properties of the QE space: its Haldane dimension[4] and the momentum decomposition.

4. Sphere-onto-Plane Mapping for Composite Particles

In Fig. 4 the exciton spectrum is shown in a wider area of L values. The accuracy of the data decreases with L, but a wide L "window" is illuminating for the systematics of states purposes. The number of states in the L_3, L_5, and L_6 branches is exactly as the momentum addition rule predicts. Every branch is a prototype of the corresponding branch in the plane geometry. The only important difference which exists between Figs. 2 and 4 is that all the branches start at $k = 0$ in Fig. 2 and at L_m in Fig. 4. To overcome it and bring both figures in a full correspondence it is necessary to generalize the Haldane prescription for the mapping between the spherical and plane geometries, $L = kR$ ($R = \sqrt{S}l$ is the sphere radius), for composite particles. The starting point of each branch on a sphere should be identified with the $k = 0$ point in the plane, and the quantum number L_m, which survives in the $S \to \infty$ limit, is related to the projection M of the internal angular momentum, $M = -L_m < 0$. M is the quantum number designating branches in Fig. 2. A relation between L and the quasimomentum k is established by the relation $L - L_m = kR$.

Implications which the generalized mapping prescription has for the selection rules for optical transitions are discused in Ref. 5.

5. Acknowledgements

We are grateful to Profs. A. L. Efros and Y.-S. Wu for suggestive discussions, and to Dr. F. G. Pikus whose expert advice maintained a considerable advancement of our computational programs. The support by NSF Grant No. DMR-9116748 and San Diego Supercomputer Center where the computations were performed is acknowledged.

6. References

* Also: L. D. Landau Institute for Theoretical Physics, Moscow 117940.

1. V. M. Apalkov and E. I. Rashba, JETP Lett. **55**, 37 (1992).
2. B. I. Halperin, Phys. Rev. Lett. **52**, 1583 (1984).
3. E. I. Rashba and M. E. Portnoi, Phys. Rev. Lett. **70**, 3315 (1993).
4. F. G. M. Haldane, Phys. Rev. Lett. **67**, 937 (1991).
5. V. M. Apalkov and E. I. Rashba, these Proceedings.

CHARGE AND SPIN EXCITATION GAP IN TILTED MAGNETIC FIELD

L.B. LIAO and D. HEIMAN
Francis Bitter National Magnet Laboratory, MIT
Cambridge, MA 02139, USA

A. PINCZUK, L. N. PFEIFFER and K. WEST
AT&T Bell Laboratories, Murray Hill, NJ 07974, USA

and

A.H. MACDONALD
Physics Department, Indiana University
Bloomington, IN 47405, USA

ABSTRACT

An inelastic light scattering study of charge- and spin-density excitations in single quantum wells was conducted in tilted magnetic fields. When the magnetic field is not perpendicular to the plane of the electron gas, the in-plane component of the magnetic field couples the inter-Landau level and intersubband transitions resulting in two magnetic field dependent transitions. We have observed two peaks near the lower hybrid energy which persists outside of the localization region near $\nu = 2$. We interpret these as the peaks associated with momentum conserving charge- and spin-density excitations. The energy gap between the charge- and spin-density excitations depends on both the strength of the depolarization shift and the tilt angle. We were able to model the system numerically and obtained quantitative agreement between the experiment and time-dependent Hartree-Fock calculation.

The observation of magnetoroton density of states in two-dimensional (2D) Landau level (LL) excitations was first reported by Pinczuk *et al.* [1] The peaks in the inelastic light scattering spectra result from critical points in the mode dispersions. When the magnetic field is perpendicular to the 2D electron gas, the density of states can be observed only near integral filling factors where the screening of the charged impurities by electrons are ineffective due to the absence of extended states near the Fermi energy.[1] The presence of an in-plane component of magnetic field will couple the field-dependent LL's with field-independent subband levels. This results in two sets of field-dependent inter-level excitations: upper hybrid (UH) and lower hybrid (LH) modes. For our samples, the LH becomes the inter-Landau (IL) transition when the tilt angle is set to zero. It is possible to observe the $q \cong 0$ excitation of the LH mode using inelastic light scattering, while the IL transition is not observable in the dipole approximation. In this paper we report the observation of the in-plane momentum-conserving peaks of charge-density excitation (CDE) and spin-density excitation (SDE) near $\nu = 2$ for various tilt angles. We find that an energy gap develops between CDE and SDE when there is an in-plane magnetic field component. (For no in-plane field the depolarization shift for the IL transition is zero, and both CDE and SDE have energies equal to the cyclotron energy as the in-plane wave vector, q, vanishes.) The size of the energy gap depends on both the tilt angle and the depolarization shift. Since the depolarization shift depends critically on the wave function in the z direction, we find that the gap cannot be accurately calculated without taking into account the finite depth of the well and self-consistent electrostatic interactions between the electrons.

Two samples of GaAs/Al_xGa_{1-x}As single quantum well with $x \cong 0.3$ were studied in detail. The samples have electron densities of 2.5×10^{11} /cm^2 and 3.6×10^{11} /cm^2 and mobilities of 3.0×10^6 cm^2/Vs and 2.8×10^6 cm^2/Vs respectively. The width of the well for both samples is 250 Å. We have studied the samples at various tilt angles, $\phi = 0°, 30°, 45°, 60°$, near $\nu = 2$. The tilt angle is defined as the angle between the normal to the sample and the magnetic field.

Figure 1a shows the representative spectra for various tilt angles. The most striking new feature we see for $\phi \neq 0$ is the presence of a new, strong peak in the spectra whose spectral position is above the single-particle energy. In order to identify this new peak, we have studied how it varies with magnetic field, tilt angle and temperature. We find that as we vary the magnetic field, the peak persists outside of the localization region and its spectral position varies with the magnetic field in a fashion similar to the single-particle energy.[7] This strongly suggests that the peak is the momentum conserving CDE excitation. We note that the shift from the single-particle energy for the peaks are on the order of 1 to 2 meV. This rules out the possibility that these peaks are due to impurities located in the center of the quantum well since such impurities will have a transition energy of approximately 7 meV for our experimental condition.[2,3] Furthermore, the intensities are not dependent on temperature in the range from 1.6 K to 15 K,[7] ruling out the possibility that these peaks are due to barrier impurity state.[4,5]

Another interesting feature we find was that the peak associated with the critical point in the dispersion of SDE broadened as we increased the tilt angle. In fact, for the lower density sample at 60 degrees tilt, there are actually two peaks associated with the spin density excitation, suggesting that one peak is due to the critical point in the density of states while the other is the momentum-conserving peak. The polarization study of the sample shows that this peak is depolarized, consistent with spin-density excitation selection rule. As we move away from the filling factor of two, we observed that the SDE peak disappeared while the CDE peak persists. This is possibly due to the changing spin characteristic of the ground state as we sweep the magnetic field. In fact, when $\nu = 1$, the only allowed spin excitation is a spin-flip excitation which has very different energy (relative to the single-particle energy) than at $\nu = 2$.

In order to explain the experimental results of tilted field inelastic light scattering, we have carried out a time-dependent Hartree-Fock calculation using a finite square well confinement potential, with the Hartree part of the potential included self-consistently. We can separate the Hamiltonian into an unperturbed part which includes the single-particle Hamiltonian and the Hartree potential and a perturbative part which includes the partial Coulomb interaction with the Hartree potential subtracted out:

$$H_o = \sum_i \frac{(\vec{p_\perp} + \frac{e}{c}\vec{A_\perp})_i^2}{2m} + \frac{p_{z_i}^2}{2m} + V(z_i) + \frac{m}{2}(\frac{eB_\parallel}{mc})^2 z_i^2 + \frac{e}{mc}\vec{\pi_i}\cdot\vec{A_{\parallel i}} + V_{Hartree}(z_i), \quad (1)$$

$$V(z_i) = \begin{cases} 0 & |z_i| < \frac{L}{2} \\ V_o & \text{otherwise,} \end{cases}$$

$$H_{e-e} = \frac{1}{2}\sum_{i\neq j}\frac{e^2}{|\vec{r_i} - \vec{r_j}|} - \sum_i V_{Hartree}(z_i). \quad (2)$$

V_o, the discontinuity at the edges of the well, is 225 meV for our samples. We have defined $\vec{p_\perp} = (p_x, p_y, 0)$, $\vec{\pi} = \vec{p_\perp} + \frac{e}{c}\vec{A_\perp}$, $\vec{A_\perp} = \frac{B_\perp}{2}(-y, x, 0)$, $\vec{A_\parallel} = B_\parallel(0, -z, 0)$, and $V_{Hartree}(z) = -2\pi Ne^2 \int_{-L/2}^{L/2} \Psi_G^* \Psi_G |x - z| dx$. We have to approach the problem in two stages. We calculate the eigenfunctions of the unperturbed Hamiltonian numerically. Then, we calculate the effect of the perturbative part. We have approached the first part of the problem by a series expansion in LL's. We assume coefficients of LL's are functions of z: $\Psi_G = \sum_{n=0}^{n_{max}-1} g_n(z)\,|n\rangle$ and $\Psi_m = \sum_{n=0}^{n_{max}-1} f_n^m(z)\,|n\rangle$, where n represents the LL index and m is the excited-state index. We substitute the series expansions into Eq. (1) and equate the coefficients of the nth LL to achieve a set of n_{max} coupled second-order differential equation:

$$\begin{aligned} E_m f_n^m(\zeta) \quad = \quad & -\frac{\hbar\omega_\perp}{2}\frac{\partial^2 f_n^m(\zeta)}{\partial\zeta} + \left(\hbar\omega_\perp(n + \frac{1}{2}) + \frac{\hbar\omega_\parallel^2}{\omega_\perp}\zeta^2\right) f_n^m(\zeta) + \\ & \hbar\omega_\parallel\sqrt{\frac{1}{2}}\zeta(\sqrt{n} f_{n-1}^m(\zeta) + \sqrt{n+1} f_{n+1}^m(\zeta)) + \\ & \left(V(\zeta) + \frac{\nu e^2}{\epsilon\ell_o}\zeta + g_{n_{max}}(\zeta)\right) f_n^m(\zeta). \end{aligned} \quad (3)$$

In Eq. (3), $f_n^0(\zeta) = g_n(\zeta)$ and ζ is the normalized length $\frac{z}{\ell_o}$ where $\ell_o = \sqrt{\frac{\hbar c}{eB_\perp}}$. The structure of the equation set is clear: each equation is coupled only to the equations of index one more and one less. In Eq. (3) we have introduced one extra variable. This variable satisfies the Poisson equation and can be written as $g_{n_{max}}(z) = V_{Hartree}(z)$. Instead of an iterative n_{max} coupled second-order differential equations problem, we have transformed the problem into a non-iterative $n_{max} + 1$ problem and avoid the instability that is often associated with iterative procedures. At integral filling factors, we can confidently apply time-dependent self-consistent Hartree-Fock calculation outlined. Figure 1b compares the measurements of the spectral position of the $q \cong 0$ peaks of CDE and SDE with the expected spectral positions at $\nu = 2$ from theoretical calculation. The experimental points as a function of tilt angle is consistent with the calculated curves for the momentum-conserving CDE and SDE excitations.

We have observed the $q \cong 0$ peaks of CDE and SDE near $\nu = 2$ in tilted magnetic fields. As $q \to 0$, CDE and SDE are no longer degenerate when we have an in-plane

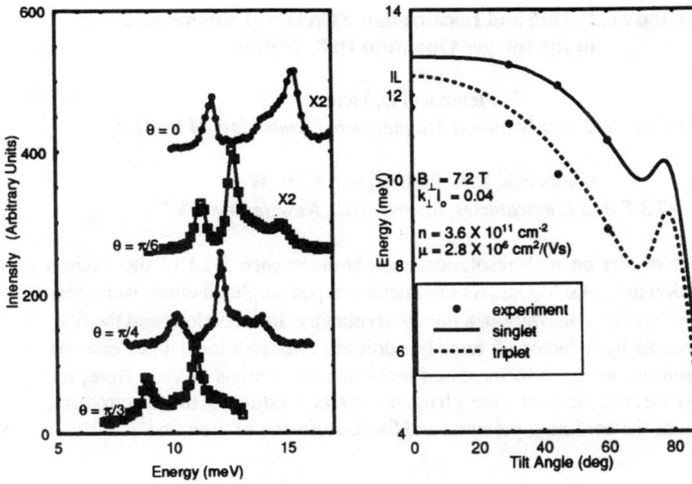

Fig. 1. Experimental data and theoretical curves for the higher density sample. The left panel (1a) shows the spectra from four tilt angles. The right panel (1b) shows the experimental points for CDE and SDE peaks as a function of the tilt angle and the theoretical curves.

component of magnetic field. The gap that develops between CDE and SDE is a function of both the tilt angle and the strength of depolarization shift. We were able to achieve quantitative agreement with theoretical calculation when we take into account the finite well depth and self-consistent electrostatic interactions between electrons.

This work was carried out with the support National Science Foundation Grant DMR-9201614. L.B.L. acknowledges the support of AT&T Bell Laboratories and NSF. The Francis Bitter National Magnet Lab is also supported by the NSF.

1. A. Pinczuk, J.P. Valladares, D. Heiman, A.C. Gossard, J.H. English, C.W. Tu, L. Pfeiffer, and K. West, *Phys. Rev. Lett.* **61** (1988) 2701.
2. N. C. Jarosik, B. D. McCombe, B. V. Shanabrook, J. Comas, John Ralston, and G. Wicks, *Phys. Rev. Lett.* **54** (1985) 1283.
3. Ronald L. Greene, K. K. Bajaj, *Phys. Rev. B* **31** (1985) 913.
4. J. M. Mercy, B. D. McCombe, W. Beard, J. Ralston, and G. Wicks, *Surface Sci.* **196** (1988) 334.
5. A. A. Reeder, J. M. Mercy, and B. D. McCombe, *IEEE Journal of Quantum Electronics* **24** (1988) 1690.
6. A. Pinczuk, S. Schmitt-Rink, G. Danan, J. P. Valladares, L. N. Pfeiffer and K. W. West, *Phys. Rev. Lett.* **63** (1989) 1633.
7. L.B. Liao *et. al. to be published.*

Exciton Scattering and Localization Effects on Luminescence in the Integer Quantum Hall Regime.

F. Plentz and D. Heiman
MIT Francis Bitter National Magnet Lab, Cambridge, MA 02139

A. Pinczuk, L.N. Pfeiffer, and K.W. West
AT&T Bell Laboratories, Murray Hill, New Jersey 07974.

We present results on high-resolution photoluminescence (PL) of the exciton recombination in GaAs/$Al_xGa_{1-x}As$ modulation-doped single quantum wells near $\nu=1$. At $\nu=1$ and σ^- polarization, a highly asymmetric line develops and the line-width increases by a factor of two. In addition, two additional lines emerge. These effects are assigned to increased scattering of excitons at $\nu=1$. Here, the unscreened electric field of the charged centers modulates the electrostatic component of the confining potential and Stark-shifts the exciton line.

Eight years after the discovery of the integer quantum Hall effect[1] (IQHE) and six after Tsui and Störmer observed the fractional quantum Hall effect[2] (FQHE) in transport experiments, the signatures of those effects on the exciton photoluminescence (PL) spectra of a two dimensional electron gas (2DEG) system where reveled.[3] After that, a number of works on the PL spectra from the 2DEG at IQHE and FQHE regimes appeared in the literature.[4,5,6] Inelastic light scattering [7,8] and time-resolved PL[9,10] also proved to be useful techniques for the study of related phenomena.

Although most of the recent experimental[10,11] and theoretical[12,13,14] work has been directed at understanding the phenomena underlying the incompressible quantum fluid at fractional filling factors and its interaction with the photoexcited e-h pairs, the exciton kinetics at $\nu=1$ still lacks a full understanding. The observed spectral anomalies, i.e, the reduction of the PL peak intensity and the plateaus in the recombination energies, have been assigned to the change in the screening response of the 2DEG as the Fermi level is moved across the localized-states region of the broadened Landau level.[3,4] It was also pointed out[4] that QHE localization should be an important component in the optical recombination.

In this paper we address the detailed structure of the PL spectra in the $\nu=1$ region. The present modulation-doped GaAs quantum wells are specially designed to have small line-widths ~ 0.2 meV. Making use of circular polarization analysis, we are now able to expose the rich spectra in the $\nu=1$ region. For σ^- polarization, we find:

a) dramatic line-width increase;
b) large asymmetry of the PL towards its lower energy side;
c) appearance of a well developed higher energy peak and a lower energy shoulder.

These effects are evidence of increased scattering and localization effects on the exciton kinetics at this filling factor.

The experiments were performed using resistive "Bitter" magnets and the sample was strain-free mounted in a two-optical-fiber apparatus[15] inside a ^3He cryostat. For the excitation we used a 780nm diode laser, and power density at the sample of ~1×10^{-4} Wcm^{-2} (3 μW total power) to avoid heating of the electronic system. A small circular polarization analyzer, centered at 800nm, was mounted in the rig before the signal recovery fiber. The PL signal was analyzed by a double 0.85 m spectrometer with 0.05 meV resolution and detected by a CCD camera.

In Fig. 1 we present PL spectra for three filling factors. Figures 1a and 1b show PL with σ^- and σ^+ polarizations, respectively. The use of circular polarization analysis is essential in resolving the fine structure of the excitonic line-shape. This is due to the fact that both polarizations are present with about the same intensity, and they do not have the same evolution as the field is increased.

At $\nu=1$ there are two remarkable changes in the σ^- PL spectra in addition to the previously observed anomalies in intensity and energy position. There is a dramatic *line-width increase*. The PL line also becomes *highly asymmetric towards its low-energy side*. It is worth noting that the σ^+ PL shows a much less pronounced increase in line-width. Furthermore, the σ^- and σ^+ spectra are broadened in opposite directions. The σ^- spectra broadens towards its *lower* energy side, while σ^+ broadens towards its *higher* energy side.

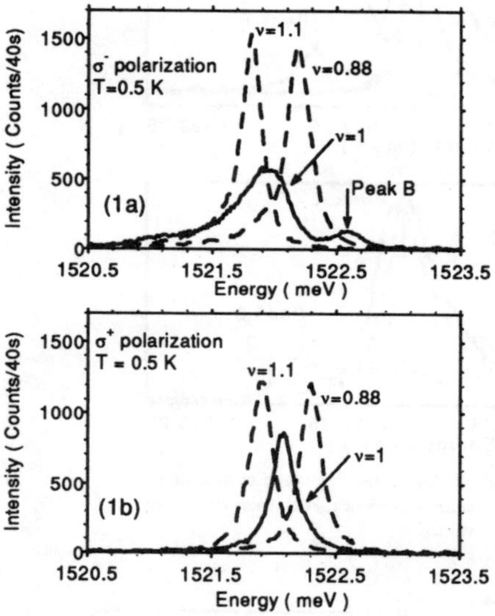

Fig. 1. PL spectra for σ^- (1a) and σ^+ (1b) polarizations across $\nu=1$. Note the huge line-width change for the σ^- PL at $\nu=1$. Note also that both polarizations show similar behavior regarding the PL peak intensity.

It is worth taking a closer look at the spectral line-shape changes in the $\nu=1$ region. This done in Fig. 2a and 2b. Figure 2a presents a comparison between the PL at $\nu=1.1$, $\nu=1.0$ and $\nu=0.88$. The energy positions and intensities of the main peak were normalized to the $\nu=1.1$ values. We can see how dramatic the changes are in the line-shape. There is not only an overall increase of the line-width, but also a large enhancement of the line asymmetry. On either side of the $\nu=1$ transition the PL line-shape is almost fully recovered, making clear that the observed effects are strictly associated to this filling factor. Figure 2b compares the spectra at $\nu=1$ and two different temperatures, T=0.5 K and T=1.2 K. It can be seen that the large asymmetry towards the low-energy side is present for both temperatures, but the line-width *decreases* as the temperature is *increased*.

In addition to those new anomalies, a high energy peak named B, and a lower energy shoulder develop in the σ^- PL. Peak B is located 0.64 meV *above* the main PL emission at $\nu=1$. Note also that peak B *decreases* in intensity as the temperature is *increased* and is almost absent at T=1.2 K. Therefore, this spectral feature does not originate from the thermal population of an upper energy state. It should be noted that the usual anomalies associated with $\nu=1$ region are still well developed at T=1.2 K. The new spectral features -- highly asymmetric line-shape, increased line-width, lower-energy shoulder, and a well-developed higher-energy peak -- are

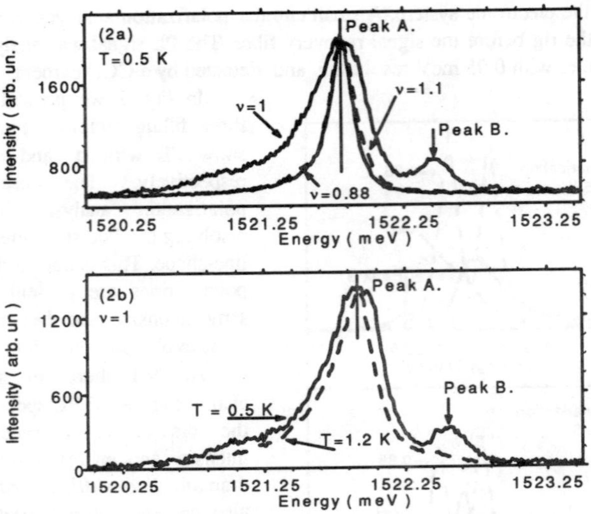

Fig. 2 PL spectra for GaAs quantum well: 2a shows the line-shape change as we move across ν=1 and 2b shows the temperature dependence. Peak energy positions and intensities were set equal to the ν=1.1 values.

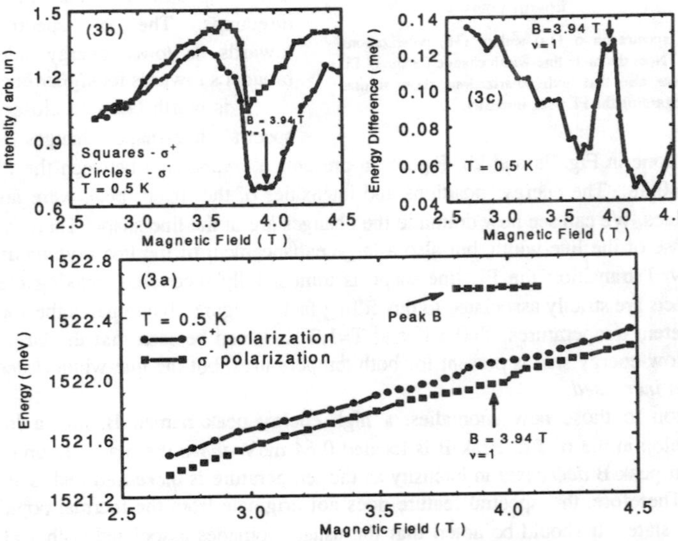

Fig. 3. Figure 3a shows PL peak energy versus magnetic field for σ⁻ and σ⁺ polarizations. Note that peak B energy position does not change with field. Figure 3b shows the peak intensity as a function of field. In Fig. 3c The energy separation between PL peaks in opposite polarizations is presented.

clearly present for temperatures below the one necessary to produce the IQHE at ν=1. This is especially true for peak B.

We present in Fig. 3a the peak energy versus magnetic field plot for both polarizations. In Fig. 3b and Fig. 3c peak intensity, and energy difference between σ^- and σ^+ main PL peak are shown as a function of field, respectively. By analyzing Fig. 3a we can observe one of the most interesting facets exposed by our measurements. The energy position of peak B is constant in the field interval across ν=1. On the other hand, the main PL emission shows the energy position anomalies associated with ν=1 as can be seen in Fig. 3a and Fig. 3c. Both polarizations have the same qualitative behavior concerning the main PL peak intensity.

In conclusion, when the filling factor reaches integer values the screening response of the 2DEG is largely reduced. Therefore, excitons are much more affected by the unscreened electric field of charged centers. This will be manifested in a small modulation of the electrostatic component of the confining potential and a Stark shift of the exciton line. The combination of these two components give rise to the observed line-shape changes. The potential fluctuations at ν=1 might also localize the photoexcited hole, giving rise to the appearance of peak B in σ^- polarization.

Acknowledgments: One of us (FP) would like to acknowledge financial support from the Brazilian funding agency CNPq. Work supported by NSF grant DMR-9201614. The Francis Bitter National Magnet lab is supported by NSF.

---

[1] K. von Klitzing, G. Dorda, and M. Pepper, Phys. Rev. Lett. 45, 494 (1980).
[2] D.C. Tsui, H.L. Störmer, and A.C. Gossard, Phys. Rev. Lett. 48, 1559 (1982).
[3] D. Heiman, B.B. Goldberg, A. Pinczuk, C.W. Tu, A.C. Gossard, and J.H English, Phys. Rev. Lett. 61, 605 (1988).
[4] B.B. Goldberg, D. Heiman, A. Pinczuk, L. Pfeiffer, and K. West, Phys. Rev. Lett. 65, 641 (1990).
[5] A.J. Turberfield, S.R. Haynes, P.A. Wright, R.A. Ford, R.G. Clark, J.F. Ryan, J.J. Harris, and C.T. Foxon, Phys. Rev. Lett. 65, 637 (1990).
[6] H.B. Buhmann, W. Joss, K. von Klitzing, I.V. Kukushkin, G. Martines, A.S. Plaut, K. Ploog, and V.B. Timofeev, Phys. Rev. Lett. 65, 1056 (1990).
[7] A. Pinczuk, J.P. Valadares, D. Heiman, A.C. Gossard, J.H. English, C.W. Tu, L.N. Pfeiffer, and K. W. West, Phys. Rev. Lett. 61, 2701 (1988).
[8] D. Heiman, A. Pinczuk, B.S. Dennis, L.N. Pfeiffer, and K.W. West, Phys. Rev. B45, 1492 (1992).
[9] M. Dahl, D. Heiman, A. Pinczuk, B.B. Goldberg, L.N. Pfeiffer, and K.W. West, Phys. Rev. B45, 6957 (1992).
[10] D. Heiman, A. Pinczuk, M. Dahl, B.S. Dennis, L.N. Pfeiffer, and K.W. West, Surf. Sci. 305, 50 (1994).
[11] I.V. Kukushkin, R.J. Haug, K. von Klitzing, and K. Ploog, Phys. Rev. Lett. 72, 736 (1994).
[12] A.H. MacDonald, E.H. Rezayi, and D. Keller, Phys. Rev. Lett. 68, 1939 (1992).
[13] V.M. Apalkov, and E.I. Rashba, Phys. Rev. B46, 1628 (1992).
[14] V.M. Apalkov, and E.I. Rashba, Phys. Rev. B48, 18312 (1993).
[15] D. Heiman, Ch. 1 in Semiconductors and Semimetals, Vol. 36, Eds. R.K. Willardson and A.C. Beer (Academic Press, 1992), p1.

STUDY OF Γ-X CROSS-OVER IN GaAs/AlAs SHORT PERIOD SUPERLATTICES BY SIMULTANEOUS APPLICATION OF HIGH PRESSURE AND PULSED HIGH MAGNETIC FIELDS

N. Miura, H. Kunimatsu, K. Uchida, S. Sasaki and T. Yagi

Institute for Solid State Physics, University of Tokyo
Roppongi, Minato-ku, Tokyo 106, Japan

The type I to type II transition due to the Γ-X cross-over was investigated in GaAs/AlAs short period superlattices under high pressure and pulsed high magnetic fields up to 40T. In $(GaAs)_{14}/(AlAs)_6$, which is of type I below 5.2kbar, the transition to type II appeared as a sudden change of the exciton diamagnetic shift in the photoluminescence spectra. At a pressure of 4.9kbar, a magnetic field-induced type I to type II transition was observed at around 20T. Near the critical pressure, the exciton state was found to possess a mixed character between type I and type II due to the Γ-X mixing.

1. Introduction

The type I to type II transition due to the cross-over of the Γ- and X-like conduction bands is one of the fundamental problems for GaAs/AlAs short period superlattices. It is generally accepted that the system is a type II superlattice when the number of monolayers of GaAs and AlAs layers is smaller than about 13 [1], because the X minima are located below the Γ minimum, whilst it is type I for a larger period. The band lineup of a type I superlattice can be converted to that of type II by applying high pressure because of the different pressure coefficient between the Γ and X minima [2]. Such a type I to type II transition can be caused also by applying high magnetic fields, because the effective mass is smaller in the Γ-like band than in the X-like band. In a previous work, we have observed a magnetic field-induced type I to type II transition in $(GaAs)_{15}/(AlAs)_{13}$ by applying high magnetic fields up to 150T [3]. The transition was observed at about 55T and 80T in two samples which have different Γ-X gaps due to strain. Since the change of the Γ-X energy gap by a unit magnetic field is not so large, we need very high magnetic fields to observe the magnetic field-induced type I to type II transition. Such a transition and the Γ-X cross-over can be studied in more detail, if we apply both high pressure and high magnetic fields simultaneously, because then we can control the initial Γ-X energy gap precisely by pressure before applying high fields. In this paper, we study the type I to type II transition in the presence of both pulsed high magnetic fields and high pressure in $(GaAs)_m/(AlAs)_n$ short period superlattices.

2. Experimental

We have built two different types of pressure cells for optical measurements under pulsed high magnetic fields. One is a diamond anvil cell made mainly from stainless steel, which allows an application of a hydrostatic pressure up to about 30kbar for a sample of about 0.4mm. In the measurements in pulsed high magnetic fields, one has to keep in mind that a considerably large eddy current is induced in the metal part of the pressure cell. It brings about a temperature rise and shielding of the magnetic fields. In fact, when we used the diamond anvil cell, the sample temperature was found to rise up to about 20K from 4.2K soon after the field pulse due to the eddy current in the clamping piston and the gasket, although we made the outer cylinder from low conductance stainless steel. We are now trying to change the material of the piston from metal to ceramics.

The other one is a piston cylinder type clamp cell made from Be-Cu, whose cross sectional view is shown in Fig. 1. In this case, the temperature rise of the sample is almost negligible during the pulse, because of the larger distance between the metal parts and the sample, and moreover, we can mount samples as large as 2mm in diameter, although the maximum pressure is limited to about 10kbar.

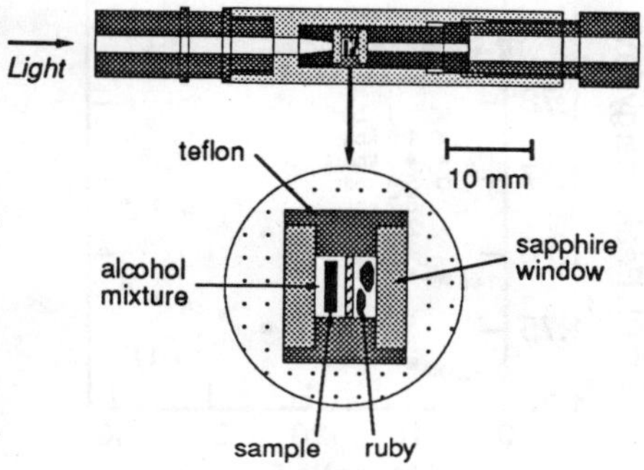

Fig. 1 High pressure clamp cell for the magneto-optical measurements in pulsed high magnetic fields.

Pulsed high magnetic fields up to 40T were generated by discharging a condenser bank of 300kJ (5kV). The duration of the field was 10ms. The samples used were a $(GaAs)_{14}/(AlAs)_6$ superlattice grown by MBE. We have measured the magneto-photoluminescence spectra at 4.2K with simultaneous application of both high pressure and pulsed magnetic fields, using an OMA system for recording the spectra and the 514.5 nm line of an Ar laser for excitation.

3. Results and discussion

$(GaAs)_{14}/(AlAs)_6$ is a type I superlattice and undergoes the transition to type II at p_c = 5.2kbar at zero field. At the ambient pressure, a sharp peak was observed at 1.743 meV ascribed to the exciton from the Γ-like state. Applying a magnetic field up to 40T, we observed a diamagnetic shift, that was a quadratic function of magnetic field at low magnetic fields but it gradually tended to a linear dependency at higher fields above 20T. The comparison of the spectra between 0T, about 20T and 40T is shown in Fig. 2 for the atmospheric pressure, and high pressures p = 4.9kbar and 7.6kbar. The exciton peak shifted to higher photon energies as the pressure was increased up to the transition pressure p_c. When a pressure higher than 5.7kbar was applied, the diamagnetic shift was completely that of the type II and much smaller than for type I. The photon energy of the exciton peak is shown as a function of magnetic field in Fig. 3(a). It is clearly seen that for p > 5.4kbar, the diamagnetic shift becomes much smaller than at lower pressures. Figure 3(b) shows the same photon energy vs. field plot in a magnified scale. At p = 5.7kbar, where the system is com-

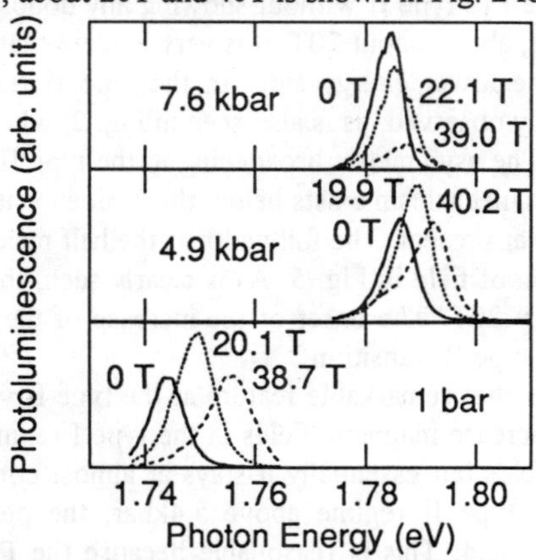

Fig. 2 Comparison of the exciton photoluminescence spectra in $(GaAs)_{14}(AlAs)_6$ between B=0 and high fields at various pressures.

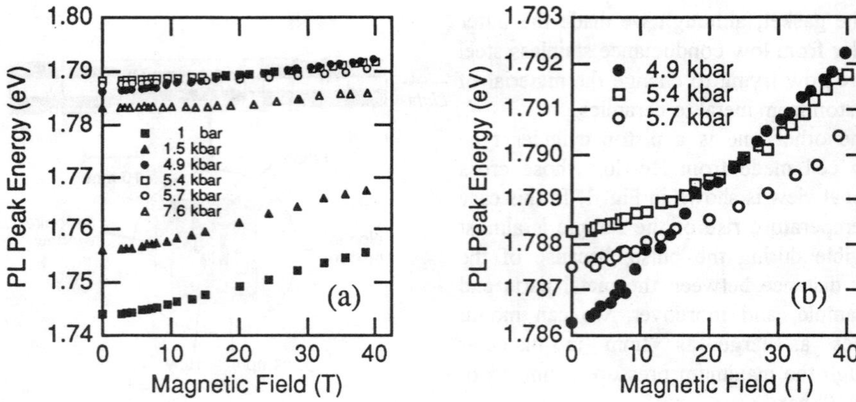

Fig. 3 (a) Photon energy of the exciton peak as a function of magnetic field at various pressures. (b) Same as (a), but with a magnified ordinate scale for high pressure data.

pletely in the type II regime, the diamagnetic shift is quite small, about 3meV at 40T, which is comparable with those in other type II superlattices with m, $n < 6$. For lower pressures, we can clearly see the similarity and the difference in the diamagnetic shift between 4.9kbar and 5.4kbar. Namely, the both graphs are almost parallel in the high field range 30-40T, whereas the shift in the B-quadratic part below 7T is much larger at $p = 4.9$kbar than at $p = 5.4$kbar. The low field diamagnetic shift at $p = 4.9$kbar is nearly the same as those for type I at $p = 1$bar and $p = 1.5$kbar, about 1meV at 7T. This implies that at $p = 4.9$kbar, the system is still type I at low fields. Therefore, we can conclude that at this pressure, the type I to type II transition occurs at some magnetic field between 7 and 40T. We can identify that transition point as a kink at around 20T.

The magneto-luminescence spectra of excitons at $p = 4.9$kbar is shown in Fig. 4 for various magnetic fields. It should be noted that the spectra exhibit a smooth transition from type I to type II without showing any double peaks near the transition field at 20T. However, above about 20T, it is very evident that the peak is broadened and have a long tail to the lower energy side. In the type II regime, the asymmetric broadening was commonly observed, as is also seen in Fig. 2, whereas the peak is symmetric in the type I regime. The asymmetric broadening in the type II regime suggests that another emission transition mechanism exists below the X-like exciton. The origin of the extra transition is not clear at present. The full width at the half maximum of the exciton peaks is plotted as a function of field in Fig. 5. As is clearly seen, the width starts increasing abruptly from about $B = 20$T. The onset of the increase of the half width is another indication of the type I to type II transition.

Another remarkable feature at the type I-type II transition is in the peak intensity. As we increase magnetic fields in the type I regime, the peak intensity increases slightly at low fields but essentially it stays at almost constant value. On the other hand, in the complete type II regime above 5.4kbar, the peak intensity decreases as we increase magnetic field. This is reasonable because the Γ-X mixture decreases as the Γ-X gap increases with increasing magnetic field. Near the transition pressure, the field dependence of the peak intensity appears to be of intermediate character between type I and type II.

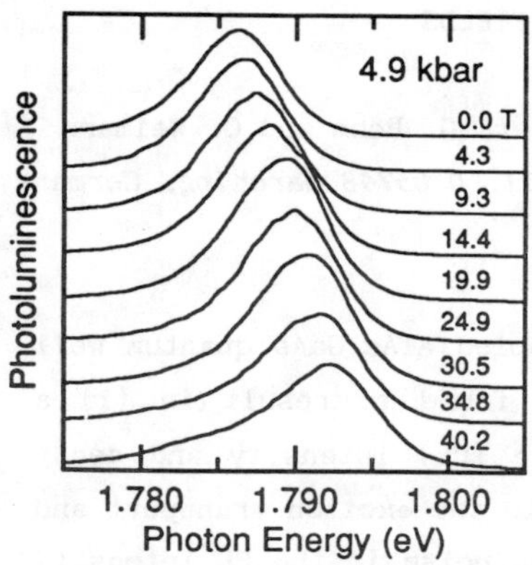

Fig. 4 Photoluminescence spectra at p = 4.9kbar for various magnetic fields.

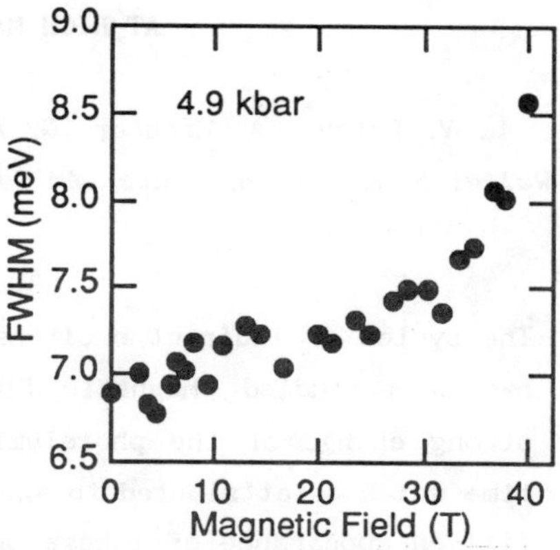

Fig. 5 Full width at the half maximum of the exciton photo-luminescence as a function of magnetic field at 4.9kbar.

The observed type I to type II transition is very gradual and not so distinct. The exciton peak smoothly changes from the Γ-like to X-like. The diamagnetic shift at p = 4.9kbar shows an intermediate magnitude between for p = 1.5kbar and for p = 5.7kbar. This indicates that the effective mass and the exciton binding energy change continuously from Γ-like to X-like. This is explained by the Γ-X mixture in the vicinity of the cross-over point.

The Γ-X cross-over of the Landau levels occurs at magnetic field B represented by the following equation,

$$\Delta E = \frac{\hbar eB}{2}\left(\frac{1}{m^{\Gamma}} - \frac{1}{m^{x}}\right) , \tag{1}$$

where ΔE is the Γ-X energy gap at zero field, and m^{Γ} and m^{X} is the effective mass at the Γ- and X-like band minima, respectively. Putting the values $m^{\Gamma} = 0.07m_0$ and $m^{X} = 0.20m_0$ [6], we obtain ΔE=10.7meV at the experimentally observed transition field B =20T. ΔE at p = 4.9kbar is about 3meV and certainly smaller than this value. Therefore, the Landau level crossing occurs at a field lower than 20T. In practice, however, the observed type I to type II transition of excitons would occur at the crossing point of exciton levels of the Γ- and X-like states, that is at higher field than that for the Landau levels [3]. More detailed calculation and experiments are required to obtain quantitative agreement.

References

1. R. Clingolani, M. Holtz, R. Muralidharan, K. Ploog. K, Reimann and K. Syassen: Surf. Sci. **228** (1990) 217.
2. G. Li, D. Jiang, H. Han Z. Wang and K. Ploog: J. Lumines. **46** (1990) 261.
3. S. Sasaki, N. Miura, T. Yagi and Y. Horikoshi: J. Phys. Soc. Jpn. **62** (1993) 2490.
4. S. Sasaki, N. Miura and Y. Horikoshi: Appl. Phys. Lett. **59** (1991) 96.
5. I. L. Spain, M. S. Skolnick, G. W. Smith, M. K. Saker and C. R. Whitehouse: Phys. Rev. B **43** (1991) 14091.
6. N. Miura, H. Yokoi, J. Kono and S. Sasaki: Solid St. Commun. **79** (1991) 1039.

EVIDENCE FOR THE EXCITON CONDENSATION IN COUPLED QUANTUM WELLS AT HIGH MAGNETIC FIELDS

L. V. Butov*, A. Zrenner, G. Abstreiter, G. Böhm and G. Weimann
Walter Schottky Institute, Am Coulombwall, D 85748 Garching, Germany

The system of indirect excitons in coupled AlAs/GaAs quantum well has been studied. Magnetic field was found to result in (i) a strong change of the photoluminescence (PL) intensity and decay time which is attributed to anomalies in the exciton transport and (ii) an appearance of a huge broad band noise in the PL intensity which is evidence for exciton condensation.

The electron-hole (e-h) interaction in the neutral e-h system has been predicted to cause the condensation of excitons into an excitonic insulator state which is analogous to the Bose-Einstein condensation in the low density limit[1]. The condensation condition can be achieved if the temperature of the excitons is below a critical temperature. The low temperature neutral e-h system can be realized in coupled quantum wells (CQWs) where the recombination rate of excitons is strongly suppressed due to the spatial separation of electrons and holes[2]. Furthermore perpendicular magnetic field increases the exciton binding energy which was predicted to improve the critical conditions for exciton condensation[3].

We have performed cw and time resolved magnetooptic experiments ($T>350$ mK, $B<14$ T) on tunable Γ-X AlAs/GaAs CQW structures in which the excitons are indirect both in real- and k-space[4]. The lifetime in this system is long enough to allow for a thermalization of the indirect excitons down to temperatures below 1 K. The active part of the structure consists of a 40 Å AlAs and 30 Å GaAs double layer, surrounded by two $Al_{0.48}Ga_{0.52}As$ barrier layers. Carriers were photoexcited by either a cw dye-laser ($\hbar\omega=1.85$ eV) or a pulsed semiconductor laser ($\hbar\omega=1.8$ eV, $\tau=4$ ns). The overlap between electrons and holes in the structure can be controlled by the external gate voltage (V_g). In the indirect regime

(V_g<0.4 V) electrons are in the AlAs QW and holes in the GaAs QW. Increasing exciton density results in a monotonic increase of the exciton energy[4] which prevents the condensation into a metallic e-h droplets[5].

In the indirect regime the radiative lifetime (τ_r) is much larger than the nonradiative lifetime (τ_{nr}) and the total recombination lifetime (τ) is determined by τ_{nr} ($\tau \approx \tau_{nr}$). The application of a magnetic field was found to result in a strong change of both I_{PL} and τ (Fig.1). With increasing $B_\perp$ both I_{PL} and τ first increase and then decrease dramatically. PL spectra at different $B_\perp$ are shown in the left insert. The decay curves are shown on the right insert. For $B_\perp$<7 T the decay is nearly monoexponential while for higher fields (at which both I_{PL} and τ drops down with $B_\perp$) it can be divided on the initial fast decay and subsequent slow decay. $B_\parallel$ results in a monotonic increase of I_{PL} and τ with nearly monoexponential decay.

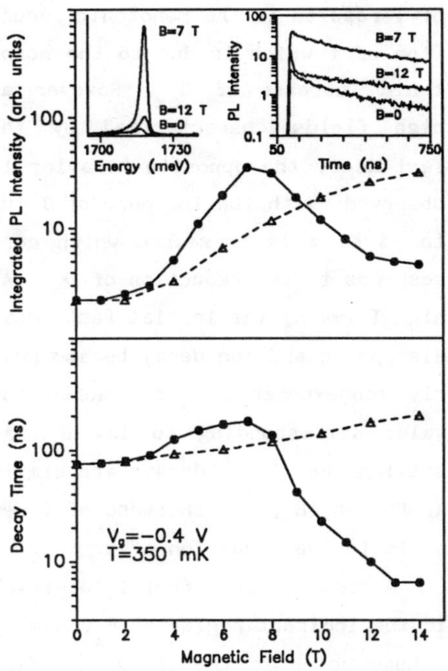

Fig.1. PL intensity and decay time at V_g=-0.4 V and T=350 mK versus $B_\perp$ (points) and $B_\parallel$ (triangles). Inserts: corresponding time integrated PL spectra (left) and decay curves (right).

The correlated variations of I_{PL} and τ means that τ_{nr} is subjected to a strong change by magnetic field. The nonradiative recombination occurs at nonradiative recombination centers (NC), the rate of recombination at NC in narrow QWs is determined by the transport of excitons to the NC[6]. We conclude therefore that magnetic field has a strong influence on the transport properties of excitons. An increase of $B_\perp$ first leads to a reduction of the exciton diffusivity (D_{ex}) and then to a strong increase. Increasing $B_\parallel$ leads to a monotonic reduction of D_{ex}.

The observed reduction of D_{ex} in magnetic field can be qualitatively explained by the coupling between the magnetoexciton motion and its internal structure resulting in the increased exciton scattering by interface roughness[4]. However this mechanism is not able to explain the dramatic increase of D_{ex} at high $B_{\perp}$. This striking anomaly in the magnetoexciton transport seems to be connected with the coherent effects in the exciton system[4]. This is confirmed by the temperature dependence of the decay process. At $B_{\perp}<7$ T the increase of T results in the monotonic reduction of τ which is due to the monotonic increase of D_{ex}. However at high fields characterized by the fast decay the opposite behavior is observed: with the increase of T (up to ~4 K) τ is increased which corresponds to the reduction of D_{ex}. At high T (~4 K) the initial fast decay disappears and the decay became nearly monoexponential, τ reaches the value corresponding to low $B_{\perp}$. At still higher T the decays are similar for any $B_{\perp}$, the increase of T results in the reduction of τ.

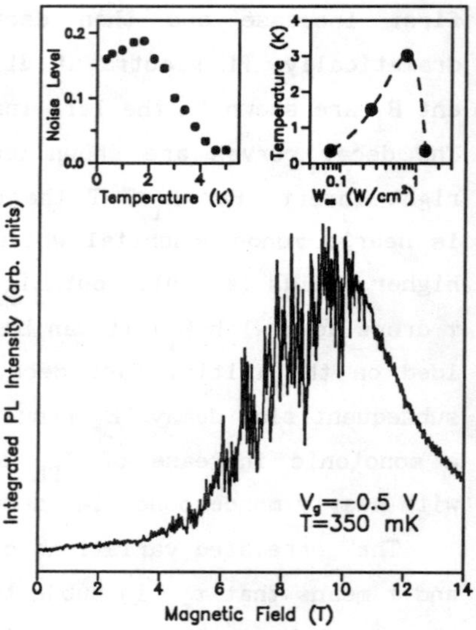

Fig. 2. PL intensity at V_g=-0.5 V and T=350 mK versus $B_{\perp}$. Left insert: temperature dependence of the noise level at V_g=-0.5 V and $B_{\perp}$=9 T. Right insert: excitation density dependence of T_c for the noise observation.

A spectacular effect is observed in the indirect regime at $B_{\perp}$, namely a huge noise in the integrated PL intensity (Fig.2). The noise has a broad band spectrum[4]. The appearance of the huge noise is strong evidence for the presence of coherence in the exciton system. The noise amplitude is known to be inversely proportional to the number of statistically independent entities in a system[7]. Large noise amplitudes therefore denote that only a small number of entities exists in the photoexcited region. We believe that the appearance of those macroscopic entities in the exciton system is given by the condensation of indirect excitons. A condensed domain can be considered as one macroscopic entity. The PL signal of condensed excitons is much higher as compared to uncondensed ones[8]. The formation and disappearance of condensed domains results therefore in a change of the total PL signal.

The conditions under which the noise is observed confirm this interpretation: (i) The noise is observed only in the indirect regime for excitons with long lifetime. (ii) The noise is observed only at low temperatures. The characteristic temperature dependence of the noise amplitude $\langle\delta I_{PL}\rangle/\langle I_{PL}\rangle$ is shown on the left insert of Fig.2. As a function of exciton density the transition temperature (T_c) first increases and then decreases when approaching plasma densities (right insert of Fig.2). The initial increase of T_c is consistent with predicted behavior for low densities[9], while at high densities (close to the plasma regime) a condensation of excitons is unlikely in the present system due to different electron and hole Fermi surfaces[1]. (iii) The noise is observed only at high $B_\perp$ which improves the condensation conditions (Fig.2). It can be also seen that the noise is strongly suppressed at the highest fields corresponding to reduced excitonic lifetime (see Fig.1) which is in agreement with the expected conditions for the exciton condensation. The possible superfluidity of the condensed excitons can also be responsible for the observed fast exciton transport at high $B_\perp$.

* Permanent address: Institute of Solid State Physics, Russian Academy of Science, 142432 Chernogolovka, Russia

REFERENCES

1. L.V.Keldysh, Yu.E.Kopaev, Fiz. Tverd. Tel. **6**, 2791 (1964).
2. Yu.E.Lozovik, V.I.Yudson, Zh. Eksp. Teor. Fiz. **71**, 738 (1976).
3. I.V.Lerner, Yu.E.Lozovik, Zh. Eksp. Teor. Fiz. **80**, 1488 (1981).
4. L.V.Butov, A.Zrenner, G.Abstreiter, G.Böhm and G.Weimann, Phys. Rev. Lett. **73**, 304 (1994).
5. V.B.Timofeev, V.D.Kulakovskii, I.V.Kukushkin, Physica **117**, 327 (1983)
6. V.N.Abakumov, V.I.Perel, I.N.Yassievich, Nonradiative recombination in semiconductors, ed. V.M.Agranovich, A.A.Maradudin, North-Holland, 1991
7. B.L.Altshuler, P.A.Lee, R.A.Webb, Mesoscopic phenomena in solids, ed. V.M.Agranovich, A.A.Maradudin, North-Holland, 1991.
8. G.E.W.Bauer, Physica Scripta **T45**, 154 (1992).
9. J.M.Kosterlitz and D.J.Thouless, J.Phys. C **5**, L124 (1972).

MAGNETIC FIELD DEPENDENT PHOTOLUMINESCENCE STUDIES OF InGaAs/GaAs STRAINED-SINGLE-QUANTUM WELLS

E. D. JONES, L. R. DAWSON, J. F. KLEM, and S. K. LYO
Sandia National Laboratories, Albuquerque, NM 87185-0350, USA

and

D. HEIMAN and X. C. LIU
Francis Bitter National Magnet Laboratory, MIT, Cambridge, MA 02139, USA

ABSTRACT

Conduction and valence-band dispersion curves determined by magnetoluminescence are presented for n-type InGaAs/GaAs strained-single-quantum well structures. The magnetic field range was 0 to 30 tesla, and the temperature varied between 4.2 and 77.4 K.

The requirement for light-hole valence-band masses for compound semiconductor optoelectronic devices has been well documented.[1,2] Light-hole masses have been achieved in layered structures by the introduction of compressive biaxial strain in the active quantum layers. The biaxial strain is realized by growing layered structures with materials with differing lattice constants, e.g., GaAs and In_xGa_{1-x} layers. However, because of interactions between the heavy-hole and light-hole valence bands, the valence-band masses are nonparabolic. For optoelectronic devices, the consequence of a valence-band nonparabolicity is that as carrier densities and/or biasing electric fields are increased, the valence-band mass increases, thereby changing lasing threshold conditions, magneto-optic parameters, etc.

Recently, it was shown that the conduction- and valence-band energy dispersion curves of modulation doped strained single-quantum-well (SSQW) and lattice-matched single-quantum-well structures can be determined[3-5] by magnetoluminescence measurements. These papers discussed the importance of the energy difference ΔE between the heavy-hole and light-hole valence bands in determining the degree of mixing of the heavy-hole and light-hole valence bands. Large values for ΔE not only give rise to small ground-state in-plane light-hole masses, but also reduce the valence-band nonparabolicity.

In this paper, we will present results of high field magnetoluminescence determinations of band structure dispersion curves of an modulation doped n-type InGaAs/GaAs SSQW structure. The sample was prepared using molecular beam epitaxy. The magnetoluminescence measurements were made in the temperature range of 4.2 and 77.4 K, and the magnetic field was varied between 0 and 30 tesla. The applied magnetic field is parallel to the growth direction, i.e., the Landau orbits are in the plane of the quantum well. For this geometry, all measurements concerning the conduction-band and valence-band dispersion curves and masses refer to their *in-plane* values. The SSQW structure (#AK242: $In_{0.15}Ga_{0.85}As/GaAs$) consisted of a single 12.5-nm-wide InGaAs quantum well. For this structure, the heavy-hole and light-hole energy difference ΔE, which includes contributions from both strain effects and quantum confinement, is about 45 meV. The 4-K 2D-carrier concentration N_{2D} and mobility μ is respectively 8.9×10^{11} cm^{-2} and 66000 cm^2/Vsec. At 77-K, $N_{2D} = 2.5 \times 10^{11}$ cm^{-2} and $\mu = 12900$ cm^2/Vsec.

For n-type structures and fairly high temperatures, e.g., 77 K, the $n_v = 0, 1, 2, 3, \ldots$ valence-band Landau levels are populated according to Maxwell-Boltzmann statistics and hence the magnetoluminescence transitions between the n_c and n_v Landau levels are *allowed* and obey the $\Delta n \equiv (n_c - n_v) = 0$ selection rule. For these temperatures, the interband luminescence transition energy is given in terms of the bandgap energy E_g and the conduction-band and valence-band cyclotron energies $\hbar\omega_{c,v} = 2\mu_\beta H / m_{c,v}$ by

$$E = E_g + (n + \frac{1}{2})(\hbar\omega_c + \hbar\omega_v), \qquad (1)$$

where $\mu_\beta = e\hbar/2m_0c$ is the Bohr Magneton, $m_{c,v}$ is the conduction-band or valence-band effective mass in terms of the free electron mass m_0, and the Landau index $n = 0, 1, 2, 3, \ldots$

For large magnetic fields, minimum laser power, and low temperatures, only the $n_v = 0$ valence-band Landau level is populated. Therefore, only the transition between the $n_c = 0$ and $n_v = 0$ Landau levels is allowed while transitions between the $n_c = 1, 2, 3, \ldots$ and the $n_v = 0$ Landau levels are zeroth-order forbidden, but are weakly allowed by higher order processes.[6,7] For this low-temperature condition, the luminescence transition energy E is given by

$$E = E_g + (n_c + \frac{1}{2})\hbar\omega_c + \frac{1}{2}\hbar\omega_v, \qquad (2)$$

where $n_c = 0, 1, 2, 3, \ldots$ From (2), the energy difference δE between the $E(n_c)$ and $E(n_c\text{-}1)$ magnetoluminescence peaks provides all of the necessary information about the conduction band energy dispersion curves.

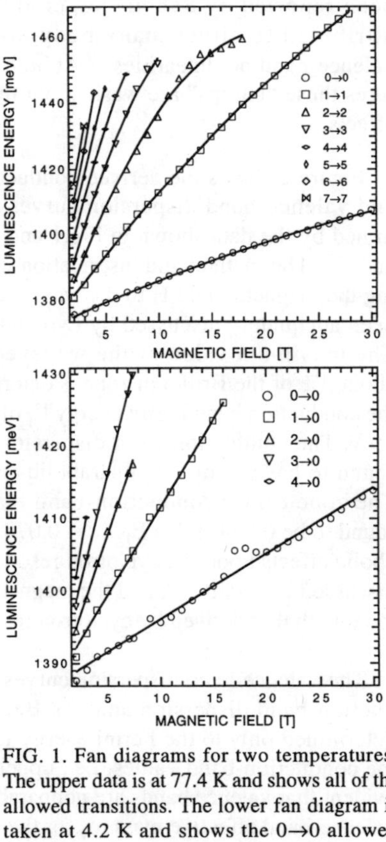

FIG. 1. Fan diagrams for two temperatures. The upper data is at 77.4 K and shows all of the allowed transitions. The lower fan diagram is taken at 4.2 K and shows the 0→0 allowed transition and the higher energy forbidden transitions.

The interband magnetoluminescence transitions for 77.4 K (upper) and 4.2 K (lower) are shown in Fig. 1. The lines are drawn through the data points in order to aid the eye. All transitions for the 77.4 K data are "allowed", while only the 0→0 transition is allowed for the 4.2 K data. The curvature in the data at 77 K for the higher energy transitions may be due to thermal population of the higher lying quantum well states in the valence band. Similar behavior was observed in p-type structures where these population effects are more apparent. The nonlinear behavior, "bumps", near 8 and 16 T in the 0→0 transition for the 4.2K occur at integer filling factors of 2 and 1. This effect has been most recently been reported by Schmiedel, et.al.[8] and are attributed to either many body effects[9] or valence-band nonlinearities.[10] At high temperatures these "bumps" are washed out by thermal effects.

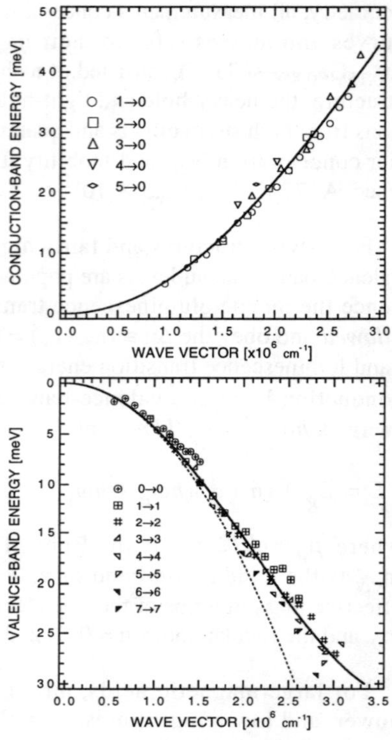

FIG 2. Magnetoluminescence determined conduction-band (upper) and valence-band (lower) dispersion curves. The solid lines are nonparabolic best-fit curves to the data. The dotted line in the lower graph is parabolic with a zone-center mass of 0.090.

Figure 2 shows the derived conduction-band and valence-band dispersion curves as determined by the data shown in Fig.1 and using (1) and (2). The method and justification used relating the magnetic field H to the wavevector $\boldsymbol{k}$ has been adequately discussed by Lyo and Jones.[11] The maximum value of the wavevector $\boldsymbol{k}$ of about 3% of the Brillouin zone is determined by the conduction band Fermi energy E_f of about 40 meV. The conduction-band dispersion curve is found to have a small nonparabolic correction. The zone center conduction-band mass m_c is found to be 0.068 and at E_f, $m_c \sim 0.075$. Nonparabolic effects upon the cyclotron resonance measured conduction-band masses have been discussed in detail[12,13], and the magnetoluminescence results presented here are in agreement with that derived by cyclotron resonance.

The valence-band dispersion curves can now be derived with a knowledge of the conduction-band dispersion and (2). Because the conduction-band dispersion curves are determined only to the Fermi energy E_f, we limit the valence-band wavevector range to the conduction-band values, i.e., upper graph of Fig 2. From lower graph in Fig. 2, it is evident that valence bands are nonparabolic. For the range of data shown in the figure, the valence-band effective mass m_v for the SSQW structure varies between 0.11 and 0.22. The solid line through the valence-band data is a best-fit nonparabolic curve to the data. The dotted line in the valence-band graph is parabolic with a zone-center mass of 0.090. As previously mentioned, the heavy-hole light-hole energy separation ΔE is about 45 meV. Magnetoluminescence determined valence-band dispersion curves for other samples (with differing values for ΔE) confirm that as ΔE increases the valence-bands become more parabolic and also that the valence-band masses become lighter.

In conclusion, we have shown that magnetoluminescence measurement techniques are powerful tools for performing single-sample simultaneous measurements of both the conduction-band and valence-band dispersion curves (and masses).

Supported in part by the Division of Material Science, Office of Basic Energy Science, U. S. DOE, No. DE-AC04-94AL8500 and a National Science Foundation Grant, DMR-9201614. The Francis Bitter National Magnet Laboratory is supported by the National Science Foundation.

References

1. G. C. Osbourn, P. L. Gourley, I. J. Fritz, R. M. Biefeld, L. R. Dawson, and T. E. Zipperian in *Semiconductors and Semimetals, 24, Applications of Multiquantum Wells, Selective Doping, and Superlattices*, edited by R. Dingle (Academic Press, NY, 1987).
2. T. P. Pearsall in *Semiconductors and Semimetals, 32, Strained-Layer Superlattices: Physics* edited by T. P. Pearsall (Academic Press, NY, 1990).
3. E. D. Jones, S. K. Lyo, I. J. Fritz, J. F. Klem, J, E. Schirber, C. P. Tigges, and T. J. Drummond, Appl. Phys. Lett. **54** (1989) 2227.
4. E. D. Jones, R. M. Biefeld, J. F. Klem, and S. K. Lyo, *Proceedings. 16th International Symposium on GaAs and Related materials*, Karuizawa, Japan 1989, Inst. Phys. Conf. Ser. No. **106** (1990) 435.
5. E. D. Jones, S. K. Lyo, J. F. Klem, J. E. Schirber, and S. Y. Lin, *Proceedings. 18th International Symposium on GaAs and Related materials*, Seattle, USA, 1991, Inst. Phys. Conf. Ser. No. **120** (1992) 407.
6. S. K. Lyo, E. D. Jones, and J. F. Klem, Phys. Rev. Lett. **61** (1988) 2265.
7. S. K. Lyo, Phys. Rev. B 40 (1989) 8418.
8. T. Schmiedel, L. P. Fu, S. T. Lee, A. Petrou, M. Dutta, J. Pamulapati, P. G. Newman, and J. Boviatsis, J. Appl. Phys. **74** (1993) 2100
9. T. Uenoyama and L. J. Sham, Phys. Rev. B **39** (1989) 11044.
10. C Delalande, J. A. Brum, J. Orgonassi, M. H. Maynadier, G. Bastard, J. C. Maan, G. Weimann, and W. Schlapp, Superlattices and Microstructures **3** (1987) 29.
11. S. K. Lyo and E. D. Jones in *Electronic, Optical, and Device Properties of Layered Structures*, Edited by J. R. Hayes, M. S. Hybertsen, and C. R. Weber, Proceedings, Fall 1990 Meeting of the Materials Research Society, Boston, MA, page 271.
12. J. Singleton, R. J. Nicholas, D. C. Rogers, and C. T. B. Foxen, Surface Science **196** (1989) 429.
13. F. A. P. Osório, M. H. Degani, and O. Hipólito, Superlattices and Microstructures **6** (1989) 107.

COHERENT OPTICAL SPECTROSCOPY IN HIGH MAGNETIC FIELDS

I. BAR-JOSEPH, S. BAR-AD and Y. LEVINSON

Department of Physics, The Weizmann Institute of Science, Rehovot 76100, Israel

ABSTRACT

We use time resolved four wave mixing to study electron scattering in modulation doped GaAs quantum wells in high magnetic fields up to 8 T. We find a decrease in scattering rates relative to the zero field behavior and a strong enhancement at the lowest Landau level which we attribute to the Fermi edge singularity.

1. Introduction

Time resolved coherent optical spectroscopy was originally developed and extensively used for the study of atomic and molecular systems.[1] The relevant time scales in these systems, micro to nano-seconds, was compatible with the resolution of available lasers two-three decades ago. The development of ultra-fast lasers, with sub-picosecond time resolution, enabled the implementation of this spectroscopic technique to semiconductors, where the typical time scales for relaxation and dephasing are in the femto to picosecond range. Indeed, time resolved coherent spectroscopy has emerged in recent years as an important experimental probe of free carriers and exciton dynamics in semiconductor quantum wells.

The underlying concept in this technique can be described as follows. Consider a short optical pulse, with a time and spectral widths, Δt and $\Delta \nu$, respectively, hitting the sample at time t_0, and excites electrons from the valence to the conduction band. This excitation creates in the medium a polarization $P(t)$, which is coherent with the laser pulse. A second pulse, which arrives at a later time t, can interact with this polarization due to the third order nonlinear susceptibility. However, because of dephasing processes, which affect different electron-hole pairs in a different manner, the polarization gradually looses its coherence with respect to the exciting pulse, with a dephasing time τ. Consequently, the measured nonlinear signal, generated by the interaction between $P(t)$ and the second pulse, exhibits a decay behavior on a time scale on the order of τ. The experimental realization of this concept is usually done in a two beam four-wave-mixing (FWM) configuration. Two laser pulses, propagating along $\mathbf{k}_1$ and $\mathbf{k}_2$, excite the system and generate a signal in the $2\mathbf{k}_2 - \mathbf{k}_1$ direction. The integrated intensity of this emitted signal decays as a function of the delay between the exciting pulses with a time constant, which is proportional to the phase relaxation of the photo-excited carriers in the sample.

The problem on which we wish to focus in this work is that of electron scattering in high magnetic field in the presence of a two-dimensional electron gas (2DEG). In fact, only a very limited number of time resolved optical experiments have been performed on a 2DEG system. Knox *et al.* have used differential absorption to study the thermalization rate of photo-excited carriers in this system.[2] Previous FWM measurement at zero magnetic field have shown that the dephasing of the induced polarization is due to electron-electron (e-e) scattering and that the scattering times are inversely proportional to the square of the excess energy above the Fermi level.[3] The application of a magnetic field changes the electronic density of states (DOS) and wave functions, and therefore is expected to affect the e-e scattering process. The zero field continuous DOS gradually changes to a discrete spectrum of degenerate Landau levels, whose occupation varies with the magnetic field strength, thus affecting the available phase space for scattering. Furthermore, the changes in the electronic wave functions, from plane waves to localized functions, modify the scattering matrix elements. Our work aimed at studying this scattering process in high magnetic field using time resolved FWM.

2. The e-e Scattering Rate at High Magnetic Field

We have studied two modulation doped (MD) samples, consisting of 30 periods of 10 nm GaAs wells and 50 nm $Al_{0.3}Ga_{0.7}As$ barriers, the center 10 nm of which were doped with silicon. The carrier concentrations in the wells at 4K, 1.8×10^{11} cm^{-2} in one sample and 10^{11} cm^{-2} in the other, were deduced from V.d.P measurements, which also gave a low temperature mobility of $4\text{-}5 \times 10^4$ cm^2/Vs for both samples (implying a Drude relaxation time of ≈ 2 ps). The excitation power densities were kept below $\approx 10^{10}$ cm^{-2}, or about 5% of the electron densities in the wells. The temporal resolution in our experiment was ~ 0.7 ps.

The FWM signal from both samples shows an identical behavior, up to a scaling with the carrier concentration of the magnetic field dependencies. Figure 1 compares the FWM signal, which is measured close to the Fermi energy at B=0, to that measured at B=6.5 T. We note that both the rise and the fall times of the signals are different. The B=0 signal is slightly asymmetric but roughly follows the autocorrelation of the laser pulse, peaking at zero time delay. Such a behavior indicates a very fast dephasing time, comparable with the pulse autocorrelation width (0.7 ps). This is in agreement with Ref. [3], where dephasing times of less than 1 ps were measured close to the Fermi energy at similar excitation densities. The high field signal rises more slowly, with a rise time which is the pulse autocorrelation width, and decays exponentially with a time constant of 1.5 ps. The same temporal behavior is measured *throughout* the lowest Landau level (LL).

The slow dephasing time implies that e-e scattering rate τ_{ee}^{-1} in high magnetic fields is lower and is not as sensitive to excess energy as compared with B=0. However, our measurements can only set an upper limit for τ_{ee}^{-1}. It is very likely that impurity scattering is the limiting process which determines the dephasing time at high magnetic field in our measurements. In fact, the measured dephasing time of 1.5 ps at 7 T is very similar to the Drude scattering time determined from the mobility measurements.

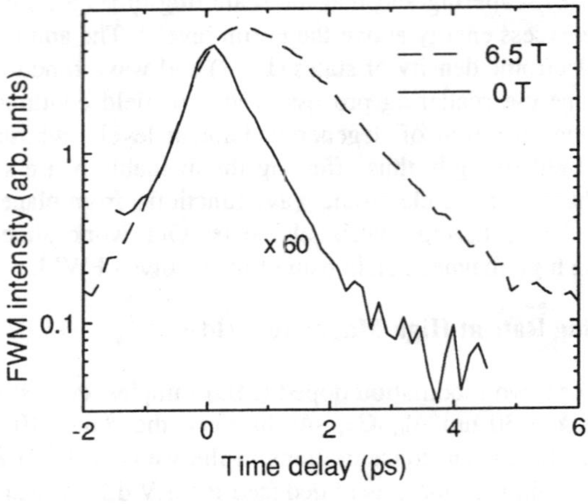

Figure 1: The temporal evolution of the FWM signal close to the Fermi energy at 0 T, and in the first Landau level at 6.5 T (note the logarithmic scale for the FWM intensity). The data shown is for the high density sample, $n=1.8\times10^{11}$ cm^{-2}.

To explain the observed reduction in scattering rate at high field we consider a 2DEG subjected to a smooth random potential U. At B=0 (and zero temperature) τ_{ee}^{-1} can be written as ~ $(g_F/p_F^2)\varepsilon^2$, where ε is the electron excess energy above the Fermi level, g_F is the DOS at the Fermi level and p_F is the Fermi momentum. The ε^2 factor reflects the number of final states which can be reached from a given initial state, since for B=0 the maximum energy transfer $\Delta\varepsilon_{max}=\varepsilon$. A smooth random potential U, with a correlation length Λ large compared to the Fermi wavelength, changes only the DOS and does not affect τ_{ee}^{-1} significantly. This situation changes drastically in strong B, and a smooth potential U, with Λ much larger than the magnetic length a_B (Fig. 2), imposes severe limitations on the e-e scattering. We shall assume for simplicity that the potential U depends only on one coordinate, $U=U(y)$. In this case the energy of an electron in the first LL is $E(y)=1/2\hbar\omega_B+U(y)$, where ω_B is the cyclotron frequency and y is the center of the electron orbit.

It can be shown[4] that at strong B, when all the electron of the 2DEG are in the first LL, the e-e interaction matrix element for inter-LL scattering, in which a test electron e_1 in the second LL is scattered by the Fermi sea into the first LL (and an electron e_2 from the Fermi sea is scattered into the second LL) is small compared to the intra-LL matrix element, when both e_1 and e_2 are in the first LL. Therefore the dominant e-e scattering is intra LL.

Consider the e-e interaction matrix element $\langle y'_1, y'_2 \mid e^2/r \mid y_1, y_2 \rangle$, where y_1 is the initial coordinate of the test electron e_1 which is above the Fermi level, y_2 is the initial coordinate of Fermi sea electron e_2, and y'_1, y'_2 are the final coordinates of these electrons. In the scattering event e_1 transfers energy to e_2. The wave functions entering the matrix element are localized near the corresponding y's within the magnetic length a_B, and hence the shift of y in the scattering event, i.e. $y_1 - y'_1$, is limited by a_B. On the other hand, it follows from a straightforward calculation of the matrix element, that when the shift of the electron $y_1 - y'_1$ is of the order of a_B, the matrix element is exponentially small for distances between interacting electrons $y_1 - y_2$ larger than a_B. *This means that electrons 1 and 2 interact only when the distance between them is of the order of* a_B. Since e_1 is above the Fermi level and e_2 is below, both have to be located near special points y^*, where the energy $E(y)$ crosses the Fermi level, and within a distance of the order of a_B from y^*.

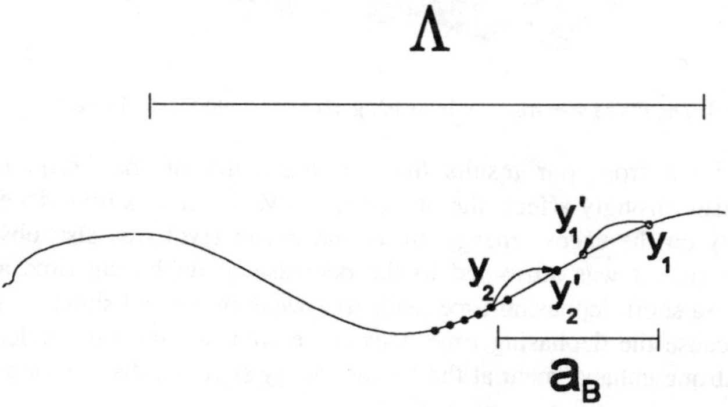

Figure 2: e-e scattering in a long range disordered potential

These considerations impose strong limitations on the available phase space for e-e scattering. One can see that the excess energy of the test electron is limited by $(\Delta\varepsilon_{max})_B = \nabla U\, a_B \approx \Gamma\, a_B/\Lambda$, where Γ is the width of the LL. This energy is small compared to the width of the LL. For larger excess energies the e-e scattering rate is exponentially small and the dephasing rate would be limited by impurity scattering.

3. The Spectrum of the FWM Signal - the Fermi Edge Singularity

Figure 3 illustrates the evolution of the FWM spectrum for the low density sample as the magnetic field is increased. It can be seen that the FWM spectra are dominated by a single peak, which at 3 T shifts to lower energy and significantly increases in intensity. These changes are correlated with the jump of the Fermi level from the second to first LL at 3 T. Thus one can conclude that the peak of the FWM signal follows the onset of the absorption.

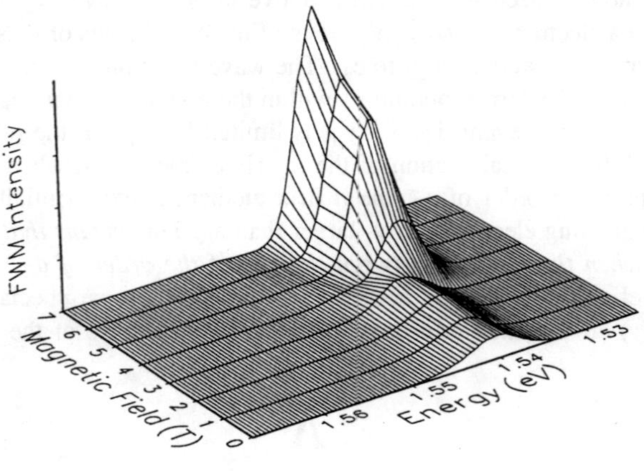

Figure 3: The FWM spectrum for increasing magnetic field for the low density sample

It is evident from our results that the proximity of the Fermi energy to the excitation energy strongly affects the observed FWM signal. A strong dependence of the FWM intensity on the excess energy above the Fermi level was also observed in zero field.[3] In that case it was attributed to the decrease of dephasing time as function of excess energy (a short dephasing time leads to a weak integrated signal). This is not the case here, because the dephasing time does not exhibit a strong dependence on excess energy. The strong enhancement at the Fermi energy suggests that the observed signal is related to a singularity at the Fermi level.

An insight into the origin of this singularity is obtained by examining the temperature dependence of the signal. Figure 4 shows the measured FWM signal at the first LL in a field of 7 T as a function of temperature. The rapid decrease of the signal can be very well fitted to an exponential decay, with a decay constant of 17K or ~ 1.4 meV. The dephasing time changes much more slowly as a function of temperature, and no significant change is observed up to 30 K. Therefore, the fast drop of the signal with temperature can not be attributed to a fast dephasing. On the other hand, these temperature and energy scales are characteristic of the decay of the Fermi edge singularity in absorption and photoluminescence measurements.[5-7] We thus arrive at the conclusion that the measured signal in our FWM experiments is the nonlinear response of the Fermi edge singularity. This finding clearly indicates that the role of the photo-generated holes should be considered if a complete understanding is desired.

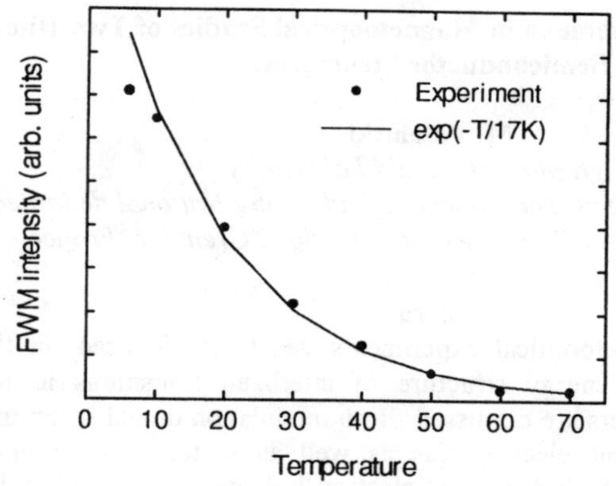

Figure 4: FWM intensity as a function of temperature for the low density sample at zero time delay and 7T. The solid line is a fit to an exponential decay.

It is surprising to note the apparent lack of any contribution to the FWM signal from light hole related transitions. This absence of light hole FWM signal is in contrast to their appearance in the absorption spectra. A possible explanation for the absence of the light hole signal could be the strong dependence of the Fermi edge singularity on the hole mass: the lighter the hole mass, the weaker is the enhancement. Another puzzling experimental result is the different ratios of the parallel polarizations to the crossed polarizations signals at the first and second LL. While in the first LL this ratio is approximately unity, in the second LL the parallel polarizations signal is much stronger.

In summary, we have shown that e-e scattering is suppressed in a strong magnetic field. We have proposed a model which explains this behavior and emphasizes the role of disorder. The strong enhancement of the FWM signal at the Fermi level was shown to be related to the Fermi edge singularity.

We wish to acknowledge Hadas Shtrikman for the MBE growth. This research was supported by the Basic Research Foundation administered by the Israeli Academy of Sciences and Humanities.

4. References

1. L. Allen and J.H. Eberly, "Optical Resonance and Two Level Atoms" (John Wiley & Sons, New York, 1975).
2. W. H. Knox *et al. Phys. Rev. Lett.* **61** (1988), p. 1290 .
3. D. S. Kim *et al.*, *Phys. Rev. Lett.* **68** (1992), p. 2838.
4. V. F.Gantmakher and Y. B. Levinson, "Carrier scattering in metals and semiconductors" (North-Holland, 1987).
5. G. D. Mahan, *Phys. Rev.* **163** (1967), p. 612.
6. M. S. Skolnick *et al.*, *Phys. Rev. Lett.* **58** (1987), p. 2130.
7. G. Livescu *et al. IEEE J. Quant. Electron.* **QE 24** (1988), p. 1677.

High Carrier Density Phenomena in Magnetooptical Studies of Two-Dimensional Semiconductor Structures.

M. Potemski
High Magnetic Field Laboratory
Max-Planck Institute fur Festkorperforschung and Centre National de la Recherche Scientifique BP 166 Grenoble Cedex 9, 38042 Grenoble, France

Abstract

The results of magnetooptical experiments which are focused on the investigation of the energy structure of interband transitions in the presence of free carriers are discussed. Both modulation doped structures with a one-component electron gas as well as systems of undoped quantum wells where high density of electron-hole pairs is created under intense optical excitation are considered. The energy diagrams of magnetooptical transitions observed in both the emission and absorption type experiments are investigated.

1. Introduction

An important application of interband magnetooptics is to investigate the dependence of the properties of two dimensional semiconductor structures on the density of free carriers.[1] The initially empty bands of thin semiconductor layer can be populated with free carriers either through modulation doping or under intense optical excitation. In the case of modulation doped structures the effect of doping has been mostly studied in luminescence experiments.[2] However, by combining of luminescence and luminescence excitation spectroscopy[3] it is possible to investigate not only the band-edge emission but also to probe the empty electronic states above the Fermi level. This type of magnetooptical experiment[4] , which are reported below, gives new information about the energy structure of interband transitions in the presence of an electron gas. In particular they show that the many body approach which successfully explains the transitions related to the lowest occupied electric subband is not applicable in the case of transitions involving empty states of higher excited subbands. The latter transitions show in contrast the two-particle excitonic correlation. In section 3, the recent results of magnetooptical studies[5] of emission spectra under intense optical excitation in undoped quantum well structures are presented. These experiments have been performed using a time-resolved technique on the nano second scale and magnetic fields up to 28T. They give new insight into our previous studies[6] and lead us to suggest a possible interpretation. It is proposed that the emission spectra of highly excited quantum wells in high magnetic fields are determined by particular properties of magnetoexcitonic states[7] which can interact only if they are built out of Landau levels with different Landau level index.

2. Energy Structure of Interband Transitions in the Presence of Free Carriers in Modulation Doped Quantum Well Structures.

A number of works[8] on luminescence experiments on modulation doped quantum wells have shown that in the range of sufficiently high concentration of two-dimensional electrons the band-edge recombination is shifted downward as a function of carrier density. This effect has been interpreted in terms of a band gap renormalization[9], i. e. in terms of the rigid collapse of the fundamental bandgap induced by Coulomb interaction between free carriers.

Recently, the interband transitions in modulation doped quantum well structure has been studied by Stepniewski et. al.[4]. In these experiments both luminescence as well as luminescence excitation spectra were measured. Transitions related to electronic states below as well as above the Fermi level have therefore been investigated. The aim of this work, which is reviewed here, was to verify to what extend one can reproduce the rich energy spectrum of interband magnetooptical transitions on the basis of the existing theories concerning many-body effects including the complexity of the valence band structure. The sample used in the experiments, schematically shown in the inset of Fig. 1., is a typical, modulation doped (on one side), 250 Å-wide, GaAs quantum well structure with an electron sheet concentration of $7.6\times10^{11}cm^{-2}$ under the conditions of the optical experiments. More precise information about sample parameters and characterization by means of transport and cyclotron-resonance measurements can be found elsewhere[4].

Typical optical spectra obtained at zero magnetic field are shown in Fig. 1. The broad band E_0H_0 observed in luminescence can be assigned to the recombination between the lowest occupied electronic E_0 and the highest heavy hole H_0 confined subbands. The sharp structure E_1H_0 observed in both luminescence and luminescence excitation is attributed to the transition between H_0 valence band subband and the first empty E_1 subband in the conduction band. Higher energy peaks are observed only in the photoexcitation spectrum and are assigned to the E_1L_0, E_1H_2, E_2H_1 and E_3H_2, transitions between subsequent conduction (E_i) and valence (L_i,H_i) subbands where L and H denote

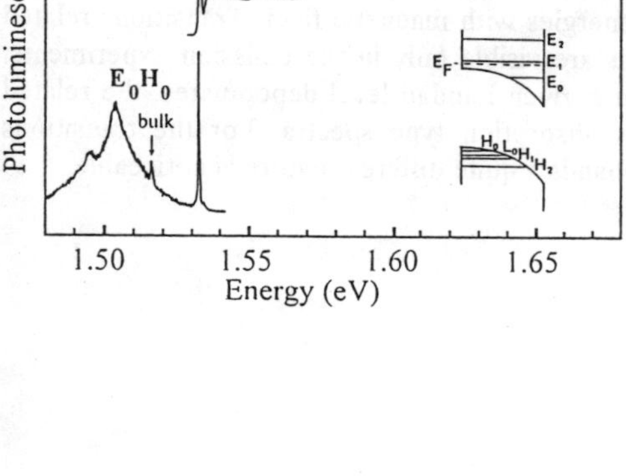

Fig. 1.
Photo-luminescence and photo-luminescence excitation spectrum of a 250 Å wide modulation doped GaAs/GaAlAs quantum well structure. The observed peaks correspond to transitions between different subbands of the conduction and valence bands. The quantum well structure is schematically presented in the inset.

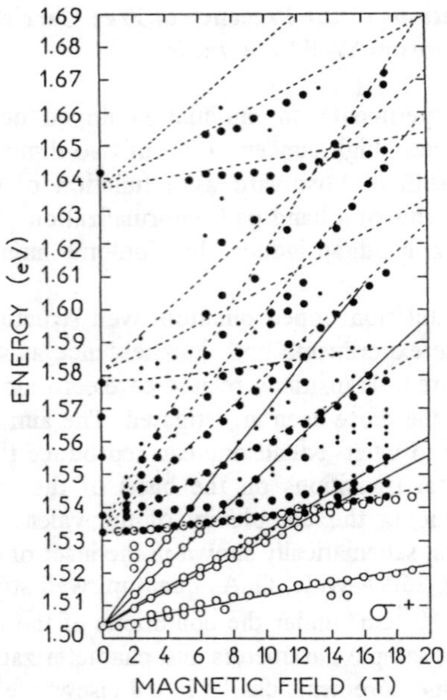

Fig. 2.
The magnetic field diagram of σ^+-transitions observed in luminescence (open circles) and luminescence excitation (closed circles) spectra

respectively, the light and heavy hole states. From Fig.1. it is clear, that the band edge emission is very broad and the zero-field spectra cannot readily be used to determine the bandgap position. Surprisingly enough these spectra show also that absorption transitions related to edges of higher excited subbands are observed as quite sharp peaks and not as a step-like singularities. In fact, the photoluminescence excitation spectrum of this modulation doped sample resemble very much the excitonic transitions observed in case of undoped quantum wells. Much more information can be obtained from magnetooptical experiments. When a sufficiently high magnetic field is applied, well resolved peaks are observed in previously featureless parts of the zero-field luminescence and luminescence excitation spectra. The measured spectra are very rich and the observed transitions exhibit a strong circular polarization. The Landau level fanchart of the dominant transitions observed for the σ^+ polarization is shown in Fig. 2. This figure shows that optical transitions observed in a magnetic field can be divided into groups related to different electric subbands. For the lowest, occupied E_0 subband one observes a nearly linear increase of the transition energies with magnetic field. Transitions related to fully occupied electronic Landau levels are visible only in the emission experiments. Upon increasing the magnetic field, when a given Landau level depopulates, the related optical transition becomes observable in absorption type spectra. For the transitions related to the higher empty electronic subbands a quite different nature is noticeable.

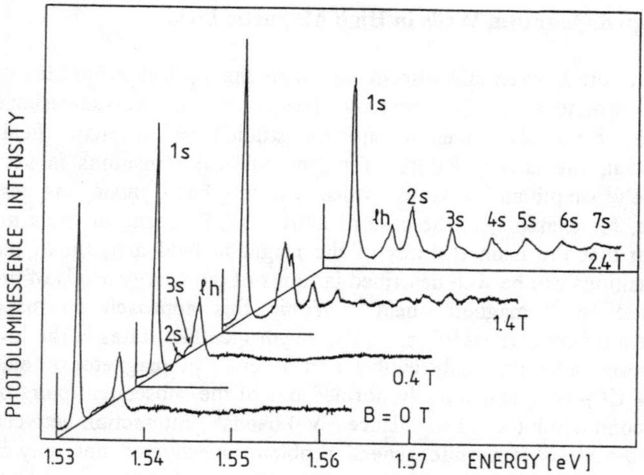

Fig. 3. Magnetic field evolution of the photoluminescence excitation spectra in the range of low magnetic fields showing the appearence of a series of magnetoexcitonic resonances related to the first excited subband

Note, for example, that the ground state within the E_1H_0 transitions shows much weaker field dependence than the lowest state of the occupied subband. In fact, all the ground states related to the excited electronic subbands show typical diamagnetic field dependence which indicates the excitonic character of these transitions. Very characteristic is the behavior of the photoluminescence excitation spectra of E_1H_0 related transitions in the low field region where a clear evolution of a series of the excited excitonic states is observed as shown in Fig3. Stepniewski et. al.[4]; have shown that the Landau level Fan chart of the transitions related to the lowest (occupied) electric subband can be extremely well reproduced in envelope function[10] calculations which take into account the complexity of the valence band structure[11] and the nonparabolicity effect[12], and in which the effect of bandgap renormalization[13] is included in the potential shape using a the functional density approach[14]. Solid lines in Fig. 2 represent the result of these calculations. It should be noted that this approach describes in a satisfactory manner all the Landau level transitions independent of whether they are occupied or empty but only if they belong to the lowest conduction band subband. The same description fails in case of transitions involving levels of higher empty conduction band subbands. It has been shown that neglecting many body effects it is impossible to reproduce the fundamental energy gap E_0H_0. In contrast when many body effects are not included the energy positions of higher subbands are well reproduced but bandgap energy is not. It seems therefore that electronic levels related to higher excited subbands do not feel the many-body contribution to the potential shape introduced by the functional density formalism. Transitions related to higher excited subband do show, however, the two-particle excitonic correlation. Dashed lines in Fig.2 represents a qualitative analysis of these transitions in terms of energy levels of two-dimensional hydrogen atom subjected to a magnetic field. This simply procedure gives satisfactory agreements with experimental data and allows for the estimation of the excitonic binding energies, which have been also reproduced in calculations[15].

3. Highly Excited Undoped Quantum Wells in High Magnetic Fields

At low excitation intensities, excitonic effects dominate the optical properties of undoped quantum well structures[16]. The excitonic resonances in two-dimensional semiconductor structures have also been intensively studied in magnetic fields experiments[17,18]. In general, the energy ladder of magnetooptical transitions is very compliated, it is greately simplified however, when valence band mixing is less pronounced. This is true, for example, in the case of narrow GaAs quantum wells for which it has been shown that the main features of the magnetic field diagram of the interband absorption transitions can be well described in terms of the energy levels of the hydrogen atom subjected to a magnetic field[18]. Within this approach interband magnetoabsorption spectra reflect a series of 1s, 2s, 3s, ...hydrogen like states in the low field region and these resonances gradually evolve into a series of magnetoexcitonic states CL_0-VL_0 CL_1-VL_1 CL_2-VL_2, respectively, formed out of the subsequent pairs of Landau levels L_i of the conduction (C) and valence (V) bands[19]. Interaction between two-dimensional magnetoexcitons is a quite general problem investigated not only in connection to interband transitions but also in the case of inter Landau level[20] or inter Zeeman level excitations[21] of a two-dimensional electron gas. Already very early theoretical works on interband magnetoexcitons[7] have shown that states which belong to the same pair of Landau levels hardly interact with each other. For, example it was shown[22] that the biexciton cannot be formed out of two magnetoexcitons in the ground states. This particular property favor the formation of the condensate form of magnetoexciton gas which was discussed theoretically in several papers[23] and which has stimulated experimental studies of highly excited quantum wells in high magnetic fields. . We present here the main results of recent experiments by Schoser et al.[5], who perfirmed experiments on GaAs-based quantum wells. In these experiments, single quantum well structures were used, the emission spectra measured within 1ns time resolution scale and the range of magnetic fields extended up to 28 T in contrast to our 20T, time integrated studies on multiply quantum well structures. The experiments under better defined experimental conditions give new insight into our previous results[6] and allow us to speculate about the possible interpretation. The results discussed here have been obtained for two different samples, one with 40 and the second woth 60 Å-wide GaAs quantum well embedded between GaAlAs barriers. The experiments were performed in a helium bath at 2K using a 23T polyhelix or a 30T hybrid magnet. In high excitation power experiments a frequency doubled Nd-YAG laser with a pulse width of 20 nsec was used as the excitation source. The luminescence signal was detected within a 4Å spectral and 1nsec time resolution scales. Optical spectra measured at B=21T are shown in Fig. 4. The upper trace represents the conventional low-power photoluminescence excitation spectrum which shows the absorption peaks related to usual magnetoexcitonic states. The evolution of the emission spectra (obtained within the 1.5 nsec-wide time-gate) during the rise of the high excitation pulse are shown below. Similar spectra observed at higher magnetic fields and also at higher laser power are presented in Fig.5. The experimental data presented in Fig.4 are in agreement with our previous observations. The energy positions of luminescence peaks observed in high excitation

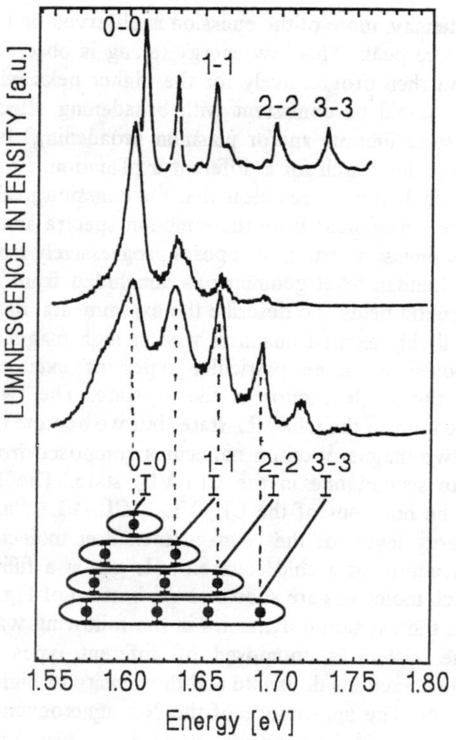

Fig. 4.
Characteristic spectra for 60Å-wide GaAs/GaAlAs single quantum well structure at magnetic field B=21T. The upper trace represents the photoluminescence excitation spectrum which shows subsequent heavy hole magnetoexcitonic states. Ground state of the light hole exciton is traced with dashed line. Two emission spectra observed within 1.5nsec time interval when increasing the high power laser pulse are shown below. At the bottom, the hypothetic excitonic molecules discussed in the text are schematically represented.

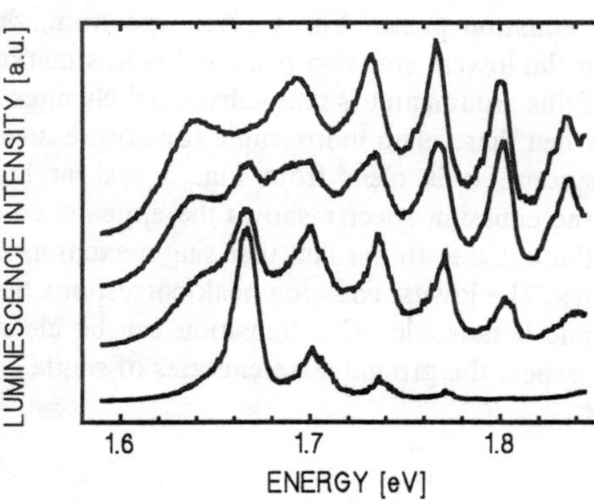

Fig. 5.
From the bottom: the evolution of the emission spectra with increasing power of the laser pulse for 40Å-wide quantum well structure at B=28T.

experiments are substantially different from the positions of magnetoexcitonic transitions measured in low power experiments. Over the broad range of excitation power the emission peaks change very weakly their energy positions. However, the results presented in Fig. 4 show more clearly another characteristic feature of the high power spectra. When increasing the excitation intensity, more of the emission is observed on the low energy side of each of the luminescence peak. This low energy tailing is observed first of all for the lowest energy peak and then progressively for the higher peks with increasing of laser power. Such behavior could be consistent with broadening effects caused by final state interaction[24](final state lifetime and/or plasmon broadening) but experimental data shown in Fig. 5 provokes the search for a different explanation. From the spectra measured at highest magnetic fields it becomes clear that the emission peaks observed at relatively low excitation powers disappear from the emission spectra and a new emission spectrum emerges. This new emission structure appears progressively from the low energy side and also shows the Landau level geometry as concluded from an analysis of measurements at different magnetic fields. To describe the experimental data, we propose to consider the system of a highly excited quantum well in high magnetic fields as a magnetoexciton gas composed of some particular types of excitonic molecules. The most simply one is just the single exciton in the 1s state. The two-magnetoexciton molecule cannot be formed out of the CL_0-VL_0 states but we assume the possibility of the local formation of the two-magnetoexciton molecules composed from the one exciton in the CL_0-VL_0 and the second one in the CL_1-VL_1 state. The N-magnetoexciton molecule is assumed to be built out of the CL_0-VL_0 , CL_1-VL_1 , and ...CL_N-VL_N states. We suppose that energy levels of the N-magnetoexciton molecule should somehow coincide with energy structure of a cold electron-hole gas at a filling factor 2N. Hypothetic energy levels of such molecules are shown at the bottom of Fig.4. Using the above picture one may imagine the excitation dynamics in the following way. At relatively low excitation powers the system is composed of different types of magnetoexcitonic molecules, however single excitons dominate and the density of higher order molecules decreases with increasin N. The appearance of the N-magnetoexciton leads to the appearance of the Nth emission peak in the spectrum and also give rise to the low energy emission tails of the existing emission peaks. For a given spectrum, this tailing is therefore the most pronounced for the lowest emission peak and is less marked for higher energy peaks. A consequence of this argumentat is that substantial changes in the emission spectrum might be expected when there is no more single (or progressively double, etc.,..) magnetoexcitons in the system. It is clear from Fig. 5 that at high excitation powers the low energy part of the emission spectra shows the appearence of the new Landau level structure which we think is due to the fact that single excitons no longer exist at very high excitation intensities. The lowest emission peak correspons then to the ground state emission of the excitonic bimolecule. This transition can be clearly observed at high magnetic fields where we expect the ground state energies of single and double molecules to be remarkably different.

4. Conclusions

Two examples of magnetooptical studies of interband transitions in two-dimensional structures containing free carriers have been presented. In case of the modulation doped quantum well it is concluded that different approaches for the description of the energy ladders are necessary depending on whether the optical transitions involve Landau levels of populated electric subband or whether they are related to empty conduction band subbands. High excitation intensity experiments on undoped GaAs quantum wells are tentatively interpreted in terms of local formation of magnetoexcitonic molecules. More data analysis and theoretical work is necessary to be able to draw firm conclusions from these experiments.

5. Acknowledgments

It is a pleasure to thank B. Etienne, C. Kutter, S. Schoser, and R. Stepniewski for fruitful collaboration and J.C. Maan, G. Martinez, and P. Wyder for many helpful discussions.

6. References

1. For a review see, e.g., *Proceedings 21st Int. Conf. Phys. Semicon., Beijing, Chine, 1992 eds. P. Jiang and H-Z. Zheng*, (World Scientific, Singapore, 1992): M. Skolnick et al., p. 41, M. Potemski et al., p. 915.
2. See, e. g., *Proc. 9th Int. Conference on Electronic Properties of Two-Dimensional Systems*, Surface Science, **263** (1992).
3. J. Orgonasi et al., *J. de Physique* **46** (1987) C5-407; R. Stepniewski et al., *Proceeding 20th Int. Conf. Phys. Semicon., Thessaloniki 1990, (eds. E.M. Anastassakis and J.D. Joannopoulos*, (World Scientific, Singapore, 1991), p. 1282.
4. R. Stepniewski et al., Submitted to *Phys. Rev*. **B15** (1994).
5. S. Schoser et al., *Proc. 22nd Int. Conf. Phys. Semicon., Vancouver 1994*, to be published.
6. M. Potemski et al., *Solid State Commun.* **75** (1990) 185.
7. I.V. Lerner and Yu.E. Lozovik, *Sov. Phys. JETP* **53(4)** (1981) 763.
8. C. H. Perry et al. in *High Magnetic Fields in Semiconductor Physics, ed. G. Landwehr*, (Springer, Berlin, 1987), p. 202; C. Delalande et al. *Phys. Rev. Lett.* **59** (1987) 2690; S. Haacke et al., *Phys. Rev*. **B45** (1992) 1736.
9. A.E. Ruckenstein and S. Schmitt-Rink, *Phys. Rev*. **B35** (1987) 7551; G.E.W. Bauer, *Phys. Rev. Lett*. **64** (1990) 60.
10. *G. Bastard in Wave Mechanics Applied to Semoconductor Heterostructures*. (Les Editions de Physique, Paris, 1982)
11. M. Altarelli in *Proc. Les Houches Winterschool on Semiconductor Superlattices and Heterojunctions* (Springer, Berlin, 1986); F. Ancilotto et al., *Phys. Rev*. **B38,** (1988) 1788.
12. U. Ekenberg, *Phys. Rev*. **B40** (1989) 7714.
13. O. Gunarson and B.I. Linqvist, *Phys. Rev*. **B13** (1977) 4274.
14. G.E.W. Bauer and T. Ando, *J. Phys. C, Solid State Phys*. **19** (1986) 1537.
15. A. B. Henriques, *Phys. Rev*. **B44** (1991) 3340.
16. C. Weisbuch et al., *Sol. State Commun*. **37** (1981) 219.
17. L. Vina et al., *Phys. Rev*. **B41** (1990) 10767, and references therein.
18. M. Potemski et al., Phys. Rev. B43, (1992) 14707.
19. J. C. Maan in *Physics and Applications of Quantum Wells and Superlattices, eds. E.E. Mendez , K. von Klitzing* (NATO ASI Series, Plenum Press, New York, 1988), p. 347.
20. C. Kallin and B.I. Halpherin, *Phys. Rev*. **B31** (1985) 3635.
21. Yu. A. Bychkov, S.V. Iordanskii, and G.M. Eliashberg , *Pis'ma Zh. Eks. Teor. Fiz.* **33** (1981) 152 [*JETP Lett*. **33** (1981) 143].
22. I.V. Lerner and Yu. E. Lozovik, *Sov. Phys. JETP* **55** (1982) 691
23. D. Paquet, T.M. Rice, and K. Ueda, *Phys. Rev*. **B32** (1985) 5208;
24. P.T. Landsberg, *Phys. Stat. Solidi* **15** (1966) 623; J.H Collet et al., *Phys. Rev*. **B40** (1989) 12296.

MAGNETO-RAMAN SCATTERING BY OPTIC AND ACOUSTIC PHONONS IN QUANTUM WELLS

T. RUF, A. CROS, J. SPITZER, G. GOLDONI*, AND M. CARDONA
Max-Planck-Institut für Festkörperforschung, Heisenbergstr. 1, D-70569 Stuttgart, Federal Republic of Germany
**Dipartimento di Fisica, Universita degli Studi di Modena, Via G. Campi 213/A, I-41100 Modena, Italy*

ABSTRACT

We study resonant Raman scattering by optic and acoustic phonons in GaAs/$Al_xGa_{1-x}As$ multiple quantum wells in high magnetic fields. The Raman efficiency exhibits pronounced magneto-oscillations at interband transitions between Landau levels from which we determine nonparabolicity and anisotropy effects of the electron effective mass.

1. Introduction

Raman scattering by longitudinal optic (LO) phonons in bulk semiconductors and multiple quantum wells (MQW's) is strongly resonant with magneto-optical interband transitions between Landau levels.[1,2] Fan plots of resonance energies vs. magnetic field B in different scattering geometries, using circularly polarized light, show sharp selection rules due to the dependence of the Raman cross section on electron-photon and electron-phonon interaction. In the acoustic-phonon regime of MQW's we find a disorder-induced continuous emission background, which also displays resonant magneto-oscillations.[3,4] We investigate both, LO-phonon and acoustic mode magneto-Raman scattering in GaAs/$Al_xGa_{1-x}As$ MQW's. From the data we determine the well-width dependence of electron effective masses and compare the results to theoretical models.

2. Optic-Phonon Scattering

Figure 1 shows experimental LO-phonon magneto-Raman resonances (triangles) in a fan plot of exciting laser energy vs. B for a symmetric GaAs/AlAs MQW with layer thicknesses of 44 Å.[2] Experiments were performed at a temperature of 10 K in backscattering Faraday configuration with σ^--polarized light for both, excitation and detection. The magnetic field was oriented along the (001) MQW growth direction. For vanishing B the fan lines below 1760 meV converge at the energy of the direct transition between the first confined electron and heavy hole subbands (≈ 1705 meV). Thus they correspond to incoming resonances of the Raman efficiency. For excitation energies above 1760 meV a different set of fan lines is observed which for zero B converges at an energy larger than the above one by about one LO phonon. These resonances are either outgoing, involving transitions between heavy hole and electron Landau levels, or incoming, and light hole valence levels participate.

Interband magneto-optical spectra have been calculated with a six-band envelope function approach.[5] Here we treat heavy and light mass valence levels separately in a 4×4-$\vec{k} \cdot \vec{p}$-model using known Luttinger parameters.[2] Electron Landau levels are described by an effective mass which is determined from fits to the experiment.[2] Coulomb effects are taken into account by subtracting the binding energy of a quasi-two dimensional exciton in the high-field limit from the calculated transition energies.[2,5] The lines in Fig. 1 represent the best theoretical description of the observed magneto-Raman resonances. Fan plots of dotted and solid lines are incoming and outgoing resonances involving heavy mass valence states. Landau indices n are given for the set of outgoing resonances. Fan lines for the same n are shifted from each other by the LO-phonon energy. Due to the strong mixing of valence levels in a magnetic field[5] and the resulting multitude of possible Raman processes,[1,2] Landau indices given in Fig. 1 represent the quantum numbers of the rather pure electron states involved in the respective transitions. Dashed-dotted fan lines in Fig. 1 are incoming resonances where light mass valence levels participate. At this particular well width the separation between heavy and light hole subbands at zero field is comparable to the LO-phonon energy. Thus fan lines of incoming light-hole related and outgoing heavy-hole related resonances are almost parallel to each other in the range of energies above 1760 meV. Both types of resonances are observed experimentally.

3. Acoustic-Phonon Scattering

Near the laser line magneto-Raman oscillations in MQW's are found at arbitrary Raman shifts.[3,6] These resonances involve processes with scattering by acoustic phonons and become allowed due to a relaxation of crystal momentum conservation along the growth direction caused by layer thickness fluctuations and interface roughness. A spread in confinement energies due to disorder, i. e., inhomogeneous broadening of interband transitions, allows Raman processes to be resonantly excited in particular parts of a MQW sample. These act as single quantum wells for which crystal momentum needs not be conserved. Such incoherent scattering leads to a continuous emission background which reflects details of the electronic structure and the electron-phonon interaction.

The smaller Raman shifts of acoustic phonons compared to those of optic modes allow us to investigate individual interband transitions where only one type of hole Landau level is involved. This is in contrast to LO-phonon magneto-Raman experiments where a multitude of different processes may contribute to an observed resonance.[1,2] A further advantage arises from the almost doubly-resonant character of acoustic-phonon Raman processes. This yields much stronger signals than the usually singly-resonant optic-phonon scattering.

We have measured acoustic-phonon magneto-Raman oscillations in a (100/103) Å $GaAs/Al_{0.35}Ga_{0.65}As$ MQW and compare the fan plots to calculations in Ref. [6]

4. Effective Masses

A theory for electron effective masses in MQW's shows that the nonparabolicity and anisotropy of the bulk conduction band has a characteristic influence on the MQW electronic structure.[7] Different energy-dependent effective masses for the calculation of confinement energies ϵ ($m_\perp$) and the in-plane dispersion ($m_\|$) are derived. In the limit of infinitely high barriers these masses are:[7]

$$m_\perp = m\,(1+\alpha'\epsilon) \tag{1}$$
$$m_\| = m\,(1+(2\alpha'+\beta')\epsilon)\,. \tag{2}$$

The parameters α' and β' describe the nonparabolicity and the anisotropy of the bulk conduction band. For GaAs we use $\alpha' = 0.642/\mathrm{eV}$ and $\beta' = 0.697/\mathrm{eV}$. The effective mass at the band edge is given by $m = 0.0665\,m_0$.[7] While this theory has been tested recently using optically-detected cyclotron resonance,[8] data from interband experiments remain scarce.[5,7]

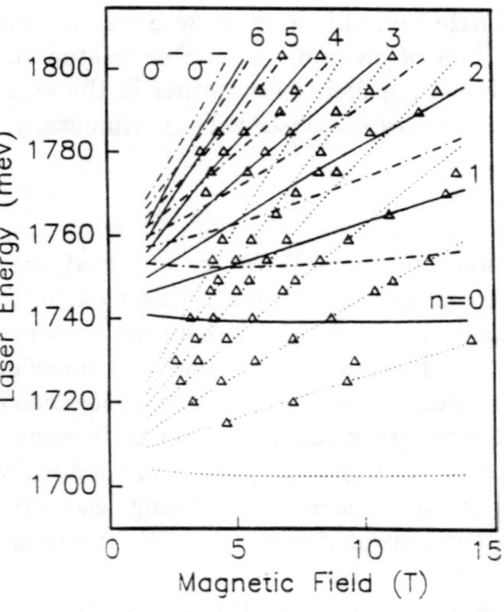

Figure 1: Magneto-Raman resonances (triangles) of a (44/44)Å GaAs/AlAs MQW compared to calculated fan lines.

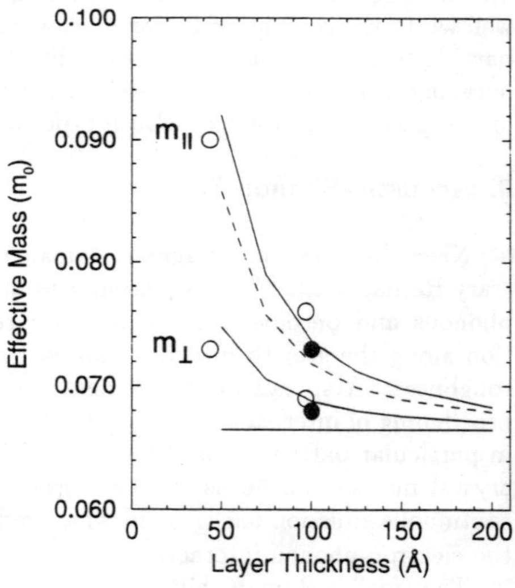

Figure 2: Well-width dependence of effective masses in GaAs.

The solid lines in Figure 2 show the layer-thickness dependence of $m_\perp$ and $m_\|$ for GaAs MQW's calculated with the above equations in analogy to Ref. [7]. For comparison m is indicated by the horizontal line. The open circles are experimental values for GaAs/AlAs MQW's determined from LO-phonon magneto-Raman scattering as outlined in sect. 2. The filled circles are effective masses obtained from acoustic-phonon resonances of the sample described in the previous section. Our interband magneto-optical experiments confirm the theoretical model, especially the larger increase of $m_{\|}$ compared to that of $m_{\perp}$ with decreasing GaAs well width. While the assumption of perfect confinement is reasonable for AlAs, smaller effective masses are observed for the sample with $Al_{0.35}Ga_{0.65}As$ barriers. These values, however, are in good agreement with the more complicated expressions for finite barrier heights given in Ref. [7].

The dashed line in Fig. 2 derives from the bulk GaAs conduction band dispersion determined by the LO-phonon magneto-Raman technique in backscattering from a (001)-oriented surface with $B \parallel (001)$.[9] It can be expressed by an energy-dependent effective mass $m_e^* = m(1 + 1.447\epsilon + 0.245\epsilon^2)$. The layer-thickness dependence is obtained according to Ref. [7]. Values of m_e^* are close to those of $m_{\|}$, which reflects the fact that cyclotron energies are determined by the effective mass of the in-plane dispersion.

5. Conclusions

Magneto-Raman scattering confirms theoretical predictions for the well-width dependence and the different enhancement of electron effective masses along and perpendicular to the MQW growth direction which are responsible for confinement and Landau quantisation, respectively.

References

1. T. Ruf, R. T. Phillips, C. Trallero-Giner, and M. Cardona, Phys. Rev. B **41**, 3039 (1990).
2. A. Cros, T. Ruf, J. Spitzer, M. Cardona, and A. Cantarero, Phys. Rev. B **50**, July 15 (1994).
3. V. F. Sapega, V. I. Belitsky, T. Ruf, H. D. Fuchs, M. Cardona, and K. Ploog, Phys. Rev. B **46**, 16005 (1992).
4. T. Ruf, J. Spitzer, V. F. Sapega, V. I. Belitsky, M. Cardona, and K. Ploog, Phys. Rev. B **50**, July 15 (1994).
5. F. Ancilotto, A. Fasolino, and J. C. Maan, Phys. Rev. B **38**, 1788 (1988).
6. G. Goldoni, T. Ruf, V. F. Sapega, and M. Cardona, to be published.
7. U. Ekenberg, Phys. Rev. B **40**, 7714 (1989).
8. R. J. Warburton, J. G. Michels, R. J. Nicholas, J. J. Harris, and C. T. Foxon, Phys. Rev. B **46**, 13394 (1992).
9. T. Ruf and M. Cardona, Phys. Rev. B **41**, 10747 (1990).

MAGNETIC TUNING OF SPATIALLY INDIRECT INTERBAND TRANSITIONS IN PARABOLIC AND RECTANGULAR QUANTUM WELLS

M. FORKEL, J. HEITZER, J. EHRLICH, U. PENNER, and G. H. DÖHLER
Institut für Technische Physik I, Universität Erlangen-Nürnberg
Erwin-Rommel-Straße 1, 91058 Erlangen, Germany

G. MACKH and W. OSSAU
Physikalisches Institut, EP III, Universität Würzburg
Am Hubland, 97074 Würzburg, Germany

K. L. CAMPMAN and A. C. GOSSARD
Materials Department, University of California
Santa Barbara, CA 93106, U. S. A.

ABSTRACT

The electronic and optical properties of two-dimensional semiconductor structures with spatially separated electron and hole subband states are investigated under the influence of in-plane magnetic fields. The probabilities for optical transitions can be tuned up or down, depending on the population of the subbands and the magnetic field. We present both results of photoluminescence measurements and of numerical calculations showing this effect.

1. Introduction

Most of the experiments on heterostructures in magnetic fields have been performed for fields perpendicular to the layers.[1] So far parallel magnetic fields have only been the subject of relatively few investigations.[2] In this paper we report on the effect of in-plane magnetic fields on the optical interband recombination in systems with spatially separated electrons and holes. This indirect band gap in real space causes a distinctly different behaviour[3] compared to the case of direct transitions.[4]

The in-plane magnetic field creates an additional parabolic potential in the direction normal to the layers. The centre of mass of the wavefunctions is shifted normal to the layers depending on the momentum perpendicular to the orientation of the magnetic field. This is equivalent to the classical picture of a Lorentz-force. Electron and hole states are driven towards each other and the overlap between their wavefunctions can be exponentially enhanced. Therefore, it is expected that the luminescence intensity can be drastically tuned by an in-plane magnetic field.[3]

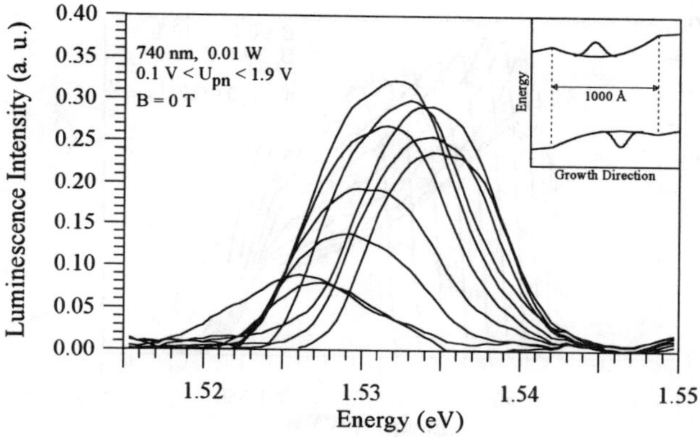

Fig. 1. Measured photoluminescence of a *p-i-n*-heterostructure for different forward voltages. The insert shows the conduction and valence band edges of the sample and the respective wavefunctions.

2. Theory

The Hamilton operator for electrons in a confining potential $V(z)$ and a magnetic field parallel to the layers of the 2-D system is given by

$$H = \frac{1}{2m_c}(\vec{p} - e\vec{A})^2 + V(z), \quad \text{with } \vec{A} = (0, -zB_x, 0). \tag{1}$$

The corresponding Schrödinger equation can be separated in a x, y-dependent and a z-dependent part. The latter one, providing the electron subband energies $E_{c,\mu}$ and envelope functions $\Phi_{c,\mu}(z)$, reads

$$\left[-\frac{\hbar^2}{2m_c}\frac{d^2}{dz'^2} + \frac{e^2B_x^2z'^2}{2m_c} + V(z)\right]\Phi_{c,\mu}(z') = E_{c,\mu}\Phi_{c,\mu}(z'), \tag{2}$$

where $z' = z + \frac{\hbar k_y}{eB_x} = z + z_0$. The x, y-dependent part is the usual plane wave equation. Note, that the solutions of Eq. (2) depend on the position of the cyclotron orbit centre coordinate z_0 with respect to the potential $V(z)$.

Eq. (2) and a similar equation for the hole subbands are solved numerically for a parabolic well[5] embedded in the intrinsic layer of a *p-i-n*-heterostructure (see insert in Fig.1). For non-vanishing photo-induced electron and hole densities Hartee- and exchange-contributions to the potential $V(z)$ have to be taken into account selfconsistently. Heavy and light hole subband mixing[6] has also been included. Fig. 2 shows calculated luminescence spectra for different magnetic fields and a fixed (photo-induced) charge carrier density $n_{\text{QW}}^{(2)} = p_{\text{QW}}^{(2)}$ and external voltage $U_{pn} = 0$. With increasing magnetic field the intensity first increases due to the enhanced overlap between the wavefunctions of electrons and holes as outlined in the introduction. At magnetic fields exceeding 5 T, however, the spectra exhibit a gradual

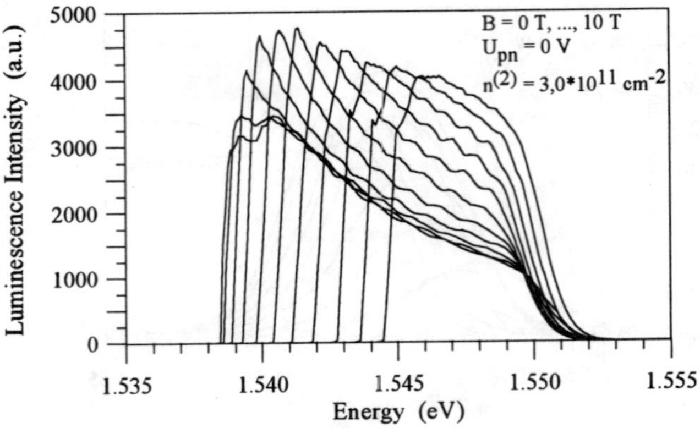

Fig. 2. Calculated luminescence spectra for different magnetic fields up to 10 Tesla.

decrease of the intensity. This reflects the fact that the spatially indirect band gap becomes now more and more indirect in momentum space, although the overlap of the wavefunctions still increases depending on the momentum parallel to the layers.

3. Experiment

Electrically controlled photoluminescence measurements without and with in-plane magnetic field have been performed on the *p-i-n*-heterostructure described in section 2 (see insert of Fig. 1). By variation of the external bias U_{pn} the electric field in the *i*-region of the sample and therefore also the spatial separation of electrons and holes can be tuned. The photoluminescence spectra obtained for different forward biases without a magnetic field (Fig. 1) clearly depict the typical tunability of peak energy and intensity of systems with spatially separated electrons and holes.[7]

The application of magnetic fields parallel to the layers results in an increase of the luminescence intensity by more than a factor of 10 for $B \leq 2\,\mathrm{T}$ (Fig. 3). For magnetic fields $B > 2\,\mathrm{T}$ the luminescence intensity is reduced drastically for the same reasons as mentioned in section 2. The extremely strong increase for $B \leq 2\,\mathrm{T}$ is attributed to a stronger spatial separation in the experiment compared to the calculation.

4. Conclusions

We have demonstrated theoretically and experimentally that magnetic fields parallel to the layers induce strong shifts of the electronic states, both in real and momentum space. The observed strong increase of the luminescence intensity at low and strong decrease at high magnetic fields is in agreement with the theoretical

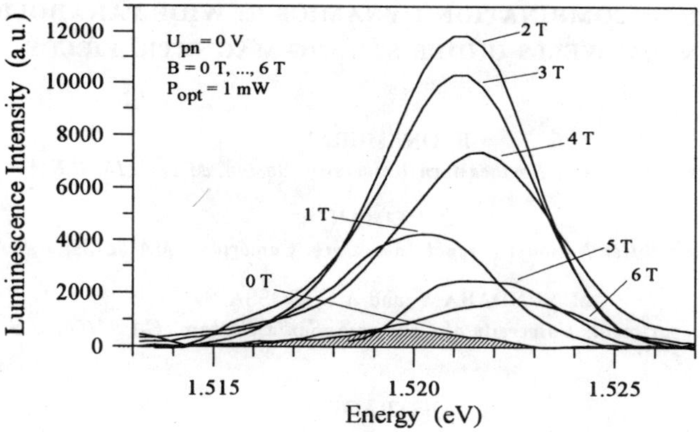

Fig. 3. Photoluminescence for different in-plane magnetic fields. The luminescence intensity increases up to $B = 2\,\mathrm{T}$. For higher fields it decreases drastically. The hatched area shows the intensity at zero magnetic field.

expectation.

Acknowledgements

We thank U. Rössler and K. Lier for helpful discussions. This work was supported by the Deutsche Forschungsgemeinschaft.

References

1. see for example *High Magnetic Fields in Semiconductor Physics II*, Springer Ser. in Sol.-State Sci., Vol. 87, ed. G. Landwehr (Springer, Berlin, 1989).
2. see for example G. Belle, J. C. Maan, and G. Weimann, *Solid State Commun.* **56** (1985) 65; J. C. Maan, in *Festkörperprobleme*, Vol. 27, ed. P. Grosse (Vieweg, Braunschweig, 1987), p. 137; J. Khurgin, *Appl. Phys. Lett.* **56** (1990) 2490.
3. G. H. Döhler, in *High Magnetic Fields in Semiconductor Physics II*, Springer Ser. in Sol.-State Sci., Vol. 87, ed. G. Landwehr (Springer, Berlin, 1989), p. 174..
4. S. K. Lyo and E. D. Jones, *Solid State Commun.* **83** (1992) 975.
5. R. C. Miller, A. C. Gossard, D. A. Kleinman, and O. Munteanu, *Phys. Rev. B* **29** (1984) 3740.
6. J. M. Luttinger and W. Kohn, *Phys. Rev.* **97** (1955) 869.
7. G. H. Döhler, *J. Vac. Sci. Technol.* **16** (1979) 851; G. H. Döhler, *IEEE* **QE-22** (1986) 1682.

EXCITON RECOMBINATION DYNAMICS IN WIDE PARABOLIC QUANTUM WELLS UNDER STRONG MAGNETIC FIELDS

H. OKAMURA*
Physics Department, Northeastern University, Boston, MA 02115, U.S.A.

D. HEIMAN
MIT Francis Bitter National Magnet Laboratory, Cambridge, MA 02139, U.S.A.

M. SUNDARAM† and A.C. GOSSARD
Materials Department, University of California, Santa Barbara, CA 93106, U.S.A.

ABSTRACT

Using time-resolved photoluminescence(PL), we have studied electron-hole recombination dynamics in n-type modulation-doped wide parabolic GaAs/AlGaAs quantum wells in magnetic fields up to B=19 T. The PL was dominated by exciton recombination due to low electron densities. At low temperatures ($T \leq 3$ K), the PL decay time showed a rapid increase with B above an onset field $\simeq 5$ T. This increase in decay time occurs in spite of an increased wave function overlap due to fields. A field-induced localization of electrons, that reduces the electron-hole correlation, is proposed as an explanation for the delayed recombination.

In the last several years, photoluminescence (PL) spectroscopy has emerged as a powerful probe in studies of two-dimensional(2D) electron systems in strong magnetic fields. In the regimes of quantum Hall effects(QHE), many anomalous behaviors have been observed in the PL spectra.[1] They include the appearance of multiple PL peaks and intensity transfer between those peaks at specific Landau level filling factors, among others.[1] These optical anomalies arise from changes in screening and chemical potential caused by the formation of the QHE states. More recently, the time-resolved(TR) PL technique has been applied to study the temporal evolution of the PL in the QHE regimes[1-5]. For example, Dahl *et al.*[2] and Heiman *et al.*[3] have found that the PL intensity minimum seen at Landau level filling factors $\nu = 1$ and $1/3$ are associated with a maximum in radiative lifetime. They suggested that the intensity minima are due to a decrease in electron-hole wave function overlap caused by a reduced screening in the QHE states. In addition, at lower filling factors $\nu \leq 1/3$, significant lifetime increases have been observed.[3,5] They have been discussed in terms of electron localizations, but the mechanisms giving rise to localizations are not yet well understood.

Modulation-doped wide parabolic GaAs/Al_xGa_{1-x}As quantum wells (WPQW) have been recently developed to create thick (≥ 1000 Å) layers of quasi-3D electron gases with a nearly flat density profile across the well[8-11]. This is in contrast to the conventional modulation-doped quantum well/interface structures, where

*Work performed at Francis Bitter National Magnet Laboratory.
†Present address: Jet Propulsion Laboratory, Pasadena, CA 91109, U.S.A.

electrons are confined within thin ($\leq$ 200 Å) layers. In a WPQW, a quantum well is grown so that the band gap changes parabolically along the growth direction, which is achieved by controlling the Al fraction x during the growth. Electrons are supplied into the well from remote donors set back from the well. Those electrons act to screen the parabolic potential, and consequently the electrons are expected to distribute uniformly, without accumulating at the well edges. Another feature of the WPQW is that there is no separation between hole and electron envelope functions. This is in contrast to the single well / interface structures, where electrons and holes have a finite separation. As a result, the electron-hole correlation in a WPQW is expected to be somewhat enhanced, making it an interesting system for optical studies as well. Fritze *et al.*[6] has found a Fermi edge singularity in the PL spectra of WPQW's.

In this work, we have studied the TR-PL from n-doped WPQW's in strong magnetic fields at low temperatures. The WPQW samples used in this work were grown by MBE using a digital alloy technique,[11] i.e., the parabolic wells consisted of 20 Å-superlattices of GaAs and $Al_xGa_{1-x}As$, with the average Al fraction changing parabolically from $x \simeq 0$ at the center of the well to $x = 0.3$ at the edges. Details of TR PL measurements in high magnetic fields can be found elsewhere.[2,7] Figure 1 shows the PL and the PL excitation(PLE) spectra of an n-doped 4640 Å WPQW sample under continuous excitation. The electron 2D density for this sample was $1.7 \times 10^{11}\text{cm}^{-2}$.[9] From the curvature of the parabolic well profile, the 3D density is estimated to be $\simeq 7\times10^{15}\text{cm}^{-3}$, with the Fermi energy $\simeq$ 2 meV. The PLE spectrum has a peak at the absorption edge, which is usually associated with exciton recombination. Upon applying magnetic fields, the PL spectrum shows a diamagnetic shift up to $B \simeq 5$ T, without Landau level splitting.[7] These observations indicate that the recombination is due to exciton. This is consistent with the low 3D density, well below the Mott critical density for exciton unbinding($\simeq 2\times10^{16}\text{cm}^{-3}$ for GaAs). Here we shall concentrate on the temporal evolution of the PL in magnetic fields, while a detailed account of the PL spectral changes in fields will be given elsewhere.[7]

Figure 2 shows the PL decay time τ as functions of temperature 0.5 K$\leq T \leq$10 K and magnetic field $B \leq$18 T. The fields were applied perpendicular to the electron layer(Faraday configuration). Significant lifetime increases are evident at low temperatures: below T=3 K the lifetime τ increases rapidly above an onset field $\simeq$5 T. At T=0.5 K, the lifetime at B=18 T is more than an order of magnitude longer than that at B=0 T. In contrast, above T=3 K, τ shows only a slight monotonic increase with B. Furthermore, τ has a peculiar temperature dependence at fields above the onset: below T=3 K, τ *increases* when T is lowered, but above T=3 K it *decreases* when T is lowered. Clearly, a different mechanism is giving rise to the decay times for $T \leq$3 K in the presence of a field. A sample with a smaller electron density($n_{2D} = 1.3\times 10^{11}$ cm^{-2}) was also studied, and a very similar lifetime increase was found with an onset at $\simeq$ 5 T.

Let us first consider effects of magnetic fields on exciton recombination rate. The oscillator strength τ^{-1} of exciton radiative recombination in a quantum well can be expressed as:[12]

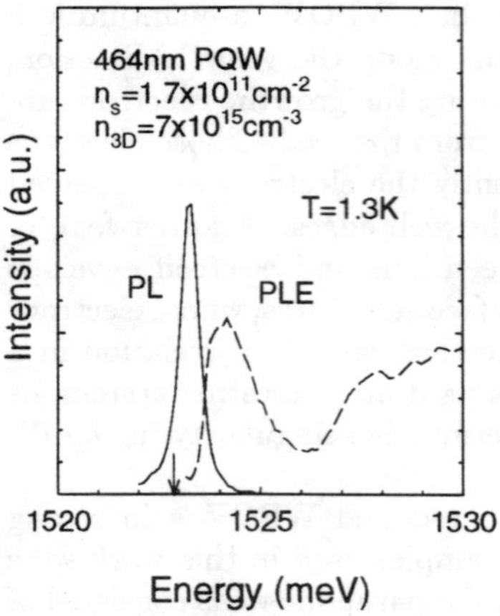

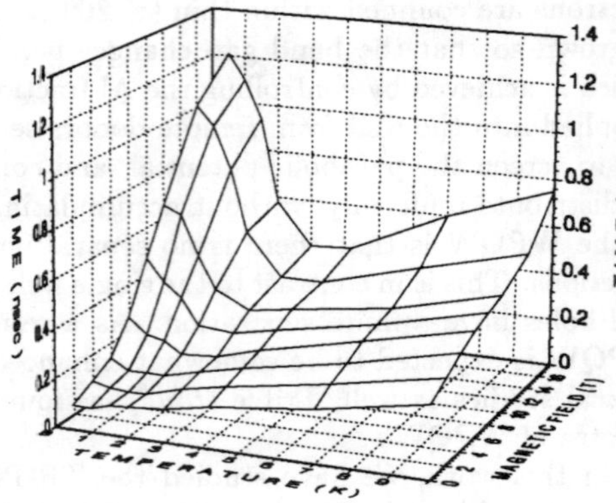

Fig. 1. (left): The PL (solid line) and the PLE (dashed line) spectra measured at T=1.3 K under continuous excitation. Fig. 2. (right): The PL lifetime as functions of magnetic field B and temperature T.

$$\tau^{-1} \propto \tau_0^{-1} \, |< \chi_e|\chi_h >_z|^2 | \, \phi_{e-h}(r=0) \, |^2 \, w(T) \tag{1}$$

where τ_0^{-1} is the squared transition matrix element, $< \chi_e|\chi_h >_z$ is the overlap of electron and hole envelopes perpendicular to the plane, $\phi_{e-h}(r)$ is the in-plane 2D hydrogenic wave function, $w(T)$ is a factor related to the homogeneous width and the thermal population of excitons. Since $w(T)$ is normally a decreasing function of temperature,[12] we expect that τ should decrease when T is lowered. Hence (1) cannot explain the T-dependence for $T \leq 3$ K, where τ *increases* when T is lowered. We now turn to the B-dependence of τ. In a B field the hydrogenic wave function of an exciton "shrinks".[13] Since this shrinkage gives a larger $| \, \phi_{e-h}(r = 0) \, |$, applying a magnetic field is expected to reduce τ, which is opposite to our observation. Therefore, the simple exciton model (1) is inconsistent with the actual data at $T \leq 3$ K in the fields. It has been reported that in a WPQW, $< \chi_e|\chi_h >_z$ can be modified in a magnetic field due to a depopulation of higher subbands.[14] This mechanism is, however, unlikely to cause the lifetime increases given that the recombination is due to excitons. Namely, once an electron is bound to a hole, whose envelope function is peaked at center of the well,[11] it will not be affected very much by a change in $\chi_e(z)$.

Since the lifetime increase at high fields becomes more enhanced at a lower temperature, a field-induced localization mechanism that slows the e-h recombination rate might be responsible. A localization of *excitons*, however, is not approppreate here since it would *enhance* the oscillator strength.[15] In that case, the internal e-h motion of an exciton is still free, even if the exciton center of mass is immobilized. Instead, to have a slower recombination rate, the e-h correlation of an exciton would have to be restricted by a localization. Since the onset field is nearly equal for samples with different electron densities, it appers to be scaled with the magnetic length(cycrotron radius) l_B, rather than the Landau level filling factor. Based on these observations, we put forward the following interpretation. Given that our sample includes $Al_{0.3}Ga_{0.7}As$ alloy, the carriers can become trapped by alloy disorder even at zero magnetic field.[16] As the applied field is increased, l_B decreases ($\simeq$ 120 Å at B=5 T), and above the onset, the electrons also can become trapped by potential fluctuations. Since both the electron and the hole are trapped, their dynamical response will be reduced; τ will increase. Similar lifetime increases versus B have been observed in PL studies of modulation-doped 2D systems.[3,5] In those 2D structures, holes have some spatial separation from electron layers,[1] hence an electron localization will lead directly to a PL lifetime increase. In contrast, there is no layer separation in our case. Our observation is unique in that a cross-over from a trapped exciton to separately trapped electron and hole might have been observed.

We thank M. Dahl, G. Favrot and L.B. Liao for useful discussions. H.O. thanks Prof. C.H. Perry for encouragement. This work is supported by NSF Grant DMR-9201614 and the National Magnet Laboratory is supported by NSF.

References

1. D. Heiman, in *Semiconductors and Semimetals*, Vol. 36, eds. C. Littler and D.G. Seiler (Academic Press, New York, 1992); A. Nurmikko and A. Pinczuk, Physics Today, June (1993) 638.
2. M. Dahl *et al. Phys. Rev.* **B45** (1992) 6953; *Surf. Sci.* **263** (1992) 638.
3. D. Heiman *et al. Surf. Sci.* **305** (1994) 50; *Physica B*, to be published.
4. S.A. Brown *et al., Surf. Sci.* **305** (1994) 42.
5. I.V. Kukushkin *et al., Phys. Rev. Lett.* **72** (1994) 3594.
6. M. Fritze *et al., Phys. Rev.* **B45** (1992) 8408; *ibid*, **B48** (1994) 15103.
7. H. Okamura *et al.*, to be published.
8. A.J. Rimberg and R.M. Westervelt, *Phys. Rev.* **B40** (1989) 3970.
9. J.H. Burnett *et al., Phys. Rev.* **B43** (1993) 12033.
10. P.F. Hopkins *et al., Appl. Phys. Lett.* **57** (1990) 2823.
11. M. Sundaram *et al., J. Appl. Phys.* **69** (1991) 2370.
12. J. Feldmann *et al., Phys. Rev. Lett.* **59** (1987) 2337.
13. Y. Yafet *et al., J. Phys. Chem. Solids* **1** (1956) 137.
14. M.P. Stopa and S. Das Sarma, *Phys. Rev.* **B47** (1993) 2122.
15. E. I. Rashba and G. E. Gurgenishvili, *Sov. Phys. Solid State* **4** (1962) 759.
16. M.S. Skolnick *et al., Phys. Rev. Lett.* **58** (1987) 2130.

Transfer of excitons between coupled CdTe quantum wells tuned by an external high magnetic field

S. Haacke[1], I. Lawrence[2], H. Mariette[2], P. Peyla[2], W.W. Rühle[3], and M. Zigone[1]

[1]Grenoble High Magnetic Field Laboratory MPIF-CNRS, B.P. 166, 38042 Grenoble Cédex 9, France and Institute for Micro- and Optoelectronics, Federal Institute of Technology, 1015 Lausanne, Switzerland.
[2]CEA-CNRS joint group "Microstructures de Semiconducteurs II-VI", Laboratoire de Spectrométrie Physique and DRFMC - Centre d'Etudes Nucléaires de Grenoble, 38054 Grenoble Cédex 9, France.
[3]Max-Planck-Institut für Festkörperforschung, Heisenbergstr. 1, 70569 Stuttgart, Germany

Abstract

Steady-state and picosecond photoluminescence experiments on CdTe/CdZnTe asymmetric double quantum wells in high magnetic fields (B≤20 T) are presented. We demonstrate the importance of inter-well transfer of excitons in this II-VI system revealed by a reduction of the excitonic lifetimes at magnetic fields where the narrow well 1s and the wide well 2s excitons coincide in energy.

1. Introduction

Asymmetric double quantum wells (a-DQWs) consist of a narrow QW of thickness L_N and a wide QW of thickness L_W which are separated by a barrier with a thickness L_B. They are widely used as model systems for the investigation of carrier transfer dynamics by time-resolved photoluminescence on a picosecond timescale[1]. As a consequence of tunneling the narrow well (NW) lifetime is considerably reduced. The calculated carrier transfer times allow to estimate the ultimate operation frequencies of tunneling devices such as resonant tunneling diodes or BRAQWET modulators[2].

In II-VI heterostructures such as CdTe/CdZnTe, the simultaneous tunneling of electrons and holes in form of excitons, rather than as single carriers is observed[3]. The importance of exciton tunneling in II-VIs when compared to GaAs/AlGaAs relies on two basic properties. First, the thin CdTe QWs investigated present exciton binding energies E_b as high as 20 meV. Second, these values of E_b are comparable with the shallow valence and conduction confinement potentials of 15-20 meV and 60-80 meV, respectively. The heavy-hole confinement potential is therefore actually a sum of the band-offset plus the Coulomb potential centered around the electron position[4]. In such a situation, the heavy-hole is believed to follow the tunneling electron which results in the observed exciton transfer.

In III-Vs, the application of an electric field tunes electron or hole wavefunctions of adjacent wells into resonance which is observed as a drastic shortening of the carrier lifetime in the NW[5]. In the present case, we will show that, using the strong diamagnetic shift of the wide well 2s exciton state, 2s(WW), resonant *exciton* tunneling can be achieved between the 1s(NW) and the 2s(WW) state in high *magnetic* fields.

2. Samples and experimental details

The MBE grown CdTe/CdZnTe a-DQWs are described thoroughly in Ref. 3. We will focus here on a sample with well widths of L_N=40 Å, and L_W=74 Å separated by a $Cd_{0.87}Zn_{0.13}Te$ barrier of L_B=109 Å. A single CdTe QW of 40 Å width and of the same barrier height show PL decay times of 147 ps, limited by non-radiative processes[3].

The photoluminescence excitation (PLE) data are recorded at T=2 K under low-power (<100 mW/cm^2) excitation by a Ti:Sapphire laser. The PL is detected with a cooled GaAs photomultiplier placed behind a 2.0m monochromator ($\Delta\lambda$<1 Å). Magnetic fields B≤ 20 T were produced by a Bitter magnet. For the time-resolved photoluminescence (TRPL) measurements the samples were excited by the 2-ps pulses of a mode-locked Ti:Sapphire laser. The excitation density estimated to be below 10^{10} cm^{-2} excitons per pulse was low enough to avoid space charge effects. The PL was dispersed in a 0.32-m monochromator (resolution 5 Å) and detected by a two-dimensional synchroscan streak camera. Exponential PL decays as fast as 7 ps can be measured. The sample temperature was held at 10 K by a cold-finger cryostat placed in the centre of a superconducting coil (B≤ 14 T). The magnetic field was applied perpendicular to the a-DQWs in Faraday configuration.

3. Experimental Results

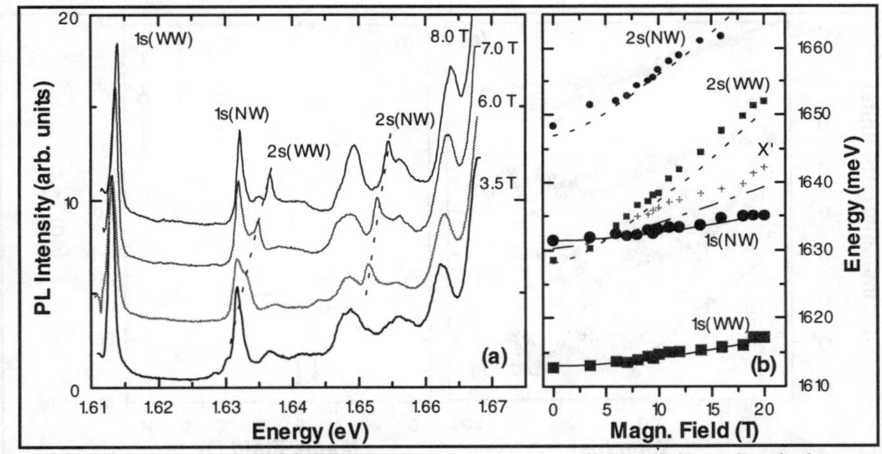

Figure I: (a): PLE spectra for B≤ 8 T excited by σ^- polarized laser light and detected at the low-energy side of the 1s(WW) σ^- emission. The 2s states of both NW and WW are indicated by a guide-to-the-eye. (b): Magnetic field dependence of the excitonic transition energies observed for the WW (squares) and for the NW (circles). Crosses indicate the indirect X'(e_W, hh_N) exciton. The lines represent results of numerical calculations.

For the sample 40/109/74 which shows a difference 1s(NW) - 2s(WW) of 2.8±0.3 meV at B=0 T PLE experiments up to B=20 T have been carried and are shown in figure I.a for exciting laser light polarized σ^- (low-energy component). The PL is detected on the σ^- polarized component of the low-energy side of the 1s(WW) exciton. The 1s states corresponding to NW and WW can be easily identified at 1.613 eV (WW) and 1.6315 eV (NW), respectively. Both peaks shift very slightly with increasing magnetic field B (< 0.2 meV/T). For B=3.5 T we observe a shoulder on the low-energy side of the 1s(NW) peak (at 1.630 eV) which evolves linearly for higher magnetic field (B>7.5 T) with a slope of roughly 1.3 meV/T (fig. I.b). This linear field dependence at higher fields strongly suggests to identify this peak as the 2s state of the WW. The 1s(NW) and 2s(WW) have crossed for B> B_C≈5 T.

The 2s(NW) exciton is identified at higher energies (E>1.65 eV) by a nearly linear shift of 1.3 meV/T (B>5 T). This transition is enhanced in PLE detected on the NW. The transition around 1.662 eV is assigned to the light-hole exciton of the CdZnTe buffer layer. The origin of the transitions at 1.648 and 1.6365 eV is not clear for the moment. For B>8 T a weak peak appears between the 1s(NW) and the 2s(WW) exciton (crosses in fig. I.b). This transition evolves slowly with B but gains in strength with increasing magnetic field. Its assignement will be clarified below.

The calculated transition energies of the excitonic states are represented in fig. Ib as lines. Numerical calculations are performed following the method of Leavitt and Little[6] adapted for the case of arbitrary magnetic fields[7]. The 1s-2s splitting of both wells and the linear slope of the 2s transitions in magnetic field are well reproduced by theory provided that an infinite in-plane mass is assumed for heavy-holes. The calculated transition energy of the « cross » exciton $X'(e_W,hh_N)$, made up of the electron localized in the WW and the heavy-hole in the NW, is represented as dahed-dotted line. Although roughly 2 meV too low in energy, its slope corresponds well to the weak transition (crosses) observed in PLE. We therefore assign this transition as the $X'(e_W,hh_N)$ exciton indirect in real-space.

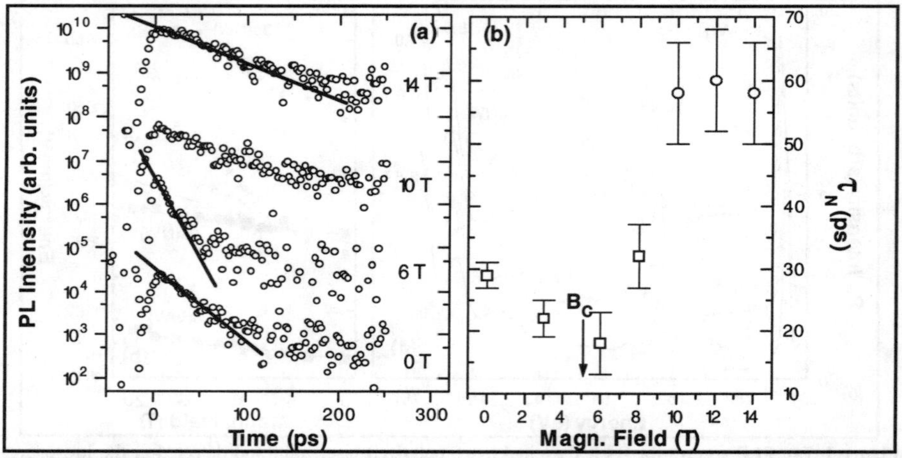

Figure II: (a): PL decay spectra of the NW for B≤14 T. An exponential decay can be observed over 1-2 decades. (b): The magnetic field dependence of the NW PL decay time τ_N indicating a minimum at B_C when the 1s(NW) and 2s(WW) coincide in energy.

The PL decay time τ_N of the NW has been measured for magnetic fields up to B=14 T(figure II). The wavelength of the σ^+ polarized laser is 745 nm which corresponds to the light-hole exciton of the CdZnTe barrier. The low-energy σ^- component of the PL is detected avoiding thereby the effect of spin relaxation on the observed PL decay times. Figure II.a and b clearly demonstrate a minimum of the NW PL decay time τ_N at B=6T close to B_C. Starting with a zero-field PL decay time of $\tau_N(B=0)=28\pm3$ ps we observe a decrease down to $\tau_N=18\pm5$ps at B=6 T. For higher fields (B>6 T) τ increases drastically up to $\tau_N=60\pm5$ ps, and remains constant for B≥10 T.

The PL decay time τ_W of the WW increases by 60 ps up to $\tau_W=225$ ps for magnetic fields B>12 T (not shown here). Moreover, an increase of the steady-state PL intensity is observed. This result is consistent with the fact that the PL decay times are limited by non-radiative processes. If the decay time τ_W was a radiative lifetime it would be expected to decrease according to an increase of the excitonic oscillator strength. Non-radiative scattering events that an exciton can undergo are apparently suppressed by the high magnetic field. To clarify this preliminary interpretation one should perform temperature-dependent measurements of the PL decay time for a given magnetic field.

4. Discussion

In analogy to the inter-well "1→2" resonances between electron states observed in GaAs/(Al,Ga)As a-DQW's[5] one tends to interpret the observed resonance induced by the magnetic field as a delocalization of the excitonic wavefunctions 1s(NW) and 2s(WW). The observed PL decay times as short as 20 ps are then due to an increase of the probability of elastic scattering (by impurities or interface defects) in this resonant situation. An LO phonon-assisted relaxation into the 1s(WW) state is not possible since the energy difference is only 18.8 meV. At high magnetic fields, the off-resonance coupling between both wells is not negligible since τ_N is as low as 60 ps and does not reach the reference value of an uncoupled QW (147 ps). The transfer of 1s(NW) excitons can occur by inelastic scattering with acoustic phonons as in the zero field case, but the excitonic wavefunction is much less delocalized.

The transfer of electrons from a direct into an indirect (« cross ») exciton has been observed in GaAs/AlGaAs[8]. Observing the cross exciton X'(e_W,hh_N) in PLE does not mean that this state is an effective channel for tunneling. The build-up of X'(e_W,hh_N) assumes a much faster transfer time for electrons than for holes. This is not compatible with an analysis of the rise-time of the WW PL recorded at B=0 T which demonstrates that electrons and holes have very comparable transfer times.

5. Conclusions

The magnetic field allows to tune the inter-well coupling in CdTe/CdZnTe a-DQWs by aligning energetically the 1s(NW) and the 2s(WW) excitons. This is demonstrated by a strong magnetic field dependence of the 1s(NW) lifetime. The magnetic field is shown to be a powerfull tool to change the coupling between the wells and to evidence that the dominant transfer proces is the tunnelin of excitons and not of single carriers.

References:

1. See for instance: K. Leo, J. Shah, E.O.Göbel, J.P. Gordon, and S. Schmitt-Rink, *Semicond. Sci. Technol.* 7, B394 (1992).
2. M. Wegener, T.Y. Chang, I. Bar-Joseph, J.M. Kuo, and D.S. Chemla., *Appl. Phys. Lett* **55** (1989) 583.
3. S. Haacke, N.T. Pelekanos, H. Mariette, A.P. Heberle, W.W. Rühle, and M. Zigone, *Journal of Crystal Growth,* **138** (1994) 831.
4. P. Peyla, Y. Merle d'Aubigné, A. Wasiela, R. Romestain, H. Mariette, M.D. Sturge, N. Magnea, and H. Tuffigo, *Phys. Rev. B* **46**, (1992) 1557.
5. D. Oberli, J. Shah, T.C. Damen, C.W. Tu, T.Y. Chang, D.A.B. Miller, J.E: Henry, R.F. Kopf, N. Sauer, and A.E. DiGiovanni, *Phys. Rev. B* **40**, (1989) 3028.
6. R.P. Leavitt and J.W. Little, *Phys. Rev. B* **42**, 11774 (1990).
7. P. Peyla, Ph. D. thesis, Université J. Fourier, Grenoble (1992).
8. A.P. Heberle, W.W. Rühle, and K. Köhler, *Phys. Stat. Sol.* **B173**, (1992) 381 and R. Ferreira,

ELECTRONIC LIFETIMES OF EXCITED LANDAU LEVELS IN GaAs/AlGaAs HETEROSTRUCTURES

W. HEISS[a], B. N. MURDIN[b], C. J. G. M. LANGERAK[b,c], G. M. H. KNIPPELS[b], I. MARAN[a], K. UNTERRAINER[a], E. GORNIK[a], C. R. PIDGEON[c], N. J. HOVENIER[d], W. TH. WENCKEBACH[d], G. WEIMANN[e]

a) Institut für Festkörperelektronik, Technische Universität Wien, 1040 Wien, Austria
b) FOM-Instituut voor Plasmafysica, Rijnhuizen, 3430 BE Nieuwegein. The Netherlands
c) Department of Physics, Heriot Watt University, Edinburgh, UK
d) Faculteit der Technische Natuurkunde, TU Delft, The Netherlands
e) Walter Schottky Institut, München, Germany

ABSTRACT

Saturation experiments in AlGaAs/GaAs heterostructures are carried out using an optically pumped far infrared laser as well as using a free electron laser. From the analysis of the cyclotron resonance at low intensity the scattering time associated with the linewidth τ_{cr} is deduced. At higher intensity the heating produces an apparent intensity dependence τ_{cr}. When sufficiently short pulses are used we find $\tau_{cr} = 15$ ps independent of the laser intensity. Therefore, for the first time a saturation of the cyclotron resonance transition in hererostructures is found which is not masked by heating effects.

1. Introduction

Contradictory results of electronic lifetimes in two- dimensional electron gas (2DEG) systems derived by saturation spectroscopy have been reported [1, 2]. Partial saturation of cyclotron resonance of a 2DEG was reported at 118.8 μm [1], using an optically pumped far infrared laser (OPFIRL); other authors, working at longer wavelengths and with sample densities higher than 1×10^{11} cm^{-2} were not able to achieve saturation up to considerably higher intensities [2, 3]. More recently saturation was achieved in high mobility samples in the spin polarised limit (filling factor < 1) using an OPFIRL[4], and also in coupled quantum well structures with $n_s < 10^{11}$ cm^{-2}, using the Santa Barbara free electron laser (FEL) [5].

In order to compare saturation measurements from two extreme regimes, i. e. steady state and transient, we have performed experiments using both an OPFIRL (steady state) and a FEL (transient). Furthermore, we present the first saturation experiments for rather high density samples ($n_s = 4 \times 10^{11}$ cm^{-2}) using the free electron laser FELIX, at Rijnhuizen.

2. Experimental

The lifetime measurements with the low density sample ($n_s = 4.2 \times 10^{10}$ cm^{-2}, $\mu = 0.9 \times 10^6$ cm^2/Vs) were performed with a high power quasi cw OPFIRL used in the pulsed mode ($\lambda = 118.8$ μm). The pulse duration (0.3 ms) was adjusted to be much longer than the expected inelastic relaxation lifetime, ensuring steady state condition during the optical excitation.

The high doped sample with $n_s = 4 \times 10^{11}$ cm^{-2} ($\mu = 1.5 \times 10^6$ cm^2/Vs) was investigated with the free electron laser FELIX, which was adjusted to deliver "macropulses" with a duration of 4 μs (repetition rate f = 5 Hz). Each macropulse consisted of a train of "micropulses", 20 ps long wih 1 ns spacing. The laser was set to a wavelength of 70 μm with a spectral width of 1.1 %. In both experiments (OPFIRL and FEL) the transmitted signal was measured with an InSb detector.

3. Experimental Results:

In a ladder of equidistant Landau levels (LL) no saturation can be achieved. In GaAs/AlGaAs heterostructures a non-equidistant spacing can be produced by the resonant polaron effect and band non-parabolicity. The polaron effect leads to a change of the effective mass of a Landau level when the energy of the LL is close to the optical phonon energy[6] (36 meV).

For a transition energy of 10.43 meV where the experiments with the OPFIRL are carried out, three LLs are well below the optical phonon edge and thus can be assumed to be equidistant. Assuming that the fourth Landau level is close to the LO phonon energy and is shifted due to the polaron effect [4], the CR saturation can be described with a three level system in the steady state. The rate equations and the normalised transmission change are given in detail in Ref. 4. Since the pulse duration of the OPFIRL is much longer than the expected relaxation time it is justified to solve the rate equations in the steady state. The equations of the three level system contain three parameters which have to be fitted to the experiment. These parameters are the inelastic relaxation lifetimes of the electrons from the n = 1 to n = 0 LLs (T_1) and from n = 2 to n = 1 (T_2), and the scattering time associated with the linewidth τ_{cr}.
The saturation experiment done with FELIX is analysed with time dependent rate equations. Furthermore, the photon energy here was 17.7 meV (70 μm) where only two LL levels are situated well below the optical phonon edge while the n = 2 LL is close to the LO phonon energy. At this photon energy the CR saturation can be described by a two level system.

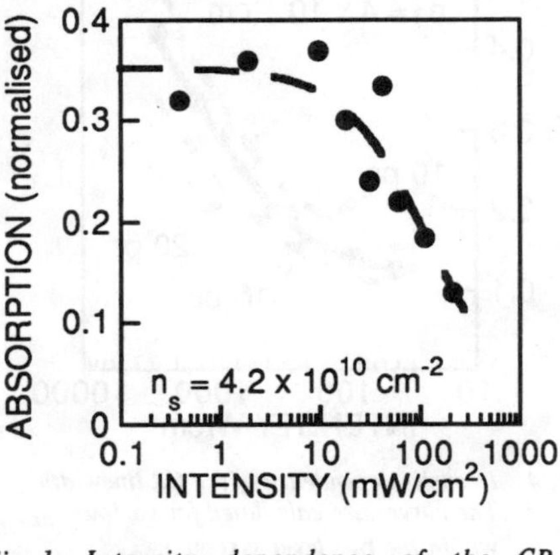

Fig. 1 Intensity dependence of the CR absorption measured with the OPFIRL at 118.8 μm (dots). The dashed line is a solution of the rate equations using $T_1 = T_2 = 4.9$ ns.

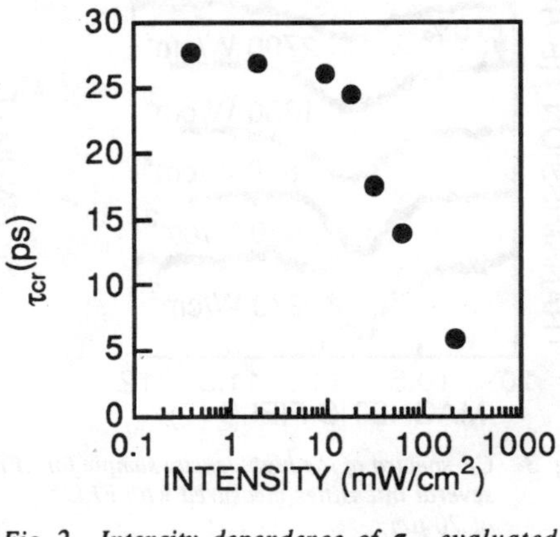

Fig. 2 Intensity dependence of τ_{cr} evaluated from a fit of the CR linewidth under the assumption T = 4.9 ns.

3.1. Steady State (OPFIRL) Measurements

In Fig. 1 the normalised absorption of the CR transition is shown as a function of the laser intensity. No change of the absorption is observed up to an intensity of 30 mW/cm^2. For higher intensities the absorption decreases significantly. In addition to the decrease of the absorption coefficient also a broadening of the CR transition line is observed.

As a first approximation we assume that τ_{cr} is independent of laser intensity and derive it from a fit of the absorption spectrum at low intensities. We find $\tau_{cr} = 27$ ps. We also assume a constant $T_1 = T_2 = T$ and Fig. 1 shows a best fit of the peak normalised absorption from the rate equations, using T = 4.9 ns (dashed curve). This fit can be improved by allowing T to vary with intensity (and is found to be inversely proportional to N_1, the carrier concentration in the n = 1 LL; see Ref. 4). Alternatively we may fix T (at 4.9 ns) and allow τ_{cr} to be intensity dependent, and thus varying the CR linewidth which results in the analysis shown in Fig. 2.

As a result these two approaches do not give consistent fits of the complete CR curves, and therefore do not lead to conclusive lifetimes. We have therefore performed the following transient experiment, which allows us to find τ_{cr} without any other fitting parameter.

3.2. Transient (FEL) Measurements

Maran et al. [4] have shown that the relaxation time T_1 is inversely proportional to the carrier concentration in the first LL ($T_1[s] = 5 / N_1[cm^{-2}]$). Extrapolating this result, we expect an minimum relaxation time of 20 ps for a sample with $n_s = 4 \times 10^{11}$ cm^{-2} (i. e. $N_1 = 2 \times 10^{11}$ cm^{-2} under conditions of a two levels system). This time allows an application of FELIX since the expected lifetime lies between the micropulse duration (20 ps) and the distance between two subsequent micropulses (1 ns). In that situation the CR absorption from the solution of the rate equations depends on the pulse duration (20 ps)

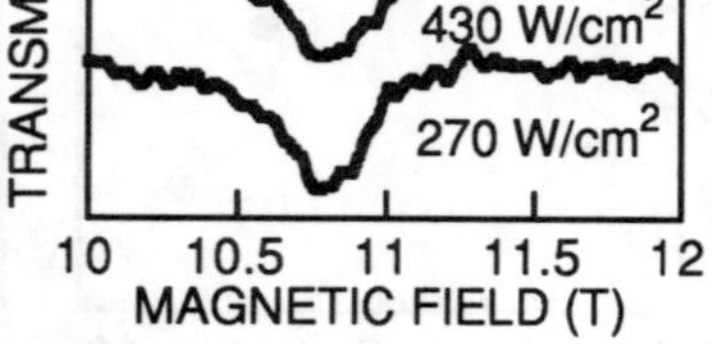

Fig. 3 CR spectra of the high density sample for several intensities measured with FELIX at 70 μm.

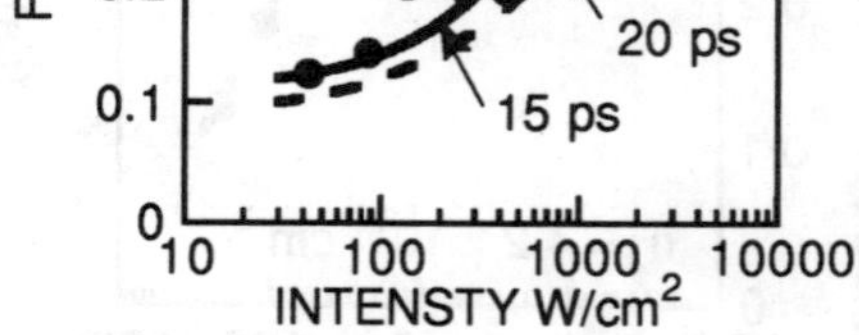

Fig. 4 Intensity dependence of the CR linewidth. The curves are calculated for various τ_{cr} within the two level system.

and is no longer influenced by the value of T_1. No assumption has to be made for T_1 for the analysis of the rate equations allowing a direct determination of the intensity dependence of τ_{cr}.

In Fig. 3 CR spectra are shown as a function of the laser intensity. The transmission of the CR transition increases for higher intensities. Furthermore, a small shift of the CR peak to higher magnetic fields is observed. The transient intensity dependence of the CR was calculated within a two level system (equations are given in Ref. 4) for a constant value of τ_{cr}. This analysis yields a (constant) characteristic time of 20 ps (i. e. equal to the micropulse duration, not due to the relaxation time T_1).

In Fig. 4 the dependence of the CR linewidth is shown in more detail. The linewidth given in this plot is the FWHM of the absorption lines shown in Fig. 3 minus the spectral width of FELIX (0.12 T). The distinct lines in Fig. 4 represent the linewidth of the CR absorption calculated for τ_{cr} = 10 ps, 15 ps, and 20 ps. By comparing these lines with the experimental data it can be seen that the intensity broadening of the CR absorption can be explained with a constant τ_{cr} of 15 ps. This is a proof that we have performed for the first time a clean saturation experiment in heterostructures not masked by heating effects.

4. Summary

Saturation experiments were performed in two regimes, the steady state (OPFIRL) and transient (FELIX). With the OPFIRL saturation can be observed in rather low density samples while for densities higher than 1×10^{11} cm^{-2} a laser with short pulses in the order of some ps has to be used to avoid heating of the sample. Using a transient analysis of the rate equations and the results of the experiments with the OPFIRL a consistent interpretation of the CR saturation experiment is shown yielding a constant value for τ_{cr} of 15 ps for a sample with $n_s = 4 \times 10^{11}$ cm^{-2}. In view of this the present system is well suited for pump-probe experiments for determination of the relaxation time T_1, which will be the subject of future work.

5. Acknowledgement

This work is partially supported by the "Fonds zur Förderung der Wissenschaftlichen Forschung, FWF", Austria (P9301 TEC) and by the "Jubiläumsfonds der österreichischen Nationalbank". It was also performed as a part of the research programme at the "Stichting voor Fundamenteel Onderzoek der Materie" (FOM) with finacial support from the "Nederlandse Organisatie voor Wetenschappelijk Onderzoek" (NWO).

6. References

1. M. Helm, E. Gornik, A. Black, G. R. Allan, C. R. Pidgeon, K. Mitchell and G. Weimann, *Physica* **134B+C** (1985) 323
2. G. A. Rodriquez, R. M. Hart, A. J. Sievers, F. Keilmann, S. L. Wright and W. I. Wang, *Appl. Phys. Lett.* **49** (1986) 458
3. Ranganathan, J. Kaminski, W. J. Li, J-P. Cheng and B. D. McCombe, *Proc. SPIE* **1678** (1992) 120
4. I. Maran, W. Seidenbusch, E. Gornik, G. Weimann and M. Shayegan, *Semic. Sci. Technol.* **9** (1994) 700
5. W. J. Li, B. D. BcCombe, J. P. Kamisnki, S. J. Allen, M J Stockman, L S Muratov, L. N.Pandey, T. F. George and W. J. Schaff, *Semic. Sci. Technol.* **9** (1994) 630
6. G. Lindemann, R. Lassnig, W. Seidenbusch, E. Gornik, *Phys. Rev. B* **28** (1983) 4693

Tilted Field Cyclotron Resonance in MegaGauss Fields using the semimetallic superlattice system InAs/GaSb

R.J. Nicholas*, Y. Shimamoto+, Y. Imanaka+, N. Miura+, N.J. Mason*, and P.J. Walker*

* Physics Dept., Oxford University, Clarendon Laboratory, Parks Rd., Oxford, OX1 3PU, U.K.

+ I.S.S.P., Tokyo University, Roppongi, Minato-ku, Tokyo, Japan

ABSTRACT

Cyclotron resonance has been studied in tilted magnetic fields up to 120T, in the semimetallic semiconductor superlattice system InAs/GaSb. This shows two effects: (i) at fields of around 40T, the resonance position is found to shift to lower magnetic fields on tilting, and (ii) at around 100T, the resonance intensity is strongly reduced due to a magnetic field induced semimetal to semiconductor transition, but returns rapidly on tilting the field. These two effects are explained in terms of the interaction of the confinement and Landau energies, which couple strongly when the fields are tilted, and influence the effective mass through non-parabolicity.

We report a study of cyclotron resonance in very high tilted magnetic fields in superlattices of the semimetallic system InAs/GaSb using the megagauss facility of the University of Tokyo. The application of very high magnetic fields has substantial effects on the properties of this system, due to the 'crossed' or 'broken-gap' alignment of the bands in which the conduction band edge of the InAs lies 140meV above the top of the valence band of the GaSb. It has been shown recently[1] by cyclotron resonance experiments that magnetic fields of order 40-50T are sufficient to induce a transition from semimetallic to semiconducting behaviour in long period InAs/GaSb superlattices, due to the additional magnetic quantization

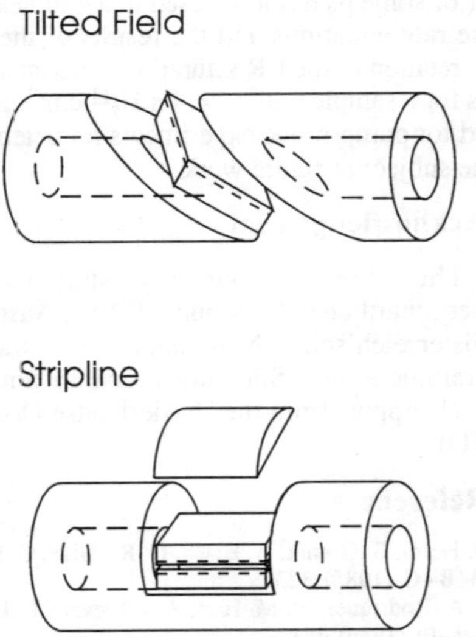

Figure 1: A schematic view of the experimental arrangement.

energy. At this point there is a very strong reduction in the intensity of the cyclotron resonance absorption. This paper describes a study of cyclotron resonance in a long period superlattice at high fields, in which tilted field experiments are performed for the first time in this field range.

The experiments consist of measurements of the transmission of infra-red radiation in the region of 10μm and 5μm from CO_2 and CO lasers, as a function of magnetic field generated using a single turn coil pulsed magnetic field system, with a maximum magnetic field of 150T, and a pulse duration of typically 5μsec. The sample arrangement consists of a pair of opposing wedges of varying angles cut in plastic tube, with the sample sandwiched between them, and a sample aperture of 1mm. In addition some measurements were performed in a stripline geometry, in which a pair of epitaxial samples are placed in the centre of the guide, with the active layers adjacent to each other. The sample arrangements are shown schematically in fig. 1.

The sample studied consisted of a 20 period InAs/GaSb superlattice, with layer thicknesses of 450Å/100Å, grown onto a 2 μm GaSb buffer layer by MOVPE on a semi-insulating GaAs substrate.

Typical experimental recordings are shown in fig. 2 for 9.64μm, for three different tilt angles (0, 45°, and 60°) and for the data taken using the stripline geometry. These all give a single resonance in the region of 40-45T, whose intensity does not change very significantly with angle, but whose position moves to a **minimum** value for tilt angles in the region of 45°. The normal expectation is that for tilted fields the resonance for a two dimensional system will follow a Bcosθ behaviour, but it is well known that once the cyclotron energy exceeds the subband spacing there will be a transition back to something close to 3-D behaviour[2,3]. In the case of InSb, which was studied in the Voigt geometry[4] with θ from 0 to 90°, it was found that for high energies the resonance reverted to an essentially bulk behaviour. It is, however, unexpected that the resonance position should shift to a <u>lower</u> field on tilting.

A second series of recordings was then taken using 5.53μm, as shown in fig. 3, in which the resonance has moved up to the region of 100T. This showed a further unexpected behaviour. The trace at 0°

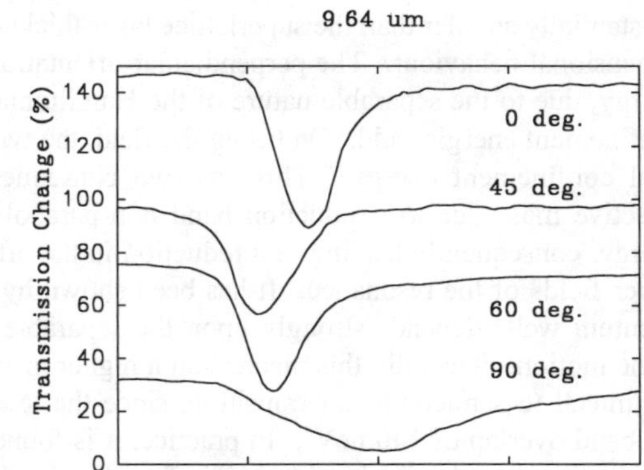

Figure 2: Typical experimental recordings for a series of different tilt angles using 9.64μm radiation.

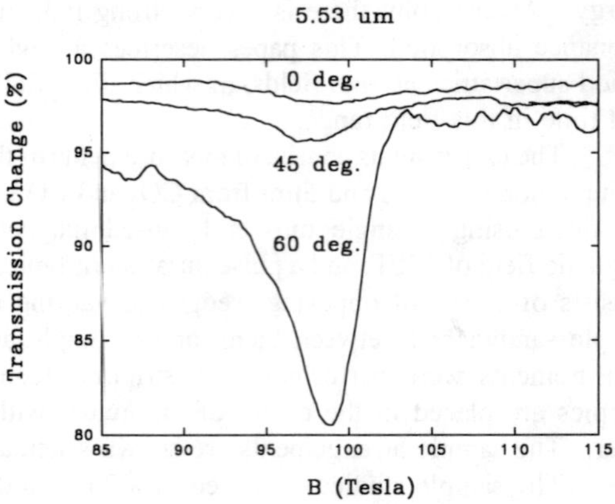

Figure 3: Typical experimental recordings for a series of different tilt angles using 5.53μm radiation.

showed a very weak resonance, in which the resonance absorption had fallen by approximately a factor of 50 compared with the data taken with 9.64μm. This is consistent with the recent study by Barnes et al[1], in which the fall in absorption was shown to be due to the uncrossing of the lowest electron and the highest hole Landau levels. Upon tilting the sample, however, there is a rapid return of the absorption strength, and the resonance again shows a weak shift down to lower fields. This behaviour is summarised in fig. 4, which shows the field positions of the resonance for the two wavelengths as a function of tilt angle, and also the carrier density per layer, as deduced from the absorption intensity, as a function of the <u>perpendicular</u> component of magnetic field (Bcosθ).

Our explanation of this behaviour is based on the mixing of the cyclotron motion with the confinement, due to the tilted fields, combined with the effects of non-parabolicity. The scale of the cyclotron motion (cyclotron radius l_c= 26Å at 100T) is substantially smaller than the superlattice layer thicknesses, leading to an essentially three dimensional behaviour. The perpendicular orientation leads to a maximum confinement energy, due to the separable nature of the Hamiltonian, since the magnetic and quantum confinement energies add. On tilting the field, the two energies mix, leading to a <u>smaller</u> total confinement energy. This has two consequences. Firstly, the increase in the effective mass due to conduction band non-parabolicity is reduced due to the smaller energy, consequently leading to a reduction in the effective mass and therefore a shift to lower fields of the resonance. It has been shown by Ekenberg[5] that non-parabolicity in quantum wells depends strongly upon the separable nature of the confinement and in-plane motion. Secondly this means that a higher magnetic field is required to induce the semimetal to semiconductor transition, since the total confinement energy must exceed the band overlap of 140meV[6]. In practice, it is found that the intensity of the resonance (proportional to the carrier density) is found to be determined by the two dimensional component of magnetic field Bcosθ, as shown in fig. 4(c).

The upward shift of the resonance position for the Voigt (stripline) geometry is not explained in this picture, and is most probably associated with the spatially dependent

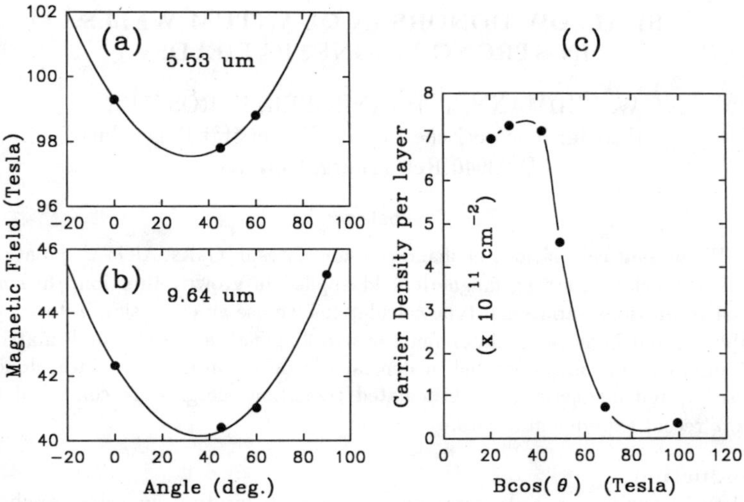

Figure 4: (a) and (b) are resonance field versus angle; (c) is electron density, deduced from absorption intensity, versus perpendicular field component Bcos(θ).

Landau level energies discussed by Merkt[7] for the Voigt geometry. Detailed modelling of the Landau energies in the Voigt geometry was not attempted, due to the rather complex self-consistent nature of the confinement.

References:

1. D.J. Barnes, R.J. Nicholas, R.J. Warburton, N.J. Mason, P.J. Walker, and N. Miura, Phys. Rev. **B49** (1994) 10474
2. S. Huant, M. Grynberg, G. Martinez and B. Etienne, Sol. State Commun. **65** (1988) 457
3. R. Merlin, Solid State Commun. **64** (1987) 99
4. J.H. Crasemann and U. Merkt, Solid State Commun. **47** (1983) 917
5. U. Ekenberg, Phys. Rev. **B40** (1989) 7714
6. D.M. Symons, M. Lakrimi, R.J. Warburton, R.J. Nicholas, N.J. Mason, P.J. Walker and M.I. Eremets, Semicond. Sci. Technol. **9** (1994) 118
7. U. Merkt, Phys. Rev. **B32** (1985) 6699

SHALLOW DONORS IN QUANTUM WELLS IN STRONG MAGNETIC FIELDS

W. WIDMANN, T. DARNHOFER, U. RÖSSLER
Institut für Theoretische Physik, Universität Regensburg
D-93040 Regensburg, Germany

Abstract

We present calculations of shallow donor states in GaAs/AlGaAs quantum wells subject to a strong magnetic field applied in growth direction. In contrast to previous variational type calculations we use an expansion in terms of subband and Landau oscillator functions to diagonalize the Coulomb matrix. Binding energies are presented in dependence on quantum well width, donor position and magnetic field. Calculated transition energies are compared to more recent experimental data.

1. Introduction

Shallow impurities in bulk semiconductors with binding energies much smaller than the bandwidth and spatial extensions much larger than the crystal lattice constant can be well described using the hydrogen model in the effective mass approximation[1]. Spectroscopic studies of impurities in an applied magnetic field show some characteristic features: the spherical symmetry of the 3-dimensional Coulomb problem is reduced to axial symmetry around the magnetic field direction[2]. For impurities in quantum wells the symmetry is already reduced by the anisotropy of the grown structure even without a magnetic field. Early theoretical studies of shallow donors in quantum wells[3] have revealed the influence of quantum confinement on the hydrogen states. These calculations and earlier ones including a magnetic field along the growth direction[4] aimed at the lowest state in the axial symmetry by using variational methods. The closely related problem of quantum well excitons in a magnetic field has been studied even earlier[5] yet with a restriction to the extreme confinement case.

In our study of shallow donors in quantum wells subject to a magnetic field we adopt a concept developped for the relative motion of the electrons in a quantum dot[6]: the diagonalization of the Coulomb matrix in a basis of subband and Landau oscillator functions. For sufficiently high magnetic fields this method can be considered as exact. More recent data from far-infrared experiments[7] together with previous ones[8] are taken to test our concept. Moreover we present calculated results concerning the dependence of the donor states on the quantum well width and on the position of the donor atom in the quantum well.

2. The method

The quantum mechanical problem of a shallow donor at position z_0 in a quantum well grown in z-direction is given by the Hamiltonian

$$H = H_0 + H_{Coul} \tag{1}$$

The first part

$$H_0 = \frac{1}{2}(\vec{p} + e\vec{A})\frac{1}{m^*(z)}(\vec{p} - e\vec{A}) + V(z) \tag{2}$$

describes the quantum well without impurities, $\vec{A} = \frac{B}{2}(-y, x, 0)$ being the vector potential and $V(z)$ the confinement potential (for rectangular quantum wells represented by step functions). The second part (using polar coordinates ρ, ϕ)

$$H_{Coul} = -\frac{e^2}{4\pi\epsilon_0\epsilon_r\sqrt{\rho^2 + (z - z_0)^2}} \tag{3}$$

is the Coulomb potential for a donor atom placed at z_0, which can be inside ($z_0 < \frac{L}{2}$) or outside the quantum well ($z_0 > \frac{L}{2}$). We do not consider dielectric mismatch. In effective atomic units H_0 can be written (ρ and z now in units of the effective Bohr radius)

$$H_0 = -\frac{\partial}{\partial z}\frac{m_e}{m^*(z)}\frac{\partial}{\partial z} + V(z) - \nabla^2_{(\rho,\phi)} + \frac{1}{4}\gamma^2\rho^2 + \gamma l_z \tag{4}$$

with $\gamma = \frac{\hbar\omega_c}{\mathrm{Ry}^*}$. The eigenvalue problem with H_0 can be solved using the basis

$$\langle\vec{r}|\nu nm\rangle = \xi_\nu(z)\mathrm{e}^{im\phi}R_{nm}(\rho) \tag{5}$$

with eigenvalues

$$E_{\nu nm} = E_\nu + (2n + m + |m| + 1)\gamma. \tag{6}$$

The eigenvalues of H, Eq. (1), are then found by diagonalizing the Coulomb matrix in the basis of the eigenstates of H_0, which is diagonal in m (due to the axial symmetry of the problem) but requires double integration over ρ and z, which cannot be performed analytically. It turns out that for increasing quantum numbers n and m numerical integration routines will soon become instable due to the oscillatory character of the eigenfunctions, Eq. (5). We make use of recursion relations for the Laguerre-polynomials in $R_{nm}(\rho)$ to formulate corresponding recursion relations for Coulomb matrixelements, which for fixed ν, ν' read

$$\begin{aligned} C(m, n'+1, n) &= \frac{1}{\sqrt{n'+1}}\Big\{\sqrt{n'+m+1}\,C(m, n', n) \\ &- \sqrt{n+m+1}\,C(m+1, n', n) + \sqrt{n}\,C(m+1, n', n-1)\Big\} \end{aligned} \tag{7}$$

where $C(m, n', n)$ stands for the Coulomb matrixelements between eigenstates $|\nu, n, m\rangle$ and $|\nu', n', m\rangle$. Eq. (7) allows to trace back all elements of a Coulomb matrix for m_0 and rank n_{max} to the matrixelements with $n = n' = 0$ and $0 \leq m \leq 2n_{max} + m_0$. Convergence was obtained to within 2 percent of all states up to 5p and $\gamma < 2$ with $n_{max} = 20$ and improves with increasing γ. It turns out that subband coupling does not become significant in the range of parameters used here. Thus, all results are obtained for $\nu = \nu' = 0$.

3. Results and Comparison with Experiment

We have performed calculations for GaAs/$Al_xGa_{1-x}As$ quantum wells with rectangular profile (effective masses $m^*(x) = (0.067 + 0.083x)m_e$, band-offset equal to 70% of the gap difference).

In Fig. 1 we show the binding energies (relative to the corresponding Landau level) of the lowest shallow donor states at the center position ($z_0 = 0$) and for $\gamma = 1$ (6.75 Tesla) as a function of the quantum well width L. The monotonous increase with decreasing L down to about 20 Å is the characteristic confinement effect, while the drop for smaller L results from penetration into the barriers. The shallower slope of the p-state binding energies can be ascribed to the different mixing of motion in z and x, y by the Coulomb interaction. The increase of the binding energy with the magnetic field is demonstrated for 1s, 2s and $2p_-$ in Fig. 2 by similar plots for different γ values. This increase is stronger for s- than for p-states which is reflected in the experimental data. For a fixed magnetic field value ($\gamma = 1$) and different widths of the quantum well we have calculated the dependence of the binding energy on the donor position z_0 (Fig. 3). The finite barrier allows for bound states even if the donor is placed outside the well ($z_0 > \frac{L}{2}$). Finally we compare calculated transition energies with experimental data for a center-doped quantum well with L=210 Å[8] and for another one with L=100 Å[7]. We find perfect agreement for higher magnetic field. The observed deviations for $B \leq 2$ Tesla are due to the limitations of the method, which is a high magnetic field concept.

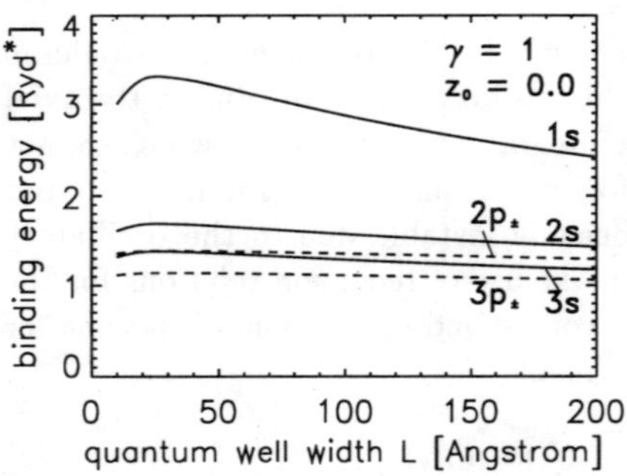

Figure 1: Calculated binding energies of 1s, 2s, 3s, $2p_\pm$ and $3p_\pm$ of a center-doped quantum well for $\gamma = 1$ in dependence on the quantum well width L.

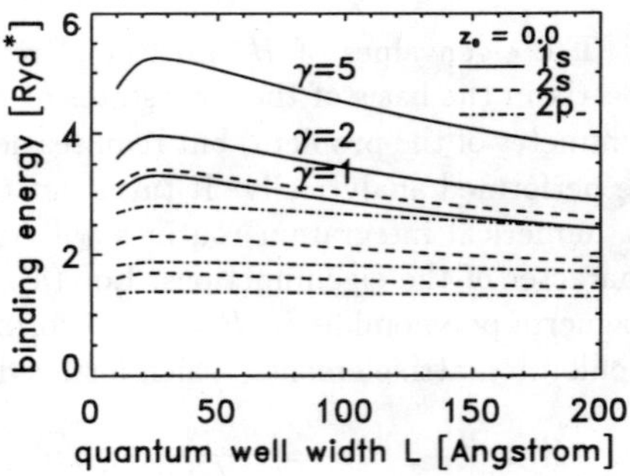

Figure 2: Same as Fig.1 but for different magnetic fields (γ=1,2,5) and 1s, 2s and $2p_-$.

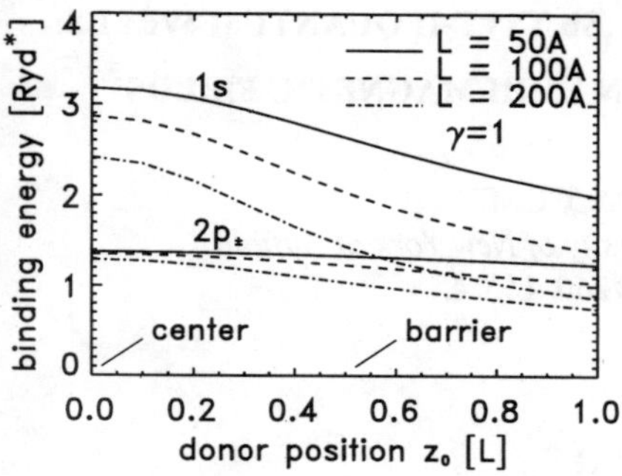

Figure 3: Calculated binding energies of 1s and $2p_{\pm}$ for different L and fixed magnetic field in dependence on the donor position z_0.

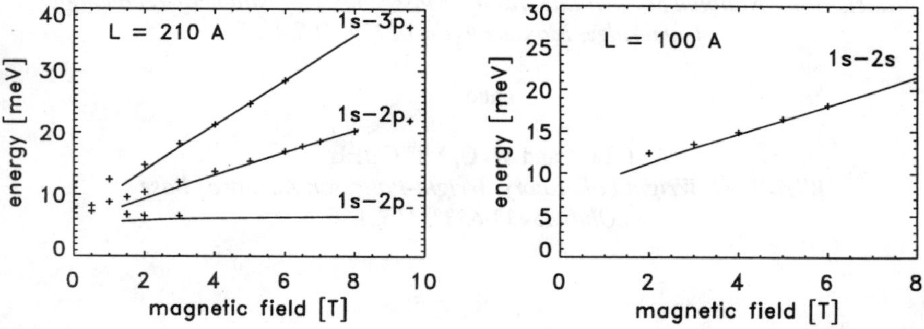

Figure 4: Comparison between calculated (solid lines) and experimental (crosses) transition energies from Ref. 8 (left) and Ref. 7 (right).

References

1. S. Pantelides, Rev. Mod. Phys. **50** (1980) 1287
   A.K. Ramdas and S. Rodriguez, Rep. Prog. Phys. **44** (1981) 1287
2. J. Simola and J. Virtamo, J. Phys. **B11**(1978) 3309
   W. Rösner, G. Wunner, H.Herold and H. Ruder, J. Phys. **B17** (1984) 29
3. G. Bastard, Phys. Ref. **B24** (1981) 4714
   C. Mailhiot, Y.C. Chang and T.C. McGill, Phys. Ref. **B26** (1982) 4449
   R.L. Greene and K.K. Bajaj, Solid State Commun. **45** (1983) 825
4. S. Chaudhuri und K.K. Bajaj, Solid State Commun. **52** (1984) 967
   R.L. Greene und K.K. Bajaj, Phys. Ref. **B31** (1985) 913
5. Yu.B. Grebenshchikov und V.V. Korneev, Sov. Phys. Solid State **19** (1977) 1255
   I.V. Lerner und Yu.E. Lozovik, Sov. Phys. LETP **51** (1980) 588
6. D. Pfannkuche and R.R. Gerhardts, Phys. Ref. **B44** (1991) 13132
   T. Darnhofer and U. Rössler, Phys. Rev. **B47** (1993) 16020
7. P.O. Holtz, B. Monemar, M. Sundaram, J.L. Merz und A.C. Gossard, Phys. Rev. **B47** (1993) 10596
8. J.M. Mercy, N.C. Jarosik, R.D. McCombe, J. Ralston und G. Wicks, J. Vac. Sci. Technol. **B4(4)** (1986) 1011

FAR-IR SPECTROSCOPY OF InAs/Al$_x$Ga$_{1-x}$Sb TYPE-II QUANTUM WELLS: ELECTRON-HOLE INTERACTION IN HIGH MAGNETIC FIELDS

J. KONO and B. D. MCCOMBE
Department of Physics, State University of New York at Buffalo
Buffalo, New York 14260, U.S.A.

and

J.-P. CHENG
Francis Bitter National Magnet Laboratory, Massachusetts Institute of Technology
Cambridge, Massachusetts 02139, U.S.A.

and

I. LO* and W. C. MITCHEL
WL/MLPO, Wright Laboratory, Wright-Patterson Air Force Base
Ohio 45433-6533, U.S.A.

ABSTRACT

We have studied far-infrared magneto-spectroscopy of a series of InAs/Al$_x$Ga$_{1-x}$Sb (x=0.1, 0.2, 0.4, 0.5, 0.8 and 1.0) quantum wells. We have observed new lines and strong oscillations of cyclotron resonance linewidth and mass with filling-factor only when electrons and holes coexist *and* magnetic field is high enough. We assign the new lines to internal transitions of excitons, and discuss the strong oscillations in terms of electron-hole correlation.

1. Introduction

Recently, much attention has been paid to spatially separated electrons and holes in high magnetic fields due to their fascinating possibilities.[1] High mobility electron-hole (*e-h*) systems with InAs quantum wells (QW's) have been fabricated in the last few years, making experimental studies possible. Although separated in one direction, electrons and holes can have large excitonic binding energies (greater than k_BT at 4.2K) in high magnetic fields.[2] We present results of far-infrared magneto-spectroscopy of InAs/Al$_x$Ga$_{1-x}$Sb QW's with x ranging from 0.1 to 1.0. When holes in Al$_x$Ga$_{1-x}$Sb coexist with electrons in InAs, we have observed new lines, whose behavior is consistent with their assignment to internal transitions of excitons composed of spatially separated electrons and holes. In addition, linewidth and mass of electron cyclotron resonance (CR) strongly oscillate with filling-factor, when holes exist. We conclude that *e-h* correlation, moreover the existence of an excitonic ground state, is responsible for the above phenomena.

2. Experimental

Six samples with different Al-composition x but with nominally the same structure were studied. The thicknesses of a top $Al_xGa_{1-x}Sb$ layer, an InAs QW layer and a bottom $Al_xGa_{1-x}Sb$ layer were 15nm, 15nm and 3μm, respectively. A 10nm-thick GaSb cap layer was grown on top of the whole structure. All the samples are type-II QW's, the existence of holes depending on x. Samples with $x > 0.3$ are semiconducting; only electrons exist in the InAs layers. Samples with $x < 0.3$ are semimetallic; holes in $Al_xGa_{1-x}Sb$ confined at the interface(s) by an electrostatic potential coexist with electrons in InAs. The electron densities in the InAs QW's range from 5.5 to 8.5×10^{11} cm^{-2}, mobilities from 1.0 to 2.0×10^5 cm^2/V-sec. Fourier-transform spectrometers were used in conjunction with superconducting magnets and a Bitter magnet to carry out the magneto-spectroscopy.

3. Results, Discussion and Conclusions

Samples with x = 0.4, 0.5, 0.8 and 1.0 always show semiconducting behavior, containing only electrons. In Fig.1, typical spectra for a semiconducting sample at several magnetic fields are displayed. Over the whole range, electron CR is the only spectroscopic feature. Due to the large nonparabolicity of conduction-band in InAs, CR splittings are clearly observed, similar to earlier reports on InAs/AlSb QW's.[3,4] The inset shows the effect of LED illumination: illumination increases the transition frequency and reduces the intensity, since it lowers the electron density due to the negative persistent photo effect.[5] All the semiconducting samples showed similar behavior, which can be understood in terms of single-particle band structure effects.

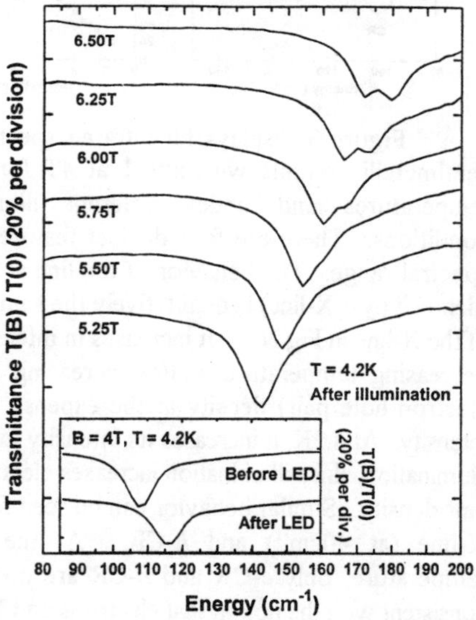

Fig.1. Spectra for a sample with x=0.5. Inset shows spectra before and after illumination at 4T.

Figure 2 shows results for a sample with x=0.2 — an intermediate sample. In the magnetic field range of the present study (2-7T), this sample shows semiconducting behavior before illumination and semimetallic behavior after illumination in magneto-transport experiments.[5] This is consistent with the far-infrared results shown in Fig. 2: before illumination, only electron CR is observed over the whole field range; after illumination, however, a new line (X-line) appears 30cm^{-1} above CR with comparable intensity, and CR shifts down in frequency and broadens. The X-line is observable only at high enough magnetic fields (> 3T) and low enough temperatures (< 40K). The results before illumination are qualitatively the same as semiconducting samples (Fig.1) and those after illumination are qualitatively the same as a semimetallic sample (Fig.3).

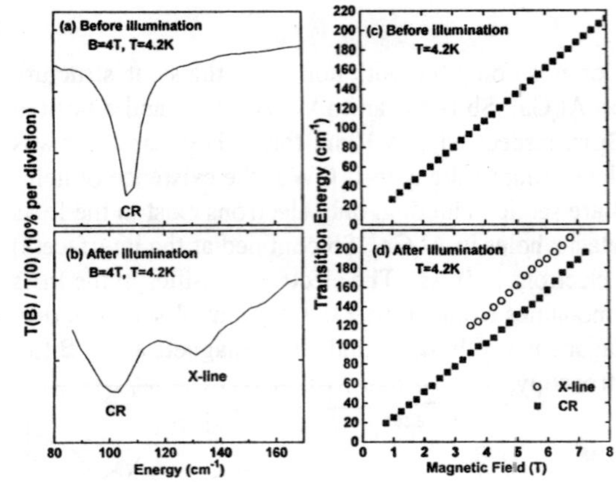

Fig. 2. Far-infrared spectra for a sample with x=0.2 (a) before and (b) after LED illumination. Transition energy vs. magnetic field is plotted in (c) before illumination and in (d) after illumination.

Figure 3 displays far-infrared spectra for a semimetallic sample with x=0.1 at 4T for several temperatures and under different illumination conditions. There are four distinct features in this spectral range. The behavior of the line at 125cm^{-1} (denoted by *e*-X-line) is qualitatively the same as that of the X-line in Fig.2(b). It increases in intensity with decreasing temperature and/or increasing hole (or electron-hole pair) density at the expense of *e*-CR intensity. At 4.2K, it increases its intensity after LED illumination, since illumination increases electron-hole pair density. Similar behavior can be seen for the *h*-X-line (at 40cm^{-1}) and *h*-CR. At the highest temperature, only *e*-CR and *h*-CR are observable, consistent with the notion that electrons and holes are thermally liberated from bound states.

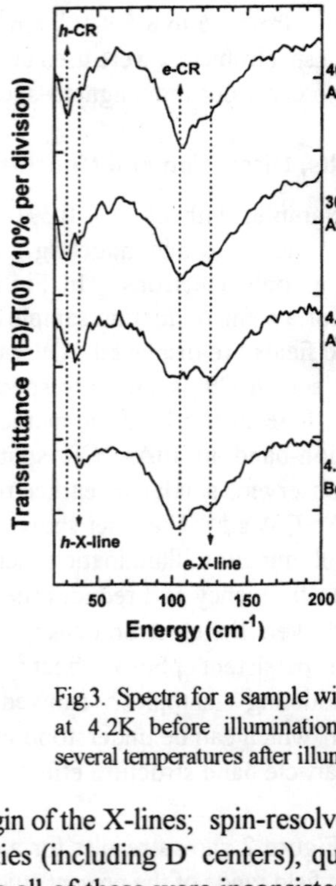

Fig.3. Spectra for a sample with x=0.1 at 4.2K before illumination and at several temperatures after illumination.

We have excluded several possibilities for the origin of the X-lines; spin-resolved CR, CR of electrons in a second subband, residual impurities (including D$^-$ centers), quasi-2D magnetoplasmons and CR of localized electrons, since all of these were inconsistent with some aspects of the experimental results, particularly with the fact that the X-lines appear only when electrons and holes coexist. On the other hand, all the observations are consistent with the assignment of the X-lines to internal transitions of excitons; the excitons are composed of spatially separated electrons and holes, and are stabilized by applying a magnetic field. The X-lines were observed only in the high-field limit, $\gamma = \hbar\omega_c / 2R^* \gg 1$, but *e*-X-line and *h*-X-line correspond to electron-like and hole-like $1s \rightarrow 2p_+$ transitions, respectively, in the low-field (hydrogenic) notation. The observation of the X-lines with no external excitation suggests the existence of an excitonic ground state. More details of the behavior of the X-lines and discussion in terms of an excitonic ground state are given elsewhere.[7]

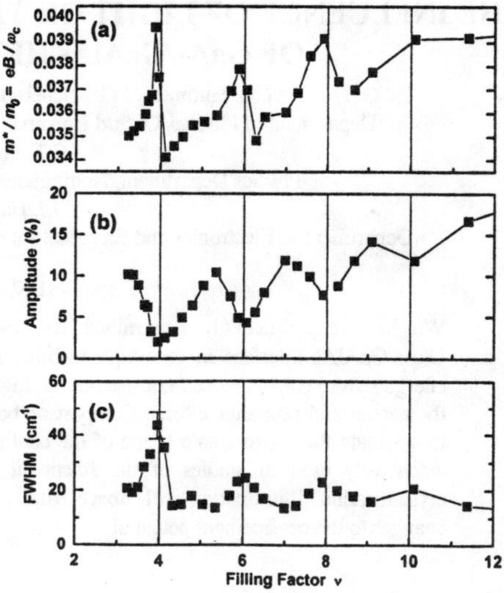

Fig.4. Oscillations of electron CR vs. filling-factor for a semimetallic sample at 4.2K after illumination: (a) CR-mass, (b) CR-amplitude, and (c) CR-linewidth.

Figs.4(a)-(c) show mass, absorption amplitude, and linewidth of *e*-CR vs. filling-factor, observed for the semimetallic sample at 4.2K after LED illumination. Pronounced CR oscillations can be seen. Mass jumps, amplitude-minima, and width-maxima occur at *even* integer filling-factors: $\nu = \ldots$ 10, 8, 6 and 4; thus the onset of nonparabolicity-induced CR splittings,[6] which we have observed for semiconducting samples, is <u>not</u> the cause of these oscillations.

Both the X-lines and the strong oscillations are *absent* for semiconducting samples; also, they disappear for semimetallic samples at high temperatures (> 50K) where the *e-h* interaction becomes relatively unimportant. Filling-factor dependent screening of impurity scatterers cannot be the cause of the oscillations in the present case, since the existence of impurities is irrelevant to whether the sample is semiconducting or semimetallic. Instead, all the experimental facts lead us to conclude that the Coulomb interaction between spatially separated electrons and holes plays a central role in the origin of the oscillations.

4. Acknowledgements

This work is supported in part by the Office of Naval Research under Grants No. N00014-89-J-1673 and N00014-91-J-1939. We are grateful to M.O. Manasreh, C.E. Stutz and R.K. Evans for sample preparation. JPC thanks IBM for support. IL is in part supported by the National Science Council of Taiwan, the Republic of China.

5. References

1. See e.g., X. M. Chen and J. J. Quinn, *Phys. Rev. Lett.* **67** (1991) 895.
2. X. Xia, X. M. Chen, and J. J. Quinn, *Phys. Rev.* **B46** (1992) 7212.
3. J. Scriba *et al.*, *Semicond. Sci. Technol.* **8** (1993) S133.
4. M. J. Yang *et al.*, *Phys. Rev.* **B47** (1993) 6807.
5. I. Lo, W. C. Mitchel, and J.-P. Cheng, *Phys. Rev.* **B48** (1993) 9118.
6. E. B. Hansen and O. P. Hansen, *Solid State Commun.* **66** (1988) 1181.
7. J.-P. Cheng *et al.*, to be presented at the 22nd Int. Conf. Phys. Semicond.

*Present address: *Department of Physics, National Sun Yat-Sen University, Koahsiung, Taiwan, 80424, R.O.C.*

THE INFLUENCE OF LIGHT ON THE CONFINEMENT POTENTIAL OF GaAs/GaAlAs HETEROJUNCTIONS

G.M. Summers, J.G. Michels, D.M. Symons, R.J. Nicholas
Department of Physics, Oxford University, Parks Road, Oxford, OX1 3PU, UK.
C.T. Foxon
Physics Department, Nottingham University, Nottingham, UK.
J.J. Harris
Department of Electronics and Electrical Engineering, University College London, UK.

Abstract

We have experimentally determined the energy levels of confined subbands at the GaAs/GaAlAs interface in very high mobility heterojunction samples. Theoretical subband energies were calculated self-consistently taking into account wave function penetration into the barrier and exchange effects. Comparison between theory and experiment has enabled us to calculate the approximate shape of the confinement potential in these samples which are extensively used in studies of the fractional quantum Hall effect and possible Wigner crystallisation. Illumination with above band gap radiation is shown to result in dramatic changes to the confinement potential.

1. Introduction

This paper addresses the question of what happens to the shape of the confinement potential in very high mobility heterojunction samples when illuminated. The interest in this topic arose because investigations of the fractional quantum Hall effect and possible Wigner crystallisation, using these and other similar samples, relied on visible radiation to alter either the carrier concentration of electrons confined at the heterojunction interface or to excite photoluminescence. It was usually assumed in previous studies that the basic shape of the conduction band profile remained relatively unchanged with the additional perturbation of light. It will be shown that this is not the case and that the behaviour of these samples indicates a delicate interplay between the depletion electric field in the GaAs, the channel density and the Fermi level of the AlGaAs barrier.

The energy levels of the confined electric subbands are determined by observing resonant coupling of the Landau levels originating from different subbands when the sample is placed in a tilted magnetic field[1-5]. Resonant coupling is apparent from a dramatic enhancement of the cyclotron resonance linewidth accompanied by shifts and sometimes splitting of the resonance. A theoretical value of the subband spacing is then calculated self-consistently and comparison with experiment allows surface depletion fields and conduction band profiles to be deduced.

2. Experiment

A series of cyclotron resonance transmission measurements were performed using a fast scan Fourier transform spectrometer and ^{3}He cryostat. The ^{3}He insert was specially modified to allow the insertion of an optical fibre down into the sample space to allow for the depopulation of the 2DES using a red HeNe laser creating power densities up to ~10^{-3}Wcm^{-2}. The samples consisted of conventional GaAs/$Al_xGa_{1-x}As$ heterojunctions, with x=0.3, and a 0.5μm thick layer of GaAs grown on a short period GaAs/AlAs superlattice. Different samples with spacer layers (800-4800Å) were used to allow a wider range of carrier densities to be studied[7].

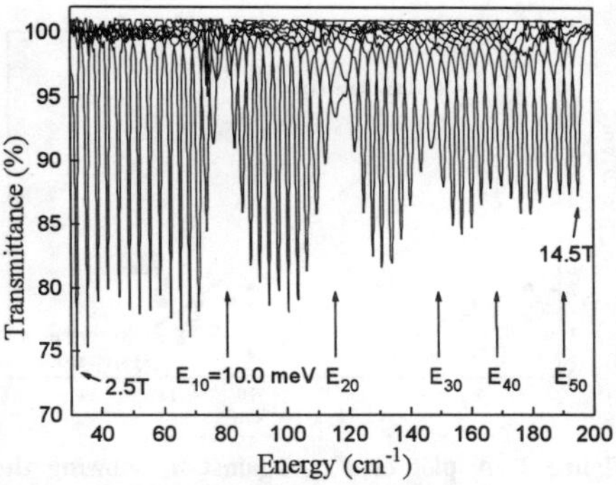

Figure 1. Cyclotron resonance spectra in sample G641 at 7° to the field direction, n_s=7×10^{10} cm^{-2}.

3. Results

Figure 1 shows a typical series of cyclotron resonance spectra. The magnetic field was incremented by 0.25 T up to 14.5 T. The carrier density determined from the area of the resonant absorption curve was 7×10^{10} cm^{-2}. The sample had been briefly illuminated with a red LED which had increased the carrier density up from an unilluminated value of 3.2×10^{10} cm^{-2}.

Well away from any possible couplings the cyclotron resonance appears as a single sharp peak at the position determined from the resonance condition: $E=\hbar eB/m^*$, with $m^*=0.07m_e$ for GaAs. As the energy of the couplings are approached, a broadening of the linewidth occurs accompanied by a decrease in the amount of transmission, and close to the lower couplings a split CR is observed. In total, couplings up to that between the zeroth and fifth subbands are observed in this data allowing $E_{10}..E_{50}$ to be measured. Only E_{10} is necessary in the calculation of the surface depletion field, although when higher couplings could be measured they supplied an additional verification of the self-consistent calculation.

Under conditions of weak continuous illumination a rather different picture is observed, illustrated in Figure 2. Again at low fields a single sharp resonance is observed and in the vicinity of the first coupling a significant enhancement of the linewidth is observed. However, above E_{10} the CR remains broadened and there is no longer any evidence for higher couplings.

The experiment was performed on the available samples under a wide variety of illumination conditions, thus allowing for a comprehensive study of the variation of subband energy spacings with n_s and illumination state.

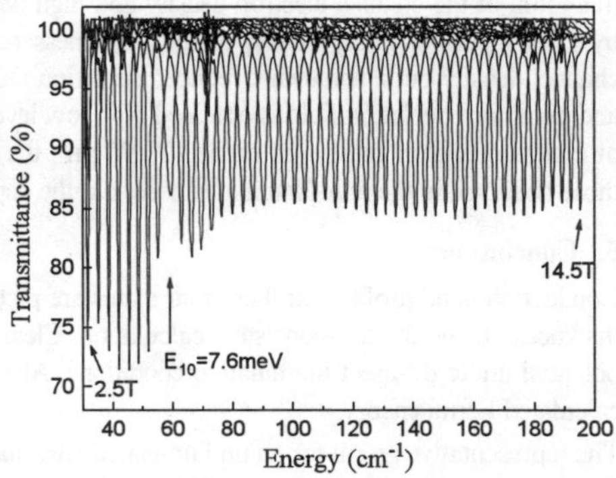

Figure 2. CR spectra observed under conditions of weak constant illumination.

4. Analysis

Using values of E_{10} and carrier density (n_s) we now calculate the effective depletion electric field (F_{depl}) arising from space charge in the GaAs. We use a self-consistent calculation of the potential profile, the electron wavefunctions and associated confinement energies. The calculation agreed with published results[6] for known values of n_s and net acceptor density in the GaAs (n_{AC}). n_{AC} was set to 5×10^{13} cm^{-3}, based on estimates made during sample growth[7,8].

An illustrative way of presenting the data is to plot F_{depl} as a function of the channel electron density, as shown in Figure 3. Evident is a progression from a minimum channel electron density and high depletion field towards a maximum channel electron density and reduced depletion field. Increasing levels of illumination then depopulate the 2DES while F_{depl} stays roughly constant. This sequence is marked by the solid lines superimposed on the data points. In this progression, four different regimes of sample behaviour as a function of illumination can be defined:

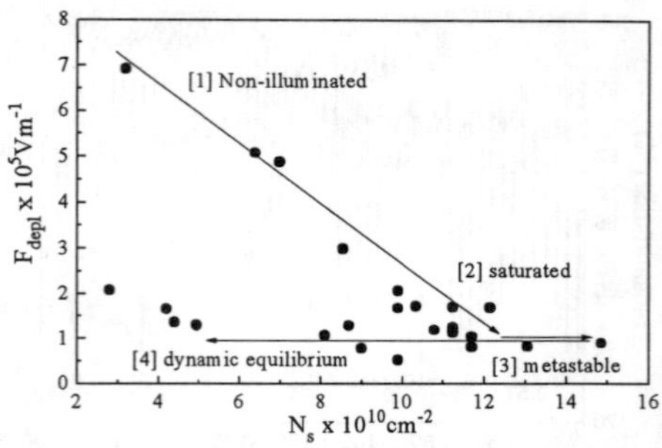

Figure 3. A plot of F_{depl} against n_s showing the relative changes as the illumination conditions are altered.

[1] non-illuminated
[2] saturated [dark]
[3] metastable low-level illumination
[4] dynamic equilibrium produced by a high level of continuous illumination.

A non-illuminated state is reached by cooling slowly under conditions of total darkness resulting in a minimum of the channel electron density and high depletion field. A short burst of illumination followed by a long period [many hours] of total darkness results in a saturated [dark] state with an enhanced channel density but a somewhat reduced depletion field. Maximum, meta-stable, channel electron density and minimum depletion field is obtained with low levels of persistent illumination. Finally, for high levels of continuous illumination, of order 10^{-3} W/cm^2, the channel electron density drops to similar values to those observed in the non-illuminated case, but the depletion field stays relatively low.

5. Conclusions

Conduction band profiles for these four states are pictured in Figure 4 as deduced from F_{depl}, spacer layer thickness, n_s and the self-consistent calculation. Clearly evident is the dramatic change in the slope of the potential under different illumination conditions. Also shown are the energy levels for E_0, E_1, and E_f, the calculated Fermi energy.

The representative graph for an unilluminated case shows the spacer layer conduction band minima set at 74 meV above the Fermi energy. The number of carriers contained in the 2DES is low and the large depletion field results in very little band bending through the 'depletion' layer - a true triangle-like potential. The level of the Fermi energy is consistent with predicted values for the D-X center[9,10] so that, as expected, almost all of the electrons are trapped at these deep donor sites.

When a very small amount of light is allowed to strike the sample for a short period of time, the value of n_s increases while F_{depl} is reduced. In the AlGaAs, the conduction band is now only 22 meV above the Fermi level, indicating that the light has ionised the electrons from the D-X states up into the shallow donor states lying just below the conduction band. These carriers are then able to tunnel through the spacer layer and into the heterojunction. 22 meV is almost the exact value of the ionisation energy for these shallow hydrogen-like donors predicted in a recent study[11].

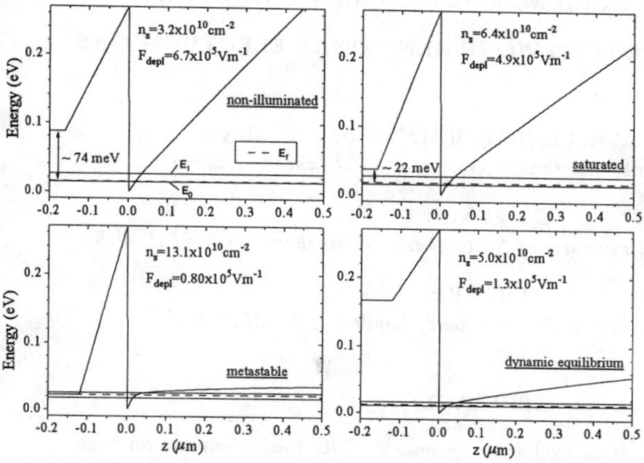

Figure 4. Plots of the conduction band profiles for the 4 itemised illumination conditions.

Low levels of persistent illumination pull the conduction band down coincident with the Fermi energy and further increase the channel electron density. The depletion field is abruptly flattened by the minimal charge remaining in the GaAs. A possible explanation for this is that the continuous illumination results in charge separation of electron-hole pairs in the AlGaAs. The electrons are repelled by the spacer layer potential, while the holes are swept into the GaAs and neutralise the acceptor charge. The fourth state caused by high levels of continuous illumination results in a decrease of the channel electron density while the depletion field remains low. The additional influx of holes, once the GaAs potential is flat, is thought to further depopulate the channel. The sample should be in dynamic equilibrium with the shape of the potential reflecting the charge distribution throughout the heterojunction.

In conclusion we can say that we have highlighted the dramatic changes in conduction band profile that can occur in GaAs/GaAlAs heterojunctions when subject to varying illumination conditions, due to the change in relative contributions to the interface field from the inversion layer and space charge layer. The potential can change from a space charge dominated triangular well to being very flat in the GaAs channel with a resulting quasi-accumulation layer at the interface.

6. References

1. Z. Schlesinger, J.C.M. Hwang and S.J. Allen Jr., *Phys. Rev. Lett.* **50** (1983) 2098.
2. G.L.J.A. Rikken, H. Sigg, C.J.G.M. Langerak, H.W. Myron and J.A.A.J. Perenboom, *Phys. Rev. B* **34** (1986) 5590.
3. R. Merlin, *Solid State Commun.* **64** (1987) 99.
4. S. Huant, M. Grynberg, G. Martinez and B. Etienne, *Solid State Commun.* **65** (1988) 457.
5. A.D. Weick, F. Thiele, U. Merkt, K. Ploog, G. Weimann and W. Schlapp, *Phys. Rev. B* **39** (1989) 3785.
6. F. Stern and S. Das Sarma, *Phys. Rev. B* **30** (1984) 840.
7. C.T. Foxon, J.J. Harris, D. Hilton, J. Hewett and C. Roberts, *Semicond. Sci. Tech.* **4** (1989) 582.
8. J.J. Harris, C.T. Foxon, K.W.J. Barnham, D.E. Lacklison, J. Hewett and C. White, *J. Appl. Phys.* **61** (1987) 1219.
9. M. Mooney, *J. Appl. Phys.* **67** (1990) r1.
10. P. Bhattacharya, *Semicond. Sci. Tech.* **3** (1988) 1144.
11. G. Oelgart, G. Grummt, G. Lippold, R. Pickenhain, R. Schwabe and L. Lehmann, *Semicond. Sci. Tech.* **5** (1990) 894.

MILLIMETRE WAVE INDUCED MAGNETIC FIELD PINNING IN A HIGH MOBILITY LOW DENSITY TWO DIMENSIONAL ELECTRON GAS

G.M.SUNDARAM, I.KULAÇ, R.MEISELS, F.KUCHAR
Institut für Physik, Montanuniversität, A-8700 Leoben, Austria

C.T.FOXON
Department of Physics, University of Nottingham, Nottingham, NG7 2RD, U.K.

J.J.HARRIS
Imperial College, Prince Consort Road, London, SW7 2BZ, U.K.

ABSTRACT

We have performed a series of cyclotron resonance (CR) measurements on high mobility low density GaAs/AlGaAs using millimetre wave radiation (60-110 GHz). Transmission measurements were made for a range of carrier densities $n_s = 3 \times 10^{10} - 4.8 \times 10^{10} \mathrm{cm}^{-2}$. The results indicate that there exists a narrow range of applied frequencies where the positions of the CR transmission minima remain fixed in magnetic field, while the incident millimetre wave frequencies are increased. The anomalous behaviour appears to be related to even Landau level filling factors. In particular, they were observed close to $\nu = 8$. We have also attempted to observe the CR of composite fermions through magnetotransport measurements.

1. Introduction

Recent interest in CR anomalies in GaAs/AlGaAs heterojunctions has brought to the forefront the importance of magnetooptical and magnetotransport measurements using millimetre waves[1-4]. However these investigations were mostly confined to infrared transmission measurements at relatively low occupancies of the Landau levels (LL). Our experiments involve the use of millimetre waves to probe both the high filling factor and $\nu < 1$ regime in a GaAs/AlGaAs heterojunction with extremely low disorder and low carrier density.

For high filling factors ν essentially a single particle picture is frequently applied whereas at low filling factors interactions play a dominant role. Jain[5] proposed a one to one correspondence between the states of the IQHE ($\nu = 1, 2, 3, ...$) and those of the $\nu = 1/3$ hierarchy $\nu = 1/3, 2/5, 3/7, ...$ resp.. The electrons of the IQHE correspond to quasiparticles, so-called composite fermions, which have two flux quanta attached and therefore see only an effective field $B_{eff} = B - B(\nu = 1/2)$. Du et al.[6] and Leadley et al.[7] have interpreted the ρ_{xx}-oscillations in the range $\nu = 1$ to $\nu = 1/3$ as Shubnikov-deHaas oscillations of the composite fermions. They found a mass of $m_{c.f.} \approx 0.6 m_e$ for the composite fermions, where m_e is the free electron mass. Scattering times of $\tau_c \approx 8ps$[7] were determined. This corresponds to an $\omega\tau_c \approx 2.5$ at 38GHz. Therefore, we have examined, in addition to the electron CR, the region around $\nu = 1/2$ for signs of the cyclotron resonance of the composite

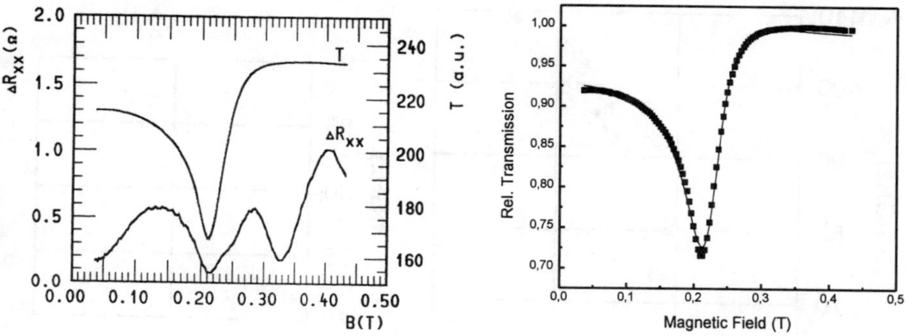

Fig. 1. Left graph: Change of longitudinal resistance due to millimetre wave irradiation ΔR_{xx} and transmission T vs. magnetic field for a frequency of 90GHz. T=1.6K. Note the correspondence between the CR and the photoconductivity minimum. Right graph: T (solid line) compared with theoretical fit (filled squares)

fermions.

2. Experiment

The sample studied was an ultra high mobility ($\mu = 3 \times 10^6 \mathrm{cm}^2/\mathrm{Vs}$) heterojunction. The measurements in the high filling factor regime were made at 1.6K with electron densities of 3.1×10^{10} and $4.8 \times 10^{10} \mathrm{cm}^{-2}$. The CR experiments were performed with modulated millimetre wave sources spanning the E (60-90 GHz) and W (75-110 GHz) bands. The applied radiation was coupled to the sample by means of an oversized Ka-band waveguide. The sample ($2.5 \times 2.5 \mathrm{mm}^2$) was placed inside the waveguide near its end, and a carbon bolometer in conjunction with a lock-in amplifier was used to measure the transmitted radiation. For all frequencies the power of the incident radiation was adjusted to give the same signal strength measured by the bolometer at $B = 0$. To examine the millimetre wave photoconductivity two additional lock-in amplifiers were used to measure the longitudinal resistance R_{xx} and its change due to the radiation ΔR_{xx}.

The experiments to investigate the composite fermion regime were performed at 0.4K at an electron concentration of $9 \times 10^{10} \mathrm{cm}^{-2}$. The CR positions of electron- and hole-like c.f. can be expected to be symmetric around $B(\nu{=}1/2){=}6.3\mathrm{T}$ at $B_{res}{=}B(\nu{=}1/2) \pm \omega m_{c.f.}/e$. ω is the angular frequency of the applied radiation.

3. Results and Discussion

Figure 1 illustrates a typical CR trace at high occupancy of the LLs. Relatively

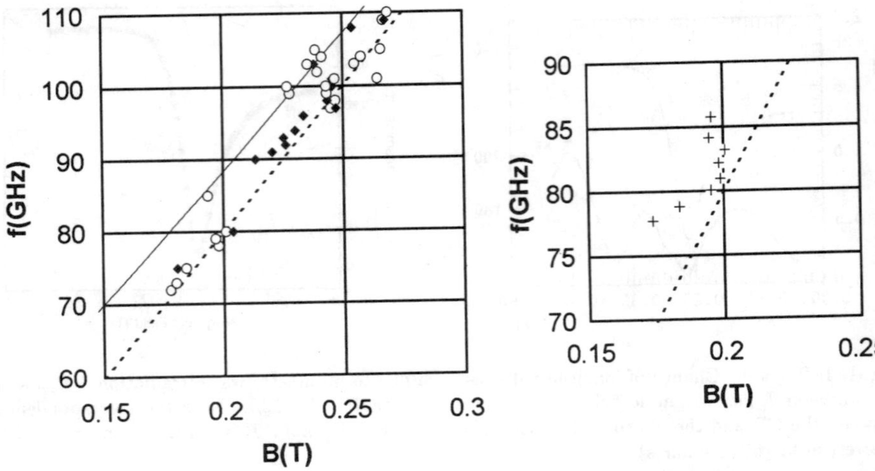

Fig. 2. Frequency of the incident microwave radiation vs. CR field for carrier concentration of (left graph) $4.8\times10^{10}\text{cm}^{-2}$ ($\nu = 8$ at $\approx$0.25T, $\nu = 10$ at $\approx$0.2T) and (right graph) $3.7\times10^{10}\text{cm}^{-2}$ ($\nu = 8$ at $\approx$0.2T). The dashed line indicates the CR, the solid line indicates the L=1 edge magnetoplasmon with a $B = 0$-frequency of 27GHz[8]. The circles and diamonds in the left graph correspond to two different measurement runs.

symmetric resonances can be obtained for some of the applied frequencies, however asymmetric resonances are also frequently encountered. In order to find the correct resonance position we have employed a theoretical fit for the transmission[9], taking into account internal interferences in the sample. Figure 2 shows a plot of the resonance position as a function of frequency. The striking feature of the data is that though the measurements were carried out at different carrier densities, a "pinning behaviour" (where the resonance position is pinned in magnetic field) is tied to the filling factor. This is especially distinct in the right graph of figure 2 at about 0.2T ($\nu = 8$) and in the left graph at 0.25T ($\nu = 8$) in the CR data. A possible explanation may be the absence of screening at even filling factors which leads to a broadening of the LLs. This would allow transitions between them for a wider frequency range for a magnetic field where ν is even. In addition, doublets of minima were observed (left graph, $n_s = 4.8 \times 10^{10}\text{cm}^{-2}$). The minima appear to be connected with the unperturbed bulk CR (dashed line) and with an edge magnetoplasmon (solid line, calculated according to a theory by Fetter[8]), respectively.

We have also looked at the photoconductivity data. A typical curve is shown for a frequency of 90GHz in figure 1. At the resonance position itself the ΔR_{xx}-signal shows a minimum. The higher field oscillations in the photoconductivity correspond to SdH-oscillations, e.g. the minimum at 0.33T to $\nu = 6$. As the photoconductivity provides a good detection mechanism for the CR we have utilized it to probe the

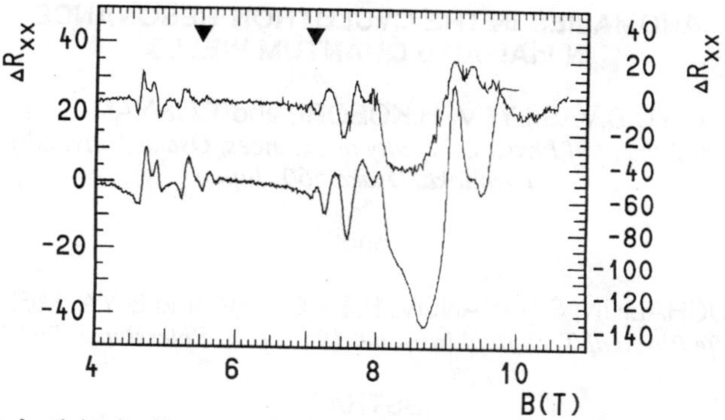

Fig. 3. Photoconductivity in the composite fermion regime around $B(\nu = 1/2) = 6.3$T. f=38GHz, T=0.4K. Upper curve: 0.8μW, lower curve 600μW. Expected c.f. resonance positions are indicated by triangles.

CR of composite fermions. Figure 3 shows traces of the photoconductivity ΔR_{xx} in the magnetic field range where CR from composite fermions would be expected. At a milimetrewave frequency of f=38GHz low (0.8μW, curve a) and high (600μW, curve b) microwave powers were applied. The expected resonance positions (assuming $m_{c.f.} = 0.6m_e$) are indicated by triangles. While strong signs of the CR of the electrons were seen at low magnetic fields (figure 1), no resonance, within the experimental precision of 1Ω, of the composite fermions was seen around $B(\nu = 1/2)$, particularly not at B=5.5T and 7.1T, where it is expected for a mass of $0.6m_e$. All the structure in ΔR_{xx} corresponds to structure in R_{xx} (without millimetre wave radiation) and does not shift with frequency (e.g. it is also seen there at 140 GHz). The signals around 4.7T and 9.4T correspond to $\nu = 2/3$ and $\nu = 1/3$.

This work was supported by the Fonds zur Förderung der Wissenschaftlichen Forschung , Austria, Project number M0050-PHY.

4. References

1. E. Batke et al., *Phys. Rev.* **B37** (1988) 3093.
2. Z. Schlesinger et al., *Phys. Rev. Lett.* **58** (1987) 73.
3. K. Ensslin et al., *Phys. Rev.* **B36** (1987) 8177.
4. R.J.Nicholas et al. *Phys. Rev.* **B39** (1989) 10955.
5. J. K. Jain, *Surf. Sci.* **263** (1992) 65.
6. R. R. Du et al., *Phys. Rev. Lett.* **70** (1993) 2944.
7. D. R. Leadley et al., *Phys. Rev. Lett.* **72** (1994) 1906.
8. A. L. Fetter, *Phys. Rev.* **B33** (1986) 5221.
9. T. A. Kennedy et al., *Sol. St. Comm.* **18** (1976) 275.

ANOMALIES IN THE CYCLOTRON RESONANCE IN InAs-AlSb QUANTUM WELLS

YU.B.VASILYEV, H.KOBORI, and T.OHYAMA,
Department of Physics, Faculty of Sciences, Osaka University, Toyonaka, Osaka 560, Japan

and

S.D.SUCHALKIN, S.V.IVANOV, P.S.KOP'EV, and B.YA. MELTSER
Ioffe Physical Technical Institute, 194021 St. Petersburg, Russia

ABSTRACT

We present novel cyclotron resonance and magneto-resistance effects in InAs-AlSb quantum wells in perpendicular and tilted magnetic fields. In perpendicular magnetic fields a pronounced splitting of the CR signal into two distinct resonances and peculiarities in magneto-transport data are attributed to a lifting of spin degeneracy due to the lack of inversion symmetry. In tilted magnetic fields an additional peak appears on the shoulder of the CR peak when a maximum of magneto-resistance occurs at integer filling factors. This phenomenon is discussed in terms of the coincidence of two Landau levels with opposite spin orientations. Our results indicate on the importance of spin-orbit coupling effects as well as electron-electron interactions in InAs-AlSb quantum wells.

1. Introduction

There has been a considerable attention to InAs quantum wells with different types of barriers[1-3]. This system is interesting because of band structure characteristics, mainly, a small band gap resulting in a very small effective mass and strong nonparabolicity of the conduction band. Here we shortly describe experimental investigations of cyclotron resonance (CR) and magneto-resistance (MR) studies in single InAs quantum wells (QW) sandwiched between AlSb barriers using perpendicular and tilted magnetic fields. The peculiarities of the QW properties allow us to observe a series of new effects associated with spin-orbit coupling and electron-electron interactions.

2. Experimental

The samples consist of a 200A QW, 100A and 50A AlSb barriers, a 100A intermediate layer of AlGaSb and a cap layer of 100A GaSb. The layer thicknesses were determined by high resolution transmission electron microscopy. All layers are undoped. Two lowest subband levels in the InAs QW have energies of 33 meV and 123 meV. The Fermi level lies between them at all investigated electron densities so that electrons occupy only the lowest one subband. CR measurements were performed using a far-infrared laser with wavelengths, λ, of 220μm, 119μm, and 84μm. The samples were wedged to avoid interference and supplied with electrical contacts allowed transport measurements. The temperature was 4.2K.

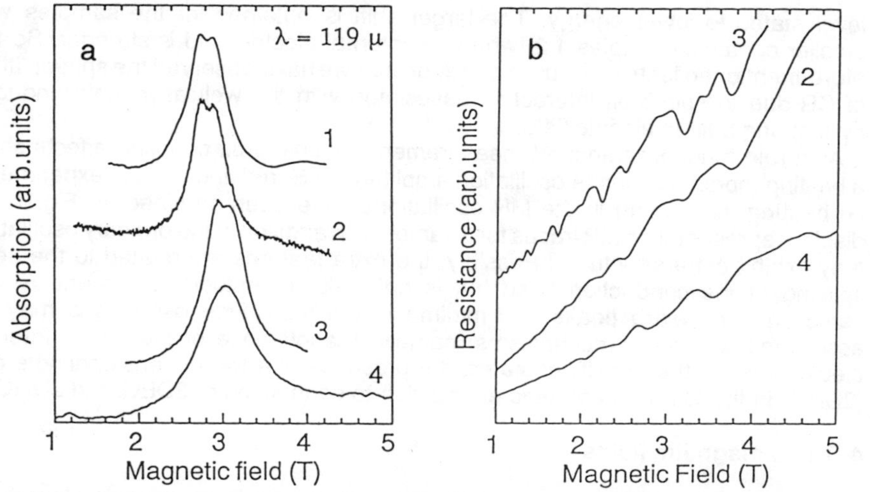

Fig.1a)Cyclotron resonance traces for different samples at λ=119μm; b)magneto-resistance traces

3. Perpendicular magnetic fields

CR spectra at λ=119μm for different samples are shown in Fig.1a. All of the samples were prepared from the same wafer although their sizes as well as the sizes of the contacts were various. Table 1 shows their parameters. Here the sample number corresponds to the curve number in Fig.1a and R is the ratio of the electric contacts sizes to the total size of the sample. It is clearly seen (Fig.1a) that the CR peaks split into two and this splitting is different for different samples. We attribute the difference to the presence of electrical contacts. For sample 1 (without contacts) and sample 2 the splitting energy ($\Delta\varepsilon$ = 0.5 meV) and resonance fields (H_c = 2.8 T) are approximately the same. For the sample 3 splitting energy ($\Delta\varepsilon$ = 0.4 meV) is smaller and for sample 4 the splitting is not observed at all (the linewidth becomes wider). Also the positions of cyclotron resonance peaks are shifted to higher magnetic field for samples 3,4 (H_c = 3.0 T) as compared with samples 1,2. These features can be explained by assuming the presence of a built-in electric field in the structures which is affected by electrical contacts. Indeed, it is known that the strong internal electric field can break inversion symmetry in the well and result in spin splitting of the conduction band[4]. The splitting is larger for the stronger built-in electric field. This is consistent with our results which indicate the influence of the contacts on the internal electric field strength: the larger contacts size, the smaller electric field and, accordingly, the smaller observed splitting. To explain the shift of the CR peak position we suppose that the strong internal electric field perturbs the Landau

Table 1

Sample	(mm × mm)	R × 100%
1	3×3	0%
2	4×4	4%
3	3×2.5	20%
4	4×1	40%

level states to lower energy. The larger shift is obtained for the samples with smaller contacts (samples 1,2) when the internal electric field is stronger. So the afore-mentioned facts allow us to conclude that we have observed the spin-splitting of CR due to spin-orbit interaction associated with the well asymmetry induced by a strong built-in electric field.

As a rule magneto-transport measurements of spin-orbit coupling effects show a beating modulation in the oscillation amplitude.[5,6] Surprisingly in our experiments no beating modulation in the MR oscillations have been obtained. In Fig.1b we display representative MR traces for 3 samples. In addition to the ordinary oscillation they exhibit extra structures. Possibly, the extra features are related to the spin-splitting of the conduction band but is not well understood at present. At the same time the extra peaks and minima in the resistance oscillations may be associated with other mechanisms, namely, the influence of the strong internal electric field on the density of states, the presence of a two dimensional hole gas (2DHG) in the GaSb, geometric effects due to connection of 2DEG and 2DHG.

4. Tilted magnetic fields

Figure 2a shows CR traces in tilted magnetic fields at the wavelength of 119μm. It is obviously seen that an additional peak appears on the shoulder of the cyclotron resonance line in the range of tilted angles between 58° and 65°. In Fig.2b we show MR oscillations as a function of total magnetic field for two angles where maxima near even filling factors are observed.

It is known that CR line splitting may occur when CR line overlap with another mode which is optically inactive in the experimental conditions: the inactive mode becomes active when the CR oscillator strength is added to the second line. The most favorable candidate for the role of this inactive mode in our experiments is spin resonance line.[7] Indeed, commonly, spin resonance is forbidden by electric dipole selection rules and does not manifest itself in the CR spectra. In tilted

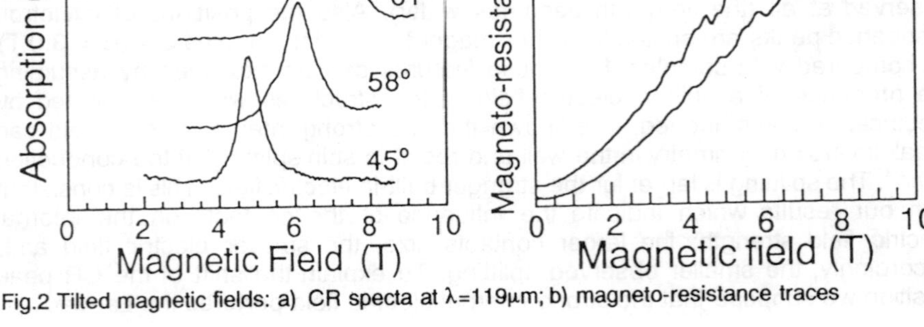

Fig.2 Tilted magnetic fields: a) CR specta at λ=119μm; b) magneto-resistance traces

magnetic fields the spin resonance energy is determined by the total magnetic field while the cyclotron resonance energy depends only on the perpendicular component of the magnetic field. So tilting the sample with respect to the magnetic field direction one can obtain the coincidence of the CR and SR lines. This situation is reached when Landau level state with spin-up orientation coincides with the Landau level state with spin-down orientation. Experimentally the overlapping is identified by an appearance of a maximum in the Shubnikov-de Haas oscillations at the even integer filling factors.[8] Comparison of the MR oscillations and CR data (Fig.2b and Fig.2a) shows that the additional peak in absorption and the maxima at even filling factors in the MR oscillations occur approximately at the same range of angles where, as we think, different Landau levels with spin-up and spin-down coincide. From the coincide condition for the angle 61° the absolute value of g*-factor 24.5 is deduced. The bulk g*-factor is -15. The enhanced g*-factor is attributed to electron-electron interaction in a two dimensional gas resulting in a population difference of spin states. The enhanced g*-factor values were so far obtained only in magneto transport measurements and never in CR studies. This is due to Kohn's theorem which says that for translationally invariant system CR is unaffected by electron-electron interactions. The fact that we measure the same enhanced g*-factor from the coincide condition by magneto-transport and CR measurements means that Kohn's theorem becomes invalid and translational invariance is broken. This can take place in systems with strong disorder, with non-parabolic conduction band, or if there are at least two types of carriers with different effective masses. All these peculiarities are inherent for InAs QWs and we are not able to identify the main origin of the translational asymmetry.

5. Conclusion

We have performed CR experiments in InAs-AlSb quantum wells in perpendicular and tilted magnetic fields. The peculiarities of the QW structure allow us to observe novel effects associated with spin-orbit coupling and electron-electron interactions.

6. Acknowledgment

One of us (Yu.V.) was supported by Japan Society for the Promotion of Science.

7. References

1. P.F.Hopkins et al., *Appl. Phys. Lett.* **58**, 1428 (1991).
2. J.Scriba et al., *Semicond. Sci. Technol.* **8**, S133 (1993).
3. M.J.Yang et al., *Phys. Rev. B* **47**, 6807 (1993).
4. Yu.A.Bychkov and E.I.Rashba, *JETP Lett.* **39**, 78 (1984).
5. J.Luo et al., *Phys. Rev. B* **38**, 10142 (1988).
6. B.Das et al., *Phys. Rev. B* **39**, 1411 (1989).
7. V.I.Fal'ko, *Phys.Rev.Lett.* **71**, 141 (1993).
8. R.J.Nicholas et al., *Phys. Rev. B* **37**, 1294 (1988).

High magnetic field properties of the negatively charged exciton (X^-) in modulation-doped quantum wells

R.T. Cox, K. Kheng and K. Saminadayar
Département de Recherche Fondamentale sur la Matière Condensée, CEA/Centre d'Etudes Nucléaires de Grenoble, 38054-Grenoble Cedex 9, France

T. Baron and S. Tatarenko
Laboratoire de Spectrométrie Physique, CNRS et Université J. Fourier, B.P.87, 38042-Saint Martin d'Hères Cedex, France

ABSTRACT

Optical absorption spectra of $CdTe/Cd_{1-x}Zn_xTe$ Multiple Quantum Wells doped with Indium donors in the $Cd_{1-x}Zn_xTe$ barriers are studied in magnetic fields to 20T. With electron concentrations of order $10^{11}cm^{-2}$ in the CdTe wells, Landau level (free electron- free hole) excitations are seen at low field. Discrete exciton states corresponding to negatively charged excitons X^- (two electrons and one hole) and to neutral excitons X appear around filling factor ν=2. Excited electron-spin triplet states of X^- are also seen, where the second electron is associated with λ=1 and λ=2 Landau levels. Some properties of the magnetoemission spectra associated with X and X^- are briefly described.

1. Introduction

Optical absorption and emission corresponding to creation and annihilation of the negatively charged exciton X^- (two electrons bound to one hole) was identified in recent studies of $CdTe/Cd_{1-x}Zn_xTe$ quantum wells[1,2,3]. Just as the "regular" or neutral exciton X can be considered to be a solid-state analogue of the hydrogen atom, the species X^- is an analogue of the H^- ion.

In a 3-Dimensional X^-, the binding energy of the second electron, E_{b2}, would be very small : only 0.055 Rydbergs for infinite hole mass (as in H^-), reducing to for example 0.030 Rydbergs for m_e/m_h= 0.5 [4]. However, as for D^- (the negatively charged donor) [5], quantum confinement greatly increases the stability of X^-, by a factor of 10 in the 2-Dimensional (2D) limit [6].

Although it has not yet been reported for GaAs quantum wells, X^- turns out to be easily observable in CdTe quantum wells, because of the large value of the Rydberg constant in CdTe (R_∞ =13.5meV, twice the GaAs value). It should also be seen in other II-VI semiconductor systems with large R_∞ : in fact, there is evidence it contributes to the optical properties of $Zn_{1-x}Cd_xSe/ZnSe$ wells of the type studied in the current blue-laser development [7].

Observation of X^- requires an excess of electrons in the quantum well. We produce this in our $CdTe/Cd_{1-x}Zn_xTe$ structures by modulation-doping : during growth of the structures by Molecular Beam Epitaxy, we place a plane of Indium donors in the $Cd_{1-x}Zn_xTe$ alloy barriers at spacing distance *s* from the edge of the well. The donor electrons transfer into the CdTe wells and form a 2D electron gas.

In the absence of excess electrons, the fundamental optical absorption and emission processes in the quantum well are the creation and annihilation of excitons, respectively :

$$h\nu \rightarrow X \qquad (1a) \qquad \text{and} \qquad X \rightarrow h\nu \qquad (1b).$$

In the modulation-doped structures, new absorption and emission transitions occur :

$$e + h\nu \rightarrow X^- \qquad (2a) \qquad \text{and} \qquad X^- \rightarrow e + h\nu \qquad (2b).$$

We will discuss the effects of a magnetic field on the absorption transitions (1a) and (2a) and very briefly its effect on the emission transitions. The samples are 10-period multiple quantum wells (MQWs) grown on $Cd_{0.88}Zn_{0.12}Te$ substrates. The well thickness L_w is 100Å. Concentrations n_e from 2 10^{10} up to about 6 $10^{11}cm^{-2}$ per well are obtained by varying the Indium concentration and/or the barrier height or the spacing-distance s. The values of n_e given here are from Capacitance-Voltage measurements or from estimates (based on Poisson's equation) of the amount of charge transfer to the wells.

The alloy substrates are transparent to light at the wavelength of the CdTe well excitons. They also impose a large strain in the CdTe wells which splits light and heavy hole states by ≈30 meV. So we need only consider heavy exciton states, a very helpful simplification.

Optical measurements to 8 and 11 T were done at 2 - 50 K in superconducting magnets at CEN-Grenoble and at Laboratoire de Spectrométrie Physique respectively. Measurements to 20 T were done in Bitter-coil magnets at the High Field Laboratory, Grenoble. Faraday optical geometry was used with B normal to the quantum well plane.

2. Optical absorption in zero-field

The 2 K absorption spectrum of *undoped* $CdTe/Cd_{1-x}Zn_xTe$ quantum wells is dominated by a sharp peak X at ≈1.6 eV corresponding to transition (1a) : creation of the 1s exciton associated with the lowest electron and heavy-hole subbands (E_1H_1 exciton).

For low modulation-doping, giving $n_e < 4\ 10^{10}cm^{-2}$ in the wells, a new excitonic absorption peak X^- (transition (2a) above) is seen below peak X in zero field, as shown in Fig.1(a) (spectrum 2K, 0T) for a sample with $n_e \approx 2\ 10^{10}cm^{-2}$. The intensity of the X^- peak increases with n_e. The separation between X and X^- peaks is the second electron's binding energy E_{b2}. It is 2.65 meV for 100Å well-width (0.20 CdTe Rydbergs).

Observation of sharp exciton peaks in the presence of an electron gas is contrary to experience from optical absorption studies of equivalent GaAs or $In_{1-x}Ga_xAs$ quantum wells, where even very dilute electron gases destabilise exciton states by the combined effect of Pauli-exclusion (the electrons occupy conduction band states needed in the exciton wavefunction) and electric screening. Such effects are seen in our CdTe wells from $n_e \approx 4\ 10^{10}cm^{-2}$.

By $n_e \approx 10^{11}cm^{-2}$, the exciton peaks X and X^- are replaced by a broad, structureless bump at the optical absorption edge, as shown in the zero-field spectrum at the bottom of Fig.2 for nominal $n_e = 1.8\ 10^{11}cm^{-2}$. This bump appears analogous to the "Fermi-Edge Singularity" (FES) features seen for GaAs quantum wells, which are attributed to the residual effects of electron-hole correlations in the many-electron one-hole system after screening and exclusion have destroyed discrete exciton states.

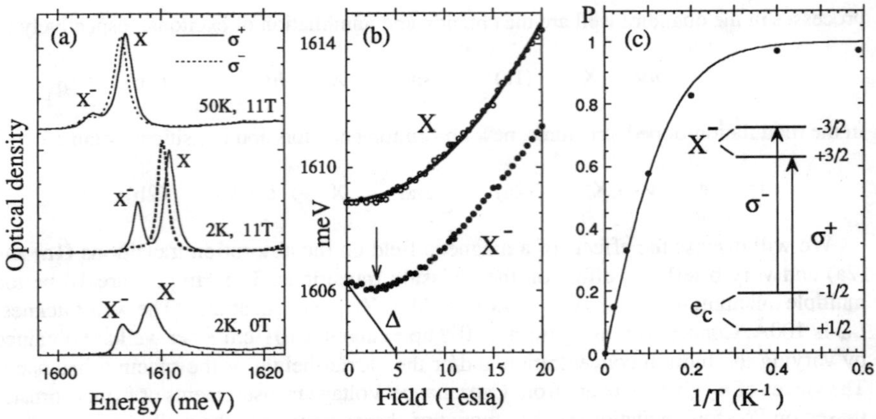

Fig.1 : Results for a lightly doped sample : ZD137, a ten period, 100Å CdTe/ 450Å $Cd_{0.85}Zn_{0.15}Te$ MQW modulation-doped at the barrier centres (i.e. s=225 Å) to give ≈2 $10^{10}cm^{-2}$ electrons in each CdTe well. 1(a) : Absorption spectra (logarithmic Optical Density) showing excitonic peaks X^- and X at zero field (unpolarised light) and at 11 T (in circularly polarised light) at 2K and 50 K. *Fig.1(b)* Parabolic field-dependence of positions of X^- and X peaks (σ+ polarisation). *Fig.1(c)* A fit to the temperature dependence of the degree of circular polarisation P at 11T using g_e=1.74; the inset show the selection rules for transitions $\Delta M=\pm1$ from the conduction electron state e_c to the (electron spin singlet) X^- state.

3. Field dependence of the absorption spectra

First, some numbers. The (neutral) exciton's binding energy, E_{b1}, is about 20meV in a 100Å wide CdTe well. The CdTe electron mass m_e is 0.096. The in-plane hole mass m_h is 0.16 if one uses the diagonal approximation with the Luttinger parameters of ref.[8]. The in-plane reduced mass μ (defined by $1/\mu=1/m_e+1/m_h$) is then 0.06 and the combined cyclotron energy $\hbar\omega_{c\mu}=\hbar eB/\mu$ is 1.94 meV/T. Defining $\gamma=(1/2)\hbar\omega_{c\mu}/E_{b1}$, we reach only $\gamma\approx1$ at our maximum field of 20T. Thus, in measurements on undoped samples, we find an essentially quadratic diamagnetic shift up to 20 T for exciton peak X.

The electron g value is g_e= -1.6 and the electronic Zeeman splitting $g_e\mu_B B$ is 0.094 meV/T. The Zeeman splitting of X and of the X^- optical *transition* (inset in Fig.1(c)) is given by $(g_e-g_h)\mu_B B$; it varies with well-width and strain[9] and for the samples considered here it is not more than ≈ 1 meV at 20T (that is, g_h is accidentally ≈ g_e).

3.1 *Lightly-doped samples:*

For lightly-doped samples that give resolved X and X^- peaks in zero field, a magnetic field induces a quadratic shift for X (see Fig.1(b)), the same shift as in an undoped sample. It induces a very similar shift for X^-.

Closer inspection shows that field increases slightly the X-X^- separation : in fact, the X^- positions go through a minimum at about 3 T, marked by an arrow in Fig.1(b). We attribute the initial negative slope of the X^- curve to the change in diamagnetic energy in transition (2a) e + $h\nu \rightarrow X^-$, which involves *charged particles*. This energy change is $(1/2)\hbar(\omega_{cX^-} - \omega_{ce})$, which is negative because $m_{X^-}=2m_e+m_h > m_e$. The difference amounts to -0.44 meV/T, marked by the straight line Δ in Fig.1(b).

The observed initial downward slope of the X^- parabola is much smaller than this. We suggest that the cyclotron motion of the centre of mass of X^- breaks down very rapidly with increasing field given that the electron cyclotron energy $\hbar\omega_{ce}$ reaches E_{b2} already at 2T. So X^- will rapidly look like a neutral exciton with a second, "outer" electron in a cyclotron orbit of energy $(1/2)\hbar\omega_{ce}$. The linear change of diamagnetic energy in transition (2a) then disappears, leaving *just the same quadratic shift* as for X.

Disorder-induced spatial localisation of the electrons (discussed later), may also contribute strongly to attenuating the initial negative diamagnetic shift of peak X^-.

We next describe the most characteristic feature of the X^- absorption, its magnetic circular dichroism. At high field and 2K, absorption transition $e + h\nu \rightarrow X^-$ is nearly 100% σ^+ polarised, see the middle spectrum in Fig.1(a).

The circular dichroism results from the selection rules for σ^+, σ^- transitions, shown in inset in Fig.1(c). In the initial state of the transition, we have conduction electron states $M=\pm 1/2$ split by $g_e\mu_B B$. In the final state, the ground state of X^-, the two electron spins are *antiparallel* and the spin of X^- is just the heavy-hole spin $M=\pm 3/2$. The two allowed absorption transitions $\Delta M=+1, -1$, seen in σ^+, σ^- respectively, are as shown. At high B and 2K, only the spin-up (+1/2) electron states are populated, so X^- is created in σ^+ only.

Fig.1(c) shows the degree of polarisation $P = (I^+ - I^-) / (I^+ + I^-)$ as a function of $1/T$ at 11 T for this sample. Assuming the absorption intensities I^+, I^- are proportional to $n_e(+1/2)$, $n_e(-1/2)$, which is valid at low n_e, and applying Boltzmann statistics (valid at filling factor $\nu \ll 1$), we can fit the values of P using $g_e = -1.74$, which is close to -1.6.

Remarkably, the X peak in Fig.1(a) is partly circularly polarised at 2K,11T. It is partly polarised σ^-, that is opposite polarisation to X^-. In an undoped sample, the transition $h\nu \rightarrow X$ is not polarised by a field. It is as if doping causes transfer of σ^+ oscillator strength from X to X^- ("oscillator strength stealing" by X^-). There is not yet any theory of this.

3.2 *Heavily-doped samples*

A magnetic field has more complex effects for the more strongly doped samples that give no resolved exciton peaks in zero-field, as discussed in ref.[1] for a sample with nominal $n_e = 1.1\ 10^{11}\text{cm}^{-2}$. In Figs 2 and 3, we show data for a new sample with nominal n_e of $1.8\ 10^{11}\text{cm}^{-2}$ which gives similar effects but at ≈50% higher fields.

First, at low fields, the Fermi Edge Singularity splits in a pattern that is characteristic of transitions creating quasi-unbound electrons and holes in conduction band and valence band Landau levels λ_c, λ_v respectively. The optically allowed transitions ($\lambda_c=\lambda_v$) have diamagnetic energy $(\lambda+1/2)\hbar(\omega_{ce}+\omega_{ch})$: the essential feature of these transitions is that they move *linearly* with field (except for peaks $\lambda=0$: in Fig.3 these curve upwards slightly, which is not yet understood). The lines $\lambda=2$, $\lambda=1$ drawn in Fig.3 yield values of the in-plane reduced mass $\mu=0.050$ and 0.048 respectively, reasonably close to the expected $\mu=0.06$.

This is classic for modulation-doped quantum wells. What is new in our CdTe wells is that a discrete peak, X^-, characterised by a *parabolic* field dependence, always appears at sufficiently high field. In Figs 2 and 3, X^- appears at ≈4.5 T, in σ^+ polarisation, and it gains strength rapidly as the field increases further. In the σ^- polarisation, the X^- peak is

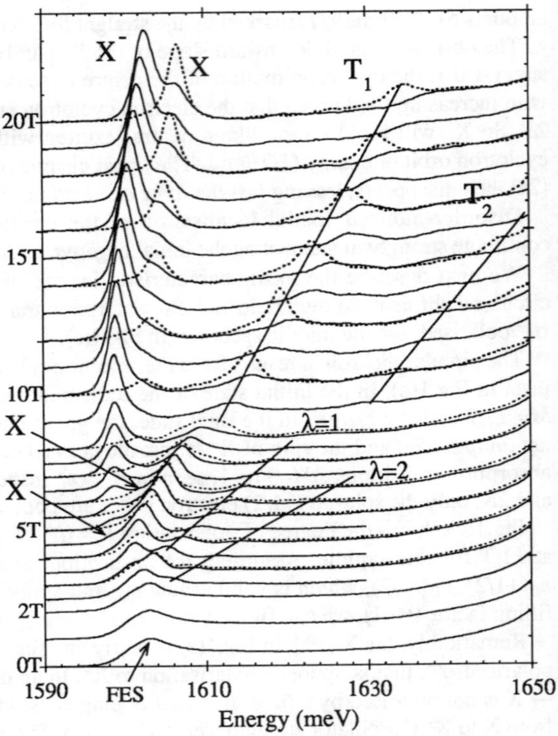

Fig. 2 : Optical density spectra in σ+ (full lines) and σ- (dotted lines) of sample ZD173, a ten period, 100ÅCdTe/ 450Å$Cd_{0.78}Zn_{0.22}Te$ MQW modulation-doped at the barrier centres (i.e. s=225Å) to give ≈1.8 $10^{11}cm^{-2}$ electrons in each CdTe well. The Fermi Edge Singularity evolves at low field into electron-hole excitations on Landau levels λ. Exciton states X^- and X appear at 4.5 and 6 T respectively (arrows). Peaks T_1 and T_2 are attributed to electron spin triplet states of X^- (section 3.3)

much weaker, as at lower n_e (section 3.1).

At higher field, peak X appears. It is stronger in σ^- than in σ^+ ("intensity-stealing" by X^- again), see Fig.2. It too gains intensity rapidly with field. The parabolic field dependence of the positions of peaks X and X^-, beginning at the threshold fields, is shown in Fig.3, and contrasts strongly with the linear dependence of the unbound e-h pair excitations λ.

The threshold field for the first appearance of the X^- peak increases with electron concentration as shown in Fig.4(a) (black points). The

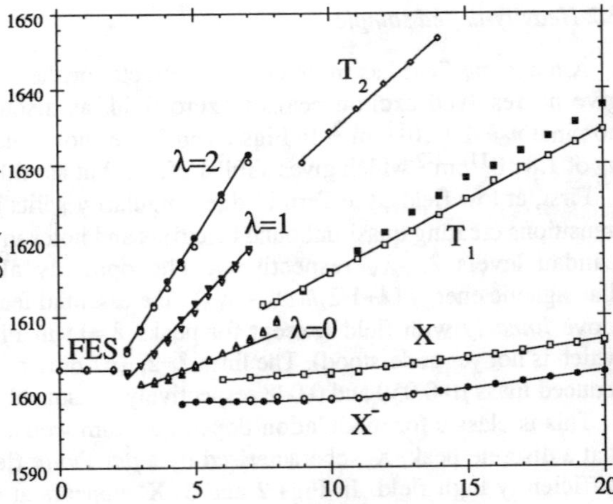

Fig.3 : Positions of main peaks in spectra of sample ZD173 (nominal n_e= 1.8 $10^{11}cm^{-2}$), from Fig.2. The black points are measured in σ+ polarisation, the white points are in σ-. The parabolic field dependence of the X and X^- peaks contrasts with the linear dependence of e-h pairs on Landau levels λ.

Figure is approximate only, because our values of n_e are not very accurate and because the thresholds are not always sharp. Nevertheless, the Figure suggests that, except at very low n_e where it exists in zero-field, the X^- species appears at field values corresponding approximately to Landau level filling factor $\nu=2$, which is marked by the line of slope $h/2e$ drawn in Fig.4(a).

Fig.4(a) also gives the threshold fields for observation of X (white points). These are more randomly distributed. In ref.[1], discussing the sample with $n_e \approx 1.1\ 10^{11}cm^{-2}$, we said that X appears at twice the threshold field of X^-, at $\nu \approx 1$. Fig.4(a) shows that this is not generally true, but X does always appear later than X^-.

In summary of this data, we observe a change from unbound e-h pair states at low field to exciton states at high field, occuring around $\nu=2$. No equivalent effect appears to have been reported for modulation-doped III-V quantum wells, where the Landau level type absorptions are seen at all fields [10]. A good theory of our results is not yet available, but we give the following, intuitive explanation. Neglecting first the electron spin splitting, we suggest that exciton states (X^- and X) appear from near $\nu=2$ for two reasons.

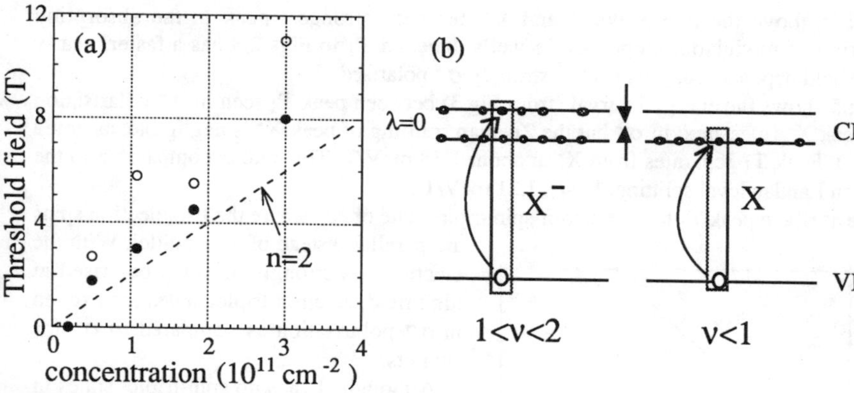

Fig 4(a): Threshold fields at which absorption peaks X^- and X are first detected, against the nominal electron concentration per well n_e for five samples. The straight line marks Landau level filling factor $\nu=2$. *Fig.4(b)* : At left, excitation of a valence band electron to an empty spin-down $\lambda=0$ conduction band level produces X^-. But X can exist only when spin-up levels begin to empty, at right. Density-of-states broadening of order $g_e\mu_B B$ or more will blur this distinction.

First, the ground state of X^- and the ground state of X (the 1s exciton) involve predominantly $\lambda=0$ conduction band states (increasingly so as γ increases). At fields such that $\nu>2$, when the $\lambda=0$ phase space is still filled with electrons, Pauli exclusion prevents excitonic binding. At higher fields, from $\nu<2$, the $\lambda=0$ space begins to empty and formation of excitonic states becomes possible.

Second, in the presence of moderate disorder, a field-induced metal-to-insulator transition ("carrier freezout") occurs between $\nu=2$ and $\nu=1$, as emphasized in recent phase-diagrams [11]. We imagine weak localisation occuring in the potential fluctuations produced by the plane of charged donor cores at distance s (e.g. $s\approx 225$Å= 3.7 donor Bohr radii for the sample of Figs 2,3). This would drastically reduce electric screening of exciton states.

The observation of X, X$^-$ in zero-field at $n_e < 4\ 10^{10}\text{cm}^{-2}$ implies that this localisation already occurs in zero field at small n_e. If the above ideas are correct, the black X$^-$ points in Fig.4(a) mark approximately a *phase boundary* between metallic and insulating regions in the concentration-field graph, with a small insulating region existing in zero-field.

Why does X appear later than X$^-$ with field ? We now include the spin splitting $g_e\mu_B B$. Then space (phase space or real space) will still be filled with spin-up λ=0 electrons between ν=2 and ν= 1. We suggest that an electron-hole excitation cannot form a neutral exciton X in a filled space of spin-up electrons, only X$^-$ can exist (Fig 4(b) at left). An exciton X can exist from ν=1 only, when the spin-up λ=0 levels start to empty (Fig.4(b) at right).

Broadening of the electronic density-of-states of the order of the electron Zeeman energy will reduce this hypothetical 2:1 ratio of the X and X$^-$ threshold fields, and indeed the ratio is not 2:1 in Fig.4(a) except for the point at nominal n_e= 1.1 10^{11}cm^{-2}.

3.3. *Electron spin triplet states of X$^-$*

Lying above the two peaks X and X$^-$, the next strongest peak in the absorption spectrum of modulation-doped CdTe wells, labelled T_1 in Figs 2,3 has a faster, nearly linear field dependence. This peak is strongly σ^- polarised.

Fig.5 shows the energy interval (from Fig.3) between peak T_1 seen in σ^- polarisation and peak X$^-$ (measured in σ^+ but the Zeeman splitting of peak X$^-$ is negligible, as noted earlier). Peak T_1 separates from X$^-$ at about 1.58 meV/T. This rate is comparable to the electron Landau level splitting, $\hbar\omega_{ce}$= 1.21 meV/T .

We attribute peak T_1 to an electron-spin triplet state of X$^-$, where the two electron spins are parallel instead of antiparallel. With the electron gas strongly M=+1/2 polarised at high field, electron triplet states are excited in σ^- polarisation as compared to σ^+ for singlets.

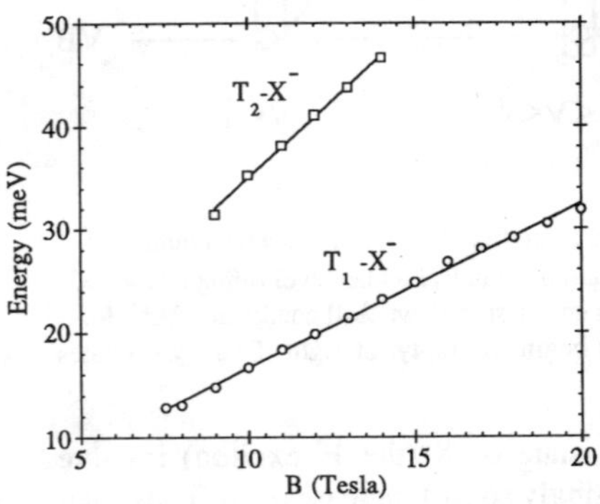

Fig.5 : Energy intervals between absorption peaks T_1 , T_2 and peak X$^-$, taken from Fig.3. Peaks T_1 and T_2, seen in polarisation σ-, are attributed to electron-spin triplet states of X$^-$, with the second electron on λ=l and λ=2 Landau levels respectively. The straight-line fits have slopes of 1.58 and 2.97 meV/T, which are to be compared with $\hbar\omega_{ce}$ = 1.21 meV/T and 2 $\hbar\omega_{ce}$ = 2.42 meV/T respectively.

Although H$^-$ has no spin-triplet states in zero-field in either 3D or 2D, such states are stabilised by a magnetic field. There is no theory of this for X$^-$ but by analogy with recent theoretical calculations of triplet states of D$^-$ [12,13], the field should create an electron spin triplet state which looks like an exciton plus an "outer" electron in the λ=1 level. As noted in 3.1, the (electron spin singlet) ground state of X$^-$ looks like an exciton plus an outer electron in the λ=0 Landau level. Thus the energy separation T_1-X$^-$ should vary appoximately as $\hbar\omega_{ce}$, as observed.

A weaker transition labelled T_2 in Figs 2,3 is seen at even higher field. Fig.5 shows that the T_2-X^- energy difference has a slope of 2.97 meV/T. The T_2-T_1 interval then increases at 1.39 meV/T, which is closer to $\hbar\omega_{ce}$= 1.21 meV/T than the T_1-X^- interval. We attribute T_2 to the next electron triplet state, with the outer electron in the λ=2 Landau level.

4. Emission spectra

Some of the interesting luminescence properties associated with X^- are as follows (more detail will be given elsewhere[7]).

Even in undoped samples, an X^- peak (transition (2b)) is always seen below the X peak (1b) and results from excitons binding with photoelectrons to form X^-. For very light modulation-doping, the X^- peak completely dominates the emission even though X may still dominate in absorption. At higher doping level, where we see a FES in absorption, the luminescence broadens. But a sharp X^- recombination peak appears with field, at about ν= 2. It is partly polarised σ^+, due to thermalisation of the X^- spin levels.

At very high field (>10 T), where the electrons are nearly all in the M=+1/2 state, a strong X emission reappears, but in polarisation σ^- only, which corresponds to the M= -1 state of X. Binding of these M=-1 excitons with the M=+1/2 electrons to form X^- (in its electron spin singlet ground state) is inhibited because it would require an electron spin-flip. This explains why the M=-1 exciton luminescence appears at high field.

Acknowledgements

We thank M. Potemski for providing the optical facilities for our 20T experiments at the Grenoble High Field Laboratory. Y. Merle d'Aubigné, A. Wasiela and W. Grieshaber contributed to the 11T spectroscopic measurements.

References

1 K.Kheng, R.T.Cox, Y.Merle d'Aubigné, Franck Bassani, K.Saminadayar and S.Tatarenko, Phys. Rev. Lett. 71, 1752 (1993)
2 K. Kheng et al, Surface Science 305, 225 (1994)
3 K.Kheng et al, J.de Physique IV, 3, Colloque C5, 95 (1993)
4 G.Munschy and B.Stébé, Phys.Status Solidi (b) 64, 213 (1974)
5 S.Huant, S.P.Nadja and B.Etienne, Phys.Rev.Lett. 65, 1486 (1990)
6 B. Stébé and A. Ainane, Superlattices Microstruct. 5, 545 (1989).
7 R.T.Cox et al , 7th Int.Conf.on Superlattices and Microstruct., Banff, 1994 (to be published)
8 Le Si Dang, G.Neu and R.Romestain, Solid State Commun. 44, 1187 (1982). (Note: the electron and hole masses that we are quoting are all polaron masses, from cyclotron resonance.)
9 J.J.Davies, R.T.Cox and Y.Merle d'Aubigné, unpublished results
10 B.B.Goldberg, D.Heiman and A.Pinczuk, Phys.Rev.Lett.63, 1102 (1989); B.B.Goldberg et al, Phys.Rev.B 38, 10131 (1988)
11 see e.g. S.Kivelson, D.-H.Lee and S.-C.Zhang, Phys.Rev.B 46, 2223 (1992). (Note: n=1,2,.. in our text corresponds to n=1/2,1,.. in this and other references which ignore spin.)
12 P.Hawrylak, Solid State Commun. 88, 475 (1993)
13 A.B.Dzyubenko and A. Yu Sivachenko, Phys.Rev.B 48, 14690 (1993)

Resonant polaron coupling of D^- centers in quantum wells

S. Huant[1] and B. Etienne[2]

1: *High Magnetic Field Laboratory, MPI-FKF/CNRS, BP 166, 38042 Grenoble Cedex 9, France*

2: *Laboratoire de Microstructures et Microelectronique, Centre National de la Recherche Scientifique, BP 107, 92225 Bagneux Cedex, France*

A resonant polaron effect has been observed for D^- centers in a 100 Å GaAs quantum-well structure. The magnitude of this effect is found to be considerably smaller than predicted theoretically. This suggests that second-order perturbation theory is unsufficient to describe the electron-phonon interaction for a two-electron state in a weakly polar two-dimensional semiconductor and/or that polaron correlation is not properly accounted for in the calculations.

1. Introduction

Since the observation that they can form readily in quantum wells (QW's)[1], D^- centers , i.e. neutral donors D^0 trapping a second electron, have been the subject of numerous recent studies, both experimental and theoretical. Works have been devoted to the mechanism of formation of D^- in QW's[2,3], on their energy spectrum in a magnetic field[4-9] and on the behavior of D^- centers in the presence of an excess of free carriers.[10,11] Here, we are interested in the nature of the electron-phonon interaction for the simple two-electron system that represents a D^- center.

In a polar semiconductor, the two D^- electrons interact with the polarization cloud of longitudinal-optical (LO) phonons which they produce to form a bound Fröhlich bipolaron. As compared to D^0 donors, this polarization cloud is enhanced because one electron not only feels its own cloud but also that of the second electron bound to the same impurity ion. This should result in a dramatic enhancement of polaron effects for D^- centers[12] which need experimental investigations. Since non-resonant polaron corrections on electronic levels usually compensate each other in transition energies, polaron effects are best probed in the resonant regime. This regime is reached in a high magnetic field when a given magneto-optical transition - here the main spin-singlet transition of D^- - approaches the LO phonon energy $\hbar\omega_{LO}$, leading to a resonant splitting of this transition. The larger the polarization cloud, the larger the magnitude of this splitting. High magnetic fields - typically 20 T in GaAs - are necessary to measure both the lower polaron branch (below $\hbar\omega_{LO}$) and the upper branch (above $\hbar\omega_{LO}$), and thus the polaron splitting.

2. Experiment

In this work, we used a 100 Å/ 200 Å $GaAs/Ga_{0.75}Al_{0.25}As$ multiple-quantum-well structure grown by molecular-beam epitaxy. This structure is planar doped with silicon donors at the middle of the wells ($N_w = 1.8 \times 10^{10} cm^{-2}$) and middle of the barriers ($N_b = 3 \times N_w$). This unbalanced douple-planar doping geometry favors the conversion of all the D^0 donors in the wells to D^- centers.[2] Far-infrared transmission spectra of this sample were taken by use of a fast-scan Fourier-transform spectrometer (Bruker IFS 113v) connected to a 21 T resistive magnet.

Typical transmission spectra (normalized to zero-field spectra) are shown in Fig. 1. At low field, two peaks are seen which are the $1s \rightarrow 2p^+$ transitions for D^0 donors in the barriers (peak C) and the $s \rightarrow p^+$ spin-singlet transition of D^- (peak B), respectively. Upon increasing the magnetic field, both peaks C and B shift to higher energies. It is found that the D^- transition lies in energy below the reststrahlen band for fields less than 16.5 T and goes above $\hbar\omega_{LO}$ for fields in excess of 19.5 T. The field dependence of the D^- peak is shown in Fig. 2. A typical anticrossing behavior is seen in the optical-phonon energy range which reveals a resonant polaron effect.

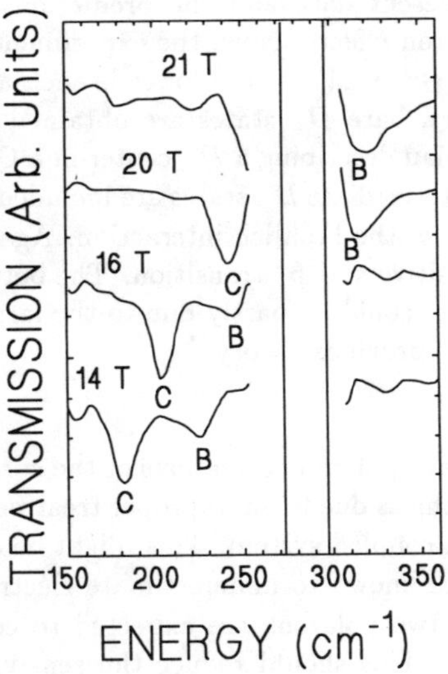

Figure 1. Transmission spectra of a D^- doped 100Å quantum-well sample showing the $1s \rightarrow 2p^+$ transition for D^0 donors in the barriers (peak C) and the spin-singlet $s \rightarrow p^+$ transition of D^- (peak B) both below and above the Reststhralen band of GaAs (represented by vertical bars).

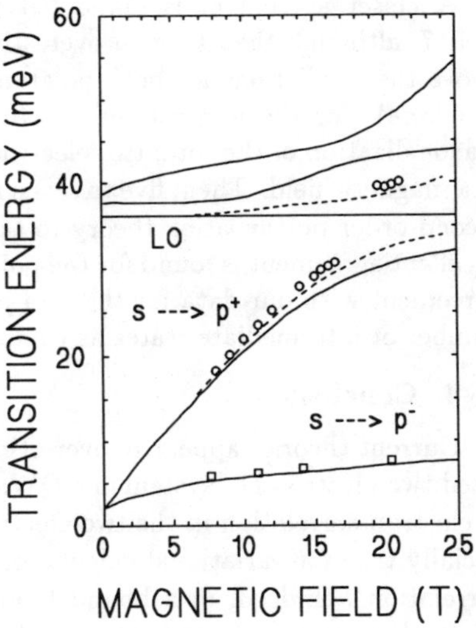

Figure 2. The field dependence of the $s \rightarrow p^+$ transition energy of D^- centers in a 100Å quantum well. The open symbols are the experimental data. The full lines are the predictions of Ref. 12 and the dashed lines are the predictions of Ref. 7.

3. Discussion

The predictions of two recent theoretical treatments of the polaron effect of D^- centers[7,12] are shown in Fig. 2. It can be seen that these theories differ considerably in the resonant regime.

In Ref. 12, a variational approach was used to compute the bare (without electron-phonon interaction) D^- states. The electron-phonon Fröhlich interaction is taken into account within an improved version of the Wigner-Brillouin perturbation theory. The resonant polaron effect is calculated in the three-level scheme (s, p^-, and p^+ singlet states of D^-). A similar analysis for D^0 donors has shown previously a very good agreement with experiment.[13,14] Also shown in Fig.2 are our former data and the predictions of the same theory for the $s \rightarrow p^-$ transition of D^- which is not sensitive to resonant polaron effects: A satisfactory agreement is found because the polaron corrections to the s and p^- states compensate each other. Such a good agreement is not found for the transition to the p^+ state. Clearly, theory greatly overestimates polaron effects in the resonant regime for this transition, i.e., the experimental data are well above ($\approx 2.5meV$) the predictions for the lower branch and substantially below ($\approx 3.5meV$) for the upper branch.

A closer agreement is found between the present data and the predictions of Ref. 7, although theory again overestimates polaron effects. Now, the experiment is above the predictions for both polaron branches ($\approx 1meV$ for the lower branch and $\approx 0.5meV$ for the upper branch). In this theory, bare D^- states are obtained by diagonalization of the total two-electron hamiltonian describing a D^- center in a QW in a magnetic field. Then, five to seven relevant intermediate D^- states are included in second-order perturbation theory to take account of the Fröhlich interaction. Again, excellent agreement is found for the polaron-insensitive $s \rightarrow p^-$ transition. The better agreement with our data for the $s \rightarrow p^+$ transition could be partly due to the larger number of intermediate states as compared to the previous theory.[15]

4. Conclusion

Current theories appear to overestimate resonant polaron corrections of the correlated two-electron D^- system in a QW. Perhaps, this is due to an improper treatment of electron correlation in the two-electron-phonon coupled system. This might be especially true for variational calculations which are known to underestimate electron correlation[4]. Indeed, in a bound bipolaron, the two polarons are expected to correlate their motion in order to avoid each other. This should reduce the sensivity of one of the D^- electrons to the polarization cloud of the other electron, thereby reducing polaron effects. Also the adequacy of second-order perturbation theory is questionable[15]. Indeed, variational calculations show that the outer D^- orbital gradually merges with the inner orbital in increasingly high fields[16]. Thus, in the high field limit, the polarization cloud around D^- is twice as large as that around D^0 (this simple picture does not hold at low field where the outer orbital is twice as much extended as the inner orbital). Since the Fröhlich electron-phonon coupling constant α ($\alpha = 0.065$ in GaAs) scales with the square of the polarization cloud, an equivalent coupling constant for D^- is $4\alpha \approx 0.25$. For such a large value, second-order perturbation theory is certainly unsufficient[17]. Therefore, it is hoped that the present work will stimulate more studies to reconcile polaron theory of D^- with experiment.

5. Acknowledgments

Helpful correspondence with A.B. Dzyubenko is gratefully acknowledged.

6. References

1. S. Huant, S.P. Najda and B. Etienne, Phys. Rev. Lett. **65**, 1486 (1990); S. Huant et al., *in Proceedings of the 20th International Conference on the Physics of Semiconductors, Thessaloniki, 1990*, ed. by E.M. Anastasakis and J.D. Joannopoulos (World Scientific, Singapore, 1990), p. 1369.

2. A. Mandray, S. Huant and B. Etienne, Europhys. Lett. **20**, 181 (1992); S. Huant, A. Mandray, G. Martinez, M. Grynberg, and B. Etienne, Surface Sci. **263**, 565 (1992).

3. D.M. Larsen, Phys. Rev. B **47**, 16333 (1993).

4. T. Pang and S.G. Louie, Phys. Rev. Lett. **65**, 1635 (1990).

5. D.M. Larsen and S.Y. McCann, Phys. Rev. B **46**, 3966 (1992).

6. A.B. Dzyubenko, Phys. Lett. A **165**, 357 (1992); Fiz. Tv. Tela **34**, 3238 (1992) [Sov. Phys. – Solid State, **34**, 1732 (1992)].

7. A.B. Dzyubenko and A.Yu. Sivachenko, Phys. Rev. B **48**, 14690 (1993).

8. S. Huant, A. Mandray, J. Zhu, T. Pang, S.G. Louie, and B. Etienne, Phys. Rev. B **48**, 2370 (1993).

9. A.B. Dzyubenko, A. Mandray, S. Huant, A. Yu Sivachenko, and B. Etienne, Phys. Rev. B (15 August 1994)

10. J.-P. Cheng, Y.J. Wang, B.D. McCombe, and W. Schaff, Phys. Rev. Lett. **70**, 489 (1993).

11. A.B. Dzyubenko and Yu. E. Hozovik, Zh. Eksp. Teor. Fiz. **104**, 3416 (1993) [JETP **77**, 617 (1993)]; P. Hawrylak, Phys. Rev. Lett. **72**, 2943 (1994)

12. J.-M. Shi, PhD Thesis, University of Antwerpen (1993), J.-M. Shi, F.M. Peeters, and J.T. Devreese, Physica B (to appear)

13. J.-M. Shi, F.M. Peeters, G.Q. Hai, and J.T. Devreese, Phys. Rev. B **44**, 5692 (1991).

14. J.-M. Shi, F.M. Peeters, and J.T. Devreese, Phys. Rev. B **48**, 5202 (1993), J.-M. Shi, F.M. Peeters, and J.T. Devreese, to be published

15. A.B. Dzyubenko, private communication.

16. S.G. Louie and T. Pang,*in New horizons in low dimensional systems*, ed. by H. Aoki et al. (Kluwer Press, 1992), p. 445.

17. J.T. Devreese and F.M. Peeters, Solid State Commun. **58**, 861 (1986).

RESONANT POLARON EFFECTS ON D$^-$-CENTERS IN GaAs/Al$_{0.3}$Ga$_{0.7}$As MULTIPLE-QUANTUM-WELL STRUCTURES

J.-P. CHENG
Francis Bitter National Magnet Laboratory, Massachusetts Institute of Technology Cambridge, Massachusetts 02139, USA

B. D. McCOMBE
Department of Physics, State University of New York at Buffalo Buffalo, New York 14260, USA

W. SCHAFF
School of Electrical Engineering, Cornell University, Ithaca, New York 14835, USA

ABSTRACT

The effects of resonant electron–optical-phonon interactions on the energy levels of a two-electron system, D$^-$-centers, in GaAs/Al$_{0.3}$Ga$_{0.7}$As multiple-quantum-well structures have been studied via far-infrared photoconductivity spectroscopy at low temperatures in high magnetic fields up to 23T. An anti-level crossing behavior has been clearly observed for the D$^-$ 1S–2P(M=+1) transition in the resonant energy region with the GaAs optical phonons. The interaction strength is enhanced in comparison with the single-electron system (D^0), and it increases with decreasing well-width.

1. Introduction

The electron–optical-phonon interaction (polaron effect) in semiconductor confined structures has been studied extensively within the last decade due to its fundamental importance in the carrier relaxation process under hot electron conditions. In comparison with bulk material, the problem of electron–phonon coupling in a confined system is much more complicated since both the electronic wave function and the phonon modes are modified.[1] Previous cyclotron resonance (CR) experiments on degenerate two-dimensional (2D) electron gases have revealed that many-electron effects are very important in determining the electron-phonon coupling strength.[2] To avoid the complications of many-electron effects (screening and occupation effects), the investigations on a single-electron system, neutral donor (D^0) impurities, have been carried out on the multiple-quantum-well (MQW) structures,[3-6] and it has been shown that the interaction strength increases with increasing confinement.[1,6]

It is of interest to study how the electron-electron correlation and the confinement effects will simultaneously affect the electron-phonon interactions in a quasi-2D system. In this work, we investigate a particularly simple many-body system, two electrons bound to a positively-charged impurity center, to form a so-called D$^-$-center. Stable D$^-$-centers can be generated in MQW structures by doping selectively in both wells and barriers so that the neutral donor (D^0) in the well can bind an extra electron donated by the barrier doping. The two electrons are indistinguishable and mutually screen each other from the positive center, leading to a very small binding energy. Several experimental[7-9] and theoretical[10-14] studies have been carried out, mostly in the energy region much lower than the GaAs optical phonons, and it has been shown that the electron-electron interaction is crucially important in determining the electronic states of a D^--center.

In this paper we report a far infrared (FIR) magneto-optical study of resonant polaron effects on D^--centers in GaAs/$Al_{0.3}Ga_{0.7}As$ MQW's at low temperatures and high magnetic fields (up to 23T). The results show that electron-electron correlation enhances the strength of electron-phonon coupling in comparison with D^0 impurities (single-electron system) in similar structures.[1, 3-6] The coupling strength increases with increasing confinement, which is consistent with previous work for the D^0 impurities.[1]

2. Experiment

The two samples used in this experiment were grown by molecular-beam epitaxy with identical structure except for the well widths: Sample 1, 100Å; Sample 2, 200Å. The GaAs wells are separated by 600Å-wide $Al_{0.3}Ga_{0.7}As$ barriers, and both well- and barrier-centers are planar doped with Si at the same nominal concentration, $2x10^{10}/cm^2$, for both samples. A total of 20 well-barrier repetitions are sandwiched between two thick (~2000Å) AlGaAs cladding layers. The whole structure was grown on a semi-insulating GaAs substrate and capped with a 100Å GaAs layer.

A capacitively-coupled photoconductivity technique was used in this experiment.[5] Two semitransparent Cr electrodes separated by a small gap were evaporated on the top surface of the sample. A low-frequency (~100Hz) ac voltage was applied between the two electrodes coupled capacitively to the resistive MQW structures, and the ac current through the sample was detected. FIR magneto-spectra were obtained by using Fourier transform spectrometers in conjunction with a 9T superconducting magnet (Buffalo) or a 23T Bitter magnet (Francis Bitter National Magnet Laboratory). All data were taken at liquid-He temperatures in the Faraday geometry (magnetic field parallel to the FIR propagation direction and normal to the sample surface).

3. Results and Discussions

At low magnetic fields (< 12T), both D^- and the D^0 transitions can be observed in the photoconductivity spectra.[9] As the field increases, the photoresponse for the D^0 transitions rapidly loses intensity, and only the D^- transitions appear at high fields. In Fig. 1 we show the spectra of the D^- singlet 1S–2P(M=+1) transition at several magnetic fields in the resonant polaron region for Sample 1. The transition (lower branch) moves to higher energy and gradually decreases in relative intensity as the field increases from 13T to 18T, and it completely disappears at 19T. Meanwhile a new line (upper branch) appears at around 15T. It increases in intensity at the expense of the

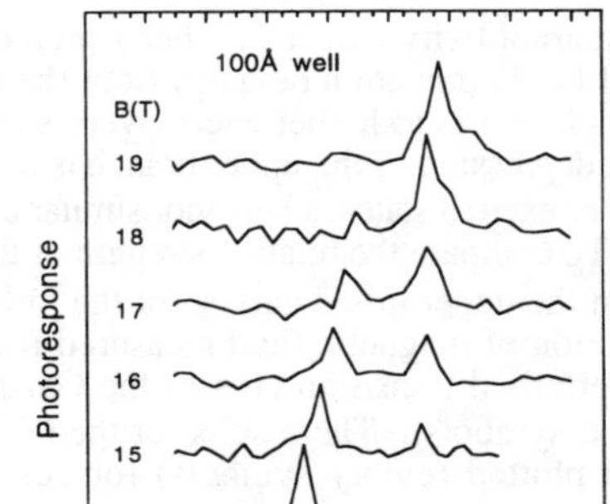

Fig. 1 Photoconductivity spectra of the D^- singlet 1S–2P(M=+1) transition for Sample 1 in the resonant region.

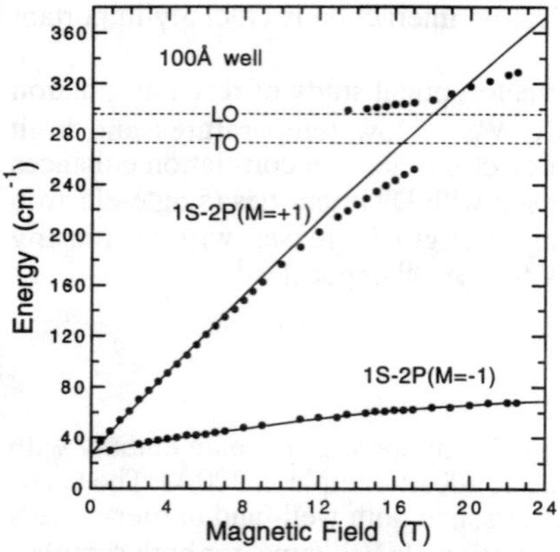

Fig. 2 The D⁻ transition energies as a function of magnetic field for Sample 1. Solid dots are experimental data, and curves are unperturbed D⁻ transitions without resonant polaron correction.

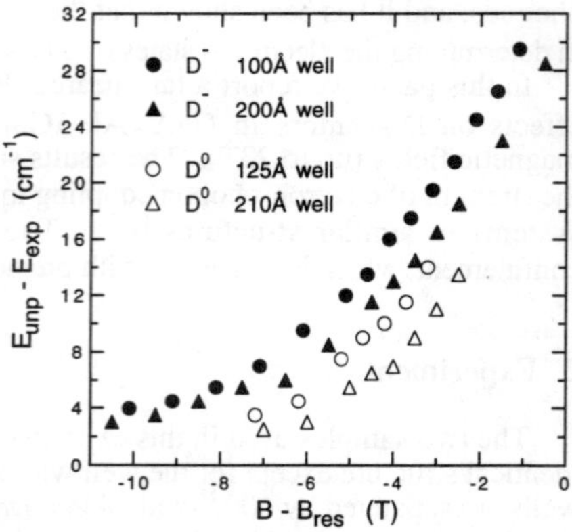

Fig. 3 Energy deviations from the unperturbed transitions in the lower branch of the resonance versus relative magnetic field for both the D⁻ and D⁰ transitions in QW with different well-widths.

lower branch and becomes the dominant line in the spectra when the field is higher than 17T. Thus the two branches exhibit typical anti-level crossing behavior in the optical phonon region.

In Fig. 2 we plot the D⁻ singlet 1S–2P(M=+1) and 1S–2P(M=-1) transition energies as a function of magnetic field for Sample 1. The solid dots are the experimental data, and the curves represent the unperturbed D⁻ transitions without resonant polaron effects. The 1S–2P(M=-1) curve is a simple smooth fit for the experimental data, and the 1S–2P(M=+1) curve is obtained by adding a CR energy, $\hbar\omega_c$, to the 1S–2P(M=-1) curve. The nonparabolicity effect has been included in the CR energy. It is clear that the 1S–2P(M=+1) transition deviates from the unperturbed curve at fields above 9T. The upper branch moves to higher energy very slowly and crosses the unperturbed curve at ~19.5T. The depression of the upper branch is due to the three-level resonant process involving the higher excited states, a behavior similar to the D^0 transitions reported in Ref. 5.

To compare the relative strength of the polaron coupling, we plot the energy deviations from the unperturbed curves for the lower branch of the D⁻ 1S–2P(M=+1) transition as a function of magnetic field measured from the resonant field, B_{res} (the field at which the unperturbed transition crosses the GaAs LO phonon energy), for both samples in Fig. 3 (solid symbols). The results for the D^0(1s-2p^{+1}) transitions with similar well widths are also plotted (empty symbols) for comparison. Systematically, the magnitudes of the deviations for the D⁻ transition are always larger than those for the D^0 transition, and the deviation increases with decreasing well width. These results indicate that the D⁻-centers show a stronger electron–phonon coupling in comparison with the D^0 impurities at the same well width, and the coupling is enhanced by confinement.

An enhanced polaron effect for D⁻-centers has been pointed out in Refs.13 and 14, based on the perturbation calculations. In this case, the two electrons bound to the positive center are not far from each other in space, and each electron interacts with the lattice field polarized not only by itself, but also by another correlated electron (bipolaron). The enhanced lattice polarization field increases the electron–phonon interaction and leads to a lower total energy. Our experimental results are in qualitative agreement with this physical picture. On the other hand, the energy splitting at B_{res} is far less than the theoretical predications,[13,14] indicating that the polaron correction is overestimated in the calculations.

4. Conclusions

Resonant electron–optical-phonon interactions have been clearly observed for a confined two-electron system, D^--centers in QW structures. The interaction is stronger for the D^--centers than the D^0 impurities (single-electron system) at the same well width, which may be attributed to the enhanced lattice polarization field due to the electron-electron correlation. The polaron coupling strength increases with decreasing QW width, indicating that the confinement increases the interaction, consistent with the previous results for the confined D^0 impurities.[1]

5. Acknowledgments

Most of this work was carried out at the Francis Bitter National Magnet Laboratory sponsored by the National Science Foundation. We are grateful to the staff for assistance. JPC thanks IBM for support, and the work in Buffalo is supported by the Office of Naval Research under Grant No. N00014-89-J-1673.

6. References

1. See, e.g., J.-P. Cheng, B.D. McCombe, G. Brozak, and W. Schaff, *Phys. Rev.* **B 48**, (1993) 17243; and references therein.
2. C.J.G.M. Langerak, J. Singleton, P.J. van der Wel, J.A.A.J. Perenboom, D.J. Barnes, R.J. Nicholas, M.A. Hopkins, and C.T.P. Foxon, *Phys. Rev.* **B 38** (1988) 13133; F.M. Peeters, X.-G. Wu, C.J.G.M. Langerak, J. Singleton, D.J. Barnes, and R.J. Nicholas, *Phys. Rev.* **B 45** (1992) 4296.
3. Y.-H. Chang, B.D. McCombe, J.-M. Mercy, A.A. Reeder, J. Ralston, and G.A. Wicks, *Phys. Rev. Lett.* **61** (1988) 1408.
4. S. Huant, W. Knap, G. Martinez, and B. Etienne, *Europhys. Lett.* **7** (1988) 159.
5. J.-P. Cheng, B.D. McCombe, and G. Brozak, *Phys. Rev.* **B 43** (1991) 9324.
6. J.-P. Cheng, B.D. McCombe, and G. Brozak, *Surf. Scie.* **267** (1992) 488.
7. S. Huant, S.P. Najda, and B. Etienne, *Phys. Rev. Lett.* **65** (1990) 1486.
8. E.R. Mueller, D.M. Larsen, J. Waldman, and W.D. Goodhue, *phys. Rev. Lett.* **68** (1992) 2204.
9. S. Holmes, J.-P. Cheng, B.D. McCombe, and W. Schaff, *Phys. Rev. Lett.* **69** (1992) 2571.
10. T. Pang and S.G. Louie, *Phys. Rev. Lett.* **65** (1990) 1635.
11. D.M. Larsen and S.Y. McCann, *Phys. Rev.* **B 45** (1992) 3845.
12. X. Xia and J.J. Quinn, *Phys. Rev.* **B 46** (1992) 12530.
13. J.M. Shi, F.M. Peeters, and J.T. Devreese, *Surf. Scie.* **305** (1994) 220; *Phys. Rev.* **B**, to be published.
14. A.B. Dzyubenko and A. Yu Sivachenko, *Phys. Rev.* **B 48** (1993) 14690.

Magnetooptics of impurity transitions in quantum wells: temperature effects and evolution from D^- centers to localized magnetoplasmons

A.B. Dzyubenko†, A.Yu. Sivachenko‡, Yu.E. Lozovik§ and A.L. Yablonskii†

†*General Physics Institute, RAS, Moscow 117942, Russia*

‡*Weizmann Institute of Science, Rehovot 76100, Israel*

§*Institute of Spectroscopy, RAS, 142092 Troitsk, Moscow District, Russia*

For double-planar doped quantum wells (QWs), we theoretically study the evolution with temperature of intensities of magneto-optical $1s \to 2p^+$ and $2p^- \to 2s$ transitions of neutral donors in the well $D^0(w)$ and in the barrier $D^0(b)$, transitions from the singlet and triplet D^- ground states and the cyclotron resonance of free electrons. For a QW with excess free electrons we study many-body effects on impurity magneto-optical transitions in the vicinity of integer filling $\nu = 1$. Theory predicts large discontinuities in energies and oscillator strengths which, however, are smeared out at temperatures T$\sim$ 4 K.

Quasi-two-dimensional D^- centers in high magnetic fields [i.e. two-electron systems formed in GaAs/GaAlAs QWs by neutral donors D^0 trapping a second electron] have been the subject of recent experimental and theoretical studies (e.g. [1-9] and references therein). In this paper we study intensities of impurity magneto-optical transitions with large oscillator strengths which can be relevant at elevated temperatures $T < 20$ K. This study is motivated by quantitative predictions of theory [1,2] that, in addition to the D^- singlet transition S, two strong triplet D^- transitions T− and T+ can be observable at elevated temperatures. Second, we study impurity magneto-optical spectra in the presence of a finite concentration of excess free electrons n_{ex} in a QW [3]. It was shown previously that for Landau level filling factors $\nu \equiv 2\pi\ell_B^2 n_{ex} \sim 1$, the picture of the transitions S and T± between the two-particle D^- states is changed to that of *collective* magnetoplasma excitations localized at the isolated Coulomb impurity [1]. In particular, this explains (see also [4,5]) the shift to higher energies observed in [3]. In this work, using a strictly-2D model in the high-field limit we investigate many-body effects on the energies and oscillator strengths of impurity transitions in the *vicinity* of integer filling $\nu = 1$.

1. Temperature effects.

To calculate the occupancies, we use a standard thermodynamic model in which the probability that a state with the energy ε_j and with N_j electrons is occupied is

$$w_j^{(\alpha)} = Z_\alpha^{-1} \exp\left[\frac{N_j^{(\alpha)}\mu - \varepsilon_j^{(\alpha)}}{k_B T}\right] \quad , \quad Z_\alpha = \sum_j \exp\left[\frac{N_j^{(\alpha)}\mu - \varepsilon_j^{(\alpha)}}{k_B T}\right] \tag{1}$$

here the index α describes the states due to either the donors in the well, in the barrier or belonging to the zero Landau levels (LLs): $\alpha =$ w, b or LL, respectively. For, e.g., $\alpha =$ w, possible occupation numbers are $N_j^{(w)} = 0, 1, 2$ and correspond to three different charged states $D^+(w)$, $D^0(w)$ and $D^-(w)$. Mean total number of electrons on levels of the kind α is $\langle N^{(\alpha)}\rangle = \sum_j w_j^{(\alpha)} N_j^{(\alpha)}$. In our calculations of occupancies we assume that a finite

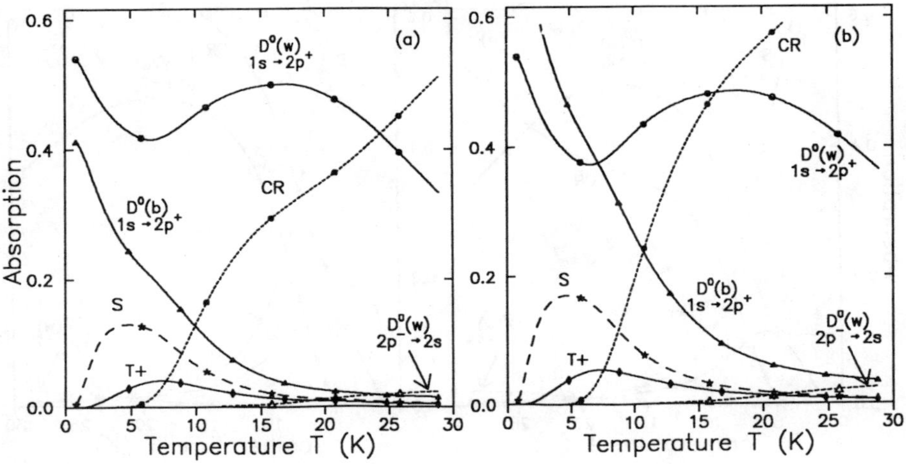

Figure 1: Magneto-optical transitions in 194 Å/196 Å GaAs/$Ga_{0.7}Al_{0.3}As$ QW at $B = 10$ T with (a) $\mathcal{N}_d^{(w)} = \mathcal{N}_d^{(b)} = 10^{10}$ cm^{-2}, $\gamma = 0.8$ and (b) $\mathcal{N}_d^{(w)} = 10^{10}$ cm^{-2}, $\mathcal{N}_d^{(b)} = 2 \times 10^{10}$ cm^{-2}, $\gamma = 0.7$. Intensities are normalized to that the singlet transition S would have at $T = 0$ provided $1s$ $D^0(b)$ state has vanishing binding energy. Intensity of the T− triplet transition (not shown) scales with temperature like that of T+ increased by $\sim 25\%$.

number of donor states split off from free LLs: four single-particle $1s$ and $2p^-$ $D^0(w)$ and $D^0(b)$ states, two lowest-lying two-electron D^- singlet $|0,0;S\rangle$ and $|0,1;S\rangle$ states and two lowest-lying D^- triplet $|0,1;T\rangle$ and $|0,2;T\rangle$ states belonging to the zero LL; see [1,2] for notations and more details of the D^- spectra. We consider magneto-optical transitions with $\Delta M_z = 1$, $\Delta N = 1$ which have large oscillator strengths in the Faraday geometry [1,2,6]: $1s \to 2p^+$ $D^0(w)$ and $D^0(b)$, $2p^- \to 2s$ $D^0(w)$ transitions, the cyclotron resonance (CR) of free quasi-2D electrons, the transition S from the D^- singlet ground state $|0,0;S\rangle$ and the two strong transitions T± from the triplet ground state $|0,1;T\rangle$. Details of calculations of eigenenergies and oscillator strengths are similar to those of Refs. 1,2,6; spin degeneracy of levels is lifted by the Zeeman energies $g^*\mu_B BS_z$ (with $g^* = -0.44$). Chemical potential μ is numerically determined from electron number balance equation

$$\mathcal{N}_0 \langle N^{(LL)} \rangle + \mathcal{N}_d^{(w)} \langle N^{(w)} \rangle + \gamma \mathcal{N}_d^{(b)} \langle N^{(b)} \rangle = \mathcal{N}_d^{(w)} + \gamma \mathcal{N}_d^{(b)} \tag{2}$$

where $\mathcal{N}_0 = (2\pi\ell_B^2)^{-1}$ is the 2D density of states in the zero LL, $\ell_B = (\hbar c/eB)^{1/2}$, $\mathcal{N}_d^{(w)}$ [$\mathcal{N}_d^{(b)}$] is the concentration (per cm^{-2}) of the donors in the well [in the barrier]; $0 < \gamma \leq 1$ phenomenologically describes the part of electrons transferred from the barrier donors into the QW. We consider QWs with parameters similar to those used in experiments [3,6,9] (and references therein): (i) with the well width $d_w = 194$ Å and the barrier width $d_b = 196$ Å, (ii) with $d_w = 194$ Å and $d_b = 600$ Å. Results of calculations of intensities are shown in Figs. 1, 2. For wider barrier QWs (Fig. 2b), $D^0(b)$ states are irrelevant, and the D^- peak S is the dominant feature at low temperatures. In contrast, for 194 Å/196 Å QWs the binding energy of the $1s$ $D^0(b)$ state exceeds that of the D^- singlet ground state; the latter, hence, should be depopulated at low temperatures. Indeed, Fig. 1 shows that the D^- peak

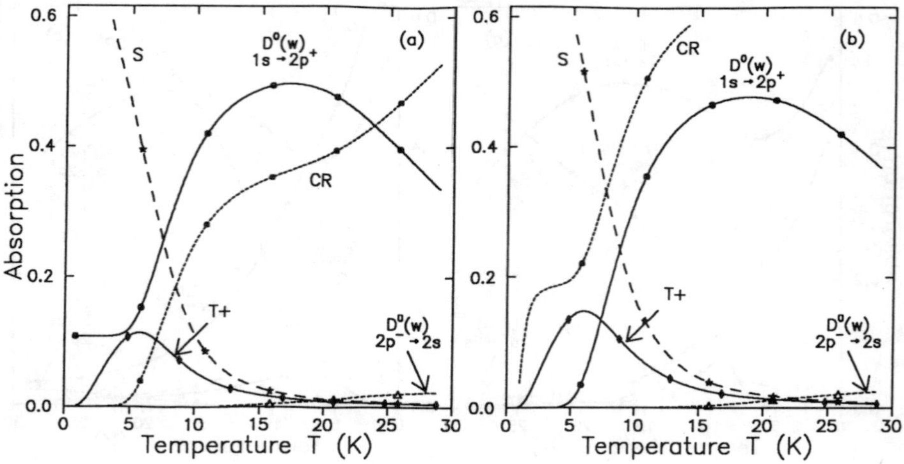

Figure 2: Same as in Fig. 1 for 194 Å/600 Å QW.

S [though having at $T \approx 4$ K comparable intensity with the $1s \to 2p^+$ $D^0(w)$ and $D^0(b)$ transitions] vanishes rapidly below $T \approx 2$ K. Note that in the limit $T \to 0$, correlations in positions of $D^+(b)$ and $D^-(w)$ ions (not included into our model) are qualitatively important and can explain the coexistence of $1s \to 2p^+$ $D^0(w)$, $D^0(b)$ and D^- singlet peaks [7] (and references therein). For a quantitative description, negative polaron corrections to D^- self-energies, which greatly exceed those for D^0 states [1,2,8] (see also [9]), should be also taken into account.

Figures 1, 2 also show that: (i) the triplet transitions T+ and T− can be observed in a rather narrow temperature region, (ii) for wider-barrier 194 Å/600 Å QWs maximal T± intensities are higher [there is no competition between D^- and $D^0(b)$] and are peaked at somewhat lower temperatures than for 194 Å/196 Å QWs. Thus, wider-barrier QWs seem to be more promising for observation of the triplet D^- transitions. Theory [1,2] predicts, however, that T+ and T− lie very close to $1s \to 2p^+$ $D^0(w)$ transition and CR, respectively. Due to this degeneracy, careful search in 194 Å/196 Å QWs [6] has not revealed new features associated with D^- triplet transitions. To find traces of the T−, we suggest to scrutinize variation with temperature [in the region 4 K $< T <$ 10 K] of the CR lineshape to which T− contributes significantly for samples with lower barrier dopings (Fig. 2a). Similarly, to reveal T+ transition we suggest to scrutinize variation with T of the lineshape of $1s \to 2p^+$ $D^0(w)$ peak for samples with higher barrier dopings, where the complete conversion $D^0(w) \to D^-(w)$ occurs at low temperatures (Fig. 2b).

2. Many-body effects.

Impurity magneto-optical transitions in the vicinity of $\nu = 1$ can be described in terms of few-particle impurity-bound magnetoexciton complexes. Indeed, consider first the filling $\nu = 1^{+0}$, when just a single electron is present in $0\uparrow$ LL (and $0\downarrow$ LL is completely filled). The initial state, in the presence of compensating background, can be thought of as a "neutral donor" state $|i\rangle = a^{\dagger}_{0\uparrow M}\,|G\rangle$ with the oscillator quantum number M ($|G\rangle$ describes

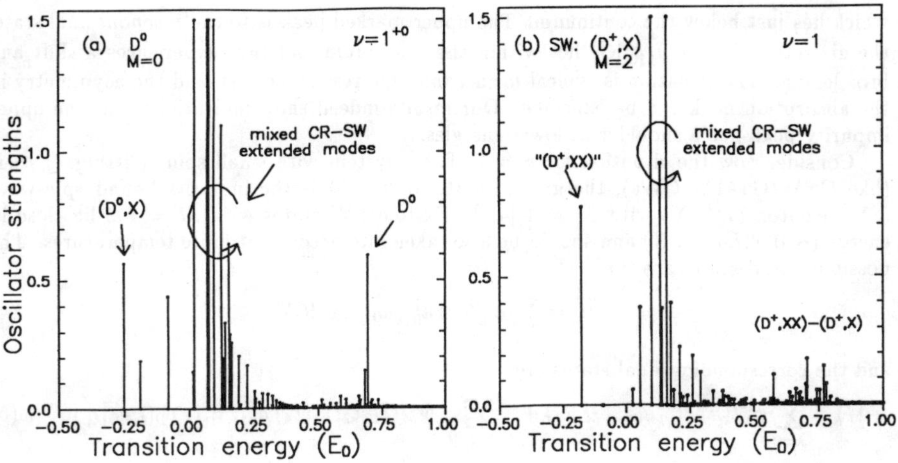

Figure 3: Oscillator strengths versus transition energies [calculated from $\hbar\omega_c$, in units of $E_0 = (\pi/2)^{1/2} e^2/\epsilon\ell_B$] for transitions (a) from the ground D^0 state $a^\dagger_{0\uparrow M}\,|G\rangle$ with $M = 0$, and (b) from the excited SW exciton (D^+, X) with $M = 2$. Prominent features associated with the impurity are marked by arrows; predominant characters of the final states are also indicated.

fully occupied $0\downarrow$ LL). At low temperatures, only transitions from the ground $M = 0$ state should be considered. The final state of the strong magneto-optical transition is

$$|f\rangle = \phi_M\, a^\dagger_{1\uparrow M}\,|G\rangle + \sum_{\ell_1,\ell_2,\ell_3} \Phi_M(\ell_1,\ell_2;\ell_3)\, a^\dagger_{0\uparrow\ell_1} a^\dagger_{1\downarrow\ell_2} a_{0\downarrow\ell_3}\,|G\rangle \tag{3}$$

and can be thought of as a mixture of the D^0 state with one electron in the $1\uparrow$ LL and the four-particle complex containing the donor ion D^+, two electrons and one hole [i.e. a magnetoexciton bound to the neutral donor (D^0, X)]. The amplitudes ϕ_M and $\Phi_M(\ell_1,\ell_2;\ell_3) \sim \delta_{M,\ell_3-\ell_1-\ell_2}$ are determined by direct diagonalization of the total Hamiltonian (cf. [10]) in which we truncate the basis set (3) including only states with $\ell_i \le \ell_{max} = 40$ [the Hamiltonian matrix to be diagonalized is then 861×861]. Qualitatively, this corresponds to dealing with a finite system of the size $R \approx \sqrt{2(\ell_{max}+1)}\,\ell_B \approx 9\,\ell_B$. The final states (3) can be either (i) localized impurity-bound states, or (ii) extended states forming a continuum. It can be shown that for an infinite system the exact lower continuum edge lies at energy $-1.25E_0$ [where $E_0 = (\pi/2)^{1/2} e^2/\epsilon\ell_B$] and corresponds to a superposition of the *extended* spin-wave (SW) exciton (see below) and the D^0 (the electron in $1\uparrow$ LL bound to D^+ with $\ell = 1$). There are also higher-lying extended states in the continuum corresponding to a superposition of magnetoplasmon (electron in $1\downarrow$ LL and the hole in $0\downarrow$ LL forming a bound pair) and the D^0 state with electron in $0\uparrow$ LL. Due to a truncation of the basis set, spurious mixing between extended magnetoplasmon modes and SW excitons occurs, and there is a redistribution of oscillator strengths from the optically-active CR mode (see Fig. 3). There are two peaks associated with the impurity transitions in Fig. 3a. The lower peak is the transition to the localized final state with the predominant character (D^0, X)

which lies just below the continuum. The upper marked peak is to the *resonant* final state: the discrete D^0 state $a^\dagger_{1\uparrow 0}\,|G\rangle$ lies within the continuum and thus experiences a shift and broadening. The situation is typical for a Fano-type resonance [11], and the asymmetry in the absorption peak can be expected. Our results indeed show indications that the upper impurity peak has a shoulder at lower energies.

Consider now the situation for $\nu = 1$. For a system with small spin splittings $g^*\mu_B B$ (like GaAs/GaAlAs QWs), the ground state for $\nu = 1$ is the impurity-bound spin-wave (SW) exciton (D^+, X) with $M = 1$ [4,5]. Excited SW states with $M = 2, 3$ lie close in energy ($\sim 0.02E_0 \approx 4$ K) and should be also taken into account at finite temperatures. The possible initial states are thus

$$|i\rangle = \sum_{k=0} \phi^{(0)}_M(k)\, a^\dagger_{0\uparrow k} a_{0\downarrow k+M}\,|G\rangle \tag{4}$$

and the corresponding final states are

$$|f\rangle = \sum_{k=0} \phi^{(1)}_M(k)\, a^\dagger_{1\uparrow k} a_{0\downarrow k+M}\,|G\rangle + \sum_{\ell_1,\ell_2,\ell_3,\ell_4} \Phi_M(\ell_1,\ell_2;\ell_3,\ell_4)\, a^\dagger_{0\uparrow\ell_1} a^\dagger_{1\downarrow\ell_2} a_{0\downarrow\ell_3} a_{0\downarrow\ell_4}\,|G\rangle \tag{5}$$

i.e., can be thought of as a mixture of the exciton (D^+, X) and biexciton (D^+, XX) bound to the ionized donor (cf. [10]); here $\Phi_M(\ell_1,\ell_2;\ell_3,\ell_4) \sim \delta_{M,\ell_3+\ell_4-\ell_1-\ell_2}$. The transition from the ground SW state (D^+, X) with $M = 1$ has been also considered in [5]; our results differ from [5] in that we find no prominent impurity peaks for $M = 1$ case. For $M = 2$ we find that only the lower-energy prominent impurity peak appears (Fig. 3b), while for $M = 3$ both the low- and high-energy peaks are present with the energies which are close to those for $\nu = 1^{+0}$ case (Fig. 3a).

Summarizing, since for $\nu = 1$ and $\nu = 1^{+0}$ the nature of the ground state is different (SW exciton (D^+, X) and D^0, respectively), at $T = 0$ large discontinuities in oscillator strengths and energies of impurity magneto-optical transitions are expected in the vicinity of $\nu = 1$. At temperatures $T \geq 4$ K, however, when transitions from excited impurity-bound SW excitons are involved, these singularities are effectively smeared out. Hence, drastic changes in the impurity magneto-optical spectra with temperature can be expected near $\nu = 1$.

Acknowledgments. A.B.D. wishes to thank L.V. Butov, S. Huant and M. Grynberg for stimulating discussions. The work of A.B.D. and A.L.Ya. is supported in part by INTAS grant No. 93–1495.

[1] A.B. Dzyubenko and A.Yu. Sivachenko, JETP Lett. **57**, 507 (1993).
[2] A.B. Dzyubenko and A.Yu. Sivachenko, Phys. Rev. B **48**, 14690 (1993).
[3] J.-P. Cheng, Y.J. Wang, B.D. McCombe, and W. Schaff, Phys. Rev. Lett. **70**, 489 (1993).
[4] A.B. Dzyubenko and Yu.E. Lozovik, JETP **77**, 617 (1993).
[5] P. Hawrylak, Phys. Rev. Lett. **72**, 2943 (1994).
[6] A.B. Dzyubenko, A. Mandray, S. Huant, A.Yu. Sivachenko, and B. Etienne, Phys. Rev. B **50**, 4687 (1994).
[7] D.M. Larsen, Phys. Rev. B **47**, 16333 (1993).
[8] J.-M. Shi, F.M. Peeters, and J.T. Devreese, to be published.
[9] S. Huant and B. Etienne, this Conference; J.-P. Cheng, B.D. McCombe, and W. Schaff, this Conference.
[10] A.B. Dzyubenko, Sov. Phys.– Solid State **31**, 1885 (1989); Solid State Commun. **74**, 409 (1990); (E) **75**, No. 6; Phys. Lett. A **173**, 311 (1993).
[11] U. Fano, Phys. Rev. **124**, 1866 (1961).

Photoluminescence from the Wigner Crystal

H.A. Fertig

Department of Physics and Astronomy,

University of Kentucky, Lexington,

Kentucky 40506-0055

The theory of photoluminescence (PL) properties of a two-dimensional Wigner crystal (WC) in a strong magnetic field is reviewed. An exact theoretical framework is developed which allows a systematic study of various quantum effects on the PL spectrum. At the mean-field level, it is shown that in principle the spectrum displays two unique signatures of the WC: (1) a "Hofstadter butterfly" spectrum with bands and gaps indicative of quantum particles in a magnetic field and a periodic potential, and (2) a characteristic upward shift upon melting. The formalism allows shakeup effects to be included as well. It is found that shakeup wipes out the Hoftadter spectrum found at the mean-field level for localized holes, although it should survive for itinerant holes. Shakeup in addition introduces PL satellite peaks, which may be directly identified with maxima in the WC collective-mode density of states. The observation of this structure distinguishes between a liquid and solid electron state.

Sixty years ago, Wigner[1] pointed out that an electron gas will undergo a zero-temperature, quantum phase transition into a crystalline phase as the density is lowered. Semiconductors are attractive systems for looking for such a phase, because one has great control over the electron densities through dopant concentrations. The intrinsic difficulty with such semiconductors is that the dopants which provide the electrons also introduce a random potential, which may spoil the long-range order associated with a crystal state. In recent years, the two-dimensional electron gas (2DEG) as realized in modulation doped semiconductors has arisen as a particularly good candidate for observing the Wigner crystal (WC), since the quality of samples of this type presently available is so high that the electron groundstate is not necessarily dominated by disorder. The possibility of observing the WC is further enhanced by the application of a strong perpendicular magnetic field, which quenches the kinetic energy, and allows the formation of a crystal state at higher densities (for which disorder effects are less important) than would be possible without it.

Experimental evidence for the WC in 2DEG's has accumulated over the last few years[2]. One probe that has produced much intriguing data is photoluminescence (PL), in which either a valence band hole[3] or a hole bound to an acceptor[4,5] recombines with an electron in the 2DEG, producing a characteristic photon spectrum. The subject of this work is to understand what type of spectrum should be expected from the system, assuming that the electronic groundstate is indeed a WC. We will see that a mean-field analysis[6] of the PL spectrum has, in principle, characteristic signatures of the WC: a "Hofstadter butterfly"[7] spectrum for the case of weak interactions between the electrons and the hole, and a characteristic shift in the PL spectrum upon melting of the crystal. When the possibility of exciting phonons and other collective excitations of the WC in the recombination process –shakeup effects – is included, we will see that the Hofstadter spectrum is lost in localized hole experiments (although we expect it to survive in itinerant hole geometries), and that it is replaced by a spectrum with several collective mode satellites.

I. Formalism

We present in detail our approach for localized hole geometries; changes required for itinerant hole experiments will be outlined below. In real experimental situations, the density of holes is extremely small compared to the density of electrons, so that one expects correlations among the holes to be unimportant in the final result for the PL spectrum. Since we are assuming that the groundstate of the electrons is a WC, we wish to choose a hole geometry that does not spoil the discrete translational invariance of the system. It is thus convenient to set up the holes in a superlattice that is commensurate with the electron lattice. For this geometry, working in the lowest Landau level (as is appropriate in strong magnetic fields), the PL intensity may be written in the form[6] $P(\omega) \propto \mathrm{Im}[R(\omega + i\delta)]$, where

$$R(\omega) = \frac{n_h \Omega}{2\pi l_0^2} \sum_{\mathbf{G}} R(\mathbf{G}, \omega) e^{-G^2 l_0^2/4},$$

and we have approximated the core-hole wavefunctions as delta-functions, n_h is the density of holes, Ω the volume of the system, and the vectors $\mathbf{G}$ are the reciprocal lattice vectors of the superlattice. The magnetic length $l_0 = \left(\frac{\hbar c}{eB}\right)^{1/2}$ will be set to unity in the remainder of this paper. $R(\mathbf{G}, \omega)$ is defined as

$$R(\mathbf{G}, \omega) = \frac{1}{g} \sum_X e^{-iG_x X + iG_x G_y/2} R_{ii}(X, X - G_y; \omega),$$

where g is the Landau level degeneracy, X is the guiding center quantum number, and

$$R_{ij}(X_1, X_2; i\omega_n) = -\int_0^\beta < T_\tau a_{X_1}(\tau) c_i(\tau) c_j^\dagger a_{X_2}^\dagger > e^{i\omega_n \tau} d\tau,$$

with $a_X^\dagger$ creating an electron in state X and $c_i^\dagger$ creating a hole in the unit cell i, and β is the inverse temperature. The quantity R_{ij} should be interpreted as a Green's function for the electron lattice, and formally has poles at the quasihole excitations of the WC. We will see that these poles merge together into narrow bands, and form well defined peaks in the PL spectrum. The equation of motion for $R_{ij}(\mathbf{G},\omega)$ may be written as

$$\begin{aligned}\frac{\partial}{\partial\tau}R_{ij}(\mathbf{G},\tau) = &< \rho(\mathbf{G}) > \delta_{ij}\delta(\tau) - \epsilon_0 R_{ij}(\mathbf{G},\tau) \\ &- n_h \sum_{\mathbf{G}'} V(\mathbf{G}') e^{i\mathbf{G}'\times\mathbf{G}/2 - G'^2/4} R_{ij}(\mathbf{G}-\mathbf{G}',\tau) \\ &- \sum_{\mathbf{G}'} \int_0^\beta d\tau' \Sigma(\mathbf{G},\mathbf{G}';\tau-\tau') R_{ij}(\mathbf{G}',\tau'),\end{aligned} \tag{1}$$

where the self-energy is implicitly defined by

$$\sum_{\mathbf{G}'} \int_0^\beta d\tau' \Sigma(\mathbf{G},\mathbf{G}';\tau-\tau') R_{ij}(\mathbf{G}',\tau') = \tag{2}$$

$$\frac{1}{\Omega}\sum_{\mathbf{q}\neq 0} v(\mathbf{q}) C_{ij}(-\mathbf{q},\mathbf{q}+\mathbf{G};\tau) e^{i\mathbf{G}\times\mathbf{q}/2 - q^2/2} + \frac{1}{\Omega}\sum_{\mathbf{q}} V(\mathbf{q}) C_{ij}(-\mathbf{q},\mathbf{G};\tau) e^{-i\mathbf{q}\cdot\mathbf{R}_i - q^2/4}.$$

In Eqs. 1 and 2, $v(\mathbf{q})$ and $V(\mathbf{q})$ are the Fourier transforms of the electron-electron and electron-hole interactions, respectively, ϵ_0 is the energy of the localized hole, and the sum over $\mathbf{G}'$ is only over reciprocal lattice vectors, while the sums over $\mathbf{q}$ are over all wavevectors. $\mathbf{R}_i$ specifies the position of the hole in the i th unit cell, and $< \rho(\mathbf{G}) > e^{-G^2/4}$ is the expectation value of a Fourier component of the electron density. The method for computing these Fourier components has been described elsewhere[6,8]. Finally, the correlation function $C_{ij}(\mathbf{p}_1,\mathbf{p}_2)$ is just the Fourier transformation of

$$C_{ij}(X_1X_2;X_3X_4;\tau) \equiv -g < T_\tau a^\dagger_{X_1}(\tau) a_{X_2}(\tau) a_{X_3}(\tau) c_i(\tau) c^\dagger_j(0) a^\dagger_{X_4}(0) > . \tag{3}$$

Eqs. 1 and 2 represent the first in an infinite series of equations relating an n particle Green's function to the $n+1$ particle Green's function[9]. To make practical the computation of the PL, one must find a sensible approximation for the self-energy in Eq. 2. We now discuss two schemes for doing this.

II. Mean Field Approximation

The simplest approximation possible for the self-energy in Eq. 2 is to use a Hartree-Fock (HF) decomposition of Eq. 3. This yields the the result $\Sigma(\mathbf{G},\mathbf{G}';\tau-\tau') = \Sigma^{HF}(\mathbf{G},\mathbf{G}')\delta(\tau-\tau')$, with $\Sigma^{HF}(\mathbf{G},\mathbf{G}') = W(\mathbf{G}-\mathbf{G}') < \rho(\mathbf{G}-\mathbf{G}') > e^{i\mathbf{G}\times\mathbf{G}'/2} + \frac{1}{2\pi}\sum_{\mathbf{q}} V(\mathbf{q}) < \rho(-\mathbf{q}) > e^{-q^2/4 - i\mathbf{q}\cdot\mathbf{R}_i}\delta_{\mathbf{G},\mathbf{G}'}$, where W is the sum of the direct and exchange Coulomb potentials[6,8]. We have investigated the results of this approximation, using an electron density of $n_s = 6\times 10^{10} cm^{-2}$, and assuming the holes are separated from the electron plane by d = 250Å. Our supercells contain up to 27 electrons. We investigate both the cases of (a) no electron-hole interaction (which is appropriate in the limit $d \to \infty$), and (b) in the presence of a Coulomb electron-hole interaction. For case (a) (no electron-hole interaction), we find a characterisic spectrum for rational filling fractions $\nu = p/q$, where p and q are integers, of p narrow peaks in the PL. An example of this is illustrated in Fig. 1. This may be understood in terms of the single particle density of states for the WC in the mean-field approximation (MFA): in this case, an individual electron moves in the periodic potential of the other electrons, and a magnetic field with q/p flux quanta penetrating each unit cell. The spectrum for this situation is the well-known "Hofstadter butterfly" of q bands, and in the MFA, p of these are filled. The PL spectrum just reflects the recombination of electrons occupying these bands. Thus, PL is a weighted measure of the electron density of states.

One unfortunate characteristic is that at the lowest filling fractions, for which the WC might be realized in real systems ($\nu < 1/5$), the gaps between the bands are extremely narrow (on the order 0.01meV); thus in practice one is likely only to resolve a single PL peak. In any case, these gaps (for the localized hole case) do not survive either electron-hole interactions or shakeup effects, so that one should not expect to observe the Hofstadter spectrum in this way. For case (b) (electron-hole interaction), we find only a single PL line, which turns out to be the contribution from a single electron in a bound state with the hole. PL from recombination of other electrons in the crystal is in principle present, but its magnitude is very small.

For both (a) and (b), we find an interesting finite temperature effect: when the crystal is brought to melting, there is a sudden shift in the spectrum to higher energies, as may be seen in the inset of Fig. 1. The interesting property of this effect is its suddenness, which is a result of the first order nature of the transition in HF. Thus, in real, weakly disordered samples, one might expect to see two peaks over a range of temperatures, one each from regions of melted and crystalline electrons, which coexist for some temperature range due to disorder effects[10]. This is in qualitative agreement with experiment[3,4], where a shift in oscillator strength between two peaks is observed over a temperature range of about 1 K.

III. Shakeup Effects

Our general formalism allows us to go beyond the MFA, to include the excitation of phonons and other collective modes in the electron-hole recombination process. More standard treatments of shakeup, using exact solutions of independent boson models, have been used to address related tunneling phenomena in a magnetic field[11]. The strength of our approach is that it allows us to treat both the tunneling electron and the lattice electrons on the same footing; i.e., the formalism does not force us to choose out a single electron and specify it as the one that will recombine with the hole. This is especially important for situations with weak electron-hole interactions, where the electron lattice at filling $\nu = p/q$ may be thought of as filling p Hofstadter bands, for which the single particle states are actually linear combinations of states localized at different lattice sites.

Beyond this, our formalism allows us to include collective modes without assuming a harmonic approximation. This last approximation is fine for phonon modes of the electron lattice near the center of the Brillouin zone; indeed, a classical approach does quite well for describing such phonons. However, for higher wavevector modes the harmonic approximation breaks down, and the collective excitations may involve very large displacements of the electrons from their lattice sites.

Our formalism allows all such effects to be accounted for. The general approach is to carry the hierarchical set of equations described in Section I to one more level. Specifically, rather that write down a HF decomposition of C_{ij}, we derive an equation of motion for it that includes the collective excitations. The result is then inserted into Eqs. 1 and 2.

The method for generating an equation of motion for C_{ij} will be described elsewhere[12]. After an arduous calculation, the result expresses C_{ij} in terms of the density-density correlation function $\chi(\mathbf{G}_1+\mathbf{q},\mathbf{G}_2+\mathbf{q};\tau) \equiv \chi_{\mathbf{G}_1\mathbf{G}_2}(\mathbf{q};\tau) = -g < T\rho(\mathbf{G}_1+\mathbf{q},\tau)\rho(-\mathbf{G}_2-\mathbf{q};0) >$. The Fourier transform of this response function contains poles at the collective mode frequencies of the system; it is thus through χ that shakeup effects are introduced. We have computed this function numerically using a generalized random phase approximation[8]. Upon substituting the resulting C_{ij} into Eq. 2, one obtains the self-energy $\Sigma = \Sigma^{HF}\delta(\tau-\tau') + \delta\Sigma$, with

$$\delta\Sigma(\mathbf{G}_1,\mathbf{G}_2,\tau_1-\tau_2) = \frac{1}{4\pi}\sum_{\mathbf{G}_1'\mathbf{G}_2'}\int_{BZ} d^2\mathbf{q}e^{i\mathbf{q}\times(\mathbf{G}_1-\mathbf{G}_2)/2-i\mathbf{G}_1\times\mathbf{G}_1'/2+i\mathbf{G}_2\times\mathbf{G}_2'/2}$$

$$\times W(\mathbf{G}_1'+\mathbf{q})W(\mathbf{G}_2'+\mathbf{q})F(\mathbf{G}_1-\mathbf{G}_1'-\mathbf{q},\mathbf{G}_2-\mathbf{G}_2'-\mathbf{q};\tau_1-\tau_2)\chi_{\mathbf{G}_1'\mathbf{G}_2'}(\mathbf{q};\tau_1-\tau_2)$$

$$+\frac{1}{(2\pi)^3}\sum_{\mathbf{G}_1'\mathbf{G}_2'}\int_{BZ} d^2\mathbf{q}e^{-(\mathbf{G}_1'+\mathbf{q})^2/4-(\mathbf{G}_2'+\mathbf{q})^2/4}V(\mathbf{G}_1'+\mathbf{q})V(\mathbf{G}_2'+\mathbf{q})$$

$$\times F(\mathbf{G}_1,\mathbf{G}_2;\tau_1-\tau_2)\chi_{\mathbf{G}_1'\mathbf{G}_2'}(\mathbf{q};\tau_1-\tau_2)$$

Here, $\int_{BZ} d^2\mathbf{q}$ represents an integral over wavevectors in the first Brillouin zone of the superlattice, and F is a generalized Green's function satisfying the equation of motion

$$\frac{\partial}{\partial\tau}F(\mathbf{G},\mathbf{G}_1,\tau-\tau_1) = \delta_{\mathbf{G}\mathbf{G}_1}\delta(\tau-\tau_1) - \sum_{\mathbf{G}'}\Sigma^{HF}(\mathbf{G},\mathbf{G}')F(\mathbf{G}',\mathbf{G}_1,\tau-\tau_1)$$

$$-\epsilon_0 F(\mathbf{G},\mathbf{G}_1,\tau-\tau_1) - n_h\sum_{\mathbf{G}'}V(\mathbf{G}')e^{i\mathbf{G}'\times\mathbf{G}/2-G'^2/4}F(\mathbf{G}-\mathbf{G}',\mathbf{G}_1,\tau-\tau_1),$$

which may be solved using methods discussed previously[6].

Examples of our calculated PL spectra are shown in Fig. 2 for filling fractions $\nu = 1/5$ and $\nu = 2/7$. For the case of no electron-hole interaction [Fig. 1(a)], at low temperature, a well-defined shakeup peak may be seen approximately 1 meV below the main PL peak; a second very weak satellite is observed approximately 1.6meV below the main peak. The origins of these peaks may be understood in terms of the phonon DOS, which is illustrated in Fig. 3. A van Hove singularity, arising from zone-edge phonons, appears as a strong double peak near 0.4meV. Two other peaks may be seen near above this. There are weak sidebands associated with each of these peaks in the PL spectrum. The precise interpretation of these peaks is unclear; however, it has been speculated that these represent vacancy-interstitial excitations[8]. We point out that it is *crucial* to use a fully quantum mechanical treatment of the collective excitations of the lattice to observe these higher order satellites; classical treatments of the phonons do not produce these unusual excitations.

As in the mean-field calculation, there is a sudden shift in the PL spectrum upon melting. The inset to Fig. 2 illustrates the PL spectrum in the melted state both without and with an electron-hole interaction. In the former case, there is no phonon sideband present. This is necessarily so, because in the melted phase, the density is uniform, and there are no collective modes in the LLL. By contrast, when an electron-hole interaction is present, there is a non-uniform electron density near the hole, allowing some local collective modes to persist even above the melting temperature. With further increase in temperature, or increased setback between the hole and the 2DEG, the oscillator strength of this mode will significantly decrease.

Fig. 2(c) illustrates our results for $\nu = 2/7$ in the absence of electron-hole interactions. As can be seen, except for a change in energy scale caused by changing the magnetic field, the lineshape is essentially identical to the case of $\nu = 1/5$. This contrasts sharply with the results found in the MFA, where without electron-hole interactions, a filling $\nu = p/q$ generally yields p distinct lines for a localized hole. While the splittings are so small in that situation that they are difficult in practice to resolve, evidently shakeup effects wipe out this structure even in principle. However, as explained below, we do expect the Hofstadter structure to survive in an itinerant hole experiment.

IV. Itinerant Holes

For itinerant holes that are relatively distant from the electron plane, for which excitonic correlations between the hole and the electrons may be ignored, the PL power may be written in the form $P(\omega) \propto \sum_{\vec{G}} \sum_j \mathrm{Im}[G(\vec{G}, \omega - \omega_j) W_h(\vec{G}, j)]$, where $G(\vec{G}, \omega)$ is the standard electron Green's function[8], and W_h is related to the hole Green's function by $G_h(\vec{G}, \omega) = \sum_j W_h(\vec{G}, j)/(\omega - \omega_j + i\delta)$. The hole Green's function may be computed directly by finding its single particle states in the periodic potential of the electron lattice. For simplicity, we have done this for a hole in its lowest Landau level. Fig. 4 illustrates the resulting PL spectrum using the MFA for the electron Green's function. One can see that as the temperature is raised, a series of peaks emerges in the spectrum. This is a result of the periodic potential felt by the hole, which has its own Hofstadter spectrum, and hence has higher bands which become thermally occupied as the temperature is raised. Unlike the bands for the electrons, which are very close together, the hole bands are on the order of .1meV apart, and presumably could be resolved in experiment. Furthermore, shakeup effects, which we saw spoil the Hofstadter spectrum for the electrons, do not affect the hole spectrum in G_h. Thus, we expect that itinerant hole experiments offer the best avenue for PL observation of the Hofstadter spectrum. Observation of such a spectrum, which is unique to particles moving in a magnetic field and a periodic potential, would give direct and dramatic evidence that the electrons have indeed formed a WC.

Acknowledgements. This work was done in collaboration with Dongzi Liu and S. Das Sarma, and was supported by NSF Grant No. DMR 92-02255. The author also acknowledges a Sloan Foundation Fellowship, and a Cottrell Scholar Award.

REFERENCES

1. E.P. Wigner, Phys. Rev. **46**, 1002 (1934).
2. For a review of recent experiments, see B. McCombe and A. Nurmikko, eds., *Electronic Properties of Two-Dimensional Systems*, (North-Holland, New York, 1994).
3. S.A. Brown et al., Surface Science **305**, 42 (1994); E.M. Goldys et al., Phys. Rev. B **46**, 7957 (1992); R.G. Clark, Physica Scripta **T39**, 45 (1991) and references therein.
4. H. Buhmann et al., Phys. Rev. Lett. **66**, 926 (1991); I.V. Kukushkin *et al*, Phys. Rev. B **45**, 4532 (1992).
5. I.V. Kukushkin et al., Phys. Rev. Lett. **72**, 3594 (1994).
6. H.A. Fertig, D.Z. Liu, and S. Das Sarma, Phys. Rev. Lett. **70**, 1545 (1993); D.Z. Liu, H.A. Fertig, and S. Das Sarma, Phys. Rev. B **48**, 11184 (1993); H.A. Fertig, D.Z. Liu, and S. Das Sarma, Surface Science **305**, 67 (1994).
7. D. Hofstadter, Phys. Rev. B **14**, 2239 (1976).
8. R. Côté and A.H. MacDonald, Phys. Rev. Lett. **65**, 2662 (1990); Phys. Rev. B **44**, 8759 (1991).
9. L.P. Kadanoff and G. Baym, *Quantum Statistical Mechanics*, (Benjamin, Reading, 1981)
10. M.C. Cha and H.A. Fertig, unpublished.

11. G. Mahan, *Many-Particle Physics*, (Plenum, New York, 1981). For a recent related application, see P. Johansson and J.M. Kinaret, Phys. Rev. Lett. **71**, 1435 (1993).

12. H.A. Fertig, D.Z. Liu, and S. Das Sarma, unpublished.

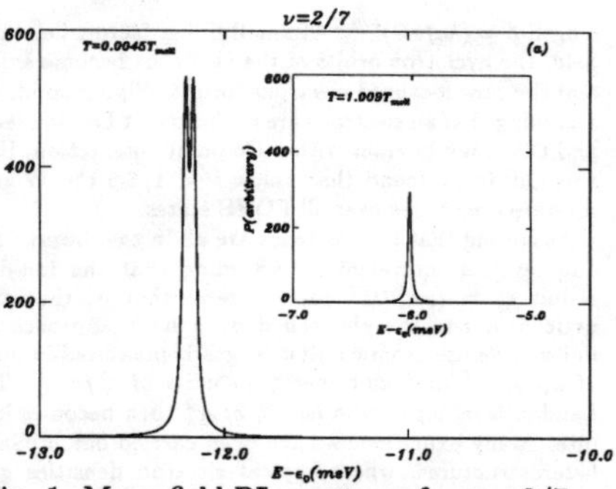

Fig. 1. Mean-field PL spectrum for $\nu = 2/7$ with no electron-hole interaction, with T below melting temperature. Inset: Same, for T above melting.

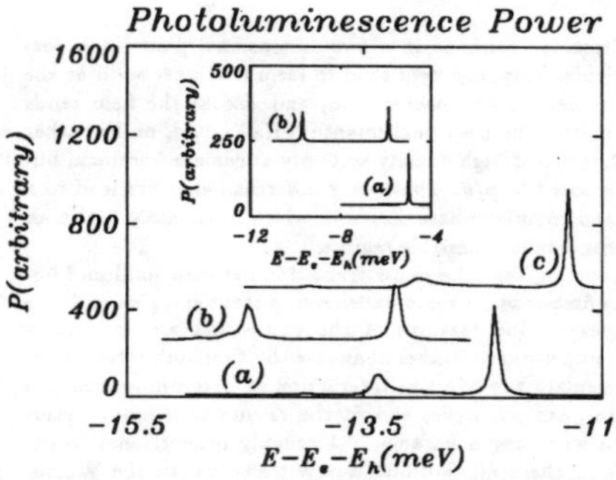

Fig. 2. PL spectra for with shakeup effects: (a)$\nu = 1/5, T = 0$, no electron-hole interaction; (b)$\nu = 1/5$, $T = 0$, with electron-hole interaction; (c)$\nu = 2/7$, $T = 0$, no electron-hole interaction. Inset: PL spectra for (a) and (b), with T just above melting temperature. Spectra are offset by 250 units.

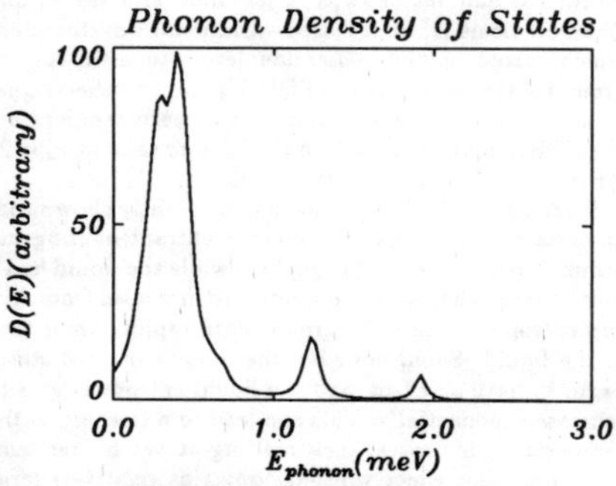

Fig. 3. Collective mode density of states for WC at $\nu = 1/5$, $T = 0$.

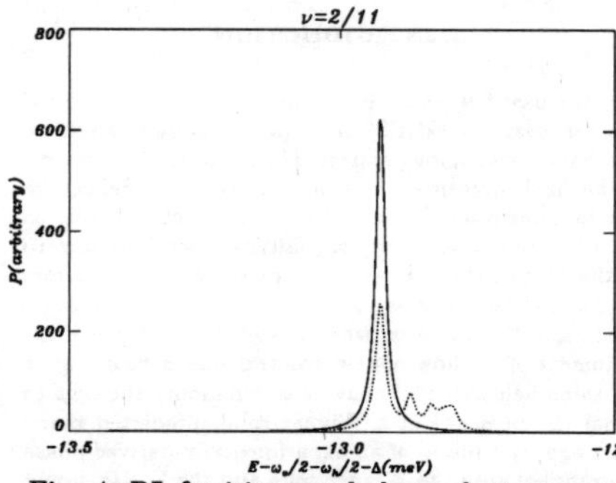

Fig. 4. PL for itinerant hole at electron filling $\nu = 2/11$, for $T \approx .005T_{melt}$ (solid line) and $T \approx .05T_{melt}$ (dotted line).
Electron and hole cyclotron frequencies given here by ω_e and ω_h, respectively.
Δ is the conduction band-valence band gap.

Electron liquid - solid phase boundary at finite density and field in heterostructures

Rodney Price*
Physics Dept., University of Maryland
College Park, MD 20742

Electrons confined in a two-dimensional plane in a perpendicular magnetic field tend to form a Wigner solid as the system density approaches zero, and also as the field tends to infinity. The fractional quantum Hall liquid, on the other hand, forms at high density and only at certain fractional filling factors $\nu = p/q$. These very different behaviors lead to a rich and complex phase diagram, which fortunately lies in an experimentally accessible regime.

Assuming that the phase transition between solid and liquid is first-order, we compare ground-state energies and free energies to find features of the phase diagram in density and temperature. Rather than use the Coulomb interaction, we calculate the effective interaction in heterojunctions and square quantum wells, and fit the results to a model interaction with length parameter λ roughly proportional to the width of the well. We find a new transition to the Wigner solid phase at high density in very wide wells, driven by the softening of the interaction at short distances. In addition, the low-lying excitations of both solid and liquid fall rapidly with well width, moving the temperature-dependent features of the phase diagram much lower in temperature. We examine these effects using a variational Monte Carlo approach.

* Work done in collaboration with Sankar Das Sarma, Xuejun Zhu, and P. M. Platzman

I. INTRODUCTION

In the past few years, experiments with 2-dimensional electron systems (2DES) in a perpendicular magnetic field have shown a long-expected behavior as the strength of the field increases. At comparatively low fields, the now-familiar fractional quantum Hall effect appears, as the longitudinal resistivity ρ_{xx} falls exponentially to zero as the temperature $T \to 0$ at fractional filling factors ν. As the field increases, however, an insulating state appears, with $\rho_{xx} \to \infty$ when $T \to 0$ [1–3]. Optical experiments [4,5] show a new spectral line developing at the same fields. This behavior is commonly thought to signal the presence of a Wigner solid, predicted many years ago, and pieces of an experimentally-derived phase diagram between the Wigner solid and the FQHE liquid can be sketched out.

There are several experimental parameters that affect the boundary between Wigner solid and FQHE liquid. The most important is the magnetic filling factor $\nu = 2\pi n\ell^2$, where n is the electron density and the magnetic length $\ell = (\hbar c/eB)^{1/2}$. At small filling factor, i.e. large field, the cyclotron orbits of the electrons become small, and the now localized electrons form a Wigner solid. By assuming that all electrons are in the lowest Landau level, and that they interact with a Coulomb interaction, Lam and Girvin [6] found that below $\nu \approx 1/6.5$ the Wigner solid predominates over all FQHE states.

Assuming that the electrons are all in the lowest Landau level is equivalent to assuming that the ion-disk radius $r_s = (\pi n)^{-1/2}$ goes to zero; that is, the magnetic field and the electron density both approach infinity. (We use atomic units: length is measured in units of $a_B = \hbar^2/me^2$ and energy in units of e^2/a_B.) The Landau level separation $\hbar\omega_c = 2/\nu r_s^2$ then becomes infinite. Many experiments have been carried out in GaAs heterostructures, where typical electron densities give $r_s \approx 1.5 - 3$ and the Landau levels are far apart compared to other energies in the system. More recently, experiments have been carried out in GaAs heterostructures using holes rather than electrons. Because of the hole's higher effective mass, $r_s \approx 5 - 20$ and Landau-level mixing becomes important. The experiments find, in fact, that the insulating state occurs at higher ν, near 1/3 rather than the $\nu = 1/5$ of the electron experiments, showing that Landau-level mixing is having an effect.

The emergence of an insulating state near $\nu = 1/3$, as observed in the experiment of [7], is consistent with recent calculations of Price, Platzman, and He [8] and Zhu and Louie [9]. Both used variational wavefunctions which mixed in higher Landau levels to estimate the ground state energy for the FQHE liquid and the Wigner solid, respectively, and found that at zero temperature the FQHE liquid gives way to the Wigner solid at $r_s \approx 22$ for $\nu = 1/3$ and $r_s \approx 15$ at $\nu = 1/5$.

Platzman and Price [10] have also recently shown that at finite temperatures a peculiar re-entrant melting can occur. Because the solid is gapless, while the liquid had a finite energy gap, as the temperature increases from zero the entropy of the solid grows more rapidly than that of the liquid. Soon, however, the liquid's excited states begin to have an effect, and the liquid entropy begins to increase exponentially. This can lead to a freezing as the temperature increases, then melting at yet higher temperature. This effect will only occur at relatively large r_s, near the zero-temperature freezing point.

Finally, the experiments are all done in real systems, which can only be considered quasi-two-dimensional. Electrons confined to the lowest subband can still move some small distance, call it L, in the z-direction. If on

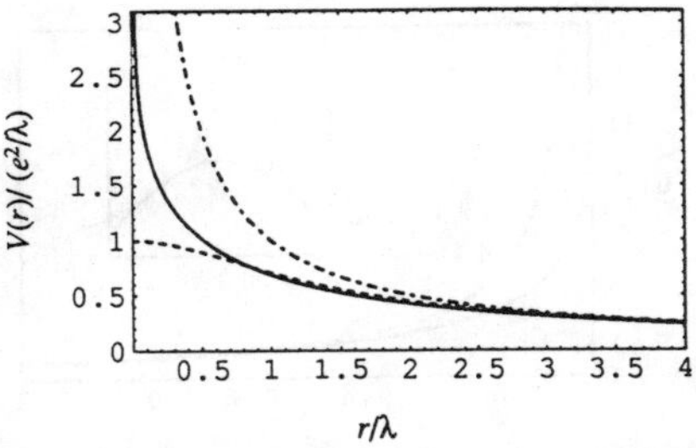

FIG. 1. The effective interaction (solid line), Coulomb interaction (dash-dot line), and model interaction (dashed line) for a GaAs heterojunction.

average a pair of electrons is separated in the z-direction by $\lambda < L$, their effective interaction at distances $r \ll \lambda$ in the xy plane becomes much softer than the Coulomb interaction, while at large distances $r \gg \lambda$, the interactions are nearly identical. This preserves the long-range character of the interaction while weakening the short-range part. Because the formation of the Wigner solid is driven by the long-range part of the interaction, while the FQHE liquid derives its energy advantage from the short-range part, the quasi-two-dimensional character of a real experimental system might be expected to favor the solid somewhat.

In fact, MacDonald and Aers [11] found that at $\nu = 1/3$, for small values of L such as those found in a GaAs heterojunction at high electron density, the ground-state energy of both liquid and solid were reduced by an approximately equal factor. Zhang and Das Sarma [12], however, found substantial corrections to the FQHE gap due to finite thickness effects in GaAs heterojunctions. At larger values of L, He *et al* [13] calculated the overlap of the Laughlin wavefunction with an exact wavefunction calculated for small systems of up to 8 electrons on a sphere. They found that at $\lambda/\ell \approx 3$ the overlap begins to deteriorate until at $\lambda/\ell \approx 7$ some other ground state clearly exists. Experimental data from wide quantum wells, such as the parabolic well of Shayegan *et al* [14], show a collapse of the FQHE gap at wide well widths, in agreement with the results of He *et al*.

In this paper, we compare ground-state energies and excited states of variational wavefunctions for the Wigner solid and the FQHE liquid as both r_s and λ are allowed to vary. We find that at small λ, such as that found in GaAs heterojunctions, the zero-temperature phase transition from liquid to solid at large r_s does not vary much from the Coulomb case [8]. The finite-temperature r_s-T phase boundary, however, is scaled down in temperature considerably, due to the softening of both phonon and magnetoroton modes. At large λ we find that the FQHE liquid gives way to the Wigner solid at low density, as expected, but most unexpectedly, we find that the Wigner solid also dominates the FQHE liquid at high density as well. Only in an intermediate range of r_s does the liquid have lower energy than the solid when λ is large.

II. THE EFFECTIVE INTERACTION

Rather than going through a full three-dimensional calculation, taking into account the shape of the confining potential in the z-direction, we will use an effective interaction in the xy plane that models the three-dimensional interaction. This is feasible because the electrons are confined to a small number of subbands, usually just one, and then can be thought of as extended rod-like charges in the z-direction which are allowed to move only in the xy plane [15]. The effective interaction between these model charges is given by

$$V_{\rm eff} = \int dz_1 \int dz_2 \frac{|\zeta(z_1)|^2|\zeta(z_2)|^2}{[r^2 + (z_1 - z_2)^2]^{1/2}}. \qquad (1)$$

Here $\zeta(z)$ is the envelope wavefunction describing quantization in the z-direction and r is the separation between electrons in the xy plane.

The confining potential in the z-direction enters into this equation only through the envelope wavefunction $\zeta(z)$. This wavefunction should be obtained in a self-consistent procedure which takes into account the many-body nature of the electrons in the xy plane. At low electron density, in quantum wells where the subband splitting is large, $\zeta(z)$ is simply the z-component of the single-particle wavefunction. When electron density becomes higher, or the well is made wider bringing the subbands closer together, $\zeta(z)$ will be modified. In GaAs quantum wells this modification can be substantial, particularly in the case of GaAs - AlGaAs heterojunctions, where the confining potential is roughly triangular near the bottom of the well. A variational wavefunction [15] describing the lowest subband is

$$\zeta_{\rm tri}(z) = \left(\frac{b^3}{2}\right)^{1/2} z e^{-bz/2} \qquad (2)$$

$$b = \left(\frac{48}{r_s^2}\right)^{1/3}. \qquad (3)$$

The strong dependence of b on r_s shows the importance of many-body effects on the envelope wavefunctions in a heterojunction.

The effective interaction $V_{\rm eff}(r)$ in this heterojunction, using this variational wavefunction, (following Zhang and Das Sarma [12]) is shown as the solid line in Figure 1. Compared to the Coulomb interaction, shown as the dash-dotted line, the central peak is much softer. We model this interaction with the model interaction, shown as the dashed line,

$$V_0(r) = \frac{1}{\sqrt{r^2+\lambda^2}} \tag{4}$$

where $\lambda = 1.5/b$. This model interaction is nearly identical to $V_{\rm eff}(r)$ at all $r/\lambda > 0.7$. Since in a heterojunction λ is a function of r_s, we can write $r_s/\lambda = 2.4r_s^{1/3} \geq 3$ for $r_s \geq 2$. Then the pair correlation function for either solid or liquid is very nearly zero at $r/\lambda \leq 0.7$, and we can safely assume that the electrons never approach each other that closely and $V_0(r) = V_{\rm eff}(r)$ over all r of interest.

In a square well, however, λ and r_s are independent parameters, as the width of the well L is to lowest order independent of the density of the electrons in the xy plane. It is only when subband mixing begins to become important that the electron density begins to affect the envelope wavefunctions $\zeta(z)$. We expect that when only the lowest subband is occupied the single-particle wavefunctions will be adequate for $\zeta(z)$, so we use the usual cosine solution for an infinite well, shown in the inset to Figure 2. The effective potential for the cosine solution cannot be evaluated explicitly, however, so we use a gaussian wavefunction

$$\zeta_{\rm sq}(z) = \frac{1}{(\pi\gamma^2)^{1/4}} e^{-z^2/2\gamma^2}, \tag{5}$$

fitted to the cosine solution. Here $\gamma = 0.277L$ gives the gaussian wavefunction shown in Figure 2. The effective potential for the two wavefunctions is almost identical. The gaussian wavefunction gives

$$V_{\rm eff}^{\rm sq}(r) = \frac{1}{\sqrt{2\pi}\gamma} e^{r^2/4\gamma^2} K_0\left(\frac{r^2}{4\gamma^2}\right), \tag{6}$$

where K_0 is a modified Bessel function. We would like to use the simpler model interaction $V_0(r)$ again, so we choose the parameter λ to fit $V_{\rm eff}^{\rm sq}(r)$ best when r is large. A least-squares fit, shown in Figure 2, yields $\lambda/L = 0.2$.

Because λ/L is small for the square well, we need a wide well if we are to investigate a system with reasonably large λ. For example, a well with $\lambda = 1$ in an electron system in GaAs is approximately 500Åwide. In these wide wells, subband mixing can become important, and the envelope wavefunctions will tend to spread out toward the edges of the well as the electrons reduce their potential energy. As the density distribution $n(z)$ spreads, we would expect the effective potential between adjacent electrons to soften. In fact, if we assume the envelope wavefunction to be constant, we find that $\lambda/L = 0.29$.

At low r_s in wide quantum wells, $n(z)$ will be modified further as it peaks near the well edges. The well then begins to resemble a highly-coupled double-layer system. To estimate the effective interaction for such a system, we used a density distribution $n(z)$ obtained from a zero magnetic field density functional calculation [16] in a square well 1100Åwide at $r_s = 2$, and found that $\lambda/L = 0.14$. The Fermi level at zero field was higher than the two lowest subbands, whereas in our high magnetic

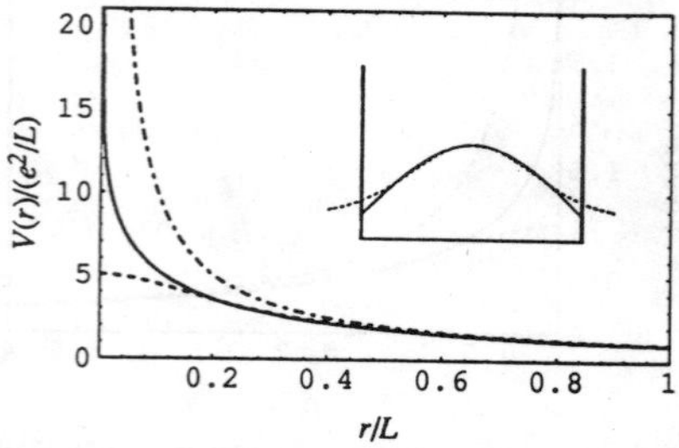

FIG. 2. The effective interaction (solid line), Coulomb interaction (dash-dot line), and model interaction (dashed line) for a square well of width L. The model interaction parameter $\lambda = 0.2L$. The lowest subband cosine wavefunction and gaussian fit are shown in the inset.

field case ($\nu = 1/3$) the filling in the lowest subband would be at most 1/3. Because the subband mixing is reduced at $\nu = 1/3$, the electron density should be more evenly distributed in the z-direction than in the zero-field case, and λ/L should be correspondingly higher.

If we choose $\lambda = 0.2L$ for the square well, $V_0(r)$ is nearly identical to $V_{\rm eff}^{\rm sq}(r)$ for $r \geq 0.2L$, and differs significantly from $V_{\rm eff}^{\rm sq}(r)$ only for $r \leq 0.1L$. The pair correlation function for both liquid and solid is small below about $r = r_s$, so we believe $V_0(r)$ to be a good description of $V_{\rm eff}^{\rm sq}(r)$ for $\lambda \leq 2$.

III. VARIATIONAL WAVEFUNCTIONS

In this paper, we determine the phase boundary between Wigner solid and Laughlin liquid by comparing ground-state energies at zero temperature, and comparing free energies at finite temperature. Because we are simply comparing energies, we are assuming that the phase transition is first-order, rather than for example, some Kosterlitz-Thouless phase transition, and we neglect the presence of any other states in the vicinity. Within these limitations we can construct a phase boundary reasonably consistent with experimental results.

At low filling factor $\nu \ll 1$ the approximation is often made that the electrons are all in the lowest Landau level. In this case, the kinetic energy of each electron is at its lowest possible energy $\hbar\omega_c/2$ and any variational wavefunction has only to minimize the potential energy of the system. We might wish to lower the potential energy by localizing the electrons, thereby keeping them farther away from each other, but the available states in the lowest Landau level are not sufficient to localize them much. The Laughlin liquid uses a different strategy by keeping the electrons at fixed angular momenta with respect to each other. Nearest neighbors are kept well apart, but almost no correlations appear beyond the nearest neighbor. This method of reducing potential energy will work best when the force law is short-ranged, and interactions with nearest neighbors are most important.

The Wigner solid, on the other hand, reduces its potential energy by localizing each electron. Since each electron is always near a particular lattice site (at least at zero temperature) correlations go on indefinitely. The interaction energy between an electron and its next-nearest neighbor, and its next-next-nearest neighbor, and so on, is reduced slightly by these correlations. Here the potential energy is reduced most effectively when the force law is long-ranged. In the lowest Landau level, the electrons come nearer to each other in the solid (at least at $\nu = 1/3$ and $1/5$) than in the liquid, but the long-range force's contribution to the potential energy is smaller.

Now if we choose to lower the potential energy as much as possible, at the expense of higher kinetic energy, we would localize the electrons as much as possible by fixing them to their lattice sites. In order to do this, we must include all Landau levels such that the electron wavefunctions become delta functions. The potential energy of this system is the lowest possible, the Madelung energy of a classical electron crystal. This Madelung energy is a lower bound for the energy of any variational wavefunction we might choose.

Although the Coulomb interaction has a singularity at the origin, it is a long-range interaction, as we can see from Wigner's original argument for the electron solid: The kinetic energy T of an electron system goes as $1/r_s^2$, whereas the potential energy V goes as $1/r_s$. Then the potential energy becomes more important as $r_s \to \infty$, and since the crystal minimizes the potential energy better than the liquid, sooner or later there must be a transition to a crystalline state. Our model interaction $V_0(r)$ removes the Coulomb interaction's singularity, except at $\lambda = 0$, but leaves the long-range part of the interaction the same. The short-range part, however, becomes much less important as λ gets large. Then the liquid will have less advantage over the solid at higher densities, and we expect to see a transition to a Wigner solid at lower r_s than we would have otherwise.

In Figure 3 we show the difference in energy of two wavefunctions, the Laughlin liquid and Lam and Girvin's correlated solid wavefunction [6], from the Madelung energy as a function of λ/r_s. (The inset to the figure shows the Madelung energy itself as a function of λ/r_s.) The Laughlin energy begins to diverge from the Madelung energy at $\lambda/r_s \approx 1$ where He *et al* [13] saw the Laughlin state's overlap with the true ground state in a small system begin to deteriorate, and the difference is large at $\lambda/r_s \approx 2.5$, where He *et al* saw virtually no overlap with the ground state. The solid energy, on the other hand, is approaching the Madelung energy closely in this range. The phonon modes get soft quickly in this range, and the zero-point energy $\sum_k \hbar\omega_k/2$ becomes small.

Both these wavefunctions are restricted to the lowest Landau level, so the energy of the system must go as

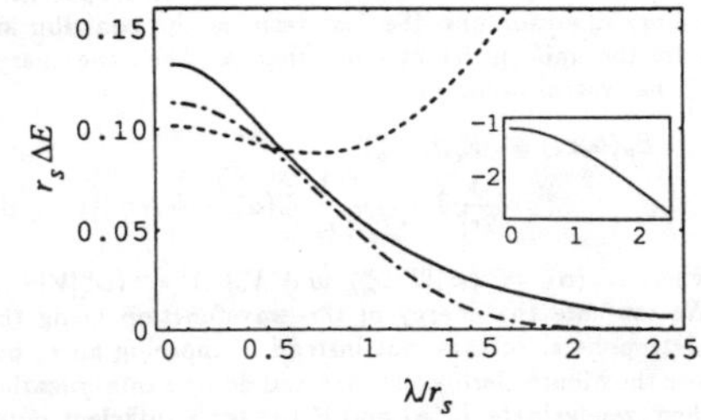

FIG. 3. The Laughlin liquid energy (dashed line) and Lam-Girvin solid energy (solid line) shown after subtracting $\hbar\omega_c/2$ and the Madelung energy. The zero-point energy for the harmonic solid in the lowest Landau level (dot-dashed line) is shown for comparison. The Madelung energy itself is shown in the inset as a function of λ/r_s.

$1/r_s$ and we cannot expect to find any phase boundary in r_s. Landau-level mixing must be introduced to find any phase transitions in r_s, so we modified the above two variational wavefunctions for liquid and solid by putting in a new factor, with new variational parameters, which mixes Landau levels.

A variational wavefunction for the liquid which interpolates in some sense between a wavefunction with the lowest possible kinetic energy, the Laughlin wavefunction, and a wavefunction with the lowest possible potential energy, in which the electrons are completely localized, might be expected to be a good variational choice. To that end, we have chosen a variational wavefunction that consists of the Laughlin wavefunction ψ_m $(m = 1/\nu)$ multiplied by a Jastrow factor $\prod_{i<j} e^{-\alpha/\sqrt{r_{ij}}}$, where r_{ij} is the distance between the ith and j^{th} particles and α is the variational parameter. When $\alpha = 0$, we recover the Laughlin wavefunction, and when $\alpha \neq 0$, the wavefunction is no longer analytic and higher Landau levels are mixed in. The Jastrow factor introduces more correlations into the wavefunction, lowering the potential energy, while introducing a kinetic energy cost.

Details of the calculation are given in [8], so we will only review it briefly here. (Note that [8] uses units of energy $e^2/2a_B$, while this paper uses atomic units e^2/a_B.) We use the spherical geometry, in which our wavefunction becomes

$$\psi_m^\alpha = \prod_{i<j}(u_i v_j - u_j v_i)^m \exp\left(\frac{-\alpha}{|u_i v_j - u_j v_i|^{1/2}}\right), \quad (7)$$

where $u_i \equiv e^{-i\phi_i/2}\cos(\theta_i/2), v_i \equiv e^{i\phi_i/2}\sin(\theta_i/2)$ are convenient spinor coordinates, and the distance between particles i and j is taken as the chord distance $r_{ij} = 2R|u_i v_j - u_j v_i|$. The Hamiltonian can be written

$$H = \frac{1}{2r_s^2}T + \frac{1}{r_s}V - \frac{1}{r_s}N^{3/2}, \quad (8)$$

where T is the kinetic energy operator, V is the potential energy operator and the last term is the contribution from the uniform background charge. Then the energy of the system becomes

$$\begin{aligned} E_\nu(\alpha, r_s) &\equiv \langle\psi_\alpha^\nu|H|\psi_\alpha^\nu\rangle \\ &= \frac{1}{2r_s^2}T_\nu(\alpha) + \frac{1}{r_s}V_\nu(\alpha) - \frac{1}{r_s}N^{3/2}, \end{aligned} \quad (9)$$

where $T_\nu(\alpha) = \langle\psi_\alpha^\nu|T|\psi_\alpha^\nu\rangle$ and $V_\nu(\alpha) = \langle\psi_\alpha^\nu|V|\psi_\alpha^\nu\rangle$. We evaluate the energy of the wavefunction using the Metropolis algorithm, but instead of choosing an r_s before the Monte Carlo code runs and doing a minimization then, we evaluate $T_\nu(\alpha)$ and $V_\nu(\alpha)$ for a sufficient number of values of α to fit a smooth curve to the points, and do the minimization for some particular r_s after the Monte Carlo code runs. In this way we avoid duplicating calculations for different r_s, and the curve fitting is used to average out some of the remaining statistical noise in the Monte Carlo estimates of $T_\nu(\alpha)$ and $V_\nu(\alpha)$.

In order to determine the value of r_s at the liquid-solid phase boundary we need a rather accurate evaluation of the solid. Lam and Girvin [6] evaluated the energy of a correlated Wigner solid wavefunction

$$\Psi = \exp\left(\frac{1}{4}{\sum_{i,j}}' \xi_i B_{ij} \xi_j\right) \prod_i \phi_{R_i}(z_i), \quad (10)$$

where $\xi_i = z_i - R_i$, $B_{ij} \equiv B(R_i - R_j)$, and

$$\phi_{R_i}(z_i) = \exp\left(-\frac{1}{4}\left[|z_i - R_i|^2 - (z_i^* R_i - z_i R_i^*)\right]\right). \quad (11)$$

Here $z_i = x_i + iy_i$ is the ith particle position and $R_i = X_i + iY_i$ is the ith lattice site. Ψ is the harmonic crystal wavefunction restricted to the lowest Landau level, and the variational parameters B_{ij} are calculated by using the values derived from the harmonic crystal. However, in order to make a reasonable comparison of solid and liquid wavefunctions, we need, as discussed above, a wavefunction which includes Landau-level mixing.

Zhu and Louie [9] have recently completed a study of the ground-state energy of the Wigner crystal including Landau-level mixing, using the Coulomb interaction. Price and Zhu have modified their code to use the modified potential (4). This calculation is similar to Lam and Girvin's, except that two more variational parameters α and β are added to the wavefunction to put in more correlations at the expense of some Landau-level mixing. They first followed Lam and Girvin's method with the exception that the gaussians in the single-particle wavefunctions (11) were "squeezed" to move the electrons farther away from each other, by making the replacement

$$\exp\left(-\frac{1}{4}|z_i - R_i|^2\right) \rightarrow \exp\left(-\beta|z_i - R_i|^2\right). \quad (12)$$

Varying the parameter β away from 1/4 introduces Landau-level mixing into the wavefunction because the single-particle wavefunctions $\phi_{R_i}^\beta(z_i)$ are no longer eigenstates of the single-particle Hamiltonian. Zhu and Louie then multiply the wavefunction by another Jastrow factor, and the wavefunction, with two variational parameters α and β, is then

$$\Psi = \exp\left({\sum_{i,j}}'\left[\frac{1}{4}\xi_i B_{ij}\xi_j - \frac{\alpha}{2}u(|z_i - z_j|)\right]\right) \prod_i \phi_{R_i}^\beta(z_i), \quad (13)$$

where

$$u(r) = \frac{1}{\sqrt{r}}\left(1 - e^{-\sqrt{r/F} - r/2F}\right), \quad (14)$$

and F is a constant chosen to optimize the pseudopotential at small r. Zhu and Louie varied B_{ij} as well as β, but found that varying B_{ij} had very little effect on the energy, as Lam and Girvin had suggested. The quantum Monte Carlo calculations were done with modified periodic boundary conditions [9], which require the addition of an unimportant phase factor to the single-particle wavefunctions (11).

IV. ZERO TEMPERATURE

Figure 4 shows our calculated ground-state energies for both the liquid and the solid, plotted as a function of λ at $r_s = 2$, 10, and 20. The energies for $r_s = 2$ resemble closely the lowest Landau level curves of Figure 3, as they should. The lowest Landau level energies are plotted there as a function of the dimensionless variable λ/r_s, which can be thought of as the ratio of the average separation in the z-direction λ and the average separation in the xy plane r_s (actually $\sim 2r_s$) of two nearby electrons. Plotting these curves as a function of λ rather than λ/r_s for $r_s = 2$, 10, and 20 would yield a set of curves similar to those in Figure 4, if we use $0 \le \lambda \le 5$. This range corresponds to $0 \le \lambda/r_s \le 2.5$ at $r_s = 2$, but only $0 \le \lambda/r_s \le 0.25$ for $r_s = 20$. The solid and liquid energies at $r_s = 10$ and 20 have been reduced somewhat, however, because of the Landau-level mixing in the variational wavefunctions. The lowest Landau level energies cross at $\lambda = 6$ at $r_s = 10$ and $\lambda = 12$ at $r_s = 20$, but the variational energies here predict a freezing transition at $\lambda \approx 4$ at $r_s = 10$ and possibly $\lambda \approx 6 - 7$ at $r_s = 20$. None of these results are particularly surprising, but they illustrate the fact that Landau-level mixing does in general tend to favor the solid somewhat.

More surprising is the close approach of the Wigner solid energy to the Madelung energy at large λ, evident at $r_s = 2$. Here Landau-level mixing is having very little effect, particularly as λ increases. The variational wavefunction for the liquid uses the fact that localizing the electrons by mixing in higher Landau levels will keep nearest-neighbor electrons farther apart, and since the

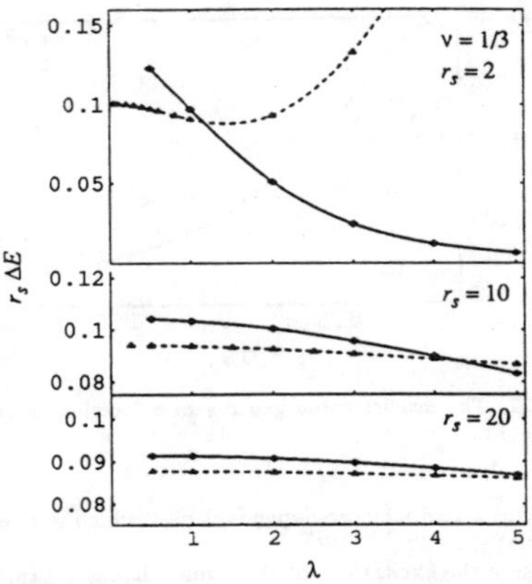

FIG. 4. The variational liquid energy (dashed line) and solid energy (solid line) shown after subtracting $\hbar\omega_c/2$ and the Madelung energy for various fixed values of r_s.

Coulomb potential rises rapidly at small r the energy savings can be significant. As the interaction softens, however, the energy savings becomes small, and as a result the liquid and solid energies at $r_s = 2$ are nearly identical to the lowest Landau level energies. Ortiz, Ceperley, and Martin [17] have recently used their fixed-phase quantum Monte Carlo method to improve the liquid energy shown here somewhat. The difference in energy is significant at $r_s = 10$ and 20, but at $r_s = 2$ the change in energy for the Coulomb interaction is only ~ 0.006.

The rapid descent of the solid energy with λ leads to a freezing transition at about $\lambda = 1.2$ at $r_s = 2$. For an electron system in GaAs with $a_B = 100$Å, this would predict a freezing transition for an approximately 1200Å wide well, brought on by the softening of the short-range part of the interaction. Keeping the well width constant and lowering the density gives a smaller λ/r_s, and the system melts once again until the increased Landau-level mixing at high r_s causes the system to freeze again. Figure 5 shows the energies of solid and liquid when λ is fixed and r_s is allowed to vary. At $\lambda = 0.5$ we find no significant lowering of the solid energy at low r_s, but at $\lambda = 1$ the solid energy at $r_s = 2$ is nearly as low as the liquid energy, and at $\lambda = 2$ and 3 the solid energy at $r_s = 2$ is much lower than the liquid energy. In all cases the liquid again becomes favorable at lower density until Landau-level mixing again causes the system to freeze.

In a heterojunction $\lambda \sim r_s^{2/3}$, so as r_s increases, the ratio λ/r_s falls as $r_s^{-1/3}$, favoring the liquid slightly. This effect partially offsets the effect of Landau-level mixing, and the crossover point is pushed slightly higher. On

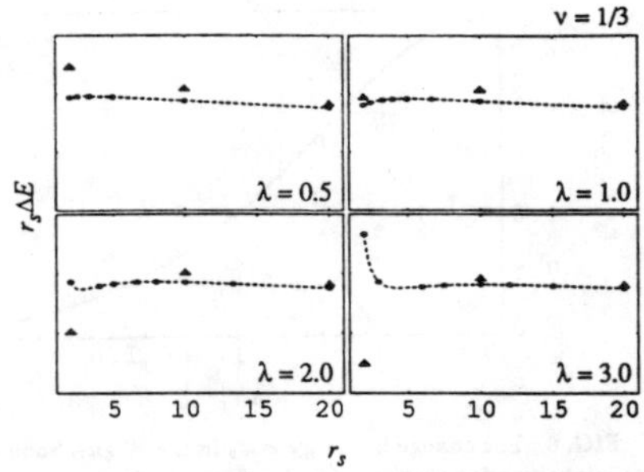

FIG. 5. The variational liquid energy (dashed line) and solid energy (triangles) shown after subtracting $\hbar\omega_c/2$ and the Madelung energy for various fixed values of λ.

the other hand, at low r_s, the slightly higher λ causes the solid energy to fall closer to the liquid energy, so the crossing is not as steep as in the Coulomb case [8]. The overall effect is small.

The effects of impurities in the system continue to be important when the Coulomb interaction is softened. The impurity potential the electrons see is slowly varying, since the impurities are offset from the well by some hundreds of angstroms. Because the solid is compressible, it can adjust to the impurity potential, while the liquid, being incompressible, to lowest order cannot. Then the solid energy will fall relative to the liquid. Simple arguments based on an elastic medium description of the Wigner solid [18] indicate that the energy gained by the solid is roughly

$$E_{\rm imp} \approx \frac{1}{2} m v_t^2 \left(\frac{r_s}{\xi}\right)^2 \tag{15}$$

where v_t is the transverse sound velocity in the absence of a magnetic field, and ξ is the correlation length for the Wigner solid. Depending on the quality of the sample we might expect, for example $5r_s \leq \xi \leq 50r_s$.

The transverse sound velocity v_t will tend to fall rapidly as the interaction softens. Using the harmonic approximation to find the phonon modes ω_t and thus v_t, we estimate the impurity energy as a function of λ/r_s as shown in Figure 6. The energy gain by the solid will be reduced, until in a heterojunction the system may freeze at $r_s \approx 15-20$. The weaker impurity effects will also have implications for the pinning of the Wigner solid formed at high λ/r_s.

V. FINITE TEMPERATURE

At finite temperature we need to know something about the excited states of the system as well as the

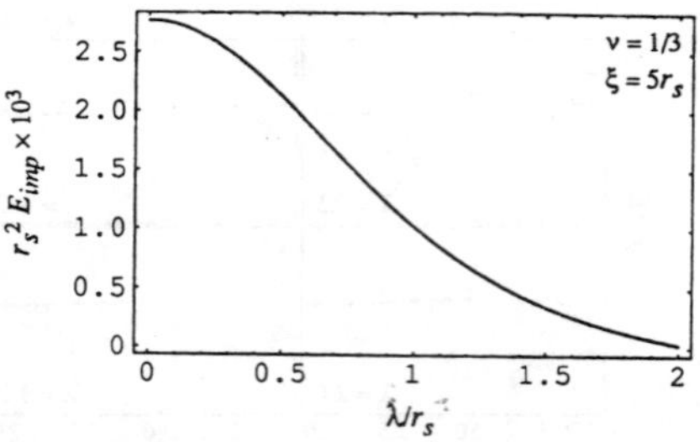

FIG. 6. The change in energy E_{imp} in the Wigner solid due to impurity effects, shown as a function of λ/r_s.

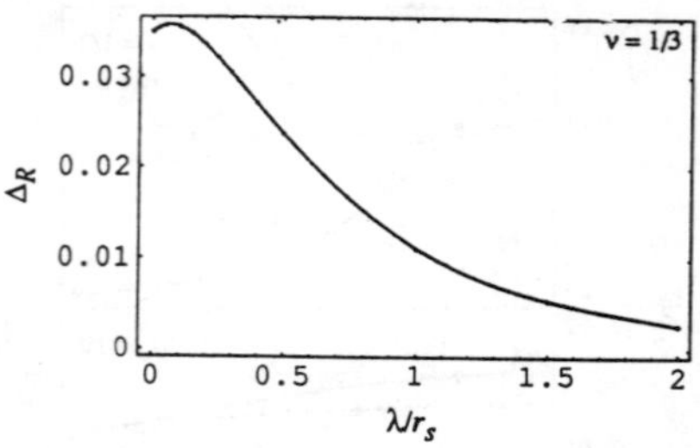

FIG. 7. The magnetoroton gap Δ_R as a function of λ/r_s for $\nu = 1/3$.

ground-state energies in order to make comparisons between the liquid and the solid. However, because the temperatures of interest are so low, we need only know the lowest-lying modes δ_k of both the solid and the liquid to calculate the free energy of either. Then the free energy is

$$F = E + T\sum_k log(1 - e^{-\delta_k/T}), \tag{16}$$

where E is the ground-state energy and the δ_k are the lowest-lying excitations of either liquid or solid. For the Wigner solid these excitations are the transverse magnetophonons ω_k^-. We evaluate the free energy F_{WS} of the solid using the ground-state energies calculated in the previous section. The magnetophonons ω_k^- were calculated using the harmonic approximation [19]. We find, as before [10], that the temperature-dependent part of the free energy for the solid goes as $T^{7/3}$ at low temperatures, although the proportionality constant in [10] is considerably lower.

On the liquid side, since we are interested primarily in the lowest-lying modes of the magnetoroton spectrum, we can approximate the magnetorotons mode with a parabola fitted to the minimum at k_R,

$$\Delta_k = \frac{(k - k_r)^2}{2m_R} + \Delta_R, \tag{17}$$

where m_R is an "effective mass" near the minimum and Δ_R is the magnetoroton gap. All three parameters are affected by the softening of the interaction, but Δ_R is most important. Figure 7 shows Δ_R as a function of λ/r_s. The roton spectra were calculated by numerically integrating (9.4.5) in [20], using the lowest Landau level structure factor $s(k)$. Rappe, Zhu, and Louie have recently calculated the magnetoroton spectrum including Landau-level mixing, and find only a 10% reduction in Δ_R at $r_s = 20$, so $s(k)$ is a good approximation for our purposes. If we assume that only the modes in the vicinity of the minimum contribute, the free energy of the liquid has a temperature-dependent part which goes exponentially, as $T^{3/2}e^{-\Delta_R/T}$.

Because the excitations of the liquid display a gap, at very low temperatures $T \ll \Delta_R$, even the lowest lying modes at the minimum k_R remain unoccupied, while the low-lying modes of the solid, which have no gap, begin to fill immediately. The free energy of the solid then falls as a power of T, while the free energy of the liquid remains constant. In this range of temperature, if we have chosen r_s such that the liquid ground state is favored only slightly, the growing entropy of the solid can cause the free energy of the solid to drop below that of the liquid, and the system freezes into a Wigner solid as temperature rises. When the temperature begins to approach the roton gap energy, however, the free energy of the liquid begins to fall exponentially as the states become occupied. Because the density of states at the roton gap energy is very large, this exponential rise is very rapid and the liquid free energy quickly falls below that of the solid, and the system melts into a Laughlin liquid once again.

In Figure 8 we plot the phase boundary at $\nu = 1/3$ in terms of r_s and T for a heterojunction with effective mass $m^* = 0.3m_e$. The phase boundary behaves as expected: if we choose r_s slightly less than the critical r_s, we can freeze the system by *raising* the temperature. Here the transition occurs at $T \approx 0.2$K, whereas using the Coulomb interaction it occured at much higher $T \approx 1$K. The dashed line shows the phase boundary found by using an estimate of the solid energy including impurities, and reducing Δ_R by a factor of 1/3. The re-entrant feature moves to lower temperature and higher r_s, but remains robust.

VI. SUMMARY

We have calculated ground-state energies and free energies of both the FQHE liquid and the Wigner solid with a model interaction rather than the Coulomb interaction,

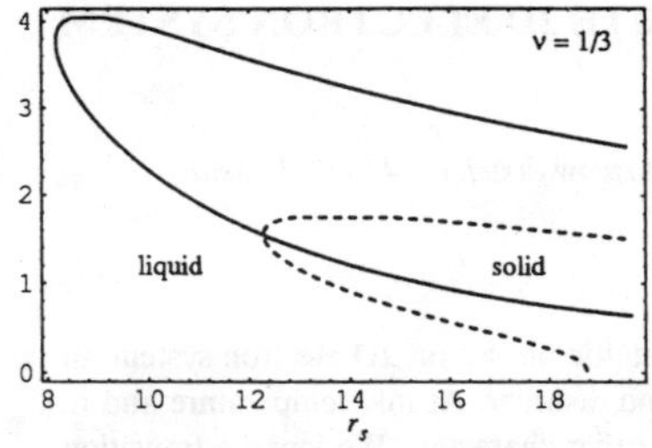

FIG. 8. The FQHE liquid, Wigner solid phase boundary at $\nu = 1/3$ in r_s and T. The effective mass of the particles is $m^* = 0.3m_e$. The dashed line is the phase boundary including impurity effects.

using a variational quantum Monte Carlo method. The model interaction includes the finite extent of the electron wavefunction in the z-direction through a length parameter λ. By comparing the energies for both liquid and solid, we obtain phase diagrams in both r_s and T. We predict a Wigner solid transition at low r_s in very wide wells, as well as another transition at high r_s. The transition at low r_s is driven by the softening of the interparticle potential in wide quantum wells, as the short-range component of the potential becomes small compared to the long-range component. At high r_s the transition is driven by Landau-level mixing.

A re-entrant transition in temperature is examined as well, caused by the magnetoroton gap in the liquid. Finite thickness effects reduce the temperature at which the transition occurs by a factor of four or more, as the magnetophonon and magnetoroton modes that drive the transition soften.

VII. ACKNOWLEDGEMENTS

The author wishes to thank P. I. Tamborenea and Song He for helpful discussions, and P. I. Tamborenea for providing him with his data. This work was done in collaboration with Xuejun Zhu, P. M. Platzman, and S. Das Sarma, and was supported by the US-ARO and the US-ONR.

[1] R. L. Willett, H. L. Stormer, *et al*, Phys. Rev. B **38**, 7881 (1988).

[2] H. W. Jiang, R. L. Willett, *et al*, Phys. Rev. Lett. **65**, 633 (1990).

[3] V. J. Goldman, M. Santos, M. Shayegan, and J. E. Cunningham, Phys. Rev. Lett. **65**, 2189 (1990).

[4] H. Buhmann, W. Joss, *et al*, Phys. Rev. Lett. **65**, 633 (1990).

[5] H. Buhmann, W. Joss, *et al*, Phys. Rev. Lett. **66**, 926 (1991).

[6] P. K. Lam and S. M. Girvin, Phys. Rev. B **30**, 473 (1984).

[7] M. B. Santos, Y. W. Suen, *et al*, Phys. Rev. Lett. **68**, 1188 (1992).

[8] R. Price, P. M. Platzman, and S. He, Phys. Rev. Lett. **70**, 339 (1993).

[9] X. Zhu and S. G. Louie, Phys. Rev. Lett. **70**, 335 (1993).

[10] P. M. Platzman and Rodney Price, Phys. Rev. Lett. **70**, 3487 (1993).

[11] A. H. MacDonald and G. C. Aers, Phys. Rev. B **29**, 5976 (1984).

[12] F. C. Zhang and S. Das Sarma, Phys. Rev. B **33**, 2903 (1986).

[13] Song He, F. C. Zhang, X. C. Xie, and S. Das Sarma, Phys. Rev. B **42** 11376 (1990).

[14] M. Shayegan, J. Jo, Y. W. Suen, M. Santos, and V. J. Goldman, Phys. Rev. Lett. **65**, 2916 (1990).

[15] T. Ando, A. B. Fowler, and F. Stern, Rev. Mod. Phys. **54**, 437 (1982).

[16] P. I. Tamborenea, personal communication.

[17] G. Ortiz, D. M. Ceperley, and R. M. Martin, Phys. Rev. Lett. **71**, 2777 (1993).

[18] P. Lee and T. M. Rice, Phys. Rev. B **10**, 3970 (1975).

[19] L. Bonsall and A. A. Maradudin, Phys. Rev. B **15**, 1959 (1977).

[20] S. M. Girvin, in *The Quantum Hall Effect*, 2nd Ed., R. E. Prange and S. M. Girvin (ed.), Springer-Verlag, N. Y. (1990).

LIQUID AND SOLID PHASES IN THE DILUTE 2D ELECTRON SYSTEM

V. M. PUDALOV
Institute for High Pressure Physics, Troitsk, Moscow district, 142092, Russia

ABSTRACT

The critical boundary between insulating and metallic phases in 2D electron system in Si was studied vs temperature, magnetic field and disorder. At mK temperature and in high mobility samples the transition has the first order character. We found a transition from the single-particle localization to collective solidification with temperature, as $T \rightarrow 0$, and with lowering disorder. The critical boundary for the collective transition exhibits quantum oscillations in magnetic field. This results in the reentrant metal–insulator transitions at fixed electron density and varying magnetic field.

1. Introduction

1.1. Metal to Insulator Transitions in 2D

The low temperature transitions between metallic and insulating phases has become the focus of theoretical and experimental interest. The problem is intimately related to the choice of the ground state for the dilute 2D electron (2DE) system, subject to electron-electron (*e-e*) interaction, magnetic field , and disorder. In the framework of the weak localization concept[1] it is predicted that all the electrons in a disordered infinite 2D system are to be localized at zero magnetic field and in the zero temperature limit. Recently, however, Azbel found[2] that, contrary to the weak localization concept, a 2DE system in a model disordered short range potential at T=0 is localized only at energies below a mobility edge. The Azbel's conclusion is evidently in agreement with the strong-localization concept[3,4], where the 2D electron system is to localize into a single-particle (SP) insulator below a mobility edge, i.e. upon increasing disorder or lowering density n_s. At all energies above this edge, the extended states do exist and provide the conduction down to the zero temperature. In any case the localization is suggested to occur when the sheet resistivity ρ exceeds $\sim h/e^2$.

1.2. Electron Solid in 2D

In the opposite disorder-free limit and strong Coulomb interaction, the 2D system is to self-localize into a collective insulating state[5], known as the Wigner lattice or electron solid (ES). The ordering potential energy in 2D $\langle E_{ee}\rangle = (\pi n_s)^{1/2}e^2/\kappa$, where n_s is the two-dimensional electron density, e is the elementary charge, and κ is the static dielectric constant of the host material. Experimentally, Wigner crystallization was observed first in an electron gas on the surface of liquid helium[6].

The electron density in this system[6,7] does not exceed 10^{10}cm^{-2}. The electron gas obeys the classical (Boltzmann) statistics, $E_F \ll k_B T$ (where $E_F = h^2 n_s/4\pi^2 m^*$ is the Fermi energy), and the ES is of the classical origin. The disordering kinetic energy is determined mainly by the thermal energy k_BT. Therefore, melting of the ES is governed by the condition $E_{ee} = \Gamma k_BT$ and occurs along the classical boundary

$$n_m = (\kappa \Gamma k_B T)^2 \pi e^4, \qquad (1)$$

where the critical value $\Gamma \approx 127$ has been found experimentally[6].

1.3. Quantum 2D Electron Solid

For the dilute 2DE system in (100) Si at milli-Kelvin temperatures, the disordering zero-point energy *exceeds* k_BT due to the low electron effective mass, $m^*=0.2m_e$, and at available electron densities $\sim 10^{11}\text{cm}^{-2}$ the 2DE system obeys the Fermi statistics $E_F >> k_BT$. Hence, the electron solid in Si must be of quantum origin. Upon increasing density the crystal is expected to melt at zero temperature due to zero-point oscillations[5]. The critical concentration for the quantum melting is:

$$n_{qm} = 1/\pi(r_c a_B)^2 \qquad (2)$$

where $a_B = \kappa h^2/(4\pi^2 m^* e^2)$ is the effective Bohr radius, 2.04 nm at the Si-SiO_2 interface, $\kappa = 7.7$, and r_c is the critical value for a dimensionless Wigner-Seitz radius[3], $r_s = 1/a_B(\pi n_s)^{1/2}$.

The ES in the quantum regime at zero magnetic field was first observed in Refs.[8,9,10] in high-mobility Si MOSFETs, and its phase boundary has been outlined vs density and disorder in the limit of T=0. M-I transition develops as a collective effect in the "high mobility" samples, with the peak mobility $\mu > 10\times 10^3$ cm^2/Vs. A number of experimental results provide evidence for the collective character of the transition, e.g.: (i) large correlation length $\xi >> (1/\pi n_s)^{1/2}$ in the ES phase, diverging towards the critical density[8], (ii) frequency spectrum of the noise[9], accompanying transport in electric field above the threshold and indicating the charge periodicity, (iii) discontinuity of the threshold electric field at the critical density, indicating the first order phase transition[11] at $T \to 0$. In the lower mobility samples or at high temperatures $T > \approx 0.6$K the M-I transition develops as a single particle effect[12].

Recently Chui and Tanatar[13] have done the numerical simulations of the quantum melting in 2D, while Kawaguchi and Saitoh[14] studied the quantum melting criterion by taking into account the phonon anharmonicity. In both calculations the quantum melting in 2D was found to occur at $r_s \approx 9$ in a good agreement with the experiments[8,9,10,15] on Si.

1.4. ES in Si and in GaAs-Al(Ga)As

The kinetic energy of electrons in GaAs-Al(Ga)As is larger (due to 3 times smaller effective mass, $m^*=0.067m_e$), and *e-e* interaction energy is lower (due to 1.7 times larger dielectric constant). As a result, the classical melting temperature $T_m = e^2(\pi n_s)^{1/2}/(\kappa \Gamma k_B)$, which is 0.877K in Si, is 1.5× lower[5] in GaAs-Al(Ga)As. Hence, the quantum melting density, estimated in accord with Eq. (2), is 20× lower. The ratio of the ordering potential to disordering kinetic energy, E_{ee}/E_0 is about .6 times larger in Si, which again favors formation of the ES in Si. Therefore, realization of the ES at zero magnetic field in GaAs-Al(Ga)As would require an unrealistically low electron density, $n_s < n_{qm} = 3\times 10^8\text{cm}^{-2}$, and the electron crystal in zero field is more likely to be observed in (100) Si.

In GaAs-Al(Ga)As, the ES, however, is induced by the strong magnetic field and forms in the range of low fractional filling factors $\nu \leq 1/3$ - $1/7$. The ES is interrupted then by the *incompressible Laughlin liquid states* at fractional fillings. This results in reentrant insulating transitions around fractional filling factors when the magnetic field is swept at fixed electron density[5]. In (100) Si the electron solid forms starting from H=0. Quantizing magnetic field bends the liquid to solid boundary and causes its quantum oscillations. This again results in the reentrant insulating state[15] interrupted by the *compressible liquid state* at integer filling, when magnetic field is swept at fixed density.

2. Experimental

The measurements were done on number of Si-MOSFET samples of different mobility, from 5×10^3 cm^2/Vs (Si-39, the most disordered sample) to 75×10^3 cm^2/Vs (Si-15). All samples were of the Hall-bar geometry[15,16]. The electron density n_s in 2D was varied by changing the gate voltage V_g. Earlier, the linear dependence $n_s(V_g)$ has been proven to remain undisturbed in high mobility samples down to the critical density of the M-I transition, and even in the ES state[17,18]. The high sample mobility is of the key importance, since it provides control of the 2D electron density to 5×10^{10} cm^{-2}.

All measurements were done by a 4-terminal dc technique, with battery operated instruments optically connected to routine electronic equipment and PC. In this way we were able to measure reproducible from day to day nonlinear resistance values up to $10^{10}\Omega$. Fig. 1 shows typical threshold current-voltage dependencies in the vicinity of n_c. The threshold electric field E_t was determined from the voltage drop extrapolated from the linear wings of the I - V curves and the known geometrical factor.

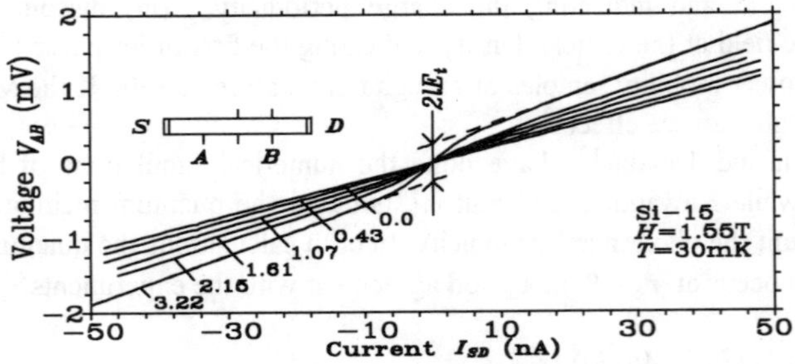

Fig. 1. Typical nonlinear current-voltage dependencies in the vicinity of the critical point, in the ES phase. Inset shows the sample geometry. Numbers designate $\delta n_c=(n_c-n_s)$ values in units of 10^9cm^{-2}.

3. M-I Transition in Low and High Mobility 2D Systems

The collective ES phase crosses over[8-10] to the SP insulator upon lowering electron density and/or decreasing sample mobility below ~10^4 cm^2/Vs. The ES shares many features of the pinned Wigner lattice, in particular, a temperature activated and nonlinear threshold electric field transport. A rough estimate for the correlation length $\xi = \Delta/eE_t$ may be constructed from the activation energy Δ and threshold electric field E_t.

The dc-conduction in the insulating phase was studied in Ref.[13] in the samples of high and low mobility. The high-μ samples exhibited the collective M-I transition, while the low-μ ones – the SP-transition. The *V-I* curves taken at the same δn_c are much sharper in the high-μ samples than in the low-μ ones. Fig. 2 (a) shows that: (i) the activation energy Δ diminishes linearly[8,9,10] with δn_c, (ii) E_t diminishes $\propto(n_c-n_s)^{1.5-1.7}$. And at last, E_t exhibits a sharp step[11] of ~100μV/cm at the very critical point. Therefore, ξ diverges $\propto(n_c-n_s)^{0.5-0.7}$ far off the transition, reaching a maximum of ~10μm.

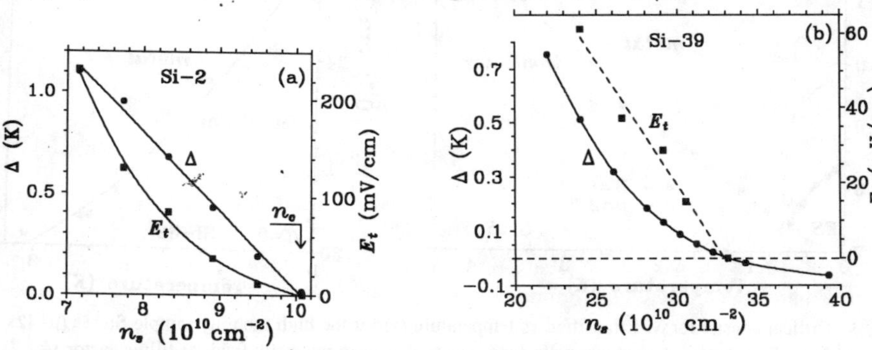

Fig. 2. Activation energy and threshold electric field in the insulating phase vs electron density: (a) in the sample Si-2 (μ=24,000cm^2/Vs), and (b) in the sample Si-39 (μ=5,000cm^2/Vs).

For the low-μ sample, the *V-I* characteristics at mK-temperature are nonlinear as well, but much smoother at the same temperature and the same deviation δn_c. Fig. 2 (b) shows that: (i) Δ diminishes first $\propto(n_c-n_s)^2$, then linearly, closer to n_c, (ii) E_t drops down linearly to the critical point, and (iii) both dE_t/dn_s and $d\Delta/dn_s$ are smaller by a factor of ~10× than those for the high-μ sample. The "correlation" length ξ ~0.2μm is about 50× lower than for the high-μ sample and does not diverge but slowly diminishes as $n_s \to n_c$. The threshold electric field is much lower than that suggested by the theory for breakdown of the localized states[20], and the reason of the discrepancy is not clear.

This comparison shows definitely different critical behavior in high and low μ samples, due to the collective transition in the former and SP localization in the latter ones. In the high-μ samples both thermal activation energy and threshold electric field agree well with the theoretical picture describing conduction in terms of the collective dislocation motion in the pinned electron lattice[5,19].

4. Competition between the Single-Particle and Collective Localization

Due to a nonzero amount of disorder, the real 2D system is always subject both to the SP and collective effects. The following questions are therefore the very topical:

(i) would both kinds of localization coexist or compete, (ii) how would they transform into each other at varying temperature, density, disorder, and other parameters.

We realized in one Si-MOSFET sample Si-15 both the collective and SP -localization at low and high temperature correspondingly, and traced the critical density for the M-I transition vs temperature. In this way we studied the interplay between SP localization and collective solidification as a *function of temperature and electron density*. In Fig. 3 (a), the measured boundary of the M-I transition is plotted vs temperature. The region above

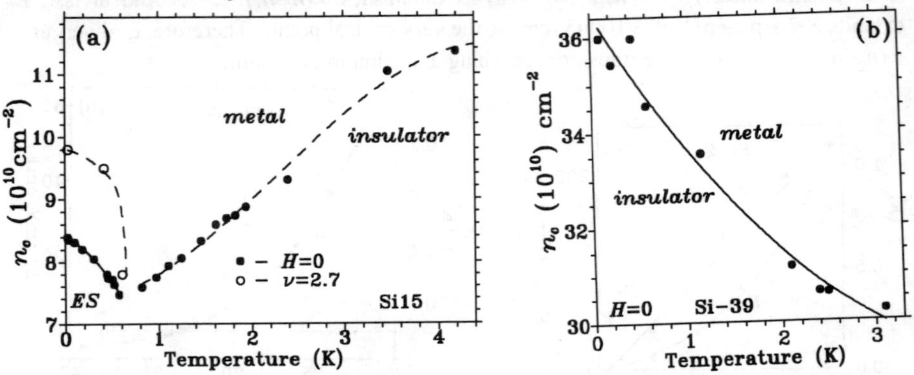

Fig. 3. Critical density for M-I transition vs temperature (a) for the high mobility sample Si-15 ($\mu=42\times 10^3\text{cm}^2/\text{Vs}$). Full dots – at zero magnetic field, empty dots – in magnetic field, at filling factor ν=2.7. (b) Critical density for the low-mobility sample Si-39 ($\mu=5\times10^3\text{cm}^2/\text{Vs}$).

the plotted boundary corresponds to the metal, while below – the insulator state. For high-mobility samples the critical density decreases slowly with lowering temperature. in the interval $4 > T > 0.8$K. The decrease in n_c becomes less for more disordered samples, and eventually for Si-39, the lowest mobility sample, the M-I boundary *rises* with lowering temperature, as seen in Fig. 3 (b). At lower temperature $T < 0.6$K in high mobility samples the M-I boundary starts sharply rising.

The quantizing magnetic field modulates the M-I phase boundary at mK-temperatures and causes its quantum oscillations[21], while the high-temperature part, at $T > 1$K, remains almost undisturbed. Empty circles in Fig. 3 (a) show $n_c(T)$ for the Landau level filling factor ν=2.7, which is the most favorable for the insulating phase[15,16,21].

With such a V-shape critical boundary, the 2DE system exhibits the M-I transition twice at varying temperature, if the density is fixed in the vicinity of n_c. The temperature driven M-I transitions, which are due to crossing the *high T* boundary at Fig. 3(a), have been studied earlier[22] and are skipped here. The temperature dependence of the diagonal resistance taken at the constant filling factor, ν=2.7, is shown in Fig. 4 in $1/T$ coordinates. Each point in this graph was found from the slope of the *I-V* curves. The plot reminds a regular temperature activated dependence of R_{xx}. The magnified fragment in Fig. 4(b), however, reveals the very unusual details: dR_{xx}/dT changes the sign in the range

$1/T$<0.8K, thus confirming the transition from an insulator to metal upon increasing temperature from zero. The dashed line connects the points of R_{xx} minima.

The M-I boundary in Fig. 3(a), terminates at $n_c(0)=8.4\times10^{10}cm^{-2}$ at zero magnetic field

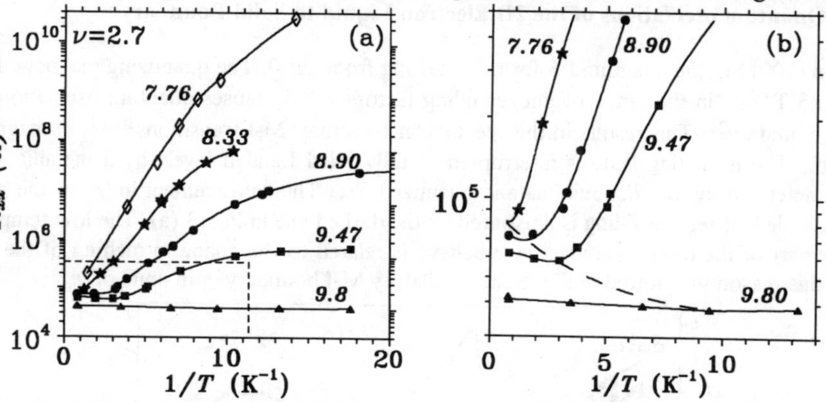

Fig. 4. (a) Diagonal resistance vs inverse temperature for 5 electron densities near n_c, and (b) the left bottom fragment of the graph. Sample Si-15. The densities are indicated in units of $10^{10}cm^{-2}$.

and T→0. This point was identified in Refs.[8,9,10] as the quantum melting critical density, n_{qm}. The M-I transition with lowering density in the high-μ samples occurs as a weak first order phase transition[11], both at H=0 and $H\neq0$. The low temperature part of the $n_c(T)$ curve, at $T < 0.6$ K, would thus correspond to the melting diagram of the ES.

At high temperature, $T > 4$ K, the M-I transition in high mobility samples doesn't show signatures of either collective transition, or first-order phase transition. It is natural therefore to interpret the transition, in the first approximation, as the single-particle localization. The critical boundary decreases with lowering temperature at 4 > T > 1K. As we mentioned, it might be the effect of *e-e* interaction[4] on the SPL. Another explanation related to the temperature dependence of the screening has also been discussed[22]. Anyway, the total diagram $n_c(T)$ reflects transformation of the SP localization (at $T >$ 4K) into the collective solidification (at $T <$ 1K).

The intermediate range of temperatures $T \approx 0.7$K is worthy of individual analysis. In this range, the SPL boundary seems to intercept the ES boundary and "sink" underneath, limiting the ES state from the low density side, as suggested in Ref.[8]. This would be natural, since lowering density at a given disorder (i.e. impurity density) has to lead to the SP localization. This possibility requires, however, a special experimental verification.

The total view of the boundary indicates the way in which two different kinds of localization affect each other. The electron solidification is favored by the strong localization (the latter pushes the ES boundary up) while the disorder is weak[25]. On the other hand, the collective effects bend the SPL boundary down starting from the high temperature, and eventually overpower the strong localization at low temperature. These novel effects of temperature driven localization and delocalization are weakening with decreasing sample mobility μ, and finally are missing in the sample Si-39 with μ=5× 10^{3}cm^2/Vs. It quantitatively agrees with the disorder phase-diagram plotted earlier[8].

5. Quantum oscillations of the 2D Electron Liquid to Solid Boundary

In (100) Si, the quantum ES forms[15] starting from H=0. The quantizing magnetic field 0 to 5 T, i.e., in the range of integer filling factors $\nu > 1$, causes quantum oscillations of the boundary[21]. This results in the spectacular reentrant M-I transitions[16,23] in magnetic field. The insulating state is interrupted at fully filled Landau levels by a metallic state characterized by low R_{xx} minima and quantized R_{xy}. The enhancement in n_c at the most favorable half-integer filling is illustrated by the dashed line in Fig. 3 (a). The low temperature part of the phase boundary we believe is related to the quantum melting of the ES. By this reason we plotted in Fig. 5 the oscillatory M-I boundary[21] in units of r_s.

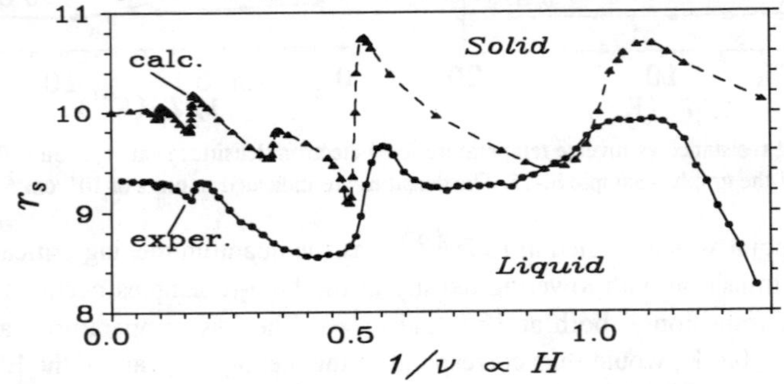

Fig. 5. The measured quantum oscillations of the critical r_s value vs magnetic field[25] (full dots), and the calculated quantum oscillations of the liquid to solid boundary[22] (triangles).

The quantum oscillations of the melting boundary are due to the oscillatory variations of the ground state energy in the liquid phase and a monotonous – in the solid phase. Thus, when the density is set in vicinity of the transition, it will repeatedly intercept the critical boundary upon varying magnetic field. This approach describes quantitatively the re-entrant M-I-transitions in magnetic field. and provides reasonable agreement with experimental data both on r_s (see Fig. 5). and on the activation energy vs magnetic field[21].

6. Summary

We reviewed the novel effects of temperature and magnetic field driven localization and delocalization in 2DE system. The global temperature dependence of the critical boundary is determined by the interplay between the single-particle and collective localization. The distinct low temperature part of the metal-insulator boundary, at $T < 0.6$K, is attributed to the 2D quantum melting, which occurs as a first order phase transition. This part of the boundary exhibits quantum oscillations in magnetic field, which lead to the reentrant metal-insulator transitions.

7. Acknowledgements.

This work has been supported by the Russian Foundation for Basic Research (grant N° 93-0214235), by the International Science Foundation (grant MUG000), by Ministry for Science and Technical Policy, and by the Nederlandse Organisatie voor Wetenschappelijk Onderzoek (grant N° 07-13-217).

8. References

1. E. Abrahams, P. W. Anderson, D. C. Licciardello, and T. V. Ramakrishnan. *Phys. Rev. Lett.* **42** (1979) 673.
2. M. Ya. Azbel', *Phys. Rev.* **B45** (1992) 4208.
3. T. Ando, A. B. Fowler, and F. Stern, *Rev. Mod. Phys.* **54** (1982) 589 .
4. H. Kamimura and H. Aoki. *The Physics of Interacting Elecrons in Disorded Systems.* (Clarendon Press, Oxford, 1989).
5. *Physics of the 2D Quantum Electron Solid,* ed. S. T. Chui (International Press, Cambridge, USA, 1994).
6. C. C. Grimes and G. Adams. *Phys. Rev. Let.* **42** (1979) 795.
7. H.-W. Jiang, A. J. Dahm. *Surf, Sci.* **98** (1980) 1.
8. V. M. Pudalov, M. D' Iorio, S. V. Kravchenko, and J. W. Campbell. *Phys. Rev. Lett.* **70** (1993) 1866.
9. M. D' Iorio, V. M. Pudalov, S. V. Kravchenko, and J. W. Campbell. *Surf. Sci*. **305** (1994) 115.
10. V. M. Pudalov, M. D' Iorio, and J. W. Campbell. *Physica B*, **194-196**, (1994) 1289.
11. V. M. Pudalov and S.-T. Chui. *Phys. Rev.* **B49** (1994) 14062.
12. V. M. Pudalov and M. D' Iorio. in proc.: *ICPS-22*, Vancouver (1994).
13. S.-T. Chui and B. Tanatar. *Bull. APS* **38** (1993) 472. To be published.
14. M. Kawaguchi and M. Saitoh. *Solid State Commun.* **88** (1993) 765.
15. V. M. Pudalov, in *Physics of the 2D Quantum Electron Solid,* ed. S. T. Chui (Intern. Press, Cambridge, USA, 1994).
16. M.D' Iorio, V. M. Pudalov, and S. G. Semenchinsky, *Phys. Rev.* **B46** (1992) 15992.
17. S. V. Kravchenko, J. A.A.J. Perenboom, and V. M. Pudalov. *Phys. Rev.* **B44** (1991) 13513.
18. V. M. Pudalov and M. D' Iorio. in proc.: *SEMIMAG-94,* Cambridge (1994).
19. S. T. Chui. *Phys.Lett.* **A180** (1993) 149.
20. Y. C. Lee, C. S. Chu, and E. Castano. *Phys.Rev.* **B27**, 6136 (1983).
21. V. M. Pudalov. *Bull. Amer. Phys. Soc.* **39** (1994) 805. To be published.
22. S. V. Kravchenko, G. V. Kravchenko, J. Furneaux, V. M. Pudalov, and M. D' Iorio, *Phys. Rev.* **B** (1994), in press.
23. M. D' Iorio, V. M. Pudalov, and S. G. Semenchinsky. *Phys. Lett.* **A150** (1990) 422.
24. V. M. Pudalov, M. D' Iorio, and J. W. Campbell. *JETP Lett* **57** (1993) 608.
25. A. G. Eguiluz, A. A. Maradudin, and R. J. Elliott. *Phys. Rev.* **B27** (1983) 4933.

Inter-Electron Coulomb Interaction and Disorder in the Activated Transport in Magnetic-Field-Induced Insulating Phase in 2D Systems

A. Fukano, S. Kawaji, J. Wakabayashi[A],
Department of Physics, Gakushuin University, Toshima-ku, Tokyo, 171 Japan

K. Hirakawa, H. Sakaki[B],
Institute of Industrial Science, University of Tokyo, Minato-ku, Tokyo, 106 Japan

and

T. Goto, Y. Koike[C] and T. Fukase
Institute for Materials Research, Tohoku University, Aoba-ku, Sendai, 980 Japan

ABSTRACT

We have measured activated magnetotransport on low-density, low-mobility two-dimensional electron systems in AlGaAs/GaAs heterostructures in Landau level filling factors of $0.1 < \nu < 0.2$. Effects of inter-electron Coulomb interaction and disorder in the activation energy are discussed comparing our results with other authors'.

1. Introduction

In recent years an insulating phase (IP) in two-dimensional electron systems (2DES) at small Landau level filling factor ν ($\nu = N_s h/eB = 2(l_B/a)^2$, N_s is the 2DES density, $l_B = (\hbar/eB)^{1/2}$, the magnetic length and $a = (\pi N_s)^{-1/2}$) in semiconductor interfaces has been a subject of intense research in connection to Wigner crystallization.[1~6] However, in the thermally activated conduction, there does not appear any indication of a phase transition between electron solid and liquid regions. Recently a new insulating phase (Hall insulator), where diagonal conductivity $\sigma_{xx} = 0$ and Hall conductivity $\sigma_{xy} = 0$ but Hall resistivity $\rho_{xy} \cong B/eN_s$, has been proposed.[7] Results of Hall effect measurements which may support the Hall insulator picture were reported.[8,9] Experimental studies of the IP in GaAs/AlGaAs heterostructures so far reported were carried out for samples which have very high electron mobilities. However, roles of disorder which affect the effects of inter-electron Coulomb interaction in the IP are not clear yet.

In a search for the $\nu = 1/7$ FQHE state in a low mobility GaAs/AlGaAs heterostructure ($N_s = 6.2\times10^{14}$ m^{-2} and $\mu = 35$ m^2/Vs) at T down to 20 mK, we found, that diagonal resistivity ρ_{xx} diverges but ρ_{xy} stays approximately at B/eN_s.[10] It is worthwhile to measure properties of activated conduction in low mobility samples to examine roles of disorder related to inter-electron Coulomb interaction in the IP. We present here experimental results of activation energy E_A in ρ_{xx} in a region of $0.1<\nu<0.2$

Present address:
A: Department of Physics, Chuo University, Bunkyo-ku, Tokyo, 112 Japan
B: Research Center for Advanced Science and Technology, University of Tokyo, Meguro-ku, Tokyo, 153 Japan
C: Department of Applied Physics, Faculty of Engineering, Tohoku University, Aoba-ku, Sendai, 980 Japan

in low mobility samples at temperatures higher than the melting temperature if these electrons could solidify at lower temperatures. We compare them with those by other authors.

2. Experimental results and discussions

The samples used were modulation-doped GaAs/$Al_{0.3}Ga_{0.7}As$ single hetero- structures which had a concentration of silicon donors and the thickness of the doped AlGaAs layer of 3×10^{23} m^{-3} and 80 nm, respectively. The AlGaAs spacer layer was 85 nm thick. The width and the length of the sample were 50 and 600 μm, respectively and potential probes were located at 1/3 and 2/3 along the sample. The electron density N_s and the electron mobility μ were 6.9×10^{14} m^{-2} and 45 m^2/Vs, respectively, in the sample A, 6.1×10^{14} m^{-2} and 20 m^2/Vs in the sample B, and 6.0×10^{14} m^{-2} and 19 m^2/Vs in the sample C. The magnetic fields were produced by hybrid magnets HM-1b and HM-2 at HFLSM, IMR, Tohoku University. The magnet HM-1b has the maximum field of 27 T combined with a dilution refrigerator and used in experiments for the sample A. The magnet HM-2 whose maximum field was 23 T combined with a ^3He refrigerator was used in experiments for the sample B and C. Magnetic field dependences of ρ_{xx} and ρ_{xy} in the sample A were measured in magnetic fields between 14 and 27 T at $T = 1$, 0.5 and 0.3 K at a current of 1 nA DC. Magnetic field dependences of ρ_{xx} and ρ_{xy} of the sample B and C were measured in magnetic fields between 9.5 and 23 T at $T = 0.90$, 0.77, 0.67 and 0.47 K

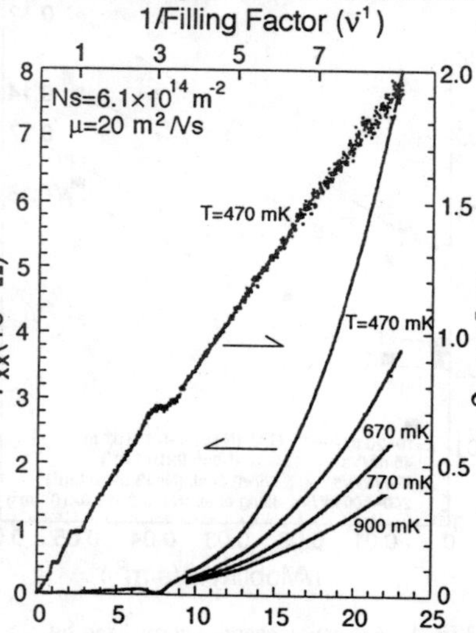

Fig. 1. Magnetic field dependence of ρ_{xx} and ρ_{xy} in the sample B.

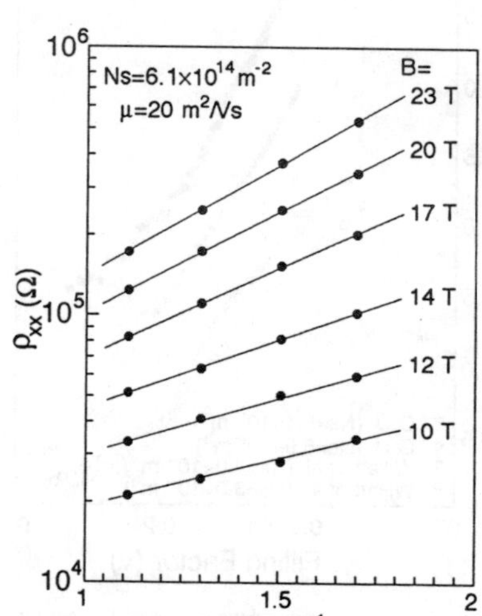

Fig. 2. Temperature dependence of ρ_{xx} in the ρ_{xy} in the sample B.

at a current of 5 nA DC. These currents were in the linear region of the current-voltage characteristics. Figure 1 shows the magnetic field dependences of ρ_{xy} and ρ_{xx} in the sample B. We obtained activation energy E_A via Arrhenius plots $\rho_{xx} = \rho_0 \exp(E_A/2T)$ as shown in Fig. 2.

In Fig. 3, activation energies of the sample A and B are shown as functions of filling factor ν. The filling factor dependence of the activation energy in the sample C is close to that of B. In Fig. 3, Willett et al.'s[11] results are also shown in comparison with our results. Electron mobility in the sample with $N_s = 3.0\times10^{14}$ m^{-2} in Willett et al.'s work[11] was $\mu\sim$ 120 m^2/Vs. Figure 3 shows that the activation energy decreases with increasing electron mobility at the same filling factor. Willett et al.'s results show clear breaks in slopes in E_A vs. ν plot at $\nu = 0.2$ where ρ_{xx} shows clear minimum in their sample with $N_s = 4.8\times10^{14}$ m^{-2}(Fig.1 in Ref. 11). Our samples also show similar breaks in their slope in E_A vs. ν plots near $\nu = 0.2$. This observation shows that though μ is low, the electron-electron interaction which causes FQHE still works in our samples. However, N_s-dependence of E_A in our samples has opposite sign compared with Willett et al.'s results at $\nu < 0.2$. When we normalize E_A in Fig.3 by Coulomb energy $e^2/\varepsilon a$, Willett et al.'s results get together almost on single curve at $\nu < 0.2$.

Jiang et al.'s results measured in high-density and very high mobility samples (N_s = $(5.3\sim11)\times10^{14}$ m^{-2} and $\mu = (200\sim800)$ m^2/Vs)[1] are also consistent Willett et al.'s results. However, though the behavior of the activation energy in the sample A appear

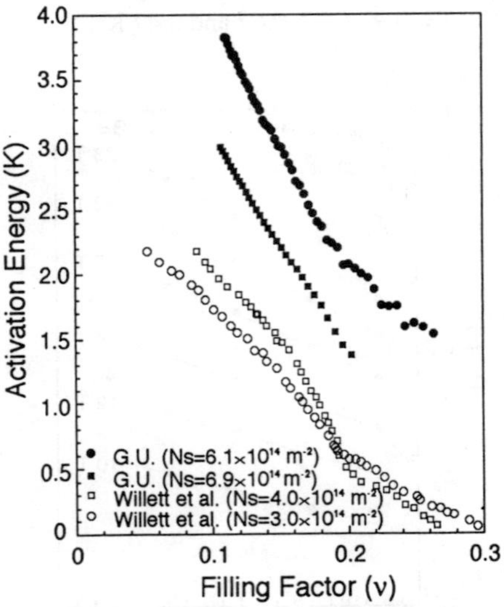

Fig. 3. Filling factor dependence of the activation energies including the results by Willett et al. in reference 11.

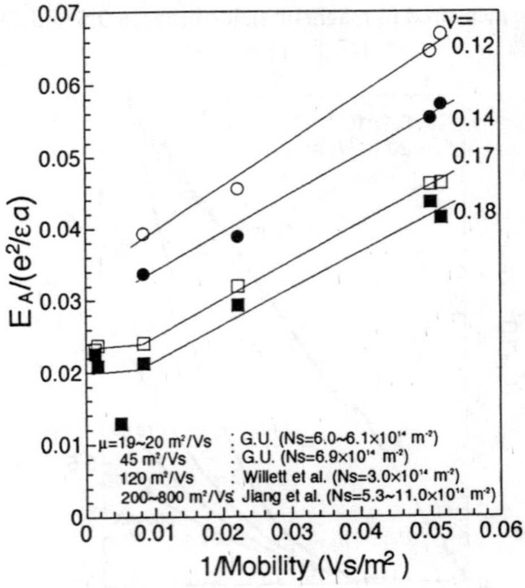

Fig. 4. Activation energies normalized by Coulomb energy $e^2/\varepsilon a$ against the inverse of electron mobility.

to be similar to these results, we consider that activation energies of our samples are not consistent with Willett et al.'s and Jiang et al's results.

We can see effects of inter-electron Coulomb interaction and disorder in E_A in Fig. 4 where the activation energies are normalized by $e^2/\varepsilon a$ and plotted against the inverse of the electron mobility. When the electron mobility is smaller than ~ 100 m^2/Vs, disorder is strong enough to destroy the electronic phase constructed by the inter-electron Coulomb interaction.

In conclusion we have measured thermally activated transport in magnetic-field-induced insulating phase in low mobility two-dimensional electron systems and found that, comparing with other authors' results, the mobility dependence of the activation energy normalized by Coulomb energy shows the inter-electron Coulomb interaction effect and the role of disorder.

3. Acknowledgments

The authors thank the staff of HFLSM, IMR, Tohoku University for the use of the high magnetic field facilities. They also thank T. Okamoto, K. Hayashi, T. Okuno and T. Ohrui for technical assistance. Part of this work was supported by a Grant-in-Aid for Scientific Research from the Ministry of Education, Science and Culture.

4. References

1. H.W. Jiang, H.L. Stormer, D.C. Tsui, L.N. Pfeiffer and K. W. West, *Surf. Sci.* **263**(1992) 18.
2. F.I.B. Williams, G. Deville, D.C. Glattli, P.A. Wright, E.Y. Andrei, R.G. Cark , B. Etienne, C.T. Foxon and J.J. Harris, *Surf. Sci.* **263** (1992) 23.
3. I.V. Kukushkin, N.J. Pulsford, K. von Klitzing, K. Ploog and V.B. Timofeev, *Surf.Sci.* **263** (1992) 30.
4. M. Heyne, M.K. Ellis, A.S. Plaut, A. Usher and K. Ploog, *Surf. Sci.* **263** (1992) 39.
5. L.W. Engel, T. Sajoto, Y.P. Li, D.C. Tsui and M. Shayegan, *Surf. Sci.* **263** (1992)44.
6. M.A. Paalanen, R.L. Willett, P.B. Littlewood, R.R. Ruel, K.W. West, L.N. Pfeiffer and D. Bishop, *Phys. Rev.* **B45** (1992) 11342.
7. S. Kivelson, D.H. Lee and S. F. Zhang, *Phys. Rev.* **B46** (1992) 2223.
8. V.J. Goldman, J.K. Wang, Bo Su and M. Shayegan, *Phys. Rev. Lett.* **70** (1993)647.
9. T. Sajoto, Y.P. Li, L.W. Engel, D.C. Tsui and M. Shayegan, *Phys. Rev. Lett.* **70** (1993) 2321.
10. J. Wakabayashi, A. Fukano, S. Kawaji, K. Hirakawa, H. Sakaki, Y. Koike and T. Fukase, *J. Phys. Soc. Jpn.* **57** (1988) 3678.
11. R. L. Willett, H.L. Stormer and D.C. Tsui, L.N. Pfeiffer, K. W. West and K.W. Baldwin, *Phys. Rev.* **B38** (1988) 7881.

A MOLECULAR DYNAMICS STUDY OF ORIENTATIONAL ORDER AND DEPINNING IN A MODEL OF THE DISORDERED ELECTRON SOLID

Min-Chul Cha and H.A. Fertig
Center for Computational Sciences and Department of Physics and Astronomy, University of Kentucky, Lexington, KY40506

ABSTRACT

We study the low-temperature depinning properties of two-dimensional classical electron solids under the influence of modulation-doped impurities by using a simulated annealing molecular dynamics method. By changing the ratio of the setback distance, d, to the lattice constant, a_0, we find that the destruction of quasi-long range orientational order occurs through disclination-dissociation near $d/a_0 \approx 1.15$. Associated with this, current-flow patterns in the presence of a depinning electric field change. In the strong disorder regime, depinning happens through string-like current flows which tear the solid, whereas in the weak disorder regime, sliding motion of patches containing many electrons is dominant. Correspondingly, the behavior of the threshold electric field as a function of d/a_0 changes as it is tuned through the two regimes.

1. Introduction

In two-dimensional systems in very strong magnetic fields where the zero-point kinetic energy is quenched the electron solid is expected to be formed.[1] Recently a considerable amount of experimental study has been devoted to the search for the magnetically induced electron solid[2] mostly in heterojuction systems. Most of these observations have been presumed to indicate the existence of a pinned electron crystal. To interpret experimental observations, however, it is necessary to understand the pinning sources and estimate the pinning forces. Random impurities which pin a electron crystal at the same time introduce defects in the lattice. The roles of the defects generated by impurities in the pinning[3] are not well understood.

In this paper, we report our study on pinning and orientational order in a model of the classical two-dimensional electron system. Our studies are most directly applicable to electrons on a helium film[4], although some of our basic conclusions should apply to heterojunction systems in strong magnetic fields, where the magnetic length is quite small compared to the interparticle spacing, so that quantum exchange effects are unimportant[5]. Disorder is introduced as modulation-doped donors randomly located on a plane separated by a setback distance d from the two-dimensional electron plane, so that the energy of the system is given by

$$E = \sum_{i \neq j} \frac{e^2}{\epsilon|\vec{r}_i - \vec{r}_j|} - \sum_{ij} \frac{e^2}{\epsilon(|\vec{r}_i - \vec{R}_j|^2 + d^2)^{1/2}} \qquad (1)$$

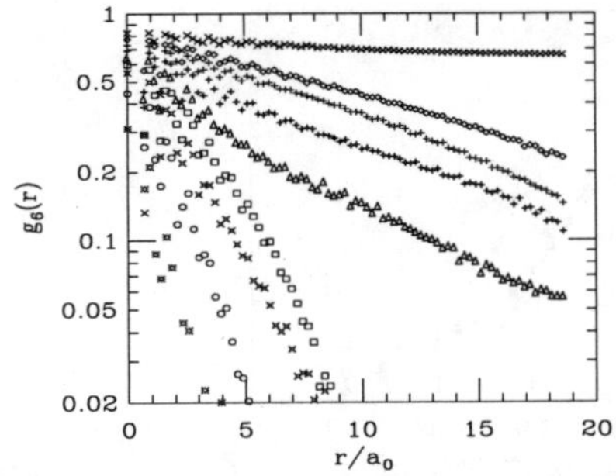

Figure 1: The orientational correlation functions for various setback distances. Different symbols represent data for samples with different setback distance. From top to bottom, d/a_0 =2.0, 1.7, 1.5, 1.3, 1.2, 1.1, 1.0, 0.9, and 0.8.

where $\{\vec{r}_i\}$ are the electron configurations, $\{\vec{R}_i\}$ are randomly quenched donor configurations on a modulation-doped plane, d is the setback distance, and ϵ is a dielectric constant.

Numerically, we use a simulated annealing molecular dynamics method[6] to find a low-temperature electron configuration, $\{\vec{r}_i\}$. The Ewald sum technique[7] is used to handle the long range interaction in computing forces with periodic boundary conditions. To minimize any mismatch of current patterns at the boundaries resulting from the fact that electrons exiting the system will need to reenter it on the opposite side of the unit cell, along the direction of the depinning field we juxtapose two square boxes whose impurity configurations are mirror-images of each other.

2. Numerical Results

From numerical simulations with up to 3200 particles, we have observed defects (*i.e.* predominantly dislocations and disclinations) generated by the impurity potential. We take $\epsilon = 13$ in the numerical simulations, and slowly lower the temperature from above the melting transition down to 20mK for a typical electron density $n = 5.7 \times 10^{10}\text{cm}^{-2}$. At this temperature we measure the orientational correlation functions and the threshold depinning electric field. In the weak disorder limit (large d/a_0), the ground state configuration contains a quasi-long range orientational order [8](*i.e.* hexatic glass), and we observed no free disclinations in the system. However, as disorder increases, isolated disclinations appear, destroying the quasi-long range orientational order[8].

Fig. 1 shows the orientational correlation functions for different values of d/a_0. We see that the quasi-long range orientational order is destroyed in the strong disorder limit (small d/a_0) and the correlation functions can be well fitted by exponential forms. Associated with this destruction of the quasi-long range orientational order, the behavior of the threshold electric field is changed as shown in Fig. 2. A break in the curve takes place approximately at $d/a_0 \approx 1.15 \pm 0.1$, which is consistent with the vanishing of orientational order. We find that this behavior is mostly due to the

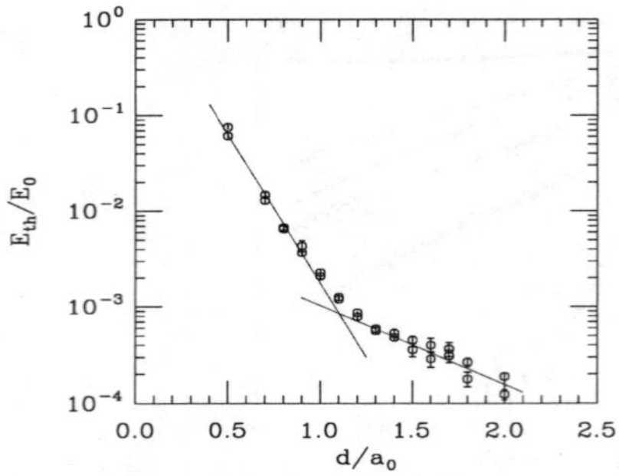

Figure 2: Depinning threshold field observed in various samples in unit of $E_0 = e/(\epsilon a_0^2)$. The dotted lines are guides to the eye.

proliferation of the disclinations. Electrons tend to flow along directions of the local symmetry of the crystal. When the system does not have long range orientational order, it is necessary for electrons to pass through regions of great strain in the lattice, where the orientation changes. These regions of strain represent bottlenecks in the electron flow. As the disorder strength is tuned and orientational order changes from quasi-long range to short range, the number of bottlenecks proliferates and there is a sharp increase in the threshold field. We also note that the threshold electric field is very sensitive to the setback distance, which might explain the very disparate values of this quantity in experiments[2].

In order to measure the threshold depinning field, we shift the positions of the particles along a chosen direction[6] by steps of $0.01a_0$ up to 1–3 lattice constants. After each shift 200 MD steps are taken to equilibrate the system, and the pinning force is obtained in the next 100 MD steps from the averaging the force on particles due to impurities. The threshold field is determined by the maximum value of the pinning force during the shifting process; error bounds may be obtained by comparing the heights of the several peaks that one sees. Fig. 3 illustrates the trajectories of electrons. One can see that in the strong disorder regime depinning happens through string-like current flows that tear the solid, whereas in the weak disorder regime sliding motion of patches containing many electrons is dominant.

In conclusion, we have observed a crossover between a weak disorder regime and a strong disorder regime in a model electron solid as we tune the level of disorder in the form of the modulation-doped donor impurities. In the weak disorder regime, we have a hexatic glass phase where a quasi-long range orientational order is present, whereas in the strong disorder regime isolated disclinations destroy the orientational order. The crossover takes place near $d/a_0 = 1.15 \pm 0.1$. Associated with this crossover, the behavior of the threshold electric field and the pattern of the depinning currents change.

3. Acknowledgements

This work was supported in part by NSF DMR 92-02255.

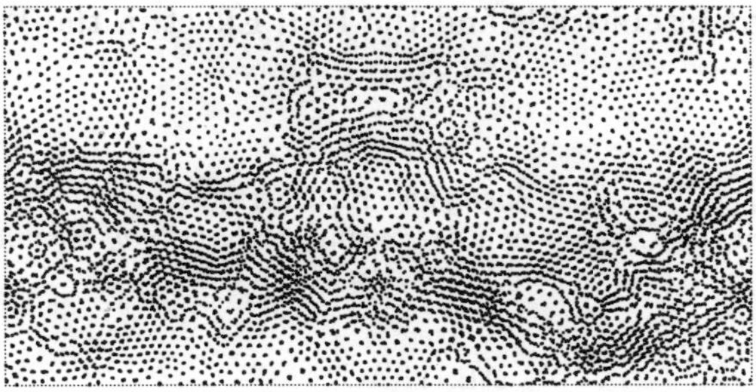

(a)

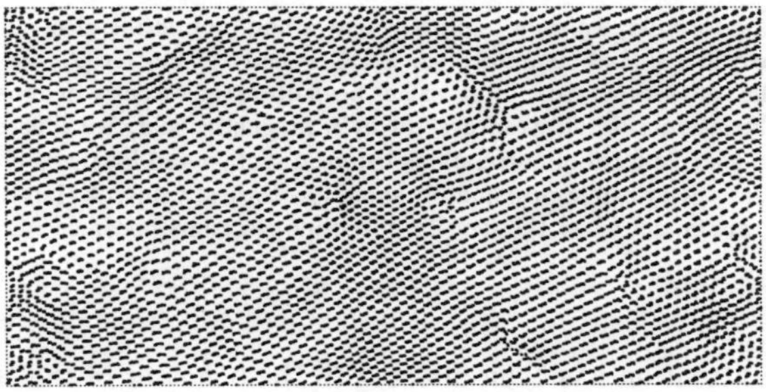

(b)

Figure 3: The trajetories of electrons as they are shifted in the case (a)$d/a_0 = 0.9$ and (b)$d/a_0 = 1.5$.

4. References

1. Y. E. Lozovik and V. I. Yudson, Pis'ma Zh. Eksp. Teor. Fiz. **22** (1975) 26 [JETP Lett. **22** (1975) 11]; Pui K. Lam and S. M. Girvin, *Phys. Rev.* **B30** (1984) 473; D. Levesque, J. J. Weis, and A. H. MacDonald, *Phys. Rev.* **B30** (1984) 1056.
2. For the review of experiments, see *Proceedings of the 10th International Conference on Electronic Properties of Two-Dimensional Systems*, Newport, RI, USA, 31 May–4 June 1993, ed. B. McCombe and A. Nurmikko (North-Holland, Amsterdam, 1993).
3. A.-C. Shi and A. J. Berlinsky, *Phys. Rev. Lett.* **67** (1991) 1926.
4. C. C. Grimes and G. Adams, *Phys. Rev. Lett.* **42** (1979) 795; H.-W. Jiang and A. J. Dahm, *Phys. Rev. Lett.* **62** (1989) 1396.
5. Kazumi Maki and Xenophon Zotos, *Phys. Rev.* **B28** (1983) 4349.
6. A. Brass, H. J. Jensen, and A. J. Berlinsky, *Phys. Rev.* **B39** (1989) 102; H. J. Jensen, A. Brass, and A. J. Berlinsky, *Phys. Rev. Lett.* **60** (1988) 1676.
7. L. Bonsall and A. A. Maradudin, *Phys. Rev.* **B15** (1977) 1959; D. S. Fisher, B. I. Halperin, and R. Morf, *Phys. Rev.* **B20** (1979) 4692.
8. D. R. Nelson and B. I. Halperin, *Phys. Rev.* **B19** (1979) 2457; D. R. Nelson, *Phys. Rev.* **B26** (1982) 269; D. R. Nelson, M. Rubinstein, and F. Spaepen, *Philos. Mag.* **A46** (1982) 105.

Correlated Interstitials and the Hall Effect in the Wigner Crystal

Lian Zheng and H.A. Fertig
Department of Physics and Astronomy, University of Kentucky, Lexington, Kentucky 40506-0055

ABSTRACT

Using the Kubo formula, we show explicitly that a non-interacting electron system can not behave like a Hall-insulator, *ie.*, a DC resistivity matrix $\rho_{xx} \to \infty$ and ρ_{xy} finite in the zero temperature limit, as has been observed recently in experiment. For a strongly interacting electron system in a magnetic field, we illustrate, by constructing a specific form of correlations between mobile and localized electrons, that the Hall resistivity can approximately equal its classical value. A Hall-insulator is realized in this model when the density of mobile electrons becomes vanishingly small.

Recently, much attention has been directed to the behavior of the Hall resistivity for a two-dimensional electron system (2DES) in an insulating phase. Magneto-transport studies of the 2DES have shown[1] that the Hall resistivity of a 2DES in a perpendicular magnetic field equals approximately the disorder-free value $\rho_{xy} \sim B/nec$ for a wide range of applied magnetic field strength, except near the QHE plateaus, where B is the applied magnetic field, n is electron density of the 2DES, e and c are the electron charge and speed of light, respectively. The fact that a large portion of the electrons may become localized by disorder scattering does not affect the value of ρ_{xy} substantially. This is in sharp contrast to the behavior of the diagonal resistivity, which is found to change from $\rho_{xx} \to 0$ in the QHE phase to $\rho_{xx} \to \infty$ in an insulating phase. This implies, in particular, the existence of a new insulating phase with $\rho_{xx} \to \infty$ and $\rho_{xy} =$ finite, the so-called Hall-insulating phase, which has generated much experimental[1,2,3] and theoretical[4,5] work recently. We will first show that the Hall-insulating behavior can not be explained with any independent-particle model. We will then explain this behavior by considering a model of correlated charged excitations.

For non-interacting electrons with an arbitrary disorder, the Kubo formula for conductivity σ_{ij} can be evaluated using the complete set of eigenfunctions of the Hamiltonian

$$\sigma_{xx}(T) \sim \sum_{nm} |J^x_{nm}|^2 \frac{\partial f(\epsilon_n)}{\partial \epsilon_n} \delta(\epsilon_n - \epsilon_m),$$

$$\sigma_{xy}(T) \sim \sum_n f(\epsilon_n) \sum_m \frac{J^x_{nm} J^y_{mn} - J^y_{nm} J^x_{mn}}{(\epsilon_n - \epsilon_m)^2}.$$

where n and m label the eigenstates of the Hamiltonian ($H = \sum_i h_i$ and $h|n>= \epsilon_n|n>$), J_{nm} is the matrix element of current operator, and $f(\epsilon)$ is the Fermi distribution function. If the lowest extended state has an energy Δ above the Fermi level, one may show from these formulas[6] that

$$\sigma_{xx}(T) \sim \sigma_{xy}(T) \sim e^{-\beta\Delta}, \quad \text{for } T \to 0.$$

The resistivity is then given by

$$\rho_{xx}(T) \sim \rho_{xy}(T) \sim \infty \quad \text{for} \quad T \to 0,$$

which shows clearly that a non-interacting 2DES is not a Hall-insulator. Several authors have described an AC form of the Hall-insulator, by calculating the zero-temperature AC conductivity and then considering the low frequency limit. For a non-interacting 2DES at $T = 0$, it may be shown quite generally[6] that $\sigma_{xx}^2(\omega) + \sigma_{xy}^2(\omega) = O(\omega^2)+$ higher power in ω; $\mathrm{Re}[\sigma_{xy}(\omega)] = O(\omega^2)+$ higher powers in ω. Then we have

$$\varrho_{xy} = \frac{\mathrm{Re}[\sigma_{xy}]}{\sigma_{xx}^2 + \sigma_{xy}^2}|_{\omega\to 0} \sim \frac{\omega^2}{\omega^2} = \text{constant}.$$

It is easy to see that this does not describe the DC conductivity since it could mislead one to conclude that every insulator is a Hall-insulator.

Next, we will argue that if correlations between conduction electrons and the localized electrons are important, then the Hall-insulating behavior may be obtained. We consider low temperature activated conduction of a 2D electron gas in an insulating phase. Let $N = N_c + N_L$ and $n = N/A, n_c = N_c/A, n_L = N_L/A$, where A is the system size and N, N_c, N_L are respectively the number of electron in the system, the number of electrons in extended states, and the number of electrons in localized states, with n, n_c, n_L the corresponding densities. A general wavefunction for a charged excitation that reduces its interaction energy with the lattice electrons by Laughlin-Jastrow type correlations may be written in the form

$$\Psi_{corrl}(z_o) = \Psi_{uncorr} \prod_{i=1}^{N_L} (z_i - z_o)^{m_i}, \tag{1}$$

where $z_i = x_i - iy_i$ are the lattice electrons in complex notation, z_o is the coordinate of an excited electron, and m_i are integers.

To show that this is a particularly low energy charged excitation, we have evaluated the energy of this wavefunction for the case $m_i = m$ for all the lattice sites, and choosing Ψ_{uncorr} to have the form of (the Hartree approximation to) a perfect crystal state with a single interstitial electron whose coordinate is given by z_o. Fig. 1 illustrates our results for this energy both for $m = 5$ and $m = 7$. The energy ϵ is defined here as the energy to add one electron to the system in this state, less the chemical potential in the groundstate; i.e., it is the energy to deform the crystal in

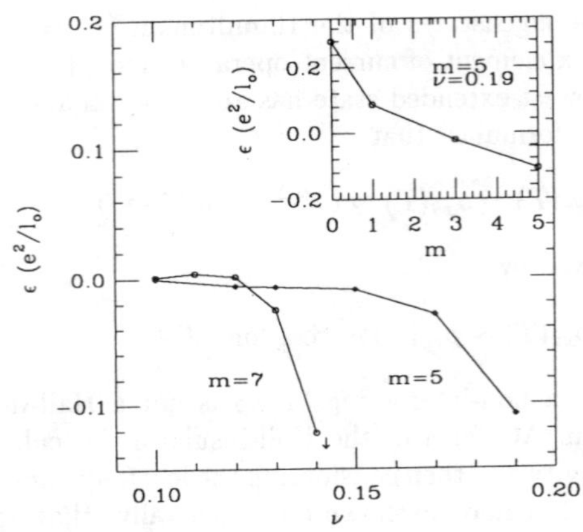

Figure 1: Energy ϵ as a function of filling factor ν for $m = 5$ and $m = 7$. Insert shows the energy at $\nu = 0.19$ for different values of m.

such a way that an interstitial is formed, without adding charge to the system. The case $m = 0$ corresponds to the usual uncorrelated interstitial. It can be shown [7] using a plasma analogy that interstitial in this wavefunction is ejected from the system if the filling factor $\nu > 1/m$, so that this is only a valid interstitial wavefunction below this limit. However, as ν approaches $1/m$ from below one can see that the energy of the interstitial is quite low, and even becomes negative close enough to this the limiting value[8]. The negative values of the ϵ imply an instability of the crystal if ν is close enough to $1/m$. The consequences of this are discussed elsewhere[7]. For our present purpose, the inset of Fig. 1 is most relevant. Here the energy is plotted for fixed filling fraction at different values of m. One can see that the energy lowers monotonically with increasing m. The magnitude of this lowering is quite large, and suggests that such correlations can be quite important in constructing excited states of the system.

Returning to the more general case of a weakly disordered crystal, Fig. 1 suggests that the values of m_i may be chosen to minimize the energy of the excitation. Since choosing the maximum values of m_i seems to minimize the energy, we should have $< m_i >= 1/\nu$, where $<>$ denotes an average over lattice sites; a larger value of $< m_i >$ would effectively eject the excited electron from the system. Using a plasma analogy, it can be shown that the localized charged excitation in Ψ_{uncorr} becomes a delocalized state Ψ_{corr}; i.e., the excited electron is no longer localized near the site of the interstitial in Ψ_{uncorr} due to the Jastrow factor. The Jastrow factor in Eq. 1 may be thought of as attaching a flux tube of strength $m_i\phi_o$ to the ith localized electron. In a mean-field description, the localized electrons will generate a fictitious magnetic field $B^{cL} = n_L\phi_o/\nu$, in addition to an applied field $B^{ext} = n\phi_o/\nu$. The excited conduction electrons are then traveling in a net field of $B^{net} = B^{ext} - B^{cL} = n_c\phi_o/\nu$. To the lowest order in the disorder potential, the motion of the conduction electrons can be described in the Drude picture as follows. Suppose there is a current

$j_x = n_c ev$ with $\rho_{xx} = 1/(n_c e\mu)$. A Hall voltage is generated to balance the Lorentz force $\rho_{xy} = E_y/j_x = (vB/c)/(n_c ev) = B/(n_c ec)$. For the strongly correlated systems $B = B^{net} = (n_c/n)B^{ext}$ so that $\rho_{xy} = B/(nec)$, regardless of the number of electrons which are localized. A Hall-insulator is realized when $n_c \to 0$. By contrast, for uncorrelated conduction electrons $B = B^{ext}$ one has $\rho_{xx} \to \infty$ and $\rho_{xy} \to \infty$ when $n_c \to 0$, which agrees with the earlier exact Kubo formula result.

Acknowledgements: The authors thank Professor A.H. MacDonald for very enlightening discussion. This work is supported by NSF through Grant No. DMR-9202255.

REFERENCES

1. V. J. Goldman, M. Shayegan, and D. C. Tsui, Phys. Rev. Lett., **61**, 881 (1988); T. Sajoto, Y. P. Li, W. Engel, D. C. Tsui, and M. Shayegan, *ibid.*, **70**, 2321 (1993). V. J. Goldman, J. K. Wang, and Bo Su, and M. Shayegan, *ibid.*, **70**, 647 (1993);
2. H. W. Jiang, R. L. Willett, H. L. Stormer, D. C. Tsui, L. N. Pfeiffer, and K. W. West, Phys. Rev. Lett., **65**, 633 (1990); Phys. Rev. B., **44**, 8107 (1991); P. F. Hopkins, M. J. Burns, A. J. Rimberg, and R. M. Westervelt, Phys. Rev. B., **39**, 12708 (1989); M. D'Iorio, V.M. Pudalov, and S.G. Semenchinsky, Phys. Lett. A., **150**, 442 (1990); V.M. Pudalov, M. D'Iorio, and J.W. Campbell, Pis'ma Zh. Eksp. Teor. Fiz., **57**, 592 (1993).
3. H. W. Jiang, C. E. Johnson, K. L. Wang, and S. T. Hannahs, Phys. Rev. Lett., **71**, 1439 (1993); T. Wang, K. P. Clark, G. F. Spencer, A. M. Mack, and W. P. Kirk, *ibid*, **72**, 709 (1994).
4. Shou-Cheng Zhang, Steven Kivelson, and Dung-Hai Lee, Phys. Rev. Lett., **69**, 1252 (1992); O. Viehweger and K. B. Efetov, J. Phys. Condens. Matter **2**, 7049 (1990); Phys. Rev. B., **44**, 1168 (1991); Y. Imry, Phys. Rev. Lett., **71**, 1868 (1993); G. Vignale, Phys. Rev. B., **48**, 11504 (1993);
5. Xiao-Fang Wang, Ziqiang Wang, and G. Kotliar, Phys. Rev. Lett., **68**, 2504 (1992);
6. Lian Zheng and H.A. Fertig, to appear in Phys. Rev. b (Rapid Communications).
7. Lian Zheng and H. A. Fertig, preprint (1994).
8. The $m = 5$ data appears to be negative in the entire range plotted. However, the magnitude of most of these values is smaller than our estimated error of the energy. Only the last two points are definitely negative within the accuracy of our calculation.

Transport and Photoluminescence studies of 2D Hole systems in the Extreme Quantum limit

P.J. Rodgers, B.L. Gallagher, M. Henini, Y. Ponomarev†, A. Usher† and G. Hill*

Department of Physics, University of Nottingham, Nottingham NG7 2RD, UK.
†Department of Physics, University of Exeter, Exeter, Devon EX4 4QL, UK.
*Department of Electronic Engineering, University of Sheffield, Sheffield S1 3JD, UK.

We present electrical transport and photoluminescence results for high mobility low density $GaAs$-$(Ga,Al)As$ 2D hole systems. We find that the longitudinal resistance, R_{xx}, shows a strongly temperature dependent insulating transition for filling factors $\nu < 1/3$ and $2/5 > \nu > 1/3$ but that the FQHE is the ground state at $\nu = 1/3$. The insulating state can be characterized by an activation energy and a threshold electric field both of which show similar re-entrant behaviour around $\nu = 1/3$. The magnitude of the threshold field and activation energy are consistent with Wigner crystal formation and at low ν the observed melting temperature of $\sim 400mK$ is close to the classical melting temperature. The photoluminescence from low density high mobility 2D quantum well p-type samples is found to show a new line for $\nu > 1/3$ which increases in strength as ν is decreased. This disappears on increasing temperature to $\sim 2K$. This is very similar to the behaviour observed in 2DEGs around $\nu = 1/5$.

In very high mobility two-dimensional (2D) electron gases there is evidence for the formation of a magnetically induced Wigner solid at a critical Landau level filling factor (ν_c) around 1/5. In the higher mass 2D hole system recent transport measurements have shown insulating transitions consistent with Wigner solid formation at a density dependent filling factor - with ν_c around 1/3 for hole densities around $5 \times 10^{10} cm^{-2}$ [1, 2]. Such lower values are consistent with the stronger Landau level mixing for holes which stablizes the Wigner state. The density dependence observed in the hole system enables one to probe areas of the r_s vs ν phase diagram (where r_s is the dimensionless mean carrier separation in units of the Bohr radius) which are not accessible in studies of electron systems.

Transport measurements where made on a low density ($6 \times 10^{10} cm^{-2}$) high mobility ($8 \times 10^5 cm^2/Vs$) p-type heterostructure at temperatures down to 85mK. The longitudinal resistivity R_{xx} exponentially diverges as the temperature $T \to 0$ for $\nu < 1/3$ and for $2/5 > \nu > 1/3$ while $R_{xx} \to 0$ as $T \to 0$ for $\nu = 1/3$ (figure 1B). This range of filling factors was studied by conventional four-terminal low frequency transport, in order to determine the activation energy, ΔE, as a function of ν, and by applying a two-terminal dc voltage bias V between the source and drain of our Hall bar sample and measuring the dc current I through the sample. Figure 1A shows some typical I-V curves for $\nu \leq 1/3$ which clearly show a very strong non-linearity which weakens with increasing filling factor. The I-V curves are very similar to those observed for high mobility 2DEGs around 1/5 [3, 4] which have been associated with the depinning of the magnetically induced Wigner solid. We define a threshold voltage V_{th} for the breakdown of the insulating phase by numerically diferentiating the I-V curves to locate the region, at high bias, where the response becomes linear and extrapolating a linear best-fit line to obtain an intercept with $I = 0$. This definition provides a consistent criterion for defining the voltage threshold regardless of the overall strength of the non-linearity and allows the effects of the Hall resistance to be eliminated.

The results of this analysis are presented in figure 1C which shows the threshold bias (E_{TH}) as function of filling factor. The magnitudes of the threshold field are consistent

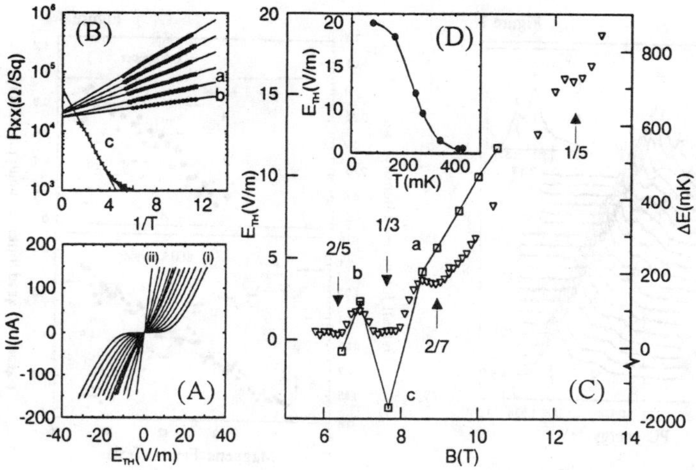

Figure 1: **A** Non-linear voltage threshold measurements at $T = 100mK$, curve (i) is taken at 12T with successive curves taken at equally spaced field intervals down to (ii) taken at 7.6T corresponding to $\nu = 1/3$; **B** Arrhenius plot of activated longitudinal resistance, curves a,b and c correspond to the indicated points in C; **C** Magnetic field dependence of threshold bias E_{TH} ($\bigtriangledown$) for $T = 100mK$ and energy gap ΔE (from $lnR_{xx} = \Delta E/kT$) of activated regions ($\square$); **D** Temperature dependence of E_{TH} at fixed field of 12T.

with the theory of Ruzin *et al* [5]. The non-linear behaviour persists to high filling factors and is seen to follow the general form of the longitudinal resistance. At the lowest temperatures, $E_{TH}(\nu)$ shows a distinct minimum at $\nu = 1/5$ despite the state being unobservable by conventional transport measurements. Also shown in figure 1C is the activation energy determined from the activated temperature dependence of the 4-terminal longitudinal resistance. Figure 1D shows the temperature dependence of E_{TH} at a fixed field of 12T, just below $\nu = 1/5$, the non-linearity is seen to rapidly decrease with increasing temperature and is seen to disappear completely above a temperature of $\sim 400mK$ which is very similar to the activation temperature derived from the extrapolation of ΔE to this filling factor. The $\nu \rightarrow 0$ melting temperature determined from a linear extrapolation of the activation data, as outlined in Rodgers *et al* [2], gives $1000mK$ which is approximately twice the calculated classical melting temperature for this system of $433mK$.

Photoluminescence (PL) measurements have been performed on two quantum well (QW) structures, having similar mobilities and densities to that described above. Both QW have a 150Å well width, one has a 600Å spacer layer and the other has 800Å. The PL was measured at temperatures down to 30mK and magnetic fields up to 17.4T, while the samples were excited at Ar^+ laser power densities 10^{-4}–$10^{-3}mW/cm^2$.

In the low-field regime the PL signal consists of two lines (see insert to figure 2) which we attribute to intrinsic recombination from the ground (line 1) and 1st excited subband (line 2) of the 2D holes with optically pumped electrons on the other side of the QW [6]. The intensity of peak 1 rises as the temperature is increased, whilst that of peak 2 falls, which rules out the interpretation of these lines as being recombination of free and bound excitons. Peak 2 shows a 1.8 times stronger linear dependence on excitation

Figure 2

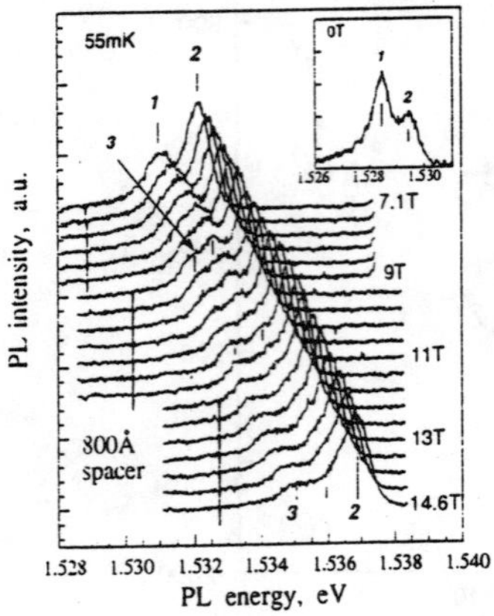

Figure 3

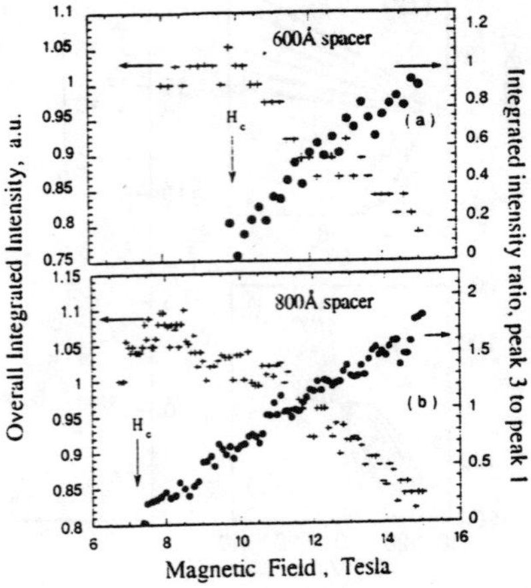

Figure 4

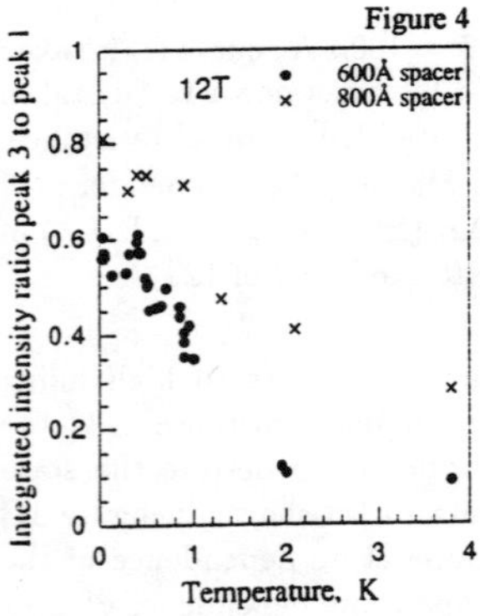

Figure 2: Photoluminescence spectra for a 150Å wide QW at high magnetic fields. Peak 1 is seen to weaken with increasing field while a new line, 3, emerges. The field is shown by following the spectral baseline (0.4T step for 9-14.6T range and 0.38T for 7.1-9T). Traces are offset values proportional to the field offset for clarity. Inset shows zero-field doublet structure at 55mK.

Figure 3: Field dependencies of overall integrated intensity and ratio of integrated intensities of peak 3 to peak 1 for two samples. Critical fields H_c are defined as the points where the ratio→ 0, which corresponds with the fields at which the overall integrated intensities begin to decrease.

Figure 4: Temperature dependencies of ratios of integrated intensities of peak 3 to peak 1 taken at 12T. Open markers are for 600Å spacer sample and filled ones are for 800Å spacer sample. Non-zero measured ratios at high temperatures are caused by thermal broadening of line 1.

power than peak 1 which supports the assertion that the former peak is associated with photocreated carriers. When samples are illuminated below the (Al,Ga)As band gap the position of line 1 shifts downwards and the separation between the two lines increases which could be explained by energy level renormalization effects caused by change in the concentration of the 2D carriers. Although the separation between the two lines is higher than the Fermi energy, one can explain the occurence of line 2 by the presence of a small photoinduced population of the excited subband together with a much stronger overlap of the wavefunctions of the excited 2D holes and conduction band electrons [7].

A full width at half maximum (FWHM) of $< 0.9meV$ for each of the peaks confirms the high quality of the interfaces in the structures [8]. Both peaks exhibit weak diamagnetic shifts at low fields associated with the formation of Wannier excitons. Reversed modulations of intensities of the peaks with magnetic field similar to those observed in the 2DEG samples have been noted although in the absence of the simultaneous transport measurements it is impossible to assign them to integer and fractional quantum Hall effect features [7, 9].

At higher fields (estimated $\nu < 0.35$) the low temperature spectra show a quite dramatic decrease of the integrated intensity accompanied by the appearance of a new line, 3, approximately $0.6meV$ below peak 1 (figure 2). This line increases in intensity with field and shows a marked temperature dependence.

One can estimate the point at which peak 3 appears by.considering evolution of the ratio of the integrated intensities of the structures 1 and 3 with field. It is interesting that the point of sharp decrease of the integrated intensity coincides with the appearance of peak 3 to within 10% of the field. Figure 3 shows the critical field H_c determined in this manner for two samples. Notably, the critical fields are different for the two samples: in the sample with the narrower spacer (higher hole density) the critical field is $10.1T$, while for the wider spacer sampler (lower hole density) it is $7.1T$. Assuming that quasi-equilibrium densities under excitation are only slightly different from those we observed in transport measurements on unilluminated samples, one can deduce that the critical filling factor at which line 3 emerges is $\nu_c \sim 0.34$, in both samples.

The temperature dependence of the new line is quite striking - it is present in the PL up to $0.9K$ and then shows an activated behaviour before disappearing at about $2.5K$ (Fig 4). It is interesting that the integrated intensity decrease exhibits the same trend.

This kind of behaviour in the magnetic quantum limit is quite similar to that observed in the luminescence experiments for Beryllium δ-doped n-type heterojunctions which was attributed to the formation of the magnetically induced Wigner solid which melts with increasing temperature and becomes favourable only below a critical filling factor (or above a critical field) [10, 11]. Although this explanation is quite tempting several others are possible for instance magnetic localisation effects persistent. Simultaneous transport and optical investigations are under way to test all those possibilities.

In conclusion we observe in a low density high mobility 2DHS heterostructure an activated insulating region reentrant around $\nu = 1/3$ which displays strong non-linear I–V response. The magnitudes of the threshold fields and the characteristic temperatures are consistent with theoretical models of a pinned Wigner solid. The photoluminescence studies of similar density and mobility quantum well 2DHS show a new low energy peak forming at $\nu \sim 0.35$ which increases in strength as ν decreases. The peak weakens with increased temperature but persists to $T \sim 2K$ which is somewhat higher than the behaviour seen in the transport measurements. The results of both transport and PL are in very good agreement with observations made in 2DEG's around $\nu = 1/5$ which have been ascribed to the formation of a magnetically induced Wigner solid.

Note added: Recently we became aware of similar PL results gained from investigation of two 200Å QW 2DHS using the same technique as described above [12]. They observed the appearance of a new low-energy peak with exactly the same characteristics as peak 3 in our investigations.

References

[1] M.B. Santos, Y.W. Suen, M. Shayegan, Y.P. Li, L.W. Engel and D.C. Tsui: Phys. Rev. Lett. **68**, 1188 (1992); M.B. Santos, J. Jo, Y.W. Suen, L.W. Engel and M. Shayegan: Phys. Rev. B. **46**, 13639 (1992).

[2] P.J. Rodgers, C.J.G.M. Langerak, B.L. Gallagher, R.J. Barraclough, M. Henini, T.J. Foster, G. Hill, S.A.J. Wiegers and J.A.A.J. Perenboom: Physica B. **184**, 95 (1993); P.J. Rodgers, C.J.G.M. Langerak, B.L. Gallagher, R.J. Barraclough, M. Henini, G. Hill, S.A.J. Wiegers and J.A.A.J. Perenboom: J. Phys. C. **5**, L449 (1993)

[3] F.I.B. Williams *et al*: Phys. Rev. Lett. **66**, 3285 (1991)

[4] H.W. Jiang, H.L. Stormer, D.C. Tsui, L.N. Pfieffer and K.W. West: Phys. Rev. B. **44**, 8107 (1991)

[5] I.M. Ruzin, S. Marianer and B.I. Shklovskii: Phys. Rev. B. **46**, 3999 (1992)

[6] Y.V. Ponomarev, A. Usher, P.J. Rodgers, B.L. Gallagher, M. Henini and G. Hill: *to be published*

[7] A.J. Turberfield, S.R. Haynes, D.A. Wright, R.A. Ford, R.G. Clark and J.F. Ryan: Phys. Rev. Lett. **65** 637 (1990); A.J. Turberfield, R.A. Ford, I.N. Harris, J.F. Ryan, C.T. Foxon and J.J. Harris: Phys. Rev. B. **47** 4794 (1993)

[8] O. Brandt, K. Kanamoto, Y. Tokudam and N. Tsukada: Phys. Rev. B. **48** 17599 (1993)

[9] B.B. Goldberg, D. Heimann, A. Pinczuk, L. Pfieffer and K. West: Phys. Rev. Lett. **65** 641 (1990)

[10] H. Buhman, W. Joss, K.v. Klitzing, I.V. Kukushkin, A.S. Plaut, G. Martinez, K. Ploog and V.B. Timofeev: Phys. Rev. Lett. **66** 926 (1991); I.V. Kukushkin, V.I. Fal'ko, R.J. Haug, K.v. Klitizing, K. Eberl and K. Totemayer: Phys. Rev. Lett. **72** 3594 (1994) [and references therein]

[11] M.K. Ellis, M. Hayne, A. Usher, A.S. Plaut and K. Ploog: Phys. Rev. B. **45** 13765 (1992)

[12] L.V. Butov, A. Zrenner, M. Shayegan, G. Abstrelter and H.C. Manoharan : Phys. Rev. B. **49** 14054 (1994)

SINGLE-ELECTRON TUNNELING IN QUANTUM DOT SYSTEMS

ROLF J. Haug
Max-Planck-Institut für Festkörperforschung
D-70569 Stuttgart, Germany

ABSTRACT

Tunneling through quantum dots is determined by the interplay between charging effects and the discrete level spectrum originating from the three-dimensional confinement. We have performed spectroscopic measurements of ground and excited states in single quantum dots by studying the linear and nonlinear transport in magnetic fields. In system consisting out of two or more quantum dots the measured conductance through the system is influenced by the charging energies of the individual dots and the coupling between the dots.

1. Introduction

Quasi zero-dimensional systems in semiconductors are called quantum dots due the confinement of the electrons in all three dimensions. The tunneling through quantum dots is influenced not only by the quantum confinement effects, but also by Coulomb charging effects, which have been intensively studied also in metallic systems during the last years[1]. These charging effects or single-electron effects are based on the quantization of the charge and the relevant energy is the charging energy $E_C = e^2/2C$ with C the total capacitance of the quasi-isolated system. If the capacitance of the enclosed system (quantum dot) is small enough, i.e. $E_c >> k_BT$ with T the temperature, one can expect to see single-electron tunneling. A single-electron-tunneling transistor was first realized in metallic systems, but very soon it became clear that many devices realized in semiconducting heterostructures could also work as single-electron transistors. Single-electron effects have been studied in vertical (perpendicular to the surface of the sample) and in lateral (parallel to the surface) semiconducting devices. Here, we will concentrate on lateral transport in AlGaAs/GaAs heterostructures.

In quantum dots the interplay between quantum confinement effects and charging effects can be studied for a small number of electrons. One of the first,detailed studies in this direction was performed by McEuen et al.[2] in measuring the gate-voltage position and amplitude of the conductance resonances observed in magnetotransport

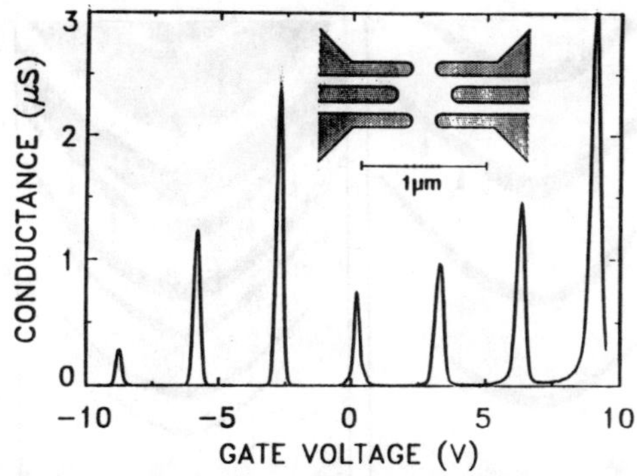

Figure 1: Conductance versus back-gate voltage measured for a quantum dot defined by the top-gate geometry as shown schematically in the inset. The voltages applied to the top gates had been kept constant (at values of about -0.7V)).

through a quantum dot. Nonlinear transport experiments allowed to observe, in addition to tunneling through the ground states of the quantum dot, also tunneling through excited states. [3–5]

Since the knowledge about transport through single quantum dots is maturing, coupled quantum dot systems are studied more intensively. In the following several experiments studying transport through single and coupled quantum dot systems in a magnetic field will be described and discussed.

2. The Devices

AlGaAs/GaAs heterostructures served as basic structures for the devices measured. The two-dimensional electron gas (2DEG) in a structure used for a device with a single quantum dot had an electron density of 3.4×10^{15} m^{-2} and a mobility of 60 m^2/Vs at a temperature of 4.2K. For the quantum dot, metallic split gate structures were evaporated on to the surface of the heterostructures. Here, for the single quantum dot we will concentrate on a structure with 6 gates enclosing an area of about 350nm diameter, as schematically shown in the inset of Fig. 1. By application of negative gate voltages (on the order of -0.7V) to the six gates a quantum dot is isolated in the middle between the gates. The electrostatic potential of the quantum dot can be changed by application of a gate voltage to a metallization on the backside of the device, in addition.

Figure 1 shows the measured conductance versus backgate voltage for the device schematically given in the inset. Sharp conductance resonances due to single-electron tunneling through the dot are observed. For positive gate voltages a finite conductance is measured between the peaks whereas for negative gate voltages co-tunneling effects (finite conductance between the peaks) are not seen. Whilst for negative gate voltages beyond -10V conductance is totally suppressed due to the large barriers, positive gate voltages strongly decrease the barrier heights and widths. Therefore,

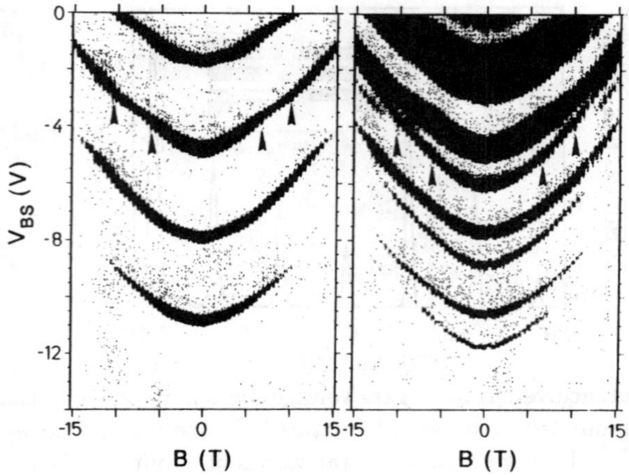

Figure 2: Differential conductance (of the quantum dot of Fig. 1) shown in a linear grey scale (black corresponds to $2\mu S$ or larger and white corresponds to $-0.1\mu S$ or smaller) as a function of backgate voltage for different magnetic fields applied parallel to the current. Left side: vanishing source-drain voltage, ac-voltage 5 μm; right side: a source-drain voltage of -0.7mV was applied in addition.

in this structure the charge of the dot can only be increased by about 10 electrons before leaving the Coulomb blockade regime. Sharp variations in the peak values of the conductance resonances are observed indicating that the quantum nature of the states in the quantum dot (of the order of 20 electrons are expected to be enclosed in the dot) dominates the conductance.

3. Transportspectroscopy of Single Quantum Dots

To investigate the different quantum states of the dot in more detail finite source-drain voltages allowing for tunneling in a finite energy range can be applied. Instead of the sharp conductance resonances as seen in Fig. 1 more complicated structures extending over a finite range in gate voltage are then observed[3]. Even regions of negative differential conductance appear[3]. Such complicated structures originate from tunnel processes through different states of the quantum dot. Whilst for small source-drain voltage and low temperature, tunneling through the quantum dot can only occur via the ground states, for a finite source-drain voltage and a correspondingly finite range of energy tunneling via excited states is possible.

A plot of the differential conductance as function of source-drain voltage and gate voltage allows not only the observation of the additional channels for tunneling due to excited states, but also a classification of these excited states[6,3].

Application of a magnetic field can further help to identify the different states observed. For small source-drain voltages transport through the dot uses the ground

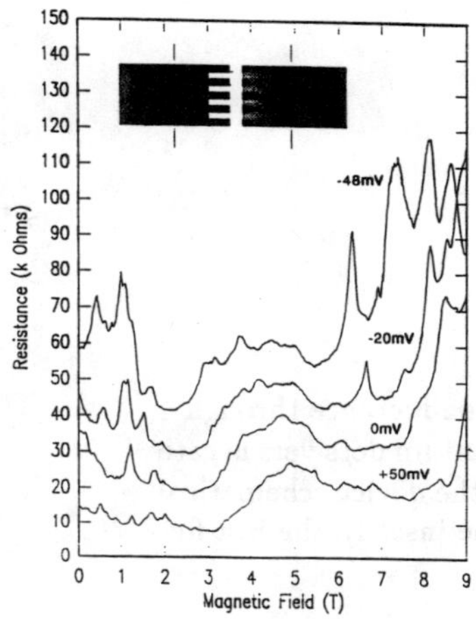

Figure 3: Resistance of the device shown in the inset (schematical picture of the gate structure used to define 4 coupled quantum dots) versus magnetic field for four different gate voltages applied at a temperature of T=0.4K.

states of the quantum dot. Varying a magnetic field influences these ground states. Changes in the nature of the ground states are reflected in the measurements in changes in the gate-voltage position of the conductance resonances. Corresponding changes can be observed in the amplitude of the conductance resonances[2,6]. By the combination of excited-state spectroscopy at finite source-drain voltages and additional spectroscopy of the ground states in a magnetic field, the evolution of ground states into excited states with magnetic field (and vice versa) can be studied[3,7] as shown in Fig. 2 (marked by arrows). The left side of Fig. 3 shows the linear conductance through the single quantum dot (as shown in Fig. 1) as a function of a magnetic field applied parallel to the current. An overall shift of peakpositions to larger gate voltages is observed for an increasing magnetic field. This strong dispersion has been explained by the influence of the magnetic field not only on the quantum dot, but also on the 2DEG in the leads[3]. In addition, some modulation of the peakamplitude and position is observed (marked for one peak with arrows)[3]. On the right side in Fig. 2 measurements of the differential conductance for a source-drain voltage -0.7mV are shown. Here, two different magnetic field dispersions are obvious, which could originate from different spin quantum numbers of the different states. These different magnetic field dispersions lead to changes in the nature of the ground states of the system, which coincide with the amplitude modulations seen in linear transport (marked by arrows on both sides).

3. Coupled Quantum Dots

Besides these investigations of single structures, more complicated systems are also studied now in more detail. When the first works about transport through strings of quantum dots were published, the nature of the interplay between quantum confinement effects and charging effects were not clear[8–10].

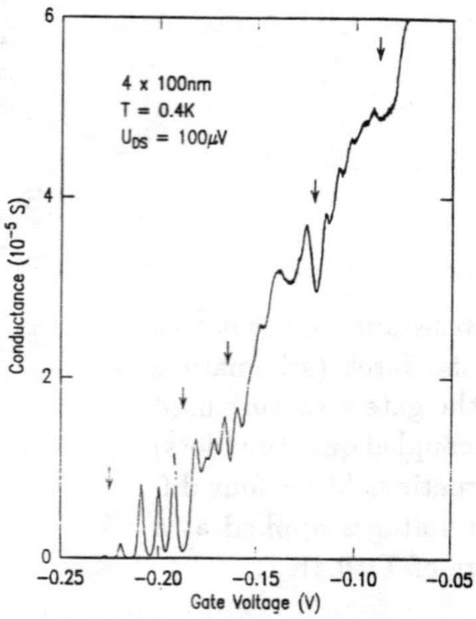

Figure 4: Conductance through 4 coupled quantum dots versus gate voltage for the device schematical shown in the inset to the last figure.

An example of a top-gate structure used for the generation of a device with 4 coupled quantum dots is shown in the inset to Fig. 3. One part of the split-gate structure has a flat end, whereas the second part ends in 5 fingers having a periodicity of 100nm[8]. Due to the presence of the gate structure on top of the heterostructure the carrier concentration in the 2DEG is strongly modulated already at zero gate voltage as can be seen (and was known from other measurements) in the resistance measured across the gate structure as a function of magnetic field (shown in Fig. 3 for different gate voltages applied to the gate). Especially, the strong resonances in the resistance at magnetic fields below 2T and above 6.2 T (corresponding to a filling factor of 2) show the quenching of the conductance due to the influence of the magnetic field on the level structure in the modulated system.

The conductance versus gate voltage for this device is shown in Fig. 4. Deep minima (marked by arrows) are separated by the occurence of 4 maxima. These measurements[8,10] have originally been interpreted in terms of transport through coherent states of ballistic quantum chains[11], which could be also interpreted as transport through molecular states. But, it is now clear that the physics of these coupled quantum dot systems is extremely rich, e.g. for intermediate coupling strenght electron-electron interaction as described by a Hartree-energy could lead to a Mott-Hubbard like metal-insulator transition[12]. For a very weak coupling between the quantum dots, i.e. around threshold of the device, the electrostatic coupling will be dominating and will lead to effects like the Coulomb blockade, which could be responsible for the sharp conductance resonances seen in Fig. 4 for the gate voltages around -0.2mV. Therefore, this gate voltage regime has been studied in more detail. The conductance versus gate voltage for different magnetic fields[8] showed that the periodicity in gate voltage of the peaks in the conductance is roughly independent of the magnetic field, one of the characteristica of charging effects, i.e. the capacitance is roughly constant. In contrast, the peak positions change and the peak amplitudes show strong variations with magnetic field, i.e the level structure of the system is affected by the magnetic field[2].

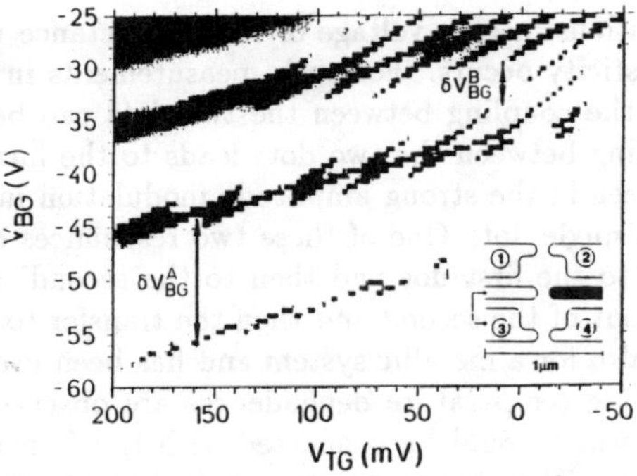

Figure 5: Conductance plotted in a greyscale (white corresponds to 0μS and black to 2μS) as function of top-gate voltage V_{TG} and backgate voltage V_{BG} for the device geometry as shown schematically in the inset. Whereas the right center-gate has been varied for this measurements, the voltages at the other gates had been kept fixed: upper two gates V_G = -0.34V, left center gate and left lower gate V_G = -0.77V, right lower gate V_G = -1.2V.

A only slighthly more complicated system in comparison to a single quantum dot is a double quantum dot, and here the coupling between the two dots can be more easily changed in a defined way than for more complex systems.

For the double quantum dot the split-gate geometry as schematically shown in the inset to Fig. 5 has been used. By application of different gate voltages (given in the figure caption) to the different parts of the split-gate device two quantum dots of different sizes are isolated in the two parts. By varying the voltage on one of the center gates (shown in dark in the inset to Fig. 5) the charges on the two dots can be changed and the coupling between the two dots is varied. Since this center gate couples to both dots, it is necessary to vary also another gate voltage, in addition, to understand the changes occuring in the system. Figure 5 shows the conductance in a grey scale measured as a function of the voltage on the central top gate V_{TG} and as a function of the backgate voltage V_{BG}. In Fig. 5 the two periodicities in gate voltage originating from the different capacitive coupling from the backgate to the two dots of different sizes are marked: A (small dot) and B (large) dot[13].

In the measurements shown in Fig. 5 the conductance of the double-quantum-dot system seems to be given by the superposition of the conductance pattern of the individual quantum dots, but some variations of the periodicity of the large quantum dot are apparent. These variations originate from the electrostatic coupling between the two dots, i.e. due to the finite capacitance between the two dots the charging of one dot influences the other. Glazman and coworkers[14] calculated the conductance through a coupled quantum dot system under conditions of single-electron tunneling and they predicted the occurence of stochastic Coulomb blockade due to the electrostatic coupling. In plotting the conductance as function of two gate voltages as done in Fig. 5 the positions in gate voltage of the conductance resonances are well defined, i.e. no stochasticity occurs. For such measurements in a two-dimensional space in gate voltages the coupling between the two dots can be studied in detail. The electrostatic coupling between the two dots leads to the formation of two conductance resonances (seen in the strong amplitude modulation in Fig. 5) instead of a single resonance in a single dot. One of these two resonances originates from the transfer of an electron to the first dot and then to the second, the other from the transfer of an electron out of the second and then the transfer to the frist one. This effect has been shown also for a metallic system and has been used for the pumping of electrons[15]. Interesting temperature dependences are observed for this coupled quantum dot system[13] which could be connected with the formation of a coherent state at low temperatures. The phase coherence in double quantum dot systems has been already started to be investigated[16]. But, the detailed temperature dependence of the conductance for such a double-quantum dot system is not fully understood, e.g. influence of co-tunneling effects etc.

5. Conclusions

Single and multiple quantum dots have been studied in linear and non-linear transport. Applications of magnetic fields help in identifying the quantum nature of the states in the quantum dots. Several experiments, starting from single quantum dots up to devices with 4 quantum dots, have been presented and discussed.

Acknowledgements

I gratefully acknowledge contributions from J. Weis, R. Blick at the Max-Planck-Institut, and K. Lee, T.P. Smith III, D. Kern, J.M. Hong, L.L. Chang, L. from IBM, Yorktown Heights. I thank K. Ploog and K. Eberl for providing MBE grown samples for some of the experiments. Discussions with D. Pfannkuche, H. Pothier, W. Häusler, D. Weinmann and others have been also invaluable. I thank K. v. Klitzing and L. Esaki for the support, encouragement and interest into this work. Part of this work has been supported by the Bundesministerium für Forschung und Technologie.

References

1. 'Special Issue on Single Charge Tunneling' in *Z. Phys. B - Condensed Matter* **85**, No. 3 (1991); 'Single Electron Tunneling and Mesoscopic Devices', eds. H. Koch and H. Lübbig, Springer Series in Electronics and Photonics, Springer Berlin (1992); 'Single Charge Tunneling', eds. H. Grabert and M.H. Devoret, Plenum Press New York (1992).
2. P.L. McEuen, E.B. Foxman, U. Meirav, M.A. Kastner, Y. Meir, N.S. Wingreen, and S.J. Wind, *Phys. Rev. Lett.* **66**, 1926 (1991); R.C. Ashoori, H.L. Störmer, J.S. Weiner, L.N. Pfeiffer, S.J. Pearton, K.W. Baldwin, and K. W. West, *Phys. Rev. Lett.* **71**, 613 (1993).
3. J. Weis, R.J. Haug, K. v. Klitzing, and K. Ploog, *Phys. Rev. Lett.* **71**, 4019 (1993); *Surf. Sci.* **305**, 664 (1994); *Semicond. Sci. Techn.* (in print); J. Weis, *PhD thesis*, Stuttgart

(1994).

4. R.J. Haug, J. Weis, K. v. Klitzing, and K. Ploog, *Physica B* **194-196**, 1253 (1994).

5. A.T. Johnson, L.P. Kouwenhoven, W. de Jong, N.C. van der Vaart, C.J.P.M. Harmans, and C.T. Foxon, *Phys. Rev. Lett.* **69**, 1592 (1992); E.B. Foxman, P.L. McEuen, U. Meirav, N.S. Wingreen, Y. Meir, P.A. Belk, N.R. Belk, M.A. Kastner, and S.J. Wind, *Phys. Rev. B* **47**, 10020 (1993).

6. J. Weis, R.J. Haug, K. v. Klitzing, and K. Ploog, Phys. Rev. B **46**, 12837 (1992); *Physica B* **189**, 111 (1993).

7. R.J. Haug, *Festkörperprobleme - Advances in Solid State Physics* **34**, (1994).

8. R.J. Haug, K.Y. Lee, T.P. Smith III, and J.M. Hong, *Proc. 20th Int. Conf. Phys. Semiconductors*, Eds. E.M. Anastassakis, J.D. Joannopoulos, World Scientific, London (1990).

9. L.P. Kouwenhoven, F.W.J. Hekking, B.J. van Wees, C.J.P.M. Harmans, C.E. Timmering, and C.T. Foxon, *Phys. Rev. Lett.* **65**, 361 (1990).

10. R.J. Haug, J.M. Hong, and K.Y. Lee, *Mat. Res. Soc. EA-* **26**, 29 (1990); *Surf. Sci.* **263**, 415 (1992).

11. S.E. Ulloa, E. Castano, and G. Kirczenow, *Phys. Rev. B* **41**, 12350 (1990); J. Brum, *Phys. Rev. B* **43**, 12082 (1991).

12. C.A. Stafford and S. Das Sarma, *Phys. Rev. Lett.* **72**, 3590 (1994).

13. R.H. Blick, R.J. Haug, J. Weis, D. Pfannkuche, K. v. Klitzing, and K. Eberl (to be published).

14. L.I. Glazman, V. Chandrasekhar, *Europhys. Lett.* **19**, 623 (1992); I.M. Ruzin, V. Chandrasekhar, E.I. Levin, L.I. Glazman, *Phys. Rev. B* **45**, 13469 (1993).

15. H. Pothier, P. Lafarge, P.F. Orfila, C. Urbina, D. Esteve, and M.H. Devoret, *Europhys. Lett.* **17**, 249 (1992).

16. U. Sivan, F.P. Milliken, K. Millkove, S. Rishton, K. Lee, J.M. Hong, V. Boegli, D. Kern, and M. deFranza, *Europhys. Lett.*, **25**, 605 (1994).

EDGE PROPERTIES OF QUANTUM DOTS IN THE FQHE REGIME

M.D. JOHNSON
Department of Physics, University of Central Florida
Orlando, FL 32816-2385, USA

ABSTRACT

This is a summary of some of what has been learned about edge states in quantum dots in strong magnetic fields, with an emphasis on the fractional quantum Hall regime. The description of edge states as chiral Luttinger liquids leads to a number of interesting consequences, including thermodynamic oscillations, edge reconstructions, and unusual tunneling properties, and these are briefly reviewed.

1. Edge States

A two-dimensional electron gas in a strong transverse magnetic field can exhibit the quantum Hall effect[1]. This is because in the bulk the electron fluid is incompressible. The physical origin of this incompressibility is quite different in the case of the integer and fractional effects (IQHE and FQHE). In the integer case the incompressibility arises from the kinetic energy gap between successive Landau levels, while in the fractional case it is a consequence of electron-electron interactions. Even so, in both cases the only low-energy excitations are edge states[2], which consequently are responsible for many of the interesting properties of quantum Hall systems. In this paper we will summarize some of what has been learned about the physics of edges in quantum dots, with an emphasis on the fractional Hall regime.

Perhaps the best-known use of edge states is the very appealing explanation they provide for quantized Hall effect itself (particularly in the integer regime) through the mesoscopic transport theory developed by Landauer and Büttiker[3,4]. This approach focuses on transport due to scattering edge states. In a two-terminal IQHE device, the current is assumed to be supplied at the ends by reservoirs which are separately in local equilibrium, with the result that right- and left-movers are occupied up to energies determined by local chemical potentials μ_R and μ_L, respectively. These states are scattered by any impurities and hence a state's current is reduced by reflection (backscattering). In an IQHE system the states on opposite edges carry current in opposite directions. Because of their macroscopic separation, backscattering between edges is highly suppressed in a strong magnetic field, and transmission is very nearly unity[5]. This gives the quantized (Hall) conductance. Obviously this explanation of quantization suffices only if current is carried almost entirely at the edges and not in the bulk. This is still not a settled question.

A deeper difficulty with this explanation is that it only works at low currents. In the Landauer-Büttiker approach it is usually assumed that the Hall voltage is much less than the inter-Landau-level spacing $\hbar\omega_c$, and consequently the currents are small. However, experiments find excellent quantization at extremely high Hall voltages on this scale (*e.g.*, 10 or 100 times $\hbar\omega_c$). It is straightforward to show that the Landauer-Büttiker approach, if applied naively to this regime, cannot give quantization[6]. As soon as $\mu_R - \mu_L$ exceeds $\hbar\omega_c$, one or more Landau levels ends up contributing more than the current $e(\mu_R - \mu_L)/h$ needed for quantization. This can be dealt with, still within the edge picture of transport, using the maximum entropy approach to nonequilibrium statistical mechanics[6]. This gives quantization at almost arbitrarily high currents for ideal systems, but the explanation suggests a need to modify (at high currents) the usual model of reservoirs as regions in local equilibrium. The electronic occupancies in this approach depend on a state's current i as well as its energy ϵ: $f = 1/(\exp[\beta(\epsilon - \xi i - \mu)] + 1)$ (where μ is a global chemical potential and ξ a constant determining the total current), rather than the Landauer-Büttiker distribution $f = 1/(\exp[\beta(\epsilon - \mu_{L,R})] + 1)$.

The existence of gapless modes at the edge of an incompressible fluid has some beautiful consequences (especially in the fractional regime) going beyond their role in transport. The simplest result is the ability to enumerate the different branches of edge excitations. For example, in the IQHE with filling factor $\nu = 1$, there is one branch of excitations on an edge. (These are obtained by multiplying the filled Landau level wavefunction by symmetric polynomials $Q(z_1, \ldots, z_n)$, where z_i are the electron coordinates.) There is also a single branch for the simpler fractions in the FQHE, such as $\nu = 1/3$ or $1/5$; these excitations can be mapped[7–11] one-to-one onto edge excitations of the integer system by multiplying the latter by a Jastrow factor.

In general an edge can support more than one branch of excitations. Wen enumerated the edge states quite generally by showing that they span representations of Kac-Moody algebras[12]. For certain cases this can be understood more simply; *e.g.*, in the IQHE with $\nu = 2$, each filled Landau level gives rise to one edge branch. A more interesting example is provided by $\nu = 2/3$. The bulk ground state at $\nu = 2/3$ is a particle-hole conjugate of the $\nu = 1/3$ bulk state. In a finite system, the $1/3$ state has an edge at which the filling factor falls from its bulk value of $1/3$ to zero. The conjugate of this has a droplet of holes in a $\nu = 1/3$ Laughlin state, with density rising to $\nu = 1$ at the edge of the hole droplet; consequently there is a second outer edge at which the density falls to zero. Each edge can support low-energy excitations, electron-like on the outer and hole-like on the inner. This picture of edges in the $2/3$ system[7] has been supported by numerical calculations, at least with sufficiently strong confining potentials at the edges[13,14].

2. Luttinger Liquids

More interesting than this simple counting is the mounting evidence that the gapless edge excitations can be described as chiral Luttinger liquids[12,7–9,15]. In the IQHE this description can be developed microscopically[15]. In the fractional QHE it arises as an effective theory of edge excitations[12,8–11]. There is also recent experimental support of this picture (Sect. 5). Wen showed by a simple hydrodynamic analogy that this should occur whenever the bulk is incompressible; and also, more formally, that this is required for consistency in the field theory when the bulk is taken finite[12,9]. The hallmark of a (here chiral) Luttinger liquid is that it permits the excitations to be bosonized, *i.e.*, mapped to an equivalent boson description[16–18]: the edge excitations can be described as fermionic particle-hole pairs, or as bosonic wave deformations in the shape of the electron droplet.

A beautiful numerical example of bosonization in the IQHE was given by Stone, *et al.*[15], who showed that the low-energy spectrum of interacting many-electron systems near $\nu = 1$ could be mapped onto a spectrum of non-interacting bosons. The simplest example of bosonization in the FQHE comes from treating $\nu = 1/3$ in the truncated pseudopotential model[5,19], in which all states lying below the interaction gap have zero interaction energy; as in the integer case, these can be counted bosonically[7,20].

It is useful to summarize some information about bosonization, here for an ordinary (chiral) Luttinger model[16–18]. This describes (here, spinless) fermions moving in one direction in one dimension, with a linear dispersion. The ground state consists of a Dirac sea with all states in $-\infty < k \leq k_F$ filled. The density operator, normal-ordered with respect to the Dirac sea, is $\hat{\rho}(x) = \frac{1}{L}\sum_q e^{-iqx}\rho_q$, where $\rho_q = \sum_k c^\dagger_{k+q}c_k$ for $q \neq 0$, and $\rho_0 = \sum_k (c^\dagger_k c_k - \theta(k_F - k))$. Here the coordinate x runs from $-L/2$ to $L/2$, and $\hat{\psi}(x) = \sum_k e^{ikx}c_k/\sqrt{L}$, with $k = 2\pi m/L$. The essential feature of the Luttinger model is that, because of the normal ordering, the density operators obey the commutation relations $[\rho_{-q}, \rho_{q'}] = \frac{Lq}{2\pi}\delta_{qq'}$. Consequently the Luttinger model can be written equivalently in terms of boson operators defined below[17,18]. Haldane showed that one-dimensional systems of interacting fermions can quite generally be described in a very similar way[18], and so called them 'Luttinger liquids.'

The description of QHE edges as chiral Luttinger liquids requires one important modification of the Luttinger model results: a branch labeled by i is characterized by a parameter ν_i related to the filling factor of the underlying electron fluid. The density operators then obey the commutation relations[12]

$$[\rho_{i,-q}, \rho_{i',q'}] = \nu_i \frac{Lq}{2\pi}\delta_{qq'}\delta_{ii'}. \tag{1}$$

Hence one can define[17,18,12] boson creation and annihilation operators $a^\dagger_q, a_q$, for $q > 0$:

$$a^\dagger_q = \sqrt{\frac{2\pi}{Lq|\nu|}}\,\rho_{q\,\mathrm{sgn}\,\nu}, \qquad a_q = \sqrt{\frac{2\pi}{Lq|\nu|}}\,\rho_{-q\,\mathrm{sgn}\,\nu}. \tag{2}$$

From Eq. (1), these satisfy $[a_q, a^\dagger_{q'}] = \delta_{qq'}$. A Luttinger liquid Hamiltonian can then be written $H_0 = \frac{v\pi}{L}\rho_0^2 + \sum_{q>0} v_q q a^\dagger_q a_q$.

An example of the power of bosonization is that it leads immediately to the QHE, even in the fractional case. Let us compute the transverse (Hall) conductance in linear response. Let y be a coordinate transverse to the edge. A scalar potential $\phi(x,y) = -Ey\cos(qx - \omega t)$ produces, in the limit $q, \omega \to 0^+$, a purely transverse electric field. For small-amplitude excitations on one edge, $y \approx$ const, and so ϕ adds

$$V = \int_{-L/2}^{L/2} e\phi(x,y)\rho(x)dx = -\frac{eEy}{2}\left(\rho_q e^{-i\omega t} + \rho_q e^{i\omega t}\right) \tag{3}$$

to the Hamiltonian. To linear order in E, the ground state of $H_0 + V$ is

$$|0'\rangle = |0\rangle + \frac{eEy}{2\hbar}\frac{1}{v_q q - \omega}e^{-i\omega t}\rho_q|0\rangle, \tag{4}$$

where $|0\rangle$ is the boson vacuum. Hence the perturbed charge density is

$$\bar{\rho}(x,t) = e\langle 0'|\rho(x)|0'\rangle = \nu\frac{e^2 Ey}{h}\frac{q}{v_q q - \omega}\cos(qx - \omega t). \tag{5}$$

From the continuity equation, the longitudinal current density is then

$$\bar{j}(x,t) = -\nu\frac{e^2}{h}\frac{\omega}{v_q q - \omega}\phi(x,y). \tag{6}$$

On the one-dimensional edge the current density j_x equals the current I. Then, taking the usual linear response order of limits, $q \to 0$ followed by $\omega \to 0$, this gives $I = \nu\frac{e^2}{h}V$, where $V = Ey$ is a constant potential applied along one edge. Thus the ability to describe an edge branch as a Luttinger liquid leads at once to the quantization of the Hall conductance, with a value determined by the bulk filling factor ν.

3. Magnetic Oscillations

A calculation[20] based on these bosonization ideas shows that a quantum dot in the fractional Hall regime should exhibit unusual mesoscopic magnetic oscillations in thermodynamic properties. In a parabolic quantum dot (with confinement potential $m^*\Omega^2 r^2/2$), the confinement energy of the many-electron states depends only on their total (z-direction) angular momentum, M_F: $E = \gamma(N + M_F)$, where $\gamma = m^*\Omega^2\ell^2$ and $\ell = (\hbar c/eB)^{1/2}$. In the truncated pseudopotential model[5,19], all states lying below the interaction gap (including the edge excitations) have zero interaction energy.

Moreover, these excitations can be bosonized—the number of N-fermion states at M_F equals the number of N-boson states at total angular momentum $M_B = M_F - mN(N-1)/2$. The simple energies, and the ability to count the number $g(N, M_B)$ of bosonic excitations, makes it possible to exactly evaluate the grand partition function $Z_G = \sum_{N,M_B=0}^{\infty} g(N,M_B)e^{-\beta(E(N,M_B)-\mu N)}$. More accurately, this gives the exact contribution of these low-energy states to Z_G, and hence the low-temperature thermodynamics. For example, the low-temperature magnetization is $M = M_1 + M_{\rm osc}\exp(-2\pi^2/\beta\gamma m)$, where M_1 is a smooth piece and

$$M_{\rm osc} = \frac{\gamma}{B}\left[\frac{4\pi N_\phi}{m\beta\gamma}\sin(2\pi(N_\phi - 1)/m) + \frac{4\pi^2}{m(\beta\gamma)^2}\cos(2\pi(N_\phi - 1)/m)\right]. \tag{7}$$

Here N_ϕ is the number of flux quanta through the dot. $M_{\rm osc}$ has a periodicity of m in flux quanta, which is m times that of noninteracting electrons. These unusual oscillations express the fact that near $\nu = 1/m$ and at constant μ, one electron is added to the correlated state in the dot for every m additional flux quanta, on average. From another viewpoint, the oscillations are a consequence of the bosonization which is possible when the edge excitations are a chiral Luttinger liquid. Since the correlated state and Luttinger liquid edges exist with the actual Coulomb interaction, so too should the periodicities.

4. Coupled Composite Edges

Next let us investigate composite edge branches when they are strongly coupled by the long-range Coulomb interaction between electrons. This applies to IQHE systems with $\nu > 1$ and to FQHE systems at fractions such as $\nu = 2/3$. The approach used here was first used by Wen[12]. Here we want to emphasize the peculiar characteristics at strong coupling, where it will be seen that one branch becomes a noninteracting Luttinger liquid exactly like those of the simple cases, and the other branch becomes neutral and disappears entirely[21]. Kane *et. al.* showed that disorder gives a reconstruction which is quite similar[22]. These results may explain the observation by Ashoori *et al.*[23] of only one edge magnetoplasmon at $\nu = 2/3$.

Consider a composite edge made up of two chiral Luttinger liquids characterized by ν_1 and ν_2. Their density operators satisfy[12] the relations Eq. (1). If the two branches are independent, then their (normal ordered) Hamiltonian is $H_0 = \sum_q \sum_{i=1}^2 \gamma_i(q){:}\rho_{iq}\rho_{i,-q}{:}$. These two branches can both be electron-like (*e.g.*, at $\nu = 2$, where $\nu_1 = \nu_2 = 1$); or instead one can be electron- and the other hole-like (*e.g.*, at $\nu = 2/3$, with $\nu_1 = 1$ and $\nu_2 = -1/3$). With noninteracting electrons (at $\nu = 2$, say), the branches are certainly independent, and the dispersion $\gamma_i(q)$ depends only on the confinement potential. At $\nu = 2/3$, numerical work shows that when only short-range interactions are included, excitations on the two edges appear independent[13], and hence can also be described by H_0 with some γ_i.

When the edge can be treated as effectively one-dimensional, the long-range piece of the Coulomb potential adds to the Hamiltonian a term[12,21] $H_1 = \sum_{q>0} V(q)\rho_q\rho_{-q}$, where $V(q)$ is the Fourier transform of the interaction potential. For the Coulomb interaction, $V(q)$ diverges logarithmically for small q as $-\frac{2e^2}{L}\log q$. In the case of a composite edge, the different excitation branches correspond to changes in electron density along strips which are close but not quite together. Then the total Hamiltonian becomes

$$H = H_0 + H_1 = \sum_{q>0} \sum_{ij=1}^{2} W_{ij}(q) \ : \rho_{iq}\rho_{j,-q} :, \tag{8}$$

where $W_{ij}(q) = V_{ij} + \gamma_{ij}\delta_{ij}$, and the V_{ij} is a matrix element giving the interaction between states in branches i, j.

It is here that the bosonization technique really comes into its own. Because the interaction piece is quadratic in boson operators, H can easily be diagonalized. Suppose (as is appropriate for $\nu = 2/3$) that $\nu_1 > 0$ and $\nu_2 < 0$ (and $|\nu_2| < \nu_1$). Use

Eq. (2) to define boson operators $a, a^\dagger$ related to ρ_{1q}, and $b, b^\dagger$ related to ρ_{2q}. The Hamiltonian is diagonalized by the Bogoliubov transformation[12]

$$a_q = \cosh\theta_q \alpha_q - \sinh\theta_q \beta_q^\dagger, \qquad b_q^\dagger = -\sinh\theta_q \alpha_q + \cosh\theta_q \beta_q^\dagger,$$
$$\tanh 2\theta_q = \frac{2\sqrt{\nu_1|\nu_2|}W_{12}}{(\nu_1-\nu_2)W_{11}+W_{22}}, \tag{9}$$

giving $H = (L/4\pi)\sum_{q>0} q(W_q^+ \alpha_q^\dagger \alpha_q + W_q^- \beta_q^\dagger \beta_q)$, where $W_q^\pm = [(\nu_1 W_{11} - \nu_2 W_{22})^2 + 4\nu_1\nu_2 W_{12}^2]^{1/2} \pm (\nu_1 W_{11} + \nu_2 W_{22})$.

Now let us show that the decoupled modes $\alpha^\dagger, \beta^\dagger$ are chiral Luttinger liquids with associated density operators $\rho_{\alpha q}, \rho_{\beta q}$. A natural way to do this is to construct edge waves using coherent states[15,12]. From the commutation relations Eq. (1), one can show that[15] the coherent state $|\psi\rangle$ has density $\bar\rho(x)$, with

$$|\psi\rangle = e^{i\eta\sqrt{Lq/2\pi}(a_q + a_q^\dagger)}|0\rangle, \qquad \bar\rho(x) = -\eta\,\mathrm{sgn}(\nu)|\nu|^{1/2}\pi^{-1} q \sin(qx). \tag{10}$$

Apply this to the α-mode coherent state $|\psi_\alpha\rangle = e^{i\eta\sqrt{n}(\alpha_q+\alpha_q^\dagger)}|0\rangle$, where $q = 2\pi n/L$. Using Eq. (10), the total charge density of this state is

$$\bar\rho_\alpha(x) \equiv \sum_{i=1}^{2} \rho_{i\alpha}(x) = -\eta\pi^{-1}(\nu_1^{1/2}\cosh\theta - |\nu_2|^{1/2}\sinh\theta) q \sin(qx). \tag{11}$$

By comparison with Eq. (10) for positive ν, one can, in the long wavelength limit, define $\nu_\alpha > 0$ with $\nu_\alpha^{1/2} = \nu_1^{1/2}\cosh\theta - |\nu_2|^{1/2}\sinh\theta$, and then construct momentum component density operators (with $q > 0$) $\rho_{\alpha,q} = \sqrt{\frac{Lq\nu_\alpha}{2\pi}}\alpha_q^\dagger$ and $\rho_{\alpha,-q}$. (Here $\theta = \theta_{q=0}$.) By construction these obey commutation relations of the form Eq. (1). More significant is that the density operator $\rho_\alpha(x) = \frac{1}{L}\sum_{q\neq 0} e^{-iqx}\rho_{\alpha q}$ indeed measures the total density $\rho_1 + \rho_2$ on the edge created by the action of $\alpha^\dagger$. (These density operators differ by a constant factor from those found by Wen[12].) The same procedure works for mode $\beta^\dagger$. When $W_{12} < W_{11}, W_{22}$, one obtains $\nu_\beta < 0$, with $|\nu_\beta|^{1/2} = |\nu_2|^{1/2}\cosh\theta - \nu_1^{1/2}\sinh\theta$, and corresponding density operators $\rho_{\beta,q} = \sqrt{\frac{Lq|\nu_\beta|}{2\pi}}\beta_q$ and $\rho_{\beta,-q}$. Note that the ρ_α and ρ_β operators commute—these are independent modes. Thus the Hamiltonian with two coupled chiral Luttinger liquid branches a, b is converted by diagonalization into a Hamiltonian with two uncoupled chiral Luttinger liquids α, β.

Now consider the limit in which the original branches are strongly coupled. In the long-wavelength limit, if the edges are very close together, $W_{ij}(q) \approx V(q)$ independent of branch indices. In this limit the boson operators α, β correspond to to Luttinger liquids characterized by $\nu_\alpha = \nu_1 + \nu_2$ and $\nu_\beta = 0$, with density operators

$$\rho_{\alpha q} = (Lq/\pi)^{1/2}(\nu_1^{1/2} a_q^\dagger + |\nu_2|^{1/2} b_q), \qquad \rho_{\beta q} = 0, \tag{12}$$

and $H = \sum_{q>0} V(q)\rho_{\alpha,q}\rho_{\alpha,-q}$. In this limit the β modes are neutral. That is, in the β mode the charge modulation on the two edges is out of phase, so that at each position x on the edge $\langle \rho_1(x) + \rho_2(x)\rangle = 0$.

Thus in the strong coupling limit, one mode forms a single chiral Luttinger liquid characterized by a total $\nu = \nu_1 + \nu_2$, with a dispersion governed by $V(q)$; and the second is neutral and has zero energy. This provides an economical explanation for the result of the experiment of Ashoori, *et al.*[23], who directly studied edge magnetoplasmons (triggered by a voltage pulse) in an etched sample at $\nu = 2/3$. The description of the edge as composite suggested there should be two branches of excitations and two corresponding signals representing counter-rotating edge magnetoplasmons. Instead, Ashoori *et al.* found only a single signal. This is easy to understand in light of the above—the voltage pulse does not couple to the neutral branch. Kane *et al.* give a similar explanation, with the neutral branch arising from disorder[22]. When the Coulomb interactions are not dominant, but still strong, the second (β) mode will be nearly neutral and provide a much weaker signal. Results very similar to the above occur when both edge branches are electron-like[21].

5. Tunneling

Finally let us touch on the tunneling properties of edge states. This is currently the most experimentally significant consequence of their Luttinger liquid nature. Wen originally calculated tunneling properties between quantum Hall edge states with the Luttinger liquid effective field theories, finding a power law current-voltage relationship as well as a power law dependence of the tunneling on temperature[12]. Similar results were obtained by Kane and Fisher[24], who also predicted a universal scaling of the transmission resonances (see also Ref. 25). Several tunneling geometries are possible. Perhaps the simplest has two quantum Hall regions able to tunnel through an intervening barrier. For Luttinger liquid edges characterized by ν, the tunneling conductance was predicted[12,24] to be go as $T^{2/\nu-2}$, *i.e.*, with an exponent characterizing the underlying incompressible bulk. For $\nu = 1/3$ the tunneling goes as T^4, quite unlike the temperature independence of $\nu = 1$.

Very recently Milliken, Umbach and Webb have performed experiments which seem to confirm the Luttinger liquid predictions in every way[26]. They use a Hall bar pinched off with a point contact. At high enough contact voltages this separates the bar into two Hall fluids weakly coupled through the contact region. At $\nu = 1/3$ they obtain a T^4 transmission conductance. By changing the magnetic field the same sample gives a temperature independent tunneling at $\nu = 1$. Moreover, the transmission resonances also satisfies the scaling predictions of halfwidth ($T^{2/3}$ at $\nu = 1/3$), as well as shape.

The Luttinger liquid properties of edges should be manifest in many other ways. For example, the universal scaling for tunneling resonances should be reflected in a scaling of energy-level splitting when electrons can tunnel between two edges in a quantum Hall system[27]. This is related to ideas of electron versus quasiparticle tunneling[28]. Numerical support of this conjecture would provide a test of fundamental Luttinger liquid ideas in a way independent of the effective field theories.

6. Acknowledgements

The understanding I have gained of edges and transport in quantum Hall systems is a result of fruitful collaborations with Allan MacDonald and Olle Heinonen which I am happy to be able to acknowledge here. This work was supported by the National Science Foundation under grant DMR-9301433.

7. References

1. R.E. Prange and S.M. Girvin (eds.), *The Quantum Hall Effect*, (Springer Verlag, New York, 1986).
2. B.I. Halperin, *Phys. Rev. B* **25**, 2185 (1982).
3. R. Landauer, *IBM J. Res. Develop.* **1**, 223 (1957); *Philos. Mag.* **21**, 863 (1970); *Z. Phys. B* **21**, 247 (1975); *Z. Phys.* **B 68**, 217 (1987).
4. M. Büttiker, *Phys. Rev. Lett.* **57**, 1761 (1986); *Phys. Rev. B* **38**, 9375 (1988); *IBM J. Res. Dev.* **32**, 317 (1988).
5. S.A. Trugman and S.A. Kivelson, *Phys. Rev. B* **31**, 5280, (1985).
6. O. Heinonen and M.D. Johnson, *Phys. Rev. Lett.* **71**, 1447 (1993).
7. A.H. MacDonald, *Phys. Rev. Lett.* **64**, 220 (1990).
8. F.D.M. Haldane, *Bull. Am. Phys. Soc.*, **35**, 254, (1990).
9. Michael Stone, *Phys. Rev. B*, **42**, 8399 (1990); *Ann. Phys.* (NY) **207**, 38 (1991);
10. Dung-Hai Lee and Xiao-Gang Wen, *Phys. Rev. Lett.*, **66**, 1765 (1991).
11. A.H. MacDonald and D.B. Murray, *Phys. Rev. B* **32**, 2707 (1985).
12. X.G. Wen, *Phys. Rev. Lett.* **64**, 2206 (1990); *Phys. Rev B* **41**, 12838 (1990); *Phys. Rev. Lett.* **66**, 802 (1991); *Int. J. Mod. Phys.* **6**, 1711 (1992).
13. M.D. Johnson and A.H. MacDonald, *Phys. Rev. Lett.* **67**, 2060 (1991).
14. Y. Meir, *Phys. Rev. Lett.* **72**, 2624 (1994).
15. M. Stone, H.W. Wyld and R.L. Schult, *Phys. Rev. B* **45**, 14156 (1992). (Ref. 1), Chapter 8.
16. J.M. Luttinger, *J. Math. Phys.* **15**, 609 (1963).
17. D.C. Mattis and E.H. Lieb, *J. Math. Phys.* **6**, 304 (1965).
18. F.D.M. Haldane, *J. Phys. C* **14**, 2585 (1981).
19. F.D.M. Haldane, in *The Quantum Hall Effect*
20. A.H. MacDonald and M.D. Johnson, *Phys. Rev. Lett.* **70**, 3107 (1993).
21. M.D. Johnson and A.H. MacDonald, preprint.
22. C.L. Kane, M.P.A. Fisher and J. Polchinski, preprint.
23. R.C. Ashoori, *et al.*, *Phys. Rev. B* **45**, 3894 (1992).
24. C.L. Kane and M.P.A. Fisher, *Phys. Rev. B* **46**, 15233 (1992).
25. K. Moon, H. Yi, C.L. Kane, S.M. Girvin and M.P.A. Fisher, *Phys. Rev. Lett.* **71**, 4381 (1993).
26. F.P. Milliken, C.P. Umbach and R.A. Webb, submitted to *Phys. Rev. Lett.*
27. M.D. Johnson and A.H. MacDonald, unpublished.
28. D.J. Thouless, *Phys. Rev. B* **40**, 12034 (1989).

STUDY OF A COULOMB ISLAND IN STRONG MAGNETIC FIELDS

O. Klein, C. de C. Chamon, D. Tang, D.M. Abusch-Magder, U. Meirav,* X.-G. Wen, and M.A. Kastner
Physics Department, Massachusetts Institute of Technology,
77 Massachusetts Ave., Cambridge MA 02139 U.S.A.

and

S.J. Wind
IBM T.J. Watson Research Center, Yorktown Heights, New York 10598 U.S.A.

ABSTRACT

The conductance resulting from resonant tunneling through a droplet of N electrons is used to measure its chemical potential μ_N. Abrupt shifts of μ_N occur when the state of the droplet changes. These are used to study its phase diagram in strong magnetic fields; we find evidence for a new phase in the spin polarized regime.

The device studied is of the type described by Meirav *et al.*[1]. It consists of a two-dimensional electron gas (2DEG) in an inverted GaAs/$Al_xGa_{1-x}As$ heterostructure with electrostatic gates above and below it. The bottom gate is a highly conducting substrate of n^+ doped GaAs. A positive bias, V_g, applied to the bottom gate varies the density of the 2DEG. On the top surface, two metallic (TiAu) gates are lithographically patterned with a double constriction. Applying a negative bias to these top gates depletes the 2DEG 100nm underneath them, confining the electrons to an island of $500 \times 500\text{nm}^2$ between the constrictions. Current flows through the resulting electron droplet via the tunnel barriers caused by the constrictions.

The conductance G of the device as a function of V_g at $B = 0$ is shown in the lower inset of Fig.1. It consists of quasi-periodic sharp peaks ($\Delta V_g = 1.2\text{mV}$), crudely described by the coulomb blockade mechanism[2]. In this model, when the bottom gate voltage is set between peaks, transport is suppressed by the charging energy $U \sim 0.66\text{meV}$ necessary to add an electron to the droplet. Each period thus corresponds to the addition of one electron to the droplet. At resonance, the electrochemical potential of the droplet, $\mu_N - e\alpha V_g$, is aligned with the Fermi energy of the leads and current flows. The value of V_g at which the peak occurs provides a measure of μ_N.

We begin by considering the effect of magnetic field on a single conductance peak. The value of the gate voltage at which the Nth conductance peak occurs is plotted as a function of magnetic field between 1 and 5T in Fig.1. The change in behavior near 1.6T results[3] from the depopulation of all but the lowest orbital Landau level (LL). The step-like behavior of the peak position above 1.6T can be thought[3] of as resulting from the transfer of electrons between the two spin-split

*Present address: Department of Physics, Weizmann Institute of Science, Rehovot 76100 Israel

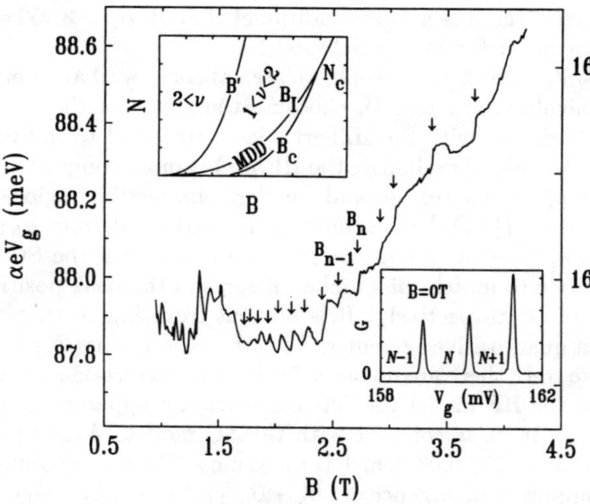

Fig. 1. Upper Inset: B-N phase diagram of the droplet. The boundaries corresponding to a change of the total spin of the droplet in the $2 > \nu > 1$ regime are omitted; the MDD domain of stability is limited on one side by $B_I(N)$, the boundary of the spin polarized phase, and on the other side by $B_c(N)$, where there is a reconstruction of the charge density. Above N_c the MDD phase is terminated. Lower Inset: Conductance through the island as a function of V_g at $B = 0$T. Main: Position of the Nth conductance peak as a function of B at $T = 100$mK. We have used a constant factor $\alpha = 0.55$ to convert V_g scale to energy [1,3]. The arrows indicate the minima of the conductance peak height. B_n is the field for the nth minimum above 1.6T, with $n = \{1, ..., 14\}$.

states of the lowest orbital LL. The number of steps above 1.6T is proportional to N, the number of electrons in the droplet, but the proportionality constant depends on the shape of the charge distribution.

We characterize the data in Fig.1 by examining the separation in B of the upward steps. We plot in Fig.2a the quantity $(B_n - B_{n-1})^{-1}$ as a function of B_n. Because each step corresponds to the flip of a spin, one may think of $(B_n - B_{n-1})^{-1}$ as being roughly proportional to the spin susceptibility. A fit to the form $y(B) = y_o[(B - B')/B']^\epsilon$ gives $B' = 1.7 \pm 0.02$T and $\epsilon = -0.41 \pm 0.06$ for our data; the solid curve in Fig.2a shows the fit.

Plotted in Fig.2c is $(B_n - B_{n-1})^{-1}$ obtained using μ_N of the semi-classical (SC) model of McEuen *et al.*[3] for a parabolic potential with cylindrical symmetry. We adjust N so that the calculated $\mu_N(B)$ has the same number of steps as observed experimentally. With N fixed, we adjust $\hbar\omega_0$, the strength of the confining potential, to match the value of $B' = 1.7$T, at which the transfer of electrons between the two spin states of the lowest LL begins in our experiment. Using this procedure, we find $N = 42$ and $\hbar\omega_0 = 1.8$meV. As seen in Fig.2c the SC model predicts values of $(B_n - B_{n-1})^{-1}$ which are roughly the same size as the measured ones. However, it does not predict the upward curvature of $(B_n - B_{n-1})^{-1}$ near B'. In the SC calculation, the last spin flip occurs at $B_I = 3.23\text{T} = 1.9B'$. Although we find a step at $2B'$ in our measurement, there is also an additional step at $B_c = 3.75\text{T} = 2.2B'$ (filled circle in Fig.1) not predicted by the SC model.

To compare our results with a more sophisticated theory, we have performed a Hartree-Fock (HF) calculation of $\mu_N(B)$, choosing the states of the symmetric gauge as the complete basis set, with the Hilbert space truncated to the two spin states of the lowest LL. Because of exchange, the HF $\rho(r)$ is more compact than the SC $\rho(r)$ with larger incompressible regions and smaller compressible regions[4].

In Fig.2b, we plot $(B_n - B_{n-1})^{-1}$ as a function B_n extracted from $\mu_N(B)$ for the HF model with $N = 27$ electrons and $\hbar\omega_0 = 2.1\text{meV}$. As with the SC model, these parameters are chosen to match the number of steps in the peak position and the experimental value of B', respectively. It is obvious from Fig.2b that the HF calculation is in excellent quantitative agreement with the experiment (Fig.2a). This is particularly impressive since there are no other fitting parameters once N and $\hbar\omega_0$ are fixed. In particular, the HF model predicts correctly the apparent divergence of $(B_n - B_{n-1})^{-1}$ near B', in clear contrast with the SC model. A fit to the HF results with $y(B)$ gives $\epsilon = -0.43 \pm 0.03$, which is the same as the experimental value within the errors. The apparent divergence of $(B_n - B_{n-1})^{-1}$ in Fig.2a suggests that because of exchange the two spin states of the lowest LL are equally occupied at B' in our droplet. This is consistent with another experimental observation: a new step in $\mu_N(B)$ is added between B' and $2B'$ for every two electrons added, implying that the two spin states of the lowest LL of our droplet are equally populated with increasing N.

The HF calculation predicts that the last spin flip occurs at $B_I = 3.15\text{T} = 1.85B'$. Thus, above B_I the droplet is spin polarized. MacDonald *et al.*[5], and Chamon *et al.*[4] showed that there exists a region in the B-N phase diagram (sketched in the upper inset of Fig.1) in which, for $N < N_c \sim 100$, the ground state (GS) of the spin polarized droplet is the maximum density droplet (MDD) with approximately constant $\rho(r)$. With increasing magnetic field the radius of the MDD decreases, the electrons get closer together, and the interaction energy eventually favors a larger area droplet. HF[4] predicts that, at B_c, the edge undergoes a reconstruction, causing[4] an abrupt upward shift of μ_N at B_c of roughly the same height as the step at B_I.

The excellent quantitative agreement between HF and the experiment for $B \leq 2B'$, strongly suggests that the MDD is formed in our experiment above $2B'$. The HF calculation predicts that the transition to the reconstructed droplet occurs at $B_c = 4.21\text{T}$ for our droplet, a value larger than the one observed experimentally. In this regard, it is important to bear in mind that although the HF energy of the MDD is exact because the MDD is an exact eigenstate of the many-body Hamiltonian[5], the HF energy of the reconstructed droplet is only variational. Therefore, the calculated value of B_c is an upper bound on the true transition field. Indeed, an exact calculation[6,7] for small N shows that the HF model overestimates B_c. The failure of HF to predict the size of the magnetic field window in which the MDD is the GS (Fig.2b) and the dependence[8] of B_c on N may indicate that correlations are playing an important role in this transition. The downward step at about 3.5T

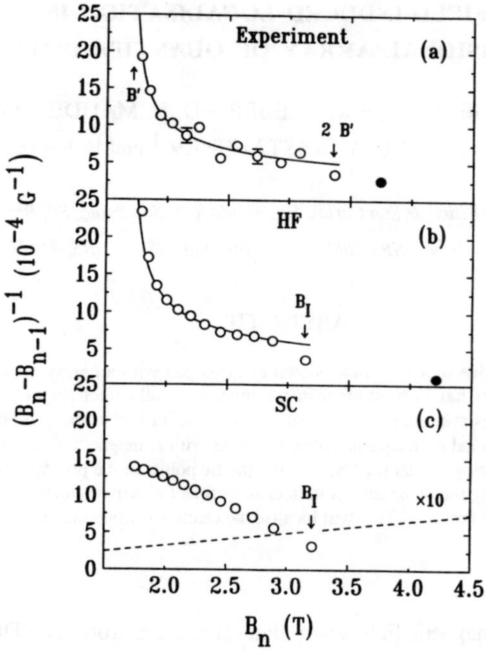

Fig. 2. (a) $(B_n - B_{n-1})^{-1}$ vs. the B_n obtained from Fig.1. The error bars are the spread of the data when the analysis is repeated for other conductance peaks on the same device. (b) and (c) results obtained with the HF ($N = 27$ and $\hbar\omega_0 = 2.1$meV) and SC ($N = 42$ and $\hbar\omega_0 = 1.8$meV) calculations. The solid line is a fit with $y(B)$ where $B' = 1.7 \pm 0.02$T and $\epsilon = -0.41 \pm 0.06$ for the experiment and $\epsilon = -0.43 \pm 0.03$ for the HF. B_I indicates the field onset of the spin polarized regime in both models. The solid circle indicates B_c. The dashed line in (c) is the constant interaction model [3] (the scale is expanded by a factor 10).

(Fig.1) is also reminiscent of features predicted to result from correlations[7].

This work was supported by NSF Grant No. ECS 9203427 and by the U.S. Joint Services Electronics Program under Contract No. DAALL03-93-C-0001.

1. References

1. U. Meirav *et al.*, Phys. Rev. Lett. **65**, 771 (1990).
2. C.W.J. Beenakker, Phys. Rev. B **44**, 1646 (1991).
3. P.L. McEuen *et al.*, Physica B **189**, 70 (1993).
4. C. de C. Chamon *et al.*, Phys. Rev. B **49**, 8227 (1994).
5. A.H. MacDonald *et al.*, Aust. J. Phys. **46**, 345 (1993).
6. D. Pfannkuche *et al.*, Physica B **189**, 6 (1993).
7. S.-R. E. Yang *et al.*, Phys. Rev. Lett. **71**, 3194 (1993).
8. O. Klein *et al.* (unpublished).

MAGNETIC FIELD-INDUCED LOCALISATION IN A TWO-DIMENSIONAL ARRAY OF QUANTUM DOTS

G. M. GUSEV,[1] ULF GENNSER,[2] X. KLEBER,[2] D. K. MAUDE,[2] J.-C. PORTAL,[2]
D. I. LUBYSHEV,[1] YU. V. NASTAUSHEV,[1] and P. BASMAJI[1]

[1] *Instituto de Física de São Carlos, Universidade de São Paulo, SP, Brazil.*

[2] *High Magnetic Field Laboratory, CNRS, 38042 Grenoble, and INSA- CNRS, 31077 Toulouse, France.*

ABSTRACT

The magnetoresistance of a two-dimensional electron gas with an array of weakly connected quantum dots has been studied in the quantum Hall effect regime. The usual Shubnikov - de Haas minima present when the Fermi level lies in the gap between two Landau levels are found to disappear above a certain critical magnetic field. We suggest that enhanced scattering occurs for levels at the saddle points in the potential between the oval antidots defining the quantum dot lattice, as well as for skipping orbits, as they pass the sharp bend of the antidots. This will localise the electrons inside the quantum dots.

1. Introduction

In the absence of a magnetic field a two-dimensional electron gas (2DEG) is localised at sufficiently low temperature. Since there are extended states in the quantum Hall effect regime, it is possible for a magnetic field to induce an insulator-metal transition. In clean, high-mobility samples, edge states, which carry the current along the boundary of the structure,[1-2] are responsible for the delocalisation. When the Fermi level lies in the gap between Landau levels, the edge states cannot scatter into the bulk, and thus the diagonal component of the resistivity, ρ_{xx}, falls to zero. The edge state scattering mechanism plays an important role for the explanation of the quantum Hall effect (QHE). Theoretical work has shown that the nature of the boundary of the 2DEG can have a significant effect on the conductance in a high magnetic field.[3,4] Scattering by 'soft' irregularities in the boundary is suppressed between the occupied edge states. Recently, it has been shown that it is technologically possible to introduce an artificial lattice of antidots, or scatterers into the 2DEG.[5] The antidots were seen to be responsible for changing the behaviour of the electrons in the small and intermediate magnetic field range. However, no influence on electron transport was found in the in the QHE regime. In this work we demonstrate, that an antidot lattice may also be responsible for a dramatic increase of the magnetoresistance in the high magnetic field quantum Hall limit. We believe that the field leads to trapping of the extended electron trajectories in the quantum dots formed between the antidots.

2. Experimental Details and Results

The samples used in this work consisted of Hall bars defined on standard AlGaAs/GaAs heterostructures with a 2DEG (initial mobility: 4 - 5 $\times$ 10^5 cm^2/Vs, electron concentration:

1.5 - 5 × 10^{11} cm^{-2}). Lattices of elliptical-shaped antidots were fabricated using electron-beam lithography and reactive plasma etching. Antidots occupy the connecting sides in a regular, quadratic array. The periodicity of the lattice is 1 μm, the length of antidots 0.8 μm, and their width 0.1 μm. A macroscopic lattice (with the size 500 μm × 200 μm) and a mesoscopic lattice (4 μm × 4 μm) were produced. Illumination of the sample slightly increases the carrier concentration while decreasing the depletion width around the antidots, thus enhancing the mobility by an order of magnitude, as the conduction is limited by the narrow channels between the tips of the antidots. Thus we are able to study the magnetoresistance as the electrically active regions around the ovals are increased (after further illumination), but with a reasonably constant carrier concentration. Only when the depletion width is fairly small do we see a significant increase in the concentration accompanying the change in ρ_{xx}. Measurements were performed in fields up to 15 T at temperatures T=50 mK-4.2 K, and up to 28 T at T = 300 mK - 1.2 K.

Without illumination, the antidot lattice acts at low temperatures as an insulator. However, after a large amount of illumination, the magnetoresistance of the native 2DEG is recovered, showing the normal Shubnikov-de Haas (SdH) oscillations. The spin-split states are resolved at filling factors $\nu \leq 6$, and some fractional filling factors are observed. Fig.1a shows the ρ_{xx} of the sample 2 as function of the magnetic field B, for different temperatures, and after some illumination. At low magnetic fields a negative magnetoresistance and commensurability oscillations due to the trapping of the classical orbits inside the quantum dots.[6] At 1.1K well-behaved SdH oscillations at higher magnetic fields clearly demonstrate, that the scattering process is not influenced by the antidot array. However, at lower temperatures the oscillations are broken up into irregular fluctuations, and furthermore, at B > 4.5 T, the resistance starts to increase linearly with the field (Fig. 1b). Further illumination pushes this linear increase towards higher magnetic fields, until, after extended illumination, the normal 2DEG behaviour of the magnetoresistance in recovered. In order to gain a better understanding of these effects, we have performed a systematic study after various illumination periods, bringing the sample from a state of high resistivity to a state of low resistivity.

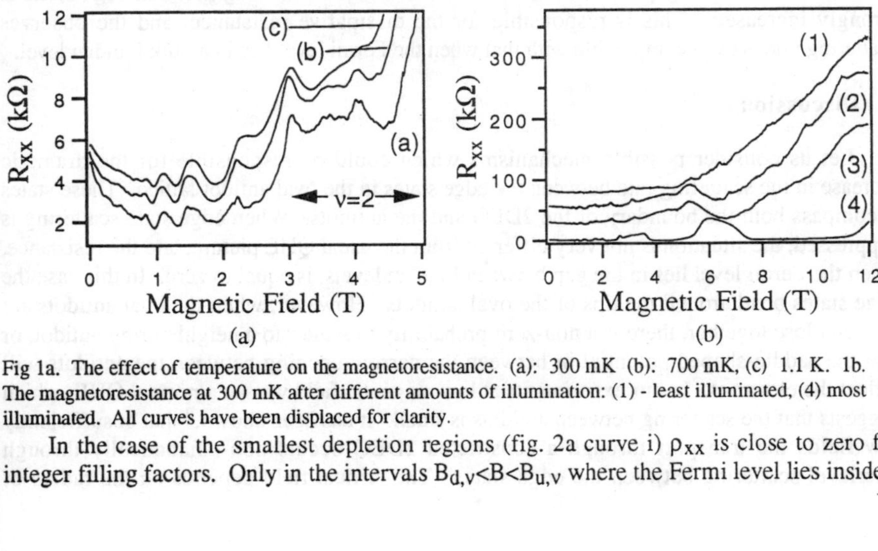

Fig 1a. The effect of temperature on the magnetoresistance. (a): 300 mK (b): 700 mK, (c) 1.1 K. 1b. The magnetoresistance at 300 mK after different amounts of illumination: (1) - least illuminated, (4) most illuminated. All curves have been displaced for clarity.

In the case of the smallest depletion regions (fig. 2a curve i) ρ_{xx} is close to zero for integer filling factors. Only in the intervals $B_{d,\nu}<B<B_{u,\nu}$ where the Fermi level lies inside a

bulk Landau level, is a maximum in ρ_{xx} seen. For example, for the second Landau level, $B_{d,2}$ = 4.2 T, and $B_{u,2}$ = 4.8 T. However, with increasing depletion region width, the minimum for $\nu = 2$ is partly transformed into a region with higher resistance. This is seen as a broadening of the resistance peak between $\nu = 1$ and $\nu = 2$; however, we note that the resistance is only increasing on the $\nu = 2$ side of the peak. A similar effect is observed at $\nu = 4$ (curve ii) and $\nu = 3$ (ii-iii). In figure 2b, with increasing depletion region width, the high resistance region at $\nu = 2$ continues to creep towards lower magnetic field, the high resistance region at $\nu = 3$ (iii) also moves towards lower fields, encompassing $\nu = 4$ and $\nu = 5$ (iv - v). $\nu = 3$ thus passes from low resistance (curve i) to high resistance (iii), back to low resistance (v) and finally again to high resistance (vi).

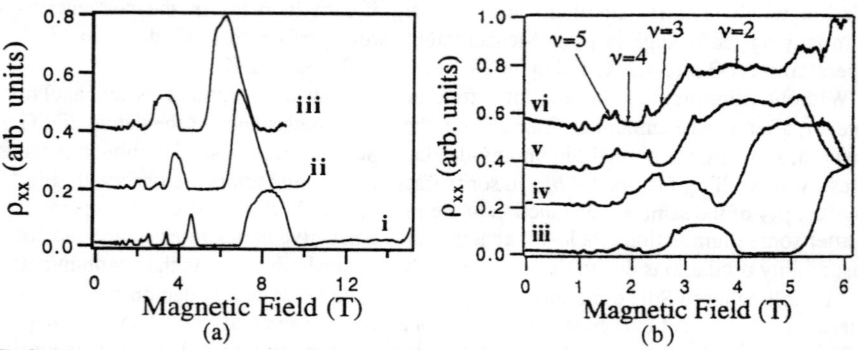

Fig 2. Magnetoresistance at T = 50 - 100 mK after successive illuminations Curve i: smallest depletion width (longest illumination); curve vi: largest depletion width. The curves have been displaced for clarity.

To summarise our experimental observations: After a large amount of illumination, the scattering between edge states is suppressed, and the situation is similar to the normal QHE regime. With increasing depletion regions we see that at certain magnetic fields, when the Fermi level lies in the gap between Landau levels, the backscattering between edge states is strongly increased. This is responsible for the dissipative resistance, and the observes scattering rate is even comparable with that when the Fermi level lies in a bulk Landau level.

3. Discussion

Let us consider possible mechanisms which could be responsible for the dramatic increase in the scattering rate between the edge states in the oval antidot lattice. These states encompass both the boundary of the 2DEG and the antidots. When edge state scattering is suppressed, the situation is not very different from the usual QHE picture, and the resistance, when the Fermi level lies in the gap between Landau levels, is equal to zero. In this case the edge states bend around the tips of the oval antidots. However, when the oval antidots are located close together, there is a non-zero probability to scatter to a neighbouring antidot, or that the saddle-shaped potential in between the narrow opening between the antidots will reflect the electrons. In our experiments at low magnetic field we see a normal QHE, which suggests that the scattering between antidots is small. It has been shown,[3] that at sufficiently low fields the transport through a constricted 2DEG occurs non-adiabatically through interlevel scattering between the edge states. The critical field depends on the radius of curvature of the constricting walls. At higher fields any scattering must therefore involve the bulk 2DEG states, or edge states around adjacent antidots, limiting the transport. As the saddle potentials between the antidots restrict the conductance at zero magnetic field, we will first consider the possibility that it plays a significant role also in the QHE regime. Estimates of the depletion region from the measured carrier concentrations and resistivity, show that the passages between the antidots can be almost closed. Scattering is expected to occur when a Landau level is pushed above the Fermi energy at the saddle potential.[3] In the case of broadened levels at the saddle point, this ejection (which does necessarily occur at the same magnetic field as a Landau level ejection from the bulk 2DEG) could explain the broad regions of scattering at different filling factors. However, at present we do not have a full picture of the energy spectra in the constrictions between the antidots.

It is also necessary to consider the scattering at the tips of the antidots. The reason for this scattering is that the electron in a skipping orbit will have a different energy as it passes by the tip, as the area, and thus the magnetic flux passing through its orbit, is changed. This can be viewed as a kink in the antidot potential around its tip, which can scatter the electrons. Since the transport is adiabatic, the electron has to be scattered away from the antidot, and it is then likely that they will be trapped in the 2DEG 'lakes' between the antidots. At higher magnetic fields, when the radius of curvature of the antidots is much larger than the cyclotron radius, this scattering mechanism will no longer be effective. It is clear, that further theoretical calculations for the specifics of the antidot lattice are necessary for a complete understanding of the scattering mechanisms. This, as well as the reproducible fluctuations which are seen on the background of the linearly increasing magnetoresistance will be reported elsewhere.[7]

4. Conclusion

We have found an increase of edge state scattering in a 2DEG with a two-dimensional array of weakly connected quantum dots at high magnetic fields. We propose that above a certain critical magnetic field, the transport in the edge states is adiabatic, so that the electrons are reflected by the saddle potential between the antidots, as well as scattered by their tips, and thus are trapped in the 2DEG islands situated between the antidots.

Acknowledgements: This work was supported by FAPESP, CNPq (Brazil) and CNRS (France). U.G. acknowledges support from SI and from NFR through the HCM program.

References

1. B. I. Halperin, *Phys. Rev.* **B 25** (1982) 2185.
2. M. Buttiker, *Phys. Rev.* **B 38** (1988) 9375.
3. L. I. Glazman and M. Jonson, *J. Phys. Cond. Matter* **1** (1989) 5547.
4. J. J. Palacios and C. Tejedor, *Phys. Rev.* **B 45** (1992) 9059.
5. K. Ensslin and P. Petroff, *Phys.Rev.* **B 41** (1990) 12307.
6. G. M. Gusev, Z. D. Kvon, Yu. V. Nastaushev, L. V. Litvin, A. K. Kalagin, and A. I. Toropov, *Superlattices and Microstructures* **13** (1993) 383.
7. G. M. Gusev, U. Gennser, D. K. Maude, X. Kleber, J.-C. Portal, D. I. Lubyshev, and P. Basmaji, *to be published*.

MAGNETOTRANSPORT OF ELECTRONS IN ARRAYS OF ANTIDOTS IMPOSED UPON $Si/Si_{0.7}Ge_{0.3}$ HETEROSTRUCTURES

D. TÖBBEN[1], M. HOLZMANN[1], S. KÜHN[2], H. LORENZ[2]
G. ABSTREITER[1], J.P. KOTTHAUS[2], and F. SCHÄFFLER[3]

[1] *Walter Schottky Institut, TU München, D-85748 Garching, Germany*
[2] *Sektion Physik, LMU München, D-80539 München, Germany*
[3] *Daimler-Benz AG, Forschungsinstitut Ulm, Wilhelm-Runge-Str. 11, D-89081 Ulm, Germany*

Electrons in arrays of antidots imposed upon $Si/Si_{0.7}Ge_{0.3}$ heterostructures are studied by magnetotransport. The lateral potential modulation is created by reactive ion etching of the semiconductor or via the field-effect by means of a modulated Schottky gate, with superlattice periods ranging between 1300nm and 500nm. The lithography is performed using laser holography. Features, which are characteristic of ballistic transport though antidot superlattices and which were so far only known from III-V semiconductor structures, are observed in the longitudinal as well as in the Hall resistivities in both types of samples. These effects are found to be weaker, but also to some extend tunable in the gated samples.

1. Introduction

Magnetotransport studies of lateral antidot superlattices [1-3] imposed upon high-mobility two-dimensional electron gases (2DEGs) have lead to the discovery of several anomalies in both the longitudinal resisitivity ρ_{xx} and the Hall resistance R_H. Theoretical studies[4] have shown that maxima in ρ_{xx} at low magnetic fields are associated with pinned orbits around groups of antidots. In smooth lateral potentials such trajectories can only exist around one or four antidots and the larger one is strongly non-circular. Additional, non-quantized Hall plateaus in this magnetic field regime stem from the same cyclotron-type orbits, while the quenching of the Hall effect around B=0 is due to chaotic channeling trajectories which tend to leave the channel in the direction opposite to the Lorentz force[5].

All of these effects were so far restricted to 2DEGs in III-V semiconductor heterostructures since the electron mobility μ has to be high enough for the elastic mean free path to be larger than the superlattice period. In this work, however, we will present magnetotransport investigations on arrays of antidots imposed upon $Si/Si_{0.7}Ge_{0.3}$ heterostructures by reactive ion etching and by electrostatic induction via the field-effect. It is shown that the sample quality is indeed sufficiently high to observe the above anomalies in ρ_{xx} and R_H.

2. Sample structure and preparation

The $Si/Si_{0.7}Ge_{0.3}$ heterostructures we used were grown by molecular beam epitaxy as described in ref. 6. Lateral antidot superlattices fabricated by reactive ion etching (RIE) were studied on samples containing a 2DEG about 82nm beneath the surface. Its mobility μ is found to be 175000cm^2/Vs at a carrier density $n_s=6.9\times10^{11}$ cm^{-2} at T=370mK. Experiments on field-effect induced antidots were performed on similar $Si/Si_{0.7}Ge_{0.3}$ heterostructures with less doping, which were also employed recently to study size effects in field-effect induced electron channels[7]. The 2DEGs in these samples are situated about 70nm beneath the surface and possess a low-temperature mobility around 170000cm^2/Vs at $n_s=4.5\times10^{11}$ cm^{-2}. The elastic mean free path is thus about 1.7μm and 1.4μm in both samples, respectively.

Directly etched antidot superlattices were generated by appropriate holographic lithography and subsequent RIE as described in ref. 8. Periods a=830nm, 620nm, and 500nm were realized. In all samples most of the doping layer was removed in the antidots. Hall bars were defined afterwards by wet chemical etching.

To fabricate a modulated gate, a 100nm Si_3N_4 film was deposited at 110°C, after the mesas were etched. It was then patterned by laser holography and selective, wet chemical etching before a Pd film was deposited as a gate. It is known from reference samples[7] that the electrons' mean free path is hardly effected underneath the nitride at zero bias while the 2DEG is already depleted via the Schottky effect wherever the Pd is in direct contact with the semiconductor. Samples were investigated with periods a=1300nm, 830nm, and 620nm.

3. Experimental results

3.1 Transport in etched antidot arrays

Fig. 1a shows the longitudinal resistances of all three antidot samples and an unstructured reference at T=370mK in the dark up to B=4T. The zero field resistance increases monotonically with decreasing superlattice period. All structured samples display two maxima at low magnetic fields, marked by two arrows for the sample with period a=620nm. These peaks shift slightly towards higher magnetic fields with decreasing period. The height of the maxima is about 7% and 20% of the zero-field value of ρ_{xx} for the two samples with the smallest periods. The maxima are less pronounced - only about 10% - for the sample with period a=830nm. However, this sample was etched slightly shallower than the other two. Therefore, the potential modulation is expected to be smaller in this specimen.

As B is increased, ρ_{xx} drops again and the onset of Shubnikov-de Haas (SdH) oscillations is observed. This regime is used to evaluate the sheet densities n_s which are found to be equal to that of the reference sample within a few percent. In all samples the positions r_c/a of the two low-field peaks are (1.51±0.10) and (0.47±0.03), respectively, with

$$r_c = \hbar\sqrt{4\pi n_s/g_s g_v}/eB \qquad (1)$$

being the cylcotron radius and $g_s = g_v = 2$ the spin and valley degeneracies of the system.

The third arrow in Fig. 1a marks a shoulder at $r_c/a \approx 0.2$ which is also observed in the sample with the smallest superlattice period. We have previously reported[8] that this shoulder vanishes with illumination and that the resistivity decreases radically while the carrier density remains almost constant. Furthermore, the smaller peak gains in intensity relative to the larger one. It is also interesting to note that the SdH oscillations and the lifting of the spin degeneracy at intermediate fields are less pronounced as compared to the reference sample. The shoulder does not appear at all in the most

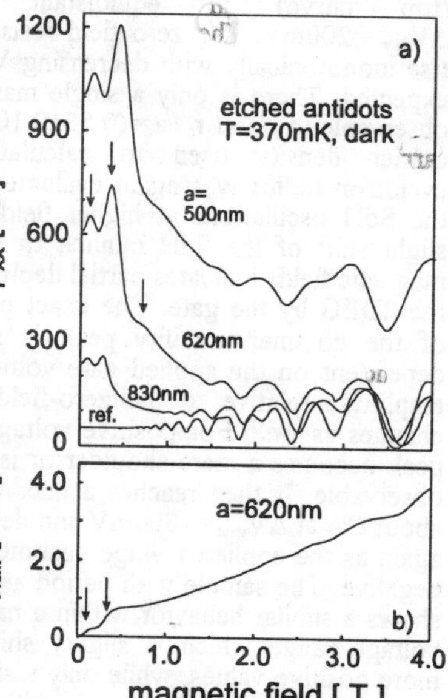

FIG. 1 a) Longitudinal resistivities of reference and etched antidot samples and b) Hall resistance of the sample with period a=620nm at T=370mK in the dark.

shallow etched sample with period a=830nm. Based on these facts we believe that charged defects, generated in the etching process, are involved. The rise of $\rho_{xx}(B=0)$ thus appears to be due to intentionally introduced scatterers, i.e. the antidots, and to defects created by the fabrication.

The Hall effect of the sample with period a=620nm is displayed in Fig.1b. While quantum Hall plateaus are clearly observable above B=2T, two additional plateaus appear at magnetic fields slightly higher than those of the two maxima in ρ_{xx}, indicated by the two arrows. Furthermore, R_H is quenched at small fields around B=0 in the sense that it is highly non-linear in this regime.

These features are well known from work on arrays of antidots in GaAs/AlGaAs structures. From the position of the two low-field peaks in ρ_{xx} we conclude that these are indeed commensurability maxima associated with pinned orbits around one or four antidots. The lateral potential is obviously rather smooth, since a commensurability maximum would otherwise not be allowed at $r_c/a \approx 3/2$ as reported by *Fleischmann et al.*[4]. Even more peaks might be observable in a steeper potential. Smoothness of the potential to some degree is also a requirement for Hall effect anomalies[5].

3.2. *Field-effect induced arrays of antidots*

Fig. 2a displays the longitudinal resistance of a sample with period a=830nm up to B=2T for gate voltages decreasing from +400mV (lowest curve) to -1400mV (top curve) in equidistant steps $\Delta V_{gate}=200$mV. The zero field resistivities rise monotonically with decreasing V_{gate} as expected. There is only a single maximum observable in ρ_{xx} at $r_c/a=(0.55\pm0.10)$. The carrier density used to calculate the cyclotron radius was again evaluated from the SdH oscillations at higher fields. The slight shift of the SdH minima to smaller magnetic fields indicates partial depletion of the 2DEG by the gate. The exact position of the commensurability peak is slightly dependent on the applied gate voltage. Its amplitude relative to the zero-field value changes as well: For positive voltages, the peak becomes a mere shoulder or is hardly observable. It then reaches a maximum of about 6% at $\Delta V_{gate}=-800$mV and decreases again as the applied voltage becomes more negative. The sample with period a=620nm shows a similar behavior within a narrower voltage range which is slighty shifted to more positive values, while only a shoulder can be observed in the sample with period a=1300nm for all gate voltages instead of a maximum.

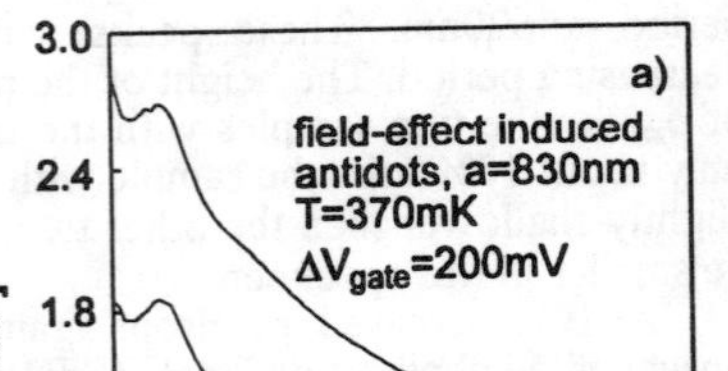

FIG. 2 a) Longitudinal resistivity and b) Hall resistance of a field-effect induced antidot superlattice with period a=830nm and V_{gate} ranging from +400mV to -1400mV at T=370mK. The Hall curves are offset vertically.

The Hall resistance of the same sample in the same range of gate voltages is shown in Fig.2b. The curves are offset vertically for clarity. The Hall trace for $\Delta V_{gate}=+400mV$ is almost perfectly linear in the low-field regime and then exhibits quantum Hall plateaus above B=1.5T. For smaller and negative V_{gate} the Hall effect becomes quenched around B=0 and a non-quantized plateau becomes more and more pronounced while the quantum Hall plateaus disappear. The arrow marks the position of the maximum in ρ_{xx} at $\Delta V_{gate}=-1400mV$.

Since only a single and less pronounced maximum is observed we conclude that the lateral potential introduced via the patterned gate is weaker and probably even smoother than in the samples prepared by RIE. It should also be remembered that the elastic mean free path in this sample is a little smaller than in the one used for RIE experiments. By applying a sufficiently high forward bias the system is turned into an almost truely two-dimensional one, the low-field peak vanishes totally, and there are no Hall effect anomalies. On the other hand, the potential modulation can also be enhanced by reverse biasing of the gate. We have found out in previous work that the mobility drops rapidly with decreasing carrier density[7]. Thus, the effect of a negative bias on both the zero field resistivity and the amplitude of the low-field maximum is two-fold: It decreases the elastic mean free path of the electrons and simultaneously broadens the lateral potential by electrostatic depletion. At very negative gate voltages, when the carrier density and thus the elastic mean free path are strongly reduced as compared to the zero bias situation, the system is so strongly disturbed that the peak amplitude decreases again. It is interesting to note that the Hall effect anomalies still become more pronounced in this regime.

4. Conclusion

In conclusion, we have demonstrated that the quality of $Si/Si_{0.7}Ge_{0.3}$ heterostructures is indeed high enough to observe typical features of ballistic transport through antidots superlattices, such as commensurability oscillations in ρ_{xx}, quenched Hall effect, and anomalous Hall plateaus. The potentials are rather smooth for both fabrication techniques and clearly less pronounced if introduced electrostatically. However, the effects of the antidot superlattices can be tuned to some extend in these samples.

Acknowledgements

We thank K. Ensslin for stimulating discussions and G. Kritzler for Si_3N_4 deposition. This work was supported financially in part by the *Siemens AG* (*SFE Mikrostrukturierte Bauelemente*) and by the *Bundesministerium für Forschung und Technologie* (Bonn, Germany) under grant NT 24137.

References

1. K. Ensslin and P.M. Petroff, *Phys. Rev. B* **41** (1990) 12307
2. D. Weiss, M.L. Roukes, A. Menschig, P. Grambow, K. von Klitzing, and G. Weimann, *Phys. Rev. Lett.* **66** (1991) 2790
3. A. Lorke, J.P. Kotthaus, and K. Ploog, *Phys. Rev. B* **44** (1991) 3447
4. R. Fleischmann, T. Geisel, and R. Ketzmerick, *Phys. Rev. Lett.* **68** (1992) 1367
5. R. Fleischmann, T. Geisel, and R. Ketzmerick, *Europhys. Lett.* **25** (1994) 219
6. F. Schäffler, D. Többen, H.-J. Herzog, G. Abstreiter, and B. Holländer, *Semicond. Sci. Technol.* 7 (1992) 260
7. M. Holzmann, D. Többen, G. Abstreiter, and F. Schäffler, *J. Appl. Phys.* (1994, in press).
8. D. Többen, M. Holzmann, S. Kühn, H. Lorenz, G. Abstreiter, J.P. Kotthaus, and F. Schäffler, submitted to *Phys. Rev. B*

MAGNETIZATION AND HEAT CAPACITY OF TWO ELECTRONS IN THREE DIMENSIONAL QUANTUM DOTS

J. H. Oh*, K. J. Chang*, G. Ihm†, S. J. Lee‡ , and T. W. Kim♮

**Department of Physics, Korea Advanced Institute of Science and Technology, Taejon 305-338, Korea*

†*Department of Physics, Chungnam National University, Taejon 305-764, Korea*

‡*Department of Physics, Korea Military Academy, Seoul 139-799, Korea*

♮*Department of Physics, Kwangwoon University, Seoul 139-050, Korea*

ABSTRACT

For three dimensional quantum dots with two electrons confined parabolically in high magnetic fields, we examine electronic heat capacities and magnetizations. The main feature is found that the combined effect of the electron-electron interactions and the hybrid-magnetoelectric quantization governs the whole energy spectrum. Thus, the ground state transition occurs in disk-like dots, while it is suppressed in football-like dots. We clearly see such behavior in heat capacity and magnetization.

I. Introduction

Quantum dots are usually modeled as a disk-like shape[1], so called, a quantum dot disk, where the lateral confinement of electrons on the x-y plane is much smaller than that along the z-direction. In contrast, three dimensional (3D) quantum dots which have been of recent interest have comparable confinement energies in all three directions.[2] Much of attentions for the quantum dot disk has been paid to investigate the electronic structure, especially, in the presence of high magnetic fields.[3–5] A key factor in these studies is the role of electron-electron (e-e) interactions which can be probed by measuring equilibrium thermodynamic quantities[4] such as heat capacity or magnetization.

In this work we consider various 3D parabolic quantum dots with two electrons and investigate their electronic magnetizations and heat capacities as a function of magnetic field.

II. Electron-electron interactions

Our model for a parabolic 3D quantum dot is characterized by two different frequencies, the confinement frequency ω_p on the x-y plane and ω_v along the z-direction. Ignoring the Zeeman spin-splittings, the energy eigenvalues in the presence of magnetic field, $\vec{B} = B\hat{z}$, are separated into two terms, the energies of the center of mass and relative motions. The former[5] is given by

$$E_{N,L} = \hbar\Omega(2N + |L| + 1) + \frac{\hbar\omega_c}{2}L + \hbar\omega_v(M + \frac{1}{2}) \tag{1}$$

where $\omega_c = eB/m^*$ is the cyclotron frequency and $N(= 0, 1, 2, \ldots)$, $L(= 0, \pm1, \pm2, \ldots)$, and $M(= 0, 1\ , 2, \ldots)$ denote the radial, azimuthal, and vertical quantum numbers, respectively, for a cylindrical geometry. The frequency $\Omega = \sqrt{\omega_p^2 + (\omega_c/2)^2}$ indicates the hybrid-magnetoelectric quantization on the x-y plane, i.e., the coupling between the electric and magnetic confinements. The energies of the relative motion are mostly determined by the e-e interactions and are specified by the quantum numbers n, l, and m.

Figs. 1-(a) and -(b) show the calculated energy dispersions as a function of magnetic field for $\hbar\omega_v = 6$ meV and $\hbar\omega_v$ =1.5 meV, respectively, with keeping $\hbar\omega_p$ = 3 meV. In this case, we choose the dielectric constant ϵ=12.4 and the effective mass $m^* = 0.0665$. The electronic structure in Fig. 1-(a) is found to be similar to that of the quantum dot disk[3] while the quantum dot in Fig. 1-(b) can be considered as a football-like dot. One of the major differences between the two figures, except for the shift of the zero point energy, is the ground state. As the magnetic field increases, the ground state of the disk-like dot in Fig. 1-(a) has a sequence of l values ($l = 0, -1, -2, \ldots$) with the other quantum numbers zero, as viewed easily by the level crossing, while in Fig. 1-(b) the l-value of the ground state is zero for the magnetic fields considered here. Previous studies[3-5] for the disk-like quantum dot demonstrated that the e-e interactions are responsible for the $l \to l-1$ transitions and the magnetic field at which this transition occurs decrease as the dot size increases because the relative strength of the Coulomb interaction increases correspondingly. This seems not to be true for the football-like dot. Although the dot size in Fig. 1-(b) is larger than that of Fig. 1-(a), this quantum dot shows no transitions. This behavior can be understood by considering the hybrid effect and the Coulomb interaction which determine the ground state transitions. Between the $l = l_0$ and $l = l_0 - 1$ states, i.e., the two lowest levels, the energy difference ΔE is separated into the hybrid and Coulomb contributions, $\Delta E = \Delta E_{hybrid} + \Delta E_{Coul}$. For the dot in Fig. 1-(a), as B increases, ΔE_{hybrid}, which is always positive, decreases more rapidly than ΔE_{Coul} which is negative.[6] Thus, ΔE can be zero at a certain magnetic field. For the football-like dot in Fig. 1-(b) which has larger size along the field direction than that in Fig. 1-(a), the change of ΔE_{hybrid} is almost negligible, while ΔE_{Coul} decreases very slowly as the magnetic field increases. Thus, the ground state transitions are removed.

III. Heat Capacity and Magnetization

The electronic heat capacities and magnetizations are calculated from the energy spectrum for the quantum dots in Fig. 1 and are plotted as a function of magnetic field in Figs. 2 and 3, respectively. At sufficiently low temperatures, the two lowest

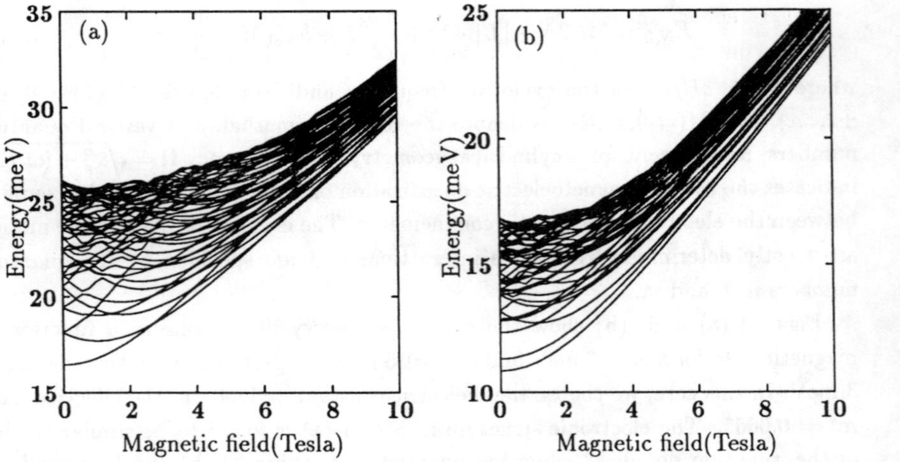

Fig. 1 Energy dispersions vs. magnetic field are plotted for (a) $\hbar\omega_v = 6$ meV and (b) $\hbar\omega_v = 1.5$ meV for 3D quantum dots with the fixed lateral confinement $\hbar\omega_p = 3$ meV.

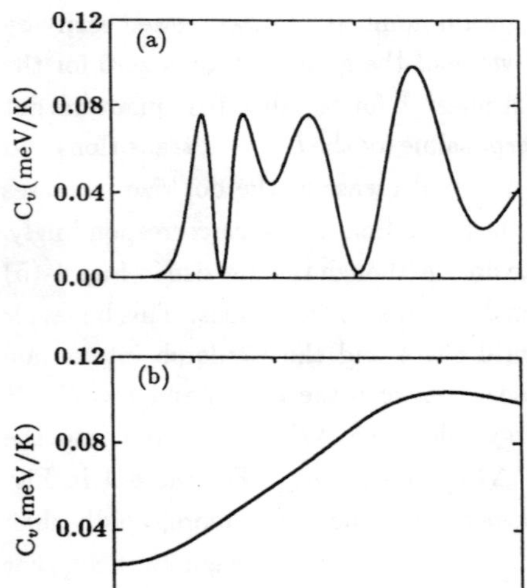

Fig. 2 Heat capacities vs. magnetic field at T=0.5 K for the quantum dots in Fig. 1.

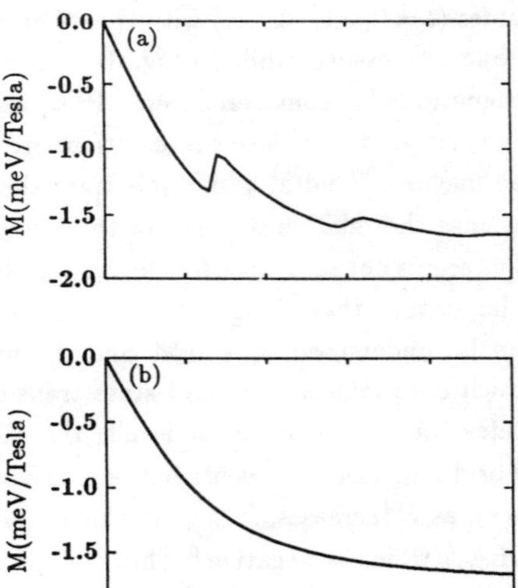

Fig. 3 Magnetizations vs. magnetic field at T=0.1 K for the quantum dots in Fig. 1.

energy levels are mainly contributed to these thermodynamic quantities and the chemical potential lies nearly in the middle of the two levels. For a quasi-0D system, it was shown[6] that the heat capacity becomes maximum when the energy separation (ΔE) between the two lowest levels is comparable to $4.80k_BT$. The heat capacity becomes exactly zero, if the contributions from other energy levels are absent, at the magnetic field where the ground state transition occurs, i.e., $\Delta E = 0$. Since ΔE is a function of B, C_v also varies with B. Thus, a remarkable difference in the heat capacities of the disk-like (Fig. 2-(a)) and football-like dots (Fig. 2-(b)) at low temperatures exists. In Fig. 2-(a), C_v shows strong oscillation with a doublet feature around the transition point, where $C_v \simeq 0$, while C_v in Fig. 2-(b) varies smoothly because of the absence of the transition points.

Now let us look at the magnetizations M, in Fig. 3. Since $M = -\partial E/\partial B$, the magnetization at very low temperatures is closely related to the slope of the ground state energy. Similarly to the heat capacity, a considerable difference is found between the two dots in Fig. 3; the one has peaks near the ground state transition points, while the other shows no peaks. At T=0, a sharp discontinuity in the magnetization is expected at the transition point since the slope of the ground state energy is discontinuous. However, at finite but very low temperatures, broadened peaks appear at higher B values from the transition point of the ground state, in which the amount of the shift is proportional to temperature as expected from the two-levels analysis.[6]

Acknowledgments

This work was supported in part by the SPRC at Jeonbuk National University, by the CMS at Korea Advanced Institute of Science and Technology, by the Ministry of Education under grant no. BSRI-93-234, and by Wharangdae Research Institute.

References

1. C. T. Liu, K. Nakamura, D. C. Tsui, K. Ismail, D. A. Antoniadis, and H. I. Smith, Appl. Phys. Lett. 55 (1989) 168.
2. B. Meurer, D. Heitmann, and K. Ploog, Phys. Rev. B 48 (1993) 11488.
3. U. Merkt, J. Huser, and M. Wagner, Phys. Rev. B 43 (1991) 7320.
4. P. A. Maksym and T. Chakraborty, Phys. Rev. Lett. 65 (1990) 108; *ibid*, Phys. Rev. B 45 (1992) 1947.
5. M. Wagner, U. Merkt, and A. V. Chaplik, Phys. Rev. B 45 (1992) 1951.
6. J. H. Oh, K. J. Chang, G. Ihm, and S. J. Lee, to be published.

THE DIAMAGNETIC SHIFT OF EXCITONS IN STRAIN CONFINED QUANTUM WIRES AND DOTS

T. H. GFROERER, M. D. STURGE
Physics Department, Dartmouth College, Hanover, NH 03755

K. KASH*
Physics Department, Case Western University, Cleveland, OH 44106

J. A. YATER,# A. S. PLAUT,@ P. S. D. LIN, L. T. FLOREZ, J. P. HARBISON
Bellcore, Red Bank, NJ 07701-7040

S. R. DAS, and L. LEBRUN
National Research Council of Canada, Ottawa, ON K1A OR6, Canada

ABSTRACT

It has been shown that excitons in quantum wells can be laterally confined by the strain potential beneath a microscopic stressor. In this paper, we report measurements made on a GaAs/AlGaAs quantum well system using 200 nm and 300 nm wide tungsten and carbon stressors. We have obtained photoluminescence spectra of these structures in magnetic fields up to 9 Telsa. The diamagnetic shifts of the quantum wire peaks are similar to those of the unstrained quantum well, indicating that the lateral extension of the exciton wavefunction is not significantly altered by the one-dimensional strain confinement. However, the tungsten quantum dot luminescence has a much smaller diamagnetic shift, suggesting that excitons in these structures are much more tightly bound.

1. Introduction

The diamagnetic shift of ground state quantum well excitons has been measured for fields both parallel[1] and perpendicular[2,3] to the growth direction of the crystal. Although the results vary somewhat quantitatively, the qualitative trend is the same: thinner quantum wells have smaller diamagnetic shifts. The reduced shift with decreasing well thickness is attributed to the diminishing size of the quantum well exciton. Recently, an enhancement of this effect has been observed for excitons that are laterally confined within the quantum well. Etched,[4] serpentine,[5] and V-groove[6,7] quantum wires all exhibit smaller diamagnetic shifts than their host quantum wells. The reduced shift in these structures is evidence that the quantum wire excitons are even smaller and more tightly bound.

For this study, we measured the diamagnetic shift of a number of strain-confined quantum wires and dots. The following procedure was used to fabricate the tungsten strain-confined structures. First, a thin layer of compressively stressed amorphous tungsten is deposited on the surface of a quantum well system consisting of a 15 nm GaAs layer between two 20 nm $Al_{.3}Ga_{.7}As$ layers. Second, portions of the amorphous tungsten are removed by reactive ion etching, leaving arrays of 200 nm and 300 nm tungsten wires and dots. Now the compressed tungsten has some space to relax, and as the microscopic wires and dots dilate, they stretch the GaAs lattice below. This lattice dilation lowers the bandgap beneath the center of the tungsten stressor, and laterally confines excitons to these regions of the quantum well.

---

* work done while at Bellcore.

\# Present address: NASA Lewis Research Center, Cleveland, OH 44135

@ Present address: University of Exeter, Stocker Road, Exeter EX4 40L

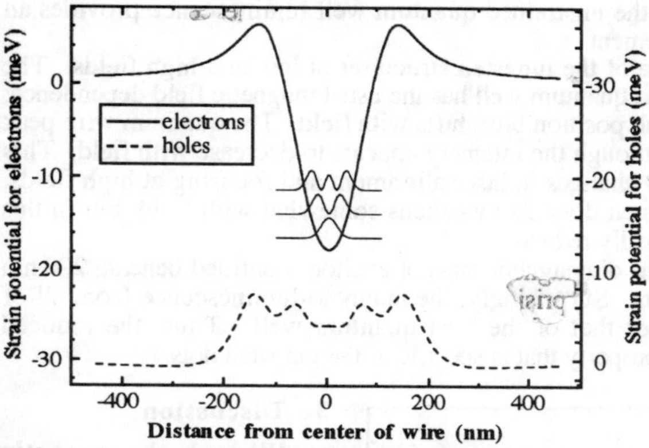

Figure 1: Strain potential as a function of lateral position beneath a 200 nm carbon stressor. The thin solid lines superimposed on the electron strain potential are the probability densities for the three lowest confined states.

The strain potential for these particular structures has not been calculated. However, a calculation has been performed for a 200 nm quantum wire where carbon, rather than tungsten, was used for the stressor material.[8] The variation of the conduction and valence bands as a function of position beneath the stressor is given in Figure 1. Since the measured stress in the planar tungsten layer is bigger than that of carbon, we expect the magnitude of the strain potential to be larger. However, if the shape of the confinement potential is comparable, we can draw two general conclusions from this calculation. First, the strain potential is nearly parabolic at the bottom of the well where the lowest confined states reside. Second, the confinement size for these states is much smaller than the actual size of the stressor.

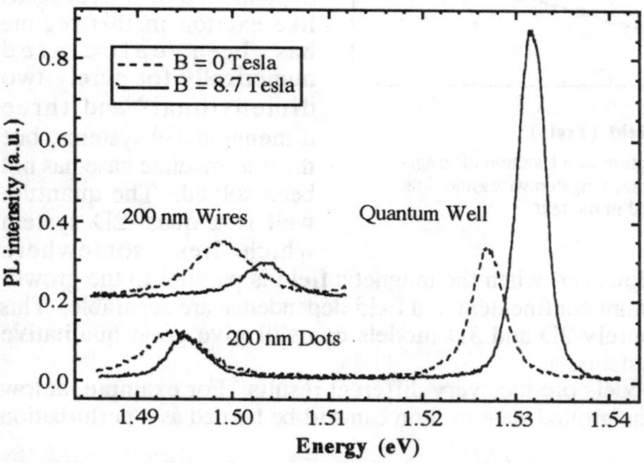

Figure 2: Photoluminescence spectra of the tungsten structures at low and high magnetic fields. The quantum wire peaks are shifted vertically for clarity.

2. Experimental

We have measured the photoluminescence spectra of the quantum wires and dots beneath 200 nm and 300 nm stressors as a function of applied magnetic field. The field was oriented parallel to the growth axis of the quantum well. The structures were excited in backscattering geometry by the 632.8 nm line of a HeNe laser with a power density of approximately 100 W/cm^2. The nominal sample temperature was 4.2K.

Strain confinement has certain advantages for this type of study. Since regions between the stressors remain relatively unstrained, the luminescence spectrum from an array of stressors includes emission from the unperturbed quantum well. Furthermore, unlike etched wires, the luminescence from strain confined quantum wires and dots is redshifted relative to the unpatterned quantum well emission so that the peak positions are easily distinguished. The diamagnetic shift of the two peaks can be obtained simultaneously, and the unstrained quantum well luminescence provides an excellent reference for the measurement.

Figure 2 shows typical spectra of the tungsten structures at low and high fields. The luminescence from the unperturbed quantum well has the usual magnetic field dependence: the intensity increases and the peak position blueshifts with field. The quantum wire peak has a similar diamagnetic shift, although the intensity appears to decrease with field. This intensity reduction may be due to changes in laser alignment and focusing at high fields. The luminescence from the tungsten dots also weakens somewhat with field, but in this case the diamagnetic shift is drastically reduced.

Recently, we have measured the diamagnetic shift of excitons confined beneath 200 nm and 300 nm carbon wires and dots. Surprisingly, the magneto-luminescence from all of these structures closely resembles that of the host quantum well. Thus, the reduced diamagnetic shift appears to be a property that is specific to the tungsten dots.

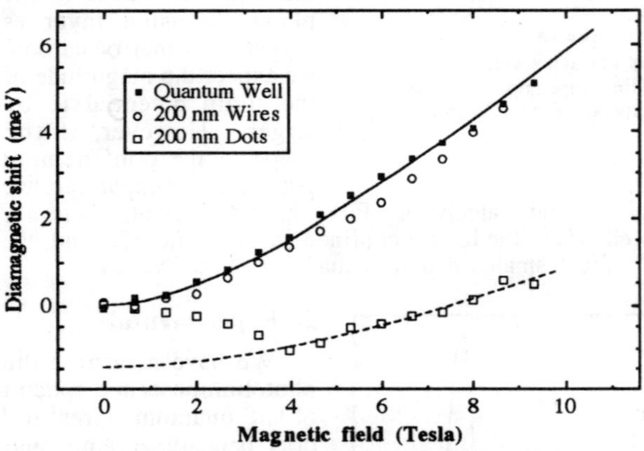

Figure 3: Photoluminescence peak position as a function of magnetic field for the quantum well and the 200 nm tungsten wires and dots. The lines are theoretical fits as described in the text.

3. Discussion

When high magnetic fields (several Tesla) are applied to excitons in GaAs, the diamagnetic shift is comparable to the Coulomb binding energy. Thus, in general, neither of these effects can be treated as a perturbation. The exact field dependence of a hydrogen-like exciton in this regime has been calculated numerically for purely two dimensional[9] and three dimensional[10] systems, but the intermediate case has not been solved. The quantum well is a quasi-2D system which lies somewhere between these two extremes. However, when the magnetic field is parallel to the growth direction of the crystal, the quantum confinement and field dependence are separable. This means that in this geometry, purely 2D and 3D models can still give good qualitative agreement with quantum well systems.

Quantitatively, these two models produce very different results. For example, at low fields, the diamagnetic shift of the ground state exciton can still be treated as a perturbation on the Coulomb interaction:

$$\Delta E = \frac{e^2B^2}{8m} <\rho^2> \tag{1}$$

Here, m is the reduced effective mass and $<\rho^2>$ is the expectation value of the square of the transverse extension of the exciton. The magnitude of this term depends strongly on the dimensionality of the system. In 3D, it is $2a_B^2$, where a_B is the exciton Bohr radius. For the 2D case, $<\rho^2> = \frac{3}{8}a_B^2$,[9] so that the low field diamagnetic shift in two dimensions is less than 1/4 of the 3D shift.

In Figure 3, the positions of the quantum well, wire, and dot peaks are plotted as a function of magnetic field for the 200 nm tungsten structures. Results for the 300 nm structures are similar. The solid line is a theoretical fit using an exact 3D hydrogen-like model for the exciton.[10] The reduced effective mass has been used as an adjustable parameter to account for the lower expectation value of the transverse exciton radius squared in a quasi-2D system. Assuming that the real reduced effective mass is $.043m_e$,[11] the expectation value of the transverse radius squared is found to be $1.3\,a_B^2$ in the unstrained quantum well.

The most striking feature of Figure 3 is the field dependence of the luminescence from the quantum dot structures. The peak position shifts to a lower energy at intermediate fields, yielding a net shift of less than 1 meV at 9 Tesla. The unusual behavior of the quantum dot luminescence with increasing magnetic field can be explained as follows. The broad peak at zero field contains luminescence from more than one confined state. At low fields, luminescence from the high energy components weakens so that the overall peak appears to redshift with field. The emission from these states decreases because they are evolving into Landau level-like excitons which shift rapidly with field. At high magnetic fields, only the ground state is observed, exhibiting a small but measurable diamagnetic shift. If this shift is fitted to the model described above (see the dotted line in Figure 3), we find $<\rho^2> = 0.3\,a_B^2$, which is considerably smaller than in the unstrained case.

4. Conclusion

We have measured the diamagnetic shift of excitons in a number of laterally confined systems. Since the quantum wire shift is very similar to that of the unstrained quantum well, one dimensional strain confinement does not appear to significantly alter the in-plane size of a quantum well exciton. However, the field dependence of the tungsten quantum dot luminescence indicates that the lateral extension of excitons in these structures is much smaller than that of the unstrained quantum well. Using a 3D hydrogen-like model for the exciton, we estimate that the expectation value of the transverse radius squared is reduced by about a factor of 4 in the 200 nm dots and a factor of 3 in the 300 nm dots. In order to evaluate the impact of these reductions on the exciton binding energy, we have used the calculation of Goff and Stébé for the excitonic ground state in a cylindrical quantum dot with finite potential barriers.[12] We find that the reductions in $<\rho^2>$ translate to exciton binding energy enhancements of approximately 60-90%.

5. References

1. N.J. Pulsford, J. Singleton, R.J. Nicholas, and C.T.B. Foxon, *J. Phys. (Paris) Colloque* **C5** (1987) 231.
2. M. Bugajski, W. Kuszko, and K. Reginski, *Solid State Commun.* **60** (1986) 669.
3. W. Ossau, B. Jakel, E. Bangert, and G. Landwehr, *Surf. Sci.* **174** (1986) 188.
4. M. Kohl, D. Heitmann, P. Grambow, and K. Ploog, *Surf. Sci.* **229** (1990) 248.
5. H. Weman, E.D. Jones, C.R. McIntyre, M.S. Miller, P.M. Petroff, and J.L. Merz, *Superlatt. Microstruct.* **13** (1993) 5.
6. Y. Nagamune, Y. Arakawa, S. Tsukamoto, M. Nishioka, S. Sasaki, and N. Miura, *Phys. Rev. Lett.* **69** (1992) 2963.
7. A.S. Plaut, K. Kash, E. Kapon, D.M. Hwang, E. Colas, *Surf. Sci.* **305** (1994) 576.
8. Yong Zhang, *Phys. Rev.* **B49** (1994) 14352.
9. A.H. MacDonald and D.S. Ritchie, *Phys. Rev.* **B33** (1986) 8336.
10. D. Cabib, E. Fabri, G. Fiorio, *Nuovo Cimento* **10B** (1972) 185.
11. L.C. Andreani and A. Pasquarello, *Phys. Rev.* **B42** (1990) 8928.
12. S. Le Goff and B. Stébé, *Phys. Rev.* **B47** (1993) 1383.

MANY BODY EFFECTS ON TUNNELING THROUGH QUANTUM DOTS: TUNNELING RATES AND ADDITION SPECTRA.

L. MARTIN-MORENO
Instituto de Ciencia de Materiales (CSIC), Universidad Autónoma, Madrid, Spain.

J. J. PALACIOS, J. H. OAKNIN, C.TEJEDOR
Departamento de Física de la Materia Condensada, Universidad Autónoma, Madrid, Spain.

G. CHIAPPE, and E. LOUIS
Departamento de Física Aplicada, Universidad de Alicante, Spain.

ABSTRACT

We present a theoretical study of addition spectra and spectral weights in quantum dots in a wide range of number of electrons and magnetic fields. Exact diagonalizations and the unrestricted Hartree-Fock approximation have been used. Our calculations reproduce qualitatively all the features, as a function of magnetic field, found in addition spectra and capacitance signals measured in single electron capacitance spectroscopy experiments. For high magnetic fields we find a "bosonic" representation of excited states comprising both edge and bulk excitations.

1. Introduction

Recent advances in nanolithography techniques have allowed the experimental study of single-electron tunneling through quantum dots (QD). A formal formulation of phenomena involving tunneling in these systems such as electronic transport in the Coulomb Blockade regime[1], or single electron capacitance spectroscopy (SECS)[2], has been put forward recently[3,4]. The main ingredients of the theory are the tunneling rates for independent electrons (SPTR) and the local density of states of interacting electrons in the dot (QDLDS) *with the dot being connected to the leads.* The calculation of the QDLDS is, in general, a very difficult problem, which can be considered as one of the paradigms of many-body theory. However, in some experimentally relevant limiting cases it is possible to describe the above mentioned phenomena in terms of, apart from the SPTR, solely the eigenenergies and the spectral weights for the *isolated* QD[4,5]. In this paper, we concentrate on the "many-body" contribution to such eigenenergies and tunneling rates (related to the spectral weights), and their dependence on both magnetic field (B) and number of electrons (N) in the QD.

More precisely, we consider a QD defined in a two-dimensional electron gas by an in-plane parabolic confining potential. The electron Hamiltonian is $H = \sum_{i,\sigma} \epsilon_{i,\sigma} c^{\dagger}_{i,\sigma} c_{i,\sigma} + \frac{1}{2} \sum_{i,j,k,l,\sigma,\sigma'} V_{ijkl} c^{\dagger}_{i,\sigma} c^{\dagger}_{j,\sigma'} c_{k,\sigma'} c_{l,\sigma}$ where the sums run over the single particle (SP) states of a two-dimensional parabolic potential defined in the xy plane, characterized by a frequency ω_0, and in the presence of a magnetic field B perpendicular to the xy

plane. $\epsilon_{i,\sigma}$ is the SP energy of a state i with spin σ. These states are labeled by the Landau index n ($0 \to \infty$), and by the third component of the angular momentum, m ($-n \to \infty$). V_{ijkl} are the Coulomb interaction matrix elements and $c^\dagger$ and c are the creation and annihilation operators. The circular gauge for the vector potential has been chosen, so both z-component of the total angular momentum (M) and z-component of the total spin (S_z) are good quantum numbers. In order to model the reported SECS experiments we have taken $\hbar\omega_0 = 5.4$ meV and standard values of the effective mass and dielectric constant in GaAs.

2. Low and Intermediate Magnetic Fields

Let us start by reviewing the low and intermediate magnetic field regimes[6], where SP states with $n > 0$ must be included in the basis set. Solid lines in Fig. I show the dependence on B of $\mu(N) = E_0(N) - E_0(N-1)$, where $E_0(N)$ is the GS energy of the N-electron system, obtained by exact Configuration Interaction calculations (exact in the sense that none of the results described below depend on the truncation of the basis set).

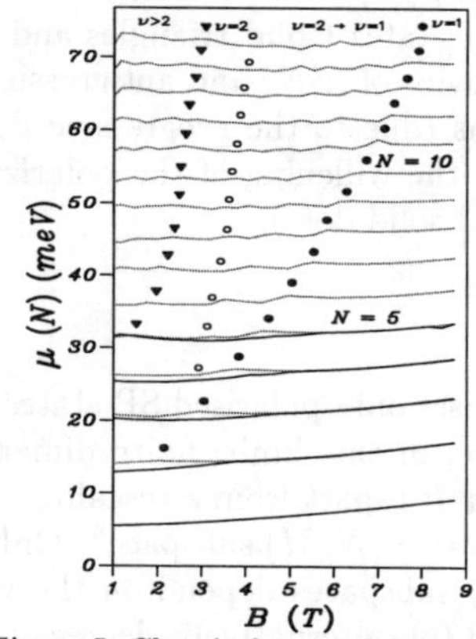

Figure I. Chemical potential as a function of B for the QD described in the text. Exact calculations (solid lines) for up to $N = 5$, and UHF calculations (dotted lines) for $N = 1$ to $N = 15$.

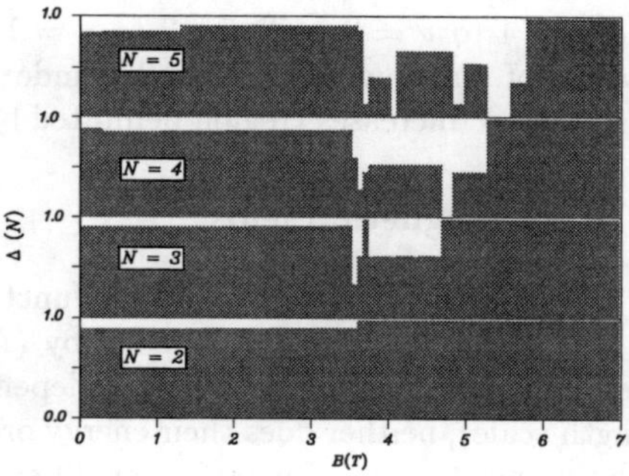

Figure II. Spectral weight up to 5 electrons as a function of B obtained by exact diagonalization.

Changes in the slope of $\mu(N)$ *vs.* B occur whenever a change in the electronic structure of the QD takes place. It is convenient to define $\nu_m = \sum_{n,\sigma} \nu_{n,m,\sigma}$ [$\nu_{n,m,\sigma}$ being the GS occupation of the SP state labelled by (n, m, σ)] as a "local filling factor", where the distance to the center increases with m. The evolution of the QD electronic structure with B is as follows: For the parameters considered, electrons at $B = 0$ form a "paramagnetic compact droplet", reminiscent of a two-dimensional state with two filled Landau levels. At large B, a "ferromagnetic compact droplet" is reached, with maximum $|S_z| = N/2$, and $\nu_m = 1$ for $0 < m < N-1$ and $\nu_m = 0$ for $m \geq N$, which is reminiscent of one filled Landau level. The polarisation transition takes place alternatively through "melted states", where the charge spreads out over a large number of SP states (usually with minimum $|S_z|$) and "compact states" with one extra spin flipped ($|S_z| \to |S_z| + 1$).

Figure II shows the dependence on B of the spectral weight at $\omega \to 0$, given by $\Delta(N) = \sum_{i,\sigma}\langle\Psi_0^{(N)}|c_{i,\sigma}^\dagger|\Psi_0^{(N-1)}\rangle\langle\Psi_0^{(N-1)}|c_{i,\sigma}|\Psi_0^{(N)}\rangle$ where $\Psi_0^{(N)}$ denotes the N-electron many-body GS wave function. Tunneling rates are directly proportional to the spectral weights[4,5]. As Fig. II shows, $\Delta(N)$ is reduced whenever the polarisation process takes place either for the N- or $N-1$- electron QD. Notice that the magnetic field interval of reduced $\Delta(N)$ increases with N, as the B interval of reduced capacitance signal in SECS experiments does. In order to verify that the B interval where polarisation occurs widens for larger N, we have performed an Unrestricted Hartree-Fock (UHF) calculation (dotted lines in Fig. I). Although UHF does not describe melted states properly (ν_m is always an integer in an angular-momentum-conserving UHF), and therefore UHF spectral weights are unreliable in the polarisation region, features related to GS energies are properly described[7] and UHF reproduces the following experimental features: a) The monotonic behavior of $\mu(N)$ after the bump due to the appearance of the stable compact $\nu = 2$ and $\nu = 1$ states (solid triangles and solid dots in Fig. I, respectively), (b) the oscillating behavior of $\mu(N)$ and anticrossings in the $\nu = 4$ to $\nu = 2$ and $\nu = 2$ to $\nu = 1$ transitions (due to the progressive depopulation of states with high Landau index), and (c) the widening of the polarization region as N increases (region delimited by open and solid dots).

3. High Magnetic Fields

At high B, the N-electron wave function comprises only polarised SP states with $n = 0$. Hilbert subspaces labelled by (N, M) have, in this limit, finite dimension. Remarkably, wave functions do not depend on ω_0 or B (apart from a rescaling of the length scale), neither does their energy ordering *within* a (N, M) subspace[8]. Only the relative positioning of energies for different (N, M) subspaces depend on the values of the parameters defining the QD. On increasing B (or, alternatively, decreasing ω_0) the N-electron GS resides in (N, M) subspaces with increasingly larger M.

As said before, for a range of parameters, the N-electron QD is in a "compact ferromagnetic state", the GS wave function, $|\Psi_{M_0}^{(N)}>$, being a single Slater determinant with total angular momentum $M_0(N) = N(N-1)/2$. We find that the GS wave function in subspaces with M close to $M_0(N)$ ($M = M_0(N) + \Delta M$ with $\Delta M \leq N$) can be accurately approximated by the variational wave function $\Phi_M^{(N)} = J_{\Delta M}^+(N)|\Psi_{M_0}^{(N)}>$, where $J_{\Delta M}^+(N) = \sum_{m=N-\Delta M}^{N-1}\alpha_{m,\Delta M}^{(N)}c_{m+\Delta M}^+c_m$, and $\alpha_{m,\Delta M}^{(N)}$ are variational coefficients (note that $J_1^+(N)$ is not obtained variationally but correspond to increasing the angular momentum of the center of mass by one unity[8]). Overlaps between the exact and variational wave functions are of the order $1-10^{-4}$. Operators $J_{\Delta M}^+(N)$ create a "hole" of integer charge in the compact droplet at the position associated with $m = N - \Delta M$. In particular, $J_N^+(N)$ creates the hole localised around the center of the droplet. Expectation values of these "bosonic" operators $J_{\Delta M}^+(N)$ satisfy a Kac-Moody algebra[7], and in the limit $N \to \infty$ coincide[7] with those describing edge excitations of a $\nu = 1$ liquid[9]. A more extended account of the properties of these operators and their effect on the spectral weight will be presented elsewhere[7].

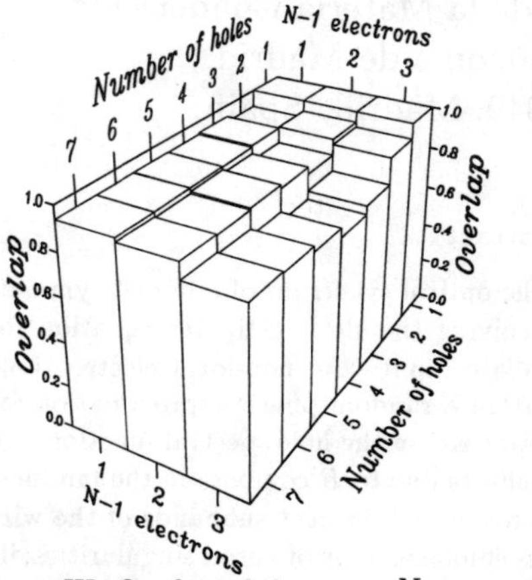

Figure III. Overlap of the exact N-electron GS's with those states created through successive application of the "bosonic" operators described in the text.

For small N ($N \leq 5$), the GS wave functions for different subspaces that become the absolute GS of the N-electron QD as B changes can be approximately obtained by successive applications of $J_N^+(N)$, i.e., $|\Psi_{M_0+LN}^{(N)}> \approx [J_N^+(N)]^L|\Psi_{M_0}^{(N)}>$, where L gives the number of holes of *integer charge* that have been created at the center of the (initially) compact state. The overlap $| < \Psi_{M_0+LN}^{(N)}|[J_N^+(N)]^L|\Psi_{M_0}^{(N)} > |^2$ between such approximate states and those obtained from the exact diagonalization is shown in Fig. III. As can be seen, the overlap is high even for a large number of applications of the operators $J_N^+(N)$.

4. Acknowledgements

We thank L. Brey for many useful discussions. This work has been supported in part by the Comisión Interministerial de Ciencia y Tecnología of Spain under contracts No. MAT 91-0905 and MAT 91-0201 and by the Commission of the European Communities under contract No. SSC-CT-90-0020.

5. References

1. M. A. Kastner, *Phys. Today*, January 1993, p. 24.
2. R. C. Ashoori *et al.*, *Phys. Rev. Lett.* **71** (1993) 613.
3. Meir and Wingreen, *Phys. Rev. Lett.* **68** (1992) 2512.
4. J. M. Kinaret *et al.*, *Phys. Rev.* **B45** (1992) 9489; *ibid.* **B46** (1992) 4681.
5. J. J. Palacios *et al.*, *Europhys. Lett.* **23** (1993) 495.
6. J. J. Palacios *et al.*, *Phys. Rev.* **B** in press.
7. Manuscript in preparation.
8. S. A. Trugman and S. A. Kivelson, *Phys. Rev.* **B31** (1985) 5280.
9. X. G. Wen, *Phys. Rev.* **B41** (1990) 12 838; M. Stone *et al.*, *Phys. Rev.* **B45** (1992) 14 156; M. Marsili, *Phys. Rev.* **B48** (1993) 18051.

Effect of a magnetic field on the optical spectra of doped quantum wires

C. Tejedor, F. J. Rodriguez and M. R. S. Tavares
Departamento de Física de la Materia Condensada.
Universidad Autónoma de Madrid.
Cantoblanco, 28049, Madrid. Spain.

Abstract

The effect of a magnetic field on the optical spectrum of a doped symmetric quantum well wire is studied by solving the Bethe-Salpeter equation for the electron-hole Green's function. We compute the non-local electron-hole Coulomb interaction in the system within a random phase approximation for the screening. The effect of the magnetic field on the hole spectral function and gap renormalization is analized. Two effects due to B compete in the luminescence spectrum: i) The symmetry is broken and distinct subbands of the wire are coupled to each other that produces enhancements of Fermi singularities; ii) Particles with different wavevectors separate from each other, which produces a decrease of the singularity. By increasing the temperature, more bands are thermally populated and singularities associated to the single particle density of states become comparable to the Fermi edge singularity.

1 Introduction

Magnetooptical experiments[1] give interesting information on the effect of B on the electron-electron interaction in low dimensional systems. Electron-electron interaction in quasi-one-dimensional (Q1D) systems, as for instance doped Quantum well wires (QWW), is strongly affected by the presence of a perpendicular magnetic field B. Due to the magnetic field, both the bare Coulomb interaction V^0 and the screening become non-local[2, 3]. Such non-locallity produces a coupling of odd and even Q1D bands due to off-diagonal terms of the screened interaction V. Therefore, properties as Fermi edge singularities (FES), very dependent on the coupling between different bands[4, 5, 6, 7, 8], are an excellent tool to study magnetic field effects on doped QWW. Existing experiments[9, 10, 11] made in doped QWW with a transverse magnetic field have not given, for the moment, much information about the behaviour of electron-electron interaction. In experiments[9] in which the system does not present a FES for $B = 0$, the field does not produce the singularity. In other experiments[10, 11] in which the QWW presents a FES for $B = 0$, a small shift and decrease of the singularity with B was observed. We present here a theoretical analisys on the conditions in which an experiment could give a better understanding of the FES as a good tool for the study of electron-electron interaction in Q1D systems with a perpendicular B.For that purpose, we calculate the non-local response function, the spectral function of the hole and the optical spectrum of QWW in the presence of transverse magnetic fields.

2 Theoretical scheme

The optical spectrum of a QWW with a transverse magnetic field is calculated along the same lines that in the case of no magnetic field. Asymmetry is the essential ingredient to couple different electron bands[8]. Such a coupling produces an enhancement of the FES when the Fermi level is slightly below the bottom of a band[4, 12, 8]. To stress the asymmetry created by the magnetic field, we use for the wire confining potentials for electrons and holes which are symmetric and coexist in space. The coupling between different bands will be due to B. We use parabolic confining potentials in the y-direction. When B is directed along the z-direction, each wavefunction (in the Landau gauge) factorizes in a planewave along x and a harmonic eigenfunction along y. States for two different values of the wavevector k see two different effective potentials which breaks symmetry. Optical spectra are computed from the linear optical susceptibility that, taking the matrix elements of the momentum operator as constant, is directly obtained from the electron-hole Green's function $G_{v,c;v',c'}(k,k',\omega)$[7]. G is calculated from the noninteracting electron-hole Green's function $G^0_{v,c}(k,-k,\omega)$ by means of a ladder approximation [7, 13]

$$G_{v,c;v',c'}(k,k',\omega) = G^0_{v,c}(k,-k,\omega)\delta_{v,v'}\delta_{c,c'}\delta(k+k') + \frac{1}{L}\sum_{v'',c'',q} G^0_{v,c}(k,-k,\omega)V_{v,c;v'',c''}(k,q)G_{v'',c'';v',c'}(q,k',\omega) \quad (1)$$

where k', k, q are 1D wavectors, the v's, c's are the indices for labelling the 1D bands, L is the wire length and $V_{v,c;v',c'}(k,q)$ the static screened electron-hole interaction. $G^0_{v,c}(k,-k,\omega)$ is obtained from the spectra of the electrons and holes. The spectral function of the electrons is well approximated by a δ-function, the only variation with the magnetic field being the dependence of the electron dispersion relation on B. On the contrary, the effective mass of the hole is so large that its dispersion relation does not change with B. However, the spectral function for the hole is not a δ-function anymore[7] and it is altered by the magnetic field.

In the lowest approximation, the hole spectral function is due to the excitation of electron-hole pairs in the conduction subband. It is obtained from the response function[14]

$$R(\omega)=\frac{1}{2\pi^2}\int dkdq\,[V_{0,0;0,0}(k,q)]^2\,[-Im\chi_{0,0}(k,k+q;\omega)] \tag{2}$$

where $\chi_{0,0}(k,k+q;\omega)$ is the electron polarizability and the interaction is obtained screening the bare Coulomb interaction V^0 by

$$V_{0,c;0,c'}(k,q)=\sum_{c'',c'''}\int dk'[\varepsilon^{-1}(k',k,q;\omega=0)]_{c,c';c'',c'''}V^0_{0,c'';0,c'''}(k',q) \tag{3}$$

The response function is obtained by inversion of the dielectric tensor computed within a random phase approximation[15]

$$[\varepsilon(k',k,q;\omega)]_{c,c';c'',c'''}=\delta_{c,c''}\delta_{c',c'''}\delta(k'-k)-\chi_{c,c'}(k',k'+q;\omega)V^0_{c,c';c'',c'''}(k'-k,q). \tag{4}$$

The polarizability $\chi_{c,c'}(k',k'+q;\omega)$ contains a diffusion term to take into account the fluctuation effects[13] as well as information on the magnetic field through the electron energies. The bare interaction $V^0_{c,c';c'',c'''}(k,q)$ contains information of the magnetic field through the electron wavefunctions and it has the form of integrals of exponentials multiplied by generalized Bessel functions which give Whittaker functions[15]. By numerical integration it is possible to get the hole spectral function. As in the case of no field, we get that $R(\omega)\to g(B)\omega$ in the asymptotic limit $\omega\to 0$ so that the spectral function is

$$A_h(-k,\omega)=\frac{2\sin(\pi g(B))\Gamma(1-g(B))}{\mu_e}\frac{e^{-|\Omega|}}{|\Omega|^{(1-g(B))}}\theta\,[\pm\Omega] \tag{5}$$

where

$$\Omega=(\omega-E_h(k)-\Delta E_h(B))/\mu_e,$$

μ_e being the electron chemical potential, $E_h(k)=-\frac{k^2}{2m_h(B)}$ the dispersion relation of the hole and $\Delta E_h(B)$ the self-energy of the hole[14] which also depends on B.

3 Results

We consider the case of a quantum wire with electrons so that the Fermi energy is $E_F=0.6E_0$ where E_0 is the binding energy of the quasi-two-dimensional exciton[7] (typically $6meV$ in GaAs quantum wells). The harmonic potential confining electrons in the wire is characterized by an intersubband energy $\omega_0=E_0$. This means approximately $400\AA$ for the wire width $l_0=\hbar\sqrt{1/m_e^*\omega_0}$ where m_e^* is the electron effective mass. Figure 1 shows both the slope g of the response function and the hole self-energy as a function of the magnetic field. The two magnitudes are computed up to second order in the series expansion of the hole Green's function[14].

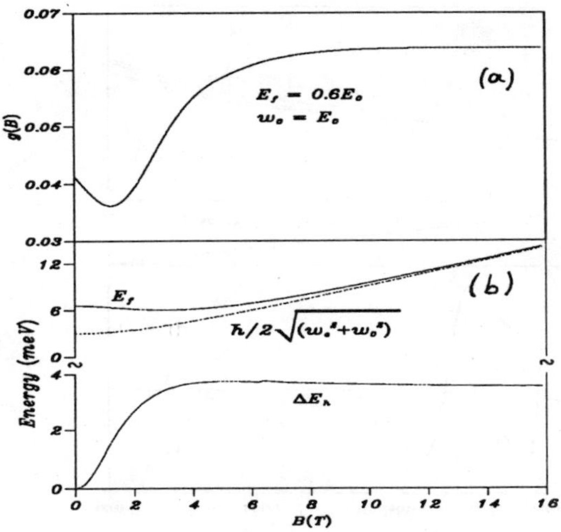

Figure 1: (a) Slope g of the response function $R(\omega)$ in the limit $\omega \to 0$ as a function of the magnetic field in a wire with $E_F = 0.6E_0$ and $\omega_0 = E_0$. (b) Band gap variation as a function of the magnetic field. Lower part shows the effect of B on the hole self-energy. Upper part shows the variation of both the conduction band bottom and the Fermi energy.

The relative variation of g is noticeable but the absolute value of that slope is so small compared with 1 that the magnetic field does not introduce changes in $A_h(-k,\omega)$ with noticeable effect on the optical spectrum. The situation is different for the self-energy of the hole. The value of this magnitude is comparable with the diamagnetic shift of the electron subbands $\hbar\omega_c/2$ in a broad range of magnetic fields. Therefore the hole self-energy has a measurable effect on the gap variation due to B.

Before discussing the optical spectra, let us show some results stressing the non-locality of the interaction. This clearly seen in the response function $[\varepsilon(k', k, q; \omega)]^{-1}_{0,0;0,0}$. In figure2 we present results for both the static (fig. 2a) and the dynamic response (fig. 2b). The non-locality is clear in the dependence on both k and q of the static response when the magnetic field is present.

In a purely one-dimensional system the FES still exists in the presence of a high magnetic field although its intensity is significantly reduced[13]. This is due to a reduction of the electron-hole interaction. Physically, for $B = 0$ electrons and holes at any wave-vector coexist in the same spatial region of width l_0. When B is turned on, each state with wave-vector k (along the wire) becomes localized around $\bar{y} = kl_B^2$ in a smaller width $l_B = \hbar\sqrt{1/(m^*(\omega_0^2 + \omega_c^2)^{(1/2)})}$. So, the effect of the magnetic field

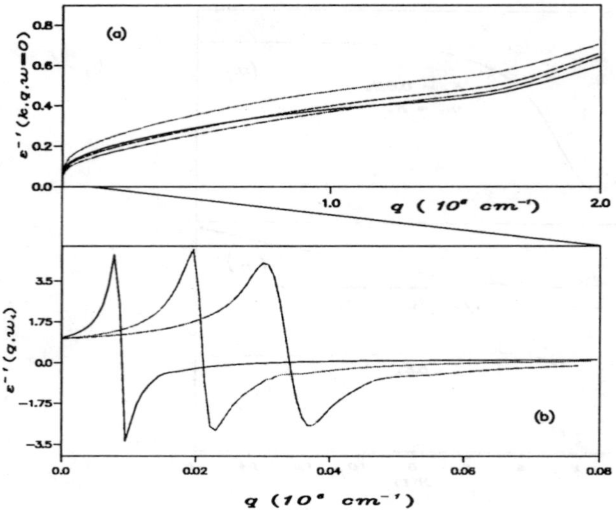

Figure 2: (a) Static response $\int dk'[\varepsilon^{-1}(k',k,q;\omega=0)]_{0,0;0,0}$ for $\omega_c = E_0/3$ with $k=0$ (dashed-dotted line), $k = k_F$ (dashed line), and $k = 2k_F$ (dotted line) compared with the local case $[\varepsilon^{-1}(q;\omega=0)]_{0,0;0,0}$ for $B=0$ (continuous line) in a wire with $E_F = 0.6E_0$ and $\omega_0 = E_0$. (b) Dynamical response $\int dk' \int dk[\varepsilon^{-1}(k',k,q;\omega_i)]_{0,0;0,0}$ for $\omega_1 = 0.3meV$ (continuous line), $\omega_2 = 2\omega_1$ (dotted line), $\omega_3 = 3\omega_1$ (dashed-dotted line).

is that electrons and holes with wave-vectors differing in δk are separated in the y-direction by a distance $\delta k l_B^2$ reducing their mutual interaction.

The situation is more complicated in actual (wider) QWW where there is more than one subband and the Fermi level can lie close to the bottom of a subband. The magnetic field changes both the intersubband spacing and the relative occupation when the Fermi level is initially above the bottom of the second subband as in the case of figure 3. The wire has $\omega_0 = 0.4E_0$ (width of roughly 600Å), $E_F = 0.6E_0$ and B is up to two Teslas. For low fields, the spectrum has two peaks ($b1$ and $b2$) related with the bottoms of the two bands. The FES is simply a shoulder due to the small coupling between the two bands (in fact is zero for $B=0$)[7, 13]. When B increases the bands become flat and the second one depopulates. The Fermi level lies in the first band slightly below the bottom of the second and the bands are strongly coupled because the system is not symmetric anymore. Therefore the FES becomes the strongest peak of the spectrum. $b2$ is a small structure produced by a few electrons thermally excited to the second band. If the field is further increased, the increasing spatial separation of electrons and holes reduces the FES.

The effect of temperature is well analized in figure 4. at the field in which the Fermi level lies just below the bottom of the second band. When the temperature is increased, the FES broadens and decreases due to the smearing of the Fermi distribution function[7, 8]. The bottom band singularity $b2$ increases because more electrons are promoted to the second band. The spectrum changes qualitatively.

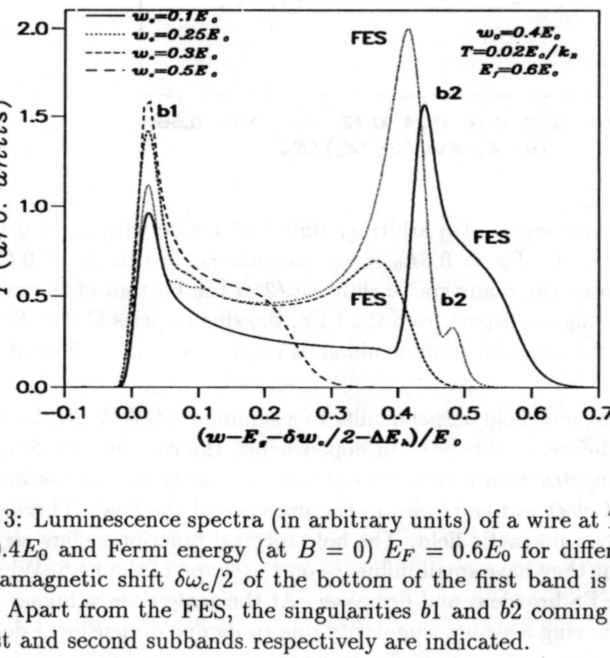

Figure 3: Luminescence spectra (in arbitrary units) of a wire at $T = 0.02E_0/k_B$ with $\omega_0 = 0.4E_0$ and Fermi energy (at $B = 0$) $E_F = 0.6E_0$ for different magnetic fields. The diamagnetic shift $\delta\omega_c/2$ of the bottom of the first band is not included in the figure. Apart from the FES, the singularities $b1$ and $b2$ coming from the bottom of the first and second subbands respectively are indicated.

Presently there is not experimental information directly comparable with our results. The experiment by Fritze *et al*[10] is the only one that in a situation similar to ours. A big difference with our system is that they work with a very wide QWW with many occupied bands. The intensity variations of the FES are difficult to follow since the Fermi level is frequently passing close to the bottom of some subband. The effects that we discuss should be much better detected in the extreme quantum limit in which a couple of bands are initially occupied.

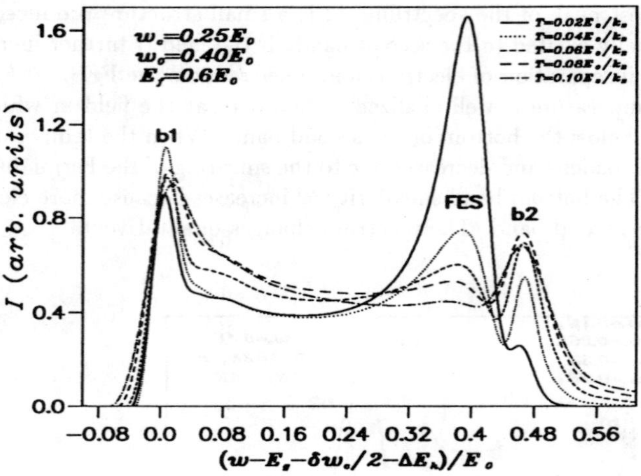

Figure 4: Luminescence spectra (in arbitrary units) of a wire with $\omega_0 = 0.4E_0$ and Fermi energy (at $B = 0$) $E_F = 0.6E_0$ in a magnetic field with $\omega_c = 0.25E_0$ for different temperatures. The diamagnetic shift $\delta\omega_c/2$ of the bottom of the first band is not included in the figure. Apart from the FES, the singularities $b1$ and $b2$ coming from the bottom of the first and second subbands respectively are indicated.

In summary, a magnetic field perpendicular to a symmetric QWW breaks the symmetry and couples different subbands. In doped wires, the coupling enhances FES. This mechanism competes with a reduction of the singularity due to the increasing spatial separation of electrons and holes in the presence of the field. Therefore, the FES is modulated by a magnetic field. The hole spectral function suffers significant variations with B but they have small influence on the shape of the FES. When temperature increases, FES broadens and decreases. At the same time, a higher band is thermally occupied having a strong singularity due to its one dimensional density of states. The combination of those two effects changes qualitatively the luminescence spectra of QWW.

This work has been supported in part by the Comision Interministerial de Ciencia y Tecnologia of Spain under contract MAT 91 0201 and the European Community under contract ESPRIT.BRA 6719. F.J.R. acknowledges to COLCIENCIAS of Colombia and M.R.S.T. to CAPES of Brazil for financial support.

References

1. *Optical Phenomena in Semiconductor Structures of Reduced Dimensions* Ed. by D.J.Lockwood and A.Pinczuk. NATO ASI Series E vol. 248 (Kluwer Academic Publishers. Dordrecht. 1993).

2. Q.P.Li and S.DasSarma, Phys. Rev. B **44**, 6277 (1991).

3. S.-R.Eric Yang and G.C.Aers, Phys. Rev. B **46**, 12456 (1992).

4. J.F.Mueller, Phys. Rev. B **42** 11189 (1990); J.F.Mueller, A.E.Ruckenstein and S.Schmitt-Rink, Mod. Phys. Lett. **5**, 135 (1990); Phys. Rev. B, **45**, 8902 (1992).

5. J.M.Calleja, A.R.Goñi, B.S.Denis, J.S.Weiner, A.Pinczuk, S.Schmitt-Rink, L.N.Pfeiffer and K.W.West, J.F.Mueller and A.E.Ruckenstein, Solid State Commun. **79**, 911 (1991).

6. P.Hawrylak, Solid St. Commun. **81**, 525 (1992).

7. F.J.Rodriguez and C.Tejedor, Phys. Rev. B **47**, 1506 (1993).

8. F.J.Rodriguez and C.Tejedor, Phys. Rev. B **47**, 13015 (1993).

9. A.S.Plaut, H.Lage, P.Grambow, D.Heitmann, K. von Klitzing and K.Ploog, Phys. Rev. Lett. **67**, 1642 (1991).

10. M.Fritze, A.V.Nurmiko and P.Hawrylak, Phys. Rev. B **48**, 4960 (1993).

11. J.M.Calleja, A.R.Goñi, A.Pinczuk, B.S.Denis, J.S.Weiner, L.N.Pfeiffer and K.W.West, to be published.

12. W.Chen, M.Fritze, A.V.Nurmiko, D.Ackley, C.Colvard and H.Lee, Phys. Rev. Lett. **64**, 2434 (1990); W.Chen, M.Fritze, A.V.Nurmiko, M.Hong and L.L.Chang, Phys. Rev. B **43**, 14738 (1991).

13. F.J.Rodriguez and C.Tejedor, Phys. Rev. B **49**, (1994).

14. G.D. Mahan, "Many particle physics" (Plenum Press. New York. 1981).

15. F.J.Rodriguez. PhD. Thesis. Universidad Autonoma de Madrid. 1994.

FIR-TRANSMISSION AND -PHOTOCONDUCTIVITY STUDIES OF SHALLOW ETCHED QUANTUM WIRES

C. M. ENGELHARDT, R. STRENZ, M. ASCHAUER, G. BÖHM, G. WEIMANN
Walter Schottky Institut, TU München, Am Coulombwall, D-85748 Garching, Germany

and V. ROSSKOPF, E. GORNIK
Institut für Festkörperelektronik, TU Wien, Gusshausstraße 25-29, A-1040 Wien, Austria

ABSTRACT

Far infrared (FIR) investigations are performed on shallow etched quantum wire arrays by transmission and photoconductivity measurements using both Fourier and FIR laser spectroscopy. The advantage of photoconductivity measurements for the investigation of small structured areas down to a few wires is demonstrated. Deviations from the plasmon dispersion are observed at high magnetic fields. In the quantum limit a deformation of the resonance compatible with the onset of spin splitting is observed, hinting at reduced Coulomb interaction in the wire system.

1. Introduction

Far infrared (FIR) spectroscopy provides a valuable tool in the investigation of mesoscopic structures. In case of electrically separated wires the FIR excites odd confined plasmon modes [1]:

$$\omega_p^2 = \frac{e^2}{2\varepsilon\varepsilon_0 m^*} n_{2D} (j+\alpha)\frac{\pi}{w}, \quad j = 1, 3, 5, ... \tag{1}$$

with the effective mass of the carriers m^*, the effective two dimensional carrier density in the wire n_{2D} and the electrical wire width w. Phenomenologically α corrects for the phase shift occurring at the reflection of the plasmon at the wire walls. In a magnetic field the plasmon should couple quadratically $\omega^2 = \omega_p^2 + \omega_c^2$ to the cyclotron resonance (CR) $\omega_c = eB/m^*$.

FIR spectroscopy with the Fourier spectrometer is performed in the low intensity limit taking spectra as a function of energy, whereas the FIR laser provides high intensities, but is bound to a few available laser frequencies; the spectra are usually measured as a function of magnetic field. The FIR transmission is measured by a bolometer or detector placed behind the sample. Photoconductivity (PC) is measured using the sample as detector. The applicability and performance of the different configurations is compared.

2. Sample preparation, experimental technique

The sample preparation starts from MBE-grown high mobility 2D electron gases in a GaAs/AlGaAs heterostructure (sample 1) and a single quantum well (sample 2). The samples are structured by holographic lithography and shallow wet chemical etching down into the modulation doping layer (see inset in Fig. 2). The filling of the wires could be tuned by above band gap illumination with a red LED. Selected sample parameters are given in table 1. From the vanishing transverse conductivity the wires appear to be electrically isolated.

Table 1: Selected sample parameters.

sample	1	2
AlGaAs spacer [Å]	240	60
Al cont. of AlGaAs	0.29	0.4
well width [Å]	–	80
n_{2D} [10^{11} cm^{-2}]	4.0	4.0
μ [cm^2/Vs] (dark)	1 500 000	25 700
n_{2D} [10^{11} cm^{-2}]		10.8
μ [cm^2/Vs] (illum.)		111 000
grating period [nm]	630	800
CR mass m^* [m_0]	0.070	0.075

The experimental setup is the standard configuration of a commercial rapid scan Fourier spectrometer or an optically pumped FIR laser connected via an oversized waveguide to a high field magnet cryostat. Unpolarized light was used. The specimens are wedged to 2.5°. The transmission spectra have been obtained by normalising the raw spectrum with another spectrum taken at a magnetic field different enough to avoid overlap of the plasmon resonances. The PC has been measured using a battery of 1.5 V with a load resistor of 5 MΩ to 1.5 GΩ. The raw spectra have been normalised by a suited transmission spectrum to account for the scaling of the PC with the FIR intensity. The FIR laser measurements employed a chopper and lock-in detection. For the PC measurements a sample current of typically 50 nA has been used. With the Fourier the transmission has been measured at 1.6 K, the PC at 0.4 K, as the signal increased strongly at lower temperatures, and the laser measurements were done at 4.2 K.

3. FIR characterisation by transmission and photoconductivity

In Fig. 1 typical spectra are shown of the unilluminated sample 1. The main resonance is attributed to the plasmon. The smaller resonance at the low energy side in a) appears at the position of the unperturbed CR. It is also visible in the structure at the high magnetic field side of c). For isolated wires the unshifted CR is not expected. But it appeared in the measurements of all the investigated shallow etched wire structures. A buried 2D gas appears unlikely. The resonance may be linked to unstructured parts of the sample. The PC spectrum d) taken with the FIR laser is in good agreement with the bolometric model of the PC [2], corresponding to the multiplication of the FIR absorption with the temperature dependence of the sample resistance and its absolute value. This mixing of different characteristics is a cause for deviations of the peak maximum from the maximum in FIR absorption. The displayed PC spectrum b) taken with the Fourier spectrometer shows a relatively high noise level as it is about to vanish at the chosen magnetic field.

Fig. 2 gives the evaluation of the resonance energies of the four configurations in a double squared plot. The plasmon resonance follows the quadratic dispersion (see above) with good agreement between the different methods. The deviations between FIR laser and the Fourier data and between the laser transmission and photoconductivity values are attributed to changes in the population of the wires. Several causes may be responsible: different cool down histories, heating of the electron system by the high FIR intensity and, for the PC measurements, heating and charging of the wires by the current.

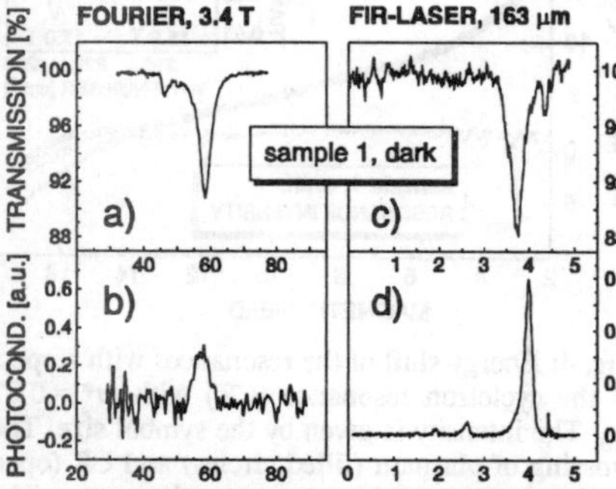

Fig. 1: a)-d) typical spectra obtained with the Fourier spectrometer and FIR laser in transmission and photoconductivity measurements.

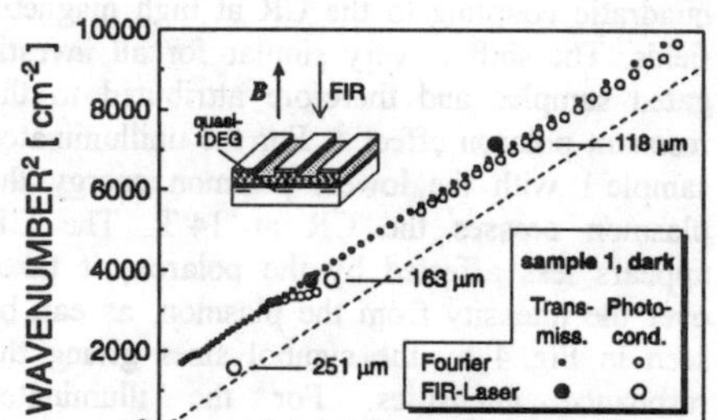

Fig. 2: Squared energy of the evaluated resonance positions vs. squared magnetic field. The dashed line corresponds to the unshifted CR with $m^* = 0.07\ m_0$. Inset: geometry of the experiment.

In Fig. 3 the evaluated intensity of the PC for the 10 ms illuminated sample 2 is compared to the two point resistance measured immediately afterwards. According to the bolometric model the maxima as well as the minima of the Shubnikov-de Haas oscillations change most with temperature, but the latter are damped in PC by the low absolute value of resistivity. Therefore the PC in the low excitation limit is still compatible with the bolometric model.

From the PC the plasmon energy ω_p of the first mode is taken from a plot like Fig. 2. The effective 2D density in the wires n_{2D} can be obtained from the maxima in the PC intensity (Fig. 3) if the distance between the peaks in $1/B$ is identified with a change of two in filling factor. The results are given in table 2. w has been calculated from Eq. 1 with $\varepsilon = 12.6$. $\alpha = -0.1$ according to Ref. [3]. The values of the strongly illuminated sample 2 are in good agreement with the result of $w = 250$ nm from Raman spectroscopy for the identical sample under continuous illumination [3].

4. Anomaly in the plasmon dispersion

Fig. 4 shows the energies of the two resonances observed in the unilluminated sample 1 in transmission with the Fourier spectrometer. The CR energy for $m^* = 0.07\ m_0$ has been subtracted to enhance visibility of small variations. The unperturbed CR almost follows the $0.07\ m_0$ trace. The plasmon shows a deviation from the quadratic coupling to the CR at high magnetic fields. The shift is very similar for all investigated samples and therefore attributed to the resonant polaron effect [4]. For the unilluminated sample 1 with the lowest plasmon energy the plasmon crosses the CR at 14 T. The CR appears less affected by the polaron, it takes over the intensity from the plasmon, as can be seen in Fig. 4 by the symbol sizes giving the resonance intensities. For the illuminated sample 1 the crossing happens at about 15 T indicating the effect to scale with either the plasmon energy or carrier density. This crossing of the two modes may be explained by an enhanced coupling of the collective plasmon excitations to the resonant polaron effect

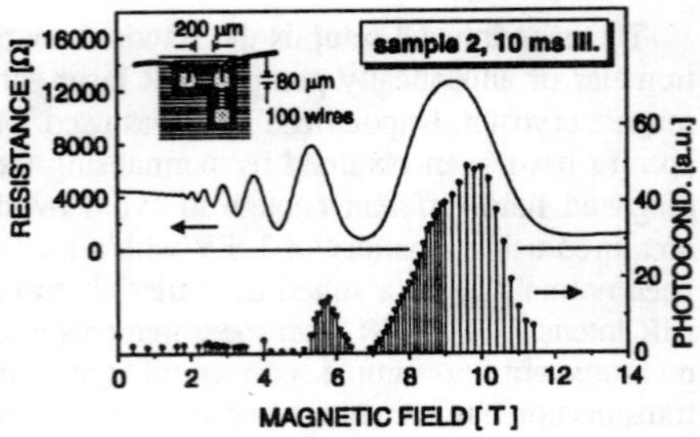

Fig. 3: Two point resistance (left scale) and strength of the photoconductivity signal (dots, right scale) of the 10 ms illuminated sample 2 at 0.4 K. Inset: Mesa structure used for the magnetotransport and photoconductivity measurements.

Table 2: Results from photoconductivity characterisation with the Fourier spectrometer.

Fourier / photoconductivity	ω_p [cm^{-1}]	n_{2D} [10^{11}cm^{-2}]	w [nm]
sample 1, dark	35.1	2.6	380
sample 1, illuminated	39.4	4.4	510
sample 2, 3 ms ill.	58.5	3.1	150
sample 2, 10 ms ill.	66.8	6.8	250
sample 2, illuminated	75.2	10.1	300

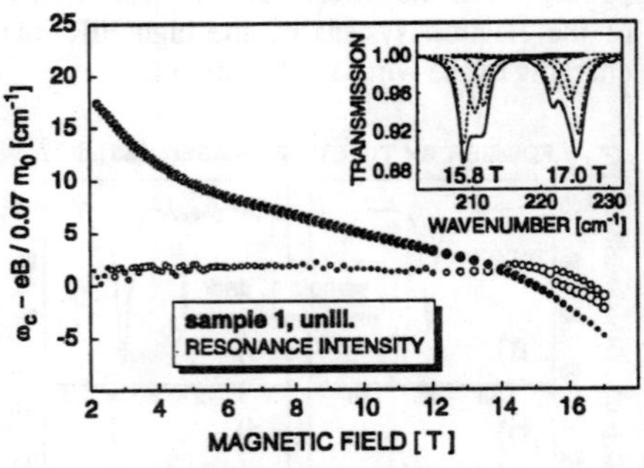

Fig. 4: Energy shift of the resonances with respect to the cyclotron resonance (CR) with $m^* = 0.07\ m_0$. The intensity is given by the symbol size. The crossing of plasmon (filled circles) and CR (open circles) is illustrated by two example spectra. The development of the asymmetry in the CR is described by two constituents (dashed lines).

analogous to the enhanced coupling to 2D intersubband transitions as observed by Kern et al. [5].

In this regime the CR appears to be deformed. It cannot be fitted by a single Lorentzian. A good fit is obtained with two lines. The energy separation of the two resonances comes out to be little above the spin splitting. The spin splitting is an effect of nonparabolicity and until now has been observed in bulk GaAs, but for 2D systems only in the extreme quantum limit [6, 7] when Coulomb coupling of the resonances is weakened [8].

Due to the limited space the data of the temperature dependence of the two constituents of the CR can not be shown. For temperatures up to 12 K an exchange of intensity between the two contributions is observed with an activation behaviour very similar to the spin splitting, with the Zeeman splitting being of the order of 5 K in the magnetic field regime under consideration. It may be an effect of the wire system to experience a reduced Coulomb coupling due to the lateral depletion between the wires.

Therefore the observation of spin splitting would by triggered in 1D systems at filling factors (here the filling factor is little below 1) much above the critical filling factors of the 2D system, which are about 1/9. To decide this question more systematic studies in this regime have to follow. As long as the origin of the crossing of the plasmon and CR is not understood in detail, the observed deformation can be part of the effect. Of course the activation behaviour of the two constituents of the CR is a somewhat stronger indication of the influence of spin splitting.

5. Summary

Four configurations of FIR-spectroscopy on quantum wires have been compared. FIR laser measurements appear to be affected by the high intensities involved. In this contribution the first demonstration of PC measurements of wires with a Fourier spectrometer is given. The PC appears at the position of the plasmon with a Lorentzian lineshape. It is most pronounced at the maxima of the two point resistance. With PC a tiny sample of $80 \times 200\ \mu m^2$ was successfully characterised in FIR. The wire widths extracted from the PC confirm the resonance to be a localized plasmon.

The plasmon energy is reduced at high magnetic fields, which is attributed to the resonant polaron effect. In the quantum limit the plasmon crosses the unperturbed CR and exchanges intensity with it. Here, the CR develops a deformation which can be described by the onset of spin splitting, which would be a hint of reduced Coulomb interaction in the wires as compared to 2D.

6. Acknowledgement

The authors gratefully acknowledge financial support by the BMFT Germany, project no. 01 BM 123 / 0.

7. References

1. W. L. Schaich and A. H. MacDonald, *Solid State Commun.* **83** (1992) 779.
2. F. Thiele, E. Batke, J. P. Kotthaus, V. Dolgopolev, V. N. Ovsyuk, G. Gusov, G. Weimann, W. Schlapp, *Solid-State Electron.* **32** (1989) 1503.
3. R. Strenz, F. Hirler, G. Abstreiter, G. Böhm, G. Tränkle and G. Weimann, Proc. ICPS 22, *World Scientific Publishing* (1994).
4. e.g.: H. Sigg, H. J. A. Bluyssen and P. Wyder, *Solid State Commun.* **48** (1983) 897.
5. K. Kern, D. Heitmann, R. R. Gerhardts, P. Grambow, Y. H. Zhang and K. Ploog, *Phys. Rev. B* **44** (1991) 1139.
6. G. M. Summers, R. J. Warburton, J. G. Michels, R. J. Nicholas, J. J. Harris and C. T. Foxon, *Phys. Rev. Lett.* **70** (1993) 2150.
7. C. M. Engelhardt, E. Gornik, M. Besson, G. Böhm and G. Weimann, *Surf. Sci.* **305** (1994) 23.
8. N. R. Cooper and J. T. Chalker, *Phys. Rev. B* **72** (1994) 2057.

RESISTANCE OF A LONG WIRE IN THE QUANTUM HALL REGIME

M. M. Fogler and B. I. Shklovskiĭ
Theoretical Physics Institute, School of Physics and Astronomy, University of Minnesota
116 Church St. SE, Minneapolis, MN 55455

ABSTRACT

We suggest a model of the dissipative transport in a long quantum wire in strong magnetic fields. This model enables us to calculate the shape of the Shubnikov-de Haas peaks in the two-probe resistance of the wire. Their shape is distinctly asymmetrical with sharp low-field and shallow high-field slopes. We use a recently developed electrostatic theory of edge states in a quantum wire, according to which the wire is split into alternating parallel strips of compressible and incompressible liquids. The resistance of the wire is determined by the equilibration in pairs of neighboring compressible strips. The equilibration rate of every pair exponentially falls off with the the width of the separating incompressible strip. The widths of different incompressible strips vary with the magnetic field in a different way. Taking into account an interplay of these dependencies enabled us to show how the peaks of magnetoresistance with peculiar asymmetrical shape appear and how the transitions between successive Hall plateaus occur with changing magnetic field.

The transport in a narrow strip of two-dimensional electron liquid (a quantum wire) in a strong magnetic field has been viewed as the propagation of current through edge channels. It is assumed that inter-edge-channel tunneling is suppressed [1,2] if the separation between edge channels exceeds several magnetic lengths. Therefore, transport in a short clean wire in high magnetic field is ballistic. The two-probe resistance (2PR) of the wire in this regime *monotonically increases* with magnetic field. Details of this dependence are determined by the structure of the edge channels. For interacting electrons the edge channels are strips of finite width located where compressible liquid is formed [3,4,5,6]. Compressible strips alternate with incompressible ones corresponding to integer filling factors. The two-probe resistance (2PR) takes quantized values in narrow ranges of magnetic field, for which the central strip is incompressible. When the central strip is compressible, 2PR takes intermediate values constituting smooth inter-plateau rises [6].

A recent experiment by Mani and von Klitzing[7] examined the 2PR of an array consisting of several hundred long narrow ($\sim 0.2 \times 100 \mu m^2$) wires. In contrast to the ballistic picture, the 2PR observed in this experiment is distinctly non-monotonous. Instead of smooth rises 2PR exhibits huge peaks between adjacent quantized plateaus. The resistance in peaks is 10-12 times greater than the quantized values. Another peculiar feature of the 2PR peaks is their strong asymmetry at high temperatures, where the low-field sides of the peaks are much more narrow than the high-field ones. The widths of low-field sides are virtually temperature independent, while the high-field sides shrink with lowering of the temperature making the peaks appear more and more symmetrical. Qualitatively similar results were also reported by Wróbel *et al.* [8].

If the wire is narrow and sufficiently long the edge channels equilibrate as electrons travel away from the probes. The electrons tend to redistribute in order to achieve an equal value of the electrochemical potential (ECP) in each channel, however suppressed those processes are. The equilibration implies not just coherent backscattering, but also *irreversible* inelastic processes; therefore, the resistance of the wire is expected to exhibit the conventional linear scaling with the length of the wire.

The comprehensive theory of the equilibration among compressible liquids can not be developed without a deep insight into the structure of the compressible liquid itself. On the other hand, the sharp exponential dependence of the inter-edge-channel equilibration on spatial separations among the strips can be taken into account rather easily. In this paper we suggest a "high-temperature" theory of the two-probe transport in a long wire based only on the variations of this suppression factors with magnetic field.

Let the wire of length L be connected to two metallic probes. Our goal is to calculate the net current I when a potential difference V is applied. Non-equilibrium currents are related to a non-uniform distribution of the ECP in the sample. In the ballistic regime the ECP has a drop across the central strip. Correspondingly, this strip carries a non-equilibrium (Hall) current $\sigma_{xy}\Delta\xi/e = (e^2/h)\nu_c V$, where ν_c is the filling factor at the center, and $\Delta\xi$ is the drop of the ECP.

In a long wire, where the transport is no longer ballistic, the drops of the ECP occur across each incompressible strip, not only across the central one (this idea was earlier put forward in Ref. 9). In this situation a generalized formula for the net current applies: $I = (e/h)\sum_i \nu_i \Delta\xi_i$, where the summation is taken over all the incompressible strips and the ν_i's are the corresponding filling factors. We neglect the variations of the ECP across compressible strips, since the equilibration within them (intra-edge-channel) is fast, whereas the inter-edge-channel equilibration is suppressed in the magnetic field, and the drops of the ECP across incompressible strips are considerable. These drops may be found to be $\Delta\xi_i = eE\nu_i l_i$, where $E = V/L$ is the longitudinal electric field, ν_i is the filling factor of a given incompressible strip, and l_i is *the equilibration length* of the pair of adjacent compressible strips: $l_i = l\exp\left\{\frac{1}{2}(w_i/\lambda)^2\right\}$. Here, λ is the magnetic length and w_i is the width of the incompressible strip. This particular form is appropriate for the single-act scattering on a short-range disorder. The parameter l of dimensionality of length is the only fit parameter in the model. Now we can give a sufficient criterion for the wire to be considered as "long": $l_i \ll L$ for all the strips. The 2PR is calculated to be $R = (h/e^2)L/\sum_i \nu_i^2 l_i$.

In the opposite limit, where all the l_i's are much bigger than L, the equilibration is ineffective, and the transport is quasi-ballistic (a "long" wire behaves like a "short" one). Naturally, $R = (h/e^2)/\nu_c$.

The following formula interpolates* between these two limits:

$$R = (h/e^2)\left(\frac{1}{\nu_c} + L\left\{\sum_i \nu_i^2 l_i\right\}^{-1}\right). \tag{1}$$

It is noteworthy that the variation of the 2PR with magnetic field is determined purely by the variations of several geometrical parameters – the widths of the incompressible strips. Our results are derived from Eq. (1) with the help of a computer program developed earlier [11]. This program calculates the one-dimensional charge distribution

*A more rigorous treatment of the problem may be found in Ref. 10.

in a quantum wire in a strong magnetic field. The confining potential defining the wire was chosen to be parabolic, and disorder potential was disregarded, since in a clean wire it may significantly affect only the equilibration, but not the distribution of the electron density.

The results of the calculations for spin-split Landau levels are presented in Fig. 1. The parameters of the simulation matched the experimental ones: the half-width of the wire $b = 1000\text{Å}$, length $L = 57\mu\text{m}$, electron density at the center $n = 1.5 \times 10^{11}\text{cm}^{-2}$. We also chose an exchange enhanced g-factor [12] of 8.0. The fit parameter $l = 2.0\mu\text{m}$ was chosen to match the experimentally observed height of the $0\downarrow$ peak. The peaks are strongly asymmetrical in shape with very sharp low-field slopes and smooth high-field ones. In order to understand the details of the shape of the peaks consider the variation of the 2PR given by Eq. (1) with the magnetic field. When the magnetic field is such that the filling factor at the center of the wire is unity, the central incompressible strip is, typically, wide [6]. Correspondingly, l_1 is very big (bigger than L) and the resistance R assumes a quantized value h/e^2(Fig. 1a). With lowering of the magnetic field, the central strip breaks in two, and a narrow compressible strip appears at the center. However, even a strip of a half the original width corresponds to $l_1 \gg L$, and R is still very close to the plateau (Fig. 1b). With a further decrease in the magnetic field, the two incompressible strips shrink. The equilibration length l_1 rapidly decreases and becomes smaller than L. The resistance grows and becomes much bigger than the quantized value (Fig. 1c). At the same time, the filling factor at the center increases and approaches two. A new incompressible strip and a new equilibration length l_2 emerge and grow. The increase of R slows down (Fig. 1d). Due to the sharp exponential dependence of the equilibration lengths on the widths of the corresponding strips, the denominator in Eq. (1) is determined by the l_i of the edge channel pair separated by the widest strip. When the width of the central strip reaches the width of peripheral ones, the resistance should exhibit a cross-over, since the new strip is expanding, while the two existing ones are shrinking. Therefore, R is at its peak value (Fig. 1e) at this magnetic field. The central incompressible strip expands tremendously fast, and very soon R assumes the next quantized value (Fig. 1f) and then the cycle is repeated again.

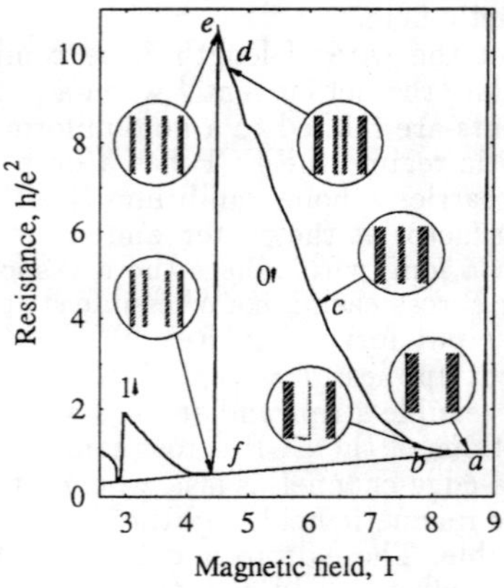

Fig. 1. The two-probe resistance as a function of magnetic field as given by our numerical simulation. The same quantity for the ballistic regime[6] is shown with a thin line for comparison. Due to the narrowness of the quantized plateaus it looks almost like a straight line. Diagrams in circles depict top view of short segments of the wire in magnetic fields corresponding to the positions pointed to by arrows. The shaded and blank areas represent compressible and incompressible strips respectively. Positions (a-f) are several stages in the evolution of the resistance peak discussed in the text.

In the experiment the low-field sides are, indeed, much steeper than the high-field ones, but not as abrupt as in the simulation. One possible explanation is the averaging of the individual peaks of several hundred distinct wires. A puzzling feature of experimental data is that a *non-zero slope* of R in the plateau areas was observed. We do not have an interpretation of this phenomenon at present.

As a conclusion we note that our model describes the dissipative transport in a long narrow wire with low disorder and at relatively high temperatures. We were able to find the two-probe resistance of the wire as a function of magnetic field by introducing a single fit parameter and without knowing details of electron tunneling and subsequent relaxation in the compressible liquid. This fit parameter affects only the heights of the peaks but not their shape. It is important to note also that we used geometrical parameters describing a clean wire and did not assume anything about the disorder potential besides its relative smallness.

The two-probe resistance in our model proved to be distinctly non-monotonous with high peaks between successive plateaus of quantized resistance. We believe that for the first time the shape of the Shubnikov-de Haas peaks is calculated explicitly in the QHE regime. We obtained the strong asymmetry of the peaks and predicted the linear scaling of their heights with the length of the wire. Both agree with the experiment. However, we were unable to interpret the observed temperature dependencies. We assume that proper treatment of this issue as well as a more accurate description of inter-edge-channel tunneling should be based on the understanding of the equilibration and localization processes in the strips of the compressible liquid.

We are grateful to K. A. Matveev for an important contribution on the early stage of the work and to M. Büttiker, D. B. Chklovskii, L. I. Glazman, K. von Klitzing, D. G. Polyakov, and I. M. Ruzin and for helpful discussions. One of us (M. F.) was supported by the NSF Grant No. DMR 9020587, and the other (B. S.) by NSF Grant No. DMR 91-17341.

References

1. M. Büttiker, *Phys. Rev.* **B38**, (1988) 9375.
2. T. Martin, S. Feng, *Phys. Rev. Lett.* **64**, (1990) 1971; *Phys. Rev.* **B44**, (1991) 9084.
3. A. M. Chang, *Solid State Commun.* **74**, (1990) 871.
4. C. W. J. Beenakker, *Phys. Rev. Lett.* **64**, (1990) 216.
5. D. B. Chklovskii, B. I. Shklovskii, L.I. Glazman, *Phys. Rev.* **B46**, (1992) 4026; **46**, (1992) 15 606 (E).
6. D. B. Chklovskii, K. A. Matveev, B. I. Shklovskii, *Phys. Rev.* **B47**, (1993) 12 605.
7. R. G. Mani, K. v. Klitzing, *Phys. Rev.* **B46**, (1992) 9877; (unpublished) 1993.
8. J. Wróbel *et al.*, in *Proceedings of the 10-th International Conference on Electronic Properties of Two-Dimensional Systems*, ed. by A. McCombe and B. Nurmikko (Elsveier, Amsterdam, 1993), p. 615.
9. R. J. Haug, K. v. Klitzing, *Europhys. Lett.* **10**, (1989) 489.
10. M. M. Fogler, B. I. Shklovskii, 1994 (to be published).
11. M. M. Fogler, E. I. Levin, B. I. Shklovskii, *Phys. Rev.* **B49**, (1994) 13 767.
12. R. J. Nicholas, R. J. Haug, K. v. Klitzing, *Phys. Rev.* **B37**, (1988) 1294.

HIGH EXCITATION MAGNETOLUMINESCENCE STUDY OF III-V SEMICONDUCTOR QUANTUM WIRES

M. Bayer, A. Forchel,
Technische Physik, Universität Würzburg, Am Hubland, D - 97074 Würzburg, Germany

T.L. Reinecke, P.A. Knipp,
Electronic Materials Branch, Naval Research Laboratory, Washington, D.C. 20375-5347, USA

Abstract

High excitation luminescence experiments on lateral subband transitions in quantum wires have been performed in magnetic fields up to B = 9.0T. Varying the orientation of B relative to the wire axis an anisotropy of the dependence of the interband transition energies on the magnetic field is observed. Applying the field normal to the quantum well plane a transition from quasi-one-dimensional to quasi-zero-dimensional behavior can be realized.

1. Introduction

Quasi-one- and quasi-zero-dimensional semiconductor structures have been intensely studied recently because of their potential both for technological applications and also for investigations of new physical phenomena. For latter investigations the application of high magnetic fields is particularly promising because the variation of the field orientation and of the field strength allow to study the geometric confinement potentials.[1–4]

Two types of samples with wire widths down to 25nm were investigated: Modulated barrier $In_{0.13}Ga_{0.87}As/GaAs$ quantum wires and deep etched $In_{0.53}Ga_{0.47}As/InP$ quantum wires. Optical excitation by an Ar^+ laser (514.5nm) was used in order to observe emission from different lateral subbands. The measurements were performed using an optical split-coil magnetocryostat. The emitted light was dispersed by a 0.25m single grating monochromator and detected either by a *Ge*-detector or by a photomultiplier.

2. Experiments in normal magnetic field

With the magnetic field normal to the quantum well plane the single particle energy levels are given by the following expression, if one uses a parabolic approximation for the geometric confinement potential:[2]

$$E = E_{n_z} + \hbar\sqrt{\omega_x^2 + \omega_c^2}(n + \frac{1}{2}) + \frac{\omega_x^2}{\omega_x^2 + \omega_c^2}\frac{\hbar^2 k_y^2}{2m} \tag{1}$$

Here the z-axis of the coordinate system is oriented parallel to the magnetic field and the y-axis is parallel to the wire axis. If the magnetic length can be varied from being much larger than the lateral dimensions of the quantum wires for low fields to being much smaller for high fields, the magnetic confinement potential is comparable to the geometric confinement potential at intermediate fields.

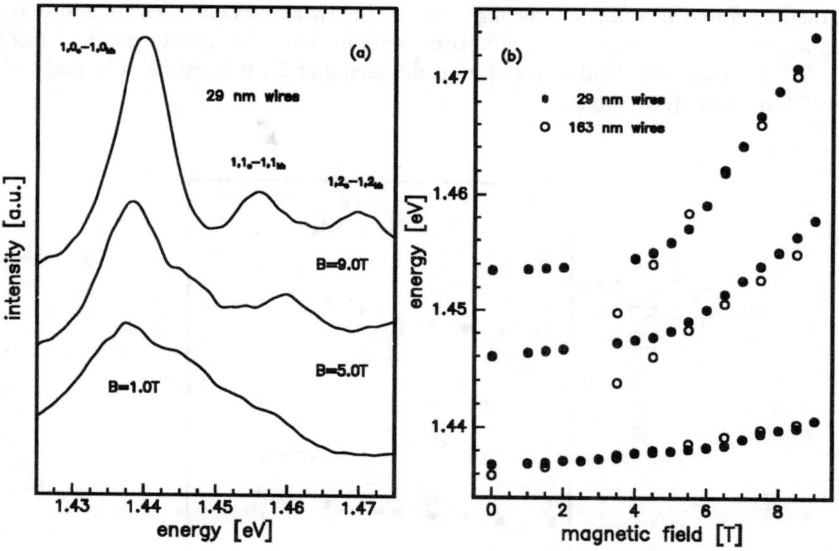

Figure 1: (a) Spectra of $In_{0.13}Ga_{0.87}As/GaAs$ quantum wires in normal magnetic field. (b) Transition energies in normal magnetic field.

From equation (1) it can be seen that at low fields the geometric confinement dominates resulting in a B^2 dependent diamagnetic shift. In contrast, at very high magnetic fields the magnetic potential dominates resulting in a linear dependence of the energies on B as expected for carriers in a quantum well in normal magnetic field. The lateral subband with index n_z, n has effectively transformed into the Landau level with index n_z, n.

This behavior is clearly reflected in high excitation spectra (Fig. 1(a)) on $29nm$ $In_{0.13}Ga_{0.87}As/GaAs$ quantum wires. Three spectral lines are observed that can be attributed to recombination between electron and hole levels that correspond to the lowest quantum well excitation, but to different lateral excitations. All of these lines shift to higher energies with increasing magnetic field but this shift is rather weak at small fields ($B < 5T$), while at high fields the transition energies rise quickly with B.

Fig. 1(b) shows the energetic positions of the transitions in the 29nm wires (full circles) in comparison with data from 163nm wide, quasi-two-dimensional wires (open circles). Especially for the higher interband transitions we observe only a weak dependence of their energies on the magnetic field for small B. This behavior changes drastically around $B = 5T$ and for higher fields the transition energies depend linearly on B as expected for carriers in quantum wells in normal magnetic field. Thus for $B > 5T$ the magnetic-field-induced zero dimensional limit is effectively realized in these 29nm wide quantum wires.

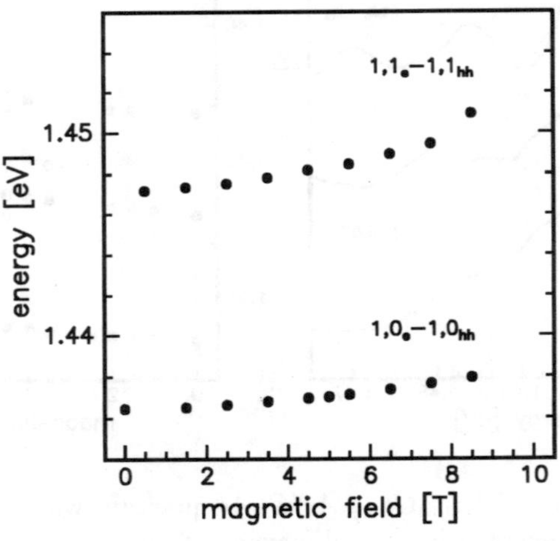

Figure 2: Transition energies in parallel magnetic field

As can be seen from Fig. 1(a), the integrated luminescence intensity of the lowest lateral transition is clearly enhanced with increasing magnetic field. This enhancement is mainly due to the increase of the density of states in the first lateral subband, that is always completely filled under high optical excitation.

According to eqn. (1) the transition to quasi-zero-dimensional behavior takes place when the cyclotron energy exceeds the lateral subband splitting. This subband splitting increases with decreasing wire width L_x and is approximately proportional to L_x^{-2}. Therefore higher magnetic fields must be applied for smaller wires in order to realize the transition. We investigated this on deep etched $In_{0.53}Ga_{0.47}As/InP$ quantum wires of different lateral dimensions, where we confirmed the dependence of the transition magnetic field defined by the equation $\hbar\omega_c = \hbar\omega_x$ on L_x^{-2}.[5]

3. Experiments in parallel magnetic field

If the magnetic field direction is parallel to the quantum wire axis both geometric confinement potentials contribute together with the magnetic confinement potential to the eigenfrequencies.

$$E = \hbar\omega_1(n_1 + \frac{1}{2}) + \hbar\omega_2(n_2 + \frac{1}{2}) + \frac{\hbar^2 k_y^2}{2m} \qquad (2)$$

$$\omega_{1/2}^2 = \Omega^2 \pm \sqrt{\Omega^4 - \omega_z^2\omega_c^2 - \omega_x^2\omega_z^2}, \qquad \Omega^2 = \frac{\omega_x^2 + \omega_z^2 + \omega_c^2}{2}$$

For all fields available in the experiments the quantum well potential dominates and only a weak dependence of the energies of the interband transitions on B is expected up to the highest fields, that was confirmed in our experiments on the $In_{0.13}Ga_{0.87}As/GaAs$ quantum wires. Fig. 2 shows the energetic positions of the two lowest lateral transitions in parallel magnetic field. Both transitions shift only weakly to higher energies over the entire field regime. The weak dependence on B was also found in the measurements with the third independent field orientation normal to the wires.

4. References

1. Y. Naganume, Y. Arakawa, S. Tsukamoto, M. Nishioka, S. Sasaki, and N. Miura, Phys. Rev. Lett. **69**, 2963 (1992).
2. M. Bayer, A. Forchel, I.E. Itskevich, T.L. Reinecke, P.A. Knipp, Ch. Gréus, R. Spiegel, and F. Faller, Phys. Rev. B **49**, 14782 (1994); this paper includes a fully numerical model to describe the lateral quantization in a magnetic field.
3. M. Oestreich, A.P. Heberle, W.W. Rühle, H. Lage, D. Heitmann, and K. Ploog, J. of Lumin. **60/61**, 390 (1994)
4. Ch. Gréus, R. Spiegel, P.A. Knipp, T.L. Reinecke, F. Faller, and A. Forchel, Phys. Rev. B **49**, 5753 (1994)
5. M. Bayer, P. Ils, A. Forchel, to be published

THE BALLISTIC TRANSPORT IN CROSSED ELECTRIC AND MAGNETIC FIELDS

TAKESHI NAKANISHI, TOMI OHTSUKI* and MOTOHIKO SAITOH

1 - 16 Department of Physics, Osaka University, Toyonaka 560, Japan

**Department of Physics, Toho University, Miyama, Funabashi 274, Japan*

ABSTRACT

The nature of the ballistic transport properties of the two-dimensional lattice in the crossed electric and magnetic fields is investigated. Combination of the electric and magnetic fields makes the electron wave functions localized on the narrow channels perpendicular to the electric field and the conductance quantization through these narrow channels is expect to occur. A tight-binding Hamiltonian is employed to describe the system which is connected with perfect leads, and numerical computations are performed on the basis of the two-terminal Landauer formula for the conductance and of the recursive Green's function technique. We find that the ballistic conductance quantization becomes sharpened by the magnetic field for the system with random potential.

1. Introduction

The electronic energy states in the 2D tight-binding model under the magnetic field applied normal to the lattice plane is known to be fairly complex as shown by Hofstadter[1]. When the normalized magnetic field ϕ/ϕ_0 is equal to the rational number p/q, where ϕ is the magnetic flux through the unit cell, $\phi_0 = ch/e$ the flux quantum and p and q are integers prime to each other, the energy eigenstates are split into q subbands. Wave functions for this tight-binding system are extended over the entire lattice. This is very different from the case of the free electrons, where the Landau states are localized around the centers of the circular motion and their energies are all degenerate for the fixed Landau index. For a finite lattice with confining walls, however, the electron states near the boundaries are appreciably modified, and the electrons migrate on the classical skipping orbits for both systems[2].

When an electric field is applied in the lattice plane on top of the magnetic field, the wave functions are confined around the system edges, because in the vicinity of the lattice edges, the triangular potential is formed by the electric field[3]. Moreover, the combination of both fields produces narrow channels in the bulk perpendicular to the electric field. The width of each channel is roughly equal to the magnetic length $\{\phi_0/(2\pi\phi)\}^{1/2}$ in units of the lattice constant. The electronic energy states in the finite system, therefore, are appreciably modified from those for the infinite system when both magnetic and electric fields are present because of the triangular potential induced by the electric field near the system edges and of the bulk effects of the electric field on the magnetically induced subband structure. We present here

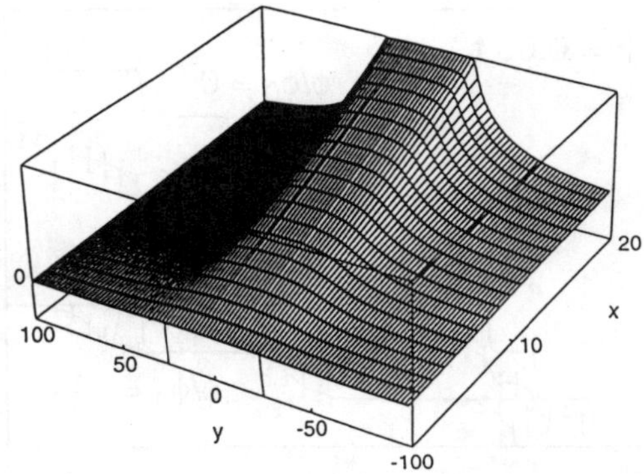

Fig. 1. The effective electric potential in the neighborhood of the sample. The sample occupies the region $0 < x < 20$, $-10 < y < 10$, and the applied voltage along the x axis is $V = 0.6\ t$ where t is the transfer energy. The condenser plates with width $D = 52$ is located at $x = 0$ and 20.

the calculations of the conductance through these narrow channels induced by both electric and magnetic fields.

2. Model and Method

We consider the 2D lattice of dimensions $L_x \times L_y$ (the lattice constant is set to be unity) and assume that the electric field is applied along the x direction by the condenser plates with width D at $x = 0$ and L_x, and the magnetic field is applied parallel to the z direction. The condenser width D is taken to be a little bit larger than L_y so that the effective potential along the y direction in the sample becomes flat. The probes to measure the conductance are attached at $y = \pm Y$ where Y is much larger than the sample width L_y. The effective electric potential in the system is determined by the Laplace equation with the boundary conditions where the electric potential is zero at $x = 0$ and $y = \pm Y$, and is equal to $V = FL_x$ for $x = L_x$ and $|y| < D/2$, where F is the electric potential drop across a unit cell. The resultant electric potential near the sample is shown in Fig. 1 for $L_x = L_y = 20$, $D = 52$ and $Y = 677$.

The Hamiltonian for the sample with random impurities is given by the following tight-binding model:

$$-t\left\{\psi(x+1,y)+\psi(x-1,y)+\psi(x,y+1)e^{2\pi i(\phi/\phi_0)x}+\psi(x,y-1)e^{-2\pi i(\phi/\phi_0)x}\right\}$$
$$+Fx\psi(x,y)+U(x,y)\psi(x,y)=E\psi(x,y), \tag{1}$$

where $U(x, y)$ is the random potential and the confining walls are assumed implicitly to be located at $x = 0$ and L_x. The corresponding boundary conditions for $\psi(x, y)$ are given by $\psi(0, y) = \psi(L_x, y) = 0$. For the calculation of the conductance, $\psi(x, +Y)$ is connected with the incident and reflected waves within the probe and

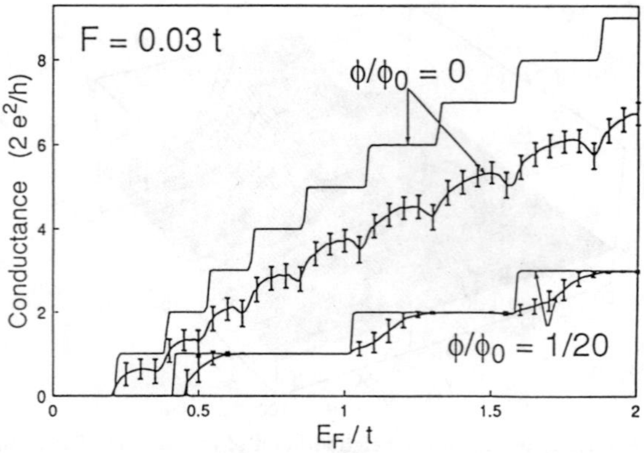

Fig. 2. The magnetic field dependence of the conductance. Thin lines show the case without the random potential and thick lines show the case without the random potential. Error bars are estimated from 1000 different samples.

$\psi(x, -Y)$ with the transmitted waves. The conductance along the y axis is evaluated as a function of the Fermi energy by the recursion method proposed by Ando[4] which is based on the Landauer formula[5,6].

3. Results and Discussion

The conductance is calculated as a function of the incident energy E_F of the electron at the probe $y = -Y$, and is shown in Fig. 2. Here the strength of the electric field is fixed to be $F = 0.03t$ and the sample size is $L_x = L_y = 20$. The conductance for the case without the random potential is plotted by thin lines and for the case with the random potential by thick lines. The strength of the random potential is distributed between $-t/2$ and $+t/2$ with the flat probability distribution. We calculate the conductance of 1000 different samples and the mean value of conductance is plotted by the thick line and the standard deviation is given by the error bars.

In the absence of the magnetic field, the conductance for the pure case shows a step-like feature with its plateau quantized in units of $2e^2/h$ This is because electrons move freely along the channel with wave vector k and the conductance for such a case is known to be quantized[7]. The step width ΔE_F for $E_F \lesssim 3t$ is characterized by the energy difference of the quantized edge states, the wave functions of which are approximated by the Airy functions[8]. When the random potential is introduced into the sample, the conductance for fixed E_F becomes smaller than the one for the pure case, and the steps become corroded, so that the step-like feature of the conductance becomes unclear. The reason for the decrease of the conductance for the impure case, i.e., the increase of the backward scattering, is that the impurities make the wave function localized in the y direction and mix the forward-going and backward-going waves and hence the transmission probability decreases.

When the magnetic field is applied, the conductance for the pure sample also shows a step-like feature with its plateau quantized in units of $2e^2/h$, and the step width is now equal roughly to the Landau level separation $4\pi t(\phi/\phi_0)$. This quantization comes from the fact that the wave function along the y direction is still characterized[2] by the wave vector k. This indicates that the electrons move freely along the y direction, and therefore the conductance is quantized. In the dirty samples, the steps of the conductance are rounded but still keep the plateaux in the middle of each step in contrast to the no magnetic field case. This is because the location of the wave functions of the forward-going and the backward-going waves are separated spatially by the distance $\sim E_F/F$, so that the mixing of these waves are reduced as compared to the case of no magnetic field. In other words, the application of the magnetic field favors the conductance quantization.

4. Conclusion

We have examined the conductance quantization of the narrow channels which are realized by the application of both electric and magnetic fields. For pure samples, the conductance quantization is clearly seen, and the step width ΔE_F reflects the energy separation either of the Airy edge states for the case of the electric field only, or of the cyclotron energy separation for the case where both electric and magnetic fields are applied. In the dirty samples, the conductance quantization is severely damaged for the case where only the electric field is present. In contrast, the quantization is only partly destroyed when both of the electric and magnetic fields are applied, i.e., the quantization of the conductance is more easily observable by the application of the magnetic field.

5. Acknowledgments

We are obliged to Professor H. Matsukawa, Dr. Y. Sakamoto and Dr. S. Yamamoto for useful discussions. One of the authors (T. N.) acknowledges the financial support from the JSPS Research Fellowships for Young Scientists.

6. References

1. D. R. Hofstadter, *Phys. Rev.* **B14** (1976) 2239.
2. B. I. Halperin, *Phys. Rev. Lett.* **57** (1982) 2185.
3. T. Nakanishi, T. Ohtsuki and M. Saitoh, *Physica* **B184** (1993) 310.
4. T. Ando, *Phys. Rev.* **B44** (1991) 8017.
5. R. Landauer, *IBM J. Res. Dev.* **1** (1957) 223.
6. R. Landauer, *Philos. Mag.* **21** (1970) 863.
7. M. Büttiker, *Phys. Rev.* **B38** (1988) 9375.
8. M. Saitoh, *J. Phys.* **C5** (1972) 914.

ENERGY–MOMENTUM RELATION FOR MAGNETOPOLARONS IN QUASI–ONE–DIMENSIONAL QUANTUM–WELL WIRES

L. Wendler and R. Kügler

Fachbereich Physik, Martin-Luther-Universität Halle, Friedemann-Bach-Platz 6, D-06108 Halle, Germany

and

R. Haupt

Institut für Festkörpertheorie und Theoretische Optik, Friedrich-Schiller-Universität Jena, Max-Wien-Platz 1, D-07743 Jena, Germany

ABSTRACT

The interaction of quasi-one-dimensional electrons and longitudinal-optical (LO) phonons in the presence of a quantizing magnetic field is calculated. Using Green's function technique, a modified Hartree-Fock approximation results in the energy-momentum relation for polarons which bends over at the phonon continuum above the renormalized ground-state. The renormalized levels as a function of the magnetic field show a minimum for nonzero magnetopolaron momentum because increasing magnetic fields decrease the kinetic energy but increase the confining energy.

Advances in growth and nanofabrication techniques have made it possible to obtain quasi-one-dimensional (Q1D) quantum-well wires (QWW) of one polar semiconductor material within another. In polar semiconductors the electronic properties are modified by *polaronic effects* in the following manner:[1,2] A quasi-particle, called *polaron*, consisting of an electron and its surrounding phonon cloud is formed. For the usual used semiconductors the coupling is weak and the *electron-phonon interaction* (EPI) leads to (i) a shift in the energy; (ii) a mass renormalization; (iii) a polaron-induced nonparabolicity and (iv) the energy-momentum relation bends over if the polaron kinetic energy approaches the energy of the LO phonon energy above the correct ground-state. The quantum confinement in two spatial directions and the quantizing magnetic field[3,4] changes these properties considerably.

The Hamiltonian of the magnetopolaron in a QWW is given in the effective-mass approximation by

$$H_p = \frac{1}{2m_e}(\mathbf{p}+e\mathbf{A})^2 + V(\mathbf{x}) + \sum_{\mathbf{q}} \hbar\omega_L (a_L^+(\mathbf{q})a_L(\mathbf{q}) + \frac{1}{2})$$

$$+\sqrt{\frac{4\pi\alpha r_p(\hbar\omega_L)^2}{V_G}} \sum_{\mathbf{q}} e^{i\mathbf{q}\mathbf{x}} \frac{1}{|\mathbf{q}|}(a_L(\mathbf{q}) + a_L^+(-\mathbf{q})). \qquad (1)$$

The first two terms represent the unperturbed electron and LO phonon system, H_e and H_{ph}, respectively and the last term is the standard Fröhlich Hamiltonian H_{ep} of the EPI[5]. Herein we assume that the electrons interact with 3D bulk LO phonons with the frequency ω_L. In Eq. (1), m_e is the effective conduction band-edge mass, α the 3D polaron coupling constant, r_p the 3D polaron radius and V_G is the volume of the crystal. For the confining potential we assume a parabolic confinement in y-direction and zero-thickness confinement in z-direction: $V(\mathbf{x}) = m_e\Omega^2 y^2/2 + V(z)$. In the unperturbed case the electron motion is described by the wave function $\langle \mathbf{x}|N, k_x\rangle = \Psi_{Nk_x}(\mathbf{x}) = L_x^{-1/2}\exp(ik_x x)\Phi_N(y)\varphi(z)$ where L_x is the length of the wire in x-direction (Born-von Kármán periodic boundary condition), $\Phi_N(y)$ is the shifted harmonic-oscillator wave function[3] and $|\varphi(z)|^2 = \delta(z)$. The corresponding eigenenergies are $\mathcal{E}_N(k_x) = \hbar\tilde{\omega}_C(N+1/2) + \hbar^2 k_x^2/(2\tilde{m}_e)$; $N = 0, 1, 2, \ldots$ where we neglect the Zeeman spin-splitting, $\tilde{\omega}_C = (\omega_C^2 + \Omega^2)^{1/2}$ is the hybrid frequency ($\omega_C = eB/m_e$: cyclotron frequency) and $\tilde{m}_e = m_e(\tilde{\omega}_C/\Omega)^2$.

To calculate the effect of the EPI on the electronic properties we use Matsubara Green's function technique. The Green's function $\mathcal{G}_{NN'}(k_x|i\nu_n)$ is given by Dyson's equation, which reads in the subband space[6]

$$\mathcal{G}_{NN'}(k_x|i\nu_n) = \mathcal{G}^{(0)}_{NN'}(k_x|i\nu_n) + \sum_{N_1,N_1'} \mathcal{G}^{(0)}_{NN_1}(k_x|i\nu_n) \times \Sigma_{N_1N_1'}(k_x|i\nu_n)\; \mathcal{G}_{N_1'N'}(k_x|i\nu_n) \qquad (2)$$

with the unperturbed single-particle Green's function

$$\mathcal{G}^{(0)}_{NN'}(k_x|i\nu_n) = \frac{\delta_{NN'}}{i\hbar\nu_n - \mathcal{E}_N(k_x) + \mu}\,. \qquad (3)$$

Herein is μ the chemical potential and $\nu_n = (2n+1)\pi/(\hbar\beta)$, $n = 0, \pm1, \pm2, \ldots$ with $\beta = (k_B T)^{-1}$. In the *Hartree-Fock approximation* the *irreducible self-energy* is given by

$$\Sigma^{HF}_{N_1N_1'}(k_x|i\nu_n) = \frac{1}{L_x}\sum_{q_x}\sum_{N_2,N_2'}\left(-\frac{1}{\beta}\right)\sum_{i\omega_n} V^{ph}_{N_1N_2N_2'N_1'}(q_x|i\omega_n)\; \mathcal{G}_{N_2N_2'}(k_x - q_x|i\nu_n - i\omega_n) \qquad (4)$$

where the EPI potential is[6]

$$V^{ph}_{N_1N_2N_3N_4}(q_x|i\omega_n) = \frac{2\omega_L/\hbar}{(i\omega_n)^2 - \omega_L^2}\, f^{LO}_{N_1N_2N_3N_4}(q_x) \qquad (5)$$

with

$$f^{LO}_{N_1N_2N_3N_4}(q_x) = 2\alpha r_p(\hbar\omega_L)^2 \int dy \int dy'\; \Phi_{N_1}(y - Y_{k_x})\, \Phi_{N_2}(y - Y_{k_x-q_x}) \times \mathbf{K}_0(|q_x(y-y')|)\, \Phi_{N_3}(y' - Y_{k_x-q_x})\, \Phi_{N_4}(y' - Y_{k_x}), \qquad (6)$$

and $\omega_n = 2n\pi/(\beta\hbar)$. $Y_{k_x} = \gamma\tilde{l}_0^2 k_x$ is the centre coordinate of the shifted harmonic-oscillator wave function with $\tilde{l}_o^2 = \hbar/(m_e\tilde{\omega}_c)$, $\gamma = \omega_C/\tilde{\omega}_C$, and $\mathbf{K}_0(x)$ is the modified Bessel function. To perform the frequency sum in Eq. (4) it is necessary to use a further approximation. Here, we assume the diagonal-approximation for $\mathcal{G}_{N_2N_2'} \approx \delta_{N_2N_2'}\, \mathcal{G}_{N_2N_2'}$

in Eq. (4). Performing the frequency sum and using the *quasi-particle approximation* (energy is determined by $Re\Sigma^{HF}_{N_1N_1'}$), we obtain:

$$\Sigma^{HF}_{N_1N_1'}(k_x|E) = \frac{1}{L_x}\sum_{q_x}\sum_{N_2}\frac{f^{LO}_{N_1N_2N_2N_1'}(q_x)}{E-\mathcal{E}_{N_2}(k_x-q_x)-\hbar\omega_L-\Sigma^{HF}_{N_2N_2}(k_x-q_x|E-\hbar\omega_L)+i\delta} \tag{7}$$

assuming a single electron at $T=0K$ and as usual $i\nu_n \to E-\mu+i\delta;\ \delta\to 0^+$. We note that for $\Sigma^{HF}_{N_2N_2}=0$ in the denominator, Eq. (7) gives the usual *Tamm-Dancoff approximation* for the self-energy. In the diagonal approximation of the Dyson equation (2) the quasi-particle energy of the Q1D magnetopolaron is given by

$$E_N(k_x) = \mathcal{E}_N(k_x) + \Sigma^{HF}_{NN}(k_x|E_N(k_x)). \tag{8}$$

Considering for the moment the energy-momentum relation for the unperturbed polaron states (n LO phonons and one electron), it is obvious that below the one-phonon energy $\mathcal{E}_0(0)+\hbar\omega_L$ there is only a finite number of subbands from zero-phonon states $\mathcal{E}_N(k_x)$. Above $\mathcal{E}_0(0)+\hbar\omega_L$ there is a continuum of states, called *phonon continuum*, for each value of the polaron momentum. The EPI couples the degenerated states, leading to a splitting of a zero-phonon state in an upper and lower branch. The upper branch is above the boundary of the phonon continuum, but the lower branch is below this line. Hence, all zero-phonon states lying below the energy $\mathcal{E}_0(0)+\hbar\omega_L$ must bend over at this line because of the anticrossing repulsion. Because the EPI renormalizes the energies, the accurate dispersion curves of all subbands must bend over at $E_0(0)+\hbar\omega_L$, i.e. the one-phonon energy $\hbar\omega_L$ above the *shifted* ground-state $E_0(0)=\mathcal{E}_0(0)+\mathrm{Re}\Sigma_{00}(0|E_0(0))$ of the system. This pinning behaviour must be described by Eq. (8). We calculate the Hartree-Fock self-energy $\Sigma^{HF}_{N_1N_1}(k_x|E)$ by setting $\Sigma^{HF}_{N_2N_2}(k_x-q_x|E-\hbar\omega_L)\to\Sigma^{HF}_{00}(0|E_0(0))$ in the denominator of Eq. (7). This approximation allows the calculation of the self-energy, gives the correct pinning behaviour and corresponds to the *improved Wigner-Brillouin perturbation theory.*[3,4]

For numerical calculation we have used a GaAs-$Ga_{1-x}Al_xAs$ QWW (GaAs: $\alpha=0.07$, $r_p=3.987nm$, $\hbar\omega_L=36.17meV$, $m_e=0.06624m_0$, m_0: bare electron mass) where the electrons are confined within GaAs.

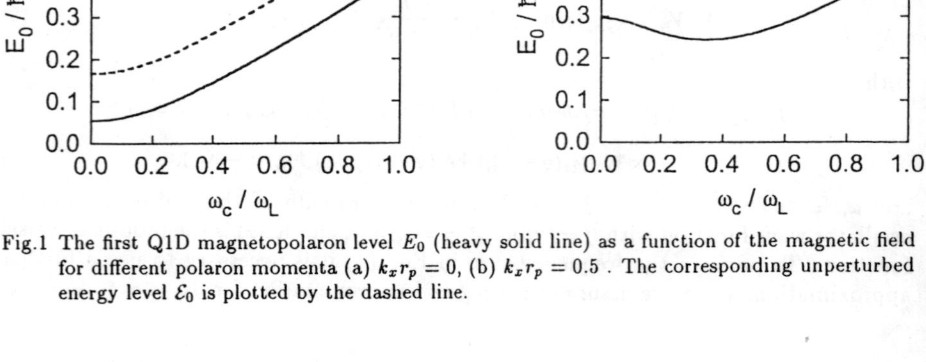

Fig.1 The first Q1D magnetopolaron level E_0 (heavy solid line) as a function of the magnetic field for different polaron momenta (a) $k_xr_p=0$, (b) $k_xr_p=0.5$. The corresponding unperturbed energy level $\mathcal{E}_0$ is plotted by the dashed line.

The calculated magnetic field dependence of the lowest subband is plotted in Fig.1 including polaron effects in the Hartree-Fock approximation. It is seen that for vanishing polaron momentum the energy $E_0 = E_0(k_x = 0, B)$ increases with increasing magnetic field. For a finite polaron momentum the curves $E_0(k_x \neq 0, B)$ have a minimum. This minimum results from the fact that with increasing magnetic field the effective mass increases ($\tilde{m}_e \to \infty$ for $B \to \infty$) and hence the kinetic energy decreases but the confining energy increases. It can be seen from Fig.2, where the energy-momentum relation of the Q1D magnetopolaron is plotted, that the magnetopolaron dispersion curve $E_0(k_x)$ bends over at $E_0(0)+\hbar\omega_L$. The bend-over region shifts to higher magnetopolaron momenta with increasing magnetic field.

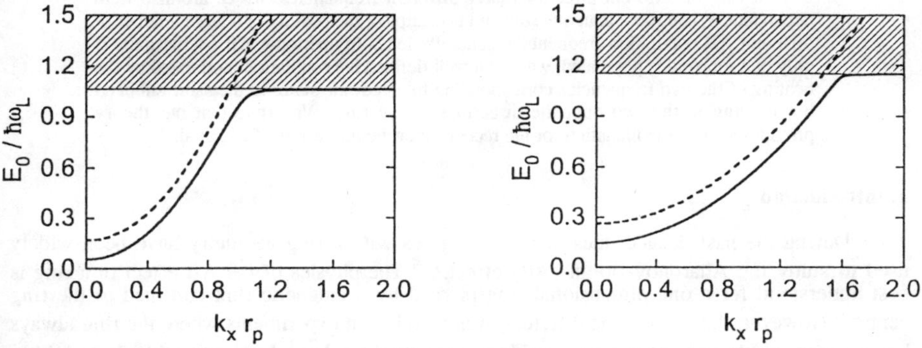

Fig.2 Energy-momentum relation for Q1D magnetopolarons in the lowest energy level: (a) $\omega_C/\omega_L = 0$ and (b) $\omega_C/\omega_L = 0.4$. The corresponding unperturbed energy level $\mathcal{E}_0$ is plotted by the dashed line. The renormalized phonon continuum is plotted by single shaded area.

In summary, we have modified the Hartree-Fock approximation of the self-energy to calculate the energy-momentum relation of Q1D magnetopolaron. The levels are shifted to lower energies and pinned at the level $E_0(0) = \mathcal{E}_0(0) + \Sigma_{00}^{HF}(0|E_0(0))$. The renormalized levels show as a function of the magnetic field a minimum for a finite magnetopolaron momentum because increasing magnetic fields decreases the kinetic, but increases the confining part of the energy of the Q1D magnetopolaron.

REFERENCES

1. C. Kuper and G. Whitfield, Eds., *Polarons and Excitons* (Oliver and Boyd, Edinburgh, 1963).
2. J. T. Devreese, Ed., *Polarons in Ionic Crystals and Polar Semiconductors* (North-Holland, Amsterdam, 1972).
3. L. Wendler, A. V. Chaplik, R. Haupt, and O. Hipólito, *J. Phys.: Condens. Matter* **5**, 4817 (1993).
4. L. Wendler and R. Kügler, *J. Phys.: Condens. Matter*, in press.
5. H. Fröhlich, *Adv. Phys.* **3**, 325 (1954).
6. L. Wendler and R. Pechstedt, *Phys. Stat. Sol. (b)* **138**, 197 (1986).

AHARONOV-BOHM EFFECT IN A RING WITH FINITE WIDTH IN THE TUNNELLING REGIME

W.-C. TAN and J. C. INKSON
Department of Physics, University of Exeter,
Exeter, EX4 4QL, UK

Exact energy spectrum and wavefunctions are obtained for a conducting ring of finite width in a perpendicular magnetic field. For the case where the ring is weakly coupled to emitter and collector, the conductance of the ring is calculated using the Landauer formula. Due to the magnetic flux piercing through the conducting region of the ring, the Aharonov-Bohm (AB) oscillations related to electronic states in different subbands and moving in different directions have different frequencies, which are also field dependent. When more than one subband is occupied, the beating of a large number of different AB oscillation frequencies generally leads to aperiodic AB oscillations. If only one subband is populated, we find a well-defined h/e oscillation modulated by the beating of the two frequencies corresponding to electronic states in the same subband but moving in the two opposite directions in the ring. We argue that our theory provides a simple explanation for the recent experimental results of Liu et al.

1. Introduction

During the past decade, mesoscopic structures with a ring geometry have been widely used to study the Aharonov-Bohm (AB) effect.[1-6] The physics of the AB effect in a ring is best understood for a one dimensional ring pierced by a magnetic flux confined to the ring centre.[5] However, the situation is different in most relevant experiments where the ring always has a finite width and subject to a uniform magnetic field.[2-3] In a ring with finite width, different electronic states have different radial wavefunctions, and will circle different amount of magnetic flux. This will result in an AB interference pattern very different from that in a one dimensional ring. Although such finite width effects have long been noticed by previous researchers,[1,2,6] a complete theoretical description has not yet been available.

In a recent experiment, Liu *et al*[2] studied the correlation between AB effects and the one-dimensional subband population in $GaAs/Al_xGa_{1-x}As$ rings in the ballistic regime. They found that, while ordered periodic AB oscillations were observed for devices having one filled subband in the ring, the AB interference patterns appear random for devices with more than one occupied subbands. They also found beating of different frequencies in the AB interference patterns *even for the case only one subband is occupied.* No conclusive explanation for such random features and beatings has been given in Ref (2).

In this paper, we propose a model potential that describes an isolated ballistic ring with finite width. Exact energy spectrum and wavefunctions can be obtained for this model with a uniform magnetic field applied perpendicular to the ring. To consider the AB oscillations, we weakly couple the ring to two leads. The conductance of the ring is calculated by using the Landauer formula. We show that the correlation between the AB effects and the subband population reported in Ref.(2) is a direct result of the finite ring width.

2. The Ring Model

We consider a two dimensional ring which is defined by a radial potential

$$V(r) = \frac{a_1}{r^2} + a_2 r^2 - V_0 \tag{1}$$

where $V_0 = 2\sqrt{a_1 a_2}$. Such a potential defines a ring with an average radius $r_0 = (a_1/a_2)^{1/4}$. For r near r_0, the transverse confining potential of the ring arm has the simple parabolic form

$V(r) \cong \frac{1}{2}\omega_0^2(r-r_0)^2$ with $\omega_0 = \sqrt{8a_2/m^*}$. The effective width of the ring at a given Fermi energy E_f is then $\Delta r \approx \sqrt{8E_f/m\omega_0^2}$.

This model is very flexible. Both the radius and the width of ring can be adjusted independently by changing parameter a_1 and a_2. V(r) can also be used to describe a quantum dot[4] (taking a_1=0)and an anti-dot (taking a_2=0). The most important advantage of the model is that it can be solved analytically. If a uniform perpendicular magnetic field B is applied, we find that the eigen values and engen wavefunction (apart from a renormalizaton constant) for the ring defined by Eq. (1) are given by[7]

$$E_{n,m} = \left[\left(n+\frac{1}{2}+M\right)\hbar\omega\right] - \frac{m}{2}\hbar\omega_c - \frac{m^*}{4}\omega_0^2 r_0^2, \quad n = 0, 1, 2, \cdots, m = \cdots, -1, 0, 1, \cdots \tag{2}$$

$$\Psi_{n,m}(r,\varphi) = r^M \exp\left[-\frac{1}{4}(r/\lambda)^2 + im\varphi\right]{}_1F_1\left(-n, M+1, \frac{1}{2}(r/\lambda)^2\right) \tag{3}$$

where

$$\omega_c = eB/m^*, \quad \omega = \sqrt{\omega_c^2+\omega_0^2}, \quad \lambda = \sqrt{h/m^*\omega}, \quad M = \sqrt{m^2 + 2a_1 m^* \hbar^{-2}} \tag{4}$$

and ${}_1F_1$ is the hypergeometric function.

The quantum numbers n and m characterise the radial motion and the angular momentum respectively. It can be shown[7] that the average radius of the (n,m)th state is $r_{n,m} = (2M)^{1/2}\lambda$ and all subband minima lie at the same m value given by $m_0 = eBr_0^2/2\hbar$. It is interesting to note that m_0 is exactly the number of quantum flux encircled by a one dimensional ring of radius r_0.[5] However, in the present case, the subband dispersions, given by Eq. (2), are non parabolic and no longer symmetric about the subband minima at non zero magnetic field.

3. Aharonov-Bohm Oscillation

To investigate the effect of finite width on the Aharonov-Bohm oscillation, we consider a ballistic ring coupled to two reservoirs (the emitter and collector). We ignore the Zeeman splitting of the electronic state because it is much smaller than the average level separation in the GaAs ring considered here. (Generally, the Zeeman splitting should be included when the magnetic field is very strong). To make the physics transparent, we assume further that the couplings between the ring and leads are very weak, and therefore the electron in one lead can reach the other one only by tunnelling through the quasi-bound states in the ring. In such case, the transmission through the ring states has the simple Breit-Wigner form[8]. Using Landauer formula, we can then express the magneto conductance as

$$G(B) = \frac{2e^2}{h}\sum_{n,m}\frac{\Gamma^e_{n,m}\Gamma^c_{n,m}}{\left(E_f - E_{n,m}(B)\right)^2 + \left(\left(\Gamma^e_{n,m}+\Gamma^c_{n,m}\right)/2 + \Gamma^i_{n,m}\right)^2} \tag{5}$$

where E_f is the Fermi energy, and $E_{n,m}$ the energies of the bound ring states given in Eq.(2). $\Gamma^e_{n,m}\left(\Gamma^c_{n,m}\right)$ and $\Gamma^i_{n,m}$ are broadening of the (n, m)th ring state caused by leaking into the emitter (collector) and inelastic scattering respectively. The elastic broadening is determined by the overlapping between the bound ring state and the extended states in the leads, which can be calculated using the wavefunctions given in Eq.(3). For simplicity, we assume here that $\Gamma^e_{n,m}$ and $\Gamma^e_{n,m}$ are a constant Γ^e. This should be a good approximation as long as only few subbands are occupied in the ring and the magnetic field is not very strong (i.e. $\omega_c < \omega_0$).

Fig.1 shows the calculated magnetoresistance ($R=G^{-1}$) for two different subband populations in the ring. To mimic the GaAs/AlGaAs ring used in the experiment of Liu et al[2], we take the ring parameter as: r_0=800nm, $\hbar\omega_0$=0.45 meV and the effective mass m*=0.067 (in unit of free electron mass). The subband populations can then be adjusted by changing the Fermi energy E_f. We have taken $\Gamma^i_{n,m} = \Gamma^i = 0.004$meV to model the thermal broadening at an experimental temperature[2] of T= 40mK. The elastic broadening is taken to be Γ^e=0.005meV.

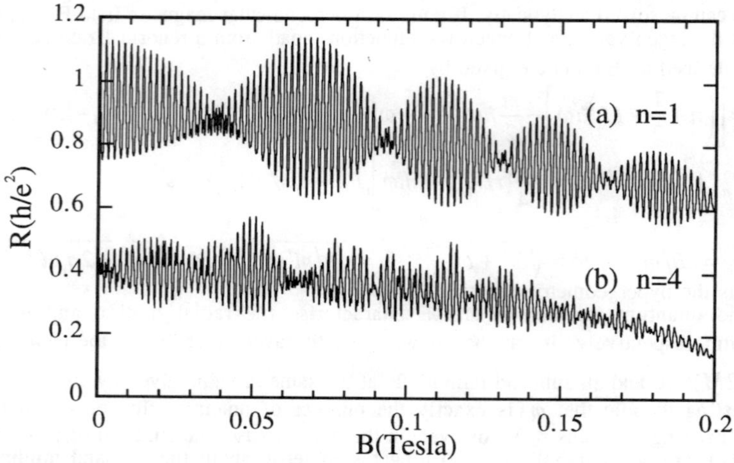

Figure 1 The calculated magneto resistance of a GaAs ring with r0=800nm at different numbers of populated subbands: (a) n=1; (b) n=4.

The curve (a) in Fig. 1 is calculated with E_f=0.5meV. In this case only one spin degenerate subband (n=1) is occupied, and $\Delta r \approx 150$ nm. In very good agreement with the experimental result of Liu et al[2], the magneto resistance in curve (a) presents the fundamental h/e AB oscillation ($f_0 \sim 500T^{-1}$) moderated by a low frequency ($\Delta f \sim 20\text{-}30T^{-1}$) oscillation (or beating).

In the present model, the explanation of such beating effect is straight forward. As shown in Eq.(2) the eigen energy of an electronic state in the ring is strongly magnetic field dependent. Whenever the eigen energy of a ring state is equal to the Fermi energy, the resonant tunnelling through the state will result in a peak in the conductance. There are two kinds of states in each subband, ones that move in the clockwise (m>m_0, $\partial E_{n,m}/\partial m > 0$) and counter clockwise (m<m_0, $\partial E_{n,m}/\partial m < 0$) directions of the ring respectively. On changing the magnetic field strength, the resonant tunnelling through clockwise and anticlockwise states will result in two nearly periodic (h/e) oscillations in the conductance. The clockwise and counter clockwise states have different average radius and encircle different amount of magnetic flux, and therefore the two oscillation frequencies f_+ (for $m > m_0$) and f_- (for $m < m_0$) are generally different. The difference between f_+ and f_- then leads to the beating pattern in curve (a).

At very weak magnetic field, the electronic states near the Fermi energy moving in different directions have nearly the same average radius and therefore $f_+ \approx f_- = f_0$. As B increases, $\Delta f = f_+ - f_-$ increases because the m>m_0 and m<m_0 states near the Fermi energy are pushed towards the outer edge and inner edge of the ring respectively. The increasing of beating frequency with increasing magnetic field is clearly shown in curve (a). However, Δf

does not always increase with increasing B. When the magnetic field is so strong that a subband is nearly depopulated, Δf will decrease with increasing B because the up shift of the subband minimum by the magnetic field effectively forces both the clockwise and anticlockwise states at the Fermi energy back to the middle of the conducting region of the ring. This occurs at a field well above that shown in Fig. 1(a).

If n subbands in the ring are populated, the oscillation in the magneto conductance will contain $2n$ frequencies corresponding to the clockwise and counter clockwise states in the n subbands. Hence, the AB oscillations will generally appear aperiodic because of the existence of a large number of different frequencies. An example of such aperiodic oscillation is displayed in curve (b) of Fig. 1, where we have taken E_f=2.0meV to give n=4 and an effective width of the ring of $\Delta r \approx$300nm. In curve (b), the magnetoresistance is dominated by apparently random fluctuations in oscillation amplitude, which is very similar to the experimental results of Liu et al[2]. The detail structures of such fluctuations are related to the width and the number of populated subbands of the ring[2], and also sensitive to the temperature and the coupling strength between the ring and leads. We will discuss such details in a future publication.

It should be pointed out that the apparently aperiodic but reproducible oscillations shown in curve (b) are different from the universal fluctuations[1] observed in mesoscopic metallic samples. The universal fluctuations are due to scattering of disordered potential, and therefore completely random. The aperiodic oscillations here, however, superposition of a number of oscillations resulting from the finite width of the ballistic ring, are non random in origin.

4. Conclusion

We have provide an analytical model for studying the AB effect in a ballistic ring of finite width. Our results reproduced the general features of the AB oscillations observed in a recent experiment[2]. Although our model calculation is in the tunnelling regime, we argue that the main results should be applicable as long as the couplings between the ring and the leads are weak (such as the set up of Liu *et al*[2]). For a complete understanding of the magneto transport in the in the strong coupling limit one needs to consider the detail scattering process at the two junctions.[5]

1. For reviews see S. Washburn and R. A. Webb, Rep. Prog. Phys. **55** (1992)1311; C. W. J. Beenakker and H. van Houten, in Solid State Physics **44**, ed. H. Ehrenreich and D. Turnbull (Academic, London, 1991); Mesoscopic Phenomena in solids, ed. B. L. Altshuler, P. A. Lee, and R. A. Webb (North-Holland, Amsterdam, 1991)
2. J. Liu, W. X. Gao, K. Ismail, K. Y. Lee, J. M. Hong, and S. Washburn, Phys. Rev. B **48**(1993)15148
3. G. Timp *et al,* Phys. Rev. Lett. **58** (1987) 2814; C. J. B. Ford *et al*, J. Phys. C. Solid State Phys. **21**(1988)L325; J. B. Pieper and J. C. Price, Phys. Rev. Lett. **72** (1994) 3586
4. G. W. J. Beenakker, H. van Houten, and A. A. M. Staring, Phys. Rev, B **44** (1991)1657
5. Y. Gefen, Y. Imry and M. Ya. Azbel, Phys. Rev. Lett. **52**(1984)129; M. Büttiker, M. Ya. Azbel Phys. Rev. A**30** (1984) 1982
6. Y. Avishai and S. A. Gredeskul, Phys. Rev. B **47** (1993) 6395.
7. W. -C. Tan and J. C. Inkson (to be published)
8. A. D. Stone and P. A. Lee, Phys. Rev. Lett. **54**(1985)1196

OPTICAL SINGULARITIES IN QUANTUM WIRES IN HIGH MAGNETIC FIELDS

J.M. CALLEJA
Universidad Autónoma de Madrid. Departamento de Física de Materiales.
Cantoblanco, E-28049 Madrid, Spain.

A.R. GOÑI
Max-Planck Institut für Festkörperforschung. Heisenbergstr. 1.
D-70506 Stuttgart, Federal Republic of Germany.

A. PINCZUK, B.S. DENNIS, J.S. WEINER, L.N. PFEIFFER and K.W. WEST
AT&T Bell Laboratories. 600 Mountain Ave. Murray Hill,
N.J. 07974 USA

1. Introduction

In this communication we report the effects of electron-electron interactions on the photoluminescence (PL) spectrum of a one-dimensional electron gas (1DEG) in an external magnetic field. The continuous change of the electronic level separation and of the electron effective mass by the field allows the study of the many-body optical Fermi-edge singularities (FES) under conditions of reduced Fermi energy and complete spin-polarization of the 1DEG.

2. Experiment

Our 1DEG is contained in quantum wires patterned in a modulation-doped single GaAs quantum well by electron beam lithography and low-energy argon bombardment[1]. The Fermi energy at zero magnetic field E_F^0 and the 1D-intersubband spacing E_{01} have been determined by inelastic ligth scattering[2] as 2.0 and 2.5 meV respectively. Consequently only the lowest electron 1D-subband is populated.

3. Results and discusion

At low fields the PL spectra (Fig.1) are dominated by a strong FES[1]. The value of E_F is obtained from the width of the emission, as indicated in the zero-field spectrum. E_F is continuously reduced with increasing field as shown in Fig.2. This is attributed to the increase of the effective mass with the field, which is well described, for a parabolic lateral potential, by the equation in the inset[3] of Fig.2.

There w_c is the cyclotron frequency. Above 2T the PL FWHM saturates around 0.6 meV due probably to wire-width inhomogeneities. The actual value of E_F has to be smaller, considering that for increasing fields the distance between Landau orbits is reduced and the electrons are accomadated in a narrower spatial region around the potential minima. The weak peak shown in Fig.1 at a 1.5 meV above the main one is attributed to the spatial modulation of the lowest Landau level by the periodic wire potential[4]. This modulated level has maxima in its density of states at the energy minima in the wires (populated by electrons) and maxima in the wire-barriers (empty or residually populated), giving rise to the observed peaks.

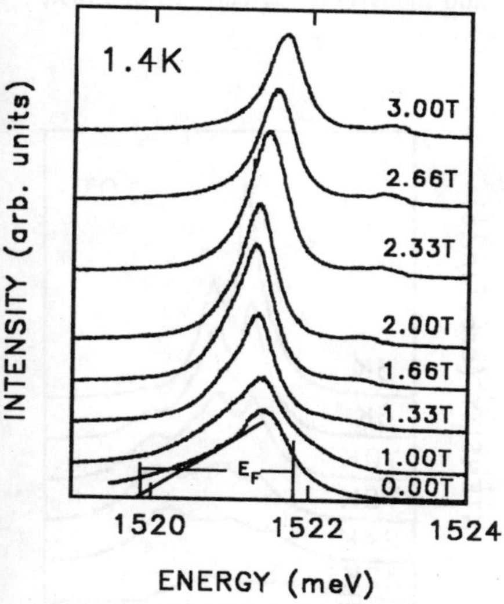

Fig.1. PL spectra for different magnetic fields at 1.4K. E_F is obtained as indicated in the zero field spectrum.

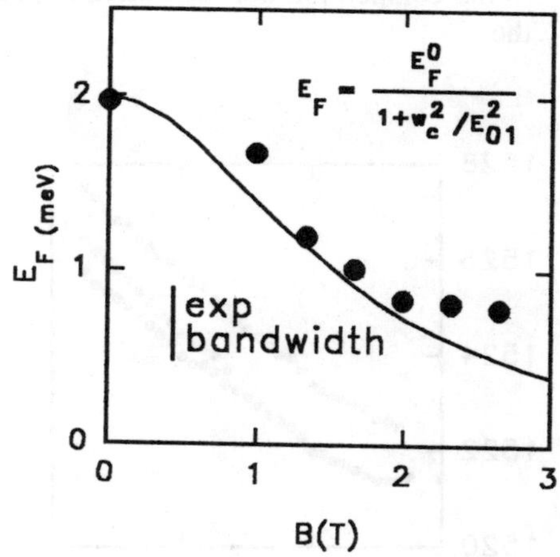

Fig.2. Experimental (dots) and theoretical (solid line) values of E_F as a function of the magnetic field.

The PL peak energies as a function of the magnetic field are shown in Fig.3. One observes that the cyclotron shift dominates over the 1D confining potential above 2T. The energy separation between the low-field PL peaks remains essentially constant until 11T, giving support to our interpretation of their origin. Above 6T the the lowest transition splits into two peaks involving different electron and hole Zeeman components with an effective g factor is close to 1. This intrinsic PL doublets correspond to the ground state in the fractional Quantum Hall effect regime observed in high mobility 2D electron gases.

At 10T the Zeeman splitting of the conduction band (0.25 meV) is expected to be bigger than E_F, so that the 1DEG is spin-polarized. Indeed, the temperature dependence of the PL spectra at 10T presented in Fig.4 shows the rapid intensity decrease characteristic of the optical FES, only for the peak of lowest energy. The higher energy component shows a much slower intensity-decrease which could be due to the competition between thermal population and many-body effects. In contrast, the

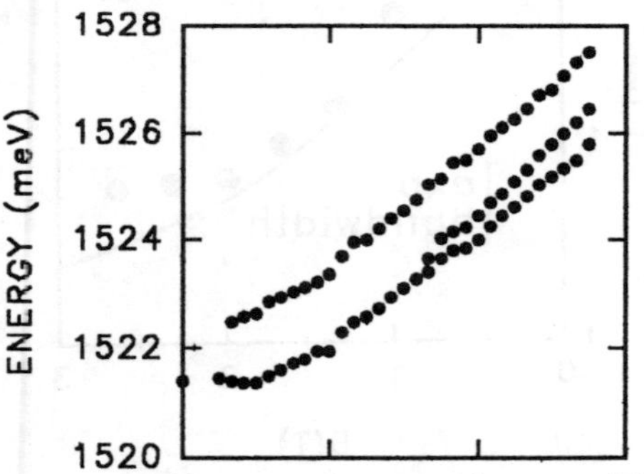

Fig.3. PL peak energies as a function of the magnetic field.

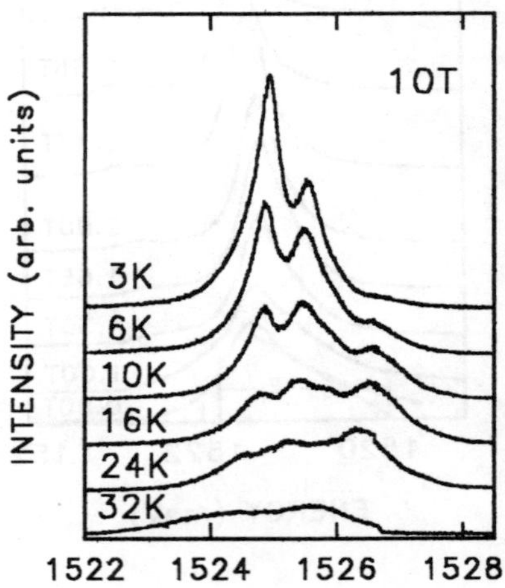

Fig.4. PL spectra at 10T for different temperatures.

peak at the highest energy, which is absent below 3K, is thermally enhanced until it reaches an intensity maximum at 24K. It shows a similar splitting, clearly observed at 16K. Above 24K the luminescent intensity decreases rapidly.

4. Conclusions

In conclusion, we have studied the optical emission of a 1DEG, whose Fermi energy can be continuously varied by an external magnetic field. The many-body enhancement of the PL intensity and its rapid thermal suppression are still observed when E_F becomes smaller than the inhomogeneous width of the optical transition. The FES is also present in the spin-polarized 1DEG. Its intensity does not depend on the magnetic field, indicating that there is no significant coupling of the electron gas to empty conduction states in the present case. The spatial modulation of the Landau levels by the periodic wire potential is also observed in the optical spectra.

5. References

1. J.M. Calleja, A.R. Goñi, B.S. Dennis, J.S. Weiner, A. Pinczuk, S. Schmitt-Rink, L.N. Pfeiffer, K.W. West, J.F. Müller and A.E. Ruckenstein. Solid State Commun. **79**, 911 (1991); Surf.Sci. **263**, 346 (1992).
2. A.R. Goñi, A. Pinczuk, J.S. Weiner, B.S. Dennis, L.N. Pfeiffer and K.W. West. Phys. Rev. Lett. **70**, 1151 (1993).
3. D. Childers and P. Pincus. Phys. Rev. **177**, 1036 (1969).
4. U. Wulf, V. Gudmundsen and R.R. Gerhardts. Phys.Rev. **B38**, 4218 (1988).
5. A. Pinczuk, B.S. Dennis, L.N. Pfeiffer and K.W. West. Phys. Rev. Lett. **70**, 3983 (1993) and references therein.

Second–harmonic generation at high magnetic fields in quasi–ballistic GaAs/AlGaAs quantum wire

J. WRÓBEL[1], F. KUCHAR[2], G. GRABECKI[1], T. DIETL[1],
K. ISMAIL[3,4], K. Y. LEE[3], R. LÖSCH[5], W. SCHLAPP[5]

[1] *Institute of Physics, Polish Academy of Sciences, Warszawa, Poland*
[2] *Institut für Physik, Montanuniversität Leoben, Austria*
[3] *IBM Thomas J. Watson Research Center, Yorktown Heights, N.Y., USA*
[4] *Faculty of Engineering, Cairo University, Giza, Egypt*
[5] *Deutsche TELECOM, Forschungs- und Technologie–Zentrum, Darmstadt, Germany*

ABSTRACT

We have measured second-harmonic generation and quantized linear conductance of high-mobility GaAs/AlGaAs quantum wire. The nonlinear signal, which is strong at low magnetic fields and also between the quantized plateaux, disappears completely when the linear conductance $G = 2e^2/h$. However, it does not disappear and is quite strong when $G = e^2/h$. We attribute this to the small spin-splitting in quantum wires, which leads to the voltage induced nonlinearities at high magnetic fields.

Already at the early stages of mesoscopic physics development it was realized that the electron transport in small samples was sensitive to the impurity distribution and to the electric field in the system. This should lead to a new type of nonlinear effects, not observed in the macroscopic conductors.[1] Strong nonlinear conductance have been measured by Webb, Washburn and Umbach[2] on submicron metal wires and rings and by de Vegvar *et al.*[3] on GaAs devices. In these experiments, however, the measurements were restricted to zero and relatively weak magnetic fields. Therefore less is known about nonlinear effects for quantum transport at high magnetic field range, where the quantum interference is suppressed and the conductance quantization is successfully described by the linear response theory approach.[4]

In this work we present experimental evidence for second-harmonic generation detected in a narrow electron channel up to the extreme quantum limit where the linear conductance plateaux, corresponding to filling factors $\nu = 2$ and $\nu = 1$, are observed. We studied the linear and nonlinear conductance in a two–terminal quasi–ballistic GaAs/AlGaAs heterojunction wire of length $L = 20$ μm and physical width $W = 0.33 \pm 0.05$ μm, in the temperature range 0.04–0.5 K and perpendicular magnetic fields up to 9 T, using dilution refrigerator cryostat. The wire was fabricated using e-beam lithography and the wet etching technique. We did not illuminate the sample after cooling-down in the dark. Assuming the parabolic confinement potential for quantum wire we determined the linear electron density $n_{\rm L} = 5.1 \pm 0.2 \times 10^6$ cm^{-1}, 1D level spacing $\hbar\omega_0 = 0.45 \pm 0.05$ meV and $N = 12$ occupied levels. The details of fabrication procedure and the method of sample parameters determination were described earlier.[5,6]

To measure conductance, a constant *ac* current I, with carefully filtered low frequency f was fed into the sample. The voltage was measured between the same leads using lock-in amplifier synchronized to f and then to $2f$ harmonic. In Fig. 1,

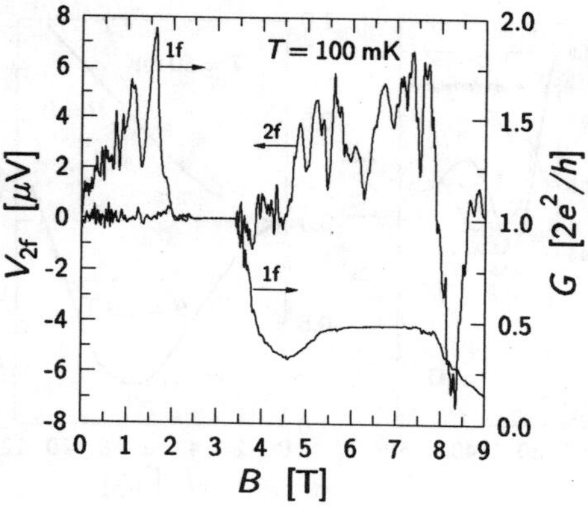

FIG. 1. V_{2f} (left y-axis) and G (right y-axis) as a function of magnetic field B, measured at $I = 10$ nA (V_{2f}) and $I = 1$ nA (G).

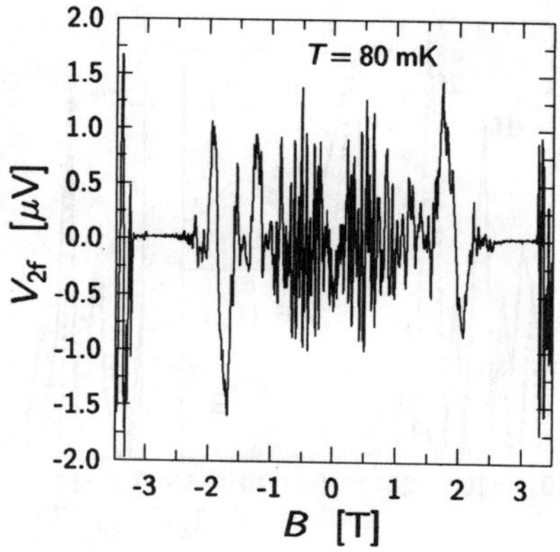

FIG. 2. $V_{2f}(B)$ measured 24 hours after recording data shown in Fig. 1. In spite of slightly larger amplitude the voltage fluctuations are reproducible with good accuracy.

the second harmonic in voltage V_{2f}, measured with $I = 10$ nA is shown as a function of magnetic field and compared to the linear conductance G (measured with $I = 1$ nA). When G exhibits the universal conductance fluctuations (UCFs), V_{2f} is strong and also fluctuates as a function of B. For higher fields, less 1D-subbans are occupied and G approaches a step-like quantized regime. A striking feature of data shown in Fig. 1 is that the horizontal "line" between 2.5 and 3.4 T belongs to the both curves. In other words, when G is quantized (with $\nu = 2$) the nonlinear signal vanishes. However, it is not the case of higher magnetic fields, where the second-harmonic signal is very strong. These features remain the same for different cool-down/warm-up thermal cycles, in spite of changes in the absolute amplitude of the second-harmonic signal.

Figure 2 shows V_{2f} measured at $T = 80$ mK for two opposite directions of the magnetic field. Up to ~ 1 T the nonlinear conductance is symmetric in B, yet at higher fields it is approximately antisymmetric. In the simple picture of non-linear transport, the V_{2f} generation may result from the left-right asymmetry in the transmission probabilities occurring in the Landauer-Büttiker formula. Therefore, the V_{2f} minimum, at say $B = +2$ T, corresponds to the observed experimentally maximum at -2 T, since the direction of edge currents flow is reversed. The question arises whether the nonlinear conductance can be analyzed in terms of linear response theory formula. This problem was recently addressed theoretically by several authors.[7,8,9] To relate our experiment to the theory we compare G and V_{2f} at some characteristic ranges of the magnetic field.

Figure 3a shows the Fourier power spectra of linear conductance and second-harmonic voltage obtained in the "UCF range" (0–0.8 T). It is evident that the second harmonic signal has some dominant frequency as a function of B. Also UCFs are rather "quasiperiodic" with a dominant peak in the Fourier spectrum. We checked by a numerical estimate that the depopulation of the 1D subbands cannot generate such oscillations in B. We think that these quasiperiodicities might originate from

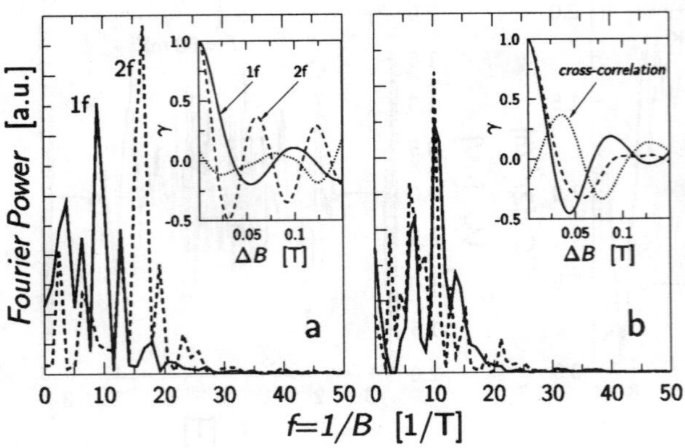

FIG. 3. Fourier power spectra for V_{2f} (- - -) and for G (—), obtained at a) 0–0.8 T and b) 3.4–4.6 T (spectra for G were calculated after substraction of the slowly varying background conductance). Insert shows the field shifted correlation coefficient γ (cross-correlation is shown with dotted line).

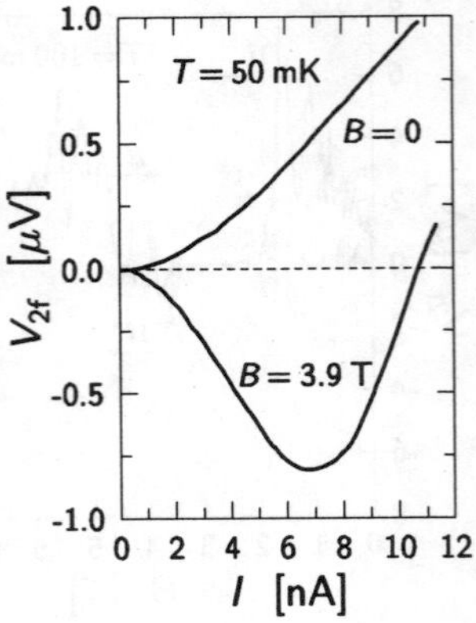

FIG. 4. V_{2f} vs current through the sample at magnetic field 0 and 3.9 T. Zero voltage level is indicated with the dashed line.

the possible spacial correlations in the scattering potential, reported already for GaAs heterostructures.[10] On the other hand, the power spectra maxima of linear and nonlinear signals occur at different frequencies . At zero magnetic field $V_{2f}(I)$ follows power law dependence at the lowest applied currents and becomes linear in I for $I > 3$ nA (see Fig.4). A similar current dependence and small cross-correlation coefficient between linear and nonlinear transport were reported in the de Vegvar *et al.* paper[3] where it was concluded that the quantum interference was responsible for higher harmonics generation at low magnetic fields.

Admittedly, for high magnetic fields the linear conductance and second harmonic signal are more closely related. When $G = 2e^2/h$ the edge currents are transmitted from "left-to-right" and "right-to-left" with the same probability (equal 1) and second-harmonic signal vanishes, provided the probabilities are not altered by large current. Recently Heinonen and Johnson[9] have argued that the Landauer-Bütiker formula $G = (2e^2/h)N$ is valid regardless of the voltage/current amplitude. Our "nonlinear conductance quantization" (for $\nu = 2$) is in agreement with their prediction.

For still higher fields we approach the "antiresonance"[6] region between two quantized plateau, where G decreases and V_{2f} is very strong. Figure 3b shows the Fourier power spectra and correlation functions calculated at $3.4 < B < 4.6$ T for first- and second-harmonic signals. Both spectra show maxima at the same frequencies and cross-correlation factor approaches 0.5. It indicates that G and V_{2f} are determined by the same physical process. Most probably the relevant mechanism is the resonant tunneling between edge currents occurring via localized states situated close to the center of the wire. On one hand, such process will create the quasiperiodic oscillations at the onset of quantized plateau, observed in linear transport[11]. On the other hand, as calculated by Ng[8], strong nonlinear effects are expected, when the voltage across the sample is large compared with the energy range where tunnelling probability changes considerably. It corresponds to our experimental situation. Furthermore, the resonant tunneling depends also on the voltage difference applied across the sample. For example, at $B = 3.9$ T $V_{2f}(I)$ behaves nonmonotonically (see Fig. 4) since the position of the nonlinear signal peaks depends on the external voltage or current.

A similar physical reasoning explains why the second-harmonic signal is not "quantized" for $B > 5$ T. The point is that G is not precisely equal e^2/h for $\nu = 1$ plateau. Furthermore, the conductance decreases with larger current and with increasing temperature. We explained it previously[6] in terms of the Kinaret and Lee[12] model which predicted the absence of exchange induced enhancement of the g-factor for quantum wires. For $I = 10$ nA the voltage across the sample is comparable to the Zeeman splitting (with bare GaAs g-factor). Therefore, one should expect voltage induced backscattering, which in turn gives rise to the breakdown of the Hall plateau and to the nonlinear effect. Moreover, such mechanism should be weakest at the center of the plateau, as observed in the experiment. It is purely 1D effect, however, since for the 2D systems the quantization is perfect, even for voltages exceeding the Landau level energy spacing.

In conclusion, we observed the presence of some characteristic frequencies in the linear and nonlinear magnetoconductance fluctuations. We detected a strong nonlinear effect in the situation of resonant tunnelling between the edge currents. Finally, we observed the second harmonic "quantization" for $\nu = 2$ but not for $\nu = 1$, when the level splitting is small.

This work was supported by grant BWF 30.405/3 and 45.283/2, Austria and grant KBN 204669101, Poland.

References

1. B. L. Al'tshuler and D. E. Khmielnitskii, JEPT Lett. **42**, 359 (1985)
2. R. A. Webb, S. Washburn, and C. P. Umbach, Phys. Rev. B **37**, 8455 (1988)
3. P. G. N. de Vegvar, G. Timp, P. M. Mankiewich, J. E. Cunningham, R. Behringer, and R. E. Howard, Phys. Rev. B **38**, 4326 (1988)
4. M. Büttiker Phys. Rev. Lett. **57**, 1761 (1986)
5. J. Wróbel, F. Kuchar, K. Ismail, K. Y. Lee, H. Nickel, and W. Schlapp, Surf. Sci. **263**, 261 (1992)
6. J. Wróbel, F. Kuchar, K. Ismail, K. Y. Lee, H. Nickel, W. Schlapp, G. Grabecki, and T. Dietl, Surf. Sci. **305**, 615 (1994)
7. P. G. N. de Vegvar Phys. Rev. Lett. **70**, 837 (1993)
8. T. K. Ng, Phys. Rev. Lett. **68**, 1018 (1992)
9. O. Heinonen and M. D. Johnnson, Phys. Rev. Lett. **71**, 1447 (1993)
10. A. M. Chang, G. Timp, J. E. Cunningham, P. M. Mankiewich, R. Behringer, R. E. Howard, and H. U. Baranger, Phys. Rev. B **37**, 2745 (1988)
11. J. K. Jain and S. A. Kievelson, Phys. Rev. Lett. **60**, 1542 (1988)
12. J. M. Kinaret and P. A. Lee, Phys. Rev. B **42**, 11768 (1990)

MESOSCOPIC PHOTOVOLTAIC EFFECT IN QUANTUM WIRES AND SMALL RINGS IN STRONG MAGNETIC FIELDS

A. A. Bykov, G. M. Gusev, Z. D. Kvon, E. B. Ol'shanetskii, S. P. Moshchenko

Institute of Semiconductor Physics,
Russian Academy of Sciences, Siberian Branch.
Prosp. Lavrent'eva, 13, 6300090 Novosibirsk, RUSSIA.

Abstract

Mesoscopic photovoltaic effect, originating under microwave radiation in GaAs/AlGaAs quantum wires and small rings has been studied in the limit of strong magnetic field ($\omega_c\tau > 1$). The microwave e.m.f. frequency dependencies were found to differ significantly in the cases of weak and strong magnetic field. Exponential decay of the e.m.f. amplitude with magnetic field in the quantum Hall regime suggests that tunnel coupling of edge current states plays a crucial role in the origination of the photovoltaic effect.

It was shown theoretically [1] and afterwards experimentally proved [2,3] that application of microwave radiation to a mesoscopic conductor produces a photovoltaic effect. The principal reason for its occurrence lies with the asymmetry of the coherent scattering of nonequilibrium electrons to the left and to the right contacts of the sample due to the random distribution of impurities. Thus, a steady photocurrent arises whose direction and magnitude are determined by a particular impurity potential. In weak magnetic fields the asymmetry factor α is of the same order as the ratio of the amplitude of universal conductance fluctuations e^2/h to the conductance G [1], $\alpha \sim e^2/hG$. In much the same way as conductance, α and related to it e.m.f. fluctuate with magnetic field and have the same correlation field value B_c as G. Along with this α is also a random function of the electron excitation energy with a correlation energy $E_c \sim h/\tau_f$, where τ_f is the time it takes an electron to cross the sample. The asymmetry factor α frequency dependence defines the frequency dependence of the mesoscopic photovoltaic e.m.f.

In strong magnetic fields the electron transport is via edge current states which in certain places may approach close enough together to become coupled by tunneling. In these circumstances the asymmetry is the consequence of asymmetrical disposition of such places. The average time it takes an electron to travel between two places of tunnel coupling becomes an important factor for the frequency dependence of α. In this paper we report about the investigation of mesoscopic e.m.f. dependence on magnetic field and frequency at high magnetic fields ($\omega_c\tau > 1$).

Experimental samples are quantum wires and submicron rings with attached ohmic voltage and current probes. The wires were created from a 2D electron gas in AlGaAs/GaAs heterostructure by means of optical lithography. The original 2D layers have the following parameters at T =4.2K: the mobility μ is 4.1×10^5 cm^2/Vs, the electron density n is 3.6×10^{11} cm^{-2}. The wires have the length $L = 1$ μm and the conducting channels' effective width $W = 0.2$ μm. The rings were fabricated by means of electron lithography with subsequent reactive ion etching from heterostructures with the following parameters: $\mu = 10^5$ cm^2/Vs, $n = (7-9) \times 10^{11}$ cm^{-2} . The effective dimensions of the rings: the width of conducting channels is $W = 0.03 - 0.06$ μm, the ring diameter is $d = 0.6 - 0.7$ μm.

The measurements were carried out at temperatures 20mK - 4.2K and magnetic fields up to 12T. The microwave radiation frequency range is 1GHz - 145GHz. Microwave radiation of frequency in the range 1-10 GHz was transmitted via a coaxial cable directly to microstructure contacts. Radiation of frequency in the range 37-145 GHz was conducted to samples by means of a waveguide. The resulting e.m.f. was measured by a lock-in amplifier on the frequency of the radiation amplitude modulation.

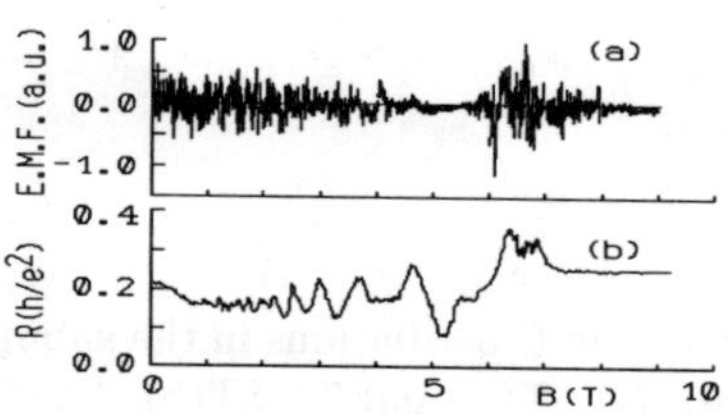

Figure 1: Microwave e.m.f. (a) and resistance (b) vs magnetic field for the submicron ring at 20mK and microwave frequency 1.6GHz.

In Fig.1 we display the magnetoresistance data for a ring structure together with the e.m.f. versus magnetic field dependence. One can easily see that there are two magnetic field intervals at which the e.m.f. falls off to zero while the resistance takes the quantised values: $h/12e^2$ and $h/4e^2$. In the investigated magnetic field range the e.m.f. has both the periodic (h/e - oscillations) and aperiodic components [4]. At the beginning of the region where the resistance takes on the value $h/4e^2$, there are, apart from h/e e.m.f. oscillations, distinct oscillations of the period $h/2e$. Fig.2 shows the e.m.f. Fourier spectra for corresponding magnetic field regions.

The e.m.f. versus magnetic field dependencies in a wire, taken at different microwave frequencies, are shown in Fig.3 together with the magnetoresistance curve. It can be seen that the e.m.f. amplitude always tends to grow smaller as the magnetic field increases. At the microwave frequency of 71 GHz the e.m.f. fluctuations are observed only in weak magnetic fields. As expected in the case of a single-connected geometry, there is no periodic component of the e.m.f. in the wire. The correlation magnetic field of the e.m.f. mesoscopic fluctuations at 9GHz is linearly proportional to the magnetic field.

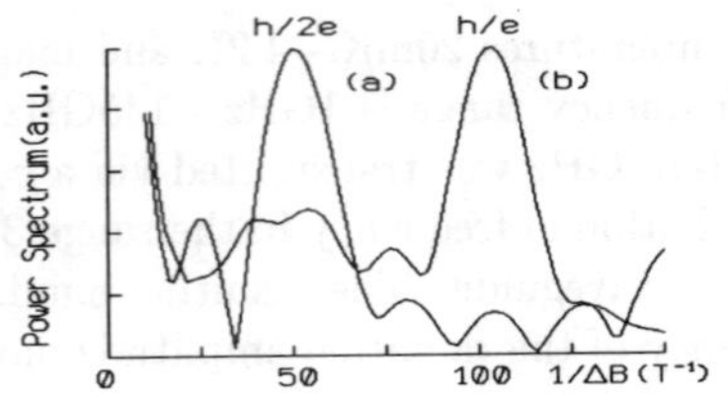

Figure 2: Fourier power spectra of the microwave e.m.f. oscillations in the submicron ring. T =20mK, the magnetic field intervals are 7-7.5T(a) and 7.5-8T(b).

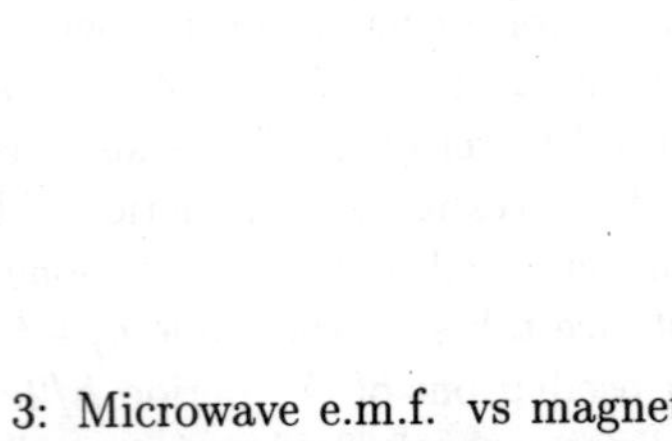

Figure 3: Microwave e.m.f. vs magnetic field (a,b,c) and magnetoresistance (d) for the quantum wire at T =4.2K and frequencies: a-9GHz, b-53GHz, c-71GHz.

Here follows a brief discussion of presented results. As stated in the beginning, we suppose that, as in the case of weak magnetic fields, in circumstances where the electron transport is via edge current states, the generation of the e.m.f. is primarily due to the asymmetry of the coherent scattering of electrons into the right and left microstructure contacts respectively. For this to be the case there must be narrow places along the channels where edge current states approach each other sufficiently close for the formation of tunnel links. The asymmetry is then a result of a microstructure imperfection: the channel width W is inhomogenous, the location of narrow places and extent of the narrowing are both random functions of the coordinate along the channel. The validity of this model is supported by a sharp reduction of the e.m.f. amplitude in the vicinity of quantum Hall plateaus. The wavefunctions of the edge states overlap less as magnetic field increases. When magnetic length l_b becomes smaller than W_{min} there is no tunnel links between edge current states any more and, therefore, no condition for the emergence of the e.m.f. This may be illustrated by the experimental dependence of the e.m.f. on magnetic field in the submicron ring. The appearance of h/2e oscillations in strong magnetic fields may be regarded as a further support for the model. The reduction of the e.m.f. amplitude with magnetic field was found to occur not only in the quantum Hall regime but also in weaker fields where Shubnicov-de Haas oscillations are already well-defined, Fig.3. In this case, the higher is the microwave frequency the less is the magnetic field value beyond which no e.m.f. fluctuations are to be found. This may be qualitatively accounted for by supposing that in the experimental conditions under consideration the asymmetry necessary for the photovoltaic effect shows up only if the microwave frequency ω satisfies the condition $\omega\tau_f l1$. The time it takes an electron to cross the sample τ_f increases with magnetic field since, quasiclassically, the electron may be thought of as moving along a spiral so that the effective length of the trajectory grows. This explains the relation between the microwave frequency and the magnetic field value at which the e.m.f. fluctuations die away. At frequency 1.6 GHz the condition $\omega\tau_f$ is satisfied in the whole range of magnetic fields so that the e.m.f. amplitude reduction in this case must be explained by other reasons. The increase of the correlation magnetic field B_c of the e.m.f. fluctuations with magnetic field under the conditions $\omega_c\tau > 1$ and $\omega\tau_f < 1$ (Fig.3a) also fits with the model of edge current states coupled by tunneling and qualitatively agrees with the results of [5].

In conclusion the experimental dependencies of the photovoltaic mesoscopic e.m.f on frequency and magnetic field in the limit $\omega_c\tau > 1$ can be qualitatively understood in terms of the electron transport via edge current states coupled by tunneling. More detailed analysis would require a theory of mesoscopic photovoltaic effect in high magnetic fields.

The research described in this publication was made possible in part by Grant N U84000 from the International Science Foundation.

1.V.I.Fal'ko and D.E.Khmelnitskyi, Zh.Eksp.Teor.Fiz. 95 (1989) 328.
2.A.A.Bykov, G.M.Gusev et al, Pis'ma Zh.Eksp.Teor.Fiz. 49 (1989) 13.
3.J.Liu, M.A.Penington and N.Giordano, Phys.Rev.B 45 (1992) 1267.
4.A.A.Bykov, Z.D.Kvon et al, Pis'ma Zh.Eksp.Teor.Fiz. 58 (1993) 538.
5.C.V.Brown, A.K.Geim et al, Phys.Rev.B 47 (1993) 10935.

MAGNETORESISTANCE OF QUANTUM WIRES FOR SELF–CONSISTENT IONIZED IMPURITY SCATTERING

M. SUHRKE,
Institut für Theoretische Physik, Universität Regensburg, D–93040 Regensburg, Germany

A. MENSCHIG, S. YU. NIKITIN*, and H. SCHWEIZER
4. Physikalisches Institut, Universität Stuttgart, Germany

Abstract

We present a quantum mechanical investigation of density of states and magnetoresistance for semiconducting quantum wires including several occupied one–dimensional subbands. As the dominating sources of scattering at low temperature we consider long–ranged ionized impurities separated from the two–dimensional electron gas (2DEG) by a spacer of finite width. In difference to the often used short–range model, the life time of a state can be much smaller than the transport relaxation time for this strongly anisotropic scattering mechanism which results in a strong broadening of the 1D subbands even in samples with high mobility. Quantitative results are given for the increase of the mobility due to the reduction of available phase space for anisotropic scattering during the approach of the one–dimensional limit.

1. Introduction

Quantum wires are investigated for some time both experimentally and theoretically since Sakaki's prediction of a further increase of the mobility in these structures due to the reduction of available phase space for scattering in one–dimensional (1D) systems[1]. However, the theoretical caculations have been done for the 1D limit (cf. e.g [2]) and it is not clear wether they apply to realistic quantum wires. In this paper we present calculations of the magnetoresistance of quantum wires of

*Permanent address: Microelectronics Department, Minsk Radioengineering Institute, Republic Byelarus (C.I.S.)

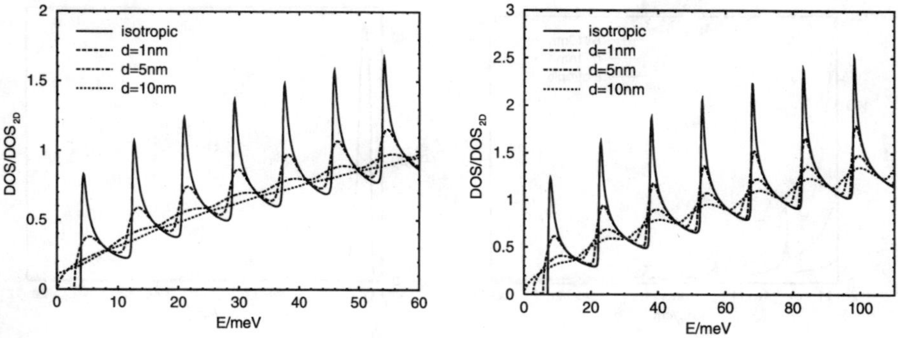

Figure 1: Normalized density of states in dependence on energy for different spacer widths d and for a wire width $W = 100nm$ and a mobility $\mu_{2D} = 94000cm^2/Vs$. The density of states for isotropic short–range scatterers is shown for comparison.

finite width for scattering by remote impurities. This represents a quite realistic model for magnetotransport simulations in real quantum wires (e.g. GaAs/AlGaAs or InGaAs/InP wires). We establish quantitative conditions for the observability of the mobility increase due to the approach of the 1D limit in dependence on the parameters of impurity scattering for realistic quantum wires.

2. Model and Theory

For simplicity, we use a parabolic shape $V(y) = \frac{m^*}{2}\omega_0^2 y^2$ for the lateral potential which confines the 2DEG located in the x y plane. This model describes well the 1D confinement for intermediate wire widths of some $100nm$ and gives simple analytic expressions for the one–particle states in a magnetic field. The magnetoelectric hybrid states are characterized by the frequency $\tilde{\omega} = \sqrt{\omega_0^2 + \omega_c^2}$ where ω_c is the cyclotron frequency. The transition from electrical to magnetical quantization occurs if ω_c becomes larger than ω_0. With increasing magnetic field the subbands are depopulated due to the increasing hybrid frequency. The wave functions at the Fermi energy are centered around the classical coordinate $y_0^n = \frac{\omega_c}{\tilde{\omega}^2} v_F^n$ where v_F^n is the subband Fermi velocity. This means that edge states are formed as soon as $y_0^n(B) > \tilde{l} = \sqrt{\hbar/m^*\tilde{\omega}}$.

Scattering is described within the self–consistent Born approximation (SCBA) in order to incorporate collisional broadening which removes the singularities of the density of states (DOS) at the subband bottoms. The screening of the impurities is taken as that in a 2DEG at zero magnetic field for simplicity. Thus, the results should apply to not too small wires and to moderate magnetic fields.

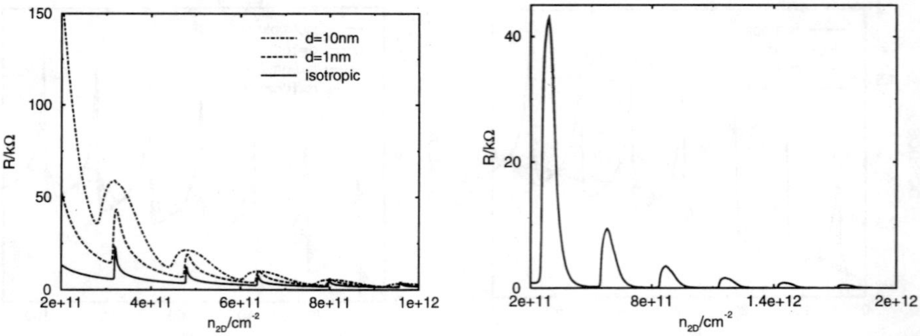

Figure 2: Absolute resistance in dependence on carrier concentration for a wire width $W = 100nm$, a wire length $L = 20\mu m$, and a mobility $\mu_{2D} = 940000cm^2/Vs$. On the left is shown the resistance without magnetic field for different spacer widths and on the right the resistance for $B = 5T$ and a spacer width $d = 10nm$.

We analyze the influence of anisotropy on the resistance of quantum wires using the Kubo formula. Due to the strongly anisotropic scattering mechanism prefering forward scattering in difference to the often used short–range model the life time of a state can be much smaller than the transport relaxation time. This results in a strong broadening of the 1D subbands even in samples with high mobility well known from the investigation of the DOS of 2D systems in a magnetic field [3, 4]. To obtain the correct transport relaxation time we have to consider vertex corrections consistently with the SCBA.

3. Results

Calculations are done for typical parameters of InGaAs/InP wires, $\mu_{2D} \sim 10^5 cm^2/Vs$ and $n_{2D} \sim 10^{12} cm^{-2}$[5].

The DOS shown in Fig.1 becomes strongly broadened with increasing anisotropy of scattering (increasing d) for fixed 2D mobility due to the rapidly growing difference between life time of a state and transport relaxation time. The broadening is weakly influenced by application of a magnetic field due the increasing separation of 1D subbands. In difference, the resistance is drasically changed by the same magnetic field due to the suppression of backscattering[6] as can bee seen from Fig.2.

Fig. 3 shows the resistance normalized by the corresponding 1D value. It shows clearly a decrease due to the reduction of available phase space in the 1D limit for anisotropic scattering. The decrease becomes more pronounced for smaller wire width because the effect increases with incrasing 1D Fermi wave vector. It is rapidly

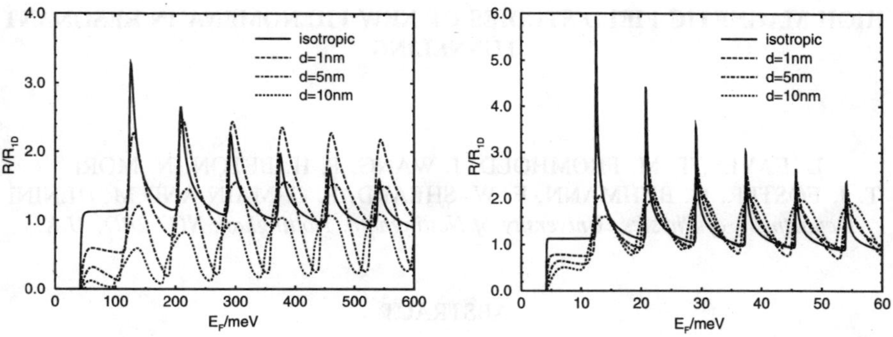

Figure 3: Normalized resistance without magnetic field in dependence on Fermi energy for different spacer widths d. The other parameters are wire width $W = 10nm$ and mobility $\mu_{2D} = 9400cm^2/Vs$ (left) and wire width $W = 100nm$ and mobility $\mu_{2D} = 940000cm^2/Vs$ (right), respectively. The resistance due to isotropic short–range scatterers is shown for comparison.

destroyed if more than one 1D subband becomes populated.

In conclusion, the predicted increase of the mobility in the 1D limit should be difficult to observe in presently available samples due to the drastical reduction of the effect with increasing number of occupied 1D subbands. Furthermore, the effect is masked by the increasing scattering strength if the 1D limit is approached by a reduction of the carrier density. A reduction of the wire width is more promising though difficult experimentally.

M. S. acknowledges financial support by the DFG.

References

[1] H. Sakaki, *Jpn. J. Appl. Phys.* **19** (1980) L735.

[2] Ben Yu-Kuang and S. Das Sarma, *Phys. Rev.* **B48** (1993) 14 388.

[3] S. Das Sarma and X. C. Xie, *Phys. Rev. Lett.* **61** (1988) 738, X. C. Xie, Q. P. Li, and S. Das Sarma, *Phys. Rev.* **B42** (1990) 7132.

[4] W. Xu and P. Vasilopoulos, (1993) preprint.

[5] S. Block, M. Suhrke, S. Wilke, A. Menschig, H. Schweizer, and D. Grützmacher, *Phys. Rev.* **B47** (1993) 6524.

[6] M. Büttiker, *Phys. Rev* **B43** (1988) 9375.

HIGH MAGNETIC FIELD STUDIES OF NEW PHENOMENA IN RESONANT TUNNELING

L. EAVES, T. M. FROMHOLD, J. WANG, P. H. BETON, N. MORI,
T. J. FOSTER, H. BUHMANN, F. W. SHEARD, P. C. MAIN AND M. HENINI
Department of Physics, University of Nottingham, Nottingham NG7 2RD, U.K.

ABSTRACT

This paper reviews some recent developments involving the use of high magnetic field magnetotunneling spectroscopy to study the electronic states of quantum wells and wires. We use this technique to probe the chaotic motion of electrons in quantum wells in tilted magnetic fields and the nature of the confined states of electrons in laterally-quantized resonant tunneling quantum wires.

1. Introduction

Magnetotunneling spectroscopy refers to the use of high magnetic fields to study energy quantization in resonant tunneling structures in which electrons tunnel into a quantum well from a doped emitter contact. This article focuses on the use of high magnetic fields to study two topical phenomena. In the first part of the paper, we describe a new dynamical system, an electron in a trapezoidal potential well, which we show is classically chaotic in the presence of a high magnetic field tilted with respect to the confining barriers. For recent reviews of quantum chaology, the study of systems exhibiting both classical chaos and quantum effects, see reference 1. When the tilt angle θ, relative to the direction normal to the barriers, is 0° or 90° the motion is regular corresponding respectively to helical orbits or skipping orbits between the barriers[2]. For intermediate tilt angles, when the cyclotron radius is sufficiently small, the orbit segments between successive collisions with the barriers, rapidly become uncorrelated giving rise to strongly chaotic motion. The second part of the paper uses high magnetic fields to demonstrate unambiguously the presence of lateral quantum confinement in small-area resonant tunneling diodes.

2. Electron States in a Quantum Well in the Regime of Classical Chaos

In our experiments the GaAs quantum well (QW) of width w = 120 nm, is enclosed between $Al_{0.4}Ga_{0.6}As$ tunnel barriers of width b = 5.6 nm surrounded by weakly n-doped (2×10^{16} cm^{-3}) contact layers[2]. Under bias, charge accumulates and a bound state is formed at the emitter-barrier interface (as shown in Figure 1(a)). At 4 K this

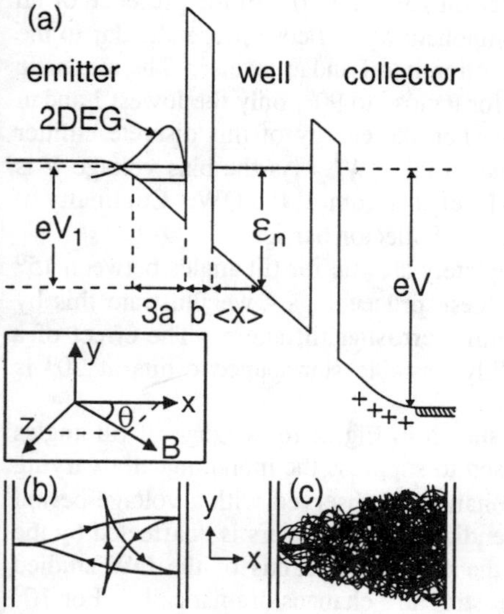

Figure 1(a) Conduction band profile under bias V Inset: tilt angle θ of magnetic field $\boldsymbol{B}$ relative to tunneling (x) direction. (b) Projection in x-y plane of star-shaped periodic orbit (c) Chaotic orbit resulting from 0.1 % change in injection velocity.

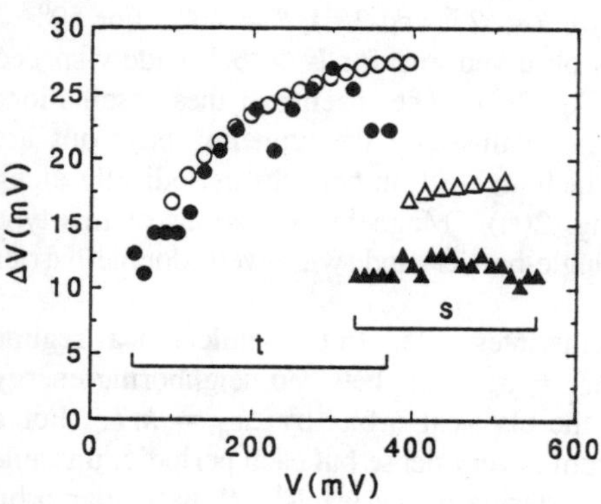

Figure 3 Variation of ΔV with V for t-series (circles) and s-series (triangles) of resonances at B = 11.4 T, θ = 20°, showing experimental values (filled circles/triangles) and theoretical values (open circles/triangles).

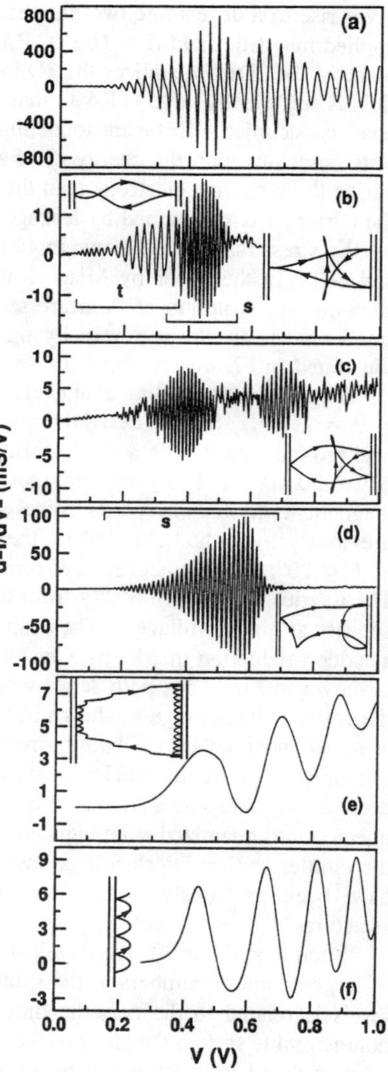

Figure 2 d^2I/dV^2 v. V plots for B = 11.4 T and tilt angle θ set to (a) 0°, (b) 20°, (c) 25°, (d) 40°, (e) 80°, (f) 90°. Voltage ranges of t- and s-series of resonances are shown in (b) and (d). Insets show periodic orbits associated with resonant structure.

gives rise to a degenerate two-dimensional electron gas (2DEG). In the presence of an applied magnetic field $\boldsymbol{B} = (B_x, 0, B_z)$, the component $B_x = B\cos\theta$ perpendicular to the plane of the 2DEG quantizes the 2D in-plane motion into Landau levels[3]. The magnetic field is sufficiently large (11.4 T) that, except for θ close to 90°, only the lowest Landau level is occupied. Resonant tunneling occurs when the energy of this discrete emitter state coincides with the energy ϵ_n of a subband in the QW[4]. As the bias voltage V is varied the tunneling electrons scan the energy-level spectrum of the QW. Continuity of the current I is maintained by leakage through the collector barrier.

We stress that the electronic motion is completely chaotic for tilt angles between 15° and 60°. In the paper by Miao et al[5] - see these proceedings - we illustrate this by showing the evolution of Poincaré sections with increasing tilt angle. The effect of a small change in electron velocity on the highly unstable star-shaped orbits at 20° is illustrated in Figures 1(b) and (c).

The resonant tunneling characteristics are shown in Fig. 2 for a range of tilt angles at $B = 11.4$ T. Second derivative plots are used to suppress the monotonically varying background. At $\theta = 0°$ a single series of resonances is observed with a voltage period $\Delta V \cong 30$ mV. In this geometry, motion perpendicular to the layers is unaffected by the magnetic field and the peaks correspond to the electric subbands of the QW studied previously[2]. As the field is tilted, the resonant structure changes dramatically. For $10° < \theta \lesssim 20°$, two distinct series of resonances are observed, labelled s and t in Fig. 2(b). The t-series emerges smoothly from the resonant structure at $\theta = 0°$ whilst the s-series appears at higher voltages. The changeover between the two series is characterized by a sudden reduction in ΔV by a factor ~ 2. The variation of ΔV with applied bias V is shown in Fig. 3 for both series when $\theta = 20°$. For $\theta \gtrsim 20°$ the s-series dominates the entire voltage range as shown in Fig. 2(c), $\theta = 25°$ and 2(d), $\theta = 40°$. For $50° < \theta < 60°$ no regular oscillatory structure is observed and for $\theta > 60°$ widely-spaced oscillations appear in d^2I/dV^2, as shown in Fig. 2(e). The spacing of these oscillations decreases steadily as θ is increased to 90°. In this case the tunneling electrons are injected into quantized skipping states, in which an electron bounces periodically along the emitter-barrier interface (see inset in Fig. 2(f)). Magneto-oscillations of this type have been previously studied for both single-barrier and wide-well double-barrier structures[4,6].

When $\theta = 0°$ or 90° the classical motion is integrable. In the semiclassical regime of large quantum numbers n, the splitting $\Delta\epsilon_n = \epsilon_{n+1} - \epsilon_n$ between neighboring energy levels is related to the periodic time T of the classical orbits by $\Delta\epsilon_n = h/T$. For a nonintegrable system the energy-level spectrum is very dense but each periodic, unstable orbit in the chaotic sea produces a regular clustering of the levels[1]. A particular orbit with period T_P leads to an oscillatory variation in the smoothed density of states with an energy period $\Delta\epsilon_P = h/T_P$. This modulation of the density of states will only be resolved in our experiments if $\Delta\epsilon_P > \hbar/\tau$, where τ is the electron scattering time. We have used $\tau \sim 0.1$ ps to take account of optic phonon emission[7] and impurity and interface scattering. We have therefore searched for periodic orbits in the QW with periods $T_P \leq 2\pi\tau \sim 0.6$ ps.

To ensure correspondence between the classical and quantum descriptions of the system the classical orbits must be accessible to the tunneling electrons. The tunneling matrix element involves the overlap of the wave functions of the QW states and emitter-2DEG states in the emitter barrier. For $\theta = 0^o$ this requires that the Landau level index for the in-plane motion is conserved[3]. For $\theta = 90^o$ both the emitter and QW states have plane-wave character in the transverse y, z directions. The tunneling matrix element only couples states with a velocity difference equal to that supplied by the Lorentz force as the electron traverses the mean distance between the 2DEG and nearside QW interface[4,6]. This requirement and that of energy conservation determine the allowed skipping states into which an electron can tunnel.

In our experiments the B-field is sufficiently large that for $\theta \leq 70^o$, electrons in the 2DEG occupy the lowest in-plane Landau level. There is thus a distribution of transverse momentum components with a mean square value $p_y^2 + p_z^2 = e\hbar B\cos\theta$. We therefore consider a set of classical electrons with a range of transverse momenta given by the above equality. On calculating the momentum change due to the Lorentz force as the electron traverses the mean distance between the 2DEG and nearside QW interface, we obtain a range of initial velocities v_y, v_z for electrons entering the well. The kinetic energy of an injected electron is obtained from the potential drop across the 2DEG and barrier, which is modelled using the Fang-Howard variational wave function[8] for the emitter bound state.

Our search for accessible periodic orbits in the QW starts with the equations for the electron trajectory in parametric form, which are obtained by analytic solution of the classical law of motion for a particle of mass $m^* = 0.067m_e$ in constant electric and magnetic fields. The times at which successive specular collisions with the well walls occur must then be calculated numerically. It may be shown that if the electron returns to the emitter barrier with no change in the in-plane velocity components v_y, v_z then it also returns to the same position and the orbit is periodic. We therefore consider 1440 initial velocity values, within the allowed range, and look for minimal change in v_y, v_z upon return to the emitter wall. Periodic orbits can be located very quickly by this technique.

For $\theta = 20^o$ we have found accessible periodic orbits in which the electron traverses the well, colliding with each barrier in turn, as in Fig. 2(b) left inset. It is only above a threshold voltage $\cong$ 100 mV that these orbits have periods < 0.6 ps. Above $V \cong 400$ mV the t-orbits are inaccessible to the injected electrons since the Lorentz force prevents electrons returning to the emitter wall after just one collision with the collector wall. A topologically distinct closed orbit is now manifest, in which the electron makes two successive collisions with the collector wall for each collision with the emitter wall. These s-orbits, which look star shaped in projection on the x-y plane, see right inset in Fig. 2(b), have periods < 0.6 ps above $V \cong 400$ mV and do not exist above a cut-off voltage $\cong$ 515 mV, when the electron has sufficiently large normal component of velocity v_x to surmount the collector barrier potential. These results are consistent with the observed sudden reduction in ΔV at $V \cong 350$ mV shown in Fig. 2(b), since the s-orbits have almost doubled periods owing to the extra loop performed in the middle of the well between the two collisions with the collector barrier. For $\theta = 25^{o}$, only s-orbits are accessible to the injected electrons and hence no sudden change in ΔV is observed, as in Fig. 2(c).

To compare these theoretical results more quantitatively with experiment we estimate the voltage period ΔV of the oscillatory structure from the calculated orbit period T_P, using a simple model of the potential distribution across the structure. The emitter Fermi level, which lies in the occupied Landau level, is close to the conduction band edge as shown in Fig. 1. We define ϵ_n to be the energy of the nth level in the well, measured from the conduction band edge at the mean position $\langle x \rangle$ for this state. If the potential drop between the emitter Fermi level and this position $\langle x \rangle$ is V_1, the resonance condition is $eV_1 = \epsilon_n$. Successive resonances correspond to an interval $\Delta V_1 = \Delta\epsilon_n/e = h/eT_P$ in the chaotic regime. The corresponding interval ΔV is obtained using $V_1/V = (3a + b + \langle x \rangle)/L$ where $3a$ is the width of the 2DEG[8], and $L \cong 187$ nm is the effective electrical width of the structure given by capacitance measurements. Taking $\langle x \rangle = ½w$ for s- and t-orbits then gives $f(V) = \Delta V_1/\Delta V \cong 0.4$ for 200 mV $\leq$ V $\leq$ 700 mV.

A comparison of experimental ΔV values with the values $\Delta V = h/feT_P$ obtained from the calculated periods T_P of the t- and s-orbits is shown in Fig. 3. For the t-orbits, ΔV increases with V as the electrons are accelerated to higher speeds in the well. The changeover from t- to s-orbits occurs at a slightly higher voltage than found experimentally and the reduction in ΔV is underestimated. We attribute these features to neglect of band nonparabolicity which would significantly increase the effective mass of the electron since its kinetic energy can reach several hundred meV . The predicted variation of ΔV with θ is also consistent with our experiments, which are described in more detail in reference 9.

3. High Magnetic Fields for Quantum Chaology in Condensed Matter Physics

Wide quantum well heterostructures, coupled with high magnetic fields, form a flexible system in which the quantum manifestations of classical chaos can be investigated. Weiss et al[10] have recently studied the chaotic motion of electrons in anti-dot superlattices in the presence of a magnetic field and have also found evidence for effects related to the de Bröglie wavelength of the carriers. Marcus et al[11] have compared the magnetoconductance fluctuation spectra in circular quantum dots and chaotic billiard stadia. Periodic components in the magnetoconductance oscillations of Bunimovich stadia[1] were attributed to unstable periodic orbits. The double barrier structure is analogous to the 2D stadia in that the chaos is generated by collisions on confining barriers which scramble the simple separable motion performed between the collisions. However, in contrast to the stadia, we can change the classical motion from regular to chaotic by altering the tilt angle of the magnetic field or the strength of the electric field between the barriers.

Prior to the recent experimental work on heterostructures, experiments on quantum chaology have focused on the ionization and spectroscopy of hydrogen-like atoms in high magnetic fields (for a review see ref. 12). To date, these spectroscopy experiments have been done at relatively modest fields (less than 10 T). However, these fields are large enough to produce classically chaotic motion at energies corresponding to the highly excited states ($n > \sim 50$) of the atoms.

On the theoretical side, it has been recognized for some time that hydrogen donors and excitons in semiconductors provide an extremely interesting system for studying chaos. Due to the screening effect of the dielectric constant and the light effective mass, shallow donors and excitons at fields of ~10 T are physically equivalent to the hydrogen atom at fields of $\sim 10^3$ T. It is therefore potentially interesting to study their spectroscopic properties at lower quantum numbers, which have not yet been explored theoretically or experimentally. The anisotropic Kepler problem[1] could be studied with or without magnetic fields for donors in Si and Ge, or in uniaxially-stressed GaAs. Other possible quantum phenomena which may have underlying chaotic classical motion include the effects on quantum magnetotransport of Landau level mixing of the valence bands of cubic semiconductors.

4. Magnetotunneling into Resonant Tunneling Quantum Wires

In recent years, several groups have investigated the effect on the tunnel current versus voltage characteristics, I(V), of confining the electron motion laterally in the plane of the well[13-19]. Such confinement produces additional quantisation, which modifies the energy-independent density of states of an unconfined quantum well into a series of discrete peaks. Resonant tunneling into the quantised states of a *one-dimensional quantum wire* or of a *zero-dimensional quantum box* should therefore give rise to additional peaks in I(V).

This section describes a new processing technique to fabricate resonant tunneling devices with sub-micron lateral dimensions. The effect of a magnetic field on the tunnel current in these devices provides unequivocal evidence for lateral quantisation. The active region of the device, grown by MBE on a semi-insulating substrate, consists of a GaAs quantum well (width $w = 9.0$ nm) formed between two AlAs tunnel barriers (thickness $b = 4.7$ nm). The active area of the device, shown schematically in Fig. 4(a), is the region of overlap of two GaAs bars, one (thickness ℓ_t) etched in the top contact layer and the other (thickness ℓ_b) in the lower contact layer. The AlAs barriers and GaAs quantum well are sandwiched in the region where the two bars overlap. The fabrication sequence, based on a series of selective wet etches, is described fully in reference 20. A notable feature of the device is the use of a freestanding GaAs bridge to provide a connection to a large ohmic contact. Optical lithography and undercut etching are used to achieve sub-micron dimensions, ℓ_t and ℓ_b can be < 0.5 μm, and typically $\ell_t < \ell_b$.

Figure 5 plots the $I(V)$ characteristics at 0.3 K for a device with $\ell_t = 0.5$ μm and $\ell_b = 1.0$ μm. In forward bias electrons flow to the top contact. From a comparison of diodes with different dimensions, we deduce that the sidewall depletion for these devices is ~ 0.2 μm for each edge and the current density is $J = 18$ nA/μm^2. The dimensions of the active conducting region are therefore ~100 nm $\times$ 600 nm so that

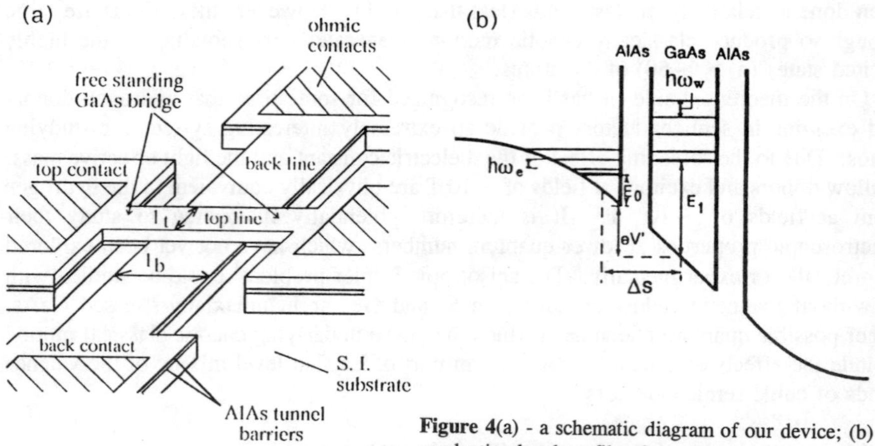

Figure 4(a) - a schematic diagram of our device; (b) - conduction band profile of our device.

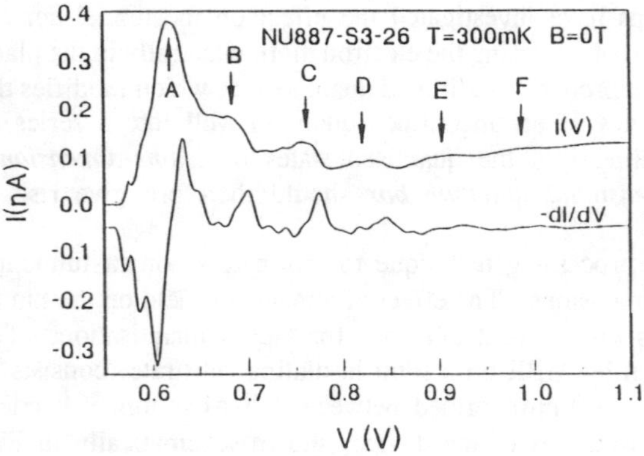

Figure 5 Low temperature (T = 0.3 K) $I(V)$ and dI/dV for a small area resonant tunneling diode.

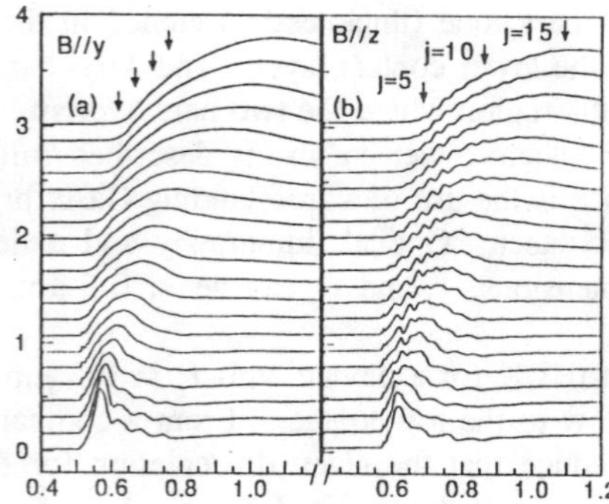

Figure 6 $I(V)$ at T = 0.3 K in the presence of a magnetic field between B = 0 T (lowest curve) and B = 10 T (top curve) in 0.5 T steps. The field is oriented perpendicular to the current flow and either perpendicular (a) or parallel (b) to the axis of the quantum wire.

electrons tunnel between 1D quantum wires formed in the emitter and the quantum well. In the following discussion the x-axis is the growth direction, and y and z are respectively perpendicular and parallel to the long axis of the wire.

Several clear peaks labelled A-F are observed in the $I(V)$ shown in Fig. 5. Peak A is the main resonance. The peak separation is $\sim$70 mV. In addition to these peaks there is a much weaker series of peaks with a separation of $\sim$20 mV. These peaks are resolved more clearly in the differential conductance (dI/dV) plot.

Figure 6 shows $I(V)$ for the same device in the presence of a magnetic field, B, oriented in the plane of the tunnel barriers, either perpendicular (Fig. 6(a)) or parallel (Fig. 6(b)) to the long axis of the quantum wire. We observe a clear anisotropy in the dependence of $I(V)$ on B for the two field orientations. This provides strong evidence that the additional structure in $I(V)$ arises from quantum confinement controlled by the lateral shape of the device, rather than donor-assisted resonant tunneling[17,21]. The origin of these peaks is discussed in detail in reference 22. For $B \parallel y$ (perpendicular to the wire, Fig. 6(a)), peaks A-F are shifted to higher voltage with increasing field. As B is increased beyond $\sim$3 T the peaks become broader and eventually cannot be resolved. In addition for $B > 4$ T a weak series of features develops in the rise to the main resonance identified by arrows in Fig. 6(a) for the $B = 10$ T curve. For $B \parallel z$ (parallel to the wire, Fig. 6(b)), peaks A-F show a complex splitting for small magnetic fields. However, for $B > 3$ T a regular series of peaks is observed in $I(V)$ with a typical spacing of $\sim$30 mV. We are able to observe more than 15 peaks in this series, as labelled in Fig. 6(b). These peaks correspond to tunneling transitions from occupied emitter states into almost equally-spaced ($\hbar\omega_w = 3.5$ meV) energy levels of the quantum well. The laterally-confining potential in the well is approximately parabolic up to energies of $\sim$50 meV.

Figure 7 shows the Fourier transforms of the laterally-confined states in the well. The states in the emitter have a similar form but are less strongly confined in real space along y, with confinement energy $\hbar\omega_e = 1.5$ meV. In k-space they are more strongly confined, and have a narrow range of k_y values. Only the first few levels are occupied at low temperatures. At zero field, the matrix element between the laterally-confined emitter state and these higher energy levels of the quantum well is almost zero, since the emitter states only couple with the lower-lying (low j) states of the well. It is remarkable that more than 15 of these levels can be observed at high magnetic field. Due to its dynamical effect on the tunneling electrons[23], the two sets of states become coupled at high magnetic field. When the electrons tunnel from the emitter through the barrier into the quantum well in the presence of a field along z, they acquire an in-plane wavevector, $k_0 = eB\Delta x/\hbar$, where Δx is the effective tunneling length[23]. Hence the range of k_y values of the emitter states is shifted to higher values, as shown by the shaded regions in Figure 7. The finite matrix element for transitions to the high j states of the quantum well is a maximum when the spread of the emitter k-states has a value close to $\pm(2(j + ½)eB/\hbar)^{½}$. At high field, there is only one peak in $I(V)$ due to tunneling via a particular laterally quantised state, as implied by the labelling $j = 1$-15 in Fig. 6(b). A full account of the transfer Hamiltonian model used to describe the data is given in reference 22. Measurements for the $B \| x$ orientation are described by Wang et al in these proceedings[24].

To conclude this section, we have shown that high magnetic fields can unambiguously demonstrate the presence of lateral confinement effects in resonant tunneling diodes and determine the form of the laterally-confining potential.

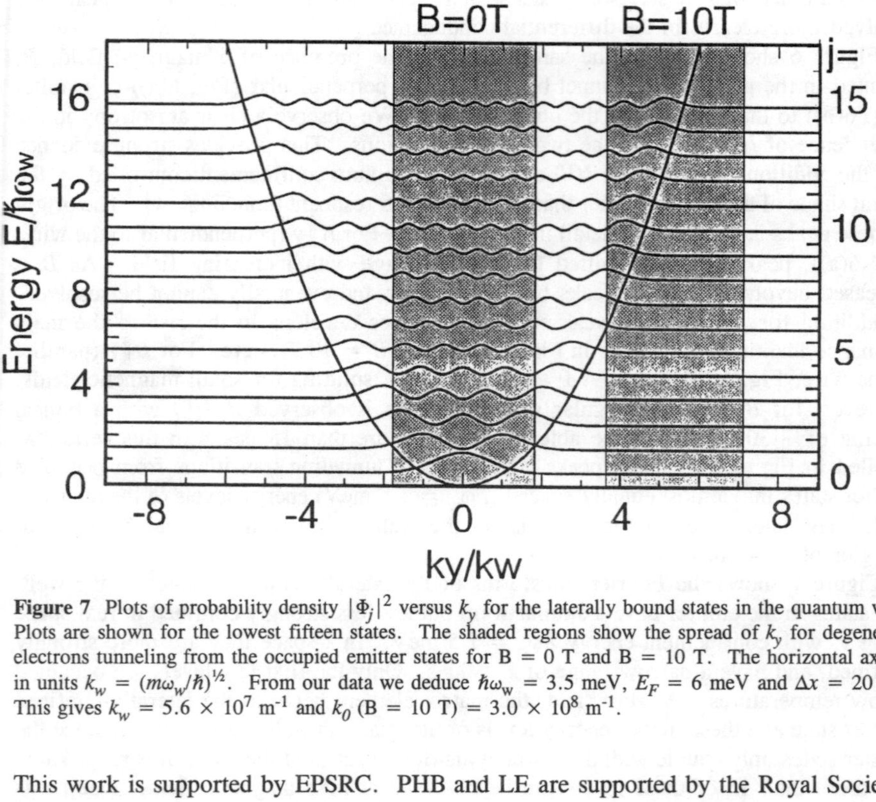

Figure 7 Plots of probability density $|\Phi_j|^2$ versus k_y for the laterally bound states in the quantum well. Plots are shown for the lowest fifteen states. The shaded regions show the spread of k_y for degenerate electrons tunneling from the occupied emitter states for B = 0 T and B = 10 T. The horizontal axis is in units $k_w = (m\omega_w/\hbar)^{1/2}$. From our data we deduce $\hbar\omega_w = 3.5$ meV, $E_F = 6$ meV and $\Delta x = 20$ nm. This gives $k_w = 5.6 \times 10^7\ m^{-1}$ and k_0 (B = 10 T) = $3.0 \times 10^8\ m^{-1}$.

This work is supported by EPSRC. PHB and LE are supported by the Royal Society.

5. References

1. M. C. Gutzwiller, *Chaos in Classical and Quantum Mechanics*, (Springer-Verlag, New York, 1990); M. V. Berry, *Proc. R. Soc. Lond. A* **413** (1987) 183.
2. M. L. Leadbeater, E. S. Alves, L. Eaves, M. Henini, O. H. Hughes, G. Hill, and M. A. Pate, *J. Phys.: Condens. Matter* **1** (1989) 4865.
3. M. L. Leadbeater, F. W. Sheard, and L. Eaves, *Semicond. Sci. Technol.* **6** (1991) 1021.

4. F. W. Sheard and T. M. Fromhold, *Acta Physica Polonica A* **82** (1992) 523.
5. J. Miao, T. M. Fromhold, F. W. Sheard, G. Edwards and L. Eaves, *Proc. 11th Int. Conf. on High Magnetic Fields in Semiconductors* (World Scientific) - these proceedings.
6. T. M. Fromhold, F. W. Sheard, and G. A. Toombs, *Surf. Sci.* **228** (1990) 437.
7. A. F. J. Levi, R. J. Späh, and J. H. English, *Phys. Rev. B* **36** (1987) 9402.
8. F. F. Fang and W. E. Howard, *Phys. Rev. Lett.* **16** (1966) 797.
9. T. M. Fromhold, L. Eaves, F. W. Sheard, M. L. Leadbeater, T. J. Foster and P. C. Main, *Phys. Rev. Lett.* **72** (1994) 2608.
10. D. Weiss, K. Richter, A. Menschig, R. Bergmann, H. Schweizer, K. von Klitzing, and G. Weimann, *Phys. Rev. Lett.* **70** (1993) 4118.
11. C. M. Marcus, A. J. Rimberg, R. M. Westervelt, P. F. Hopkins, and A. C. Gossard, *Phys. Rev. Lett.* **69** (1992) 506; R. A. Jalabert, H. U. Baranger and A. D. Stone, *Phys. Rev. Lett.* **65** (1990) 2442.
12. M. L. Du and J. B. Delos, *Phys. Rev. A* **38** (1988) 1896 and 1913.
13. M. A. Reed, J. N. Randall, R. J. Aggarwal, R. J. Matyi, T. M. Moore and A. E. Wetsel; *Phys. Rev. Lett.* **60** (1988) 535.
14. M. Tewordt, V. J. Law, M. J. Kelly, R. Newbury, M. Pepper, D. C. Peacock and J. E. F. Frost; *J. Phys. Condens. Matter* **2** (1990) 8969.
15. S. Tarucha, Y. Hiarayama, T. Saku and T. Kimura; *Phys. Rev. B* **41** (1990) 5459.
16. B. Su, V. J. Goldman, M. Santos and J. E. Cunningham; *Appl. Phys. Lett.* **58** (1991) 747.
17. M. W. Dellow, P. H. Beton, C. J. G. M. Langerak, T. J. Foster, P. C. Main, L. Eaves, M. Henini, S. P. Beaumont and C. D. W. Wilkinson; *Phys. Rev. Lett.* **68** (1992) 1754.
18. P. Guéret, N. Blanc, R. Germann and H. Rothuizen; *Phys. Rev. Lett.* **68** (1992) 1896.
19. A. Ramdane, G. Faini and H. Launois; *Z. Phys. B - Condensed Matters* **85** (1991) 389.
20. J. Wang, P. H. Beton, N. Mori, L. Eaves, H. Buhmann, L. Mansouri, P. C. Main, T. J. Foster and M. Henini; submitted for publication.
21. J.-W. Sakai, T. M. Fromhold, P. H. Beton, L. Eaves, M. Henini, P. C. Main, F. W. Sheard and G. Hill; *Phys. Rev. B* **48** (1993) 5664.
22. J. Wang, P. H. Beton, N. Mori, L. Eaves, H. Buhmann, L. Mansouri, P. C. Main, T. J. Foster and M. Henini; submitted for publication.
23. R. K. Hayden, D. K. Maude, L. Eaves, E. C. Valadares, M. Henini, F. W. Sheard, O. H. Hughes, J. C. Portal and L. Cury; *Phys. Rev. Letts.* **66** (1991) 1749.
24. J. Wang, P. H. Beton, N. Mori, L. Eaves, P. C. Main, T. J. Foster and M. Henini; *Proc. 11th Int. Conf. on High Magnetic Fields in Semiconductors* (World Scientific) - these proceedings.

HIGH-FIELD MANGETOTUNNELING SPECTROSCOPY OF VALENCE BAND DISPERSION IN STRAINED SiGe QUANTUM WELLS

A. ZASLAVSKY*
Div. of Engineering, Brown University, Box D, Providence, RI, 02912
and
T. P. SMITH III, D. A. GRUTZMACHER, S. Y. LIN, and T. O. SEDGWICK
IBM T. J. Watson Research Center, P.O. Box 218, Yorktown Heights, NY, 10598

ABSTRACT

We employ magnetotunneling measurements in fields up to 30 T to probe the valence band Landau level spectrum and anisotropic, nonparabolic in-plane dispersion in strained SiGe quantum wells. The double-barrier *p*-Si/$Si_{1-x}Ge_x$ resonant tunneling devices used in the measurements exhibit excellent tunneling characteristics with well-resolved heavy- and light-hole current peaks. The low-lying Landau level spectrum is mapped out using parallel (B || I) field data, where we observe satellite peaks corresponding to tunneling with $\Delta n = 1, 2$ changes in Landau index n. Since Landau index-nonconserving tunneling is forbidden in bands with isotropic dispersion and scattering is not a viable mechanism in our devices, we attribute the satellite peaks to Landau level mixing arising from valence band anisotropy. The anisotropic band dispersion is probed in the transverse (B ⊥ I) configuration, where we map out the heavy-hole nonparabolicity as a function of crystallographic orientation. From a simple two-band model we extract a heavy-hole nonparabolicity factor that is twice as large in the <110> direction as in the <100>.

1. Introduction

Resonant magnetotunneling in double-barrier resonant tunneling structures (DBRTS) is an effective technique for studying in-plane dispersion and Landau level spectra in quantum wells, particularly in indirect semiconductors where optical techniques are difficult to apply. In this report we present magnetotunneling results on scientifically and technologically interesting strained *p*-type $Si_{1-x}Ge_x$ quantum wells, with their complex, anisotropic and nonparabolic dispersion.

The essential features of DBRTS magnetotunneling spectroscopy are easy to appreciate within the sequential tunneling model,[1] where tunneling into the well is governed by conservation of energy E and transverse momentum $k_\perp$. This model, which successfully predicts the threshold, lineshape and $k_\perp$ conservation-induced cut-off of resonant tunneling

*Work done at IBM T. J. Watson Research Center.

peaks, is easily extended to include magnetic field effects as long as the dispersion $E(k_\perp)$ of the tunneling carriers is parabolic and isotropic -- e.g. electrons in III-V heterostructures. In the parallel field geometry ($B_\parallel \parallel I$), emitter and well states undergo Landau quantization and Landau index n must be conserved.[2,3] The resonant peak then acquires a weak regular structure due successive Landau level alignment, while additional features at a higher bias than the resonant peak can appear if inelastic processes lift the Landau index conservation.[4,5] Conversely, the main effect of a transverse field $B_\perp$ is peak broadening and shift to higher bias due to $\Delta k_\perp = eB_\perp \langle z \rangle / \hbar$ change in transverse momentum as the carrier tunnels a distance $\langle z \rangle$ from the emitter into the well.[6,7] In combination with a self-consistent treatment of the potential distribution over the DBRTS,[3] the incorporation of magnetic field-altered conservation laws into the sequential tunneling model produces remarkable agreement with experimentally observed I(V,B) characteristics.[8,9]

If the in-plane dispersion $E(k_\perp)$ of the tunneling carriers is more complex, magneto-tunneling characteristics become more difficult to interpret. Clearly, a changing in-plane mass due to band nonparabolicity can result in an irregular Landau level spectrum, with level separation $\hbar\omega_c$ changing as a function of $B_\parallel$. Further complications arise if different interacting bands -- the heavy-hole (HH) and light-hole (LH) bands in p-type quantum wells, for example -- enter into picture: theoretically calculated hole Landau spectra then have a complicated structure with strong nonlinearities and level crossings.[10] Finally, if $E(k_\perp)$ is anisotropic with respect to crystallographic directions, then the $I(V,B_\perp)$ resonant peak shift will clearly depend on $B_\perp$ orientation.[11]

All these complications are present in the valence band dispersion of strained $Si_{1-x}Ge_x$ quantum wells, which have attracted much attention for potential electronic and opto-electronic devices compatible with silicon technology. For this reason, since their first fabrication in 1988,[12] $Si/Si_{1-x}Ge_x$ DBRTS have been studied extensively. We have employed magnetotunneling spectroscopy on state-of-the-art p-$Si/Si_{1-x}Ge_x$ DBRTS to elucidate the valence band dispersion and Landau level spectrum in strained $Si_{1-x}Ge_x$ quantum wells. In $I(V,B_\parallel)$ measurements,[13] we have mapped the lowest-lying Landau level structure and observed strong index-nonconserving ($\Delta n \neq 0$) tunneling that could not be attributed to scattering,[14] but rather to elastic tunneling allowed by anisotropy-induced Landau level mixing. In $I(V,B_\perp)$ measurements, we have followed up previous work[15] in improved samples, mapped out the anisotropy of lower-lying HH_0 and LH_0 subbands, and extracted a single, anisotropic nonparabolicity parameter to describe the HH_0 in-plane dispersion.[16]

2. Samples

Our strained p-Si/$Si_{1-x}Ge_x$ DBRTS were grown by atmospheric pressure CVD on Si substrates. The growth sequence and complete DBRTS parameters have been published elsewhere:[13] here we note that tunneling occurs from a $Si_{1-x}Ge_x$ emitter into a $Si_{1-x}Ge_x$ well of width W = 23, 35, or 46 Å through 50 Å Si barriers (Ge content $x = 0.25$). The strain in the emitter splits the HH and LH band edges by $\Delta E \approx 40$ meV, so only the heavy hole states are occupied at $T = 4.2$ K (the temperature of all measurements); these heavy holes then tunnel into the HH and LH quantum well subbands. The I(V) curves of the three structures at B = 0 are shown in Fig. 1, with resonant peak identification obtaining from subband energies calculated using interpolated band-edge masses.[13] The peak-to-valley ratios are very good for p-type structures, reaching 4:1 for LH_0 peaks. Due to the remoteness of the LH bands, the HH_0 subband should be reasonably parabolic on the scale of emitter Fermi energy $E_F \sim 10$ meV, leading to expected threshold V_{th} and peak V_p bias values marked by arrows in Fig. 1.[13] For the LH_0 peaks an analogous procedure is hindered by the extreme non-parabolicity of LH_0 in-plane dispersion.

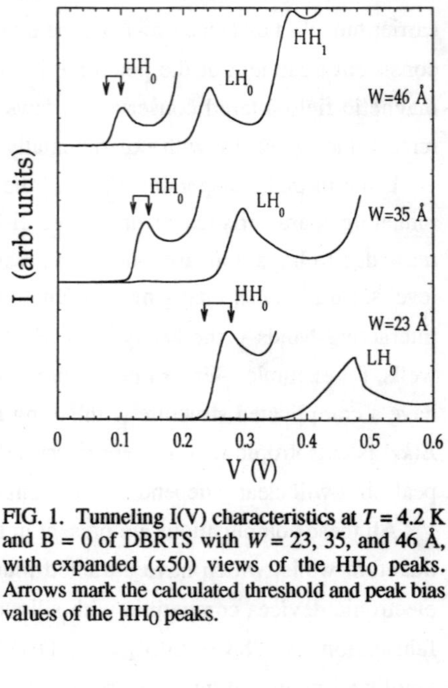

FIG. 1. Tunneling I(V) characteristics at $T = 4.2$ K and B = 0 of DBRTS with W = 23, 35, and 46 Å, with expanded (x50) views of the HH_0 peaks. Arrows mark the calculated threshold and peak bias values of the HH_0 peaks.

3. Parallel field geometry: Landau spectrum and level mixing

The I(V,$B_\parallel$) characteristics of the HH_0 peak in the $W = 35$ Å DBRTS for $0 \le B_\parallel \le 30$ T are shown in Fig. 2. While the peak remains at the same V_P and exhibits no discernible fine structure,[3] at higher bias there appears a satellite peak that increases with $B_\parallel$ and dominates the I(V,$B_\parallel$) at large $B_\parallel$. Further, for $B_\parallel \ge 15$ T another, weaker satellite appears in between (see Fig. 2). The structures with W = 23 and 46 Å exhibit analogous behavior.

The HH_0 and satellite peak positions in the W = 23 and 35 Å DBRTS for $B_\parallel \le 30$ T are

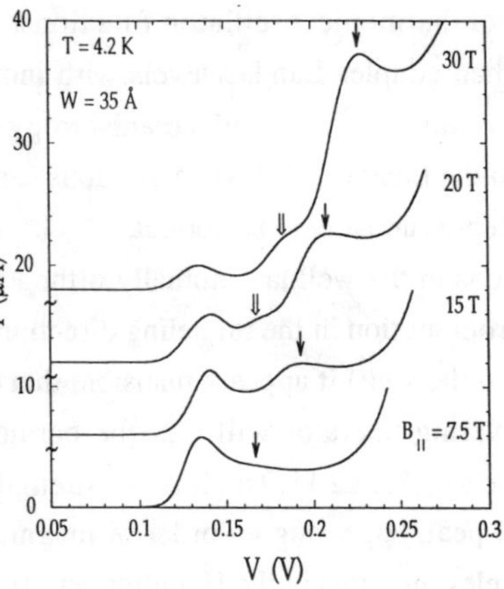

FIG. 2. $I(V,B_{\parallel})$ characteristics of W = 35 Å DBRTS at $B_{\parallel}$ = 7.5, 15, 20, and 30 T (curves displaced for clarity). Open and solid arrows mark the $B_{\parallel}$-induced Δn = 1, 2 satellite peaks.

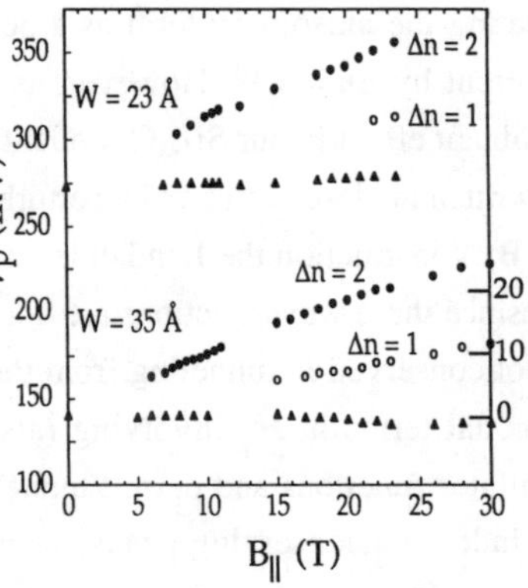

FIG. 3. HH_0 (triangles) and satellite (Δn = 1, 2 marked by open and closed circles) peak positions *vs.* $B_{\parallel}$. At right is the conversion to Landau level separation energy for the W = 35 Å structure.

summarized in Fig. 3. Since the satellite peaks occur at higher bias than the HH_0 peak, they must arise from index-nonconserving ($\Delta n \neq 0$) processes labeled in Fig. 3. By self-consistently converting the bias positions of these satellite peaks to energy we obtain the separation between the lowest three Landau levels in the well, as shown in Fig. 3 for the W = 35 Å structure. The strength of the satellite peaks, particularly the Δn = 2, is surprising and unexpected. An obvious mechanism for $\Delta n \neq 0$ tunneling is impurity or interface scattering, which relaxes index conservation.[14] However scattering-assisted features are unlikely to exceed the main resonant peak in intensity and, furthermore, Δn = 2 tunneling requires a larger change in $k_\perp$ than Δn = 1 and should hence produce weaker features, whereas the opposite is experimentally observed.

Alternatively, the $\Delta n \neq 0$ satellites could arise from Landau level mixing which allows elastic tunneling with $|\Delta n|$ = 1, 2. The simplest appropriate Hamiltonian H_{ij} for the relevant hole bands in a $B_{\parallel}$ field can be expressed as a 4x4 matrix in the $|J\ m_j\rangle$ basis, with J = 3/2 and $m_j = \pm 3/2$ and $\pm 1/2$ for HH and LH respectively.[17-19] The strain contributes to diagonal H_{ii} elements splitting the HH and LH bands by ΔE, and all H_{ij} elements can be expressed in terms of harmonic oscillator raising and lowering operators a and a_+, Luttinger valence band parameters γ_i, and momentum in the tunneling direction k_z.[17] If the in-plane anisotropy term is neglected the Landau level envelope functions for the quantum well can be written as four-component superpositions of harmonic oscillator functions.[18,19] Treating the anisotropy term as a perturbation then couples Landau levels with indices different by $\Delta n = 4$.[19] However, as shown in the next section, in-plane anisotropy is a dominant effect in our $Si_{1-x}Ge_x$ quantum wells, so the Landau level wavefunctions cannot be written in closed form and a perturbation treatment is unlikely to be correct.

By construction the Landau level wavefunctions in the well are mutually orthogonal, but since these wavefunctions are not separable from motion in the tunneling direction (k_z is not conserved in tunneling from the emitter into the well) it appears plausible that off-diagonal terms in H_{ij} involving raising and lowering operator will mix the harmonic oscillator functions and permit tunneling with $\Delta n = 0, \pm 1, \pm 2$.[13] The relative strength of the index-nonconserving peaks, with the $\Delta n = 2$ peak appearing an order of magnitude stronger than $\Delta n = 1$, appears consistent with the relevant terms in the Hamiltonian: $\Delta n = 2$ mixing arises from $H_{13} \sim a_-^2$ and $H_{31} \sim a_+^2$ anisotropic terms, while $\Delta n = 1$ mixing comes from $H_{12} \sim k_z a_-$ and $H_{21} \sim k_z a_+$ terms that are weaker because the resonant peaks occur when the subbands in the well are nearly aligned with the top of the valence band in the emitter, making the k_z of the tunneling holes small. Obviously, a more rigorous description of the elastic index-nonconserving tunneling mechanism requires a real calculation of the Landau level envelope functions in the emitter and well, in principle similar to calculations performed for quantum well subbands in Refs. 20, 21.

4. Transverse field geometry: in-plane nonparabolicity and anisotropy

The $I(V,B_\perp)$ characteristics of the HH_0 peak in the $W = 35$ Å DBRTS for $B_\perp = 25$ T are plotted in Fig. 4 as a function of $B_\perp$ orientation with respect to the principal crystallographic axes. As expected, the resonant peak broadens and shifts to higher bias. The variation of the peak shift ΔV_p with $B_\perp$ orientation is pronounced: at 25 T it ranges from $\Delta V_p \approx 65$ mV (when $B_\perp \parallel$ <110>) to $\Delta V_p \approx 110$ mV (when $B_\perp \parallel$ <100>). The variation of the LH_0 peak shift is comparable to the HH_0 result:[16] anisotropy follows the crystal symmetry, but with the opposite sign -- when $B_\perp \parallel$ <100>, ΔV_p is largest for the HH_0 peak and smallest for the LH_0 peak. Hence, for the HH_0 subband, a simple geometrical construction[6,15,16] yields the curvature of $E(k_\perp)$ to be steeper in the <100> direction, making the effective heavy-hole mass m^* away from the $k_\perp = 0$ band edge lighter along the <100> than the <110> axes.

The HH_0 and LH_0 peak shift positions in the $W = 35$ Å structure are summarized in

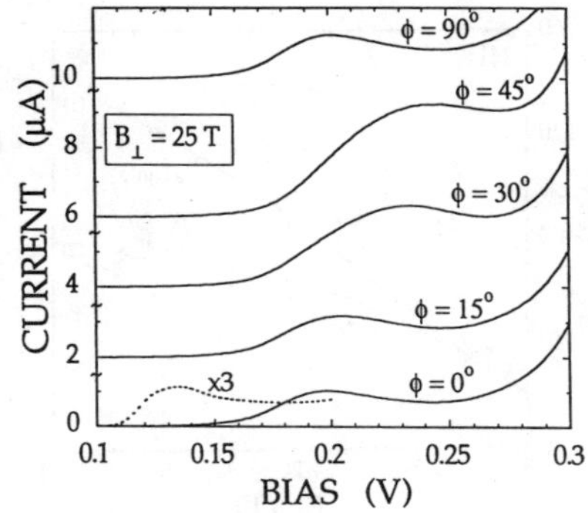

FIG. 4. I(V,B$_\perp$) data on the HH_0 peak in the W = 35 Å DBRTS at 25 T *vs.* angle φ between B$_\perp$ and the <110> direction; dashed line is B$_\perp$ = 0 trace.

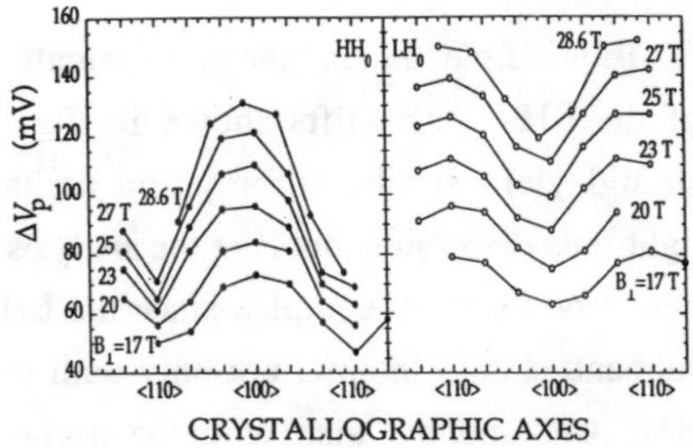

FIG. 5. B$_\perp$-induced voltage shifts ΔV_p *vs.* B$_\perp$ magnitude and orientation in W = 35 Å DBRTS. Lines connecting data points are guides to the eye.

Fig. 5 (DBRTS with W = 23 and 46 Å exhibited analogous behavior). The $\Delta V_p \sim B_\perp^2$ dependence obtained from E and $k_\perp$ conservation laws in materials with parabolic in-plane dispersion[6] does not hold for any B$_\perp$ orientation, signifying non-parabolic $E(k_\perp)$ in all directions. Since the HH_0 subband is expected to interact most strongly with the LH_0 subband, a first analysis the anisotropy and nonparabolicity of the heavy holes can proceed from a two-band model of in-plane dispersion:[22]

$$E(k_\perp) = \hbar^2 k_\perp^2/2m^*(k_\perp) = \hbar^2 k_\perp^2/2m_0^*(1 + \alpha k_\perp^2) \tag{1}$$

where m_0^* is the in-plane mass at $k_\perp = 0$ obtained from interpolated band parameters; the anisotropic nonparabolicity factor $\alpha \equiv \alpha(\phi)$. Then, by calculating the overlap of the well states described by Eq. (1) with the occupied HH emitter states for a given $\Delta k_\perp = eB_\perp \langle z \rangle/\hbar$ and applied bias V, one can compute the predicted ΔV_p shifts for various B$_\perp$ orientations as a function of $\alpha(\phi)$ and the material DBRTS parameters.[16] The results are shown in Fig. 6, where the observed ΔV_p for B$_\perp$ ∥ <110> and <100> are compared to the calculated values for α = 0, 3.5, and 8.5x10 $^{-18}$ m^2. Clearly, the α = 0 (parabolic band) curve diverges strongly from the data for B$_\perp$ > 15 T, while the cited values of $\alpha(\phi)$ give reasonable agreement with the experimental results. Similar values for $\alpha(\phi)$ were obtained in measurements on the W = 23 Å DBRTS,[16] consistent with the two-band model prediction[9] that the nonparabolicity factor should be inversely proportional to the HH_0-LH_0 subband energy separation, which is rather similar in the W = 35 Å and 23 Å DBRTS (see the bias separation between HH_0 and LH_0 peaks in Fig. 1).

In principle, an analogous examination of the LH_0 peak shifts shown in Fig. 5 should yield similar information on the light-hole dispersion, but here the analysis would be far more complex since the LH_0 subband should interact not only with the HH_0 subband but also with the higher-lying subbands, becoming strongly nonparabolic and not amenable to the simple approximation of Eq. 1.

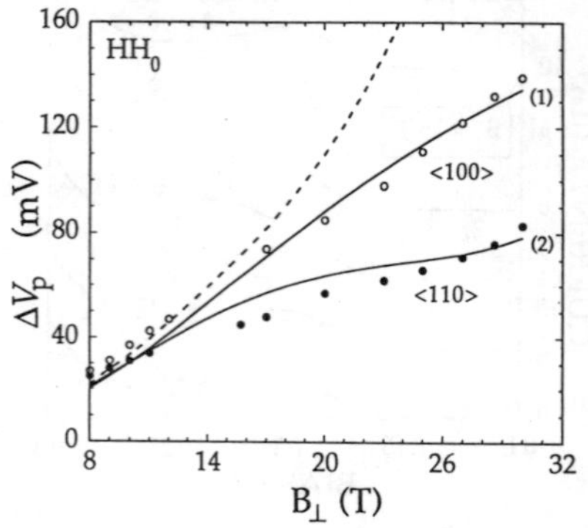

FIG. 6. HH_0 peak shift ΔV_p *vs.* $B_\perp$ ‖ <110> and <100> in W = 35 Å DBRTS. Solid lines are calculations with α = 3.5×10^{-18} m^2 (1) and 8.5×10^{-18} m^2 (2), dashed line is α = 0 calculation.

5. Conclusions

Our magnetotunneling measurements on high-quality *p*-Si/$Si_{1-x}Ge_x$ double-barrier resonant tunneling structures have yielded experimental data on the valence band dispersion in strained SiGe quantum wells. In the parallel field geometry we have mapped out the low-lying Landau level structure using the unexpectedly strong Landau index-nonconserving tunneling, while sample rotation in transverse fields was used to probe the nonparabolicity and anisotropy of the in-plane dispersion. While numerical band structure calculations comprise a well-established field with an extensive literature, calculations on new materials like SiGe are nontrivial because some of fundamental parameters are unknown (e.g. Luttinger parameters interpolated from the Si and Ge values) and standard approximations (e.g. neglect of in-plane anisotropy) may not be appropriate. For this reason we hope that the experimental data presented in this paper will stimulate further theoretical activity and comparison with numerical calculations.

6. Acknowledgments

The authors gratefully acknowledge the assistance of numerous collaborators, including but not limited to Y. H. Lee, K. Milkove, S. S. Iyer and C. Stanis (IBM); D. A. Syphers (Bowdoin College); and J.-P. Cheng and others on the Francis Bitter National Magnet Laboratory. Work at IBM was partially supported by ONR (contract N0001492-C-0017).

6. References

1. S. Luryi, *Appl. Phys. Lett.* **47**, (1985) 490.
2. E. E. Mendez, L. Esaki and W. I. Wang, *Phys. Rev.* **B33** (1986) 2893.
3. V. J. Goldman, D. C. Tsui and J. E. Cunningham, *Phys. Rev.* **B35** (1987) 9387.
4. V. J. Goldman, D. C. Tsui and J. E. Cunningham, *Phys. Rev.* **B36** (1987) 7635.
5. M. L. Leadbeater, E. S. Alves, L. Eaves, M. Henini, O. H. Hughes, A. Celeste, J. C. Portal, G. Hill, and A. Pate, *Phys. Rev.* **B39** (1989(3438.
6. L. Eaves, K. W. H. Stevens and F. W. Sheard, in *The Physics and Fabrication of Microstructures*, ed. M. J. Kelly and C. Weisbuch (Springer, New York, 1986), p. 343.
7. R. A. Davies, D. J. Newson, T. G. Powell, M. J. Kelly, and H. W. Myron, *Semicond. Sci. Technol.* **2** (1987) 61.
8. A. Zaslavsky, D. C. Tsui, M. Santos, and M. Shayegan, *Phys. Rev.* **B40** (1989) 9829; A. Zaslavsky, Yuan P. Li *et al.*, *Phys. Rev.* **B42** (1990) 1374.
9. S. Ben Amor, J. L. Rascol, K. P. Martin, R. J. Higgins, R. C. Potter, and H. Hier, *Phys. Rev.* **B41**, 7860 (1990).
10. U. Ekenberg and M. Altarelli, *Phys. Rev.* **B32** (1985) 3712.
11. J. P. Eisenstein, T. J. Gramila, L. N. Pfeiffer, and K. W. West, *Phys. Rev.* **B44** (1991) 6511.
12. H. C. Liu, D. Landheer, M. Buchanan, and D. C. Houghton, *Appl. Phys. Lett.* **52** (1988) 1809.
13. A. Zaslavsky, D. A. Grützmacher, S. Y. Lin, T. P. Smith III, R. A. Kiehl, and T. O. Sedgwick, *Phys. Rev.* **B47** (1993) 16036.
14. G. Schuberth, G. Abstreiter, E. Gornik, F. Schäffler, and J. F. Luy, *Phys. Rev.* **B43** (1991) 2280.
15. U. Gennser, V. P. Kesan, D. A. Syphers, T. P. Smith III, S. S. Iyer, and E. S. Yang, *Phys. Rev. Lett.* **67**, 3828 (1991).
16. A. Zaslavsky, T. P. Smith III, D. A. Grützmacher, S. Y. Lin, T. O. Sedgwick, and D. A. Syphers, *Phys. Rev.* **B48** (1993) 15112.
17. J. M. Luttinger, *Phys. Rev.* **102** (1956) 1030.
18. D. A. Broido and L. J. Sham, *Phys. Rev.* **B31** (1985) 888.
19. S.-R. Eric Yang, D. A. Broido, and L. J. Sham, *Phys. Rev.* **B32**, 6630 (1985).
20. L. C. Andreani, A. Pasquarello, and F. Bassani, *Phys. Rev.* **B36**, 5887 (1987).
21. R. Winkler and U. Rössler, *Phys. Rev.* **B48**, 8918 (1993).
22. G. C. Osbourn, J. E. Schirber, T. J. Drummond, L. R. Dawson, B. L. Doyle, and I. J. Fritz, *Appl. Phys. Lett.* **49** (1986) 731.

MAGNETO-TRANSPORT IN TRIPLE BARRIER RESONANT TUNNELING DIODES

M. R. Deshpande[1], M. A. Reed[1], L. D. Macks[2], S. A. Brown[2], R. P. Starrett[2], R. G. Clark[2], C. J. L. Fernando[3] and W. R. Frensley[3]

[1] *Dept. of Electrical Engineering and Applied Physics, Yale University, New Haven, CT 06520 USA*
[2] *National Pulsed Magnet Laboratory, University of New South Wales, Sydney 2052, Australia*
[3] *School of Engineering & Computer Science, University of Texas, Richardson, TX 75083 USA*

We have studied magneto-transport of two triple barrier resonant tunneling structures with slightly different barrier / well widths designed for simultaneous (triple) resonance of the quantum well energy levels and the lowest quasi-2D emitter state. For sample A we observe both emitter-to-well and well-to-well transitions which are used to probe correlated electron ground states. There is strong evidence that sample B exhibits a triple resonance.

1. Introduction

Resonant tunneling has been used extensively to probe both emitter and well states in double barrier resonant tunneling structures (DBRTS). In particular, magneto-transport is an effective method for the investigation of elastic and phonon assisted tunneling[1-3], and charge accumulation in the quantum well.[4]

We report low temperature transport measurements to 48T of GaAs double quantum well – AlAs triple barrier structures designed for a triple resonance involving the lowest quasi-2D state in the emitter and bound states in *both* quantum wells. The two structures are detailed in Table 1. The triple resonance is achieved in forward bias, defined as electron injection into the wider well first.

Fig. 1 shows the calculated conduction band profile for structure A at 305mV where the bound state in well 1 is resonant with the lowest quasi-2D state in the emitter; the well 2 state is at higher energy, just off triple resonance. The model used is a self-consistent Poisson solution, treated in a finite-temperature Thomas-Fermi approximation[5]. The calculated energies of the quantum well states in sample A (referenced to the Fermi level) at zero applied bias are 47meV for well 1 and 101meV for well 2. Calculated voltage/energy conversion ratios, α, for the position of the quantum well states in sample A are α_1=3.56mV/meV for well 1, α_2=2.34 mV/meV for well 2, and α_{12}=6.74 mV/meV for the difference between well states. The situation for sample B is similar.

In a magnetic field applied parallel to the tunneling current, the emitter and quantum well energies are quantized into Landau levels with index n. The selection rule Δn=0 can be broken by interaction with a LO phonon[3] or by other mechanisms[1], causing a series of inter-Landau level transitions. In our double well structures three distinct resonant processes are possible: i) emitter to well 1 (e$\rightarrow$1); ii) well 1 to well 2 (1$\rightarrow$2), and iii) emitter to well 2 (e$\rightarrow$2). The e$\rightarrow$2 process is very weak but the remaining two processes each may have a 'no-LO' inter-Landau level fan as well as GaAs and AlAs LO-phonon assisted transitions and associated inter-Landau level fans.

	SAMPLE A	SAMPLE B	
EMITTER	0.5 μm	0.5 μm	n+ GaAs Si: 1.5 x 10^{18}cm^{-3}
	100 Å	100 Å	GaAs graded doping
	150 Å	150 Å	GaAs undoped
BARRIER	40 Å	40 Å	AlAs undoped
WELL1	80 Å	65 Å	GaAs undoped
BARRIER	15 Å	20 Å	AlAs undoped
WELL 2	50 Å	50 Å	GaAs undoped
BARRIER	40 Å	40 Å	AlAs undoped
	150 Å	150 Å	GaAs undoped
	100 Å	100 Å	GaAs graded doping
COLLECTOR	0.5 μm	0.5 μm	n+ GaAs Si: 1.5 x 10^{18}cm^{-3}
SUBSTRATE			n+ GaAs (001)

Table 1. Sample growth details.

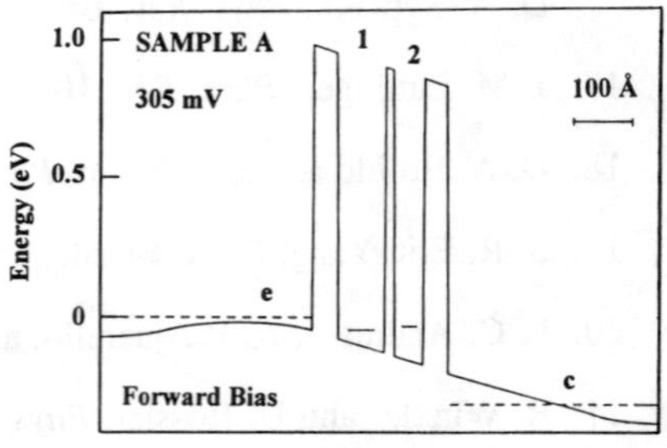

Fig. 1. Calculated band diagram for sample A at 305mV. Dashed lines mark the Fermi level.

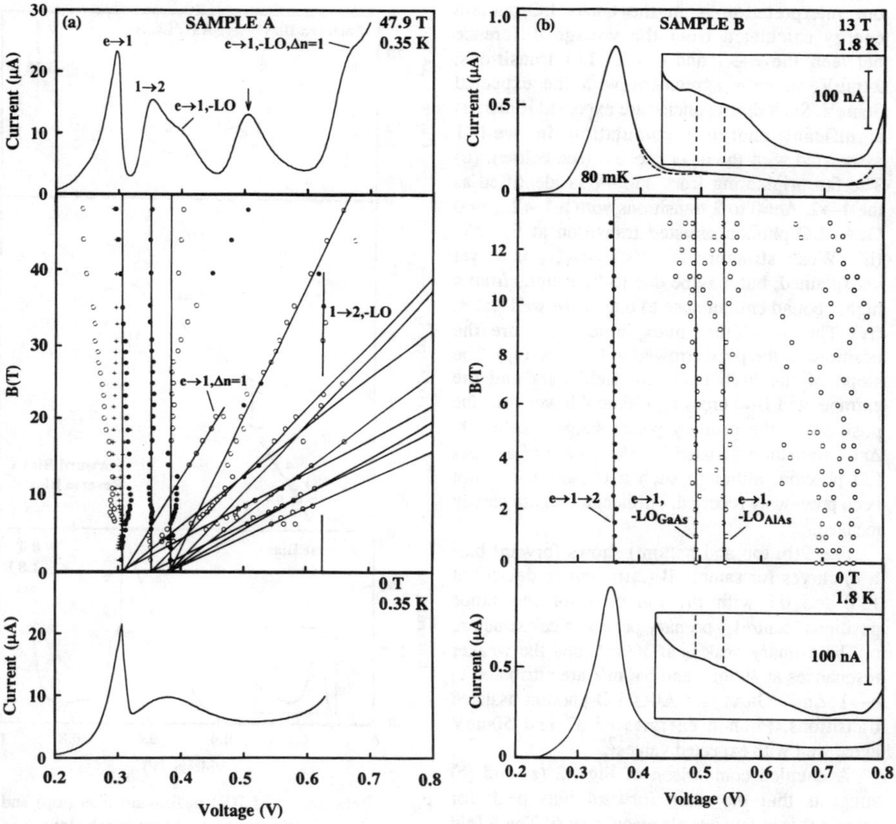

Fig. 2. Relationship of I(V) at high magnetic field (top) and at B=0T (bottom) to the peak position fan chart (centre) for (a) sample A at 0.35K and (b) sample B at 1.8K. Insets (b, top and bottom) show weak high bias I(V) structure. Dashed line (b, top) shows minor changes to I(V) for sample B at 80mK.

2. Current – Voltage Characteristics

High magnetic field measurements were conducted on sample A in a ^{3}He cryostat and 60T, 20ms pulsed magnet using a fast (2ms) voltage ramp centred on the peak field[6]. Further measurements were carried out on both samples using fast and slow voltage ramps in steady magnetic fields.

Fig. 2(a) (top and bottom) shows forward bias I(V) curves for sample A (64μm square device) at 47.9T and 0T. Fig. 2(a) (centre) shows the position of the resonances as a function of pulsed magnetic field. The •, o data represent strong and weak resonances for increasing bias[6]. Hysteresis is observed only for the primary resonance as shown by + data for decreasing bias. Comparison of fast and slow ramp data indicate that this hysteresis has an intrinsic time dependence.

The most reasonable fit to the complex inter-Landau level transition fans for sample A is shown in Fig. 2(a, centre). A detailed analysis has been presented elsewhere[6] and is summarised as follows: (i) Originating from the primary resonance at 305mV is the fan of e→1, Δn=0 to 3 transitions and from 381mV is the fan of GaAs LO-phonon assisted e→1, Δn=0 to 5 transitions. The only major discrepancy in our interpretation is in the GaAs LO-phonon energy calculated from the voltage difference between the e→1 and e→1, -LO transitions, 21meV, in poor agreement with the expected 36meV. Such discrepancies are expected if there is significant charge accumulation in well 1 associated with the e→1 process (see below). (ii) The fan originating from 350mV is identified as the 1→2, Δn=0 to 2 transitions with a 1→2, Δn=0 GaAs LO-phonon assisted transition at 630mV. (iii) Weak structure for $V<V_{primary}$ is as yet unexplained, but may be due to transitions from a higher bound emitter state to one of the well states. (iv) The only other unexplained data are the positions of the peak arrowed in Fig 2 (a, top). The slope of the high magnetic field data and the extrapolated B=0 crossing (36meV lower than the position of the primary peak) suggest an e→1, Δn=1 transition assisted by *absorption* of a GaAs LO phonon, although such a transition has not been previously reported. Further studies are clearly required.

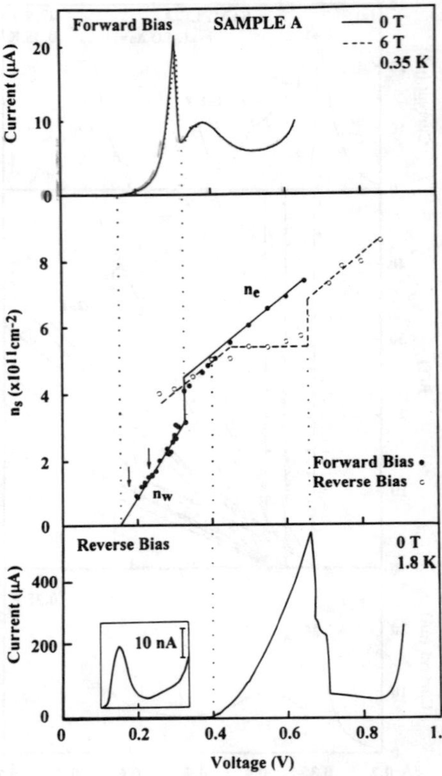

Fig. 3. Relationship of I(V) in forward bias (top) and reverse bias (bottom: inset shows weak low bias structure) to determined electron densities (centre) where solid line and dashed line are guides to the eye.

Fig. 2(b, top and bottom) shows forward bias I(V) curves for sample B (2μm square device) at 13T and 0T with the fan chart of resonance positions (centre) – primary peak ●, weak structure o. The primary peak is at 360mV and the weaker resonances at 495mV and 540mV are attributed to e→1, Δn=0 GaAs and AlAs LO-phonon assisted transitions. Phonon energies of 37 and 50meV agree well with expected values[2].

A detailed comparison of Figs. 2 (a) and (b) suggests that the main forward bias peak for sample B is in fact a triple resonance: (i) The 5-fold increase in peak to valley ratio at B=0T and 50-fold increase in peak current *density* from sample A (64μm device) to sample B (2μm device) indicates more direct transmission through structure B on resonance. (ii) The analysis of the fan chart for sample A distinguishes e→1 and 1→2 transitions. A separate 1→2 transition is not present in sample B. (iii) For sample B the LO phonon resonances are relatively weak and there is no evidence for significant charge accumulation suggesting enhanced transmission through the structure. This is reinforced by the observation of GaAs and AlAs phonon resonances of comparable height in sample B; in sample A charge accumulation in well 1 (see below) causes GaAs phonons to be dominant. (iv) The larger resonance width in sample B is not inconsistent with a process in which three energy levels must be aligned.

A reverse bias I(V) curve at 0T for sample A (64μm device) is shown in Fig. 3 (bottom); sample B shows similar structure. Weak low bias structure (Fig. 3 bottom, inset) is due to a C→1, Δn=0 transition and the strong resonance to a C→2, Δn=0 transition.

3. Shubnikov – de Haas (SdH) Oscillations

Under constant DC bias both samples show SdH oscillations in I(B) measurements[6] which can be used to calculate a 2D electron density (n_s)[4]. The same n_s values were obtained in measurements of differential conductance, g, using AC modulation/lock-in techniques. Fig. 3 shows n_s(V), obtained from I(B) and g(B) data for sample A, related to forward and reverse bias I(V). Forward bias data (●) show an increase from zero at the threshold for tunneling (~0.15V); at the primary resonance there is a distinct change in slope and a discontinuity. The data are similar to those for an asymmetric DBRTS[4] where there is charge accumulation in the well: we identify $n_s(V)$ data for $V<V_{primary}$ with charge accumulation in well 1 (n_w). At higher biases, charge in well 1 is ejected and a 2D emitter accumulation layer forms (n_e). The decrease of dn_s/dV corresponds to a shift of the 2D electron system (2DES) responsible for the SdH oscillations from well 1 to the emitter, resulting in a decreased capacitance of the structure.

In reverse bias, $n_s(V)$ data for sample A are similar to forward bias n_e data but exhibit a constant density in the region of the main resonance, in general agreement with the emitter density in Ref 4. No oscillations are observed from well electrons as emitter oscillations dominate[4].

All strong $\Delta n=0$ peaks for sample A exhibit SdH-like oscillations in peak position and current. Fig. 4 shows peak current oscillations for the primary forward bias peak (arrowed in inset) for increasing bias. Maxima in the peak current are strongly suggestive of quantum Hall effect (QHE) states associated with a fixed electron density. The assignment of Landau level filling factors $\nu=4$, 2, 1, 2/3 shown gives $n_s=4.5\times10^{11}cm^{-2}$. At elevated temperature (Fig. 4, T~2K) the $\nu=2/3$ peak decreases in magnitude whereas the integer filling factor peaks are unchanged, providing support for the correlation of the $\nu=2/3$ peak in our data with a fractional QHE state. A density of $4.5\times10^{11}cm^{-2}$ is in reasonable agreement with n_e at ~305mV (Fig. 3) but for $V<V_{primary}$ SdH oscillations are observed only from the well. It is not clear experimentally whether the Fig. 4 data ($V=V_{primary}$) are related to emitter or well charge.

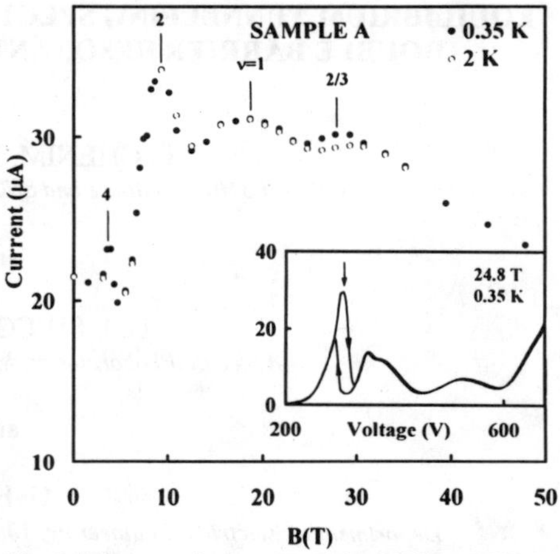

Fig. 4. Primary peak current v. B for 0.35K and 2K in sample A. Inset: high B hysteresis.

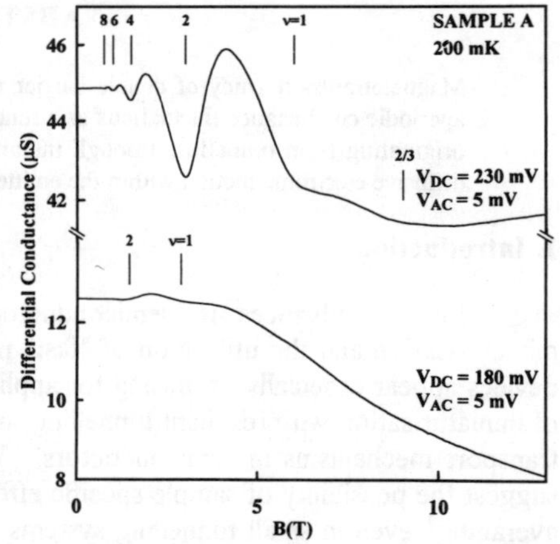

Fig. 5. g(B) for two well 1 densities (arrowed in Fig. 3) in sample A.

Due to the absence of SdH oscillations from the emitter at low bias, sample A provides a method for studying tunneling into a low density 2DES (provided by charge accumulation in well 1) in the extreme quantum limit ($\nu<1$). We have mounted sample A in a dilution refrigerator and measured g(B) at mK temperatures over a wide range of biases. Fig. 5 shows clear minima in g(B) at integer QHE states and some additional structure beyond $\nu=1$. Further measurements, to beyond $\nu=1/5$, are in progress.

References

1. E. E. Mendez, L. Esaki, and W. I. Wang, Phys. Rev. B **33** (1986) 2893.
2. V. J. Goldman, D. C. Tsui, and J. E. Cunningham, Phys. Rev. B **36** (1987) 7635.
3. M. L. Leadbeater et al., Phys. Rev. B **39** (1989) 3438.
4. L. Eaves et al., Solid State Electronics **32** (1989) 1101.
5. M. A. Reed et al., Appl. Phys. Lett. **54** (1989) 1256.
6. L. D. Macks et al., Physica B (in press).

EQUILIBRIUM TUNNELLING SPECTROSCOPY OF IMPURITY STATES IN DOUBLE BARRIER RESONANT TUNNELLING STRUCTURES.

L. GHENIM, J. C. PORTAL
LPS-CNRS, F31077 Toulouse and SNCI-CNRS, F38042 Grenoble, FRANCE

and

D. L.SIVCO, A.Y. CHO
AT&T Bell Laboratories, Murray Hill, NJ 07974, USA

and

G. HILL
Department of Electrical Engineering, University of Sheffield, Sheffield S1 4DU, UK

ABSTRACT

Magnetotransport study of double barrier resonant tunnelling structures demonstrates aperiodic conductance fluctuations as a function of magnetic field, at low temperatures, originating from tunnelling through impurity states within the well, and concomitant diffusive electronic motion within the emitter and collector contacts.

1. Introduction

Further advances in semiconductor technology will invariably involve both miniaturisation and the utilisation of "fast" physical processes. Small resonant tunnelling devices appear especially promising for applications because they combine the advantages of miniaturisation with resonant tunnelling, which is known to be one of the fastest charge transport mechanisms in semiconductors. Yet, previous studies of mesoscopic systems suggest the possibility of sample specific effects originating from the absence of ensemble averaging[1] even in small tunnelling systems. Indeed, Ricco and Azbel[2] have identified a role for impurity levels located in the barrier-well-barrier region of double barrier resonant tunnelling structures (DBRTS); the relative importance of these mechanisms in limiting performance of small tunnelling devices remains to be clarified.

Here, we examine the role of impurity states in DBRTS with equilibrium tunnelling spectroscopy. We report vanishing-bias, perpendicular magnetotransport measurements in (3D) AlGaAs/GaAs DBRTS where the well thickness was chosen narrow (20Å) in order to place the quasi bound state far above the chemical potential of the emitter and collector at equilibrium. The experiments reveal reproducible aperiodic conductance fluctuations, with rms amplitude of the order of e^2/h, which have a mesoscopic origin like the universal conductance fluctuations (UCF) observed in disordered metals. However, here, the effect is attributed to diffusive electronic motion within metallic contacts, during the tunnelling process.

2. Experiment

These molecular-beam-epitaxy grown DBRTS consist of heavily doped (n(Si) = 2.5×10^{18} cm^{-3}) $Al_{0.16}Ga_{0.84}As$ emitter and collector, 100 Å thick $Al_{0.4}Ga_{0.6}As$ undoped barriers and an undoped GaAs well. Based on previous studies,[3] we estimate the Hall mobility $\mu(4.2\ K) = 1000\ cm^2V^{-1}s^{-1}$ for a carrier density $n = 1 \times 10^{18}$ cm^{-3} in the emitter and collector. The two terminal resistance was measured under constant current conditions as a function of magnetic field in the ($B//I$) configuration, in 0.8 mm mesas, using lock in techniques at temperatures as low as $T = 0.285$ K. The reported experimental results highlight observations of conductance fluctuations at high magnetic fields in the two terminal measurement.

3. Results and discussion

Figure 1 and 3 show the conductance g units of e^2/h as a function of magnetic field B over the high ($1.7 \leq T \leq 15$ K) and the low ($0.285 \leq T \leq 1.9$ K) temperature range, after subtraction of a smooth background. Here, the sample was thermally cycled to room temperature in the interim between measurements in He^3 and He^4 systems. At $T = 0.285$ K, the data show aperiodic fluctuations in the conductance over the entire range of B, which were reproducible in the sample so long as the sample was not warmed above 100 K. However, these magnetofingerprints varied from sample to sample, and changed upon cycling the sample to room temperature (see figures 1, 3 and 4). The root-mean-square of the conductance fluctuations assumed a temperature independent value at low temperatures ($T < 1$ K), and decayed as a weak power law in T, i.e. rms(g(T)) ~ $(T)^{-0.25}$, at higher temperature (figure 2). A Fourier analysis demonstrated a complicated power spectrum and it did not show any apparent periodicity in B or in $1/B$. All these features suggest that these magnetofingerprints are not associated with a tunnelling effect between the emitter and collector Fermi levels via the Landau Levels of the 2D confined state in the well. Indeed, these features suggest that the magnetofingerprints are representative of sample specific disorder in each specimen. Specific similarities to universal conductance fluctuations[4] observed in non-DBRTS based mesoscopic systems are: (a) they are aperiodic, (b) the rms amplitude of

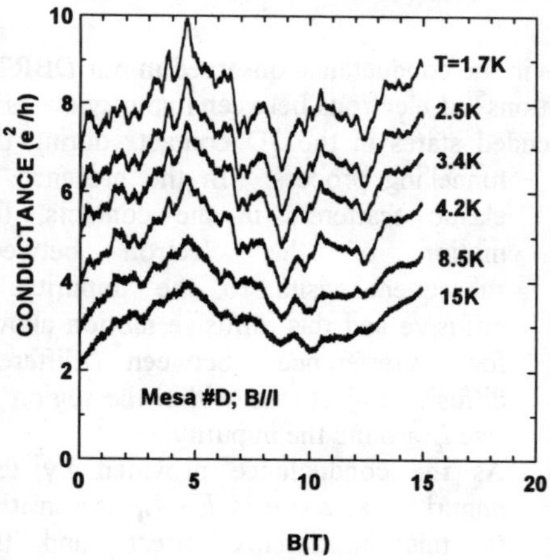

Figure 1: Conductance fluctuations as a function of B measured at several temperatures ($1.7 \leq T \leq 15$ K). The curves have been offset for the sake of clarity.

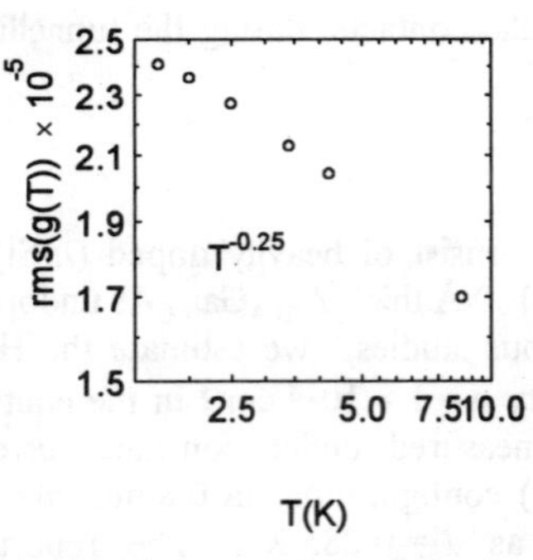

Figure 2: The rms value of the fluctuations as a function of T for the data of Fig. 1.

fluctuations is of the order of e^2/h (~ 0.55 e^2/h for sample #D and 0.45 e^2/h for sample #B), (c) the effect depends weakly upon temperature at low T, (d) it is a sample specific effect which requires ($g \geq e^2/h$). In order to determine the relevance of coherent effects in our samples, we have obtained the electronic phase coherence length l_c by evaluating the correlation function (see Fig. 5) in magnetic field from the experimental data shown in figure 1. Here, the half-width of the correlation function reflects the correlation field B_c, which corresponds to the magnetic field for which one has a flux quanta through an area of order l_c^2. Such an analysis suggests that B_c (0.285 K) = 0.18 T, which implies a coherence length $l_c = (h/eB_c)^{1/2} = 0.15$ µm, that is much smaller than the thermal length of the emitter, $l_T = 0.8$ µm, and much larger than the elastic mean free path of the emitter, $l_e = 0.01$ µm, i.e., $l_e < l_c < l_T$. Furthermore, the correlation function (figure 5) is consistent with the expected dimension-dependent power-law decay as $C(\Delta B) \propto (B_c/\Delta B)^{4-d}$ within the UCF picture as well as the weak temperature dependence of the fluctuations observed in our system as $T^{(d-4)/4}$ in the high temperature range $T > 1.7$ K, where l_T and l_c are smaller than the size of the sample for the 3D case.

Thus, we suggest that the fluctuations in the conductance observed in our DBRTS structures, arise due to multiple virtual transitions[5] of electrons between impurity states in the barrier-well-barrier region, and the extended states in the 3D contacts during the tunnelling process. In the presence of elastic scatterers in the contacts, the motion of the electron between subsequent visits to the impurity is diffusive and this diffusive motion allows for interference between different diffusive trajectories within the region of size l_c around the impurity.

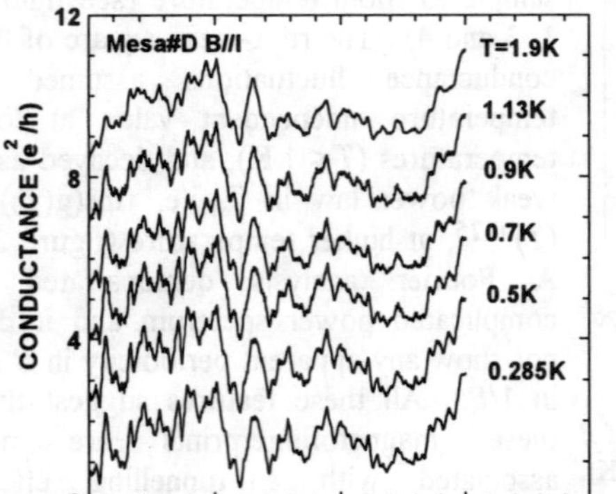

Figure 3: Conductance fluctuations as a function of B for 0..285 ≤ T ≤ 2 K after thermal cycling.

As the conductance provided by few impurities at energies $E \sim E_F$ is sensitive to this interference effect, and the application of a magnetic field changes the interference pattern, one expects B-dependent variations in the probability of resonant tunnelling and, therefore, conductance fluctuations with magnetic field. In order for this picture to be valid and to explain the transport phenomena involved in our samples, the presence of impurities in the barrier well-barrier region at energies close to E_F is critical. In addition, it is necessary that electrons are "trapped" in the DBRTS for times t much longer than the characteristic elastic lifetime of the emitter. These conditions appear to hold for the samples examined here.

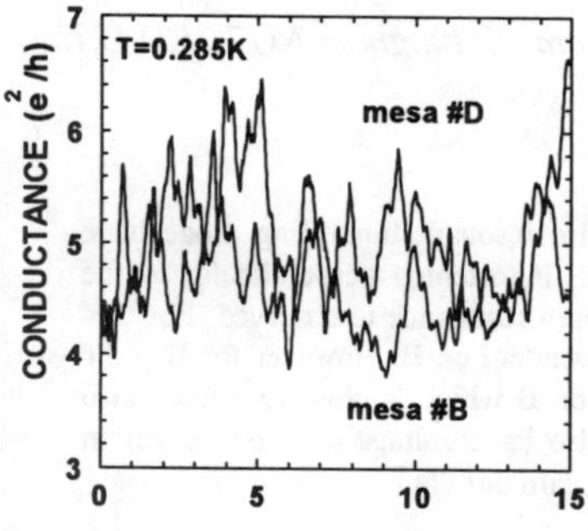

Figure 4: Magnetoconductance fluctuations of sample #B and #D at $T = 0.285$ K

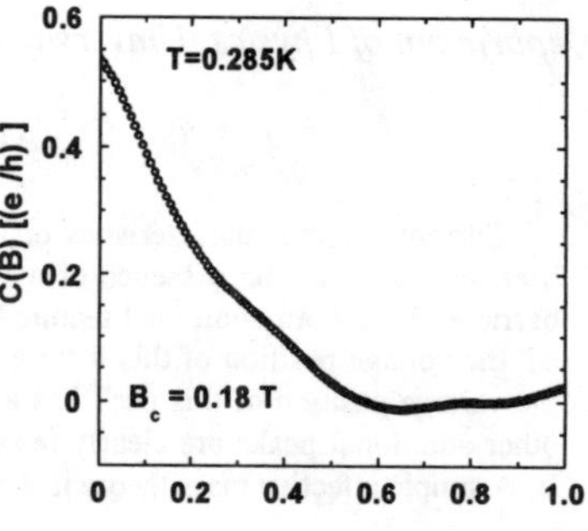

Figure 5: Autocorrelation function in magnetic field of the conductance fluctuations of sample #D at $T = 0.285$ K

4. Conclusions

In conclusion, we have observed reproducible aperiodic conductance fluctuations versus magnetic field ($\sim e^2/h$) in double barrier resonant tunnelling structures. We have attributed these fluctuations to quantum coherence effects originating from tunnelling between the highly disordered diffusive metallic electrodes and the impurities located in the barrier-well-barrier region.

5. Acknowledgements

This research was supported in part by the CNRS and the NATO

6. References

1. S. Washburn and R. A. Webb, *Adv. Phys.* **35**, 375-422 (1986).
2. B. Ricco and M. Ya. Azbel, *Phys. Rev. B* **29**, 1970 (1984).
3. N. Chand, T. Henderson, J. Klem, W. T. Masselink, R. Fisher, Y. C. Chang and H.Morkoc, *Phys. Rev. B* **30**, 4481 (1984).
4. P. A. Lee, A. D. Stone and H. Fukuyama, *Phys. Rev. B* **35**, 1039 (1987).
5. I. V. Lerner and M. E. Raikh, *Phys. Rev. B* **45**, 14036 (1992).

TRANSITION FROM LATERALLY CONFINED STATES TO LANDAU LEVELS IN A 1D RESONANT TUNNELLING DIODE

J.WANG, P.H.BETON, N.MORI, L.EAVES, P.C.MAIN, T.J.FOSTER AND M.HENINI

Department of Physics, University of Nottingham, Nottingham NG7 2RD U.K.

ABSTRACT

Current-voltage characteristics of a quantum wire resonant tunnelling diode have been measured in the presence of a magnetic field, B, oriented perpendicular to the barriers. At B=0 an additional feature beyond the main resonance is observed. For B < 3T the voltage position of this feature is weakly dependent on B. However for B > 3T the voltage position of this peak has a dependence on B which is close to linear. Two other additional peaks are clearly resolved which also have voltage positions linear in B. A simple effective mass theory is developed to explain our data.

Resonant tunnelling via the quasi-bound states formed in the quantum well of double barrier GaAs/AlAs diodes gives rise to strong peaks in the current-voltage characteristics, I(V), [1]. Additional peaks are observed in the I(V) of resonant tunnelling diodes (RTD) which have sub-micron lateral dimensions. It was originally proposed that these additional peaks were due to resonant tunnelling via a set of laterally quantised states in the quantum well [2]. However it has been shown that additional peaks in I(V) can also be caused by the incorporation of intentional [3] or unintentional [4] donors in the quantum well. More recently we have shown that the peaks in I(V) due to lateral quantisation and donors may be distinguished by a magnetic field, B, applied perpendicular to the current [5]. In this paper we discuss the influence of B applied *parallel* to the current. For this orientation we expect to observe a clear transition from a low field regime, in which the electrostatic confinement dominates, to a high field regime in which magnetic confinement dominates. The transition occurs when the cyclotron energy $\hbar\omega_c \approx \hbar\omega_0$, the characteristic energy of the confining potential. In previous work such a transition has not been observed indicating that the additional structure observed in I(V) for B=0 for those particular experiments was not due to lateral quantum confinement [6,7]. In the present work we do observe this transition confirming the observation of 1D lateral confinement at B=0.

The active region of the device used for our experiments is shown in Fig. 1a. It is fabricated from a GaAs/AlAs heterostructure using a novel optical lithography and wet etching process [8]. The GaAs quantum well width is 9.0nm and the AlAs barrier width is 4.7nm. The physical device dimensions $\ell_t \times \ell_b \approx 0.5 \times 1.0\ \mu m^2$ but the conducting area is much smaller due to sidewall depletion and is estimated to be $\approx 0.06 \times 0.6\ \mu m^2$ (the depletion edge is shown in Fig. 1a as a dashed line). We have shown previously for this device that the motion in the y-direction is quantised, and that the confining potential is parabolic [5]. Thus a set of 1D subbands is formed in the emitter and quantum well of the device, with free electron motion assumed for the z-direction. For B=0 the levels in the emitter and well (denoted by the indices i and j respectively) have an energy separation of $\hbar\omega_e$ and $\hbar\omega_w$. Note that our device is asymmetric so that $\hbar\omega_e \neq \hbar\omega_w$. Below we discuss only the polarity in which electrons flow from the top contact to the lower contact. For this polarity $\hbar\omega_e > \hbar\omega_w$ and for this diode only the lowest (i = 0) state is occupied in the emitter.

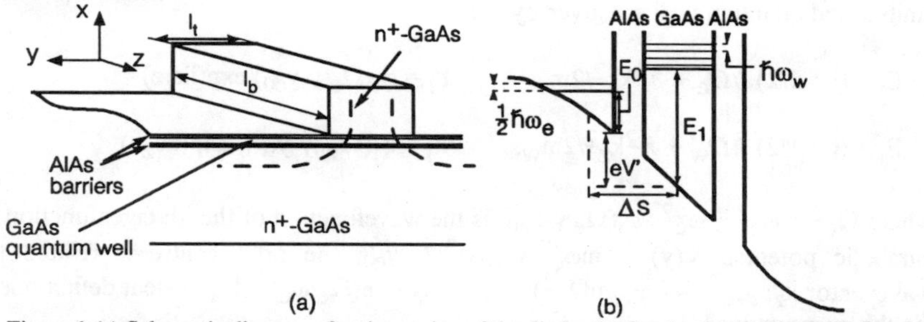

Figure 1 (a) Schematic diagram of active region of the diode; (b) Conduction band profile for B=0T. E_0 is the confinement energy of the accumulation layer formed in the emitter and E_1 is the energy of the quasi-bound state formed in the quantum well for a large area device.

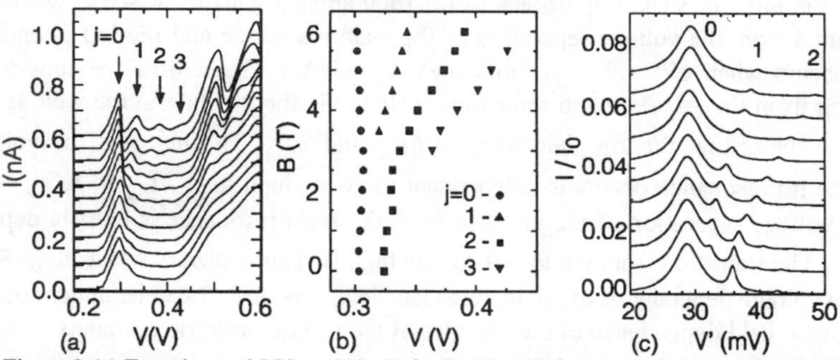

Figure 2 (a) Experimental I(V) at 300mK for B=0T - 6T from bottom to top with 0.5T step; (b) Fan diagram shows the voltage dependence of j=0, 1, 2 & 3 peaks on magnetic field; (c) Calculated I(V).

Figure 2(a) shows the low temperature I(V) for this device for B=0 - 6T. For B=0 several peaks are observed in I(V): (i) the main resonance (V=0.29V) due to tunnelling between the i=0 emitter state and the j=0 state in the well; (ii) an additional peak in I(V) (V=0.32V) which is due to to tunnelling between the i=0 state in the emitter and the j=2 state in the quantum well; (iii) peaks due to GaAs and AlAs LO phonon-assisted tunnelling [9] which occur at V=0.49V and V=0.6V respectively. For low magnetic field the j=2 peak is only weakly affected by B. However for B > 3T the voltage position has a strong, approximately linear, dependence on B (see Fig.2b). In this high field regime we observe other peaks (j=1, 3 in Fig. 2b) which also have a linear dependence on B. These peaks closely resemble the inter-Landau level transitions which are observed for a large area diode [10]. However for large area diodes these peaks do not continuously develop from a peak in I(V) which is observed even for B=0.

The Schroedinger equation for an electron in a parabolic potential and a magnetic field $\mathbf{B} = (B,0,0)$ is separable and may be readily solved in the Landau gauge $\mathbf{A} = (0,0,By)$. The energies and wavefunctions (y and z components only) of the quantised states in the emitter and quantum well are given by

$$E_i = (i + 1/2)\,\hbar\Omega_e + \hbar^2 k_z^2/2m_e, \qquad \Psi_i = \varphi_i(\Omega_e;y-y_{e0})\exp(ik_z z)$$

$$E_j = (j + 1/2)\,\hbar\Omega_w + \hbar^2 k_z^2/2m_w, \qquad \Psi_j = \varphi_j(\Omega_w;y-y_{w0})\exp(ik'_z z)$$

where $\Omega_e^2 = \omega_e^2 + \omega_c^2$. $\varphi_i(\Omega_e;y-y_{e0})$ is the wavefunction of the ith eigenfunction of the parabolic potential $V(y) = m\omega_e^2(y-y_{e0})^2/2$. y_{e0}, the orbit centre is related to the wavevector k_z, $y_{e0} = -(\hbar\omega_c/m\Omega_e^2)k_z$, and $m_e = m(\Omega_e/\omega_e)^2$. Equivalent definitions apply for the quantum well.

Resonant tunnelling occurs between state i and j when the voltage across the device is adjusted so that these levels are aligned in energy, i.e. $eV' = E_1 + E_j - (E_o + E_i)$, where V' is defined in Fig. 1b. V' is proportional to the total applied voltage, $V = \alpha V'$ where $\alpha \sim 6$ is estimated from the voltage separation of the main resonance and the LO phonon peak which occurs when $eV' = \hbar\omega_{LO}$ (36.5 meV for GaAs). Henceforth we consider only tunnelling from the i=0 state and refer to tunnelling via the jth state in the well as the jth resonance. Note that for low field $\hbar\Omega_e \approx \hbar\omega_e$ and $\hbar\Omega_w \approx \hbar\omega_w$ and the voltage at which the jth resonance occurs is independent of B. At high field $\hbar\Omega_e \approx \hbar\Omega_w \approx \hbar\omega_c$, and the voltage separation of the jth peak from the main resonance is linearly dependent on $j\hbar\omega_c$. The transition from the low B to the high B regime occurs when $\hbar\omega_w \approx \hbar\omega_c$. Thus the voltage dependence of the features labelled j = 0, 1, 2, 3 can be understood from the magnetic field dependence of the energies of the magnetoelectric subbands.

The current flowing from state i to j, $I_{ij} \propto |M_{ij}|^2$, where M_{ij} is the overlap integral [11]

$$M_{ij} = \int_{-\infty}^{\infty} \phi_i(\Omega_e;y-y_{e0})\phi_j(\Omega_w;y-y_{w0})dy$$

For the special case, $\hbar\omega_e = \hbar\omega_w$, the states in the emitter, ϕ_i, and well, ϕ_j, form an orthogonal set: $M_{ij} = \delta_{ij}$, and we would not expect to observe any additional peaks in I(V). For our asymmetric device $\hbar\omega_e > \hbar\omega_w$. However, for B=0, $M_{ij} = 0$ when (j-i) is an odd integer, that is tunnelling between states with different parity is not permitted. Thus we do not observe the j = 1 or j=3 peaks at B = 0. This parity conservation is broken for B > 0 because for this case the orbit centres of the intial and final states are not the same, i.e. $y_{e0} \neq y_{w0}$ (note that k_z is conserved in the tunnelling process). The j=1 and 3 resonances should therefore be observed only for B > 0 in accordance with our data.

From the voltage position of the j = 2 peak at B = 0 we estimate that $\hbar\omega_w$ = 3.5meV, very close to the value deduced from the effect of a perpendicular magnetic field [5]. We have calculated I as a function of V' for various magnetic fields using a theory based on the Bardeen transfer Hamiltonian [11]. The results for $\hbar\omega_w$ = 3.5meV are shown in Fig. 2c and reproduce the main features of our experiment. In particular the transition from low field to high field behaviour discussed above is observed at approximately the same field (B ~ 3T) as for our data.

For very high field $\hbar\Omega_e$, $\hbar\Omega_w \rightarrow \hbar\omega_c$, and $y_{e0} \approx y_{w0}$ and we would expect that all peaks other than j = 0 to have vanishing amplitude. This is not observed in our data - for example the j=1 peak becomes progressively stronger at high field. This behaviour is similar to that observed for a large area diode in which the additional peaks are attributed to scattering, which is not taken into account in our calculations.

We have observed additional peaks in I(V) due to resonant tunnelling via magneto-electric subbands. Resonances which are parity forbidden for B=0 are permitted and observed for B > 0. A transition from a low field regime dominated by lateral confinement to a high field regime dominated by magnetic confinement is identified.

Acknowledgements: This work was supported by the UK SERC (now EPSRC) and by the ESPRIT Partners Basic Research Project. PHB and LE thank the Royal Society, and NM the British Council, for financial support.

References

1. L.L. Chang, L. Esaki and R. Tsu, *Appl. Phys. Lett.* **24** 593 (1974).
2. M. A. Reed, J.N. Randall, R.J. Aggarwal, R.J. Matyi, T.M. Moore and A.E. Wetsel, *Phs.Rev.Lett.* **60** 535 (1988).
3. M.W. Dellow, P.H. Beton, C.J.G.M. Langerak, T.J. Foster, P.C. Main, L. Eaves, M. Henini, S.P. Beaumont and C.D.W. Wilkinson, *Phys. Rev. Lett.* **68** 1754 (1992).
4. J-W. Sakai, T.M. Fromhold, P.H. beton, L. Eaves, M. Henini, P.C. Main, F.W. Sheard and G. Hill, *Phys. Rev.* **B48** 5664 (1993).
5. J. Wang, P.H. Beton, N. Mori, L. Eaves, H. Buhmann, L. Mansouri, P.C. Main, T.J. Foster and M. Henini, preprint
6. P.H. Beton, M.W. Dellow, P.C. Main, L. Eaves and M. Henini, *Phys.Rev* **B49** 2261 (1994).
7. M. Tewordt, L. Martin-Moreno, V.J. Law, M.J. Kelly, R. Newbury, M. Pepper, D.A. Ritchie, J.E.F. Frost and G.A.C. Jones, *Phys. Rev.* **B46** 3948 (1989).
8. J. Wang, P.H. Beton, N. Mori, H. Buhmann, L. Mansouri, L. Eaves, P.C. Main, T.J. Foster and M. Henini, preprint
9. V.J. Goldman, D.C. Tsui and J.E. Cunningham, *Phys. Rev.* **B36** 7635 (1987).
10. M.L. Leadbeater, E.S. Alves, L. Eaves, M. Henini, O.H. Hughes, A. Celeste, J.C. Portal, G. Hill and M.A. Pate, *Phys. Rev.* **B39** 3438
11. N. Mori, P.H. Beton, J. Wang and L. Eaves, preprint.

THE TRANSITION TO CHAOS IN A QUANTUM WELL SUBJECTED TO A STRONG TILTED MAGNETIC FIELD

J. MIAO, T.M. FROMHOLD, F.W. SHEARD, G. EDWARDS and L. EAVES
Department of Physics, University of Nottingham, Nottingham NG7 2RD, U.K.

ABSTRACT

The transition from regular to chaotic classical electron dynamics in a wide potential well in a high tilted magnetic field is investigated. We calculate the quantised energy level spectrum for the well and show that in the regime of classical chaos the nearest-neighbor energy level spacings obey Wigner statistics.

1. Introduction

When the Hamiltonian of a system is integrable, the classical orbits are stable and regular. Quantisation of these orbits using the Bohr-Sommerfeld procedure provides a clear link between the classical and quantum descriptions of the system. However, if the Hamiltonian is nonintegrable, the classical orbits are highly sensitive to small changes in the starting conditions for the motion and the system is chaotic. In this regime, the Bohr-Sommerfeld quantisation scheme is inapplicable[1,2]. The correspondence between a chaotic classical system and its quantum analogue is therefore of fundamental interest.

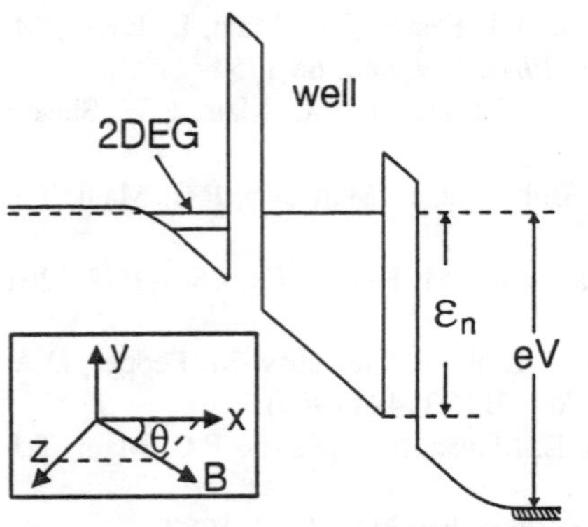

Fig. 1 Schematic conduction band diagram of the RTD under bias V showing the 2DEG in the LH contact and the energy ϵ_n of the nth level in the well. Inset: angle θ of the magnetic field relative to the tunneling (x) direction.

The link between the classical and quantum pictures of chaotic electron dynamics has been explored in recent experiments on GaAs/AlGaAs resonant tunneling diodes (RTDs) containing a 120 nm wide potential well with a strong magnetic field $\boldsymbol{B}$ tilted at an angle θ to the tunneling direction[3,4], as shown in Fig. 1. At $\theta = 0^{\circ}$ or 90°, the classical motion in the well is regular and separable. For a range of intermediate angles, the classical orbits are chaotic at the high values of B (> 10 T) used in the experiments.

In this paper we use Poincaré sections to characterise the transition to chaotic classical motion in the potential well of the RTDs and investigate the manifestations of this transition in the energy level spectrum.

2. The Transition to Chaos

Electrons are injected into the potential well from a two-dimensional electron gas (2DEG) in the left hand (LH) contact (Fig. 1). The injection energy and electric field F in the well are found using a simple model for the potential variation through the device[3] which gives 2DEG sheet densities comparable with the measured values.

To study the changeover from regular to chaotic classical motion in the potential well we have calculated electron trajectories for a range of starting velocities at the LH barrier consistent with the injection energy. Fig. 2 shows Poincaré sections generated by plotting the lateral velocity components (v_y, v_z) each time the electron bounces off the LH barrier. When $\theta = 0^o$, the classical motion is regular. The electrons execute helical traversing orbits and collide with the two barriers in turn. Because the electrons perform cyclotron motion about the magnetic field, the lateral velocity components lie on a series of concentric circles whose radii depend on the starting velocities chosen. When $\theta = 3^o$, the lateral velocity components still lie on well-defined families of curves for certain starting velocities. However, these islands of stable regular orbits are surrounded by a sea of chaos within which the electron trajectories are highly irregular. Comparison of Figs. 2(b) and 2(c) shows that the size of the chaotic sea increases with increasing θ until it fills all of the phase space when $\theta = 24^o$ (Fig. 2(d)), and the system exhibits strongly chaotic behavior.

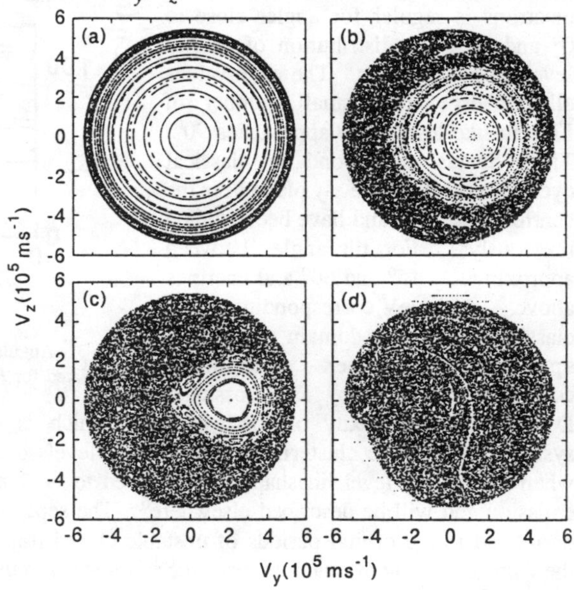

Fig. 2 Poincaré sections showing (v_y, v_z) for 22500 collisions with the LH barrier generated from 225 starting velocities with B = 11.4 T, V = 0.4 V and θ = (a) 0°, (b) 3°, (c) 6°, (d) 24°.

To illustrate how the onset of chaotic classical motion influences the quantum mechanical energy level spectrum, we have calculated the energy levels ϵ_n for a wide 120 nm potential well with impenetrable walls when B = 11.4 T and $F = 2.1\text{x}10^6$ V m^{-1} (corresponding to 0.4 V bias voltage on the structure of Ref. 3). In the gauge given by the vector potential $A = (0, xB\sin\theta - zB\cos\theta, 0)$, $p_y = -i\hbar\partial/\partial y$ is a constant of the motion and the Hamiltonian reduces to a real two-dimensional symmetric matrix within a set of basis functions consisting of products of simple harmonic oscillator states along the z-direction and sinusoidal states along the x-direction. In Fig. 3 we show the angular variation of the lowest 100 energy eigenvalues ϵ_n obtained using a basis size of 2500 which gives accurate energy levels below $\sim$ 1 eV except for θ close to 90°.

When the classical motion of the electrons is regular for angles close to 0° and 90°, the distribution of energy levels is also regular. The eigenvalues all converge to a small number of Landau-like levels as θ approaches 90°. These levels correspond to classical cyclotron orbits in the x-y plane near the centre of the well and have been studied previously[5]. For tilt angles between approximately 15° and 60° and energies above $\sim$ 150 meV corresponding to the classically chaotic domain, the level spectrum becomes extremely complicated with many anti-crossings. Despite the complexity of the spectrum, which is characteristic of nonintegrable systems[1,2], periodic clustering of the energy levels occurs. This is most easily seen when the energy level lineshape is broadened to $\sim$ 7 meV to account for LO phonon emission and will be described elsewhere[6]. The separation of neighboring clusters can be related to the orbital periods of unstable closed trajectories in the well and controls the voltage spacing of quantum resonances observed in the current-voltage characteristics of the RTDs[3,4].

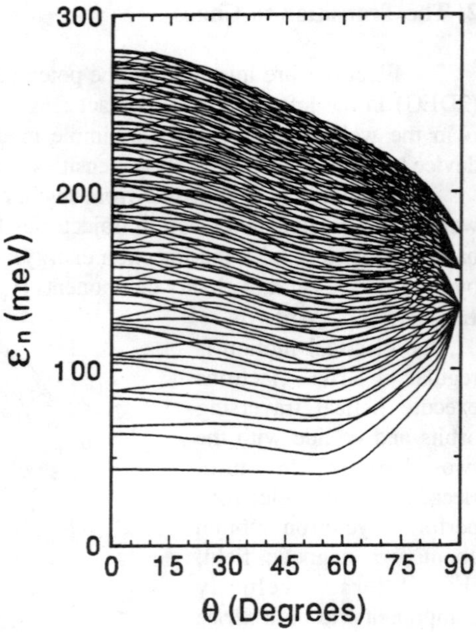

Fig. 3 Angular variation of energy levels ϵ_n calculated for B = 11.4 T, F = 2.1 x 10^6 V m^{-1}.

Calculations of the energy levels of a narrow 15 nm GaAs/AlGaAs well in a tilted magnetic field have also been reported[7]. However the corresponding classical motion, which is regular when the electrons are bound in the well, was not considered.

3. Energy Level Statistics

The long-range periodic modulation of the density of levels, described above, occurs on an energy scale determined by specific unstable closed orbits in the well. This modulation therefore depends on the parameters of the well and on the magnetic field strength and applied bias voltage. However, the energy level spectrum also exhibits behavior which is thought to be universal for nonintegrable systems. The density of

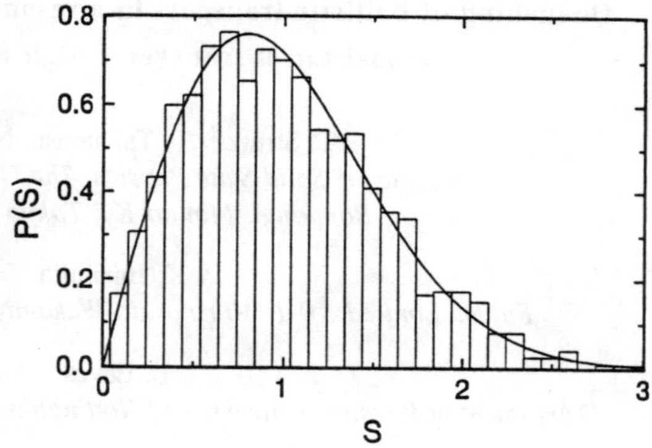

Fig. 4 Histogram: probability distribution of scaled level spacings S. Solid curve: Wigner probability density $P(S)$.

levels can be expressed as the sum of the local mean density $D(\epsilon) \propto \epsilon^{1/2}$, where ϵ is the electron energy in the well, and a term which describes fluctuations about the mean. To study the statistical properties of the fluctuations, we have scaled the energy levels so as to produce a constant average density, with unity mean level spacing[2]. The histogram in Fig. 4 shows the probability distribution of the separation S between adjacent scaled energy levels. This distribution closely resembles the Wigner probability density function $P(S) = (\pi S/2)\exp(-\pi S^2/4)$ shown by the solid curve. This function characterises the statistical distribution of energy level spacings calculated for an ensemble of real symmetric matrices with elements chosen at random from a Gaussian spread[1,2] (GOE statistics). The scaled energy levels of numerous nonintegrable systems have been found to obey similar Wigner-like statistics, which are thought to be a universal quantum signature of classically chaotic systems[1,2,8].

4. References

1. M.C. Gutzwiller, *Chaos in Classical and Quantum Mechanics* (Springer-Verlag, New York, 1990).
2. L.E. Reichl, *The Transition to Chaos* (Springer-Verlag, New York, 1992).
3. T.M. Fromhold, L. Eaves, F.W. Sheard, M.L. Leadbeater, T.J. Foster and P.C. Main, *Phys. Rev. Lett.* **72** (1994) 2608.
4. T.M. Fromhold, M.L. Leadbeater, L. Eaves, T.J. Foster, F.W. Sheard, P.C. Main, P.J. McDonnell and A. Fogarty, Semicond. Sci. Technol. **9** (1994) 488.
5. T.M. Fromhold, F.W. Sheard and G.A. Toombs, in *Proc. 20th Int. Conf. on the Physics of Semiconductors*, eds. E.M. Anastassakis and J.D. Joannopoulos (World Scientific, Singapore, 1990), p. 1250.
6. T.M. Fromhold et al, to be published.
7. G. Marx, K. Lier and R. Kümmel, in *High Magnetic Fields in Semiconductor Physics III (Springer Series in Solid-State Sciences 101)*, ed. G. Landwehr (Springer-Verlag, Berlin Heidelberg, 1992), p. 180.
8. M.V. Berry, *Proc. R. Soc. London* A **413** (1987) 183.

Quenching of ballistic transport in a resonant-tunneling hot-electron transistor at transverse high magnetic field

T. Strutz,† T. Takamasu, N. Miura
Institute for Solid State Physics, The University of Tokyo, Roppongi, Minato-Ku, Tokyo 106, Japan

K. Imamura
Fujitsu Limited, 10-1, Morinosato-Wakamiya, Atsugi 243-01, Japan

L. Eaves
Department of Physics, University of Nottingham, Nottingham NG7 2RD, UK

Base and collector currents have been studied in an InAlAs/InGaAs resonant-tunneling hot electron transistor at 4.2 K in pulsed magnetic fields up to 42.5 T. For transverse fields, the base/emitter current ratio increases initially with an approximate parabolic dependence and then saturates at high fields. Using a classical model for electron transport under a magnetic field perpendicular to the current, values for the mean free path and scattering time in the base are obtained that agree well with theoretical values for electron-electron scattering.

I. Introduction

Ballistic transport[1] in thin semiconductor layers has been receiving considerable attention and has been employed in quantum electron devices such as a resonant-tunneling hot electron transistor (RHET).[2] Nonequilibrium transport has been theoretically investigated by Monte Carlo simulations.[3, 4] It was found that major scattering mechanisms in an RHET base doped to a level of $1x10^{18}$ cm^{-3} are electron-electron scattering, coupled plasmon-LO-phonon scattering, and ionized impurity scattering.[3, 4] Experimental evidence of scattering effects, limiting hot electron transport through a thin layer, is still small. In some works, small transverse magnetic fields were applied resulting in a 25 % reduction of the ballistic electron peak in the energy spectrum.[5] In this study, scattering in the base of an RHET is investigated by using high magnetic fields.

II. Experiment

A schematic cross section of the InAlAs/InGaAs RHET and its energy band diagram are shown in Fig. 1. Electrons are injected as hot and monoenergetic electrons from the emitter into the base by resonant tunneling through a quantum well structure. The thick collector barrier prevents equilibrium electrons in the base from reaching the collector.

All layers have been grown on an InP substrate using molecular beam epitaxy (MBE) at a constant growth temperature of 480°C. The doping material was Si. The layered structure was processed into a three-terminal device with self-aligned emitter and base electrodes.[6] The RHET base consists of a 50 nm thick layer of n-InGaAs doped at a concentration of $1x1o^{18}$ cm^{-3}.

---

† present address: RCAST, University of Tokyo, 4-6-1 Komaba, Meguro-ku, Tokyo 153, Japan

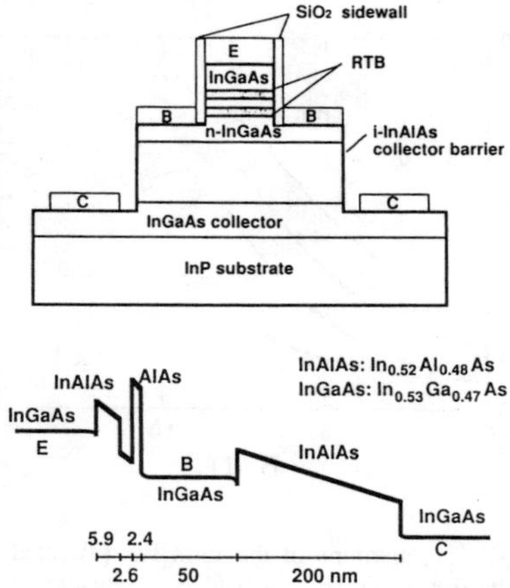

Fig. 1: Schematic cross section and energy band diagram of the RHET under bias.

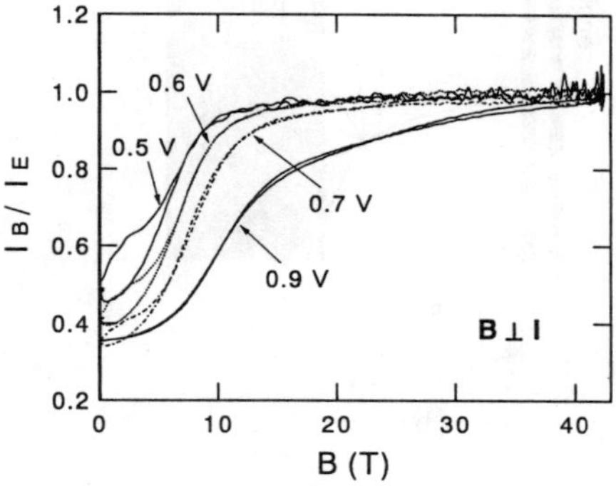

Fig. 2: Base current I_B normalized on the emitter current I_E as a function of transverse magnetic field B for different emitter base voltages below the peak voltage.

Keeping the emitter collector voltage V_{EC} constant at 2.5 V, the base current I_B and the collector current I_C were measured in common emitter configuration, whilst sweeping the magnetic field. The field was oriented perpendicular to the current direction. The sample was immersed in a liquid helium bath. The pulsed magnetic field was provided by discharging a 200 kJ condenser bank into a solenoid coil wound with Cu wire that has been reinforced by Nb-Ti filaments. The field pulse had a duration of about 10 ms.

III. Results and Discussion

The RHET exhibits an I_C-V characteristics typical of resonant tunneling. Its field dependence has been discussed in Ref. 7. Data for the base current normalized on the emitter current are plotted in Fig. 2 for different emitter base voltages V_{EB} as a function of magnetic field. The base/emitter current ratio increases initially with an approximate parabolic dependence and then saturates at high field.

Ballistic electrons, injected into the base region after resonant tunneling, experience the Lorentz force and are subjected to cyclotron motion. If the magnetic field B is increased, the curvature of the trajectory increases until the electrons are forced into skipping orbits. In the following model it is assumed that the cyclotron radius R_c is larger than the base width b (for example at $V_{EB} = 0.9$ V, this is valid for fields up to about 9 T). Then, with increasing field the path length and net transfer time through the base is prolonged, thus, the probability of each electron to undergo scattering increases. This model is schematically illustrated in Fig. 3. The length s of the path is

$$s = R_c J = R_c \sin^{-1}(b/R_c). \qquad (1)$$

The cyclotron radius is given by $R_c = m^* v_o/eB$, where m^* is the effective mass in the base, v_o the velocity with which electrons are injected into the base, e the electron charge and B

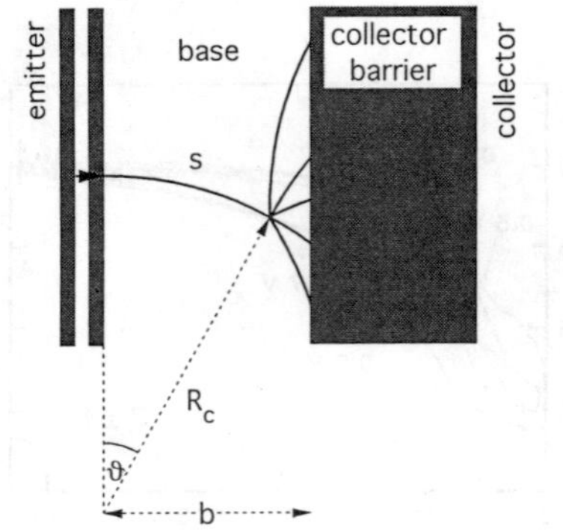

Fig. 3: Schematic drawing for model explaining electron transport and scattering in the RHET base under a magnetic field perpendicular to the current direction.

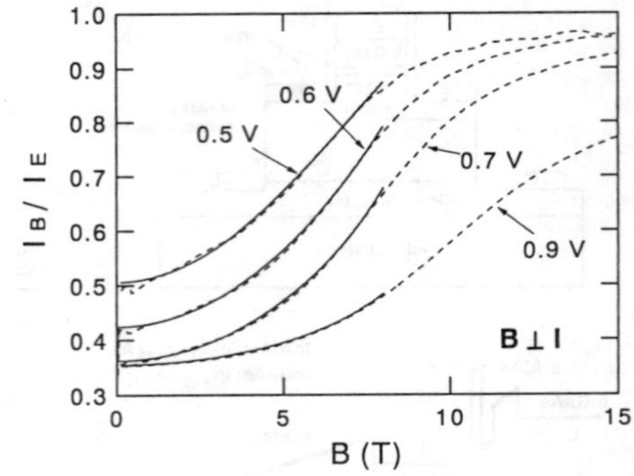

Fig. 4: Functions of the form Eq. (3) fitted to the data of I_B/I_E at a field perpendicular to the current direction (plain lines). Data (dashed lines) are average of up and down sweep data.

the magnetic field. After having undergone scattering, the initial ballistic hot electrons in the base lose their energy and are eventually thermalized in the base contact. Therefore, the base current can be described by

$$I_B = I_E\{1 - k\exp(-s/l_s)\} \tag{2}$$

where l_s is the mean free path. The parameter k accounts for the fact that those electrons that are only once or a few times scattered might still have sufficient energy to tunnel through the collector barrier, and hence, cannot contribute to the base current. Combining Eqn. (1) and (2) we obtain

$$I_B/I_E = 1 - k\exp\left\{-\frac{\frac{m^* v_o}{eB}\sin^{-1}\left(\frac{ebB}{m^* v_o}\right)}{l_s}\right\}. \tag{3}$$

Taking an effective mass $m^*/m_e = 0.042$ for $In_{0.53}Ga_{0.47}As$, Eq. (3) was fitted to the data taking k, v_o and l_s as adjustable fitting parameters (Fig. 4). The data are well described with Eq. (3) at fields smaller than about 8 T. At higher fields the assumption $R_c > b$ does not hold and Eq. (3) does not describe the data any longer.

The parameters obtained from the fit are shown in Fig. 5. Velocities are in the order of 2×10^6 m/s and do not vary so much with bias. For a peak voltage of about 1 V an injection velocity of 2.9×10^6 m/s is estimated which is in good agreement with the experiment. The mean free path obtained for a bias of $V_{EB} = 0.9\ V$ is 46 nm. This is in good agreement with studies in GaAs, where from the dependence of the hot electron spectrum on the base width, a mean free path of 30-50 nm was estimated. [5] It also agrees well with a theoretical value obtained in a Monte Carlo simulation for hot electron transport in InAlAs; for a doping level of 10^{18} cm^{-3} a mean free path of 48 nm was calculated.[4]

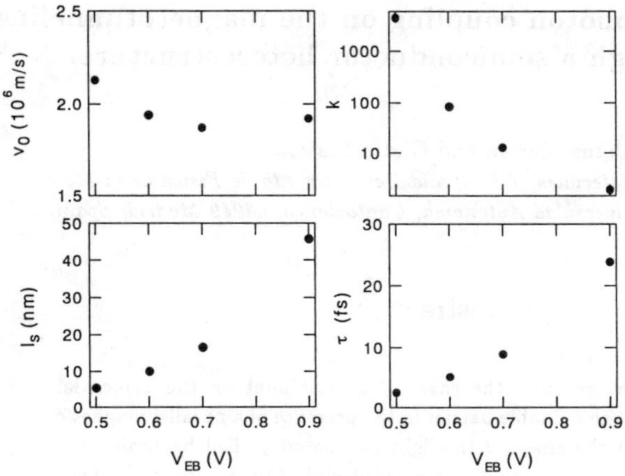

Fig. 5: Parameters obtained by fitting Eq. (3) to I_B/I_E: velocities v_o, mean free path l_s, and the fitting parameter k. The scattering time $\tau = l_s/v_o$ is included as well.

Taking the values for velocity and mean free path, one can obtain a scattering time which is in the order of 0.01 ps which also is in agreement with calculations[4]. This time is much shorter than the electron phonon scattering time for hot electrons (about 0.16 ps).[8] Therefore, our experimental results agree with the theoretical prediction that scattering in a base region doped to a level of $1x10^{18}$ cm^{-3} is not dominated by phonon scattering, but by electron-electron scattering.[4]

In conclusion, from the field dependence of the RHET currents at transverse magnetic field, electron velocity and mean free path in the base were determined. The scattering time is in the order of 0.01 ps, i.e. much shorter than expected for electron-phonon scattering (about 16 times). Our finding is in agreement with theoretical studies predicting that electron-electron scattering is the dominant scattering mechanism.

The RHETs were fabricated under the management of the FED sponsored by NEDO.

References:

1. M. Heiblum, M. I. Nathan, D. C. Thomas and C. M. Knoedler, *Phys. Rev. Lett.* **55**, 2200 (1985).
2. N. Yokoyama, K. Imamura, S. Muto, S. Hiyamizu and H. Nishi, *Jpn. J. Appl. Phys.* **24**, L853 (1985).
3. H. Ohnishi, N. Yokoyama and A. Shibatomi, *IEEE Trans. Electron Devices.* **36**, 2335 (1989).
4. A. P. Long, P. H. Beton and M. J. Kelly, *J. Appl. Phys.* **62**, 1842 (1987).
5. A. P. Long, P. H. Beton and M. J. Kelly, *Semicond. Sci. Technol.* **1**, 63 (1986).
6. K. Imamura, M. Takatsu, T. Mori, T. Adachihara, H. Ohnishi, S. Muto and N. Yokoyama, *IEEE Trans. Electron Devices.* **39**, 2707 (1992).
7. T. Strutz, T. Takamasu, N. Miura and K. Imamura, *Physica B.* **184**, 254 (1993).
8. J. A. Kash, J. C. Tsang and J. M. Hvam, *Phys. Rev. Lett.* **54**, 2151 (1985).

Effect of electron-photon coupling on the magnetotunneling current through a semiconductor heterostructure.

Jesus Iñarrea and Gloria Platero
Instituto de Ciencia de Materiales (CSIC) and Departamento de Fisica de la Materia Condensada C-III, Universidad Autonoma, Cantoblanco, 28049 Madrid, Spain.

Abstract

We have analyzed theoretically the effect of a laser light on the sequential tunneling current through a double barrier in the presence of a parallel magnetic field. We observe that the effect of the light can be controlled by tuning the ratio between the cyclotron and the photon frequency. The modification of the number of Landau levels in the well contributing to the current due to the light is discussed.

The analysis of resonant tunneling through semiconductor heterostructures has a lot of interest from a fundamental point of view as well as for its applications as microelectronic devices [1]. The application of light modifies the transport properties of these resonant devices [2–4]. In a recent work [5,6] we have developed a model to analyze the sequential tunneling current in the presence of a photon field through a double barrier (DB). In this paper we analyze the effect of a photon field on the sequential magnetotunneling current through a DB . The scattering that electrons suffer into the well due to impurities, phonons or surface roughness, implies that there are electrons which tunnel sequentially from the emitter to the well, they suffer memory loss in the well and they cross coherently in a next step the collector barrier, therefore for sequential tunneling the Landau level (LL) index conservation takes place from the first barrier to the well and from there to the collector independently. In order to study the sequential magnetotunneling, before switching on the light, we have calculated in the framework of the Transfer Hamiltonian formalism the current through the first and the second barriers separately, J_1, and J_2. These currents are related to the Fermi level in the well E_w , i.e., to the amount of charge stored into the well. In this model we obtain selfconsistently the Fermi level and the sequential current when the currents through the first and the second barriers result to be equal. We calculate the current density J_1 for the electrons to cross the first barrier from the emitter to the discrete state into the well:

$$J_1 = \frac{e^2 B}{2\pi m^*}(N_t - \Delta)\frac{k_w(E_z)T_1(E_z)}{(w2 + 1/\alpha_e + 1/\alpha_c)} \quad (1)$$

where w_2 is the well width and α_e (α_c) the wave vector of the emitter (collector) barrier. $T_1(E_z)$ is the transmission coefficient through the emitter barrier, evaluated at the energy of the resonant state in the well E_z and $k_w(E_z)$ is the z component of the wave vector into the well, Δ is the fraction of the total LL occupied in the well and N_t the total number of LL available to tunnel for an applied bias (Δ runs between 0 and N_t). For the second barrier, we apply exactly the same formalism and we obtain the current through the second barrier J_2. The sequential magnetotunneling current is obtained when both currents J_1 and J_2 are equal and it determines the LL which are occupied within the well. Before switching on the light the electrons in the emitter have just one way to get into the well : from an emitter state which is resonant with the well state. Once the light is applied to the structure, there is a coupling between the photons and the tunneling electrons. There are three ways to tunnel through the emitter barrier to the well: a direct way corresponding to an emitter state which resonates with the well state. The second (third) one is through and absorption (emission) process from an emitter state which is found at one photon energy below (above) the resonant well state. From these contributions we get in the J_1 expression the sum of three terms: the direct term has in its expression a $\delta[E_z - E_R]$, the absorption term a $\delta[E_z-(E_R-\hbar w)]$, and the emission term a $\delta[E_z - (E_R + \hbar w)]$. The tunneling current through the emitter barrier is given by:

$$J_1 = \frac{e^2 B T_1(E_z)}{2\pi m^*} \times [\frac{(N_{td}-\Delta)k_{w,0}}{w_2 + (1/\alpha_{b,0}) + (1/\alpha_{d,0})} \frac{1}{1 + k_1/k_0|C^{(1)}_{1,0}|^2 + k_{-1}/k_0|C^{(1)}_{-1,0}|^2} +$$
$$\frac{(N_{ta}-\Delta)k_{w,-1}}{w_2 + (1/\alpha_{b,-1}) + (1/\alpha_{d,-1})} \frac{|C^{(1)}_{1,-1}|^2}{k_{-1}/k_0 + |C^{(1)}_{1,-1}|^2 + k_{-2}/k_0|C^{(1)}_{-1,-1}|^2} +$$
$$\frac{(N_{te}-\Delta)k_{w,1}}{w_2 + (1/\alpha_{b,1}) + (1/\alpha_{d,1})} \frac{|C^{(1)}_{-1,1}|^2}{k_1/k_0 + |C^{(1)}_{1,1}|^2 k_2/k_0 + |C^{(1)}_{-1,1}|^2}] \tag{2}$$

where N_{ta} (N_{te}) is the number of LL available to tunnel due to absorption (emission) processes and N_{td} is the number of LL available to tunnel without interaction with light (direct tunneling). $C_{i,j}$ are the coefficients of the wave functions :

$$\Psi_0(t) = \alpha[\Phi_D(k_0) + C_{1,0}\Phi_D(k_1)e^{-iwt} + C_{-1,0}\Phi_D(k_{-1})e^{iwt}]e^{-iw_0 t} \tag{3}$$
$$\Psi_1(t) = \alpha'[\Phi_D(k_1)e^{-iwt} + C_{1,1}\Phi_D(k_2)e^{-2iwt} + C_{-1,1}\Phi_D(k_0)]e^{-iw_0 t} \tag{4}$$
$$\Psi_{-1}(t) = \alpha''[\Phi_D(k_{-1})e^{iwt} + C_{1,-1}\Phi_D(k_0) + C_{-1,-1}\Phi_D(k_{-2})e^{2iwt}]e^{-iw_0 t} \tag{5}$$

where $\Phi_D(k) = \Phi_0(k)\sum_{n=-\infty}^{\infty} J_n(\beta_k)e^{-inwt}$, being $\Phi_0(k)$, the electronic double barrier eigenstate with no photon field present in the sample, and $\beta_k = \frac{eF<P_z>}{m^*\hbar w^2}$. Here, α, α' and α'' are normalization constants ,Ψ_0 corresponds to a state at one photon energy lower (higher) than Ψ_1 (Ψ_{-1}), and the coefficients $C_{i,j}$ are given

in ref [6].For those coefficients, the first subscript is referred to the interaction with light processes i.e., '1'('-1') means absorption (emission) process, whereas the second subscript is referred to the state i.e., the subscript "0" means the reference state and the subscripts "1" and "-1" mean states one photon energy above and below that reference state respectively. The expression for the current through the second barrier J_2 can be obtained in the same way as J_1.
We have analyzed the effect of an external laser in the far infrared regime on the magnetotunneling current through a GaAs-AlGaAs DB. The magnetic field is applied in the current direction (z) and the light propagates linearly in the interface plane and is polarized in the z direction. In fig.1a the sequential magnetocurrent as a function of the bias is represented for B=5T. We observe six LL which contribute to the current. The photon frequency is 4.32 meV (one half of the cyclotron frequency). The current difference between the case with light minus the case with no applied light is shown in figure 1b. In fig. 1c. the current difference has been drawn separately for each LL for the same case as in fig. 1b. One can explain schematically the main features observed for each LL (fig. 1c): for small bias the first LL in the well is higher in energy than the Fermi energy in the emitter. As the bias increases there are electrons close to the Fermi energy which absorb a photon and tunnel resonantly from the first LL in the emitter with LL index conservation ,therefore the threshold bias for the current is smaller than the corresponding one for no light present (it moves twice the photon energy) and there is a positive peak in the current difference . For higher bias the first LL in the well crosses the emitter Fermi energy and the current difference becomes negative abruptly due to the fact that the electrons have the possibility to absorb a photon reducing the number of electrons efficient to tunnel resonantly. As the bias increases and the energy of the resonant state for the first LL lies one photon higher than the bottom conduction band of the emitter the electrons have a probability to emit a photon below the bottom of the conduction band and the resonant current is reduced (it gives the small negative contribution to the current difference for high bias). Once the well state crosses the bottom of the conduction band there are electrons in the emitter which can emit a photon and tunnel resonantly, therefore there is a positive peak in the current difference and the current cut off moves to higher bias . If one now changes the photon frequency to the same value as the cyclotron one for the same magnetic field (5T) the current difference changes dramatically, the main contribution comes from the peak at the threshold bias and there is an additional narrow structure in this region of bias (fig. 2a). For higher bias the additional features to the current difference are much smaller in intensity than in the previous case (fig. 1b) . The reason for this difference is not only the change of the bias threshold and cut off of the current due to the difference of photon frequencies but also the fact that when the ratio of the cyclotron frequency to the photon frequency is one, there are absorption and

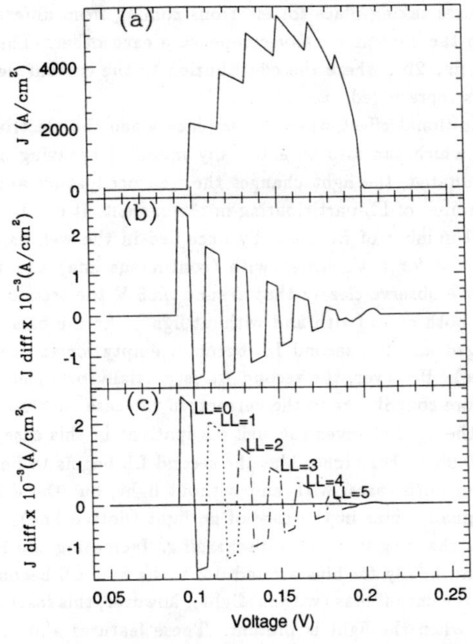

Figure 1. a) Sequential magnetotunneling current density as a function of voltage. ($F = 1.10^4 V/m, \hbar w = 4.32 meV, B = 5 Tesla$). b) Sequential magnetocurrent difference as a function of voltage between photo assisted current and current without light present. Same configuration as in a). c) Same as in b) for each landau level separately.

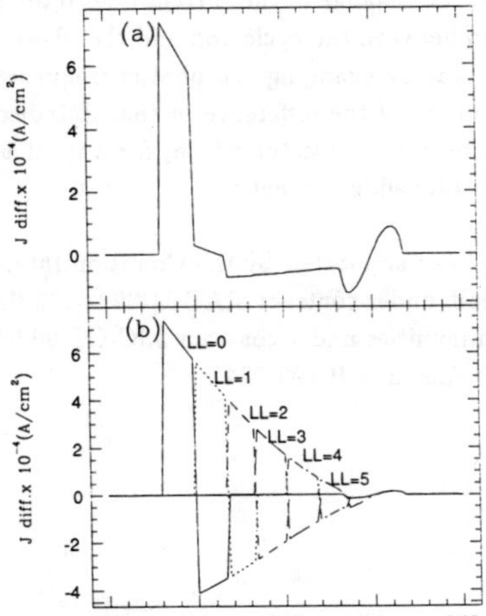

Figure 2. a) Sequential magnetotunneling difference as a function of voltage, between photo assisted current and current without light present, for $\hbar w = \hbar w_{cyclotron}$. ($F = 1.10^4 V/m, \hbar w = 8.65 meV, B = 5 Tesla$). b) Same as in a) for each landau level separately.

emission processes taking place for electrons coming from different LL whose contribution to the current density compensate each other. This feature can be observed in fig. 2b , where the contribution to the current density coming from each LL is represented.
There is an additional effect which takes place when the electrons tunnel sequentially and which can also be externally modified applying a laser: if the sample is illuminated, the light changes the well occupation and therefore it modifies the number of LL participating in the current. It can be seen in fig.3a where the total number of LL partially occupied in the well as a function of the bias is plotted for both cases: with (continuous line) and with no light (dotted line). We observe clearly that around 0.05 V the second LL begins to be occupied in both cases (with and without light). As the bias increases, the well is discharged and the second LL becomes empty for the case with light (continuous line). However, the second LL is partially occupied up to higher bias and therefore contributes to the current, in the case with no light present. It means that the light changes the well occupation: in this case it discharges the well at V=0.06 V. For higher bias the second LL begins to be occupied for the same bias in both cases, with and without light, but the discharge takes place now for smaller bias in the case of no light (dotted line) : the effect of the light now is charging instead of discharging. Increasing the bias the same effect occurs, and finally for bias around 0.2 V the first LL becomes empty as corresponds to the cut off bias (without light), however, this feature takes place for higher bias when the light is present. These features above, happen also with respect to the number of LL in the emitter which contribute to J_1. It can be visualized in fig.3b where the total number of LL in the emitter producing current is drawn as a function of the bias.
In conclusion, the effect of the light on the sequential magnetotunneling current has been analyzed. The change of the current due to the light can be controlled by tuning the ratio between the cyclotron and the photon frequency. Other interesting effect is that by changing the photon frequency and intensity of the laser is possible to control the difference in the LL index in the emitter minus the same magnitude in the collector which, for a fixed bias contributes to the sequential magnetotunneling current.

This work has been supported by the Comision Interministerial de Ciencia y Tecnologia of Spain under contract MAT 94-0982-c02-02 , by the Comission of the European Communities under contract SSC-CT 90 0201 and by the Acción Integrada Hispano-Alemana HA93-034.

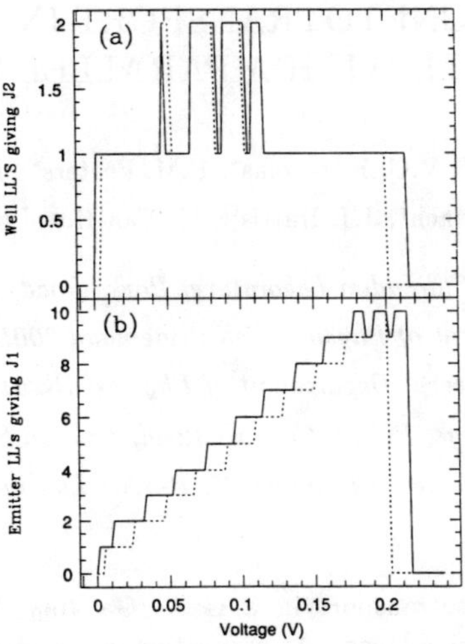

Figure 3. a) Total number of Landau levels into the well contributing to the current as a functio of voltage with (continuous line) and without (dotted line) light. ($F = 5.10^6 V/m, \hbar w = 7.meV, B = 6Tesla$). b) Total number of Landau levels in the emitter contributing to the current as a function of voltage, with (contionuous line) and without (dotted line) light. Same configuration as in a).

References

1 E.E.Mendez, L.Esaki and W.I.Wang, Phys. Rev. B **33** 2893 (1986)

2 T.C.L.G.Sollner, W.D.Goodhue, P.E.Tannenwald, C.D.Parker and D.D.Peck, Appl. Phys. Lett. **43** 588 (1983)

3 V.A.Chitta, R.E.M.de Bekker, J.C.Maan, S.J.Hawkworth, J.M.Chamberlain, M.Henini and G.Hill,Surf.Sci **263** 227 (1992)

4 H.C.Liu, Phys. Rev. B **43** 12538 (1991)

5 J.Iñarrea,G.Platero and C.Tejedor,Semicond. Sci. Technol.,**9**,515 (1994).

6 J.Iñarrea , G.Platero and C.Tejedor,Phys. Rev. B, in press.

HIGH FIELD MAGNETOTRANSPORT IN A $Ga_{0.8}In_{0.2}As$ QUANTUM WELL WITH A PARALLEL Si δ-LAYER

M. van der Burgt[°†], V.C. Karavolas[*], F.M. Peeters[*], J. Singleton[°], R.J. Nicholas[°], F. Herlach[†], J.J. Harris[□], M. Van Hove[‡] and G. Borghs[‡]

[°] *University of Oxford, Clarendon Laboratory, Parks Road, Oxford OX1 3PU*

[†] *K.U.Leuven, Department of Physics, Celestijnenlaan 200D, B-3001 Leuven*

[*] *University of Antwerp, Department of Physics, B-2610 Antwerpen*

[□] *Imperial College, Prince Consort Road, London SW7 2BZ*

[‡] *IMEC vzw, Kapeldreef 75, B-3001 Leuven*

We investigated magnetotransport in a $GaAs/Ga_{0.8}In_{0.2}As/Ga_{0.75}Al_{0.25}As$ heterostructure with a Si δ-layer. As a function of density drastic changes in the transport data are observed, especially at fields beyond 30 T. This is interpreted as due to: 1) magnetic freeze-out of carriers in the δ-layer, and 2) quantization of the electron motion in the δ-layer.

Due to the deeper well depth for two dimensional (2D) electrons in $GaAs/Ga_{1-x}In_xAs/$-$Ga_{1-y}Al_yAs$ structures compared to those in $GaAs/Ga_{1-x}Al_xAs$ heterojunctions it is possible to accommodate a higher density of 2D electrons. High electron concentrations, $n_s > 1\times10^{16}$ m^{-2}, can be provided by including a Si δ-layer in the $Ga_{1-y}Al_yAs$ layer. The resulting high conductivities in these systems makes them attractive for use in high-electron mobility transistors.

The ionisation of the DX centres in Si doped $Ga_{1-x}Al_xAs$ after illumination can lead to parallel conduction in modulation doped heterostructures. Although this is often regarded as an undesired effect degrading the performance of a device we will show that it can give rise to interesting results in sufficiently high magnetic fields.

We present transport measurements in pulsed magnetic fields up to 45 T on a $GaAs/Ga_{0.8}In_{0.2}As/Ga_{0.75}Al_{0.25}As$ structure with a Si doped δ-layer in the $Ga_{1-y}Al_yAs$ layer. Different features, ranging from the Shubnikov-de Haas (SdH) oscillations of a single subband to magnetic freeze-out and the quantum Hall effect (QHE) in two separate layers, are observed as the density is changed. A complete description of conduction valid for arbitrary magnetic fields is unavailable; however we have combined two different models to provide a good theoretical description of the experimental data.

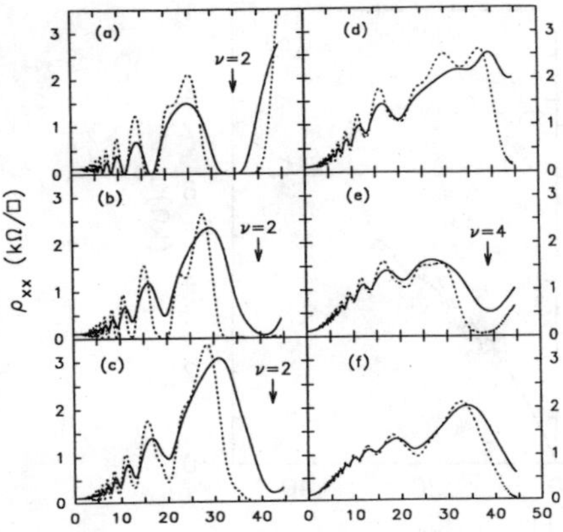

Figure 1: The resistivity ρ_{xx} at 4.2 K (full line) and 1.4 K (dotted line) at six different Hall densities: n_H = 1.69 (a), 1.87 (b), 1.91 (c), 1.95 (d), 2.05 (e) and 2.25 × 10^{16} m^{-2} (f).

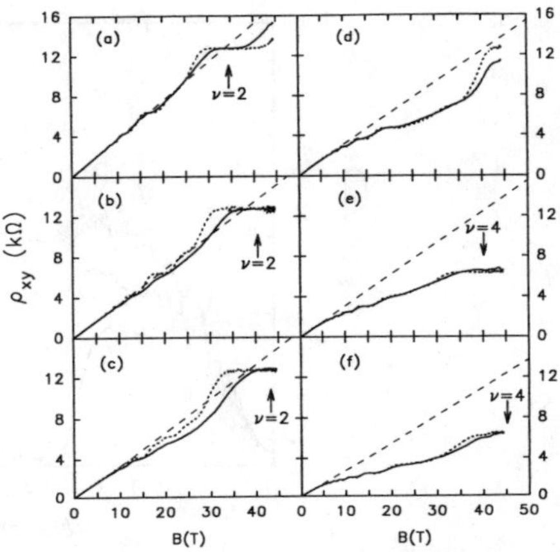

Figure 2: The Hall resistivity ρ_{xy} at 4.2 K (full line) and 1.4 K (dotted line) at the same densities as in fig. 1. The dashed line represents the classical Hall effect.

The transport experiments were done with a pseudomorphic GaAs/$Ga_{0.8}In_{0.2}As$/-$Ga_{0.75}Al_{0.25}As$ structure grown by molecular beam epitaxy. It consists of a 130 nm thick $Ga_{0.8}In_{0.2}As$ layer grown on GaAs, followed by a 5 nm wide $Ga_{0.75}Al_{0.25}As$ spacer layer, a Si δ-layer with a Si concentration of 5×10^{16} m^{-2}, a 30 nm thick Si doped (5×10^{23} m^{-3}) $Ga_{0.75}Al_{0.25}As$ layer and n^+GaAs cap layer. The sample was mesa etched into a 55 μm wide Hall bar. The electron density in the sample studied can be increased by illuminating the sample with a red LED. The density is $n_s = 1.67 \times 10^{16}$ m^{-2} and the mobility is $\mu = 3.2$ m^2/Vs when the sample is cooled to 4.2 K in the dark.

Both the Hall effect ρ_{xy} and the longitudinal magnetoresistance ρ_{xx} were measured at temperatures of 1.5 and 4.2 K using dc currents in the range 1–10 μA. Some specific problems regarding dc transport measurements of 2D semiconductors in pulsed magnetic fields are discussed in a separate paper in these proceedings [1].

An overview of the high field transport data is given in figures 1 and 2. The density indicated in the figure captions is the Hall density n_H; it is obtained from the low field slope of the Hall effect: $\rho_{xy} = B/n_H e$. Note that the Hall density only equals the total density when only one type of carrier is involved in the conduction process [2].

At the dark value of the density ($n_H = 1.69 \times 10^{16}$ m^{-2}, figs. 1(a) and 2(a)), the sample shows plateaux in ρ_{xy} and corresponding minima in ρ_{xx}. This is the QHE originating from one occupied subband in the $Ga_{0.8}In_{0.2}As$ well.

Increasing the density by 5 % up to $n_H = 1.75 \times 10^{16}$ m^{-2} results in the population of the second subband in the $Ga_{0.8}In_{0.2}As$ layer giving rise to a second frequency in

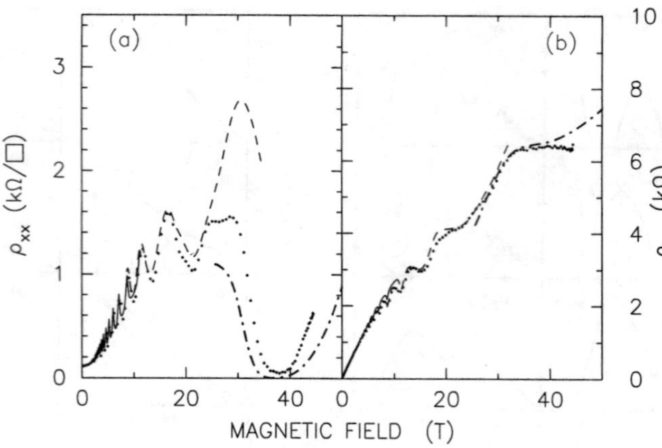

Figure 3: The magnetoresistivity (a) and the Hall resistivity (b) at 1.4 K for $n_H = 2.05 \times 10^{-16}$ m^{-2} (case (e) of fig. 1 and fig. 2) both experimental (dots) and theoretical for the three different regimes (low fields (full line), intermediate fields (dashed line) and high fields (dashed-dotted line)) with $\Gamma = 4.90\sqrt{B}$ meV, $\lambda = 0.8\sqrt{B}$ meV for the $Ga_{0.8}In_{0.2}As$ quantum well and $\Gamma = 13.90\sqrt{B}$ meV, $\lambda = 2.3\sqrt{B}$ meV for the Si δ- layer.

the SdH oscillations (not shown). From the self-consistent solutions of the Poisson-Schrödinger equations for the conduction band and the theoretical modelling of the resistivities it is clear that the second subband does not contribute significantly to conduction at higher densities. As pointed out in a previous paper [3] parallel conduction in the Si δ-layer becomes important at higher densities.

In fig. 1(b) the electron density in the δ-layer is relatively low ($n_\delta \approx 0.7 \times 10^{16}$ m^{-2}) [3]. At 1.5 K these carriers freeze out onto the Si impurity states at relatively low magnetic fields. This means that the rising background in ρ_{xx} at 4.2 K disappears at 1.5 K.

The rising background in ρ_{xx} becomes stronger when the density is increased further as is seen in fig. 1(c). Lowering the temperature to 1.5 K at this density does not bring the low field minima down to zero again but ρ_{xx} drops steeply back to the base line at fields above 30 T. The Hall effect (fig. 2(c)) starts to deviate from the classical line $\rho_{xy} = B/n_H e$ at 4.2 K but approaches the classical value again when the temperature is lowered to 1.5 K. This behaviour of ρ_{xx} and ρ_{xy} is attributed to the magnetic freeze out of the carriers in the δ-layer for $B > 30$ T.

Illuminating further to $n_H = 1.95 \times 10^{16}$ m^{-2} results in an extra shoulder at 25 T and a new minimum in ρ_{xx} at $B = 34$ T (fig. 1(d)). In the Hall effect (fig. 2(d)) the onset of a plateau, which has a value of $\rho_{xy} = h/2e^2$ at 1.5 K, is still visible at 45 T. At 4.2 K the onset of the plateau is at significantly lower values then $h/2e^2$ indicating that the freeze out has become weaker at 4.2 K and $B \leq 45$ T.

When the density in the δ-layer is sufficiently high, the carriers in the δ-layer can be treated as a second two dimensional electron gas showing the QHE at high fields. This behaviour is apparent from figs. 1(e) and 2(e). A filling factor of two in the δ-layer and in the $Ga_{0.8}In_{0.2}As$ well results in a total filling factor of $\nu = 4$ as is clear from the $\rho_{xy} = h/4e^2 = 6.5$ kΩ plateau and the strengthening of the high field ρ_{xx} minima. The total electron density at this point is $n_s \approx 3.6 \times 10^{16}$ m^{-2}. A further increase in density (figs. 1(f) and 2(f)) results in a shift of these structures to higher fields without any drastic changes.

For the theoretical modelling we made use the of the simple principle that conductivities are additive for parallel conducting layers [2]. The conductivity in the separate layers (i.e. in the first and second subband in the $Ga_{0.8}In_{0.2}As$ well and in the δ-layer) are modelled using different models. For $\omega_c\tau_q \leq 1$ (ω_c is the cyclotron frequency and τ_q is the single particle scattering time) we made use of the model of Isihara and Smrčka [5] which is based on the assumption of a constant density of states with a small sinusoidal oscillatory part superimposed on it. This model describes the SdH oscillations in ρ_{xx} for small magnetic fields. For higher fields i.e. $\omega_c\tau_q \geq 1$ the model of Englert [4] is applied where we assume a gaussian DOS with width Γ. Another gaussian with width $\lambda < \Gamma$ is superimposed on this DOS which allows us to discriminate quantitatively between localized and extended states in a Landau level. Typically $\omega_c\tau_q \sim 1$ at $B = 3$ T in the quantum well, and at $B = 30$ T in the δ-layer. This leads us to divide the magnetic-field range into three regimes (fig. 3). At very low fields the Isihara model is valid for both layers, at intermediate fields the Englert model is used for the $Ga_{0.8}In_{0.2}As$ quantum well while the δ-layer is described by the Isihara model, and at high fields the Englert model is applied for both layers. When the electron concentration in the δ-layer is sufficiently low and magnetic field is high enough the electron wavefunction becomes localized around a single donor impurity leading to magnetic freeze-out. This explains the transition observed at $B > 30$ T for the cases (c) and (d) in figs. 1 and 2.

MvdB and VCK acknowledge the support of the HCM programme of the EC. FMP is funded by of the Belgian National Fund for Scientific Research.

References

[1] M. van der Burgt, F.M. Peeters, J. Singleton, R.J. Nicholas, F. Herlach, J.J. Harris and C.T. Foxon, in these proceedings.

[2] M.J. Kane, N. Apsley, D.A. Anderson, L.L. Taylor and T. Kerr, J. Phys. C: Solid State Phys. 18, 5629 (1985).

[3] M. van der Burgt, A. Van Esch, F.M. Peeters, M. Van Hove, G. Borghs and F. Herlach, Physica B 184, 211 (1993).

[4] T. Englert, in: Application of High Magnetic Fields in Semiconductor Physics, ed. G. Landwehr (Springer-Verlag, Berlin, 1983) p. 87.

[5] A. Isihara and L. Smrčka, J. Phys. C: Solid State Phys. 19, 6777 (1986).

HIGH FIELD MAGNETOTRANSPORT IN Si-δ-DOPED GaAs LAYERS

R. Bogaerts*, P.M. Koenraad+, A.F.W. van de Stadt+, H. Leopold*, L. Van Bockstal*, F. Herlach* and J.H. Wolter+

* *Katholieke Universiteit Leuven, Celestijnenlaan 200D, B-3001 Leuven, Belgium*
\+ *Technische Universiteit Eindhoven, Den Dolech 2, P.O. 513, 5600 MB Eindhoven, Nederland*

We have carried out a magnetotransport investigation of Si-δ-doped GaAs layers with high carrier concentrations (above $2{\cdot}10^{12}\,\mathrm{cm}^{-2}$), in pulsed magnetic fields up to 50 T and at 4.2 K. The analysis of our measurements focuses on the observation of multi-subband effects in the "low filling factor" magnetic field regime, where pronounced multi-subband quantum Hall plateaux are observed. Simulations of this effect demonstrate that the plateaux can either be completely suppressed or deviate markedly from the quantized values.

1. Introduction

Semiconductor structures containing extremely narrow "δ"-doping profiles, with the dopants confined to a single or a few monolayers of the host semiconductor lattice, form a relatively new class of twodimensional (2D) semiconductor systems in which several 2D subbands are occupied. For Si-δ-doped layers in GaAs, the formation of 2D electron subbands has been corroborated by the observation of different, independent series of SdH oscillations in perpendicular magnetic fields[1,2]. In addition, for layers with low carrier concentrations ($\sim 7{\cdot}10^{11}\,\mathrm{cm}^{-2}$), the quantum Hall effect has been observed at filling factor ν=2, corresponding to the occupation of the lowest, spin-split Landau level of the lowest 2D subband[3,4]. Magnetic fields up to 50 T make it possible to study the quantum Hall effect in δ-doped layers with higher carrier concentrations, at filling factors corresponding to the occupation of more than one 2D subband. In addition, in this range of magnetic fields, strong multi-subband 2D effects, such as strong oscillations of the Fermi level (with a frequency proportional to the total carrier concentration[5]) or oscillations of the individual subband carrier concentrations, may interfere with the SdH effect.

As an extension of a magnetotransport study of Si-δ-doped GaAs layers in d.c. fields up to 20 T [2,6], we have carried out magnetotransport measurements in pulsed fields up to 50 T. The samples have donor concentrations ranging from $2{\cdot}10^{12}\,\mathrm{cm}^{-2}$ to $8{\cdot}10^{12}\,\mathrm{cm}^{-2}$, and donor distribution widths ranging from 20 Å to 110 Å. We shall discuss the results for three representative samples, for which the growth parameters and the electron concentrations are given in Table 1. Note that the electron concentration of the highest occupied subband for samples W53 and W81 (the 4th and the 3^{d} subband respectively) could not be determined from SdH measurements. The occupation of these subbands has been established in subband depopulation measurements in a parallel magnetic field, and is in agreement with theoretical calculations (see below).

The details of the growth are given by Koenraad et al.[6]. The doping layers are enclosed by a 2.5 μm thick GaAs buffer layer on a semi-insulating (100) GaAs substrate, and a 1 μm thick GaAs top layer.

2. Experimental results and analysis

The experimental results (Fig.1) show different series of SdH oscillations. Above 30 T, pronounced quantum Hall plateaux and deep magnetoresistance minima develop for samples W53 and W81, corresponding to a filling factor ν=6 and ν=2, respectively.

	d_{don} Å	T_{growth} °C	n_{don}	n_{2Dtot}	n_0	n_1	n_2	n_3
W53	20	480	8	5.3	3.50	1.23	0.51	–
W104	80	530	8	5.7	3.22	1.66	0.70	0.16
W81	20	480	2	1.8	1.43	0.35	–	

Table 1. Growth parameters and total and individual subband carrier concentrations (in 10^{12} cm^{-2}) obtained from d.c. magnetic field SdH measurements[2,6].

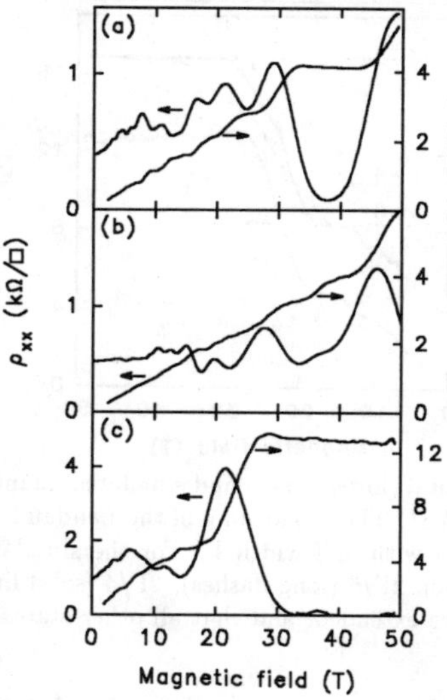

Fig. 1. The Hall effect and the perpendicular magnetoresistance at 4.2 K for samples (a) W53, (b) W104 and (c) W81.

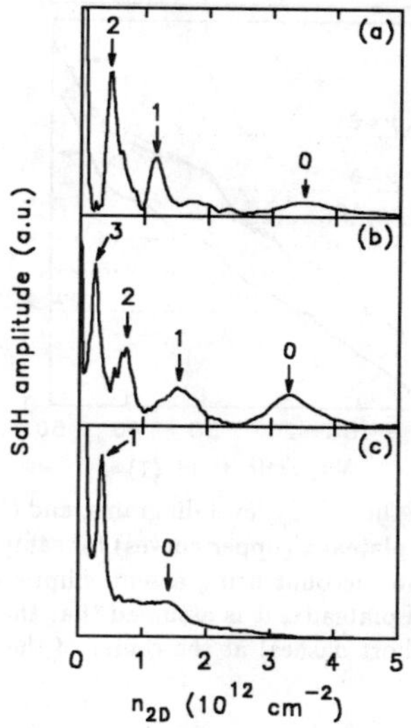

Fig. 2. FFT spectra of the SdH oscillations for samples (a) W53, (b) W104 and (c) W81.

2.1. Shubnikov-de Haas effect

To obtain FFT spectra of the SdH oscillations, a classical background magnetoresistance was first subtracted from the ρ_{xx} curves. This background signal was calculated with the classical Drude expressions for a multi-band electron system with isotropic energy bands. The values of the different Drude parameters (the carrier concentrations and the transport mobilities for each subband) were either obtained from the analysis of previous measurements[2,6], or from a fit of the weak field magnetoresistance and Hall effect to the Drude expressions.

The FFT spectra of the resulting $\Delta\rho_{xx}$ curves show distinct frequency peaks corresponding to the electron densities of the occupied subbands (Fig. 2). Typical strong multi-subband effects, such as the variation of the SdH frequencies as a function of the magnetic field or the appearance of sum frequencies, are not observed. This indicates that the broadening of the Landau levels inhibits strong fluctuations of the individual carrier densities or of the Fermi level over the entire magnetic field range. The latter observation suggests that the position of the 2D subband edge energies is not strongly dependent on the magnetic field; such a dependence results from a redistribution of

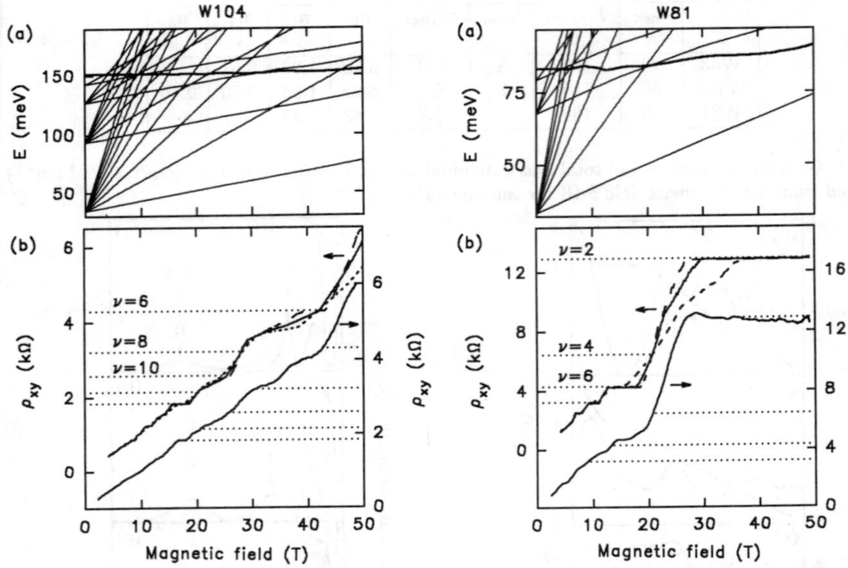

Fig. 3. (a) the energy level diagrams, and (b) the experimental (lower curve) and simulated quantum Hall effect plateaux (upper curves) for samples W104 and W81. The broadening of the Landau levels is taken into account using a semi-elliptic density of states with half width Γ. For the simulation of the Hall plateaux, it is assumed that the states in a region $2\Gamma/6$ (long dashes), $2\Gamma/4$ (solid lines) or $2\Gamma/2$ (short dashes) at the center of the Landau level are extended, and that all other states are localized.

the electrons over the different subbands and the self-consistency of the potential well.

To investigate this situation quantitatively, the Landau level energy structure has been calculated self-consistently, including the effect of redistribution of the electrons. The density of states was assumed to be semi-elliptic with half width Γ. The effective mass was taken to be the band edge effective mass of GaAs. Fig. 3a shows the resulting energy diagrams for samples W104 and W81, with respectively 4 and 3 occupied subbands. For typical values of the quantum mobilities ($500\,\mathrm{cm}^2/\mathrm{Vs}$ up to $5000\,\mathrm{cm}^2/\mathrm{Vs}$[2,6]) the changes in the carrier densities and the jumps of the Fermi level are weak and smooth; this is in accordance with the conclusions drawn from the FFT spectra.

2.2. *Quantum Hall effect*

The Landau level fan diagrams have been used to simulate the quantum Hall effect for a system with more than one occupied subband. To account qualitatively for the existence of localized states in the tails of the semi-elliptic density of states, the states within an energy range Γ/x from the Landau level center are assumed to be extended. The value of x is initially chosen to be 4. The details of the simulation will be described elsewhere.

The results of the calculations for samples W104 and W81 are shown in Fig. 3b (upper curves). The simulations do not exactly reproduce the experimentally observed Hall effect, but the same general tendencies are present. These are similar to the tendencies observed in d.c. magnetic field experiments on systems with two occupied subbands[7], and can be understood by considering the degeneracy of different Landau levels at the Fermi level: (i) at some filling factors, the observed quantum structure corresponds to strong and exactly quantized Hall plateaux. Beside the obvious example of the ν=2 plateau for W81, a pronounced plateau is observed at ν=14 for W104. Other Hall plateaux, such as the ν=4 plateau for W81, are completely suppressed. This occurs when different Landau levels cross the Fermi level simultaneously. (ii) If extended states are partially involved in the conduction, quantum Hall plateau-like features appear either below or above the exactly quantized value (e.g. the ν=6 plateau for W104). The degree to which the plateaux deviate from the quantized values can be tuned in the simulations by increasing (decreasing) the energy range of extended states, which enhances (reduces) the degree of deviation.

3. Conclusions

We have studied multi-subband 2D effects in Si-δ-doped GaAs layers with relatively high carrier concentrations. For these systems, pulsed magnetic fields up to 50 T provide the possibility to investigate phenomena in the low filling factor regime, such as the quantum Hall effect.

Over the entire magnetic field range, the broadening of the Landau levels is found to inhibit large fluctuations of the carrier densities and of the Fermi level. As a consequence, the SdH spectra show well-resolved frequency peaks, corresponding to the individual subband carrier concentrations. The quantum Hall effect for multi-subband occupation has been observed and modelled. The development of Hall effect plateaux is governed by the interplay of Landau levels corresponding to different 2D subbands. This can result in a complete suppression of the Hall plateaux or in marked deviations of the plateaux from the exactly quantized values.

Acknowledgements

R.B. is supported by the Onderzoeksraad of the K.U. Leuven. L.V.B. is a Senior Research Assistant of the Belgian Nationaal Fonds voor Wetenschappelijk Onderzoek.

References

[1] A. Zrenner, H. Reisinger, F. Koch, K. Ploog, *Proceedings of the 17th International Conference on the Physics of Semiconductors, San Fransisco*, eds. J.P. Chadi and W.A. Harrison (Springer-Verlag, 1984), p. 325.

[2] P.M. Koenraad, B.F.A. van Hest, F.A.P. Blom, R. van Dalen, M. Leys, J.A.A.J. Perenboom, and J.H. Wolter, *Physica B* **177** (1992) 485.

[3] A. Zrenner, F. Koch, J. Leotin, M. Goiran, and K. Ploog, *Semicond. Sci. Technol.* **3** (1988) 1132.

[4] G. Gillmann, P. Bois, E. Barbier, B. Vinter, D. Lavielle, M. Stohr, S. Najda, A. Briggs, and J.C. Portal, *Semicond. Sci. Technol.* **3** (1988) 620.

[5] R.J. Nicholas, *Physics and Applications of Quantum Wells and Superlattices, Nato ASI Series B* **170** (1989) 217.

[6] P.M. Koenraad, F.A.P. Blom, C.J.G.M. Langerak, M.R. Leys, J.A.A.J. Perenboom, J. Singleton, S.J.R.M. Spermon, W.C. van der Vleuten, A.P.J. Voncken, and J.H. Wolter, *Semicond. Sci. Technol.* **5** (1990) 861.

[7] Y. Guldner, J.P. Vieren, M. Voos, F. Delahaye, D. Dominguez, J.P. Hirtz, and M. Razeghi, *Phys. Rev.* **B33** (1986) 3990.

MAGNETIC FREEZE-OUT INDUCED TRANSITION FROM THREE- TO TWO-DIMENSIONAL MAGNETOTRANSPORT IN Si-δ-DOPED InSb LAYERS GROWN ON GaAs

R. Bogaerts, A. De Keyser, L. Van Bockstal, F. Herlach,
Department of Physics, Katholieke Universiteit Leuven, B-3001 Leuven, Belgium

V.C. Karavolas , F.M. Peeters,
Department of Physics, Universitaire Instelling Antwerpen, B-2610 Antwerpen, Belgium

G. Borghs
Interuniversitair Micro-Elektronica Centrum, Kapeldreef 75, B-3001 Leuven, Belgium

Pulsed field magnetotransport measurements up to 50 T have been carried out on Si-δ-doped InSb layers grown on GaAs. At low magnetic fields, a complex 3D transport behaviour is observed. Magnetic freeze-out of the bulk carriers at about 25 T results in a 3D-2D transition with 2D anisotropic behaviour at high magnetic fields. In this field range, the Hall effect shows a saturation towards a ν=1 quantum Hall plateau.

1. Introduction

The growth and characterization of heteroepitaxial InSb layers on GaAs is a relatively new research topic[1] which may be important for future applications, such as InSb-based opto-electronic devices integrated in GaAs. If heteroepitaxial InSb layers contain an additional 2D channel, such as a δ-doped layer, different types of 3D parallel conduction in the InSb layer[1,2] may mask the transport in the 2D channel. Previous publications on Si-δ-doped layers in InSb report measurements on samples without parallel 3D conduction[1]; e.g. this has been prevented by increasing the degree of compensation using p-type Be-doping[3]. For our MBE-grown δ-doped samples, no attempt has been made to achieve compensation of the InSb layers. These samples consist of a semi-insulating GaAs substrate, followed by a 0.5 μm GaAs buffer layer and a 3 μm i-InSb layer, grown at a temperature of 410 °C. The Si-δ-doping layer is situated 40 nm beneath the InSb surface, with a thickness of about 40 Å and a Si concentration of $\sim 3.3 \cdot 10^{18}\,cm^{-3}$ for sample G1145 and $\sim 3.8 \cdot 10^{18}\,cm^{-3}$ for sample G1284.

2. Experimental results

Four-terminal magnetotransport measurements were carried out in a perpendicular magnetic field up to 50 T and at temperatures ranging from 4.2 to 1.6 K. At 4.2 K, the magnetoresistance was measured at different angles θ between the magnetic field and the normal to the sample surface, including $\theta = 90°$. For these measurements, the sample current path was kept perpendicular to the magnetic field (transverse configuration). In addition, the magnetoresistance in the longitudinal configuration ($B \parallel I$) was measured. Figs. 1 and 3 show some results of the measurements on sample G1145, in low and high magnetic fields respectively. In Figs. 2 and 4, the experimental results for sample G1284 are shown.

3. Discussion

3.1. Low magnetic field behaviour

The ρ_{xx} SdH oscillations observed at low magnetic fields are independent of the angle θ, thus clearly due to bulk electrons. The 3D SdH maximum at 5 T in the transverse

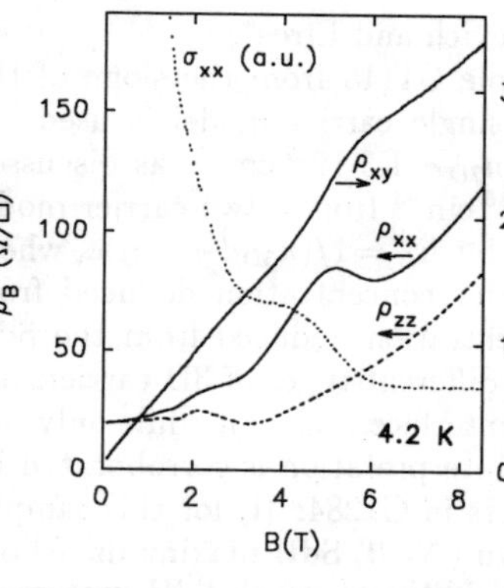

Fig. 1. The transverse magnetoresistance and the Hall effect (solid curves), and the longitudinal magnetoresistance (dashed curve) up to 8 T for sample G1145. In addition, the magnetoconductivity σ_{xx} is shown (dotted curve).

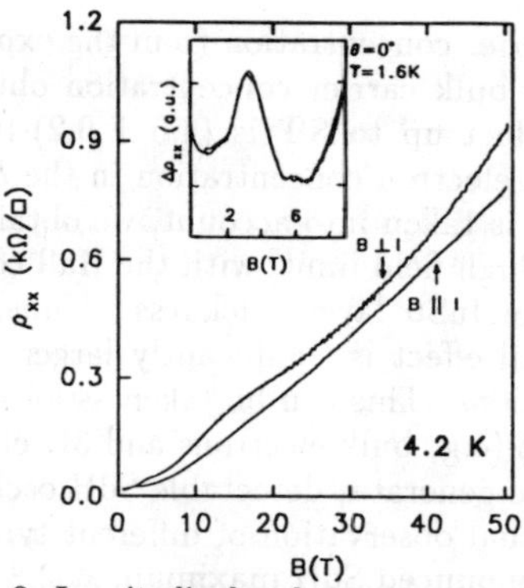

Fig. 2. Longitudinal and transverse magnetoresistance in a parallel magnetic field for sample G1284. The inset shows the low field SdH signal in a perpendicular magnetic field at 1.6 K. A second order polynom was subtracted from the data to emphasize the oscillations.

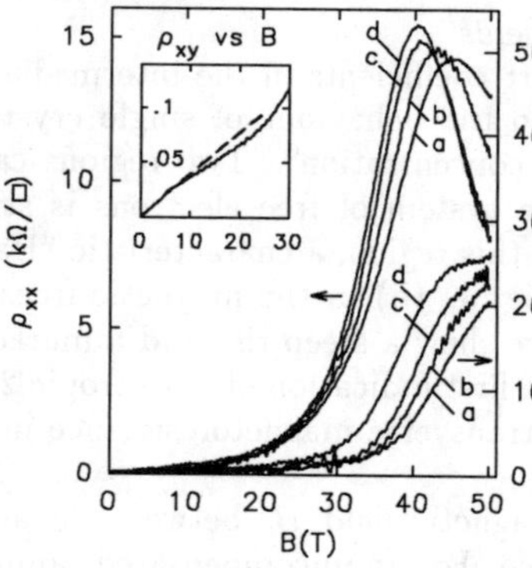

Fig. 3. The magnetoresistance and the Hall effect for sample G1145 up to 50 T in a perpendicular magnetic field at the temperatures 4.2 K (a), 2.8 K (b), 2.1 K (c) and 1.4 K (d). The inset shows the Hall effect up to 30 T.

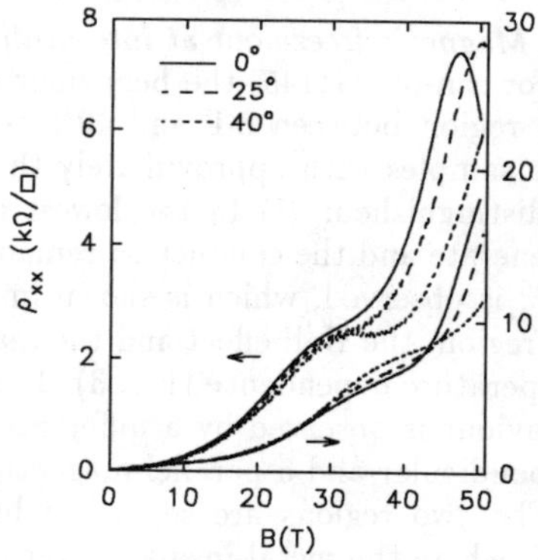

Fig. 4. Angular dependence of the Hall effect and the transverse magnetoresistance of sample G1284 at 1.6 K. For $\theta \neq 0°$, the Hall amplitude is divided by $\cos\theta$ to correct for the reduction of the Hall voltage.

magnetoresistance of sample G1145, is found to be absent in the longitudinal measurements. This is consistent with the behaviour of the SdH oscillations in n-type InSb single crystals[4], where the SdH maximum corresponding to the (N=0) spin down Landau level appears in the transverse magnetoresistance and in the Hall effect, while it is absent in the longitudinal measurements. From the expression given by Gurevich and Efros[5] for the position of the (N=0) maximum in σ_{xx}, a bulk electron concentration n_{3D} of $(2.6 \pm 0.1)\cdot 10^{16}\,\mathrm{cm}^{-3}$ is obtained. A calculation of n_{3D} from the frequency of the SdH oscillations (deduced from the magnetic field positions of the (N=2) and the (N=1) SdH maxima) yields a value $(3.2 \pm 0.2)\cdot 10^{16}\,\mathrm{cm}^{-3}$. The discrepancy between both values for n_{3D} probably arises from the broadening of the Landau levels, which shifts the SdH maxima to a lower magnetic field. This leads to an underestimate of the carrier concentration from the expression of Gurevich and Efros[5].

The bulk carrier concentration obtained for sample G1145 from the slope of the Hall effect up to 8 T is $(4.6 \pm 0.2)\cdot 10^{16}\,\mathrm{cm}^{-3}$, if a single carrier model is used. If the 2D electron concentration in the δ-doped layer ($n_{2D} \sim 1.3\cdot 10^{12}\,\mathrm{cm}^{-2}$, as discussed below) is taken into account, we obtain $n_{3D} \geq 4.1\cdot 10^{16}\,\mathrm{cm}^{-3}$ from a two-carrier model in the high field limit, with the Hall coefficient given by $R_H = 1/(n_{3D}d + n_{2D})e$, where d is the InSb layer thickness. Thus, the bulk carrier concentration deduced from the Hall effect is significantly larger than the concentration deduced from the SdH oscillations. This can be taken as an indication that different types of 3D carriers are present (e.g. bulk electrons and 3D electrons in the interface region[2]), while only one of these generates detectable SdH oscillations. This interpretation is corroborated by the actual observation of different types of 3D carriers in G1284: (i) for this sample, the pronounced SdH maximum at 2.8 T is a spin down (N=0) SdH maximum. Above the field position of this maximum, we observe an additional, weak SdH maximum that is independent of θ (Fig. 2). (ii) The parallel, transverse magnetoresistance curve of G1284 shows a maximum at about 15 T, which is absent in the longitudinal measurements (Fig. 2). This maximum thus behaves as a maximum generated by a (N=0) spin-down Landau level of a second type of InSb 3D electrons.

3.2. *Magnetic freeze-out at intermediate magnetic fields*

For sample G1145, the behaviour of the transport coefficients in the intermediate field region between 8 T and 40 T is comparable to the behaviour of single crystal InSb samples with approximately the same carrier concentration[6]. Two regions can be distinguished: (i) In the lowest field region the system of free electrons is still degenerate and the conduction remains metallic. In this region, a characteristic "Hall dip"[6] is observed, which is shown in the inset of Fig. 3. (ii) In the magnetic freeze-out region, the Hall effect and the magnetoresistance show a steep rise and a marked temperature dependence (Fig. 3). In this region, the first indication of anisotropic 2D behaviour is observed by a difference between the transverse magnetoresistance in a perpendicular and a parallel magnetic field.

The two regions are separated by a critical magnetic field B_c between 25 and 30 T, where the metal-insulator transition takes place. For an uncompensated sample ($K = N_{acc}/N_{don} = 0$) Neuringer[7] has obtained the relation $B_c \propto N_{don}^{0.84}$ between the critical field and the donor concentration. Ishida et al.[8] and Robert et al.[9] have studied partially compensated samples ($K \sim 0.5$) and found that the former equation is valid if N_{don} is replaced by N_{don}-N_{acc}. Extrapolating the results of Refs. [9,10] to the value N_{don}-$N_{acc} = n_{3D}$ observed for sample G1145, we obtain $B_c = 24$ T for n_{3D} deduced from the SdH oscillations, and $B_c = 29$ T for n_{3D} deduced from the Hall effect. These values can be reconciled with the data of Fig. 3. The 3D-2D transition in sample G1145 is thus accompanied by magnetic freeze-out of the 3D carriers. This is confirmed by theoretical simulations of the magnetoresistance and the Hall effect for this sample.

The nature of the transition in sample G1284 appears to be different. The anisotropic behaviour of the transverse magnetoresistance sets in at much lower magnetic fields than for sample G1145. Furthermore, the Hall effect and the transverse magnetoresistance for $\theta \neq 90°$ exhibit an additional feature at 30 T. This apparently marks the transition point between isotropic and anisotropic behaviour for $\theta \leq 40°$ (Fig. 4). The magnetoresistance characteristics that are indicative of magnetic freeze-out, and in particular the dramatic, temperature-dependent rise of the magnetoresistance, are not observed.

3.3. 2D high magnetic field behaviour

The transverse magnetoresistance, measured with the magnetic field perpendicular to the sample, shows a maximum between 40 and 50 T, and decreases strongly for higher magnetic fields (Figs. 3 and 4). These characteristics are shifted towards higher magnetic fields if the angle θ is increased. The Hall effect of sample G1145 shows a saturation towards a Hall plateau. We attribute this behaviour to 2D transport in the Si-δ-doped layer, and interpret it as a manifestation of a quantum Hall plateau accompanied by a resistance minimum at a filling factor ν=1. An estimate of the 2D electron concentration for G1145, assuming that the ν=1 minimum in ρ_{xx} occurs at 55 T, yields $n_{2D} \sim 1.3 \cdot 10^{12}\,\mathrm{cm}^{-2}$.

4. Conclusions and perspectives

Low temperature magnetotransport measurements performed on heteroepitaxial Si-δ-doped InSb layers unambiguously show (i) that a low resistivity 3D parallel path is present in these samples and masks the conduction in the δ-doped layer at low magnetic fields, (ii) that high magnetic fields induce a transition from bulk-carrier dominated transport to 2D carrier dominated transport.

The complex behaviour of the parallel conduction in the InSb layer merits further study. An interesting possibility is the presence of a high carrier density parallel path close to the GaAs/InSb interface, as was observed by Song et al.[2], who estimated the thickness of the interface layer to about 1000 Å. For InSb layers with slab-doped regions of this thickness, Droopad[10] has observed the formation of 2D subbands. Hence, for a system as described by Song et al., the possibility arises that parallel conduction occurs in a quasi-2D channel instead of a 3D channel.

Acknowledgements

R.B. is supported by the Onderzoeksraad of the K.U. Leuven. L.V.B. is a Senior Research Assistant and F.M.P. a Senior Research Associate of the Belgian Nationaal Fonds voor Wetenschappelijk Onderzoek.

References

[1] R.A. Stradling, in *Condensed Systems of Low Dimensionality*, eds. J.L. Beeby et al. (Plenum, 1991), p. 141.

[2] S.N. Song, J.B. Ketterson, Y.H. Choi, R. Sudharsanan, and M. Razeghi, *Appl. Phys. Lett.* **63** (1993) 964.

[3] M.J. Yang, W.J. Moore, R.J. Wagner, J.R. Waterman, C.H. Yang. P.E. Thompson, and J.L. Davis, *J. Appl. Phys.* **72** (1992) 671.

[4] S.T. Pavlov, R.V. Parfen'ev, Yu.A. Firsov, and S.S. Shalyt, *Soviet Physics JETP* **21** (1965) 1049.

[5] L.E. Gurevich and A.L. Efros, *Soviet Physics JETP* **16** (1963) 402.

[6] V.J. Goldman, M. Shayegan, and H.D. Drew, *Phys. Rev. Lett.* **57** (1986) 1056.

[7] L.J. Neuringer, *Proceedings of the 9th International Conference on the Physics of Semiconductors* (Moskou, 1968), p. 715.

[8] S. Ishida and E. Otsuka, *J. Phys. Soc. Japan* **43** (1977) 124.

[9] J.L. Robert, A. Raymond, R.L. Aulombard, and C. Bousquet, *Phil. Mag.* **B42** (1980) 1003.

[10] R. Droopad, S.D. Parker, E. Skuras, R.A. Stradling, R.L. Williams. R.B. Beall, and J.J. Harris, *Springer Series in Solid State Sciences* **89** (1989) 199.

QUANTUM TRANSPORT OF δ-DOPED GaAs STRUCTURES

ZHENG GUOZHEN, WEI YAYI, GUO SHAOLING, TANG DINGYUAN
National Laboratory for Infrared Physics, Shanghai Institute of Technical Physics, Academia Sinica, Shanghai, 200083, China

and

PENG ZHENGFU, ZHANG YUNQIANG
Nanjing Institute of Electronic Devices, Nanjing, 210016, China

ABSTRACT

By using the MBE technique, some GaAs films have been fabricated which contain single Si δ-doped layer, three δ-doped layers and Si δ-doped $Al_xGa_{1-x}As/GaAs$ heterostructure. Transverse magnetoresistance, longitudinal magnetoresistance and Hall resistance have been measured at low temperature from 0.3 K to 4.2 K and in high magnetic field up to 10 T. Transverse magnetoresistance oscillation and "diamagnetic" longitudinal magnetoresistance oscillation have been observed. Based on the experimental results, the transport behavior of three kind of samples have been discussed.

1. Introduction

We know that there are two major differences between δ-doped 2DEG and heterostructure 2DEG. First, even at moderate δ-doped concentrations the electrons populate already several excited subbands. At a doped concentration of 4×10^{12} cm^{-2} four subbands are occupied. The second difference is the considerable delocalization of the carrier in the dopant plane[1]. So the carrier mobility of the sample is quite low. This makes the observation of quantum transport quite difficult[1,2].

In this paper, single Si δ-doped GaAs Sample, three-layers δ-doped GaAs sample and Si δ-doped $Al_xGa_{1-x}As$ heterostructure were grown by using MBE method. Electric parameters of the samples have been measured. We also try to discuss their transpot behavior and coupling effect between heterojunction 2DEG and δ-doped 2DEGs.

2. Sample Preparation and Experimental Results

The Schematic plot of structures of two kinds of samples is shown in Fig. 1. They were grown by MBE method. Dopant density of the δ-doped layer is 1.09×10^{13} cm^{-2}, that of each δ-doped layer for three δ-doped layers is equal. The width of δ-doped layer is measured by mercury $C-V$ probe. It is smaller than 7 nm in the sample.

Ohmic contact electrodes were obtained by In diffusion under the protection of hydrogen gas. The depth of electrodes was controlled by the duration time of In diffusion. $I-V$ measurement shows that these electrodes are really ohmic.

Magnetoresistances were measured in L^{3}He cryostat system. 10L^{3}He gas was

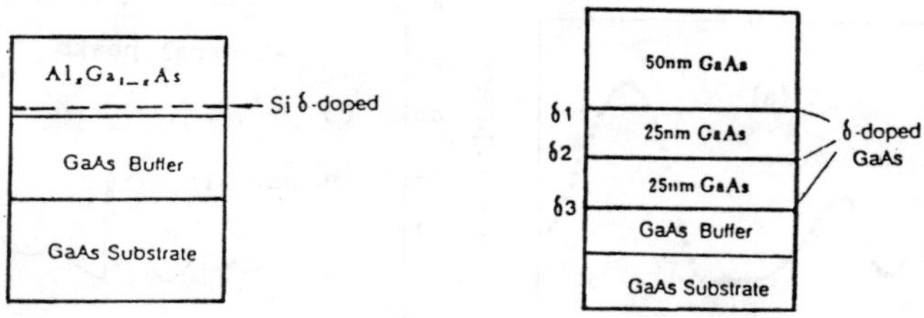

Fig. 1 Depth–profiles of structure of two kinds of samples

cooled down in sample pot. By changing the sorb temperature we can get the lowest temperature, which is 0.3 K. The sample was placed vertical or parallel to the magnetic field.

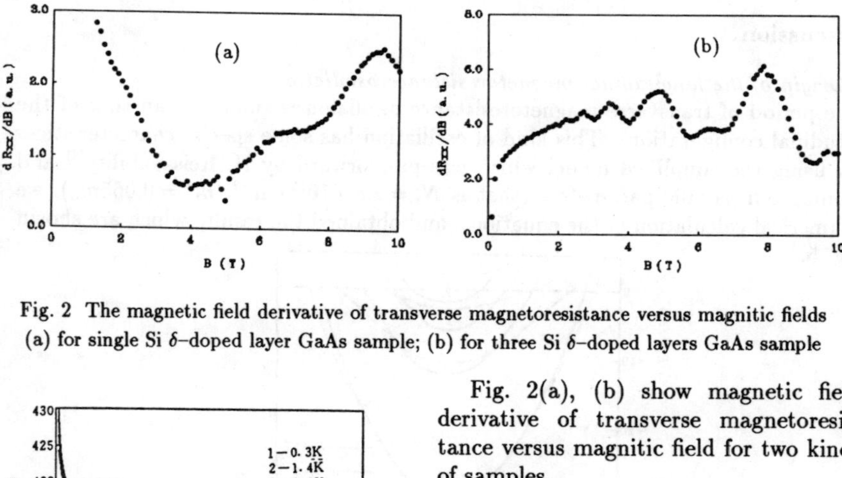

Fig. 2 The magnetic field derivative of transverse magnetoresistance versus magnitic fields (a) for single Si δ–doped layer GaAs sample; (b) for three Si δ–doped layers GaAs sample

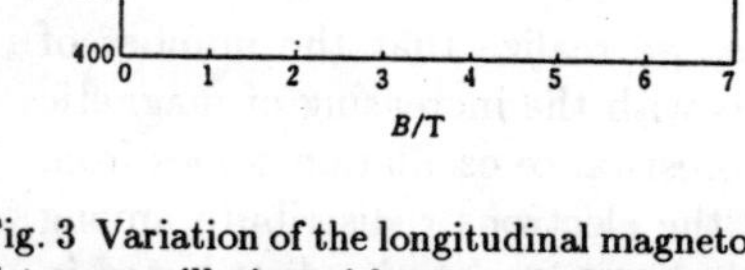

Fig. 3 Variation of the longitudinal magnetoresistance oscillation with magnetic field at 0.3 K to 4.2 K temperature region for three δ–doped layers GaAs sample

Fig. 2(a), (b) show magnetic field derivative of transverse magnetoresistance versus magnitic field for two kinds of samples.

The longitudinal magnetoresistance oscillation of three δ–doped layers GaAs sample is shown in Fig. 3. The longitudinal magnetoresistance oscillation is very clear in the experimental temperature region till 4.2 K. its period is larger than that of transverse configuration, and it remains unchangeable from 0.3 K to 4.2 K. Fig. 4(a) and 4(b) show SdH oscillations for normal and anomalous samples.

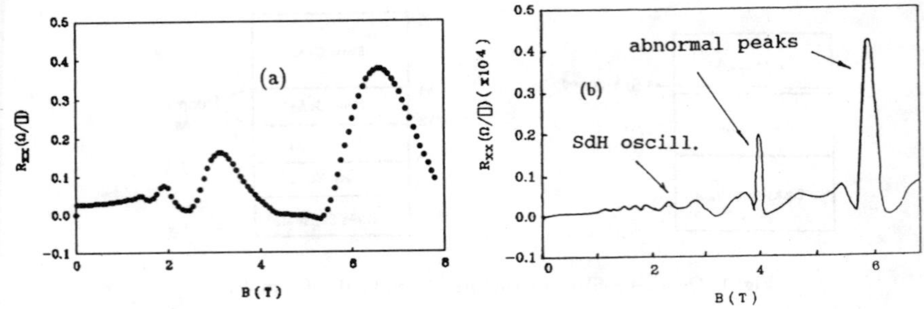

Fig. 4 Magnetoresistance oscillations for Si δ–doped $Al_xGa_{1-x}As/GaAs$ heterojunction
(a) normal peak; (b)anomalous peak

3. Discussion

3.1. Origin of the longitudinal magnetoresistance oscillation

The period of transverse magnetoresistance oscillaion is smaller than that of the longitudinal configuration . This kind of oscillation has some special characteristics.

By using the simplified model which was put forward by H. Reissuability[4] and combining our sample parameters (that is $N_s = 2.8 \times 10^{12}\,\mathrm{cm}^{-2}$, $m^* = 0.065m_0$), we did numerical calculation of the equations, and obtained the results which are shown in Fig. 5.

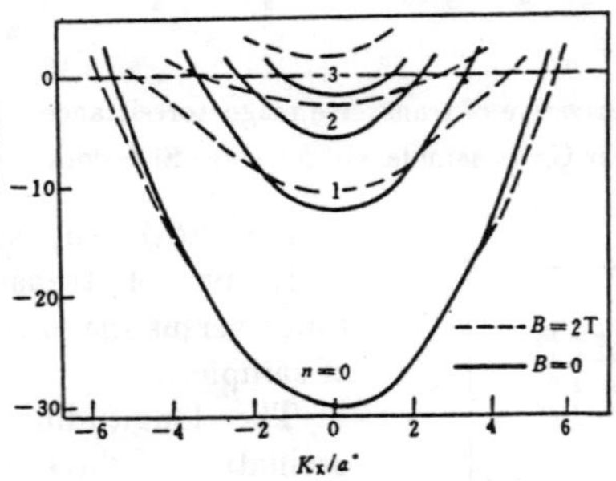

Fig. 5 Variation of the subbands under applied magnetic field

According to the theoretical calculation results, we realize that the number of subbands which are occupied by electrons decreases with the increasing of magnetic field. From this point of view, longitudinal magnetoresistance oscillation comes from the changing of subbands structures which makes the electrons redistribute among subbands. With the increasing of magnetic field electrons tend to be distributed in lower subbands. When one subband bottom comes through the Fermi level under the influence of magnetic field, the scattering of the carrier is the strongest, and this also corresponds to the oscillation peak.

3.2. Consideration of the Sample Structure

By using In diffusion in the sample, we make three δ–doped layers in parallel conduct.

The magnetoresistance oscillation in three layers δ–doped sample is obviously stronger than that of single layer δ–doped sample. Under the applied magnetic field, subband in each δ–doped 2DEG passes through the Fermi level simultaneously. It increases the intensity of oscillation. the oscillation curves in Fig. 2(b) are very strong.

3.3. Coupling effect between heterojunction 2DEG and δ–doped 2DEG

As shown in Fig. 4(b) the anomalous peaks are quite higher than normal oscillation peaks. This novel phenomenon has been observed in the sample for many times. We suggest that it comes from coupling effect between heterojunction 2DEG and δ–doped 2DEG.

3.4. Comparison between Transverse and Longitudinal Magnetoresistance Oscillations

When the sample temperature increases from 0.3 K to 4.2 K, transverse magnetoresistance oscillation gradually disappeared, but longitudinal magnetoresistance oscillation don't reduce obviously with the temperature. Based on this difference we may deduce that the mechanism of longitudinal and transverse oscillation is completely different. It means that electron distribution in subbands is not obviously changed with sample temperature from 0.3 K to 4.2 K.

Period of longitudinal magnetoresistance oscillation is larger than that of transverse configuration, and it isn't equal to $(1/B)$. This means although longitudinal magnetic field can change the electron distribution among subbands, it is weaker than that of transverse configuration.

4. Conclusion

Quantum well 2DEG formed by Delta–doped has two characteristics. One is high carrier density, the other is low mobility. These characteristics make Si δ–doped GaAs structures have many special transport behaviors. Some questions are still to be answered. On other aspect δ–doped structures offer a new kind of sample which may be used to study quantum interferences.

5. References

1. K. Ploog, *J. Crystal Growth*, **81**:(1987) 304–311.
2. O. Mezrin *et al.*, *Superlattice and Microstructure*, **10** (1991) 107.
3. S. Hiyamizu, *Semiconductors and Semimetals*, **30** (1990) 53.
4. H. Reisinger *et al.*, *Surface Science*, **170** (1986) 397–401.

QUANTUM MOBILITIES AS A FUNCTION OF THE SUPERLATTICE PERIOD IN Si ATOMIC LAYER DOPED GaAs

A.B.Henriques, V.Bindilatti and N.F.Oliveira Jr
Instituto de Física da Universidade de São Paulo
C.P.20516, 01498-970 São Paulo, Brazil

ABSTRACT

The quantum mobilities of electrons in the various minibands of GaAs with a periodical doping with silicon are studied as a function of the superlattice period, by analyzing their Shubnikov–de Haas oscillations in perpendicular fields of intensity up to 14 Teslas. The samples studied had a nearly constant sheet carrier density (2.5×10^{12} cm^{-2}), and a SL period varying in the range 40 Å– 830 Å. The magnetoresistance oscillations associated with individual minibands were separated from the rest of the spectrum by using Fourier transform techniques. The quantum mobilities were then determined from a Dingle plot for each miniband. The quantum mobilities decrease sharply for periods below 500 Å for the E2 miniband and below 200 Å for the E1 miniband. This behaviour is correlated with the overlap of the carrier spatial distribution and the impurity sheet.

1. Introduction

Modern epitaxial growth techniques allow researches to localize a sheet of Si atoms within a few monolayers of the host GaAs crystal [1]. In GaAs with a periodical Si layer doping, the electrons released from the shallow donors are confined by a space–charge potential, whose periodicity equals the spacing between the doping layers, and a superlattice (SL) is formed. The carrier distribution among the SL minibands can be obtained from a study of the transverse Shubnikov–de Haas spectra of the SL. Each magneto–resistance oscillation period is assigned to an extremal cross–section of the Fermi surface, which is calculated by a self–consistent simultaneous solution of Poisson and Schrödinger equations, wherein the overall carrier density is the single adjustable parameter [2,3]. The lineshape of the Shubnikov–de Haas spectrum of an atomic layer doped superlattice also encodes quantitave information about the broadening of the energy levels due to the Coulomb scattering by the ionized impurities. A given SL sample in general will contain several populated minibands with a dimensionality ranging from two (2D) to three (3D). Minibands with a 2D character contribute to the magnetoresistance oscillations with an amplitude whose field dependence can be approximated by an exponential dependence on $1/B$:

$$\text{Amplitude} \sim \exp(-\alpha x) \tag{1}$$

where $\alpha = \frac{2\pi}{\mu_Q}$, μ_Q is the quantum mobility, $x = 1/B$, and B is the magnetic field strength. By plotting the logarithm of the amplitude of the magnetoresistance oscillations due to an isolated miniband against the inverse magnetic field (the so called *Dingle plot*), the miniband's μ_Q can be estimated. For samples with widely spaced impurity sheets (SL's with a large period), in which the wells are disconnected and all minibands are of a two–dimensional character, it is found that the quantum mobilities of the carriers increase as we move to more energetic minibands, and the ratio between the μ_Q for adjacent minibands is ~ 3 [4,5]. Yamada and Makimoto also estimated the Hall mobility to be larger than μ_Q by a factor of ~ 1.4 [4].

When the period of the SL is made shorter, the interaction between adjacent wells becomes increasingly important, and a given miniband acquires an increasing 3D character. For 3D minibands, it is well known from standard textbooks that the amplitude of the oscillations also obey Eq.1. For minibands with a character which is intermediate between 2D and 3D, the field dependence of the amplitude of the oscillations can be approximated by (1) only if the width of the miniband is less than the broadening of the energy levels at the Fermi energy for this particular miniband [6]. This is usually the case with GaAs with a periodical doping with Si [6]; thus, the Dingle plot for an individual miniband will yield its quantum mobility μ_Q, which is the subject of the present work.

2. Results and discussion

The samples were grown on (100) GaAs semi-insulating substrates by MBE. The intended doping level was $N_d = 2.5 \times 10^{12}$ cm^{-2}. Shubnikov–de Haas measurements were made using a four contact geometry on approximately square samples with contacts in the corners, at a fixed temperature of 4.2 K, in magnetic fields up to 14 Teslas.

For each sample, the magnetoresistance oscillations associated with individual minibands were separated from the rest of the spectrum by using Fourier transform techniques. Figure 1 illustrates this procedure for a SL with a 400 Å period : (a) shows the original spectrum, (b) the inverse field Fourier Transform (full line) and its decomposition into individual peaks (dashed lines) by applying a band-pass filter, (c) the individual oscillations obtained through an inverse Fourier transformation and (d) the Dingle plot. The quantum mobilities for minibands E1 and E2 are plotted as a function of the superlattice period in Figure 2. It is seen that the quantum mobility decreases sharply below a SL period of ~500 Å for the E2 miniband and below a SL period of ~200 Å for the E1 miniband.

The origin of the decreasing mobility when the SL period is made made shorter can be understood from the calculated spatial distribution of the free electrons for each miniband, shown in Figure 3. The calculations included the conduction band non-parabolicity and the exchange–correlation correction, as described in Ref.2. The impurities were taken to be distributed by a Gaussian of full width at half maximum of 15 Å for all samples. The vertical dashed lines in Fig.3 enclose most of the Si atoms in the respective SL's. It can be seen that when the SL period is made shorter, the fraction of electrons within the impurity slab increases. This implies that the mean distance between the free electrons and the impurities decreases, which leads to a more efficient ionized impurity scattering process, and hence a decreasing μ_Q.

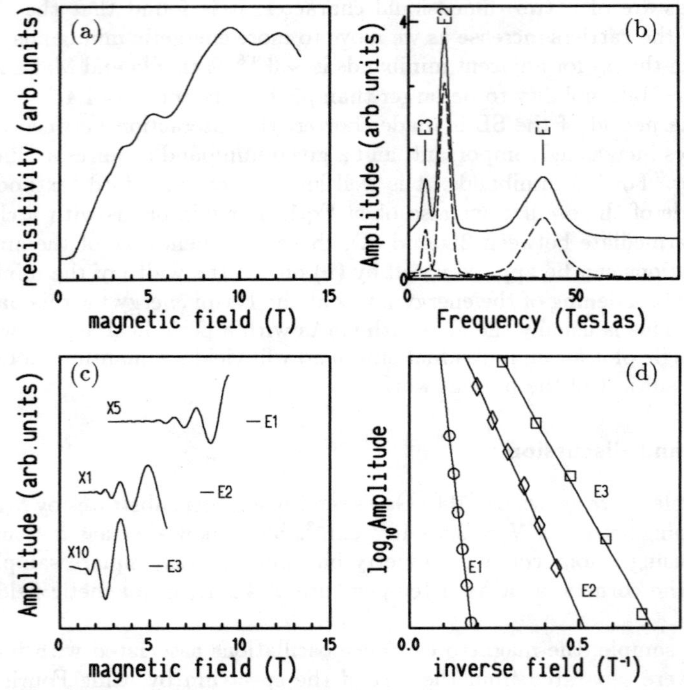

Fig.1: (a)-Original spectrum (b)-Fourier transform
(c)-individual oscillations (d) - Dingle plot.

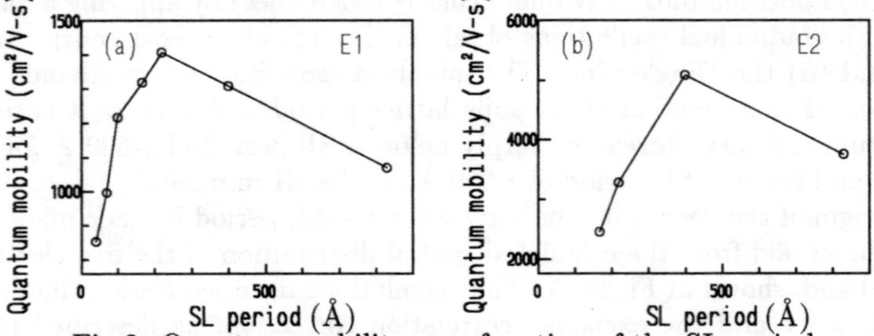

Fig.2: Quantum mobility as a function of the SL period
(a)-miniband E1; (b)-miniband E2.

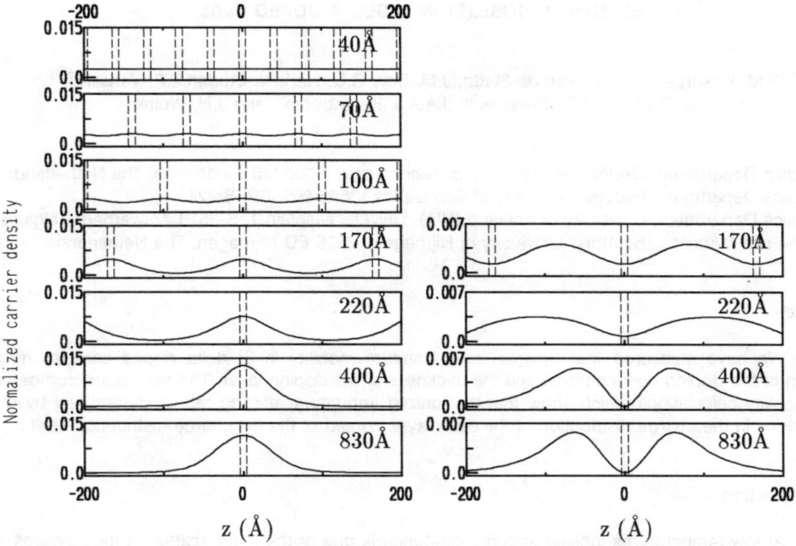

Fig.3: Free carrier density, normalized to an interval of 400 Å, in the E1 (left) and E2 minibands (right).

3. Acknowledgements

The authors are grateful to S.M.Shibli for preparation of the samples. A.B.H. acknowledges support by CNPq, CAPES and FAPESP.

4. References

1. E.F.Schubert, *Doping in III–IV Semiconductors*, (Cambridge University Press, Cambridge, 1993).
2. A.B.Henriques and L.C.D.Gonçalves, Semicond. Sci. Technol. **8**, 585 (1993).
3. A.B.Henriques, V.N.Morgoon, P.L. de Souza, V.Bindilatti, N.F. Oliveira Jr and S.M.Shibli, Phys. Rev. B **49**, 11248 (1994).
4. S.Yamada and T.Makimoto, Appl. Phys. Lett. **57**, 1022 (1990).
5. P.M.Koenraad, B.F.A. van Hest, F.A.P.Blom, R.van Dalen, M.Leys, J.A.A.J.Perenboom and J.H.Wolter, Physica B **177**, 485 (1992).
6. A.B.Henriques, Phys. Rev. B (1994), in press.

ELECTRON MOBILITY IN Si DELTA DOPED GaAs

P.M. Koenraad*, A.F.W. van de Stadt*, J.M. Shi*, G.Q. Hai#, N. Studart#, P. Vansant†, F.M. Peeters†, J.T. Devreese†,*, J.A.A.J. Perenboom°, and J.H. Wolter*

* Physics Department, Eindhoven University of Technology, 5600 MB Eindhoven, The Netherlands
# Physics Department, Federal University of São Carlos,13565-905, SP, Brazil
† Physics Department, University of Antwerp (UIA), Universiteitsplein 1, B-2610 Antwerpen, Belgium
o High-Field Magnet Laboratory, University of Nijmegen, 6525 ED Nijmegen, The Netherlands

abstract

We have measured the transport and quantum mobility in Si delta doped samples as a function of the doping concentration and the thickness of the doping layer. The results are compared with mobility calculations which show that the ionized impurity scattering rate is determined by the fluctuations in the charge distribution of the delta layer instead of the full charge distribution itself.

0. Introduction

At low temperatures ionized impurity scattering is one of the main scattering mechanisms in most semiconductor structures. Ionized impurity scattering has been studied extensively in GaAs/AlGaAs heterostructures. The strong improvement, in the last decade, of the mobility at low temperatures in GaAs/AlGaAs heterostructures originated from the successful reduction of the ionized impurity scattering is these structures. In contrast with these high mobility heterostructures are the well-known delta doping layers. In delta doping layers ionized impurity scattering is by far the most important scattering mechanism due to the strong overlap between the ionized dopant atoms and the free carriers. In delta doping layers normally more than one subband is populated. The population of multiple subbands will influence the scattering via screening and intersubband scattering. Thus delta doping layers are ideal structures for the study of ionized impurity scattering in the strong interaction regime, and for the study of the effect of multiple subband occupation on the ionized impurity scattering.

1. Experimental Methods and Analysis

We have studied ionized impurity scattering in delta doping layers by measuring both the transport and quantum mobility in Si delta doped GaAs layers. The quantum mobility and transport mobility in the case of a two dimensional electrongas are defined by the relations:

$$1/\mu_q = (m^*/e)\int W(\theta)d\theta \qquad \text{and} \qquad 1/\mu_t = (m^*/e)\int W(\theta)(1-\cos(\theta)d\theta \qquad [1]$$

where $W(\theta)$ is the probability for scattering over an angle θ. The quantum mobility is determined by the averaged scattering time, i.e. all scattering events have the same weight. The so-called transport mobility is determined mostly by the back-scattering events. The ratio of the transport to the quantum mobility thus gives information about the angle dependence, $W(\theta)$, of the scattering proces. For instance for phonon scattering the scattering probability is isotropic and therefore the ratio is equal to 1. For ionized impurity scattering in GaAs/AlGaAs heterostructures the ratio varies from 3 to 20 depending on the separation of the ionized impurities from the electron gas [1].

The transport mobility is obtained from the classical magneto-resistance in low magnetic fields. The classical magneto-conductivity is described by the Drude expressions

$$\sigma_{xx} = \Sigma_i n_i\mu_{t,i}e/(1 + \mu_{t,i}^2B^2) \qquad \text{and} \qquad \sigma_{xy} = \Sigma_i n_i\mu_{t,i}^2Be/(1 + \mu_{t,i}^2B^2) \qquad [2]$$

The quantum mobility is obtained from the amplitude dependence of the Schubnikov-de Haas (SdH) oscillations in high magnetic fields. Typically a magneto-resistance measurement on a delta doped structure contains multiple oscillation periods where each period belongs to a populated subband. The individual oscillation periods are resolved by Fourier Transformation. In the frequency domain we filter the peak corresponding to a single subband and perform an Inverse Fourier Transform to obtain the SdH oscillations of a single subband. The SdH oscillation amplitude of an individual subband is described by

$$\Delta \rho_{xx} \ (:) \ X/\sinh(X) \ \exp(-\pi/\mu_{q,i}B) \ \cos(\pi h/eB - \pi) \qquad [3]$$

where $X = 4\pi^2 m^* k_B T/heB$. The quantum mobiltiy is than obtained from so-called Dingle plots, i.e. plots where $\sinh(X)/X \ \Delta \rho_{xx}$ is plotted as function of $1/B$.

2. Results

In table 1 we present the results of the transport mobility measurements which where obtained from the classical part of the magneto-resistance measurements by the analysis method of Beck and Anderson [2], Their analysis method could only be used in the low doping concentration regime because we where not able to obtain reproducible results when three or more subbands are populated. Our results show a better mobility in the higher subbands. It seems reasonable to explain this by the smaller overlap of the wavefunction with the ionized impurities in the higher subbands, see Fig. 1.

N_{DON} (10^{12} cm^{-2})	2.0	3.5
N_{2D} (10^{12} cm^{-2})	1.8	3.2
l = 0	1500	1250
l = 1	7100	4800

Table 1: Transport mobility (cm^2/Vs) in the populated subbands.

In table 2 we present the results of the defemination of the quantum mobility in sample where the doping layer thickness was 2 nm or less. Firstly it is clear that the mobility increases with the subband number. Again the lowest subband, which is populated most heavily, has the lowest mobility. In the higher subbands the mobility is better. The table also clearly shows that the mobility decreases as function of the doping concentration. This drop might be due to an increasing overlap between scatterers and electron gas as a results of the stronger confining potential at higher doping concentrations.

N_{DON} (10^{12} cm^{-2})	2.0	2.5	3.5	5.0	8.0
N_{2D} (10^{12} cm^{-2})	1.8	2.2	3.2	4.5	5.6
l = 0	670	582	500	463	370
l = 1	4400	3390	2040	2020	1640
l = 2	-	-	-	4080	2450

Table 2: Quantum mobility (cm^2/Vs) in the populated subbands.

The ratio of the transport to the quantum mobility is about a factor of two. This is agrees with the results of Das Sarma et al [2] who calculated that the ratio should be close to 2.5 for GaAs/AlGaAs heterostructures where the ionized doping atoms overlap with the electrongas.

The sample with an electron concentration of 5.5 10^{12} cm^{-2} was annealed at 800 °C for various durations ranging from 3 to 300 s. Due to the annealing the thickness of the doping layer increased from 2 nm to 20 nm. The doping layer width was determined from a fit of the measured population with the calculated population where the width of the doping layer was the fitting parameter. We observed that the width could be described by a normal diffusion proses, σ = (2Dt). In fig 2 the mobility as a function of the thickness of the doping layer is plotted. The results show that for an increasing thickness of the doping layer that the mobility drops to the same value for all subbands and that the mobility in the lowest subband is nearly independent on the doping layer thickness. Also in this case the overlap of scatterers and electron gas seems to be the main reason for the mobility drop in the higher subbands.

3. Theory

The mobility in the individual subbands has been calculated by Mezrin and Shik [3]. They calculated the transport mobility in the populated subbands for delta doped structures with different doping concentrations having a mathematical delta doping layer of zero thickness. If we compare their quantum mobility in the lowest subband with our experimental results for the structures with different doping concentrations we get a perfect fit. However their results did not show the quantum mobility in the higher subbands.

We have also calculated [4] the quantum and transport mobility. We used wavefunctions that where obtained from the selfconsistent solution of the Poisson and Schrödinger equations. Although we observe a very good fit for the lowest subband, in accordance with results of Mezrin and Shik, we find that the quantum mobility in the higher subbands is nearly equal to the mobility in the i = 0 subband.

In studies of the mobility in GaAs/AlGaAs heterostructures is was proposed by several authors [5-7] that ionized impurity

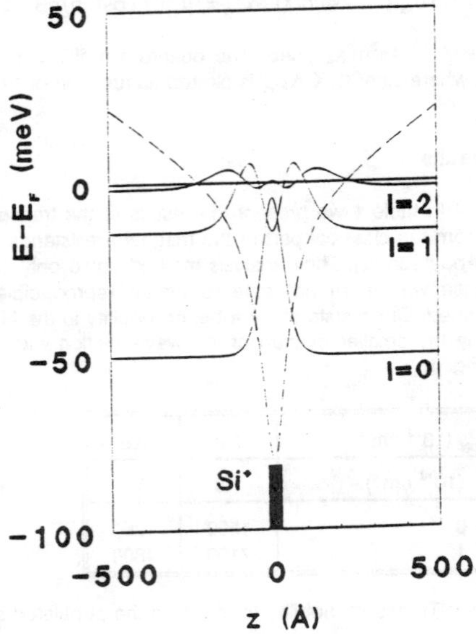

Fig1 : Charge probability distribution in Si delta doped GaAs.

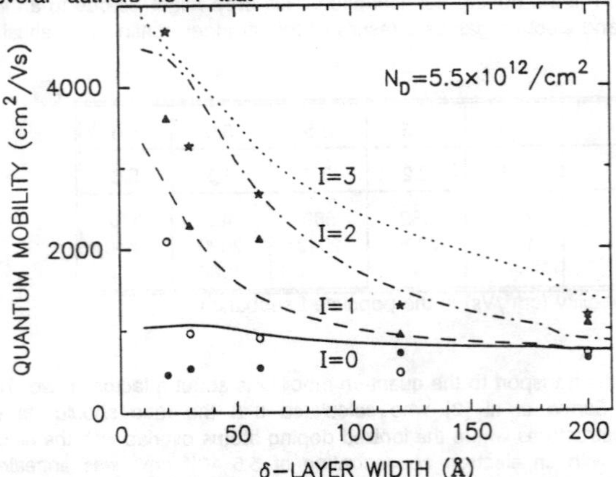

Fig 2: Quantum mobility as a function of the doping layer thickness. The lines show the calculated quantum mobilities.

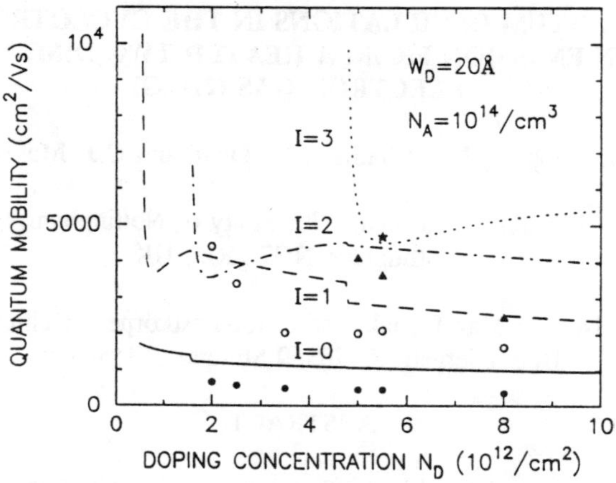

Fig 3: Quantum mobility as a function of the doping concentration. The lines show the calculated quantum mobilities.

scattering is not determined by the full distribution of the ionized impurities but by the fluctuations in this distribution. Following these arguments the mobility can increase strongly. The reason for taking the fluctuations in the charge distribution lies in the fact that when the confined states are calculated, by solving the coupled Poisson and Schrödinger equations, the averaged distribution of the charged impurities is used. Thus when the scattering is calculated on the same impurity distribution each ionized doping atom would be used twice. In Fig. 2 and Fig. 3 we show the calculated mobility we have obtained by taking the fluctuations in the charge distribution as the scattering potential. In contrast with the calculation on the full distribution we now have a better mobility in the higher subbands. Overall we obtain a satisfactory agreement between the experimental results and the calculated mobilities.

4. Conclusions

Measurements of the quantum and transport mobility show that the ionized impurity scattering rate is determined by the overlap of the ionized impurities and the electron gas. The overlap decreases with the subband number leading to a better mobility in a higher subband. In an individual subband the overlap increases with the doping concentration resulting in a lower mobility at higher doping concentrations. By increasing the thickness of the doping layer the mobility in all subbands reduces to the mobility in the lowest subband due to strong overlap in all subbands. Comparison with mobility calculations shows that the scattering rate for ionized impurities is determined by the fluctuations in the charge distribution of the delta doping layer instead of the full charge distribution.

Acknowledgements : JD and FMP acknowledge the sponsorship of the Belgian National Sience Foundation projectno 2.0093.91.

References

[1] S. Das Sarma and F.Stern, Phys. Rev. B **32**, 8442 (1985)

[2] W.A. Beck and J.R. Anderson, J. Appl. Phys. **62**, 541 (1987)

[3] O.A. Mezrin amd A.Y. Shik, Superlatt. Microstruc. 10, 107 (1990) and O.A. Mezrin, A.Y. Shik, and V.O. Mezrin, Semicond. Sci. Technol. 7, 664 (1992)

[4] G.Q. Hai, N. Studart, F.M. Peeters, J.T. Devreese, P.M. Koenraad, A.F.W. van de Stadt, and J.H. Wolter, acc. Int. Conf. "Physics of Semiconductors", August, 1994, Vancouver, Canada

[5] P.J. van Hall, T. Klaver, and J.H. Wolter, Semicond. Sci. Technol **3**, 120 (1988)

[6] R. Lassnig, Solid State Comm. **65**, 765 (1988)

[7] E.F. Schubert, L. Pfeiffer, K.W. West, and A. Izabell, Appl. Phys. Lett. **54**, 1350 (1989)

QUANTUM OSCILLATIONS IN THE CYCLOTRON PHONON EMISSION FROM A HEATED TWO-DIMENSIONAL ELECTRON GAS (2DEG)

J. Cooper*, L.J. Challis, F.F. Ouali and C.J. Mellor

Physics Department, University of Nottingham,
Nottingham NG7 2RD, UK

* Now at Max-Planck-Institut für Festkörperforschung
Heisenbergstr. 1, 70569 Stuttgart, Germany

ABSTRACT

Oscillations have been observed in the phonon intensity I_N emitted normal to a heated 2DEG bridge in a Si-mosfet in a quantizing magnetic field B as the sheet density is varied at constant power input. The oscillations are attributed to changes with filling factor in the angular distribution of the emission caused by variations in screening of the electron-phonon interaction.

1: Introduction

Phonon emission from a heated magnetically quantized 2DEG can occur from both intra- and inter-Landau level transitions. Its spectrum should include acoustic cyclotron phonons and their harmonics, $\omega = j\omega_c$ (j=1,2,3... and $\omega_c = Be/m^*$) and optic phonons at ω_{opt} together with lower frequency acoustic phonons from inter-Landau level transitions plus low frequency phonons from intra-Landau level transitions. When $k_BT_e \ll \hbar\omega_c$ (T_e is the electron temperature) the emission should be predominantly at $\omega \ll \omega_c$, except at the current entry and exit corners[1], and, when E_F is moved, oscillations should occur in the phonon emission from the bulk of the 2DEG because of the Shubnikov de Haas oscillations in ρ_{xx}. However when $k_BT_e \sim \hbar\omega_c$, the dominant emission should be cyclotron phonons mainly from transitions between the n and (n-1) Landau levels either side of E_F although contributions should also occur from higher and lower levels at ω_c and its harmonics. Optic phonons should also contribute and become dominant when $k_BT_e \gg \hbar\omega_c$. To investigate these changes and their variations with the position of E_F we have measured the phonon intensity I_N normal to an area of strongly heated magnetically quantized 2DEG in a Si mosfet.

2: Experimental

To reach $k_BT_e \sim \hbar\omega_c$ for modest power input, a $15\times180\ \mu m^2$ bridge was etched into the $1\times3\ mm^2$ gate of an (001) Si mosfet of 1K peak mobility $1.2 m^2V^{-1}S^{-1}$. For low currents and zero magnetic field, ~80% of the power is dissipated in the bridge and this proportion rises with increasing current. Since the 0.4 mm substrate is transparent to phonons of frequency $\upsilon \leq 1500$GHz ($\upsilon=\omega/2\pi$) at B=6T, where $\upsilon \sim 900$GHz, the cyclotron phonons emitted by the bridge travel ballistically across the substrate in directions close to the [001] normal due to the combined effects of in-plane momentum conservation[2] and phonon focussing while their harmonics are strongly scattered by isotopes into other focussing directions. The phonon intensity I_N along [001] is measured from the rise in temperature ΔT relative to that of a reference contact of a 0.25×0.25mm^2 metal contact attached to the substrate opposite the bridge: 90% of this temperature rise is due to the ballistic phonon intensity I_N. The power input to the device P ($1<P<200$ μW) and so total phonon intensity is kept constant to $<0.05\%$ while the sheet density n_s and so E_F are varied by the gate voltage so that changes in ΔT are due to changes in I_N. In this power range, FIR measurements in zero field on samples processed from the same wafer indicate that the electron temperature T_e varies from 15 to 30 K[3] at $n_s=1.5\times10^{16}$ m^{-2}. Although these temperatures may change in magnetic fields, we take them as a guide for the present work.

3: Results

Fig 1 shows ΔT together with the source-drain resistance R_{sd} for B=6T as a function of n_s for powers of 1μW and 100μW (data at other powers and fields are given elsewhere[4]. At 1μW, ΔT and hence I_N are in-phase with R_{sd} and so ρ_{xx} with minima when E_F is midgap. This behaviour is similar to that seen in 2DEG devices without bridges[2] and attributable to variations in the relative proportions of the emission from the bridge and from the current entry and exit points of the device. However, at 100 μW, ΔT has maxima close to the points where E_F is midgap. The phase-change occurs at 5 μW and oscillations of this phase persist up to the highest power used of 200 μW. At these higher powers, the magnitude of the oscillations $\delta(\Delta T)$ (peak to peak) and ΔT both increase approximately linearly with power over the range studied[4] so that $\delta(\Delta T)/\Delta T$ remains approximately constant for a particular field and filling factor with, for example, a value of $\sim0.3\%$ for B=6T and i=12 which decreases with decreasing B.

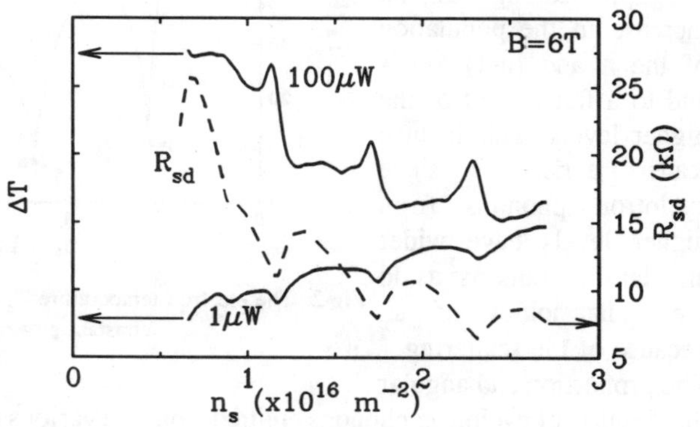

Fig 1: Variation of ΔT with sheet density for powers of 1 and 100 μW. Also shown R_{sd} measured at a power of 1 μW.

4: Theoretical Analysis

We associate the phase change in ΔT to the change in the predominant emission from intra to inter-level level processes. Since the efficiency of cyclotron phonon emission is determined by the strength of the screening, T_e oscillates with filling factor as shown by the solid curve in Fig 2 which is calculated using the theoretical expressions of Toombs et al[2] for B=6T but including screening assuming a Thomas-Fermi model. The screening length q_s is given by [5]

$$q_s(E_F) = \frac{e^2}{2\epsilon_0\epsilon_r}\frac{4\pi m^*}{\hbar^2}\left(1 - C\cos(\frac{2\pi E_F}{\hbar\omega_c})\right)$$

where C is a fitting parameter, so the screening is least when E_F is midgap.

Oscillations also occur in the populations of the Landau levels, with a fall in temperature when E_F is midgap, leading to an increase in the population of the n and (n-1) levels and to a fall in that of the higher levels. This in turn leads to a rise in I_N since cyclotron phonons from higher levels have wider angular distributions[2] as do the harmonics of ω_c because of the scattering.

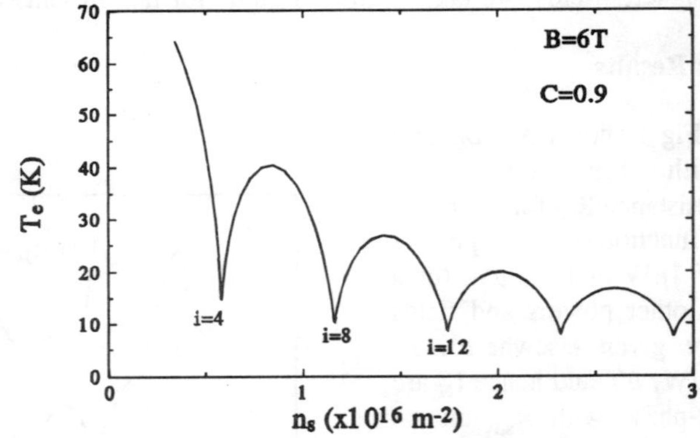

Fig 2: The electron temperature T_e calculated as a function of n_s at a constant power for C=0.9.

The proportion and angular distribution of cyclotron phonons emitted from the various inter-Landau level transitions can be calculated for each T_e and the corresponding value of ΔT for the geometry used can then be obtained by weighting the distribution by a function $W(\theta)$ (θ is the angle the phonon wavevector makes with the normal). This function is determined from calculations of the temperature distribution along the opposite face as a function of angular distribution allowing for phonon focussing[6]. Making the rather rough approximation that the scattered phonons of $2\omega_c$ and above make no contribution to ΔT we can then calculate the variation of ΔT with filling factor as shown in Fig 3 for C=0.9.

The size of the oscillations varies approximately linearly with C and a value of C= 0.03 is needed to obtain agreement with experiment. This however neglects thermal smearing which is significant so q_s must in fact be replaced by $\int q_s(E)(-dF/dE)\,dE$ (F is the Fermi function)[5] and for Gaussian linewidths with Γ=200 GHz ($\sim 0.2\omega_c$ at 6T), the effective values of C are 0.05, 0.006 and 0.001 at 15, 20 and 25K respectively. The magnitude is clearly too small at the higher temperature although is in line with observations at around 15K. However, the rapid fall with temperature is inconsistent with the approximately constant value observed experimentally and no improvement is obtained by varying Γ. So while this simple model explains, qualitatively, several of the main experimental features a more sophisticated treatment would be needed to obtain quantitative agreement.

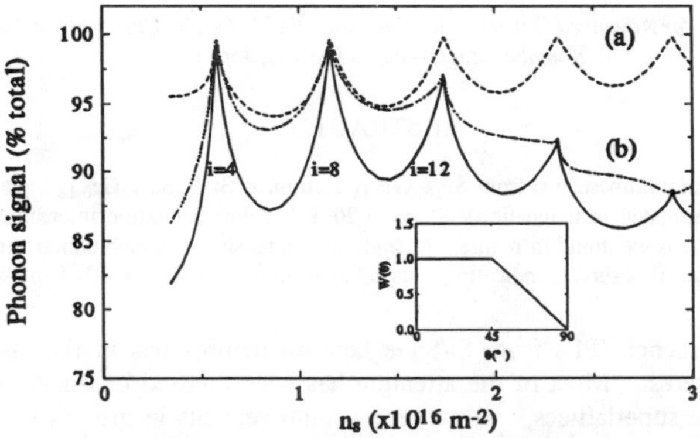

Fig 3: The proportion of the total phonon intensity reaching the detector calculated as a function of n_s for C=0.9. The two broken curves show the effects of (a) the omission of phonons of frequency $j\omega_c$ (j=2,3..) and (b) the effect on I_N of the cyclotron phonons of including $W(\theta)$. The solid curve shows the combined effects.

5: Acknowledgments

We are very grateful to Dr K A Benedict, Prof W Dietsche and Prof A Y Shik for valuable discussions and to the Science and Engineering Research Council, the European Commission and NATO for financial support.

6: References

1. P. A. Russell, F. F. Ouali, N. P. Hewett and L. J. Challis Surf. Sci. **229**, 54 (1990)
2. G. A. Toombs, F. W. Sheard, D. Neilson and L.J. Challis, Solid State Commun. **64**, 577 (1987)
3. A V Akimov, L J Challis and C J Mellor, Physica **B169**, 563 (1991), also F F Ouali, PhD thesis, University of Nottingham (1991)
4. J. Cooper, PhD thesis, University of Nottingham (1992)
5. T. Ando, A.B. Fowler and F. Stern, Reviews of Modern Physics, **54**, 550 (1982)
6. A.G. Every, N.P. Hewett, L.J. Challis and J.Cooper, Phonons 89, eds S. Hunklinger, W. Ludwig and G. Weiss (World Scientific, Singapore), 1358 (1990)

PHOTOLUMINESCENCE FROM A SiGe / STRAINED Si / SiGe QUANTUM WELL IN A HIGH MAGNETIC FIELD.

ULF GENNSER,[1] F. COPPINGER,[1] M. POTEMSKI,[1] D.K. NAYAK,[3] and J.-C. PORTAL[1,2]

[1] *Laboratoire des Champs Magnétiques Intenses, CNRS, F-38042, France*

[2]*INSA-Toulouse, 31077, France*

[3]*Research Center for Advanced Science and Technology (RCAST), The University of Tokyo, 4-6-1 Komaba, Meguro-ku, Tokyo 153, Japan.*

ABSTRACT

The photoluminescence from $Si_{0.82}Ge_{0.18}$ / strained Si / $Si_{0.82}Ge_{0.18}$ quantum wells is investigated in magnetic fields up to 20 T. At low excitation intensities, the recombination is excitonic in nature. At high laser intensities, recombination through an excited state is observed, indicating a partial formation of an electron - hole plasma.

Photoluminescence (PL) from Si/SiGe heterostructures has in the last few years attracted much interest. Most of the attention has been focused on short-period strain-symmetrized Si/Ge superlattices,[1] where recent improvements in growth technology has made it possible to reduce the dislocation density in relaxed SiGe buffer layers,[2] thus enabling the observation of luminescence from the quasi-direct transitions.[3] However, it has been found that well-resolved photoluminescence peaks can also be obtained from strained SiGe quantum wells (QWs).[4] In these structures, the holes are confined in the SiGe layer. It is believed that the linewidth of the observed transitions, of the order of a few meV, is limited by the alloy fluctuations.[5-7]

The observation of PL from strained Si layers, in which the strain results in a staggered type II band-alignment with confinement of the electrons in the Si QW, has only recently been reported.[8] Similar to strain-symmetrized superlattices (but unlike strained $Si_{1-x}Ge_x$ QWs grown on Si substrates) strained Si QWs necessitate a high-quality, step-graded, fully relaxed $Si_{1-x}Ge_x$ buffer layer grown on top of the Si substrate. In very thin Si quantum wells, the excitonic wavefunction penetrates considerably into the SiGe barriers, where the alloy randomness allow for a no-phonon (NP) recombination to be observed. The quantum confinement effect has been clearly established from the shift of the PL peaks with the well-width (10 - 21 Å), demonstrating monolayer control of the interfaces. The width of the observed peaks was $\approx$ 4 meV, indicating a very high quality of the top region of the SiGe buffer.

This narrow linewidth suggests the possibility to observe the Landau level spectra. This can give an important indication on how well-controlled the growth technology is, but is also of scientific interest, as it would allow for the study of recombination processes in type II quantum wells in an elemental semiconductor system. We present here the first investigation of the magneto-optical properties of a strained Si quantum well between fully relaxed $Si_{0.82}Ge_{0.18}$ barriers.

The samples were grown by gas-source molecular beam epitaxy (MBE). Each consisted of five Si QWs of equal width, separated by 350 Å $Si_{0.82}Ge_{0.18}$ barriers, and grown on top of a step-graded SiGe buffer layer and a 2.5 µm fully relaxed $Si_{0.82}Ge_{0.18}$ buffer cap layer. The details of the growth have been reported elsewhere.[8] Two samples were studied, with 9 Å and 12 Å wide Si quantum wells, respectively. PL spectroscopy was performed at T = 1.5 K in magnetic fields up to 20 T, using the 514.5 nm line of an argon-ion laser for excitation. The sample was mounted with the growth axis parallel to the magnetic field. A single fibre, chosen to have weak absorption in the near-infrared spectra, was used for both the excitation and collection light.

In fig. 1a the spectra at B = 0 T and B = 20T, using a moderate excitation intensity ≈ 0.5 W/cm^{-2}, for the sample with a 12 Å wide Si QW is shown. A similar behaviour is observed for the sample with 9 Å wide quantum wells. As has been shown in previous work,[8] the peaks around 1010 meV and 1070 meV correspond to transitions from the relaxed buffer layer, whereas the peak at 1040 meV can be assigned to a no-phonon (NP) transition from the quantum well, and the peak at 985 meV is the transverse optical phonon replica (TO) of the NP transition. A small blue-shift, and an increase in the intensity is seen at high magnetic fields for the QW related transitions. The shift is identical for the X^{NP} and X^{TO} lines, whereas no shift is seen for the buffer layer transitions.

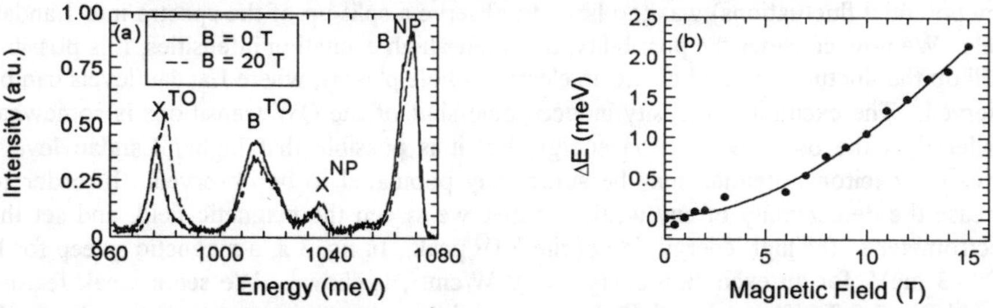

Fig. 1a. PL spectra of the sample with a 12 Å wide, strained Si QW, at B = 0T and B = 20 T. Fig. 1b. Shift of the X^{TO} transition as a function of magnetic field. Filled circles: experimental values; line: excitonic fit, using , Ry = 15 meV and $m_{exc} = 0.084\ m_0$.

In fig. 1b we plot the shift of the X^{TO} energy, ΔE, as a function of applied magnetic field. We fit the X^{TO} energy shift to the theory of a two-dimensional exciton in a magnetic field,[9] using the bulk exciton binding energy, Ry, and the excitonic effective mass, m_{exc}, as fitting parameters. The fit is fairly insensitive to variations in both the binding energy and the effective mass. However, PL experiments for strained SiGe/Si QWs show only a small deviation of the binding energy from that obtained from bulk Si.[10] In fig. 1b the best fit is shown for this bulk value, Ry = 15 meV, and gives a value of $m_{exc} = 0.084$. Due to the strain split of the Δ-valleys the electron effective mass in the Si quantum well should be taken as the transverse effective mass, m_{tr}. The exciton effective mass is then given by $1/m_{exc} = (1/m_{tr} + \gamma_1)$, where γ_1 is one of the Luttinger valence band parameters.[11] Using interpolated values for the $Si_{0.82}Ge_{0.18}$ alloy[5] gives $m_{exc} = 0.090$, in reasonable agreement with the fitted value. However, the uncertainty in the model must be pointed out. As the heterostructure is of type II, the charge separation will tend to lower the exciton binding energy. Also, due to the narrow well, parts of the electron wavefunction will penetrate into the SiGe barriers, were the six-fold degeneracy of the Δ-valley will lead to a heavier electron effective mass.

The increase of the QW recombination intensity with increasing magnetic field can qualitatively be understood in terms of the wavefunction of the electron - hole relative motion. This is squeezed in the direction of the magnetic field, for relative distances less than the Bohr radius.[12] Due to the electron confinement in a quantum well of width smaller than the exciton radius, we expect a larger overlap between the holes and electrons in a high magnetic field.

Photoluminescence excitation spectroscopy on SiGe QWs have shown that the excitons are bound to shallow (≈ 5 meV) wells due to fluctuations in the alloy composition.[4] Similar conclusions have been drawn from excitation intensity dependence of the PL[7] from Si/Ge short-period superlattices,[3] and from SiGe quantum wells,[7] where blue shifts in recombination energies have been described in terms of the higher excitation intensities filling up the shallow potential fluctuations binding the excitons. In our samples the X^{NP} and X^{TO} transitions show a large dependence on the excitation intensity, I_{exc}: a blue shift of ≈ 6 meV is observed for a 25-fold increase in I_{exc}. However, in these type II structures space-charge effects may also play a role in this shift.

At low intensities, we found that the excitonic binding energy (including random alloy potential fluctuations) was too large to observe a split-up of the spectra into Landau levels. We now consider the possibility, that under high excitation intensities, it is possible to fill up the fluctuations, and create an electron - hole plasma, where Landau levels can be observed. The excitation intensity-induced blue-shift of the QW transitions is somewhat smaller than the observed exciton energy, but it is possible that higher Landau levels, above the exciton potential, may be sufficiently populated to be observed. In order to increase the detectability of any weak features, we sweep the magnetic field, and set the spectrometer to the high-energy side of the X^{TO} peak. In fig. 3 a, a magnetic sweep for E = 986.3 meV, for an excitation energy ≈ 10 W/cm^{-2}, is shown. We see a weak feature around B = 7.5 T. Through repeated sweeps at different energies, we trace out the E - B diagram of the observed feature. This is shown in fig. 3 b, together with the shift of the central X^{TO} peak. The observed feature, which we have labeled L, has a linear dispersion

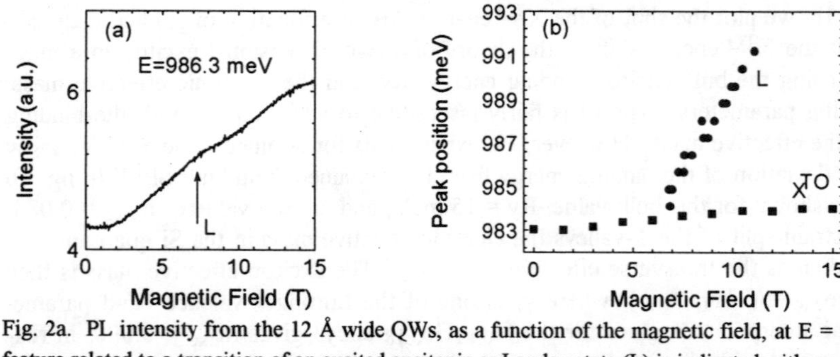

Fig. 2a. PL intensity from the 12 Å wide QWs, as a function of the magnetic field, at E = 986.3. The feature related to a transition of an excited excitonic or Landau state (L) is indicated with an arrow. Fig. 2b. The feature L, obtained from similar field sweeps as in fig. 2a, at different energies, vs. the magnetic field. Also shown in the fig. is the peak position of the X^{TO} transition.

too large to be that of the lowest excitonic level or Landau level. Thus, it cannot be explained assuming, for example, a recombination involving the light-hole inside the Si layer (where it is separated from the heavy hole due to the strain). Instead, the large magnetic field dependence indicates that an excited state is involved. Fitting it to the 2s, or n=1 Landàu level, gives a reduced effective mass = 0.11 m_0., which is in reasonable agreement with the value found for the exciton at low excitation energy, especially when considering the uncertainties in the model, mentioned above. We therefore believe, that under high illumination intensities, it is possible to populate the local fluctuations in the SiGe alloy sufficiently to partly create an electron-hole plasma. The shift of the X^{TO} peak with the magnetic field does now only represent the average distribution of Landau levels in the almost full potential fluctuations, and can no longer be adequately be described by a hydrogenic level.

In summary, we have observed the PL from a strained Si quantum well surrounded by SiGe barriers. At low excitation intensities, we observe bound exciton recombinations, that can adequately be described by the theory for a hydrogen like exciton in a magnetic field. The excitons may be bound to local composition fluctuations in the SiGe alloy. At higher excitation intensities, it is possible to populate these fluctuations, to create an electron - hole plasma, where we are able to observe an excited Landau level. With increasing advances in the SiGe growth technology, we believe that the SiGe material system may be suitable for further high magnetic field investigations.

Acknowledgements: U.G. acknowledges support from SI and from NFR through the HCM program.

References

1. See f. ex. E. Kasper and F. Schäffler, in *Semiconductors and Semimetals*, ed.T. P. Pearsall (Academic Press, London, 1991).
2. F. K. Legoues, B. S. Meyerson, and J. F. Morar, Phys Rev. Lett. **66** (1991) 2903; E. A. Fitzgerald, Y.-H. Xie, M. L. Green, D. Brasen, A. R. Kortan, J. Michel, Y.-J. Mii, and B. E. Weir, Appl. Phys. Lett. **59** (1991) 811.
3. U. Menczigar, G. Abstreiter, J. Olajos, H. Grimmeiss, H. Kibbel, H. Presting, E. and Kasper, Phys. Rev. **B 47** (1993) 4099.
4. J. C. Sturm, H. Manoharan, L. C. Lenchyshyn, M. L. W Thewalt, N. L. Rowell, J.-P. Noèl, and D. C. Houghton, Phys. Rev. Lett. **66** (1991) 1362.
5. D, J. Robbins, L. T. Canham, S. J. Barnett, A. D. Pitt, and P. Calcott, J. Appl. Phys. **71** (1992) 1407
6. S. Fukatsu and Y. Shiraki, Appl. Phys. Lett. **63** (1993) 2378.
7. L. C. Lenchyshyn, M. L. Thewalt, J.C. Sturm, P. V. Schwartz, E. J. Prinz, N. L. Rowell, J.-P. Noèl, and D. C. Haughton, Appl. Phys. Lett. **60** (1992) 3174.
8.D.K. Nayak, N. Usami, S. Fukatso, and Y. Shiraki, Appl. Phys. Lett. **63** (1993) 3509.
9. A. H. MacDonald and D. S. Ritchie, *Physical Review* **B 33** (1986) 8336.
10. S. Fukatso, H. Yoshida, A. Fujiwara, Y. Takahashi, Y. Shiaraki, and R. Ito, Appl. Phys. Lett. **61** (1992) 804.
11. N. O. Lipàri and A. Baldereschi, Phys. Rev. **B 3** (1971) 439.
12. S. K. Lyo, Phys. Rev. **B 48** (1993) 2152.

Si-SiGe 2DE&HS TRANSPORT IN PULSED HIGH MAGNETIC FIELDS

F.F. Fang[1*], R.B. Dunford[2], R.G. Clark[2], R. Newbury[2], R.P. Starrett[2], V.A. Stadnik[2], A.V. Skougarevsky[2], R.H. McKenzie[2], J.O. Chu[1], K.E. Ismail[1] and B.S. Meyerson[1]

[1] *IBM T.J. Watson Research Center, Yorktown Heights, NY 10598, USA*
(Research Staff Member Emeritus)*

[2] *National Pulsed Magnet Laboratory, University of New South Wales, Sydney 2052, Australia.*

Low temperature (mK) magneto-transport measurements for p- and n-type modulation-doped Si-SiGe heterostructures have been extended to high magnetic fields. For the **2DHS in p-SiGe** two insulating phases (IP) are identified, in regions $1<\nu<2$ and $\nu<1$, with reentrance to the QHE regime at $\nu=1$ (B~15T). The IP centred at $\nu=1.5$ has the characteristics of a Hall insulator but is unanticipated by the recently proposed global phase diagram. Wigner crystallisation is not expected at this filling factor in this sample. The dominant IP in the region $\nu<1$ strengthens significantly around $\nu=0.5$ (~30T), in line with the global phase diagram. For the high mobility **2DES in n-Si**, a two-valley system, signatures of the FQHE in the region $\nu<1$ (one valley occupied) usually observed in GaAs are replicated out to $\nu=2/5$ (B ~ 50T). For $1<\nu<2$ however (two valleys occupied) prominent FQHE states are absent, similar to the case of double layer GaAs. Valley splitting is found to depend only on the normal field component.

1. Introduction

Recent advances in preparing SiGe layers on (100)Si substrates by MBE or UHV-CVD[1,2] exploit pseudomorphically-strained layers on strain-relieved substrates or intermediate layers. For n-type samples, the 2D electron system is formed in a biaxially tensile stressed silicon layer (~10nm thick) clad between strain-relieved SiGe layers. The 2D electrons are transferred from phosphorous doping in the top-clad SiGe alloy, with an undoped spacer extending 10nm from Si-SiGe interface. For our p-type samples, holes are confined to 2D in a SiGe layer (~40nm thick) clad between the Si substrate and a top Si cap layer, which is modulation-doped with boron (10nm spacer). The SiGe layer is under compressive biaxial stress. The Ge fraction of the alloy is ~ 12 to 15 atomic percent.

In this paper, we extend previous magneto-transport results[3,4] to focus on the low temperature properties of these strained layers in intense magnetic fields. Pulsed field experiments to 50T were conducted in a ^{3}He system. For steady fields to 16T, measurements were performed in a dilution refrigerator.

2. 2DHS in p-type SiGe

Figs. 1-3 show magneto-resistance and non-linear I-V data for a p-$Si_{0.88}Ge_{0.12}$ sample with hole density $3.6x10^{11}cm^{-2}$ and mobility 8000 cm^2/Vs. The ρ_{xx} and ρ_{xy} traces at 30mK in the region $\nu>1$ are shown in Figs. 1a and 1b and a temperature evolution of ρ_{xx} for the region $3>\nu>2$ is shown in Fig. 1c. Both strain and confinement act to lift the doubly degenerate valence bands; the $m_j=\pm3/2$ heavy hole band has the lowest energy for holes. The energy level scheme in these samples is non-standard, due to spin splitting being comparable to the cyclotron energy ($g_\perp^*$ ~ 6-8, m^* ~ 0.34-0.4 m_e)[4]. Together with Landau level (LL) broadening, this results in spin splitting not being resolved (except for the lowest spin state of the lowest LL), since spin-up and spin-down energies associated with adjacent Landau levels are nearly degenerate. The QHE is then only observed at odd-integer filling, as shown for $\nu\geq3$ in Fig. 1.

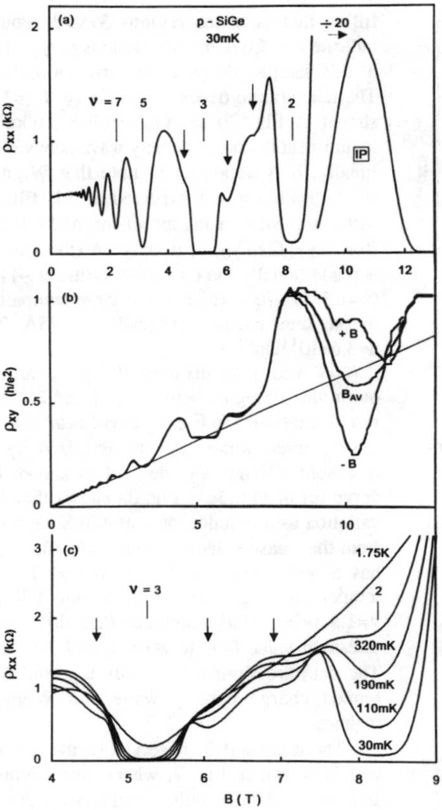

Fig. 1 (a) Longitudinal and (b) Hall resistivities ρ_{xx}, ρ_{xy} at T=30mK for the 2DHS in p-SiGe ($n=3.6\text{x}10^{11}\text{cm}^{-2}$). An insulating phase (IP) is identified in the region 2>ν>1. (c) T - dependence of ρ_{xx} in the region around ν=2 and 3.

For 3>ν>1 there are three features of note in Fig. 1: (i) ρ_{xx} structure develops close to ν=7/3 and 8/3 (arrowed) and a ρ_{xy} plateau forms at 3/7 h/e^2 (but which does not line up in field with the ν=7/3 ρ_{xx} minimum). (ii) At ν=2 a ρ_{xx} minimum forms but ρ_{xy} jumps to h/e^2, with no plateau at $h/2e^2$. The position of the ρ_{xx} minimum labelled as ν=2 in Fig. 1a is actually 0.6T higher in field than the correct ν=2 position as judged by the odd-integer sequence to lower field. At higher field the Hall resistance tends to return towards the classical value (B/ne) indicated by the straight line in Fig.1b; the details depend on the polarity of the magnetic field (+B, -B) with the average value B_{AV} (which removes ρ_{xx} admixture effects) reasonably close to the classical line. (iii) An anomalously large ρ_{xx} peak is observed in the region 2>ν>1 centred at ν=1.5 which increases from ρ_{xx} ~ 5kΩ at T=1.75K to ~40kΩ below 100mK, as shown in Fig. 2. The detailed temperature dependence of ρ_{xx} at ν=1.5 is contrasted with that at ν=2.0 (ρ_{xx} minimum) and ν=2.1 (next largest ρ_{xx} peak to lower field) in Fig. 2b depicting insulating, QHE and metallic character respectively.

The ρ_{xy} jump to h/e^2 at ν=2 and ρ_{xx} minimum at a field value slightly higher than the expected ν=2 position can be understood as resulting from the unusual energy level degeneracy in p-SiGe. At exactly ν=2 filling with unresolved spin states of adjacent Landau levels, the Fermi energy E_F will be in overlapping extended states of the 0↑, 1↓ levels and ρ_{xx} will be finite. If extended states are sufficiently narrow, a small increase in magnetic field will move E_F into localised states with only the lowest Landau level (0↓) extended states below E_F resulting in $\rho_{xx} \rightarrow 0$ and $\rho_{xy} \rightarrow h/e^2$. If this unusual Landau level degeneracy persists at higher fields, one would expect a very broad ν=1 plateau which is not the case in our data due to the observed onset of an insulating phase centred at ν=1.5, for which the single particle density of states picture above is unlikely to be appropriate. The IP requires separate explanation.

The IP at ν=1.5 is characterised by finite ρ_{xy} and an order of magnitude increase in ρ_{xx} as the temperature is lowered, as shown in Figs. 1 and 2, characteristic of the recently proposed Hall insulator[5]. However the global phase diagram[5] that situates the onset of this insulating phase on a plot of sample disorder versus inverse filling factor excludes such a phase being observed in a transition between υ=1 and 2 QHE states[6]. Several points may be relevant. The IP is centred at υ=1.5 which may indicate that extended states are important (if the LL degeneracy is lifted). The observation of structure in ρ_{xx} and ρ_{xy} at fractional

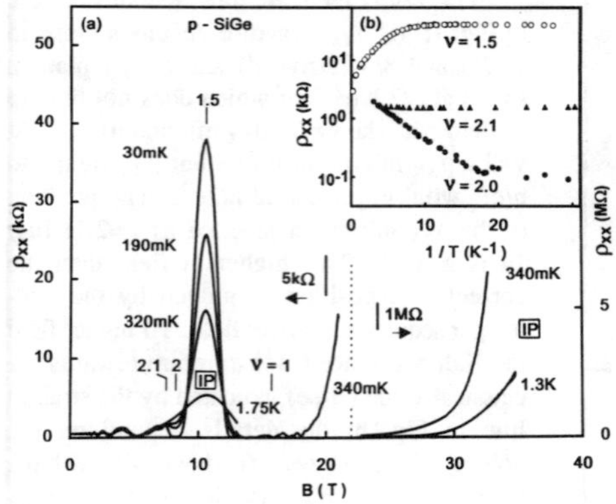

Fig. 2 (a) T - dependence of ρ_{xx} in the IP regions 2>ν>1 and ν<1.
(b) Detailed T - dependence of ρ_{xx} at ν=1.5, 2.0 and 2.1 illustrating insulating, QHE and metallic behaviour respectively.

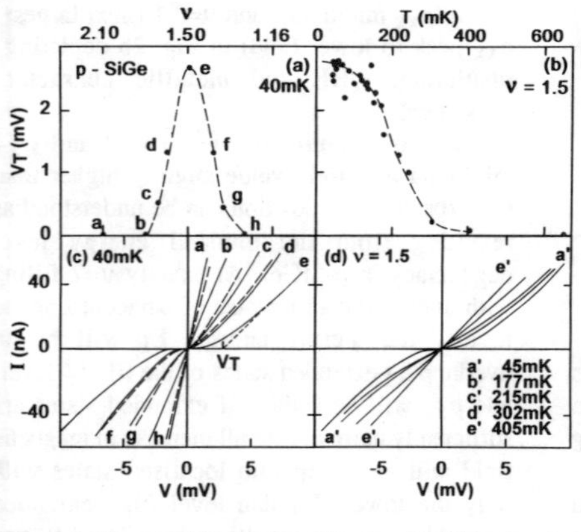

Fig. 3 Threshold voltages V_T for the IP centred at ν=1.5 are shown (a) as a function of ν at T=40mK and (b) as a function of T at ν=1.5, derived from non-linear I-V shown in (c), (d).

filling factors in the region 3>ν>2 could indicate that Coulomb interactions play a role in this sample, despite the low mobility. The temperature dependence of ρ_{xx} at ν=1.5 shown in Fig. 2b saturates below 100mK, reminiscent of charge density wave behaviour. Finally, it is important to note that Wigner crystallisation is not expected at this filling factor in p-SiGe; using $m^*=0.4m_e$ and ε=12.2 (for $Si_{0.88}Ge_{0.12}$) r_s is only ~6 (too small) and additionally the cyclotron radius at ~11T (ν=1.5) is ~70Å which is too large compared to the inter-particle separation ~ 95Å for $n=3.6\times10^{11}cm^{-2}$.

I-V measurements have also been carried out in this IP region between ν=1 and 2. The results are shown in Fig. 3. Non-linear effects are observed which are characterised by a threshold voltage V_T, defined as shown in curve (e) of Fig. 3c. Fig. 3a shows the V_T variation as a function of ν at 40mK derived from the measurements shown in Fig. 3c; V_T has a peak value of 2.7mV at ν=1.5. A temperature-dependent study of the I-V at ν=1.5 (Figs. 3b,d) indicates that this peak threshold value falls to zero around 300mK. The data are similar to results reported for pinned charge density wave and Wigner regimes.

The dominant IP observed in the region ν<1 is shown in Fig. 2, where data beyond 16T were taken in pulsed fields with 20ms duration. Whilst ρ_{xx} increases below ν=1, a prominent divergence sets in around 30T (ν~0.5) at 340mK, which is quenched with increasing temperature (see the 1.3K data shown for comparison). This is not inconsistent with the predicted Hall insulator region of the global phase diagram. Measurements to lower temperature and of ρ_{xy} in this regime are underway to examine this observation, using a dilution refrigerator configured for pulsed field operation.

3. 2DES in n-type Si

For n-type modulation-doped Si-SiGe the tensile strain and confinement effects give rise to a splitting of the conduction band into two subbands with energy difference ~200meV, so that intersubband scattering is suppressed. For the lowest subband, the energy of the two valleys corresponding to momentum along ±z are lowered while the other four are raised. There is consequently a fourfold degeneracy for each Landau level, two for spin and two for valley. The carrier concentrations in these samples are 4–7.5x10^{11}cm^{-2} with typical mobility ~1.5x10^{5}cm^{2}/Vs (see Ref. 4 for a review).

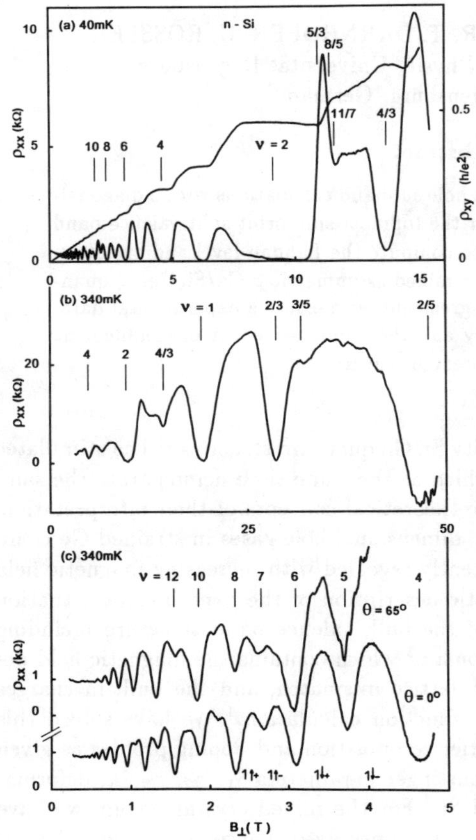

Fig. 4 Transport data for the 2DES in n-Si taken (a) in steady fields at 40mK and (b) pulsed fields at 340mK. (c) Comparison of low field ρ_{xx} data at 340mK in normal magnetic fields (θ=0°) and 65° tilt angle.

Transport data for a sample with n=4.2x10^{11}cm^{-2} are shown in Fig. 4. Parallel conduction observed in this sample has been subtracted from the ρ_{xx} data in Figs. 4a,b; error in the simple B^2 background subtraction at highest field leads to the ν=2/5 ρ_{xx} minimum lying below zero. Parallel conduction effects have not been subtracted from the Hall trace in Fig. 4a, where some distortion of the higher field ρ_{xy} data is evident. The important observation is that for ν<1 (Fig. 4b) where one valley only is occupied, FQHE structure normally observed in GaAs is replicated, namely ρ_{xx} minima at ν=2/3, 3/5 and 2/5. In contrast, for the region 2>ν>1 (two valleys occupied), whereas a deep ρ_{xx} minimum is observed at ν=4/3 there is no structure at ν=5/3, normally a strong FQHE state. Whilst it is tempting to speculate that odd numerator states are absent (with a view that ν=4/3 corresponds to 2/3 filling in both valley 1 and valley 2 for example), the only other weak ρ_{xx} structure in this region occurs more accurately at ν=11/7 than 8/5 and similarly the ρ_{xy} plateau is closer to 7/11 h/e^2. It is conceivable however that valley occupation plays an important role in the many-body FQHE states. Fig. 4c shows a comparison of the ρ_{xx} traces in normal magnetic fields and 65 degree tilt angles. Spin splitting at ν=6,10,14,... which depends on total magnetic field is clearly enhanced, whereas valley splitting at ν=5,7,9,... is the same strength in the tilted field. This provides experimental evidence that valley splitting depends only on the normal component of the magnetic field.

References

1. F.K. LeGoues, B.S. Meyerson and J.F. Morar, *Phys. Rev. Lett.* 66 (1991) 2903.
2. S.F. Nelson *et al.*, *Appl. Phys. Lett.* 61 (1992) 64.
3. F.F. Fang, P.J. Wang, B.S. Meyerson, J.J. Nocera and K.E. Ismail, *Surf. Sci.* 263 (1992) 175.
4. F.F. Fang, *Surf. Sci.* 305 (1994) 301.
5. S.Kivelson, D.H. Lee and S.C. Zhang, *Phys. Rev.* B46 (1992) 2223.
6. An IP has also been observed between the 1/3 and 2/5 FQHE states in p-GaAs which is at variance with the global phase diagram; H.C. Monoharan and M. Shayegan, preprint.

QUANTUM RESONANCES OF HOLES IN STRAINED Ge/Si_xGe_{1-x} QUANTUM WELLS

M. MERKLER, R. WINKLER, T. DARNHOFER, U. RÖSSLER
Institut für Theoretische Physik, Universität Regensburg
D-93040 Regensburg, Germany

Abstract

We have combined self-consistent hole subband calculations with a diagonalization of a 6×6 **k·p** Hamiltonian for the topmost spin-orbit split valence band including strain and magnetic field to compute the Landau level and cyclotron resonance spectra. In application to strained asymmetric *p*-Ge/Si_xGe_{1-x} quantum wells we find remarkably good agreement with existing experimental data. Spin-splitting due to the asymmetry and the combined effect of confinement and strain are responsible for the observed features.

1. Introduction

The progress in growth of high-quality Si/Ge quantum structures[1] has stimulated detailed spectroscopic measurements which at the same time demonstrate the sample quality and provide data to test the theoretical concepts of their interpretation. Cyclotron resonance experiments of two-dimensional hole gases in strained Ge channels between Si_xGe_{1-x} barriers have recently revealed with increasing magnetic field increasingly complex spectra.[2] A realistic description of the experimental situation requires to consider the peculiarities of the bulk valence band structure including the split-off band in a 6×6 **k·p** Hamiltonian,[3] which contains the magnetic field dependence and the biaxial strain due to lattice mismatch, and the built-in charges by a self-consistent multiband-envelope function calculation.[4] We have solved this problem for the sample geometries, barrier composition and doping profiles as given in Ref. 2 with bulk band parameters (Luttinger parameters γ_1, γ_2, γ_3, κ, deformation potentials) well known for Ge and Si.[5] For the mixed crystal system we have made a nonlinear interpolation following the concepts of Lawaetz[6] in view of qualitative ab-initio calculations for Si_xGe_{1-x}.[7] Thus our calculation is parameter-free, but there exist some uncertainties on the experimental side concerning the width of the *p*-channel and a partial relaxation of the biaxial strain.[8] Our results show a pronounced nonlinear dependence of Landau levels and crossings of pairs of spin-states and provide a remarkably good agreement with the experimental results.

2. The theoretical concept

The subband states in a *p*-Ge channel between barriers of Si_xGe_{1-x} with different composition on either side derive from the topmost bulk valence bands, which can be properly described (including biaxial strain) by a 6×6 **k·p** Hamiltonian

$$\mathcal{H}_{6\times6} = \mathcal{H}_k + \mathcal{H}_\varepsilon\,, \tag{1}$$

with

$$
\mathcal{H}_k = \begin{pmatrix}
P+Q+Z_{3/2} & S & R & 0 & -S/\sqrt{2} & -R\sqrt{2} \\
S^* & P-Q+Z_{1/2} & 0 & R & Q\sqrt{2} & S\sqrt{3/2} \\
R^* & 0 & P-Q+Z_{-1/2} & -S & S^*\sqrt{3/2} & -Q\sqrt{2} \\
0 & R^* & -S^* & P+Q+Z_{-3/2} & R^*\sqrt{2} & -S^*/\sqrt{2} \\
-S^*/\sqrt{2} & Q\sqrt{2} & S\sqrt{3/2} & R\sqrt{2} & P+\Delta_0+Z'_{1/2} & 0 \\
-R^*\sqrt{2} & S^*\sqrt{3/2} & -Q\sqrt{2} & -S/\sqrt{2} & 0 & P+\Delta_0+Z'_{-1/2}
\end{pmatrix}, \quad (2)
$$

where

$$
\begin{aligned}
P &= \frac{\hbar^2}{2m}\gamma_1(k_x^2+k_y^2+k_z^2), \qquad Q = \frac{\hbar^2}{2m}\gamma_2(k_x^2+k_y^2-2k_z^2) \\
Z_M &= 2\mu_B B(\kappa M + qM^3), \qquad Z'_M = 2\mu_B B\,\kappa M \\
R &= -\frac{\hbar^2}{2m}\sqrt{3}\left(\gamma_2(k_x^2-k_y^2)+2i\,\gamma_3\{k_xk_y\}\right), \quad S = -\frac{\hbar^2}{m}\sqrt{3}\,\gamma_3\,(k_x-ik_y)k_z
\end{aligned} \quad (3)
$$

and $\mathcal{H}_\varepsilon$, the strain Hamiltonian, being the same expression but with

$$
P_\varepsilon = -D_d\, tr\varepsilon, \qquad Q_\varepsilon = \frac{1}{3}D_u(\varepsilon_{xx}+\varepsilon_{yy}-2\varepsilon_{zz}). \quad (4)
$$

and the other terms set equal to zero. The Luttinger parameters γ_1, γ_2, γ_3, κ, q in (3) and the deformation potentials D_d and D_u as well as the components of the strain tensor depend on z according to the different materials of the layered structure.

The samples are δ-doped in the upper barrier with a spacer of width d_s between the doping layer and the p-channel. The hole concentration is taken as determined by magneto-transport experiments at low temperature.[2] The corresponding Hartree potential $V_H(z)$, which is to be superimposed to the valence and spin-split-off band with their offsets, is calculated by a self-consistent solution of the 6-band envelope function problem and Poissons equation for zero magnetic field. This is done in k-space by solving integral equations with quadrature methods.[4] We take into account the lowest 13 subbands of which 8 can be classified according to their dominant spinor component as HH and 5 as LH.

For finite magnetic field we introduce Landau raising and lowering operators[3]

$$
a^+ = \frac{\lambda_c}{\sqrt{2}}(k_x+ik_y), \quad a = \frac{\lambda_c}{\sqrt{2}}(k_x-ik_y) \quad (5)
$$

according to which each subband splits into a fan chart of Landau levels (see e.g. Fig. 1). In diagonalizing the Hamiltonian for $\vec{B} \neq 0$ we consider 19 Landau levels in each of the 13 subbands to obtain convergence for the Landau levels shown in Fig. 1.

3. Results and Comparison with Experiment

The Landau levels emerging from the two lowest heavy hole subbands (Fig. 1) show a nonlinear dependence on the magnetic field characteristic for nonparabolic bands. For the lowest levels the Landau level quantum number ($N = 0,1,2$) and the hole spin ($\pm$ for $\pm\frac{3}{2}$) of the dominant eigenvector component are indicated. The crossing of spin up and spin-down levels and the anticrossings for some of the higher levels are a result of subband coupling in the 6×6 Hamiltonian. The heavy solid line indicates the position of the Fermi level. The energies for dipole-allowed transitions are calculated from the spectrum and converted into cyclotron masses $m_{CR} = \frac{\hbar eB}{\Delta E}$. By taking into account the occupation factors defined by the magnetic field dependent Fermi energy we find the data shown in Fig. 2 with the transitions between pairs of Landau levels as indicated. For low magnetic field the cyclotron masses converge to separate values for m_{CR} of about 0.13 and 0.15, which correspond to transitions $N\pm \rightarrow N+1,\pm$ between pairs of spin-up or spin-down levels. This spin-splitting of cyclotron resonance is a consequence of the asymmetry of the quantum well. For increasing magnetic field the dipole-transitions (and corresponding cyclotron masses) show a more complex behavior related to the irregular slopes of Landau levels with small N (Fig. 1) as initial and final states for dipole transitions. This situation resembles the quantum resonances investigated some time ago in p-Ge under uniaxial strain.[9]

Using Fig. 2 we have generated absorption spectra by assigning to each dipole-transition a Lorentzian with phenomenological broadening but taking into account dipole-matrixelements and occupation factors (Fig. 3). The results of Fig. 1–3 have been calculated for sample 2 of Ref. 2, with a p-channel width of 168 Å, spacer thickness $d_s = 100$ Å, hole concentration $p = 11.5 \cdot 10^{11}$ cm^{-2} and composition $x = 0.6$ (0.7) in the lower (upper) barrier. The channel width of 168 Å was obtained from TEM micrographs and differed from the nominal width of 125 Å.[2] As we learned in the course of our calculations, the strain is partially relaxed to an unknown extend. We have assumed 65% of the maximum strain due to lattice mismatch.

The calculated cyclotron masses (Fig. 2) and transmission spectra (Fig. 3) reproduce the essential features of the experimental spectra and show quantitative agreement with the low field cyclotron masses from the spin-split heavy hole Landau levels. The additional cyclotron resonance at high magnetic fields can be clearly assigned to a $0^- \rightarrow 1^-$ transition, although its mass (transition energy) turns out to be too small (large) in comparison with the experimental finding. This transition is not found in samples with narrower p-channels.

References

1. G. Abstreiter, Physica Scripta Vol. T49 (1993) 42

2. C.M. Engelhardt, D. Többen, M. Aschauer, F. Schäffler, G. Abstreiter, and E. Gornik, Solid State Elelctronics **37** (1994) 949
3. H.-R. Trebin, U. Rössler, and R. Ranvaud, Phys. Rev. **B20** (1979) 686
4. R. Winkler and U. Rössler, Phys. Rev. **B48** (1993) 8918
5. Landolt-Börnstein, New Series III/22a Semiconductors, ed. O. Madelung, Springer Berlin, Heidelberg 1987
6. R. Lawaetz, Phys. Rev. **B4** (1971) 3460
7. M. Rieger and P. Vogl, Phys. Rev. **B48** (1993) 4276
8. C.M. Engelhardt, private communication
9. J.C. Hensel and K. Suzuki, Phys. Rev. **B9** (1974) 4219

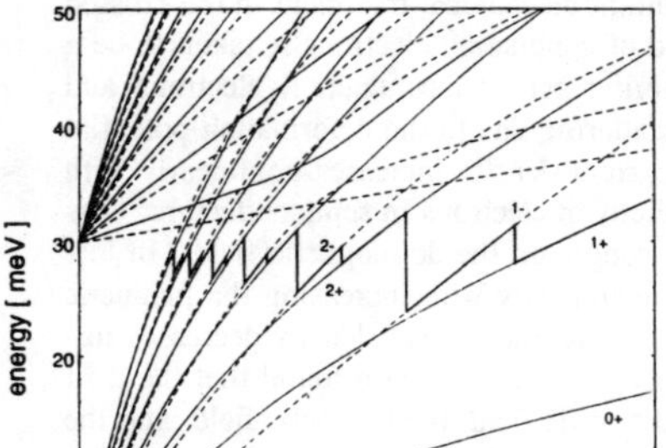

Figure 1: Fan chart of Landau levels deriving from the two lowest heavy hole subbands for the sample 2 of Ref. 2. The heavy solid line is the Fermi energy.

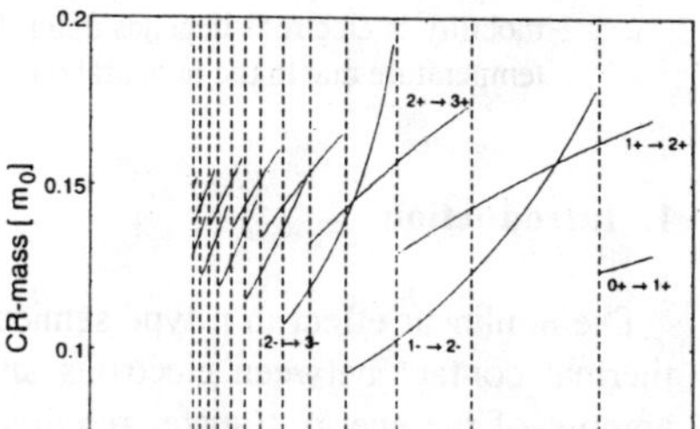

Figure 2: Cyclotron masses calculated from dipole transition energies of Fig. 1. The transitions are indicated by refering to the dominant components of the eigenvectors.

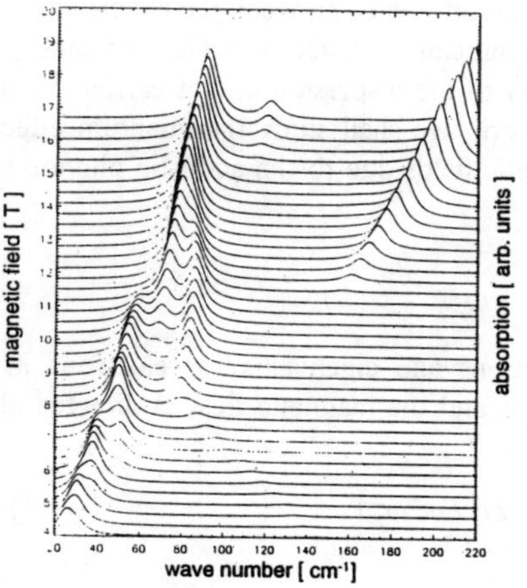

Figure 3: Absorption spectra for different magnetic fields calculated with transition energies from Fig. 1 and dipole matrixelements.

QUANTUM EFFECT OF HIGH DC MAGNETIC FIELDS ON HOT-ELECTRON SCATTERING BY ACOUSTIC PHONONS IN n-TYPE GERMANIUM

CHHI-CHONG WU [1] and CHAU-JY LIN [2]

[1] *Institute of Electronics,* [2] *Department of Applied Mathematics*
National Chiao Tung University
Hsinchu, Taiwan

ABSTRACT

The quantum effect of a dc magnetic field on hot-electron scattering in n-type germanium has been investigated using a quantum mechanical treatment in the crossed electric and magnetic fields. The energy band of conduction electrons is taken to be a nonparabolic model and the dominant scattering between conduction electrons and lattices is coming from the acoustic phonon scattering due to the deformation-potential coupling. Results show that the dissipative current density increases nonlinearly with the applied electric field due to the nonparabolicity of electrons in semiconductors. It is also shown that the current density depends strongly on the dc magnetic field. In low temperatures the current density decreases monotonically with increasing the magnetic field, while in high temperatures such as T = 300 K, the current density decreases initially and then increases slowly with the magnetic field. We also found that the drift mobility of electrons depends upon the applied electric field, the magnetic field, and the temperature due to the nonparabolicity of electrons in semiconductors.

1. Introduction

The nonlinear effect in n-type semiconductors at high electric fields will occur when the thermal contact between electrons and the host lattice can no longer handle the large amount of the energy transfer required to keep electrons in the thermal equilibrium with the host lattice. As the electric field increases, the difference between the electron temperature and the lattice temperature becomes significant. A factor which influences the nature of the current flow is the nonparabolicity of the dispersion law of carriers in III-V compound semiconductors.[1] In the present work, we shall study the quantum effect of hot electrons in the crossed electric and magnetic fields due to the acoustic phonon scattering in n-type germanium.

2. Theory

For the nonparabolic model, the eigenfunctions and eigenvalus for electrons in the applied electric field **E** directed along the x-axis and the magnetic field **B** directed along the z-axis are given by

$$\Psi_{kn}(\mathbf{r}) = \exp(ik_y y + ik_z z)\Phi_n\left[x - (\hbar c / eB)k_y + eE / m^* \omega_c^2\right] \qquad (1)$$

and

$$E_{\mathbf{k}n} = -\frac{1}{2}E_g\left\{1-\left[1+\frac{4}{E_g}\left\langle(n+\tfrac{1}{2})\hbar\omega_c+\frac{\hbar^2k_z^2}{2m^*}+\frac{\hbar cE}{B}k_y-\frac{e^2E^2}{2m^*}\right\rangle\right]^{1/2}\right\} \tag{2}$$

respectively, where $m^*=(m_\ell^* m_t^{*2})^{1/3}$ is the effective mass of electrons with the transverse effective mass m_t^* and the longitudinal effective mass m_ℓ^*,[2] E_g is the energy gap between the conduction and valence bands, $\omega_c = |e|B/m^*c$ is the cyclotron frequency of electrons, Φ_n is the harmonic-oscillator wave function, and k_y and k_z are the components of the electron wave vector in the y and z directions, respectively.

As hot electron transport in semiconductors, there exists a scattering between hot electrons and acoustic phonons. From Liouville's theorem

$$\frac{d\boldsymbol{\rho}}{dt}=\frac{\partial\boldsymbol{\rho}}{\partial t}+\frac{1}{i\hbar}[\boldsymbol{\rho},H]=0 \tag{3}$$

where $\boldsymbol{\rho}$ is the density operator. The Hamiltonian H can be written as

$$H = H_o + V_s(\mathbf{r}) \tag{4}$$

where H_o is the unperturbed Hamiltonian which has the eigenvalues as shown in Eq. (2), and $V_s(\mathbf{r})$ is the scattering potential between electrons and acoustic phonons in semiconductors. The current density J_x can be obtained as [1,3]

$$J_x = \frac{2\pi|e|}{\hbar}\sum_{\mathbf{kk}'nn'}(f_{\mathbf{k}'n'}-f_{\mathbf{k}n})\left|\langle\mathbf{k}'n'|V_s(\mathbf{r})|\mathbf{k}n\rangle\right|^2\delta\left(E_{\mathbf{k}'n'}-E_{\mathbf{k}n}\pm\hbar\omega_{\mathbf{q}}\right) \tag{5}$$

where $f_{\mathbf{k}n}$ is the Maxwell-Boltzmann distribution for a nondegenerate semiconductor.

When the dominant interaction between hot electrons and acoustic phonons is of the deformation-potential coupling, $V_s(\mathbf{r})$ is given by [4]

$$V_s(\mathbf{r})=\left(\frac{k_BT}{2\rho v_s^2V}\right)^{1/2}E_d\exp(i\mathbf{q}\bullet\mathbf{r}) \tag{6}$$

where E_d is the deformation potential due to the electron-phonon interaction, v_s is the sound velocity, ρ is the density of material, and V is the volume of the crystal. From Eqs. (1), (2), (5), and (6), one can obtain the current density as

$$J_x = C\left[\sum_{\ell=0}^{\infty}a_\ell^{3/2}\exp(\alpha_\ell)\right]^{-1}\left\{q_m-(k_BT/\hbar v_s)\left[1-\exp(-\hbar v_sq_m/k_BT)\right]\right\}\sum_{n=0}^{\infty}\{a_n^{5/2}\exp(\alpha_n)$$
$$\times[(2n+1)P_n(E,B,T)+(n+\tfrac{1}{2})^2Q_n(E,B,T)]+\tfrac{1}{2}(n+1)(n+2)a_{n+2}^{5/2}\exp(\alpha_{n+2})Q_{n+2}(E,B,T)\} \tag{7}$$

while the drift mobility of electrons can be obained from

$\mu_x = J_x/(n_o eE)$ (8)

where n_o is the electron concentration of semiconductors, $q_m = \pi/a$ is the cutoff Debye wave number with lattice constant a, a_n , C, and α_n are given as follows

$$a_n = \left[1 + 4\hbar\omega_c (n + \tfrac{1}{2}) / E_g\right]^{1/2} \tag{9}$$

$$C = 3n_o c k_B T E_d^2 E / 2\rho v_s^2 V q_m^3 \hbar^3 B \tag{10}$$

$$\alpha_n = -E_g a_n / 2k_B T + m^* c^2 E^2 / 2k_B T B^2 a_n. \tag{11}$$

P_n and Q_n are functions of E, B, and T with the quantum state n.[5]

3. Numerical Results and Discussion

As a numerical example, we consider the hot-electron scattering by acoustic phonons in n-Ge. The relevant values of physical parameters for n-Ge are taken to be [2,6] : $n_o = 2.4 \times 10^{13}$ cm^{-3}, $\rho = 5.3267$ gm/cm^3, $\varepsilon = 16$, $m^* = 0.2224 m_o$ (m_o is the mass of free electron), $E_g = 0.72$ eV, $E_d = 19$ eV, $v_s = 3 \times 10^5$ cm/sec, and a = 5.64613 Å. In Fig. 1, we plot the current density as a function of electric field in n-Ge at B = 3 kG. It is shown that the current density increases nonlinearly with the electric field. In lower temperatures, the current densisty increases monotonically with the electric field. However, at T = 300 K

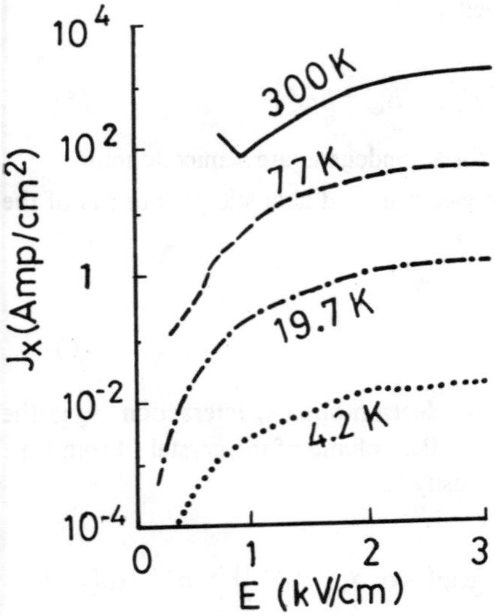

Fig. 1 Current density in n-Ge as a function of E at B = 3 kG.

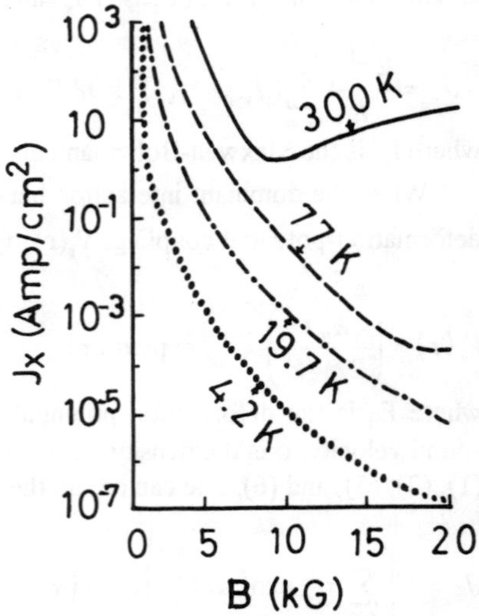

Fig. 2 Current density in n-Ge as a function of B at E = 2.4 kV/cm.

there exists a minimum point in the lower electric field. In Fig. 2, we plot the current density as a function of the magnetic field at E = 2.4 kV/cm in n-Ge. It shows that at low temperatures the current density decreases monotonically with the magnetic field and at high temperatures a minimum point could be observed. In Fig. 3, the drift mobility of electrons is plotted as a function of the electric field in n-Ge. In low temperatures such as T = 4.2 K, the drift mobility increases monotonically with the electric field. As temperature increases, there can be observed a minimum point in the lower electric fields. In Fig. 4, the drift mobility in n-Ge has the same property as shown in Fig. 2 for the current density in n-Ge. As the temperature at T = 300 K, a minimum point can be observed. As temperature decreases, the minimum point will be shifted to high magnetic fields and finally be vanished. From these numerical results, it can be seen that both current density and drift mobility of electrons appear nonlinearly due to the nonparabolicity of electrons.

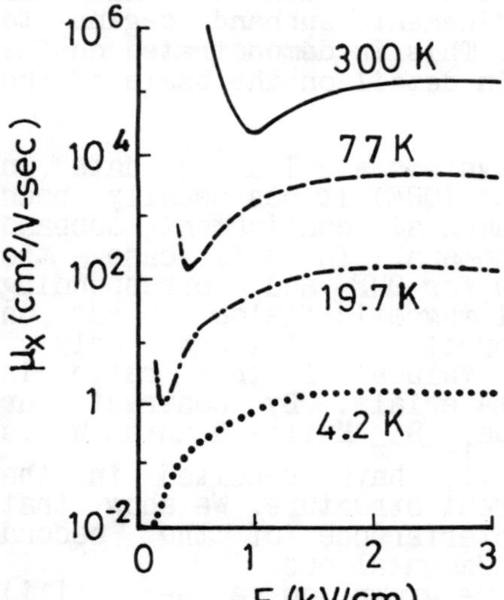

Fig. 3 Drift mobility in n-Ge as a function of E at B = 3 kG.

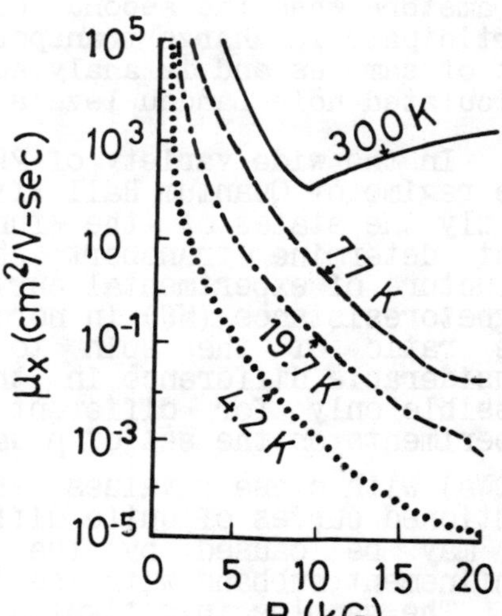

Fig. 4 Drift mobility in n-Ge as a function of B at E = 2.4 kV/cm.

4. References

1. C. C. Wu and C. J. Lin, *Semicond. Sci. Technol.* 7, (1992) **B**367.
2. S. M. Sze, *Physics of Semiconductor Devices* (Wiley-Interscience, New York, 1981).
3. H. F. Budd, *Phys. Rev.* **175**, (1968) 241.
4. B. R. Nag, *Theory of Electrical Transport in Semiconductors* (Pergamon Press, Oxford, 1972).
5. C. C. Wu, NSC Report in Taiwan, China (unpublished 1994).
6. C. C. Wu and C. J. Lin, *J. Appl. Phys.* **73**, (1993) 8319.

THE INFLUENCE OF THE SECOND CONFINEMENT SUBBAND ON THE QHE IN p-Ge/$Ge_{1-x}Si_x$ MULTI-QUANTUM WELLS

Yu.G.Arapov, N.A.Gorodilov, V.N.Neverov, M.V.Yakunin
Institute for Metal Physics, Ural Branch of RAS, S.Kovalevskaya Str.18, Ekaterinburg GSP-170, 620219 Russia

and

A.L.Chernov, O.A.Kuznetsov, L.K.Orlov, R.A.Rubtsova
NIFTI, Nizhny Novgorod State University, 603600 Russia

We show that the structure of experimental curves for the Quantum Hall Effect and corresponding magnetoresistance of p-Ge/$Ge_{1-x}Si_x$ multi-quantum wells is changed radically and becomes very sensitive to the well width and material parameters when the second confinement subband begins to participate in charge transport. This is demonstrated on the set of samples and is analyzed in detail on the basis of the calculated hole Landau levels.

In the wide variety of 2D systems studied to date in the regime of Quantum Hall Effect (QHE) it has usually been solely the states of the fundamental confinement subband that determine transport phenomena. In this case the structure of experimental curves for QHE and corresponding magnetoresistance (MR) in normal magnetic fields depends on the ratio of the spin to orbital splittings only[1]. Considerable difference in the values of this ratio is possible only for different materials. By contrast our experiments on the set of p-Ge/$Ge_{1-x}Si_x$ Multi- Quantum Wells (MQWs) with close x values ($x\cong 0.1$) have resulted in the mentioned curves of quite different structure. We show that it may be caused by the interference of the second confinement subband with the fundamental one.

The samples investigated were grown on a Ge (111) substrate by the Chemical Vapor Deposition method[2]. The widths of the alloy and Ge layers were approximately the same within one sample. The middle parts of $Ge_{1-x}Si_x$ layers were Boron doped leaving spacers approximately 1/4 thick of these layers on both sides. Due to the lattice mismatch the structure is strained and the strain is mainly located in the Ge layers. Sample characteristics are described in the Table along with some evaluated parameters. The hole concentrations $p=(3\div 8)\cdot 10^{11}\,cm^{-2}$ as determined from the Hall effect and SdH oscillations. X-ray and Raman measurements gave the average composition and the strain along the structure. The strain and Fermi energy E_F were also estimated from the hole masses deduced from the temperature dependence of the weak field SdH oscillation amplitudes[3].

Sample	Number of periods	Hall mobility(4.2K) $cm^2/V \cdot s$	Well width W,Å	Strain 3D gap δ,meV	E_F,meV	pW^2	$\frac{E_F-E_1}{E_2-E_1}$
1006-1	90	14000	125	20	8.0	0.4	0.3
1124b3	27	10000	200	8	7.0	1.0	0.7
1003-2	90	15000	220	21	7.2	1.4	1.0
1123a6	15	14000	235	10	7.0	1.4	1.0

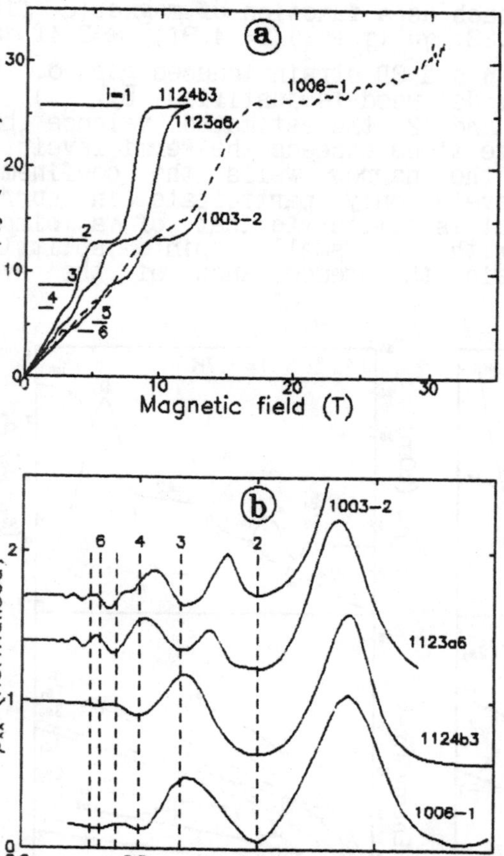

Fig.1: Comparison of magnetotransport results for various samples

QHE and MR were measured on the double cross samples in pulsed magnetic fields up to 35T and up to 13T in stationary fields at T=0.1÷4.2K. The Hall resistances calculated from the measured data to give the result for one period reveals distinct QHE plateaux $\rho_{xy}= h/ie^2$ for i=1,2 within an accuracy of several percents for each sample (Fig.1a). The correctness of the identification of the i=1 plateau is also proved by the lack of other plateaux in much higher fields. The obtained coincidence of the number of 2DHG sheets with the number of periods means that (1) all Ge layers participate in the current transfer equivalently and (2) a single 2DHG is connected with the Ge layer rather than two 2DHGs that might be awaited in triangular deeps near each potential wall due to the well bottom bending.

The curves below the i=2 plateau have quite diverse structure. For the samples with the narrowest layers (1006 and 1124) the next to the plateau with i=2 in the lower fields comes the one with i=4 and other even-numbered plateaux dominate. But for the samples with the wider layers the distinct plateau with i=3 exists while the i=4 plateau is absent. This difference is still more pronounced in the diagonal magnetoresistance ρ_{xx} which exhibits the same succession of the relative minima amplitudes as the corresponding ρ_{xy} plateau widths (Fig.1b).

Using a simple model of an infinitely deep square quantum well[4] we have calculated the hole Landau levels in the Kohn-Luttinger approach as a function of magnetic field for Ge parameters γ_1=13.38, $\gamma\equiv(\gamma_2+\gamma_3)/2$= 4.97, æ=3.41 and for various well widths W and 3D strain induced gaps δ. The simplicity of the well model used is justified by (1) the single 2DHG in the well and (2) the estimated valence band offset that at least five times exceeds the Fermi level[2].

For samples with the narrow wells the confinement fundamental subband levels only participate in current transfer and their layout is similar to that of a simple nondegenerate band with a small spin splitting[1] (Fig.2,left). This explains the predominance of the even-numbered peculiarities.

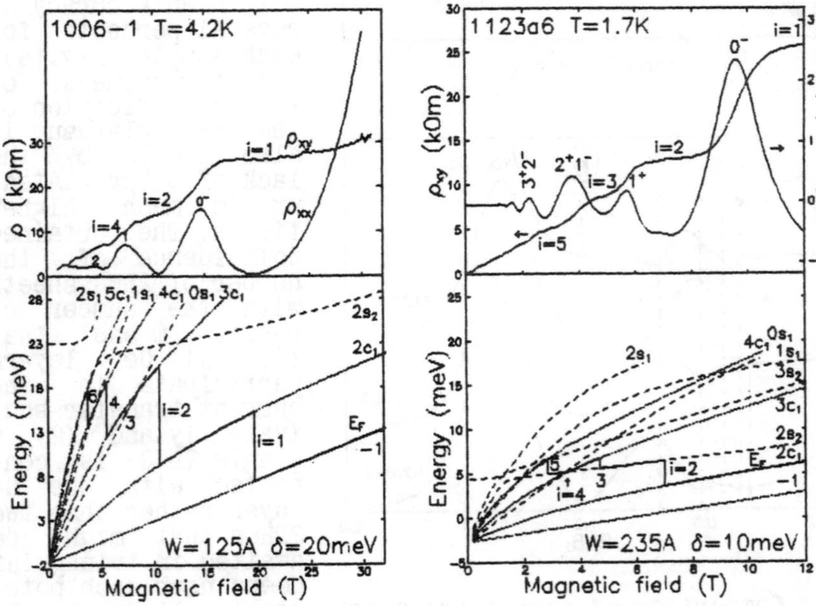

Fig.2: Magnetotransport data and hole Landau levels

For samples with the wide enough wells the Fermi level reaches the area where the second subband magnetic levels interfere with those of the fundamental subband (Fig.2,right). This results in a shift of the peculiarities' numbering while the occurring magnetic level crossings lead to the complications in the experimental data. If the Fermi energy transits between two levels in the vicinity of such a crossing then the corresponding peculiarity drops out from the experimental data since the localized states between these two levels vanish (see i=4). The trace of the Fermi energy motion as a function of field indicates that its step heights do correspond to relative peculiarity amplitudes on the experimental curves confirming that our simple model used in the calculations is sufficient to describe the main features.

The values of pW^2 and of the ratio of the Fermi energy to the second confinement level energy $(E_F-E_1)/(E_2-E_1)$ that characterize the participation of the second confinement subband in charge transport are also given in the Table. The comparison of these values with the structure of experimental curves shows a good correlation.

Since the energy distances between the confinement subbands depend strongly on the potential well shape and width and on the material parameters the structure of the curves measured is highly sensitive to these parameters provided that the second subband participate in transport. In this case the fitting of the calculated relative peculiarity amplitudes to the experimental data may be used to control these parameters. To illustrate this the initially absent peculiarities (e.g.i=4 on Fig.2,right) might be revived by different ways: by the well width diminishing that would move the second subband up, by the Fermi level shift with the carriers generation (irradiating of the sample), by tilting the magnetic field etc.

Acknowledgements: This work is partly supported by the ISF(Soros) and Russian Fund for Fundamental Investigations. We thank A.Germanenko, G.Harus, G.Minkov, N.Shelushinina and I.Tsidilkovski for helpful discussions.

References

1. R.W.Martin, R.J.Warburton, R.J.Nicholas *et al.*, *Proc.20th Int. Conf. Phys. Semiconductors* 1990 Greece, v.**3**, p.909.
2. B.A.Aronzon, N.K.Chumakov, J.Leotin, J.Gallibert *et al.*, *Superlatt. and Microstruct.* **13**, 159 (1993).
3. Yu.G.Arapov, N.A.Gorodilov, O.A.Kuznetsov, *et al.*, *Fizika i Tehnika Poluprovodnikov (Rus.)* **27**, 1165 (1993).
4. A.V.Germanenko, G.M.Minkov, E.L.Rumyantsev, O.E.Rut, *Adv. Mater. Opt. Electron.*, **3**, 57 (1994).

THE NOVEL STATE OF A TWO-DIMENSIONAL ELCTRON GAS NEAR SILICON SURFACE IN HIGH MAGNETIC FIELDS

G.M.Gusev[1,2] , Z.D.Kvon[1]
[1]Institute of Semiconductor Physics, Novosibirsk, Russia
[2]Instituto de Física de São Carlos, Universidade de São Paulo, SP, Brasil

V.T.Dolgopolov, A.A.Shashkin
Institute of Solid State Physics, Chernogolovka,Russia

The anomalous behaviour of diagonal ρ_{xx} and Hall ρ_{xy} components of the resistivity tensor of a two-dimensional (2D) electorns near the high index silicon surface tilted from the exat (100) surface has been investigated in magnetic fields 10T-16T and temperatures T=1.5K-0.02K. The observation of features of ρ_{xx} and ρ_{xy} of this system was reported for a first time in our paper[1]. It was shown that these features arose for for electron concentrations $N_s > N_s^{\Delta}$ (N_s^{Δ} - electron concentration when minigap begins) and were caused by the transformation of energy spectrum of 2D electrons in hihg magnetic fields owing to influence of the one-dimensional (1D) superlattice potential. However measurements in [1] was carried out only at T>1.5K and couldn t get the complete information about the transformed Landau level and related transport properties.

The experimental samples were MOS-transistors fabricated on on silicon surfaces tilted at 9.5° from the exact (100) surface around [011] direction. The maximum mobility at 1.5K was 2 10^3 cm^2 /Vs.

Fig. 1 show the dependencies of ρ_{xx} and ρ_{xy} versus the Landau level filling factor ν for 16T and for T=0.02K. One can see that for $N_s < N_s^{\Delta}$ ρ_{xx} and ρ_{xy} exibit the usual behaviour observed in 2D electron gas near Si surface in the quantum Hall regime: wide minima in ρ_{xx} , where it equals zero, and plateaus in $\rho_{xy}(\nu)$ dependence corresponding to gaps between neighboring Landau levels and spin splitting, and much weaker valley-splitting minima. However, quite other picture is observed for $N_s > N_s^{\Delta}$ - the spin-splitting minima in $\rho_{xx}(\nu)$ and plateau in $\rho_{xy}(\nu)$ almost disappears whereas the valley splitting minima becomes deeper. With the further increasing of ν maximum arises instead of a spin-splitting minima. Simultaneously the plateau in the ρ_{xy} is replaced by a wide minimum. Comparative analyses of the results shows that the value of this minima depends on the behaviour of ρ_{xx}: when the depth of the spin splitting minima is comparable witn valley splitting minima the minimum in the Hall component of the resistivity is absent. With decreasing of spin-splitting minima depth the minimum in ρ_{xx} appears and reaches the largest when the spin splitting minima is replaced by a maximum. The same behaviour is observed in ρ_{xx} and ρ_{xy} as functions of the magnetic field B .

The temperature dependence of $\rho_{xx}(B)$ is absent in the anomalous "spin" maximum and this dependence is relatively weak in valley-splitting minima. Insert to Fig.1 shows ρ_{xx} versus T in one of the same valley minima. One can see for T>0.4K the ordinary activation dependence is observed, but for T<0.4K activated energy begins to decrease and for T<0.1K pratically metallic behaviour of ρ_{xx} sets in.

We can make three conclusions from the results discribed above. First, for $N_s > N_s^{\Delta}$ Landau level of 2D electrons near high index Si surface are transformed and is not determined as the level consisting of four sublevels which separated by spin and valley splitting, because superlattice potential entangles these sublevels. It results in strong decreasing of spin splitting and its total disappearance with increasing of minigap. It should be noted that the value of minigap increses with increasing of electron concentation. Second, the states of the transformed Landau levels have a quite other character than the ordinary Landau level states in IQH regime. In a usual situation Landau level states are divided into localized (when Fermi level is placed between Landau sublevels) and delocalized (when Fermi level crosses Landau subleves). But in our case as the temperature dependence of ρ_{xx} in fig.1 shows all states of the transformed Landau levels are delocalized. The behaviour of ρ_{xy} is anomalous and is not described by the common equation $\rho_{xy} = B/e*c*N_s$. Discord between this equation and the experimental value of ρ_{xy} in the minima reachs 30%. It indicates the anomalous properties of delocalized states of the transformed Landau level. They are likely to result from the formation of the "hole" Fermi surface in 2D multivalley electron system under the influence of the high magnetic field and the potential of 1D superlattice. Thus, the described experimental results enable to make the conclusion that the novel state of a 2D electron gas near the Si hihg index surface in high magnetic fields, arisen under the influence of 1D superlattice, performs unusual properties requiring further experimental and theoretical reseaches.

G.M.G.is supported by FAPESP (Brazil).

Reference

1.. Gusev G.M., Kvon Z.D., Cheremnykh P.A. J.Phys.C, **18,** (1985), L1083

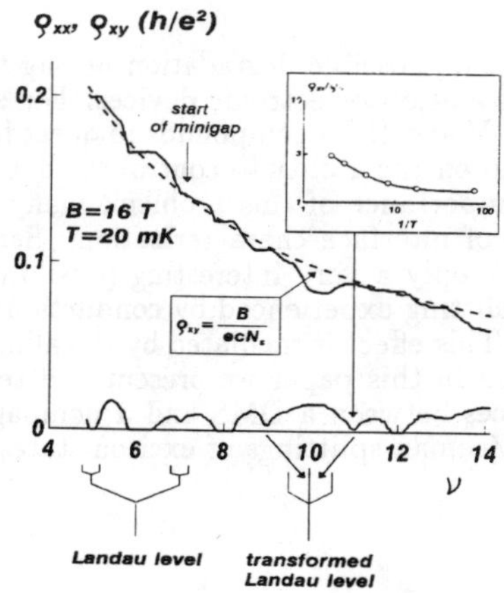

Fig.1 The dependence of ρ_{xx} and ρ_{xy} on filling factor of Landau levels ν.

INTERFACE CHARACTERIZATION BY ZEEMAN EFFECT IN STRUCTURES WITH SEMIMAGNETIC SEMICONDUCTORS

J.A. GAJ[1], J. CIBERT[2], G. FEUILLET[3], W. GRIESHABER[2], G. KARCZEWSKI[4], P. KOSSACKI[1], Y. MERLE d'AUBIGNÉ[2], NGUYEN THE KHOI[1], A. WASIELA[2], and T. WOJTOWICZ[4]

[1] *Institute of Experimental Physics, Warsaw University, 69 Hoża, 00-681 Warszawa, Poland*

[2] *Laboratoire de Spectrométrie Physique, Université Joseph Fourier Grenoble I – CNRS, BP87, 38402 Saint Martin d'Hères, France*

[3] *Département de Recherche Fondamentale - Matière Condensée, SP2M-PSC Centre d'Etudes Nucléaire de Grenoble, BP85, 38041 Grenoble, France*

[4] *Institute of Physics, Polish Academy of Sciences, Al. Lotników 32/46, 02-668, Warszawa, Poland*

ABSTRACT

We describe a method of interface characterization based on Zeeman effect measurements of exciton states confined in structures with barriers containing magnetic ions. It enables one to study the shape of the interface profile, the influence of the growth conditions or thermal treatment on the interface and provides a perspective to decouple the influence of interface mixing from truly 2D magnetic effects in studies of ultra thin magnetic layers. Influence of the parameters of the model on the obtained results and practical applicability of the method are discussed.

1. Introduction

The quality of interfaces and its possible degradation belong to the key issues in development of new electronic and optoelectronic devices. For example in the current competition between the III-V and II-VI compounds in quest for production of a blue laser the success may depend on the ability to control the degradation of laser structures. Understanding the importance of this problem, many researchers constantly look for new efficient ways of interface characterization. Semimagnetic or Diluted Magnetic Semiconductors[1] supply a very interesting possibility to study interfaces. It is related to giant spin splitting experienced by conduction- and valence electrons in external magnetic field[2]. This effect is mediated by the alinement of magnetic ions present in those compounds. In this paper we present and test a proposed recently method of studying interfaces between a DMS and a nonmagnetic semiconductor[3], based on measurements of Zeeman splittings of exciton states localized in a nonma-

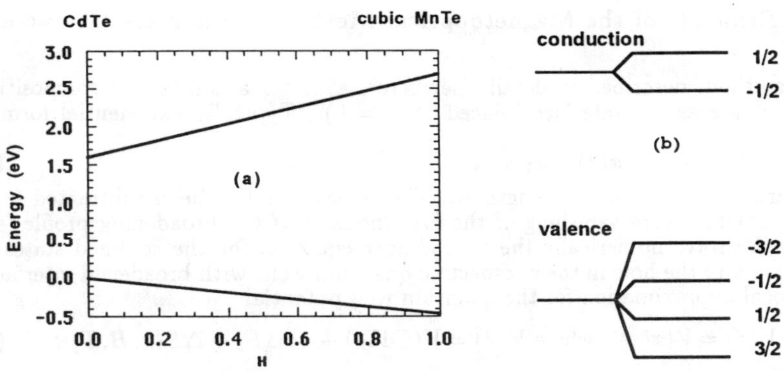

Fig. 1. Energy bands in the CdTe/CdMnTe system (a) and their splitting scheme in magnetic field (b).

gnetic quantum well with DMS barriers. A carrier (exciton) in such a well has been shown to be a perfect tool for interface studies: its Zeeman effect will originate from the interface region, where its wavefunction overlaps with magnetic ions, making possible an efficient influence of their magnetization on the excitonic Zeeman effect. The method has been developed for the CdTe/CdMnTe system but its principle applies to any similar system with magnetic ions present only in the barriers. To illustrate the possibilities of the method, we present some experimental data obtained on suitable molecular beam epitaxy (MBE) grown structures with CdTe wells and CdMnTe barriers. We used samples from two sources: Centre d'Etudes Nucléaire de Grenoble and Polish Academy of Sciences at Warsaw. Experimental details have been given elsewhere[6].

2. CdTe-MnTe System

Zinc blende cadmium manganese telluride single crystals can be obtained by a modified Bridgman method up to about 70% Mn mole fraction and by MBE for any composition. Its energy gap increases with the Mn content; according to recent results (e.g., those of B. Kuhn-Heinrich et al[4]), the gap difference is divided between the valence and conduction bans in the proportion 30/70 (Figure 1). Cadmium manganese telluride is one of the first DMS where giant spin splittings of excitonic states have been observed and explained[5]. In this large gap material a particularly simple model is applicable, where the direct influence of the magnetic field on the effective mass states is neglected and only the giant spin splittings resulting from ion-carrier $sp-d$ interaction are included in a mean field approach. Those splittings are proportional to the macroscopic magnetization of the material. The magnetization itself, resulting from a complicated combination of ion-ion interactions in the external magnetic field, is difficult to predict theoretically with a reasonable precision, and often is treated as a parameter to be determined from the experiment. The resulting splittings of bands are schematically represented in Figure 1b.

3. The Principle of the Magnetooptical Method of Interface Characterization

The method, described in detail elsewhere[6], assumes a continuous composition profile $x(z)$ across the interface (placed at $z = 0$). Typically exponential form is used,

$$x(z) = x_1 + (x_2 - x_1)\exp(-z/l_0 - 1) \tag{1}$$

parameterized by intermixing length l_0. The constant 1 has been subtracted from the exponent to assure vanishing of the first moment of the broadening profile (see Ref. 6). We solve numerically the Schrödinger equation for the confined states of the electron and the hole in their respective quantum wells with broadened interfaces using a local approximation for the quantum well potential

$$V(z) = V(x(z)) \text{ where } V(x) = V(\text{CdTe}) + xa\Delta E_g + \Delta E(x, B, T) \tag{2}$$

where a is the relative band offset (typically 0.7 and 0.3 for the conduction- and valence bands respectively), $\Delta E_g = 1.563\text{eV}$ is the band gap difference between the (extrapolated) cubic MnTe and CdTe and $\Delta E(x, B, T)$ is the Zeeman splitting for the respective band, assumed identical to the bulk value measured for composition x (in the simplest version of the model). Then, including a calculated exciton binding energy, the computed results are compared with experimental data, allowing to fit the intermixing length l_0 and (if necessary) the quantum well width.

4. Bulk Zeeman Splittings

In order to know the bulk Zeeman splitting values , we parameterized a large set of experimental data in the composition range up to $x = 0.67$ (see Ref. 4). The heavy hole exciton splitting as a function of magnetic field can be at our experimental conditions (pumped He temperature, fields up to 5T) represented with a good accuracy by a modified Brillouin function (see Ref. 3)

$$\Delta E_{ex} = \Delta_E^S B^S \left(\frac{g\mu_B S_B}{k_B (T - T_0)} \right) \tag{3}$$

where B^S is the Brillouin function, $S = 5/2$, $g = 2$, while Δ_E^S and T_0 depend on the crystal composition x and obey expressions given in Ref. 4. Using this empirical description, one can compute bulk magnetooptical splittings for any composition up to about 0.7 as a function of magnetic field at pumped helium temperatures. For example Figure 2 shows heavy hole exciton splitting at 1T and 2K as a function of composition.

Especially useful is the splitting related to the Mn mole fraction, i.e., normalized per Mn ion. We may deduce from the curve that by diluting Mn ions from, e.g., 70% to 5%, we increase the contribution of each ion to the Zeeman splitting by about two orders of magnitude! This explains, why the method is so sensitive to the interface broadening: the dilution will increase drastically the Zeeman effect of the confined exciton, which, as it was said before, is sensitive primarily to the Mn ions in the interface region.

5. Influence of the Parameters of the Model on Simulation Results

Figure 3 shows Zeeman splitting calculated for a 20 monolayer CdTe quantum well with $Cd_{0.68}Mn_{0.32}Te$ barriers as a function of magnetic field for various values of

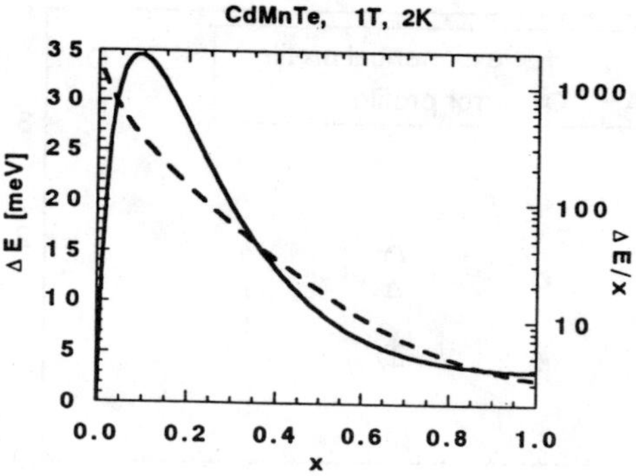

Fig. 2. Heavy hole exciton Zeeman splitting calculated using Eq. 3: absolute – cont. line, and divided by Mn mole fraction x – dashed

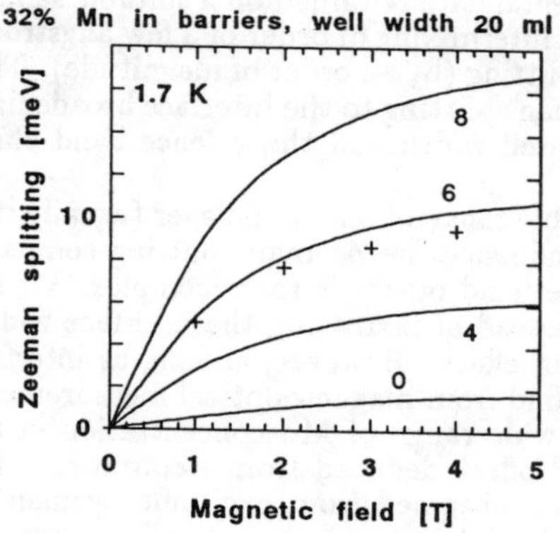

Fig. 3. Zeeman splitting of the ground state of heavy hole exciton as a function of magnetic field: experimental (points) and calculated for indicated values of intermixing length in angstroms (lines)

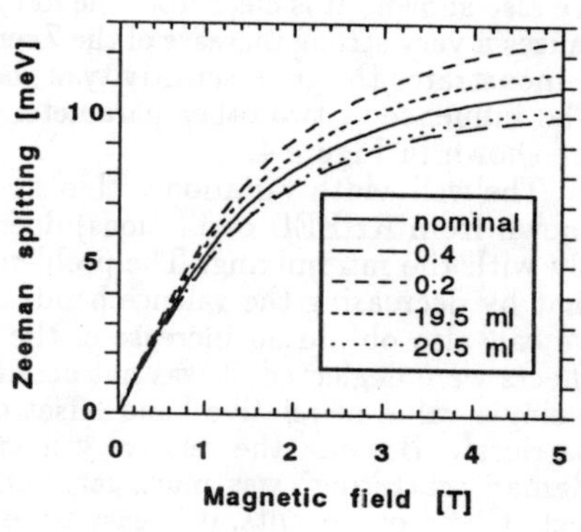

Fig. 4. Zeeman splitting of the ground state of heavy hole exciton calculated as a function of magnetic field. Nominal: 32% Mn in barriers, well width 20 ml, relative valence band offset 0.3, intermixing length 6 Å; other lines show indicated deviations from the nominal values.

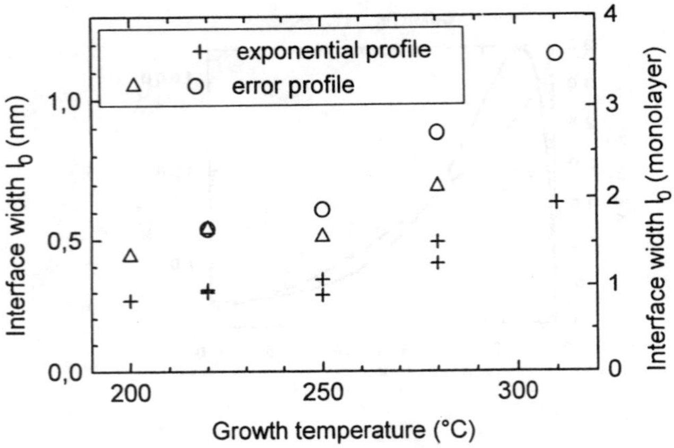

Fig. 5. Intermixing length determined from the Zeeman effect for a series of quantum wells as a function of growth temperature. Barrier composition $x = 0.2$, quantum well width 20 monolayers.

intermixing length. For comparison, experimental data obtained on a suitable sample are also shown. It is clear from the figure that intermixing of order of a few angstroms causes a very strong increase of the Zeeman splitting (by an order of magnitude). This demonstrates the great sensitivity of the Zeeman splitting to the interface broadening. The influence of two other parameters: the well width and the valence band offset are shown in Figure 4.

The well width variation within a reasonable range of one monolayer (usually it is known from RHEED oscillations) does not influence the Zeeman splitting comparably with the intermixing. The problem of the band offsets is more complex. We see that by decreasing the valence band offset instead of increasing the interface width, we can also obtain an increase of the Zeeman effect. However, as long as interface effects were neglected, it was not possible to find from magnetooptical measurements a single value of relative band offset over a wide range of Mn concentration in the barriers[7]. Besides, the relative valence band offset deduced from electron spin flip Raman scattering[8] was much larger than that obtained from excitonic Zeeman effect. Given our results, it is easy to explain the difference: interface mixing strongly enhances the Zeeman splitting of the band states. Depending whether the experiment detects primarily the valence band splitting (excitonic magnetooptics) or the conduction band splitting (spin flip Raman scattering) an analysis neglecting interface effects will tend to artificially decrease the offset of the carrier in question in order to increase its sensitivity to the barrier splitting. Therefore without intermixing one can expect an overestimation of the value of relative valence band offset for spin flip measurements and its underestimation for excitonic Zeeman splitting measurements. We can get rid of this difficulty by estimating the band offsets on the ground of some other measurements, such as light-heavy hole splitting[3].

6. Examples of the Application of the Method.

Figure 5 shows experimental results obtained on a series of samples grown at different temperatures compared to simulation results. The intermixing increases

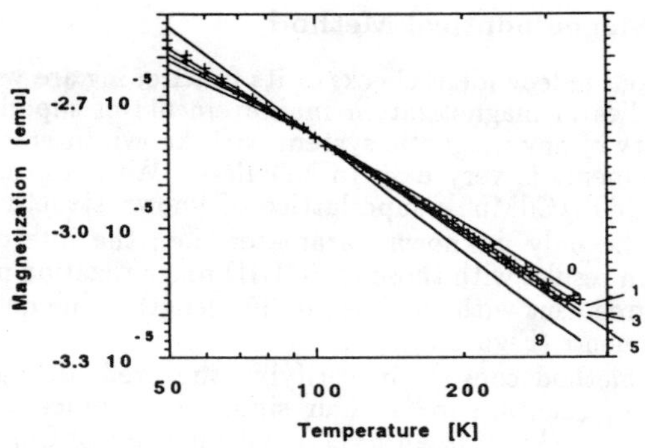

Fig. 6. Magnetization of a CdTe/CdMnTe superlattice calculated for indicated values of intermixing length in Å (lines) compared to SQUID measurements (points). The origin of the y axis chosen arbitrarily for each curve (uncertainty in the diamagnetic contribution from the substrate).

with growing substrate temperature in the MBE process. It is interesting to note that an opposite tendency is observed in the line widths: they are bigger for samples grown at lower temperatures[9] It means that depending on the method, we detect various aspects of the interface imperfection: roughness on the scale of exciton size in the linewidth, or width of the interface composition profile in Zeeman experiments. This shows, how important is using various interface characterization methods to obtain a complete image.

Another example of the application of the magnetooptical method is to study a thermal treatment on the Zeeman splittings of excitons confined in quantum wells[10]. After an annealing by RTP (Rapid Thermal Processing) for 15 sec at a nominal temperature of 450 centigrade, a strong increase of the Zeeman effect is observed, equivalent to an increase of the intermixing width from below 4 Å to about 12 Å. These experiment opens a way for precise studies of influence of RTP processes on interfaces in semiconductor structures.

Our method, when applied to especially designed structures, can also be used to study separately the normal (CdMnTe on CdTe) and the inverted (CdMnTe on CdTe) interfaces[11]. Asymmetric quantum wells were grown for this purpose, where only one barrier (normal or inverted) was of CdMnTe, whereas the other one (respectively inverted or normal) was of nonmagnetic CdZnTe alloy. We found that the Zeeman splitting due to the inverted magnetic interface is much stronger (by a factor of three to four) than that due to the normal interface. These results can be explained using the exponential profile of the same width for both interfaces. The asymmetry comes from the asymmetry of the exponential profile: the tail of low Mn concentration gives a strong enhancement of the Zeeman effect in case of the inverted interface, whereas in the case of the normal one the rapid onset of the exponential profile does not produce such results.

7. New Tests of the Magnetooptical Method

As for any new method, independent checks of its functioning are very important. One of the possibilities lies in magnetization measurements of superlattices. High temperature susceptibility of any magnetic system with known interactions between elementary magnetic moments is very easy to calculate. We performed a series of such calculations for a CdTe/CdMnTe superlattice of known structure and barrier composition[12], varying the only unknown parameter, i.e., the intermixing length. Comparing the simulation results with those of SQUID magnetization measurements, we found a very good agreement with the intermixing length value determined from magnetooptical measurements (Figure 6).

Another test of the method consists in verifying the prediction about Zeeman splitting in Mn-containing quantum wells. Our simulations indicate that the Zeeman splitting in a quantum well of $Cd_{0.86}Mn_{0.14}Te$ with barriers of $Cd_{0.3}Mn_{0.7}Te$ does not depend significantly on the intermixing. To test this prediction we performed magnetoreflectivity measurements on a sample containing three quantum wells: a 12 monolayer CdTe well and two $Cd_{0.86}Mn_{0.14}Te$ wells (12 and 16 monolayers thick) before and after an RTP treatment[10]. The results confirmed fully the tested prediction: the Zeeman splitting in the CdTe well increased after an RTP treatment by about 50%, whereas for the two $Cd_{0.86}Mn_{0.14}Te$ wells remained unchanged with a good precision.

8. Unresolved Problems and Limitations of the Method

The exact shape of the interface profile has not been determined with certainty. At present the exponential form seems to be the most probable, but we cannot exclude corrections in the future. Further studies are necessary to elucidate that question, e.g., using magnetic measurements, ESR techniques, high resolution transmission electron microscopy, etc.

Another problem that has not been discussed here, is the influence of so called intrinsic interface magnetic correction[6]. It is well known that in alloys such as $Cd_{1-x}Mn_xTe$ with large x the paramagnetism is strongly reduced by the short range exchange interaction between the Mn spins (see Ref. 1 and Figure 3). Thus the Mn spins sitting at the CdTe interface having less Mn neighbors should exhibit a larger magnetization than the spins lying deeper inside the CdMnTe layer. A simple way to take those effects into account by introducing an effective composition corresponding to the local average number of magnetic neighbors[6] predicts in case of very thin magnetic barriers the intrinsic magnetic correction comparable to the total measured value. Such estimations seem to work quite well when confronted with experimental data[13]. Nevertheless, more serious studies of interface magnetism are desirable. Systematic studies of the influence of the parameters of the model[6] allowed to determine the most useful range of the quantum well width as up to about 50 Å and that of the barrier compositions as above 0.2.

9. Conclusions and Perspectives

Zeeman effect of excitons confined in quantum wells with DMS barriers has been shown to be very sensitive to the interface broadening. This sensitivity allows to study very small variation of the interface width under influence of thermal treatment, of modification of the growth conditions etc. While the sensitivity of the method has been proven beyond any doubt, some questions remain concerning, e.g., the precise shape of the interface profile, the absolute values of the intermixing length, or the exact contribution of the intrinsic interface magnetic properties due to missing magnetic neighbors of the ions at the interface. The method is obviously limited to ternary systems where one of the components is magnetic – only the CdTe/CdMnTe system has been tested so far. Furthermore, the Zeeman effect based method detects only the interface roughness on the atomic scale (dilution), whereas larger scale roughness may be studied using other methods. It is therefore important to use various methods to obtain a complete characterization of an interface.

It seems that the most important area of future application of the magnetooptical method will be in studies of processes related to thermal treatment such as creation and migration of defects in semiconductor structures.

1 J. K. Furdyna, *J. Appl. Phys.* 64 R29 (1988) and references therein.

2 see e.g. J. A. Gaj in *Semiconductors and Semimetals*, ed. J. K. Furdyna and J. Kossut (Academic, Boston 1988) vol 25 p 275.

3 J. A. Gaj, C. Bodin-Deshayes, P. Peyla, J. Cibert, G. Feuillet, Y. Merle d'Aubigné, R. Romestain and A. Wasiela, Proc. XXI ICPS Beijing 1992, ed. Ping Jiang and Hou-Zhi Zheng, World Scientific 1992, p. 1936.

4 B. Kuhn-Heinrich, W. Ossau, T. Litz, A. Waag and G. Landwehr, to be published.

5 J. A. Gaj, R. Planel, G. Fishman, *Sol. State Commun.* 29, 435 (1979).

6 J. A. Gaj, W. Grieshaber, C. Bodin-Deshayes, J. Cibert, G. Feuillet, Y. Merle d'Aubigné, and A. Wasiela, to be published in *Phys. Rev.* B

7 B. Kuhn-Henrich, W. Ossau, M. Popp, E. Bangert, A. Waag and G. Landwehr, Proc of XXI Int. Conf. Phys. 29. Semicond. Beijing 1992, ed. Ping Jiang and Hou-Zhi Zheng, World Scientific 1992, p. 923

8 M. P. Halsall, D. Wolverson, J. J. Davies, D. E. Ashenford and B. Lunn, *Sol. State Comm.*, 86 15 (1993)

9 W. Grieshaber, C. Bodin, J. Cibert, J. Gaj, Y. Merle d'Aubigné, and A. Wasiela, to be published in *Appl. Phys. Lett.*

10 P. Kossacki, Nguyen The Khoi and J. A. Gaj, G. Karczewski, T. Wojtowicz, E. Janik, A. Zakrzewski, M. Kutrowski and J. Kossut, to be presented at VII Int. conf on Superlattices, Microstructures and Microdevices, Banff 1994

11 W. Grieshaber, C. Bodin, J. Cibert, A. Wasiela, Y. Merle d'Aubigné, J. Gaj and G. Feuillet, to be published in Proc. of MRS Spring Meeting (1994).

12 P. Kossacki et al., to be published

13 W. Grieshaber, J. Cibert, A. Wasiela, Y. Merle d'Aubigné, J. A. Gaj, and G. Feuillet, to be presented at 22th Int. Conf. Phys. Semicond., Vancouver 1994

MAGNETIC POLARONS IN SEMIMAGNETIC QUANTUM WELLS AND SUPERLATTICES

D.R.Yakovlev

Physikalisches Institut der Universität Würzburg, 97074 Würzburg, Germany
and
A.F.Ioffe Physico-Technical Institute, Russian Academy of Sciences, 194021 St.Petersburg, Russia

We review the optical study of exciton magnetic polarons in semimagnetic quantum wells and short-period superlattices based on $Cd_{1-x}Mn_xTe$. A reduction of the dimensionality of the system from 3D to 2D realized by the decrease of the quantum well width is shown to affect strongly the stability of exciton magetic polarons, which is in agreement with theoretical predictions. The modification of the magnetic properties of the Mn-system in the quasi-two-dimensional barrier layers of superlattices is found to be stronger than that at the interfaces of $CdTe/Cd_{1-x}Mn_xTe$ single quantum wells.

1. Introduction

The magnetic polaron (MP) formation is associated with the local magnetic ordering of spins of magnetic Mn-ions due to their exchange interaction with carrier spins. It has been intensively studied in bulk semimagnetic semiconductors[1,2] and the observation of two-dimensional magnetic polarons (2DMP's) in quantum well (QW) structures has been reported[3,4]. In $Cd_{1-x}Mn_xTe$ the exchange field induced by heavy-holes on the Mn-ions exceeds the electron exchange field by several times. As a consequence, for the exciton magnetic polaron formation the structure of the hole wave function is of importance and the electron input can be neglected. Two types of magnetic polarons with respect to the conditions of primary localization of carriers have been considered: (i) bound magnetic polarons if carriers are initially bound to a Coulomb impurity and (ii) free magnetic polarons (FMP) formed from carriers not bound to Coulomb impurities. The stability of FMP's has been investigated by Kasuya et al.[5]. Their analisys shows that the polaron energy is determined by a balance of two terms of opposite sign: a positive kinetic energy due to carrier localization, and a negative energy due to the exchange potential well. Theoretical estimations predict that the stability conditions for FMP's are hardly realized in bulk semimagnetic semiconductors and that the primary localization of carriers has a key importance for the MP stability[6,7]. The analisys of the FMP stability in the systems of different dimensionality (one-, two- and three-dimensional systems have been compared) shows that the reduction of dimensionality strongly favors the FMP formation and that the self-trapped polaron state is always stable in the one-dimensional case[6,7].

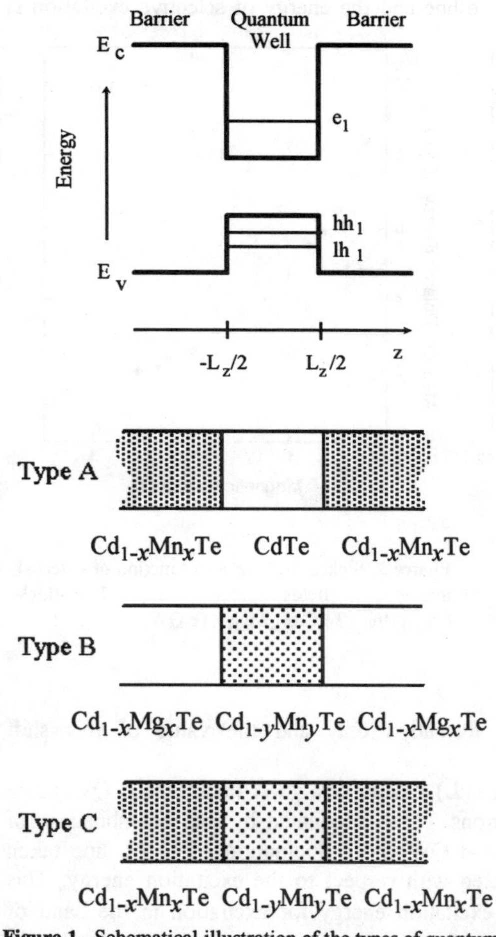

Figure 1. Schematical illustration of the types of quantum well structures under study.

In this paper we review the optical study of exciton magnetic polarons in semimagnetic QW's and SL's. Different types of QW structures ($CdTe/Cd_{1-x}Mn_xTe$, $Cd_{1-y}Mn_yTe/Cd_{1-x}Mg_xTe$, $Cd_{1-y}Mn_yTe/Cd_{1-x}Mn_xTe$) containing Mn-ions in the barriers or/and in well layers allow to study the properties of MP's across the transition from the 3D- to 2D regimes. We show that the reduction of the dimensionality of the system down to 2D favors the MP formation. Properties of 2DMP's in QW's are compared with properties of MP's in SL's with pronounced minibands. Apart from the studies of the intrinsic properties of MP's, MP's could be used as a sensitive tool for the investigation of the magnetic properties of the Mn-system. Evidences of it modification at the interfaces between nonmagnetic and semimagnetic semiconductors have been found in QW's and SL's.

2. Magnetic Polarons in Quantum Wells

We have studied and compared the MP formation in three different types of semimagnetic QW structures grown by molecular-beam epitaxy on (100)-oriented CdTe substrates.[4] These types are specified in Fig. 1, where dark patterns denote the presence of Mn-ions. Structures of $CdTe/Cd_{0.74}Mn_{0.26}Te$ (type A) have nonmagnetic wells embeded in semimagnetic barriers. Structures of types B and C contain semimagnetic wells of $Cd_{0.9}Mn_{0.1}Te$. In the type B structure the wells are confined by nonmagnetic $Cd_{0.62}Mg_{0.38}Te$ barriers and in type C by semimagnetic $Cd_{0.75}Mn_{0.25}Te$ barriers. For all types of structures we have studied a set of single QW's with well widths (L_z) varying between 12 Å and 300 Å. Energies of the exciton magnetic polarons have been determined by the method of selective excitation[3]. Excitons were excited selectively in the band of localized states at energies where the spectral diffusion due to phonon-assisted tunneling does not occur during the exciton lifetime. In this case the Stokes shift between the luminescence line and the energy of selective excitation is

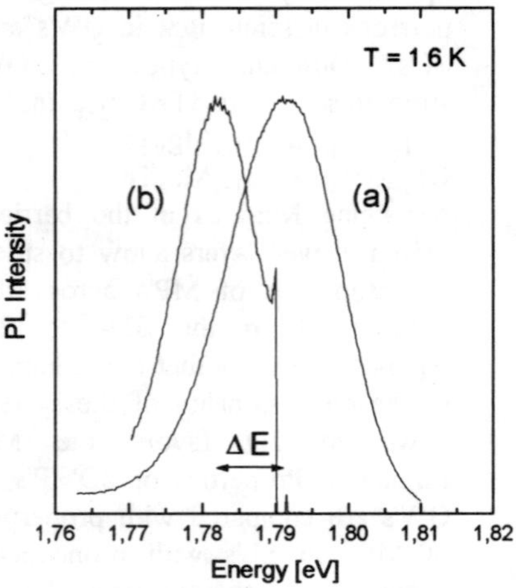

Figure 2. PL spectra of a 75-Å-thick $Cd_{0.9}Mn_{0.1}Te/Cd_{0.4}Mg_{0.6}Te$ QW (type B) taken under nonresonant excitation at 2.41 eV (a) and selective excitation at 1.792 eV (b).

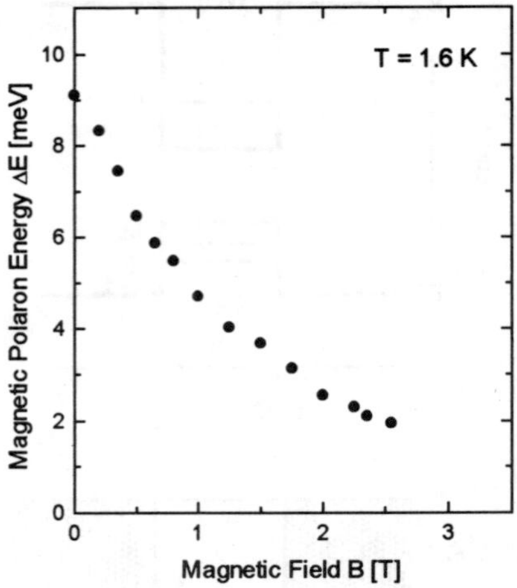

Figure 3. Stokes shift ΔE as a function of external magnetic fields for a 75-Å-thick $Cd_{0.9}Mn_{0.1}Te/Cd_{0.4}Mg_{0.6}Te$ QW.

contributed by the magnetic polaron formation only and the value of this shift corresponds to the MP energy.

Fig.2 displays photoluminescence (PL) spectra of a 75-Å-thick type B QW taken under nonresonant and selective excitations. PL is dominated by the recombination of heavy-hole excitons localized on alloy and QW width fluctuations. The PL line taken under selective excitation is Stokes-shifted with respect to the excitation energy. This Stokes shift ΔE is independent of the excitation energy for excitation in the band of localized states and is effectively suppressed in external magnetic fields (see Fig.3) and with temperature increase up to 15 K. That let us consider the shift ΔE as the MP energy.[3]

In Fig.4 we summarize the MP energies versus well width for different types of QW structures. In type A structures the MP energy is 25 meV in 6-Å-thick QW's and decreases strongly with increasing well width. In this type of QW's only the part of the wave function penetrating into the barriers contributes to the MP formation and this part of the wave function is drastically reduced for thicker QW's. As a consequence the MP energy decreases strongly with increasing well width and MP's are not formed in QW's thicker than 30 Å. The MP energies in type B structures are constant on the level of 9 meV in the range of well widths from 18 to 100 Å, but are not detectable in QW's thicker than 200 Å. In type B QW's only the part of the wave function situated in the well layer

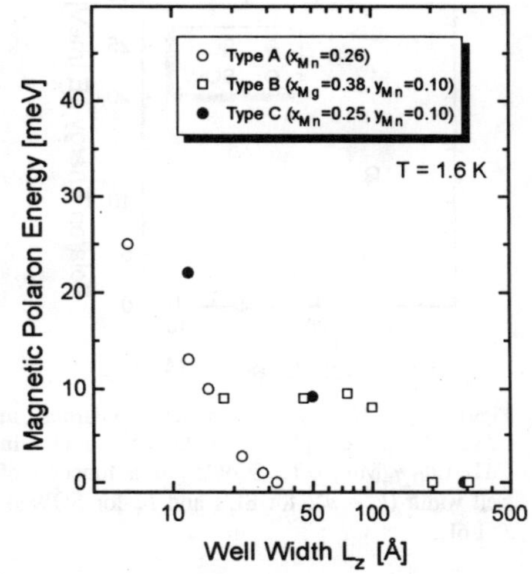

Figure 4. Energy of exciton magnetic polarons as a function of well width in QW structures of different types.

contributes to the MP formation. The part of the wave function penetrating into the barrier is obviously very small since the MP energy in independent of L_z for studied well widths up to 100 Å. In the third type of structures (type C) we have found MP energies of 22 meV and 9.5 meV for QW widths of 12 Å and 50 Å, respectively. Like in the other two types of structures MP's are not formed in a 300-Å-thick QW. In QW's of type C the main features of types A and B are combined. Consequently, the relation of the contributions from the Mn-spins situated in the barriers and in the well should differ strongly for different QW widths. For $L_z > 30$ Å the Mn-spins in the barriers give no contribution to the MP energy due to small penetration of the hole wave function in the barrier layers. In these QW's polarons in type C have the same energies as in type B structures. The large polaron energy of 22 meV found in 12-Å-thick type C QW demonstrate the joint contribution of the Mn-spins from barrier and well layers to the MP energy.

The difference in the MP energies for 100- and 200-Å-thick QW's of types B and C evidences the effect of the reduction of the dimensionality of the system on the stability of magnetic polarons. In the 300-Å-thick QW's of types B and C we have found no indication of MP formation like it has been found in bulk $Cd_{1-x}Mn_xTe$[2], where the content x=0.1 is just on the edge below which no polaron formation is detected. The experimentally established dependence of the MP energy on the dimensionality of the system is in qualitative agreement with the results of numerical calculations[6].

3. Magnetic Polarons in Short-Period Superlattices

In this part we report on the study of magnetic polarons in the minibands of short-period superlattices (SL's), which could be considered as a quasi-3D case. Properties of the superlattice MP's are compared with the properties of the 2DMP's in single quantum wells (SQW's). Experiments have been performed on $CdTe/Cd_{0.76}Mn_{0.24}Te$ superlattices grown by molecular-beam epitaxy on (100)-oriented CdTe substrates after a 0.6-µm-thick buffer layer of $Cd_{0.76}Mn_{0.24}Te$. All superlattices have equal thicknesses of well (l_z) and barrier (l_b) layers, and SL's with periods $d = l_z + l_b = 2l_z$ of 24, 40 and 60 Å have been studied. A structure with a 12-Å-thick $CdTe/Cd_{0.74}Mn_{0.26}Te$ SQW has been used for comparison with the 24-Å-period SL.

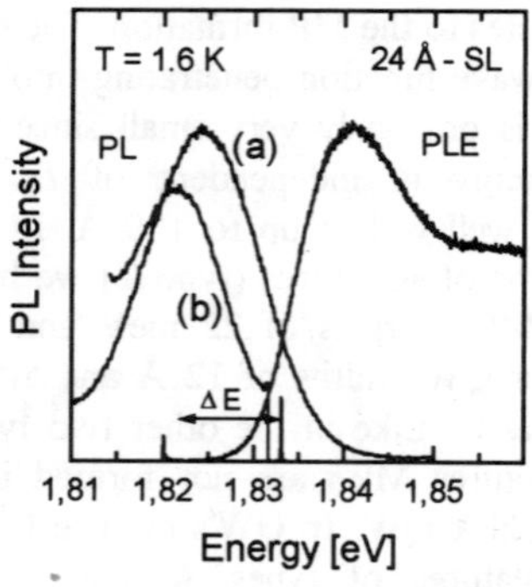

Figure 5. Photoluminescence (PL) and PL excitation (PLE) spectra of a 24-Å-period $CdTe/Cd_{0.76}Mn_{0.24}Te$ SL. Excitation energies: (a) 2.41 eV, (b) 1.833 eV.

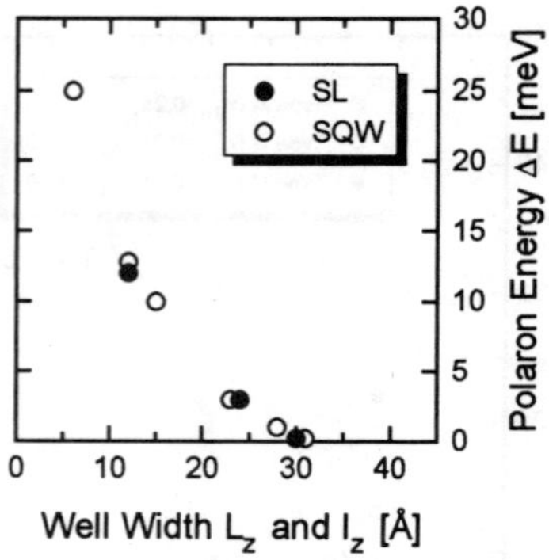

Figure 6. Energy of magnetic polarons in $CdTe/Cd_{0.76}Mn_{0.24}Te$ superlattices and in $CdTe/Cd_{0.74}Mn_{0.26}Te$ SQW's[3] as a function of well width ($l_z = d/2$ for SL's and L_z for SQW's). T=1.6K.

In Fig. 5 photoluminescence and PL excitation (PLE) spectra of a 24-Å-period $CdTe/Cd_{0.76}Mn_{0.24}Te$ SL are shown. The maximum of the PLE spectrum corresponds to the energy of free excitons. The PL line (a) is shifted to lower energies with respect to the PLE maximum reflecting that recombination is dominated by localized excitons. Under selective excitation of localized excitons the PL line exhibits a pronounced shift ΔE=12 meV from the excitation energy (line (b)). The value of the shift is independent of the energy of selective excitation. This shift is fully suppressed in magnetic fields of 3 T and with temperature increase above 10 K (see Figs.7,8). In analogy with results for SQW's[3] we associate the shift ΔE with the energy of the exciton magnetic polaron.

In Fig.6 the MP energies determined for SL's of different periods are shown and compared with polaron energies in SQW's. The polaron energy in the SL's decreases with period increase and is not detectable for $d \geq 60$ Å (i.e. $l_z \geq 30$ Å). That let us conclude that in SL's the hole wave function is not involved into the barrier layers in the process of polaron formation and is centered in the CdTe well layers. The same conclusion has been drawn previously for SQW's[3]. In QW's the holes are affected by the Coulomb potential of strongly confined electrons,[8] which induces additional localization of the holes in the QW layers. Electrons in the SL minibands do not induce the Coulomb potential which enhances the localization for the holes, but even in this case the hole wave function is still centered in the CdTe well layers. We suggest that this is due to a considerable valence band offset (about 30%), which can not been overcome by the exchange energy gained in the polaron formation process.[9]

As the next step we have performed a comparative study of the MP formation in a 24-Å-period SL and in a 12-Å-thick SQW. The polaron dynamics and suppression of the MP energy at external magnetic fields (Fig.7) and temperature (Fig.8) have been measured and the following experimental facts have been found: (i) At 1.6 K the polaron energies in the SL and SQW are very close to each other (12 and 13 meV, respectively);

(ii) The dynamics of the polaron formation determined from the time-evolution of ΔE after a 5-ps excitation pulse[2] is very similar in both structures and gives 75 ps for the MP formation time both in the SL and SQW. Exciton lifetimes determined from the decay of integral PL intensity are 150 ps and 170 ps for the SL and SQW, respectively; (iii) At magnetic fields applied parallel to the structure growth axis (Faraday geometry) MP's in the SL are suppressed two times faster than in the SQW (Fig.7); (iv) The anisotropy of the MP suppression in the fields applied parallel and perpendicular to the structure axis in the SQW is considerably larger than in the SL (Fig.7); (v) The polaron energy in the SL decreases rapidly with temperature increase and disappears at $T > 6$ K, whereas the polaron shift in the QW vanishes for $T > 16$ K only (Fig.8). Generally speaking, the same polaron energies found at 1.6 K in the SL and SQW does not mean that the structure and properties of these MP's are identical.

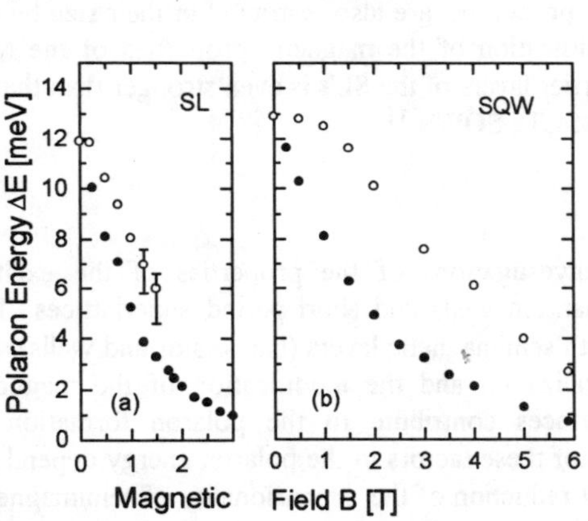

Figure 7. Polaron energy as a function of external magnetic fields applied parallel (closed circles) and perpendicular (open circles) to the structure growth axis for a 24-Å-period CdTe/$Cd_{0.76}Mn_{0.24}Te$ SL (a) and a 12-Å-thick CdTe/$Cd_{0.74}Mn_{0.26}Te$ SQW (b). $T = 1.6$ K.

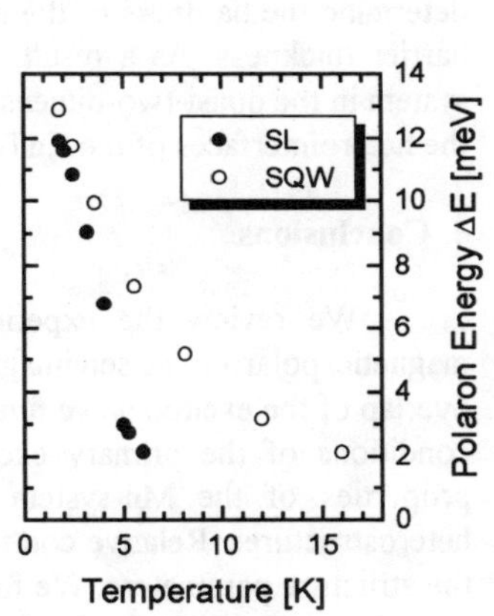

Figure 8. Polaron energy as a function of temperature in a 24-Å-period CdTe/$Cd_{0.76}Mn_{0.24}Te$ SL and in a 12-Å-thick CdTe/$Cd_{0.74}Mn_{0.26}Te$ SQW.

The anisotropic behavior of the MP energy in external magnetic fields is due to the different suppression mechanisms actual for the Faraday and Voigt geometries. In QW's and SL's at zero field the moments of the heavy-hole excitons are oriented along the growth axis since the light-hole states are lifted to higher energies. This results in the orientation of the polaron moments along the structure axis as well. Magnetic fields applied in the Faraday geometry do not change the orientation of the Mn-spins inside the polaron, but align the external Mn-spins and reduce the exchange energy gained by the MP formation. In the Voigt geometry the MP suppression requires: either (a) to reach fields, which are strong enough to provide a magnetic energy of the Mn-spins exceeding the value of their exchange interaction with the heavy-hole spin, or (b) to reach magnetic fields which induce a strong mixing of the light- and heavy-hole states and change the "best" quantization axis of excitons from the initial quantization axis to the field direction.

The relative contributions of the mechanisms (a) and (b) depend on the structure parameters such as the polaron energy, the value of the Zeeman splitting and the value of the light- and heavy-hole splitting. In the compared SL and SQW different anisotropies of the polaron suppression are mainly due to the different values of the splitting between the light-hole and heavy-hole exciton states which are 5 meV and 32 meV in the SL and SQW, respectively (details will be published elsewhere).

Another striking experimental result is that the MP's in the SL are suppressed much more effective in magnetic fields and with temperature increase than in the SQW. It reflects stronger paramagnetic properties of the Mn-spins participating in the MP formation in the short-period SL in comparison with the SQW.[10] In the 24-Å-period SL the barrier layers are 12 Å wide, i.e. 5 monolayers only. That means that the majority of the Mn-ions is located at or next to the interface layers. The large Mn-clusters, which determine the hardness of the magnetic properties, are also restricted in their size by the barrier thickness. As a result, the modification of the magnetic properties of the Mn-system in the quasi-two-dimensional barrier layers of the SL's is even stronger than that at the heterointerfaces of the CdTe/$Cd_{1-x}Mn_xTe$ SQW's.[11]

4. Conclusions

We review the experimental investigations of the properties of the exciton magnetic polarons in semimagnetic quantum wells and short-period superlattices. The overlap of the exciton wave function with semimagnetic layers (barriers or/and wells), the conditions of the primary exciton localization and the modification of the magnetic properties of the Mn-system at interfaces contribute to the polaron formation in heterostructures. Relative contributions of these factors in the polaron energy depend on the structure parameters. We found that reduction of the dimensionality of semimagnetic layer from 3D to 2D favors the magnetic polaron formation, as it has been theoretically predicted. Using magnetic polaron as a local "probe" of magnetic properties of the Mn-system we shown that the modification of these properties in the quasi-two-dimensional barriers of superlattices is even stronger that at the interfaces of SQW's.

Acknowledgements - I am indebted to A.Waag for providing the high quality quantum well and superlattice structures. I would like to thank G.Mackh, R.Hellmann, B.Kuhn-Heinrich, W.Ossau, G.Landwehr, and E.O.Göbel for fruitful collaboration, as well as K.V.Kavokin and I.A.Merkulov for helpful discussions. This work has been supported in part by the Volkswagen Foundation.

References

1. P.A.Wolff, in: *Semiconductors and Semimetals* Vol.25, ed. by J.K.Furdyna and J.Kossut, p.413, Academic Press, London (1988)

2. G. Mackh, W. Ossau, D. R. Yakovlev, A. Waag, G. Landwehr, R. Hellmann, and E. O. Göbel, *Phys. Rev. B* **49**, 10248 (1994)

3. D. R. Yakovlev, W. Ossau, G. Landwehr, R. N. Bicknell-Tassius, A. Waag, S. Schmeusser, and I. N. Uraltsev, *Solid State Commun.* **82**, 29 (1992)

4. G.Mackh, W.Ossau, D.R.Yakovlev, A.Waag, T.Litz, and G.Landwehr, *Solid State Commun.* **88**, 221 (1993)

5. T.Kasuya, A.Yanase and T.Takeda, *Solid State Commun.* **8**, 1543 (1970)

6. C.Benoit a la Guillaume, *Phys.Stat.Sol.(b)* **175**, 369 (1993)

7. A.V.Kavokin and K.V.Kavokin, *Semicond.Sci.Technol.* **8**, 191 (1993)

8. A.V.Kavokin, V.P.Kochereshko, G.R.Pozina, I.N.Uraltsev, D.R.Yakovlev, G.Landwehr, R.N.Bicknell-Tassius, and A.Waag, *Phys.Rev.B* **46**, 9788 (1992)

9. B.Kuhn-Heinrich, W.Ossau, T.Litz, A.Waag, and G.Landwehr, *J.Appl.Phys.* **75**, 8046 (1994)

10. W.J.Ossau and B.Kuhn-Heinrich, *Physica B* **184**, 422 (1993)

11. D.R.Yakovlev, *J. de Physique IV, Colloque C5*, **3**, 67 (1993)

MAGNETIC FIELD TUNABLE CARRIER CONFINEMENT IN $ZnSe/Zn_{1-x}Fe_xSe$ QUANTUM WELLS

T. SCHMIEDEL, National High Magnetic Field Laboratory
Florida State University, Tallahassee, FL 32306

W.Y. YU, and A. PETROU, Department of Physics, SUNY at Buffalo
Buffalo, NY 14260

H. ABAD, and B.T. JONKER, Naval Research Laboratory, Washington DC 20375

J. WARNOCK , IBM Research Division, Thomas J. Watson Research Center,
Yorktown Heights, NY 10598

ABSTRACT

The formation of a spin superlattice in a $ZnSe/Zn_{0.96}Fe_{0.04}Se$ quantum well structure has been investigated from a study of the exciton energies and relative intensities as function of magnetic field. At low fields the lower energy exciton component has a type-II character; at higher fields the trend reverses due to the increased conduction band splitting of the $Zn_{0.96}Fe_{0.04}Se$ barriers and the structure exhibits spin superlattice behavior. The experimental results are described quantitatively by a variational model which incorporates the tunable confining potentials as well as the electron-hole Coulomb attraction.

1. Introduction

Von Ortenberg introduced the term "spin superlattice" to describe heterostructures composed of alternating diluted magnetic semiconductor (DMS) and non-magnetic layers[1]. The bandgaps of the constituent layers are equal within a few meV, and as a result at zero magnetic field no confining potentials due to band discontinuities are present. The application of an external magnetic field causes large (tens of meV) band splittings in the DMS layers[2]; in contrast, the non-magnetic layers have negligible magnetic band splittings. The magnetic field induced band offsets between the magnetic and non-magnetic layers creates spin-dependent potentials which tend to segregate the electrons and holes according to their spin state. The m_j=-1/2 (+1/2) electrons and m_j=-3/2 (+3/2) holes are localized by band discontinuities in the magnetic (non-magnetic) layers of the heterostructure. The energy of the lower σ_+ (-3/2,-1/2) excitonic interband transition exhibits a large red shift as function of magnetic field B because both participating carriers are confined in the DMS layers. In contrast, the energy of the upper σ_- (+3/2,+1/2) exciton component does not depend on B for magnetic fields high enough to localize the m_j=+1/2 electrons and m_j=+3/2 holes in the non-magnetic layers. Both excitonic transitions are type-I i.e. spatially direct and thus have comparable intensities. Spin superlattice behavior as outlined above has been realized in $ZnSe/Zn_{0.99}Fe_{0.01}Se$[3] and $ZnSe/Zn_{0.99}Mn_{0.01}Se$[4] quantum wells. In $ZnSe/Zn_{0.9}Fe_{0.1}Se$[5] structures the band offset at zero field is significant (tens of meV) and most of it is distributed in the conduction band. In this system the magnetic field induced band discontinuities result in a type-II σ_+ (-3/2,-1/2) interband transition with the m_j=-1/2 electron in the ZnSe wells and the m_j=-3/2 hole in the DMS barriers. The upper σ_- (+3/2,+1/2) transition on the other hand remains type-I with both carriers confined in the ZnSe wells.

In this work we have investigated a $ZnSe/Zn_{0.96}Fe_{0.04}Se$ quantum well structure which shows an intermediate behavior. The zero field conduction band offset is large enough so that the lower σ_+ exciton component exhibits type-II character at low magnetic fields. The Coulomb attraction between the m_j=-1/2 electron and the m_j=-3/2 hole prevents a complete type-II transition which would pull the (-3/2,-1/2) exciton apart. As the magnetic field is increased the magnetic splitting of the conduction band in the $Zn_{0.96}Fe_{0.04}Se$ layers overwhelms the zero field offset, and the structure becomes a spin superlattice with the σ_+ and σ_- transitions confined in the $Zn_{0.96}Fe_{0.04}Se$ and ZnSe layers, respectively.

2. Experimental

We have used two samples for this work. Sample 1 is a 100Å/100Å $(ZnSe/Zn_{0.96}Fe_{0.04}Se)_4$ multiple quantum well structure. Sample 2 is a 952 Å $Zn_{0.96}Fe_{0.04}Se$ epilayer that allows us to determine the magnetic band splittings of the DMS layers in sample 1. Both samples were grown by molecular beam epitaxy (MBE) on semi-insulating GaAs (100) substrates from elemental source ovens at 300°C with a rate of approximately 0.25 μm/hr. The thickness and magnetic ion concentration in each sample were measured following growth using x-ray fluorescence techniques. The energies and intensities of the interband transitions were determined from the reflectivity spectra.

3. Results and Discussion

The energies of the σ_+ (-3/2,-1/2) and σ_- (+3/2,+1/2) heavy hole exciton transitions from sample 1 are plotted as function of magnetic field B in fig.1. They exhibit the pronounced asymmetry which has been observed in $ZnSe/Zn_{0.99}Fe_{0.01}Se$ spin superlattices[3] and $ZnSe/Zn_{0.9}Fe_{0.1}Se$ type-I/type-II heterostructures[5]. The lower σ_+ transition has the large red shift which is characteristic of DMS materials. The upper σ_- component on the other hand shows a small blue shift for B<4 tesla; for B>4 tesla its energy does no change with field.

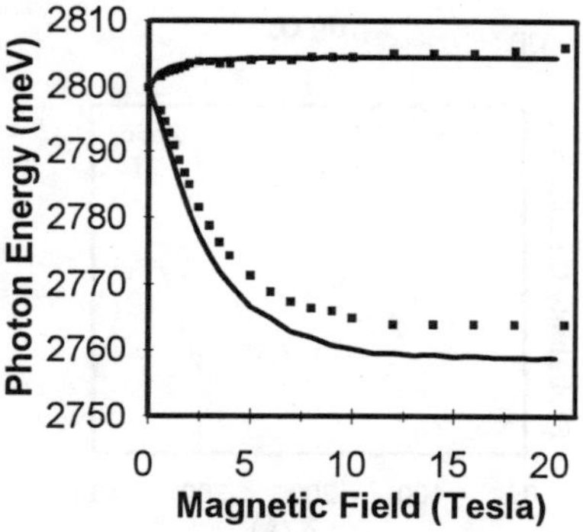

Fig. 1 Energies of the σ_+ and σ_- heavy hole exciton transitions plotted versus magnetic field

In fig.2 we plot the intensity ratio I(+)/I(-) of the lower and the upper heavy hole exciton components versus B for sample 1. The ratio initially decreases with increasing magnetic field between 0 and 1.75 tesla. For B>2 tesla the ratio increases monotonically with field. This behavior is quite distinct from that of a spin superlattice in which the ratio I(+)/I(-) initially increases with B and then saturates to 1. The observed difference is attributed to the fact that the $ZnSe/Zn_{0.96}Fe_{0.04}Se$ structure has a non-zero conduction band offset at B=0. A variational model has been developed which describes excitons in type-I or type-II heterostructures in which the electrons and holes are weakly confined[6]. This model has been used to fit the data from sample 1 shown in fig.1 and fig.2. The only adjustable parameters are the band offsets and the zero field exciton energy. The best fits to the data were obtained using a conduction and valence band offset of 5 meV. The deviation of the calculated from the experimental energies for the lower σ_+ (-3/2,-1/2) exciton component could be due to a small difference in the Fe concentration x between sample 2 and the DMS layers of sample 1. The process of carrier segregation can best be understood by examining the calculated electron and hole wavefunctions. In fig.3 the square of the electron (heavy hole) wavefunction is plotted as function of the distance along the growth axis using the solid (dashed) line. Above (below) the wavefunction plot we give a schematic diagram (not to scale) of the conduction (heavy hole) band alignment. The band discontinuities tend to confine the electrons (holes) in the ZnSe ($Zn_{0.96}Fe_{0.04}Se$) layers of the structure. The holes are predominantly localized in the $Zn_{0.96}Fe_{0.04}Se$ layers due to their larger effective mass. The lighter electrons on the other hand are distributed almost evenly throughout the structure. The electron-hole Coulomb attraction is responsible for the weak maxima of the electron probability distribution at the centers of the DMS layers,

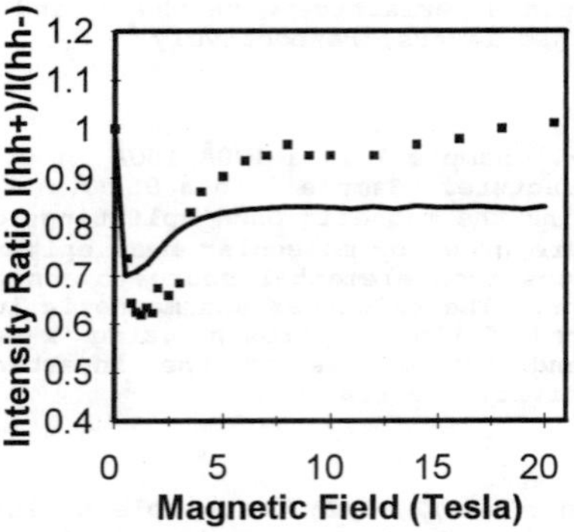

Fig.2 Intensity ratio I(+)/I(-) of the heavy hole exciton components plotted as function of magnetic field

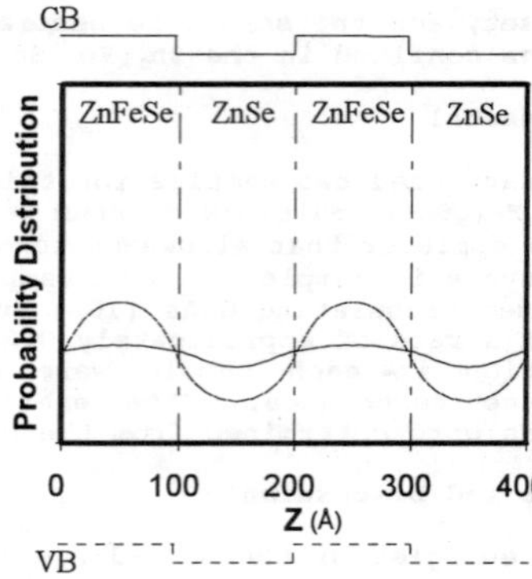

Fig.3. Square of the electron and hole wavefunctions plotted versus distance along the growth axis at zero field

even though the conduction band offsets tend to confine the electrons in the ZnSe wells.

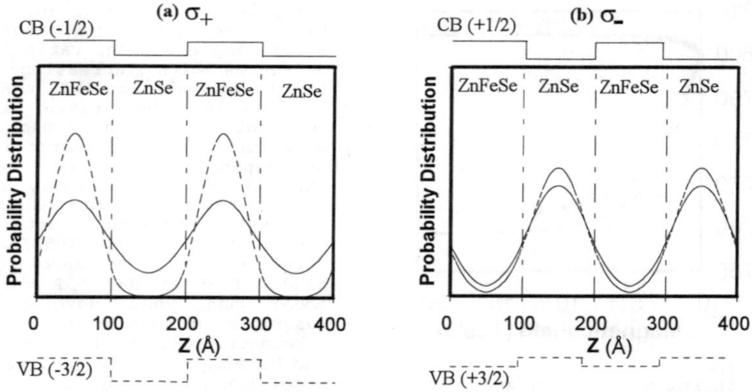

Fig.4. Square of the electron (solid line) and hole (dashed line) wavefunctions plotted as function of the distance along the growth axis at B=0.5 tesla (a) m_j=-1/2 electrons, m_j=-3/2 holes (b) m_j=+1/2 electrons, m_j=+3/2 holes

The electron and hole wavefunctions at B=0.5 tesla are shown separately for the σ_+ and σ_- transitions in fig.4a and 4b, respectively. The confining potentials for the m_j=+1/2 electrons and m_j=+3/2 holes are equal to 5.9 meV and 0.2 meV and tend to confine both types of carriers in the ZnSe wells (see fig.4b). The confining potential for the m_j=-1/2 electrons decreases to 4.1 meV while that of the m_j=-3/2 hole increases to 10.2 meV. Thus the m_j=-3/2 holes are almost entirely localized in the DMS layers while a large fraction of the m_j=-1/2 electron wavefunction remains in the ZnSe wells (see fig.4a). The electron-hole overlap for the σ_+ transition is significantly smaller than that of the σ_- exciton and leads to the reduction of the I(+)/I(-) ratio with increasing B in the 0-1.75 tesla field range. We note that the type-II band alignment for the σ_+ transition is not strong enough to pull the exciton apart. For B>2 tesla the trend reverses and the ratio I(+)/I(-) increases with B reaching a value close to unity.

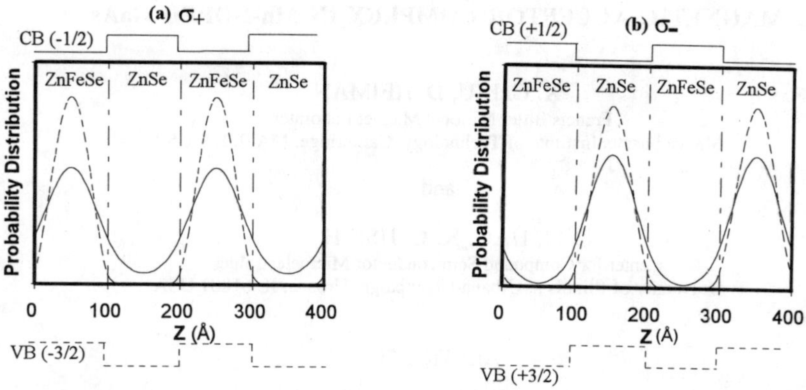

Fig.5. Square of the electron (solid line) and hole (dashed line) wavefunction plotted as function of distance along the growth axis at B=20 tesla (a) m_j=-1/2 electrons, m_j=-3/2 holes (b) m_j=+1/2 electrons, m_j=+3/2 holes

This is because the magnetic splittings of the conduction band in the $Zn_{0.96}Fe_{0.04}Se$ layers become larger than the zero field offset. The magnetic field at which the transition from type-II to spin superlattice occurs is roughly marked by the position of the minimum in I(+)/I(-) shown in fig.2. In fig. 5 we give the electron and hole wavefunctions and the band discontinuities at B=20 tesla. These are: 1.5 meV for the m_j=-1/2 electrons; 43.2 meV for the m_j=-3/2 holes; 11.5 meV for the m_j=+1/2 electrons; 33.2 meV for the m_j=+3/2 holes. The resulting potentials segregate the carriers according to their spin.

4. Conclusion

We have studied the transition from a weakly type-II to a spin superlattice behavior in a $ZnSe/Zn_{0.96}Fe_{0.04}Se$ quantum well structure. This work demonstrates the interplay between the Coulomb attraction which tends to keep the m_j=-1/2 electrons and m_j=-3/2 holes together and the type-II potentials which act to pull the carriers apart.

5. Acknowledgments

Work at SUNY was supported by NSF Grant No. DMR 9223054 and ONR/DOD under the MFEL program. The work at NRL was supported by ONR.

6. References

1. M. von Ortenberg, *Phys. Rev. Lett.*, **49**, (1982), 1041.

2. J.K. Furdyna, *J. Appl. Phys.*, **64**, (1988), R29.

3. W.C. Chou, A. Petrou, J. Warnock, and B. T. Jonker, *Phys. Rev. Lett.*, **67**, (1991), 3820.

4. N. Dai, H. Luo, F.C. Zhang, N. Samarth, M. Dobrowolska, and J.K. Furdyna, *Phys. Rev. Lett.*, **67**, (1991), 3824.

5. X.C. Liu, A. Petrou, J. Warnock, B.T. Jonker, G.A. Prinz, and J.J. Krebs, *Phys. Rev. Lett.*, **63**, (1989), 2280.

6. J. Warnock, B.T. Jonker, A. Petrou, W.C. Chou, and X. Liu, *Phys. Rev.*, **B48**, (1993), 17321.

MAGNETIC ACCEPTOR COMPLEX IN Mn-DOPED GaAs

X. C. LIU, D. HEIMAN
Francis Bitter National Magnet Laboratory
Massachusetts Institute of Technology, Cambridge, MA 02139 USA

and

J. HAO, K. C. HSIEH
Center for Compound Semiconductor Microelectronics
University of Illinois at Urbana-Champaign, Urbana, IL 61801 USA

ABSTRACT

Optical studies of Mn-doped GaAs have been carried out in magnetic fields up to 30T. Energy splittings of Mn-acceptor-related luminescence are compared with those of hydrogenic shallow acceptors present in the same samples. Larger energy splitting and lower saturation fields from the Mn-related transitions are observed. In particular, the senses of the dominant circular polarization of photoluminescence transitions are found to be opposite among the two sets of transitions. These results reveal that the exchange between the Mn d-shell and its bound hole is antiferromagnetic, making this magnetic acceptor complex (MAC) a spin-1 entity. The measured g-factor for this Mn MAC is in agreement with an AF coupled model calculation.

The study of diluted magnetic semiconductors (DMS) has generated great interest in recent years.[1,2] The large exchange interactions between spins of band carriers (electrons and holes) and localized magnetic moments of transition metal ions are responsible for several spin-dependent phenomena including giant Zeeman splittings and large Faraday rotations. In addition, impurity-bound carriers form bound magnetic polarons (BMP)[3] that induce sizeable magnetization clouds in their vicinity. The total moment of the BMP depends on the magnitude of the exchange interactions, the size of the Bohr radius of the impurity carrier, and the concentration of the magnetic spins. In DMS materials based on II-VI compounds, the BMP formation generally involves an extended impurity orbit and a large number of magnetic moments.

Another type of exchange-coupled system can be found in transition metal doped III-V semiconductors. One example is the magnetic acceptor complex (MAC) in GaAs:Mn. The manganese impurity consists of a negatively charged Mn^{2+} ($3d^5$) core[4] and a hole with a binding energy of ~113 meV and Bohr radius ~10Å.[5-7] The hole is expected to interact with the attractive Mn^{2+} core state through exchange interaction.[4]

It has been demonstrated recently[8] that highly Mn-doped GaAs (up to ~ 5×10^{19} cm^{-3}) can be produced by diffusion technique with appropriate choice of Mn sources and As over-pressure. In this paper, we present results of photoluminescence (PL) studies on a series of GaAs:Mn materials with relatively low Mn concentrations in high magnetic fields and at low temperatures. (Results on highly Mn-doped GaAs materials will be presented in a separate publication.[9]) Circular polarization-PL of the Mn MAC related transitions were measured and compared to those from shallow acceptors. The Mn-related transitions exhibited larger spin-splittings and lower polarization saturation magnetic fields than their non-magnetic counterparts. In particular, the senses of their polarizations showed opposite signs. Our results confirm the earlier suggestions[10,11] that the exchange interaction between the bound hole and the d-shell is antiferromagnetic (AF), making Mn MAC a spin-1 system. From the energy splitting, the effective g-factor of the Mn MAC is also obtained and its larger value is consistent with calculation of the AF model.

The GaAs:Mn samples were made by diffusion technique on undoped semi-insulating GaAs substrate materials, using sealed quartz ampoule technique with solid MnAs as the diffusion source. Details of the diffusion procedures have been published previously.[8] Capacitance-voltage (C-V) electrochemical etch profiler or secondary ion mass spectroscopy techniques were used to determine the Mn concentration profile that usually includes several plateaus of order ~μm each. Within each plateau the Mn concentration was constant with uncertainty typically within ~10%. In order to get the desired concentration for PL measurements, some of the sample surface was etched away. Magneto-PL measurements were done using a two-fiber optic probe[2] and 514.5nm excitation.

For samples with relatively low Mn concentrations ($< 1 \times 10^{18}$ cm^{-3}), the zero-field PL spectra exhibited well-defined Mn-related transitions[5,7] at ~1406meV. They are

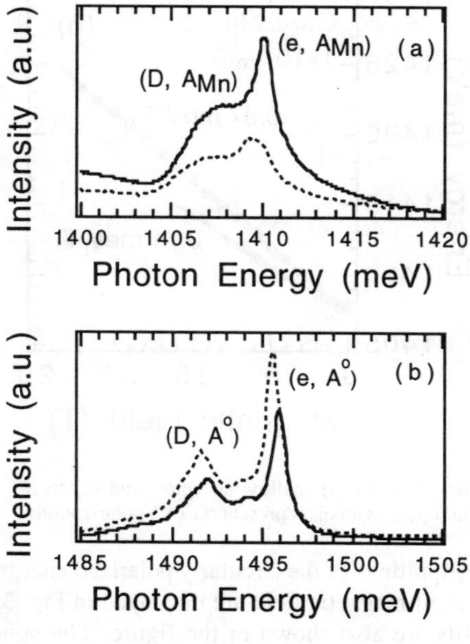

Fig. 1 PL spectra related to (a) Mn-acceptor, and (b) Shallow acceptor. Solid (dashed) lines are analyzed with σ_+ (σ_-) polarizers, at B=3T and T=4.2K.

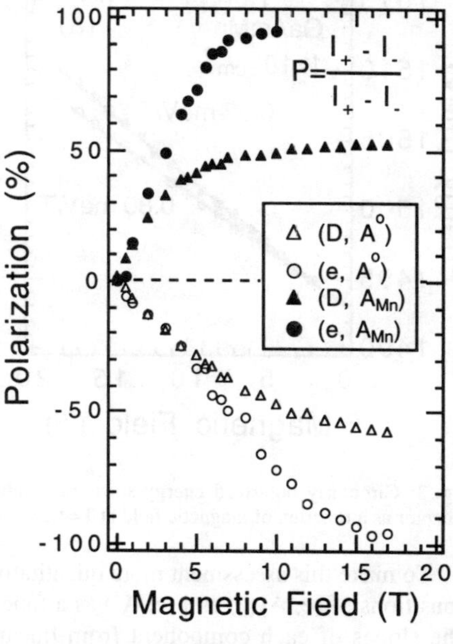

Fig. 2 Circular polarizations of the PL features shown in Fig. 1 as a function of magnetic field.

attributed to recombinations of a hole with electrons from the conduction band (e, A_{Mn}) and shallow donors (D, A_{Mn}). PL recombination also originated from residual shallow acceptors (A°). The conduction band (e) and shallow donor (D) to A_{Mn} and A° transitions were studied and compared in magnetic fields, as exemplified in Fig. 1. The A°-related transitions exhibit primarily σ_- polarization, while the ones associated with Mn are dominated by σ_+. In both cases, the σ_+ components appear at higher energies. Fig. 2 shows the circular polarization of the peak intensities as a function of magnetic field. Although they exhibit the same general field-dependent behavior (saturating in magnetic field), they are quantitatively different. In particular, the senses of the dominant polarization for the Mn related transitions are opposite to that for shallow acceptors.

The primary difference between the two types of acceptor-bound holes is that the ones in the Mn MACs are also affected by the exchange interaction with the Mn d-shell. The opposite sign in their polarizations indicates that there exists a large AF exchange interaction[10,11] between the Mn bound hole and the Mn d-shell that forces the hole spin (j=3/2) to be antiparallel to the Mn d-shell spin (S=5/2), making the Mn MAC a spin-1 entity. The AF coupled Mn MAC exhibited an enhanced g-factor as evidenced from the polarization behavior shown in Fig. 2, in which the circular polarization of the Mn MAC saturated at much smaller magnetic field than that for the shallow acceptors.

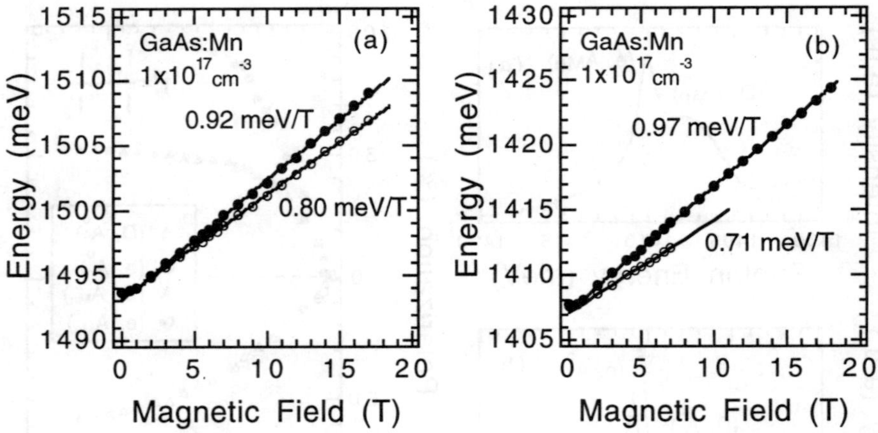

Fig. 3 Circularly polarized energy shifts of conduction band to (a) shallow acceptor, and (b) to Mn acceptor as a function of magnetic field at T=4.2K. Solid (open) symbols represent σ_+ (σ_-) polarizations.

To make this assessment more quantitative, splittings of the circularly polarized energy transitions of (e, A_{Mn}) and (e, A°) as a function of magnetic field are displayed in Fig. 3. The slopes of each component from linear fits are also shown in the figure. The spin splitting of the (e, A_{Mn}) transition is apparently larger than that of the (e, A°). Using a g-value of -0.44 for the electron, we obtain an effective g-factor of ~0.84 for the j=3/2 hole on a shallow acceptor, in reasonable agreement with the value quoted in the literature.[4] On the other hand, a g-factor of 2.47 is determined for the Mn MAC. The large g-value, although not identical to the one measured by ESR experiment,[4] is nevertheless an indication of enhancement caused by the exchange interaction within the Mn MAC.

The effective Lande g-factor of the AF-coupled angular momenta for the Mn MAC (S and j) can be calculated according to the formula similar to spin-orbit coupling. For the ground state (J=1): $g_1 = \frac{7}{4} g_{Mn} - \frac{3}{4}\kappa$, where g_{Mn} =2 is the Mn g-factor, and κ is the hole g-factor that can be taken as the measured value, 0.84, for the shallow acceptor hole. From this, the total g-factor for the Mn MAC ground state is ~ 2.87, which compares favorably to the measured value discussed above for the Mn MAC ground state.

At very high magnetic field, the ground state of the Mn MAC is expected to transform from the spin-1 to a spin-2 configuration. A measurement of the magnetic field at which the level crossing occurs could yield a value for the exchange constant within the Mn MAC. The level crossing is anticipated as a change in the slope of the (e, A_{Mn}) transition energy as a function of the magnetic field.

In summary, we have performed polarization magneto-PL measurements on lightly Mn-doped GaAs materials. The PL results (polarization and energy shifts) reveal that the bound hole and the Mn d-shell within the Mn MAC are AF coupled. The ground state of the Mn MAC in GaAs forms a spin-1 entity.

Acknowledgments

This work was supported by the National Science Foundation through grant DMR-9121452. The Francis Bitter National Magnet Lab is supported by the National Science Foundation.

References

1. For a review, see e.g., *Diluted Magnetic Semiconductors,* eds. J. K. Furdyna and J. Kossut, *Semiconductors and Semimetals*, Vol. **25** (Academic, New York, 1988).
2. D. Heiman, in *Semiconductors and Semimetals,* Vol. **36** (Academic, New York, 1992) p. 1.
3. P. A. Wolff, in Ref. 1.
4. J. Schneider, et.al., *Phys. Rev. Lett.* **59** (1987) 240; and Th. Frey, et.al., *J. Phys. C: Solid State Phys.* **21** (1988) 5539.
5. T. P. Lee and W. W. Anderson, *Solid State Commun.* **2** (1964) 265.
6. D. A. Woodbury and J. S. Blakemore, *Phys. Rev.* **B8** (1973) 3803.
7. M. Ilegems, R. Dingle, and L. W. Rupp Jr., *J. Appl. Phys.* **46** (1975) 3059.
8. C. H. Wu and K. C. Hsieh, *J. Appl. Phys.* **72** (1992) 5642.
9. X. C. Liu, D. Heiman, J. Hao, and K. C. Hsieh, *Proc. Int'l Workshop Semimagnetic (Diluted Magnetic)Semiconductors, Linz, Austria, Sept. 28-30, 1994*
10. A. Petrou, et.al., *Solid State Commun.* **55** (1985) 865.
11. Ya. Karlik, et.al., *Sov. Phys. Solid State* **24** (1982) 2022.

MAGNETO-OPTICAL STUDY OF $CdTe/Cd_{1-x}Mn_xTe$ MULTIPLE QUANTUM WELLS IN THE MEGA-GAUSS FIELD RANGE

Shinji KURODA, Kazutoshi KOJIMA, Kôji KOBAYASHI and Kôki TAKITA
Institute of Materials Science, University of Tsukuba, Tsukuba, Ibaraki 305, JAPAN

Kazuhito UCHIDA and Noboru MIURA
Institute for Solid State Physics, University of Tokyo, Roppongi, Minato-ku, Tokyo 106, JAPAN

ABSTRACT

Magneto-absorption measurement of $CdTe/Cd_{1-x}Mn_xTe$ multiple quantum wells (MQW's) with Mn-content of $x \simeq 30$ % was performed under ultra-high magnetic fields up to 150 T. Observed Zeeman shift energies of the absorption edge of the $Cd_{1-x}Mn_xTe$ layers and the heavy-hole exciton in the quantum well were analysed by comparing with theoretical calculations.

1 Introduction

$CdTe/Cd_{1-x}Mn_xTe$ multiple quantum wells (MQW's) are modulated structures consisting of diluted magnetic semiconductors. In this system, confinement potential of the quantum well depends strongly on the magnetization of the $Cd_{1-x}Mn_xTe$ barrier layer due to the *sp–d* exchange interaction[1]. It is of interest to study magneto-optical properties under high magnetic fields, since the application of high magnetic fields are expected to change electronic states in the quantum well significantly through the large Zeeman splitting in the $Cd_{1-x}Mn_xTe$ barrier layer.

In a previous paper, we reported magneto-optical studies of $CdTe/Cd_{1-x}Mn_xTe$ MQW's with Mn-content of about 30 % under magnetic fields up to 40 T using a non-destructive pulsed magnet[2]. In the present work, we study the excitonic transitions under ultra-high magnetic fields up to 150 T through the magneto-absorption measurement on the same MQW samples using the single-turn coil system.

2 Experimental Methods

The procedures of the sample preparation are the same as described in Ref.[2,3]. Samples were grown on a GaAs substrate in the following order by the MBE technique; a CdTe buffer layer (~ 2000 Å), a $Cd_{1-x}Mn_xTe$ cladding layer (~ 8000 Å), an MQW structure consisting of 50 cycles and finally a $Cd_{1-x}Mn_xTe$ capping layer (~ 2000 Å). After the growth, the substrate was removed by selective etching for the optical transmission measurement.

Absorption spectra was measured under magnetic fields in the mega-gauss range using the single-turn exploding coil system. The detection system consisting

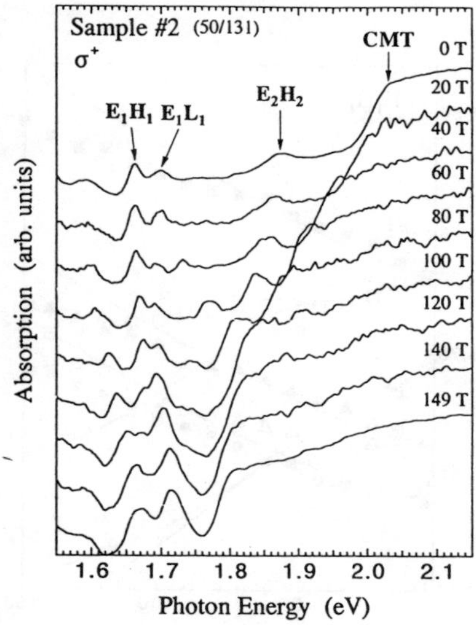

Figure 1 : Absorption spectra of Sample #2 (L_W = 50 Å, L_B = 131 Å, x = 0.27) at various fields for the σ^+ polarization.

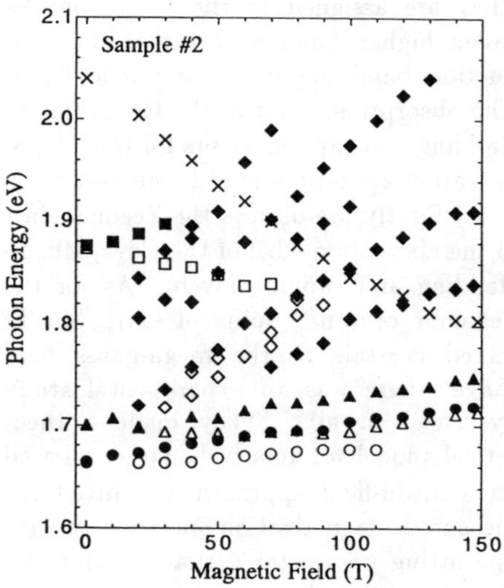

Figure 2 : The magnetic field dependence of absorption peak positions of Sample #2 for the σ^+ (open symbols) and σ^- (closed symbols) polarizations. See text for the notation of symbols.

of a streak camera, an image intensifier and a CCD camera enables us to obtain the change of optical spectra during the sweep of a pulsed magnetic field. The direction of the magnetic field was perpendicular to the sample surface and light from a Xe flush lamp was transmitted along the magnetic field (Faraday configuration). The temperature at the measurement was typically 16 ~ 18 K.

3 Experimental Results and Discussion

Figure 1 describes a typical example of absorption spectra under magnetic fields up to 150 T. This is the result on an MQW with L_W = 50 Å, L_B = 131 Å and $x = 0.30$ (Sample #2) measured for the σ^+ polarization. Peaks located at 1.66, 1.70 and 1.87 eV are assigned to the absorption of heavy-hole exciton of the first subband (E_1 H_1), light-hole exciton of the first subband (E_1 L_1) and heavy-hole exciton of the second subband (E_2 H_2), respectively. An upheaval at about 2.0 eV is due to the absorption of the $Cd_{1-x}Mn_xTe$ cladding and capping layers. It can be seen that these peaks shift with magnetic fields due to the enhanced Zeeman splitting.

In Fig. 2, observed peak energies are plotted against magnetic field for the σ^+ (open symbols) and σ^- (closed symbols) polarizations. Circles, triangles and squares denote peak positions of the E_1 H_1, E_1 L_1 and E_2 H_2 excitons, respectively. Absorption peaks which appear with the application of magnetic fields are plotted by diamonds.

They are assigned to the transition between higher Laudau states of the conduction band and the heavy-hole band. The absorption edges of the $Cd_{1-x}Mn_xTe$ cladding and capping layers for the σ^+ polarization are represented by crosses.

Firstly, we discuss the Zeeman shift of the absorption edge of the $Cd_{1-x}Mn_xTe$ cladding and capping layers. As for the behavior of Mn^{2+} spins of $Cd_{1-x}Mn_xTe$ mixed crystals in the mega-gauss field range, there was an experimental study by Isaacs et al[4]. They made a theoretical model for relatively large x based on a mean-field approach and fitted the measured magnetization successfully with one fitting parameter ζ that accounts for the interaction between Mn^{2+} spins beyond the nearest neighboring position. Although the temperature of their measurement $T \sim 10$ K is lower than our case, we calculate spin alignment $\langle s_z \rangle$ of the $Cd_{1-x}Mn_xTe$ layers with the same paramter ζ as chosen for the fitting to their data. The calculated $\langle s_z \rangle$ agrees well with the observed Zeeman shift of the $Cd_{1-x}Mn_xTe$ absorption.

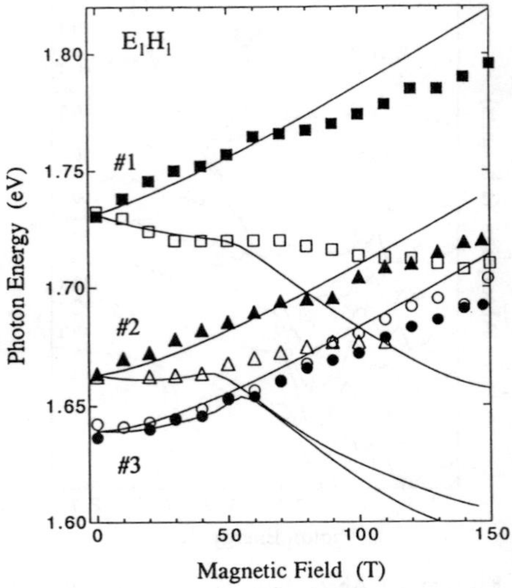

Figure 3 : The magnetic field dependence of the $E_1 H_1$ exciton energies of Samples #1 ~ #3 for the σ^+ (open symbols) and σ^- (closed symbols) polarizations. Squares, triangles and circles denote the results of Sample #1 ($L_W = 34$ Å, $L_B = 155$ Å, $x = 0.30$), #2 and #3 ($L_W = 67$ Å, $L_B = 163$ Å, $x = 0.32$). Solid lines represent calculated exciton energies by a variational approach.

Next we turn to the analysis of the Zeeman splitting of the heavy-hole exciton in the quantum well. Similarly to the previous paper[2], we made a calculation of $E_1 H_1$ exciton energies by variational approach[5]. In the calculation, Mn^{2+} spin alignment $\langle s_z \rangle$ in the barrier layer is assumed to be given by the above estimation. In Fig. 3, calculated and observed exciton energies are compared for three samples with different well and barrier widths. In the figure, open and closed symbols denote observed exciton energies for the σ^+ and σ^- polarization and solid lines denote calculated energies when the valence-band offset ratio Q_V is assumed to be 20 %.

As for the σ^- polarization, the overall agreement between the calculation and the observation is reasonable except for the higher-field region ($H \gtrsim 50$T). A small deviation in the lower-field region ($H \lesssim 50$T) is considered to originate from an anomolous behavior of Mn^{2+} spins located near the QW heterointerfaces, as discussed in the previous paper[2]; that is, the Zeeman splitting of excitons confined in the well layer are much influenced by the alignment of Mn^{2+} spins at the heterointerfaces, whose magnetization follows Brillouin function as in $Cd_{1-x}Mn_xTe$ with low Mn-content because of the reduced antiferromagnetic coupling[6] and/or interdiffusion of Mn during the growth process. Though the reason for the deviation in the higher-field region ($H \gtrsim 50$ T) is not clear, it might be attributed to the uncertainty in the estimation of $\langle s_z \rangle$ at high fields.

On the other hand, the calculated lines for the σ^+ polarization deviates significantly from the observed peak positions. A rapid decrease of the calculated energies above about 50 T is due to the theoretically expected type I → II transition. In the experiment, however, such a distinct feature is not found in the field dependence of the exciton energies. It is to be noted that this discrepancy is not resolved by choosing another value of Q_V; for an estimated value of $\langle s_z \rangle$, which amounts to 1.8 ~ 1.9 at $H = 150$ T, the type I → II transition is expected to take place below the highest field if Q_V is less than about 50 %. For this discrepancy about the type I → II transition, some pictures such as the compensation by decrease of the excitonic binding energy[7] were suggested. However neither of them is considered to be applicable to the present case and further studies will be necessary for the satisfactory explanation.

4 Summary

We have performed the magneto-absorption measurement of CdTe/$Cd_{1-x}Mn_xTe$ MQW's with $x \sim 0.3$ under very high magnetic fields up to 150 T. We have found that the Zeeman shift energies of the absorption edge of the $Cd_{1-x}Mn_xTe$ cladding and capping layers are in good agreement with an estimated value of $\langle s_z \rangle$ by a mean-field model. For an analysis of heavy-hole exciton energies in the quantum well, we have compared them with the calculation by a variational approach. As a result, the agreement of exciton energies for the σ^- polarization was found to be good in the overall feature. On the other hand, for the σ^+ polarization, the observed exciton energies deviate largely from the result of calculation; in the calculation exciton energies decrease rapidly due to the field-induced type I → II transition, while no distict feature appear in the field dependence of observed exciton energies.

Acknowledgment

This work is supported in part by Iketani Science and Technology Foundation.

References

1. J. K. Furdyna, J. Appl. Phys. **64** (1988) R29.
2. S. Kuroda, K. Kojima, K. Kobayashi, Y. Shirai, K. Takita, K. Uchida and N. Miura, Physica B *in press.*
3. S. Kuroda, Y. Shirai, K. Kojima, K. Uchida, N. Miura and K. Takita, Jpn. J. Appl. Phys. **32** (1993) Suppl. 32–3 364.
4. E. D. Isaacs, D. Heiman, X. Wang, P. Becla, K. Nakao, S. Takeyama and N. Miura, Phys. Rev. B **43** (1991) 3351.
5. K. -S. Lee, Y. Aoyagi and T. Sugano, Phys. Rev. B **46** (1992) 10269.
6. W. Ossau and B. Kuhn-Heinrich, Physica B **184** (1993) 422.
7. A. Wasiela, P. Peyla, Y. Merle d'Aubigné, J. E. Nicholls, D. E. Ashenford and B. Lunn, Semicond. Sci. Technol. **7** (1992) 571.

MAGNETIZATION STUDIES OF EuTe/PbTe SUPERLATTICES IN HIGH MAGNETIC FIELDS

J.J. CHEN and M.S. DRESSELHAUS
Department of Physics, MIT
Cambridge, MA 02139, USA
and
E.J. McNIFF and G. DRESSELHAUS
Francis Bitter National Magnet Laboratory, MIT
Cambridge, MA 02139, USA
and
G. SPRINGHOLZ and G. BAUER
Johannes Kepler Universität
Linz, Austria

ABSTRACT

MBE-grown EuTe(ξ)/PbTe(η) superlattices (SLs) with ξ EuTe (111) monolayers alternate with η PbTe monolayers in each SL cell, such that $\eta = 3\xi$ (ξ=2,3,4,5) were studied by a vibrating sample magnetometer for field strengths up to 10 tesla. Magnetic saturation effects were observed at 1.2 K. The linear region of the M vs. H curves was analyzed by a mean-field model, and approximate phase diagrams in the $H - T$ plane are produced for four SLs, each with a different ξ value.

1. Introduction

High quality EuTe(ξ)/PbTe(η) SLs were grown along the [111] direction by MBE. Detailed growth conditions and X-ray characterization results are given elsewhere.[1] Bulk EuTe is a type II Heisenberg antiferromagnet which has ferromagnetically ordered spins in the (111) planes, and spins in the adjacent (111) planes are antiferromagnetically aligned. PbTe is a non-magnetic semiconductor. EuTe/PbTe SLs belong to a general set of Heisenberg antiferromagnetic SLs. Compared to other SLs in that set, EuTe/PbTe does not have the complications of interfacial coupling that FeF_2/CoF_2[2] and CoO/NiO[3] have; and it has a simpler magnetic structure when compared to ZnTe/MnTe (AF III).[4] Furthermore, EuTe/PbTe is the first Heisenberg antiferromagnetic SL that has been studied by direct magnetization measurements.[5] EuTe also has the lowest transition temperature (T_c=9.6 K) of all the antiferromagnets mentioned above. This allows us to investigate the entire range of $M(H)$ up to saturation with magnetic fields attainable in the laboratory.

2. Theory

In a mean-field model, $M(H)$ for EuTe should follow a straight line (see Fig.1a) for fields from zero to H_F where $M(H)$ saturates at M_S.[6] This model applies to systems for which magnetic anisotropy can be neglected, and to cases of weak anisotropy of

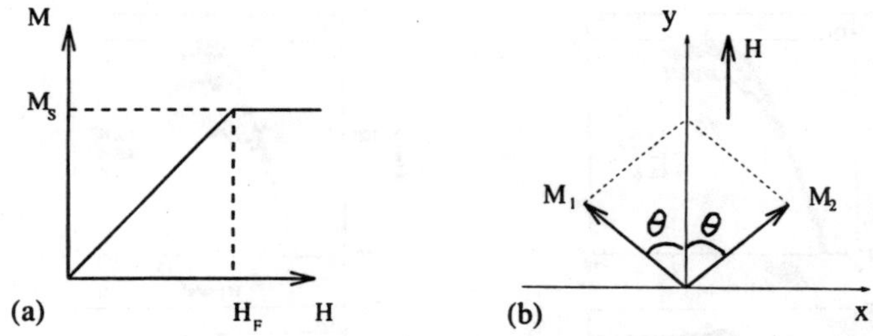

(a) (b)

Figure 1: (a) $M(H)$ curve of antiferromagnets, (b) directions of sublattice (adjacent (111) planes) magnetizations M_1 and M_2 with respect to H pointed along the positive y direction.

the "easy plane" type for $H>\sqrt{H_A H_E}$, in which H_A and H_E are, respectively, the in-plane anisotropy field and the exchange field. Bulk EuTe belongs to the latter category, with H_A=8 gauss (within the (111) planes) and H_E=3.6 tesla.[7] EuTe in the SLs retains its bulk properties insofar as $M(H)$ is independent of field direction when H is applied parallel to the (111) planes. Also neutron diffraction studies show that the spins are lying in the (111) planes for $T<T_c$.[8]

Since EuTe is an "easy plane" type antiferromagnet, the spins will rotate to become $\perp H$ when $H\sim\sqrt{H_A H_E}\sim$500 gauss. Thus we label the phase for $0<H<\sqrt{H_A H_E}$ as antiferromagnetic. In this regime, the magnetizations of the neighboring (111) planes are equal and antiparallel to each other. For $\sqrt{H_A H_E}<H<H_F$, EuTe is in the canted-spin phase, where the magnetizations of the neighboring planes are equal in magnitude and make an equal but nonzero angle θ with H (Fig. 1b). The linear increase of $M(H)$ in this phase corresponds to the spins opening up until they align along the field direction. After H exceeds H_F, the system enters the paramagnetic phase. In a mean-field model,[6] the slope of $M(H)$ in the canted-spin phase can be calculated by considering the energy per atom for nearest-neighbor (n.n) and next nearest-neighbor (n.n.n) interactions,

$$E = \frac{(-6J_1S^2 - 6J_2S^2)\cos 2\theta - 6J_1S^2}{2} - 2HS\mu_B\cos\theta \tag{1}$$

where J_1 and J_2 are the exchange couplings for the n.n and n.n.n interactions, μ_B the Bohr magneton, S=7/2 for a Eu ion, and H is the external field. Minimizing the energy of Eq.(1) with respect to the angle θ, we obtain

$$\cos\theta = \frac{-H\mu_B}{(6J_1+6J_2)}S \tag{2}$$

and the magnetic moment per atom in the field direction is then

$$M = 2S\mu_B\cos\theta = \frac{-2{\mu_B}^2 H}{(6J_1+6J_2)}. \tag{3}$$

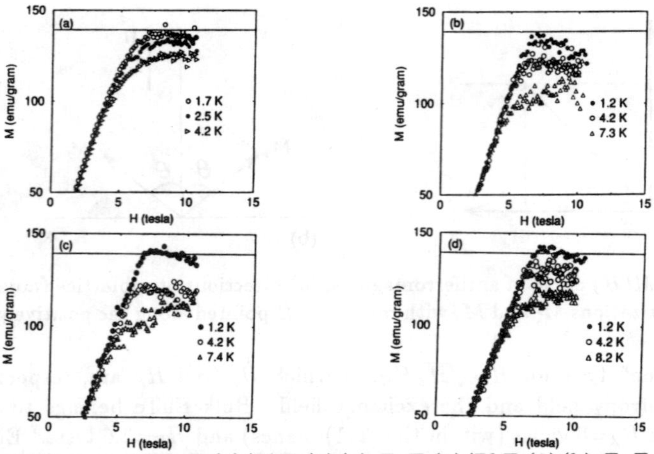

Figure 2: $M(H)$ curves for EuTe(2)/PbTe(6) (a), EuTe(3)/PbTe(9) (b), EuTe(4)/PbTe(12) (c), and EuTe(5)/PbTe(15) (d) SLs.

The coefficient of H in Eq.(3) is the slope of the M vs. H curve. With J_1/k_B and J_2/k_B given as 0.04 ± 0.01 K and -0.15 ± 0.01 K, respectively,[5] and multiplying by 2.2×10^{21}/g, we find the slope of $M(H)$ to be 44 emu/tesla gram. We will later compare our observed $M(H)$ slopes for the SLs to this calculated value.

3. VSM Measurement and Discussion

The $M(H)$ data were taken using a vibrating sample magnetometer (VSM). The field H is applied parallel to the (111) planes, and each SL that was measured contains approximately 50 micrograms of EuTe. Thus the magnetization signal is very small, and there is scatter in the measured $M(H)$ due to increased instrumental noise at high fields. Here in Figs. 2a-d, for the purpose of clarity, we only show $M(H)$ curves at three temperatures T per SL, though we studied $M(H)$ at many settings for each SL. All magnetization data are normalized to emu/gram, and the solid lines in Fig. 2 represent the saturation magnetization for EuTe (139 emu/gram). The mass of EuTe in the SL is obtained by multiplying the area of the SL by 5.4×10^{14} Eu/cm^2 and the number of (111) EuTe layers present in the SL.

The transition from the canted-spin state to the paramagnetic state is a second-order phase transition accompanied by a discontinuity in the differential $M(H)$. However, because of noise in the $M(H)$ data, the derivative is not well defined. Instead, we chose to fit the $M(H)$ curve for $H > 0.5$ tesla by a straight line, and we define the point where the $M(H)$ data start to deviate from the linear fit as the transition point from the canted-spin phase to the paramagnetic phase. In this manner, approximate boundaries for four different SLs were obtained and the results are plotted in relation to the bulk EuTe phase diagram[7] in the H-T plane (see Fig. 3). The phase boundaries

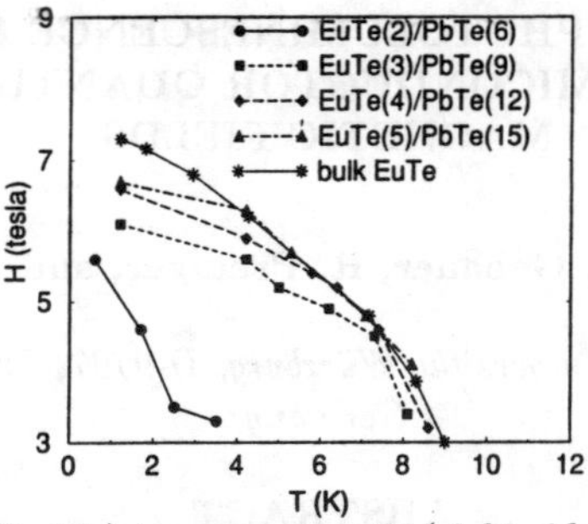

Figure 3: Canted-spin (left side) to paramagnetic (right side) boundaries for EuTe/PbTe SLs and bulk EuTe. Values of T_c (Ref. 5) are given in Table I.

Table 1: Experimental slopes of $M(H)$ for EuTe(ξ)/PbTe(η) SLs for various values of ξ

# of EuTe monolayers ξ	2	3	4	5	∞
T_c (K)	4.2	8.3	9.4	9.6	9.6
$\partial M/\partial H$ (emu/tesla gram)	24.0	21.5	20.4	20.4	15.5

in the H-T plane move to lower field and temperature values as ξ decreases from 5 to 2, consistent with decreasing T_c values (see Table I).

The slope $\partial M/\partial H$ obtained by the mean-field calculation is on the same order of magnitude as the experimental values. The measured slopes of the $M(H)$ curves in the canted-spin phase are temperature independent for each SL, but decrease as ξ increases (Table I). This is consistent with our simple formulation, where we would expect the slope to decrease as ξ increases, due to the increase in the energy per atom for larger ξ. More importantly, the present study suggests there is negligible effect on the strengths of the exchange couplings J_1 and J_2 when EuTe is reduced from the bulk down to 2 atomic monolayers in thickness.

The MIT authors acknowledge NSF grant #DMR-92-01878 for financial support. The Linz authors acknowledge support from FWF #8446, Vienna.

1. G. Springholz and G. Bauer, *Appl. Phys. Lett.* **62** (1993) 2399.
2. C.A. Ramos *et al.*,*Phys. Rev. Lett.* **65** (1990) 2913.
3. J.A. Borchers *et al.*, *Phys. Rev. Lett.* **70** (1993) 1878.
4. P. Klosowski *et al.*, *J. Appl. Phys.* **70** (1991) 6221.
5. J.J. Chen, Z. H. Wang, M. S. Dresselhaus, G. Dresselhaus, G. Springholz and G. Bauer, *Solid State Electronics*, **37**, Nos 4-6, (1994) 1073.
6. E.L. Nagaev, *Physics of Magnetic Semiconductors*, (Mir Publishers Moscow 1983).
7. N.F. Oliveira, Jr., S. Foner, and Y. Shapira, *Phys. Rev. B* **5** (1972) 2634.
8. T.M. Giebultowicz *et al.*, to be published.

TIME–RESOLVED PHOTOLUMINESCENCE FROM DILUTED MAGNETIC SEMICONDUCTOR QUANTUM WELLS IN MAGNETIC FIELDS

M. Dahl, W. Gebauer, R. Prediger, and A. Waag

Physikalisches Institut, Universität Würzburg, D-97074 Würzburg, Am Hubland, Germany

ABSTRACT

The effect of an external magnetic field on exciton decay time in $(Cd, Mn)Te/CdTe$ quantum wells is investigated by means of time–resolved photoluminescence. A reduction of the effective decay time in magnetic fields is reported. We relate this observation to the suppression of magnetic polaron formation in the external magnetic field.

1. Introduction

Diluted magnetic semiconductors (DMS) owe their unique properties to the strong exchange interaction between free carriers and localized magnetic ions. For several DMSs high quality growth of heterostructures has been achieved in the last couple of years. Among the new effects studied in low dimensional systems is the formation of quasi 2D magnetic polarons (MP) in $CdTe/(Cd, Mn)Te$ quantum wells (QWs)[1]. Theoretical studies have shown that the existence of free 2D MPs is unlikely for DMS at temperatures $T \geq 1.5K$[2]. An initial localization is necessary to provide stability for the MP complex. In narrow QWs the initial localization can be provided by fluctuations of the well width, in mixed crystals compositional disorder may support polaron formation. An important tool to study polarons is the observation of photoluminescence (PL) from excitons where the constituents, are part of a MP. The dynamics of these exciton magnetic polarons (EMPs) is only partly understood. As shown by Mackh et al.[3] the exciton lifetime determines wether the MP reaches its equilibrium magnetisation.

In the following we present time–resolved PL data on $CdTe/(Cd, Mn)Te$ and $(Cd, Mn)Te/(Cd, Mg)Te$ QWs in external magnetic fields. We demonstrate that the suppression of polaron formation i.e. the suppression of the contribution of a magnetic potential to exciton localization leads to an increase in exciton oscillator strength.

2. Experimental details

Our samples were grown by molecular beam epitaxy (MBE). The typical sample structure consists of a set of isolated quantum wells (QWs) separated by thick barrieres (500Å to 1000Å). A buffer layer of barrier material with typical thickness

2000$\AA$ was grown on (001)–oriented CdTe substrates. The structures were capped by 1000$\AA$ to 2000$\AA$ barrier material. Two types of samples were investigated: (i) QW structures with nonmagnetic CdTe layers embedded between barriers of the dilute magnetic semiconductor (Cd,Mn)Te. (ii) QW structures with wells formed by (Cd,Mn)Te and the barriers of (Cd,Mg)Te. The latter is nonmagnetic and its high quality MBE growth has been achieved recently. The sample was placed in the bore of a superconducting solenoid magnet ($B \leq 16\,T$) and immersed in superfluid Helium ($T \approx 1.8K$). Optical access was achieved by optical fibers[4]. The time–resolved PL data were measured using time correlated single–photon counting technique. Diode lasers ($\lambda = 635\,nm,\ 655\,nm,\ 675\,nm$) were applied to generate excitation pulses with repetition rates $f = 100\,MHz$ and an average power of $P = 100\,\mu W$. A $f = 0.25\,m$ spectrometer was used for dispersion, a small side–on photomultiplier for detection of the radiation. The overall time response of the setup is typically $300\,ps$. Using numerical deconvolution we achieve a time resolution of $\approx 60\,ps$ for the PL decay measurements. Time-resolved PL spectra can be measured by setting time windows of width $\Delta T \geq 20\,ps$ at variable delays relative to the laser pulse.

The simple phenomenological model used in the deconvolution procedure to draw times from the PL decay data is characterised through an effective lifetime τ_{eff} and a rise time τ_r:

$$I_{PL}(t) = I_0 \frac{\eta}{(\tau_{eff} - \tau_r)}(e^{-t/\tau_{eff}} - e^{-t/\tau_r}) \qquad (1)$$

τ_{eff} corresponds to the sum of the radiative $1/\tau_{rad}$ and nonradiative $1/\tau_{non}$ decay probabilities. η is the PL efficiency given by the ratio τ_{eff}/τ_{rad}.

For the discussion of the observations presented below we use the following relations given by Feldman et al.[5] for the radiative lifetime τ_{rad} and oscillator strength F^{2D} of 2D excitons:

$$\tau_{rad} = \frac{\tau_{eff}}{\eta} \propto (F^{2D})^{-1} = f_0^{-1}\frac{A_X}{A_c} \qquad (2)$$

A_X is the area of the 2D exciton and A_c is the area covered by the motion of the exciton center of gravity. f_0 is the dipol moment between the Bloch states of the valence and conduction band.

3. Results and Discussion

Fig.1 shows effective PL decay times τ_{eff} versus well width L_z for several CdTe QWs with various manganese concentrations ($0.09 \leq x \leq 0.21$) of the $Cd_{1-x}Mn_xTe$ barrier material. The decay data were measured at the spectral position of the cw PL emission peak. All values determined for the rise time were below the resolution ($\leq 50\,ps$) of our setup. The common feature for all samples is a strong increase of τ_{eff} for well widths less then $L_z = 50\,\AA$. The largest τ_{eff} reaches $520\,ps$ and was found for the narrowest and deepest QW ($L_z = 6\,\AA$, x=0.21). For $L_z > 50\,\AA$ we observe decay times of $\approx 150\,ps$ and a weak L_z dependence which is in agreement with the findings of Pohlman et. al.[6] An increase of τ_{eff} for narrow wells has been reported for III–V QWs; it can be explained by the reduced electron–hole overlap for narrow

QWs where the penetration of the wave functions into the barrieres is important.

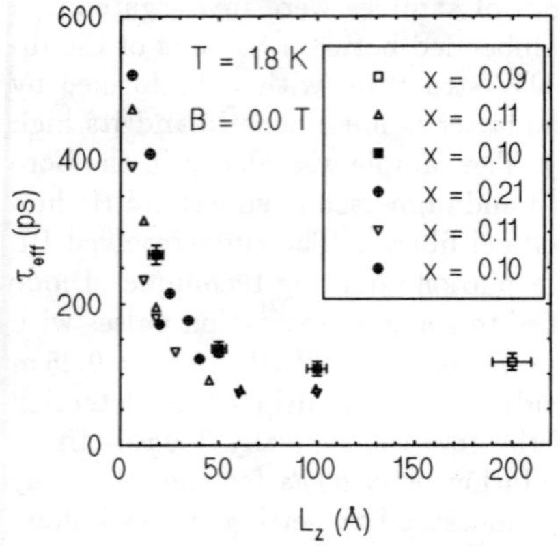

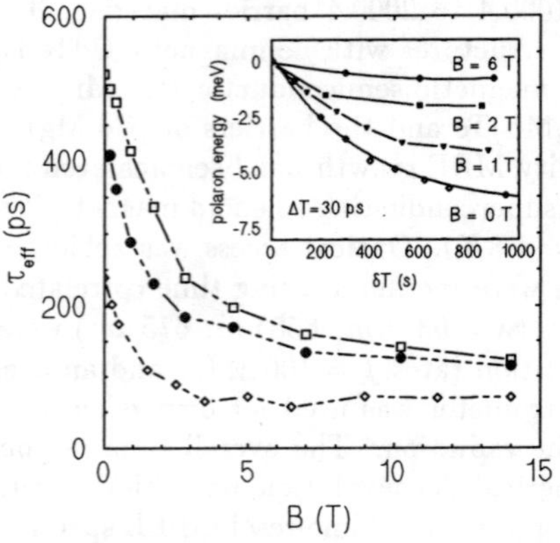

Fig.1 (left): Well width dependence of the PL decay time for $Cd_{1-x}Mn_xTe/CdTe$ quantum wells.

Fig.2 (right): Magnetic field dependence of τ_{eff} for 3 quantum wells: (□) : $L_z = 6\,\mathring{A}, x = 0.21$; (•) : $L_z = 15\,\mathring{A}, x = 0.21$; (◇) : $L_z = 18\,\mathring{A}, x = 0.09$. Inset: polaron shift versus delay time for different magnetic fields.

We calculated the L_z dependence of the radiative recombination time using a variational approach to the exciton problem. Using resonable parameters the result is about a factor 3 too small to explain the observations. In Fig.2 examples of the magnetic field dependence of τ_{eff} are shown for 3 narrow $CdTe/(Cd,Mn)Te$ QWs. (The orientation of the field is normal to QW.) For all samples we observe qualitatively the same behaviour: A rapid decrease of the decay time at small fields is followed by saturation. In the inset of Fig.2 we present the result of a time–resolved PL experiment. The peak position of the PL emission from a $L_z = 15\,\mathring{A}$ well of a $CdTe/Cd_{0.79}Mn_{0.21}Te$ sample is plotted for different magnetic fields versus delay time. The time dependend shift of the peak position is an indication of magnetic polaron formation since it can be completely suppressed by fields above $\sim 3T$. An external B field only modifies the magnetic contribution to localization while the disorder and well width fluctuation induced part is not or only weakly affected. The magnetic field induced supression of the MP energy has been studied by Yakovlev et al.[1] by means of resonant PL excitation with qualitatively similar results. We believe that the magnetic field dependence of τ_{eff} is caused by polaron formation. In (Cd,Mn)Te the dominant polaronic contribution to EMPs is through the exchange interaction of the hole with the Mn ions. Since we are concerned with heavy–hole excitons, localization of the hole is very efficient in reduction of the area for the center of gravity motion A_c of the EMP. Polaron formation in $(Cd,Mn)Te$ should therefore according to Eq.(2) lead to an increase in the radiative lifetime. The relation between the experimentally determined effective lifetime τ_{eff} and τ_{rad} is given through the PL efficiency η. The latter can by determined from the integrated PL intensity. In Fig.3 we present an example of the magnetic field dependence of τ_{eff} and τ_{rad} for

$(Cd, Mn)Te/(Cd, Mg)Te$ QWs. The magnetic field dependence of τ_{eff} is similar to $(Cd, Te)/(Cd, Mn)Te$ QWs however, the largest change in τ_{eff} is found for the wide wells. This is could be an effect of the reduction of the polaron energy due to penetration of the wavefunction into the non-magnetic barrier.

4. Conclusion

We showed that the exciton decay time in narrow $CdTe/(Cd, Mn)Te$ QWs is strongly reduced in external magnetic fields. The effect can be understood by the suppression of magnetic contributions to exciton localization and the related increase in oscillator strength. The field induced increase in oscillator strength is also observed in $(Cd, Mn)Te/(Cd, Mg)Te$ QW structures.

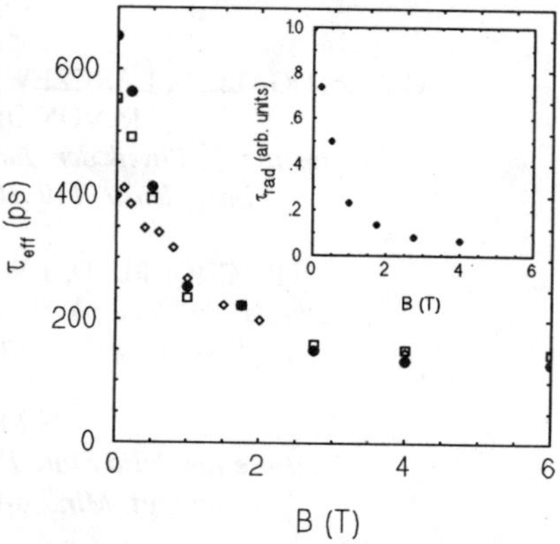

Fig.3: Magnetic field dependence of the effective PL decay time for $Cd_{0.9}Mn_{0.1}Te/Cd_{0.64}Mg_{0.36}Te$ quantum wells. □ $L_z = 300Å$, • $L_z = 100Å$, ◇ $L_z = 45Å$. Inset: Magnetic field dependence of the radiative lifetime for the $L_z = 100Å$ well.

Acknowledgement

The authors are in debt to Prof. Dr. G. Schaack for his interest and support. This work was supported by the Deutsche Forschungsgemeinschaft.

References

1. D.A. Yakovlev, W. Ossau, G. Landwehr, R. N. Bicknell–Tassius, A. Waag, S. Schmeusser, I. Uraltsev, Solid State Commun. **82** (1992) 29
2. A.V. Kavokin and K.V. Kavokin, Semicond. Sci. Technol. **8** (1993) 191
3. G. Mackh, W. Ossau, D.R. Yakovlev, A. Waag, G. Landwehr, R. Hellmann, E.O. Göbel, Phys. Rev. **B49** (1994) 10248
4. D. Heiman, X.L. Zheng, S. Sprunt, B.B. Goldberg, and E.D. Isaacs, SPIE **1055** (1989) 96
5. J. Feldmann, G. PeterE.O. Göbel, P. Dawson, K. Moore, C. Foxon, R.J. Elliot, Phys. Rev. Lett. **59** (1987) 2337
6. A. Pohlmann, R. Hellmann, E.O. Göbel, D.R. Yakovlev, W. Ossau, A. Waag, R.N. Bicknell–Tassius, G. Landwehr, Appl. Phys. Lett. **61** (1992) 2929

MAGNETO-RESONANCES IN EPITAXIAL Hg(Fe)Se LAYERS

O.PORTUGALL, M.BARCZEWSKI, S.LUTHER, N.PUHLMANN, M.VON ORTENBERG
Institut für Physik der Humboldt-Universität zu Berlin
Invalidenstr. 110, 10115 Berlin, Germany

R.SCHÖNFELD, T.WIDMER, D.SCHIKORA
Institut für Halbleiterphysik und Optik der TU Braunschweig
Mendelssohnstr. 3, 38106 Braunschweig, Germany

N.MIURA
Institute for Solid State Physics, University of Tokyo
Roppongi, Minato-ku, Tokyo 106, Japan

ABSTRACT

The results of infrared-transmission measurements on epitaxial layers of Hg(Fe)Se in pulsed magnetic fields up to 150 T and static fields up to 12 T are presented. The dominant resonant structures are assigned to the cyclotron resonance and a new type of transitions which cannot be observed in bulk material.

1. Introduction

The narrow gap compound Hg(Fe)Se maintains a special position among semimagnetic semiconductors: the pinning of the Fermi energy at the Fe^{2+}-impurity level above the conduction band edge yields a strongly correlated electronic system[1] and - as a consequence - electric response properties which are particularly sensitive to modulations of the density of states[2]. Naturally, the application of high magnetic fields discloses a profound source of interesting phenomena in this respect. The corresponding *external* modulation of the density of states can be furthermore complicated by the *internal* quantization in nanostructured samples of reduced dimensionality. In the case of Hg(Fe)Se such samples are now available for the first time[3]. The present report is focussed on infrared cyclotron resonance measurements on HgSe in fields up to 150 T and farinfrared low field measurements up to 12 T on HgSe and Hg(Fe)Se.

2. Experimental

As a first step undoped HgSe samples were grown by the hotwall technique on BaF_2 or GaAs with a thin ZnTe buffer layer. More recently the MBE technique was employed to produce Hg(Fe)Se-layers on ZnTe/GaAs-substrates. The quality of the samples was checked in the conventional way by x-ray analysis and magnetotransport studies. The reproducibility of all physical properties was found to be remarkably good.

The infrared and farinfrared transmission measurements were performed in the Faraday geometry with the principal direction parallel to the [111]-crystal axis for the case of HgSe on BaF_2 and parallel to the [100]-orientation for HgSe on ZnTe/GaAs. For the high field measurements on HgSe the destructive single turn coil method was used to generate up to 150 T. The experiments on Hg(Fe)Se were performed in a conventional 15 T superconducting magnet. A CO_2-laser was used as radiation source for the high field measurements, an optically pumped molecular gas laser in the low field regime.

3. Results and Discussion

Typical results for the high field infrared transmission of epitaxially grown HgSe-layers are displayed in fig. 1 and 2. Despite the different substrate materials, the spectra in both figures can be characterized by three common features, namely, a transmission edge around 30 T and two groups of resonant structures at higher fields.

The set of one or two peaks between 40 and 70 T corresponds to effective masses between $0.055m_o$ and $0.060m_o$ and must therefore be associated with the cyclotron resonance of charge carriers in HgSe bulk states. The pair of resonances in

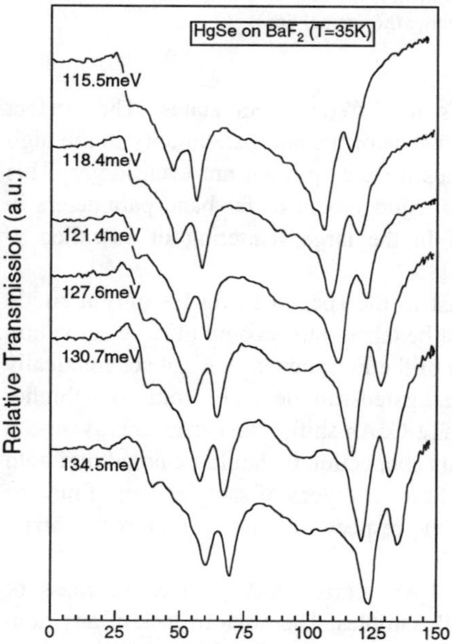

Fig.1: Radiation energy dependence of the infrared transmission of HgSe on BaF_2.

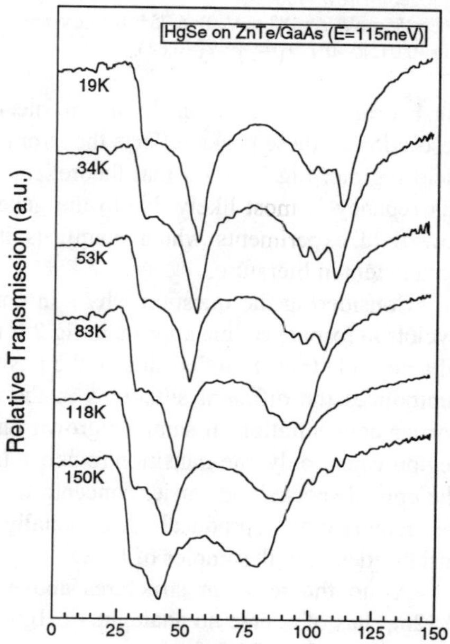

Fig.2: Temperature Dependence of the infra-red transmission of HgSe on ZnTe/GaAs.

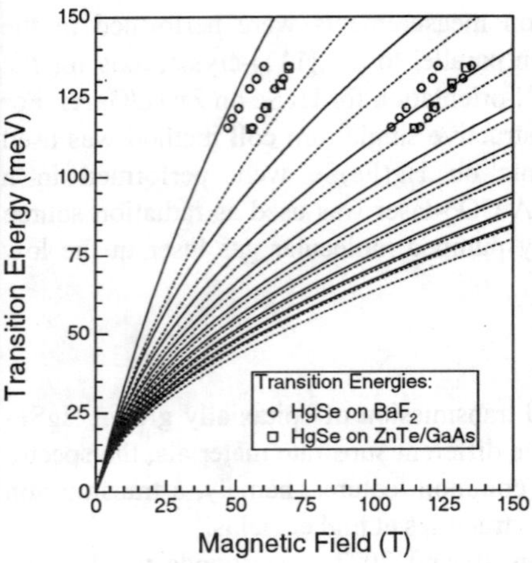

Fig.3: Experimental peaks and cyclotron resonance transitions calculated with the Pidgeon-Brown model.
(E=255meV, Δ=383meV, P=6.75*10^{-7}meV cm, F=-0.01, k=-0.1, γ_1=-γ_2=-γ_3=0.01).

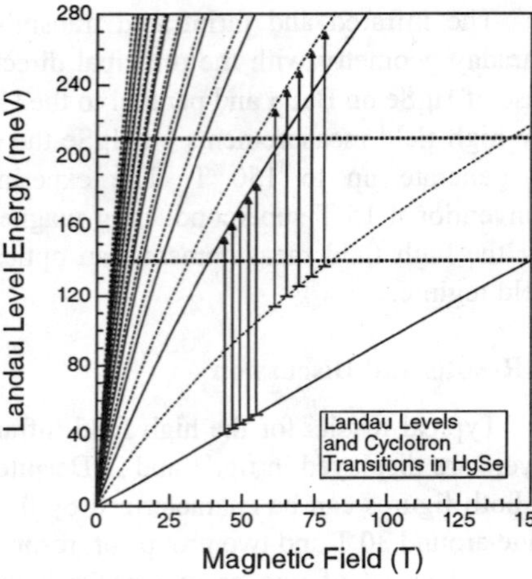

Fig.4: Cyclotron resonance transitions between Landau levels in HgSe. The number of possible resonances depends on the position of the Fermi energy (horizontal lines).

fig.1 originates from transitions of electrons in different spin states. The perfect resolution of these peaks reflects the strong influence of the non-parabolicity in the high field regime. Fig.3 shows that theoretical values for the splitting are even larger. This discrepancy is most likely due to the general misjudgement of far band parameters in low field experiments which manifests itself in the large scattering of so-called γ-parameters in literature.

Considering the question why - in contrast to the spectra in fig.1 - only a single cyclotron resonance line appears in fig.2 it must be taken into account that the activation of a possible transition depends on the position of the Fermi energy. Fig.4 schematically reproduces the different situations in the investigated samples. The somewhat higher carrier concentration in samples grown on ZnTe/GaAs shifts the Fermi energy into a region where only one transition remains. In this connection it should be noted that both the optical spectra and carrier concentrations of the two types of samples were found to be remarkably reproducable, especially with respect to the wellknown carrier instabilities in bulk samples of HgSe.

As to the resonant structures above 90 T the corresponding effective mass of $0.12m_o$ has of course no analogue in HgSe bulk samples. The obvious lack of any non-parabolic shift in fig.3 furthermore excludes other conduction band transitions such as spin-flips or combined spin-flips as a possible source for these peaks. Alternatively, the formation of a twodimensional electron gas due to the lattice mismatch between

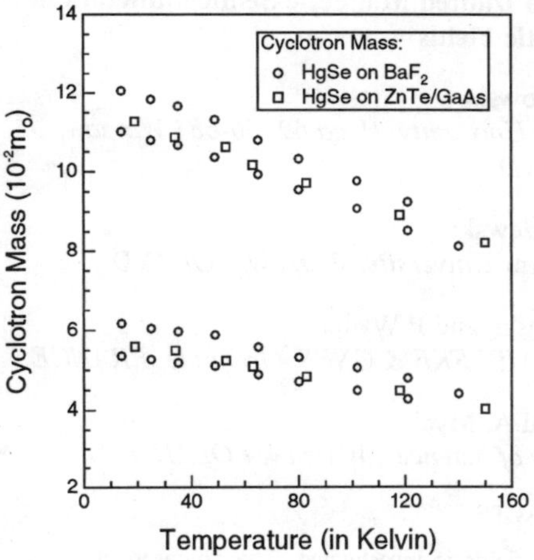

Fig.5: Temperature Dependence of effective masses corresponding to the resonant structures observed in HgSe.

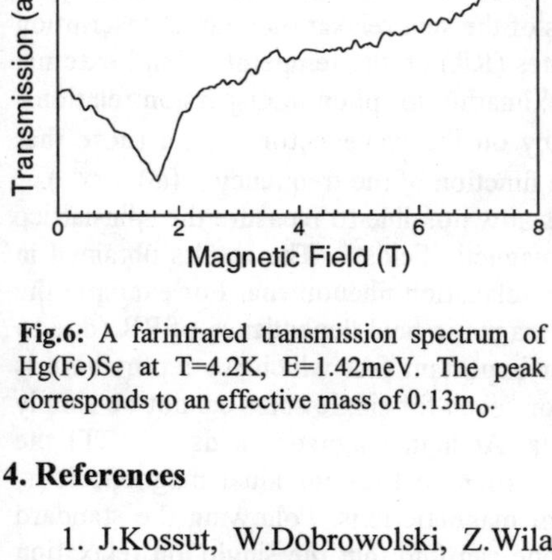

Fig.6: A farinfrared transmission spectrum of Hg(Fe)Se at T=4.2K, E=1.42meV. The peak corresponds to an effective mass of $0.13m_0$.

epitaxial layers and substrates may therefore be considered. In this case, however, the sample characteristics should be sensitive to the particular substrate material. Especially the different lattice constants and thermal expansion coefficients of BaF_2 and ZnTe/GaAs should be imposed on the epitaxial layer. Fig.5 on the other hand visualizes that the temperature dependencies of spectra from both sample types are the same so that an involvement of twodimensional electrons in the observed characteristics remains doubtful.

The application of magnetic fields in the megagauss range leaves of course the possibility for new phenomena due to either the field strength itself or inductive effects of the pulses. However, fig.6 shows that a broader but similar structure corresponding to 0.13 m_0 also appears in DC field measurements on Hg(Fe)Se were a transmission window for submillimeter waves exists between 800 µm and 1000 µm in the otherwhise opaque sample. The corresponding structure is therefore most likely related to a new intrinsic mechanism of the HgSe-host material which is yet to be identified. The fact that this feature has not been detected in DC-field measurements on HgSe may be related to the lower mobility of these samples. Such discrepancy between Hg(Fe)Se and HgSe is typical for bulk samples and has been reconfirmed in Shubnikov-de Haas measurements on epitaxial layers of both materials.

4. References

1. J.Kossut, W.Dobrowolski, Z.Wilamowski, T.Dietl, K.Swiatek, *Semicond. Sci. Technol.* **5** (1990) 260.
2. M.von Ortenberg, *Semicond. Sci. Technol.* **8** (1993) 16.
3. D.Schikora et al., to be published in *Proc. of the 22nd ICPS, Vancouver 1994.*

Investigation of Spin-Lattice Relaxation in Diluted Magnetic Semiconductors in High Magnetic Fields

A.M. Witowski
Institute of Experimental Physics, Warsaw University, Hoza 69, 00-681 Warsaw, POLAND

W. Bardyszewski
Institute of Theoretical Physics, Warsaw University, Warsaw, POLAND

Ch. Kutter, M. Potemski, and P.Wyder
High Magnetic Field Laboratory Grenoble, MPI FKF & CNRS, Grenoble, FRANCE

B. Witkowska, and A. Mycielski
Institute of Physics, Polish Academy of Sciences, Warsaw, POLAND

ABSTRACT

The spin-lattice relaxation at high magnetic fields is investigated. The nonresonant technique allows us to study of the relaxation rates at arbitrary magnetic fields. The role of non-Debye phonons is shown. The conclusions are based on the experimental data obtained for CdMnTe and HgCdMnTe at low (2K) temperature and in the range of magnetic fields up to 24T.

The spin-lattice relaxation (SLR) is conventionally studied using the EPR microwave techniques[1]. In these experiments even at helium temperatures the involved energies are smaller than or comparable with the thermal energy kT. It is important to note that spin flip energies are also much smaller than the Debye energy and, as far as the direct processes are considered , the phonons involved have small wave vectors q. In consequence, in the case of conventional studies of the spin relaxation a good description of the observed dependence of the relaxation rates (RR) on the temperature and external magnetic field was obtained[1] using Debye approximation for phonon dispersion relations, i. e., linear dependence of the phonon frequency on the wavevector: $\omega \sim q$ (note that within this model the phonon density is a square function of the frequency:.$\rho(\omega) \sim \omega^2$).

Applying new experimental technique, it is now possible to measure the spin-lattice relaxation rates at all arbitrary available static magnetic fields[2-4]. The results obtained in such experiments show new features in the spin relaxation phenomena. For example the presence of spin-pairs has been observed in the magnetic field dependence of RR (due to the SLR in exchange coupled pairs[5,6]). Another important fact which we report here is that the field dependence of spin lattice relaxation, even for single spins cannot be simply interpreted in the range of high magnetic fields. At high magnetic fields (> 12T) the relaxation is mainly due to the spin-lattice relaxation of the individual magnetic ions, especially in materials with a small admixture of magnetic ions. Following the standard method[1,7] and using the Debye approximation one can find that the single ion relaxation via one-phonon process (direct process) leads to the magnetic field dependence of the relaxation rates proportional to the third power of the magnetic field strength B. Unfortunately this dependence is not experimentally observed. The dependence is rather of the type B^5 (Ref. 6). Therefore we suspect that at such high magnetic fields the phonons with the energy corresponding to the spin splitting energy no longer can be described by the Debye model (i.e. linear dispersion).

Apparently $\rho(\omega)$ strongly depends on the phonon dispersion relation. The magnetic field dependence of the relaxation rates is sensitive to the applied model of this relation and in the present paper we calculate SLR rates using more realistic description of the phonon dispersion relation (e.g. Ref. 8):

$$\omega = C * |\sin(qa/2)| \quad (1),$$

where ω is acoustic phonon frequency, a is the lattice constant, and C can be considered as the maximum value of ω (in the case of CdTe [9] it is about 1THz - 4.2 meV). With the presented dispersion relation one can calculate the density of phonon states (DoPS). Using a two level approximation, which at low temperatures well describes the relaxation of Mn^{2+} spins in high magnetic fields (above 10T) [10], we obtained the formula describing relaxation rates as a function of temperature and magnetic field. It describes very well the magnetic field dependence of SLR rates for CdMnTe[6] at 4.5K and at high magnetic fields (see Fig. 1). Of course, the presented approach is not valid for phonons with wave vectors far from the center of the Brillouin Zone, (BZ) where the dispersion relation is much more complicated.

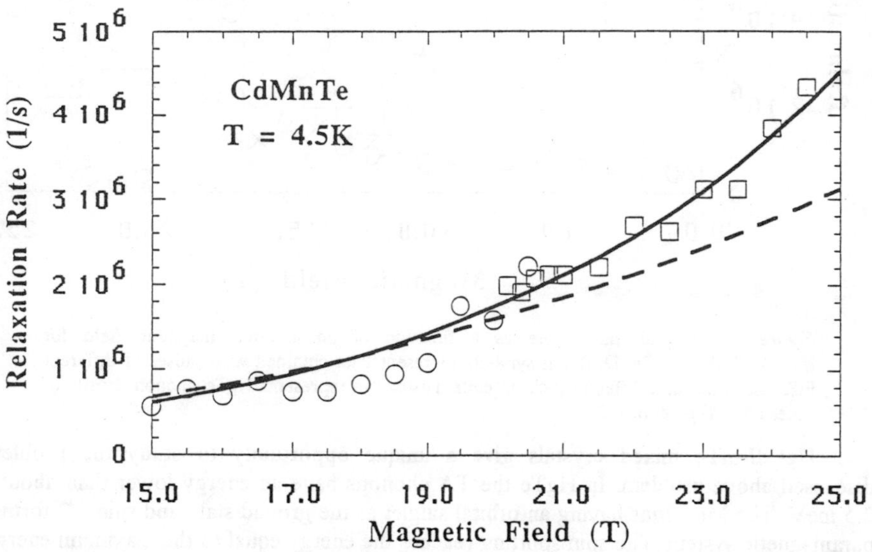

Figure 1. The magnetic field dependence of the spin-lattice relaxation rates measured for $Cd_{0.99}Mn_{0.01}Te$ (squares) and $Cd_{0.998}Mn_{0.002}Te$ (circles) - after *Strutz et al. (92)*, and calculated using Debye approximation (broken line), and presented model (solid line).

Low field SLR investigations show that a long wavelength phonons are very effective in the relaxation process. It is not obvious that phonons with very large wave vectors (edge of the BZ) will contribute equally effective to the SLR. They have short wavelengths and low symmetries. One may study this problem experimentally in the range of magnetic fields where the spin splitting becomes comparable with and/or greater than the energy of phonons with the wave vectors from the edges of BZ (e. g. using the nonresonant technique[3]). The nonresonant method allows studies of SLR as a function of the magnetic field in a broad range of fields (up to 24T). In this method[3] the lattice is heated with the nonresonant far infrared pulse. The spins then warm up to reach thermal equilibrium with the lattice changing sample magnetization. The magnetization changes are then detected with a pick-up coil.

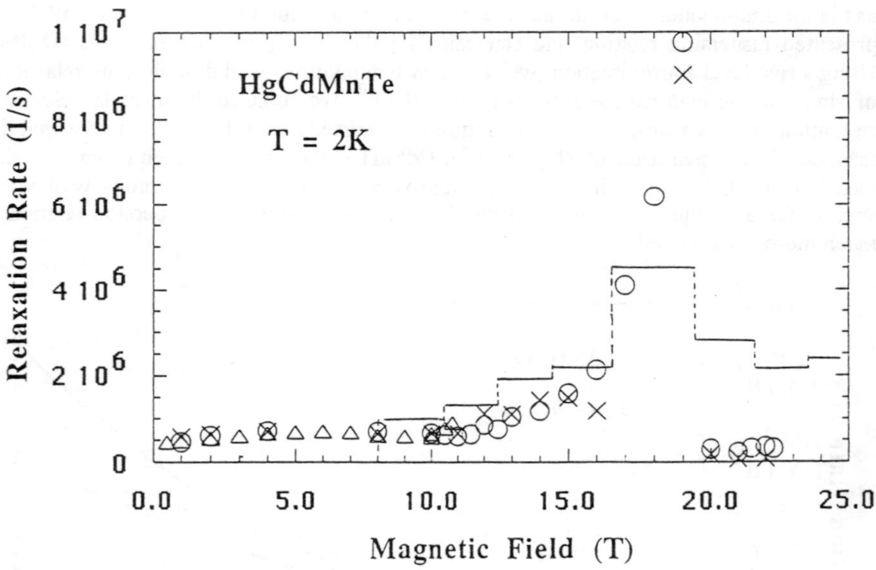

Figure 2. The relaxation rate as a function of an external magnetic field for $Hg_{0.685}Cd_{0.3}Mn_{0.015}Te$. Different symbols represent data obtained with pulses of different FIR laser lines and different pick-up coils. Histogram represents HgTe phonon density of states after *Kepa et al.(82)*.

HgCdMnTe mixed crystals give a unique opportunity to study the problem discussed above problem. In HgTe the TA phonons have an energy lower than about[11] 2.5 meV. The Mn^{2+} ions having an orbital singlet as the ground state and spin 5/2 form a paramagnetic system. The spin splitting reaches the energy equal to the maximum energy of TA phonons at about 20T (g factor of 2). Since HgTe is a zero gap semiconductor, therefore cadmium is added to open the energy gap. An open gap allows the FIR radiation to pass through the crystal, which is essential in our experimental method[3] (homogenous heating of the lattice). Free carriers can create additional paths of spin-lattice relaxation, therefore their concentration should be as small as possible. This is achieved by proper annealing of the material (HgCdMnTe).

The measurements were performed at 2K for a crystal with 30% of cadmium and 1.5% of manganese, which was n-type (n ~ 10^{15} cm^{-3}). At T = 2K one can expect to observed relaxation due to the direct process which should be therefore strongly affected by the phonon density of states (e.g. Ref.1). Our results show a rapid increase of the relaxation rates (RR), when the splitting becomes comparable with the energy of phonons from the BZ edge (the peak in the PDoS) and later an abrupt reduction of RR for splitting bigger than the maximum of the TA phonon energy. The quite large relaxation rates at low fields are probably due to the presence of free carriers. At high magnetic fields (above 14T) the direct relaxation of Mn spins to the lattice becomes the dominant mechanism of the SLR. The relaxation rates given by this mechanism are related to the phonons density of states. Therefore the magnetic field dependence of the RR reflects structures of the PDoS.

From the presented data it is clear that the TA phonons with wave vectors from the edge of the Brillouin Zone play a very important role and are very effective in the process spin-lattice relaxation. At high magnetic fields, the spin - TA phonon interaction become the dominant spin-lattice relaxation path. The phonon density of states is therefore visible in the magnetic field dependence of the relaxation rates. In some cases, where the involved phonons do not come from the center of the Brrillouin Zone, the high magnetic field data cannot be described, if the Debye approximation of the phonon subsystem is used. In contrast the experiment is well understood with simple model which falls beyond this approximation.

Acknowledgment

One of us (A.M.W.) acknowledges the financial support from the European Community Grant (No ERB-CIPA-CT-92-0211) for the stay in Grenoble.. We would also like to thank Prof. M. Grynberg for valuable discussions. We are indebted to Dr. W. Dobrowolski and E. Grodzicka for sample characterization.

References

1. K. J. Standley and R. A. Vaughan, *Electron Spin Relaxation Phenomena in Solids*, (Adam Hilger, London 1969).
2. M.A. Butler, S.J. Martin, and R.J. Baughman, *Appl. Phys. Lett.* **49** (1986) 1053.
3. T. Strutz, A. M. Witowski, and P.Wyder, *Rev. Sci. Instrum.* **64** (1993) 1853.
4. Ch. Kutter, et al. to be published.
5. Xiaomei Wang, M. Dahl, D. Heiman, P.A. Wolff, and P. Becla, *Phys. Rev.* **B46** (1992) 11216.
6. T. Strutz, A.M. Witowski, and P.Wyder, *Phys. Rev. Lett.* **68** (1992) 3912.
7. K.W.H. Stevens, *Rep. Prog. Phys.* **30** (1967) 189.
8. Ch. Kittel, *Introduction to Solid State Physics* (John Wiley & Sons, New York, 1966).
9. J. M. Rowe, R. M. Nicklow, D. L. Price, K. Zanio, *Phys. Rev.* **B10** (1974) 671.
10. A.M. Witowski, *Solid State Commun.* **86** (1993) 347.
11. H. Kepa, T. Giebultowicz, B. Buras, B. Lebech, and K. Clausen, *Physica Scripta* **25** (1982) 807.

TRANSITION METAL IMPURITIES AND BAND OFFSETS: $Zn_{1-x}Mn_xSe$

V.R. GALAKHOV, T.P. SURKOVA, E.Z. KURMAEV, V.I. SOKOLOV
Institute of Metal Physics, Ural Division of Russian Academy of Sciences
620219 Yekaterinburg, GSP-170, Russia

Chr. ZUBRÄGEL, M. NEUMANN
Univarsität Osnabrück, F/B Physik
Barbarastrasse 7, 49069 Osnabrück, Germany

Yu.N. IVANOV
Institute of Solid State Physics
142432 Chernogolovka, Russia

ABSTRACT

Absorption, reflectivity and photoemission spectra have been used for the estimation of valence band offsets for the compounds with a different Mn content. The partial Mn3*d* derived valence band spectra are analyzed also.

1. Introduction

It was established[1,2] for some compounds that the energy differences between 3*d* transition metal (TM) energy levels in the series of isovalent II-VI and III-V semiconductors are equal to the respective band offsets. The results of systematic consistent investigations of this interesting phenomenon are presented in this report.

2. Results and Discussion

We choose $Zn_{1-x}Mn_xSe$(Ni) compounds with the different Mn content that ensures both a moving of energy bands (due to Mn) and the 3*d* impurity level in an energy gap (due to Ni).[3] For all samples the Ni concentration is $\sim 5.4 \cdot 10^{17}$ cm^{-3}. The characteristics of the investigated samples are listed in Table 1.

Figure 1 presents optical absorption spectra (at T = 5 K) for several samples at the range of Ni photoionization band. The spectra are given for the $\alpha^{2/3}$ vs $\hbar\varpi$ plot, which should be a straight line for the photoionization process with an intercept giving the impurity ionization energy $\hbar\varpi_{th}$.[4] The structure in the range 1.6 – 1.8 eV is due to: a) Ni^{2+} $(3d^8)$ $^3T_1(F) \rightarrow {}^1T_2(G)$ internal transitions, b), c), uncontrollable Co^{2+} $(3d^7)$ $^4A_2(F) \rightarrow {}^4T_1(P)$ internal transitions.

Assuming that the difference between TM (namely, Ni) ionization energies is equal to a respective band offset[1,2] one can get the values for valence band offsets (VBO) for $Zn_{1-x}Mn_xSe$(Ni) solid solutions through the composition change. These values changing from ~20 mev for x = 0.03 to ~40 meV for x = 0.19 are in a

Table 1: Composition x, free exciton energies E_{FE} and threshold energies $\hbar\varpi_{th}$ for the $Ni^{2+} \rightarrow Ni^{+}$ photoionization process in $Zn_{1-x}Mn_xSe(Ni)$ Crystals at 5 K.

x	E_{FE}(eV)	$\hbar\varpi_{th}$
0.00	2.802	1.838
0.13	2.798	1.820
0.16	2.840	1.814
0.18	2.900	1.812
0.19	2.906	1.794
0.37	—	1.816

good agreement with one of <20 meV derived from the magnetic field studies of $ZnSe/Zn_{0.77}Mn_{0.23}Se$ quantum wells.[5] But it is almost order of magnitude less than was reported at Ref. 6. At the last paper the VBO for $Zn_{0.81}Mn_{0.19}Se$ crystal has been determined from photoemission spectra as a difference between the binding energies of the Zn3d core level for the ternary and parent binary compounds.

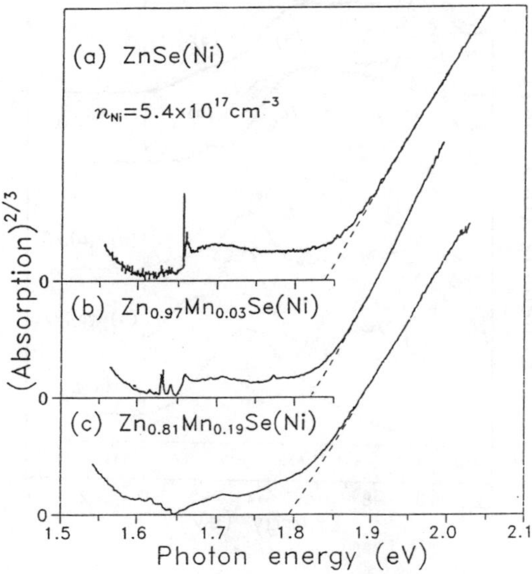

Figure 1. Absorption spectra near the photoionization threshold of $Ni^{2+} \rightarrow Ni^{+}$ for different concentration x: a) x = 0.0, b) x = 0.03, c) x = 0.19.

The similar resonant photoemission measurements (Fig. 2) at a set of our crystals showed that within the limit of the experimental error (±0.05 eV) the position of the valence band top of the samples with a different x does not change in the reference to the Zn3d maxima, and this is an agreement with our optical data. Thus, one can conclude that VBO are likely to be small through a Mn composition change, and the observed band gap change is mostly caused by the shift of the conduction band.

The resonant photoemission spectra obtained at the storage ring BESSY in Berlin show that the main peak at 3.5 – 3.7 eV is attributed mainly to the d^5L^{-1} final state originating from $L \to d$ charge screening of d hole (L^{-1} denotes a hole in the ligand p level). The satellite region (5 – 9 ev) is more d^4 like states and the region of the valence band is due to d^5L^{-1} states.

When we compare our experimental results for $Zn_{1-x}Mn_xSe$(Ni) with calculations[7] for the model cluster $[MnTe_4]^{6-}$ it is obvious that the semiconductor shows charge transfer type behaviour rather than Mott-Hubbard behaviour.

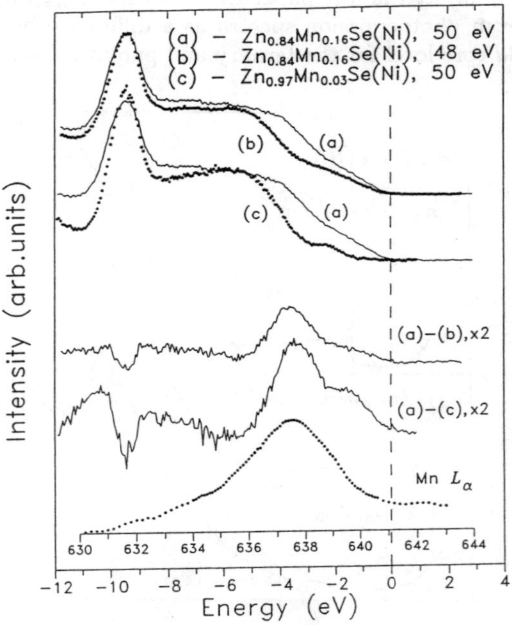

Figure 2. Energy-distribution curves of $Zn_{0.84}Mn_{0.16}Se$(Ni) obtained: a) at the Mn3p – 3d resonance $\hbar\varpi = 50$ eV, solid line); b) at the antiresonance ($\hbar\varpi = 48$ eV, points); c) energy-distribution curve of $Zn_{0.97}$Mn0.03Se(Ni) at $\hbar\varpi =$ 50 eV (points); a)–b) the difference curve between the spectra taken on and off the resonance for $Zn_{0.84}Mn_{0.16}Se$(Ni); a)–c) difference curve for $Zn_{0.84}Mn_{0.16}Se$(Ni) and $Zn_{0.97}Mn_{0.03}Se$(Ni) at $\hbar\varpi = 50$ eV. At the bottom the Mn$L\alpha$ X-ray emission spectrum for $Zn_{0.90}Mn_{0.10}Se$ is shown (Mn2p – 3d transition).

Acknowledgements

This work was supported by the Deutsche Forschungsgemenschaft and the German Bundesministerium for the Science and Technology (BMFT, Project 05 5MPAAB 7) and by the Soros Foundation Grant.

References

1. J.M. Langer, C. Delerue, M. Lannoo, and H. Heinrich, *Phys. Rev.B* **38**, 7723 (1988).
2. J. Tersoff and W. Harrison, *Phys. Rev. Lett.* **58**, 2367 (1987).
3. S.A. Permogorov, L.N. Tenishev, V.I. Sokolov, and T.P. Surkova, *Defects and Diffusion Forum*, ed. by N.T. Bagraev (Trans. Tech. Publications, Switzerland, 1993) V. 103 – 105, p. 193
4. J.M. Noras and J.W. Allen, *J. Phys. C* (Solid State Phys.) **13**, 3511 (1980).
5. Y. Hefetz, J. Nakahara, A.V. Nurmikko, L.A. Kolodziejski, R.L. Gunshor, and S. Datta, *Applied Phys. Lett.* **47** 989 (1985).
6. R. Weidemann, H.-E. Gumlich, M. Kupsch, H.-U. Middelmann, and U. Becker, *Phys. Rev. B* **45**, 1172 (1992).
7. L. Ley, M. Taniguchi, J. Ghijsen, R.L. Johnsson, and A. Fujimori, *Phys. Rev. B* **35**, 2839 (1987).

QUANTUM ASPECTS OF LANDAU EMISSION BY MAGNETOELECTRIC PHOTOEFFECT OF InSb AT HIGH MAGNETIC FIELDS

Takeshi MORIMOTO, Meiro CHIBA, Shizumasa UEDA
Institute of Atomic Energy, Kyoto University, Uji, Kyoto 611, Japan

Giyuu KIDO and Terufumi HAMAMOTO
Institute for Materials Research, Tohoku University, Sendai 980, Japan

Quantum effects on the interband Landau emissions have been investigated for n-InSb in a magnetic field H up to 22 T. At 20 K two dominant emissions become stimulated with a reduction in the virtual energy gap ε_g of ~ 25 meV in the cw excitation under the crossed fields E^* and H. The band parameters are determined from the emission spectroscopy.

As an inverse phenomenon of interband magneto-optical absorption[1] (IMO), notable quantum effect on stimulated emission has recently been revealed in bulk n-InSb in quantizing high magnetic fields H ($//\hat{z}$) up to 7 T at low temperatures, when passing a current J ($//\hat{x}$) across the magnetic field.[2,3] A large number of electrons and holes are then considered to be excited nonequilibriumly by the effective transverse electric field E_y^*, as given by $E_y^* \sim JH / n^* |e| c$, which is induced in the configuration $\boldsymbol{J} \perp \boldsymbol{H}$ and becomes much larger than the longitudinal electric field E_x in high magnetic fields.[4,5] Here, n^* is the effective concentration of electrons.

At the quantum limit, almost all electrons and holes thus excited, will populate in the lowest Landau levels, $l = l' = 0$, l and l' being Landau's orbital quantum numbers of electrons and (heavy) holes. The peak frequencies, $\omega_\pm$, of the strongest emissions $\alpha_1(0_+ \rightarrow 0')$ and $\alpha_2(0_- \rightarrow 0')$ obey the relation (1) with the selection rule, $\Delta l = 0$:[2,3]

$$\hbar\omega_\pm = \varepsilon_g + \frac{1}{2}\hbar\omega_{c1}\left[1 \pm \frac{m_1^* g_1}{2m_o} \right] + \frac{1}{2}\hbar\omega_{c2,h} - |e| E_y^* l^*. \tag{1}$$

Here, $E_y^* \sim JH / n^* |e| c$, ω_{c1} and $\omega_{c2,h}$ are the cyclotron frequencies of electrons and holes, respectively, ε_g is the energy gap, m_1^* is the effective mass of electrons, g_1 is the effective g-factor, m_o is the rest mass of electrons, and l^* is the mean free path of lucky electrons[6] drifting along $-y$ direction until the annihilation by recombination in the E_y^* field. The last term in Eq.(1) is added to interpret the virtual decrease in the energy gap under the crossed fields (E_y^*, H). Here we report further investigations carried out using the facilities of high field laboratory of Tohoku University.

The experiments were made on non-doped n-InSb whose electron concentration n_o and the mobility μ at 80 K were 5×10^{14} cm^{-3} and 3.7×10^5 cm^2/Vs, respectively. The dimensions were 4.06 mm long, 3.18 mm wide and

0.22 mm thick, and the surfaces were etched by CP-4A solution so that the surfaces did not form the Fabry-Pérot cavity. The magnetic field was applied along the [110] direction. A constant dc current J was electrically chopped with a 50% duty factor.

Figure 1 is the emission spectrum for H = 12 T at 19 K, which indicates remarkable line narrowing in the $\alpha_1(0_+ \rightarrow 0')$ and $\alpha_2(0_- \rightarrow 0')$ emissions within the resolution of the monochromator ($\Delta\lambda$ = 0.08 μm). This emission spectrum is much sharper than the spectra[2)] observed first for H < 7 T. The line narrowing observed here would manifest occurrence of the stimulated emission brought about by divergingly high value of the gain $g(\omega,H)$ at the quantum limit at $\omega = \omega_\pm$ at the band edge.[7)] This is endorsed by the characteristic V-I curves showing sharp increase in the voltage drop across the sample associated with breaking out of the stimulated emission.[2,8)]

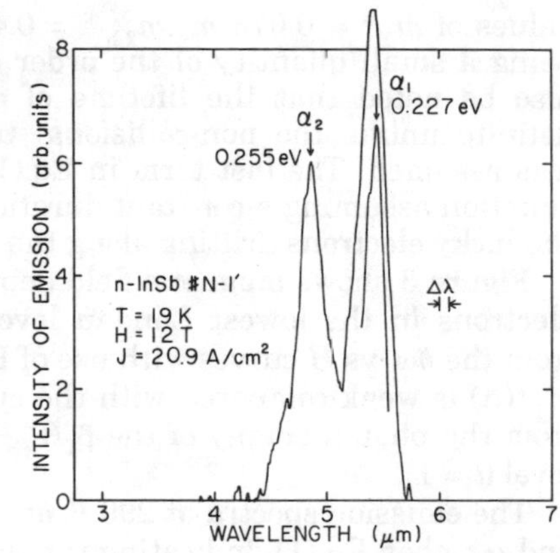

Fig.1. Emission spectrum for H = 12 T at 19 K. J = 20.9 A/cm^2.

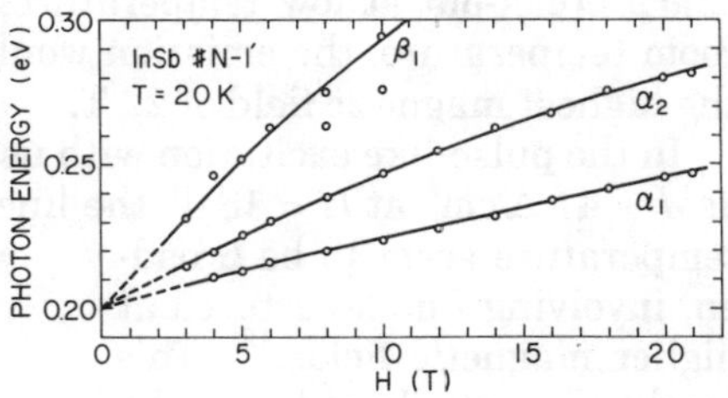

Fig.2. Photon energy of emission peaks vs magnetic field at 20 K.

Figure 2 is the dispersion curves of the photon energy of the emission peaks vs magnetic field at 20 K. Slight effect of non-parabolicity,[1)] so that deviation from Eq.(1), is seen at H > 6 T. From the extrapolation of the curves to zero magnetic field we find a value of the virtual energy gap to be 0.220 eV at 20 K in the cw excitation using relatively low current density as J ~ 20 A/cm^2, whereas the value of ε_g for InSb is 0.225 eV at low temperatures. The decrease in the virtual energy gap in the crossed fields (E_y*, H) can be interpreted from the last term in Eq.(1): assuming the reasonable values of E_y* ~ 250 V/cm[5)] and l* ~ 10^{-4} cm, we have a value of $-\Delta\varepsilon = -|e|E_y{}^*l^*$= -0.025 eV, in agreement with the observed value of the decrease in the virtual energy gap. Note that the value of E_y* can be estimated from the expression, $E_y{}^* \sim JH / n^*|e|c$, to be 226 V/cm for J = 21 A/cm^2 at H = 12 T, taking a value of $n^* = 7 \times 10^{14}$ cm^{-3}.

On the other hand, Aronov's decrease in the energy gap,[9,10)] $-\Delta\varepsilon = -(1/2)(m_1{}^*$

+ $m_{2,h}$*) $(cE_x / H)^2$, is estimated to be -1.84 x 10^{-7} eV using the observed values of $m_1^* = 0.016\ m_o$,$m_{2,h}^* = 0.42\ m_o$ and E_x = 45 V/cm[8] at H = 12 T, being a small quantity of the order of 10^{-5} of the observed value. It should also be noted that the lifetime of electrons and hence the value of l^* is definite unlike the non-collisional treatments[9,10] where no recombination was assumed. The last term in Eq.(1) can be derived from th effective mass equation assuming a constant duration length, l^*, of the effective field E_y^* on the lucky electrons drifting along the -y direction.[11]

Figure 3 shows magnetic field dependences of the effective mass $m_1^*(\alpha)$ of electrons in the lowest Landau level and the effective g-factor g_1, derived from the $\hbar\omega$ vs H curves with use of Eq. (1). The magnetic field dependence of $m_1^*(\alpha)$ is weak compared with the cyclotron effective mass $m_1^*(\beta)$ estimated from the photon energy of the $\beta_1(1_+ \rightarrow 0_+)$ emission involving higher Landau level (l = 1).

The emission spectra at 290 K are shown in Fig. 4. The emission peaks, α_1 and α_2, obey Eq.(1), indicating tunability of the interband Landau emissions at the band edge (k_z = 0). However, the line widths of the emissions, are fairly broader than the spectral resolution ($\Delta\lambda$ = 0.096 μm), in contrast to the spectra at low temperatures. The current dependence of the emitted intensities did not indicate sharp increase with a critical value J_c, in contrast to the sharp break-out at low temperatures.[2,3,8] Therefore, in the cw-operation at room temperature, the emission would not be regarded as stimulated even at the highest magnetic field of 22 T.

In the pulse-like excitation with use of moderately higher current densities, as $J \sim 47$ A/cm^2 at H = 18 T, the line widths of the emission spectra at room temperature seem to be broaden, involving fine structures at higher magnetic fields.[12] This would be attributed to the existence of holes excited to higher Landau levels (l' = 1, 2), being still non-degenerate under such a moderately high excitation. Then, the higher order transitions obeying the selection rules, $\Delta l = \pm 1$ and Δl = 2, will be allowed[10] under the presence of strong crossed fields (E^*, H), which cause the fine structures beside the main peaks α_1 and α_2.

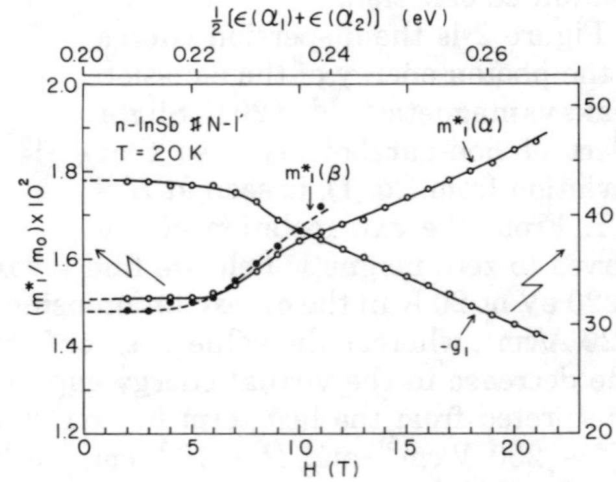

Fig.3. Magnetic field variation of the electron effective masses $m_1^*(\alpha)$ and $m_1^*(\beta)$, and of the effective g-factor g_1 at 20 K. $m_1^*(0) = 0.015\ m_o$. $g_1(0) = -43.3$.

On the contrary, no fine structures were observed in the emission spectra at low temperatures even at the highest magnetic field. At low temperatures the excess electrons and holes are easily degenerate in the lowest magnetic energy levels, and the stimulated emission takes place between the lowest Landau levels ($l = l' = 0$). This will be the reason why Eq.(1) becomes the good approximation for the emission peaks α_1 and α_2 even in higher magnetic fields $H > 10$ T despite the complicated structures of the magnetic energy levels in the valence bands.

At room temperature, the zero-extrapolation of the $\hbar\omega$ vs H curves gives a value of $\varepsilon_g(0)$ to be 0.18 eV, being the same value as that of ε_g in the absence of external fields.[13] At room temperature, the value of E_y^* is smaller than 1/20 of that of low temperatures owing to the large value of n^*. Then the contribution of the last term in Eq.(1) will be negligibly small.

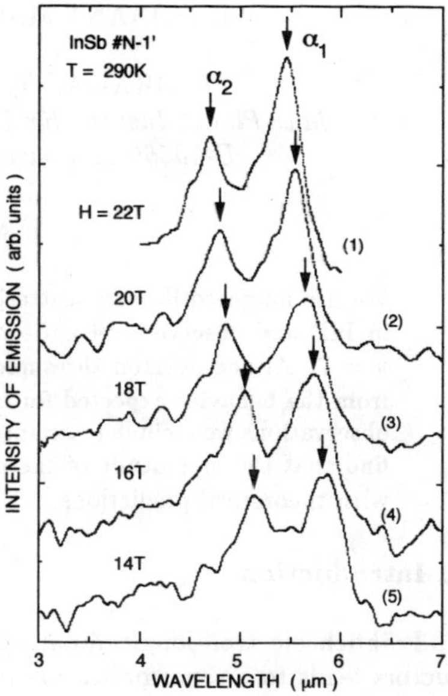

Fig.4. Emission spectra at 290 K. (1) J = 18.5 A/cm^2. (2) J = 19.3 A/cm^2. (3) J = 20.2 A/cm^2. (4) J = 21.2 A/cm^2. (5) J =22.9 A/cm^2.

1. For a review see, B. Lax and K. J. Button, in *Physics of Solids in Intense Magnetic Fields*, ed. E. D. Haidemenakis (Plenum, New York, 1969), p. 145.
2. T. Morimoto and M. Chiba, J. Phys. Soc. Jpn. **60** (1991) 2446.
3. T. Morimoto, M. Chiba, G. Kido and A. Tanaka, Semicond. Sci. Technol. **8** (1993) S417.
4. For the classical case see, T. Morimoto and M. Chiba, Infrared Phys. **29** (1989) 371.
5. See also, M. Toda and M. Glicksman, Phys. Rev. **140** (1965) A1317.
6. See for example, F. Capasso, in *Semiconductors and Semimetals*, Vol. 22, Part D, ed. R. K. Willardson and A. C. Beer (Academic Press, New York, 1985), p.1.
7. T. Morimoto and M. Chiba, Semicond. Sci, Technol. **7** (1992) B652.
8. T. Morimoto, M. Chiba, S. Ueda, G. Kido and T. Hamamoto, to be presented in the 22nd Int. Conf. Physics of Semiconductors (1994).
9. A. G. Aronov, Sov. Phys.-Solid State **5** (1963) 402.
10. For a review see, W. Zawadzki, Surface Sci. **37** (1973) 218.
11. T. Morimoto, Bull. Inst. Atomic Energy, Kyoto Univ. **85** (1994) 57.
12. T. Morimoto, M. Chiba, S. Ueda, G. Kido and M. Inoue, Physica B **184** (1993) 123.
13. See also, F. R. Kessler, P. Paul and R. Nies, Phys. Stat. Sol. (b) **167** (1991) 349.

RESONANT MAGNETO-POLARONS IN InP

F. IIKAWA,* T. RUF, AND M. CARDONA
Max-Planck-Institut für Festkörperforschung, Heisenbergstr. 1, D-70569 Stuttgart, Federal Republic of Germany

ABSTRACT

We use magneto-Raman scattering to investigate resonant magneto-polarons in InP and observe level anti-crossings for Landau quantum numbers up to $n = 7$. At the polaron threshold experimental fan lines deviate significantly from the behavior expected for isolated two-level systems. To account for the observations we include resonant coupling with neighboring Landau levels. We find that the magnitude of the polaron-splitting depends on n, in agreement with theoretical predictions.

1. Introduction

Fröhlich electron-longitudinal optic (LO) phonon interaction in III-V semiconductors leads to a renormalization of cyclotron energies and anticrossings appear at the threshold where the $|n = 0,\ 1\,\mathrm{LO}\rangle$-state coincides with $|n,\ 0\,\mathrm{LO}\rangle$ Landau levels. In InP these resonant magneto-polarons constitute a weak effect due to the coupling constant $\alpha = 0.11$ being smaller than one. However, they have been observed in modulated piezo-reflectivity and magneto-Raman experiments.[1] Magneto-Raman profiles measure the LO-phonon scattering intensity vs. magnetic field B for fixed laser exciation energy $h\nu_L$. Fan plots of resonance energies vs. B make detailed studies of electronic structure and electron-phonon interaction possible.

2. Results

Figure 1 shows one-LO magneto-Raman profiles in InP for different $h\nu_L$ measured in $\bar{z}(\sigma^-, \sigma^+)z$ circular polarizations of incident and scattered photons. The experiments were performed at 10 K in Faraday backscattering configuration. In the range of small $h\nu_L$ the magneto-Raman profiles consist of a series of sharp peaks labeled a, b, d, and f. Their intensities decrease with increasing $h\nu_L$ and new resonances, labeled c, e, and g appear in between. For the largest $h\nu_L$ shown, the latter peaks dominate the spectra. They are, however, significantly broader than those at the smallest $h\nu_L$.

Figure 2 shows a fan plot of magneto-Raman resonances vs. B for a wide range of excitation energies. For vanishing B the fan lines converge to the direct band gap of InP increased by the LO-phonon energy $\hbar\omega_{LO}$. They are therefore attributed to outgoing resonances of the Raman efficiency for which the energy of scattered photons matches with that of interband magneto-optical transitions.[1]

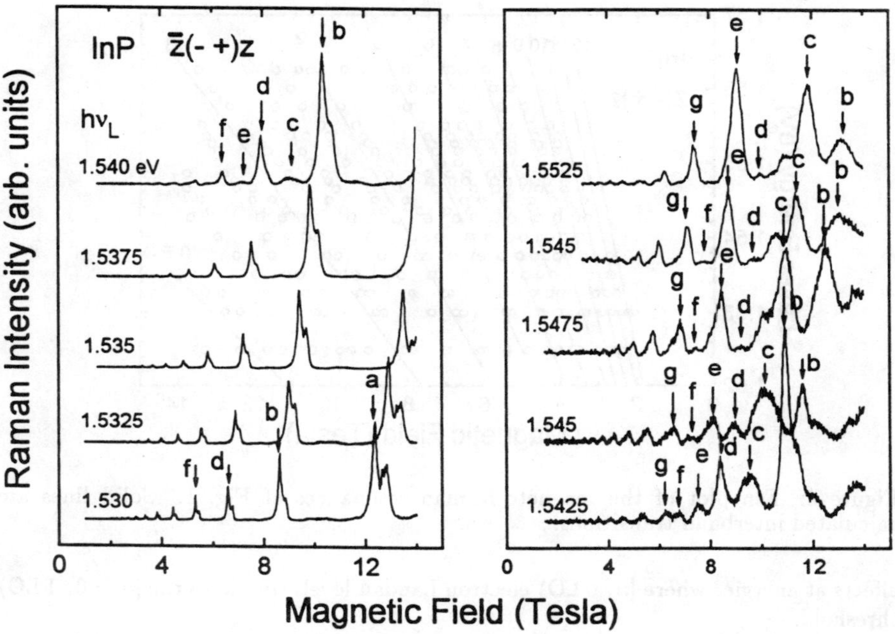

Figure 1: LO-phonon magneto-Raman profiles in InP for different laser excitation energies $h\nu_L$ around the magneto-polaron threshold.

3. Discussion

The solid lines in Fig. 2 represent the theoretical energies of interband magneto-optical transitions between light mass valence and conduction Landau levels.[1] The hole levels are strongly mixed and the indices n given next to each fan line are those of the electron states. An exciton correction depending on both n and B, has been subtracted from the transition energies which are calculated in a one-particle picture.[1] The energy of one LO-phonon was added to the theoretical fan lines to account for the outgoing character of the resonances.

The agreement between theory and experiment is satisfactory only for the largest and smallest $h\nu_L$. The assignment of the resonances in Fig. 1 to calculated fan lines is as follows: Peak a corresponds to the $n = 2$ resonance; peaks b and c scale with the fan line for $n = 3$; resonances d and e are associated with $n = 4$; peaks f and g belong to $n = 5$. Several resonances with smaller intensities are also included in Fig. 2. Pronounced deviations from almost linear behavior are found in the intermediate regime. For $h\nu_L$ around 1.545 eV in Fig. 2 anticrossings between pairs of resonances associated which each fan line occur. We attribute this to resonant magneto-polaron

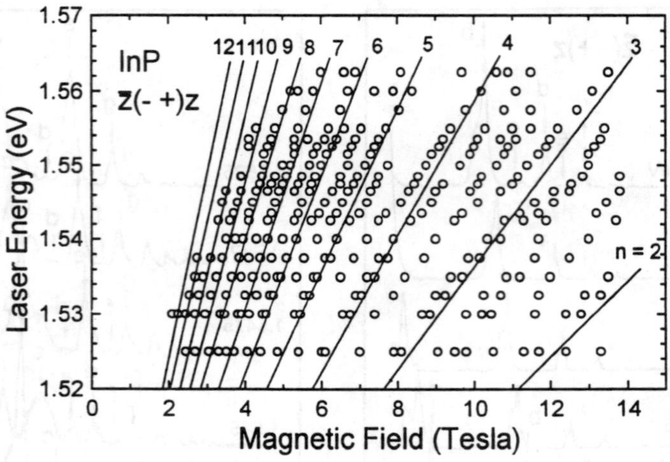

Figure 2: Fan plot of the magneto-Raman resonances of Fig. 1. Solid lines are calculated interband transitions.

effects at energies where $|n,\ 0\,\mathrm{LO}\rangle$ electron Landau levels couple to the $|n = 0,\ 1\,\mathrm{LO}\rangle$ threshold.

However, the observed anticrossings are different from those expected for two-level coupling for which upper and lower branches at resonance should be pinned to the $|n = 0,\ 1\,\mathrm{LO}\rangle$ level.[2] Renormalized interband transition energies should deviate strongly from the respective $|n,\ 0\,\mathrm{LO}\rangle$ fan lines and quickly disappear outside the resonance region. As can be seen in Figs. 1 and 2, experimental magneto-Raman resonances extend over a wide range of energies and magnetic fields beyond the magneto-polaron regime.

To visualize this effect Fig. 3 shows electron Landau levels obtained from Fig. 2 by subtracting theoretical hole energies and $\hbar\omega_{LO}$ from the observed resonances and correcting for excitonic effects.[1] We subtract the hole energy for that Landau index n from which a particular transition is derived in the limit of large or small $h\nu_L$. Open and closed circles are for $\bar{z}(\sigma^-, \sigma^+)z$ and $\bar{z}(\sigma^+, \sigma^-)z$ configurations, respectively. We find that the lower branches of levels $|n,\ 0\,\mathrm{LO}\rangle$ and the upper branches of levels $|n-1,\ 0\,\mathrm{LO}\rangle$ coincide.

This behavior can be understood if resonant coupling of the $|n,\ 0\,\mathrm{LO}\rangle$ states with neighboring Landau levels is allowed. In a phenomenological treatment of coupled two-level systems we diagonalize a larger matrix where all Landau levels interact with the $|n = 0, 1\,\mathrm{LO}\rangle$-state via a coupling constant $V(n)$. We use $V(n) = (\alpha/2n)^{2/3}\cdot\hbar\omega_{LO}$ as suggested by calculations of resonant magneto-polarons in the three-dimensional case.[3] The continuous lines in Fig. 3 are renormalized Landau levels obtained this way. They agree well with the experiment. We cannot describe these results using a constant $V(n) = \alpha \cdot \hbar\omega_{LO}$.

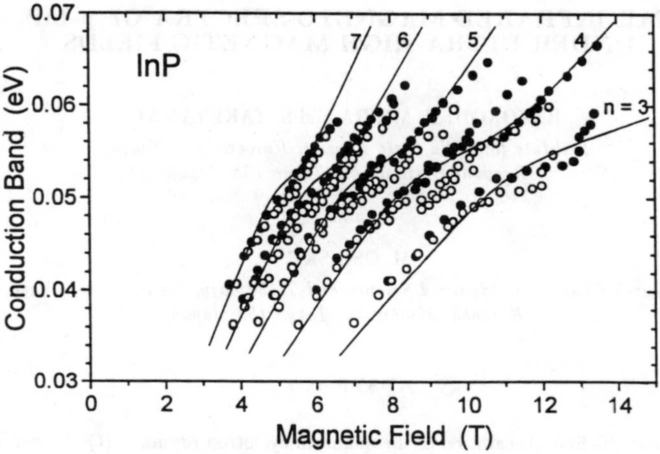

Figure 3: Electron contributions to the fan lines of Fig. 2.

4. Conclusions

Magneto-Raman scattering allows the study of resonant magneto-polarons in InP. Observed fan lines cannot be explained with a simple two-level model: Coupling with neighboring Landau levels has to be taken into account. In agreement with theoretical predictions, the polaron coupling constant depends on the Landau level index. The magnitude of magneto-polaron anticrossings decreases with n.

Acknowledgments

We thank E.J. Thrush for the InP sample and H. Hirt, M. Siemers, and P. Wurster for technical assistance. F. I. acknowledges Conselho Nacional de Desenvolvimento Cientifico e Tecnológico for financial support. We thank M. P. Chamberlain for a critical reading of the manuscript.

References

[*] Permanent Address: Instituto de Física "Gleb Wataghin" - UNICAMP, CP 6165, CEP: 13083-970, Campinas, SP, Brazil.

1. T. Ruf, R. T. Phillips, A. Cantarero, G. Ambrazevičius, M. Cardona, J. Schmitz, and U. Rössler, Phys. Rev. B **39**, 13378 (1989).
2. P. Pfeffer and W. Zawadzki, Solid State Commun. **57**, 847 (1986).
3. F. M. Peeters and J. T. Devreese, Phys. Rev. B **31**, 3689 (1985).

FAR INFRARED MAGNETO-SPECTRA OF p-InP UNDER ULTRA-HIGH MAGNETIC FIELDS

H. YOKOI[†], N. MIURA and S. TAKEYAMA[‡]

Institute for Solid State Physics, University of Tokyo, Roppongi, Minato-ku, Tokyo 106, Japan

and

M. OSIŃSKI[*]

Research Center for Advanced Science and Technology, University of Tokyo, Komaba, Meguro-ku, Tokyo 153, Japan

ABSTRACT

We report the first observation of the quantum cyclotron resonance(QCR) and impurity cyclotron resonance(ICR) transitions in p-type InP(Zn). The measurements were made under ultra-high magnetic fields up to 150 T produced in the single-turn coil system at 119, 28.0 and 16.9 μm and $9-202$ K. The band parameters γ_1, γ_2, γ_3, κ, Δ and P in the Pidgeon-Brown model are determined to be 1.4, -0.35, -0.069, -1.44, 0.108 and 8.14×10^{-8} eV cm. We obtain the excitation energy of 29 ± 2 meV for the acceptor states.

1. Introduction

The III-V semiconductor InP is an important material for optical devices. Despite that the electrical band structure of InP has been investigated intensively, the valence band parameters determined by several authors show a large discrepancy each other[1]. This is because of the scarecity of the experimental studies. Though an observation of the classical cyclotron resonance transitions at the temperature of 110K has been reported[2], the quantum cyclotron resonance(QCR) transitions have never been observed yet. It is difficult to measure cyclotron resonance transitions at low temperatures in p-type InP because the holes freeze out and the impurity scattering becomes prominent. The purpose of this work is to investigate the valence band structure of InP through the measurement of QCR transitions under ultra-high magnetic fields, where it is possible to satisfy the quantum limit condition, $\omega_c\tau > k_BT$, even in the region around 100 K. Here, ω_c, τ, k_B and T are cyclotron angular frequency, average relaxation time, the Boltzman constant and temperature, respecively. Study on the impurity cyclotron resonance(ICR) transition of the acceptor(Zn) levels in InP is also reported.

2. Experimentals

A sample of p-type InP doped with Zn was prepared in the liquid-encapsulated Czochralski method. Its surface direction, thickness, carrier density and mobility are

(001), 353 μm, 1.8 $\times 10^{16}$ cm^{-3} and 150 cm^2/Vs at room temperature, respectively. Pulsed ultra-strong magnetic fields were generated in the single turn coil system with a fast capacitor bank(40 kV, 100 kJ)[3]. A magnetic field up to 150 T and of 3.5 μsec in rise time was available with a copper coil of 10 mm in the inner diameter. The magnetic field was measured with a pickup coil inserted in the center of the field coil. A liquid helium flow-type cryostat made of regin was used to control the sample temperature. Samples and cryostats were not destroyed in each pulse. Water vapor lasers were used as light sources at 119 μm, 28.0 μm and 16.9 μm wavelengths. The strength of transmission light was measured with photocurrent-type Ge(Ga) and Ge(Cu) detectors. Every experiment was performed in the Faraday configuration($\boldsymbol{B}//\boldsymbol{k}//[001]$). The lefthand circularly polarized light was used in the measurements at 119 μm wavelength. Below 100K, the absorption peaks became small as temperature decreased. At 12 K, no absorption structure was observed. Figure 1 shows the magnetic field dependence of the absorption strength at three wavelengths and various tempratures. The trace (a) presents a result measured at 119 μm and 74 K where the quantum limit condition is satisfied, i.e. the ratio $\hbar\omega_c/kT\ (\equiv \eta)$ is 1.63. Several absorption

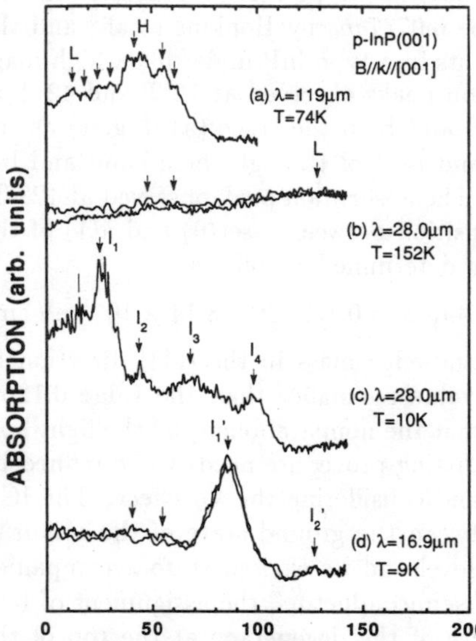

Fig.1 Far-infrared magneto-absorption spectra of p-InP at several wavelengths and temperatures. We have assigned the peaks L and H in the traces (a) and (b) as QCR transitions of the light hole band and the heavy hole band, respectively. The peaks I_{1-4} and $I_{1,2}$' in the traces (c) and (d) can be assigned as ICR transitions.

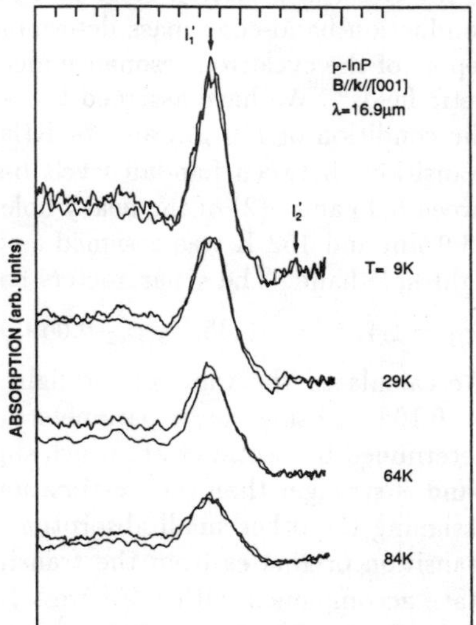

Fig.2 Tempearture variation of infrared magneto-absorption spectra of p-InP at 16.9 μm. The absorption peaks (I_1' and I_2') are assigned as ICR transitions.

peaks assigned as QCR transitions were obserbed. At 28 μm and 152 K, QCR transitions were also observed (trace (b) in Fig. 1). The absorption spectra at lower temperatures were quite different at 28.0 μm from those at 119 μm. As temperature decreased, these peaks disappeared and new peaks grew larger. The trace (c) in Fig. 1 shows a result at 28.0 μm and 10 K. At 16.9μm, two absorption peaks were observed to become larger as temperature decreased (Fig. 2). Considering the temperature dependence of the absorption strength, we have assigned the absorption peaks observed at lower temperatures as ICR transitions. QCR transitions were not observed up to 150 T at 16.9 μm and lower temperatures than 84 K.

3. Discussions

We have determined the γ parameters of the energy band of InP in the framework of the Pidgeon-Brown model[4]. In this paper, we are following the notation of Pidgeon and Brown for the band parameters and Landau levels. In the procedure, we set $E_g = 1.423$ eV, $\delta = 0.108$ and $F = q = 0$. We took into account the results of the classical cyclotron resonance measurement in p-type InP by Leotin et al., the conduction band-edge mass determined to be $0.0777\, m_0$ by Hopkins et al.[5] and the report of the cyclotron resonance measurements in n-type InP under ultra-high magnetic fields[6]. We have assigned the absorption peaks observed at 12 T and 42 T in the condition of 119 μm and 74 K(labeled L and H in the trace (a), Fig. 1) as the transitions between Landau levels b-set(0) and b(1) of the light hole band and between b(1) and b(2) of the heavy hole band. The absorption peak observed at 128 T, 28.0 μm and 152 K was assigned as the transition between a-set(0) and a(1) of the light hole band. The γ parameters have been determined as follows,

$$\gamma_1 = 1.4,\ \gamma_2 = -0.35,\ \gamma_3 = -0.069,\ \kappa = -1.44,\ \Delta = 0.108,\ P = 8.14 \times 10^{-8}\ \text{eV cm}.$$

We calculated the value of the light hole band-edge mass in the [111] direction to be $0.108m_0$ using the γ parameters. This value is smaller than the value $0.12m_0$ determined by Leotin et al., which suggests that the nonparabolicity of the light hole band is stronger than their estimation. The fitting procedure needs to be refined by assigning the other small absorption peaks and considering the k_H effect. The ICR transition originates from the transition between the ground state of the impurity state accompanied with the lowest Landau level and its excited state accompanied with a higher Landau level. In p-type III-IV semiconductors, the assignment of ICR is more complex than in n-type ones because of the degeneracy at the top of the valence band. In this paper, we will determine briefly the excitation energy of the impurity state at zero field. In Fig. 3, the ICR and cyclotron resonance transition energies are plotted vs. magnetic fields. We obtained two excitation energies, 27.4 and 30.8 meV, from linear extrapolation of field dependence of the ICR transitions. The result gives a rough estimate of the excitation energy of the acceptor(Zn) state as 29 ± 2 meV. This value is fairly reasonable, considering that no absorption structure was observed at 119 μm(10.4 meV) and 12 K and that the ionization energy of the acceptor level determined to be 47.3 meV through the luminesence measurement in

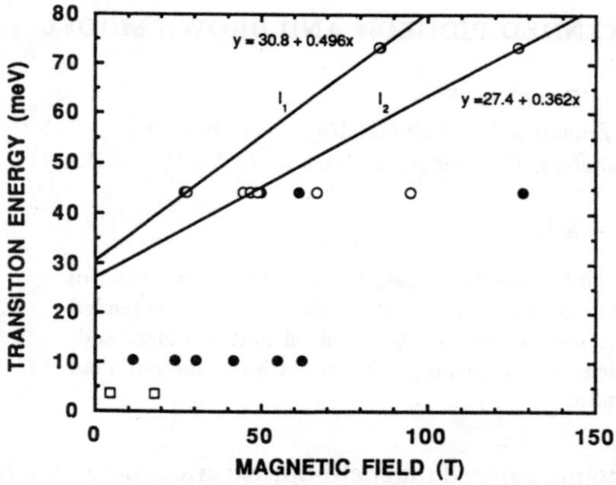

Fig.3 Transition energies vs. resonance magnetic fields in p-type InP. ○: ICR transitions(this work), •: QCR transitions(this work), □: classical cyclotron resonances[2].

Zn-doped InP[7].

In conclusion, the ultrahigh magnetic field fascility enables us the first observation of the QCR and ICR transitions in p-type InP(Zn). The determined γ parameters in the Pidgeon-Brown model explains not only our results but also the result of the classical cyclotron resonance reported by Leotin et al. fairly well. The excitation energy of the acceptor impurity state is determined to be 29 ± 2 meV from linear expolaration of the field dependence of the ICR transitions.

4. References

† Present address: National Institute of Materials and Chemical Research, 1-1 Higashi, Tsukuba, Ibaraki 305, Japan.

‡ Present address: Faculty of Science, Himeji Institute of Technology, Harima Science Garden City, Hyogo 678-12, Japan.

* Present address: Center for High Technology Materials, University of New Mexico, Albuquerque, New Mexico.

1. M. Osiński, in *CRC Critical Reviews in Solid State and Materials Sciences* (CRC Press, 1989), Vol.15, Issue 4, pp.327-343.
2. J. Leotin, R. Barbaste and S. Askenazy, Solid State Commun. **15** (1974) 693.
3. N. Miura, T. Goto, K. Nakao, S. Takeyama, T. Sakakibara, T. Haruyama and T. Kikuchi, Physica B **155** (1989) 23.
4. C. R. Pidgeon and R. N. Brown, Phys. Rev. **146** (1966) 575.
5. M. A. Hopkins, R. J. Nicolas, P. Pfeffer, W. Zawadzki, D. Gauthier, J. C. Portal and M. A. DiForte-Poisson, Semicond. Sci. Technol. **2** (1987) 568.
6. S. P. Najda, H. Yokoi, S. Takeyama, N. Miura and P. Pfeffer, Phys. Rev. B **44** (1991) 1087.
7. A. M. White, P. J. Dean, K. M. Fairhurst, W. Bardsley, E. W. Williams and B. Day, Solid State Commun. **11** (1972) 1099.

IRREDUCIBLE SPHERICAL TENSOR FORMULATION OF MAGNETO-OPTICAL TRANSITIONS IN THE EXTENDED PIDGEON AND BROWN MODEL

G. FAVROT
Physics Department and Francis Bitter National Magnet Laboratory, Massachusetts Institute of Technology, Cambridge, Massachusetts 02139, U.S.A.

ABSTRACT

Irreducible spherical tensor methods have been used to describe the rotation of the effective mass Hamiltonian for zincblende-type semiconductors in the extended Pidgeon and Brown model. The results are used to obtain a better understanding of the selection rules and orientation dependence of the intra-conduction-band magneto-optical transitions in InSb.

Inversion-asymmetry- and warping-induced magneto-optical transitions in zincblende-type semiconductors (and warping, only, in diamond-type semiconductors) have been the subject of experimental and theoretical studies over the past 40 years.[1,2] In the present work the irreducible spherical tensor formalism, introduced into effective mass theory by Lipari and Baldereschi,[3,4,5] has been applied to the problem in order to clarify the mathematics and to provide a better understanding of the selection rules and orientation dependence of the transitions. The theory is used to obtain the magnetic field dependence of intra-conduction-band (cb) transitions in InSb in the extended Pidgeon and Brown (PB) model.[6,7] The results are compared with previous experimental results.[8]

Treatment of the problem requires use of a coordinate frame $\xi\eta\zeta$ which is rotated relative to the crystallographic x, y and z directions.[9,10,11] The $\hat{\zeta}$ direction is chosen to coincide with the direction of the applied magnetic field $\mathbf{B}$. The LK basis states[12] may be quantized along $\hat{\zeta}$ because each degenerate set of LK basis states transforms with respect to the point group T_d like basis states of the *full* rotation group $R(3)$, and not just the group T_d.[13] This causes the matrix operators of the PB model to form sets which transform like irreducible spherical tensors[3] with respect to arbitrary $R(3)$ rotations and reflections. The kinetic momentum operator $\mathbf{k} \equiv \mathbf{p} + \frac{e}{c}\mathbf{A}$, where $\mathbf{A}$ is the vector potential, can be multiplied in different powers to form irreducible spherical tensor operators of arbitrary rank. The PB Hamiltonian H_{eff} contains products of parts involving matrix operators and operators made from $\mathbf{k}$ and $\mathbf{B}$.[13]

The effective Hamiltonian may be divided into parts which have spherical $R(3)$, octahedral (O_h) and antisymmetric tetrahedral (T_d) symmetry.[13] The latter are antisymmetric with respect to inversion. The spherical terms contain combinations of irreducible tensor operators which form irreducible tensors of rank zero, like k^2, $\mathbf{B} \cdot \mathbf{J}$ or $K^{(2)} \cdot J^{(2)}$.[3,4,5] The octahedral (warping) terms in the PB model may be made irreducible of rank 4, as shown for the valence band block of H_{eff} by Lipari and Baldereschi. The antisymmetric (inversion-asymmetry) terms in the PB model are irreducible of rank 3.

When the direction of $\mathbf{B}$ is chosen, the terms in H_{eff} which were spherical become axial, with symmetry group $C_{\infty h}$. This part of H_{eff} commutes with the operator $\hat{N}1_F + F_\zeta$, where $\hat{N}$ is the Landau level number operator for a spinless electron, 1_F is the unit 8×8 matrix, and F_ζ is the diagonal matrix whose elements are the m_J values of the LK basis states.[13] The eigenfunctions of the axial part of H_{eff} have a simple form and conserve the total angular momentum along ζ (spin plus orbital) associated with $\hat{N}1_F + F_\zeta$. Conservation of this quantity in the axial approximation was discussed by Yafet,[14] by Trebin *et al.*[13] and recently by Rashba and Sheka.[2] (The valence band case had been considered by Luttinger.[15]) We use λ to denote the eigenvalue of $(\hat{N} - \frac{1}{2})1_F + F_\zeta$, so that $\lambda = 0$ for the lowest energy level in the cb, and call this the 'axial quantum number'.

The fourth rank octahedral terms in H_{eff} transform like the irreducible cubic harmonic $x^4 + y^4 + z^4 - \frac{3}{5}r^4$, and have the irreducible tensor form[4]

$$T_C^{(4)} \equiv \tfrac{1}{2}[T_4^{(4)} + T_{-4}^{(4)} + \tfrac{\sqrt{70}}{5}T_0^{(4)}].$$

The third rank antisymmetric terms transform like the function xyz and have the form

$$T_A^{(3)} \equiv T_2^{(3)} - T_{-2}^{(3)}.$$

The antisymmetric tetrahedral invariants in H_{eff}, which were not considered by Lipari and Baldereschi,[3,4] are irreducible of rank $\ell = 3$, and can be expressed in irreducible tensor components. Using Weiler's designation of the band parameters,[7] and with $\mathbf{v} = \sqrt{3}\mathbf{T}$ of Trebin *et al.*[13] some of these are

$$H(G) = -iG[\mathrm{v}_x\{k_y, k_z\} + c.p.] = -\sqrt{3}G[\mathrm{v}^{(1)} \otimes K^{(2)}]_A^{(3)},$$

$$H(N_2) = -\sqrt{2}N_2[(k_x^2 - k_y^2)\sigma_z \mathrm{v}_z + c.p.] = -\sqrt{6}N_2[K^{(2)} \otimes (\sigma \mathrm{v})^{(2)}]_A^{(3)}$$

$$H(C) = (C/\sqrt{3})[(k_x\{J_x, J_y^2 - J_z^2\} + c.p.] = -(2C/\sqrt{3})[k^{(1)} \otimes J^{(3)}]_A^{(3)}.$$

$H(G')$, $H(C')$ and $H(N_3)$ are similar to $H(G)$.

Rotation of the antisymmetric terms may be carried out with the Cartesian form of the invariants by the Euler angle transformation method of Rashba[9], Gopalan[16], and La Rocca[10,11]. La Rocca's description is the easiest to follow, but neglects to make the connection of the antisymmetric coefficients $(\kappa\lambda; \mu\nu)$ with symmetric coefficients $(\kappa\lambda\mu)$ introduced by Rashba and Sheka,[9] but expressible in a simpler form. These coefficients are expressed in terms of four functions of the Euler angles,[11] F_0 through F_3. Rotation of the octahedral invariants is accomplished in a similar fashion, as was done by Sheka and Zaslavskaya[17] and recently by Obuchowicz and Wlasak.[18] The symmetrical coefficients $(\kappa\lambda\mu\nu)$ which are obtained for this case may be expressed in terms of five functions of the Euler angles which we denote W_0 through W_4. The rotation of H_{eff} may be simply expressed when the irreducible spherical tensor form is used. The spherically symmetric terms have the same form in all rotated frames. For the O_h case note that the $\ell = 4$ warping terms contain only the $m = 0$ and ± 4 spherical tensor components in the xyz frame. The $\ell = 3$ antisymmetric T_d terms contain only the $m = \pm 2$ components. If we express H_{eff} in a rotated coordinate frame, all 9 of the $\ell = 4$ components will occur, in general, and all 7 of the $\ell = 3$ components will occur. The general expressions are

$$T_C^{(4)} = \sum_m D_m^{(4C)}(\alpha,\beta,\gamma)T_m^{(4)\prime}, \qquad \text{and} \qquad T_A^{(3)} = \sum_m D_m^{(3A)}(\alpha,\beta,\gamma)T_m^{(3)\prime},$$

where $$D_m^{(4C)}(\alpha,\beta,\gamma) \equiv \tfrac{1}{2}[D_{4,m}^{(4)}(\alpha,\beta,\gamma)^* + D_{-4,m}^{(4)}(\alpha,\beta,\gamma)^* + \tfrac{\sqrt{70}}{5}D_{0,m}^{(4)}(\alpha,\beta,\gamma)^*]$$

and $$D_m^{(3A)}(\alpha,\beta,\gamma) \equiv [D_{2,m}^{(3)}(\alpha,\beta,\gamma)^* - D_{-2,m}^{(3)}(\alpha,\beta,\gamma)^*],$$

and $T_m^{(4)\prime}$ and $T_m^{(3)\prime}$ represent the spherical tensor components in the rotated frame, *i.e.* with k_+ defined as $k_\xi + ik_\eta$, *etc.* Here $\xi\eta\zeta$ are the coordinates in the rotated frame, specified with respect to the xyz frame by the Euler angles $\alpha\beta\gamma$. The matrices $D_{m'm}^{(\ell)}(\alpha,\beta,\gamma)$ are the representation matrices[19] for the ℓth representation of $R(3)$. The operator $T_m^{(\ell)\prime}$ raises the axial quantum number λ by m when $\mathbf{B}$ is along $\hat{\zeta}$.

The functions $D_m^{(3A)}(\alpha,\beta,\gamma)$ are proportional to the angular functions $F_m(\alpha,\beta,\gamma)$ defined by La Rocca *et al.*, and originally derived by Rashba and Sheka[9] for $\gamma = 0$. The functions $D_m^{(4C)}(\alpha,\beta,\gamma)$ are the analogous angular functions for the warping terms in H_{eff}, and are proportional to the functions W_m, for $m \neq 0$, and to $W_0 - \frac{3}{5}$ for $m = 0$. The γ-dependence in both cases is simply $\exp(im\gamma)$.

The zeros of the angular functions which occur for specific $\hat{\zeta}$ directions, specified by α and β, are easily understood. The cubic harmonic $x^4 + y^4 + z^4 - \frac{3}{5}r^4$ with O_h symmetry is symmetric with respect to C_4 rotations about the [001] directions, C_3 rotations about the [111] directions, and C_2 rotations about the [110] directions. Coordinate rotations by an angle γ about $\hat{\zeta}$ cause the mth component of the cubic harmonic in the $\xi\eta\zeta$ frame to be multiplied by $e^{im\gamma}$. For this term to be invariant with respect to C_4 rotations, m can only be 0 or ± 4. Thus all of the angular functions W_m associated with warping with $m \neq 0$ or ± 4 must vanish when $\hat{\zeta}$ is along [001] or a similar direction. Similarly, the functions must vanish when $m \neq 0$ or ± 3 when $\hat{\zeta}$ is along a [111] direction, and they must vanish when $m \neq 0$, ± 2 or ± 4 when $\hat{\zeta}$ is along a [110] direction. Similarly, the cubic harmonic xyz is antisymmetric with respect to C_4 rotations about [001], symmetric with respect to C_3 rotations about [111], and antisymmetric with respect to C_2 rotations about [110]. This implies that the angular functions F_m must vanish if $m \neq \pm 2$ when $\hat{\zeta}$ is along a [100] direction, if $m \neq 0$ or ± 3 when $\hat{\zeta}$ is along a [111] direction, and if $m \neq \pm 1$ or ± 3 when $\hat{\zeta}$ is along a [110] direction.

The vanishing of the angular functions along the high-symmetry directions is exactly what is required to produce the selection rules obtained for magneto-optical transitions by Trebin *et al.*[13] when $\mathbf{B}$ is along those directions.

The present formulation has been used to obtain the magnetic field dependence of the absorption coefficients of the $2\omega_c(\pi)$, $2\omega_c + \omega_s(\sigma_\perp)$ and $3\omega_c(\pi)$ intra-conduction-band transitions[7,8] at magnetic fields near 10 T where nonparabolic effects are important, using the form of the interaction Hamiltonian given by Grisar *et al.*[20] A controversy regarding values of the band parameters which contribute to inversion asymmetry complicates the comparison of theory and experiment. However, the basic effect of nonparabolicity is to reduce the absorption strength relative to the extrapolated low-field result. Our experimental values for the absorption[8] are of the order of 50% higher for $2\omega_c$ and $2\omega_c + \omega_s$ and 80% higher for $3\omega_c$ than the theoretical values obtained from high-field calculations.

I am grateful to R. L. Aggarwal, O. W. Shih and H. J. Jiménez-González for many useful discussions, to E. S. Bobrov for translation of the article by Sheka and Zaslavskaya, and to L. P. Byrne for encouragement in this work.

1. D. G. Seiler, C. L. Littler and M. H. Weiler, in *Semiconductors and Semimetals*, eds. R. K. Willardson, A. C. Beer and E. R. Weber (Academic, New York, 1992), vol. 36, p. 293.
2. E. I. Rashba and V. I. Sheka, in *Landau Level Spectroscopy*, eds. G. Landwehr and E. I. Rashba (North-Holland - Elsevier, Amsterdam, 1991), p.131.
3. N. O. Lipari and A. Baldereschi, *Phys. Rev. Lett.* **25** (1970) 1660.
4. A. Baldereschi and N. O. Lipari, *Phys. Rev. B* **8** (1973) 2697.
5. A. Baldereschi and N. O. Lipari, *Phys. Rev. B* **9** (1974) 1525.
6. C. R. Pidgeon and R. N. Brown, *Phys. Rev.* **146** (1966) 575; C. R. Pidgeon and S. H. Groves, *Phys. Rev.* **186** (1969) 824.
7. M. H. Weiler, R. L. Aggarwal, and B. Lax, *Phys. Rev. B* **17** (1978) 3269.
8. G. Favrot, R. L. Aggarwal and B. Lax, *Solid State Commun.* **18** (1976) 577.
9. E. I. Rashba and V. I. Sheka, *Soviet Phys.-Solid State* **3** (1961) 1257, 1357.
10. G. C. La Rocca, N. Kim and S. Rodriguez, *Phys. Rev. B* **38** (1988) 7595.
11. G. C. La Rocca, S. Rodriguez and F. Basani, *Phys. Rev. B* **38** (1988) 9819.
12. J. M. Luttinger and W. Kohn, *Phys. Rev.* **97** (1955) 869.
13. H.-R. Trebin, U. Rössler, and R. Ranvaud, *Phys. Rev. B* **20** (1979) 686.
14. Y. Yafet, in *New Developments in Semiconductors*, eds. P. R. Wallace, R. Harris, M. J. Zuckermann (Noordhoff, Leyden, 1973), p. 441.
15. J. M. Luttinger, *Phys. Rev.* **102** (1956) 1030.
16. S. Gopalan, J. K. Furdyna and S. Rodriguez, *Phys. Rev. B* **32** (1985) 903.
17. V. I. Sheka and I. G. Zaslavskaya, *Ukr. Fiz. Zh.* **14** (1969) 1825. (In Ukrainian and in Russian.)
18. A. Obuchowicz and J. Wlasak, *Acta Phys. Pol. A* **79** (1991) 329. (The expression for β_4, here, is incorrect. See ref. 17.)
19. M. Tinkham, *Group Theory and Quantum Mechanics*, (McGraw-Hill, New York, 1964), p. 110.
20. R. Grisar, H. Wachernig, G. Bauer, J. Wlasak, J. Kowalski and W. Zawadzki, *Phys. Rev. B* **18** (1978) 4355.

THE WAVE–FUNCTION MIXTURE OF SUBBAND ELECTRON LANDAU LEVELS IN HgCdTe

JUNHAO CHU, R. SIZMANN*, KUN LIU, F. KOCH*
National Laboratory for Infrared Physics, Shanghai Institute of Technical Physics, Academia Sinica, Shanghai 200083, P. R. China
**Physics Department E–16, Technical University of München D–85784 Garching, Germany*

ABSTRACT

The wave–function mixture, which arises from the contribution of the k–linear term of the spin–orbit interaction of the surface electrons in narrow gap semiconductors, of subband electron Landau levels $|n,+\rangle$ and $|n+1,-\rangle$ in n–type inversion layer of p–type HgCdTe metal–insulator–semiconductor (MIS) heterostructure has been investigated both experimentally and theoretically. The spin resonance of subband electrons has been measured, which is attributed to the transitions of $\Delta n=-1, \Delta s=0$ and $\Delta n=0, \Delta s=-1$ due to the wave–function mixture of subband landau levels.

1. Introduction

Because the k–linear term of the spin–orbit interaction of the surface electrons causes a large spin–splitting effect on subband electrons in narrow gap semiconductors[1,2], an additional contribution to the landau level separation has to be taken into account in the calculation of landau levels, thus the wave–function mixture between the landau levels $|n,+\rangle$ and $|n+1,-\rangle$ will happen and make spin resonance of subband electrons more complicated. The quantitative description of the subband Landau levels and the wave–function mixture becomes possible due to the recent development in the study of the subband model for inversion layer electrons[3,4]. In this paper the expressions of the wave–function mixture of the subband landau levels for $Hg_{1-x}Cd_xTe$ with $x=0.21$ and $N_A=8\times10^{16}\,cm^{-3}$ and $x=0.234$ and $N_A=4\times10^{17}cm^{-3}$ are derived from the Kane model with the consideration of spin–orbit interation. The subband parameters in the expressions are derived from the experimental results of capacitance spectroscopy, SdH oscillation and cyclotron resonance by using the subband model. From the quantitative analysis of the wave–function mixture between the Landau levels $|n,+\rangle$ and $|n+1,-\rangle$, one can see clearly that the spin resonance optical transition will occur between the subband electron states with $\Delta n=0, \Delta s=-1$ and $\Delta n=-1, \Delta s=0$. The transition with $\Delta n=-1, \Delta s=0$ is the first observation to our knowledge.

2. Subband Landau Levels

For the subbands of n–type inversion layer of p–type HgCdTe, in the presence of magnetic field ($B \parallel k_z$), without considering the k–linear term of spin–orbit interation, the Landau levels of subband electrons can be derived, being

$$E_{n,\pm} = -\frac{E_{g,\text{eff}}}{2} + \sqrt{\frac{E_{g,\text{eff}}^2}{4} + E_g\left[\hbar\omega_c\left(n+\frac{1}{2}\right) + \frac{g\mu_b B}{2}\right]} \quad (1)$$

Where $E_{g,\text{eff}} = E_g + 2E_i(1-j)$ is the effective energy bandgap, which stands for the average value of the energy separation from the valence band to the subband, E_i is the subband energy. In the electric quantum limit, only the ground state subband is occupied, $i = 0$. $j = z_{av}/z_i$ is a characteristic parameter, which is introduced for describing the wave–function distribution of subband electrons, thus the electron gas can be regarded as distributing on the average at the position of z_{av} from the surface. If the k–dependent term $(\nabla U \times K)\cdot\sigma$ of spin–orbit interation[2] is considered, the subband electron Hamiltonian can be written as

$$H = \begin{pmatrix} \boldsymbol{H}_1 & iA\sqrt{n} \\ -iA\sqrt{n} & \boldsymbol{H}_2 \end{pmatrix} \quad (2)$$

where $\boldsymbol{H}_1$ and $\boldsymbol{H}_2$ are the subband electron Hamiltonians in the magnetic field without considering the k–dependent term of spin–orbit coupling. The eigenvalues of $\boldsymbol{H}_1$ and $\boldsymbol{H}_2$ are $E_{n,+}$ and $E_{n,-}$. Here $A = 2\hbar\alpha E_z/\lambda_c$, λ_c is the cyclotron radius. E_z is the electric field in the inversion layer, α is the coupling coefficient. The quantity $\hbar\alpha E_z$ characterizes the intensity of spin–orbit coupling. It can be estimated from Refs. [5] and [6].

The energy eigenvalues obtained after perturbation is considered are

$$E_{\substack{n,- \\ n-1,+}} = \frac{E_{n-1,+} + E_{n,-}}{2} \pm \sqrt{\frac{(E_{n,-} - E_{n-1,+})^2}{4} + A^2 n} \quad (3)$$

The wave–function can be expressed as

$$\begin{cases} \Psi_{n,-} = \sqrt{\dfrac{1+\sqrt{1+C}}{2\sqrt{1+C}}}\,|n,-\rangle + i\sqrt{\dfrac{-1+\sqrt{1+C}}{2\sqrt{1+C}}}\,|n-1,+\rangle \\ \Psi_{n-1,+} = \sqrt{\dfrac{1+\sqrt{1+C}}{2\sqrt{1+C}}}\,|n-1,+\rangle + i\sqrt{\dfrac{-1+\sqrt{1+C}}{2\sqrt{1+C}}}\,|n,-\rangle \end{cases} \quad (4)$$

Where $c = 4A^2 n/(E_{n,-} - E_{n-1,+})^2$. By using Eqs. (3) and (4), the energy positions and wave–functions of the Landau levels of the subband in narrow gap semiconductors can be derived.

3. Spin Resonance

The spin resonance of subband electrons is the most important experiment for the study of the wave–function mixture. However, this is very difficult to be observed, since the intensity of the spin resonance signal is almost 20 times smaller than that of the cyclotron resonance for subband electrons. In order to investigate this project, we have measured the subband electron spin resonance with photon energy of 10.5 meV for p–$Hg_{1-x}Cd_xTe$ MIS structure samples with $x = 0.21$, $N_A = 8 \times 10^{16}\,cm^{-3}$, and $x = 0.234$, $N_A = 4\times10^{17}\,cm^{-3}$ at different N_s. The sample preparation and the details of experimental for magneto–optical measurements have been reported in Ref. [3]. A far infrared laser beam perpendicular to the surface of the sample and parallel to the magnetic field is incident onto the sample. In the reflective configuration[3], by measuring the differential reflectance ΔR between flat band voltage and the applied gate voltage, the resonance peak can be observed with sweeping magnetic field. By changing the direction of magnetic field, one can switch the cyclotron active (CRA) mode to its inactive (CRI) polarization[7]. The measured spin resonance spectroscopy for photon energy of 10.5 meV at different N_s for p–HgCdTe with $x = 0.21$ and $N_A = 8 \times 10^{16}\,cm^{-3}$ is shown in Fig. 1. The resonance peaks are corresponding to the transitions $1^+ \to 1^-$ at $N_s = 2\times10^{11}\,cm^{-2}$, $2^+ \to 2^-$ at $N_s = 6.5\times10^{11}\,cm^{-2}$, $3^+ \to 3^-$ at $N_s = 4.8\times10^{11}\,cm^{-2}$ and $4^+ \to 4^-$ at $N_s = 6.5\times10^{11}\,cm^{-2}$, respectively. In these cases, the positions of the Fermi energy are obtained to be 32.5 meV, 45 meV, 64 meV and 77 meV above the bottom of the subband, respectively. The corresponding magneto–optical transitions are described by arrows in the Landau level fan chart in Fig. 2.

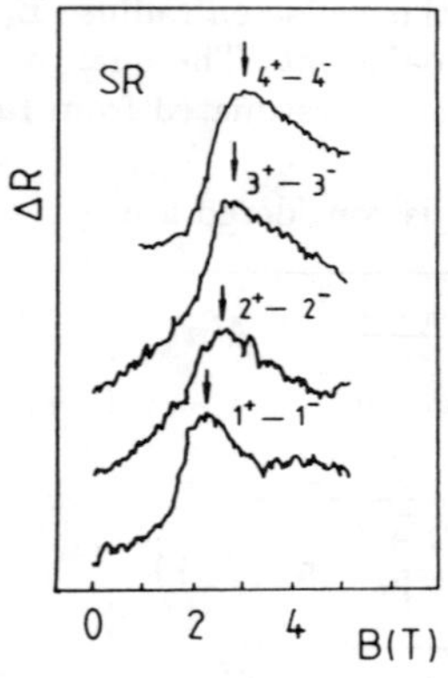

Fig. 1 Spin resonance spectroscopy for photon energy of 10.5 meV at different N_s for p–HgCdTe with $x = 0.21$ and $N_A = 8\times10^{16}\,cm^{-3}$

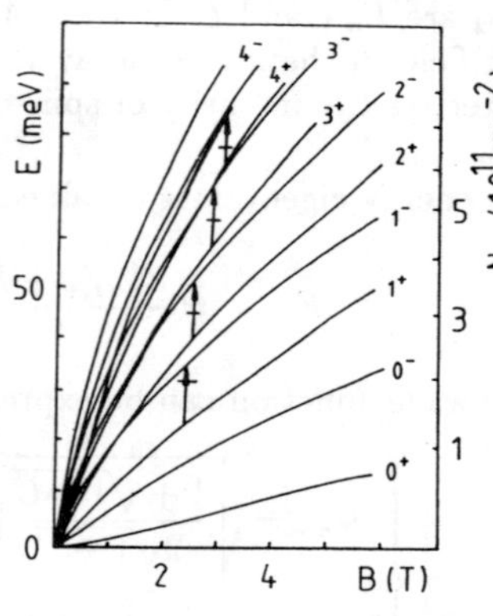

Fig. 2 The obtained Landau level fan chart and the corresponding magneto–optical transitions (shown in arrows)

4. Wave–Function Mixture

The origin of the spin resonance transition can be analyzed from the subband level electron wave–function coupling of $|n,-\rangle$ with $|n-1,+\rangle$. As can be seen that the states of Landau level $|n,+\rangle$ include two components of $|n,+\rangle$ and $|n+1,-\rangle$, the states $|n,-\rangle$ include also two components of states $|n,-\rangle$ and $|n-1,+\rangle$ due to the wave–function mixture of Landau levels with alternating Landau numbers and opposite spin numbers. They can be expressed as

$$\Psi_{n,-} = C_{n,-}|n,-\rangle + i\sqrt{1 - C_{n,-}^2}\,|n-1,+\rangle \tag{5}$$

$$\Psi_{n,+} = C_{n,+}|n,+\rangle + i\sqrt{1 - C_{n,+}^2}|n+1,-\rangle \tag{6}$$

The variation of $|C_{n,\pm}|^2$ versus magnetic field can be calculated from Eq. (4). Fig. 3 shows the calculated results for the sample with $x = 0.234$, $N_A = 4 \times 10^{17}\,\text{cm}^{-3}$. It is obvious that $|n,\pm\rangle$ is constructed not only by the component of states $|n,\pm\rangle$, but also by that of $|n \pm 1, \mp\rangle$ apart from 0^-. The more larger the Landau number, the stronger the wave–function mixture is. The effect of wave–function mixture decreases as the magnetic field increases.

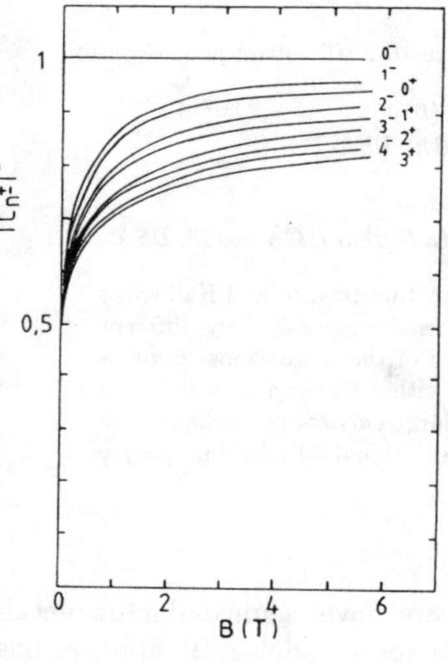

Fig. 3 The calculated $|C_{n,\pm}|^2$ for the sample with $x = 0.234$ and $N_A = 4\times 10^{17}\,\text{cm}^{-3}$

Perusal of Eq. (5) reveals that there are two contribution from the states of Landau level $|n,+\rangle$ to $|n,-\rangle$. The first contribution is the transition from the first term $|n,+\rangle$ of Eq. (6) to the first term $|n,-\rangle$ of Eq. (5). It arises from the wave–function mixture of the conduction band and the valence band, which has been investigated for MCT bulk material by McCombe[8] long ago. The second contribution includes two possible transitions from the first term $|n,+\rangle$ and second term $|n+1,-\rangle$ of Eq. (6) to the second term $|n-1,+\rangle$ and the first term $|n,-\rangle$ of Eq. (5), respectively, which arise from the transitions of $\Delta n = -1$, $\Delta s = 0$. This is the first observation of such a transition. The quantitative description of the intensity of the above transitions will be presented elsewhere.

In conclusion, the wave–function mixture of subband Landau levels and spin resonance of subband electron for MCT have been investigated from both theory and experiment.

5. References

1. F. J. Ohkawa and Y. Uemura, *J. Phys. Soc. Jap.*, **37** (1974) 1325.
2. Y. A. Bychkov and E. I. Rashba, *J. Phys.*, **C17** (1984) 6039.
3. J. H. Chu, Z. Y. Mi, R. Sizmann and F. Koch, *Phys. Rev.*, **B44** (1991) 1717.
4. Kun Liu, Junhao Chu, Biao Li, Dingyuan Tang, *Acta Physica Sinica*, **43** (1994) 267.
5. U. Rossler, F. Malcher, G. Lommer, *Proc. of the Int. Conf. of Application of High Magnetic Field in Semicond. Physics*, Ed. Landwehr, (Wurzburg 1988), p157.
6. J. H. Chu, R. Sizmann, and F. Koch, *Science in China*, **A33** (1990) 515.
7. R. Sizmann, Junhao Chu, R. Wollrab, F. Koch, J. Ziegler and H. Maier, *19th Int. Conf. on Physics of Semicod.*, (Warsaw, Poland, 1988), p471.
8. B. D. McCombe, R. J. Wagner, *Phys. Rev. Lett.*, **4** (1971) 1285.

QUANTUM OSCILLATIONS IN THE HALL EFFECT OF THIN $Sc_{1-x}Er_xAs$ EPITAXIAL LAYERS BURIED IN GaAs

R. Bogaerts, A. De Keyser, F. Herlach,

Department of Physics, Katholieke Universiteit Leuven, B-3001 Leuven, Belgium

F.M. Peeters,

Department of Physics, Universitaire Instelling Antwerpen, B-2610 Antwerpen, Belgium

F. DeRosa, C.J. Palmstrøm,

Bellcore, Red Bank, N.J. 07701-7040, USA

D. Brehmer, and S.J. Allen, Jr.

Department of Physics, University of California Santa Barbara, CA 93106, USA

$Sc_{1-x}Er_xAs$ epitaxial layers buried in GaAs show magnetoresistance and Hall effect oscillations containing Shubnikov-de Haas (SdH) frequencies generated by different parts of the Fermi surface. The analysis of the phases of the oscillations confirms the attribution of the different series of oscillations to either electron or hole Fermi surface pockets. Oscillations due to lower mobility charge carriers are enhanced in the Hall effect, which results in the observation of an additional SdH hole frequency for a 20 nm $Sc_{.43}Er_{.57}As$ sample.

1. Introduction

Semimetallic $Sc_{1-x}Er_xAs$ layers buried in GaAs are novel semiconductor-metal-semiconductor heterostructures opening perspectives for technological applications such as metal base transistors[1]. The composing rare earth monoarsenide compounds have a complicated band structure[2], which can be studied experimentally by SdH measurements in high magnetic fields. Apart from SdH oscillations in the magnetoresistance, we have studied SdH-like quantum oscillations in the Hall effect of $Sc_{1-x}Er_xAs$ layers with a thickness ≤ 20 nm. The measurements were carried out in pulsed magnetic fields up to 50 T, on a $Sc_{.43}Er_{.57}As$ layer of 71 monolayers (ML) (20 nm) and a $Sc_{.46}Er_{.54}As$ layer of 20.5 ML.

2. Experimental results and interpretation

The magnetotransport measurements were carried out at 4.2 K, with the magnetic field perpendicular to the sample surface. To improve the resolution of the oscillations, the background magnetoresistance and Hall resistance were suppressed by means of an additional amplifier with a filter window of 1 kHz-100 kHz. To retain the low field SdH oscillations with a high resolution, data sets were composed of several SdH traces with a different maximum field. Fig. 1 shows five representative signals for the quantum structure superimposed on the classical background curves.

The frequency spectra of the observed oscillations (Fig. 2) are interpreted in terms of the Fermi surface model shown in Fig. 3, which is based on theoretical band structure calculations for GdAs[2]. The calculations predict a semimetallic band structure with three valence bands degenerated at the Γ-point of the first Brillouin zone. One of these bands shifts down if spin-orbit interaction is taken into account. The exchange interaction between the magnetic Er ions and the mobile charge carriers induces a splitting of the energy levels of the spin up and spin down charge carriers, which results in a splitting of the SdH frequencies for each part of the Fermi surface.

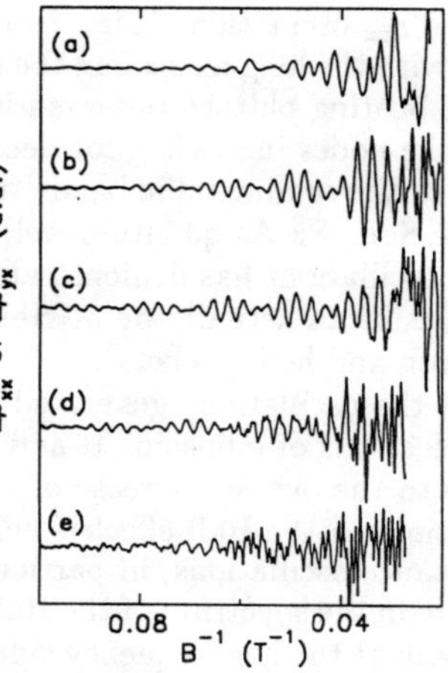

Fig. 1. Quantum oscillations superimposed on the classical background Hall effect and magnetoresistance curves: (a) $\Delta\rho_{xx}$ for 20 ML ErAs, (b) $\Delta\rho_{xx}$ for 20.5 ML $Sc_{.46}Er_{.54}As$, (c) $\Delta\rho_{yx}$ for 20.5 ML $Sc_{.46}Er_{.54}As$, (d) $\Delta\rho_{xx}$ for 71 ML $Sc_{.43}Er_{.57}As$, and (e) $\Delta\rho_{yx}$ for 71 ML $Sc_{.43}Er_{.57}As$.

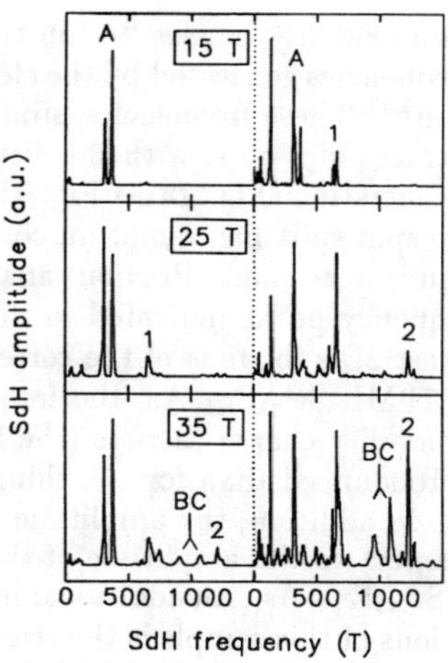

Fig. 2. Maximum Entropy Method frequency spectra of the oscillations in the magnetoresistance (left column) and in the Hall resistance (right column) of 71 ML $Sc_{.43}Er_{.57}As$, for measurements up to 15 T, 25 T and 35 T. The labels of the frequency peaks correspond to the different parts of the Fermi surface model of Fig. 3.

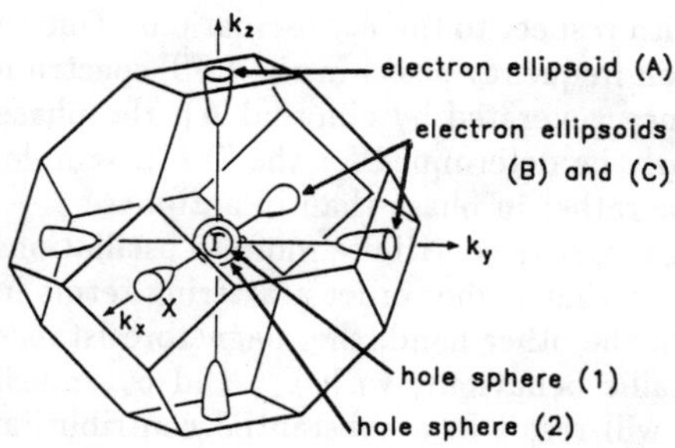

Fig. 3. A simplified model for the Fermi surface of $Sc_{1-x}Er_xAs$, consisting of three electron ellipsoids and two hole spheres. The third hole sphere, centered at Γ and corresponding to the spin-orbit split-off valence band, is not shown. Due to the exchange interaction between the mobile charge carriers and the magnetic Er ions, the energy levels of the spin-up and the spin-down charge carriers are split, resulting in a splitting of the Fermi surface (this is not shown) and a splitting of the SdH frequencies (cfr. Fig. 2).

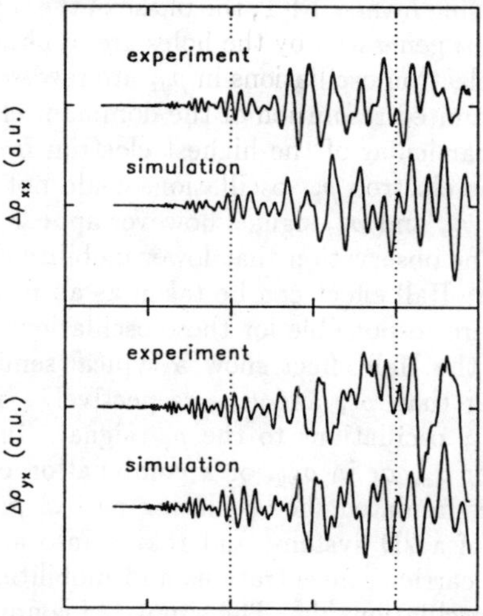

Fig. 4. Comparison of the experimentally observed oscillations in the Hall effect and the magnetoresistance with the simulated oscillations for 20.5 ML $Sc_{.46}Er_{.54}As$.

The dominant frequencies "A" in the spectra of the ρ_{xx} oscillations (Fig. 2) are spin split frequencies generated by the electron ellipsoid with the long axis along the growth direction[3,4]. These frequencies produce the regular beating pattern observed in most of the traces of Fig. 1, with the distance between the nodes increasing for decreasing Er concentration (Fig. 1a → Fig. 1b). In addition to the semimetallic band structure and the spin splitting, quantum confinement in the $Sc_{1-x}Er_xAs$ quantum well has to be taken into account. Previous analysis of the ρ_{xx} oscillations has demonstrated that the frequency peaks indicated in the spectra of Fig. 2 can actually be attributed to the lowest 2D subbands of the corresponding electron and hole pockets[5].

For 71 ML $Sc_{.43}Er_{.57}As$, the frequency peaks of the oscillations generated by the lower mobility charge carriers (the holes and the electrons of ellipsoids B and C) are more pronounced than for the thinner sample due to the overall increase of the mobilities. In addition, the amplitude of these oscillations in the Hall effect is enhanced with respect to the amplitude of the magnetoresistance oscillations, in particular for 71 ML $Sc_{.43}Er_{.57}As$. An additional interesting feature in the spectrum of the Hall effect oscillations of this sample is the strong frequency peak at the low frequency side of the electron frequencies of ellipsoid A. If this low frequency peak is attributed to a spherical Fermi surface, the corresponding 3D carrier concentration is $0.09{\cdot}10^{20}$ cm^{-3}, which represents only a small fraction of the total carrier concentration ($n \approx p \approx 3.3{\cdot}10^{20}$ cm^{-3}). Thus, it is possible that the frequency is generated by a slightly occupied hole band, which is the spin-orbit split-off band of the theoretical band structure calculation. If this is the case, the absence of a low frequency peak for the thinner sample is explained by the depopulation of the hole band as a consequence of quantum confinement.

Using inverse FFT, the phase of the separate oscillations was determined. The oscillations generated by the holes are in phase for both samples. For the 20.5 ML sample, the electron oscillations in ρ_{yx} are reversed with respect to the ρ_{xx} oscillations. Due to the limited resolution of the dominant electron frequency peaks in the FFT spectrum (in particular of the highest electron frequency generated by ellipsoid A), the phase of the electron ρ_{yx} oscillations could not clearly be determined for the 71 ML sample. The ρ_{xx} and ρ_{yx} signals however appear to be rather in phase than in antiphase.

The observation that lower mobility charge carriers contribute more to oscillations in the Hall effect can be taken as an indication that higher order scattering terms in σ_{xy} are responsible for these oscillations[6]. On the other hand, the magnetoresistance and the Hall effect show a typical semimetallic behaviour, with ρ_{xx} and σ_{xx} much larger than ρ_{yx} and σ_{xy}, respectively. This will result in a substantial contribution of σ_{xx} oscillations to the ρ_{yx} signal. In order to investigate whether oscillations in either σ_{xx} or in σ_{xy}, or a combination of both, lead to the observed oscillations, we have simulated the oscillating part of ρ_{xx} and ρ_{yx}, using the expressions for σ_{xx} and σ_{xy} of a 2D system[7] and taking into account one 2D subband of each carrier type. The carrier concentrations and mobilities were taken from a previous analysis of the ρ_{xx} oscillations[3,4,5]. The simulated conductivity tensor components were converted to ρ_{xx} and ρ_{yx}, and the background was subtracted. The simulated oscillations plotted directly versus the magnetic field show a striking resemblance to the experimentally observed oscillations (Fig. 4). The relative amplitude of the hole oscillations with respect to the main electron oscillations is somewhat different, but other features of the experimental signals are perfectly reproduced.

For 20.5 ML $Sc_{.46}Er_{.54}As$, the oscillations show the correct phase relations: the hole oscillations are in phase, while the electron oscillations in ρ_{yx} are inverted with respect to the oscillations in ρ_{xx}. A selection and inverse FFT of the electron and hole frequencies in the spectrum of the simulated signals indicates that the simulated oscillations of 71 ML $Sc_{.43}Er_{.57}As$ follow the same behaviour. For the electrons, this is different from the observed behaviour. When the oscillating components of σ_{xy} are omitted in the simulation for 71 ML $Sc_{.43}Er_{.57}As$, the phases of both the electron and hole oscillations are the same for ρ_{yx} and ρ_{xx}, since $\rho_{xx} \approx 1/\sigma_{xx}$ and $\rho_{yx} \approx \sigma_{xy}/\sigma_{xx}^2$, and only σ_{xx} oscillates. Thus, at least for 71 ML $Sc_{.43}Er_{.57}As$, the initial simulation overestimates the magnitude of the oscillations in σ_{xy}. Probably, the contributions of the oscillations in σ_{xx} and σ_{xy} to the oscillations in ρ_{yx} are of comparable magnitude for this sample. For the electrons, these contributions have different phases, and the competition between them leads to a deterioration of the electron signal. This is exactly what is observed (Fig. 2). For 20.5 ML $Sc_{.46}Er_{.54}As$, which has lower carrier concentrations, the magnitude of σ_{xx} is smaller, while σ_{yx} has increased. This results in a reduction of the oscillating σ_{xx} contribution to ρ_{yx}. By contrast, the higher order scattering terms in σ_{xy} are larger. Therefore, the resulting ρ_{yx} oscillations are more "σ_{xy} like" than "σ_{xx} like".

3. Conclusions

Apart from SdH oscillations in the magnetoresistance, we have observed quantum oscillations in the Hall effect of thin $Sc_{1-x}Er_xAs$ epitaxial layers buried in GaAs. Simulations of these oscillations are qualitatively in good agreement with the experimental results and show (i) the relative enhancement of the oscillations of lower mobility charge carriers, such as the holes, and (ii) the competition of the antiphase σ_{xx} and σ_{xy} contributions to the Hall effect electron oscillations. This can result in a deterioriation of the electron signal as experimentally observed for 71 ML $Sc_{.43}Er_{.57}As$.

For 71 ML $Sc_{.43}Er_{.57}As$, the enhancement of the amplitude of the hole oscillations in the Hall effect results in the observation of an additional SdH frequency, which is attributed to the hole sphere corresponding to the spin-orbit split-off valence band.

Acknowledgements

R.B. is supported by the Onderzoeksraad of the K.U. Leuven. F.M.P. is a Senior Research Associate of the Belgian Nationaal Fonds voor Wetenschappelijk Onderzoek.

References

[1] T. Sands, C.J. Palmstrøm, J.P. Harbison, V.G. Keramidas, N. Tabatabaie, T.L. Cheeks, R. Ramesh, and Y. Silberberg, *Materials Science Reports* **5** (1990) 99.
[2] A. Hasegawa and A. Yanase, *J. Phys. Soc. Jpn.* **42** (1977) 492.
[3] S.J. Allen, Jr., F. DeRosa, C.J. Palmstrøm, and A. Zrenner, *Phys. Rev.* **B43** (1991) 9599.
[4] R. Bogaerts, A. Van Esch, L. Van Bockstal, F. Herlach, F.M. Peeters, F. DeRosa, C.J. Palmstrøm and S.J. Allen, Jr., *Physica B* **184** (1993) 232.
[5] R. Bogaerts, A. De Keyser, F. Herlach, F.M. Peeters, F. DeRosa, C.J. Palmstrøm, D. Brehmer and S.J. Allen, Jr., *Solid-State Electronics* **37** (1994) 789.
[6] R.G. Mani, J.R. Anderson, and W.B. Johnson, *J. Phys. Chem. Solids* **48** (1987) 687.
[7] A. Isihara and L. Smrčka, *J. Phys. C* **19** (1986) 6777.

Anisotropic electron-phonon interaction observed in the magnetophonon resonance of black phosphorus

T. Strutz†
RCAST, The University of Tokyo, 4-6-1 Komaba, Meguro-ku, Tokyo 153, Japan

N. Miura
Institute for Solid State Physics, The University of Tokyo, 7-22-1 Roppongi, Minato-ku, Tokyo 106, Japan

Y. Akahama
Himeji Institute of Technology, 2167 Shoshya, Himeji, 671-11 Hyogo, Japan

G. Kido††
Institute for Materials Research, Tohoku University, Katahira 2-1-1, Aoba-Ku, Sendai, Japan

Transverse magnetophonon resonance was studied in single crystals of black phosphorus at 77 K using a continuous magnetic field up to 27 T. With the field oriented parallel to the b-axis, magneto-oscillations were observed for currents parallel to the a-axis and c-axis. Anisotropy in the electron phonon interaction is demonstrated by comparing the magneto-oscillations for both current directions.

I. Introduction

Magnetophonon resonance (MPR)[1] studied in a variety of semiconductors is a powerful tool to investigate effective mass, phonon energies, and electron-phonon interaction. MPR occurs when carriers are resonantly scattered between any two Landau levels close to $k_Z = 0$ by longitudinal optical (LO) phonons causing magneto-oscillations in the magnetoresistivity[1]. In the simplest case the resonance condition is given by

$$\hbar\omega_{LO} = N\,\hbar\omega_C,\ N=1,2,3..., \tag{1}$$

where $\hbar\omega_{LO}$ is the LO phonon energy at the Γ point, N the harmonic number, and $\omega_C = eB/m^*$ the cyclotron frequency with effective mass m^*. One of the systems that has been extensively studied is tellurium. Miura et al.[2] have found that in tellurium, which has six optical phonon modes, the different phonon modes contribute to the MPR depending on whether the current is parallel or perpendicular to the c-axis. This study has been extended by Maeda et al.[3] Like tellurium, black phosphorus (black-P) is a p-type semiconductor with a narrow energy gap of about 0.3 eV.[4] With four atoms per primitive cell, black-P has three acoustic and nine optical phonon dispersion curves, three more than

---

† also Institute for Solid State Physics, The University of Tokyo.
†† present address: National Research Institute for Metals, Sengen, Tsukuba City, Ibaraki 305, Japan.

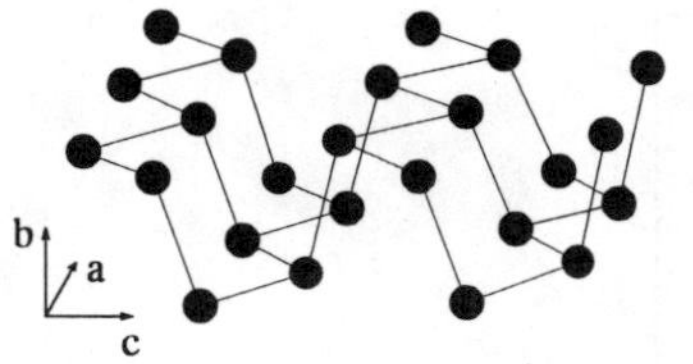

Fig. 1: Part of a puckered layer of phosphorus atoms in the structure of black-P and definition of the crystal axes.

tellurium.[5] Black-P has an orthorhombic puckered layered structure (Fig. 1). The electronic properties as well as the phonon energies are both very anisotropic as it is, for example, reflected in the transport properties.[6, 7]

Recently we reported on the first observation of MPR in black-P at temperatures $T \geq 77$ K using pulsed fields up to 40 T.[8] The magnetic field was oriented parallel to the a-axis and the current was parallel to the c-axis. In this study we report results for the magnetic field oriented parallel to the b-axis using continuous fields. Samples with different crystal orientations were used in order to probe MPR dependence on crystal orientation. A large anisotropy in the MPR spectra for currents parallel to the a-axis and c-axis has been found.

II. Experiment

Single crystals of black-P have been prepared under high pressure (10 kbar) and at high temperatures ($T \leq 900$ C) with crystal orientations along the a-axis and c-axis.[9] The crystal orientations of the samples were checked by x-ray Laue back reflection measurements. The samples had dimensions of about 8 mm x 1.5 mm x 0.7 mm. Ohmic contacts along the crystal axis direction were made on newly cleaved surfaces using gold wire and silver paste. The resistivity along the crystal-axis was measured by using the standard DC four-point contact method, driving a constant current along the crystal-axis. As first results had shown,[8] MPR is strongest at 77 K in the temperature range from 77 K to 200 K. Therefore, the samples have been immersed in a liquid nitrogen bath. The magnetic field up to 27 T was produced by a hybrid magnet providing for low noise level measurements.

III. Results and Discussion

The transverse resistivity of black-P samples at 77 K was measured for the magnetic field parallel the b-axis. The inset of Fig. 2 shows the sample resistance ρ for the c-axis sample. Clearly magneto-oscillations are observed. The relative change of the resisitivity due to MPR is about 3×10^{-4} at 23 T. By subtracting the part of the magnetoresistance $\Delta\rho = \{\rho(B)-\rho(0)\}/\rho(0)$ that increases linearly with increasing field[8], one obtains the trace indicated with $\Delta\rho_{MPR}$ shown in Fig. 2. For transverse MPR the maxima in the resistivity correspond to the resonance field positions.[1]

By taking the second derivative $d^2\rho/dB^2$, the magneto-oscillations can be more easily analyzed. This is shown in Fig. 3 for current orientations parallel to the a-axis and c-axis (a-axis and c-axis sample respectively). Maxima in $-d^2\rho/dB^2$ correspond to maxima in

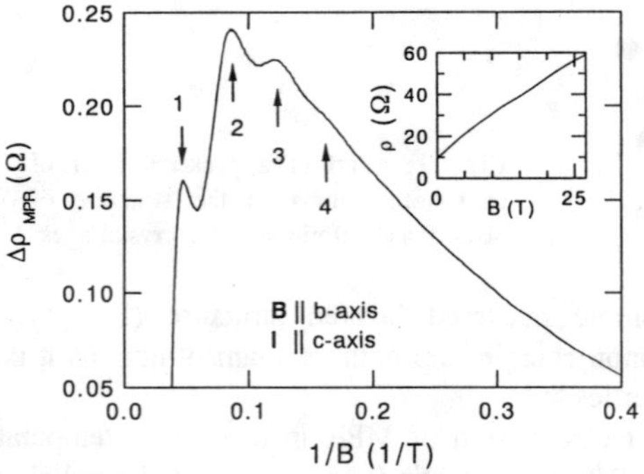

Fig. 2: Oscillating part $\Delta\rho_{MPR}$ of the transverse magnetoresistance (see text) in black-P for **B** ‖ b-axis observed in a c-axis sample. The inset shows the total sample resistance ρ.

$\Delta\rho_{MPR}$. For the current parallel to the c-axis, strong oscillations are seen with beating between different phonon frequencies. Clearly oscillations (6 maxima) with a frequency of 1/24.3 T^{-1} are observed. For the current parallel to the a-axis, the magneto-oscillations look very different with relatively more contributions of higher frequency phonons.

Since several phonon modes contribute to the MPR spectra of Fig. 3, Fourier transformations of $-d^2\rho/dB^2(1/B)$ were performed in order to distinguish between different phonon modes. Fig. 4 shows the result. The difference between both sample orientations is clearly recognizable in the Fourier spectra. For the current parallel to the c-axis there is a strong contribution at about 25 T, whereas for the current parallel to the a-axis the Fourier component at 25 T is relatively weak. However, for the current parallel to the a-axis there is a strong contribution between 35 and 60 T which in the case for the current parallel to the c-axis is weaker. Moreover, both samples show some Fourier components at 90 T.

The Hall coefficient of black-P is positive in a large temperature range from 4.2 K to 550 K.[6] At 77 K black-P is well below the temperature range for intrinsic conduction which starts above 220 K. It has an effective acceptor concentration $(N_A\text{-}N_D)$ in the order of 2–5x10^{15} cm^{-3}. Therefore, MPR is given by the scattering of holes due to LO phonons.

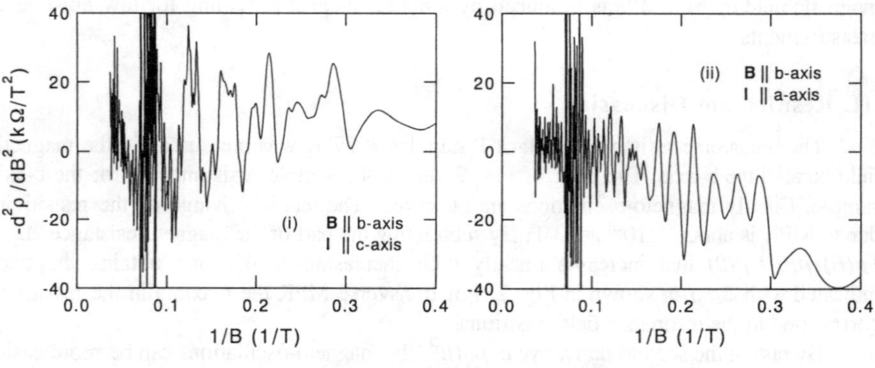

Fig. 3: Second derivative of the transverse magnetoresistance for **B** ‖ b-axis observed in a (i) c-axis and (ii) a-axis sample at 77 K.

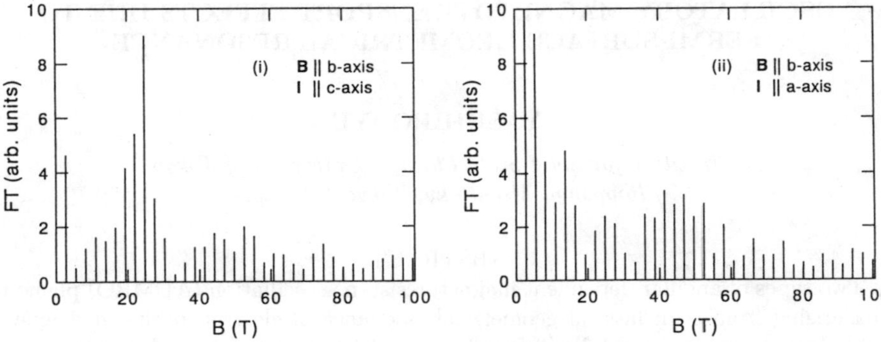

Fig. 4: Fouriertransformation of $-d^2\rho/dB^2$ for a (i) c-axis and (ii) a-axis sample with **B** || b-axis.

Taking the effective cyclotron mass $m_{ac}/m_0 = 0.238$ for holes,[10] one obtains for the MPR with fundamental resonance field of 24.3 T a phonon energy of 95 cm^{-1}. This is smaller than the energy of the lowest LO phonon of 138 cm^{-1}.[11] The resonance fields observed at about 35 T, 50 T, 60 T, and 90 T can be explained by scattering of the B_{1u} phonon mode (138 cm^{-1}, calculated resonant field $B_{res} = 35.2$ T), the B_{1g} phonon mode (197 cm^{-1}, $B_{res} = 50.3$ T), the B_{3g}^1 phonon mode (233 cm^{-1}, $B_{res} = 59.5$ T) and the A_g^1 phonon mode (365 cm^{-1}, $B_{res} = 93.2$ T) respectively.[11] The different contributions for different current directions clearly demonstrate the anisotropy in the electron phonon interaction.

In summary, magnetophonon resonance was observed in black phosphorus at 77 K using a continuous magnetic field. Different phonon modes inducing magnetophonon resonance were identified. Fourier transformation of the MPR spectra reveals the anisotropy in the electron phonon interaction.

The authors are grateful to T. Todo for doing the Laue back reflection analysis.

References:

1. R. J. Nicholas, *Prog. Quant. Electr.* **10**, 1 (1985).
2. N. Miura, R. A. Stradling, S. Askenazy, G. Carrere, J. Leotin, J. C. Portal, and J. P. Ulmet, *J. Phys. C* **5**, 3332 (1972).
3. Y. Maeda, N. Miura, M. Sakata, and E. Ohta, *J. Phys. Soc. Jpn.* **53**, 3120 (1984).
4. A. Morita, *Appl. Phys. A* **39**, 227 (1986).
5. C. Kaneta, H. Katayama-Yoshida, A. Morita, *Solid State Commun.* **44**, 613 (1982).
6. Y. Akahama, S. Endo, and S. Narita, *J. Phys. Soc. Jpn.* **52**, 2148 (1983).
7. M. Baba, F. Izumida, A. Morita, Y. Koike, and T. Fukase, *Jpn. J. Appl. Phys.* **30**, 1753 (1991).
8. T. Strutz, N. Miura, and Y. Akahama, *Physica B* (1994) in press.
9. S. Endo, Y. Akahama, S. Terada, S. Narita, *Jpn. J. Appl. Phys.* **21**, L482 (1982).
10. S. Takeyama, N. Miura, Y. Akahama, S. Endo, *J. Phys. Soc. Jpn.* **59**, 2400 (1990).
11. S. Sugai and I. Shirotani, *Solid State Commun.* **53**, 753 (1985).

OSCILLATORY MAGNETOTRANSPORT EFFECTS DUE TO FERMI SURFACE GEOMETRICAL RESONANCE

YASUHIRO IYE

Institute for Solid State Physics, University of Tokyo
Roppongi, Minato-ku, Tokyo 106 Japan

ABSTRACT

Two types of angular dependent magnetoresistance oscillation (ADMRO) phenomena arising from semiclassical geometrical resonance of electron orbits in magnetic fields have been uncovered for different types of Fermi surface topology. One, that arises from extended orbits circling around a weakly corrugated cylindrical Fermi surface, has been observed in three different systems, *i.e.* (BEDT-TTF)-based organic conductors, graphite intercalation compounds, and semiconductor superlattices. The second type is found in a low temperature phase of a (BEDT-TTF)$_2$KHg(SCN)$_4$ salt and is attributed to commensurability with the Brillouine zone, of open orbits on a quasi-one dimensional Fermi sheet.

1. Introduction

Electron transport in high magnetic fields is an everlasting source of new physics, that keeps this series of Conference always exciting. Historically, magneto-quantum oscillatory phenomena such as de Haas-van Alphen effect have played a key role in the development of metal physics. Analyses of magnetotransport phenomena furnish much information on the nature of charge carriers — Fermi surface, effective mass *etc.* —, and their scattering mechanisms. During the past few years, a family of novel magnetotransport phenomena that originate from commensurability of electron cyclotron orbits with periodic potentials, either natural or artificial, have been uncovered. One example is the so-called Weiss oscillation effect in a two-dimensional electron gas subjected to a one-dimensional potential modulation, in which magnetoresistance oscillation occurs as a function of field strength.[1] A closely related phenomenon occurs in a quantum antidot array, i.e. a two-dimensional electron gas in a two-dimensionally modulated potential.

Another group of comensurability magnetotransport phenomena, to be discussed in this paper, is the angular dependent magnetoresistance oscillation (ADMRO) effect, in which magnetotransport exhibits oscillatory behavior as a function of magnetic field angle with respect to the crystalline axes. As will be shown below, there exist two distinct types of ADMRO effect, which we call "extended orbit ADMRO" and "open orbit ADMRO". The former arises from a cylindrical Fermi surface of a quasi-2D system, while the latter is associated with an open Fermi sheet of a quasi-1D system.

2. Extended Orbit ADMRO

2.1 Model

We consider a quasi-two dimensional electron gas with a free-electron-like band in the conducting plane and a narrow tight-binding band in the third direction.

$$\varepsilon(\vec{k}) = \frac{\hbar^2}{2m^*}(k_x^2 + k_y^2) - 2t \cos(ck_z). \tag{1}$$

For a small value of t, the Fermi surface is a cylinder whose radius k_F is weakly modulated along the k_z-direction with period $2\pi/c$, as shown in Fig. 1a. When magnetic field is applied at a special angle, θ, satisfying the following condition first pointed out by Yamaji[2]

$$ck_F \tan\theta = (n - \frac{1}{4})\pi \qquad (n = 1, 2, 3 \ldots), \tag{2}$$

the crosssections of the Fermi surface cut by a plane perpendicular to the magnetic field collapse into a single value.

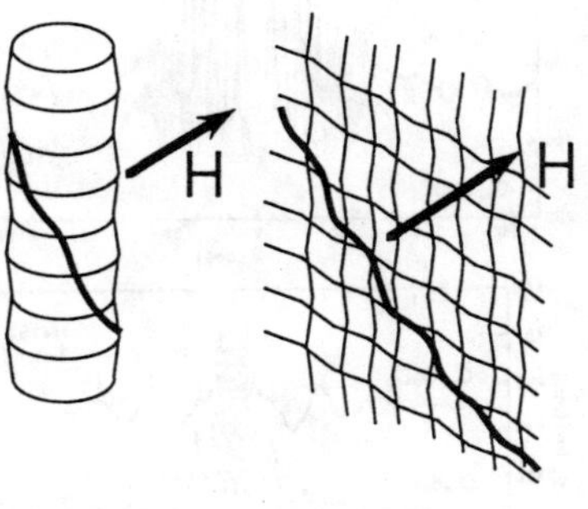

Fig.1 (a) Extended orbit on a quasi-2D Fermi cylinder, and (b) Open orbit on a quasi-1D Fermi sheet.

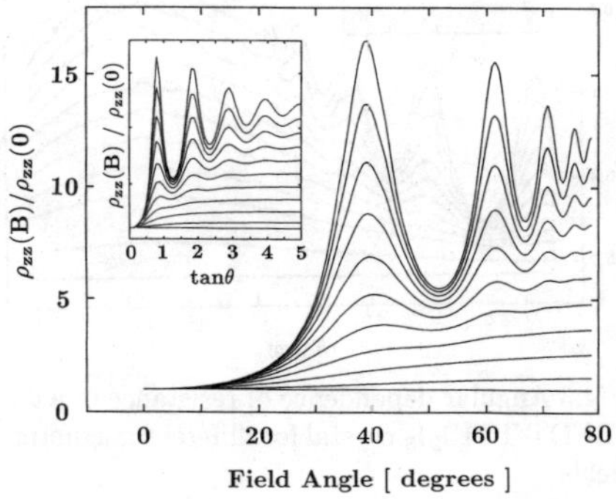

Fig.2 Angular dependence of ρ_{xx} calculated by Eq.(3).

In a semiclassical picture of electron motion in magnetic fields, this correspond to the condition that the k_z-component of the group velocity, $v_z = \frac{1}{\hbar}\frac{\partial\varepsilon(\vec{k})}{\partial k_z}$, vanishes upon averaging over each extended orbit as depicted in Fig. 1-(a). This leads to a suppression of the electron motion in the c-direction, and hence an enhancement of the c-axis resistivity ρ_{zz}. Yagi *et al.*[3] calculated the relevant conductivity tensor components by a semiclassical model based on the Shockley tube integral formalism. The conductivity along the c-axis is written as

$$\sigma_{zz} = \sigma_{zz}^0\big(J_0^2(ck_F tan\theta) + 2\sum_{n=1}^{\infty}\frac{J_\nu^2(ck_F tan\theta)}{1 + (n\omega_0\tau cos\theta)^2}\big). \tag{3}$$

Here, $\sigma_{zz}^0 = 2e^2mt^2c\tau/\pi\hbar^4$ is the c-axis conductivity in the absence of magnetic field, $J_n(z)$ is the Bessel function of the n-th order, and $\omega_0 = eH/mc$ is the cyclotron frequency at $\theta = 0$. The angular dependence of $\rho_{zz}(\theta)$ calculated by this simple model is shown in Fig.2. The calculation indicates that the corresponding structures is absent in $\rho_{xx}(\theta)$.

2.2 ADMRO in (BEDT-TTF)-salts

These are the systems in which the ADMRO effect was first discovered[4,5]. A crystal of BEDT-TTF (*bis*-ethylendithio tetrathia fulvalene) salt is formed by alternate stacking of conducting BEDT-TTF layers and insulating anion layers. Figure 3 shows the θ-dependence of resistivity of a θ-(BEDT-TTF)$_2$I$_3$ sample at different magnetic fields. It is seen that the angular positions of the resistivity peaks are independent of the field strength — a feature that distinguishes ADMRO from usual SdH oscillations. Inset shows a replot of the data for $H = 6.5T$ as a function of *tan* θ. The peak positions are expressed as *tan* $\theta = 0.42(n - 0.26)$.

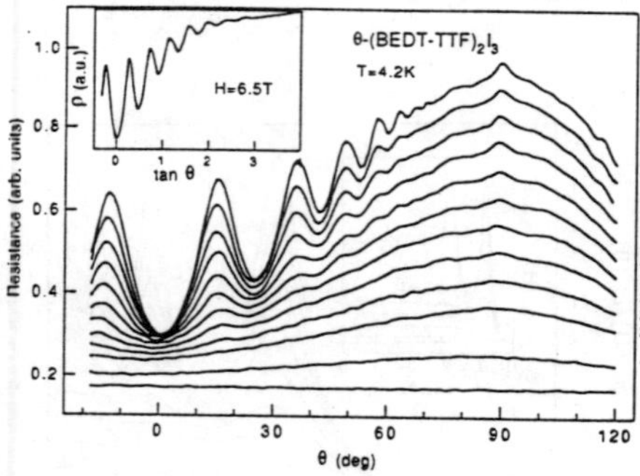

Fig.3 Angular dependence of resistance of a θ-(BEDT-TTF)$_2$I$_3$ crystal for different magnetic fields.

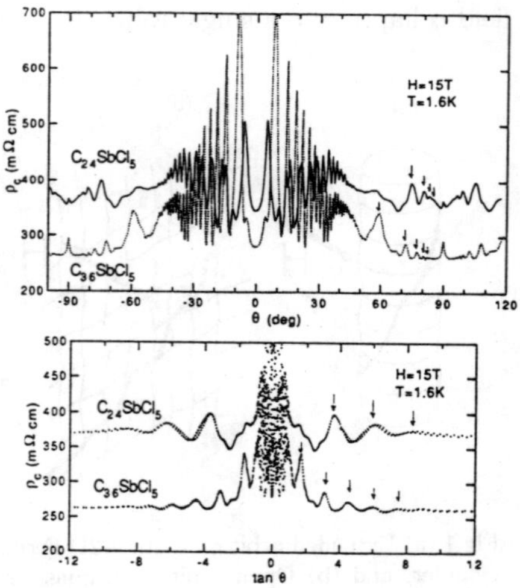

Fig.4 Angular dependence of ρ_c of stage 2 and 3 SbCl$_5$-GICs. The lower panel shows replot as a function of tanθ.

2.3 ADMRO in SbCl$_5$-Intercalated Graphite

Semimetallic graphite is made metallic by intercalation either with donor species such as alkali metals, or with acceptor species such as Br_2, HNO_3 and $SbCl_5$. We chose SbCl$_5$-graphite intercalation compound (GIC), which is most air-stable among GICs, as a candidate for experimental persuit of the ADMRO effect[6]. Figure 4 shows the angular dependences of ρ_{zz} for a stage 2 and a stage 3 SbCl$_5$-GIC at $H = 15T$ and $T = 1.6K$. The rapid oscillations in the angular range $|\theta| < 50°$ are the usual Shubnikov-de Haas (SdH) oscillations. A series of distinct peaks at higher angles marked by arrows are the ADMRO effect. (The sharp peak at $\theta = 90°$ arises from an open orbit for that particular field direction.)

The lower panel of Fig.4 is a replot of the data as a function of $\tan\theta$. The peak positions can be expressed as $\tan\theta = 2.52(n - 0.56)$ for the stage 2 sample, and $\tan\theta = 1.43(n - 0.85)$ for the stage 3 sample. These values of the periodicity in $\tan\theta$ correspond to $k_F = 0.098$ Å^{-1} and $k_F = 0.137$ Å^{1}, respectively. These values of the Fermi cylinder radius are in good agreement with those estimated from the dHvA frequencies of the strongest SdH oscillations for $H \parallel c$-axis.

The ADMRO phenomenon in $SbCl_5$-GICs corroborates the following features expected from the theoretical model:

(1) The peak positions are independent of the field intensity, and can be observed down to much lower fields than the SdH oscillations, as shown in Fig. 5-(a).
(2) The ADMRO features are observable up to much higher temperatures than the SdH oscilations, as shown in Fig. 5-(b).
(3) The ADMRO effect occurs only in ρ_{zz}. The angular trace of ρ_{xx} shows the SdH oscillations but no ADMRO features.
(4) Being a purely transport phenomenon in nature, the ADMRO effect does not manifest itself in thermodynamic quantities. The angular traces of magnetization (magnetic torque) shown in Fig.5-(c) have many dHvA oscillations but no ADMRO-related features.

(a)

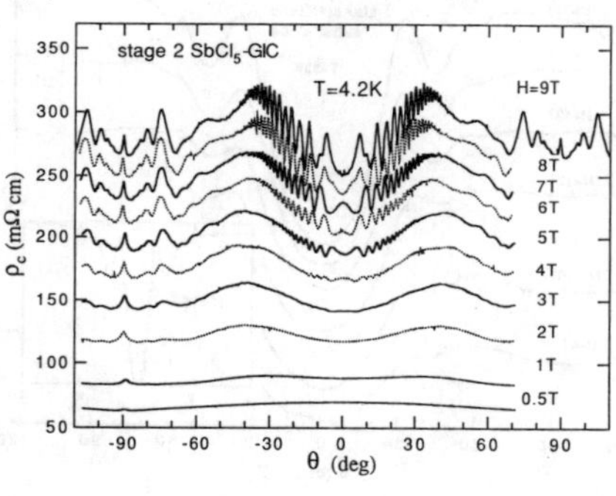

(b)

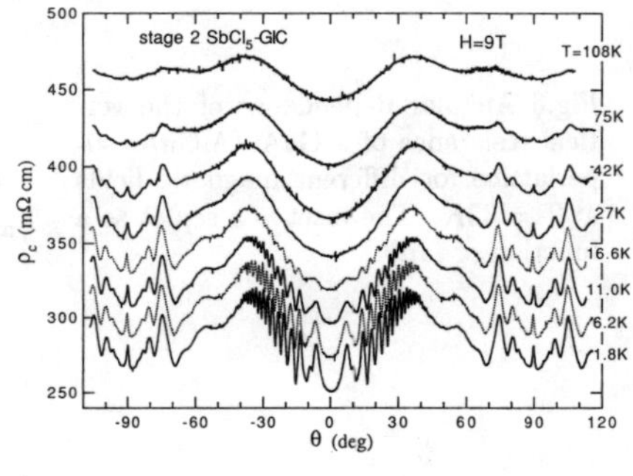

(c)

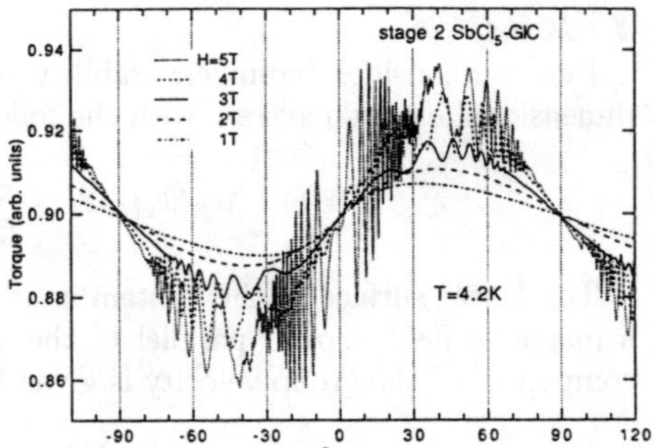

Fig.5 (a) $\rho_c(\theta)$ of $C_{24}SbCl_5$ for different magnetic fields. (b) $\rho_c(\theta)$ for 9T at different temperatures. (c) Magnetic torque of the same sample.

The ADMRO phenomenon in the $SbCl_5$-GICs is by and large explained within the framework of the simple model outlined in the preceeding subsection. There is, however, one feature that shows deviation from the simple model. The "phase factor" of the ADMRO for the stage 2 and 3 $SbCl_5$-GICs are -0.56 and -0.85, respectively. These values are in substantial deviation from the $-\frac{1}{4}$ in Eq.(2). Preliminary calculations indicate that the observed change in the phase factor may be explained by appropriate incorporation of higher order (odd) harmonic components in the k_z-dispersion.

2.4 ADMRO in Semiconductor Superlattices

Growth of semiconductor superlattice samples by molecular beam epitaxy enables us to construct designed Fermi surfaces. Figure 6 shows the angular traces of vertical resistance of a GaAs/AlGaAs superlattice sample whose superlattice periodicity and doping are so chosen to give a cylindrical Fermi surface suitable for the ADMRO effect[7]. Unlike the metallic systems, due to low carrier density (*i.e.* small Fermi surface) and light effective mass, only a few Landau levels are left below the Fermi level in the field range of interest. As a result, the SdH and ADMRO periodicity are close to each other. In fact, at $T = 4.2K$, they are difficut to separate. The data shown in Fig. 6 are therefore taken at $T = 33K$ to eliminate the SdH component.

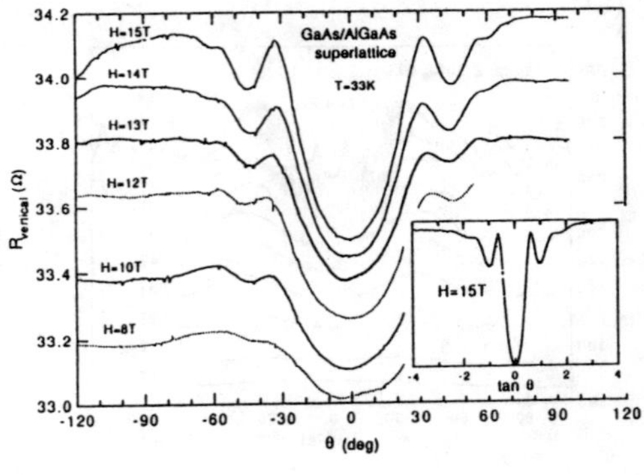

Fig.6 Angular dependence of the vertical resistance of a GaAs/AlGaAs superlattice for different magnetic fields at $T = 33K$. The inset is a replot as a function of $tan\theta$.

3. Open Orbit ADMRO

3.1 Model

Here we consider "commensurability resonance" of open orbits[8,9] in a quasi-one dimensional electron system with the following dispersion relation,

$$\varepsilon(\vec{k}) = \hbar v_F(|k_x| - k_F) - \sum_{m,n} t_{m,n}\ cos\ (mbk_y + nck_z). \tag{4}$$

The Fermi surface of this system is an open sheet as sketched in Fig. 1-(b). For a magnetic field applied parallel to the yz-plane at angle θ from the z-axis, the y-component of the group velocity is given by

$$v_y(\vec{k}, t) = \sum_{m,n} \frac{t_{m,n}mb}{\hbar}\ sin\ \left(G_{m,n}x(t) + mbk_y^0 + nck_z^0\right), \tag{5}$$

where $G_{m,n} = eB(mb\cos\theta - nc\sin\theta)/\hbar$, and $x(t) = sgn(k_x)v_F t$. v_z is given by a similar expression, and $v_x = sgn(k_x)v_F$. For a general value of θ, each electron trajectory is ergodic, *i.e.* it travels through the whole first Brillouin zone in the limit of $\tau \to \infty$. The time-averaged values of v_y and v_z then vanish, and electrons cannot carry current in the yz-plane. For special values of θ satisfying the condition, $\tan\theta = pb/qc$ (p, q: integers), the electron trajectories become commensurate with the reciprocal lattice, leading to non-vanishing v_y and v_z, and hence a finite conductivity. This type of geometrical resonance was first considered by Lebed and Bak[8]. Osada *et al.*[9] made an explicit calculation and comparison with their experimental result for $(TMTSF)_2ClO_4$.

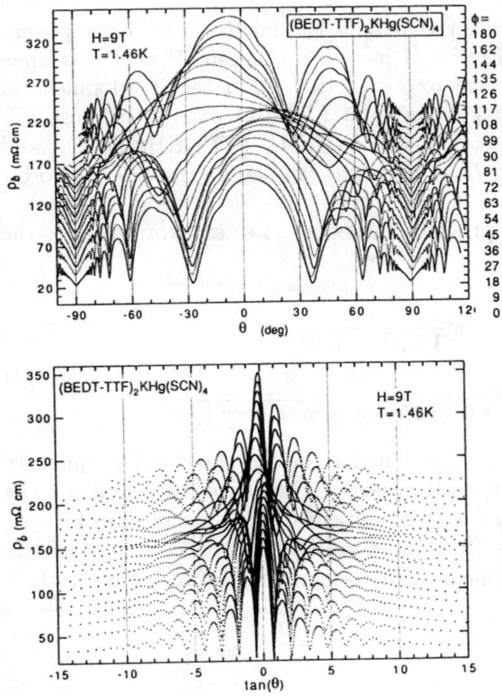

Fig.7 $\rho_b(\theta)$ of α-$(BEDT\text{-}TTF)_2KHg(SCN)_4$ at 9T and 1.46K for different azimuthal angle, φ. The lower panel shows the replot as a function of $\tan\theta$.

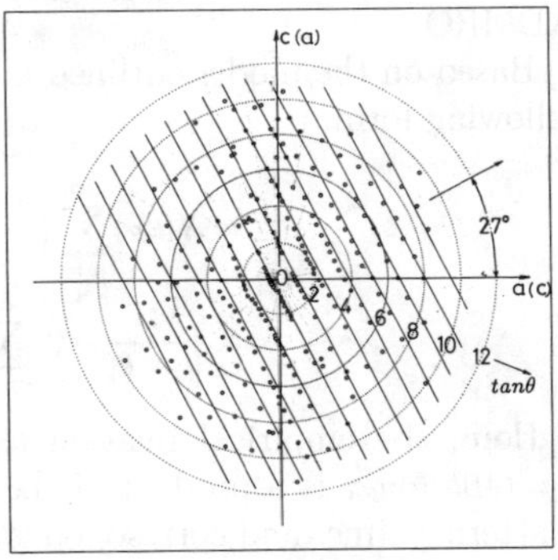

Fig.8 Polar projection plot of the positions of the dips in Fig.7. The data points align on equally spaced straight lines rotated by 27° from one of the principal crystalline axis.

3.2 ADMRO in α-(BEDT-TTF)$_2$KHg(SCN)$_4$

Figure 7 shows traces of ρ_{zz} of a crystal of α-$(BEDT\text{-}TTF)_2KHg(SCN)_4$ (z is perpendicular to the BEDT-TTF molecular layer, and is the least conducting direction) as a function of the polar angle θ and the azimuthal angle φ. The same set of data are replotted as a function of $\tan\theta$ in the lower panel of Fig. 7. Although also periodic in $\tan\theta$, the ADMRO in this system is distinct in many respects from the extended orbit ADMRO discussed in the previous section:

(1) ρ_{zz} shows dips, instead of peaks, as a function of θ.

(2) The periodicity changes drastically as a function of the azimuthal angle φ, and the ADMRO disappears at a certain angle $\varphi = \varphi_0$.
(3) The oscillation pattern is asymmetric about $\theta = 0$.

The positions of the resistance dips in the traces of Fig. 7 are marked with dots in Fig. 8, which is a ploar projection plot with the radial axis representing $tan\ \theta$. The fact that the dots align themselves on equally spaced parallel lines indicates that the ADMRO in this system is associated with an open Fermi sheet. The positions of the resistance dips can be summarized by the following expression;

$$tan\ \theta\ cos\ (\varphi - \varphi_0) = 1.27\nu - 0.55 \qquad (\nu = \ldots, -2, -1, 0, 1, 2, \ldots). \tag{6}$$

The factor $cos\ (\varphi - \varphi_0)$ reflects the fact that only the magnetic field component along the direction $\varphi_0 = 27 \pm 2°$ is effective. I has been suggested[11] that this open Fermi sheet is formed by Fermi surface reconstruction upon the development of magnetic order below $T \approx 8K$. For $T > 8K$, the ADMRO pattern is quite different The ADMRO in the paramagnetic phase of (BEDT-TTF)$_2$KHg(SCN)$_4$, and that in (BEDT-TTF)$_2$(NH)$_4$Hg(SCN)$_4$ (no orderin) are similar to the extended orbit ADMRO.

Based on the model outlined in the previous subsection, we can express σ_{zz} by the following form;

$$\begin{aligned} \sigma_{zz} &= N(\varepsilon_F) \sum_{n,m}^{\infty} \left(\frac{e\ t_{m,n}}{\hbar}\right)^2 n^2 c^2 \frac{\tau}{1 + (G_{m,n} v_F \tau)^2} \\ &\propto \frac{1}{|cos\ \theta|} + 2 \sum_{\nu=-\infty}^{\infty} \frac{t_{\nu,1}^2}{1 + \tilde{\tau}^2 [(1.27\nu - 0.55) cos\ \theta - sin\ \theta]^2} \end{aligned} \tag{7}$$

Here, the empirical relation $tan\ \theta = 1.27\nu - 0.55$ is incorporated. The quantity $\tilde{\tau} = (eBb/\hbar) v_F \tau$ is a product of the open orbit cyclotron resonance frequency and the scattering time, and corresoponds to $\omega_c \tau$ for a closed orbit. The first term in Eq. (7), introduced phenomenologically to reproduce the overall θ-dependence of ρ_{zz}, reflects the magnetoresistance approximately linear in the z-component of magnetic field.

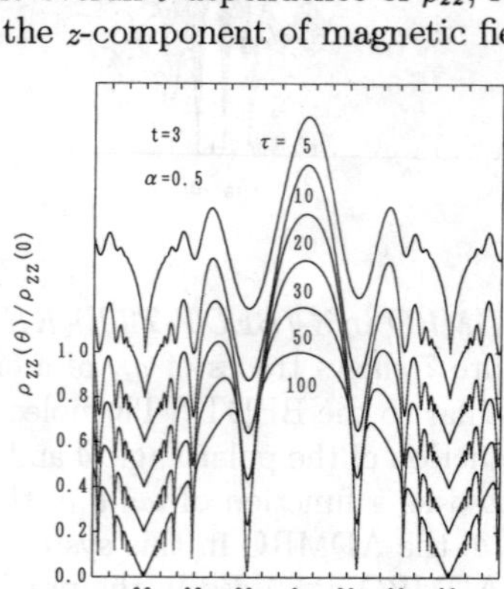

Fig.9 $\rho_{zz}(\theta)$ calculated by Eq.(7), reproducing the experimental traces in Fig.7.

Figure 9 shows examples of the $\rho_{zz}(\theta)$ curves calculated by Eq. (5). Here, the ν-dependence of $t_{\nu,1}$ is simulated by a functional form $t_{\nu,1} = t(1.27\nu - 0.55)^{\alpha}$. The model therefore contains three parameters, t, α and $\tilde{\tau}$. It is seen that the observed angular dependence of ρ_{zz} is very well reproduced by a suitable choice of these parameters.

4. Conclusion

We have seen that oscillatory magnetoresistance periodic in *tanθ* is ubiquitous among low dimensional electronic system. They are manifistation of "commensurability resonance" of electron orbital motion in high magnetic field, and can be interpreted by semiclassical models based on extremely simple models for the band structure and a constant relaxation time approximation.

Acknowledgements

The author appreciates pleasant collaboration with the co-authors of Ref. 5,6,7 and 10. This work has been supported in part by Grant-in-Aid (#04452037) for Scientific Research from the Ministry of Education, Science and Culture, Japan, and by the University-Industry Joint Research Project "Mesoscopic Electronics".

References

1. R.R.Gerhardts, D.Weiss and K.von Klitzing: Phys. Rev. Lett. **62**, 1173 (1989); R.W.Winkler, J.P.Kotthaus and K.Ploog: Phys. Rev. Lett. **62**, 1177 (1989).
2. K.Yamaji: J. Phys. Soc. Jpn. **58**, 1520 (1989).
3. R.Yagi, Y.Iye, T.Osada and S.Kagoshima: J. Phys. Soc. Jpn. **59**, 3069 (1990).
4. M.V.Kartsovnik, P.A.Kononovich, V.N.Laukhin and I.F.Shegolev: Pis'ma Zh. Eksp. Theor. Fiz. **48**, 498 (1988). [translation: JETP Lett. **48**, 541 (1988).]
5. K.Kajita, Y.Nishio, T.Takahashi, W.Sasaki, R.Kato, H.Kobayashi and Y.Iye: Solid State Commun. **70**, 1189 (1989).
6. Y.Iye, M.Baxendale, V.Z.Mordkovich: J. Phys. Soc. Jpn. **63**, 1643 (1994).
7. R.Yagi, Y.Iye, Y.Hashimoto, T.Odagiri, H.Noguchi, H.Sakaki and T.Ikoma: J. Phys. Soc. Jpn. **60**, 3784 (1991).
8. A.G.Lebed' and P.Bak, Phys. Rev. Lett. **63**, 1315 (1989); K.Maki: Phys. Rev. **B45**, 5111 (1992).
9. T.Osada, S.Kagoshima and N.Miura: Phys. Rev. **B46**, 1812 (1992).
10. Y.Iye, R.Yagi, N.Hanasaki, S.Kagoshima, H.Mori, H.Fujimoto and G.Saito: J. Phys. Soc. Jpn. **63**, 674 (1994).
11. M.V.Kartsovnik, A.E.Kovalev, V.N.Laukhin and S.I.Pesotskii: J. Phys. I France **2**, 223 (1993).

FERMI SURFACE STRUCTURE IN THE LOW-DIMENSIONAL ORGANIC CONDUCTORS $(BEDT\text{-}TTF)_2X$

S.J. Klepper,[a] J.S. Brooks,[b] G.J. Athas,[b] C.E. Campos,[b] S. Valfells,[b], P. Sandhu,[b] M. Tokumoto,[c] N. Kinoshita,[c] T. Kinoshita,[c] Y. Tanaka,[c] H. Anzai,[d] J. van Bentum,[e] J.A.A.J. Perenboom,[e] D. Heiman,[a] X. Liu,[a] and U. Gennser[a]

[a] *F.B. National Magnet Laboratory, MIT, Cambridge, MA 02139, USA*
[b] *Boston University, Department of Physics, Boston, MA 02215, USA*
[c] *Electrotechnical Laboratory, Tsukuba, Ibaraki, 305, Japan*
[d] *Himeji Institute of Technology, Himeji, Hyogo 671-22, Japan*
[e] *High Field Magnet Laboratory, University of Nijmegen, Nijmegen, The Netherlands*

ABSTRACT

We have used a variety of experimental probes to determine the Fermi surface (FS) structure of the low-dimensional organic conductors of the type $(BEDT\text{-}TTF)_2X$. Materials of the family α-$(BEDT\text{-}TTF)_2MHg(SCN)_4$ (M = K, Rb Tl, NH_4) have both quasi-2D closed hole orbit and quasi-1D open orbit portions of the FS, which are revealed in electron transport measurements. Small differences in the band structure of these charge-transfer salts have a dramatic effect on their electronic transport properties. We have used three techniques to study the FS and the electronic structure of these molecular crystals: measurements of angular magnetoresistance (MR) oscillations (AMRO), of MR under uniaxial stress, and of photoluminescence. These methods allow us to refine our FS model and to better understand electron transport in these molecular crystals.

1. Introduction

Low-dimensional organic conductors of the type $(BEDT\text{-}TTF)_2X$, or $(ET)_2X$ for short, show a rich variety of low temperature transport behavior ranging from superconductivity to a density-wave ground state, the details of which are quite sensitive to the Fermi surface (FS) structure. Figure 1 shows the magnetoresistance (MR) of $(ET)_2KHg(SCN)_4$, which has a large MR with a maximum near 10 T, and which has a sharp drop in the MR background at a characteristic "kink" field $H_K = 24$ T.[1] This data was obtained in the dc field of the Hybrid III magnet at FBNML, using Holmium pole pieces to focus the field to a maximum of 38.7 T. The Shubnikov-de Haas (S-dH) oscillations do not change from the fundamental frequency of 670 T above H_K, but their harmonic content and amplitude are altered going through this high field transition. There is also hysteresis in the value of the kink field; H_K is approximately 2 T lower on the downsweep than on the upsweep. This high field behavior can be modeled in terms of domain-like behavior of the FS.[2] This low temperature transport behavior is dependent on details of the FS of the organic material, and in particular, on details of the interactions between the closed- and open-orbit FS. We have recently worked to refine three techniques to better measure the behavior of the FS in this and similar organic materials, in order to understand the origins of these low temperature transport phenomena.

2. Angular Magnetoresistance Oscillations

The first method is to measure the angular magnetoresistance of the $(ET)_2X$ materials. Angular magnetoresistance oscillations (AMRO) result from both the open- and closed-orbit FS.[3] These oscillations, seen by rotating the angle φ between the axis perpendicular to the highly conducting planes, and the magnetic field, are anisotropic with respect to the in-plane angle θ. AMRO measured in α-$(ET)_2MHg(SCN)_4$ at fields up to 30 T display evidence for the high-field phase transition in this material, changing from sharp minima below the "kink" field H_K, to Yamaji-type oscillations above H_K.[4] The inset to Fig. 2 shows the AMRO for the mixed-dimensional $(ET)_2RbHg(SCN)_4$; note the broad central maxima at $\varphi=0°$ (field perpendicular to the conducting planes) and the minima in resistance at $\varphi=90°$. In contrast, the AMRO for the two-dimensional organic θ-$(ET)_2I_3$, shown in the inset to Fig. 3, has a sharp maximum at $\varphi=90°$ and a minimum at $\varphi=0°$.

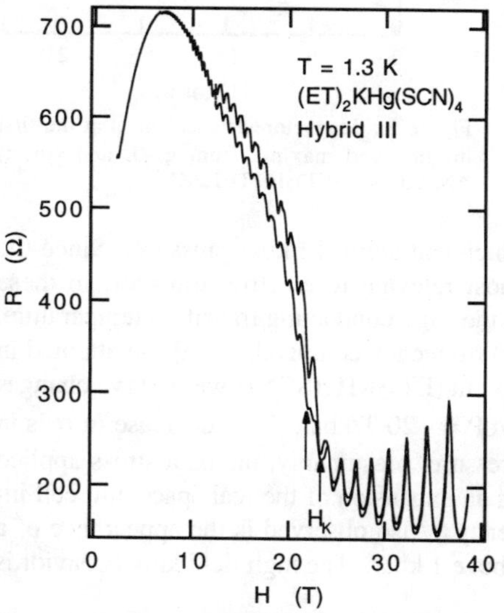

Figure 1. Low temperature magnetoresistance of $(ET)_2KHg(SCN)_4$, showing the kink field H_K.

θ-$(BEDT\text{-}TTF)_2I_3$ has a superconducting transition at 3.6 K; the AMRO in this material are anisotropic in the in-plane angle θ, but are in agreement with the Yamaji model for AMRO in a two-dimensional conductor. High field data ($H > 15$ T) in θ-I_3 show the presence of magnetic breakdown. Figures 2 and 3 show the sharp difference in these materials in the MR measured at angles φ corresponding to the first maxima and the first minima away from $\varphi=0°$ (field $\perp$ to the planes). These differences point to different transport mechanisms at work at these extrema.[2] Also of note is the clear presence of S-dH oscillations and a superconducting transition in I_3 (Fig. 3). The AMRO and the SdH oscillations at different orientations allow us to map out the FS of the different organics.

3. Effect of Pressure and Uniaxial Stress

Our second measurement involves the application of uniaxial stress to specially prepared samples, using a gas-driven piston arrangement to vary the pressure applied perpendicular to the conducting planes. This method allows us to alter the band structure of a sample *in situ*, which extends previous work using hydrostatically pressurized samples.[5] The uniaxial stress probe enables us to apply pressures up to 4 kbar on a

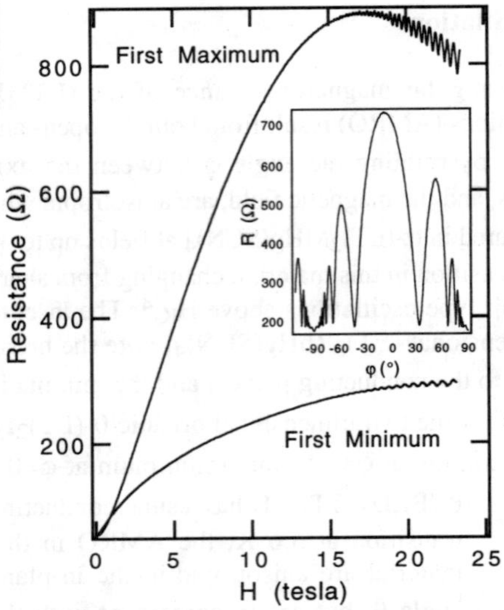

Figure 2. Magnetoresistance for φ at the first minima and maxima from φ=0, and (inset) AMRO for α-$(ET)_2RbHg(SCN)_4$ (T=1.2K).

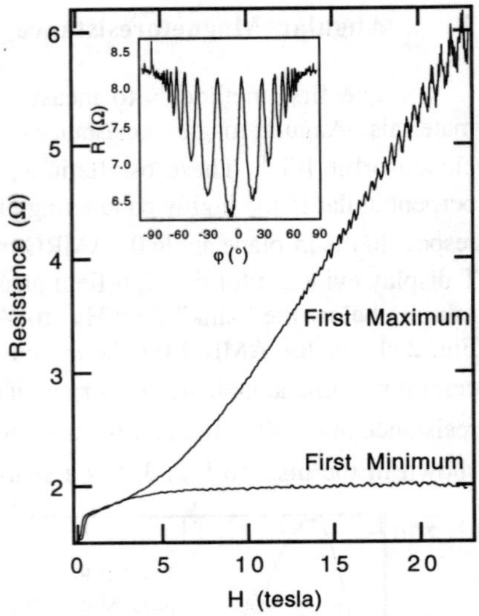

Figure 3. Magnetoresistance for φ at the first minima and maxima from φ=0, and (inset) AMRO for θ-$(ET)_2I_3$ (T=1.2K).

sample, in contrast to the pressure clamp, which can apply 12 kbar pressure. Since the uniaxial stress is applied in the direction most relevant to electron transport in these materials, it is adequate to achieve changes in the superconducting transition temperature, kink field H_K, and Shubnikov-de Haas (S-dH) frequency comparable to those attained in the higher pressure hydrostatic pressure clamp. In $(ET)_2KHg(SCN)_4$ we observe changes in the fundamental S-dH frequency α of dα/dP = - 20 T/kbar. This decrease in α is in contrast to the increase seen with uniform pressure; presumably, uniaxial stress applied perpendicular to the conducting planes causes an expansion of the real space unit cell in-plane, with resultant decrease in S-dH frequency. Also observed is the appearance of a slow frequency of about 50 T, which sets in above 1 kbar. The high field kink behavior is destroyed with uniaxial stress above 1.5 kbar.

4. Optical Studies - Photoluminescence

Our last measurement is of the photoluminescence spectra of single crystals of the low-dimensional organic conductors. These optical measurements serve to probe the electronic structure of the organic crystals; changes with temperature or field may play a role in the FS changes seen in transport measurements.[6] Figure 5 presents data for $(ET)_2Cu(NCS)_2$, which is a superconductor, and for $(ET)_2KHg(SCN)_4$, whose low temperature ground state has already been discussed, both at 1.9 K. There is no luminescence in $(ET)_2Cu(NCS)_2$ in the 1450-2650 meV range for a variety of laser lines

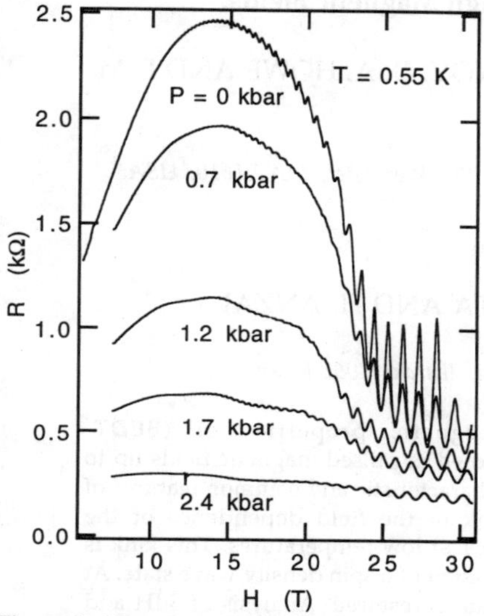

Figure 4. Magnetoresistance at various uniaxial pressures for $(ET)_2KHg(SCN)_4$. Note the disappearance of kink field H_K above 2 kbar.

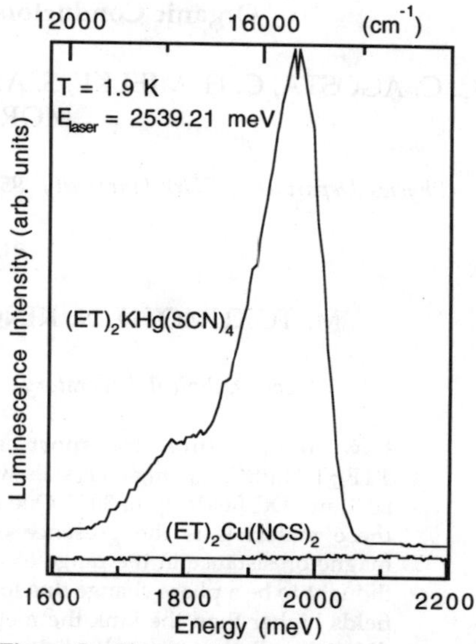

Figure 5. Photoluminescence spectra for two organic conductors.

(2335-2707 meV), and for temperatures between 1.9 - 25 K (above and below T_c), or for any polarization ∥ or ⊥ to the ET planes. This is in contrast to the strong peak seen for the KHg material as well as for neutral ET crystals and other members of the $(ET)_2MHg(SCN)_4$ family, which all display an asymmetric luminescence peak around 1.97 eV with a vibronic shoulder extending to 1.75 eV. More work needs to be done in order to understand possible excitonic processes which may be responsible for this emission.

Acknowledgments

This work was carried out at Francis Bitter National Magnet Laboratory (supported by the NSF); work at Boston University is supported by NSF DMR-92-14889. C.E.C. is supported by a fellowship from Junta Nacional de Investigação Cientifica e Technológica.

References

1. J.S. Brooks, *et al.*, *Phys. Rev. Lett.* **69** (1992) 156, and references therein.
2. G.J. Athas, *et al.*, to be published.
3. M.V. Kartsovnik, *et al.*, *J. Phys. I France* **2** (1992) 223, and references therein.
4. J. Caulfield, *et al.*, *J. Phys.:Condensed Matter* **6** (1994) L155.
5. S.J. Klepper, *et al.*, *Phys. Rev. B* **48** (1993) 9913.
6. C.E. Campos, *et al.*, *Physica B* **194-196** (1994) 1995.

Fermiology and Field-Induced Phase Transitions of Low-Dimensional Organic Conductors in High Magnetic Fields.

C. C. AGOSTA, C. H. MIELKE, S. A. IVANOV, D. A. HOWE AND F. M. MORGAN

Physics Department, Clark University, 950 Main St., Worcester, MA 01610, USA.

and

M. TOKUMOTO, N. KINOSHITA AND H. ANZAI

Electrotechnical Laboratory, Tsukuba, Ibaraki 305, Japan.

Low temperature transport and magnetic properties of (BEDT-TTF)$_2$TlHg(SCN)$_4$ single crystals were studied in pulsed magnetic fields up to 50 T and DC fields up to 30 T. One of the characteristic and common features of these materials is the presence of a kink in the field dependence of the magnetoresistance in the range of 22 to 26 T at low temperatures. This kink is thought to be a phase change due to suppression of a spin density wave state. At fields higher than the kink the metallic state is restored. Analysis of SdH and dHvA oscillations reveals changes in the electron system that occur at the phase boundary. The object of our studies is to clarify these changes via angular and field sweeps through and beyond the phase boundary.

1. Introduction

The search for higher T_c organic superconductors has given researchers an interesting array of non superconducting metallic organic compounds. Many of these novel materials are likely to answer fundamental questions regarding the competition of superconductivity with other ground states. As an example, the absence of the superconducting state in (ET)$_2$MHgSCN)$_4$ [where M = Tl, K, Rb] is believed to be related to the onset of a spin density wave (SDW) [1] below 8 K at zero magnetic field. This SDW is not present in the compound with M = NH$_4$ which becomes a superconductor at ~1.2 K. During magnetic field sweeps the evidence for the presence of this SDW includes a large initial magnetoresistance (MR) followed by a region of negative MR (called the kink) as the SDW gap is closed by the high magnetic field. Recently, sharp negative spikes have been observed[2] in the MR near the kink, or drop in the MR. These new features occur at the boundary between the SDW and the high field state that is assumed to be metallic. In this paper we explore the nature of the MR beyond the SDW phase boundary by making use of the pulsed magnetic field facilities at both the National High Magnetic Field Laboratory (NHMFL, 50 T) in Los Alamos and the High Field Low Temperature Laboratory at Leuven, Belgium. This information along with the fixed field angular sweeps and magnetization measurements carried out at the Francis Bitter National Magnet Laboratory (FBNML) helps describe some of the changing physical parameters of these novel conducting systems.

2. Experimental

In the transport measurements of the $(ET)_2TlHg(SCN)_4$ single crystals, the standard four lead method was used. The lead configuration was set up with the current flowing along the b axis in the pulsed field experiments. Currents of 1 to 100 μA (DC and rms AC) were passed through the samples in all of the experiments. Both DC and AC (f =200-1000 kHz) methods were used for the pulsed field measurements. 20 μm gold wires were attached to the crystals using gold paste while some single crystals had approximately 1000 angstroms of gold evaporated to the contact surface. The typical crystal size was approximately 1 x 1 x 0.1 mm. Magnetization measurements were carried out by use of the cantilever magnetometer method.

All experiments were performed in pumped liquid ^3He cryostats that were built entirely of plastic to avoid eddy current heating and forces associated with currents induced by the rapidly changing magnetic field.

3. Results and Discussion

The presence of a closed orbit in the Fermi surface (FS) of $(ET)_2TlHg(SCN)_4$ gives rise to SdH and dHvA oscillations. Several transport experiments have been done within the SDW phase determining the fundamental frequency of this orbit to be 650(±10)[3] tesla. FS reconstruction and complex magnetic breakdown orbits have been the topic of recent papers[4,5], suggesting that the high temperature high field phase FS is quite different from the low temperature (below 8 K) low field FS.

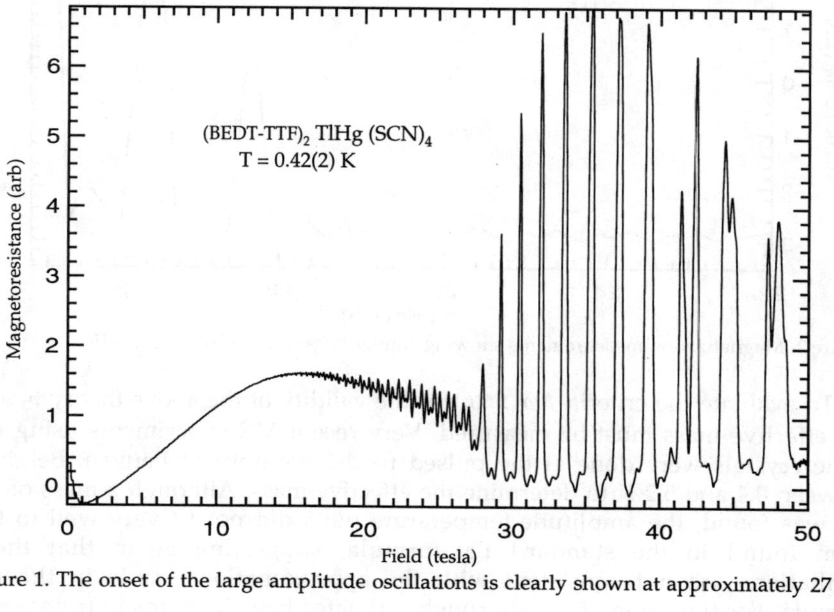

Figure 1. The onset of the large amplitude oscillations is clearly shown at approximately 27 T.

Much of our data on the TlHg containing salt has yielded large anharmonic oscillations in fields above the kink (see figure 1). This change in character suggests that the effective mass, Dingle temperature and number of carriers may be different above the resistance kink. Simulations of these systems based on the Lifshitz-Kosevich (LK) formula consistently yield symmetric traces of the oscillating magnetization, with sharp peaks in the direction of both high and low resistance. The particular shape of the observed oscillations in $(ET)_2TlHg(SCN)_4$ above the kink suggest that they may be a result of the model proposed by Chaikin et al[6] or Kang[7] where the separation of the Landau levels in the electron spectrum exceeds the width of the Landau levels. If we assume that the largest contribution to the width of the Landau levels is due to the warping of the Fermi surface in the b direction, then the criteria for observation of this effect is $\hbar\omega_c > 2t_b$ where t_b is the transfer energy bandwidth along the b axis. In this high field region, square root singularities would be found at the Landau band edges, causing double oscillations of larger amplitude that the oscillations produced before the gaps between the Landau bands formed. Qualitatively our data is consistent with the above theory. As further support for this data we show earlier results in Fig. 2 of the magnetization of $(ET)_2TlHg(SCN)_4$ measured in DC fields. The shape of the magnetization oscillations and the large amplitude change at 27 T is similar to the MR data.

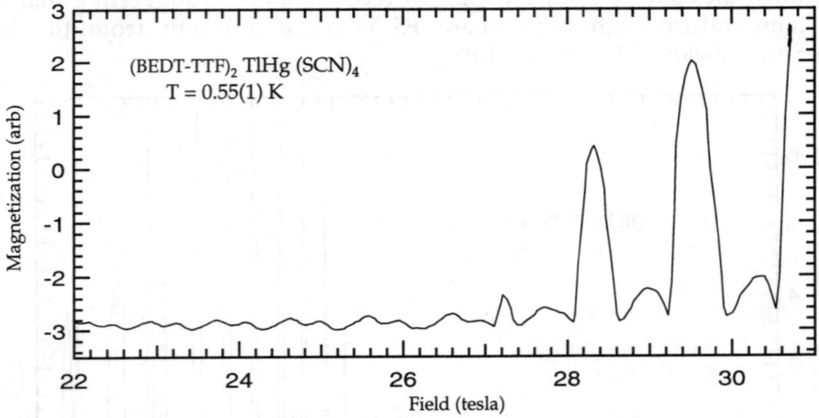

Figure 2. Magnetization measurements showing similar behavior as observed in MR.

To evaluate the criteria $\hbar\omega_c > 2t_b$ for the validity of the above theory t_b and the effective mass must be measured. Very recent MR experiments using the same crystals were done at the pulsed field laboratory at Leuven, Belgium between 0.4 and 5.2 K to determine the effective mass. Although a mass of 1.2 m_e was found, the amplitude-temperature plots did not fit very well to the form found in the standard LK formula, suggesting again that these oscillations are not consistent with the approximations made in the LK formula. For the value of t_b only rough estimates have been made from room temperature conductivity, yielding the value 5 meV.[8] Using these numbers for the effective mass and t_b the above criteria is met at fields above 86 T. Given the uncertainties in the parameters this is not far from the field where we see these large anharmonic oscillations.

Conclusions

We have observed SdH and dHvA oscillations in $(ET)_2TlHg(SCN)_4$ beyond the SDW phase boundry and suggested a possible novel mechanism based on previous theories for the high magnetic field limit. Further analysis and experiments will certainly bring to light the details of the mechanisms of electron transport in these materials and could eventually shed light on the nature of superconductivity in related materials.

Acknowledgments

We would like to thank J. Goettee, E. Haanappel, and D. Rickel, for their assistance at the NHMFL in Los Alamos, Igne Deckers for her assistance at the pulsed field laboratory at Leuven, Belgium, the FBNML for their use of the Hybrid II magnet, N. D. Kushch, V. N. Laukhin and E. B. Yagubskii for the samples used in the magnetization measurements and the Airforce URI for support (grant #F49620-92-J-0525).

References

1 J. S. Brooks et al. *Physical Review Letters*, **69**, 156, (1992).

2 C. C. Agosta, S. A. Ivanov, C. H. Mielke, D. A. Howe, M. Antia and F. M. Morgan, *Solid State Communications*, submitted May 1994.

3 A. E. Kovalev, M. V. Kartsovnik, and N. D. Kushch, *Solid State Communications* **87** No. 8, 705, (1993).

4 M. V. Kartsovnik, A. E. Kovalev, and N. D. Kushch, *J . Phys. I France* **3**, 1187, (1993).

5 J. S. Brooks et al., Preprint.

6 P. M. Chaikin, T. Holstein, and M. Ya. Azbel. *Philosophical Magazine B*, **48** No. 5, 457, (1983).

7 W. Kang, *Solid State Commmunications*, **78**, 25, (1991)

8 H. Mori et. al., *Bull. Chem. Soc. Jpn.*, **63**, 2183 (1990).

MILLIMETER WAVE PHOTOCONDUCTIVITY IN THE MAGNETIC-FIELD-INDUCED INSULATOR PHASE OF KISH GRAPHITE

T. Takamasu and N. Miura
Institute for Solid State Physics, University of Tokyo
Roppongi 7-22-1, Minato-ku, 106 Tokyo, Japan

ABSTRACT

Millimeter wave photoconductivity was measured in the magnetic-field-induced insulator phase of kish graphite. The change of the magnetoconductance due to the radiation of the millimeter wave was found in a limited range of magnetic field just above the critical field of the phase transition. The shape of the photoconductance can be explained in terms of the increase of the conduction electrons excited by the millimeter wave across the BCS like energy gap.

1. Introduction

Transverse resistance of single crystal graphite shows a drastic increase at a critical magnetic field at low temperatures. Theoretically, this anomalous change is explained by magnetic-field-induced charge density wave(CDW) and formation of energy gap[1]. Above a critical field, a large number of electrons fall below the energy gap and form a CDW along c-axis. Only remaining electrons above the energy gap can contribute to the conductance. In a previous paper[2], we focused on the dynamics of the CDW in terms of the high frequency measurements of magnetoresistance and found the similar frequency dependence to that of pinned CDW in conventional one dimensional systems. On the other hand, the existence of the gap and its energy has not been confirmed directly up to now. Theoretically, the energy gap analogous to the BCS state of superconductors should exist depending on the temperature and the magnetic field. To confirm the existence of this energy gap, optical measurement using millimeter wave is suitable. However, it is not so straight forward to analyze the absorption signal of the single crystal graphite because the absorption due to plasma and the cyclotron resonances make the optical signal complicated. In this report, we present photoconductance measurements using millimeter waves of photon energies close to the energy gap.

2. Experimental

Samples used are kish graphite. As shown in figure 1, four terminals for the conductance measurements are attatched to the sample with silver paste and placed in the staninless steel light pipe for the millimeter wave radiation. A Carcinotron millimeter wave generator was used to radiate samples with f =150GHz millimeter

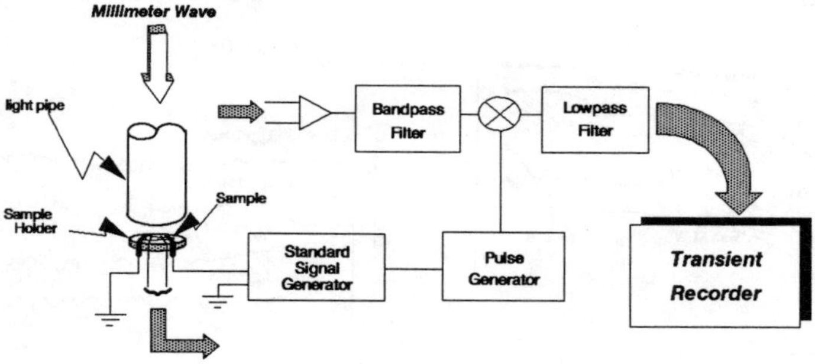

Figure 1: Schematic diagram of the measuring circuit.

wave. Magnetoconductance was measured using high frequency a.c. technique to get a high signal to noise ratio and avoid the millimeter wave rectifier effect at the voltage contacts under pulsed high magnetic fields up to 40T. The measurement frequency was chosed to be 200KHz to avoid the effect of sliding motion of the CDW which was well studied in the previous paper[2]. Photoconductance was determined as a difference between magnetoconductance traces with and without the radiation of millimeterwave. Because the critical field of the phase transition largely depends on the temperature, we limited the measurements only at the two most stabilized temperatures $T=$ 4.2K and 1.5K.

3. Result and Discussion

Figure 2 shows typical recorder traces of magnetic field dependece of resistance of kish graphite at 4.2K and 1.5K with and without millimeter wave radiation. As shown in the figure, the effect of the radiation of millimeter wave is very weak: the overall feature of the magnetoresistance stays unchanged. However, if we take the difference between the conductances with and without millimeter wave radiaion, small but clear photoconductance effect appears. We show in figure 3 the photoconductance signal $\Delta\sigma$ at T=4.2K with different power levels of the millimeter wave as a parameter. Here, the incident power of the millimeter wave is indicated as a line current I_{line} of the Carinotron generator whose output power is almost propotional to the line current. The absolute power of the radiation at I_{line}=20mA is estimated to be less than 1mW, which is weak enough not to raise the lattice temperature of the sample. It is clear that the magnetoresistance changes only in a limitted range just above the critical magnetic field $B_{cr}(T=4.2\text{K})=34\text{T}$. However $\Delta\sigma$ increases with increasing millimeter wave power, the shape of the $\Delta\sigma$ remains unchanged. This fact suggests that the photoconductance effect is not due to change of lattice temperature but to the photoexcited electrons by the millimeter wave.

In Fig.4, we show the photoconductance $\Delta\sigma$ at T=1.5K together with that at

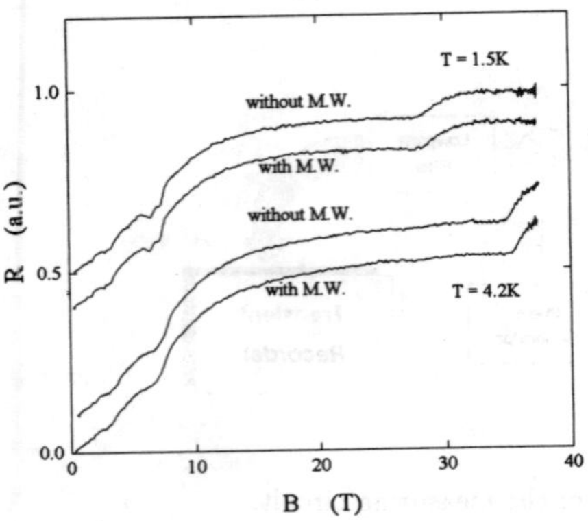

Figure 2: Magnetic field dependence of resistance of kish graphite at T=1.5K and 4.2K.

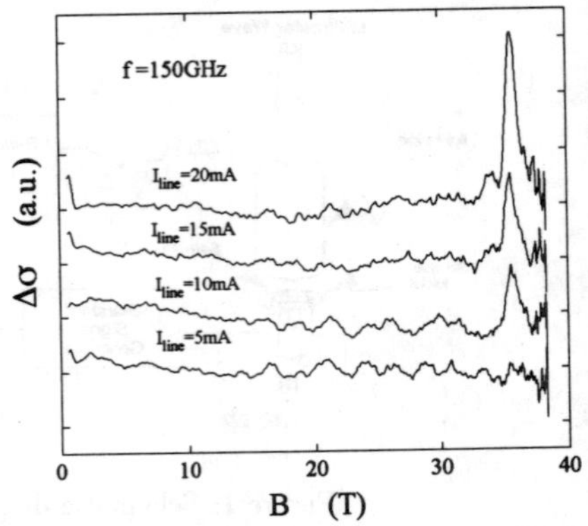

Figure 3: Magnetic field dependence of the photoconductance $\Delta\sigma$ with the incident power of the millimeter wave as a parameter.

4.2K. At T=1.5K, the peak of $\Delta\sigma$ shifts to a lower field according to the lower critical field $B_{cr}(T\text{=1.5K})$=27T, and becomes broader than that at 4.2K.

In the theoretical work of Yoshioka and Fukuyama[1], the CDW is formed along the c-axis with the energy gap. The abrupt increase of the magnetoresistance in the c-plane is cnsidered to be due to the decrease of the number of conduction electrons[4]. The present signal of the photoconductance may originate from the photoexcited electrons across this energy gap. From the analogy to the BCS theory of superconductors, the temperature(T) and magnetic field(H) dependence of the energy gap of the suggested CDW [1] 2Δ is considered as,

$$\Delta(H,T) = \Delta(H,0)\sqrt{1-(\frac{T}{T_c(H)})^2} \tag{1}$$

$$\Delta(H,0) = 1.76k_B\alpha\exp(-1/\zeta H) \tag{2}$$

where α and ζ can be empirically determined from the experiment to be 94.9K and $8.93\times10^{-3}\mathrm{T}^{-1}$, respectively[3]. According to the above equation, we show the magnetic field dependence of the energy gap at several lattice temperatures in Fig. 5 together with the energy of the used millimeter wave $\hbar\omega_0$. Electrons forming the CDW can be excited by the radiation of the millimeter wave when $\hbar\omega_0$ is larger than 2Δ. On the other hand, the density of states at the top and bottom of this energy gap can be assumed to be constant because the width of the related Landau subband is much larger than $\hbar\omega_0$. Thus we can estimate the photoconductance as,

$$\Delta\sigma \propto \hbar\omega_0 - 2\Delta. \tag{3}$$

In Fig. 4, we show the calculated $\Delta\sigma$ for T=4.2K and 1.5K by dotted lines. With

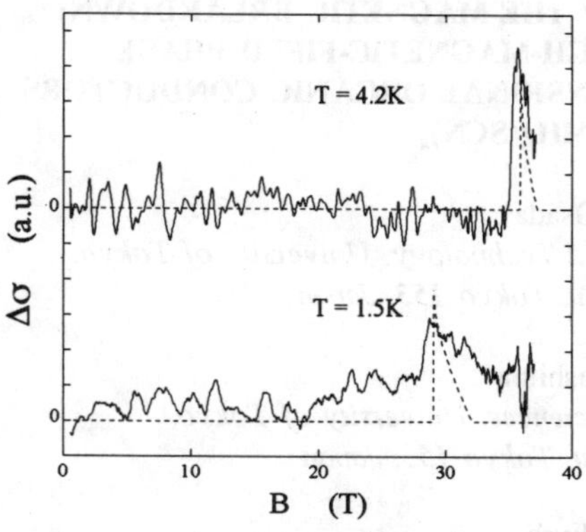

Figure 4: Photoconductance $\Delta\sigma$ at different temperatures 1.5K and 4.2K. The dotted line represents the calculated $\Delta\sigma$ from the equation(3).

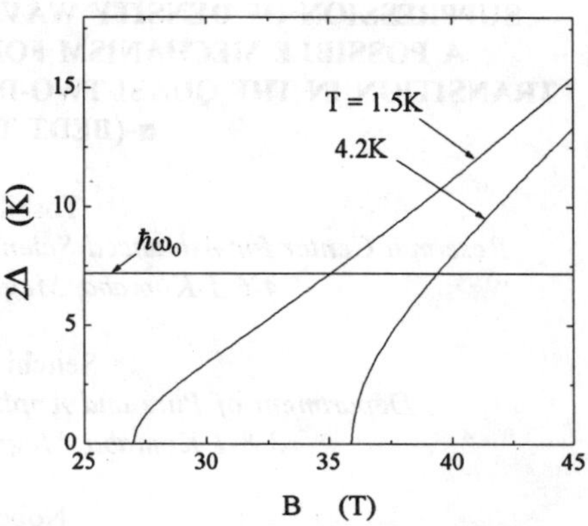

Figure 5: Magnetic field dependence of the calculated energy gap 2Δ.

these curves, the experimental results are well reproduced including the asymmetry and the temperature dependence of the photoconductance peak. This fact strongly suggests that the energy gap of the insulator phase can be described in analogous to the BCS gap of the superconductors. The width of the photoconductance peaks are about two times broader than the calculated $\Delta\sigma$. We may suggest the prefactor of the equation (3) should be larger than the predicted value.

3. Summary

In summary, we demonstrated the direct evidence of the energy gap in the magnetic-field-induced insulator phase of kish graphite in the millimeter wave photoconductance. The shape of the photoconductance can be well fitted with the calculation considering the BCS like dependence of the energy gap in this phase.

References

1. D. Yoshioka and H. Fukuyama, *J. Phys. Soc. Jpn.* **50** (1981) 725.
2. T. Takamasu, H. Ochimizu and N. Miura, *Physica* **B184** (1993) 327.
3. Y. Iye, P. Tedrow, G. Timp, M.S. Dresselhaus, G. Dresselhaus, A. Furukawa and S. Tanuma, *Phys. Rev.* **B25**(1982) 5478.
4. K. Sugihara, and J.A. Woollam, *J. Phys. Soc. Jpn.* **45**(1978)1891.

SUPPRESSION OF DENSITY WAVE BY THE MAGNETIC BREAKDOWN: A POSSIBLE MECHANISM FOR HIGH-MAGNETIC-FIELD PHASE TRANSITION IN THE QUASI-TWO-DIMENSIONAL ORGANIC CONDUCTORS α-(BEDT-TTF)$_2$MHg(SCN)$_4$

Toshihito Osada
Research Center for Advanced Science and Technology, University of Tokyo, 4-6-1 Komaba, Meguro-ku, Tokyo 153, Japan

Seiichi Kagoshima
Department of Pure and Applied Sciences, University of Tokyo, 3-8-1 Komaba, Meguro-ku, Tokyo 153, Japan

Noboru Miura
Institute for Solid State Physics, University of Tokyo, 7-22-1 Roppongi, Minato-ku, Tokyo 106, Japan

ABSTRACT

In order to explain the high-magnetic-field phase transition of the two-dimensional organic conductors α-(BEDT-TTF)$_2$MHg(SCN)$_4$ (M=K,Rb,Tl), we have studied the density wave instability of the two-band two-dimensional magnetic breakdown system. It was found that the density wave transition is oscillatorily suppressed synchronizing with the Shubnikov-de Haas oscillations. We also report new features in the high-magnetic-field transition under tilted magnetic fields.

1. Introduction

The unidentified low-temperature electronic phase observed in organic compounds α-(BEDT-TTF)$_2$MHg(SCN)$_4$ (M=K, Rb, Tl) have been one of the most popular current subjects in physics of organic conductors. These compounds are quasi-two-dimensional conductors with a pair of open Fermi surfaces (FS's) and one closed FS in the two-dimensional (2D) **k**-space. This unidentified low-temperature phase becomes unstable in high magnetic fields, and the system undergoes the phase transition back to the normal phase[1-3]. Recent magnetotransport studies have revealed this low-temperature phase as a kind of density wave (DW) state caused by the nesting of the open FS's[4-6].

One unexplained problem of this phenomenon is the cause of the high-field phase transition, that is, the reason why the DW state is suppressed by strong magnetic fields. According to the "standard model" for the field-induced DW in the single band system, the DW phase should be enhanced by magnetic fields. This discrepancy suggests that the importance of the interband coupling in the present two-band system.

Another problem is the coexistence of the DW and the magnetic breakdown (MB). Recently, the Shubnikov-de Haas (SdH) oscillations corresponding to the large MB orbit have been observed in the low-temperature phase[7]. It seems a contradiction that the SdH oscillations indicating the MB is observed in the DW phase where the open FS's are erased by the nesting.

2. Density Wave in Magnetic Breakdown System

In order to answer above questions, we have studied the DW state in the MB system. We consider the two-band 2D electron system with open and closed FS's as shown in Fig.1. First, we obtain the electronic states of this MB system quantum-mechanically, and then study the DW instability for the obtained electronic states. The details of this model will be published elsewhere[8].

If the interband coupling due to the MB can be ignored, we can consider the electronic states under magnetic fields for each band independently. Around the Fermi level, the energy spectrum of the closed orbit band is quantized to the discrete Landau levels with no dispersion on the center coordinate x_0, and that of the open orbit band is quantized to a set of subband with a linear dispersion on x_0. The interband coupling due to the MB opens the "MB gaps" at the cross points of the Landau levels and the linear subbands in the x_0 dispersion. It also opens the smaller MB gaps at the cross points of two linear subbands. These MB gaps separate the original continuous spectrum into a set of subbands, which correspond to the Landau levels of the large MB orbit.

Figure 2 shows the numerically calculated magnetic-field-dependence of the magnetic energy spectrum ("fan chart") of the two-band MB system. The MB gaps appear in the

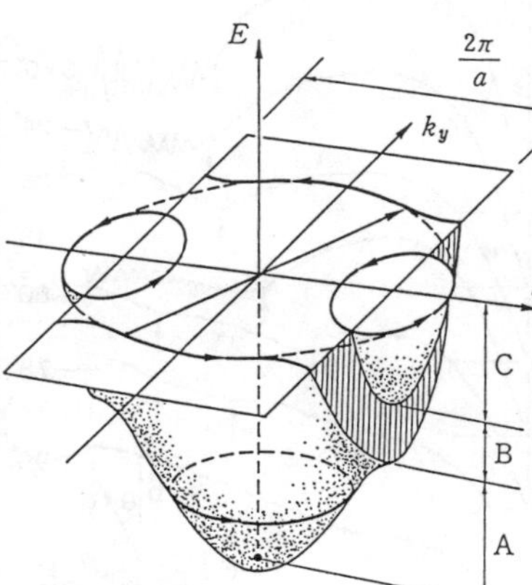

Figure 1
Band model for the 2D MB system. Fermi level is located in the energy region C, so that there are a pair of open FS's and one closed FS. MB between these two type of FS and perfect nesting of open FS's are assumed.

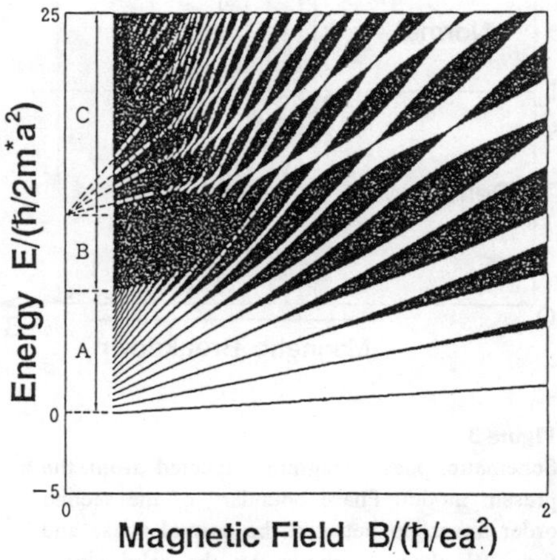

Figure 2
Fan chart of 2D MB system calculated numerically. In the energy region C, the Landau levels of the closed orbit and the linear subbands of the open orbit are mixed up, and MB gaps appear.

continuous spectrum in the energy region C where the Fermi level is located. At high magnetic fields, the spectrum is reconstructed into a set of Landau levels of the MB orbit.

Next, we study the DW instability for the electronic states under magnetic fields. The DW instability can be studied by evaluating the nonlocal static susceptibility $\chi_0(\boldsymbol{q})$. $\chi_0(\boldsymbol{q})$ is calculated from the analytically obtained eigenstates and the energy spectrum[8]. From the model calculation, we can obtain the following results: When the Fermi level is far from the MB gap, $\chi_0(\boldsymbol{q})$ has a sharp peak at $\boldsymbol{q}=(\pi/a, 2k_F)$ which is the nesting vector at the zero magnetic field. This peak gives the DW transition under the magnetic field. On the other hand, when the Fermi level is located around the MB gap, this $\chi_0(\boldsymbol{q})$ peak is strongly suppressed since the electronic state around the MB gap contains the component which does not contribute the nesting. When the Fermi level passes through one MB gap, one Shubnikov-de Haas (SdH) oscillation occurs. Therefore, the DW instability is oscillatorily suppressed synchronizing with the SdH oscillations. This suppression becomes stronger at high magnetic fields because the MB gaps become larger. So, the DW transition is perfectly suppressed at sufficiently high fields.

The expected phase diagram is schematically shown in Fig.3. The phase boundary of the second order transition between the normal phase and the DW phase oscillates against the magnetic field synchronizing with the SdH oscillations in the normal phase. Correspondingly, in the DW phase, the DW order parameter also oscillates synchroniz-

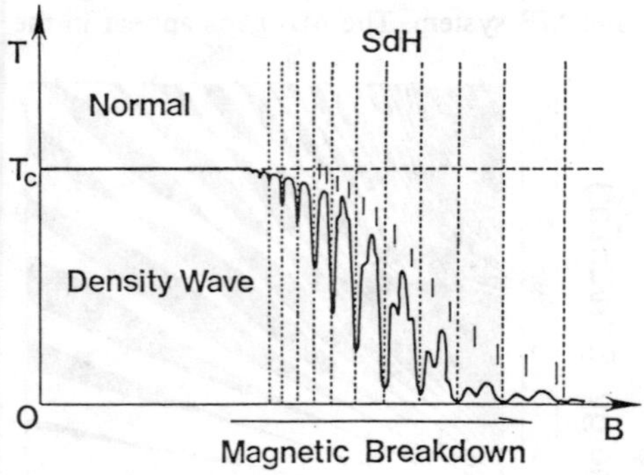

Figure 3
Schematic phase diagram expected from the present model. Phase boundary of the second order transition between the normal phase and the DW phase is shown by the solid curve. Vertical broken lines indicate the SdH oscillation field corresponding to the closed orbit. The short vertical segments indicate the SdH oscillation fields corresponding to the MB orbit. The DW transition is oscillatorily suppressed synchronizing with the SdH oscillations.

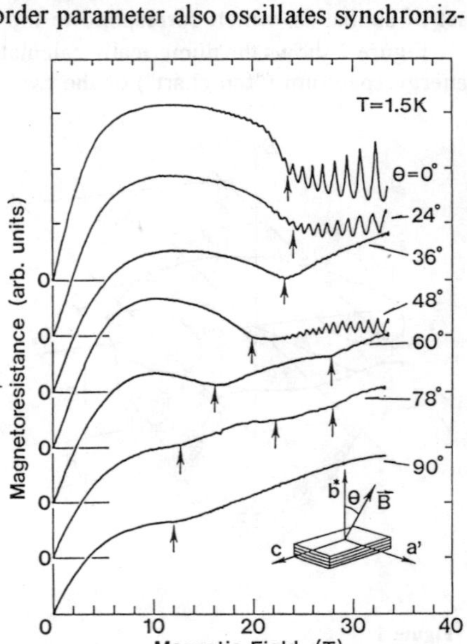

Figure 4
Magnetoresistance of $(BEDT\text{-}TTF)_2KHg(SCN)_4$ under tilted magnetic fields. The "kink" structures are indicated by arrows.

ing with the SdH oscillations of the normal phase. In the DW phase, this order parameter oscillation could be observed as the "SdH oscillations".

3. High-Magnetic-Field Transition under Tilted Magnetic Fields

The present model well explains the above-mentioned problems on the low-temperature phase of $(BEDT\text{-}TTF)_2MHg(SCN)_4$ under magnetic fields normal to the 2D plane. In the framework of this model which treats the pure 2D system, the phenomena under the tilted field are scaled by the normal component of the magnetic field $B\cos\theta$.

Figure 4 shows the magnetoresistance traces of $(BEDT\text{-}TTF)_2KHg(SCN)_4$ measured under tilted magnetic fields. The field direction were tilted in the b^*-a' plane. The high-field transition from the DW phase to the normal phase occurs at the "kink" field indicated by arrows. The background magnetoresistance and the SdH amplitude show the oscillatory angular dependence. The "kink" field does not obey the expected cosine scaling.

The real $(BEDT\text{-}TTF)_2MHg(SCN)_4$ system is not a pure 2D system but a quasi-2D one with a finite interlayer coupling as demonstrated by the angular dependent magnetoresistance oscillations[4-6]. In addition, the Zeeman effect, which was not considered in the model, works in the real system. These effects could possibly cause the deviation from the cosine scaling.

We also found the appearance of new "kink" structures above the former "kink" field in tilted magnetic fields (θ=60°, 78° in Fig.4). It is not clear whether these new structures correspond to successive phase transitions or a kind of oscillations. At the present stage, the origin of these structures is unknown.

It seems that the behaviors under tilted magnetic fields are not simply understood. To understand these behaviors, it will be necessary to take into account the three-dimensionality, the Zeeman effect, or something new.

4. References

1. T.Osada, R.Yagi, A.Kawasumi, S.Kagoshima, N.Miura, M.Oshima, and G.Saito, *Phys. Rev.* **B41** (1990) 5428.
2. T.Sasaki and N.Toyota, *Solid State Commun.* **82** (1992) 447.
3. J.S.Brooks, C.C.Agosta, S.J.Klepper, M.Tokumoto, N.Kinoshita, H.Anzai, S.Uji, H.Aoki, A.S.Perel, G.J.Athas, and D.A.Howe, *Phys. Rev. Lett.* **69** (1992) 156.
4. M.V.Kartsovnik, A.E.Kovalev, and N.D.Kushch, *J. Phys. I(France)* **3** (1993) 1187.
5. Y.Iye, R.Yagi, N.Hanasaki, S.Kagoshima, H.Mori, H.Fujimoto, and G.Saito, *J. Phys. Soc. Jpn.* **63**, (1994) 674.
6. T.Sasaki and N.Toyota, *Phys. Rev.* **B49** (1994) 10120.
7. S.Uji, H.Aoki, J.S.Brooks, A.S.Perel, G.J.Athas, S.J.Klepper, C.C.Agosta, D.A.Howe, M.Tokumoto, N.Kinoshita, Y.Tanaka, and H.Anzai, *Solid State Commun.* **88** (1993) 683.
8. T.Osada, S.Kagoshima, and N.Miura, in preparation, and to be published in *Synthetic Metals* (*Proceedings of ICSM'94*, Seoul, Korea, 1994).

A STUDY OF THE RAPID OSCILLATIONS OF $(TMTSF)_2ClO_4$ IN DIFFERENT REGIONS OF THE H–T PHASE DIAGRAM

C.C. AGOSTA, D.A. HOWE, M.A. ANTIA, S.A. IVANOV, C.H. MIELKE and F.M. MORGAN.
Clark University, Worcester, MA 01610, USA

ABSTRACT

Several experiments have been performed at temperatures from 0.5 K to 25 K at both Clark University's 30 T pulsed field and the National High Magnetic Field Lab (NHMFL) at Los Alamos where we used fields up to 50 T. Cooling rates through the anion ordering transition were varied from 200 mK/min to 5 mK/min. From our data, we are able to compare the transport properties of crystals of $(TMTSF)_2ClO_4$, both before and after what is known as the "reentrant" transition, and at temperatures above and below the onset of the field-induced spin density wave (FISDW). Based on these observations, we are able to draw conclusions about different mechanisms for the rapid oscillations and shed new light on the details of the phase diagram, particularly in the high temperature and high field limit.

1. Introduction

The compound $(TMTSF)_2ClO_4$ is one of the most studied among the molecular crystals (more correct term for organic crystals).[1] It is a quasi one-dimensional organic metal which has an exceptionally rich phase diagram including a cascade of FISDW phases and superconductivity at temperatures below 1.3 K.[2] It is one of a family of superconducting $(TMTSF)_2X$ salts, where X = TaF_6, SbF_6, AsF_6, PF_6, and ReO_4 (the Bechgaard salts). While all of the salts previously mentioned only become superconducting under pressure, $(TMTSF)_2ClO_4$ was discovered as the first ambient pressure organic superconductor.

The magnetic field - temperature (H–T) phase diagram of this compound has been worked on for more than a decade now and is relatively well defined, at least in fields up to 28 T, as is seen, for example, in the paper by Kang *et al*.[3] The main task of this work was to clarify details concerning the rapid oscillations and the high field part of the phase diagram.

2. Experimental

The typical procedure for running a sample in the pulsed field was to begin the experiment at around 40 K, and then choose a cooling rate. The sample was cooled at this rate down to approximately 15 K trough the ordering transition at around 24 K, and then brought down to 4.2 K. Then the field dependencies of magnetoresistance (MR) were measured at temperatures down to 1.5 K. Experiments at NHMFL were done in magnetic fields up to 50 T and temperatures down to 0.7 K.

Leads were made of 20 μm diameter gold wire, attached to the sample with silver paint. Some samples had gold pads evaporated on to them. Typical contact resistances were on the order of 10 Ω.

3. Results and Discussion

One of the problems we were interested in was the change in the spectral content of the rapid oscillations as a function of temperature in magnetic field range corresponding to the last FISDW phase. The cooling rate of the sample we analyze below was 6 mK/min, and the pulses were done at temperatures from 1.5 K to 9 K.

A Fast Fourier Transform (FFT), shows that a 2nd harmonic is present in the oscillations. The amplitude of the 2nd harmonic is much smaller than that of the 1st, and changes with the temperature approximately the same way that the 1st harmonic does (Fig. 1) while the sample is in the SDW phase. However, once temperatures above 5 K are reached and $(TMTSF)_2ClO_4$ is no longer in the SDW phase, the 2nd harmonic disappears. The amplitude of the 1st harmonic also starts to decrease, but never goes away. This fact indicates that 1st and 2nd harmonics may have different origins.

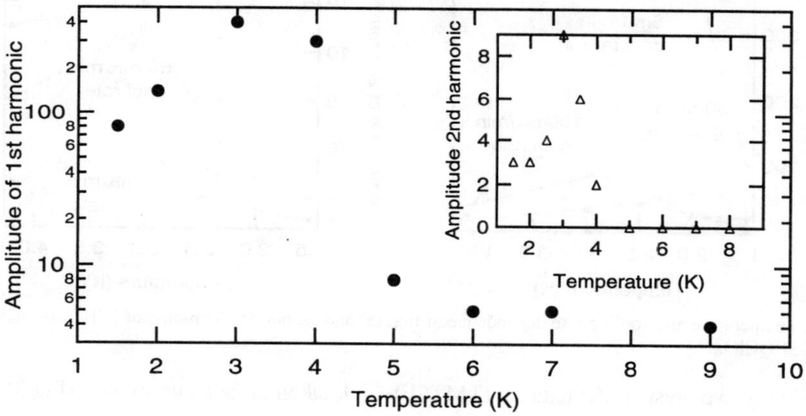

Fig. 1. Amplitude of first and second harmonics with temperature. Notice that the amplitude of the 2nd harmonic completely disappears above 4 K, while the amplitude of the 1st harmonic simply decreases by an order of magnitude.

Another variable that affects the amplitude of the oscillations is the cooling rate or degree of anion ordering. For this analysis we directly measured the amplitude of the oscillations rather than using an FFT to calculate a power spectrum so we would not lose all the phase information. In the case of a sample very slowly cooled through the ordering transition, it is difficult to measure amplitudes of 1st and 2nd harmonics separately. But, if a sample is cooled relatively fast there are two obvious sets of peaks of the same frequency, "A" and "B" that appear. These two sets of oscillations were seen first by Ulmet *et al.*[4] and have been discussed in other papers.[5,6] The "B" peaks are usually of equal or smaller amplitude and not in phase with "A" peaks, which makes them easily distinguishable. In the FFT the "B" peaks appear as a 2nd harmonic of the fundamental frequency. Thus, the 2nd harmonic in very slow cooled samples may be closely correlated with the "B" peaks observed in less well cooled samples. We have also noticed that the phase shift between the "A" and "B" peaks tends to decrease with decreasing the cooling rate (or increasing the degree of ordering in the material) causing the "B" peaks to be more difficult to see in a well ordered sample.

Fig. 2 shows that the "A" peaks' amplitude increases with increasing the degree of anion ordering while the "B" peaks' amplitude decreases. The different behavior of amplitudes of the "A" and "B" peaks as a function of degree of anion ordering and disappearance of "B" peaks outside the SDW phase probably indicate a different physical origin of these oscillations. It seems to be reasonable to think that the mechanism for "B" oscillations is similar to the oscillations in $(TMTSF)_2PF_6$[7] while the "A" oscillations have an origin, specific to $(TMTSF)_2ClO_4$ due to its anion ordering transition. There is a very extensive amount of literature concerned with proposed mechanisms of oscillations in the $(TMTSF)_2X$ family, although none of these explanations is entirely consistent with our data or propose two simultaneous mechanisms.[8,9,10]

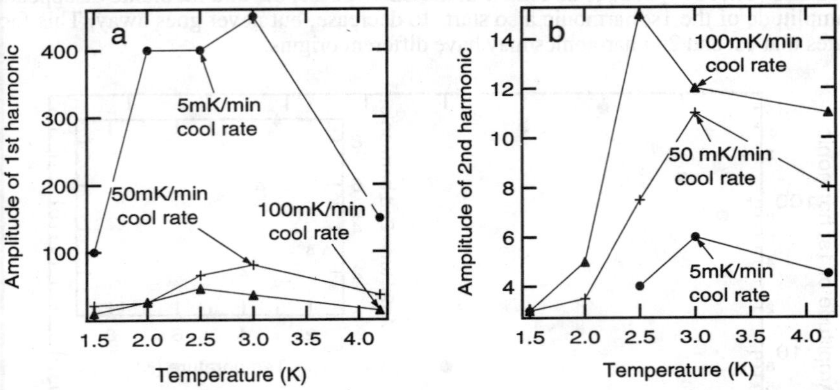

Fig. 2 Temperature and cooling rate dependence of first (a) and second (b) harmonic of SdH* oscillations in $(TMTSF)_2ClO_4$.

Finally, we present MR data for $(TMTSF)_2ClO_4$ taken in fields up to 50 T (Fig.3).

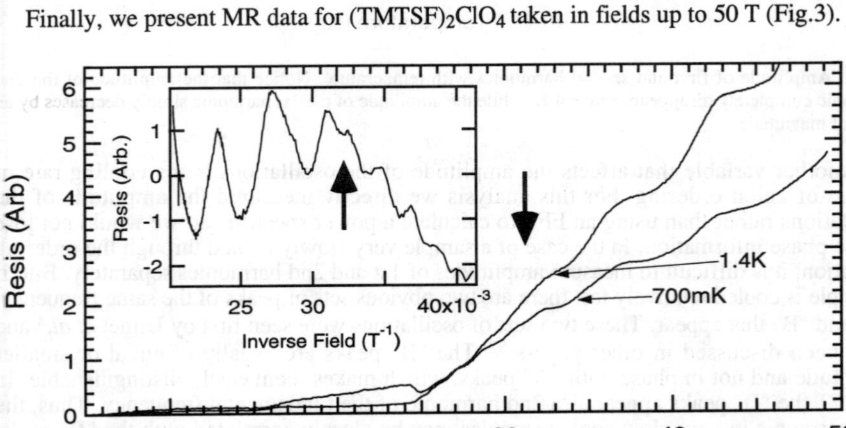

Fig. 3 MR of $(TMTSF)_2ClO_4$ sample at 1.4 K and 0.7 K. Inset shows high field part of MR at 0.7 K with linear background subtracted, plotted *vs.* inverse field. Arrows indicate field where the character of MR changes on 0.7 K data.

Resistance after the last FISDW starts to grow, revealing metallic behavior. The inset on Fig. 3 shows the high-field part of the MR at T= 0.7 K with a linear background subtracted, plotted *vs.* inverse field to make the change in the character of oscillations at approximately 31.2 T more visible. At the lower field the frequency of oscillations is about 500 T. At 31.2 T oscillations change phase and the amplitude, and the frequency halves. The origin and even the existence of this transition needs to be studied more thoroughly, but a transition in a similar part of the phase diagram was seen by Yu *et al.*[11] at about 30 T and 2.5 K. Our experiments may reveal a similar transition.

4. Conclusion

MR measurements of $(TMTSF)_2ClO_4$ show different behavior of two types of oscillations observed in the final FISDW phase as a function of temperature and degree of ordering. Measurements in magnetic field up to 50 T seem to indicate the change in the character of oscillations and background MR at 31.2 T and 0.7 K.

5. Acknowledgments

We thank J.S. Brooks (supported by NSF DMR 9214889 grant) for assistance in conducting the experiments and P.M. Chaikin for help in getting the samples. We are also grateful to NHMFL personel and A. Lacerda and D. Rickel in particular.

The work was supported by AFOSR grant #F49620-92-J-0525.

6. References

1. J.S. Brooks, private communication.
2. K. Bechgaard, K. Carneiro, F. B. Rasmussen, and M. Olsen, *J. Am. Chem. Soc.,1981*, **103**, 2440.
3. W. Kang, S. T. Hannahs, and P. M. Chaikin, *Physical Review Letters*, **70**(1993)
4. J. P. Ulmet, A. Khmou, P. Auban, and L. Bachere, *Solid State Communications,* **58** (1986), 753.
5. M.J. Naughton and G. Montambaux, *ICSM*, Tubingen (1990).
6. X. Yan, M.J. Naughton, R.V. Chamberlin, L.Y. Chiang, S.Y. Hsu, and P.M. Chaikin, *Synthetic Metals,* **27** (1988), B-145.
7. J.P. Ulmet, P. Auban, A. Khmou, S. Ashkenazy and A. Moradpour, *J.Physique Lett.* **46** (1985) L535.
8. A. G. Lebed, P. Bak, *Phys. Rev. B, Rapid Communications,* **40** (1989), 11,433.
9. M. Heritier, G. Montambaux, P. Lederer, *J. Physique Lett..*, **45**, (1984), L-943.
10. T. Osada, N. Miura, and G. Saito, *Solid State Communications,* **60** (1986), 441.
11. R.C. Yu, L. Chiang, R. Upsani, and P.M. Chaikin, *Phys. Rev. Letters,* **65** (1990), 2458.

MAGNETIC-FIELD-INDUCED ELECTRONIC PHASE TRANSITION IN NEUTRON-IRRADIATED GRAPHITE

H. Yaguchi, Y. Iye, T. Takamasu, N. Miura and T. Iwata*

Institute for Solid State Physics, University of Tokyo, Roppongi, Minato-ku, Tokyo 106, Japan

**Advanced Science Research Center, Japan Atomic Energy Research Institute, Tokai-mura, Ibaraki, 319-11, Japan*

ABSTRACT

The effect of neutron irradiation on a magnetic-field-induced electronic phase transition in graphite has been investigated. It is found that the transition temperature is lowered as concentration of defects introduced by irradiation is increased. The suppression of the transition temperature can be interpreted in terms of the combined effect of the Fermi level shift and pair-breaking effect.

1. Introduction

The electron system in graphite undergoes an electronic phase transition at high magnetic fields and low temperatures. Abrupt increase in magnetoresistance appears at a certain value of magnetic field applied along the *c*-axis.[1] The onset field strongly depends on temperature and the transition temperature T_C at a given field B is empirically expressed as

$$T_C = T^* \exp(-B^*/B), \qquad (1)$$

corresponding to a BCS-like formula for a mean-field-type transition, where T^* and B^* are fitting parameters.[1] Yoshioka and Fukuyama (YF)[2] have proposed a $2k_F$-type instability as a possible model of this transition. In the field range of interest B>20T, only the lowest electron (n=0) and the lowest hole (n=−1) Landau subbands are populated. Under such a condition, there are various possible $2k_F$-type instabilities. YF assigned the lowest field transition (which we call α) to a charge density wave (CDW) instability associated with the $2k_F$ for the (n=0, σ=1) subband. However, the other $2k_F$-type instabilities cannot be ruled out as candidates for the relevant transition. Recent experiments extended to higher fields revealed a few additional features, one of which (β) seems to be of nature similar to α.[3]

In the present work, we have investigated neutron-irradiated graphite to obtain further information on the transitions α and β. Neutron irradiation creates vacancies, some of which act as acceptors.[4] It has been found, in an earlier study, that ionized impurities suppress T_C, which can be understood as a manifestation of the pair-breaking effect.[5]

2. Experimental

The samples used in the present study were flakes of single crystal graphite (Kish graphite) which were irradiated at JAERI JRR-4 with fast neutrons. The samples used in this study are listed in Table 1. Samples #1, 2 and 3 are in the increasing order of neutron dose. Sample #1a was irradiated to the same dose as #1 and then annealed at 600°C for ~1hour, to reduce the amount of defects. Typical dimensions of the samples were ~2mm in length, ~1mm in width, and a few tens of μm in thickness. For the resistance measurement, a phase sensitive detection technique was employed with an alternating current of typically ~300μA at 200kHz. High magnetic fields were generated by using a pulse magnet connected to a capacitor bank. The maximum field was ~40T and the pulse duration was ~10ms. Temperatures down to ~0.4K were attained by use of a ^{3}He insert.

Table 1. List of neutron-irradiated graphite samples. Criterion for the fast neutron is E>1MeV. The values of τ_0, $\tau_{n=0}$ and $\tau_{n=-1}$ are obtained from the fitting of Eq.(2), and those of τ_{tr} correspond to the resistivities at 4.2K. The values of T^* and B^* are estimated from the Fermi level shift for the two cases when the transition occurs in the n=0 and n=−1 Landau subband, respectively.

	Dose (10^{16} /cm^2)	Annealing	τ_0	$\tau_{n=0}$ (10^{-12}sec)	$\tau_{n=-1}$	τ_{tr}	T^*(K), B^*(T) (n=0)	T^*(K), B^*(T) (n=−1)
#1a	4	600°C, 1hr.	12	12	12	5.5	191, 128	191, 128
#1	4	------	3.9	2.9	4.5	1.8	112, 105	282, 142
#2	8	------	3.0	2.2	2.9	1.1	92.1, 96.2	311, 144
#3	12	------	2.0	1.6	1.7	0.76	63.6, 82.1	357, 146

3. Result and Discussion

Figure 1(a) shows traces of magnetoresistance in the low-field region for the irradiated samples and a pristine graphite sample. The effect of irradiation appears in the following way. The Shubnikov-de Haas (SdH) dips associated with the spin-split n=1 electron Landau subbands shift towards lower fields with increasing neutron dose. This reflects the decrease of the electron Fermi-surface volume by introduction of acceptors.[5] Judging from these SdH dips, the increase in hole concentration is approximately 20-30 % in Samples #1 to #3. On the other hand, the Fermi level of Sample #1a is nearly recovered.

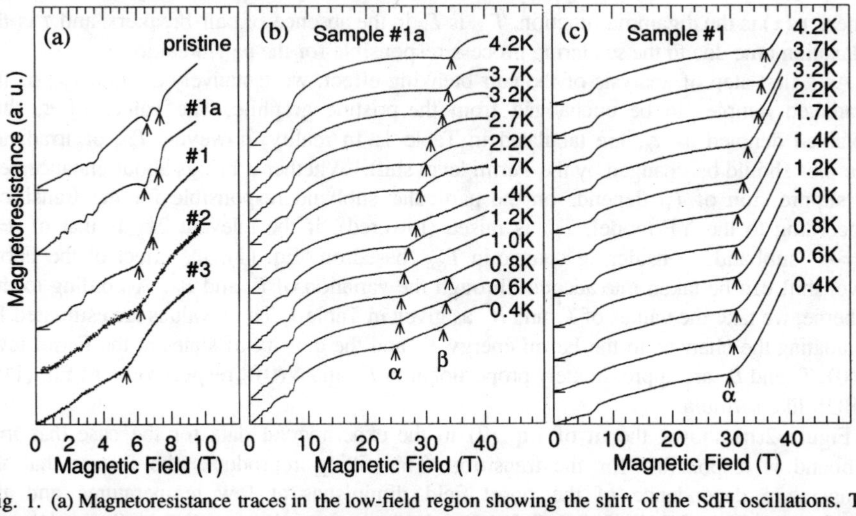

Fig. 1. (a) Magnetoresistance traces in the low-field region showing the shift of the SdH oscillations. The dips associated with the spin-split n=1 electron subbands are indicated by arrows. (b), (c) Magnetoresistance traces up to ~40T at different temperatures for two samples #1a and #1.

Figures 1(b) and (c) show the magnetoresistance traces at different temperatures for Samples #1a and #1. While in the trace for Sample #1a are seen the two transitions of α and β, in the trace for #1 only one transition α. Samples #2 and #3 also go through only the α transition in the temperature-field range of the present study. The onset field is shifted towards higher field as neutron dose is increased. In other words, T_C is lowered with increase of dose.

In Fig. 2(a), the onset points for α are plotted as log$T$-1/B, in the scope of Eq. (1). It should be mentioned that the data points below 1.0K are not used in this plot, because eddy current heating hampered reliable thermometry at such low temperatures. The introduction of acceptors by neutron irradiation has two aspects in affecting the high field transition. One is lowering of the Fermi level. The other is increase of carrier scattering. If the observed change in T_C is caused solely by the Fermi level shift, the relation between T_C and B is expected to follow the relation Eq. (1). However, the data systematically deviate downward from the functional form of Eq. (1) at lower temperatures. Besides, for the irradiated specimens, fitting of Eq. (1) yields a quite large value of T^*. It is not reasonable because T^* in Eq. (1) of a BCS-like formula should be on the order of the Fermi energy. This suggests that the increased carrier scattering plays a significant role in the suppression of T_C. The effect of impurities on a mean-field-type pairing transition such as CDW state or excitonic phase can be theoretically treated as pair-breaking effect. Its mathematical formalism is identical with the case of magnetic impurities in a superconductor.[6] T_C in the presence of pair-breaking process is given by

$$\ln\left(\frac{T_C}{T_{C0}}\right)=\psi\left(\frac{1}{2}\right)-\psi\left(\frac{1}{2}+\frac{\hbar}{2\pi k_B T_C \tau}\right), \quad (2)$$

where $\psi(x)$ is the digamma function, T_{C0} is T_C in the absence of pair-breakers, and τ is the relaxation time due to the scattering process responsible for the pair-breaking.

As a first step of analysis of the pair-breaking effect, we tentatively assume T_{C0} of the irradiated samples to be unchanged from the pristine graphite. The values of τ, thus obtained denoted as τ_0, are tabulated in Table 1. In reality, however, T_{C0} of irradiated samples should be changed by the Fermi level shift. Whether it brings about enhancement or suppression of T_C depends on the particular subband responsible for the transition. According to the YF model, T_{C0} is raised (lowered), if the relevant $2k_F$ is that of n=0 (n=−1) subband. Considering change in T_{C0} based on Eq. (1), the effect of the Fermi level shift can be taken into account through the variation of T^* and B^*. According to this scheme, we take the values of T^* and B^* as given in Table 1. These values are estimated by evaluating the changes in the Fermi energy E_F and the density of states at the Fermi level $N(0)$. T^* and B^* are approximately proportional to E_F and $N(0)^{-1}$ respectively, as Eq. (1) is a BCS-like formula.

Figure 2(b) shows the fit of Eq. (2) to the experimental data for the case that n=0 subband is responsible for the transition. This fitting reproduces the feature that the temperature dependence of the onset field diminishes at low temperatures and the extrapolated onset field at T=0 is finite. For the case of n=−1, as well as n=0, the data can be fitted to Eq. (2). The values of τ are obtained from the fitting for the above two cases, and, denoted as $\tau_{n=0,\,-1}$, given in Table 1. Reflecting that the Fermi level shift has the opposite effect on the subbands n=0 and n=−1, each value of $\tau_{n=0}$ is smaller than that of $\tau_{n=-1}$. In either case of n=0 and n=−1, the observed suppression of T_C can be explained in terms of the Fermi level shift and the pair-breaking effect. It is noted that the values of τ are for the n=0 case are somewhat closer to the transport relaxation time τ_{tr}, than those for the n=−1 case. The original aim of the present study was to determine which subband is responsible for the transition. The above comparison of the pair-breaking time τ with the transport relaxation time τ_{tr} may be taken as an evidence to favor the n=0 subband. However, considering that the scattering processes responsible for τ and τ_{tr} are not necessarily the same, a definite conclusion must await further experimental information.

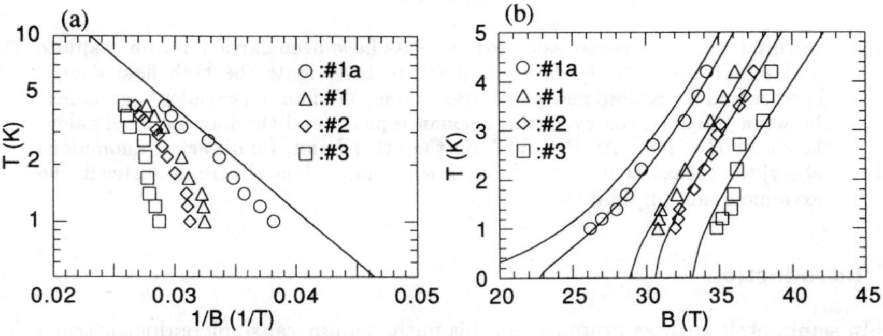

Fig. 2. Phase diagram in the magnetic field-temperature plane. (a) The left panel shows the plot as log$T$-1/B. Solid line represents fit of Eq. (1) for pristine graphite. (b) The right panel shows the plot of the onset points with the fit of Eq. (2) in the case that the relevant $2k_F$ is that of n=0 subband. The left most curve represents fit of Eq. (1) for pristine sample.

4. Conclusion

We have investigated the magnetic-field-induced electronic phase transition in neutron-irradiated graphite. The transition temperature is found to be lowered with the increase of concentration of defects introduced by irradiation. The suppression of the transition temperature can be interpreted in terms of the combined effect of the Fermi level shift and the pair-breaking effect. This provides experimental support for the interpretation of the CDW instability.

Acknowledgement

We are grateful to K. Takahashi and Y. Shimamoto for fruitful discussions. One of the authors (H. Y.) appreciates support from the Japan Society for Promotion of Science.

References

1. Y. Iye, P. M. Tedrow, G. Timp, M. Shayegan, M. S. Dresselhaus, G. Dresselhaus, A. Furukawa and S. Tanuma, *Phys. Rev.*, **B25** (1982) 5478.
2. D. Yoshioka and H. Fukuyama, *J. Phys. Soc. Jpn.*, **50** (1981) 725.
3. H. Yaguchi, Y. Iye, T. Takamasu and N. Miura, *Physica B*, **184** (1993) 332.
4. R. O. Dillon, I. L. Spain and J. W. McClure, *J. Phys. Chem. Solids*, **39** (1978) 1071.
5. Y. Iye, L. E. McNeil and G. Dresselhaus, *Phys. Rev.*, **B30** (1984) 7009.
6. J. Zittartz, *Phys. Rev.*, **164** (1967) 575.

INFRARED MAGNETOSPECTROSCOPY IN GRAPHITE UNDER PULSED ULTRAHIGH MAGNETIC FIELDS

Y. Shimamoto, H. Nojiri and N. Miura

Institute for Solid State Physics, University of Tokyo, Roppongi, Minato-ku, Tokyo 106, Japan

Abstract

Infrared magneto-transmission experiments have been carried out on graphite under high magnetic fields up to 400T to investigate the high field energy spectra and to explore new phase transitions. We found a excellent agreement between the observed cyclotron resonance peaks and the Landau level calculation at $B < 60$T. At $B > 60$T, on the other hand, we observed anomalous absorption peaks, which are related to some anomalous electronic states in the extreme quantum limit.

1. Introduction

In semimetals such as graphite and bismuth, semimetal-semiconductor transition is induced by sufficiently high magnetic fields. Moreover, various types of field induced electronic phase transitions (excitonic phase, CDW or SDW phase) are expected to occur at high magnetic fields and low temperatures.

Under strong magnetic fields ($B//c$-axis), graphite has four one-dimensional Landau subbands ($n = 0$ and $n = -1$ with spin up and down levels) remaining under the Fermi level. At low temperatures, an anomalous increase of magnetoresistance has been reported, which is explained by the CDW instability associated with the one-dimensional character of the energy spectrum from the calculation of Yoshioka *et.al.*[1] However, it is not clear what kind of the phase transition is responsible for it or which subband is related to the transition. Nakao[2] proposed that in a very high field at around 100T, a semimetal-semiconductor transition may take place, and when it occurs, the Fermi level would coincide with the crossing point of the two subbands.

To clarify the details of the high field phase and to explore new electronic phase transitions, we have studied the infrared magnetospectra related to the $n = 0$ and $n = -1$ Landau subbands under high magnetic fields up to 180T using the single turn coil technique. We also extended the measurements of the infrared magnetospectroscopy up to 400T generated by the electromagnetic flux compression method.

2. Experimental

The samples used are single-crystal graphite (kish graphite). The samples are cleaved from kish graphite ingots and the typical thickness of the samples is a few microns. Ultrahigh pulsed magnetic fields up to 180T (pulse width $\sim 6\mu$s) were generated by the single turn coil method.[3] We also made the measurements using the electromagnetic flux compression method. The recent progress in generating ultrahigh magnetetic fields up to 550T and in measuring the infrared magnetospctroscopy at low temperatures enables us to study the higher field properties of several materials such as semiconductor, semimetals and so on.[4]

As the incident light source, we used a CO_2 laser pumped far-infrared laser for 119, 306μm, a H_2O laser for 17, 28μm, a CO_2 laser for 9.2-10.6μm, a CO laser for 5.3-5.7μm and a He-Ne Laser for 3.39μm.

3. Results and Discussion

Figure 1 shows typical traces of the magnetic field dependence of the transmission at various photon energies ranging from 3 to 370meV. We can see two kinds of striking structures in these data. One is the cyclotron resonance peak indicated by "CR". The other is the magneto-plasma peak appearing as the transmission maxima indicated by arrows. Figure 2 illustrates the k_z-dispersion of the Landau levels when a magnetic field of 50T is applied parallel to the c-axis. The experimentally observed cyclotron resonance peaks correspond to the transition between the Landau levels (0,1), (-1,1) and (1,2) at the K-point. However, the hole cyclotron resonance peak of the (-1,1) transition is too broad to be distinguished from the electron resonance of the (0,1) transition even at $B \sim 50$T with $\lambda = 10.6\mu$m. This hole resonance can be seen at $B \sim 90$T with $\lambda = 5.317\mu$m. The difference in resonance width between electrons and holes is caused by the difference of the plasma frequency. In the present case, $\omega_{\mathrm{pe}} < \omega_{\mathrm{ph}}$. Here, ω_{pe} and ω_{ph} are plasma frequncies of electrons and holes, respectively.

On the other hand, another sharp structure can be explained by considering the electron-hole coupled plasma effect as pointed out by Nakamura *et.al.* [5] from the mag-

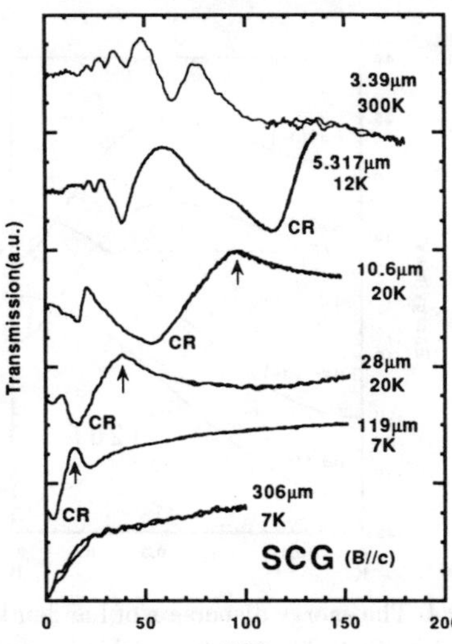

Fig.1 Magnetic field dependence of transmission at various photon energies (3-370meV).

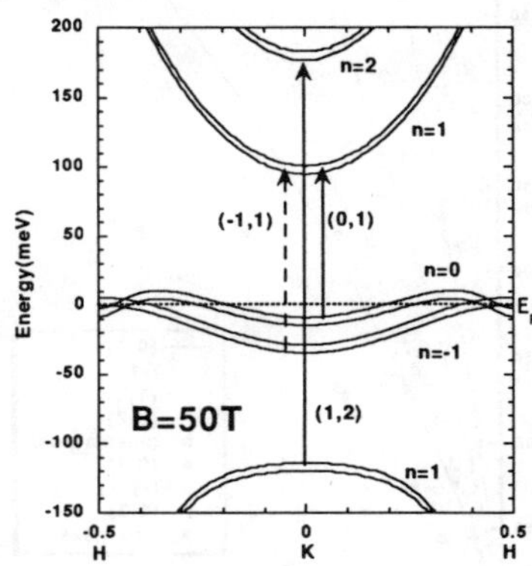

Fig.2 The k_z-dispersion of the Landau levels at $B = 50$T. The possible intraband and interband transitions are shown by arrows.

netoreflection experiments up to 40T. From the analysis using the classical dielectric constant in semimetals, the observed plasma-mode which has a superlinear dependence on magnetic field is derived from the condition $\epsilon = 0$, when graphite can not be regarded as an ideal compensated semimetal with parabolic bands. We can determine the relation between the magnetic field and the photon energy where the dielectric anomaly occurs, by considering both ω_{pe} and ω_{ph} as fitting parameters. The best fit for the data of $\lambda = 5-28\mu$m provides values of the plasma frequency, $\omega_{pe} = 0.25$eV and $\omega_{ph} = 0.60$eV. However, it was not possible to explain the data for $\lambda = 119\mu$m using these values. Quantum mechanical treatments for the dielectric constant including the contribution of the interband transition are required.

Next, we compare the observed cyclotron resonance with the Landau level calculation based on the SWM model.[2] This model expresses the electron dispersion relation in terms of seven prameters ($\Delta, \gamma_i, i = 0-5$) that is related to tight-binding parameters. We selected the most plausible parameters for the Landau level calculation. The parameters used are $\gamma_0 = 3.16$, $\gamma_1 = 0.39$, $\gamma_2 = -0.019$, $\gamma_3 = 0.28$, $\gamma_4 = 0.044$, $\gamma_5 = 0.038$, $\Delta = -0.008$ [2,6] in units of eV. Figure 3 shows both the observed (0,1), (-1,1), (1,2), (0,2) transition energies and results of the calculation as a function of magnetic fields. For $B < 60$T, the agreement between the experimental data and the calcultion is excellent. For $B > 60$T, however, the effective masses determined by both the (0,1) and (-1,1) transition energies are smaller than expected from the calculation. The calculation predicts that the $n = 0$ ($\sigma+$) Landau subband crosses the Fermi level at around 70T, and the $n = 0$ ($\sigma-$) Landau subband crosses it at around 120T in the carrier compensated condition. Figure 4 shows the k_z-dispersion of the Landau levels when a magnetic field of 120T is applied. The electronic states shown in Fig.4 corre-

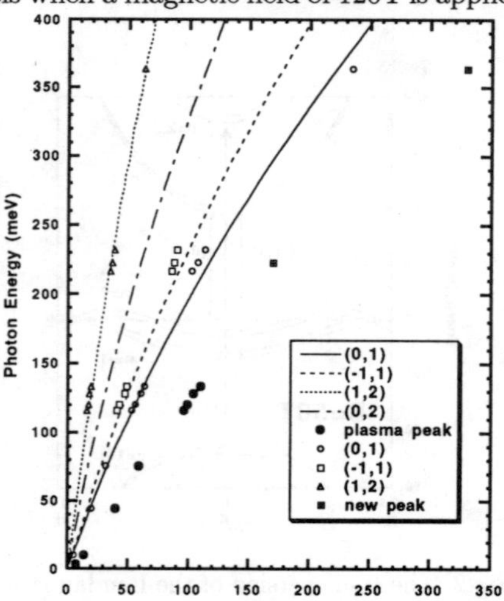

Fig.3 The relation between the photon energy and the magnetic field where the cyclotron resonance peaks and the plasma peaks appear.

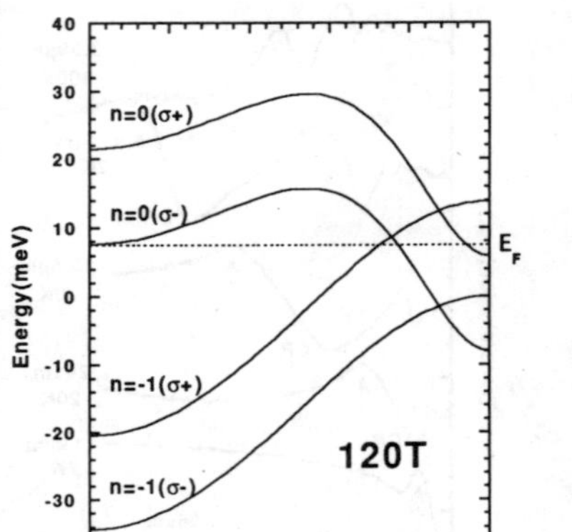

Fig.4 The energy dispersion of Landau levels at $B = 120$T. The Landau level index and the spin direction ($\sigma\pm$) are shown for each level.

sponds to the extreme quantum limit where only one Landau level remains below the Fermi level. In this regime, the cyclotron resonance between the (0,1) Landau subbands should disappear and a new (1,0) transition should appear in this field range ($B = 70 - 120$T). The (1,0) transition energy is almost equal to the (0,1) transition energy. The discrepancy between the experimental data and the calculated results at $B > 60$T may be caused by the anomalous electronic states where the $n = 0$ Landau subband exceeds the Fermi level. An alternative possibility is that the compensation of electrons and holes breaks down, and the extra electrons are created, thus the Fermi level increases. The photon energy of $\lambda = 5.317\mu$m (232meV) is just comparable to the plasma frequency of the electrons (=250meV) as mentioned above. The strong plasma effect may cause the (0,1) transition peak shift to the low field side, reflecting a sharp increase of the transmission at the plasma edge.

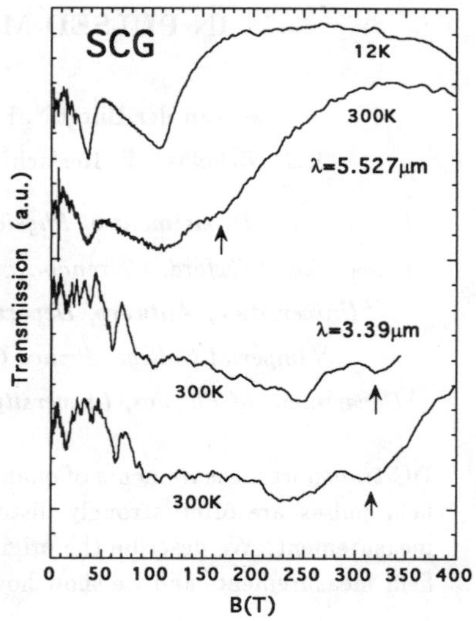

Fig.5 Transmission spectra at $\lambda = 3.39, 5.527\mu$m up to 400T. The unidentified new absorption peaks are shown by arrows.

To investigate the $n = 0$ subbands at higher fields, we have measured the transmission spectra at $\lambda = 3.39, 5.527\mu m$ under ultrahigh magnetic fields up to 400T using the electromagnetic flux compression method. These magnetospectra are shown in Fig.5. We observed the cyclotron resonance peak at $B = 240$T for $\lambda = 3.39\mu m$, which may be related to the (0,1) or (1,0) transition. We found new cyclotron resonance peaks at $B = 170$T for $\lambda = 5.527\mu m$ and at 320T for $\lambda = 3.39\mu m$. The origin of these peaks is not clear yet, but this fact suggests that the high field energy spectrum can be regarded as a different spectrum from the prediction of the calculation.

References

1. D. Yoshioka and H. Fukuyama, *J. Phys. Soc. Jpn.* **52**(1983)725.
2. K. Nakao, *J. Phys. Soc. Jpn.* **40**(1976)761.
3. N. Miura, T. Goto, K. Nakao, S. Takeyama, T. Sakakibara, T. Haruyama and T. Kikuchi, *Physica* **B155**, (1989)23.
4. H. Nojiri, T. Takamasu, S. Todo, T. Haruyama, K. Uchida, H. A. Katori, T. Goto and N. Miura , to be published in *Physica* **B**.
5. K. Nakamura, G. Kido and N. Miura, *Solid. State. Commun.* **47** (1983)349.
6. E. Mendez, A. Misu and M. S. Dresselhaus, *Phys. Rev.* **B21**(1980)827.

MEASURING QUANTUM HALL RESISTORS IN PULSED MAGNETIC FIELDS

M. van der Burgt[†°], F.M. Peeters[*], J. Singleton[°],
R.J. Nicholas[°], F. Herlach[†], J.J. Harris[□], and C.T. Foxon[‡]

[†] *K.U.Leuven, Department of Physics, Celestijnenlaan 200D, B-3001 Leuven*
[°] *University of Oxford, Clarendon Laboratory, Parks Road, Oxford OX1 3PU*
[*] *University of Antwerp, Department of Physics, B-2610 Antwerpen*
[□] *Imperial College, Prince Consort Road, London SW7 2BZ*
[‡] *Department of Physics, University of Nottingham, Nottingham NG7 2RD*

DC-transport measurements of quantum Hall (QH) resistors in short magnetic field pulses are often strongly distorted due to the transient nature of the measurement. We describe the origin of the observed differences from steady field measurements and we show how to obtain the correct resistivity values.

Non-destructive pulsed magnetic fields, in which a reinforced magnet is energised for a short duration (typically 10-50 ms), are nowadays commonly used to provide fields up to 50 T. The short duration of the pulses however, often causes complications in dc-transport measurements. This is especially clear from resistivity measurements on two dimensional semiconductor samples showing the quantum Hall effect (QHE).

The first experiments on the QHE in pulsed magnetic fields were reported by Horstman et. al. [1]. The authors measured flat plateaux in the Hall resistance ρ_{xy} of GaAs/Ga$_{1-x}$Al$_x$As heterojunctions in magnetic field pulses with a decay time of 20 ms from the peak field to half the peak field value. However, during the fast rise of these pulses strong deformations were seen in the plateaux in ρ_{xy} [2]. Similar deformations were reported by Mallett et. al. [3], and interpreted as an effect of induced currents in the sample. Mallett et. al. [3] also reported a typical shift in the ρ_{xx} resistivity minima: the minima occurring during the down sweep of the pulse were seen at lower fields then the minima measured during the up sweep.

By analysing a series of dc-transport experiments on GaAs/Ga$_{0.67}$Al$_{0.33}$As heterojunctions in pulsed magnetic fields we show that the origin of these discrepancies is solely due to capacitive effects distorting the measured signal and the applied sample current. This led us to devise a few simple methods which allow one to reconstruct the signals as they would appear in a steady field experiment. Although the present work focuses on the measurement of QH resistors, it is applicable to any dc-tranport measurement of high resistivity materials in pulsed magnetic fields.

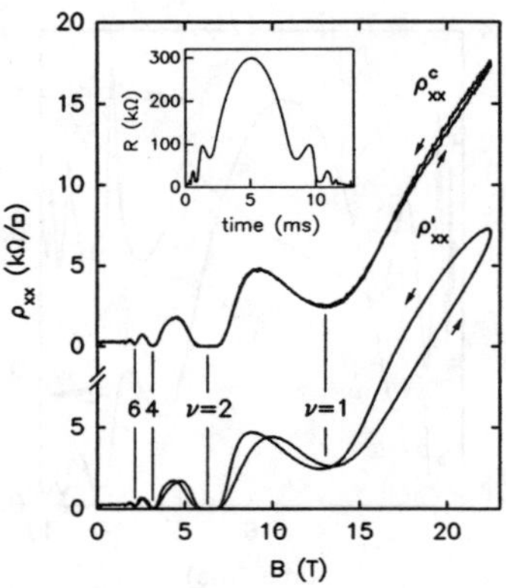

Figure 1: The measured resistivity ρ'_{xx} and the corrected resistivity ρ^c_{xx} calculated using eq. (1) with the two terminal resistance R shown in the inset and $C = 1100$ pF.

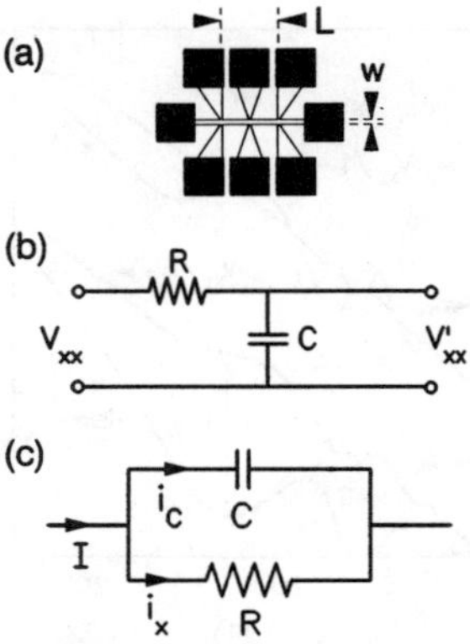

Figure 2: (a) The geometry of the sample with $w = 75$ μm and $L = 775$ μm. The electrical equivalent for (b) the resistivity measurements and (c) the Hall measurements.

For the experiments we used magnetic field pulses up to 25 T with a duration of 10–15 ms (half period of a damped sine wave). Details of the pulsed field set-up and the magnets can be found in refs. [4] and [5] respectively. We discuss the results of a GaAs/$Ga_{.67}Al_{.33}As$ heterojunction with a mobility of 9 m^2/Vs and an electron density of 2.9×10^{15} m^{-2} (4.2 K). The sample geometry is shown in fig. 2(a). The sample is mounted on a 1.4 m long probe which is immersed in liquid helium. Twisted pairs of 50 μm copper wire are connected to the sample contacts. The measured intra-pair and inter-pair capacitances are of the order of 200 and 100 pF respectively. The complete dc-transport measurement set-up is described in ref. [3].

The resistivity ρ'_{xx} measured during a 25 T pulse is shown in fig. 1. The signal measured during the trailing edge of the 15 ms pulse appears at lower fields than the signal measured during the leading edge. The difference increases with magnetic field and is maximal around $B = 20$ T.

This result can be understood when we consider the electrical Thévenin equivalent of the measurement circuit which is shown in fig. 2(b). Here V_{xx} is the voltage over the sample contacts which would be measured in an ideal experiment and V'_{xx} is the voltage which is actually measured. The latter is shifted in phase by the RC-filter formed by the time dependent sample resistance R and the capacitance C of the wiring. The relation between the two voltages is

$$V_{xx} = V'_{xx} + RC\frac{dV'_{xx}}{dt}. \quad (1)$$

When the second term in this equation is not negligible, a distorted signal is measured. After measuring R, the resistance between the contacts used for the V_{xx} measurement, and using C as an adjustable parameter, eq. (1) can be solved to find the correct

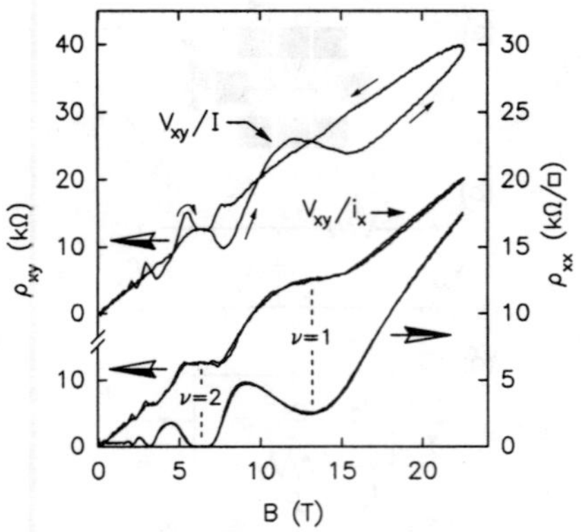

Figure 3: The measured Hall resistance V_{xy}/I, the corrected value V_{xy}/i_x and the doubly corrected resitivity ρ_{xx} (using i_x from fig. 4 and $C = 500$ pF in eq. (1)).

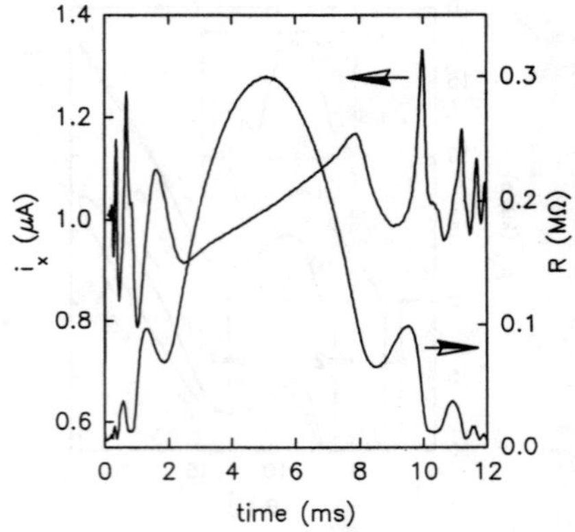

Figure 4: The effective sample current i_x calculated from eq. (2) with the resistance R as shown, $C = 870$ pF and $I = 1.03$ μA.

voltage V_{xx}. The result of this reconstruction is shown in fig. 1. The inset of fig. 1 shows the two terminal resistance R. The corrected resistivity ρ^c_{xx} reproduces almost completely for the up and down sweep of the magnetic field pulse. Only a narrow loop remains at fields above 16 T.

A typical measurement of the Hall resistance V_{xy}/I is shown in fig. 3. Here V_{xy} is the measured Hall voltage and I the constant, externally applied sample current. Very striking in these data are the strong distortions of the quantized Hall plateaux: e.g. strong deformations are seen both on the low and the high magnetic field side of the filling factor $\nu = 2$ plateau (the $\nu = 1$ plateau is not fully developed at 4.2 K). The sign of the deformations is different for the leading and the trailing edges of the magnetic field pulse. No deformations of the $\nu = 2$ QHE plateaux are seen in the central region of the plateau where $\rho_{xx} = 0$. The shape of the deviations does not depend on the sign of the magnetic field or the current direction, indicating that the observed effect is not a consequence of induced currents as suggested in ref. [3].

Again the deformations of the Hall plateaux can be understood using a simple electrical model as shown in fig. 2(c). Here R represents the two terminal resistance measured along the current path of the sample and C is the capacitance in parallel with R. When an external constant current I is applied to this system the current splits into two parts: one part, i_x flows through the sample and the other part, i_c, (dis-) charges the capacitor C. The current i_x is determined by the following differential equation:

$$\frac{di_x}{dt} = \frac{I - i_x(1 + C\frac{dR}{dt})}{RC}. \qquad (2)$$

Figure 4 shows the solution of this equation: the time dependent R is shown together with the correct sample current i_x. The net sample current i_x is not constant but has a sharp maximum where R drops and a strong minimum when R rises again. To find the correct Hall resistance we have to divide the measured Hall voltage, V_{xy}, by i_x instead of I. As shown in fig. 3 this results in the correct Hall resistance V_{xy}/i_x, which reproduces well for the up and down sweep of the field pulse. The capacitance C is again taken as a fitting parameter. The value for the capacitance, $C = 870$ pF, obtained in this way represents an effective capacitance connected to all the sample contacts. This acounts for the fact that C is usually a factor 4 or 5 higher than any capacitance measured directly on the sample holder (100–200 pF).

Of course, the varying net sample current i_x also has an effect on the measurement of ρ_{xx}. But since ρ_{xx} varies relatively quickly with field in the regions where i_x shows the strongest variations, this effect is less pronounced in the ρ_{xx} measurements. The ρ'_{xx} measurement from fig. 1, corrected for both the RC-phase shift and the varying sample current, is shown in fig. 3. The narrow loop which was still present in ρ^c_{xx} in fig. 1 has been eliminated now. Note that the RC-phase shift also has a small effect on the measurement of the Hall effect, but since the variations with field in ρ_{xy} are much less pronounced compared to those in ρ_{xx} and, for the sample geometry used, the two terminal resistance measured across the Hall bar is much smaller than the resistance measured along the bar, the RC-phase shift is not pronounced in the Hall measurement.

Kido et al. [6] described a set-up in which the effects of the experimental capacitances are electronically eliminated. We built such a set-up but the distortions were only found to be reduced rather then totally eliminated.

We can conclude that the observed distortions in pulsed field measurements of a GaAs/$Ga_{0.67}Al_{0.33}As$ QH resistor are a consequence of the varying sample resistance in combination with the capacitances associated with the experimental setup. We also demonstrated how to reconstruct the correct resistivity values.

MvdB acknowledges the support of the HCM programme of the EC. FMP is a research associate of the Belgian National Fund for Scientific Research.

References

[1] R.E. Horstman, E.J. van den Broek, J. Wolter, A.P.J. van Deursen and J.P. André: in *Proceedings of the 17th International Conference on the Physics of Semiconductors*, San Francisco, CA, 1984, eds. J.D. Chadi and W.A. Harrison (Springer Verlag, New York, 1985) p. 295.

[2] A.P.J. van Deursen, private communication.

[3] J.R. Mallett, P.M.W. Oswald, R.G. Clark, M. van der Burgt, F. Herlach, J.J. Harris and C.T. Foxon, Springer Series in Solid State Physics 101 (1992) 277.

[4] F. Herlach, L. Van Bockstal, M. van der Burgt and G.Heremans, Physica B 155 (1989) 61.

[5] L. Van Bockstal, G. Heremans and F. Herlach, Meas. Sci. Technol. 2 (1991) 1159.

[6] G. Kido, N. Miura, H. Ohno and H. Sakaki, J. Phys. Soc. Jpn. 51 (1982) 2168.

INSTRUMENTATION FOR SEMICONDUCTOR PHYSICS RESEARCH AT THE AUSTRALIAN NATIONAL PULSED MAGNET LABORATORY

R.G. Clark, R. Newbury, R.P. Starrett, L.D. Macks, A.V. Skougarevsky and A.G. Davies
National Pulsed Magnet Laboratory, University of New South Wales, Sydney 2052, Australia.

The National Pulsed Magnet Laboratory (NPML) provides access to high pulsed and steady magnetic fields, with particular emphasis on combining high fields with temperatures down to the millikelvin regime and a range of physical probes for semiconductor physics studies. NPML provides both short pulse (10-20ms) fields to above 60T and long pulse (1s) fields to 30T in addition to superconducting magnets (16T and 18T). The control arrangements and environment for combination of the 7kV, 0.8MJ thyristor-switched capacitor bank pulsed system and low temperature experiments have many novel features related to high performance, reproducibility and safety.

The NPML facility comprises three distinct experimental areas, the layout of which is shown in Fig. 1(a). The pulsed field laboratory is located centrally, with adjacent laboratories providing steady field capability. A schematic diagram of the 0.8MJ pulsed field circuitry is shown in Fig. 1(b). The design and construction of the pulse power supply and long pulse coil was carried out by Oxford Instruments Ltd., UK in a joint venture with UNSW. The selectable (1-32mF) capacitor bank is charged by a 7kV power supply in 60s. The pre-selected voltage is accurately maintained by trickle charge circuitry and the bank can be held on charge for up to 120s prior to firing to allow for final tuning of measurement apparatus (e.g. lasers) immediately prior to the shot. The bank is discharged via a parallel combination of thyristor switching stacks. Coil selection, polarity reversal and emergency dump of the bank charge are achieved via pneumatic switches. The entire pulsed system is fully computer controlled by purpose-written software.

System earthing and isolation has been designed for optimum signal to noise performance and user safety. All high voltage circuitry, with the exception of the 7kV power supply, is located within a blast-proof enclosure and is completely decoupled from the control console and system/analysis computers, located on a mezzanine level outside the enclosure, by a bundle of 60 fibre optic (FO) cables. The reinforced enclosure is designed to contain a worst-case projectile in the event of system or coil failure, 2kg (e.g. metal busbar) at 300m/s. A video camera is used to monitor the pulse room from the external control area. The enclosure has a number of protected wall penetrations at ground and mezzanine level for experimental access. A screened cabinet within the enclosure houses a four channel, 5MHz transient recorder for pulsed field data capture, together with amplification, low frequency electronics for thermometry, timing and fast I-V electronics. Data are downloaded to an on-line analysis computer located on the mezzanine level via a separate FO link, ensuring complete electrical isolation between the high voltage enclosure and operator interface. Full details of individual components of the pulsed system have been set out in Ref. 1.

New coil technology has been developed in collaboration with KU Leuven using copper-silver wire insulated with Kapton, to produce non-destructive pulsed coils with >20mm bore designed to achieve fields ~ 65T. The maximum non-destructive field attained at NPML during commissioning trials in 1992 is a 69.6T, 8ms pulse in a 12mm bore size, using high performance Cu coils. A cross section showing details of our CuAg pulsed coils is shown in Fig. 1(c), together with recent coil test data (percentage change in coil inductance v. field) for a 17mm bore coil (full circles) and a 22mm bore coil (open circles). Shot repetition rate to B_{max} is ~ 30 mins. The CuAg windings (2.4mm x 4.0mm section, tensile strength 930MPa, conductivity 82% IACS) are impregnated with glass-fibre reinforced Stycast and are externally reinforced with wet-wound (epoxy resin) 0.3mm-thick stainless steel ribbon. Carbon fibre has also been used in place of the internal glass fibre reinforcement for some coils.

The system earthing arrangements are shown in Fig. 2 (top) and are particularly relevant to obtaining high quality transport data in intense fields. Performance has been confirmed by sensitive measurements in

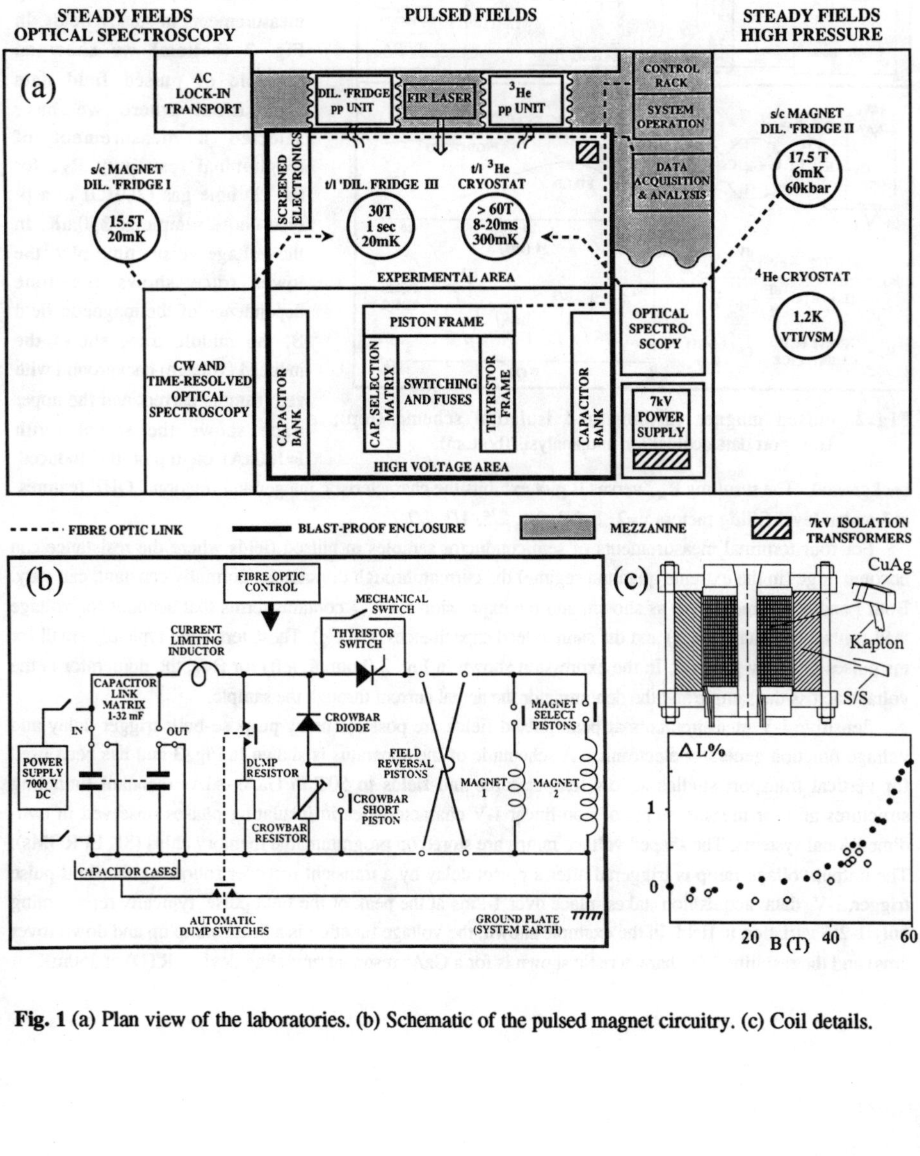

Fig. 1 (a) Plan view of the laboratories. (b) Schematic of the pulsed magnet circuitry. (c) Coil details.

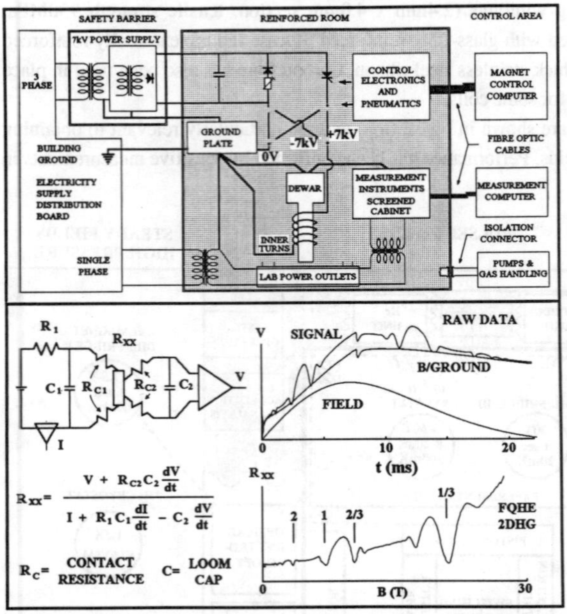

Fig.2 Pulsed magnet earthing and isolation scheme (top); Transport data acquisition and analysis (bottom).

the fractional quantum Hall effect (QHE) and extreme quantum regimes of two-dimensional electron and hole systems formed in n- and p-type GaAs-AlGaAs heterojunctions and n-and p-type Si-SiGe heterostructures, which have been compared with overlap measurements in steady fields. In Fig. 2 (bottom) we show an example of pulsed field data acquisition, where we have selected a measurement of longitudinal resistivity R_{xx} for the 2D hole gas (2DHG) in a p-type GaAs sample at 350mK. In the voltage versus time plot, the lower trace shows the time dependence of the magnetic field B, the middle trace shows the induced (voltage) background with zero sample current and the upper trace shows the signal (with I=100nA) on top of the induced background. The resulting R_{xx} versus B plot exhibits the characteristic integer and fractional QHE features at Landau level filling factors ν=2, 1, 2/3, 3/5, 2/5, 1/3, 2/7.

For four terminal measurements of semiconductor samples in pulsed fields where the resistance can become large (in the extreme quantum regime) the current through the sample, normally constant, can vary. Both I and V are measured, as shown, and the expression for R_{xx} contains terms that account for voltage lead contact resistance (R_{C2}) and the sample lead capacitances (C_1, C_2). These terms are typically small for ms pulses (~ a few percent). In the expression shown in Fig. 2 (bottom, left) for R_{xx}, the numerator is the voltage across the sample and the denominator the actual current through the sample.

Sensitive I-V measurements at peak pulsed fields are possible using purpose-built trigger delay and voltage function generator electronics. A schematic of the apparatus is shown in Fig. 3 and has been used for vertical transport studies at low temperatures and fields to 50T in GaAs-AlAs resonant tunneling structures and for measurements of non-linear I-V characteristics in insulating phases observed in two-dimensional systems. The shaped voltage ramps are stored on programmable memory chips (8K EPROMs). The output voltage ramp is triggered after a preset delay by a transient recorder following the field pulse trigger. I-V data acquisition takes place over 1-2ms at the peak of the field pulse, typically representing only 1-2% variation in field. In the example shown, the voltage function is a linear ramp up and down (over 2ms) and the resulting I-V characteristic shown is for a GaAs resonant tunneling device (RTD) at 350mK.

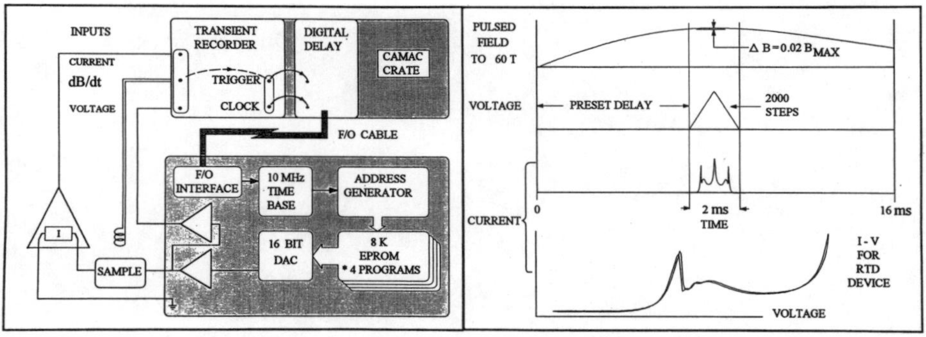

Fig. 3 I-V electronics for pulsed field studies (left); Timing sequence for I-V measurement (right).

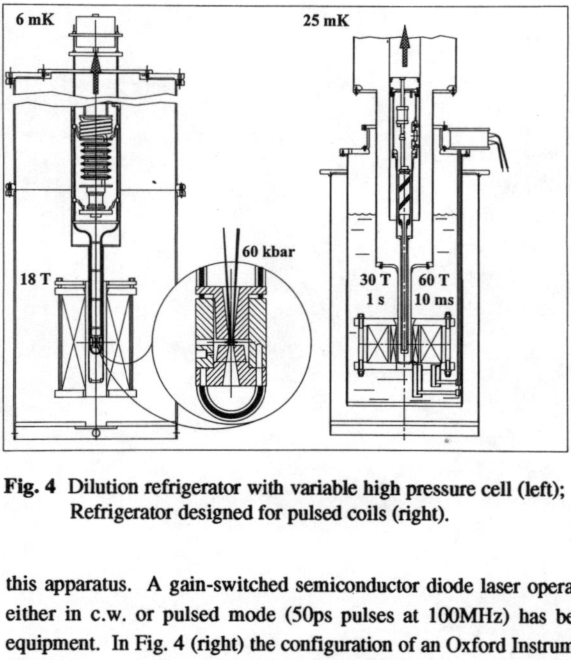

Fig. 4 Dilution refrigerator with variable high pressure cell (left); Refrigerator designed for pulsed coils (right).

Fig. 4 shows the general arrangement of two special dilution refrigerators available at NPML. An *in situ* variable miniature diamond anvil pressure cell has been interfaced to a separate top-loading 6mK dilution refrigerator/18T magnet facility for high steady field, hydrostatic pressure studies to beyond 60kbar at mK temperatures (Fig. 4 left). The solid-argon-loaded BeCu cell (inset) is driven by a stainless steel cable mechanism critically heat sunk at all temperature stages of the cryostat, tensioned by a hydraulic ram at the cryostat top-plate. Studies of low dimensional semiconductors by optical and transport probes are possible with this apparatus. A gain-switched semiconductor diode laser operating at 750nm with mW output power either in c.w. or pulsed mode (50ps pulses at 100MHz) has been constructed and interfaced to this equipment. In Fig. 4 (right) the configuration of an Oxford Instruments dilution refrigerator in our 30T, 1s and 60T, 10ms pulsed coils is shown. This refrigerator, with fibreglass tails and non-metallic dilution unit is currently being installed in our laboratory.

Reference

1. R.G. Clark et al, Proc. 4th ISSP Int. Symposium on 'Frontiers in High Magnetic Fields', Tokyo, 1993; Physica B in press.

AUTHOR INDEX